PORT ENGINEERING

PORT ENGINEERING

Planning, Construction, Maintenance, and Security

EDITED BY

Gregory P. Tsinker

JOHN WILEY & SONS, INC.

This book is printed on acid-free paper. ♾

Published by John Wiley & Sons, Inc., Hoboken, New Jersey
Published simultaneously in Canada

For general information on our other products and services or for technical support, please contact our Customer Care Department within the United States at (800) 762-2974, outside the United States at (317) 572-3993 or fax (317) 572-4002.

Wiley also publishes its books in a variety of electronic formats. Some content that appears in print may not be available in electronic books. For more information about Wiley products, visit our web site at www.wiley.com.

Library of Congress Catagloging-in-Publication Data:

Tsinker, Gregory P.
Port engineering : planning, construction, maintenance, and security / Gregory P. Tsinker.
p. cm.
ISBN 0-471-41274-0 (cloth : alk. paper)
1. Harbors—Design and construction. 2. Harbors—Maintenance and repair. I. Title.
TC205 .T7523 2004
627′.2—dc21

2003004785

Printed in the United States of America

10 9 8 7

To my wife, Nora, for her
help and inspiration

CONTENTS

PREFACE

The subject of port engineering is very broad. In fact, this is a blend that encompasses the array of engineering disciplines—civil, structural, geotechnical, hydraulic, naval architecture, and others—not to mention the socioeconomic and environmental aspects, knowledge of which is required to produce a sound design of a modern port or marine terminal.

In previous work published in the past 10 to 15 years, I have attempted to address in depth some important subjects related to design and construction of port-related structures and facilities. However, many port-related issues, such as dredging, maintenance, environmental problems, security, and many others, still need to be addressed and presented in a reader-friendly format. This volume has been written by a group of experts in their respective fields of expertise to fill a niche in the existing literature on port and harbor engineering and to provide port designers and operators, particularly those concerned with the design of a port- and harbor-related marine structures, with state-of-the-art information and commonsense guidelines to port planning, design, construction, and maintenance of port facilities.

Contributors to this book have been motivated by the desire not only to share their experiences with their colleagues in the profession, but also to provide those intended to join the field of port engineering with essential knowledge of the subjects related to successful port design, construction, and operation. This is important, because unfortunately, all too frequently, good and less costly engineering solutions are not used, due to a lack of familiarity on the part of the designer. On the other hand, use of a textbook solution to solve the problem may also be counterproductive. Every project is site specific, thus no engineer should be content merely to follow another's designs but should study such designs and use them as a starting point for developing his or her own ideas that are best suitable for the particular site conditions.

It must be understood that engineering in general and port engineering in particular is the profession in which knowledge of the mathematics, economics, and physical sciences gained from previous experiences must be applied to the particular project with thorough understanding that in a great many practical cases the problem cannot be solved with mathematical precision; the use of mathematical formulation of a particular problem must be practical and compatible with the data available. The design procedures and guidelines contained in the book are intended to point out the complexity of the particular problem and to illustrate factors that should be considered and included in an appropriate design scenario. They should not be used indiscriminately, particularly not for the detailed design, and should always be combined with good engineering judgment. The book is designed to serve as a guide and a reference for practicing engineers but is also aimed at engineering students and others seeking to enter the field of port engineering.

This volume represents the effort of 13 experts from around the world, to whom I extend the deepest gratitude for their contributions. It was a pleasure to work with this group of distinguished experts, who committed resources and hundreds of hours of their time to this project.

Deepest gratitude is extended to many individuals and organizations whose opinions and published works have been used and cited in this book. Among these stands out work by Dr. V. Buslov previously published and reprinted in Chapter 6. Many thanks go also to Miss A. Avolio, who typed part of the manuscript and dealt ably with many difficulties in the process. I also wish to thank my publisher, John Wiley & Sons, for full cooperation.

As usual, special thanks go to my wife, Nora, for her moral support, help, patience, and encouragement.

GREGORY P. TSINKER, Ph.D., P.E.

CONTRIBUTORS

Santiago Alfageme, Moffatt & Nichol Engineers, 104 West 40th Street, 14th Floor, New York, NY 10018; e-mail: salfageme@moffattnichol.com

Alberto Bernal, B y A Estudio de Ingenieria, S.L. Ferrer del Rio, 14, 28028 Madrid, Spain; e-mail: abernal@iies.es

Christopher M. Carr, Han-Padron Associates, 22 Cortlandt Street, New York, NY 10007; e-mail: ccarr@han-padron.com

Michael A. Gurinsky, Engineering Consultant, 7-09 Elaine Terrace, Fair Lawn, NJ 07410; e-mail: mgurinsky@optonline.net

John Headland, Moffatt & Nichol Engineers, 104 West 40th Street, 14th Floor, New York, NY 10018; e-mail: jheadland@moffattnichol.com

Michael A. McNicholas, Phoenix Management Services Group, Inc., 4758 West Commercial Boulevard, Fort Lauderdale, FL 33319; e-mail: mike.ptu@faxcom.com

Constantine D. Memos, School of Civil Engineering, National Technical University of Athens, Heroon Polytechniou 5, 15780 Zografos, Greece; e-mail: memos@hydro.ntua.gr

Dennis V. Padron, Han-Padron Associates, 22 Cortlandt Street, New York, NY 10007; e-mail: dpadron@han-padron.com

Bill Paparis, Han-Padron Associates, 22 Cortlandt Street, New York, NY 10007; e-mail: bpaparis@han-padron.com

Krystian W. Pilarczyk, Dutch Public Works Department (RWS), Hydraulic Engineering Division, P.O. Box 5044, 2600 GA Delft, The Netherlands; e-mail: k.w.pilarczyk@dww.rws.minvenw.nl

Robert E. Randall, Center for Dredging Studies, Ocean Engineering Program, Civil Engineering Department, Texas A&M University, College Station, TX 77843; e-mail: r-randall@tamu.edu

Eric D. Smith, Moffatt & Nichol Engineers, 104 West 40th Street, 14th Floor, New York, NY 10018; e-mail: esmith@moffattnichol.com

Gregory P. Tsinker, Tsinker & Associates, Inc., 2941 Piva Avenue, Niagara Falls, Ontario, Canada L2J 2X3; e-mail: gtsinker@niagara.com

PORT ENGINEERING

INTRODUCTION

Gregory P. Tsinker
Tsinker & Associates, Inc.
Niagara Falls, Ontario, Canada

As a very important link in a total transportation chain, ports have long been the gateway for goods and people to flow into cities and nations. They serve as an interchange between the land and sea transport and are used for assembling and breaking down commodities moved in bulk or in containers. Ports contributed to the stable development of industries and improved the lives of people by securing links with trade partners around the globe. Historically, port developments and their evolution have begun, motivated by both economic and technological pressures resulting from the global industrial revolution. To a great extent this has been influenced by the changing nature of ships and by the demand that greater volumes of cargo be handled at ports more rapidly. The latter stimulated development of more and more efficient methods and technologies for handling and hauling of miscellaneous cargoes.

The revolutionary shift to new technologies occurred in the 1950s with introduction of specialized container ships built to transport large freight containers. Initially, containers have been handled by conventional quay-edge cranes. The first specially designed container crane was introduced in 1959, and ever since, container handling cranes have grown in size and handling capacity. The need for efficient handling of containers stimulated development of new equipment, such as straddle carriers, heavy-lift forklift trucks, gantry cranes, special tractors, and others. These are discussed extensively in Chapter 2.

Furthermore, the roll on–roll off method of handling containers has been developed and used extensively. This method allows containers, but also cars, tracks, trains, and so on, to roll on ships via large stern or side ramps. The introduction of container systems for transporting goods resulted in dramatically reduced ship turnaround time, speed, and efficiency and safety of handling all types of containerized cargoes. This new technology drastically changed the approach to port planning. In most ports, a previously very effective pier system was disused as general cargo operations have been moved to the usually remote, high-volume container facilities with their large paved container storage areas and relatively few berths. These modern specialized ports and terminals tie directly into upland staging areas (marshaling yards) with multimodal links to several cities, a region, or the entire country.

The dramatic surge in port activity begun at the end of the twentieth century and expected to continue in the twenty-first century was responsible for the strong demand for space required for port expansion. Traditionally, ports developed in natural harbors and formed the nuclei for many cities. Today, ports and marine terminals are built wherever they can be justified economically: on land or when land is not available, in existing waterways, as in the case of the port of Los Angeles, where the 234-hectare Pier 400 has recently been built from dredged and other fill materials.

The need for large open areas to accommodate a modern container facility has induced ports to move to the periphery of cities, often on poor-quality land. The latter usually presents a challenge to port designers and has been an area of major controversy, related mostly to dredging and disposal of the contaminated dredged soils. Alternatives to dredging have been found in constructing offshore island ports and moving up-river shallow-draft ports down river, to deeper waters.

Furthermore, ports are not only looking for more space on land and/or existing waterways and offshore; they are also concerned with deepening their basins and approach channels to accommodate larger vessels. Most major ports now are aimed at having water depths up to 15 to 18 m to accommodate a new generation of post-Panamax vessels. The growth in vessel sizes and the development of new vessel technologies has a significant impact on port development. For example, the first generation of containerships had a capacity of around 1100 TEU (20-foot equivalent units). They were typically less than 200 m long with beams of approximately 25 m and a draft of less than 9 m.

Thanks to rapid growth in container shipping markets, the first generation of containerships was followed by second-generation vessels with capacities of 2700 TEU. These ships are typically 210 to 220 m long, having beams up to 27 m and a draft of about 10 m. Basically, the major constraint on container vessel developments was the need to keep within Panamax limits, which effectively limited the carrying capacity of containerships to less than 4000 TEU. This was however, exceeded in the late 1980s by the advent of vessels with a carrying capacity in excess of 4000 TEU. The higher capacity of these vessels was achieved by structural changes in the ships' hulls, but the vessels themselves remained within the Panamax category with an increased draft of up to 11.5 m.

In the 1990s the Panamax limit was finally exceeded by a new generation of vessels. At this time, vessels with a capacity above 6000 TEU have been introduced. Now that the Panamax barrier has been overcome, new vessels with a capacity above 8000 TEU have appeared, and future vessels are on the drawing boards with planned capacities of 12,000 to 15,000 TEU. These vessels of 120,000+ dwt (dead-weight tonnage) will be approximately 380 to 400 m long, with a beam of 60 m and a draft up to 14.8 m, and will travel across the oceans at a speed of some 23 to 25 knots. These latest and increasingly larger carriers will have a significant impact on the demands for new terminal facilities. These will be constructed for the increased depths in approach channels and alongside extended-length berths. Furthermore, these terminals will need increased maneuvering areas, and to ensure rapid vessel turnaround, the vessels will be assisted by tugboats during their berthing and unberthing maneuvers. These terminals will certainly need marshaling yards capable of dealing with a large volume of containers.

The increased size of vessels has resulted in demand for larger heavy-lift quay-side container cranes to handle the outreach of the giant post-Panamax vessels. A new generation of cranes with an outreach of 60+ m having a lifting capacity of 50 to 70 metric tons are now operational in several ports around the world. The high productivity of these cranes is due to significant advances in crane technology, which includes use of twin-lift spreaders capable of handling two 6.1-m (20-ft) boxes together, and faster hoist and trolley speeds that allow for a faster cycle time of handling containers. Most recently, in Japan, the world's first container crane with a base isolating system that allows for crane safety during an earthquake event has been introduced.

Land-side container stacking equipment includes straddle carriers, rubber-tired gantries, rail-mounted gantries, and overhead bridge cranes, supported by forklift trucks, reach stackers, tractor trailer units, and other specialized vehicles. Details on this equipment are provided in Chapters 1 and 2. Recent technical developments in quay-side container handling equipment followed by similar developments in land-side container handling technology, which now is able to operate with much greater stacking heights and width, producing much higher stacking density than before. Most recently, an overhead crane capable of stacking nine high and 10 rows across have been introduced in the port of Singapore. This and other cranes usually operated remotely with a high degree of built-in automation. Furthermore, the wider, nonstandard containers are already the reality, and the trend for the use of larger containers will continue. New container sizes will inevitably have a great impact on the design of new containerships and container handling technologies.

The future development of container terminals is envisioned by many experts as a fully automated process in which "equipment with brains" will be capable of handling container flows in both directions automatically. These most radical and far-reaching changes in container terminal technology will lower personnel requirements and operating costs, increase control, and speed the flow of goods through ports. Dramatic developments in vessels and container handling technologies will result in concentrating maritime freight worldwide on hub megaports that meet the logistical requirements posed by the ever-increasing size of container ship. Megaports able to receive the largest vessels, strategically located around the world, will be serviced by feeder traffic from small ports. Ultimately, this will appreciably reduce the investment in construction and maintenance of smaller ports.

The continuing worldwide demand for energy and raw materials is creating a strong need for specialized liquid and dry bulk material ports. Studies carried out worldwide indicate that the economics of bulk transportation in large quantities is such that the unit cost per tonne of transported material is reduced considerably with the use of larger ships able to carry superloads to terminals for further distribution. Consequently, deepwater facilities able to handle a large deep-draft vessel, such as supertankers for transporting crude oil and liquid natural gas and superbulkers to transport dry bulk materials, such as coal and iron ore, have been constructed at many locations in the world. The key factor in the operation of these terminals has been utilization of very high capacity loading and unloading technology.

More than 100 years following introduction of the first tank-mounted vessels, the procedure and method of liquid bulk handling have not changed in principle, but technical improvements in the area have been spectacular. The capacity of petroleum liquid bulk carriers (tankers) in the 1940s has reached 22,000 dwt, and at the present time, 500,000-dwt tankers ply the oceans. Today, several shipyards in Europe and Asia have the capacity to build 1,000,000-dwt tankers.

Over the last decade there has been steady increase in the production, transport, and consumption of liquid natural gas (LNG). According to some experts, the trade in LNG will grow steadily in the next decade with the expansion of existing production facilities and the development of new ones. The upsurge in LNG marine transportation has created an unprecedented demand for new LNG carriers and specialized terminals at both the point of production and the point of reception. The world's principal LNG exporters at present are Indonesia [28.5 × 10^6 metric tons (tonnes)], Algeria (19 × 10^6 tonnes), and Malaysia (15 × 10^6 tonnes), which together produce approximately 70% of the world's total LNG production. New LNG exporters and users are emerging constantly, all requiring new marine terminals with sufficient area on land for storage and processing facilities. These should be located in reasonable proximity to the marine terminal. Ideally, these facilities will be located at naturally protected deepwater locations or those protected by breakwaters.

Prudent developers of these facilities should consider the possibility of future expansion in terms of a larger vessel to be accommodated there and/or additional berths to be added. This is particularly important for new startup facilities serviced initially by a smaller vessel. In this case, access to the port, the size of its basin, and the capacity of its marine structure must all be able to cater for the larger vessels that will be required in the future. The relatively small increase in overall initial capital cost for such a facility, however, will pay off handsomely. LNG carriers now in service vary in capacity from approximately 20,000 m^3 to 137,500 m^3. In the future the capacity of the largest LNG carriers is expected to be approximately 130,000 to 150,000 m^3. Hence, new terminals will be required to serve vessels of 65,000 to 70,000 dwt around 300 m in length, with drafts of 12 to 13 m.

Similar developments have occurred in the transportation of dry bulk materials. Bulk carriers have lagged but tracked tanker growth; similarly, tankers may be expected to grow in the future. The use of large and very large deep-draft ships for transporting liquid and dry bulk materials and material hauling innovations have changed the nature of the modern bulk handling port. Ports actually become and continue to be highly specialized terminals able to handle specific cargoes at very high rates: for example, loading of 20,000 tonnes/h and more of dry bulk and 220,000 m^3 of crude oil per day.

Deep-draft vessels need deepwater ports. It has been learned, however, that the conventional approach to construction of such ports, involving dredging of large quantities of sometimes contaminated sediments, can be prohibitively expensive. The solution has been found in the construction of offshore fixed or floating marine facilities not protected from the effects of environmental forces such as waves and currents. At these facilities, the low berth occupancy due to rough sea conditions has been compensated for by a very high rate of material handling on calm days. These facilities have been constructed far enough offshore, where sufficiently deep water is found and no maintenance dredging is needed. In some instances the terminals have been moved as far as 2 km and more offshore and have been linked to the shore either by submarine pipelines or by a bridgelike trestle designed to support pipelines or conveyor systems and to provide access to the terminal for lightweight vehicular traffic. In some cases, particularly in heavily populated areas where local residents object to the construction of conventional trestles as an unacceptable "visual pollution," submarine tunnels have been constructed as a solution to the problem.

The modern cargo handling port is developed as an important link in a total transportation system, and planners of such multimodal systems seek to optimize the total network, not just one of its components. Construction of a new port, or expansion or modernization of an existing port, is usually carried out to increase port capacity and its effectiveness. Traditionally, this has focused on the sea, and consequently, construction of new berths and modernization and expansion of existing ones became prime areas of interest. However, as urban coastal areas, particularly in developed countries, have expanded substantially over the last five decades while international trade has increased and continued to expand, making the world more and more economically interdependent, the port land-side capacity to transfer cargo from wharf to end user has become increasingly critical. In some densely populated areas, the available transportation network (e.g., highway and rail) is limited to moving a certain amount of cargo and cannot be expanded further. Under these conditions there is no logic in increasing the existing port capacity unless the land-side transportation infrastructure is equally capable of moving the increased volume of cargo through the land-based transportation network.

Furthermore, in the future, exporters of petroleum products and raw materials will seek to increase the added value of their trade, which will tend to reduce tonnages of raw commodities and increase those in partially or fully processed materials. Thus increasingly, refined petroleum products, chemicals, aluminum products, steel products, sawn and processed timber products, processed agricultural commodities, and so on, will be the cargoes rather than the basic raw commodities. These are more valuable and readily damaged cargoes which require more careful handling and storage. This trend is already quite visible in Southeast Asian countries, including Malaysia and Thailand.

An important development in the maritime industry that occurred about three decades ago was the virtual disappearance of true passenger liners; the cruise industry has emerged to replace them. Today, the passenger trade exists basically on local lines, on inland waterways, and between coastal ports. Cruise vessels have continued to look like liner vessels, notwithstanding that speed of sailing is no longer of paramount importance. Future cruise vessels are expected to be larger, slower-moving floating resorts. A new generation of cruise vessels up to 150,000 dwt will be in service before 2005. This will require that existing cruise terminals adapt to these giant vessels in terms of both passenger handling and ship accommodation, where vessel draft usually is the first consideration.

Globalization and further modernization of existing ports and construction of new terminals together with the importance of sustainable development are very significant for the maritime industry. One important challenge awaiting the maritime sector of the global transportation system is implementation of efficient environment protection programs. Sound environmental management will play a vital role in future ports, from their construction through their operation and maintenance. Features of this process include but are not limited to the following: identification and classification of all potential environmental impacts that can result from the growth of transport in general and maritime sector in particular; comprehensive integrated environmental management of port construction, operation, and maintenance processes; monitoring and enforcement of environmental laws; public participation and effective communication; and cooperation with related industries.

Implementation of all of the above will be aimed at reducing the negative effects caused by intensive use of vessels: leaking of oil and ballast water and damping of dry litter; the nuisance related to noise produced by working equipment, vehicles, and the transport system; negative effects due to space consumption and visual intrusion into surrounding landscape; and atmospheric pollution produced by fuel engines of all kinds. In most cases of port modernization or construction of a new terminal, approach channels and basins must be dredged to increase basin size and depth of water. Sometimes this operation results in alterations of coastal morphology, effects on wave climate and water current, reduction or improvement in water quality, effects on sea fauna, and similar conditions.

The dredging activity may also result in a short-term increase in the level of suspended sediments at the site of dredging and the site of disposal of dredge material. However, severe damage to the environment can occur especially as a result of dredging contaminated sediments. These sediments must be prevented from spreading in the form of suspended particles to the ambient environment, and those excavated must be treated before their disposal on land or sea. Several recently published guidelines by PIANC contain information on the subject of management of contaminated soils. References to these and other guidelines and regulations concerned with port construction, operation, and maintenance are provided in Chapters 11 and 12.

Finally, it should be noted that in the foreseeable future sustainable development of maritime transportation is seen as inseparable from the global war on terrorism of all kinds. This includes but is not limited to fighting against direct threats to maritime shipping and ports worldwide from miscellaneous radical terrorist organizations and ever-increased smuggling of drugs and human beings, cargo theft, piracy, and the like. Targets of these criminal elements include terminals, vessels, cargo, containers, equipment, and personnel. In addition to concerns related to the rising number of incidents of violent piracy and waves of stowaways/refugees in the past several years, ports in the post–9/11/01 period must also face the looming specter of terrorist attacks involving weapons of mass destruction or of a vessel being used by terrorists as an instrument of destruction. To deter or deny these threats effectively, ports must develop a security strategy that identifies potential security threats, defines critical assets and information, integrates security resources and capabilities, and ensures successful design, implementation, and management of world-class seaport security programs. Some basic in-port security programs currently in place in ports worldwide are discussed in Chapter 13.

1

PORT PLANNING

Constantine D. Memos
National Technical University of Athens
Zografos, Greece

1.1 INTRODUCTION

Port development can refer either to the creation of a new port or to the expansion of an existing one, usually aimed at increasing its capacity or upgrading port operations. The issue of port development is examined at three different levels: national, local, and port terminal. Complete study of the above can be a complicated procedure since it presupposes a contribution by many specialists of various disciplines. The analysis laid out in the following pages derives from the discipline of a civil engineer specialized in port planning who has undertaken the task of conceiving and designing the pertinent elements, in most cases as part of an interdisciplinary team charged with the overall port development planning. In designing at the port or terminal level, aspects pertaining to the maritime aspects of ports are also dealt with. Such issues include the general layout of breakwaters and quays and the design of entrances and maneuvering areas.

1.2 PORT PLANNING AT THE NATIONAL LEVEL

1.2.1 National Port Policy

Until recently, ports in many countries have usually been developed as part of local port development programs. Such programs normally do not take into consideration the corresponding plans of other ports within the country, a factor that would have resulted in better coordination for increased national benefit. Indeed, in many cases, instead of attempting to achieve mutual complementing of aims, undue competition tends to develop between ports within the same country. In government-owned ports this situation can result in uneconomical investment of national capital in competing projects, and moreover, in loss of opportunities to attract a portion of international maritime traffic.

The competitive tendencies relate to the foreign trade of the country, foreign goods in transit, and goods being transshipped: the international flows that evidence potential for development as opposed to internal transports, which have more-or-less preset movement patterns. These trade flows can be defined as follows:

- *Foreign trade flows* relate to the exports and imports of a country, and consequently, have their origin or destination in that country.
- *Goods in transit* are those goods in international flow whose land transport leg uses the territory of the country and one of its ports.
- *Goods being transshipped,* where both origin and destination are located outside the country but both of whose transport modes are marine. Consequently, in this flow only the specific ports of the country are used, not overland transport.

The latter two flows in general make up the target of the competition between ports in a country.

Given that major ports constitute integral elements of the transport network of a country, it is evident that some sort of framework for centralized coordination of port development efforts is required at a national level. A significant service that such coordination would produce refers to determination of the most suitable ports for attracting transit or transshipment movement on a national level. This acquires particular significance nowadays, where such cargo movement is conducted mainly in containers, and the corresponding port installations are very costly.

In more general terms, the existence of a national port policy could broadly define the role of each port in a country, so that in the context of the national economy, the available funding can be employed as productively as possible. Depending on a country's development and its tendency for privatization, the allocation of roles to each port may be conducted in such a manner as to permit a large percentage of these ports to be released from national coordination and to undertake their own development.

1.2.2 Definitions of Port Functions

Today, the port has acquired its standing within the intermodal transport system by constituting a nodal point between two transport modes. In seaports, one mode concerns maritime transport; in river ports, this mode concerns river transport. The nodal linkage between two different modes of transport should be functional, permitting efficient and secure movement of passengers, cargo, and vehicles. A civil port is a passenger, cargo, or combined port depending on the traffic that it serves. In a combined part, both passengers and cargo provide a significant percentage of the traffic. Of course, specialized ports exist, such as marinas (for harboring pleasure craft), fishing ports, and naval military bases.

There are two basic methods of loading and unloading cargo to vessels. They are *lift on–lift off* (Lo-Lo), which refers to the loading and unloading method, employing either the vessel's gear or quay-side cranes, and *roll on–roll off* (Ro-Ro), which refers to the loading and unloading method conducted by horizontally moving equipment. Vessels allowing this type of loading and unloading are equipped with a loading ramp that permits the movement of cargo handling equipment and other vehicles (trucks, forklifts, straddle carriers, tractors, etc.) between quay and vessel.

At cargo ports, the type and packaging of cargo products determine the manner of loading and unloading as well as of other operations. Thus, the following basic categories of port terminals can be identified, each having varying equipment and operational features:

- *General cargo terminals.* These arc terminals equipped with conventional cranes, which handle cargo in all types of packaging compatible with cranes. The packaging could be parcels, sacks, pallets, or containers. The latter should not, however, constitute a major percentage of the traffic, because otherwise a specialized container terminal would be required to improve throughput performance.
- *Container terminals.* In this case, containers are handled using special loading/unloading, transfer, and stacking equipment. They are typified by extensive yard areas for container stowage.
- *Multipurpose terminals.* These terminals combine a variety of functions in a single terminal, where containers, but also conventional general cargo or other packaged products, can be handled.
- *Ro-Ro terminals.* Here cargo is transferred within a roll on–roll off system, with loading and unloading of cargo by horizontally moving lorries, forklifts, tractors, and so on.
- *Bulk cargo terminals.* At these terminals, liquid or dry bulk cargo without packaging is handled. Usually, pumping machinery with suitable piping or grab cranes is used at these terminals.

The main quantity that may be affected by a suitably implemented national port policy lies in international cargo flow. Consequently, the initial and basic step in formulating a country's port system includes the determination of those ports that will undertake to serve the flows of foreign trade, transshipment, or transit. These flows operate more-or-less independently of one another, and thus for simplification of the analysis, may be studied individually.

The basic criteria to be considered in developing a proposition as to the roles of a country's ports may be classified into the following four groups:

1. The national and regional development policies of the country
2. The transportation infrastructure of the hinterland and its prospects
3. Existing port capacity and potential for development
4. Cargo forecasts for each port

After each of the three independent international flows has been examined, the findings should be pooled, to define the core of the country's port system. Thus, the role of each port that participates in international cargo flow will be specified and the basic cargo throughputs can be determined. Considering these throughput values, and factoring in the national flows, master plans can be drawn up for individual ports.

Apart from international cargo flow, other aspects of the overall port development study are usually examined. Although these are not of primary significance in the formulation of the core of a national port system, they do have a role in evaluation of the main subsystems and in developing the final proposal. Such aspects include:

- Special bulk cargoes, such as coal, cement, petroleum products, grains
- Industrial ports
- Shipbuilding and ship repair
- Free zones
- Coastal shipping
- Passenger movement

1.3 PORT PLANNING AT THE INDIVIDUAL PORT LEVEL

1.3.1 Port Development Planning

1.3.1.1 Port Development and Master Planning. The master plan of a port allocates the land within the port to the various uses required, describes the projects needed to implement the plan, and gives an indicative implementation scheme by development phase. These phases are related directly to the projected port traffic which has to be monitored closely. When in due course a decision is reached to proceed with implementation of a development scheme, this should be integrated smoothly with, or derive from, the master plan for the port. Therefore, it is important that a master plan exist, and drafting one should be among the primary concerns of port management. Of course, a variety of continuously varying factors have a bearing on such a plan, ranging from statistical data on port traffic to international treaties. For this reason, the plan should be revised regularly, at least every five years. Moreover, if during the design of a particular development phase the need arises for a review of the plan, this should be conducted concurrently, if possible, to ensure compatibility with the other functions and operations of the port. However, the lack of a master plan at a particular port should not delay the making of decisions for small-scale immediate improvement, although it is recommended that at the first opportunity an effort should be made to draft a master plan for the port.

1.3.1.2 Long-Term Planning. In the event that a national ports plan does not exist, the consultant should proceed with drafting a master plan, after studying the following components of long-term planning:

1. The role of the port—in particular:
   a. The servicing of its inland area as regards foreign trade
   b. The support that the port may offer to the region's commercial and industrial development
   c. The attraction of transiting and transshipment traffic
2. The responsibility of the port for the construction of both port and land works. Frequently, more than one agency becomes involved: for example, when a port area is serviced by a railroad.
3. The land use in the area and the potential for expansion of the port. It is important that there be general agreement between interested parties over the proposed expansions and land use so that the resulting master plan meets with wide acceptance.
4. The policy for financing the port development, which may be formulated on the basis of its own resources and/or through a state grant.

In general, in modern port development the basic requirement is for large expanses of land to ensure productive operation of the individual terminals. Therefore, a careful examination of point 3 assumes particular importance.

1.3.1.3 Medium-Term Planning. As stated, each port development scheme should be incorporated in the master plan and should proceed to implementation following the results of an appropriate feasibility study. The latter study should refer individually to each independent section of the overall development proposal, such as a container terminal or a bulk cargo terminal. Thus, under a positive but reduced yield from the overall proposal, the risk of concealment of a nonproductive section is avoided. The drafting of a port development plan calls for the conduct of the following special studies:

1. Analysis of the functionality of the port as regards the services offered in conjunction with capacity
2. Designs, with budgets
3. Operational design, with budget
4. Financial and financing study

In large port development projects it is customary to reexamine the organization and management of the port operating agency and to recommend organizational improvements on a small or larger scale. It is possible that many of the ports in a country do not warrant a development effort beyond maintenance of existing structures or appropriate modification, such as to serve fishing vessels or pleasure craft. Such modifications are nowadays met quite frequently, since old ports, traditionally being part of the core of their town, cannot easily incorporate large land expanses needed in modern port layouts. Also, environmental and social issues do not allow in many cases major expanses of an old port site. The requirement that the citizenship should be granted free access to the waterfront of their city is gradually being respected by more and more authorities. Nevertheless, the problem of what to do with the old port installations is a complex one, where both the needs of the local community and the benefits of the relevant port authority should be accommodated. As noted above a common trend is to change the character of a past commercial port into a marina or fishing vessels refuge. There are also examples (London, Marseille, etc.) where old ports were completely refurbished into commercial or recreational zones, some of them arousing controversial discussions among town-palnners.

Moreover, since ports interact in many ways with the surrounding township, port master planning should take into account, apart from strictly engineering issues, such aspects as social, economic, and environmental constraints and should easily fit within the relevant town and regional plans. This frequently calls for a compromise between the requirements of the port and the local authorities.

1.3.2 Principles of Port Design

1.3.2.1 Guiding Principles. If the undertaking involves the development of an existing port, before proceeding with development plans it would be prudent to make efforts to (1) increase productivity and (2) improve existing installations. Factors that contribute to increasing productivity in an existing port are improvements in loading and unloading practices, to the overall operation of the port terminals, and to modernization of cargo handling and hauling equipment. As pointed out, the expansions that may be required additionally to the improvements above should be incorporated in the master plan of the port and should be implemented within a time horizon in order to constitute productive projects according to the pertinent feasibility studies.

Particularly as regards the individual terminals within a port, the respective capacity calculations are based on different factors, depending on the nature of each terminal as follows:

1. In conventional cargo terminals, the required number of berths is determined first, to keep vessels' waiting time below a specified limit, determined by economic and other criteria.
2. In container terminals, the land area required for the unobstructed movement of cargo flow is calculated.
3. For specialized bulk cargo terminals, the cargo flow during loading and unloading

has to be calculated first, to ensure that vessels will be serviced within acceptable periods of time.

As arrival times of commercial vessels at ports cannot adhere to an exact schedule, enabling ready scheduling of requisite berthing and eliminating waiting time, to determine the number of berths a compromise is usually made between two extreme situations: on the one hand, the minimization of vessel waiting time, and on the other, the maximization of berth occupancy.

1.3.2.2 Port Costs. Two factors constitute port costs: investment cost, which does not depend on traffic, and operating cost, which does. If the cost were to be expressed per unit of cargo throughput, the relation between cost and traffic volume is depicted as in Figure 1.1. A ship's cost in port is also made up of two constituents: the cost of the vessel's waiting time and the cost of the ship while berthed. The ship's total port cost curve expressed as above is shown in Figure 1.2. The sum of the port cost and the cost of the ship in port provides a total cost, as shown in Figure 1.3.

Traffic corresponding to point *B* in Figure 1.3 is less than that at point *A*. This means that the optimum traffic volume for a port is lower when the total cost is taken into account than when either the total port cost or the total vessel cost is considered. Of course, the difference between *A* and *B* depends on vessel types, which determines the corresponding vessel cost curves.

A measure often used to describe the level of service offered to vessels is the ratio of waiting time to service time. It is generally recommended that this ratio be lower than, say, 20%, but there is a danger here of showing an improvement of service provided through a unilateral increase in service time. This is why for the purposes of evaluation, absolute values of total vessel waiting time at the port are also required.

1.3.2.3 Traffic Fluctuations. Even a proportionally small but persistent increase in the traffic of a port may very quickly cause congestion in a port lacking in reserve spaces; the congestion will cause a reduction in the productivity of serviced vessels, which aggravates the problem further. The increase in traffic may be

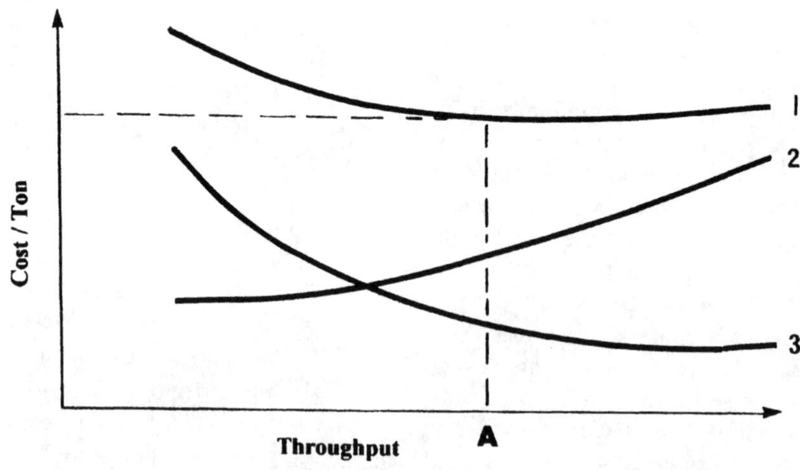

Figure 1.1 Port cost as a function of cargo throughput. 1, Port's cost; 2, cost of operation; 3, capital cost.

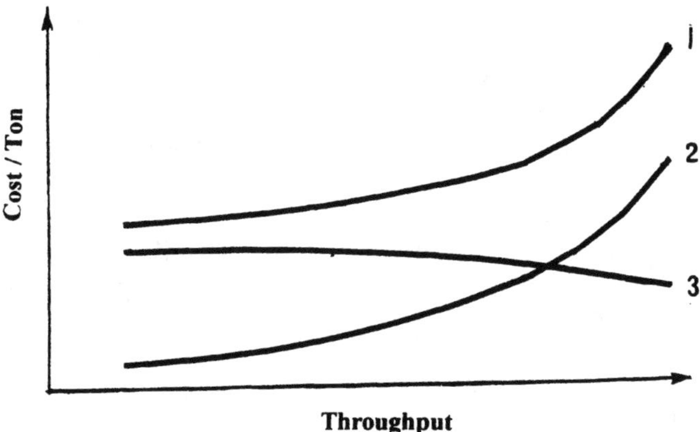

Figure 1.2 Cost of ship in port. 1, Ship cost in port; 2, cost of waiting; 3, cost of berth.

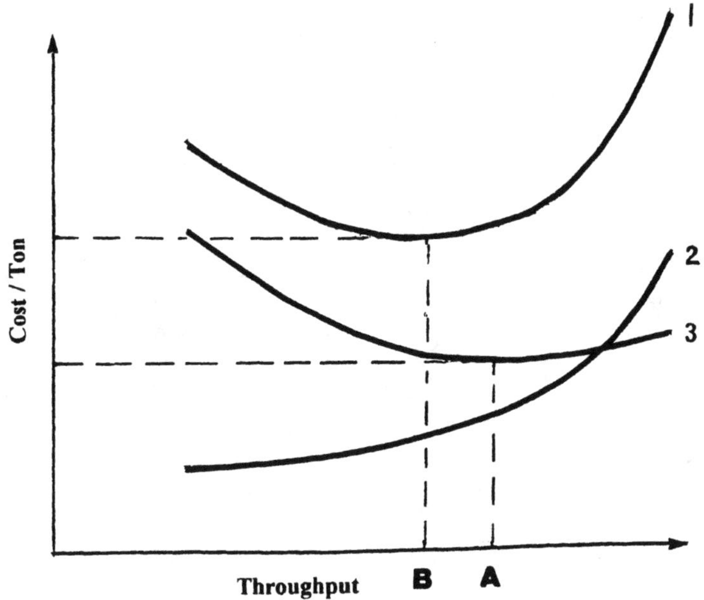

Figure 1.3 Total vessel–port cost curve. 1, Total cost; 2, cost of vessel; 3, cost of port.

caused by a new shipping line, larger cargo volumes, more frequent or occasional vessel calls, and so on. Even a change in the packing method of a product of large throughput may affect the efficiency and productivity of a port adversely. It is assumed that the problems created by a steady increase in traffic will be met in good time through the implementation of suitable development projects based on the master plan of the port.

The fluctuations around more-or-less regular average traffic may be faced by a carefully designed emergency plan according to which old quays, anchorages, and so on, on reserve, which are not used as vessel servicing positions, may be brought into operation. Usually, the reserve capacity of a port consists of inexpensive installations, which, however, give rise to a high cost of operation. These reserves should be allocated equally among all the port's sections. Other means of a temporary increase to port capacity could be an improvement in cargo handling: for example, an increase in work gangs per vessel serviced, the hiring of additional mobile cranes or other equipment, or the use of lighters for loading and discharging on two sides.

The size of the cargo to be taken for planning purposes should be selected carefully so that potential fluctuations may be absorbed with some acceptable increase in vessel waiting time. As regards high-cost installations and vessels, a method of smoothing the peaks in waiting time is that of *serving by priority,* according to which, when the vessel arrives at a predetermined time, it will have guaranteed access. The more such agreements between ports and liner operators are signed, the greater the smoothing of the traffic curve.

1.3.2.4 The Optimum State. The chief benefit from investments in port projects is the possibility if reducing total vessel time at a port. Despite the fact that ships are the first party to benefit, in the medium term both the port and the country benefit overall from the development of ports. From a practical point of view, optimization of the waiting time–quay use issue may result in a 75% occupancy factor for a group of, say, five general cargo berths, which produces a wait of half a day, for an average service time of 3.5 days. This means that over a long period of time: 55% of vessels will berth immediately, 10% of vessels will wait for 2 days, and 5% of vessels will wait for 5 days. It can be deduced from the above that the fact that some vessels experience excessive waiting times does not necessarily mean that the port is congested.

1.3.2.5 Grouping of Installations. Depending on the type of cargo traffic and on the equipment required, berthing positions and other installations are grouped in more-or-less independently operating areas of a port. This grouping implies specialization in the type of cargo traffic being served in each port section. Thus, better utilization is achieved: for example, in wharf depths and quicker servicing of vessels and cargoes. However, there are also disadvantages to grouping port installations. Basically, the flexibility obtainable by the greater number of berths is reduced. This offers a more productive exploitation of both water and land spaces.

Implementing a sort of grouping therefore should proceed when conditions are ripe: for example, when there is high traffic or when a good number of berths are required. An intermediate stage of providing a multipurpose terminal serving two (or even three) types of movement may be interposed prior to the final stage of specialized port terminal. This terminal will require cargo handling equipment capable of handling more than one type of cargo. Such equipment may be more expensive, so the servicing of vessels and of cargoes may not attain the efficiency of specialized terminals, but there is more than acceptable utilization of equipment and in general of the entire installation of a multipurpose terminal. A multipurpose terminal should retain some flexibility so that in the future it may be converted into a specialized terminal when conditions permit.

1.3.3 Cargo Volume Forecasts

1.3.3.1 Scope. Cargo volume forecasts for a port provide estimates of:

- The types and quantities of the various goods to be moved through the port

- Packing by type of cargo
- The number of vessel calls corresponding to the quantities above

If a national ports policy has been drawn up, the magnitudes above will already be known; otherwise, forecasts are made individually for the specific port under consideration. There is potentially great uncertainty in forecasts, and therefore the planning should accommodate flexibility to enable adaptation to meet future traffic. The parameters considered in cargo volume forecasts include:

- Population and national product
- Regional development programs
- The transport network and its projected future
- Coastal shipping
- Diversion of a portion of the traffic to other harbors

It is customary to hold interviews with government and local authorities, the shipping community, and interested parties to gain an understanding of the present and future traffic patterns. An independent review of global commercial and trade prospects that play a major role in traffic forecasts should also be conducted. Usually, the dependence of the results on the parameters is estimated on the basis of sensitivity control of the various calculations. Thus, in addition to the central forecast, we frequently include both an optimistic and a downside forecast, based on the corresponding growth scenario. An important port function involves monitoring the accuracy of the forecasts by comparing them with actual traffic.

1.3.3.2 Cargo Flow Combination. Usually, forecasts of significant cargo flows are conducted by type of cargo and by route. Bulk cargoes should be distinguished by type of cargo; container and Ro-Ro cargoes are distinguished by type of vessel performing the carriage. Ro-Ro cargoes may consist of (1) containers, (2) vehicles, (3) general cargo, and (4) products of intermediate unitization.

Container cargoes are calculated in 20-foot equivalent units (TEU), inclusive of empty containers. Forecasts should provide for some increase in the number of products accepting containerization. The net weight of the TEU ranges from 5 to 18 tons, depending on the stowage factor of the cargo within the container. For instance, if this factor were 2.8 m^3/ton, the net weight per TEU would amount to 10.4 tons. The results of cargo projections by cargo type and route should be reformulated by cargo category (e.g., dry bulk cargo). The total probability of a complex flow depends on the partial probabilities of the constituent flow forecasts and on the degree of their interdependence. It is advisable to analyze the flows of products with intense seasonal fluctuation separately and then to add them to the other flow forecasts.

1.3.3.3 National Transshipment. To estimate the transshipment flows either originated from or directed to a national port, and of the corresponding quays required, the alternative cargo flows between ports A and B should be examined. The latter implies that the required volume of cargo could be delivered to each port either directly, or the total volume of cargo directed to both locations will be delivered to one port only, from which it will be transshipped to the other location either by land, or by sea using smaller ships (coasters). For details, consult the proceedings of a United Nations conference (1978).

1.3.4 Port Productivity

The productivity of a port is the measure of its ability to move cargo through it within a unit of time under actual conditions. It is known that cargoes undergo various stages of handling while in port. For example, imported goods undergo the following handling procedures:

- Discharging while a vessel is berthed
- Transport to storage area and stowage
- Removal from storage and transport to area of transshipment or to means of overland transport
- Loading onto means of overland transport
- Departure from the port

Obviously, the total productivity of a port is determined by the lowest partial productivity of each link in the cargo handling chain. The conditions prevailing at the port at any given moment, such as weather conditions, human resources, and condition of machinery, affect the productivity of the partial procedures considerably. Consequently, a substantial time range representative of prevailing conditions has to be assumed for the evaluation.

The cargo handling practices pursued in each port have a decisive bearing on productivity, and any attempt at their improvement should also factor-in a period of adjustment of these practices to the new machinery and handling methods. Generally, a reference to any measure of productivity should be correlated with its corresponding time period. If this involves an extensive time period, on the order of several months, productivity may be reduced to half its value achieved in a short period of time (e.g., 1 hour). This may apply to all the particular procedures and handling of cargo flows within the port. For instance, over a short period of time, say a few hours, the container discharge efficiency at the dockside phase may amount to 750 TEU per day per berthing position, whereas over a period of several months the corresponding output for the same berth may drop to 400 TEU per day. Obviously, the long-term efficiency rate is important in the design of port installations.

Since the total efficiency of a commercial port or terminal is determined by the lowest productivity of the partial handling leg, every intervention for increased productivity should be directed initially at the least efficient procedure, with the purpose of balancing it out with handling legs of higher efficiency. The following are the most typical pairs of consecutive cargo handling legs in port cargo handling procedures:

- *Dock loading and unloading:* transport from quay to storage area, or vice versa
- *Transport from storage area to means of overland transport:* flow of means of transport to and from inland areas

An efficiency equalization between each of the constituent parts of a cargo handling pair should be achieved, measured on an hourly (or even daily) basis. Equalization should also be effected between the pairs themselves, although over a greater time period, that of a week, during which the cargoes remain in the storage areas, where the various checks and other procedures are conducted. This requirement for efficiency equalization ensures smooth functioning of the storage areas, thus averting the risk of congestion.

Efficiency increase may be achieved by intervention in three areas: (1) human resources, (2) technical matters, and (3) management and procedures. Intervention in the first area involves mainly an improvement in working conditions; in the second area, equipment renewal, better maintenance, and backup provisions; and for the third area, procedure simplification, imposition of a maximum time limit for cargo to remain at the storage areas, and so on.

It should be noted that an increase in productivity of a terminal by L % does not reduce vessel servicing time by the same percentage, but rather by $L/(1 + L)$, as is easily deduced by the definition of loading/unloading productivity at the quay (= cargo loaded or unloaded/vessel servicing time). The efficiency of a port terminal is affected by the quantity of cargo to be loaded to and unloaded from a vessel. It has been found that a large quantity of homogeneous products increases productivity, but usually this is not considered in the relevant calculations.

1.3.5 The Master Plan

1.3.5.1 Port Categories. From a construction point of view, ports may be classified into the following categories.

Artificial Ports. Artificial ports are those constructed along a shoreline by means of earth fill or excavation (Figure 1.4). In both cases these ports have to be protected from the adverse effects of waves and currents. In the former case (Figure 1.4*a*) the land part of a port is created by means of earth fill, and in the latter case (Figure 1.4*b*) the port basin is created artificially by means of excavation of land adjacent to the shoreline. The geometry of the excavated basin depends on port size and mode of operation. The excavated harbor is joined with the sea via an approach channel. The entrance to this channel is usually protected from waves and current by means of breakwaters and dikes. For more information on excavated harbors, readers are referred to Memos (1999).

Ports Constructed in a Natural Harbor. Examples are shown in Figure 1.5. Significant factors to be considered in opting for one of the foregoing types of port is availability of land, land fill material, soil quality, depth of water, environmental conditions, and others.

1.3.5.2 Port Location. Traditionally, ports are situated in a location central to the urban area they serve. The port is thus surrounded by urbanized area, and both further development of the port and access to it are rendered difficult. This situation restricts expansion of the port required to meet modern demands. In most cases, a feasibility survey for relocation of the port outside the city will have to be conducted. The prerequisites for such relocation are (1) secure maritime approaches, (2) ample availability of land area, and (3) satisfactory access by land.

For an initial new site evaluation, an extensive list of data to be collected is usually drawn up. Some of the items included are:

- Uses and ownership of the land
- Topography and access
- Existing utilities and structures at the site
- Wind and rainfall data
- Hydrographic information
- Geotechnical data, including potential sources of construction materials
- Environmental assessment of the area

During the initial site evaluation, some aspects of the project that may affect its development should be investigated. These may include necessary permissions and ownership implications, dredging and spoil disposal requirements, environmental constraints, and so on. In cases of inability to relocate, an alternative to be examined is that of establishing additional land facilities inland such as an inland depot.

1.3.5.3 Design Criteria. During the master planning stage of a project preliminary design criteria should be proposed covering aspects such as types of operations to be undertaken (e.g., containers, transit and transshipment flows, import/export; design vessel, operating equipment).

1.3.6 General Layout of Port Works

1.3.6.1 Guiding Principles. The arrangement of port works should be such as to ensure easy berthing of vessels, secure efficient cargo loading and unloading, and safe passenger embarkation and disembarkation operations. Specifically, easy access of vessels to a port should

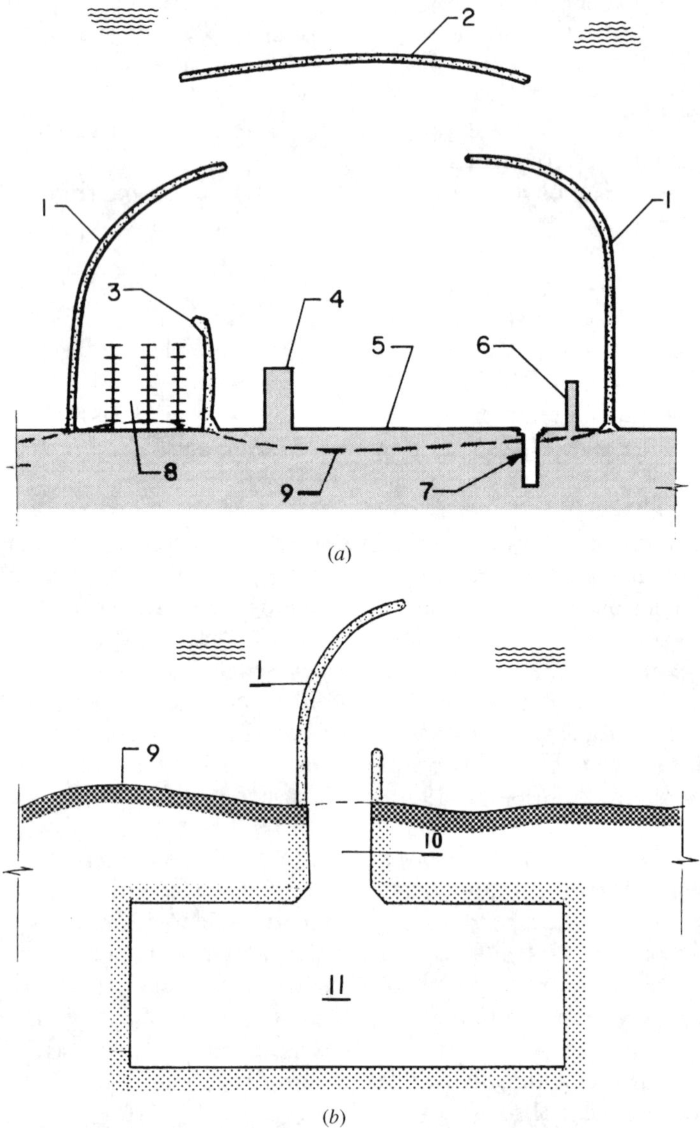

Figure 1.4 Conceptual arrangements of artificial ports: (*a*) created by earthfill; (*b*) created by excavation. 1–3, Breakwaters, 4, pier; 5, marginal wharf; 6, outfitting pier; 7, dry dock; 8, marina; 9, existing shoreline; 10, approach channel; 11, excavated basin.

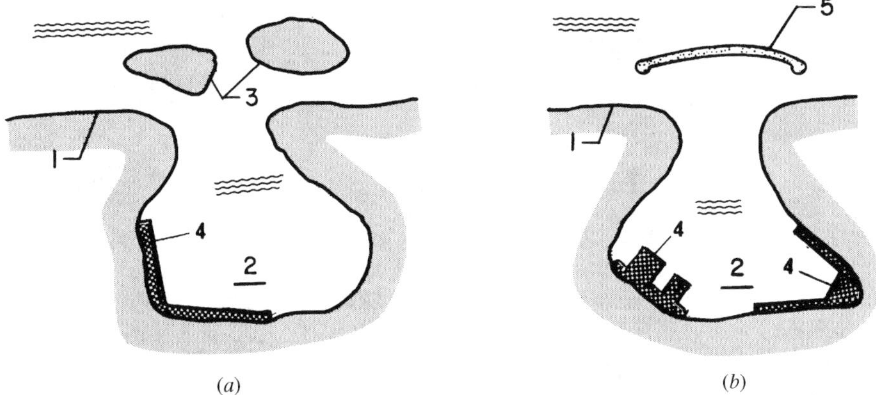

Figure 1.5 Ports constructed in natural harbors. (*a*) Entrance to the harbor is naturally protected by existing islands. (*b*) Entrance to the harbor is protected by the breakwater. 1, Coastal line; 2, harbor; 3, existing island; 4, port facilities; 5, breakwater.

be ensured through an appropriate navigation channel, a suitably designed port entrance, an adequate maneuvering area, and avoidance of undesirable erosion or deposition of material in and around the harbor area.

Factors to be considered in drafting a well-designed layout of port works include winds, waves, and currents and also the transportation of deposits in the study area. The existence of river or torrent mouths in the vicinity of the works has to be considered seriously in choosing the location and arrangement of the harbor.

The disturbance of harbor basins is a significant parameter, and low agitation should be achieved through a suitable arrangement of harbor structures. Specifically, the appearance of reflection and resonance phenomena within the harbor should be avoided through the use of absorbing beaches and suitable geometry of the structures that delineate it. The problem of excess wave agitation should be explored in either a physical or a mathematical model in order to arrive at an optimum layout of port works. Such models may also be used to optimize the constituent elements of the port, such as the port entrance.

Several of the subjects above may be tackled successfully by providing for an outer harbor that functions as a relief zone for the incoming waves, thus producing easier port-entry conditions. Next comes a closer examination of the most important elements that have a direct impact on the general layout of the principal port structures. For issues related to the navigation channels that serve ports, readers are referred to Chapter 10.

1.3.6.2 Port Entrance. The port entrance demands careful consideration to ensure quick and safe entry of vessels in the harbor. The orientation and width of the entrance should reconcile two opposing criteria. For reasons of comfortable navigation, the harbor entrance should communicate directly with the open sea and should be as wide as possible. On the other hand, the narrower and more protected the entrance, the smaller the degree of wave energy and deposits that penetrate the harbor basin, resulting in more favorable conditions for attaining tranquility of the in-harbor sea surface.

It is recommended that orientation of the entrance be such that vessels entering the harbor have the prevailing wind to the fore. Transverse winds and waves create difficult conditions for steering a vessel through the critical phase of entering the harbor basin, and a layout of port works that would permit frequent occurrences of such situations should be avoided.

Naturally, in most cases, the designer is obliged to compromise, as mentioned above. Obviously, the designer should avoid placing the entrance in the zone of wave breaking because of the difficulties to vessel maneuvering that may arise. Frequently, the entrance is formed by a suitable alignment of the protection works, whose structure heads are suitably marked with navigation lights. In the event that it is not possible to avoid transversal winds and waves, it is recommended that calm conditions at the harbor entrance be created by means of extending the windward breakwater to a satisfactory length beyond the entrance, at least to the length of one design vessel. In such cases it is advisable that the superstructure of the outward port structure be raised so that the wind pressures on the sides of the incoming vessel are reduced.

To attain the calmest possible conditions at the harbor entrance area, it is recommended that the external works in its vicinity be formed with sloping mounds so that wave energy in the entrance area can be absorbed. Breakwaters with a vertical front near the entrance may cause difficult navigation conditions there, because of the reflected and semistationary wavetrains created in that region. Moreover, in designing the layout of the harbor arms that bound the entrance, care should be taken that any sedimentation of deposits in the area be reduced. For significant projects, study of the entrance usually culminates in a physical model in which optimization of the arrangement is effected by conjoining all the relative requirements.

The width of the harbor entrance is defined in terms of the smallest length vertical to the entrance axis for which the minimum required draft applies. The depth at the entrance is generally determined by the maximum draft of the design vessel to be served. This figure should be taken beneath the lowest low water so that the harbor will always be accessible. In areas with a large tidal range in which the sea level can fluctuate by several meters, the question arises as to whether it is necessary to ensure accessibility to the port at all times. To meet such a requirement would signify an increase in the dredge depth equal to the range in tidal level. Alternatively, it could be accepted that the entrance be equipped with gates and that the port not be accessible during certain low-tide periods. Because such periods are foreseeable, as relying mainly on precise astronomical predictions, and because they are of relatively small duration, this solution is not to be rejected offhand, particularly if the harbor is accessible by means of long access channels. Vessels wait in the open sea up to the time when the channel is navigable for a specific vessel. Obviously, the internal harbor works of a tidal harbor will be compatible as regards drafts, with the planned navigation channel drafts suitably increased by a factor to compensate for the tidal increase during the open phase of the harbor. Thus, the vessels may always be safely afloat as long as they are in the harbor. Such a solution for periodic operation of the port entrance and channel has shortcomings, of course, because of vessel delays and other harbor malfunctions. Consequently, a careful cost–benefit analysis should be conducted prior to deciding the extent to which the port will be of free or of limited navigability. Such problems do not arise in ports with relatively small tidal fluctuations.

A safety factor of about 15% of the design vessel draft is sufficient for purposes of defining the minimum entrance depth. Alternatively, a margin of about 1.5 to 2.0 m over the draft of a loaded vessel gives a safe water depth at the port entrance. The width of a free entrance usually ranges between 100 and 250 m, depending on the size of the port. It is recommended that width be at least equal to the length of the design vessel the port is to serve. Thus, for small harbors it is possible to specify

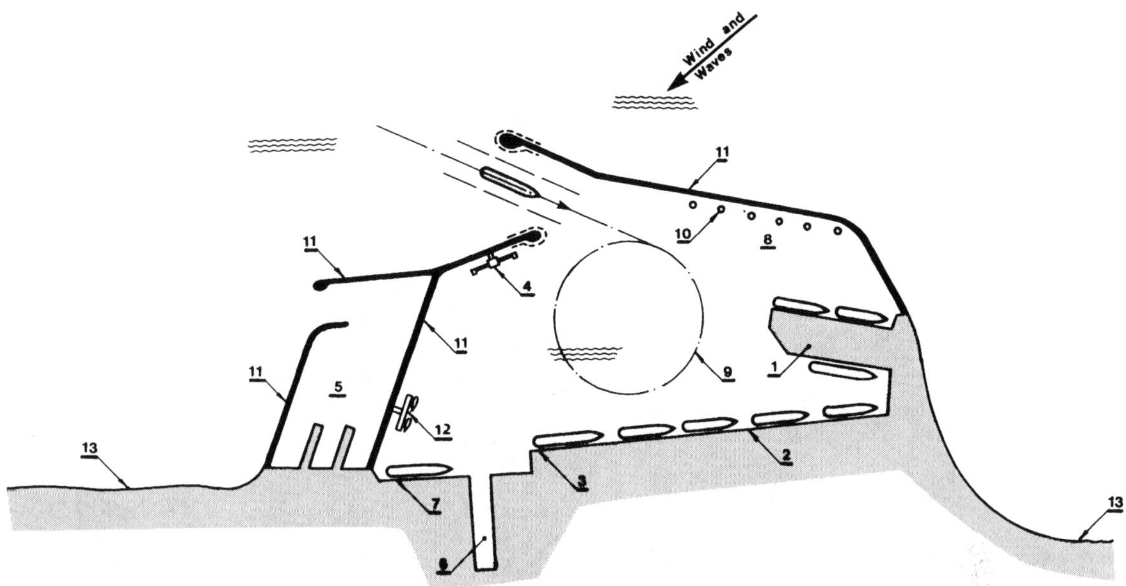

Figure 1.6 Layout of a large multipurpose artificial port. 1, General cargo terminal; 2, container terminal; 3, passenger terminal; 4, oil berth; 5, fishing port; 6, dry dock; 7, ship repair area; 8, anchorage area; 9, maneuvering circle; 10, mooring dolphins; 11, breakwater; 12, tugboat berth; 13, coastal line.

entrance width to be as low as, say, 50 m. The corresponding width of a closed port is significantly smaller than the sizes above. For more information, readers are referred to Tsinker (1997) and Chapter 9.

1.3.6.3 Maneuvering Area. When a vessel enters the harbor basin, its speed needs to be reduced to proceed with anchoring and berthing maneuvers. In practical terms, these maneuvers may be conducted at a normal speed of 8 to 11 knots over a length of 2 to $3L$, L being the vessel length, although larger distances may be required for larger vessels with modern hydrodynamic shapes. A significant consideration in determining the required length for minimizing speed is the vessel's fittings in maneuvering equipment, as well as the type of propeller; if the latter is of variable pitch, the distance can be reduced to $1.5L$. The maneuvering area is located either in the outer harbor, situated between the port entrance and the main port, or in the main harbor basin closest to the entrance.

Apart from reducing speed during an initial stage of straight movement, the vessel conducts maneuvers for positioning itself appropriately for the berthing position, which has been determined beforehand. This expanse of sea, called the *maneuvering area* or *circle,* should have dimensions calculated on the basis of the harbor's design vessel. If the port is sufficiently large, more than one maneuvering area may be designed and located at intervals of about 1 km. Figure 1.6 depicts the layout of a large artificial port with a maneuvering circle.

The diameter of the maneuvering circle required is affected directly by the type of rudders and propellers with which a vessel is equipped, whether or not tugboats will be employed, or whether anchors or wrapping dolphins will be used. For unfavorable maneuvering conditions, no tugs, and vessels with only one rudder, a $4L$ diameter is required, whereas in favorable conditions with modern navigation systems, a $3L$ diameter may suffice. Instead of a circle, maneuvering requirements may be satisfied by an ellipse with $3L$ and $2L$ axes, the main axis being lengthwise of the vessel's course. If maneuvers are conducted with the aid of tugboats, the minimum diameter of the maneuvering circle may be reduced to $2L$. A corresponding decrease is also achieved if the vessel is fitted with a second rudder or a lateral propeller, usually a bow thrust.

During towage, a vessel's engines usually are stopped or are in excellent synchronization with the tugboats. Furthermore, if a vessel has the ability to use bow and stern anchors or wrapping doplhins, the diameter of the maneuvering circle may reach the minimum dimension of $1.2L$.

In the maneuvering area, the sea surface is generally calmer than that at the entrance, and it is advisable that the lateral currents in this area be weaker than approximately 0.15 m/s. Furthermore, the reduction in available draft due to squat is insignificant in the maneuvering circle. Consequently, the required draft in the maneuvering area may be somewhat smaller than that at the entrance. In most cases, a safety margin of about 1.5 m below the maximum draft of the design vessel is sufficient.

To avoid accidents, the maneuvering area should be surrounded by a safety zone from fixed structures or vessel moorings. It is accepted that the width of this zone is a minimum of $1.5B$, where B is the design vessel's beam, and in any case it should be above 30 m. More information is given in Chapter 9.

1.3.6.4 Vessel Anchorage and Mooring. Perhaps the most significant role of a harbor is to provide shelter to vessels and to protect them from waves, currents, and strong winds. Once ships enter port, they generally use one or more anchors for their maneuvers, and while they are preparing for their berthing, mooring lines are also used, tied to the dock bollards. It may be necessary to immobilize vessels before entry into port, either while waiting for a free berth or for the tidal water to rise above the critical level at the entrance channel. This is achieved either by using the ship's anchors or by using suitable mooring buoys or dolphins located in the waiting area. Detailed information on anchors and anchorage area is provided in Chapters 7 and 8.

1.3.6.5 Wave Agitation in the Port Basin. It was mentioned previously that the basic function of a port is provision of a protected anchorage for vessels and the facilitation of quick and safe loading and unloading operations and embarkation and disembarkation of passengers. Therefore, the absence of disturbing waves in the basin that would impede the smooth functioning of the port is mandatory. The study of disturbances in a harbor basin should take as input the prevailing wave pattern and provide as output the percentage of time during which the port, or individual sections of it, cannot be operational. As stated earlier, the main factor causing an interruption in the operation of a port, and indeed one that demands careful examination, is that of wind-generated waves. Apart from penetration through the entrance, wave transmission and overtopping at breakwaters should be considered in determining surface agitation in a basin.

It follows that planning the layout of port structures is of crucial importance in attaining the necessary tranquility of the sea surface in a harbor basin. That is why particular attention must be paid to this problem in the course of studying the layout of port works. A satisfactory answer may be obtained by laboratory testing of various designs in a physical model. In these tests, wave disturbance is recorded at suitably selected locations in the harbor basin, as well as resulting movements of berthed vessels. The acceptable limits of these movements are determined depending on the loading and unloading method and the type of cargo handling equipment being used.

Apart from physical models, a good deal of information can be obtained from mathematical models, which can be developed to various degrees of accuracy. In this case, the wave heights in sections of the harbor basin are determined under various environmental conditions and degrees of absorption of the solid boundaries, although it is exceedingly difficult to simulate vessel movements. Wavelengths of the incident wave field have a particularly significant effect on vessel behavior; certain wavelengths produce dangerous conditions, as noted below when we discuss disturbance due to long oscillations. Any examination of port basin tranquility does, of course, include an assessment of the cost of the port works required to obtain each degree of basin calmness.

Long Oscillations. Apart from wind-generated waves, a range of other natural factors can disturb a harbor basin, although to a lesser extent. Many of these have to do with extreme events, such as storms and seismically created waves. In such cases, many harbors do not offer satisfactory shelter to vessels, which prefer to sail out to the open sea to avoid sustaining or causing damage in port.

Among these factors, those most significant as to continuous effects on harbor basins and therefore on ships' operations can generically be termed *long oscillations* (*seiches*). In effect, these refer to trapped oscillations with periods in excess of 30 s caused by changes in atmospheric pressure, long waves caused in the open sea by barometric lows, surf beats, edge waves, and so on. A serious problem arises when the harbor basin's geometry favors the development of resonance at the frequencies of the free oscillations prevailing in the region. In such cases, the flow velocity at the nodes of the oscillation of the free surface may reach 0.5 m/s even though the vertical surface excursions may generally be small. Long waves with periods usually in the region of 1 to 3 min place stresses on docked vessels, particularly when this involves larger ships with taut mooring lines. The phase velocity of these long waves in relatively shallow harbor waters is given approximately by $(gd)^{1/2}$, d being the uniform depth of water. Consequently, for a harbor basin with a rectangular plan of dimensions $L \times W$ with an entrance on the W (width) side, the resonance period of standing waves, T_L, along the two directions will be

$$T_L = \frac{4L}{n(gd)^{1/2}} \qquad n = 1, 3, 5, \ldots \qquad (1.1)$$

with a node of the standing wave at the entrance and an antinode at the opposite end of the harbor basin, and

$$T_w = \frac{2W}{n(gd)^{1/2}} \qquad n = 1, 2, 3, \ldots \qquad (1.2)$$

with antinodes at both opposing docks.

A basic means of avoiding resonance in a new harbor is the design of harbor basins with such geometry that the frequencies above are far from the usual frequencies of long waves in the region. The latter may be traced through the use of recording devices of surface elevation not sensitive to high-frequency waves. In cases where the harbor evidences complex geometry, the typical resonance modes are determined through mathematical models, or even through physical models in some cases, in a way similar to examination of the disturbance due to wind waves. As known, low-frequency waves may penetrate harbor basins without undergoing significant reduction of their amplitude. That is why any attempt toward a better layout of the protection works and of the entrances will be fruitless with regards to the elimination of long waves.

Recommendations for Improving a Port Basin's Tranquility. It is obvious that a basic element in designing a port is to achieve the lowest possible disturbance in the harbor basin, particularly close to berths. For this reason, it is recommended that the following factors be examined:

1. Provision for an adequate extent of the outer harbor area and of all the harbor basins, for dispersion of wave energy penetrating the harbor
2. Provision for spending beaches in suitable locations of the harbor, especially those attacked directly by waves entering the basin.
3. Provision for absorbent wharves with suitable design for dissipating disturbing wave agitation. It is recommended that this type of work be checked through physical modeling because the phenomena of conversion of wave energy, expelling of air, upward loading of the crown slab of the quay, and so on, are sufficiently complex and do not easily lend themselves to analysis through mathematical modeling.

In any case, the usefulness of absorbing quay walls is debatable, chiefly because of the wave reflection caused by berthed vessels at their sea side, a fact that reduces the efficiency of these structures considerably.

1.3.6.6 General Layout of Protection Works

Layout of Main Structures. Works whose function is to ensure the calmest possible conditions within harbor basins and along quays, particularly from wind-generated waves, are termed *harbor protection works*. These may include the following:

1. Breakwaters, usually constructed either connected to the shore or detached. Shore-connected breakwaters are classified as *windward* or *primary* and *leeward* or *secondary*. The former protect the harbor from the main wave direction, and the latter protect from waves of secondary directions. Frequently, leeward breakwaters are partially protected by windward breakwaters.
2. Jetties, usually arranged in pairs to form entrances to harbors located inward from the shoreline or in rivers. Paired jetties may also increase the flow speed and thus prevent sedimentation.

Figures 1.4 through 1.6 depict certain common arrangements of outer port works, depending on the type of harbor. The free end of protection works is called the *structure head,* and the remainder is the *structure trunk*. The effect of harbor works to be constructed on the transport regime of sediments in the region is particularly important. Quite often, port works are located in the surf zone, where the largest percentage of sediment transport takes place. Consequently, the effect of these works on coastal erosion or deposition may be quite significant. The phenomena usually caused by harbor protection works as regards sedimentation is a concentration of deposits upstream of the windward breakwater, erosion of the shore downstream of the leeward breakwater, sedimentation in the vicinity of the harbor entrance and the approach channel, and others (Figure 1.7).

The solution to such types of problems is not an easy matter, and in many cases recourse to the method of sand bypassing is considered to minimize the dredging required for mainte-

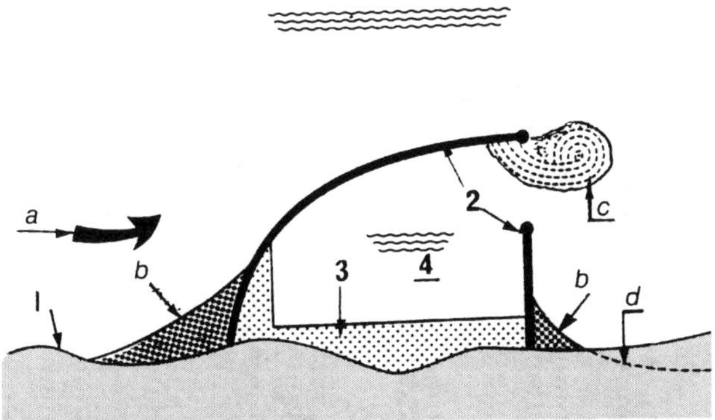

Figure 1.7 Effects of harbor works on coastal sedimentation. a, Longshore littoral transport; b, accretion; c, deposition; d, erosion; 1, natural shoreline; 2, breakwater; 3, landfill for port construction; 4, artificial harbor.

nance of drafts. The general idea in designing the layout of protection works should be to favor the transfer of sediment to deeper waters, where they are less harmful. Application of this general rule is not always easy, of course; that is why port designers usually resort to laboratory tests of the general arrangement of a harbor's defense works.

The protection structures are in principle laid out such as to provide the space required for a calm harbor basin, maneuvering areas, and necessary safety margins. Following that, an examination is conducted to ascertain the degree to which a large portion of the outer works is located in the wave-breaking zone. Selected values of wave heights are examined and the required modifications to the layout of the works are made so that the works are placed outside the breaking zone of the crucial design waves. This is done to reduce wave loads on the relevant structures and consequently, their cost. An important step follows: that of forming the harbor entrance in accordance with the guidelines of Section 1.3.6.2. Another point that relates to the shape of the breakwaters refers to the avoidance of angles to the open sea smaller than 180°, to evade a concentration of wave energy, with adverse effects on the structure's integrity.

Finally, the possibility of water renewal should be investigated, to reduce pollution of harbor basins to the minimum possible. It is not easy to suggest arrangements that can attain this target. As regards intervention in the harbor's protection works, the matter is usually handled by providing openings across the body of the structure, to facilitate water circulation. However, for these openings to be effective, they should be of sufficient width, which of course results in allowing significant disturbance into the harbor basin. Also, undesirable sediments may enter the harbor and be deposited if the openings extend down to the seabed. Therefore, in most cases the openings are not extended at depths beyond the surface layer in which the wind-generated water circulation generally takes place, to prevent the transfer of heavy sediments that occurrs at the lower part of the water column.

1.3.6.7 General Layout of Inner Port Works

Geometric Elements. The arrangement of berths and docking installations follows the principles noted in Section 1.3.6.5. Layouts that favor enhancement of long oscillations should be avoided, and it is also recommended that spending beaches be placed in suitable locations in the harbor basin. The geotechnical properties of the seabed in the project area play a significant role in deciding the general layout of the inner works. If a rocky seafloor is present, it is usually advisable to place the line of wharfs close to their final depth, to avoid expensive excavations of the rocky bed. If the latter is soft, the location of the wharfs is determined by, among other factors, a detailed technical and economic comparison of reclaiming versus dredging.

It has been pointed out that maneuvering surfaces should have a security distance of between 30 and 50 m from any vessels docked at the planned berths. Figures 1.4*a* and 1.6 give the main elements of a harbor's inner works. As a general rule, the plan must ensure that the shape of the docks provides for better use of the harbor basin and easier navigational conditions for vessel maneuvers, and that the functioning of dock equipment and machinery is not hampered. Furthermore, to keep pollution of harbor basins to a minimum, placing docks and basins in recessed positions of a harbor should be avoided, because the renewal of water there is weak. If narrow piers are planned (e.g., only for the mooring of small vessels), it is advisable to examine the possibility of designing them on piles with openings for facilitation of water circulation. The development of a port over time is generally associated with a required strip of land parallel to the berths. Previously, this strip was planned to be about 50 m wide; later, adapting to technological development in cargo handling, this was increased to 100 and 200 m. A result of this change was a tendency to shift from narrow piers that created a zigzag layout of docks to straight quay lines parallel to the shore, which ensures large land areas.

The linear dock arrangement, however, takes up a far greater length of coast, which frequently is very expensive, or not feasible to acquire for other reasons. In such cases, wide piers are used to increase quay length. Their width can be 300 m or more, and they may be placed at a small angle to the shoreline if this would have the benefit of protecting them from waves and provide better operational conditions.

Quay length is determined by the particular method of docking and by the number of berths. Alongside berthing for a vessel of length L requires a quay length of $b = L + 30$ to 40 m or $b = 1.2L$. For Ro-Ro stern (or bow)-to-shore berthing, the required quay length b is determined by the vessel's beam B and is roughly $b \approx 1.2$ to $1.5B$. The minimum depth h of the sea at the quay is determined by the design vessel's maximum draft d_{max}. A safety factor for this value (i.e., pilot's foot) in the region of 1 m should be added to cover for any heaving motion due to wave disturbance. Thus $h \approx d_{max} + 1$ m. The dimensions usually recommended for seaport docks are illustrated in Figure 1.8. Other inner installations apart from berthing quays, such as dry docks, slipways, and maintenance quays, should be situated independent of the customary loading and unloading quays and as much as possible in protected areas of the harbor.

Connections with Inland Areas. It has already been mentioned that the nature of a modern cargo port resembles more a cargo handling hub within a combined transport system than a sea transport terminal point. Consequently, a basic element in the smooth operation and development of a terminal are the port's inland connections. These connections, through which nonsea transport of goods to and from the port is effected, may be road or rail accesses, artificial or natural inland navigable routes, airlines, or oil product pipelines. Road, rail, and river connections (to which we refer later) can also connect a port with specialized cargo concentration terminals located in suitable inland depots. These stations serve to smooth out the peaks in demand and supply of goods to a port

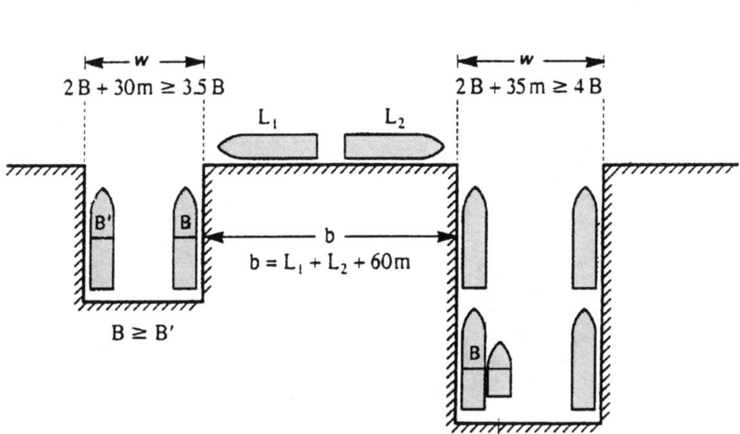

Figure 1.8 Main dimensions of sea docks.

that has limited storage areas. Figure 1.9 depicts some general arrangements of such connections.

The provision of inland storage areas forming part of a port is a modern tendency pronounced in container transport, which creates the need for larger backup areas and also a need for boxes to stay in port for a shorter time. The transport of goods between port and inland depots is thus carried out quickly and efficiently, in contrast with the traditional servicing of all destination points directly from a port without intermediate transshipment. In addition to being effected by road, the connection between port and inland depot may be by rail, particularly when the distance is great. In the latter case, the loading of trains, when this involves imports, may be effected at a small distance from the port, where the goods are forwarded through a system of wheeled trailers fed from the port, as shown in Figure 1.10. In each case, the traditional arrangement in general cargo terminals in which rail (or road) vehicles approach the docks for immediate loading and unloading of cargo through the use of dock cranes is being abandoned. The main reason for this development is that loading/unloading vehicles obstruct dock operations, in addition to the frequent inability to coordinate ship–train operations, resulting in vessel delay. Two alternative handling options are available in this respect: (1) the full cargo can be forwarded inland via port sheds, or (2) "direct" loading/unloading to and from rail or road vehicles can be retained but conducted at some distance from the docks. The second alternative demands an additional fleet of tractors and platforms to link docks with transshipment areas to means of overland transportation. This alternative solution is depicted in Figure 1.11 together with the traditional arrangement, which, as mentioned, is gradually being abandoned by many ports.

The tendency to shift land transportation away from docks is even more prevalent in container or Ro-Ro port terminals. Inland connections are allowed only to reach a delivery and receiving area, which in container terminals is generally located near the container freight station (for details, see Section 1.4.3). In most cases, road access to ports is appealing, particularly for small and moderate distances.

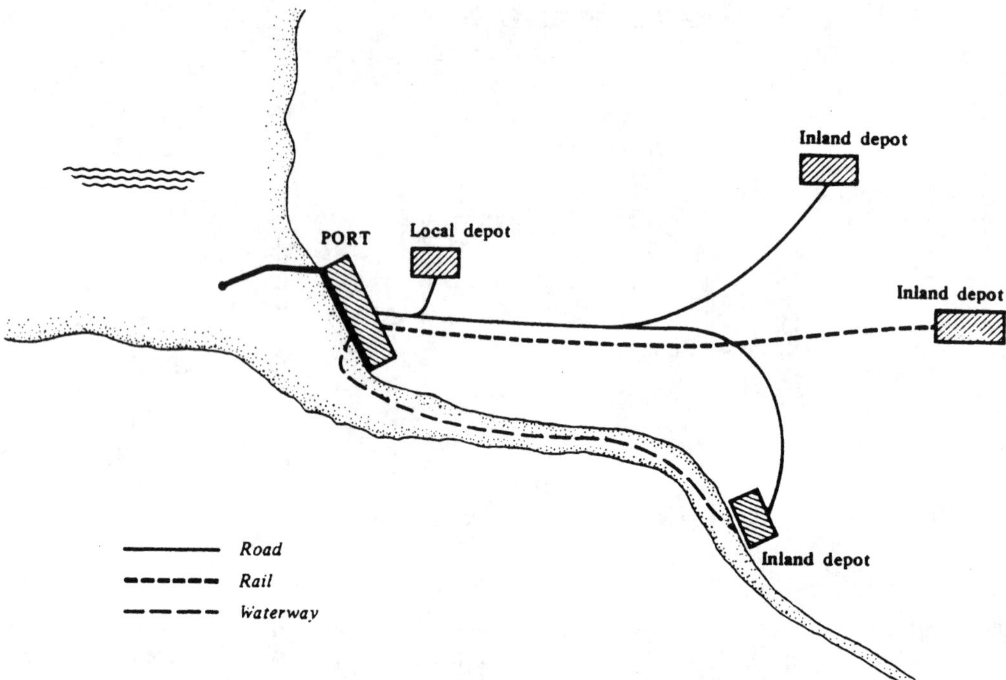

Figure 1.9 Connection of a port with an inland cargo collection terminal. (From United Nations, 1978.)

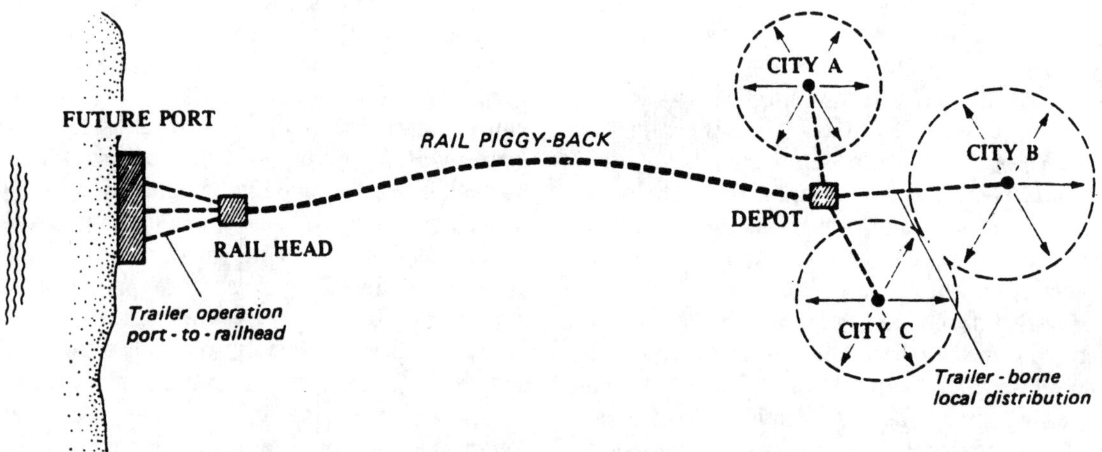

Figure 1.10 Combination of road–rail connections of the port with the inland depot. (From United Nations, 1978.)

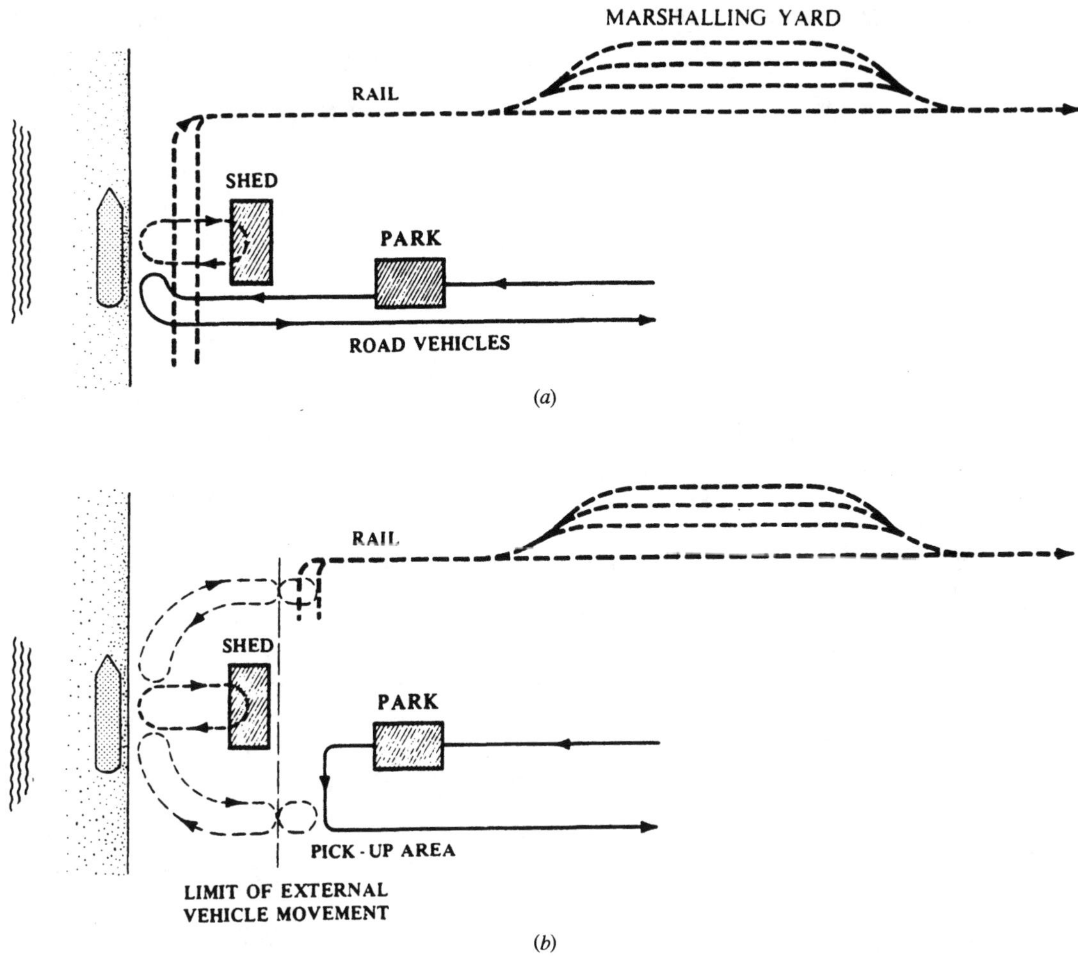

Figure 1.11 Restricting the approach of vehicles to the docks: (*a*) traditional approach; (*b*) alternative approach. (From United Nations, 1978.)

The variety and types of road vehicles render them versatile, and in conjunction with a dense road network in many regions, make them suitable for "door-to-door" service. Rail connection at ports offers security, speed, and economical transport of bulky goods over large distances.

Many ports throughout the world are constructed at the mouths of navigable rivers or canals, to connect them with other areas by means of inland navigable routes. Connections by inland navigation offer economy and are particularly suitable for the transport of bulk cargoes and for supporting combined transports between river ports and seaports that serve barge-carrying vessels.

Additional Points to Be Considered. Several issues of general application to the layout of land installations of a port are listed below.

1. The conventional berthing positions for general cargo require a smaller draft at the quay (usually 7.70 to 10 m) than those required for containers or bulk cargo.
2. Much larger land areas are required in terminals where containers are to be handled.
3. Care should be taken in drawing up the land use so that smells from bulk cargoes carried by prevailing winds do not damage the environment.
4. Security issues should be examined, particularly as regards flammable materials or explosives.
5. Product compatibility should be examined for cargoes adjacent to their respective handling areas. For instance, pairing coal with grains is incompatible, as is pairing grains with fertilizers.
6. The overall traffic pattern in the land area at a port should be examined, to avoid potential congestion or a need for bridging.

1.4 PORT PLANNING AT THE TERMINAL LEVEL

1.4.1 Port Development

1.4.1.1 Phases of Port Development. The course of development of a port or port terminal usually undergoes phases, which also indicate its age. Evolution from a traditional break-bulk cargo port to a specialized unitized cargo port may be gradual. However, it is distinguishable into qualitative changes that take place in specific periods throughout the overall life of the port. These phases are as follows:

Phase 1: Traditional General Cargo Flow. A port with break-bulk or packaged bulk cargo terminals, such as for bagged grains or petroleum in barrels.

Phase 2: Break-Bulk Cargoes. When break-bulk cargo flow exceeds an economically acceptable limit, these cargoes are transported in bulk form and the port develops a special bulk-cargo terminal. At the same time, the break-bulk berths are increased, to accommodate the higher demand.

Phase 3: Unit Loads. Unit loads start being carried on conventional vessels in small quantities in units such as palettes, containers, or packaged lumber. At the same time, break-bulk cargo flows, particularly those of bulked break-bulk cargoes, start diminishing to levels that require separation of cargo terminals for various cargo categories.

Phase 4: Multipurpose Terminal. Unitized cargoes on specialized vessels start appearing in quantities that do not yet require development of a specialized terminal. Thus, a multipurpose terminal is created in which break-bulk cargo traffic is diminished, although unitized cargo is also handled. At the same time, the specialization of dry bulk cargo terminals continues.

Phase 5: Specialized Terminal. With an increase in unit loads beyond certain levels, specialized cargo terminals are created for handling containers, packaged lumber, and Ro-Ro. The multipurpose terminal of phase 4 is converted into a specialized terminal, with the addition of specialized cargo handling equipment. Break-bulk general cargo is reduced further.

It should be noted that in normal situations, the transition from phase 3 to phase 5 should progress through phase 4, so as to provide an opportunity to the port to increase unitized cargo traffic to volumes that will enable economically feasible development of a specialized terminal in phase 5. Moreover, in the event that a port has entered phase 3 of its development, care should be taken to avoid creating additional general cargo berths.

1.4.1.2 Review of Existing Port Installations. The examination of existing installations should precede any decision to expand old, or to construct new, port terminals. The purpose of such a study is to identify any functional difficulties that would detract significantly from the theoretical productivity of the marine and land sector of the port terminal. In many cases, improved organization of the component operations of the port terminal produces a significant increase in its productivity. In addition to an improvement in the terminal's organizational structure, there is the possibility of introducing structural changes and upgrades of port installations, which will usually necessitate a considerable expenditure. It should be noted that in many cases, technological developments and changes in packaging and cargo handling methods frequently render the upgrading of existing installations a difficult and complicated task. At the same time, the existence of spare capacity is always a desirable feature in a modern port able to accommodate peaks in cargo flows, albeit with reduced productivity. Thus in cases where the recommended installation upgrade marginally covers the expected demand, it is recommended that old installations be placed on standby to cover unforeseen requirements and that expansion of an existing, or construction of a new, port terminal be opted for.

1.4.1.3 General Cargo Terminal. The first phase in a design for expansion of an existing break-bulk cargo terminal or for the creation of a new one involves diligent collection and analysis of statistical data regarding the existing terminal's output. This analysis will also determine the "age" of the existing terminal—in other words, the degree to which the owners of this break-bulk cargo terminal are prepared to see it evolve into a multipurpose terminal or even into a specialized container or bulk-cargo terminal. This decision will be based on the percentages of the flows and the unit loading that conventionally packaged cargoes assume over time.

Analysis of these data will also reveal whether berth productivity falls short of theoretical values. In this case, and particularly if significant vessel waiting times are observed, the cause of the reduced output should be looked into carefully. Usually, a standard efficiency rating per berth with a high degree of break-bulk cargo traffic is 100,000 tons per year, whereas if unitized cargoes constitute 30 to 40% of the traffic, this productivity figure may rise to more than 150,000 tons per year.

1.4.1.4 Bulk Cargo Terminal. To decide on the expansion of a bulk cargo terminal, the data from the existing terminal have to be considered. Just as in the case of break-bulk terminals, the purpose of this examination is to determine whether the lower productivity of the terminal is due to malfunctioning or to increases in traffic volume. In ore-exporting terminals, the latter case may be due to improvements in mining technology or to discoveries of new deposits. The study should focus on such issues as coordination between the various phases of product movement, on lags, if such exist, during which no product is available for loading on the vessel, and on the method of cargo movement over land. The findings of this examination will lead to a decision either to improve the operational procedures and the equipment of the existing terminal, or to create an additional bulk cargo terminal.

1.4.2 General Cargo Terminal

Despite the fact that the general cargo terminal is becoming increasingly scarce, the main factors pertinent to its organization and operation are presented below, so they may also be used in the study of a multipurpose terminal.

1.4.2.1 *Vessel Waiting Time.* It is generally accepted that arrivals of general cargo vessels follow a Poisson distribution. According to this, the probability $P(n)$ for n vessels to arrive in port within a specified period—usually 1 day—is

$$P(n) = \frac{(N)^e e^{-N}}{n!} \tag{1.3}$$

where N is the average number of arrivals per day over a long time period. The observation above is equivalent to saying that the distribution of the time intervals t between successive arrivals is negative-exponential:

$$P(t) = e^{-t/T} \tag{1.4}$$

where T is the average of these intervals over a large time period. On the basis of existing data it is estimated that the time periods t for servicing of berthed vessels follow an Erlang distribution with $K = 2$. The Erlang distribution is expressed by the formula

$$P(t) = e^{-Kt/T} \sum_{n=0}^{K-1} \frac{(Kt/T)n}{n!} \tag{1.5}$$

where T is the average servicing time. Within reasonable accuracy, queue theory can provide values of vessel waiting time for various degrees of utilization of the system. In the case of the general cargo terminal, assumptions are made of random arrivals and distribution of servicing times according to an Erlang2 distribution. This in fact corresponds to an $M/E_2/a$ queue, where M denotes the Poisson distribution of arrivals and a is the number of berths.

1.4.2.2 *Berth Occupancy.* The occupancy rate of a group of berths expresses the percentage of time that berth positions are occupied by ships being serviced. The effect of berth occupancy on waiting time depends on the probability distributions of arrivals and of servicing times as well as on the number of berths available to the sector of the port being examined. With regard to a general cargo terminal, an $M/E_2/n$ queue is usually assumed, as stated above. The effect that the grouping of berthing places on vessel waiting times can be seen through the congestion factor, defined below, values of which are contained in Table 1.1. In general, a larger number of berths enables greater occupancy rates for the same waiting periods.

For the sake of demonstration, let us assume 10 general cargo berths and an average of two vessel calls per day headed for these berths. If the average servicing time is 3.5 days, the occupancy factor k_0 is

$$k_0 = \frac{2 \times 3.5}{10} = 0.70$$

in which case the congestion factor k_0', which in average terms expresses waiting time as a percentage of servicing time, amounts to 6% or 0.2 day. Now, if the total of these berths is divided into two independently operating groups, with one vessel call per day per group, the occupancy rate remains the same, while the congestion factor is tripled, to 19%. Table 1.1 provides an approximation of the waiting time for the queue above expressed as a percentage of the average servicing time as a function of the number of berths and of their occupancy.

The optimum berth use depends on the cost ratio between berths and vessels. The values given in Table 1.2 give occupancy factors generally recommended for a 1:4 cost ratio, depending on the number of berths of the general cargo terminal. It should be noted that the

Table 1.1 Congestion factor in queue $M/E_2/n$

Occupancy Factor	Number of Berths														
	1	2	3	4	5	6	7	8	9	10	11	12	13	14	15
0.10	0.08	0.01	0	0	0	0	0	0	0	0	0	0	0	0	0
0.15	0.13	0.02	0	0	0	0	0	0	0	0	0	0	0	0	0
0.20	0.19	0.03	0.01	0	0	0	0	0	0	0	0	0	0	0	0
0.25	0.25	0.05	0.02	0	0	0	0	0	0	0	0	0	0	0	0
0.30	0.32	0.08	0.03	0.01	0	0	0	0	0	0	0	0	0	0	0
0.35	0.40	0.11	0.04	0.02	0.01	0	0	0	0	0	0	0	0	0	0
0.40	0.50	0.15	0.06	0.03	0.02	0.01	0.01	0	0	0	0	0	0	0	0
0.45	0.60	0.20	0.08	0.05	0.03	0.02	0.01	0	0	0	0	0	0	0	0
0.50	0.75	0.26	0.12	0.07	0.04	0.03	0.02	0.01	0.01	0.01	0	0	0	0	0
0.55	0.91	0.33	0.16	0.10	0.06	0.04	0.03	0.02	0.02	0.01	0.01	0.01	0	0	0
0.60	1.13	0.43	0.23	0.14	0.09	0.06	0.05	0.03	0.03	0.02	0.02	0.01	0.01	0.01	0.01
0.65	1.38	0.55	0.30	0.19	0.12	0.09	0.07	0.05	0.04	0.03	0.03	0.02	0.02	0.02	0.02
0.70	1.75	0.73	0.42	0.27	0.19	0.14	0.11	0.09	0.07	0.06	0.05	0.04	0.03	0.03	0.03
0.75	2.22	0.96	0.59	0.39	0.28	0.21	0.17	0.14	0.12	0.10	0.08	0.07	0.06	0.05	0.05
0.80	3.00	1.34	0.82	0.57	0.42	0.33	0.27	0.22	0.18	0.16	0.13	0.11	0.10	0.09	0.08
0.85	4.50	2.00	1.34	0.90	0.70	0.54	0.46	0.39	0.34	0.30	0.26	0.23	0.20	0.10	0.16
0.90	6.75	3.14	2.01	1.45	1.12	0.91	0.76	0.65	0.56	0.50	0.45	0.40	0.36	0.33	0.30

Source: United Nations Conference on Trade and Development (UNCTAD), 1978.

Table 1.2 Recommended occupancy factors

Number of Berths	Occupancy Factor k_0 (%)	Congestion Factor k_0' (%)
1	40–50	50–75
2	50–60	26–43
3	53–65	14–30
4	56–65	11–19
5	60–70	9–19
6–10	62–75	2–21
>10	70–85	0–26

higher factor values are more fitted for $E_2/E_2/n$ queues, which are more applicable to container terminals.

1.4.2.3 Number of Berths. The key parameter in the design of a general cargo port terminal is that of the number of berths. This parameter depends mainly on the annual cargo throughput of the terminal and on the predetermined level of vessel servicing to be offered by the terminal. The latter depends on the corresponding waiting periods discussed previously. The number of berths n can be expressed as

$$n = \frac{Q}{24k_0 qprN} \tag{1.6}$$

where Q is the annual cargo flow estimate (tonnes), k_0 the berth occupancy factor, q the average tonnage handled by one gang per hour (calculated from statistical data of this or a similar port), p the fraction of time during which the berths are operational (e.g., if the total daily working hours per berth is 16 over 6 days per week, this factor would be $16 \times 6/24 \times 7 = 0.572$), r the average number of gangs concurrently loading or unloading an average-sized vessel (depends on cargo type and vessel size), and N the days of berth operation in a year (days when berths are in a position to receive vessels, e.g., not closed for maintenance).

The number of berths, n, may also be expressed approximately as a function of cargo throughput, Q, expressed in units of 100,000 tons per year as follows:

$$n \simeq \frac{Q}{k_0} \tag{1.7}$$

where k_0 is the occupancy factor. Having determined the number of berths in the terminal, berth length is then calculated on the basis of the length of the design vessel to be calling at the terminal. Berth length is generally taken to be 20% above the design vessel length. Wharf width should typically include free sea space of at least two design vessel widths. The productivity per running meter of a general cargo berth usually ranges from 600 to 1200 tons of cargo per year for average occupancy. Where container units are handled by conventional quay cranes or by vessel gear, this output may reach 1600 tons per year.

1.4.2.4 Storage Area. A small portion of the total throughput of a general cargo terminal is either loaded directly to or discharged directly from land transportation means without requiring storage at the terminal. The other cargo is stored for a period of time in sheds, open areas, or warehouses. The required cargo storage area A (thousands of m^2) can be expressed as a function of known parameters, by adopting the following simple relation:

$$A = \frac{1.7}{365}\left(\frac{QD}{dH}\right)\left(1 + \frac{p}{100}\right) \tag{1.8}$$

where Q is the annual tonnage to be stored (thousands of tons; this refers to the portion of total cargo flow that requires storage); D the average storage duration (days; it is assumed on the basis of existing statistical data); d the cargo density [tons/m^3; this may be calculated using the stowage factor (in m^3/ton), typical values of which are shown in Table 1.3]; H the average stowage height (m; depends on type of cargo, its packing, and stowage means; an average value is 2 to 3 m; the smaller the stowage height, the larger the storage areas, but simpler mechanical means are required for cargo handling; for this reason, comparisons should be made between various alternatives); and p the peak factor, multiplies the average area required to accommodate cargo flow peaks (usually, this increase is between 25 and 40%).

The factor 1.7 in eq. (1.8) covers the extra space required because of the splitting of consignments into smaller units and accommodates areas not used for stacking, such as corridors and offices. Assuming a rectangular shape of the storage area, the dimensions of the shed may be calculated to have a width of roughly half the length. In any case the width should be above 40 to 50 m.

In ports, cargo is stored in sheds, warehouses, or in the open. Sheds usually are steel frame constructions at ground level, situated

Table 1.3 Typical cargo densities

Cargo	Stowage Factor (m^3/ton)	Cargo Density (tons/m^3)
Bagged cement	1.0	1.00
Plaster, bagged	1.2	0.83
Sand, bagged	0.5	2.00
Animal feed, bagged	1.5	0.67
Bagged coffee	1.8	0.56
Citrus fruits	2.5	0.40
Cotton bales	2.7	0.37
Bagged flour	1.3	0.77
Grapes	3.9	0.26
Frozen fish (boxed)	2.1	0.48
Paper rolls	2.5	0.40

lengthwise and relatively near the wharves and used for cargo storage over a short period of time. Conversely, since they are not part of the fast-track cargo handling chain, warehouses are usually situated behind the sheds so as not to take up valuable space near the berths. Cargo that is to remain in port for a substantial period of time is stored there. Such situations arise when the port owners wish to engage in the warehousing of goods: for instance, goods requiring ripening or separation and repackaging for direct ex-warehouse sale. Contrary to sheds, warehouses may be multistoried buildings, although single-storied warehouses are more practical. A typical layout of a general cargo terminal for three berths is shown in Figure 1.12.

1.4.2.5 Sheds. The basic requirements for a port shed are as follows:

1. To be of sufficient width, which should extend at least to 40 to 50 m
2. To have as few columns as possible within the storage area
3. To have sufficient ventilation and lighting
4. To have a smooth and durable floor surface
5. To have an adequate number of large sliding doors, with easy handling
6. To save floor space by placing offices at a higher level
7. To be constructed so as to enable expansion or other envisaged modifications

The shed floor should be adequately sloped to enable drainage. Usually, such a slope is specified up to 1:40 for purposes of good functioning of handling equipment and stacking stability. The shaping of this slope may be combined with the construction of a loading platform lengthwise to the land side of the shed, to an approximate height of 1 m. A loading platform is needed to connect the shed with inland areas by road and by rail, if such a connection exists. Rail tracks are laid embedded so as not to protrude from the floor surface. If it is not possible to create a permanent platform as indicated above, mobile loading ramps may be employed. In this case, the shed floor may be shaped with a double slope, with a watershed along the lengthwise axis of the shed.

The width of the area between the shed and the berth (apron) is about 20 to 30 m. Traditionally, conventional portal cranes placed on rail tracks alongside the quay have been used in this zone, and railcars approached this zone to load and unload directly from the dock cranes. Experience has shown that it is difficult to load and unload railcars satisfactorily, with the result that cargo handling efficiency is reduced. Currently, the practice of approaching general cargo berths by rail has been abandoned, and cargo flow is effected through sheds and warehouses.

A further development in the dockside zone is the increasingly reduced presence of dock cranes on rails. Many such cranes, which in the past were characteristic of general cargo terminals, are now being replaced by versatile heavy mobile cranes supplemented, if possible, by a vessel's gear. Apart from loading and unloading heavy unitized cargo at the dockside, these cranes, with an approximate 20-ton lifting capacity, may assist operations in other areas of a terminal. In general, the number of cranes and their lifting capacity depend on the type and volume of cargo and its method of handling at the port. The overall width of the land zone required to sustain all cargo handling operations in a modern general cargo port terminal should extend 200 m from the quay line.

1.4.2.6 Cargo Handling. Following unloading by cranes of general cargo onto a dock, transporting and stacking it in sheds follows. A reverse course applies in the case of cargo export. Transfer to and from a shed may be effected in the following ways: (1) use of a

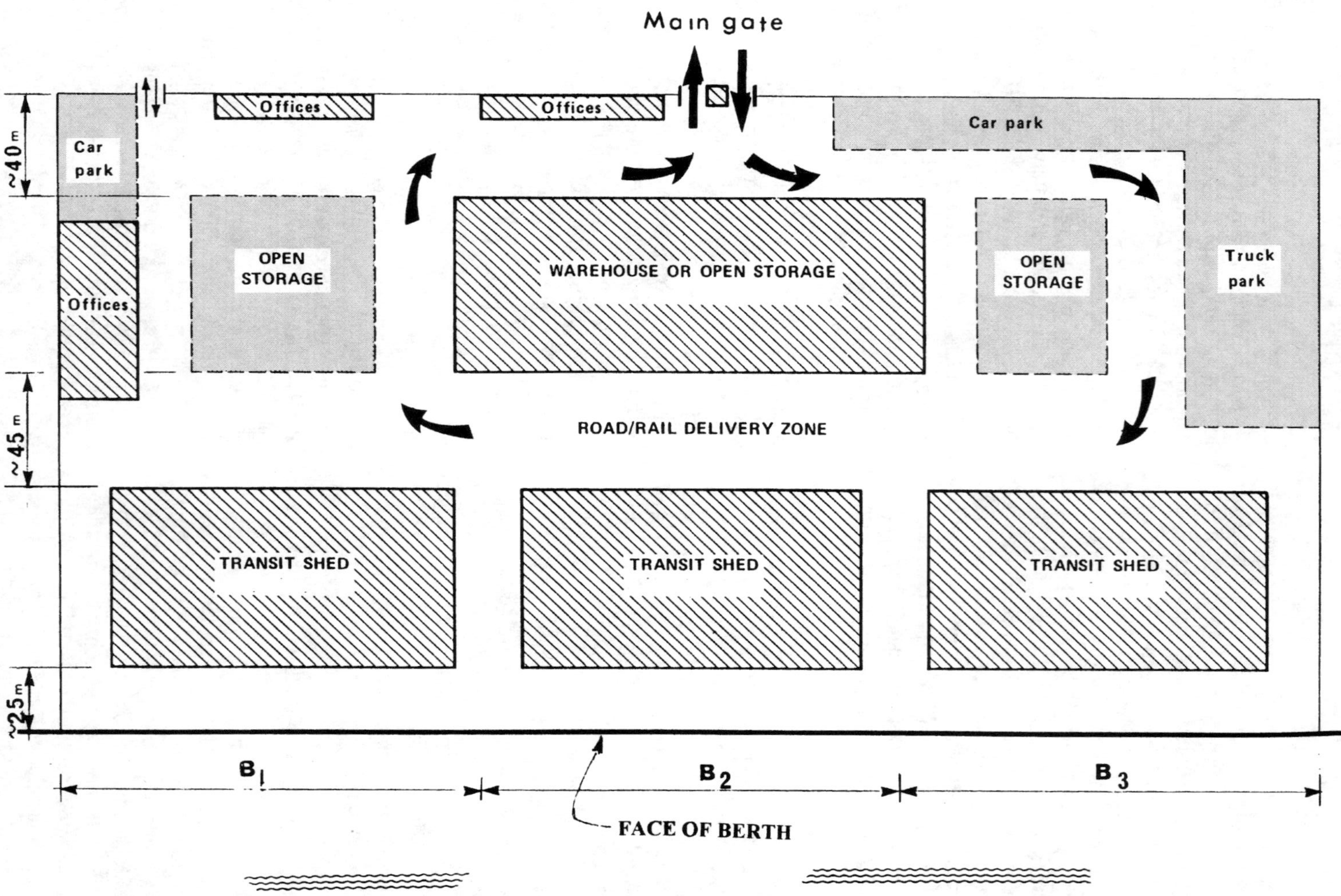

Figure 1.12 Typical layout of a general cargo terminal.

tractor–trailer combination and (2) use of heavy forklift trucks. Cargo unloaded by dockside cranes can be placed directly on trailers that are transported back and forth by tractors. Under normal working conditions, a tractor may service three or four trailers. If forklifts are used instead of tractors and trailers, the cranes discharge the cargo directly onto the dock floor for forklifts to pick up. Cargo stacking at a shed is effected by means of forklifts, while in open storage areas it is performed either by forklifts or by 10-ton mobile cranes. In the absence of statistical data, the cargo handling equipment required at break-bulk cargo terminals may be calculated by means of the following approximate norms:

- Number of loading and unloading gangs per vessel: $3\frac{1}{2}$ for oceangoing vessels; $1\frac{1}{2}$ for feeder vessels
- 3 forklifts per gang, or 2 tractors and 8 trailers per gang
- 0.8 forklift and 0.4 stacking crane per gang

Furthermore, for equipment an extra 20 to 25%, and for trailers an extra 5%, is required for repair and maintenance purposes.

1.4.3 Container Terminal

1.4.3.1 Cargo Unitization. One of the most significant developments in maritime transport was the establishment some 40 years ago of the container as a cargo packaging unit. Over the past 30 years the amount of goods shipped in containers increased at a rate of about 7% per year (i.e., more than double the growth in the world economy and 50% over the expansion in world trade). In the container terminal, increased throughput productivity is attained in addition to other advantages, such as canceling the need for extensive sheltered storage areas, security, and standardization in equipment dimensions and required spaces.

Containers are transported mainly in specialized vessels, classified into "generations" depending on their size. Typical dimensions of modern container ships are given in Chapters 2 and 10. Most container ships are capable of crossing the Panama Canal (Panamax-type vesssels), allowing 13-box-wide storage across the deck. During the 1990s post-Panamax vessels appeared, having capacities exceeding 8000 TEU with drafts of 14.5 m. These vessels have beams of 43 m, allowing 17-box-wide deck storage. It has been announced that in 2004 two containerships of 9800 TEU will enter trans-Pacific service. Engineers consider that there is no technical constraint to building a ship of 15,000 or even 18,000 TEU, the latter size being imposed by the shallowest point in the Malacca Strait in Southeast Asia, allowing a draft of 21 m. Such megaships might have a length of 400 m and a beam of 60 m, giving 24-box-wide deck storage. Table 1.4 shows the principal dimensions of some of the new generation vessels together with the projected 12,500 TEU capacity vessel. This latter Ultra Large Container Ship (ULCS) was found to be of an optimal size by a study carried out by Lloyds Register of Shipping and Ocean Shipping Consultants. These gradually increasing dimensions of new vessels have a significant impact on the geometric requirements of ports' layout. Thus berths of up to 400-m long with water depths down to 16 m become increasingly the norm for modern container terminals. Also, gantry cranes should be able to cope with increased beams and the capacity of handling equipment should be compatible with larger consignments. Containers can be stacked in the hold or four high on the ship's deck. Difficulties arise with large stacking heights as regards container fastening and other aspects.

The container ships mentioned above are oceangoing vessels and in many cases avoid making frequent calls at nearby ports. Thus, smaller, intensively utilized feeder vessels are employed in short distances for the collection

Table 1.4 New generation container ships

Vessel Name	Launch Date	Dead-Weight Tonnage	TEU	LOA (m)	Beam (m)	Draft (m)
Hyundai Admiral	1992	59,000	4,411	275	37.1	13.6
NYK Altair	1994	63,163	4,473	300	37.1	13.0
APL China	1995	66,520	4,832	276	40.0	14.0
Ever Ultra	1996	63,388	5,364	285	40.0	12.7
Hajin London	1996	67,298	5,302	279	40.4	14.0
Regina Maersk	1996	82,135	6,418	318	42.8	12.2
NYK Antares	1997	81,819	5,798	300	40.0	14.0
Sovereign Maersk	1997	104,696	8,736	347	42.9	14.5
ULCS		120,000	12,500	400	60	14.8

or distribution of cargoes from a region (e.g., the eastern Mediterranean). These feeder vessels are of 30 to 350 TEU capacity and usually have no lifting gear. Loading and unloading are conducted by means of a single dockside gantry crane, with a corresponding reduction in output. These feeder vessels are usually Ro-Ro or combined type. Table 1.5 lists the main dimensions of typical feeder vessels.

Because of the container terminal's specialization and the large investment involved, a minimum level of cargo volume is required to render the investment profitable. This throughout depends on individual conditions and ranges typically around 70,000 TEU annually. It is characteristic that the investment cost per TEU for an annual traffic rate of 20,000 TEU is triple that of the corresponding cost for 80,000 TEU. Containers are of simple rectangular shape, as shown in Figure 1.13. Table 1.6 lists the typical dimensions of various container sizes. It is estimated that in the future the trend toward greater container length, in the region of 45 ft, and a weight of over 35 tons will gain momentum.

1.4.3.2 Cargo Handling. Practice has shown that the actual productivity of container terminals is significantly lower than the theoretical productivity. An average daily productivity per berth used to be in the region of 450 TEU for many small container terminals, whereas large modern terminals can achieve up to 2000 movements, as in the port of Singapore. A concept of narrow docks has been proposed, where a vessel could be served by cranes at both sides, thus achieving high productivity, on the order of 300 movements per hour per berth.

Table 1.5 Typical dimensions of feeder vessels

Feeder Vessel Type	Dead-Weight Tonnage	TEU	Length (m)	Beam (m)	Draft (m)
Ro/Ro	4580	176	130	17	6.25
Lo/Lo	1260	106	77	13	3.70
Combined	2080	111	87	14	4.70
Combined	6500	330	115	19	7.40

Figure 1.13 Steel container.

Loading and unloading operations are carried out by means of powerful dock gantry cranes that can attain an output of 25 to 30 TEU per hour, although usually their average productivity is lower. The operational life of a typical gantry crane extends to 15 years and 2,000,000 operating cycles. Some typical gantry crane dimensions are:

- Lifting capacity 30–50 tons
- Rail gauge 15–40 m
- Maximum lifting height above dock 25 m
- Maximum depth beneath dock 15 m
- Maximum seaward overhang 25–40 m
- Landward overhang 5–25 m

Large quayside gantry cranes may serve vessels up to 18 container rows, while several terminals around the world are already operating gantries capable of serving vessels 22-boxes wide with outreaches more than 60 m, serving super post Panamax vessels. Among the advances in gantry technology the twin-lift spreaders are worth mentioning, capable of handling two 20-ft boxes simultaneously. The critical operating parameter of a dock gantry crane is its output, which should be as high as possible to reduce vessel berthing time. For this reason, methods of making the loading/unloading cycle at the dock independent of the transport cycle of the boxes to open-air storage are employed, to attain a continuous supply to and removal of containers from the dock gantry crane. Extensive land areas, required for storage of containers forwarded through a terminal, constitute the distinguishing characteristic of specialized container terminals. In the case of

Table 1.6 Selected container sizes

ISO Type	TEU	External Dimensions (m)	Maximum Lifting Capacity (tons)	Cubic Capacity (m^3)
1C (20 ft)	1	6.05 × 2.435 × 2.435	20	29.0
1A (40 ft)	2	12.190 × 2.435 × 2.435	30	60.5
1B (30 ft)	$1\frac{1}{2}$	9.125 × 2.435 × 2.435	25	45.0
1D (10 ft)	$\frac{1}{2}$	2.990 × 2.435 × 2.435	10	14.1

imports, containers are transferred from docks to the stacking yard, for pickup a few days later for overland transport. The reverse procedure applies for exports. The simplest handling procedure involves the use of container chassis such as the one depicted in Figure 1.13.

The procedure followed in the case of imported containers involves the following stages:

- Loading of the container by dock gantry crane onto a container chassis
- Transport of container chassis by tractor to the storage area
- Chassis and container retained in storage area until delivery

Unloaded container chassis are parked in a dedicated lot. In a storage area, containers may be handled by straddle carriers, miscellaneous rubber-tired high-lift (front loader) high-reach stackers, and so on. For details, consult Chapter 2. Loaded containers may be stacked to a maximum height of three or four, depending on the type of equipment used. Empty containers may be stacked six or seven high. Representative examples are shown in Figures 1.14 through 1.16.

The minimum width of corridors between container rows in a linear layout is approximately 1.20 m, to enable access by a straddle carrier's legs. Circulation lanes are provided at regular intervals, forming a road network for the use of straddle carriers and other vehicles. These lanes have a minimum width of 12 m when they have to allow for turning of the rubber-tired straddle carrier, and 5.5 m in other cases. Usually, free gaps about 0.80 m wide are also allowed between the smaller surfaces of adjacent containers to facilitate handling, inspection, and so on.

This handling system may be simplified as regards the variety of equipment. Thus, tractors and chassis may be replaced by rubber-tired straddle carriers so that the latter also carry out the transport of containers from docks to the storage area. However, using straddle carriers for long distances does not put them to optimum use. Other disadvantages of these vehicles include the problem of requiring frequent maintenance and repairs and providing limited visibility to the operator; on the other hand, they are exceedingly versatile machines. Recent technical developments in straddle carriers include the incorporation of twin spreader systems, similar to those used in quayside gantry cranes.

Another method of cargo handling in the stowage area is through use of special gantry cranes with a 45-ton lifting capacity that can stack containers four, or even five, high (Figure 1.14). These gantry cranes, usually called *portainers,* may move on rails, spanning about 20 container rows. They can also be fitted with tires, in which case they have a smaller span, in the region of six or seven container rows and smaller stacking capacity; usually three to four container height. Portainers on tires are, however, more versatile and capable of being applied to various operations.

Stowage gantry cranes are preferred in container terminals with large throughput, particularly export traffic, and are amenable to adaptation for automated applications in container placing and identification. It is noted that information technologies are applied increasingly in most operations that take place in modern container terminals, not only in box stacking. A recent attempt toward full automation between dockside and yard was manifested in the design of dockside and stacking gantries with overlapping reaches.

Yard gantry cranes may also be used to move containers between open-air storage and rail or road vehicles. The handling systems above may be combined to suit the requirements of any particular port terminal. It is evident that with exports, a higher stacking height

Figure 1.14 Containers stacked in storage area by gantry crane.

can be accepted than in the imports section because of the reduced probability for additional maneuvers to reach an underlying container in the stack.

Overhead cranes were recently introduced in Singapore port. These are capable of stacking nine boxes high spanning ten rows across. They are operated remotely, having a high degree of automation built in.

New ideas on container storage are also being considered to replace the method of placing the boxes on the ground with automated racking systems.

1.4.3.3 Storage Yard. Containers remain in open-air storage areas for a few days until they are forwarded to either sea or land transport. Indicative average values of waiting time for imported containers is roughly 6 days, and 4 days for containers destined for export, while empty containers usually remain in port about

Figure 1.15 Containers stacked by high-reach stacker with telescopic boom.

10 to 20 days. The required container storage area depends on the stowage method and available equipment. Table 1.7 lists the area required per container, including space for access to the corresponding handling equipment.

The vehicle access lanes at the container terminal should have a width of 3.5 m for trucks or trailers, 5.5 to 7.0 m for straddle carriers, and 5 m for side loaders. In 90° bends, the widths above become 6, 12 to 15, and 7.5 m, respectively. Front-loading forklifts require an access lane of width equal to the length of the containers handled, increased by a safety margin of approximately 1.0 m on each side.

The performance of various transport and stacking equipment may be calculated by the time it takes to stack (or to remove) a container

Figure 1.16 Container storage area; typical linear container stacking configuration.

Table 1.7 Gross storage area requirements

Stacking Method	Container Height (no. containers)	Storage Area (m^2/TEU)
Trailer	1	65.0
Straddle carrier	3	10.0
	4	7.5
Gantry crane	3	10.0
	4	7.5
	5	6.0
Forklifts, side loaders	2	19.0
	3	13.0

and by the average speed of the vehicle. Stacking time ranges from 0.5 to 1 min for straddle carries, 1 to 2 min for forklifts, and 2 to 4 min for side loaders. Average speed ranges from 450 to 500 m/min for trucks, tractors, and side loaders, to 400 to 430 m/min for straddle carriers, and 300 to 350 m/min for forklifts.

The storage area, E, in hectares required in a container terminal may be calculated using the relation

$$E = \frac{QD}{3560} \frac{e}{f} \left(1 + \frac{p}{100}\right) \tag{1.9}$$

where Q is the number of containers handled annually (thousands of TEU), D the average container waiting time (days), e the area required per TEU (m^2; taken from Table 1.6 on the basis of the maximum possible height), f the ratio of average to maximum stacking height, and p the peak factor (%).

The working surface of an open-air storage yard is designed according to the type of container equipment used. It could be either paved or simply gravel covered. Usually, heavy forklifts impose stricter requirements on road surfaces than do tractors or straddle carriers. The rolling zones of portainers on tires are usually reinforced. The U.K. guidelines indicate the need for a minimum thickness of bituminous surfacing of 18 cm to avoid reflective cracking due to the cement-bound base. Bituminous surfacing is relatively inexpensive, but it can be damaged by corner castings in the container storage area. Cast-in-situ concrete is more expensive, inflexible, but generally hard-wearing. The other options include gravel, reinforced concrete plinths with gravel or other infill, and block paving. Gravel is the cheapest option, but it tends to spread onto adjacent readways, to get stuck in corner castings of boxes, and to render slot marking difficult. Block paving is relatively expensive but is being accepted as the most flexible surfacing for storage yards, since it allows lifting and relaying of damaged sections.

The yard surface should display a 1:40 to 1:50 gradient for efficient runoff of rainwater. However, a yard surface should ideally be horizontal for box stacking, so a compromise of about 1:100 gradient is generally used. Continuous slot drains or individual catch pits provided along roadways collect runoff and discharge it to outfall pipes. The terminal-included yard and gates should be amply illuminated to ensure efficient round-the-clock operations. Lighting is generally provided by high-mast columns, typically 30 to 50 m high. Layout of columns should be considered carefully to avoid risk of collision or taking up vital space in the storage area, achieving at the same time a more-or-less uniform illuminance. Fire-fighting facilities in the form of fire hydrants should also be provided throughout the terminal, including the storage yard. Hydrants can be in pillars or in pits, the latter case requiring a standpipe to be attached before hoses can be connected. A typical paved surface storage yard is shown in Figure 1.16.

1.4.3.4 Container Freight Station and Other Areas. A percentage of the containers handled at a container stripping terminal pass through a special shed, where chartering, container repacking, stuffing, and cargo reallocation operations are conducted. This shed, called a *container freight station* (CFS), should have a capacity calculated on the basis of 29 m^3 per TEU. The CFS's design area, S (in thousands of m^2), can be estimated by the formula

$$S = \frac{QD}{365}\frac{29}{h}(1 + r)\left(1 + \frac{p}{100}\right) \qquad (1.10)$$

where Q is the annual CFS container throughput (thousands of TEU), D the average duration of stay (days), h the average stacking height (m), r the access factor (accommodating space for lanes, maneuver areas, etc.), and p the peak factor (%).

Along the two long sides of the shed, containers and trailers are served, respectively, to facilitate repacking operations. Trucks can park outside or even within the station. The CFS is usually located at the rear of open-air storage areas of the terminal. It is possible, however, in case the land required is not available within the terminal, to plan for this installation at a distance from the port, and to maintain an exclusive connection with it. This arrangement is preferred, for example, when an expansion of an existing port within an urban area would otherwise be required in an area where obtaining additional space normally presents a problem. Figure 1.17 indicates the two arrangements in question.

In addition to open-air storage areas and container freight stations, other spaces are needed to cover requirements, such as maneuvering for land vehicles (road or rail), personnel parking, customs, administration building, refrigerated containers, storage of hazardous or flammable materials, and maintenance workshops. These additional installations amount to about 2 to 3 ha per berth.

1.4.3.5 Berths. Another parameter required for the design of container terminals is the number of berths required. To estimate this number, the number of berth-days needed annually, D, is calculated initially using the relation

$$D = \left(\frac{T}{HPm} + \frac{1}{12}\right)C \qquad (1.11)$$

where T is the ship's cargo to be loaded or unloaded (TEU), H the vessel working time per day, P the average quantity of TEU handled hourly per crane (including work stoppages or breakdowns), m the cranes per berth (allowing for an efficiency factor as follows: 1 crane/berth: $m = 1.0$; 2 cranes/berth: $m = 1.9$; 3 cranes/berth: $m = 2.4$; 4 cranes/berth or more: 80% efficiency per crane), and C the annual number of vessels calling at the container terminal.

It should be pointed out that the real-life data of crane productivity vary significantly between ports. However, a design figure of 120,000 TEU per crane per year can be used for initial planning purposes. To convert the annual number of berth-days into the number of berths required for the terminal, an optimum level of vessel servicing has to be determined, after having analyzed the corresponding waiting queue.

For specialized container terminals, the assumption is usually made that the time intervals between successive vessel arrivals do not follow the negative exponential distribution applicable to general cargo terminals (see Section 1.4.2.2), but rather, follow an Erlang distribution, with $K = 2$, because here there is some regularity of container ship arrival compared to that of general cargo vessels. It is further assumed that vessel servicing time follows an E_2 distribution as well. Table 1.8 gives the average waiting time (congestion factor) for an $E_2/E_2/n$ queue as a percentage of servicing time for various degrees of berth use (occupancy). Using the data of Table 1.8, the choice of the suitable number of berths for a container terminal is calculated by eq. (1.11) through trial and er-

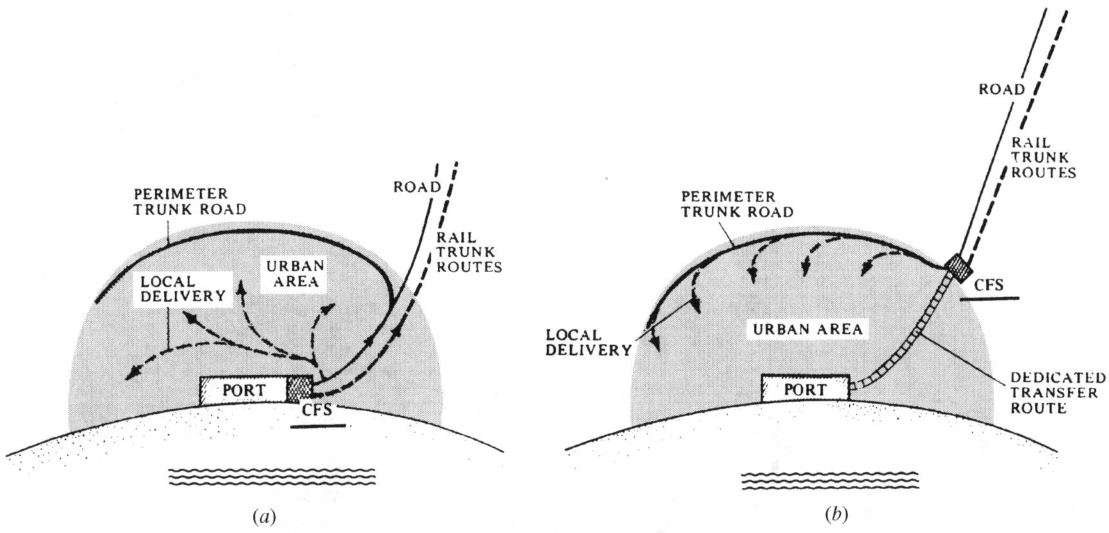

Figure 1.17 Location of a CFS within (*a*) and outside (*b*) a port.

Table 1.8 Congestion factor in queue $E_2/E_2/n$

	Number of Berths							
Occupancy	1	2	3	4	5	6	7	8
0.01	0.02	0.00	0.00	0.00	0.00	0.00	0.00	0.00
0.15	0.03	0.01	0.00	0.00	0.00	0.00	0.00	0.00
0.20	0.06	0.01	0.00	0.00	0.00	0.00	0.00	0.00
0.25	0.09	0.02	0.01	0.00	0.00	0.00	0.00	0.00
0.30	0.13	0.02	0.01	0.00	0.00	0.00	0.00	0.00
0.35	0.17	0.03	0.02	0.01	0.00	0.00	0.00	0.00
0.40	0.24	0.06	0.02	0.01	0.00	0.00	0.00	0.00
0.45	0.30	0.09	0.04	0.02	0.01	0.01	0.00	0.00
0.50	0.39	0.12	0.05	0.03	0.01	0.01	0.01	0.00
0.55	0.49	0.16	0.07	0.04	0.02	0.02	0.01	0.01
0.60	0.63	0.22	0.11	0.06	0.04	0.03	0.02	0.01
0.65	0.80	0.30	0.16	0.09	0.06	0.05	0.03	0.02
0.70	1.04	0.41	0.23	0.14	0.10	0.07	0.05	0.04
0.75	1.38	0.58	0.32	0.21	0.14	0.11	0.08	0.07
0.80	1.87	0.83	0.46	0.33	0.23	0.19	0.14	0.12
0.85	2.80	1.30	0.75	0.55	0.39	0.34	0.26	0.22
0.90	4.36	2.00	1.20	0.92	0.65	0.57	0.44	0.40

Figure 1.18 Small craft harbor at Monaco.

ror after estimating the days at berth required annually. Typically, a key performance indicator for a container terminal is the number of TEUs handled per annum per linear meter of quay. Based on data from major international container terminals, a design figure of 1000 TEU per annum per linear meter of quay may be used for the initial planning of well-equipped facilities.

1.4.4 Marinas

1.4.4.1 Basic Design Criteria. Marinas provide harboring and supply and repair services for pleasure boats. Recently, marine tourism and other recreational activities, such as amateur fishing and sailing, have increased rapidly worldwide, with a corresponding increase in pleasure craft and in a requirement for mooring spaces. To be classified as a fully developed marina, a harbor should satisfy certain criteria that extend beyond the provision of mooring slots. These services include water and bunker supply, availability of a repair unit, vessel lifting and launching arrangements, a supplies and provisions outlet, and vessel dry berthing. An example of a fully developed small craft harbor is shown in Figure 1.18.

Pleasure boats fall mainly into two categories: motor-powered and sailboats. Boats of these categories differ with regard to the geometric characteristics necessary for designing the moorings and in general, all the elements of a marina. The percentage of participation of

Table 1.9 Typical design parameters of pleasure boats

Length (m)	Number of Vessels (%)	Powerboats (%)	Sailboats (%)	Draft (m)		Beam (m)	
				Powerboats	Sailboats	Powerboats	Sailboats
0–5	50	40	10	0.80	1.40	2.20	1.80
5–9	30	21	9	1.00	2.00	3.60	3.00
9–12	10	5	5	1.20	2.40	4.10	3.40
12–15	7	4	3	1.040	2.080	4.80	3.90
15–20	3	2	1	1.660	3.40	5.30	4.40
Total	100	72	28				

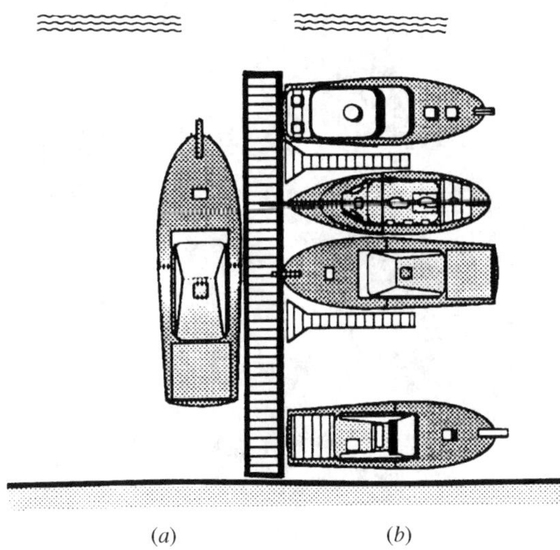

Figure 1.19 Typical moorings.

each category in the total number of vessels to be serviced in the marina depends primarily on the country and the marine region involved. Over time, these percentages vary in accordance with the development of this type of recreation as well as other parameters. A typical allocation of pleasure boats into the two categories above and five length classes is given in Table 1.9, where the figures refer to typical dimensions of the largest vessel in each class.

1.4.4.2 Dock Layout. Marinas possess docks, often floating docks, for vessel berthing, which may be either parallel (Figure 1.19*b*) or perpendicular (Figure 1.19*a*) to the quay line. Perpendicular berthing is effected either with light buoys, fixed or dropped anchors, or through the use of fingers. Fingers placed perpendicular to the main dock form single or double boat slips. Usually, single boat slips are for the use of relatively large boats; smaller boats are accommodated in double boat slips. Figure 1.20 indicates a mooring method in a double boat slip. For purposes of economy, the length of a finger may be designed to be smaller than that of the largest boat by a percentage depending on the size of the boat to be served. The ratio of finger length to the largest boat length may be a minimum (according to British Standards) of $\frac{3}{4}$ for boats up to 10 m, $\frac{7}{8}$ for lengths up to 15 m, and 1.0 for larger boats. Obviously, it is advisable that this reduction in length be applied in comfortable navigating conditions and low environmental loads, such as wind and waves.

Navigation channels within a harbor basin should be sufficiently wide to permit the necessary maneuvers. For comfortable conditions, this width should be $2L$ for motorboats and $2.5L$ for sailboats, where L is the length of the design boat. In sheltered areas and favorable conditions, the channel width can be reduced to $1.75L$ or even $1.5L$, measured between fixed or movable obstacles, such as between fingers or moored boats. The width of boat slips B de-

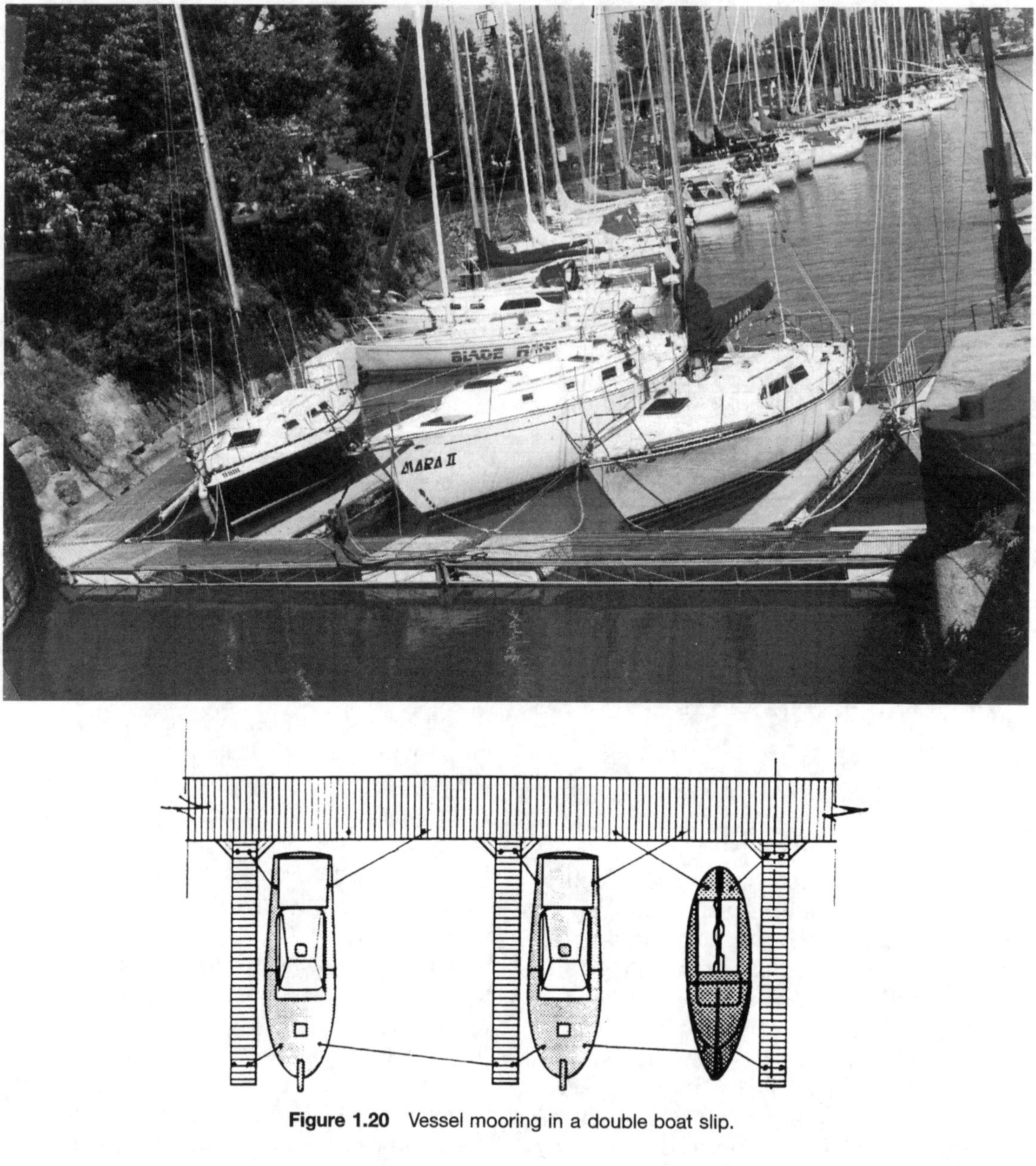

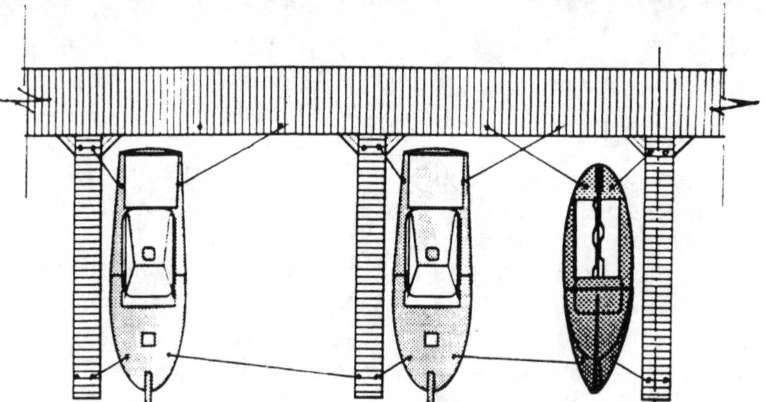

Figure 1.20 Vessel mooring in a double boat slip.

pends directly on the beam of the maximum boat W to be served. For a single boat slip it is $B = W + 2C_1$; for a double boat slip it is $B = 2W + 3C_2$ where C_1 and C_2 are the respective safety clearances. These depend on boat size, and according to the American Society of Civil Engineers (ASCE, 1994), they are as given in Table 1.10.

Usable water depth at slips and channels should be maintained at 0.50 to 1.00 m greater than the maximum draft of vessels using the marina. In moorings without fingers, a common type of mooring arrangement in the Mediterranean, safety clearances between moored boats are maintained at 0.5 m for boats up to 7.5 m long, 0.75 m for boats up 12 m, and 1.0 m for larger boats. Finger width lies around 0.9 m for finger lengths between 9 and 11 m and 1.2 m for lengths between 12 and 15 m.

The width of floating docks at which the fingers are connected at right angles depends on the total length of each dock, which is related directly to the number of people using them. The figures in Table 1.11 are typical dock widths for marinas of rather high-level specifications. Access between floating docks and fixed marginal quays is achieved by means of articulated ramps, as shown in Figure 1.21. These ramps are usually hinged on the fixed quay while the other end, resting on the floating dock, is fitted with a connecting plate rolling on the floor of the floating dock. The maximum longitudinal ramp slope is 1:4 ($m = 4$), the usable ramp width $W = 1.20$ m, and the rail height $H_r = 1.10$ m above the walking surface.

Table 1.10 Safety clearances in boat slips

Boat Length (m)	C_1 (m)	C_2 (m)
7.3	0.46	0.41
9.7	0.60	0.51
12.2	0.76	0.61
15.2	0.91	0.71
24.3	1.06	0.81

Table 1.11 Floating dock width

Dock Length (m)	Dock Width (m)
Up to 100	1.5
100–200	1.8
Above 200	2.4

1.4.4.3 Floating Docks. Floating docks are commonly adopted to ensure the availability of mooring slots in marinas, because of the relatively small loads they receive from berthed vessels and operation loads. They are made up of floats on which passageway decking, usually wooden, is fitted. The floats may be either full or hollow, and they are basically constructed of expanded polystyrene, fiberglass, or plain concrete. Floating docks are anchored through gravity anchors and chains or by vertical piles that prevent horizontal movement. An example of a gravity anchor is depicted in Figure 1.22. Gravity anchor design calculations are made using the customary methods for floating bodies. In these methods, boat impacts and wind forces on berthed boats have to be considered. Dock fingers are lighter constructions designed similar to floating docks. For more recent information on mooring systems for recreational craft, readers are referred to data from the Permanent International Association of Navigation Congresses (PIANC, 2002).

1.4.4.4 Marina Services. A well-organized marina possesses a range of facilities and equipment for its users.

Freshwater Supply. Water pipes—generally, those of the local water supply network—run the length of the docks and supply water to vessels through appropriate outlets. Usually, fire hydrants are provided in a water supply network. They are positioned at approximately

Figure 1.21 Articulated access ramp to a floating dock.

50-m intervals and are equipped with a 1.5-in. flexible hose kept at special firefighting points. Fire hydrants are attached to water mains of relatively large diameter, typically 2 in. or more. As water is not the most suitable fire-fighting means for a fire caused by fuel or an electrical short-circuit, there is a tendency to replace conventional fire hydrants with chemical fire-extinguishing equipment located appropriately in the marina.

Roughly 1-in.-diameter pipes are needed for water supply of adequate pressure, excluding firefighting service, to serve up to 50 mooring slots. Flexible pipe sections are placed at crossings between floating elements and at shore connections to absorb the corresponding movements. Pipelines exposed to the sea are made of plastic or steel to avoid corrosion. Measurement of water consumption can be made centrally for the marina as a whole or individually at outlet points. The points of water supply are frequently combined with the power supply within special pillars.

Power Supply. Power supply sockets should be provided along the length of docks to provide an electric current of 20, 30, or 50 A at 120 or even 230 V. Typically, every vessel exceeding 6 m in length should have access to the relative power outlet. Cabling is arranged in special ducts or suspended lengthwise along docks, to satisfy safety regulations. Grounding is provided by means of returns to shore. The marina lighting network is arranged in parallel with that of the power supply. The lighting fixtures are either incorporated in the supply

Figure 1.22 Raw iron gravity anchor used for anchoring of floating docks.

points or are mounted on independent poles preferably 3 m in average height.

Telephone Connection. The telephone system offered by each marina depends on the needs of the particular situation and in conjunction with cost, on the level of services offered. There have been systems of full coverage, with suitable supply points at each mooring position, and others with a telephone switchboard and paging or with the more accessible method of card-operated phones. In any event, the development of cellular telephony has nearly eliminated the need for providing telephone service to marinas.

Waste Disposal and Sewerage. An increasing number of pleasure boats possess systems for disposal of their accumulated waste by means of pumping. It would be useful to provide, preferably on a fixed dock, appropriate intakes and conduits connected to the local sewerage network. For solid waste, garbage dumpers are placed at suitable locations, accessible to garbage trucks.

Storage Lockers. Many marinas provide lockers for the storage and safekeeping of provisions, equipment, and so on, close to the moorings. These lockers may be combined with the water or power supply stands described above.

Bunker Supply. A bunkering point can be situated on an appropriate berth of the marina, connected to shore storage tanks. Pumps with measuring devices are located on this dock. Care must be taken to avoid accidents, such as fuel leakage into the marina basin. Bunker supply points are usually combined with installations for receiving slops and removal of chemical substances from boats' tanks. Frequently, design of the fuel supply is assigned to companies engaged in marina bunkering.

Cleats and Fenders. Along the length of docks, cleats or light bollards are to be provided at suitable intervals. In the case of alongside berthing, these will be located at either end of the berthing place, with one more in the middle for vessels exceeding 10 m. Cleats are manufactured of rustproof alloys or of hardwood. Boats can also be tied fast on piles, placed for this purpose along lines parallel to the docks, thus delimiting the boundaries of the navigation channels within the marina. In addition, floating dock guide piles may also be used for mooring purposes. Fenders of fixed or floating docks constitute serious equipment for the safety of both vessels and marina installations. Various types of fenders are used, such as continuous rubberform alongside a dock, single tires hanging vertically on the sides of the dock, or vertical wooden or plastic fenders for soft contact.

Vessel Lifting and Launching Installations. Boat lifting and launching procedures are a significant part of an organized marina. A large variety of lifting arrangements could be used as required. The commonest arrangements for vertical lifting in marinas are the travel lift, the fixed jib crane with horizontal boom, the special forklift, and the monorail. The travel lift (Figure 1.23) is equipped with a crane mechanism mounted on a steel frame usually fitted with rubber tires. It travels along and above the water surface of a boat slip so that it can be placed above the boat to be lifted. Travel-lift frames can be open at one end for servicing sailboats. Lifting a vessel is done using appropriate nylon slings.

A fixed jib crane (Figure 1.24) with a horizontal boom is placed in an appropriate location in a marina and at such a distance from the dock as to avoid damage from a potential collision with the dock wall of boats being lifted. The transfer of significant point loads from a crane on the quay wall should be taken into consideration in the design of the latter.

A special forklift possesses a vertical stem that enables the forks to reach below the bottom of the boat to be lifted. The forklift approaches the dock, alongside which a suitable retaining bar has been fixed to avert accidents. A safety margin between the movable parts of the forklift and the vertical dock wall should also be factored into the design. These forklifts may be used for boat storage during the winter layup period. An example of multilayered winter storage of pleasure boats is depicted in Figure 1.25.

Finally, monorails are easy-to-use installations since the conveyor holding the vessel moves by remote control. The conveyor is suspended over rails running centrally along the length of the monorail. The monorail is placed transversally to the dock and extends over the sea by means of a protruding beam to enable vertical lifting and relaunching of vessels. Figure 1.26 indicates the approximate relation between the length and weight of motor-driven craft and sailboats, from which the required lifting capacity of the marina's equipment can be estimated.

The commonest method of launching relatively small boats, which normally constitute the majority of vessels, is by use of launching ramps. These are slopes extending above and below sea level with nonskid surfaces formed by means of deep, gently sloped grooves of sufficient width. The vehicles that are to pull out or launch boats approach these ramps laterally with special trailers and make use of the wire rope that holds the vessel. A submarine

Figure 1.23 Travel lift frame for launching and retrieving pleasure boats.

horizontal gravel mound is provided to stop a vehicle from falling into the sea in the event of an inability to brake. The ramp width is a minimum of 5 m. A sufficient expanse for parking vehicles pulling boat-bearing trailers should be provided for in a suitable location close to the ramp. Moreover, this area should also contain a space for rinsing seawater off the vessel, the trailer, and the boat. Runoffs should be collected for treatment because it usually contains oil, mud, and so on, that should not be allowed to flow back freely into the harbor basin. Embarkation and disembarkation docks and berths for boats waiting their turn to be lifted should be situated near the launching slip. In areas with weak tides, small floating ramps may be used for relatively small vessels. Table 1.12 summarizes the basic characteristics of the primary vessel lifting and launching systems.

Auxiliary Buildings and Installations. A well-organized marina should contain a number of auxiliary buildings and networks that should be arranged and designed according to the needs they are to serve. The following are the most important such buildings and installations:

- *Marina administration building.* This structure houses the administration, accounts, inquiries, telephone switchboard, and so on.
- *Harbor master's building.* This structure is used to house the navigation and security

Figure 1.24 Fixed jib crane.

services. It may be combined with the administration building.

- *Boat repair shop*. This building or area constitutes a point of attraction for many pleasure boats. It may be designated only for small or for larger vessels, in which case the arrangement for vessel lifting and transfer to the boat repair shop is designed accordingly. A range of equipment from simple wheeled carriers to powerful lifts and rails are used for the transport of vessels to and from the repair shop.
- *Repair and maintenance building*. This structure is used for land equipment and machinery. Usually, this building is combined with the vessel repair shop if a shop is provided.
- *Provisions kiosk*. All types of consumables and durable goods related to operation of the marina may be supplied through a shop in the marina, as part of the administration building or otherwise.
- *Sanitation areas*. Approximately one toilet for each 15 mooring places should be provided at intervals of less than 300 m.
- *Road network, utilities networks, and lighting*. These are designed as for urban areas.
- *Entrance gate and fencing*. Security is always a sensitive issue in marinas, and special care should be given to protection from theft and vandalism. Fencing of the marina land area and safeguarding of its perimeter contribute a great deal.

Figure 1.25 Winter storage for pleasure boats.

- *Parking lots.* Attention should be paid to ensure adequate parking space for marina users, with clear signposting and unobstructed traffic flow. A typical parking place with a trailer occupies an area of 3 m by 12 m.

Boat Dry Stacking. A good number of marinas provide shore areas for laying-up vessels ashore. Dry stacking of boats is preferred by many users because of the improved maintenance achieved (washing with sweet water, etc.), but it adds extra capacity to the marina. Under normal circumstances, dry storage is provided for vessels smaller than 2 tons, but if the marina possesses the appropriate mechanical equipment, much larger vessels can be laid-up ashore. Table 1.13 lists typical dimensions and weights of pleasure boats for dry berthing.

The majority of small sailing boats, under 4.5 m, are placed by hand, keel upward, on special shelves, after their mast has been removed. Motor vessels under 7 m are placed on shelves by forklift, keel downward (Figure 1.25). The stacking areas may be open-air or sheltered. Larger vessels, both sailboats and motor vessels, are usually placed on special trailers which are drawn by their owner's vehicle from and to the storage area. When the storage is done on scaffolding, marina personnel undertake handling of the vessels. The lifting and launching equipment methods referred to previously are employed. Moreover, special arrangements can be used that combine lifting

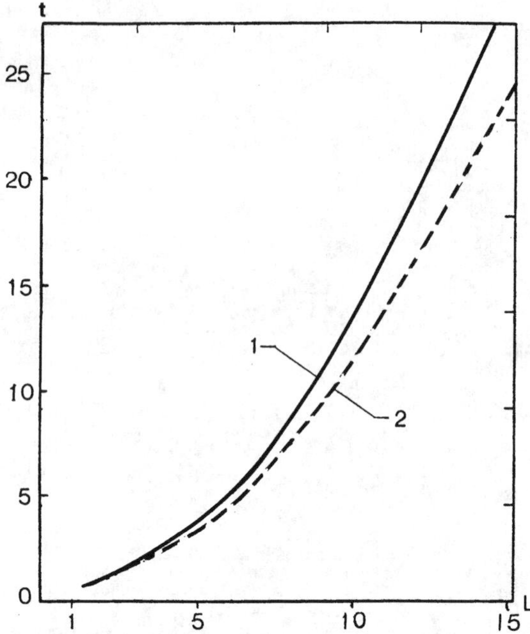

Figure 1.26 Approximate relation between length and weight of pleasure boats. 1, Motorboats; 2, sailboats. *Note: L* in meters, *t* in tons. (Adapted from U.S. Army Corps of Engineers, 1974.)

and launching with transport and stowage at the dry berthing positions. Such an arrangement may include a forklift suspended from a gantry crane operating in a covered vessel slip. The layup slots are arranged appropriately on scaffolds along the wet slip perimeter.

One of the advantages of laying-up ashore is that a marina requires a far shorter quay length than that of conventional mooring arrangements, amounting to approximately 15% of the latter. The total required marina area is smaller than the corresponding surface for wet berthing. For instance, a 200-vessel marina of an average 6.5-m-long vessel with 22-m-wide navigation channels requires roughly the surfaces denoted in Table 1.14 when it uses exclusively wet or dry berthing.

Marina Water Renewal. Marina basins frequently suffer from seawater pollution deriving from the marina area and also directly from craft using the marina. Pollution of the surrounding region may result from wastewater or stormwater effluents discharging in the marina basin and from surface water that carries a sig-

Table 1.12 Principal vessel lifting and launching system characteristics

No.	System	Lifting Capacity (tons)	Number of Vessels Transferred Daily	Turnover Cycle[a] (min)	Appropriate for Large Tide Fluctuation
1	Dry dock	Adequate	1–2	20–60	Yes
2	Slipway	Adequate	1–6	20–60	Yes
3	Lifting platform	Adequate	1–10	20–50	Yes
4	Ramp and tractor/trailer	5	100–250	3–8	No
5	Crane and trailer	15	20–50	20–40	Yes
6	Monorail	20	30–80	10–30	Yes
7	Forklift	2	100–250	3–8	No (special accessory required)
8	Travel-lift with straps	250	~50	10–20	Yes

Source: Adapted from PIANC (1980).

[a]Lifting, landing of vessel, and return of equipment to its original position.

Table 1.13 Typical dimensions of vessels for dry stacking

Boat Class	Beam (m)	Length (m)	Height (m)	Weight (tons)
I	<2.40	<5.40	0.90–1.50	<1.25
IIa	2.40^{-}	4.80–6.30	1.20–1.80	0.75–1.75
IIb	2.40^{+}	5.40–7.20	1.50–2.10	1.75–2.75
IIIa	>2.40	6.30–7.80	1.65–2.40	2.25–3.25
IIIb	>2.40	7.50–8.70	2.10–2.70	3.00–4.25

Source: Dry Stack Marinas, Florida.

Table 1.14 Typical space requirements in a small marina (thousands of square meters)

	Berthing	
Marina Surfaces	Dry	Wet
Land	9.5	5.1
Sea	2.5	13.2
Total	12.0	18.3

nificant polluting load. Boats may give rise to pollution through effluents from washing, garbage, oils, and so on. In each case the possibility exists to avoid seawater pollution through appropriate design of networks in the surrounding region by not allowing discharges within the port and by providing for the collection of garbage and other refuse from the boats, as mentioned above. At the same time, pertinent regulations governing protection of the marine environment have to be enforced.

In any case, frequent renewal of marina water is desirable to avoid potential eutrophication due to the lingering pollution. For this reason, marinas with two sea entrances have an advantage as regards their ability to enhance some streaming motion, which boosts the exchange of marina waters with offshore seawater. Usually, an effort is made to invigorate these streams by leaving openings at key locations across the protection structures. It is obvious that the problem becomes even more acute in regions with a weak tide, such as the Mediterranean. It has been estimated that the water quality begins to be unacceptable when the period of water renewal exceeds roughly 10 days. In severe cases, when no other method of coping with a problem is available, recourse can be taken to mechanical mixers, which are positioned in the marina basin to create artificial water circulation, thus renewing the polluted water. For detailed information associated with small craft marina design, construction, and operation, readers are referred to a work by Tobiasson and Kolmeyer (1991).

1.4.5 Fishing Ports

1.4.5.1 Main Features. Annual world sea fishing products amount to approximately 100,000,000 tons, with China providing one-fifth of the catch. Of this quantity, 28% is converted into fishmeal, the balance being consumed by people (29% fresh fish, 12% canned, 8% cured, and 23% frozen). Fishing ports serve professional fishing vessels and demonstrate a series of particularities which differentiate them from other commercial ports. These particular characteristics are summarized below.

The services that a fishing port is required to provide to fishing vessels are not limited to safe mooring to discharge the catch. The port should also be able to provide a suitable number of places for safe anchorage to fishing vessels during long periods of inactivity. Due to the nature and duration of the stay of such fishing vessels at port, the mooring types and rules of safe clearances determining the berthing positions of vessels are less strict than those for a commercial port.

In addition to being a refuge, a fishing port should possess small to medium-sized shipbuilding and repair facilities. This is because in addition to conducting purely repair work, fishing vessels conduct their regular maintenance work while in port. Thus, fishing ports should provide all the necessary means to ensure a minimum level of maintenance of the fleet they serve. Similarly, there are significant differences between the land zone of a fishing port and that of a conventional commercial port. For a fishing port, there is the systematic conduct of commercial activity regarding the catch, with the frequent presence of industrial units for processing and packaging. Consequently, the nature of a fishing port expands and it no longer acts as a hub in a combined transport system as is the case with conventional ports. Rather, it evidences the features of a commercial and industrial zone, and its land area is set out accordingly.

Moreover, it should be noted that in most fishing ports no exporting sector exists, and consequently, only unloading of vessels is carried out at the docks. In line with the specific requirements and characteristics above, a fishing port may include, in addition to loading wharves and mooring positions, the following elements:

- Repair docks
- Launching ramps
- Repair workshops
- Open-air spaces for drying nets and repairing nets and vessels
- Provisioning and equipment stores
- Sheds for storage of ships' gear
- A sheltered area for cleaning and sorting the catch
- A sheltered area for exhibiting the catch and for conducting the relevant commercial transactions
- Offices and ancillary areas
- Fish processing and packaging units
- Refrigerators for maintenance of the catch
- An ice-making unit
- Fuel, power, fire safety, and water supply networks
- Open-air areas for fish drying

There are a large variety of fishing vessels, and therefore the periods when vessels are away at sea for fishing vary accordingly. Vessels fall under the following categories:

I. Small vessels up to 30 gross registered tons (GRT), capable of putting out to sea for 1 day. These vessels are usually not equipped with refrigerating equipment.

II. Medium-sized vessels between 30 and 150 GRT, with a fishing autonomy of about 1 week. These vessels are equipped with a refrigerated hold.

III. Deep-sea vessels over 150 GRT, equipped with refrigeration and deep-freeze installations. Times out at sea for this category usually extend to 1 month. Such vessels may reach the 2000-GRT size.

IV. Large specialized industrial vessels.

Figure 1.27 shows the type of packing and respective processing stages of the catch toward consumption corresponding to the categories of fishing vessels above.

1.4.5.2 Design Criteria for Marina Installations. In each case, the design vessel determines the scale of a port and its constituent

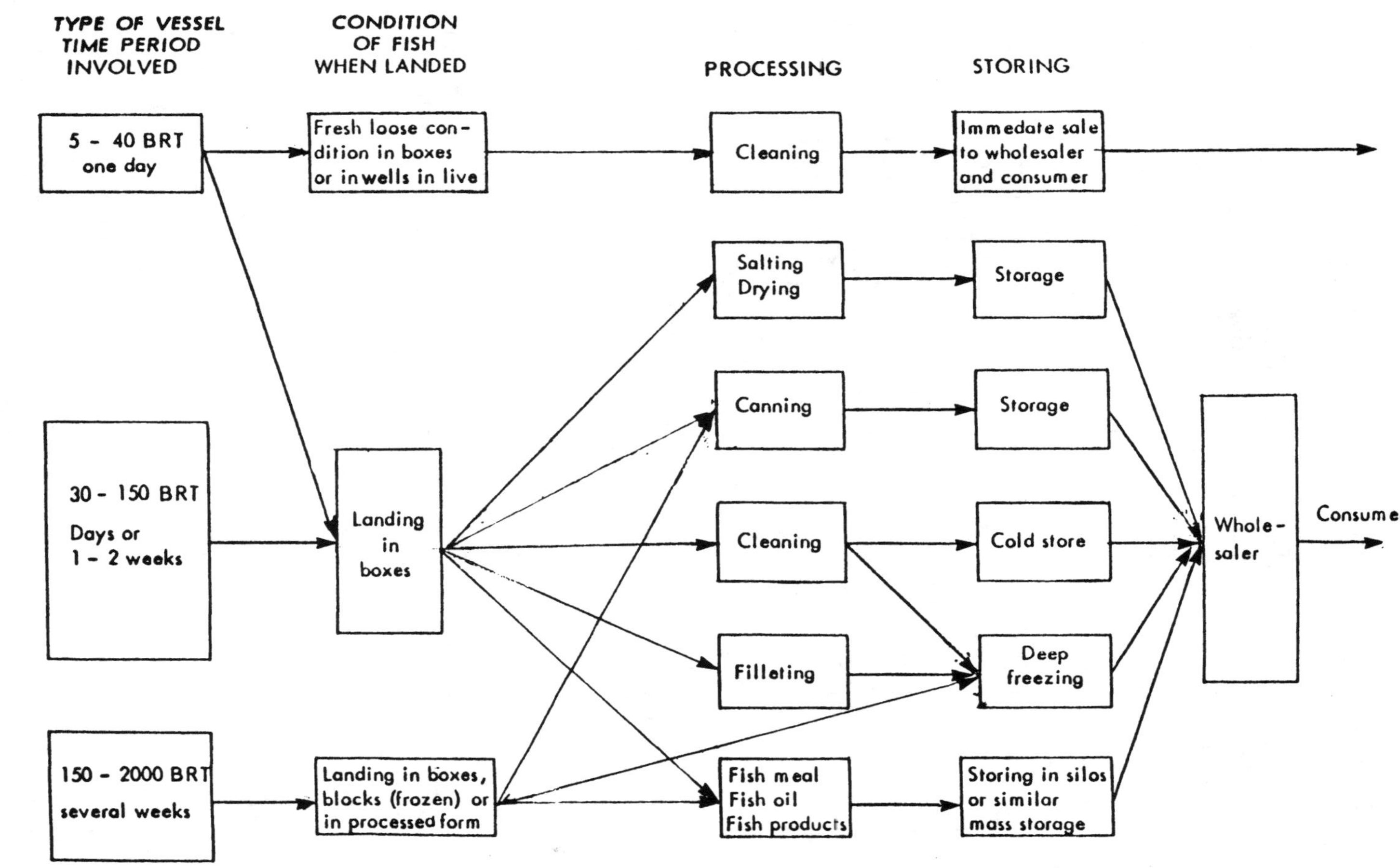

Figure 1.27 Processing methods of catch and distribution stages. (After Bruun, 1981.)

elements. Thus, depending on vessel size, the entrance width of a port usually ranges between 20 and 120 m. Table 1.15 lists typical dimensions of fishing vessels falling under the categories listed in Section 1.4.5.1. Based on the earlier discussion, an indicative fishing cycle for each vessel category is given in Table 1.16. The total duration of the cycle consists of days at sea and days in port for unloading and provisioning, from which an estimate of the required moorings can be made.

Repetition of the fishing cycle within the year depends on climatic conditions, the pertinent regulations determining the fishing period, local conditions, and repair and maintenance requirements. Category III or IV vessels usually need two months annually for such work, while smaller vessels take up a significant portion of their overall time for repairs and maintenance. These percentages may vary according to region; thus the allocation of overall time by vessel category listed in Table 1.17 is purely indicative and should always be adapted to local conditions. In a fully developed fishing port, the functions in the second to fifth columns in Table 1.16 are conducted in different sections of the port. Of course, there are situations where the functions, such as the second and third columns, may be combined in the same location without the need to move the vessel around.

Table 1.15 Typical dimensions of fishing vessels

Vessel Category	Length (m)	Draft (m)	Beam (m)
Ia	<7	<1.0	<3.5
Ib	7–10	1.0–1.5	3.5–4.0
II	10–20	1.5–2.5	4.0–6.0
IIIa	20–30	2.5–3.5	6.0–7.0
IIIb	30–60	3.5–5.0	7–10
IV	60–170	5.0–8.5	10–24

Table 1.16 Indicative duration of fishing cycle

Vessel Category	Days at Sea	Unloading and Provisioning (days)	Duration of Cycle (days)
I	1	1	2
II	6	4	10
III	35	5	40
IV	45–100	~8	50–110

Fishing vessel arrivals at port adhere to a more-or-less given pattern with peaks at certain periods of the year. Indicative occupancy factors of the landing quays may be in the range n = 0.4 to 0.7, depending on vessel size. A rough way of calculating the number of unloading berths is to consider that about 15% of the number of vessels using the port should be able to find a free unloading berth at any time. The functions in the third to fifth columns in Table 1.17 require additional berthing facilities since such functions are normally conducted in locations other than those housing the unloading operations. Consequently, to calculate the number of these positions, it is necessary to determine occupancy factors n just as in the unloading sector. Table 1.18 lists several values of factor n for the various vessel categories and port functions. The factor n = 1.0 in the fourth column reflects the fact that the said "function" actually is the idle time of an obligatory stay in port.

Fishing vessels usually are secured alongside or in a tight arrangement stern to shore along straight docks. There are ports with a sawlike arrangement of unloading docks (e.g., Esbjerg in Denmark), to increase the number of vessels being served. In the case of a simple straight dock, the requirements for the water area relevant to the mooring type shown in Figure 1.28 can be accepted. Depending on the vessel category and its function, two (or more) rows of vessels moored side by side could be considered. For reasons of safety, this increase

Table 1.17 Allocation of fishing vessel time (days per year)

Vessel Category	Days at Sea	Unloading of Catch and Loading of Provisions	Bunkering/ Provisioning and Associated Idle Time	Idle Time and Small-Scale Repairs and Maintenance	Major Repairs and Maintenance	Number of Fishing Cycles per Year
I	140	70	75	75	5	140
II	170	85	30	70	10	28
IIIa	250	20	15	65	15	7
IIIb	250	20	15	60	20	7

Table 1.18 Indicative occupancy factors

Vessel Category	Unloading of Catch and Loading of Provisions	Bunkering/Provisioning and Associated Idle Time	Idle Time and Small-Scale Repairs and Maintenance	Major Repairs and Maintenance
I	0.7	0.7	1.0	0.8
II	0.6	0.6	1.0	0.7
IIIa	0.5	0.5	1.0	0.6
IIIb	0.4	0.4	1.0	0.5
IV	0.4	0.4	1.0	0.5

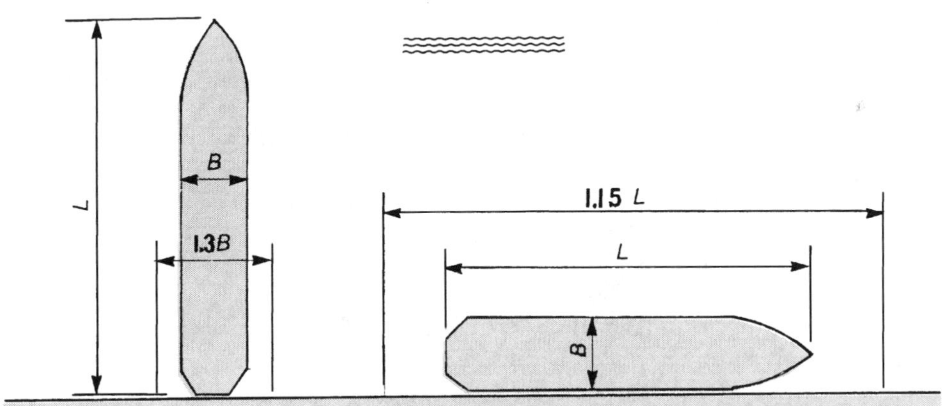

Figure 1.28 Mooring types of fishing vessels.

in number of vessel mooring places should not exceed a factor of about 50%. Table 1.19 gives indicative values of the hold capacity of fishing vessels.

Fishing vessel provisioning involves primarily fuel, water, and ice. The quantities of fuel and water required are estimated on the basis of the capacity of the respective tanks of the vessel. Some indicative values of the latter are given in Table 1.20.

Table 1.19 Net capacity of fishing vessels

Vessel Category	Length (m)	Hold Capacity (m^3)	Dead-Weight Tonnage
Ia	<7	1.5	0.8
Ib	7–10	4.5	2.5
II	10–20	25	15
IIIa	20–30	85	55
IIIb	30–60	400	250
IV	60–170	500–3500	300–2200

Table 1.20 Vessel tank capacities

Vessel Category	Length (m)	Fuel (tons)	Water (tons)
Ia	<7	0.3	0.2
Ib	7–10	0.8	0.5
II	10–20	10	5
IIIa	20–30	50	12[a]
IIIb	30–60	300	20[a]

[a] Additional seawater supply.

Category III and IV vessels usually have their own refrigeration installations and do not require stocking of ice. Vessels of the other categories need about 3 tons of ice on average per day during the fishing season. Unloading of the catch is effected in a manner related to packing type, hence by size of vessel. Usually, the vessel's own lifting gear, 3- and 6-ton mobile cranes, and corresponding forklifts suffice for the unloading and forwarding of catch to the cleaning sheds. Unloading by conveyor belts applies to catch packaged in boxes or crates. Given the tendency for improved packing of the merchandise during the voyage, particularly in the larger fishing vessels, the use of conveyor belts is becoming increasingly popular.

1.4.5.3 Land Installations. As stated in Section 1.1, the land installations of a fishing port are diverse and differ from those of ports for other commercial purposes. When a fishing port is fully developed, its land installations include the auction shed, the central building with cleaning and sorting areas, an exhibition area and auction room, a packing room with ice, refrigerators for overnight or longer storage of the catch, deep-freeze stores, salted or dried fish stores, weighing rooms, packaging material stores, and auxiliary installations (offices for administration, sellers, buyers, etc.). Depending on the particular situation, the cold display for auctioning may be replaced by a display of the catch in ambient conditions (PIANC, 1998).

The dimensions of an auction shed depend mainly on whether the display of the fish relies on a sample or on the totality of the catch. In the latter case, the building is located adjacent to the unloading zone of the dock, whereas in the former, it could be located farther inward of the port, at the same time being smaller than in the preceding case. Some basic criteria of the individual functions taking place under roof are listed below, to assist in the preliminary design of a shed with full view of the catch:

- Washing and sorting — 15–30 tons/m^2 annually
- Exhibit and sale — 1–15 tons/m^2 annually
- Weighing and arrangement — 7–15 tons/m^2 annually
- Storage in freezer — Capacity for 2–3 days' production
- Packaging plant — 6–12 tons/m^2 annually
- Access corridor — 8–16 tons/m^2 annually
- Auxiliary installations — May be installed in a mezzanine or on the ground floor, requiring 15 to 20% of the overall building

The typical overall building width ranges from 40 to 80 m. Frequently, a separate shed is provided for cleaning and storage of containers of the catch. The washing area for the containers requires about 1 m^2/ton per year, while the storage area varies depending on the specific packing type—a representative value being 0.2 m^2/ton of annual product handling. The wastewater from the washing of both the catch and the packaging containers should be conducted through floor grilles to a suitable treatment installation prior to final disposal. The floor should slope around 1:75 to facilitate surface drainage.

Repair and maintenance work may be provided by a series of installations, ranging from the simplest ramps to the most complex shiplift or dry dock facilities. A lifting arrangement providing ease of application on relatively small vessels is the Syncrolift, equipped with a vertical lifting platform supported by four legs at both sides (Tsinker, 1995). Repair/maintenance installations may use the longitudinal or transverse transport system for moving vessels to/from their respective dry berth for repair or maintenance.

REFERENCES AND RECOMMENDED READING

Agerschow, H., H. Lundgren, T. Sorensen, T. Ernst, J. Korsgaard, L. R. Schmidt, and W. K. Chi, 1983. *Planning and Design of Ports and Marine Terminals*, John Wiley and Sons, New York.

ASCE, 1969. *Report on Small Craft Harbors,* No. 50, American Society of Civil Engineers, Reston, VA.

———, 1994. *Planning and Design Guidelines for Small Craft Harbors,* American Society of Civil Engineers, Reston, VA.

———, 2001. *ASCE Proc. Specialty Conference Ports '01,* Norfolk, VA.

Bruun, P., 1981. *Port Engineering,* 3rd ed., Gulf Publishing, Houston, TX.

Brunn, P. 1989–1990. *Port Engineering,* 4th ed. (Vols. 1&2), Gulf Publishing Company, Houston, TX.

Chapon, J., 1966. *Travaux Maritimes,* Eyrolles, Paris.

Dally, H. K. (ed.), 1983. *Container Handling and Transport,* CS Publications, Worchester Park, Surrey, U.K.

E.C. COST 330, 1998. *Telinformatic Links between Ports and Partners,* Final Report of the Actions, European Comission, Brussels, Belgium.

European Commission, DG Environment, 2001. *Assessment of Plans and Projects Significantly Affecting Natura 2000 Sites. Methodological Guidance on the Provisions of Article 6(3) and Article 6(4) of the Habitats Directive 92/43/66,* Brussels.

European Sea Ports Organisation, 1995. *Environmental Code of Practice,* Brussels, Belgium.

Frankel, E. G., O. G. Houmb, and G. Moe, 1981. *Port Engineering,* Gulf Publishing Company, Houston, TX.

Herbich J. B. (ed.), 1992. *Handbook of Coastal and Ocean Engineering,* Vol. 3, Gulf Publishing Company, Houston, TX.

Hershman, W. (ed.), 1988. *Urban Ports and Harbor Management: Responding to Change along U.S. Waterfronts,* Taylor and Francis, New York.

IMO, 1991. *Port Logistics,* Compendium for Model Course 5.02, International Maritime Organization, London.

Knapton J., and A. Meletiou, 1996. *The Structural Design of Heavy Duty Pavements for Ports and Other Industries,* The British Precast Concrete Federation, 3rd edtion, London.

Memos, C., 1999. Lecturers on Harbor Works, "Symmetry" Publ., Athens, Greece.

PIANC, 1980. *Dry Berthing of Pleasure Boats,* Suppl. Bull. 37, Permanent International Association of Navigation Congresses, Brussels, Belgium.

———, 1997. *Review of Selected Standards for Floating Dock Designs,* Suppl. Bull. 93, Permanent International Association of Navigation Congresses, Brussels, Belgium.

———, 1998. *Planning of Fishing Ports,* Suppl. Bull. 97, proceedings of Specialized ASCE Conferences on Ports, PIANC Congresses, Harbor Congresses, and the other specialized conventions. Permanent International Association of Navigation Congresses, Brussels, Belgium.

———, 1999. *Environmental Management Framework for Ports and Related Industries,* PECO, Brussels.

———, 2002. *Mooring Systems for Recreational Craft,* Rep. WG10, Permanent International Association of Navigation Congresses, Brussels, Belgium.

Takamatsu, T., M. Yosui, and H. Sanada, 2002. *A Port and Harbor Vision to Connect People's Lives with the Sea and World,* PIANC Proceedings, 30th Congress, Sydney, Australia.

Tobiasson, B. O., and R. C. Kolmeyer, 1991. *Marinas and Small Craft Harbors,* Van Nostrand Reinhold, New York.

Tsinker, G., 1995. *Marine Structures Engineering: Specialized Applications,* Chapman & Hall, New York.

———, 1997. *Handbook of Port and Harbor Engineering: Geotechnical and Structural Aspects,* Chapman & Hall, New York.

UNCTAD, 1984. *Port Development: A Handbook for Planners in Developing Countries,* United Nations, Geneva.

UNCTAD, 1992. *Développment et amélioration des ports, Les principes et l'organisation modernes des ports* (*Development and Amelioration of Ports, Principles of Modern Port Management and Organization*), UN-Unctad, Conseil du commerce et du développment, Commission des transport maritimes, groupe intergouvernemental spécial d'experts des ports, GE.92-50027/1038C, Geneva.

UNCTAD, 1996. *Sustainable Development Strategies for Cities and Ports,* UN-Unctad, IAPH & IACP, Geneva.

UNCTAD, 2001. *Review of Maritime Transport,* United Nations, Geneva.

United Nations, (Food and Agriculture Organization), 1970. *Fishing Ports and Markets,* Fishing News Books, Oxford.

———, 1978. Conference on Trade and Development, U.N., New York.

U.S. Army Corps of Engineers, 1974. *Small-Craft Harbors: Design, Construction, and Operation,* SR 2, USACE, Washington, DC.

2

PORT-RELATED MARINE STRUCTURES

Dennis V. Padron and Bill Paparis
Han-Padron Associates
New York, New York

2.1 INTRODUCTION

Starting in the 1960s, the evolution of ports has increased exponentially. Prior to that time, the general cargo ship was the most prominent vessel for shipping nonbulk cargo, but in 1956 the container concept was introduced, and by the 1960s, specialized container ships replaced general cargo ships as the dominant vessel for shipping nonbulk and nonspecialized cargo. The first container carrier was a converted T2 tanker approximately 160 m long, with a loaded draft of approximately 9 m. By 2001, fully cellularized container vessels of length over 335 m with a draft of more than 14 m, and able to carry more than 8000 TEU (20-foot equivalent units) were under construction, with even larger vessels planned. Similarly, in the early 1950s the largest tankers afloat were less than 40,000 dead weight tons (dwt), with a length of approximately 215 m and a draft of approximately 11 m. But by the 1970s, tankers in excess of 500,000 dwt, with a length of approximately 430 m and a draft of approximately 26 m, were in operation, although these vessels proved to be uneconomical, and the maximum tanker size leveled off at approximately 350,000 dwt. The growth in dry bulk carriers followed a trend similar to that of tankers but did not exceed approximately 350,000 dwt.

In a modern port, specialized terminals are provided to accommodate the special requirements of particular types of vessels and cargoes. For purposes of this chapter these terminals are classified as container, general cargo, Ro-Ro (roll on–roll off), liquid bulk, dry bulk, and passenger. Of course, there are other types of terminals for specialized purposes, such as passenger ferries, rail or automobile ferries, and cryogenic liquids, but these terminals represent a relatively small percentage of port-related marine structures.

2.2 CARGO HANDLING EQUIPMENT

2.2.1 Containers

Container handling equipment can be placed in two very general categories: (1) equipment to move containers between the vessel and the terminal, and (2) equipment to move containers within the terminal. Within each of these categories, a great variety of equipment is in use.

At the wharf, containers can be handled to and from ships by gantry container cranes, jib cranes, multipurpose cranes, and even mobile cranes. For purpose-built container terminals handling specialized container ships, only gantry container cranes (also known as quay-side gantry cranes, portainers, or simply, container cranes) are used. Transfer of containers between the wharf and the container yard may be handled by tractors towing trailers, straddle carriers, heavy-duty front-end loaders, or automated guided vehicles. Within the container yard, stacking and unstacking may be carried out by straddle carriers, rubber-tired or rail-mounted gantry cranes, or a variety of lift-truck designs, while receipt–delivery operations may also involve those equipment types as well as tractor–trailer systems.

UNCTAD (1987) defines six types of container handling systems, each a combination of equipment types working together to perform the shoreside handling function:

1. The *tractor–trailer system,* in which containers are both handled and stored on over-the-road chassis or terminal trailers, which are moved around the terminal by heavy-duty tractor units
2. The *straddle carrier direct system,* in which wharf transfer, stacking, and other duties are performed by straddle carriers
3. The *straddle carrier relay system,* in which straddle carriers are responsible for in-yard stacking and unstacking,

while wharf transfer and other movements are performed by tractor–trailer sets or other equipment

4. The *yard gantry system,* where the container yard is equipped with rubber-tired or rail-mounted gantry cranes for stacking and unstacking, with tractor–trailer units for wharf transfer and other movements
5. The *front-end loader system,* either performed entirely by heavy-duty lift trucks of one sort or another or with other equipment for wharf transfer
6. *Combination systems,* consisting of various hybrid combinations of straddle carriers, yard gantry cranes, and other equipment, with more than one type of stacking equipment in use at a time, each carrying out a function to which it is best suited

Container cranes (Figure 2.1) are the most distinctive feature of a container terminal. The important factors for the selection of container cranes are:

- Height of lift, as determined by the ability to lift a container over the top of the stacked containers on board a vessel at its highest position
- Outreach, to the most distant outboard container stack on the vessel
- Back reach, to handle containers or vessel hatch covers behind the crane
- Clearance between the crane legs to handle the maximum length container or hatch cover
- Wheel loads as determined by track gauge, number of wheels, and spacing between wheels

Figure 2.1 Super post-Panamax container cranes at Port Salalah, Oman. (Courtesy of Han-Padron Associates and Salalah Port Services.)

- Overall length of the carriage between bumpers, to permit close working of two cranes on alternate hatches
- Lifting capacity, including spreader devices
- Lifting and trolley travel velocity
- Operating control, including programmability
- Wind capacity, instrumentation, alarms, and tie-down provisions
- Powering requirements and method

The largest container cranes currently in use have an outreach that can service vessels 22 containers wide, with a lifting capacity of 65 tonnes or more, and operating on a 30.5-m-gauge track. The straddle carrier (Figure 2.2) is a wheeled frame that lifts and transports a load within its framework. Despite problems with the earliest generations of straddle carriers, they are the most popular form of container lifting device. There are two main types of straddle carrier: twin-engined low-mounted types, and single-engined top-mounted types.

Figure 2.2 Straddle carrier.

The overall height varies between about 9 m (for rising-arch and smaller fixed-arch designs) and 15 m for those stacking one-over-three and 18 m for those stacking one-over-four.

Yard gantry cranes were developed from industrial overhead cranes. There are two distinct types: rubber-tired gantry cranes (RTGs), frequently referred to as transtainers, run on heavy-duty pneumatic-tired wheels, and rail-mounted gantry cranes (RMGs), run on steel wheels over fixed rails. Although these two types of transfer cranes serve the same function, there are distinct differences between them in addition to their wheels.

Rubber-tired gantry cranes (Figure 2.3) span several rows of containers and one truck lane and can stack containers up to six high (one-over-five). RTGs are essentially container-yard stacking devices and are used in combination with other container handling equipment, usually tractor–trailer sets, for the wharf transfer operation. The motive power for rubber-tired gantries is diesel or diesel–electric. The containers within the storage area are normally arranged in long rows parallel to the wharf, with about 30 TEU slots per row, which is thus approximately 180 m long. There are usually five or six rows of containers per block, plus the truck lane, giving a total block width of 23 to 27 m. At the ends of the blocks are roadways about 20 to 30 m wide, and an extra roadway space is provided between adjacent blocks. The wheels of the rubber-tired gantries can turn through 90° so that the equipment can be moved from one storage block to another as required to meet operational needs.

Rail-mounted gantry cranes function in a manner similar to rubber-tired gantry cranes and are used in combination with tractor–trailer sets for the wharf transfer operation. It is important to distinguish between two distinct types of rail-mounted gantry cranes: those used for stacking in the container yard and those used on receipt and delivery operations at rail terminals. The latter are generally small gan-

Figure 2.3 Rubber-tired gantry cranes (RTGs) at Port Salalah, Oman. (Courtesy of Han-Padron Associates and Salalah Port Services.)

tries, spanning perhaps two or three rail tracks and a roadway. Container-yard rail-mounted gantry cranes generally have large spans, some spanning 60 or 90 m. Stacking height generally is one-over-six, and lifting capacity is over 40 tonnes. Rail-mounted gantries are generally electrically powered, by cable or by their own local generators.

Terminal tractors and their trailers or chassis are among the most common equipment types in container terminals. The tractor is basically a heavy-duty motive power unit, or "prime-mover," fitted with one of a variety of coupling devices for attaching rapidly to an over-the-road chassis or a yard trailer. The tractor also has a prominent role in the handling of containers to and from Ro-Ro vessels, either mounted on their over-the-road chassis or lifted from their storage position below deck onto low-bed yard trailers.

Over the years tractor design has evolved to become specialized for port work. The cab has been moved to an off-center position to provide easy access to the trailer coupling device, brake cable connections, and so on, and its seating and control layout have become modified for easy reversing, with all-round driver vision a particular feature.

Lift trucks have become a prominent means of handling containers in terminals. In front-end loader configuration (Figure 2.4), lift trucks can employ top, bottom, or side-lift

Figure 2.4 Lift truck.

Figure 2.5 Reach stacker.

spreaders to lift containers and can also use a fork attachment that engages with "fork pockets" on the base of some containers. Higher-capacity trucks are usually fitted with a fixed-length or telescopic frame that attaches to the side or top of the container at two or four points. The top-lift attachment is a "spreader" similar to those used on straddle carriers and gantry cranes. Other machines comparable to front-end loaders, employing very different designs and lifting principles, are the side loader and the reach stacker (Figure 2.5). The side loader is a truck with a mounted lifting device which lifts a container onto the side of the chassis, in which position the box is carried to its storage position or pickup point. Reach stackers are tractorlike vehicles having a lifting boom located within the wheelbase and resemble mobile cranes, from which they were largely developed.

2.2.2 General Cargo

The nature of cargo defined as general cargo varies dramatically (e.g., bagged coffee on pallets, reels of newsprint, bales of textiles, drums of chemicals, machinery, steel ingots), and the equipment used to handle the cargo, both in the vessel's holds and on the wharf, is selected based on the characteristics of the particular cargo. Most general cargo vessels are equipped with lifting gear at each deck hatch, and this gear is used to transfer cargo between the ship and the wharf or pier. Substantially higher loading and unloading rates can be achieved by rail-mounted revolving, level-luffing cranes traveling on the wharf or pier deck. These cranes operate such that the boom can be raised or lowered (luffed) without changing the height of the hook. They are usually mounted on portal frames that span over truck or train traffic on the wharf or pier. Where throughputs are low, rubber-tired or tracked mobile cranes of various capacities are used to transfer cargo between the ship and the wharf.

Heavy lifts or other special cargo are often handled by floating cranes or derricks. Ports usually have only one such piece of equipment available since it is not used often and can be moved where needed. The floating crane will usually operate on the outboard side of the vessel and transfer the cargo to a barge, from which it is rehandled.

Within the hold of the vessel, cargo is often handled by a forklift, which is lifted into the hold by a shore-side crane. Forklifts are also used for handling general cargo on the wharf or pier deck. They are available in capacities ranging from 2 or 3 to 40 tonnes or more, and can be fitted with special handling devices for particular cargoes.

Rubber-tired mobile cranes, some with hydraulically extensible booms, are also used on the wharf deck for handling cargo. They are generally not suitable for horizontal travel of more than 50 or 60 m and are used most often to transfer cargo onto other equipment, such as trucks or railcars. Other specialized equipment, such as straddle carriers of various sizes and capacities, are used to handle particular types of cargoes.

2.2.3 Roll on–Roll off Handling

The principal reason for utilizing a Roll on–Roll off (Ro-Ro) type of operation is that it eliminates the need for terminal cargo handling equipment, other than tractors, with or without chassis or low-bed yard trailers. These are the same units as those described above under container handling equipment except that those units usually feature tractors with two-wheel drive, whereas four-wheel-drive tractors are usually used for Ro-Ro work, where they are required to climb and descend ramps, often under wet and slippery conditions. The Ro-Ro tractor generally has a manual gearbox, for maximum power, while "level-ground" tractors normally have automatic gearboxes. Since a clean exhaust is essential for working between decks of a Ro-Ro vessel, these tractors are fitted with catalytic exhaust cleaners and are designed to produce as little noise as possible.

2.2.4 Liquid Bulk Cargo

Liquid bulk cargo is transferred between the vessel and the terminal, in either direction, utilizing one or more of several available transfer systems. These systems are designed to accommodate the characteristics of the liquid, the required flow rates, and the horizontal and vertical operating envelope of the vessel's manifold. Two general types of cargo transfer systems are in use, rigid metal articulated loading arms and flexible hoses. Hybrids of these systems are also in use.

2.2.4.1 Loading Arms. Loading arms are all-metal articulated units used to connect the stationary terminal manifold to the moving vessel manifold. A loading arm is constructed of three tubular sections (riser, inboard arm, and outboard arm) connected by multiaxis swivel joints. The riser is fixed to the pier or wharf deck and is connected to the terminal piping system as illustrated in Figures 2.6 and 2.7. The inboard arm is connected to the top of the riser and the outboard arm is connected to the inboard arm and to the vessel manifold. To minimize the power required to maneuver the loading arm and to desensitize the arm to changes in its position within the operating envelope, counterweights are incorporated to balance the inboard and outboard arms.

As discussed in Padron (1998), loading arms with a diameter of 8 in. or less are usually light enough to be maneuvered manually. Loading arms with diameters greater than 8 in. usually

Figure 2.6 Articulated loading arms at Curaçao refinery. (Courtesy of Han-Padron Associates.)

require a hydraulically powered system for arm maneuvering. These larger loading arms are equipped with a control console and usually a portable pendent controller or radio-controlled unit for maneuvering the arms into the desired position. The pendant controller or radio-controlled unit enables the operator to control the movements of the arm remotely while on the pier or onboard the vessel.

A loading arm may be equipped with an emergency release coupling (ERC). This coupling is made up of two adjacent block valves with a short spool piece between them. In case of a vessel surge and/or drift, which would cause the vessel manifold to move beyond the operating envelope of the loading arm, the ERC valves will close simultaneously, and then the spool piece between the valves separates. Loading arms may also be equipped with an electronic range monitoring and alarm system to monitor the extreme points reached by the arm.

2.2.4.2 Hoses. Flexible hoses are used for cargo transfer at terminals that accommodate smaller vessels or have lower throughput. Generally, hoses are less costly than loading arms and, similar to loading arms, can accommodate a wide range of vessel sizes. However, the diameter of hoses is normally limited to 10 in., due to the difficulty of handling larger sizes, while loading arms may be as large as 24 in. in diameter.

At small terminals that serve a narrow range of vessels, primarily barges, hoses run from the manifold on the pier or wharf deck directly to the vessel manifold. At large terminals that serve a wide variety of vessels (barges and tankers), the operating envelope of the vessel manifold covers a wide range. At these terminals, the hoses are supported on a tower, as illustrated in Figure 2.8, and their position is adjusted continuously to accommodate the vessel manifold position. Hoses are sensitive to kinking and cannot be bent beyond their minimum bend radius. To maintain this minimum radius, a saddle or hose "bun" is often used to support and handle the hoses. Most modern hose towers are equipped with a hydraulic crane to facilitate handling and transfer of the hoses to and from the vessel manifold.

2.2.4.3 Hybrids. Sometimes, hybrids of loading arms and hose tower systems are used. One such hybrid is an articulated rigid piping system supported by a tower similar to a hose tower (Figure 2.9). With this system, both the inboard and outboard segments are pipe. Similar systems are in use where the inboard segment is pipe and the outboard segment is hose.

2.2.5 Dry Bulk Cargo

Other than self-unloading ships, dry bulk cargo is loaded or unloaded either by a fixed-tower loader and unloader or by a moving or traveling loader/unloader. Terminals with a fixed-

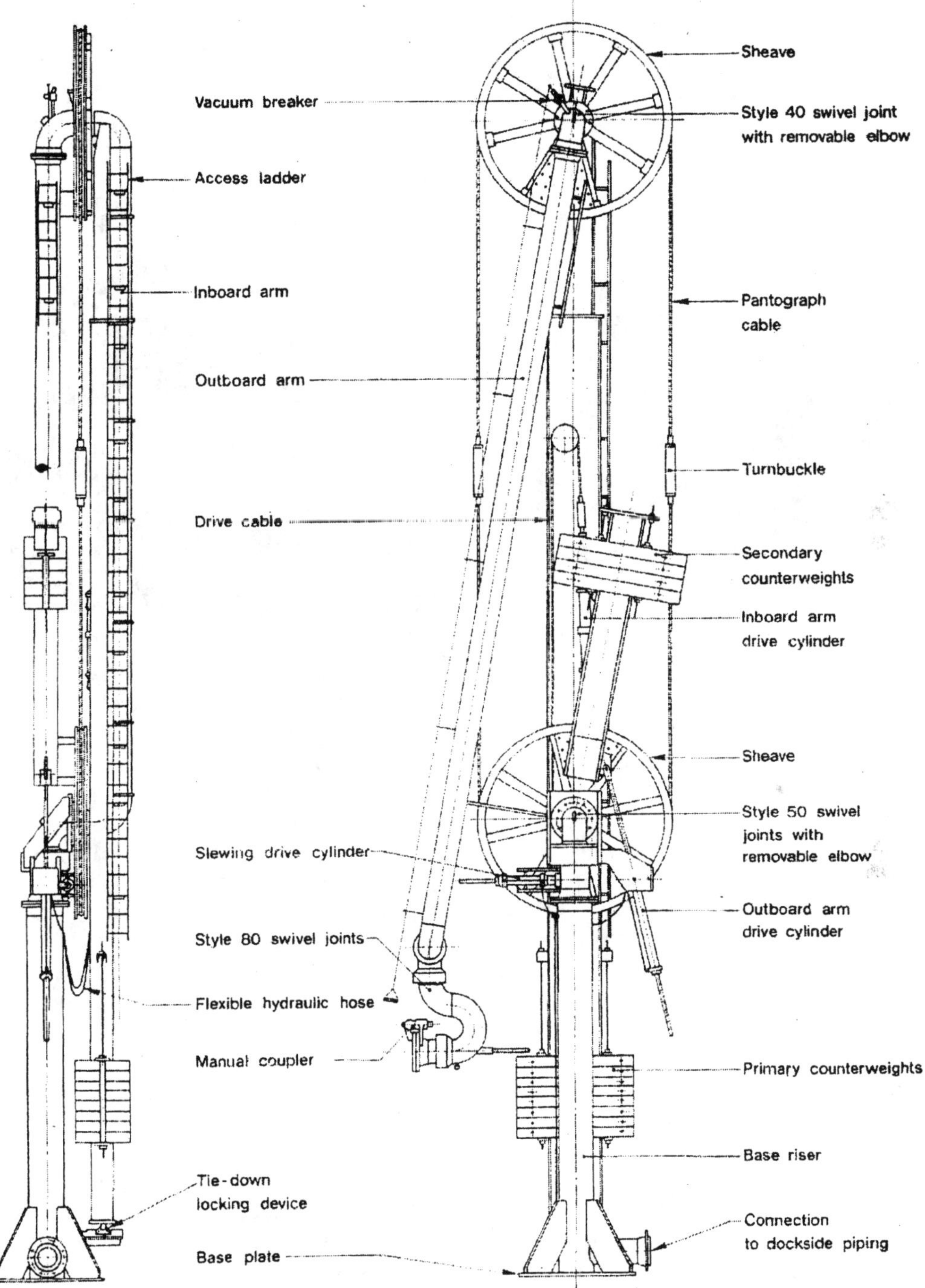

Figure 2.7 Articulated loading arm. (Courtesy of FMC.)

Figure 2.8 Hose tower at East Boston, Massachusetts, petroleum product terminal. (Courtesy of Han-Padron Associates.)

Figure 2.9 Articulated rigid piping loading system supported by a tower at Everett, Massachusetts. (Courtesy of Han-Padron Associates.)

tower loader/unloader require moving the ship along the berth, whereas the moving loader/unloader will move from hold to hold with the ship staying in a fixed position. Figure 2.10 illustrates a fixed loader.

A ship loader generally has a side boom carrying a conveyor that receives cargo from the main conveyor bringing the cargo from the storage yard and loads the ship at the outboard end through a chute. Telescoping chutes are used at certain facilities to minimize dust emissions as material drops from the chute into the ship's hold. Many different types of moving loaders are in use, and they are generally classified as traveling, radial, and linear, with many variations of each type.

A traveling loader moves on rails along the pier or wharf, with a "tripper" attached to transfer the cargo from the main conveyor on the pier or wharf to the conveyor on the traveling loader boom. A traveling loader for loading coal into 150,000-dwt vessels is illustrated in Figure 2.11.

A radial loader is pivoted at the point where the main conveyor transfers the cargo onto the loading boom. The outboard end of the loader rides on a circular track, and the loader boom shuttles in and out to reach the vessel's holds as the loader travels along the circular track. To service very large vessels, radial loaders are used in pairs (Figure 2.12).

A linear loader is also fed from the main conveyor at a single point, but it travels along a straight track on the pier or wharf deck, with a shuttling conveyor connecting the main conveyor to the loader boom shuttling conveyor. The operating principle of a linear loader is illustrated in Figure 2.13.

Unloading of dry bulk ships is usually accomplished by one of the following methods:

Figure 2.10 Fixed loader and continuous bucket-ladder unloader at lower Mississippi River coal transshipment terminal. (Courtesy of Han-Padron Associates and International Marine Terminals. Photograph by Commercial & Industrial Photographers.)

- Self-unloader bulk carriers with feeders and a collecting conveyor below the holds, feeding a long boom conveyor for discharging to shore
- Clamshell unloaders mounted on gantries that travel on rails parallel to the pier face or attached to traveling or stationary slewing cranes on the pier (Figure 2.14)
- Bucket-ladder unloaders, which consist of an endless line of buckets connected by wire ropes or chains and suspended from a hinged boom (Figure 2.10)
- Pneumatic unloaders, which use a vacuum system to transfer relatively light, fine-grained bulk cargoes
- Screw conveyor unloaders, which feature a screw conveyor at the outboard end of a boom conveyor, usually mounted on a traveling gantry (Figure 2.15)

A conveyor system is used to transfer bulk material between the berth and the storage stock piles. Dry bulk materials are usually stored in large open stock piles or in covered

Figure 2.11 Traveling ship loader at lower Mississippi River coal transshipment terminal. (Courtesy of Han-Padron Associates and International Marine Terminals.)

buildings or silos if the product must be protected from the weather. Conveyor systems connect the stock piles with the mooring facility on one end and the inland transport system on the other. Material coming into the storage area via one of the main conveyors is usually placed in the pile by a rail-mounted traveling slewing stacker or an overhead conveyor system, as illustrated in Figure 2.16, often supplemented by dozing to increase storage capacity. For smaller storage volumes, a radial stacker, supplemented by dozing, is used (Figure 2.17).

Reclaiming materials from stock piles can be accomplished by mobile or rail-mounted bucket or clamshell reclaimer, rail-mounted bucket wheel reclaimer, combination stacker–reclaimer (Figure 2.18), or an underground hopper and tunnel system. For smaller installations, reclaiming may be done by payloaders feeding stationary or movable hoppers.

2.3 VESSEL CHARACTERISTICS

When planning and designing port-related marine structures, it is critical to define the appropriate design vessel or vessels. The vessel dimensions and characteristics are key to the appropriate design of the facilities that will service the vessels. The design vessel may be an actual vessel or a composite of the characteristics of a number of vessels. The size and

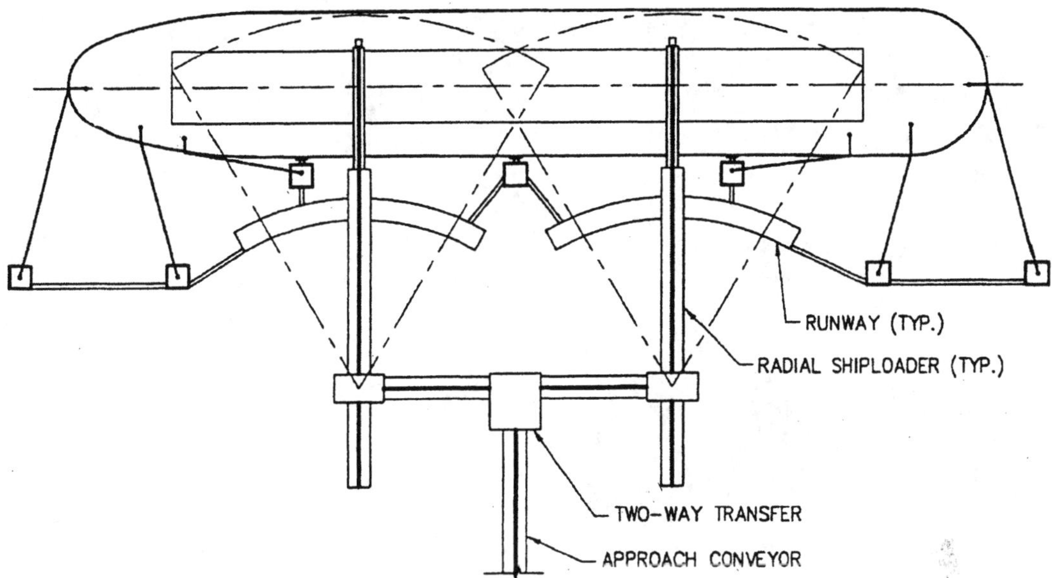

Figure 2.12 Dual radial ship loaders.

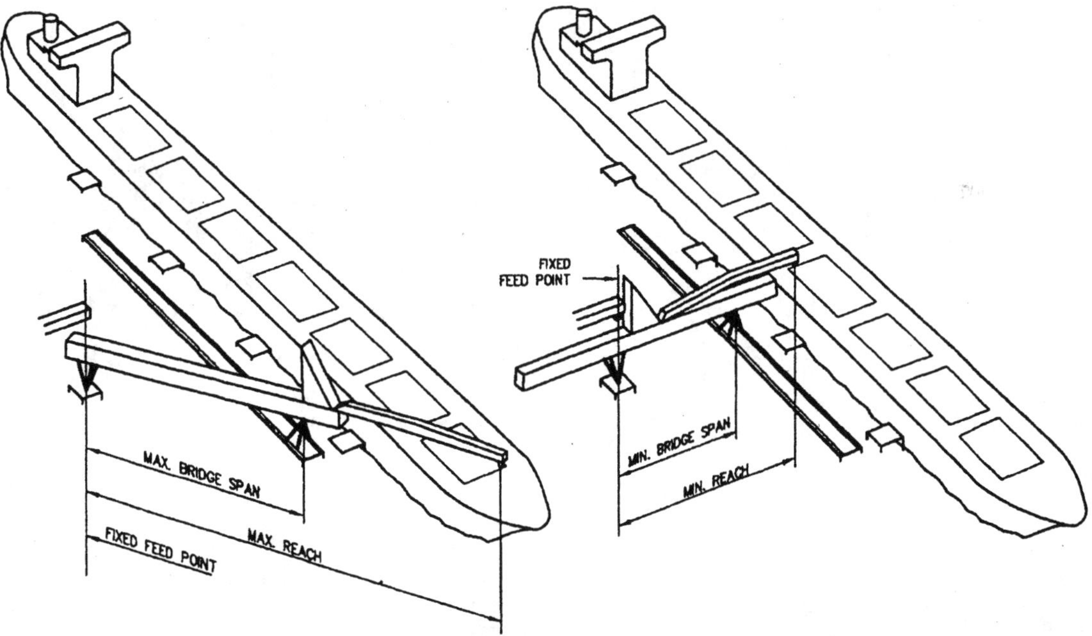

Figure 2.13 Linear ship loader.

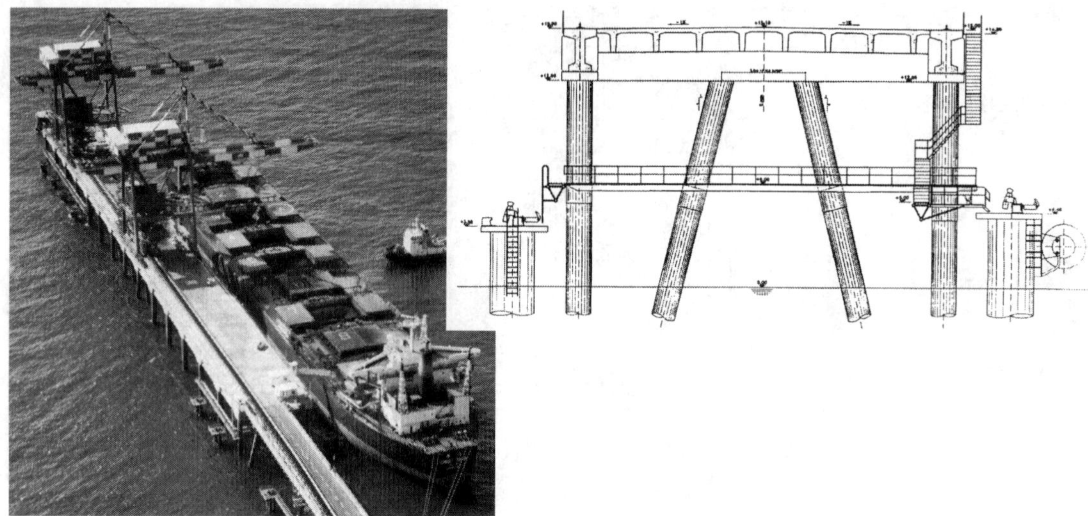

Figure 2.14 Traveling clamshell unloaders at Ashkelon, Israel. (Courtesy of Yaron-Shimoni-Shacham.)

Figure 2.15 Screw conveyor unloaders. (Courtesy of Siwertell.)

Figure 2.16 Overhead stacking and tunnel reclaim system. (Courtesy of Han-Padron Associates.)

characteristics of the design vessel should be based on forecasts of future trends in shipping. Port structures normally have a relatively long useful life, so it is important to consider the nature of the vessels likely to be calling in the future.

Figure 2.17 Radial stacker at the Meramec Power Plant, Missouri. (Courtesy of Energy Associates.)

The basic characteristics of container ships, general cargo ships, auto carriers, tankers, bulk carriers, and passenger ships are provided in Tables 2.1 through 2.6, respectively. The dimensions given in the tables are approximate and may vary considerably among different vessels in the same size class. For more information on vessel characteristics the reader is referred to Chapter 1, PIANC (2002), and Cork and Holm-Karsen (2002).

2.4 TYPES OF STRUCTURES

2.4.1 Wharves

A wharf or quay (the terms are used interchangeably) is a marine structure for berthing vessels which is constructed essentially parallel to the shoreline. It is usually contiguous with the shore and is generally created by constructing a wall or other type of retaining structure, and dredging in front of the structure to create sufficient water depth for the vessels expected to call at the facility, while filling behind the structure to raise the ground surface to an elevation compatible with the intended upland use of the facility.

There are numerous concepts that are suitable for the design of wharves or quays. These concepts fall into two broad categories: those providing a fully closed structure along the face of the wharf and those with an open structure profile. Each has advantages and disadvantages, and selection of the optimum concept for a particular situation will depend on the conditions that exist at the location. In some situations, the fact that closed wharf structures reflect wave energy while open wharf structures tend to dissipate wave energy may be an important consideration.

In the following sections, various generic types of wharf structure concepts are discussed. Each time one of these structures is designed to suit a particular scenario, it will undoubtedly incorporate modifications and features not discussed or illustrated, to satisfy the conditions and objectives of the particular project. Readers should not consider the concepts discussed and illustrated to be a comprehensive compilation of available concepts, but rather, an identifica-

Figure 2.18 Stacker/reclaimer at lower Mississippi River coal transshipment terminal. (Courtesy of Han-Padron Associates and International Marine Terminals.)

Table 2.1 Container ship characteristics

TEU	Dead-Weight Tonnage	Loaded Displacement (tonnes)	Length (m)	Beam (m)	Loaded Draft (m)
400	6,500	9,000	115	19	6.4
800	15,000	20,000	180	27	8.8
1,200	22,000	30,000	200	30	10.4
1,600	30,000	42,000	215	31	11.3
2,000	36,000	50,000	260	32	11.6
2,500	44,000	65,000	275	32	11.9
3,000	51,000	75,000	290	32	13.1
3,500	56,000	80,000	275	39	12.5
4,000	60,000	85,000	280	39	12.8
5,000	65,000	95,000	285	40	13.1
6,000	80,000	115,000	305	41	13.7
8,000	105,000	150,000	335	46	14.0

Table 2.2 General cargo ship characteristics

Dead-Weight Tonnage	Loaded Displacement (tonnes)	Length (m)	Beam (m)	Loaded Draft (m)
1,000	1,600	65	10	4.3
2,000	3,000	80	13	4.9
5,000	7,100	110	16	6.7
10,000	13,600	135	20	8.5
20,000	25,800	165	24	9.8
40,000	49,300	200	29	11.6
60,000	71,900	225	32	12.8
80,000	94,000	245	35	13.7
100,000	116,000	260	39	15.9

Table 2.3 Auto carrier characteristics

Number of Cars	Dead-Weight Tonnage	Loaded Displacement (tonnes)	Length (m)	Beam (m)	Loaded Draft (m)
5,600	18,000	33,000	195	32	9.8
6,000	26,000	42,000	210	32	10.4
6,200	28,000	45,000	200	32	11.9

Table 2.4 Tanker characteristics

Dead-Weight Tonnage	Loaded Displacement (tonnes)	Length (m)	Beam (m)	Loaded Draft (m)
1,000	1,400	60	9	4.0
2,000	2,800	75	11	5.0
5,000	7,000	100	14	6.4
10,000	13,000	140	17	7.9
20,000	25,000	180	22	9.5
40,000	50,000	215	29	11.0
60,000	75,000	245	32	13.1
80,000	100,000	260	37	13.7
100,000	125,000	285	41	14.6
120,000	150,000	295	42	16.5
150,000	180,000	300	44	17.1
200,000	240,000	310	47	18.9
250,000	300,000	325	50	20.4
300,000	356,000	340	53	22.0

Table 2.5 Bulk carrier characteristics

Dead-Weight Tonnage	Loaded Displacement (tonnes)	Length (m)	Beam (m)	Loaded Draft (m)
5,000	7,000	110	16	6.7
10,000	13,000	140	18	7.9
20,000	24,000	180	23	9.8
40,000	53,000	220	30	11.3
60,000	78,000	245	32	11.9
80,000	100,000	260	39	12.5
100,000	125,000	270	42	13.1
120,000	150,000	280	43	14.9
150,000	180,000	290	45	17.4

Table 2.6 Passenger ship characteristics

Gross Registered Tonnage	Loaded Displacement (tonnes)	Length (m)	Beam (m)	Loaded Draft (m)
30,000	30,000	230	28	9.5
40,000	35,000	265	30	10.0
50,000	45,000	300	31	10.4
60,000	55,000	310	33	10.6
70,000	65,000	315	34	11.0
80,000	75,000	315	35	11.6

tion of the generic types of concepts that have been used over the years. It is the designer's responsibility to modify, combine, and otherwise use these generic concepts as a basis for beginning the process of creating a design that will be most appropriate for the project at hand.

2.4.1.1 Closed Wharves. Figure 2.19 shows an example of a closed wharf concept. This concept utilizes a continuous, straight, heavy-duty steel sheet pile bulkhead. In this arrangement, the bulkhead is tied back near its top to resist the soil pressure acting behind it. This is a very common type of wharf construction technique and has been used widely around the world for many years. Wharves have also been constructed similarly substituting concrete sheet piles for steel sheet piles. However, handling and installation impose significant limitations on the use of concrete sheet piles, and they are not used widely for this type of structure.

The lateral support for the bulkhead illustrated in Figure 2.19 is provided by a tie rod and pile-supported deadman system. A number of alternative methods of providing lateral support are also used, depending on the local bathymetric, geotechnical, and other conditions. For example, a deadman that utilizes passive earth pressure may be used in place of a pile-supported deadman. Alternatively, a tie-back system featuring rock anchors or soil

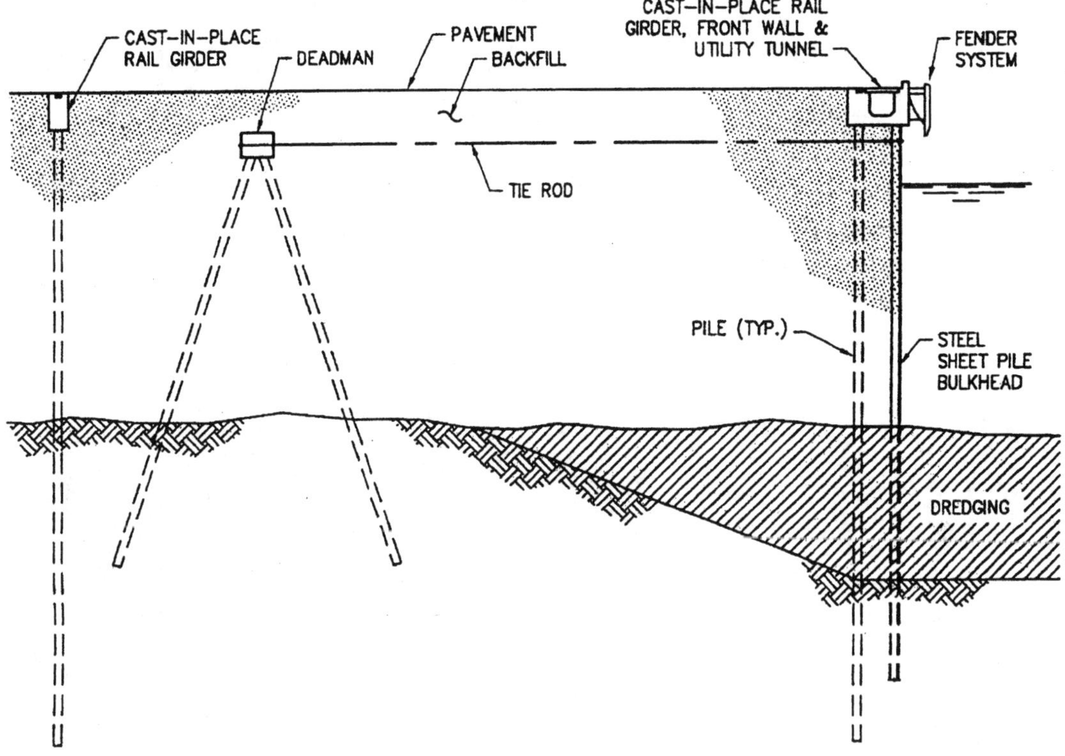

Figure 2.19 Steel sheet pile closed-face wharf.

anchors may be substituted for the tie rod and deadman system. In deepwater applications, tie rods at two or more elevations may prove to be cost-effective.

Another steel closed type of wharf features individual cells formed of flat steel sheet piles interconnected to provide a continuous structure (Figure 2.20). A cast-in-place concrete header cap provides the multiple functions of (1) maintaining sheet pile integrity along the critical edge of the wharf, (2) providing a continuous straight wharf face for fender system mounting, (3) providing a trench for housing of utilities, and (4) serving as the waterside crane rail girder, with the addition of supporting piles, for wharves that will support rail-mounted cargo handling cranes. The cell geometry and weight provide lateral stability for the wharf without the need for tieback systems or other restraining devices.

A closed wharf of concrete construction is illustrated in Figure 2.21. This wharf concept features large, prefabricated concrete caisson units positioned on a foundation bed of dense gravel or crushed stone. The caissons may be designed as closed units to be floated in position and lowered to the seabed by controlled flooding, or they may be designed as open units to be lifted and lowered into position by a crane. As a variation of the lifted type caisson, the rear wall may be omitted and the cross walls modified, resulting in a buttressed L-shaped unit in cross section (Figure 2.22). Eliminating the rear wall reduces the lifting

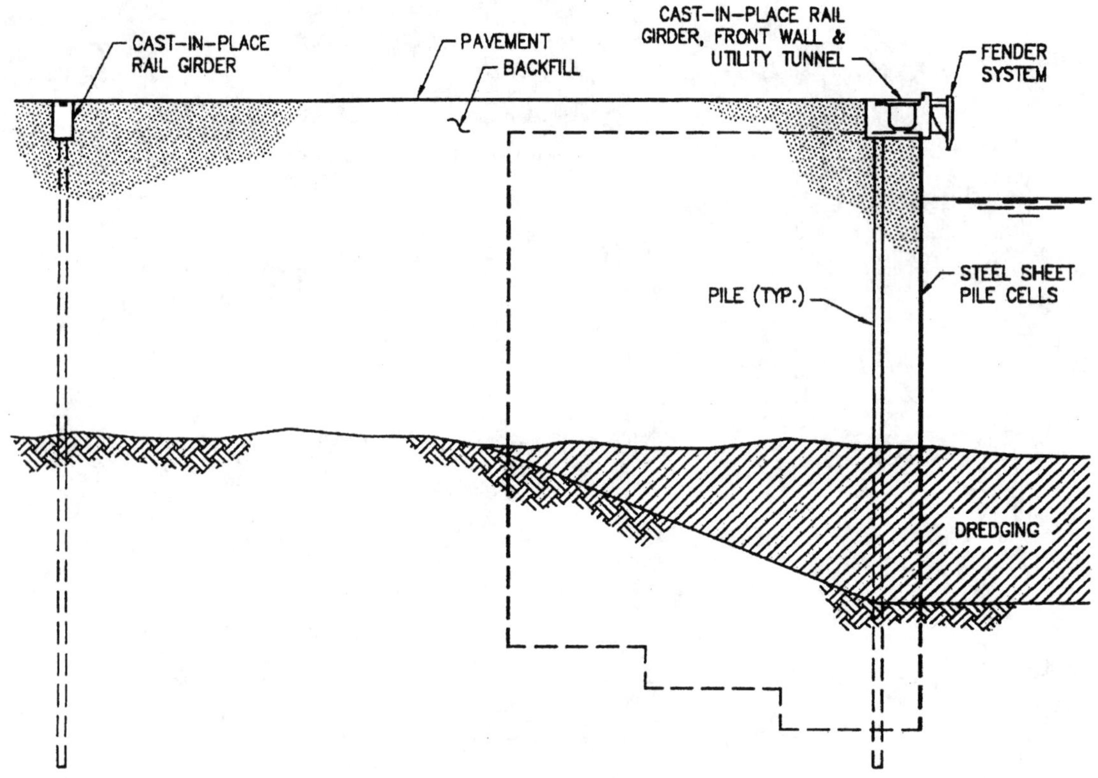

Figure 2.20 Steel sheet pile cell closed-face wharf.

weight and thus allows longer units to be handled by a crane of given capacity. Once in position, the caissons or L-sections are filled with sand or gravel to increase their weight and provide the required resistance to overturning and sliding. After backfilling and compaction of fill material is complete, the tops of the caisson units are fabricated of reinforced cast-in-place concrete, followed by the gantry crane rail support beam if required, and the cantilever section of the superstructure. Scour protection along the toe of the caisson units is usually provided.

Another type of closed concrete wharf is the block gravity wall concept (Figure 2.23). After dredging, a foundation bed of dense gravel or crushed stone is placed, followed by installation of large prefabricated concrete block units. Both solid and hollow blocks may be considered. Hollow blocks can be made larger than solid blocks without increasing their weight. Therefore, for the same weight as a solid block, the hollow block occupies a larger surface area of wall. This means that fewer blocks would be lifted with heavy lift equipment during construction. The hollow vertical cavities are filled with either crushed stone or tremie concrete, which can be installed using less costly land based equipment. The blocks are sufficiently large and heavy to withstand lateral loads resulting from soil pressure combined with surcharge live loads, berthing and mooring loads, and seismic events. As with the caisson type of construction, scour protection is usually pro-

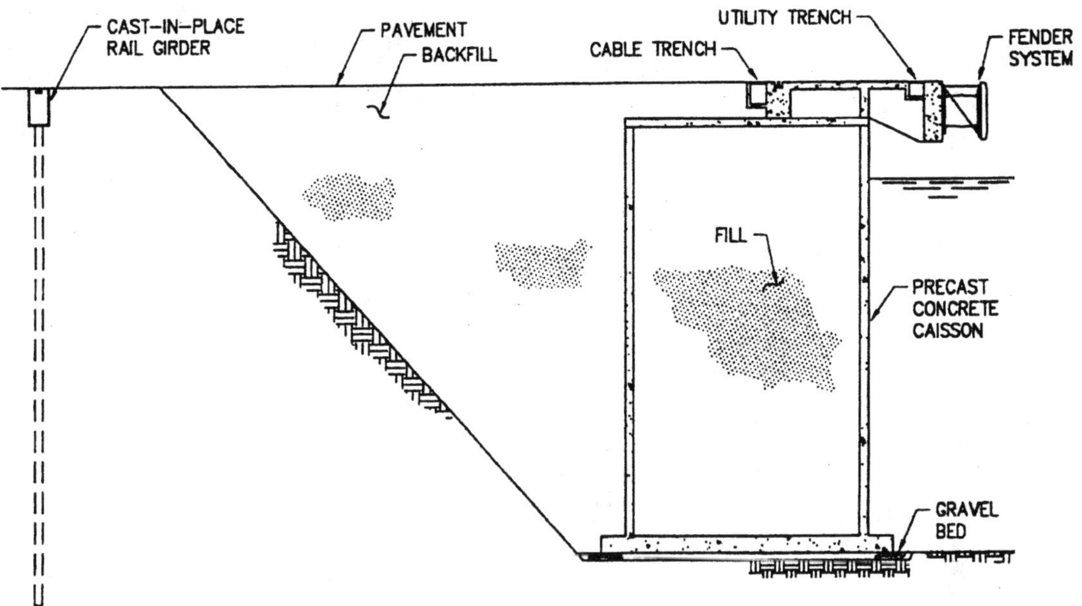

Figure 2.21 Precast concrete caisson closed-face wharf.

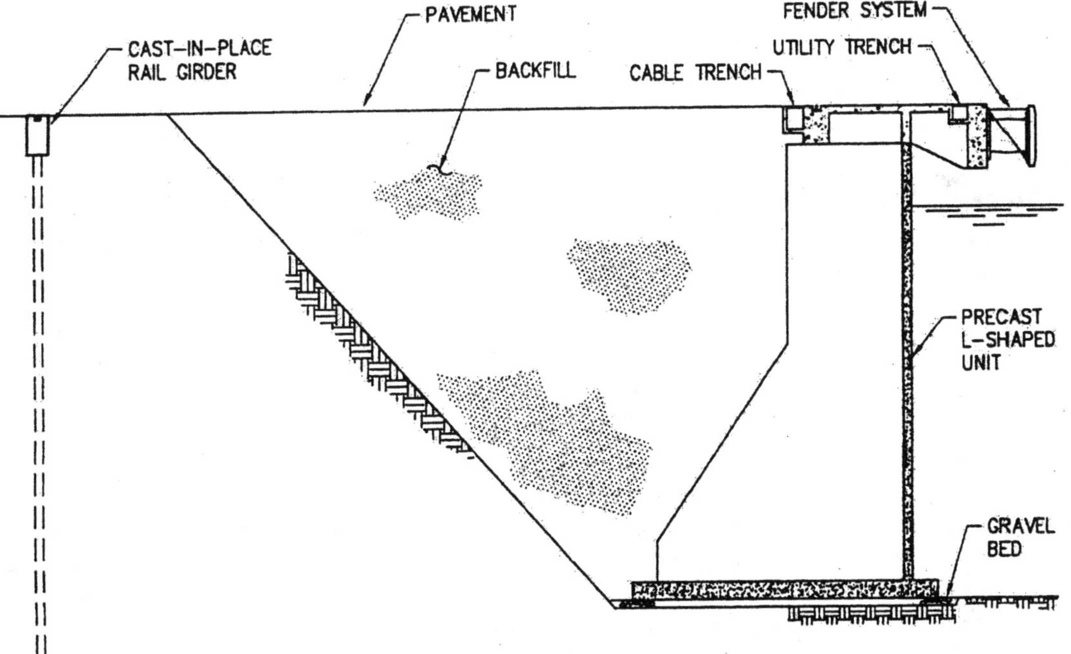

Figure 2.22 Precast concrete L-shaped unit closed-face wharf.

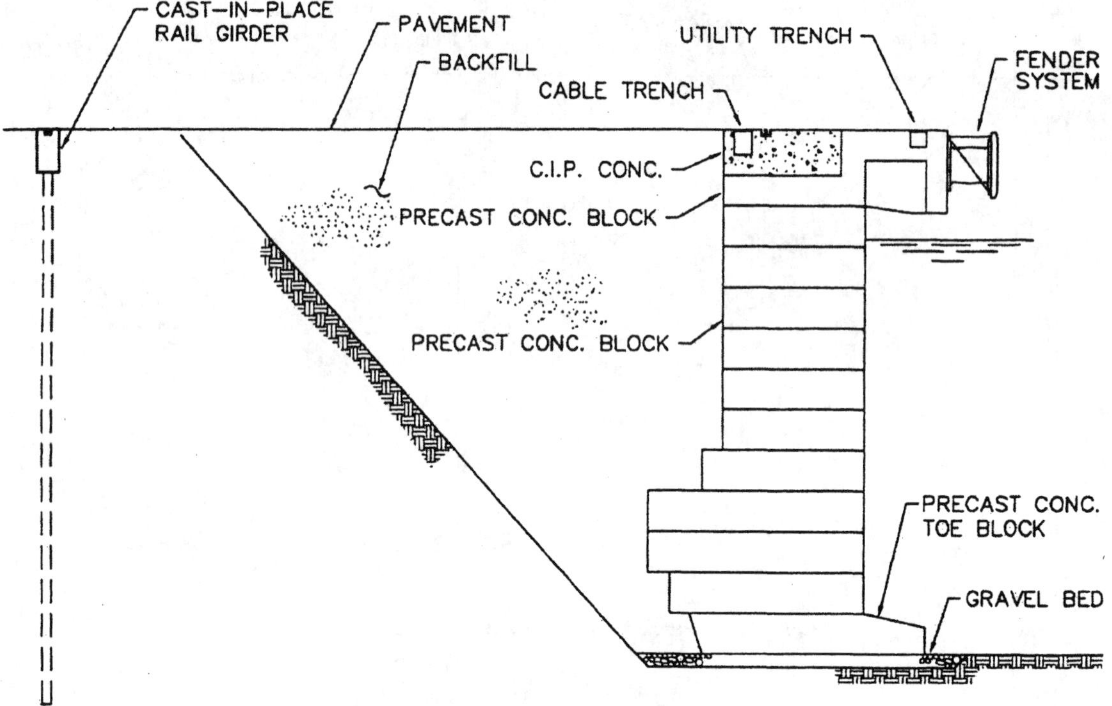

Figure 2.23 Precast concrete block closed-face wharf.

vided at the toe. For more information consult Tsinker (1997).

2.4.1.2 Open Wharves. An open type of wharf design is shown in Figure 2.24. This is a relieving, or low-level platform, a concept that is often cost-effective for deeper-water applications where relatively large vessels are to be accommodated. Along the inshore edge of the platform, short lengths of sheet piling of either steel or precast concrete are used as a cutoff wall. The pile-supported platform structure then extends from the face of the sheet piling to the face of the wharf. The soil pressure acting on the sheet piling is resisted by batter piles built into the platform structure as shown. The backfill material on top of the platform provides the required dead weight to ensure that the vertical piles are not subjected to significant tension forces as a result of the uplift component from the batter piles. In addition, the backfill tends to distribute heavy concentrated loads on the wharf pavement, such as wheel loads from cargo handling equipment, so that the concrete deck slab need not be designed to accommodate these concentrated loads. Thus, even though the deck slab must support more dead load than the deck slab of the high deck platform described below, it may not necessarily be of heavier construction. The backfill also permits relatively easy adjustments to accommodate future changing deck operational requirements for the facility.

Figure 2.25 illustrates an open wharf concept, referred to as a *high deck platform*, that is similar to the relieving platform concept, except that the concrete deck structure is positioned at the operating level of the wharf and

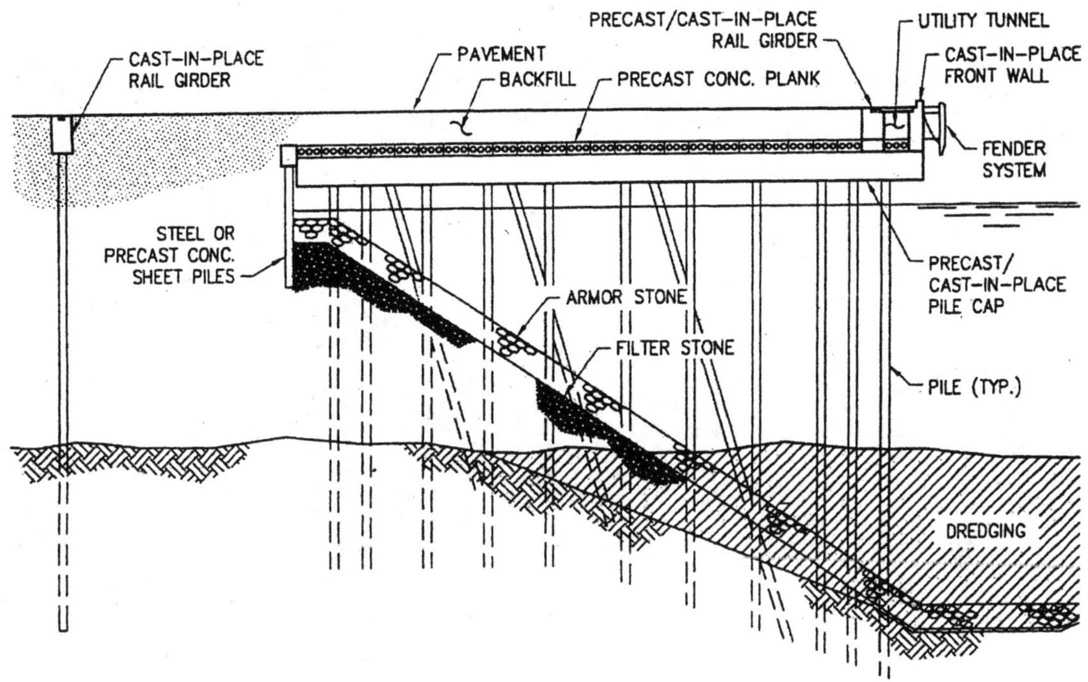

Figure 2.24 Relieving platform open-face wharf.

the backfill above the deck has been eliminated. This concept offers the advantage of a reduced dead load to be supported by the pile system, as compared to the relieving platform, but has the disadvantage of not readily allowing future changes to accommodate changing operational requirements.

Of course, for each of the alternatives illustrated, there are many possible variations and refinements. For example, lateral stability for the relieving platform concept (Figure 2.24) or high deck platform (Figure 2.25) can be provided by batter piles at the cap of the sheet piles (Figure 2.26), rather than by batter piles under the deck. Alternatively, lateral stability may be provided by any of a number of different tie rod and deadman systems, or rock or soil tieback anchor systems, as described above for a sheet pile bulkhead. In another variation of the high deck concept, the width of the deck can be increased and the underdeck slope extended to meet the deck slab (Figure 2.27), thus eliminating the sheet pile cutoff wall. This eliminates most of the lateral earth pressure acting on the structure and permits the use of vertical piles alone to resist horizontal forces acting on the structure. High deck platforms supported solely on vertical piles are particularly attractive in areas subject to high seismic activity because of their relatively light weight (compared to low-level platform structures) and the relatively flexible nature of the vertical pile system (compared to a batter pile system). For more information consult Tsinker (1997).

2.4.2 Piers

A pier is a shore-connected marine structure for berthing vessels, which may be of several types of configurations, principally categorized as finger, T-head, and L-shaped. A finger pier is generally oriented more or less perpendicular

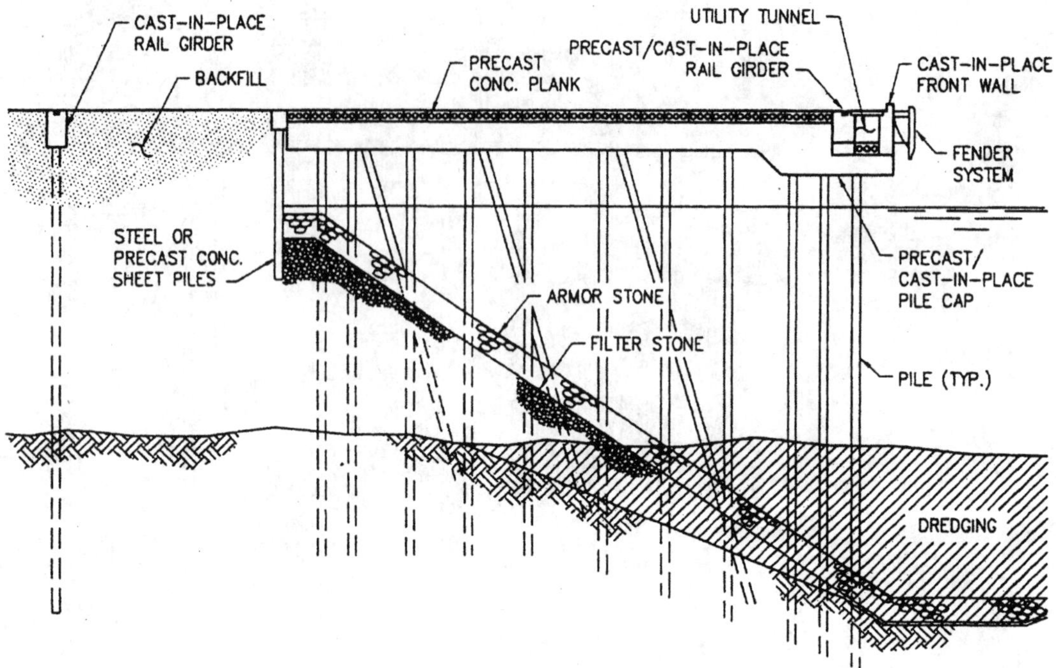

Figure 2.25 High-level platform open-face wharf.

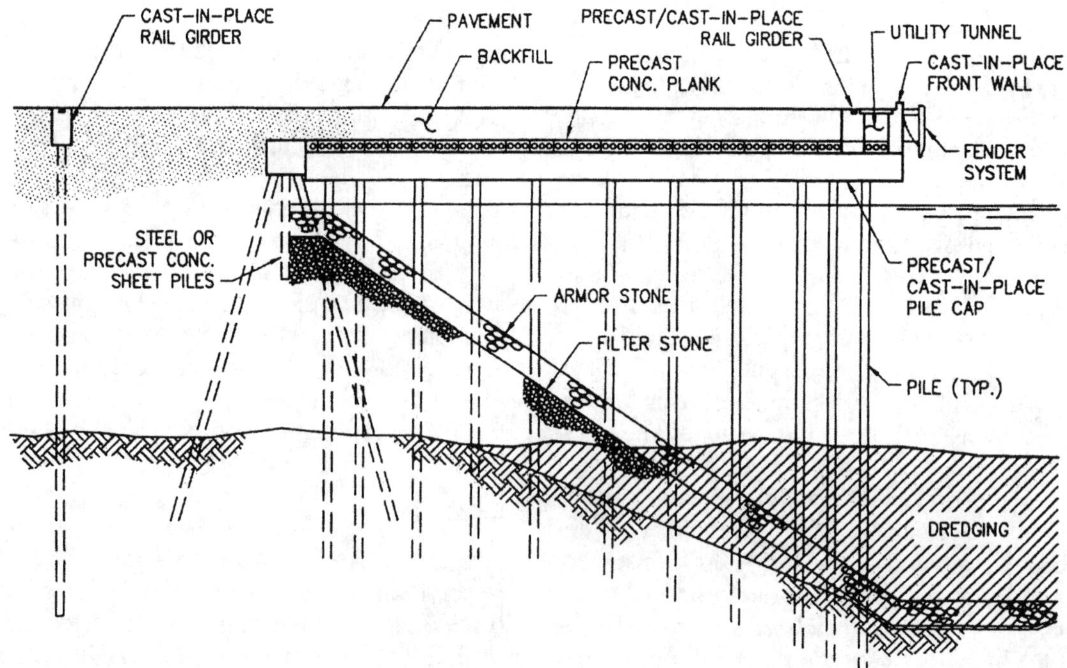

Figure 2.26 Batter pile anchored relieving platform open-face wharf.

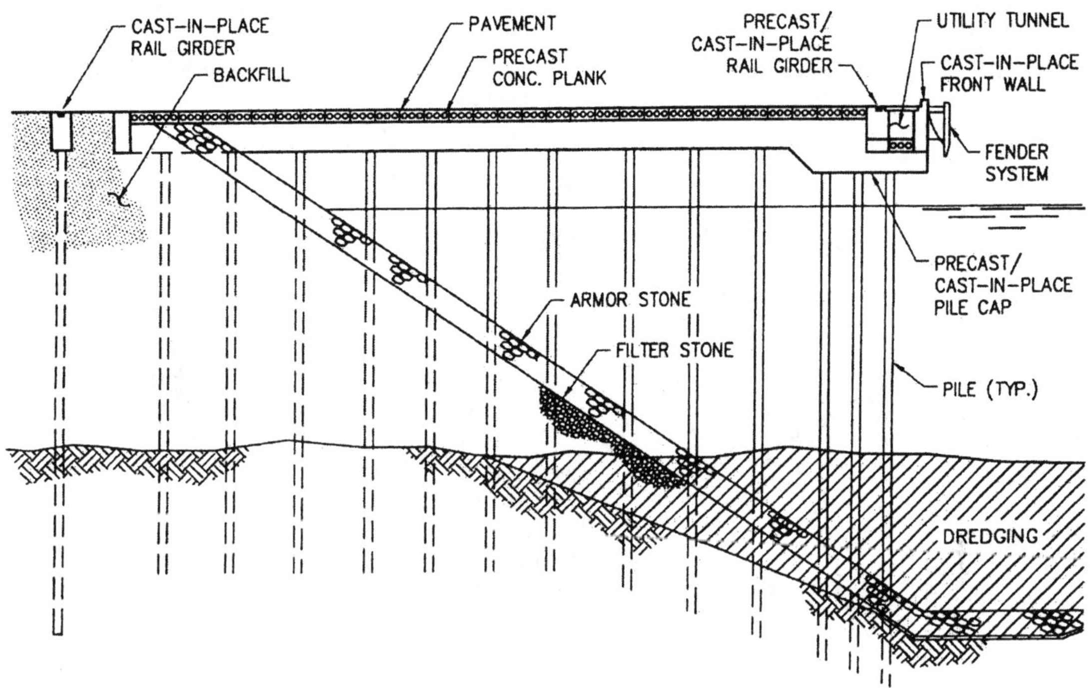

Figure 2.27 Wide high-level platform open-face wharf.

to the shoreline and usually provides two-sided berthing. A T-head pier is generally oriented essentially parallel to the shoreline, with an access trestle that connects to the shoreline at a point near the center of the pier. It generally provides for berthing only on the offshore side, but can provide two-sided berthing if the T-head is of sufficient length. An L-shaped pier is similar to the T-head pier, but the access trestle connects one end of the pier to the shoreline so that both sides of the pier may readily be used for berthing.

Although a pier may be constructed as a closed structure, similar to the closed type of wharf structures described above, most piers are constructed as open structures. Open piers are supported on piles and may usually be categorized as those supported by vertical piles only (Figure 2.28), those supported by a combination of vertical and batter piles (Figure 2.29), or those supported by batter piles only (Figure 2.30). Of course, many variations of these configurations are used. Also, in some situations a jacket substructure, similar to the concept used for offshore platforms in the petroleum industry, may be the preferred choice for the pier substructure rather than the free-standing pile concepts illustrated. Under certain conditions, particularly where the tidal range is extreme or where the pier is to be used only for a relatively short period of time, a floating type of pier construction may be the most appropriate solution.

A vertical pile pier is particularly effective in locations subject to significant seismic activity. The vertical piles resist horizontal loading (berthing impact, mooring forces, environmental forces, and seismic loading) through the

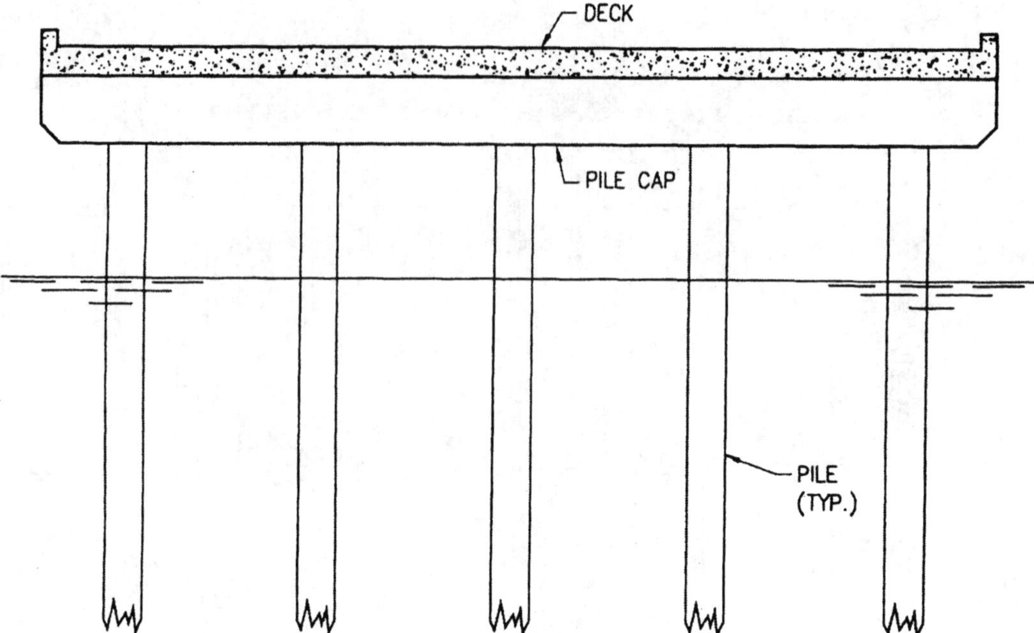

Figure 2.28 Vertical pile-supported pier.

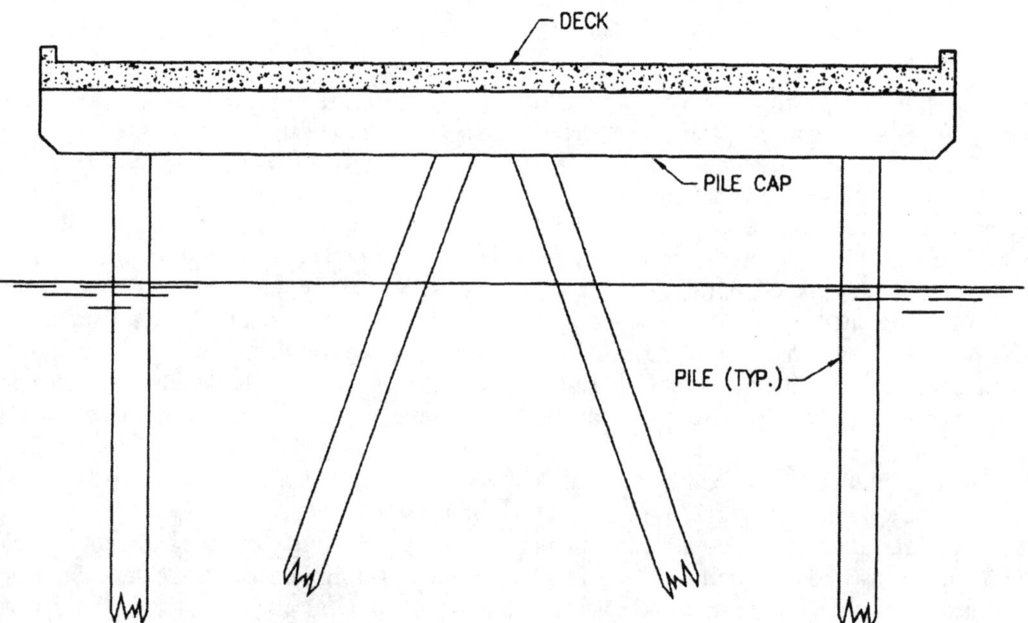

Figure 2.29 Vertical and batter pile-supported pier.

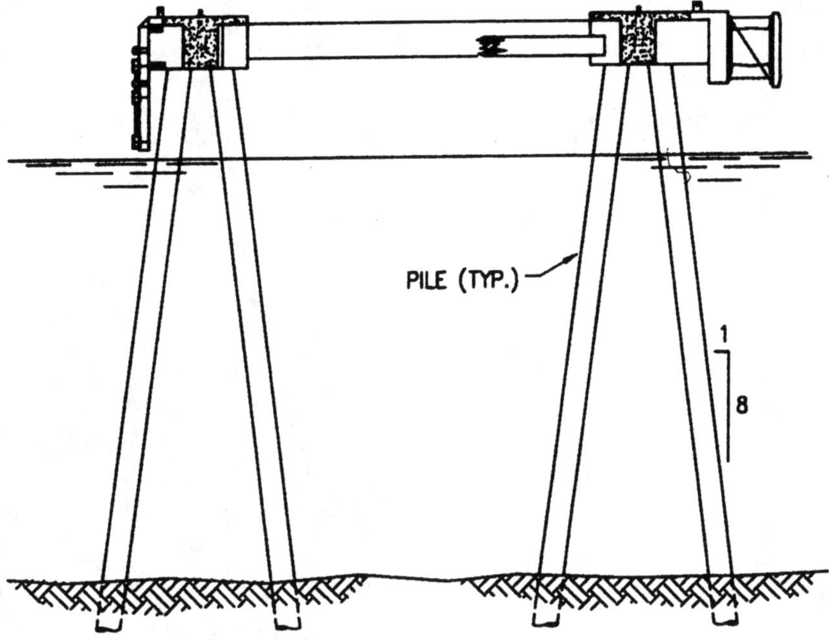

Figure 2.30 Batter pile-supported pier.

frame action of the piles and pile cap of each bent. The load is resisted primarily by means of bending of the piles and lateral pile–soil interaction. Since the flexibility of the piles in bending is significantly greater than in axial loading (as for batter piles), the magnitude of the seismic loading is less than for the batter/vertical pile pier. Also, the vertical piles resist horizontal forces equally in all directions, whereas batter piles resist horizontal forces only in the direction of the batter. Thus, batter piles must be provided in at least two directions, usually at right angles, to resist seismic loading, which can occur in any direction.

A batter/vertical pile pier is usually more cost-effective where seismic activity is relatively low and the major horizontal loading is primarily in one direction (berthing impact and transverse mooring forces). Horizontal loading is resisted primarily by the axial stiffness of the batter piles, which is more cost-effective than resisting this loading through bending of vertical piles for loading applied along one line of action. This type of structure is stiffer than the vertical-pile-only structure, resulting in higher seismic loading. The batter/vertical pile pier structure is also usually more cost-effective than the vertical-pile-only type, where the seabed soil is weak for a substantial depth so that laterally loaded piles are relatively ineffective.

An all-batter pile system is, in effect, a combination of the two concepts discussed previously. The inclination of the batter piles is usually very slight, resulting in a relatively flexible structure, making this concept attractive in regions of high seismicity. The flexibility of the structure also results in the distribution of berthing impact forces over a great many pile bents, and consequently, results in a cost-effective pier structure.

2.4.3 Dolphins

Often, for berths intended to serve liquid bulk or dry bulk vessels, a series of individual structures, referred to as *dolphins,* is more cost-effective than a continuous wharf or pier structure. A typical liquid bulk berth would be made up of a group of individual structures consisting of breasting dolphins, mooring dolphins, loading platform, interconnecting walkways, and an approach trestle (Figures 2.31 and 2.32). Berths for dry bulk vessels are generally similar to those for liquid bulk vessels, except that a runway suitable for the bulk cargo loader or unloader is provided in place of the loading platform. A typical dry bulk berth arrangement, in which a radial loader is utilized, is shown in Figures 2.33 and 2.34.

Breasting dolphins support the fender system and usually are equipped with mooring fittings to secure the vessel's spring mooring lines. As a general rule, the breasting dolphins of tanker berths are arranged symmetrically

Figure 2.32 Sea island tanker berth at Kharg Island, Iran.

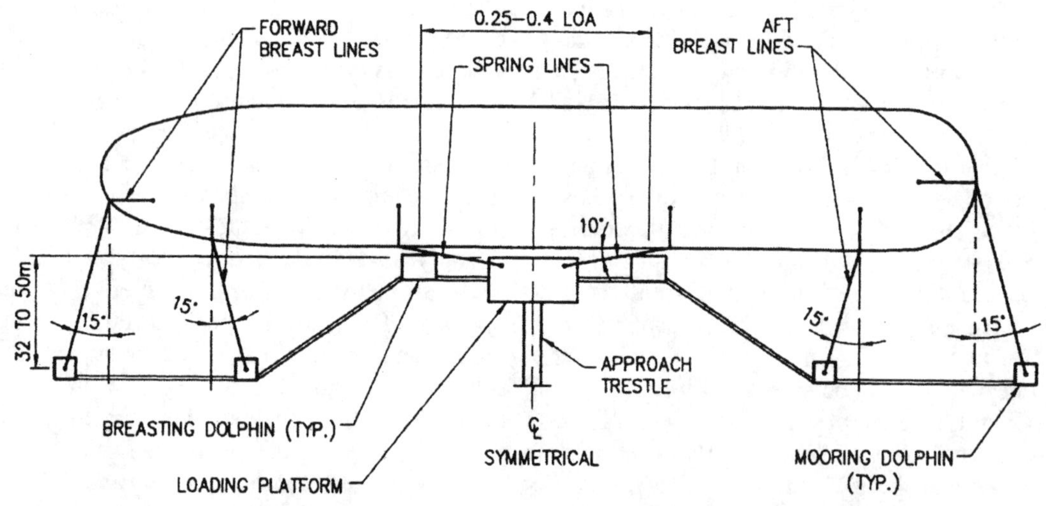

Figure 2.31 Tanker berth layout. (After OCIMF.)

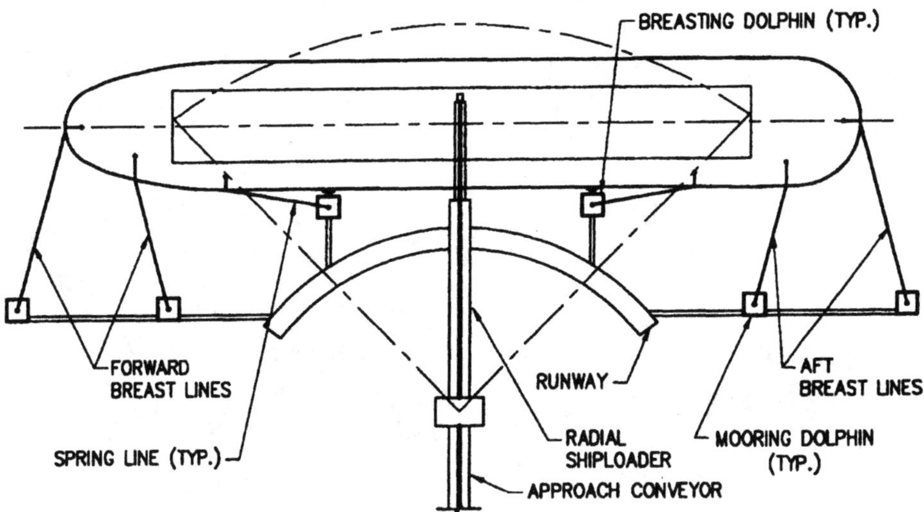

Figure 2.33 Typical dry bulk berth layout.

about the center of the loading platform piping manifold, spaced approximately one-third the length overall (LOA) of the design vessel. Wider spacing of the breasting dolphins will minimize the energy of berthing impacts and decrease mooring line loads resulting from yawing of the moored vessel. However, the maximum spacing should not exceed 0.40 times the LOA of the vessel, to ensure that the breasting dolphins are within the parallel sides of the vessels. Also, the spacing should not be less than 0.25 times the LOA to ensure the stability of the moored vessel. When a berth must accommodate a wide range of vessel sizes, two pairs of breasting dolphins may be required so that the spacing between the dolphins remains within these limits.

The discussion above regarding the arrangement of the breasting dolphins for a liquid bulk terminal also generally applies to a dry bulk terminal with a traveling loader or unloader of sufficient range so that the bulk carrier does not have to be shifted along the length of the berth to permit the loader or unloader to reach all the vessel's holds. For dry bulk berths with a stationary loader or unloader, or a movable loader or unloader with limited reach, the vessel must be shifted during the loading or unloading process. In this case, the breasting dolphins should be of sufficient number and in such locations that the principles described above regarding location of the dolphins are achieved at all positions of the vessel.

Two primary types of structures are used for breasting dolphins: rigid and flexible. Rigid structures deflect very little under design load, and the tanker berthing energy is absorbed by a resilient fender system. Flexible dolphins deflect under impact, and both the structure and resilient fender absorb the impact energy, thus requiring a substantially smaller resilient fender.

Mooring dolphins support mooring fittings to secure the vessel's head, stern, and breast mooring lines. Generally, mooring dolphins are designed as rigid structures, with batter piles

Figure 2.34 Radial loader coke loading berth at St. Croix, Virgin Islands. (Loader not yet constructed.) (Courtesy of Han-Padron Associates and Misener Marine Construction.)

providing the necessary horizontal load resisting mechanism. However, under certain conditions vertical monopile structures have proven to be a cost-effective mooring dolphin.

The arrangement of the dolphins, and consequently the arrangement of a vessel's mooring lines and the location of the fender units, can have a significant effect on the magnitude of the forces acting on the mooring system and structures. By arranging the mooring and breasting points efficiently, the forces in the mooring system are minimized, improving the safety of the berth. For relatively small vessels, the holding power of the mooring lines usually provided on the vessel is large relative to the environmental forces acting on the vessel. However, for large vessels this is not the case, and it is essential to provide an efficient mooring arrangement. It is most efficient to moor the vessel within its own length, as recommended by the Oil Companies International Marine Forum (OCIMF, 1997) and illustrated in Figure 2.31.

Sound mooring principles, as described by OCIMF, are as follows:

- Mooring lines should be arranged as symmetrically as possible about the midship point of a vessel.
- Breast lines should be oriented as perpendicular as possible to the longitudinal cen-

terline of a vessel and as far aft and forward as possible.

- Spring lines should be oriented as parallel as possible to the longitudinal centerline of a vessel.
- The vertical angle of the mooring lines should be kept to a minimum.
- Mooring lines of the same size and type (material) should be used for all leads, especially for all leads in the same service (i.e., breast lines, spring lines, etc.).
- If synthetic tails are used on wires, the same size and type of tail should be used on all lines run out in the same service.
- Mooring lines should be arranged such that all lines in the same service are approximately the same length between a vessel's winch and the shore mooring fitting.

Access to the breasting and mooring dolphins may be provided by walkways or by launch, or in some situations the mooring dolphins may be located on shore. For more information on this subject matter consult Tsinker (1997).

2.5 STRUCTURE COMPONENTS AND MATERIALS

Modern port-related structures are comprised of relatively few types of components; generally either steel, concrete, or combinations of these materials. Although timber has been used extensively for port-related structures in the past, particularly in the United States, its use has been reduced dramatically and it will not be used widely in the future for substantial structures. Therefore, timber is not addressed here as a structural material.

The major components of most port-related structures can be classified as piles, bulkheads, pile caps, decks, fenders, and mooring fittings. Each of these components and the materials from which they are fabricated are discussed below.

2.5.1 Piles

Piles supporting port-related structures are fabricated of steel or concrete, selection of which can be controversial. It will depend not only on the site conditions and local availability of materials, but also on the preferences of the designer, because steel and concrete both have advantages and disadvantages. Steel piles are easier to handle than concrete piles in that they are lighter and not subject to cracking during handling. They can also be cut off readily if they cannot be driven to the anticipated tip elevation, or they can readily be lengthened by a welded splice if driven to a greater embedment than anticipated. Concrete piles have a lower material cost and can be extremely durable if fabricated, handled, and installed properly.

Steel piles are usually either pipe sections or H-sections. Pipe sections are normally preferred to minimize the surface area exposed to corrosion and to eliminate corners where coatings are thin and subject to damage. H-sections should be avoided in a marine environment unless there is a compelling reason for their use.

Steel pipe piles may be driven either open or closed ended. When driven open ended, an internal soil plug will develop when the internal skin friction exceeds the end bearing of the plug. Once the plug is fully developed, an open-ended pile has the same capacity as a closed-ended pile. Where a steel pipe pile is to be driven through a rock underdeck slope, or to bedrock, it may be preferable to drive it closed ended. The end closure may be a flat plate or a cast steel conical point. The conical point centers the reaction force when the hard stratum is reached and is particularly useful when the piles are being driven to a sloping bedrock.

Some designers prefer to fill pipe piles with concrete, and some even require the removal of the internal soil plug in piles driven open ended in order to fill the entire pile with concrete, presumably to prevent internal corrosion and/or to increase the capacity of the pile. However, in most cases, neither of these objectives is necessary. It has been demonstrated that once the small amount of oxygen present inside a pile is used up in creating an insignificant layer of rust on the interior of the pile, no further corrosion takes place. Also, steel pipe piles are always designed on the basis that the steel pipe will support the design load. Therefore, any slight increase in structural capacity that may result from filling the pile with concrete serves no purpose.

Steel piles in seawater are subject to corrosion, particularly in the splash zone, and measures have to be taken to ensure their durability. One method is to increase the thickness of the steel beyond that required for structural purposes and to allow the steel to corrode. The amount of additional thickness, the corrosion allowance, will depend on the desired life of the structure and the corrosion rate of steel at the location of the structure. Unfortunately, the corrosion rate can be highly variable, depending on such parameters as temperature, salinity, current velocity, and abrasion from suspended sediments or ice.

Usually, the preferred solution is a corrosion-preventing coating. Steel coatings should be selected for durability in seawater immersion. The most commonly used steel pile coating is coal tar epoxy. Often, a combination of corrosion allowance and coating is used. The use of plastic resin sleeves bonded to pipe piles in the splash zone prior to installation may also be given consideration. Concrete jackets applied in the splash zone after installation have also sometimes been used as a corrosion prevention measure.

Steel H-piles may be preferred where small soil displacement is required so that the piles can penetrate hard layers. Because of this characteristic, an H-pile is often used as a "stinger" on the end of a concrete pile that must be driven through hard layers. The length of the stinger is selected such that the entire stinger is below the finished mudline to avoid the corrosion problems of H-piles in seawater. The tip of an H-pile or stinger may be provided with a cast steel shoe when particularly hard driving or boulders are anticipated.

A cathodic protection system installed in conjunction with a high-quality coating system on the immersed portions of a steel structure is an economic necessity for long design life. When high-quality coating systems are employed, a sacrificial anode system with a long life span is usually the recommended choice over an impressed current system, due to ease of installation and very low maintenance. However, numerous impressed current systems have been installed and operated very successfully. A sacrificial anode system can readily be designed for a 20- or 30-year life and requires virtually no maintenance. With any type of cathodic protection system, a set of test stations should be installed and clear maintenance and operation manuals made available, along with mandates to promote periodic testing.

Concrete piles are probably used most extensively for port-related structures. They are usually either of solid square or octagonal cross section up to a maximum 900 mm across, or hollow cylindrical up to 2000 mm in diameter, and are normally prestressed to prevent cracking during handling and driving. The larger square and octagonal piles are usually cast with a hollow circular core to reduce the weight of the pile. When subject to comprehensive quality control during fabrication, handling, and driving, they can be relatively maintenance free, even in a severe marine environment.

The material cost of concrete piles is generally lower than for equivalent steel piles, but they are heavier and more difficult to handle and are more difficult to splice or extend should site conditions so necessitate. Also, they have more restrictive length limitations than steel piles since handling stresses and weight become excessive for very long piles. When long concrete piles are driven into soft soils, high-tension stresses that can cause cracking of the concrete may develop as tension waves from the hammer blows reflect back from the pile tip. The pile must be adequately prestressed and/or reinforced to prevent these tension stresses from creating hairline cracks which will negatively affect the durability of the pile.

Cast-in-place concrete piles are used occasionally in port-related marine structures. They can be constructed in almost any reasonable size and in most soil conditions. A hole is drilled in the foundation soil and kept open either by the installation of a steel casing or by the use of drilling mud. Reinforcing is placed in the hole, properly centered, and concrete is pumped into the hole by the tremie method. Cast-in-place concrete piles are particularly useful where pile-driving vibration or noise cannot be tolerated or where the soil displacement caused by driven piles may damage adjacent facilities.

Several types of piles made of composite plastic materials or of steel pipe encased in plastic materials have been developed but are not yet used widely for structural applications. More information on this subject matter is provided in Chapter 6.

2.5.2 Sheet Pile Bulkheads

A bulkhead is a waterfront structure provided to permit an abrupt change in ground elevation so that the ground level on one side is substantially higher than the ground or seabed level on the other side. By this definition, all the wharf structures illustrated in Section 2.4.1 can be classified as bulkheads. In this section we cover only sheet pile bulkheads of the type used at or near the face of the wharf, as illustrated in Figure 2.19. Most such bulkheads are tied-back walls constructed of steel sheet piling, but concrete sheet piling is also used widely. Figure 2.35 illustrates conceptually four basic tied-back sheet pile bulkhead configurations.

In the past, timber was used extensively for bulkheads and is still in use today. However, it is expected that timber bulkheads will not be widely used in the future for major marine structures, and therefore, timber bulkheads are not considered here.

Tied-back steel sheet pile bulkheads are the most widely used. Steel sheet piles are readily available in a wide range of shapes and weights, and they can be combined readily with steel H-piles or pipe piles for deepwater applications where greater bending strength is required.

Reinforced concrete sheet piles, either prestressed or not, are also used for construction of bulkheads. Although concrete sheet piles are more difficult to handle and install than steel sheet piles, the relatively maintenance free nature of good-quality concrete can make them an attractive alternative. Usually, concrete sheet piles are fabricated as flat panels, with tongue-and-groove edges, although many variations are in use. Prestressing tends to minimize cracking and thus improve the durability of the sheet piles.

In applications where the height of the bulkhead is such that available steel sheet pile sections do not have adequate bending strength, the bulkhead wall may be made up of a combination of steel H-piles or pipe piles, with pairs of steel sheet piles between. The H-piles or pipe piles are designed to provide virtually the full required bending strength and are driven to a depth sufficient to provide the required lateral restraint. The sheet piles restrain the soil between and transfer the soil pressure to the H-piles or pipe piles, and are driven only a short distance below the design seabed elevation. This type of bulkhead is often referred to as a *combi-wall*. The H-piles may be fabricated from conventional H-piles with interlocks

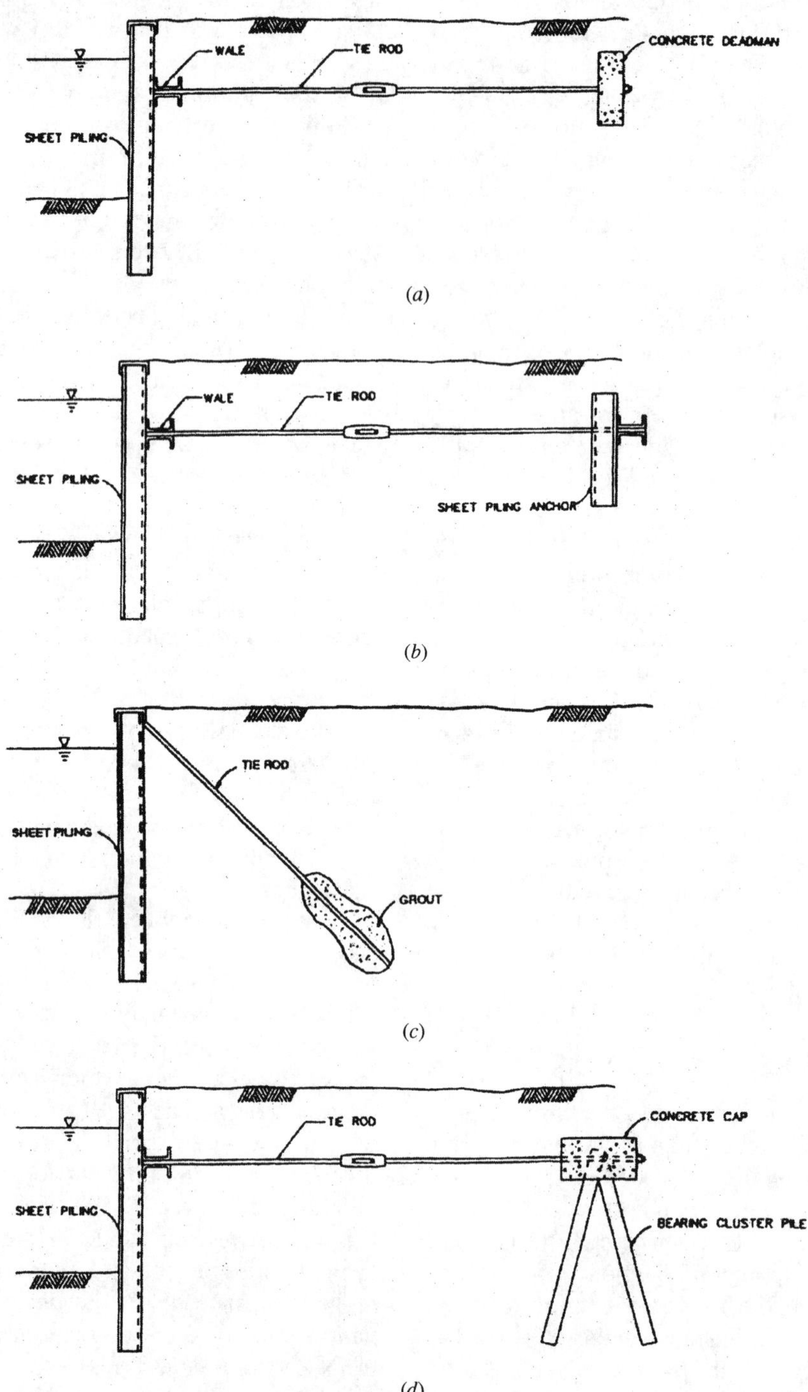

Figure 2.35 Various types of anchored sheet pile bulkheads: (*a*) tie rods and dead man; (*b*) tie rods and anchor wall; (*c*) tiebacks with grout anchor; (*d*) tie rods and A-frame. (From ASCE, 1996.)

cut from sheet piles welded along the edges of one flange, or they may be patented rolled sections with the interlocks integral with the flanges. Figure 2.36 illustrates one type of patented H-pile combi-wall system. Combi-wall pipe piles are fabricated from conventional pipe piles, with sheet pile interlocks welded along diametrically opposite sides (Figure 2.37).

Most tied-back bulkheads are anchored near the waterline by means of a wale and tie rods secured to a deadman located beyond the failure wedge of the soil being retained by the bulkhead. However, there are a virtually unlimited number of variations of this type of bulkhead anchoring system that have been used over the years to suit the circumstances of a particular situation.

Typically, for steel sheet pile bulkheads, the tie rods are secured to the bulkhead by means of a wale extending the length of the bulkhead on either the exterior or the interior face, immediately above the waterline. The wale may be of steel construction, usually back-to-back channels spaced sufficiently far apart for the tie rod to pass through, or a variety of other steel shapes. Sometimes a precast concrete member is used as an exterior wale because of its durability. Although the wale is often located on the exterior face of the bulkhead, in applications where vessels will berth against the bulkhead, or where projections from the bulkhead cannot be tolerated for other reasons, the wale may be installed on the interior face. In this case, the connection between the bulkhead and the wale is a tension connection and must be detailed carefully to ensure that the connecting bolts do not pull through the sheet piles.

Figure 2.36 H-pile combi-wall bulkhead system. (Courtesy of Arbed.)

In some situations, particularly for high bulkheads, tie rods at two or more levels may be cost-effective. In this case, one of the tie rod levels may be below the waterline. Since the cost of installing tie rods under water can be quite high, a substantial savings in bulkhead sheet pile material is required to justify the cost of the lower level of tie rods.

The tie rods are usually round steel bars, often in two segments connected by a turnbuckle to facilitate tightening the rods. For larger bulkheads, the tie rods are usually fabricated with upset threaded ends to reduce the weight and cost of the tie rods. Although most installations utilize solid round steel tie rods, other types of tie rods, such as prestressing strands or wire rope, are also used. All tie rods must be protected against excessive sag caused by soil settlement, and from corrosion. Prestressing strands and wire rope are particularly susceptible to corrosion.

The deadman anchorage system for the tie rods may be any of a number of different concepts, including precast or cast-in-place concrete plates, a continuous concrete wall, individual panels made up of several steel sheet piles, a continuous steel sheet pile wall, or steel or concrete piles of various shapes and in var-

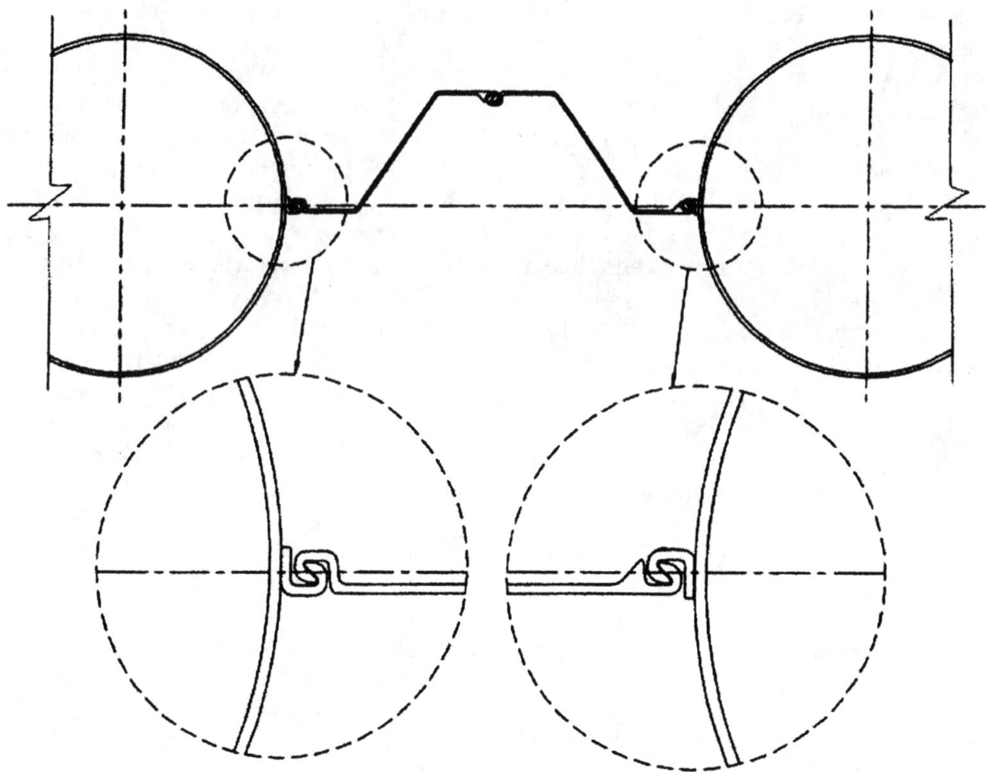

Figure 2.37 Pipe pile combi-wall bulkhead system. (Courtesy of Arbed.)

ious configurations, most often in A-frame pairs. Often, some convenient structure, such as the land-side crane rail support beam at a container terminal, may be modified cost-effectively to serve as the deadman.

In some instances, a soil or rock anchor system may be preferable to a tie rod and deadman system. The soil or rock anchor system applies the tensile load directly into a strong soil stratum or bedrock and is installed by drilling a hole, with or without a casing, inserting the anchor, and grouting the annular space. Although these types of anchorages are more costly than tie rods, the elimination of the deadman and the minimal disruption of operations at existing facilities may offset the higher cost of the soil or rock anchors in some situations, particularly where existing facilities or site limitations make installation of a tie rod and deadman system impractical.

The sheet pile bulkhead is usually provided with some type of cap. The cap may be as simple as a steel channel section of a depth slightly greater than the flange depth of the sheet piles, with flanges oriented downward, and bolted to the sheet piles. Alternatively, the cap may be cast-in-place reinforced concrete of a configuration most suitable for the particular situation. Sometimes, a concrete cap is also used to secure the outboard end of the tie rods, in place of a wale. For more information consult Tsinker (1997).

2.5.3 Pile Caps

In considering pile caps, there are generally three practical alternatives: (1) cast-in-place concrete, (2) precast concrete (Figure 2.38), or (3) a precast/cast-in-place concrete combination consisting of a U-shaped precast shell filled with cast-in-place concrete (Figure 2.39). An advantage of the precast/cast-in-place combination is the elimination of costly over-water formwork. Also, the concrete closest to the water, in this case the precast section, is denser and of higher strength and the precasting process provides for better quality control in the curing process and the placement of rebar, ensuring consistent concrete cover. These factors provide a considerably more durable structure in the harsh marine environment, with reduced maintenance requirements. Casting in place the interior of the cap reduces the lifting weight and permits a simple connection between the pile and the cap.

For port-related marine structures supported on steel piles, steel pile caps are sometimes used, particularly in locations where good-quality cast-in-place concrete is not readily available. In such cases, usually rolled or built-up wide-flange shapes are utilized as the pile cap, with appropriate stiffeners provided at the pile connections and as necessary to support the deck structure properly.

2.5.4 Decks

The decks of modern port-related structures are constructed of concrete, either cast-in-place or precast, or some combination of cast-in-place and precast. Generally, precast concrete is denser and of higher strength than cast-in-place concrete, and the precasting process provides for better quality control in the curing process and the placement of reinforcing.

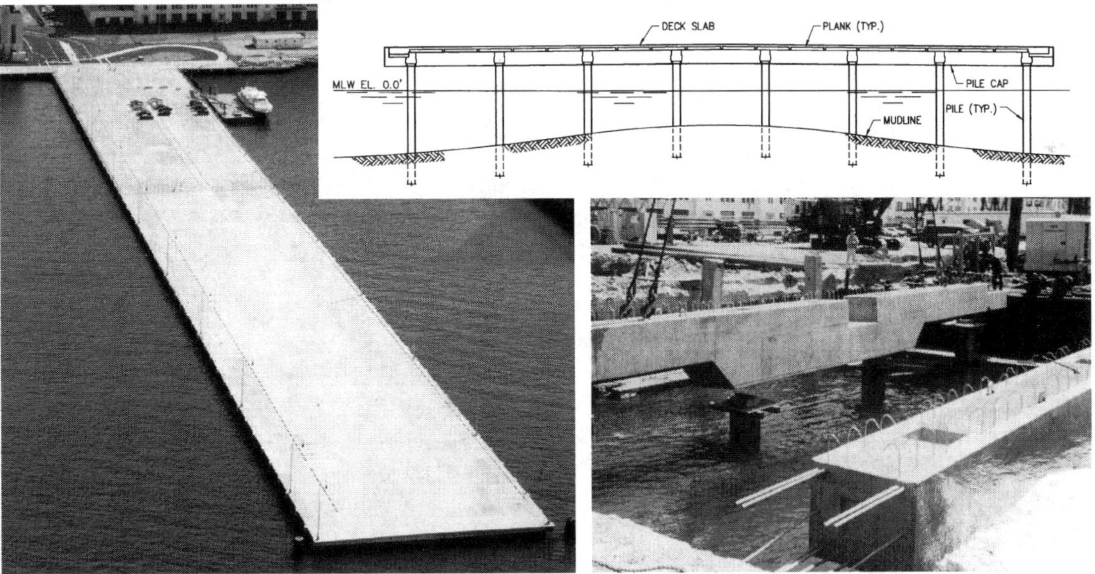

Figure 2.38 Solid precast concrete pile caps used for pier at Brooklyn Army Terminal, New York. (Courtesy of Han-Padron Associates.)

Figure 2.39 U-shaped precast shells for pile caps of pier at Naval Weapons Station Earle, New Jersey. (Courtesy of Han-Padron Associates.)

For cast-in-place concrete decks, the design should be robust, featuring thick slabs and few, if any, beams or girders. Thin slabs, as well as the corners of beams and girders, are prone to premature deterioration due to corrosion of the reinforcement. Many modern cast-in-place concrete deck designs completely eliminate the use of pile caps, beams, and girders, and are based on flat slab design principles.

The individual elements of precast concrete decks may be simple solid planks, prestressed or nonprestressed, or hollow elements of various configurations, usually prestressed. Often, standard precast, prestressed concrete bridge deck sections (Figure 2.40) may prove cost-effective.

Probably the most widely used deck construction consists of precast concrete elements with a cast-in-place concrete topping. The cast-in-place topping distributes concentrated loads among adjacent individual precast elements and over the thinner portions of hollow elements, ties the entire deck structure together, and provides a convenient location for installing negative moment reinforcing steel to create a continuous structure.

Concrete decks should be designed in accordance with the requirements of the American Concrete Institute (ACI, 1995). The elements should satisfy the crack control requirements for severe exposure. Great care is required in specifying the concrete mix to include the proper additives (fly ash, silica fume, etc.) and admixtures to ensure a dense and durable product. Similarly, the specifications must ensure the use of appropriate chloride-free aggregates and water with low mineral content. Design details must provide adequate cover over the reinforcing bars. The use of epoxy-coated reinforcing bars has proven to be cost-effective in many locations. However, regardless of how carefully the concrete is specified, it is essential that the construction be controlled properly.

When there is not a substantial amount of fill between the concrete deck and the pave-

Figure 2.40 Standard precast, prestressed concrete bridge box sections used for pier deck, Naval Weapons Station Earle, New Jersey. (Courtesy of Han-Padron Associates.)

ment, concentrated wheel loads or crane outrigger loads will control the design of the deck. Particular care in designing for concentrated loads in decks featuring hollow precast concrete sections is required.

The number of expansion joints provided in the deck should be kept to a minimum. Historically, expansion joints have resulted in maintenance problems, and they should be provided only where an analysis of the deck and pile system indicates that they are necessary to relieve thermal stresses, particularly bending stresses in piles resulting from thermal expansion or contraction of the deck. The required location of expansion joints will depend on the relative stiffness of the deck and piles, but for most structures it may be expected that the required spacing will be between 180 and 300 m.

2.5.5 Fenders

The purpose of a marine fender system is to prevent damage to both the vessel and the wharf or pier structure during the berthing process and while the vessel is moored. As a vessel approaches a structure it possesses kinetic energy by virtue of its displacement and motion. As the vessel contacts the structure and is brought to a stop, this kinetic energy must be dissipated. For very small vessels, the kinetic energy is quite low and there are a number of mechanisms acting at the time of impact that

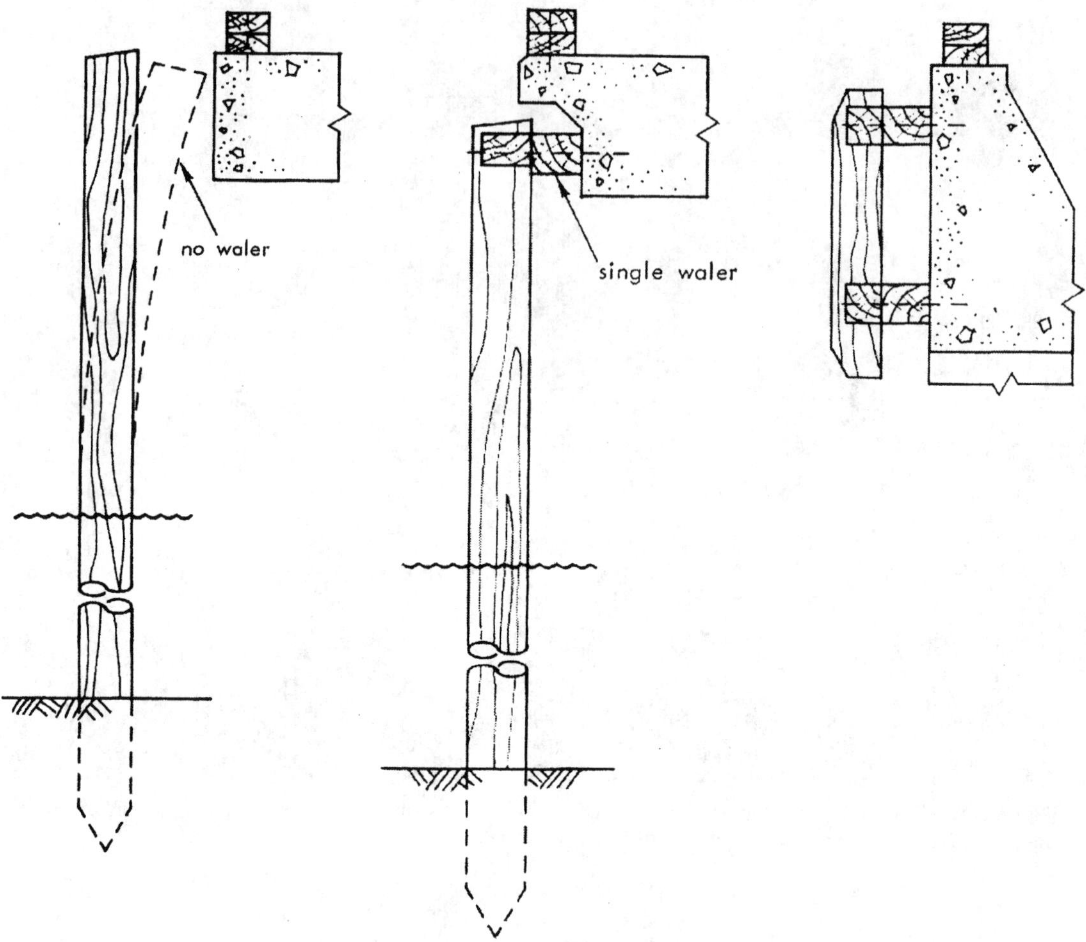

Figure 2.41 Typical timber fender systems.

will dissipate small amounts of energy. However, for larger vessels it is necessary to consider the berthing impact and provide a fender system with sufficient energy absorption capacity to prevent damage.

Timber fender systems are the most widely used for small vessels, but are also used on wharves and piers for large vessels. Timber fender systems absorb energy by bending and local crushing of the wood fibers. The energy absorption capacity is quite low, and when a high-energy impact occurs, the timber will usually break, absorbing a considerable amount of additional energy. Timber fender systems are usually designed to allow easy replacement of individual elements. There are almost an infinite number of variations of timber fender arrangements in use. Figure 2.41 illustrates just a few.

Often, there is a need to provide a significant amount of energy absorption capacity in the fender system. One of the most common early

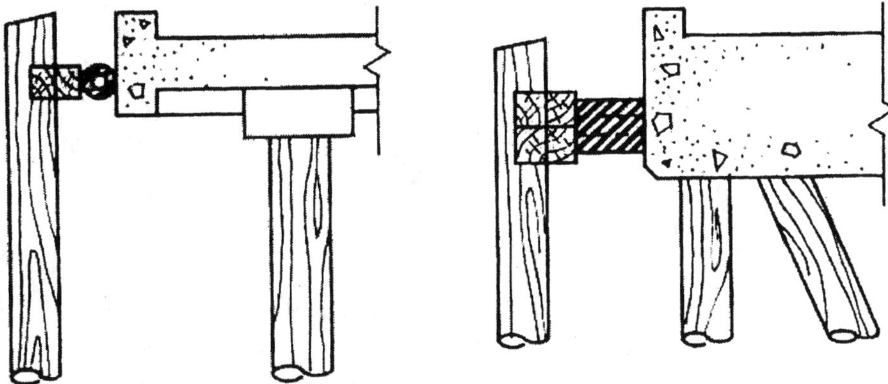

Figure 2.42 Typical timber fender systems with rubber elements.

ways of doing this was to provide an extruded rubber element between the traditional timber fender and the wharf or pier structure (Figure 2.42). In fact, this type of system is in wide use today, but the energy absorption capacity is still quite small.

Early attempts to develop fender systems with large energy absorption capacity led to the development of gravity fender systems. These fender systems are based on the principle of converting the ship's kinetic energy into potential energy by raising a heavy weight. Gravity fender systems, however, have the disadvantage of requiring a substantial, and therefore costly, structure to support them, and they usually require a significant amount of maintenance.

A number of other types of fender systems, utilizing buoyancy, steel springs, and other energy absorption mechanisms, were developed over the years. However, with the development of rubber fender units, these other types of fender systems have ceased to be used for new installations.

The first of the modern rubber fender units capable of absorbing high levels of energy was the raykin fender (Figure 2.43). The raykin fender works on the principle of absorbing energy by shearing and compressing rubber elements that are bonded to steel plates. Quite a few raykin systems were installed in the 1960s, but in the 1970s they were, for the most part, displaced by other types of rubber fenders that are less costly per unit of energy absorption capacity and require less maintenance.

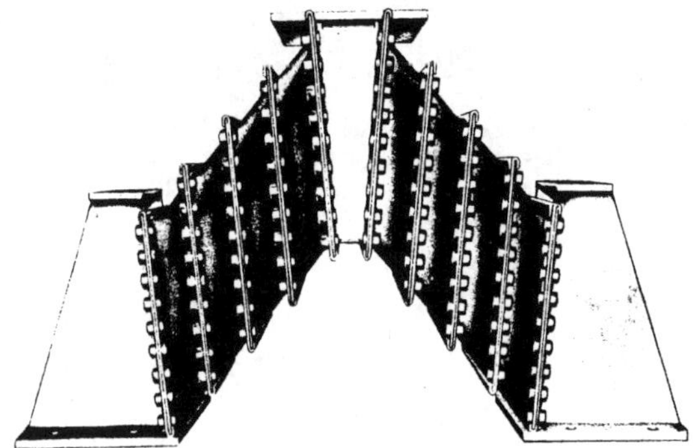

Figure 2.43 Raykin fender unit.

There are a great many types of rubber fender units on the market today. Each has different characteristics and certain advantages and disadvantages. The fender systems most widely used in new installations include:

- Buckling (Figure 2.44)
- Pneumatic (Figure 2.45)
- Foam-filled (Figure 2.46)

Figure 2.44 Buckling fender units.

Figure 2.45 Pneumatic fender units.

- Side-loaded rubber (Figure 2.47)
- Shear (Figure 2.48)
- Flexible pile (Figure 2.49)

The fenders illustrated in the figures represent only a few of the many variations of each type of fender system that are supplied by various fender manufacturers.

Figure 2.50 illustrates the reaction and deflection characteristics of these types of fenders. In this example, all the fenders have the same design reaction and the same energy absorption capacity. The energy absorption capacity is equal to the area under the reaction/deflection curve. It is evident from the figure that buckling fenders require considerably less deflection to absorb the design energy for a given maximum reaction. This characteristic has made them quite popular, but it does have some drawbacks. The maximum reaction occurs during almost every berthing, even with vessels smaller than the maximum design vessel. Also, many buckling fenders cause rather high contact pressure against a ship's hull and a panel is often needed to distribute and reduce this pressure. These types of fenders are susceptible to significant reductions in energy absorption capacity when subjected to impacts not perpendicular to their face.

Floating pneumatic and foam-filled fenders have similar reaction/deflection relationships. As can be seen from Figure 2.50, they must be larger than the corresponding buckling types and thus require greater reach of the cargo han-

Figure 2.46 Foam-filled fender units.

Figure 2.47 Side-loader rubber fender units.

Figure 2.48 Shear fender units on a barge fleeting dolphin. (Courtesy of Han-Padron Associates.)

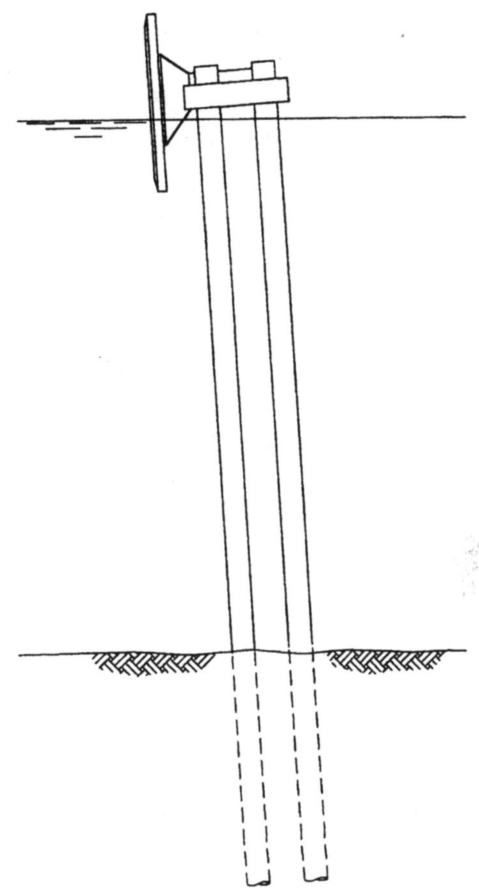

Figure 2.49 Flexible pile breasting dolphin.

dling equipment. These fenders have low hull contact pressures, eliminating the need for a panel between ship and fender. Also, reactions at or near the level of the maximum design reaction will occur very few times during the life of the facility. These types of fenders are relatively "soft" and are particularly attractive where vessels experience relatively large motions due to wave action while berthed.

Large side-loaded cylindrical fenders are very popular where energy absorption requirements are not too high. Their relatively low cost makes them an attractive alternative despite their relatively high hull pressures and the difficulty of mounting them securely.

Shear fenders may also be attractive where large energy absorption capacities are not required. A number of shear fenders can be mounted on a single panel, so a wide range of energy absorption capacities can be achieved. These fenders have an essentially linear reaction/deflection relationship, similar to that of the flexible pile type.

The flexible pile fender is often attractive where soil conditions are suitable because it combines the functions of fender and breasting structure. The energy-absorbing capacity of a pile is a function of its length, and consequently, this type of fender system is particularly attractive in deepwater applications.

The objective of the fender system designer is to select the optimum fender system for a particular situation by evaluating all the factors. The circumstances in which fender systems are installed vary greatly, and there is no single type of fender system that is most suitable in all situations. The optimum system can usually be defined as the system that results in the least cost, over the useful life of the facility, considering both initial capital cost and annual maintenance and repair costs, including fender

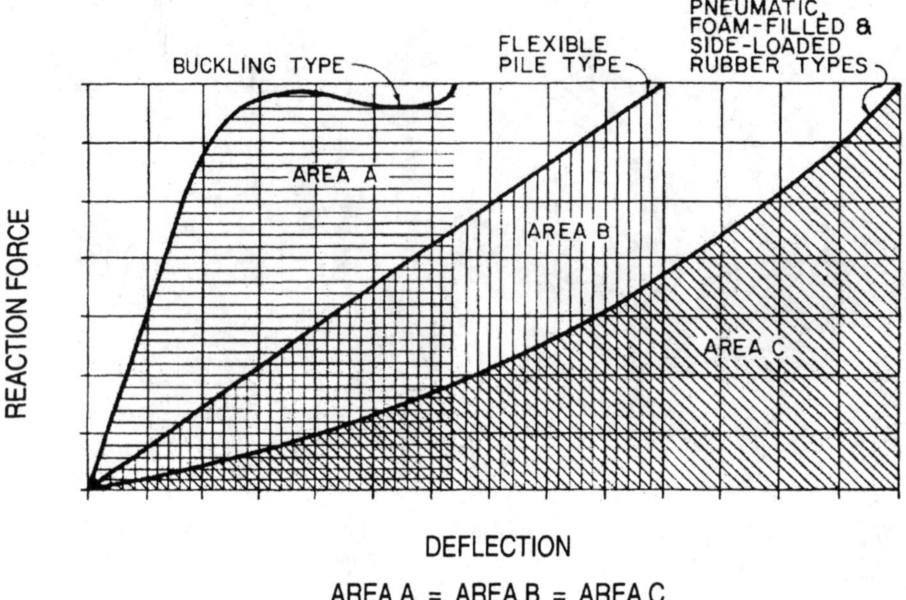

Figure 2.50 Reaction–deflection characteristics of various types of fender units.

system, berthing structure, and ship repair costs. For detail guidelines of fender system design consult PIANC (2002).

2.5.6 Mooring Fittings

Mooring fittings are provided on piers and wharves to secure ships' mooring lines. Different types of mooring fittings, with a wide range of capacities, are available. The use of mooring fittings of different capacities on the same pier or wharf face should be avoided to preclude the possibility of a high-capacity mooring line being inadvertently secured to a low-capacity fitting.

Where flammable products are handled, mooring fittings and metal components of rub rails, guard curbs, gallows, gangways, platform surfaces, and so on, which may come in contact with a vessel's wire mooring lines at any time (including during deployment or release), should be insulated electrically from the mooring structure.

2.5.6.1 Bollards. Bollards are short single- or double-column cast steel fittings of various configurations and capacities and are normally filled with concrete. Double-column bollards are often referred to as bitts. Conventional bollards generally are 600 to 1200 mm high and have one or two horns to prevent the mooring lines from slipping off when secured at a vertical angle. Modern low-profile bollards are generally less than 600 mm high, with a flared top to prevent the mooring line from slipping off. Low-profile bollards are more efficient than conventional bollards because the point of load application is closer to the deck, where the load is resisted. Bollard capacities generally range from approximately 30 tonnes to more than 200 tonnes. Figure 2.51 illustrates several different types of bollards.

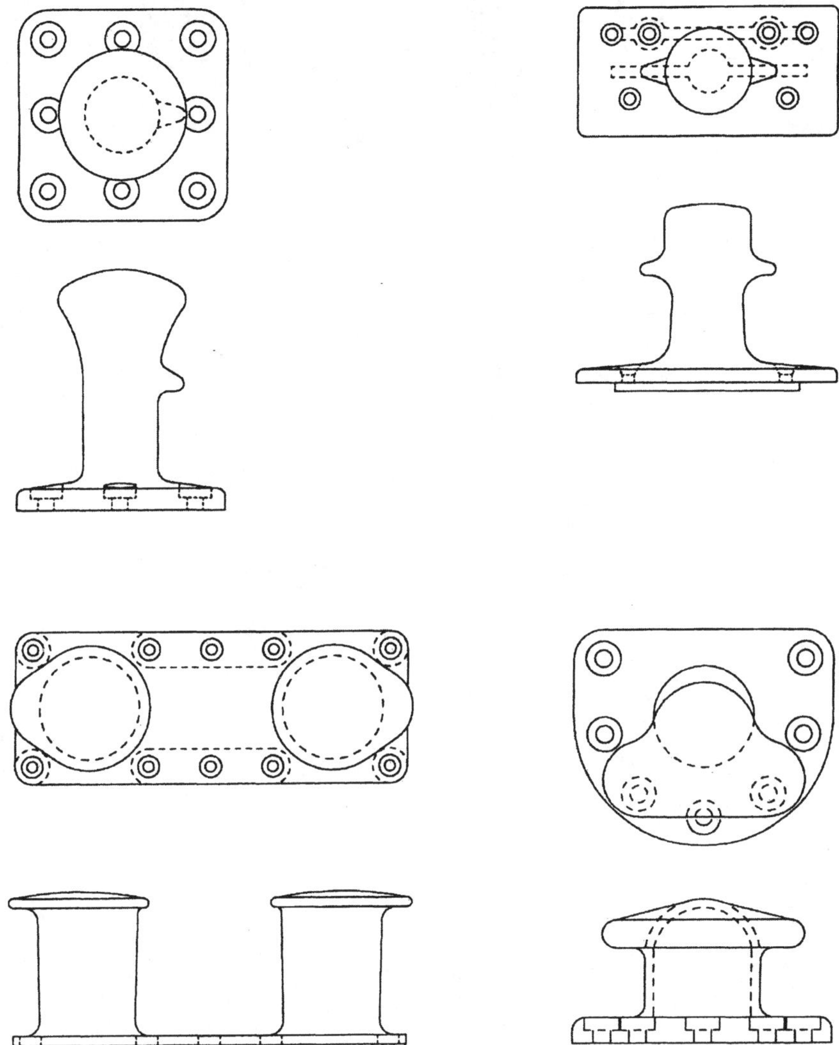

Figure 2.51 Various types of bollards.

2.5.6.2 Cleats. Cleats are relatively low capacity cast steel fittings of various configurations, but all have two projecting arms for securing mooring lines of smaller vessels. Cleats are intended for tugs, barges, and similar small vessels and normally have capacities not exceeding 20 tonnes. They should not be used in combination with higher-capacity mooring fittings. Figure 2.52 illustrates typical styles of cleats.

2.5.6.3 Quick-Release Hooks. Quick-release mooring hooks are preferred to bollards for securing the mooring lines of large tankers

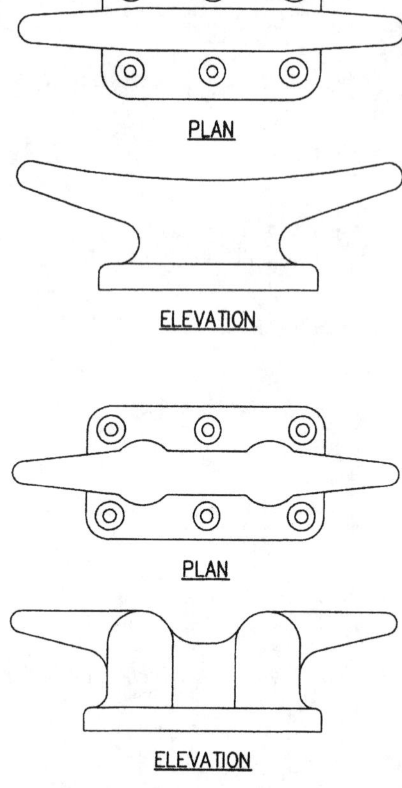

Figure 2.52 Two types of cleats.

or other vessels that must be able to depart a berth very quickly in case of an emergency. Quick-release hooks should be provided for berths that accommodate tankers or bulk carriers of 25,000 dwt and larger. Quick-release hooks should be equipped with a remote release system for added safety. To avoid accidental release of the hooks, the release should be a two-step activity, requiring (1) lifting a cover or pulling a pin, and (2) pulling a lever. Only one mooring line should be attached to any hook. Where more than one line is required to be connected to a point, multiple hooks should be provided. Typically, two hooks are provided at each mooring point for a berth that accommodates tankers or bulk carriers up to 75,000 dwt, with a third hook added for the breast, head, and stern mooring lines for vessels larger than 75,000 dwt.

As many as six hooks can be mounted on a single assembly, and each hook can have a capacity ranging from 65 to 220 tonnes. Figure 2.53 illustrates a single-hook assembly and a triple-hook assembly with an integral capstan. Quick-release hooks can be fitted with load cells which, along with the interconnecting wiring and display and alarm console, form a

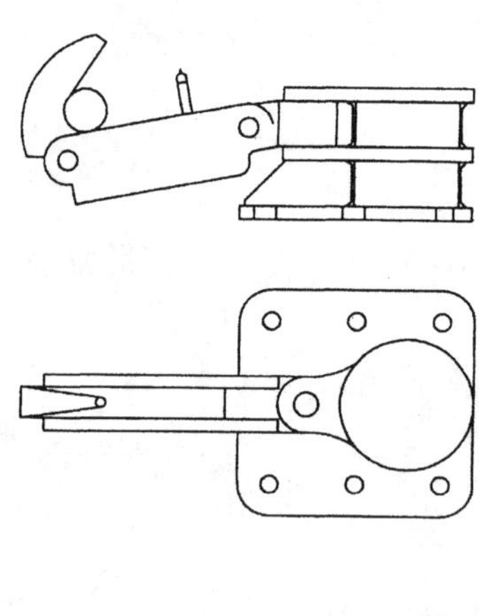

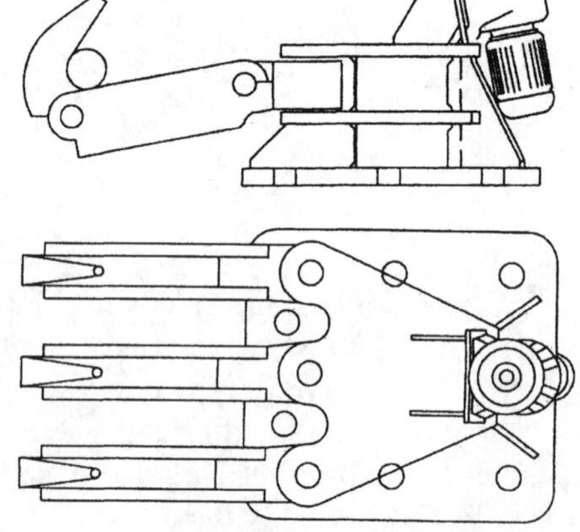

Figure 2.53 Quick-release hooks.

system to indicate and record mooring line loads. The loads can be monitored manually or automatically during environmental conditions which may cause excessive loading in the mooring lines, and the information can be used by operating personnel in deciding when it is safe for the vessel to remain in the berth.

2.5.6.4 Capstans. Capstans are used for hauling the heavy mooring lines of large vessels ashore, using lighter messenger lines. Capstans should be located at each mooring point, especially where wire mooring lines are used. The capstan may be mounted on a quick-release mooring hook assembly (Figure 2.53), or immediately adjacent to the mooring fitting such that the eye of the mooring line can be slipped over the fitting as the line is being hauled ashore. The capstan should have a minimum pulling capacity of 2 tonnes at a speed of 25 m/min with a minimum 15-hp motor. The motor should be of the reversing type to allow for unwrapping of a seized messenger line and should be operated by a foot pedal to allow a single operator to control operation of the coiling rope with both hands. The capstan should be equipped with a positive braking system so that there will be no slippage under full load.

2.6 DESIGN LOADS

To ensure safe and effective operation of port-related marine structures, it is necessary to reliably define the loading that the structures will be subjected to during their service life. For these types of structures the various loadings are usually categorized as:

- Dead loads
- Vertical live loads
- Mooring loads
- Berthing loads
- Seismic loads

The manner in which these loadings are combined, and the allowable stresses utilized for each load combination, must also be defined. Much of the following discussion is derived from the *Marine Oil Terminal Engineering Standards* developed for the California State Lands Commission (2002). For more information on this subject matter the reader is referred to Tsinker (1997).

2.6.1 Dead Loads

Dead loads consist of the weight of all components of the structure as well as the weight of all permanent attachments, such as piping, mooring equipment, gangway structures, light poles, railings, sheds, and stationary cargo transfer equipment. When estimating the dead load, the designer should consider both the present intended use of the structure and possible future uses.

It is worth noting that under normal circumstances, the dead load of a port-related marine structure constitutes a relatively small percentage of the total load acting on the structure. Therefore, precision in estimating the dead load is usually not necessary. The designer may make simplified conservative estimates of the dead load without significantly affecting the cost-effectiveness of the design.

The unit weights of the construction materials used for the assessment of dead loads should be based on the actual construction weights. If this information is not available, the unit weights listed in Table 2.7 may be used.

2.6.2 Vertical Live Loads

Vertical live loads consist of the weight of all movable equipment, cargo stored on or moving across the structure, wheel loads of trucks, mobile cranes and other rubber-tired cargo handling equipment, outrigger float loads from

Table 2.7 Unit weights for assessing dead loads

Material	Unit Weight (kN/m^3)
Rolled steel or cast steel	77
Cast iron	71
Aluminum alloys	28
Timber (untreated)	6–8
Timber (treated)	7–9
Concrete, reinforced (normal weight)	23–25
Concrete, reinforced (lightweight)	14–19
Asphalt paving	24

mobile cranes, wheel loads on the rails of steel-wheeled cargo handling equipment, and any other moveable equipment for the facility. Uniform live loads can vary from as much as 48 to 58 kPa for container terminal wharves and piers, to as little as 5 to 7 kPa for structures intended to support only personnel and small vehicles. Loading on the rails supporting large container gantry cranes can vary from 290 kN/m to as much as 730 kN/m or more. The concentrated load from the outrigger float of a large mobile crane can be 100 tonnes or more.

When considering uniform live loads it is not necessary to consider impact loading. However, when considering live loads from wheeled vehicles, impact can be an important consideration. Generally, for rubber-tired vehicles, an impact factor of 15% is applied when designing slabs, beams, and pile caps. If there is 0.5 m or more of pavement and fill material between the tire and deck slab, it is not necessary to consider impact in the design of the slabs, beams, and pile caps. It is also not necessary to consider impact for the design of structural elements below the pile caps. For steel-wheeled cargo transfer equipment, an impact factor of 20% is used when designing the rail support beams, but need not be applied to piles or other substructure supporting the rail support beams.

Because of the wide variation in vertical live loads, it is essential for the designer to fully define the live loads applicable to the particular structure. This definition should consider not only the current use of the structure, but also potential future uses.

Although usually not a controlling factor, vertical uplift forces should also be considered. These forces can result from buoyancy and wave "slamming" for open structures that have relatively low level decks. Since most open structures with low-level decks have substantial fill on the deck between the pavement surface and the deck structure, these uplift forces are usually not a concern. But in a situation where an open structure is in a location within the port that is relatively exposed to wave action, these forces should be considered.

2.6.3 Mooring Loads

The forces acting on a moored vessel arise from the following sources: winds, currents, wind waves, waves from passing vessels, tidal variations, and seiche. The methodologies for computing these forces are described below. For more information on this subject matter consult Chapters 7 and 8.

2.6.3.1 Wind Loads. Environmental loads induced by wind on vessels while in a fixed mooring position are generally computed by considering two load cases.

- *Survival condition:* 25- or 50-year return period, with 30-s-duration wind speed.
- *Operational condition:* wind speed established in terms of an operational wind rose. When operational wind speed is exceeded, cargo transfer operations are terminated.

The 30-s-duration wind speed should be determined based on data that give the annual maximum wind speed and direction for each year for a substantial number of years. Data for eight wind directions (45° increments) should ideally be obtained such that the maximum wind speed can be calculated for each 45° sector. If data for only one direction are available, it is usually assumed that this wind speed applies to all directions unless it can be demonstrated that the wind speed in other directions should be less. If 30-s-duration wind speed data are not available, the data should be adjusted to a 30-s wind, as described below. A 25- or 50-year return period, depending on the nature and importance of the facility, should be used to establish the design wind speed for each direction. The methods outlined by the American Society of Civil Engineers (ASCE, 1998a) may be used to calculate the design wind speeds from statistical material.

A wind velocity of 30 s duration measured at an elevation of 10 m above the water surface should be used in the wind load equations. For wind velocities obtained at a different elevation, adjustments to the equivalent 10-m velocity can be made using the formula

$$V_w = v_w \left(\frac{10}{h}\right)^{1/7} \tag{2.1}$$

where V_w is the 10-m wind velocity (knots), v_w the wind velocity at elevation h (knots), and h the elevation above the water surface (m).

The available wind duration should be adjusted to a 30-s wind duration using the formula

$$V_{t=30\,\mathrm{s}} = \frac{v_t}{c_t} \tag{2.2}$$

where v_t is the velocity over a given time period and c_t is the windspeed conversion factor (Figure 2.54). If wind data are available over land only, the following formula should be used to convert the wind speed from over-land to over-water conditions:

$$V_w = 1.10 V_L \tag{2.3}$$

where V_w is the over-water wind speed and V_L is the over-land wind speed.

OCIMF's (1997) guidelines for VLCC's or the *British Standard Code of Practice for Maritime Structures* (British Standards Institution, 1984) may be used to compute the wind loads for all vessels shaped as a tanker. For barges and other vessels with configurations different than those of tankers, the wind loads may be calculated based on the guidelines of ASCE (1998a). Alternatively, wind loads for any type of vessel may be calculated using the guidelines in the U.S. Department of the Navy's *Mooring Design Manual* (1998).

2.6.3.2 Current Loads.

Environmental loads induced by currents on vessels while in a fixed moored position should be considered. Where current velocities are low, published sources of current velocities are sufficient. However, where current velocities are high, the published data should be supplemented by site-specific data. These data should be obtained by real-time measurements over an extended period of time.

An average current velocity (V_c) over the draft of the vessel should be used in computing the current forces and moments. If the vertical current velocity profile is known, the definition of the average current velocity over the draft of the vessel can be obtained from the equation

$$V_c^2 = \frac{1}{T \int_0^T (v_c)^2 \, ds} \tag{2.4}$$

where V_c is the average current velocity (knots), T the draft of vessel (m), v_c the current

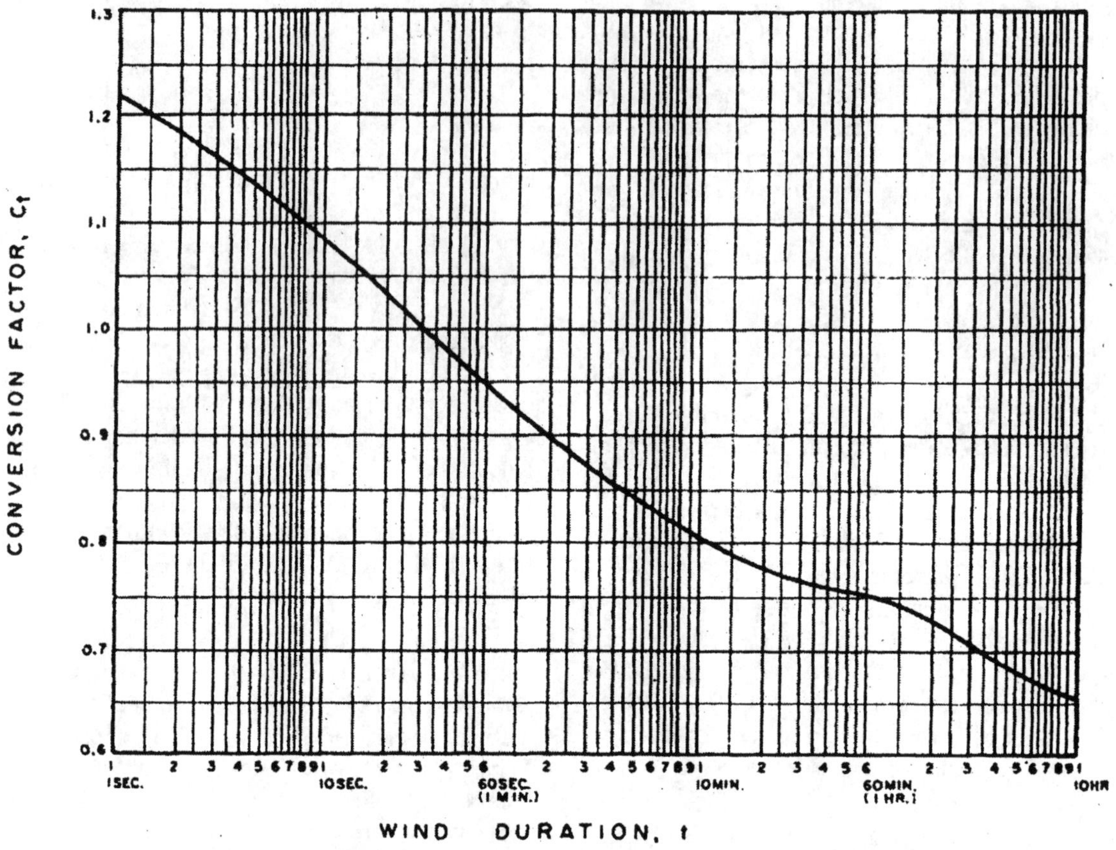

Figure 2.54 Wind speed conversion factor.

velocity as a function of depth (knots), and *s* the water depth measured from the surface (m). If the velocity profile is not known, the velocity at a known water depth should be adjusted by the factors provided in Figure 2.55 to obtain the equivalent average velocity over the draft of the vessel.

OCIMF's (1977) guidelines or the British standard (British Standards Institution, 1984) may be used to compute the current loads for all vessels shaped as a tanker. For barges and other vessels with configurations different from that of tankers, the current loads may be calculated based on the guidelines of ASCE (1998a). Alternatively, current loads for any type of vessel may be calculated using the guidelines in the U.S. Department of the Navy's (1998) document.

2.6.3.3 Wind Wave Loads. Most port structures are located in relatively sheltered waters such that typical wind waves can be assumed not to affect the moored vessel when the wave period is less than approximately 4.0 s. When wave periods exceed 4.0 s, a simplified dynamic mooring analysis should be carried out.

Wind-generated waves can cause high loads on vessels that are moored to fixed berths. The

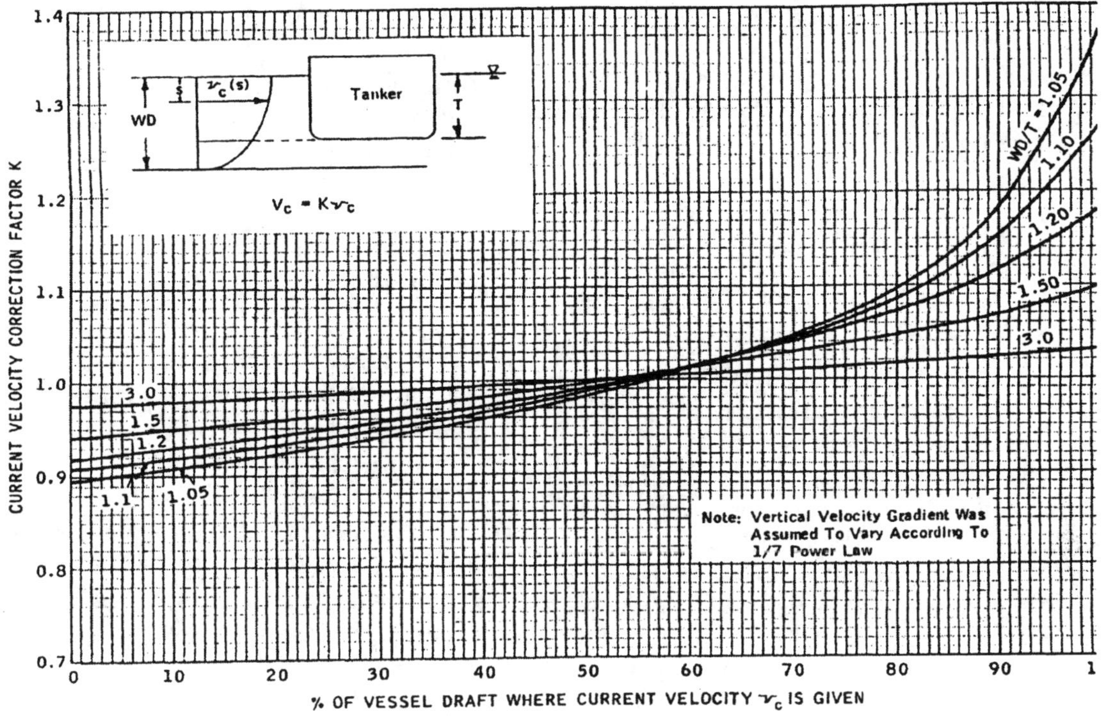

Figure 2.55 Current velocity correction factor.

associated structural loads result primarily from first-order wave-induced vessel motions. The major design objective is to provide a sufficiently flexible fender system to accommodate wave-induced vessel motions within the allowable deflection limits of the fender units. The relation between the natural frequency of a vessel mooring system and the exciting wave frequency is a major factor affecting the magnitude of the loads. If the two frequencies are similar, resonance may occur that will result in an amplification of the vessel motion, which in turn causes higher mooring and breasting loads on the structure.

To estimate transverse wave-induced vessel motion and the associated fender reactions, the vessel and its mooring system may be analyzed as a forced linear spring–mass system. The wave excitation force is the harmonic forcing function for the system, and the mooring system is analyzed as a linear spring system using an effective spring constant for the fender system, based on the linear range of the fender units.

The horizontal water particle accelerations should be calculated for the various wave conditions, taken at the middepth of the loaded draft. The water particle accelerations should then be used to calculate the wave excitation forces to determine the static displacement of the vessel. The Froude–Krylov method may be used to calculate the wave excitation forces by conservatively approximating the vessel as a rectangular box with dimensions similar to the actual dimensions of the vessel. The excitation force computed assumes a 90° incidence angle with the longitudinal axis of the vessel, which will result in forces that are significantly greater than the forces that will actually act upon the vessel from quartering seas. A load reduction factor may be used to account for the design wave incidence angle from the longitudinal axis of the ship. The overall excursion of the vessel should be determined for each of the wave conditions by calculating the dynamic response of the linear spring–mass system. The corresponding fender reactions should be calculated from the fender unit load-excursion curves.

2.6.3.4 Loads from Passing Vessels. The force generated by a passing vessel is a complex problem that has been largely ignored. However, the effect of a passing vessel on a moored vessel may be significant. The force created by the passing vessel is due to pressure gradients associated with the pattern of flow that accompanies the passing vessel. Pressure is generated at the bow and stern of the passing vessel and suction is generated at port and starboard. These pressure and suction fields cause the moored vessel to sway, surge, and yaw, and these motions impose forces in the mooring lines.

Passing vessel analysis should be conducted when the following conditions exist:

- The clear distance between moored vessel and passing vessel is less than 150 m.
- The combined speed of vessel and current is greater than 5 knots.

To calculate sway and surge forces as well as yaw moment on a moored vessel due to a passing vessel, the following parameters should be established:

- *Passing distance.* Clear distance between passing and moored vessels should be estimated. This distance is normally established based on the width of the channel and the vessel traffic pattern.
- *Passing vessel traveling speed.* The passing vessel speed should be estimated.
- *Operational wind and current conditions.* These may be assumed when calculating forces due to a passing vessel.

In addition, the following should be considered when calculating the sway, surge, and yaw motions of the moored vessel:

- *Size of passing vessel.* Force due to passing vessel increases with increasing size of the passing vessel.
- *Ratio of length of moored vessel to length of passing vessel.*
- *Ratio of midship section areas of the moored and passing vessels.*
- *Underkeel clearance of the passing vessel.* Force due to passing vessel increases with decreasing underkeel clearance.
- *Draft condition of the passing vessel.* A loaded vessel generates more force due to a passing vessel than a vessel in ballast.
- *Mooring line condition.* Pretension in the mooring lines produces less mooring line loads than do slack lines.
- *Current condition.* When current is present, the speed of the vessel is relative to the moving water. Thus, moving against the current will increase the force, and moving with the current will decrease it.

There are very few references for calculating the forces due to a passing vessel. A simple theoretical analysis presented by Wang (1975) may be used to evaluate the surge and sway forces and yaw moment of a passing vessel. Wang developed graphs of nondimensional surge and sway forces and yaw moment as a function of the separation between the moored and passing vessels and the effect of water depth.

The following passing vessel positions should be investigated:

- The passing vessel is centered on the moored ship. This position produces maximum sway force.
- The midship of the passing vessel is fore or aft of the centerline of the moored ship by a distance of 0.33 times the length of the moored ship. This position produces maximum surge force and yaw moment at the same time.

For calculating the mooring line forces, the following conservative assumptions can be used:

- The surge force is assumed to be resisted only by the spring lines. Either the bow or stern lines will resist a small portion of the longitudinal force, but their contribution can be neglected.
- When the sway force pushes the moored ship against the berth, it is assumed to be resisted equally by the fender units contacted.
- When the sway force pulls the moored ship away from the berth, it is assumed that the breast lines will resist this force.
- Depending on the direction, yaw moment is assumed to be resisted either by bow or stern breast lines and fender unit at the compressed side.
- The surge force and yaw moment are in phase and are out of phase with the sway force.

2.6.3.5 Seiche Loads. The penetration of long-period/low-amplitude waves into a harbor can result in resonant standing-wave systems when the wave forcing frequency coincides with a modal natural frequency of the harbor. Such long-period waves can occur due to seiche. The resonant standing waves can result in large surge motions if the resonant frequency is close to the natural frequency of the vessel and mooring system.

The standing-wave system created by seiche is characterized by a series of nodes and antinodes. *Antinodes* are regions of maximum vertical (water particle) motion and minimum horizontal motion. Conversely, *nodes* are regions of maximum horizontal motion and minimum vertical motion. For a moored vessel that is located in a nodal region, the large horizontal water particle motions can result in large surge motions of the vessel, depending on the natural frequency of the vessel and mooring system. Resonant vertical motions occurring at antinode locations typically do not present problems for conventional mooring systems.

Seiche typically has wave periods ranging from 20 s to several hours, with wave heights in the range 30 to 120 mm. The following procedures should be used in evaluating the effects of seiche.

- Calculate the natural oscillating period of the basin. The basin should be idealized as a rectangular basin, either closed or open at the seaward end so that the formula provided in the U.S. Department of the Navy's *Harbors Design Manual* (1984) can be applied. The formula give the wave period and wave length for different modes. The first three modes should be considered in the analysis.
- Determine the location of the moored ship with respect to the antinode and node of the first three modes to determine the chances for resonance.
- Determine the natural period of the vessel and mooring system. The calculation should be based on the stiffness of the mooring system in surge and the total mass of the system. The surge motion of the moored ship is estimated by analyzing the vessel motion as a harmonically forced linear single-degree-of-freedom spring–mass system.
- Vessels are generally berthed parallel to the channel, therefore, only longitudinal

(surge) motions need be considered. The loads on the mooring lines (spring lines) are then determined from the computed vessel excursions and mooring line stiffnesses.

2.6.4 Berthing Loads

Berthing loads are quantified in terms of energy transfer from the kinetic energy of the vessel into potential energy dissipated by the fender system. The reaction generated by the fender system usually imposes a substantial horizontal loading on the structure. In Section 2.5.5 we describe the various types of marine fender systems in common use, and in Section 2.8.6 we describe how berthing energies are calculated.

The longitudinal and vertical component of the berthing force should be calculated using appropriate coefficients of friction between the vessel and the fender. In lieu of specific data, the values listed in Table 2.8 may be used for typical fender/vessel materials.

Longitudinal and vertical components can be determined by

$$F = \mu N \tag{2.5}$$

where F is the longitudinal or vertical component of berthing force, μ the coefficient of friction of contact materials, and N the maximum berthing force.

Table 2.8 Coefficients of friction between vessel and fender

Contact Materials	Friction Coefficient
Timber to steel	0.4–0.6
Urethane to steel	0.4–0.6
Steel to steel	0.25
Rubber to steel	0.6–0.7
UHMW[a] to steel	0.1–0.2

[a] Ultrahigh-molecular-weight plastic rubbing strips.

2.6.5 Earthquake Loads

Port structures located in seismically active areas should be designed to resist earthquake motions considering the relationship of the site to active faults, the seismic response of soils at the site, and the dynamic response characteristics of the entire structure. The required level of sophistication in developing the earthquake input motions depends on the design parameters for earthquake motions.

In the absence of a site-specific response spectrum, the seismic forces should be calculated in accordance with the requirements of the American Association of State Highway Transportation Officials (AASHTO, 1996). The AASHTO method considers the interrelationship of factors such as the location of the site relative to active faults, the effect of the overlying soil on the earthquake motion, and the dynamic response characteristics of the structure. When applying the AASHTO method to the design of port structures, the following should be considered:

- The weight of the structure should include the total dead load and a portion of the design live load, generally between 10 and 20%, depending on the nature of the facility.
- The depth of overburden to "rocklike" material should be determined from borings or other geological data.
- The maximum expected acceleration at the site, expressed as a percentage of g at bedrock, should be estimated. This earthquake coefficient may be calculated based on site-specific conditions, or it may be estimated based on the AASHTO procedures.
- The period of vibration of the structure should be calculated to determine the seismic force.
- The framing factor should be taken as 1.0 for framing featuring both vertical and bat-

ter piles and 0.8 for framing with only vertical piles.

- The structure should be designed to resist seismic forces acting nonconcurrently in the direction of each of the principal axes of the structure.

Detail discussion of this subject matter is provided in Chapter 3.

2.6.6 Earth Pressure

The lateral active and passive earth pressure acting on the port structure should be evaluated in accordance with standard geotechnical engineering procedures. In determining lateral earth pressures, the effects of seismic events and the pressures due to water-level differentials should be considered.

2.6.7 Ice Forces

In addition to the weight of accumulated ice on the structure, the forces exerted by floating ice should be considered. The AASHTO standard procedure for calculating the dynamic impact ice forces generally provide conservative results. The AASHTO value of the effective crushing pressure of ice during impact is 400 psi, but actual effective pressure may be only half of this value. When considering static pressure, freshwater ice will exert less pressure on a structure than seawater ice of the same thickness. For freshwater ice, pressures of 100 to 200 kPa may be assumed, while for sea ice, pressures of 275 kPa to as much as 1000 kPa may be assumed. These are maximum values and relate to crushing of the ice. However, the total force acting on the structure cannot exceed the wind and current driving force acting on the ice floe. Where ice conditions are severe, the structure should be designed in a configuration that causes the ice to fail in bending rather than crushing against the structure. For detailed information on ice loads the reader is referred to Tsinker (1995, 1997).

2.6.8 Loading Combinations

Typical port-related structures are comprised of components of different materials that require the use of both load factor and allowable stress designs on an individual basis. For example, concrete piles and decks are designed using load factor design but also checked for service loads (cracking), whereas the mooring fittings, bolts, and so on, are checked using allowable stress design. The selection of mooring lines is controlled by safety factors against the minimum breaking strength of the line.

The various loads to be considered are defined below, along with a general reference to the type of component to which each design methodology applies. Each component of the structure should be analyzed for all applicable combinations that produce the maximum stresses, bearing, and/or uplift in the component. Table 2.9 presents the load factors that should be used for load factor design. Table 2.10 presents the load factors that should be used for allowable stress design.

2.6.8.1 *Dead Load.* Upper- and lower-bound values are applied for the normal condition to check the maximum moment and shear with minimum axial load.

2.6.8.2 *Live Load.* Approximately 10 to 20% of the live load is typically considered for the earthquake load case. Also, some reasonable percentage of the maximum live load should be considered during berthing.

2.6.8.3 *Buoyancy Load.* Typically, port structure decks are not low enough to be subjected to buoyancy or wave slamming forces. However, in the event that portions of the structure are subjected to these forces, an appropriate uplift force should be applied.

Table 2.9 LRFD load factors for load combinations

Load Type	Normal Condition	Mooring Condition	Berthing Condition	Earthquake Condition
Dead load	1.4[a]	1.2	1.2	$1 \pm k$
Live load	1.7[b]	1.7[b]	0.1–0.2	0.1–0.2
Buoyancy	1.3	1.3	1.3	
Wind on structure	1.3	1.3	1.0	
Current on structure	1.3	1.3	1.0	
Ice on structure	1.3	1.3	1.0	
Earth pressure	1.6	1.6	1.6	1.0
Mooring load		1.3		
Berthing load			1.7	
Earthquake load				1.0

[a]Reduce the load factor for dead load (*D*) to 0.9 to check components for minimum axial load and maximum moment.
[b]The load factor for live load (*L*) may be reduced to 1.3 for the maximum outrigger float load from a truck crane.

Table 2.10 Allowable stress load factors for load combinations

Load Type	Normal Condition	Mooring Condition	Berthing Condition	Earthquake Condition
Dead load	1.0	1.0	1.0	$1 \pm k$
Live load	1.0	1.0	0.1–0.2	0.1–0.2
Buoyancy	1.0	1.0	1.0	
Wind on structure	1.0	1.0	1.0	
Current on structure	1.0	1.0	1.0	
Ice on structure	1.0	1.0	1.0	
Earth pressure	1.0	1.0	1.0	1.0
Mooring load		1.0		
Berthing load			1.0	
Earthquake load				1.0

2.6.8.4 Wind and Currents on the Structure. Wind and currents on the vessel are included in the mooring loads. The wind and current loads acting directly on the structure are therefore additional loads that can act simultaneously with the mooring or berthing loads.

2.6.8.5 Earth Pressure. The lateral soil pressure acting on port structures should be considered.

2.6.8.6 Mooring Loads. Multiple mooring load cases may be required, depending on the combination of environmental loads (wind and current) on the vessel considered in the mooring analysis for both operational and survival conditions. In addition, mooring loads caused by passing vessels should be considered in the mooring analysis for the operational condition.

2.6.8.7 Berthing Load. The berthing load should be based on the maximum fender system reaction. The structural capacity due to the berthing demand should be established based on allowable concrete or steel strains in the structural components, similar to the seismic load. Because berthing is an everyday occurrence, a load factor has been applied to this load.

2.6.8.8 Earthquake Loads. No load factors should be assigned to earthquake loads where a performance-based seismic analysis methodology is being used. This type of analysis requires that the actual force demand be limited to defined concrete and steel strains. For deck and pile evaluation, two cases of dead load (upper and lower bound) should be considered in combination with the seismic load. The upper- and lower-bound dead load values are expressed in relation to the peak ground acceleration (PGA) such that the load factor becomes $(1 \pm k)DL$, where $k = 0.5\text{PGA}$.

2.7 GEOTECHNICAL ISSUES

The basic principles of geotechnical engineering apply to port-related structures. However, in addition to the normally encountered geotechnical issues, some considerations more or less unique to these types of structures are encountered frequently. These issues are addressed below. Also, since a meaningful geotechnical analysis cannot be carried out unless reliable information on the geotechnical conditions that exist at the site are available, a recommended comprehensive geotechnical investigation program for port-related marine structures is also provided.

The discussions of axially and laterally loaded piles and shallow foundations are based on the American Petroleum Institute (API, 2000). The discussions of lateral earth pressure, slope stability, and liquefaction are based on Das (1993). The discussion of settlement is based on Winterkorn and Fang (1991). For a more detailed discussion of these issues, readers are referred to these publications.

2.7.1 Axially Loaded Piles

The following discussion of pile capacities has been developed for steel pipe piles. However, the basic principles are applicable to steel H-piles and concrete piles, with appropriate adjustments to account for the different characteristics of these other types of piles.

The design pile penetration for axially loaded piles should be sufficient to develop adequate capacity to resist the maximum computed axial bearing and pullout loads, with an appropriate factor of safety. The allowable pile capacities are determined by dividing the ultimate pile capacities by the following factors of safety:

- Operating conditions 2.0
- Extreme conditions 1.5
- Seismic conditions 1.5

The ultimate bearing capacity of pipe piles, Q_d, should be determined by the equation

$$Q_d = Q_f + Q_p = fA_s + qA_p \qquad (2.6)$$

where Q_f is the skin friction resistance (kN), Q_p the total end bearing (kN), f the unit skin friction capacity (kPa), A_s the side surface area of pile (m^2), q the unit end bearing capacity (kPa), and A_p the gross end area of pile (m^2).

Total end bearing, Q_p, should not exceed the capacity of the internal plug. In computing pile loading and capacity, the weight of the pile–soil plug system and hydrostatic uplift should be considered.

For piles driven open-ended, the shaft friction, f, acts on both the inside and outside of the pile. The total resistance is the sum of the external shaft friction, the end bearing on the pile wall annulus, and the total internal shaft friction or the end bearing of the plug, whichever is less. For piles considered to be plugged, the bearing pressure may be assumed to act over the entire cross section of the pile (i.e., once the total internal friction exceeds the total end bearing less that acting on the annulus, the pile can be considered to be plugged). At this point, it is assumed that additional soil cannot enter the pile, and the internal friction is therefore limited by the end bearing of the plug. For unplugged piles, the bearing pressure acts on the pile wall annulus only. Whether a pile is considered plugged or unplugged may be based on static calculations. For example, a pile could be driven in an unplugged condition but act plugged under static loading.

2.7.1.1 *Cohesive Soils.* For pipe piles in cohesive soils, the shaft friction, f (kPa), at any point along the pile may be calculated by the equation

$$f = \alpha c \tag{2.7}$$

where α is a dimensionless factor and c is the undrained shear strength of the soil at the point in question. The factor α can be computed by the equations

$$\alpha = \begin{cases} 0.5\psi^{-0.5} & \text{for } \psi \le 1.0 \\ 0.5\psi^{-0.25} & \text{for } \psi > 1.0 \end{cases} \tag{2.8}$$

with the constraint that $\alpha \leq 1.0$, where ψ is c/p'_o for the point in question and p'_o is the effective overburden pressure at the point in question (kPa).

For a pile end bearing in cohesive soils, the unit end bearing, q (kPa), may be computed by the equation

$$q = 9c \tag{2.9}$$

2.7.1.2 *Cohesionless Soils.* For pipe piles in cohesionless soils, the shaft friction, f (kPa), may be calculated by the equation

$$f = Kp_o \tan \delta \tag{2.10}$$

where K is the coefficient of lateral earth pressure, p_o the effective overburden pressure (kPa) at the point in question, and δ the friction angle between the soil and the pile wall.

For open-ended pipe piles driven unplugged, it is usually appropriate to assume K as 0.8 for both tension and compression loadings. Values of K for full displacement piles (plugged or closed end) may be assumed to be 1.0. The values indicated in Table 2.11 may be used for selection of δ if other data are not available. It should be noted, however, that the values provided in Table 2.11 are applicable for siliceous sands only and cannot be used for carbonate sands, as their use would generally result in an overestimation of pile capacity. Appropriate values for carbonate sands must be based on site-specific experience. For long piles, f may not indefinitely increase linearly with the overburden pressure, and it may be appropriate to limit f to the values given in Table 2.11.

For a pile end bearing in cohesionless soils, the unit end bearing, q (kPa), may be computed by the equation

$$q = p_o N_q \tag{2.11}$$

where p_o is the effective overburden pressure (kPa) at the pile tip and N_q is the dimensionless bearing capacity factor. Recommended values of N_q are indicated in Table 2.11.

2.7.1.3 *Axial Pullout Loads.* The ultimate pile pullout capacity may be equal to or less than, but should not exceed, Q_f, the total skin friction resistance. The effective weight of the pile, including hydrostatic uplift and the soil plug, should be considered in the analysis to determine the ultimate pullout capacity.

Table 2.11 Design parameters for cohesionless siliceous soils

Density	Soil Description	Soil-Pile Friction Angle, δ (deg)	Limiting Skin Friction Values (kPa)	N_q	Limiting Unit End Bearing Values (MPa)
1. Very loose	Sand	15	47.8	8	1.9
Loose	Sand–silt				
Medium	Silt				
2. Loose	Sand	20	67.0	12	2.9
Medium	Sand–silt				
Dense	Silt				
3. Medium	Sand	25	81.3	30	4.8
Dense	Sand–silt				
4. Dense	Sand	30	95.7	40	9.6
Very dense	Sand–silt				
5. Dense	Gravel	35	114.8	50	12.0
Very dense	Sand				

Table 2.12 *t–z* relationship for noncarbonate soils[a]

Clays		Sands	
z/D	t/t_{max}	z	t/t_{max}
0.0016	0.30	0.000	0.00
0.0031	0.50	0.250	1.00
0.0057	0.75	∞	1.00
0.0080	0.90		
0.0100	1.00		
0.0200	0.70–0.90		
∞	0.70–0.90		

[a] z, local pile deflection (mm); D, pile diameter (mm); t, mobilized soil-pile adhesion (kPa); t_{max}, maximum soil-pile adhesion or unit skin friciton capacity (kPa).

2.7.1.4 Soil Reaction. The axial resistance of the soil is provided by a combination of axial soil-pile adhesion or load transfer along the side of the pile and end bearing resistance at the pile tip. The plotted relationship between mobilized soil-pile shear transfer and local pile deflection at any depth is described using a *t–z* curve. Similarly, the relationship between mobilized end bearing resistance and axial tip deflection is described using a *Q–z* curve. In the absence of more definitive criteria, the *t–z* curves listed in Table 2.12 are recommended for noncarbonate soils.

Relatively large pile tip movements are required to mobilize the full end bearing resistance. A pile tip displacement up to 10% of the pile diameter may be required for full mobilization in both sand and clay soils. In the absence of more definitive criteria, the curve listed in Table 2.13 is recommended for both sands and clays.

Table 2.13 *Q–z* relationship for noncarbonate soils[a]

z/D	Q/Q_p
0.002	0.25
0.013	0.50
0.042	0.75
0.073	0.90
0.100	1.00

[a] z, axial tip deflection (mm); D, pile diameter (mm); Q, mobilized end bearing capacity (kN); Q_p, total end bearing (kN).

2.7.2 Laterally Loaded Piles

For laterally loaded piles, the lateral resistance of the soil near the surface is significant to pile design, and the effects on this resistance due to scour and soil disturbance during pile installation should be considered.

2.7.2.1 Clay. For static lateral loads, the ultimate unit lateral bearing capacity of soft clay, p_u, has been found to vary between $8c$ and $12c$, except at shallow depths where failure occurs in a different mode due to minimum overburden pressure. Cyclic loads cause deterioration of lateral bearing capacity below that for static loads. In the absence of more definitive criteria, p_u may be assumed to increase from $3c$ to $9c$ as X increases from 0 to X_R according to

$$p_u = \begin{cases} 3c + \gamma X + J\dfrac{cX}{D} & (2.12) \\ 9c \quad \text{for } X \geq X_R & (2.13) \end{cases}$$

where p_u is the ultimate resistance (kPa), c the undrained shear strength for undisturbed clay soil samples (kPa), γ the effective unit weight of soil (kN/m^3), X the depth below the soil surface (mm), J a dimensionless empirical constant with values ranging from 0.25 to 0.5 having been determined by field testing, D the pile diameter (mm), and X_R the depth below the soil surface to the bottom of the reduced resistance zone (mm). For a condition of constant strength with depth, the equations above are solved simultaneously to give

$$X_R = \frac{6D}{(\gamma D/c) + J}$$

Where the strength varies with depth, the equation may be solved by plotting the two equations (i.e., p_u vs. depth). The point of first intersection of the two equations is taken to be X_R. In general, minimum values of X_R should be about 2.5 pile diameters.

Lateral soil resistance–deflection relationships for piles in soft clay are generally nonlinear. The p–y curves for the short-term static load case may be generated from Table 2.14.

For the case where equilibrium has been reached under cyclic loading, the p–y curves may be generated from Table 2.15.

For static lateral loads the ultimate bearing capacity, p_u, of stiff clay ($c > 96$ kPa), as for soft clay, would vary between $8c$ and $12c$. Due to rapid deterioration under cyclic loadings, the ultimate resistance will be reduced to something considerably less and should be considered in cyclic design. While stiff clays also have nonlinear stress–strain relationships, they are generally more brittle than soft clays. In

Table 2.14 ***p–y* relationship under static loading for soft clay**[a]

p/p_u	y/y_c
0.00	0.0
0.50	1.0
0.72	3.0
1.00	8.0
1.00	∞

Table 2.15 ***p–y* relationship under cyclic loading for soft clay**[a]

$X > X_R$		$X < X_R$	
p/p_u	y/y_c	p/p_u	y/y_c
0.00	0.0	0.00	0.0
0.50	1.0	0.50	1.0
0.72	3.0	0.72	3.0
0.72	∞	$0.72\ X/X_R$	15.0
		$0.72\ X/X_R$	∞

[a] p, actual lateral resistance (kPa); y, actual lateral deflection (mm); y_c, $2.5\varepsilon_c D$ (mm); ε_c, strain that occurs at one-half the maximum stress on laboratory undrained compression tests of undisturbed soil samples.

developing stress–strain curves and subsequent p–y curves for cyclic loads, good judgment should reflect the rapid deterioration of load capacity at large deflections for stiff clays.

2.7.2.2 *Sand.* The ultimate lateral bearing capacity for sand has been found to vary from a value at shallow depths, p_{us}, to a value at deep depths, p_{ud}, as determined by the following equations. At a given depth the equation giving the smallest value of p_u should be used as the ultimate bearing capacity:

$$p_{us} = (C_1 H + C_2 D)\gamma H \qquad (2.14)$$

$$p_{ud} = C_3 D \gamma H \qquad (2.15)$$

where p_u is the ultimate resistance (kN/m) (s = shallow, d = deep); C_1, C_2, C_3 the coefficients determined from Figure 2.56 as function of ϕ', the angle of internal friction of sand (deg); H the depth (m); D the average pile diameter from surface to depth (m); and γ the effective soil weight (kN/m^3).

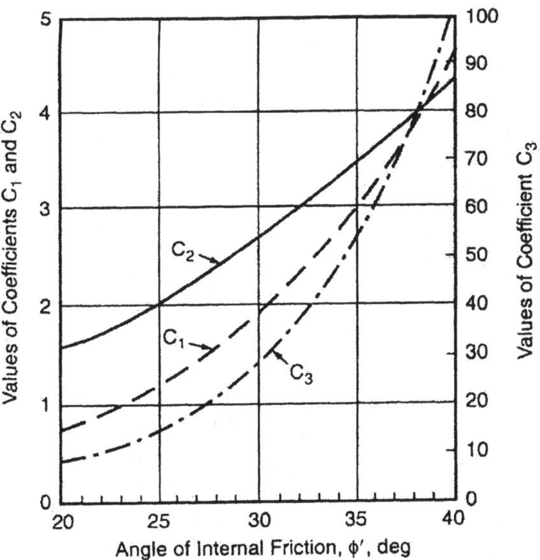

Figure 2.56 Coefficients as function of ϕ'.

The lateral soil resistance–deflection (p–y) relationships for sand are also nonlinear and in the absence of more definitive information may be approximated at any specific depth, H, by the following expression:

$$P = Ap_u \tanh\left(\frac{kH}{Ap_u}\,y\right) \qquad (2.16)$$

where A is a factor to account for a cyclic or static loading condition, evaluated as $A = 0.9$ for cyclic loading and $A = [3.0 - 0.8(H/D)] \geq 0.9$ for static loading, p_u the ultimate bearing capacity at depth H (kN/m), k the initial modulus of subgrade reaction (kN/m^3) determined from Figure 2.57 as a function of the angle of internal friction ϕ', H the depth (m), and y the lateral deflection (m).

2.7.3 Shallow Foundations

Shallow foundations are those foundations for which the depth of embedment is less than the minimum lateral dimension of the foundation element. The design of shallow foundations should include, where appropriate to the intended application, consideration of the following:

- Stability, including failure due to overturning, bearing, sliding, or combinations thereof.
- Static foundation deformations, including possible damage to components of the structure and its foundation or attached facilities.
- Dynamic foundation characteristics, including the influence of the foundation on structural response and the performance of the foundation itself under dynamic loading.

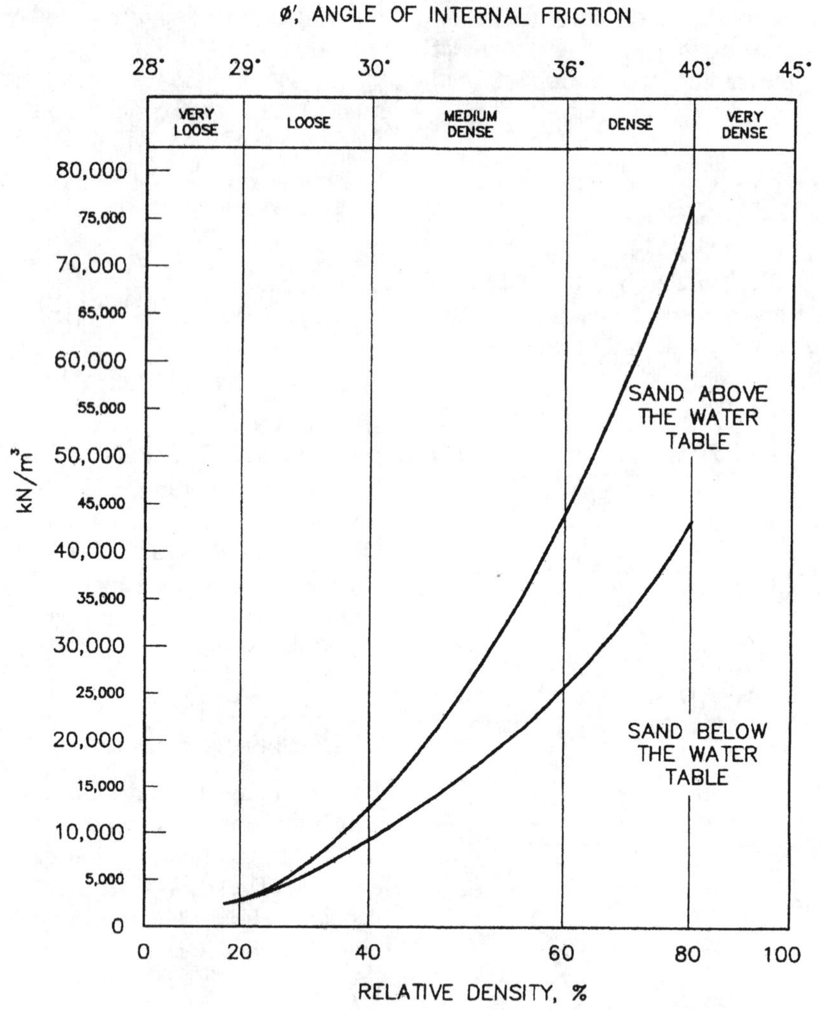

Figure 2.57 Relative density.

- Hydraulic instability such as scour or piping due to wave pressures, including the potential for damage to the structure and for foundation instability.

The foundation must be designed to provide a minimum factor of safety against failure. This minimum factor of safety is applied against the calculated ultimate bearing capacity or ultimate bearing pressure. The recommended factors of safety are discussed in Sections 2.6 and 2.8. The values recommended are based on the assumption that a thorough geotechnical investigation has been performed. If such is not the case, these values should be increased accordingly.

2.7.3.1 *Undrained Bearing Capacity* (ϕ = 0).

The maximum gross vertical load that a footing can support under undrained conditions is

$$Q = (cN_cK_c + \gamma D)A' \qquad (2.17)$$

where Q is the maximum vertical load at failure; c the undrained shear strength of the soil; N_c a dimensionless constant (5.14 for ϕ = undrained friction angle = 0); K_c a correction factor that accounts for load inclination, footing shape, depth of embedment, inclination of base, and inclination of the ground surface; γ the total unit weight of the soil; D the depth of embedment of the foundation; and A' the effective area of the foundation depending on the load eccentricity.

A method for determining the correction factors and the effective area is given below. Curves showing the numerical values of N_q, N_c, and N_γ as a function of ϕ are shown in Figure 2.58. Two special cases of eq. (2.17) are frequently encountered. For a vertical concentric load applied to a foundation at ground level where both the foundation base and ground are horizontal, the equation for two foundation shapes is as follows:

1. *Infinitely long strip footing:*

$$Q_0 = 5.14cA_0 \qquad (2.18)$$

   where Q_0 is the maximum vertical load per unit length of footing and A_0 is the actual foundation area per unit length.

2. *Circular or square footing:*

$$Q = 6.17cA \qquad (2.19)$$

   where A is the actual foundation area.

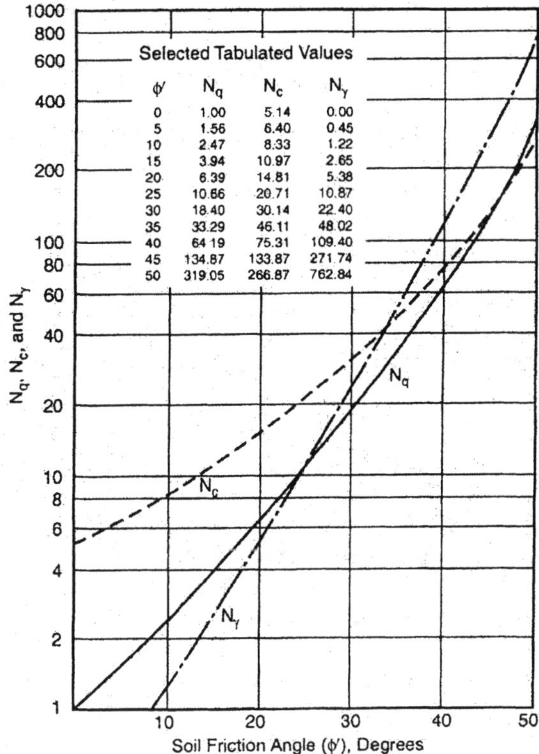

Figure 2.58 Recommended bearing capacity factors.

2.7.3.2 *Drained Bearing Capacity.*

The maximum net vertical load that a footing can support under drained conditions is

$$Q' = (c'N_cK_c + qN_qK_q + \tfrac{1}{2}\gamma' BN_\gamma K_\gamma)A' \qquad (2.20)$$

where Q' is the maximum net vertical load at failure; c' the effective cohesion intercept of the Mohr envelope; $N_c = (N_q - 1)\cot\phi'$, a dimensionless function of ϕ' (the effective friction angle of the Mohr envelope); K_c, K_q, and K_γ are correction factors that account for load inclination, footing shape, depth of embedment, inclination of base, and inclination of the ground surface, respectively (the subscripts c,

q, and γ refer to the particular term in the equation); $q = \gamma' D$, where D is the depth of embedment of the foundation; N_γ an empirical dimensionless function of ϕ' that can be approximated by $2(N_q + 1) \tan \phi$; γ' the effective unit weight; $N_q = (\exp [\pi \tan \phi]) [\tan^2(45° + \phi'/2)]$, a dimensionless function of ϕ'; B the minimum lateral foundation dimension; A' the effective area of the foundation, depending on the load eccentricity. Curves showing the numerical values of N_q, N_c, and N_γ as a function of ϕ' are given in Figure 2.58. A description of the K factors is given below.

Two special cases of eq. (2.20) for $c' = 0$ (usually, sand) are frequently encountered. For a vertical, centric load applied to a foundation at ground level, where both the foundation base and ground are horizontal, the equation for two foundation shapes is as follows:

1. *Infinitely long strip footing:*

$$Q_0 = 0.5\gamma' B N_\gamma A_0 \qquad (2.21)$$

2. *Circular or square footing:*

$$Q = 0.3\gamma' B N_\gamma A \qquad (2.22)$$

2.7.3.3 *Sliding Stability.* The limiting conditions of the bearing capacity equations, with respect to inclined loading, represent sliding failure and result in the following equations:

1. *Undrained analysis:*

$$H = cA \qquad (2.23)$$

where H is the horizontal load at failure.

2. *Drained analysis:*

$$H = c'A + Q \tan \phi' \qquad (2.24)$$

2.7.3.4 *Effective Area.* Load eccentricity decreases the ultimate vertical load that a footing can withstand. This effect is accounted for in bearing capacity analysis by reducing the effective area of the footing according to empirical guidelines.

Figure 2.59 shows footings with eccentric loads, the eccentricity, e, being the distance from the center of a footing to the point of action of the resultant, measured parallel to the plane of the soil–footing contact. The point of action of the resultant is the centroid of the

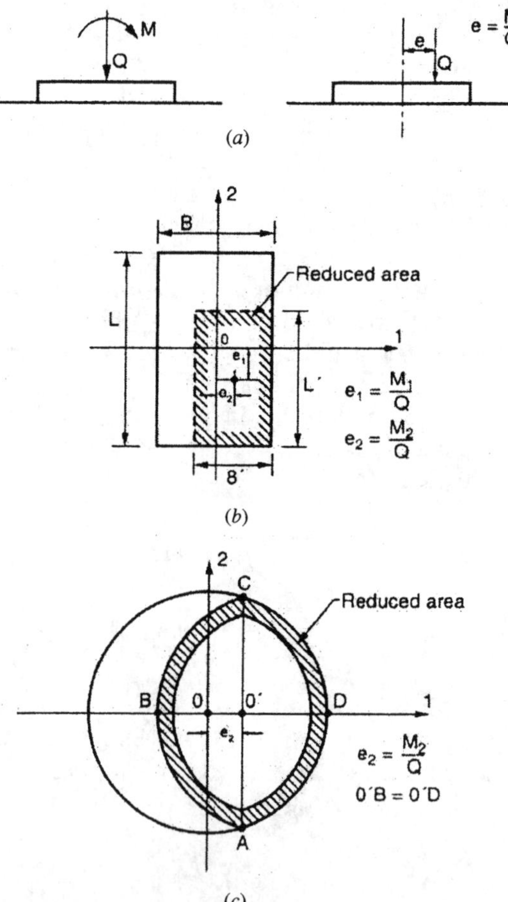

Figure 2.59 Eccentrically loaded footings: (*a*) equivalent loadings; (*b*) reduced area, rectangular footing; (*c*) reduced area, circular footing.

reduced area; the distance e is M/Q, where M is the overturning moment and Q is the vertical load.

For a rectangular base area (Figure 2.59b), eccentricity can occur with respect to either axis of the footing. Thus, the reduced dimensions of the footing are

$$L' = L - 2e_1$$
$$B' = B - 2e_2 \quad (2.25)$$

where L and B are the foundation length and width, respectively, the prime denotes effective dimensions, and e_1 and e_2 are eccentricities along the length and width.

For a circular base with radius R, the effective area is shown in Figure 2.59c. The centroid of the effective area is displaced a distance e_2 from the center of the base. The effective area is then considered to be two times the area of the circular segment ADC. In addition, the effective area is considered to be rectangular with a length/width ratio equal to the ratio of line lengths AC to BD. The effective dimensions are therefore

$$A' = 2s = B'L'$$
$$L' = \left(2s\sqrt{\frac{R+e}{R-e}}\right)^{1/2} \quad (2.26)$$
$$B' = L'\sqrt{\frac{R-e}{R+e}}$$

where

$$s = \frac{\pi R^2}{2} - \left(\sqrt[e]{R^2 - e^2} + R^2 \sin^{-1}\frac{e}{R}\right)$$

Examples of effective areas as a function of eccentricity are shown in Figure 2.60 in a dimensionless form. No data are available on other foundation shapes. Intuitive approximations must be made to find an equivalent rectangular or circular foundation when nonstandard shapes are encountered.

2.7.3.5 *Correction Factors.* The correction factors K_c, K_q, and K_γ are usually written

$$K_c = i_c s_c d_c b_c g_c$$
$$K_q = i_q s_q d_q b_q g_q \quad (2.27)$$
$$K_\gamma = i_\gamma s_\gamma d_\gamma b_\gamma g_\gamma$$

where i, s, d, b, and q are individual correction factors related to load inclination, foundation shape, embedment depth, base inclination, and ground surface inclination, respectively. The subscripts c, q, and γ identify the factor (N_c, N_q, or N_γ) with which the correction term is associated.

The recommended correction factors for N_c and N_q that account for variations in loading and geometry not considered in the theoretical solutions are obtained from the expressions for N_c and N_q as suggested by DeBeer and Ladanyi [as cited by Visic in Winterkorn and Fang (1991)]. Thus, the appropriate correction factor for the N_c term can be determined once it is shown for the N_q term. Most expressions for correction factors for N_q and N_γ are determined empirically. Following are the expressions recommended for the correction factors.

Inclination Factors. For $\phi > 0$,

$$i_q = \left(1 - \frac{H}{Q + B'L'c\cot\phi}\right)^m$$
$$i_\gamma = \left(1 - \frac{H}{Q + B'L'c\cot\phi}\right)^{m+1} \quad (2.28)$$
$$i_c = i_q - \frac{1 - i_q}{n_c \tan\phi}$$

For $\phi = 0$,

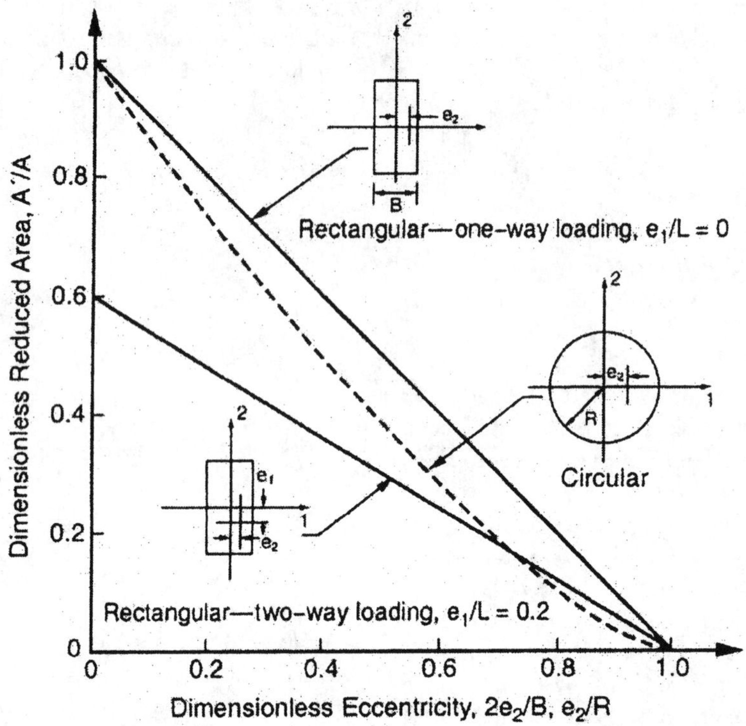

Figure 2.60 Area reduction factors, eccentrically loaded footings.

$$i_c = 1 - \frac{mH}{B'L'cN_c}$$

where H is the projection of the load resultant on the plan of the footing, m is a dimensionless function of B'/L', and θ is the angle between the long axis of the footing and H. The general expression for m is

$$m = m_L \cos^2\theta + m_B \sin^2\theta$$

where

$$m_L = \frac{2 + L'/B'}{1 + L'/B'} \quad \text{and} \quad m_B = \frac{2 + B'/L'}{1 + B'/L'}$$

Shape Factors. Rectangular:

$$s_c = 1 + \frac{B'}{L'}\frac{N_q}{N_c}$$

$$s_q = 1 + \frac{B'}{L'} \tan\phi \tag{2.29}$$

$$s_\gamma = 1 - 0.4\frac{B'}{L'}$$

Circular (centric load only):

$$s_c = 1 + \frac{N_q}{N_c}$$

$$s_q = 1 + \tan\phi \tag{2.30}$$

$$s_\gamma = 0.6$$

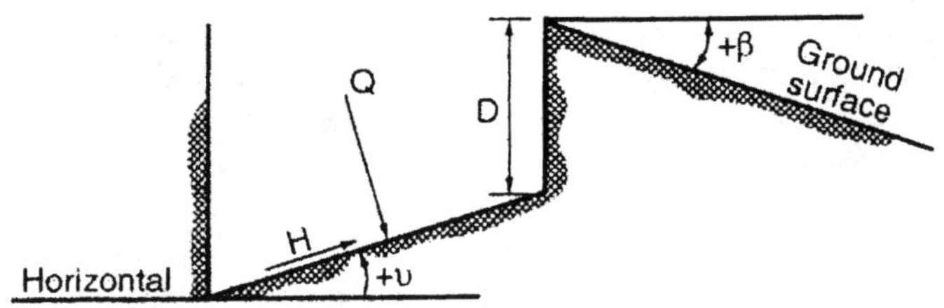

Figure 2.61 Definitions for inclined base and ground surface. (After Vesic.)

For an eccentrically loaded circular footing, the shape factors for an equivalent rectangular footing are used.

Depth Factors

$$d_q = 1 + 2 \tan \phi \, (1 - \sin \phi)^2 \frac{D}{B'}$$

$$d_{\tilde{a}} = 1.0 \qquad (2.31)$$

$$d_c = d_q - \frac{1 - d_q}{N_c \tan \phi}$$

It should be emphasized that the effect of foundation embedment is very sensitive to soil disturbance at the soil–structure interface along the sides of the embedded base. Where significant disturbance is expected, it may be prudent to reduce or discount entirely the beneficial effect of overburden shear strength.

Base and Ground Surface Inclination Factors. Base inclination: For $\phi > 0$,

$$b_q = b_\gamma = (1 - \upsilon \tan \phi)^2$$

$$b_c = b_q - \frac{1 - b_q}{N_c \tan \phi} \qquad (2.32)$$

For $\phi = 0$,

$$b_c = 1 - \frac{2\upsilon}{N_c}$$

Ground slope: For $\phi > 0$,

$$g_q = g_\gamma = (1 - \tan \beta)^2$$

$$g_c = g_q - \frac{1 - g_q}{N_c \tan \phi} \qquad (2.33)$$

For $\phi = 0$,

$$g_c = 1 - \frac{2\beta}{N_c}$$

where υ and β are base and ground inclination angles in radians. Figure 2.61 defines these angles for a general foundation problem.

2.7.3.6 Dynamic Bearing Capacity. For footings on sand, the results of laboratory testing indicate that the bearing capacity changes as a function of the loading rate. This change in bearing capacity corresponds to a change in the friction angle of the soil. If soil strength parameters using the appropriate strain rate are known from laboratory testing of the soil in question, they should be used to estimate the friction angle for seismic loading. However, if these data are not available, it can conservatively be assumed that for shallow footings on dense sand, the dynamic friction angle is

$$\phi_d = \phi_s - 2 \qquad (2.34)$$

For footings on loose submerged sands, liquefaction effects must be investigated, as described in Section 2.7.7. For footings on saturated clays, it has been found that the undrained cohesion of the clay increases with increase in strain rate. Consequently, in evaluating seismic loading there is no need to decrease the bearing capacity of the footing, and in fact, an increase may be appropriate.

2.7.4 Lateral Earth Pressure

Lateral earth pressure as an applied loading and as a load-resisting mechanism is a consideration in the design of gravity-stabilized bulkheads and tied-back bulkheads.

2.7.4.1 Earth Pressure at Rest. The at-rest earth pressure may be appropriate in the analysis of stiff, heavy, gravity-stabilized bulkheads, and even stiff tied-back bulkheads where the backfill has been compacted and minimum lateral movement can be expected. The lateral earth pressure coefficient at rest can be calculated as follows:

$$K_0 = 1 - \sin\phi \tag{2.35}$$

2.7.4.2 Active Earth Pressure. Active earth pressure is typically assumed for cantilevered retaining walls and sheet pile walls that retain backfill. The lateral active earth pressure coefficient can be calculated according to the Coulomb theory as follows:

$$K_a = \frac{\sin^2(\alpha + \phi)}{\sin^2\alpha\,\sin(\alpha - \delta)} \times \left[1 + \sqrt{\frac{\sin(\phi + \delta)\,\sin(\phi - \beta)}{\sin(\alpha - \delta)\,\sin(\alpha + \beta)}}\right]^{-2} \tag{2.36}$$

where α is the angle of the wall with the horizontal, ϕ the friction angle of soil, δ the soil–wall friction angle (from Table 2.16), β the angle of the backfill with the horizontal.

Table 2.17 provides values of K_a for a vertical bulkhead and selected values of ϕ, δ, and β.

2.7.4.3 Passive Earth Pressure. The passive earth pressure is applicable in front of bulkheads. The lateral passive earth pressure can be calculated according to the Coulomb theory as follows:

$$K_p = \frac{\sin^2(\alpha + \phi)}{\sin^2\alpha\,\sin(\alpha - \delta)} \times \left[1 - \sqrt{\frac{\sin(\phi + \delta)\,\sin(\phi - \beta)}{\sin(\alpha - \delta)\,\sin(\alpha + \beta)}}\right]^{-2} \tag{2.37}$$

Table 2.18 provides values for K_p for selected values of ϕ, δ, and β.

2.7.4.4 Effect of Tidal Variations and Drained/Undrained Conditions. In tidal conditions, the effect of differential water pressure between the front and back sides of the bulkhead should be considered. It is recommended that a differential water head of one-half the normal tidal range be considered in the design, but no less than 0.5 m. Analyses should always be carried out for both the drained (long-term) and undrained (short-term) condition for clayey soils. A drained analysis is applicable for long-term static design loads, while an undrained analysis is applicable immediately and shortly after construction, and in a seismic event.

2.7.4.5 Seismic Effects: Active Earth Pressure. The active earth pressure coefficient when considering seismic effects for cohesionless soil can be calculated by use of the Mononobe–Okabe equation as follows:

$$K_{AE} = \frac{\cos^2(\phi - \theta - \beta)}{\cos\theta\,\cos^2\beta\,\cos(\delta + \beta + \theta)} \times \left[1 + \sqrt{\frac{\sin(\phi + \delta)\,\sin(\phi - \theta - i)}{\cos(\delta + \beta + \theta)\,\cos(i - \beta)}}\right]^{-2} \tag{2.38}$$

Table 2.16 Friction angles between various foundation materials and soil or rock

Interface Materials	Friction Angle,[a] δ (deg)
Mass concrete or masonry on the following:	
Clean sound rock	35
Clean gravel, gravel–sand mixtures, coarse sand	29–31
Clean fine to medium sand, silty medium to coarse sand, silty or clayey gravel	24–29
Clean fine sand, silty or clayey fine to medium sand	19–24
Fine sandy silt, nonplastic silt	17–19
Very stiff and hard residual or preconsolidated clay	22–26
Medium stiff and stiff clay and silty clay	17–19
Steel sheet piles against:	
Clean gravel, gravel–sand mixture, well-graded rock fill with spalls	22
Clean sand, silty sand–gravel mixture, single-size hard-rock fill	17
Silty sand, gravel, or sand mixed with silt or clay	14
Fine sandy silt, nonplastic silt	11
Formed concrete or concrete sheet piling against:	
Clean gravel, gravel–sand mixtures, well-graded rock fill with spalls	22–26
Clean sand, silty sand–gravel mixture, single-size hard-rock fill	17–22
Silty sand, gravel or sand mixed with silt or clay	17
Fine sandy silt, nonplastic silt	14
Masonry on masonry, igneous and metamorphic rocks	
Dressed soft rock on dressed soft rock	35
Dressed hard rock on dressed soft rock	33
Dressed hard rock on dressed hard rock	29
Masonry on wood (cross-grain)	26
Steel on steel at sheet-pile interlocks	17
Wood on soil	14–16[b]

Source: Bowles (1988).

[a] Single values ±2°. Alternative for concrete on soil is $\delta = \phi$.

[b] May be higher in dense sand or if sand penetrates wood.

where

$$\theta = \tan^{-1}\frac{k_h}{1 - k_v}$$

$$k_h = \frac{\text{horizontal component of earthquake acceleration}}{g}$$

$$k_v = \frac{\text{vertical component of earthquake acceleration}}{g}$$

Typical values of K_{AE} for $k_v = 0$ and $\beta = 0°$ are provided in Table 2.19.

The results of laboratory tests that have been conducted indicate that the resultant pressure P_{AE} acts at a distance $\overline{H}$, which is somewhat greater than $\frac{1}{3}H$, measured from the bottom of the wall (Figure 2.62). For design purposes, Seed and Whitman (1970) have proposed the following procedure for determination of the line of action of P_{AE}.

1. Calculate P_A.
2. Calculate P_{AE}.
3. Calculate $\Delta P_{AE} = P_{AE} - P_A$. The term ΔP_{AE} is the incremental force due to an earthquake.

Table 2.17 Active earth pressure coefficients for vertical bulkheads

δ (deg)	φ (deg) 26	28	30	32	34	36	38	40
				For β = 0°				
0	0.390	0.361	0.333	0.307	0.283	0.260	0.238	0.217
16	0.349	0.324	0.300	0.278	0.257	0.237	0.218	0.201
17	0.348	0.323	0.299	0.277	0.256	0.237	0.218	0.200
20	0.345	0.320	0.297	0.276	0.255	0.235	0.217	0.199
22	0.343	0.319	0.296	0.275	0.254	0.235	0.217	0.199
				For β = 5°				
0	0.414	0.382	0.352	0.323	0.297	0.272	0.249	0.227
16	0.373	0.345	0.319	0.295	0.272	0.250	0.229	0.210
17	0.372	0.344	0.318	0.294	0.271	0.249	0.229	0.210
20	0.370	0.342	0.316	0.292	0.270	0.248	0.228	0.209
22	0.369	0.341	0.316	0.292	0.269	0.248	0.228	0.209
				For β = 10°				
0	0.443	0.407	0.374	0.343	0.314	0.286	0.261	0.238
16	0.404	0.372	0.342	0.315	0.289	0.265	0.242	0.221
17	0.404	0.371	0.342	0.314	0.288	0.264	0.242	0.221
20	0.402	0.370	0.340	0.313	0.287	0.263	0.241	0.220
22	0.401	0.369	0.340	0.312	0.287	0.263	0.241	0.220

4. Assume that P_A acts at a distance of $\frac{1}{3}H$ from the bottom of the wall (Figure 2.62).
5. Assume that ΔP_{AE} acts at a distance of $0.6H$ from the bottom of the wall (Figure 2.62); then

$$\overline{H} = (P_A \tfrac{1}{3}H + \Delta P_{AE}\, 0.6H)/P_{AE} \qquad (2.39)$$

The lateral earth pressure theory developed above is for retaining walls with dry soil backfills. However, for quay walls, the hydrodynamic effect of the water must be considered. This is usually done according to the Westergaard (1933) theory. Based on this theory, the total dynamic water force on the seaward side per unit length of the wall, $P_{1(w)}$ is

$$P_{1(w)} = \tfrac{7}{12}\, k_h \gamma_w h^2 \qquad (2.40)$$

The location of the resultant water pressure, $\bar{y}$, is

$$\bar{y} = 0.6h \qquad (2.41)$$

Matsuo and O'Hara (1960) have suggested that the increase in pore water pressure on the landward side is approximately 70% of that on the seaward side. Thus,

$$p_2 = 0.6125 k_h \gamma_w h^{1/2} y^{1/2} \qquad (2.42)$$

Table 2.18 Passive earth pressure coefficients for vertical bulkheads

δ (deg)	φ (deg) 26	28	30	32	34	36	38	40
				For β = 0°				
0	2.561	2.770	3.000	3.255	3.537	3.852	4.204	4.599
16	4.195	4.652	5.174	5.775	6.469	7.279	8.230	9.356
17	4.346	4.830	5.385	6.025	6.767	7.636	8.662	9.882
20	4.857	5.436	6.105	6.886	7.804	8.892	10.194	11.771
22	5.253	5.910	6.675	7.574	8.641	9.919	11.466	13.364
				For β = 5°				
0	2.943	3.203	3.492	3.815	4.177	4.585	5.046	5.572
16	5.250	5.878	6.609	7.464	8.474	9.678	11.128	12.894
17	5.475	6.146	6.929	7.850	8.942	10.251	11.836	13.781
20	6.249	7.074	8.049	9.212	10.613	12.321	14.433	17.083
22	6.864	7.820	8.960	10.334	12.011	14.083	16.685	20.011
				For β = 10°				
0	3.385	3.713	4.080	4.496	4.968	5.507	6.125	6.841
16	6.652	7.545	8.605	9.876	11.417	13.309	15.665	18.647
17	6.992	7.956	9.105	10.492	12.183	14.274	16.899	20.254
20	8.186	9.414	10.903	12.733	15.014	17.903	21.636	26.569
22	9.164	10.625	12.421	14.659	17.497	21.164	26.013	32.602

where p_2 is the dynamic pore water pressure on the landward side at a depth y. The total dynamic pore water force increase, $P_{2(w)}$, per unit length of the wall is

$$P_{2(w)} = 0.4083 k_h \gamma_w h^2 \tag{2.43}$$

During an earthquake, the force on the wall per unit length on the seaward side will be reduced by $P_{1(w)}$ and that on the landward side will be increased by $P_{2(w)}$. Thus, the total increase of the force per unit length of the wall is equal to

$$P_w = 0.9917 k_h \gamma_w h^2 \tag{2.44}$$

2.7.4.6 *Seismic Effects: Passive Earth Pressure.* The passive earth pressure coefficient when considering seismic effects for cohesionless soil can also be calculated by use of the Mononobe–Okabe equation as follows:

$$K_{PE} = \frac{\cos^2(\phi + \beta - \theta)}{\cos\theta \cos^2\beta \cos(\sigma - \beta + \theta)} \times \left[1 - \sqrt{\frac{\sin(\phi + \sigma)\sin(\phi + i - \theta)}{\cos(i - \beta)\cos(\sigma - \beta + \theta)}}\right]^{-2} \tag{2.45}$$

Table 2.19 Active earth pressure coefficients (seismic effects)

			ϕ				
k_h	δ	i	28°	30°	35°	40°	45°
0.1	0°	0°	0.427	0.397	0.328	0.268	0.217
0.2			0.508	0.473	0.396	0.382	0.270
0.3			0.611	0.569	0.478	0.400	0.334
0.4			0.753	0.697	0.581	0.488	0.409
0.5			1.005	0.890	0.716	0.596	0.500
0.1	0°	5°	0.457	0.423	0.347	0.282	0.227
0.2			0.554	0.514	0.424	0.349	0.285
0.3			0.690	0.635	0.522	0.431	0.356
0.4			0.942	0.825	0.653	0.535	0.442
0.5			—	—	0.855	0.673	0.551
0.1	0°	10°	0.497	0.457	0.371	0.299	0.238
0.2			0.623	0.570	0.461	0.375	0.303
0.3			0.856	0.748	0.585	0.472	0.383
0.4			—	—	0.780	0.604	0.486
0.5			—	—	—	0.809	0.624
0.1	$\phi/2$	0°	0.396	0.368	0.306	0.253	0.207
0.2			0.485	0.452	0.380	0.319	0.267
0.3			0.604	0.563	0.474	0.402	0.340
0.4			0.778	0.718	0.599	0.508	0.433
0.5			1.115	0.972	0.774	0.648	0.552
0.1	$\phi/2$	5°	0.428	0.396	0.326	0.268	0.218
0.2			0.537	0.497	0.412	0.342	0.283
0.3			0.699	0.640	0.526	0.438	0.367
0.4			1.025	0.881	0.690	0.568	0.475
0.5			—	—	0.962	0.752	0.620
0.1	$\phi/2$	10°	0.472	0.433	0.352	0.285	0.230
0.2			0.616	0.562	0.454	0.371	0.303
0.3			0.908	0.780	0.602	0.487	0.400
0.4			—	—	0.857	0.656	0.531
0.5			—	—	—	0.944	0.722
0.1	$2\phi/3$	0°	0.393	0.366	0.306	0.256	0.212
0.2			0.486	0.454	0.384	0.326	0.276
0.3			0.612	0.572	0.486	0.416	0.357
0.4			0.801	0.740	0.622	0.533	0.462
0.5			1.177	1.023	0.819	0.693	0.600
0.1	$2\phi/3$	5°	0.427	0.395	0.327	0.271	0.224
0.2			0.541	0.501	0.418	0.350	0.294
0.3			0.714	0.655	0.541	0.455	0.386
0.4			1.073	0.921	0.722	0.600	0.509
0.5			—	—	1.034	0.812	0.679
0.1	$2\phi/3$	10°	0.472	0.434	0.354	0.290	0.237
0.2			0.625	0.570	0.463	0.381	0.317
0.3			0.942	0.807	0.624	0.509	0.423
0.4			—	—	0.909	0.699	0.573
0.5			—	—	—	1.037	0.800

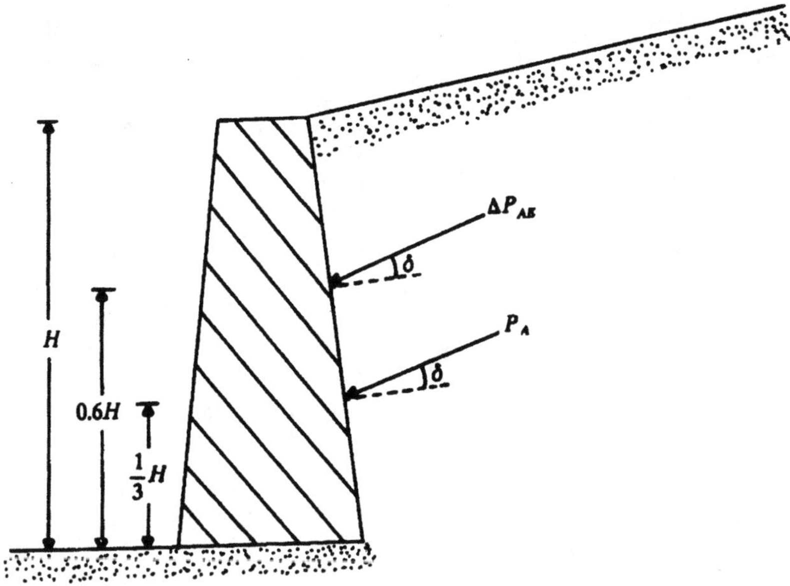

Figure 2.62 Point of application of resultant active earth pressure.

where

$$\theta = \tan^{-1} \frac{k_h}{1 - k_v}$$

Note that this equation has been derived for dry cohesionless backfill. Figure 2.63 shows the variation of K_{PE} for values of soil friction angle ϕ and k_h (with $k_v = i = \beta = \delta = 0$).

For more discussion on earthquake effects on soil pressures and waterfront structure design the reader is refered to Chapter 3.

2.7.5 Slope Stability

Slope stability is a consideration in the design of breakwaters, dikes, and other earthen structures. In carrying out slope stability analyses of slopes consisting of or founded on clayey material, both drained (long-term) and undrained (short-term, $\phi = 0$) analyses should be carried out to identify the most critical case. The undrained analysis would apply immediately and shortly after construction, and in a seismic event. The drained analysis applies when considering the static design loads on the slope.

In the design of earth slopes, the factor of safety is generally taken into account by dividing the sum of the resisting moments by the sum of the moments tending to cause failure. The safety factors to be used in design should be selected based on the sensitivity of the structure to movement. For example, a breakwater is not particularly movement sensitive, whereas a dike below a pile-supported wharf is quite sensitive to movement, as movement of the slope will result in lateral loads being exerted on the piles that support the wharf. In consideration of this, the following factors of safety are recommended for movement-sensitive structures:

- Operating conditions 1.5
- Extreme conditions 1.2
- Seismic conditions 1.2
- Temporary structures 1.3

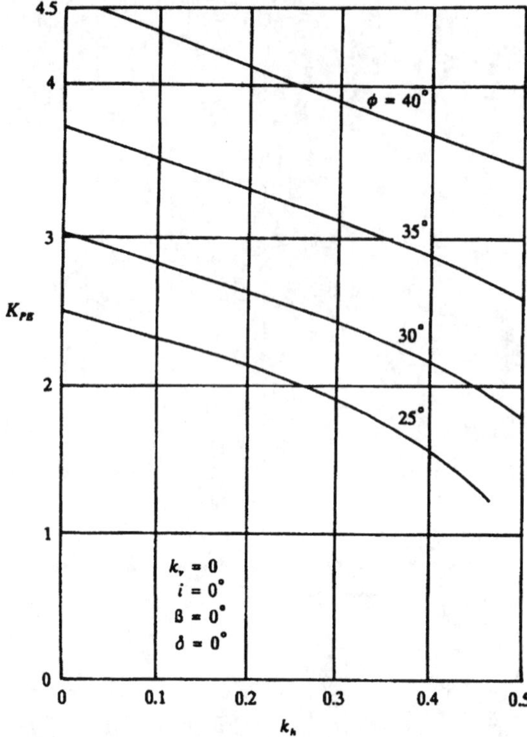

Figure 2.63 Variation of K_{PE} with soil friction angle and k_h. (After Davies et al., 1986.)

and for movement-insensitive structures:

- Operating conditions 1.4
- Extreme conditions 1.1
- Seismic conditions 1.1
- Temporary structures 1.2

The values above are based on the assumption that a thorough geotechnical investigation has been performed. If this is not the case, it is recommended that these values be increased accordingly.

2.7.5.1 Bishop Method of Analysis. The most general and widely used method of slope stability analysis is the Bishop method of slices. This method can account for several different layers of soil with different values of cohesion and friction angle. It does require one to know or estimate the pore water pressure. Figure 2.64 illustrates application of the Bishop method. In essence, the mass of soil in question is divided into a number of vertical slices, and the forces acting on each slice are determined by considering the equilibrium equations of that particular slice. The equilibrium of the entire mass of soil is then determined by summing the forces on all the slices.

Once the factor of safety for a particular trial circle has been determined, additional circles must be evaluated until the minimum factor of safety is determined. Such calculations can be carried out most efficiently through the use of a computer program developed specifically for analyzing slopes.

2.7.5.2 Dynamic Stability of Slopes. The analysis of the stability of slopes under seismic events is generally carried out using a pseudo-static method of analysis. This is accomplished by applying a lateral force equal to $k_h \times W$ on the slope, where k_h is the average coefficient of horizontal acceleration and W is the weight of the mass of soil. The magnitude of k_h is generally assumed to be 0.67 of the peak ground acceleration. The analysis is then carried out in a manner similar to the static analysis described above (i.e., using the Bishop method, and calculating the factor of safety as the summation of the resisting moments divided by the summation of the overturning moments).

A second method, which has been used more commonly in recent years, is the Newmark sliding block method. Using this method, by applying the time history of the earthquake to the slope in question, the displacement of the slope can be determined for a factor of safety of 1.0 by integration. Engineering judgment can then be used to evaluate whether the calculated displacement is considered acceptable for the structure in question. Application of the Newmark sliding block method is illustrated by Figures 2.65 and 2.66.

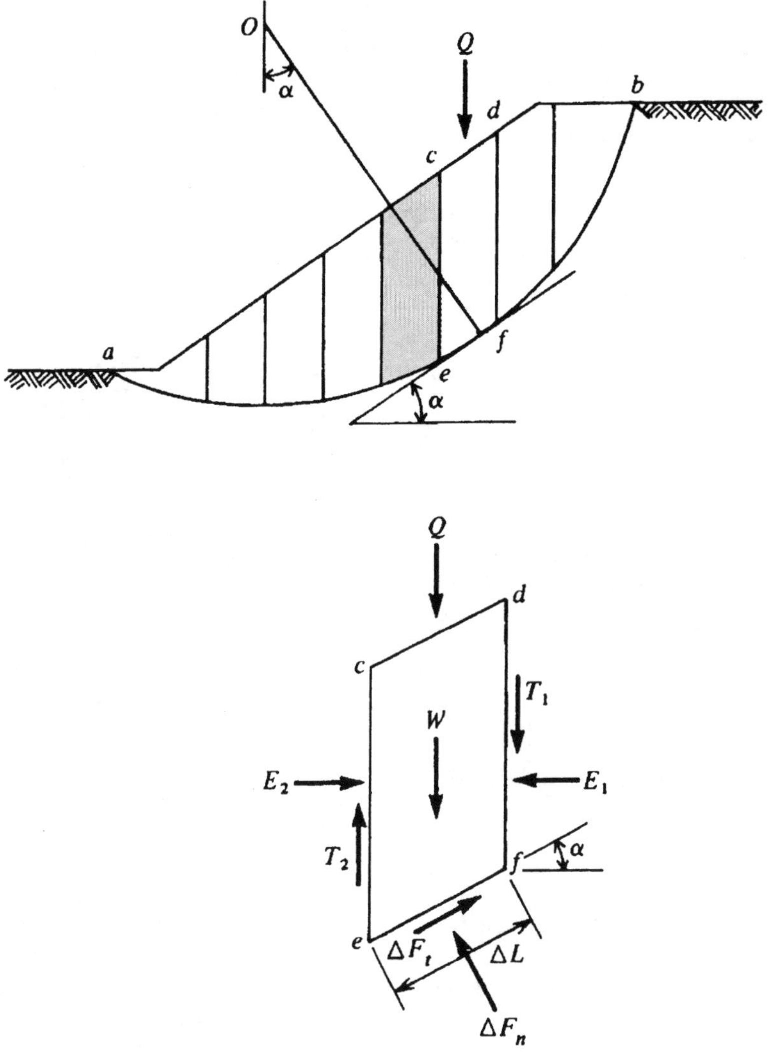

Figure 2.64 Bishop method of slope stability analysis.

2.7.6 Settlement

Settlement is a consideration in the design of earthen and gravity stabilized structures, as well as entire upland areas of terminals, particularly if the area is reclaimed. The total settlement can be calculated as the summation of the elastic settlement, consolidation settlement, and secondary compression settlement.

2.7.6.1 *Components of Total Settlement.* The elastic settlement occurs immediately upon application of load. It can be calculated as follows:

$$S_e = \frac{CpB(1 - \mu^2)}{E} \tag{2.46}$$

where S_e is the immediate elastic settlement, C

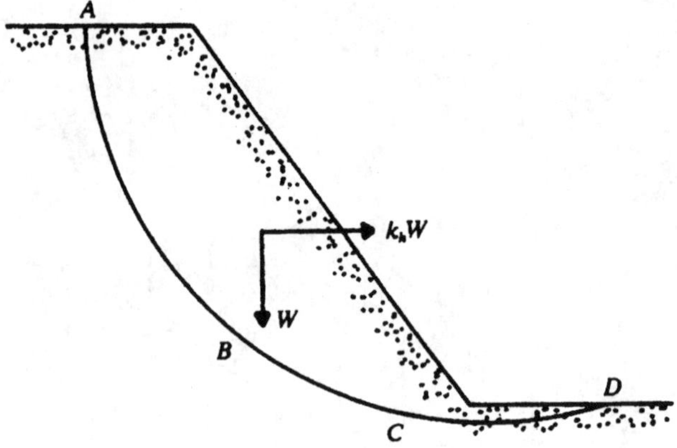

Figure 2.65 Soil slope.

the shape/rigidity factor, p the uniformly distributed load, B the foundation least width or diameter, μ is Poisson's ratio of soil, and E is Young's modulus of soil. Table 2.20 provides values of C.

In reality, most sites consist of layered soils. For a discussion of the effect of layering, the reader is referred to Winterkorn and Fang (1991).

2.7.6.2 Stress Distribution. To calculate settlements it is necessary to determine the stress distribution in the soil under a particular loading. For a point load the stress distribution is calculated to be

$$\sigma_z = I_z \frac{P}{z^2} \tag{2.47}$$

where σ_z is the vertical stress; I_z the influence value, which is a function of the geometry; P the point load; and z the depth from the surface.

I_z due to a point on the surface of an elastic half space is illustrated by Figure 2.67. Similarly, for a uniformly distributed load, the stress distribution is calculated to be

$$\sigma_z = I_z p \tag{2.48}$$

where p is the uniformly distributed load. I_z at various points within an elastic half space under a uniformly loaded circular area and for various loadings on rectangular areas can be found in Winterkorn and Fang (1991).

Cases arise where an earthen embankment or breakwater is to be constructed on a compressible layer. Similarly, there are cases where dredging will occur adjacent to a particular area (e.g., dredging may be required adjacent to a wharf that had been constructed by reclaiming land). The effect of such breakwaters or dredging can be taken into account by the use of influence diagrams provided in Winterkorn and Fang.

2.7.6.3 Consolidation Settlement. Consolidation settlement is due to the dissipation of pore water pressures, which begins at some time after a load is placed on a foundation. For purposes of the discussion herein, it will be assumed that the loaded area is large compared to the thickness of the soil layers, so that vertical strains predominate, and strains in other directions can be neglected.

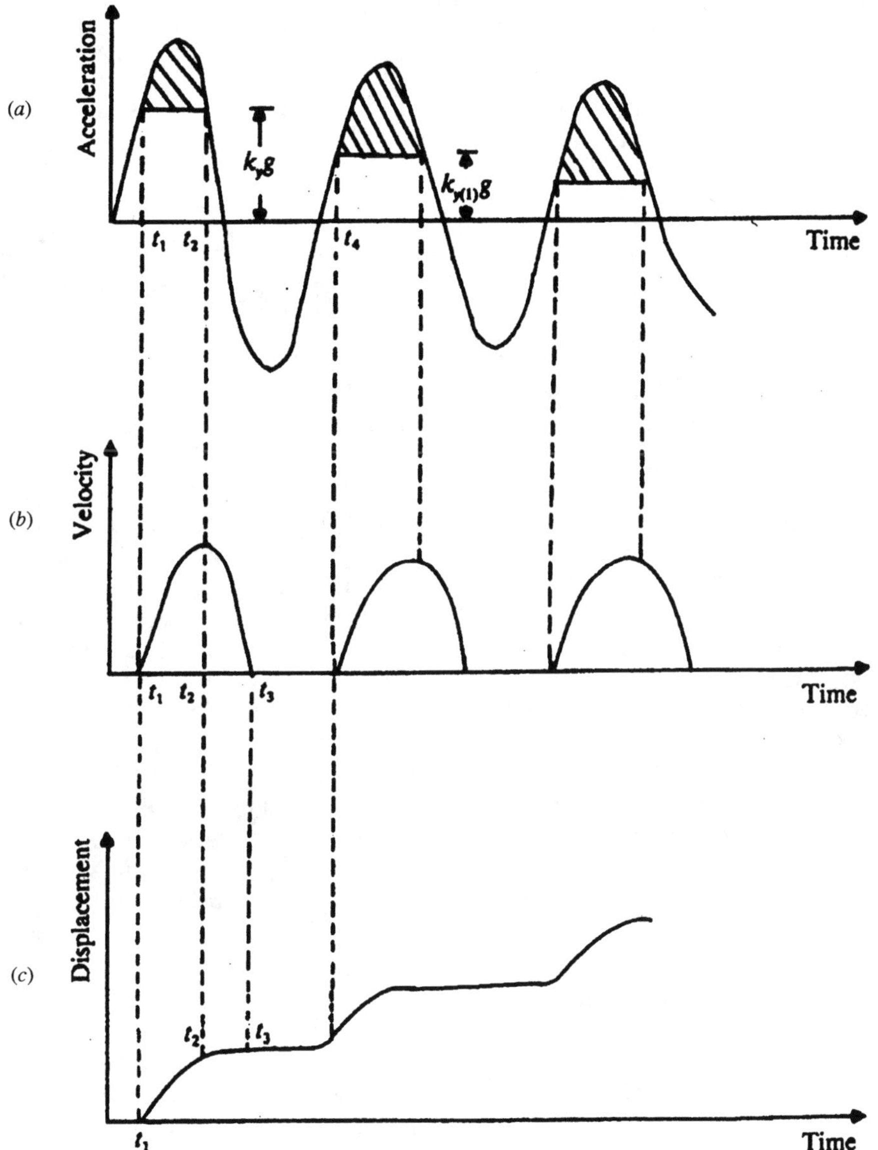

Figure 2.66 Integration method to determine downslope displacement.

Table 2.20 Values of shape/rigidity factor

Shape	Center	Corner	Middle of Short Side	Middle of Long Side	Average
Circle	1.00	0.64	0.64	0.64	0.85
Circle (rigid)	0.79	0.79	0.79	0.79	0.79
Square	1.12	0.56	0.76	0.76	0.95
Square (rigid)	0.99	0.99	0.99	0.99	0.99
Rectangle					
length/width ratio					
1.5	1.36	0.67	0.89	0.97	1.55
2	1.52	0.76	0.98	1.12	1.30
3	1.78	0.88	1.11	1.35	1.52
5	2.10	1.05	1.27	1.68	1.83
10	2.53	1.26	1.49	2.12	2.25
100	4.00	2.00	2.20	3.60	3.70
1,000	5.47	2.75	2.94	5.03	5.15
10,000	6.90	3.50	3.70	6.50	6.60

To calculate one-dimensional consolidation settlements, it is necessary to determine the relationship between the in situ void ratio and effective vertical stress. However, sample disturbance will result in a difference between laboratory and field testing as illustrated by Figure 2.68 for normally consolidated and overconsolidated clays, respectively.

The loading history for a point in a *normally consolidated deposit* is shown on a void ratio versus log pressure diagram in Figure 2.68*a*. The field virgin compression curve is indicated as a solid line down to the point that represents the in situ conditions, for which the overburden pressure, σ_0', is equal to σ_p', the preconsolidation pressure. An additional large increment of load on this deposit will produce a void ratio change corresponding to the dashed continuation of the field virgin compression curve. Due to disturbance effects, however, the effective consolidation pressure for a specimen brought into the laboratory is reduced, as shown in the figure, even though the void ratio remains constant. When the specimen is reloaded in the laboratory, a void ratio decrease occurs due to the disturbance effect, and the solid laboratory curve shown in the figure results.

In the case of an *overconsolidated clay* (Figure 2.68*b*), the in situ stress history is represented by the solid field virgin compression curve to the point at which the maximum past pressure, σ_p', was reached, after which the load was reduced to the existing overburden pressure, σ_0'. The solid rebound curve shows the in situ void ratio versus log pressure relation during the stress release. Subsequent reloading in the field would produce the dashed field recompression curve, which when the preconsolidation pressure was exceeded, would rejoin the field virgin compression curve. Again the effect of disturbance is to reduce the effective consolidation pressure at constant void ratio, leading to the laboratory curve shown as a solid line in the figure.

Thus, to predict settlements in the field, it is necessary to reconstruct the field compressibility curve from that observed in the laboratory. This is done by the method developed by Schmertmann (1955). For calculation purposes, it is usually convenient to divide the compressible strata into layers which are sufficiently thin that they can be considered approximately homogeneous. For a normally consolidated clay, the settlement ΔS_{ci} of the ith layer can be as-

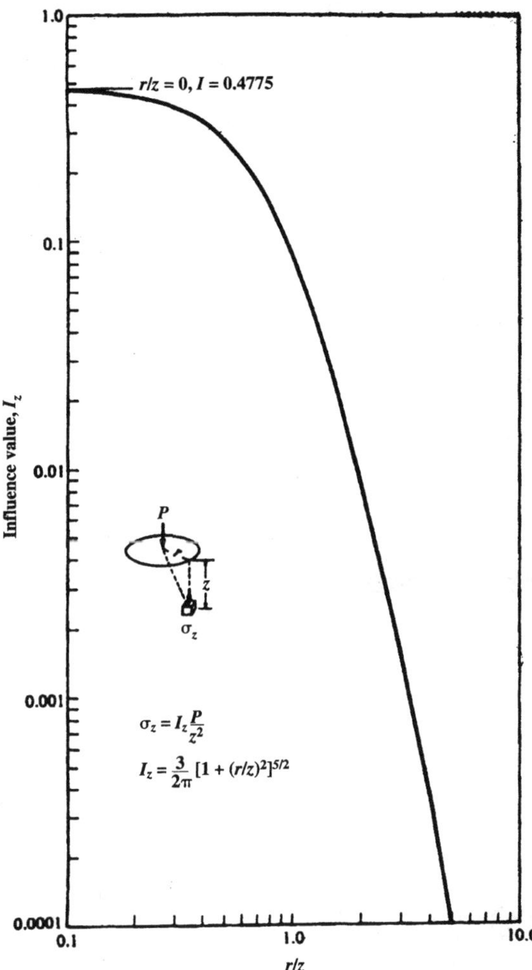

Figure 2.67 Influence diagram for vertical normal stress due to point load on surface of elastic half-space.

sumed to result from the void ratio change Δe_i at the middle of the ith layer:

$$\Delta S_{ci} = \frac{\Delta e_i}{1 + e_{0i}} H_i \tag{2.49}$$

where e_{0i} is the initial void ratio at the middle of the ith layer and H_i is the layer thickness. The total settlement is the sum of the settlements of the n individual layers:

$$S_c = \sum_{i=1}^{n} \frac{\Delta e_i}{1 + e_{0i}} H_i \tag{2.50}$$

It is generally necessary to use increments of uniform thickness when dividing a compressible stratum into layers. With increasing depth, the error introduced by assuming an average σ_0' and $\Delta\sigma'$ throughout the depth of a layer decreases because the rate of change of the applied pressure increment with respect to depth decreases as the depth increases. Thus, it is usual to increase the thickness of the layers subdividing an approximately homogeneous stratum as the depth increases.

If the stratum is normally consolidated and compression occurs entirely along the virgin compression curve, the void ratio change for each layer is

$$\Delta e_i = C_{ci} \log \left(\frac{\sigma_0' + \Delta\sigma}{\sigma_0'} \right)_i \tag{2.51}$$

in which σ_0' is the initial effective overburden pressure at the middle of the layer, $\Delta\sigma$ is the increment in vertical stress at the middle of the layer, and C_{ci} is the corrected in-situ value of the compression index.

When the compressible material is overconsolidated, the calculation of void ratio change depends on the magnitude of the stress increment. If the increase in effective stress $\Delta\sigma'$ is greater than $(\sigma_p' - \sigma_0')$, the consolidating soil will undergo both recompression and virgin compression. The total void ratio change can be calculated as the sum of the two components. The first, $\Delta e'$, is recompression due to the change in stress from the present overburden pressure, σ_0', to the preconsolidation pressure, σ_p':

$$\Delta e' = C_e \log \frac{\sigma_p'}{\sigma_0'} \tag{2.52}$$

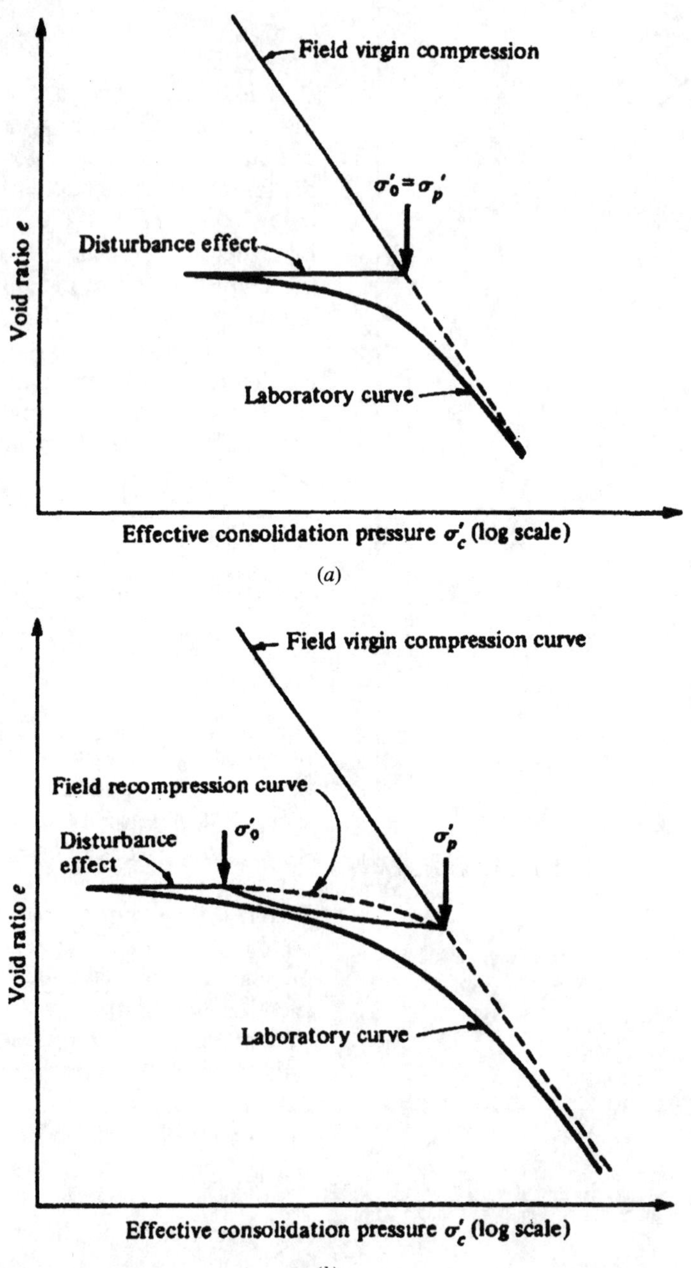

Figure 2.68 Effect of sample disturbance on compressibility: (*a*) normally consolidated clay ($\sigma'_p = \sigma'_0$); (*b*) overconsolidated clay ($\sigma'_p > \sigma'_0$).

The second component is the virgin compression, $\Delta e''$, due to the stress increment from σ'_p to $(\sigma'_0 + \Delta\sigma')$:

$$\Delta e'' = C_c \log \frac{\sigma'_0 + \Delta\sigma'}{\sigma'_p} \tag{2.53}$$

The total void ratio change is thus

$$\Delta e = C_e \log \frac{\sigma'_p}{\sigma'_0} + C_c \log \frac{\sigma_0 + \Delta\sigma}{\sigma'_p} \tag{2.54}$$

If the preconsolidation pressure is not exceeded, that is, if $\Delta\sigma' \leq (\sigma'_p - \sigma'_0)$, the soil will undergo recompression only. The change in void ratio is then

$$\Delta e = C_e \log \frac{\sigma'_0 + \Delta\sigma'}{\sigma'_0} \tag{2.55}$$

The total consolidation settlement can be calculated by summing the consolidation settlements for each layer.

The consolidation process involves expulsion of water from the soil being compressed. The preceding discussion has been concerned with determination of "ultimate" consolidation settlement (i.e., that which has occurred after all excess pore water pressure has dissipated and the soil is in an approximate equilibrium state). At any time between application of the load producing consolidation and the time at which essentially ultimate or 100% consolidation has occurred, the progress of settlement can be described by the average degree of consolidation. For a discussion of the time rate of settlement, the reader is referred to Winterkorn and Fang (1991).

2.7.7 Liquefaction

At sites susceptible to significant seismic risk, the potential for liquefaction of the soil supporting port-related marine structures, as well as the soil retained by these structures, should be evaluated. The phenomenon of liquefaction is associated primarily with medium- to fine-grained saturated cohesionless soils and is a result of an increase in the pore water pressure.

2.7.7.1 Potential Factors Affecting Liquefaction. The liquefaction potential of a soil deposit depends primarily on the following five factors:

1. *Relative density.* For a given value of peak pulsating stress, initial liquefaction and failure, which is defined as 20% double-amplitude strain, occur simultaneously for loose sand. However, as the relative density increases, the difference in the number of cycles of straining to cause initial liquefaction and failure increases.
2. *Confining pressure.* For a given initial relative density and peak pulsating stress, the number of cycles to cause initial liquefaction or failure increases with the confining pressure. This is true regardless of the relative density.
3. *Peak pulsating stress.* For a given initial void ratio (i.e., relative density) and number of cycles of load application, the variation of peak pulsating stress for initial liquefaction with confining pressure is linear.
4. *Number of cycles of pulsating stress application.* For a given vertical stress, as the number of cycles increases, the peak pulsating shear stress decreases.
5. *Overconsolidation ratio.* For a given relative density and number of cycles causing liquefaction, the ratio of shear stress to vertical stress decreases with a decrease in K_0, which is dependent on the overconsolidation ratio.

2.7.7.2 Zone of Liquefaction. There are five general steps for determining in the field the zone where soil liquefaction due to an earthquake can be initiated:

1. Establish a design earthquake.
2. Determine the time history of shear stresses induced by the earthquake at various depths of the sand layer.
3. Convert the shear stress-time histories into N equivalent stress cycles. These can be plotted against depth, as shown in Figure 2.69.
4. Using laboratory test results, determine the magnitude of the cyclic stresses required to cause initial liquefaction in the field in N cycles (determined from step 3) at various depths. Note that the cyclic shear stress levels change with depth due to change of σ_v. These can be plotted with depth as shown in Figure 2.69.
5. The zone in which the cyclic shear stress levels required to cause initial liquefaction (step 4) are equal to or less than the equivalent cyclic shear stresses induced by an earthquake is the zone of possible liquefaction, as shown in Figure 2.69.

2.7.7.3 Relation between Maximum Ground Acceleration and Relative Density for Liquefaction. A simplified procedure was developed by Seed and Idriss (1971) to determine in the field the relation between the maximum ground acceleration due to an earthquake and the relative density of a sand deposit for the initial liquefaction condition. It has been

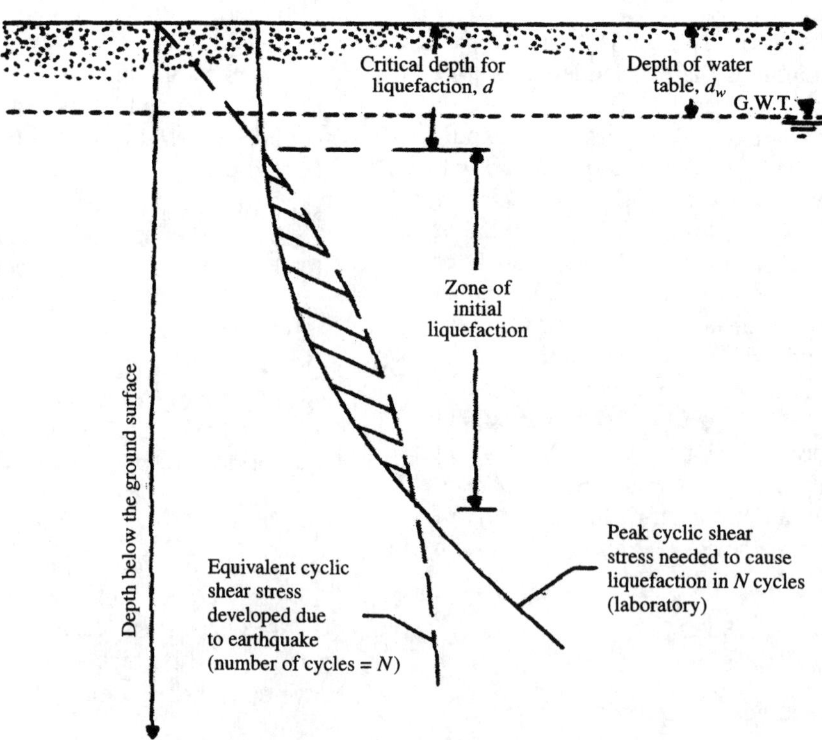

Figure 2.69 Zone of initial liquefaction.

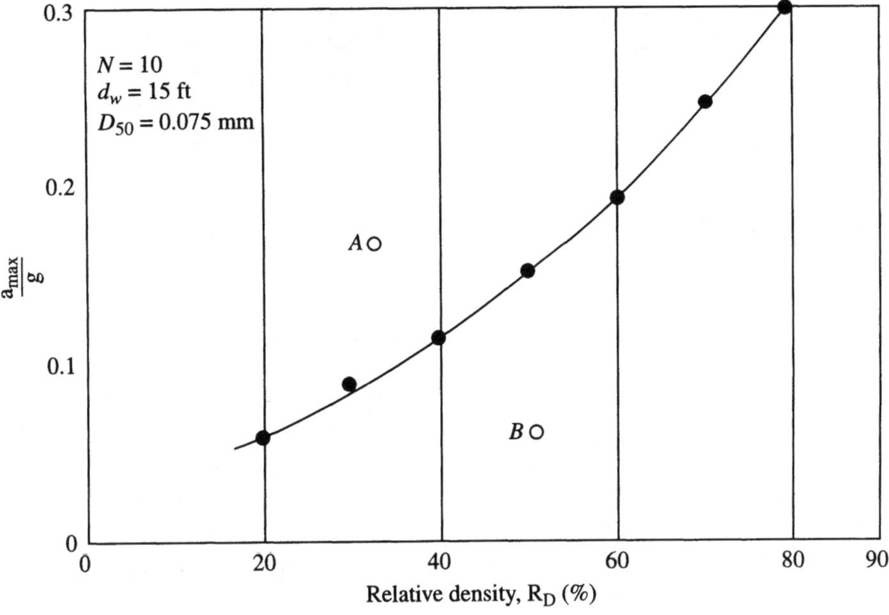

Figure 2.70 Plot of a_{max}/g versus relative density.

shown that the maximum shear stress determined from the shear stress-time history during an earthquake can be converted into an equivalent number of significant stress cycles. According to Seed and Idriss, one can take

$$\tau_{av} = 0.65\tau_{max(modif)} = 0.65C_D\left(\frac{\gamma h}{g}a_{max}\right) \quad (2.56)$$

The corresponding number of significant cycles, N, for τ_{av} is given in the following table:

Earthquake Magnitude	N
7	10
7.5	20
8	30

The following equation gives the correlation of laboratory results of cyclic triaxial tests to the field conditions.

$$\left(\frac{\tau_h}{\sigma_v}\right)_{\text{field }[R_{D(2)}]} = C_r\left(\frac{\sigma_d/2}{\sigma_3}\right)_{\text{triax }[R_{D(1)}]}\frac{R_{D(2)}}{R_{D(1)}} \quad (2.57)$$

This equation, along with the equation for τ_{av}, can be used to determine the relationship between a_{max} and R_D. One can then determine the variation of a_{max}/g with $R_{D(2)}$, where $R_{D(2)}$ is the relative density in the field.

Figure 2.70 shows a typical plot of a_{max}/g versus the relative density. For this given soil (i.e., given D_{50}, d_w, and N significant stress cycles), if the relative density in the field and a_{max}/g are such that they plot as point A in Figure 2.70 (i.e., above the curve showing the relationship of a_{max}/g), liquefaction would occur. On the other hand, if the relative density and a_{max}/g plot as point B (i.e., below the curve), liquefaction would not occur.

This type of diagram could be prepared for various combinations of D_{50}, d_w, and N. Since for liquefaction the range of D_{50} is 0.075 to 0.2 mm and the range of N is about 10 to 20, one can plot graphs for these critical combinations (i.e., $D_{50} = 0.075$ mm, $N = 20$; $D_{50} = 0.2$ mm, $N = 10$) which would serve as a useful guide in evaluation of the liquefaction potential in the field.

2.7.8 Geotechnical Investigations

It is essential for the successful design of any port-related marine structure for the designer to have reliable information on the geotechnical characteristics of the soil on which the structure will be founded and the soil that will be retained by the structure. The following provides guidelines and recommendations for carrying out a thorough marine geotechnical investigation. It is intended to ensure that all potentially critical geotechnical data are accounted for in planning the investigation program. Of course, the requirements for any particular project may dictate the need for some different or additional data, and the designer must thoroughly evaluate the specific needs for that project.

Prior to initiating a geotechnical investigation program, a literature search of all previous investigations should be conducted, including boring programs, geologic survey maps at or near the project site, and pile-driving records for nearby structures. Based on information obtained from these previous investigations, as well as the size and type of structure to be constructed, the scope of the investigation should be established.

2.7.8.1 Number of Borings. The need for and number of borings is related to the structure size and purpose and the uniformity of the soil strata as well as the cost of obtaining the borings and the cost of the project. In many cases with deepwater or remote sites, the cost of taking borings is excessive. Less costly vibracores, which can be obtained subsea with submersible equipment to a depth of 6 to 12 m, may provide adequate information. However, in general, the *minimum* number of borings in an investigation should be as follows:

- For a continuous structure, one boring for each 30 to 60 m of overall length.
- For a pier or sea island made up of individual structures such as breasting and mooring dolphins, one boring at each of the individual structures.
- For a long approach trestle to a pier, one boring for each 60 to 90 m, including one at the onshore end.
- For dredging, one boring for every 25,000 m^2 of area to be dredged. For large dredging projects, where conditions are anticipated to be relatively uniform throughout, vibracores may be substituted for some of the borings.

2.7.8.2 Depth of Borings. Minimum boring depths should be as follows:

- All borings for piled structures, such as a wharf, pier, loading platform, trestle, and mooring or breasting dolphin, should extend at least 6 m into competent soil below the estimated pile tip elevation, but not less than 20 m below the adjacent seabed (taking into account future dredging), and at least two borings should extend 15 m deeper. Should rock be encountered at higher levels, it should be cored for a depth of at least 3 m.
- For dredging, borings should extend a minimum of 1.5 m below the dredged depth, or 1 m below the depth to which the dredging contractor will be paid.

2.7.8.3 Field Sampling and Testing.

Field sampling and testing should be conducted as follows.

General Requirements

- Depending on the soil, sampling and/or field testing is required at every change of strata and at 1.5-m intervals within a given stratum throughout the full depth of all borings. Where undisturbed sampling is combined with successive standard penetration tests (SPTs), the interval between the two may be 1 m.
- SPTs should employ a standard split-spoon sampler, driven into the soil a distance of 460 mm by a 63.5-kg hammer free-falling 760 mm. The number of blows for each 150 mm of penetration should be observed and recorded.
- From at least one boring in each area, samples should be preserved to permit laboratory tests for soil classification over the full depth of the boring.
- For projects involving the stability of a slope and/or placement of fill, measurements of pore water pressure should be taken during the drilling. Piezometers should be constructed and monitored for sufficient time to characterize the groundwater conditions. In addition, special considerations are required in selecting the type and construction method of piezometers for taking measurements in impermeable materials such as silt and/or clay.

Cohesive Soils

- Undisturbed samples should be taken with a Shelby thin-walled sampler or piston sampler when highly undisturbed samples are required. Since undisturbed samples are costly, their number should be limited if possible. However, where accurate determination of soil strength is critical, the number should be increased. Tests on undisturbed samples are the primary means for estimating the shear strength and compressibility of cohesive soils. They are also used to confirm the results of in situ testing, which is described below.
- If mixed soils are found on initial sampling whereby difficulty exists with obtaining good undisturbed samples, or samples that are likely to give unrealistic test results, SPTs should be made immediately following undisturbed sampling. SPTs should also be considered for use in combination with undisturbed sampling on the basis that if adequate correlation of the two can be obtained in a representative number of borings, SPTs may be used alone in any supplemental borings to reduce time and costs.
- For projects where an accurate estimation of shear strength of the soil is critical, and where undisturbed samples are difficult to obtain, in situ vane shear tests should be considered. The results of the tests provide a direct measurement of the undrained shear strength of the soil. However, for soft clays, the speed of the test has an influence on the results, and it is consequently necessary to apply a correction factor.
- For a large project, where many borings would normally be required, soundings may be substituted for some borings, assuming that a representative number of borings indicates fairly uniform conditions throughout the area. In this case, soundings using a Dutch cone or similar type of penetrometer can be used to provide a log of penetration resistance versus depth. The results must first be correlated to nearby conditions.
- Handheld shear vane and pocket penetrometer tests should be made to obtain

rapid estimates of soil strength in the field. The vane shear and/or pocket penetrometer tests should be made at not over 1.5-m intervals in, or adjacent to, a representative number of borings in each area to develop shear strengths. The in situ vane shear test can also be used to estimate remolded shear strength, although the results are not as reliable as lab vane shear tests. Where such soils start at the surface, they should be tested continuously in the top 5 m adjacent to a representative number of borings for which undisturbed sampling has been done at 1.5-m intervals. At any supplemental locations, the amount of undisturbed sampling and vane and pocket penetrometer testing may be reduced to suit conditions.

- Pressuremeter tests should be made to obtain in situ lateral response data of silts and clays. Tests should be at depths pertinent to the design considerations for the structure. The tests are used to determine in situ load deformation and can be incorporated directly into *p–y* analyses.

Cohesionless Soils

- Primarily, only SPTs should be employed for cohesionless soils. However, if feasible, disturbed samples can be tested in the laboratory, either triaxially or in a direct simple shear or direct shear device, to obtain a correlation of density and internal friction angle with SPT blow counts.

Rock

- All coring should be done with N-series double-tube core barrels and be limited to a representative number of borings in each area.
- Where rock is encountered and needed for pile support, cores should be taken to a depth of 3 m into sound rock or 6 m into fractured rock, but not more than 6 m for any combination thereof.
- Where rock is encountered but not needed for pile support, cores should be taken to a depth of 3 m regardless of the rock character.
- If encountered rock is limestone, at a depth where the possible presence of solution cavities could affect the design of foundations, the geotechnical investigation program should be adjusted as appropriate to identify such cavities to the extent practical.
- Regardless of rock depth, complete testing of representative overburden soils is required. Supplemental borings, required only for added definition of overburden soils or rock levels may terminate at the rock surface. However, where added definition of overburden soils at supplemental locations is not required, SPTs, or other acceptable less costly and time consuming means than conventional borings, may be used to obtain supplemental rock-level data. The latter types should be used only when no interface media, such as boulders or large gravels, exist to affect measurement accuracy significantly.

2.7.8.4 Laboratory Testing. Laboratory tests on representative soil samples may include the following, depending on the needs of the project.

Cohesive Soils

- Water content
- Specific gravity
- Bulk unit weight
- Atterberg limits
- One-dimensional consolidation
- Unconfined compression

- Shear strengths from vane shear and/or triaxial test [Vane shear tests are useful for developing remolded shear strengths. Triaxial tests can be unconsolidated-undrained (UU), consolidated–undrained (CU), or consolidated–drained (CD). The type of test should be selected so as to match actual site loading conditions. The unconsolidated–undrained (UU) test is typically most suitable for the design of piling.]
- Strength sensitivity to remolding
- Strength sensitivity to free water contact, as may apply following any preaugering for piles
- Shear strengths from cyclic loading, as applied to design of flexible dolphins under recurring lateral loading
- pH and soluble salts content of the soil moisture
- Basic mineralogy of soil particles, but only if this is a factor affecting design
- Carbonate and organic carbon content

Cohesionless Soils

- Water content
- Specific gravity
- Natural and relative density, if feasible
- Grain size analysis from wet sieving
- Permeability, if feasible
- pH and soluble salts content of the soil moisture
- Carbonate and organic carbon content

Note: If the soil is sand and undisturbed samples cannot be attained as desired for triaxial tests at natural density, development of natural and relative density, and permeability, consideration should be given to the feasibility of making reconstituted samples for these purposes.

Rock. (Observations may result from from laboratory tests or inspection, as applicable.)

- Unconfined compressive strength of sound rock
- Quality of weathered rock
- Type and amount of soil inclusions in fractured rock and apparent effect on strength

2.7.8.5 Final Report. A final report detailing the results of the investigation should be issued after completion of all field and laboratory testing. The report should be in two parts.

Part 1: Data. The following elements should be included, as applicable:

- A sketch showing the actual locations of all borings and other field tests. These should be referenced to survey markers and the local coordinate system identified. The sketch should also include a plot of the coordinate system, seabed contours, and the general location in outline form of proposed facilities. The report should clearly state how the depth of water was established at the boring. In addition, a description of how the reference to local datum was established should be provided.
- A log of each boring showing the following:
  - Depth to seabed, top of each stratum, and final depth of penetration, referenced to local datum
  - Descriptive identification of the soil in each stratum based on the Unified Soil Classification System
  - Time at start and completion of the boring
  - Method of boring advancement, including borehole diameter
  - Depth and resulting SPT blow count for each 150 mm of 450 mm total penetration

    - Size of SPT split-spoon, and weight and free-fall height of hammer
    - Depth of undisturbed sampling, sampler size and type, and indication of whether pushed or driven
    - Depth, size, and weight of casing pipe, hammer weight and drop, and cumulative progressive blow count
    - Use and depth of drilling
    - Elevation where any drilling fluid is lost, and casing elevation at time
    - Observed water/mud levels at beginning of each day
    - Size of rock core samples, speed of rotation, and rate of penetration
    - Values of vane shear tests
    - Values of pocket penetrometer tests
- A log of vane shear test values for tests made outside borings
- Generalized profiles of basic soil strata change throughout the site
- The results of all laboratory tests, together with copies of all laboratory consolidation and shear strength data sheets
- A general description of field procedures, equipment, and testing techniques.

Part 2: Recommendations. The following elements should be included, as applicable:

- A general description of site soil conditions, together with basic conclusions regarding the types of foundations required
- A geologic description of the site, together with any indications of seismic faults and/or solution cavities at or near the site
- A design soil profile applicable for each structure, and identification of its range of applicability
- Recommendations for the design of foundations and other works affected by soil conditions that include the following as applicable:
    - Suitable pile types covering vertical and batter piles for the range of potential anticipated loads
    - The formulas applying to development of skin friction and end bearing for compression and tension on piles
    - Minimum pile penetrations with respect to any particular stratum that must be reached
    - The formula applying to reduction in rating due to downdrag on piles from fill
    - The formula applying to reduction in ratings for friction piles in groups
    - Requirements for jetting, preaugering, or socketing piles into rock
    - Requirements for pile tip reinforcement
    - Pile-driving wave equation analysis results, including required hammer energy
    - Criteria for pile installation (i.e., penetration and/or set and/or cumulative blow count)
    - Recommendations for, and locations of, static and dynamic pile load tests and/or pile-driving tests
    - Formula to cover the point of effective fixity for piles subject primarily to axial loading
    - Pile t–z and Q–z curves for each soil type within each design soil profile, if required
    - Subgrade moduli of deformation and formulas for development of pile lateral load capacity (includes p–y curves for each soil type within each design soil profile; formulas should be provided for both static and cyclic loadings, if applicable)
    - For gravity-stabilized structures that are founded at or near the seabed, the bearing and settlement characteristics of the soil

- For revetments and breakwaters incorporating armor stone or concrete armor units, the filtration and piping requirements of natural soils into and through the coarser-grained structure
- Geosynthetics as applied to seabed and backfill stabilization, reinforcement, drainage, and separation applications
- Allowable dredged slope inclinations to maintain adequate passive resistance against structures

2.8 DESIGN OF STRUCTURES

Standard structural design principles apply to the design of port-related marine structures. Also, the standard design codes, such as the American Institute of Steel Construction (AISC, 2000), ACI (1995), and AASHTO (1996), among others, are applicable. However, these types of structures are often of unique configurations, may be subject to some extremely heavy loading conditions, are exposed to severe environmental conditions, and frequently must be constructed utilizing floating or other marine construction equipment, all of which must be addressed in the design process.

The circumstances in which a particular port-related marine structure is to be installed are often unique, either with regard to loading conditions, environmental conditions, bathymetry/topography, or any number of other aspects of the particular scenario. It is essential for the designer, in attempting to create the optimum structure for that scenario, to utilize the concepts and examples presented herein, or elsewhere, as the point at which the creative process begins, not at which it ends. Designers should not be reluctant to impart creativity and ingenuity to the design process in order to develop structural solutions that will best achieve the objectives of the project.

Port-related marine structures are generally designed for a useful life of between 20 and 50 years, with a useful life of 30 years probably being the most commonly selected. Of course, the useful life of almost any structure may be extended virtually indefinitely through appropriate maintenance and periodic rehabilitation. For purposes of this discussion, useful life is defined as the period of time during which the structure will be able to serve its intended function while receiving only routine periodic maintenance and without major repairs or rehabilitation. Detailed discussion of standard structural design procedures is found in Tsinker (1997).

2.8.1 Gravity-Stabilized Bulkheads

As discussed in Section 2.4, wharf structures may be classified as closed types and open types. Closed wharf structures may be further classified as gravity stabilized and tied back. Gravity-stabilized structures derive their stability primarily from their weight, combined with the interaction of the structure surfaces with the soil on which it is supported and the soil it is restraining. Within the gravity-stabilized classification there are three basic structural concepts in general use: cellular, caisson, and block. Of course, there are numerous variations and combinations of these three basic concepts.

Often, because of their heavy weight and the distribution of this weight to the supporting soil at or near the bottom of the structure rather than on deep foundations, gravity-stabilized bulkheads are particularly attractive where relatively strong seabed soils are present. Assuming that seabed soils are suitable for a gravity-stabilized bulkhead, the type of bulkhead that will be most suitable for a particular situation depends on numerous factors, the most important of which are relative cost and availability of materials, relative cost of labor versus materials, and the experience of the construction contractors in the region that will perform the construction.

Although the specific design procedures for each type of gravity-stabilized bulkhead will differ, the same basic design considerations apply to all. All loads and load combinations acting on the bulkhead are defined as discussed in Section 2.6, and then the structure is evaluated for sliding stability, overturning stability, foundation bearing capacity, and global stability. Usually, a gravity-stabilized bulkhead is analyzed as a two-dimensional structure consisting of the typical bulkhead cross section. For each of the four stability evaluations, the live load should be applied in the manner that results in the lowest factor of safety. For example, when considering sliding stability, the live load should be applied behind the structure but not on the structure. But when considering foundation bearing capacity, the live load should be applied both behind the structure and on the structure.

The design process for a gravity-stabilized structure is usually initiated by selection of the basic dimensions of the typical cross section. The optimum width/height ratio for this type of structure will depend on the geotechnical properties of the foundation soil, the characteristics of the soil retained by the bulkhead, and the type of bulkhead construction. Typically, the width/height ratio of port-related gravity-stabilized bulkheads is in the range 0.5 to 0.8.

When considering sliding stability, the driving forces usually consist of lateral earth pressure, hydrostatic pressure, and mooring forces. The resisting force consists of the friction between the bottom of the structure and the foundation soil, derived from the buoyant weight of the structure and any fill material acting integrally with the structure. Normally, vertical live load acting directly on the structure is not considered. For structures where the bottom may be embedded in the foundation soil, the passive pressure acting on the embedded portion of the structure also resists sliding. The minimum factor of safety against sliding is usually taken as 1.5 for normal loading and 1.3 for extreme loading and for temporary structures.

When considering overturning stability, the driving and resisting forces are essentially the same as when considering sliding stability. The minimum factor of safety against overturning is usually taken as 2.0 for normal loading and 1.5 for extreme loading and for temporary structures.

When considering foundation bearing capacity, in addition to the driving forces considered for sliding and overturning stability, the vertical live load acting directly on the structure is also included. There are several generally accepted theories and procedures for calculating the ultimate foundation bearing capacity for a structure subjected to substantial horizontal and vertical loading. One of the most widely accepted is that presented in *Recommended Practice for Planning, Designing and Constructing Fixed Offshore Platforms: Working Stress Design,* issued by the API (2000) and discussed in Section 2.7.3. The minimum factor of safety against foundation bearing capacity failure is usually taken as 2.0 for normal loading and 1.5 for extreme loading and for temporary structures.

Global stability refers to the ability of a large soil mass, in which the bulkhead structure is embedded, to resist failure in shear along a basically circular slip surface. Global failure results in a downward and outward movement of the soil mass with the embedded bulkhead structure. There are several accepted manual and computer-based procedures available for analyzing global stability. Typical practice is first to use a relatively simple and conservative procedure to determine if global stability may be a controlling consideration. If so, a more precise procedure will be used. The minimum factor of safety against global failure is usually taken as 1.5 for normal loading and 1.3 for extreme loading and for temporary structures.

2.8.1.1 Cellular Structures. Cellular bulkhead structures are gravity-stabilized structures formed of flatweb steel sheet piles interconnected to create essentially watertight cells. The two most commonly used forms of cellular bulkhead structures are the circular type (Figure 2.71*a*), and the diaphragm type Figure 2.71*b*). In the more widely used circular type, the cells are positioned side by side and are usually interconnected by additional sheet piles forming arcs of smaller diameter between the cells along both sides of the bulkhead. Sometimes the arcs on the interior face of the bulkhead are eliminated. The arcs are generally connected to the cells at a point on the circumference located between 30 and 45° from the longitudinal axis of the bulkhead. The junction points require special prefabricated sheet piles in the cell. These sheets have an additional interlock welded along their length for the connection of an end sheet pile of the arc. The cells and arcs are filled with granular material such as sand or gravel, and a cast-in-place concrete cap provides a continuous wharf edge.

A less commonly used variation of the cellular structure is the diaphragm type. This type of cellular structure is made up of equally spaced parallel rows of flat web steel sheet piles, perpendicular to the line of the wharf, with circular arcs at each end. The end sheet piles in each row are specially fabricated to provide an interlock for the end sheet pile of the arc on each side of the row. The diaphragm type of cellular structure presents a flatter face than that of the circular cell type.

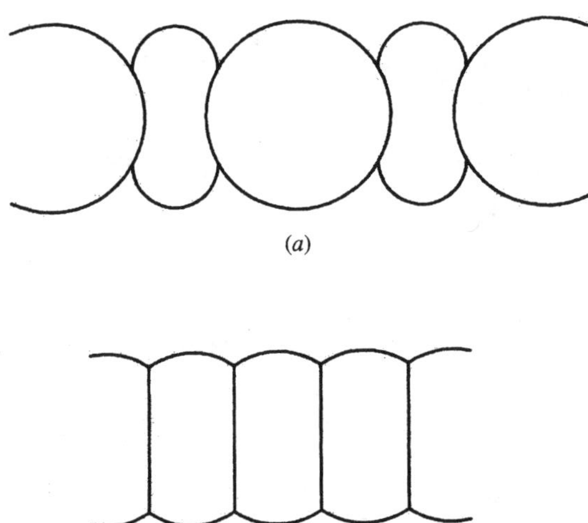

Figure 2.71 Cellular bulkhead structures: (*a*) circular cells; (*b*) diaphragm cells.

Each cell of the circular cell structure is independently stable and may be filled without regard to adjacent cells. However, diaphragm-type cell structures cannot tolerate substantial differences in the height of the fill in adjacent cells, requiring careful sequencing of the filling operation, and thus requiring more construction time.

Cellular structures derive their strength through a combination of the shear strength of the granular fill material and friction in the sheet pile interlocks resulting from loop tension within the cell sheet piles. Sometimes the above water portion of each of the sheet pile interlocks is welded to supplement friction resistance to shear deformation of the cells. The integrity of the interlocks of the cells is very important because the cells are subject to substantial hoop tension forces.

Circular cell structures are less susceptible than the diaphragm type to progressive failure in the event of a rupture of one cell. Since each circular cell is independently stable, a rupture of one cell will not affect adjacent cells. This is not the case for the diaphragm type, where the straight walls cannot tolerate substantial differential lateral loading.

The location of the outboard edge of the cells relative to the wharf face must be given careful consideration. For wharves that will serve vessels with a bulbous bow, particularly those wharves that do not have a continuous fender system, adequate clearance between the bulbous bow and the outboard edge of the cells must be provided. For example, a wharf serving large container vessels, which have bulbous bows, may have a fender system consisting of discrete fender units spaced between 10 and 20 m on centers. If a berthing vessel approaches at the maximum berthing angle, contacts the wharf midway between two fender units, and fully compresses these fender units, the vessel's bulbous bow may project a substantial distance inboard of the wharf face. The outboard edge of the cells must be set sufficiently far back from the wharf face to prevent contact. This can be accomplished by cantilevering the deck structure beyond the outboard edge of the cells, or by mounting the fender units on projections from the wharf face to increase the stand-off distance.

Individual circular sheet pile cells are also frequently used as freestanding structures for such applications as breasting, mooring, and turning dolphins, and fender structures to protect bridge piers from ship impact. In the latter application, the sheet pile cell will absorb the impacting vessel's kinetic energy, thus bringing it to a stop, by a combination of shear deformation of the cell, rupturing of individual sheet piles, raising the center of gravity of the vessel, friction between the vessel hull and the sheet piles, and a variety of other energy dissipation mechanisms. The cell is destroyed in the process, but substantial damage to the bridge pier and the vessel is prevented (Tsinker, 1995).

Cellular bulkhead structures consist of two very different materials, steel and soil, resulting in a complex interaction that makes a rational design approach very difficult. Although various theories have been suggested to derive analytical solutions for the stresses in a cell, most designers rely heavily on past practice and experience.

Before a design can be initiated, the necessary controlling dimensions must be set. The height of the bulkhead must be established based on the dredged depth and the elevation of the upland area. After the height of the bulkhead is established and the pertinent physical properties of the underlying soils, together with the cell fill, are determined, a tentative equivalent width, B, is chosen. The equivalent width of the cellular structure is defined as the width of an equivalent rectangular section having a section modulus equal to that of the actual bulkhead. The stability of the cellular structure must then be analyzed considering horizontal sliding, overturning, bearing capacity, global stability, slipping between cell fill and sheet piles, shear failure on centerline of cell, horizontal shear, and interlock tension.

2.8.1.2 Caisson Structures. Caisson bulkhead structures are gravity-stabilized structures consisting of large, prefabricated concrete units, generally positioned on a prepared foundation bed of gravel or crushed stone. The caissons are usually filled with sand to increase their weight and improve their stability. The caissons may be designed to be floated into position or lifted into position, depending on local conditions and the availability of equipment and fabrication facilities.

The size and weight of caissons that are to be lifted is limited by the availability of lifting equipment. Caissons that are to be floated into position are generally much larger than lifted caissons. A basic design objective for either type of caisson is to minimize its fabricated weight. For the lifted caisson, by minimizing the weight, the size of the caissons can be maximized for a given limiting lifting weight. This will minimize the number of units required to construct a bulkhead of a given length, and thus minimize the cost. For the floated caissons, minimizing the weight facilitates fabrication and launching. Generally, the minimum thickness of caisson walls is approximately 30 mm, and the minimum thickness of the bottom slab is approximately 60 mm, to provide adequate concrete cover for the reinforcing steel.

Once the basic dimensions of the caissons are determined to satisfy the sliding and overturning stability, foundation bearing capacity, and global stability requirements, the elements of the caisson, including interior and exterior walls, bottom slab, and top slab if so equipped, must be designed in accordance with standard reinforced concrete design procedures. In addition, for floated caissons, the structures must be designed for the stresses imposed during launching and while afloat and under tow. Also, the caissons must be designed for buoyant stability throughout the launching, towing, and installation cycle. In this regard, the basic principles of buoyant stability must be considered as applicable to ship design. The launching and installation phases are usually particularly critical with regard to stability, as is the caisson's stability if damaged.

2.8.1.3 *Block Structures.* Block bulkhead structures are probably the oldest type of bulkhead structures, with granite blocks being used for bulkheads more than 2000 years ago. Modern block structures utilize concrete blocks, either solid or hollow and of various sizes and shapes.

The basic design principles of other gravity-stabilized bulkheads with regard to sliding and overturning stability, foundation bearing capacity, and global stability requirements also apply to block structures. In addition, the stability of the block structures at each layer of blocks must be assured. If the blocks are not keyed vertically, a coefficient of friction between the underwater concrete blocks of 0.5 may be assumed.

Block wall construction is quite heavy and exerts relatively high pressure on the foundation soil. For this reason, block bulkhead structures are cost-effective where the seabed is rock or dense sand. Where the seabed is rock, a gravel or crushed stone bedding is provided as a leveling course on which to set the toe blocks. Where the seabed is sand, decomposed rock, or other weaker soil, the grading and thickness of the bedding material must be carefully designed to distribute the pressure sufficiently under the toe of the wall and to prevent the loss of fines from the soil below the bedding.

The size of the individual blocks is established based on stability requirements and the lifting capacity of available equipment. It is often attractive to cast the blocks with large voids, so that larger blocks, and thus fewer blocks for a given length and height of bulkhead, will be required. Subsequently, the voids are usually filled with tremie concrete to increase the weight and stability of the blocks. Reinforcing bars may be inserted in the concrete-filled voids to create a monolithic structure, but this is usually not cost-effective in other than the top row or two of blocks to distribute concentrated berthing and mooring loads, and other concentrated loads, throughout a large length of the bulkhead.

The blocks should be of sufficient size so that each row is only one block deep. The blocks are normally keyed, both vertically and horizontally, to maximize wall integrity, and are sized and positioned to maximize stability and minimize foundation bearing pressure. Special attention must be paid to the design of the toe of the bottom row of blocks to ensure that tension stresses in the concrete are minimized and that the toe is appropriately reinforced.

The design of the uppermost concrete block and the cast-in-place capping that is usually provided must be such that it minimizes loading on the wall and is suitable for the intended use of the bulkhead. Usually, the uppermost concrete block is cantilevered beyond the rear of the wall to reduce the lateral earth pressure acting on the wall from the fill above the block and from the design live load. The uppermost block is also often cantilevered beyond the front face of the wall to provide adequate clearance between the hulls of berthing vessels and the bulkhead. The cast-in-place capping structure is designed to provide a straight and level finished appearance for the wall. It also usually accommodates the fenders and mooring fittings and any below-grade utility services, as illustrated in Figure 2.72. It should be designed to distribute the concentrated berthing and mooring forces along a substantial length of the bulkhead.

2.8.2 Tied-Back Bulkheads

Tied-back bulkheads are generally sheet pile walls, usually of steel but also of concrete, or walls made up of a combination of steel sheet piles and either steel H-piles or steel pipe piles

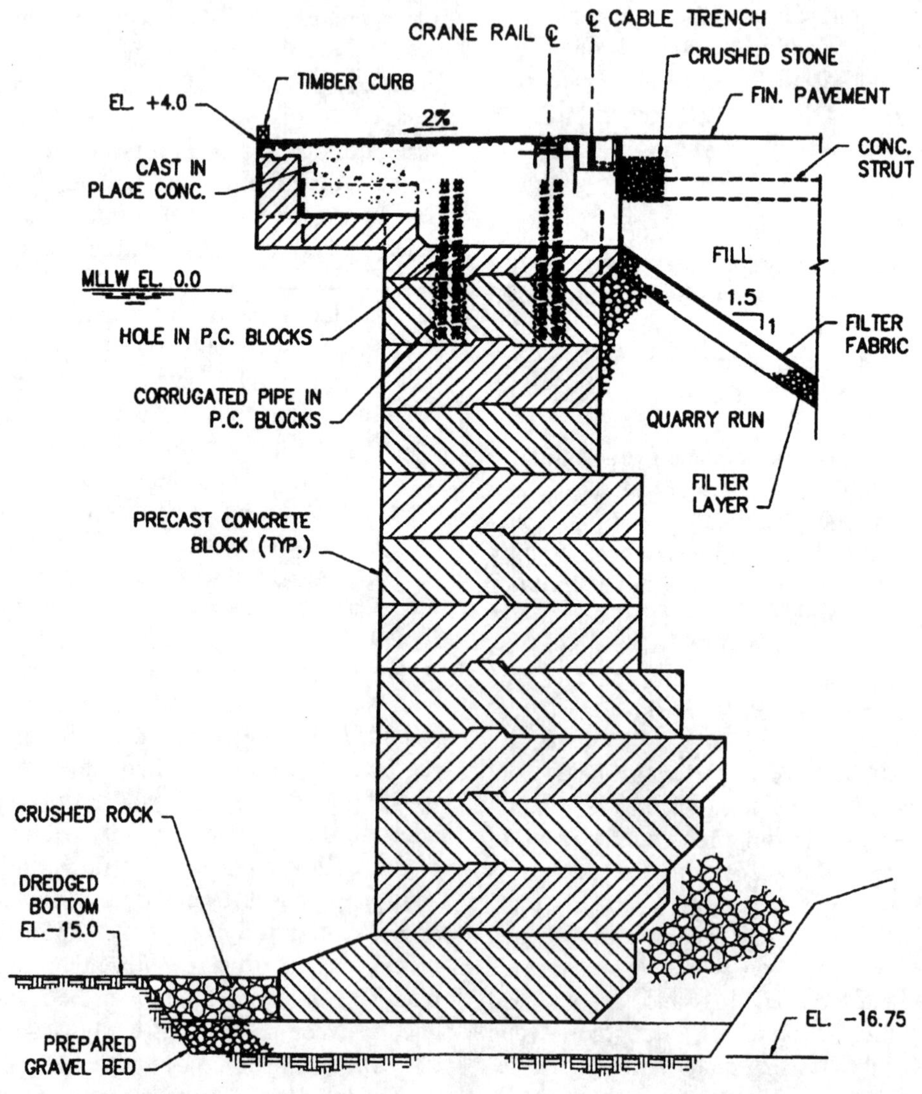

Figure 2.72 Concrete block bulkhead at the Port Salalah Container Transshipment Terminal, Oman. (Courtesy of Han-Padron Associates.)

for increased bending strength. Other types of tied-back bulkheads, such as reinforced concrete walls constructed by the slurry wall method, may be suitable in special circumstances.

The theory for the design of tied-back sheet pile bulkheads is well established and there are numerous publications available that present in great detail the various earth pressure theories and design procedures in use. Also, several excellent computer programs have been developed and are commercially available to assist in the design of sheet pile bulkheads. The following discussion is a brief summary of the design procedures. The reader is referred to ASCE (1996), Tsinker (1997), Briand and Lim (1999), or other publications for a more detailed treatment of the subject.

The earth pressure acting on the bulkhead is a function of the properties of the soil and the configuration and movement of the structure. The active, at-rest, and passive earth pressures are of concern in the design of the bulkhead. The active pressure accompanies outward movement of the bulkhead, the at-rest pressure is applicable for rigid bulkheads where the backfill has been compacted, and the passive state accompanies inward movement.

There are two well-known classical earth pressure theories: Rankine theory and Coulomb theory. Each furnishes expressions for active and passive pressures for a soil mass at the state of failure. In either case, the at-rest lateral earth pressure is taken as

$$K_0 = 1 - \sin \phi \tag{2.58}$$

Rankine theory is based on the assumption that the bulkhead introduces no changes in the shearing stresses at the surface of contact between the bulkhead and the soil. However, since the friction between the bulkhead and the soil has a significant effect on the vertical shear stresses in the soil, the lateral stresses on the bulkhead are different from those assumed by Rankine theory.

Most of the error inherent in Rankine theory can be avoided by using Coulomb theory, which considers the changes in tangential stress along the contact surface due to friction. However, Coulomb theory assumes that the failure surface is a plane, which is not the case. For the active case the error introduced is small, but for the passive case the error can be large and is unconservative. If the angle of wall friction is low, the failure surface is a log-spiral curve along the surface of least resistance. This failure surface effect is taken into account in log-spiral theory.

In addition to lateral earth pressure, the bulkhead must also support the imposed live loads acting on the deck of the structure. For a uniform live load, the conventional theories of earth pressure are utilized. For live loads over a limited area, or for concentrated live loads, the theory of elasticity, modified by experiment, as summarized by Winterkorn and Fang (1991), are used. In addition to live loads, the bulkhead may have to be designed to resist wave forces, berthing and mooring forces, and ice thrust. For bulkheads constructed in regions subject to seismic activity, the increase in lateral pressure due to the dynamic forces transferred from the surrounding soil, and from liquefaction of the soils, must be considered. For a discussion of the seismic design of these structures, the reader is referred to publications by the U.S. Army Corps of Engineers (1992), ASCE (1998b), the International Navigation Association (PIANC, 2001), and Chapter 3.

The design of sheet pile bulkheads requires several successive operations: (1) evaluation of the forces and lateral pressures, (2) determination of the required depth of piling penetration, (3) computation of the maximum bending moments in the piling, (4) computation of the stresses in the bulkhead and selection of the appropriate piling section, and (5) the design of the anchorage system.

Tied-back sheet pile bulkheads derive their stability from passive pressure on the front of the embedded portion of the bulkhead and from the tie rods or anchors near the top of the bulkhead. For bulkheads up to a height of appropriately 10 to 12 m, with a single row of tie rods, conventional steel sheet piling can be used. For higher bulkheads, high-strength steel sheet piling, combinations of steel sheet piling with steel H-piles or pipe piles, or an additional tier of tie rods is required. The overall stability of tied-back bulkheads, and the stresses in the members, depend on the relative stiffness of the piling, the depth of piling penetration, the relative compressibility and strength of the soil, the amount of anchor yield, and so on. In general, the greater the depth of penetration, the lower the resulting flexural stresses.

There are a number of different methods in use for the design of tied-back bulkheads, as well as various computer programs to determine the forces and moments in the structure. The analysis for granular soils is treated differently from the analysis for cohesive soils. As discussed above for gravity-stabilized bulkheads, a tied-back bulkhead must also be checked for global stability. In this case, the potential circular slip surface will probably pass immediately below the tip of the sheet piles.

When designing the tie rods, it should be considered that the actual force in the tie rods may be substantially greater than the values calculated. This is due to the fact that the actual pressure distribution on the bulkhead may be somewhat different from that assumed for the calculation, the flexibility of adjacent anchorages may be greater, the tie rods may sag, or there may be a repeated application and removal of heavy surcharges. Generally, an increase of 25 to 30% of the calculated force is used for designing the tie rods. Where the tie rods may be prone to sagging due to settlement of the soil, they may be supported by light piles at 7- to 10-m intervals, or they may be installed inside conduits having an inside diameter somewhat larger than the settlement anticipated.

The horizontal reaction from the sheet piles is transferred to the tie rods by a wale. It normally consists of two steel channels placed with their webs back to back in the horizontal position and spaced with a sufficient distance between their webs to clear the tie rods. Structurally, the preferred location for the wales is on the exterior face of the bulkhead, where the sheet piles will bear against the wales. However, the wales are often placed on the interior face of the bulkhead to provide a clear exterior face. In this case, every sheet pile is bolted to the wale. Figure 2.73 illustrates a typical wale and tie rod detail for a wale on the interior face of the bulkhead.

The stability of a tied-back sheet pile bulkhead depends on the stability of the anchor system to which the tie rods are secured. The reaction of the tie rods may be carried by any of a wide variety of anchorages. For an anchorage system to be effective, it must be located outside the potential active soil failure zone developed behind the bulkhead. Its capacity is also impaired if it is located in unstable ground or if the active failure zone prevents the development of full passive resistance of the system.

Short steel sheet piles driven in the form of a continuous wall may be used as a deadman to anchor tie rods. The tie rods are connected with a wale system similar to that for the bulkhead and resistance is derived from passive pressure developed as the tie rod pulls against the deadman. Tie rod connection to the deadman should ideally be located at the point of the resultant earth pressures acting on the deadman.

Intermittent concrete panels or continuous concrete walls, either cast in place or precast, may also be used as the deadman for the tie

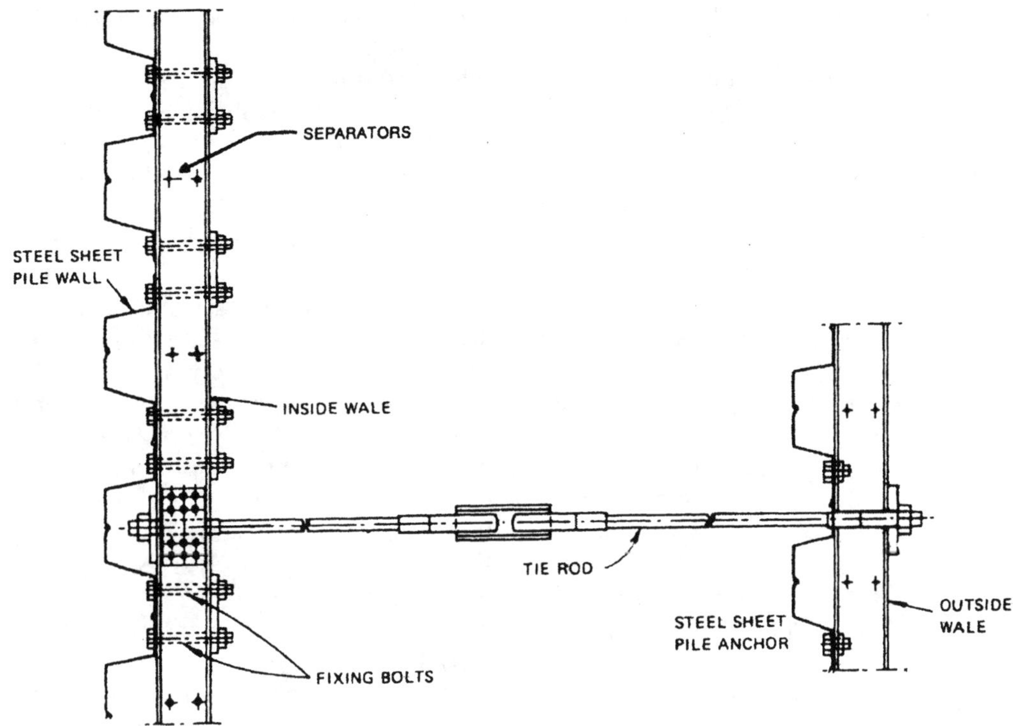

Figure 2.73 Typical interior wale and anchor rod details.

rods. Care must be exercised to see that the deadman does not settle after construction. This is generally not a problem in undisturbed soils, however, where the deadman is located in unconsolidated fill, light piles may be needed for support. Also, the soil within the passive wedge of the deadman should be compacted to at least 90% of maximum density unless the deadman is forced against firm natural soil.

Piles arranged as A-frames can often be used effectively to anchor tie rods. If only two piles form each frame, it is necessary to connect the frames with a continuous reinforced concrete cap. The tie rods can then be attached to the concrete cap. However, if three piles are used, each frame can support a tie rod through the center pile and act independently. The pile angled toward the bulkhead will be in compression, while the pile or piles angled away from the bulkhead will be in tension.

In some situations, a tieback anchor may be used in place of the tie rod and deadman system. A tieback anchor is a structural element that uses an anchor grouted into the soil or rock behind the bulkhead to secure a tendon that applies a restraining force to the bulkhead. A tieback is installed by drilling a hole or driving a casing and grouting a bar or strand tendon in place. Tiebacks differ from tie rod and deadman anchors in that the tieback installation does not require excavation behind the bulkhead. In congested ports, disruption behind the wall can cause unacceptable loss of operating capacity. The savings from avoiding disruption of structures and loss of operating capacity may more than offset the higher cost of the tiebacks.

Tieback design includes the evaluation of tieback feasibility; selection of tieback type; estimation of tieback capacity; selection of a corrosion-protection system; and selection of a testing program. An initial evaluation should be made to determine whether or not it is feasible to install tiebacks at a particular site. Normally, tiebacks can be installed in sandy and gravelly soils with a standard penetration resistance greater than 30 blows per meter, or in rock. The unbonded length of the tieback should be designed to place the anchor zone beyond the potential failure surface of the soil retained by the bulkhead. Overall length should be considered, as the cost of any tieback system varies directly with the length of the tieback. Tiebacks normally are installed at angles up to 45° from horizontal. If a tieback is installed steeper than 45°, the majority of its load will be applied vertically to the bulkhead, making it relatively inefficient.

The normal design load for tiebacks is between 450 and 1200 kN. Tendons of this capacity can be installed economically with a minimum of heavy equipment. Soil type and installation method are the key elements in designing the appropriate tieback capacity. Tieback capacity depends on the length, size, and shape of the grouted anchor, the size and type of tendon, the relative density of the soil, the in situ strength of the soil or rock, the method used to clean the hole, and the method of grouting.

Proof, performance, and creep tests are performed to verify that the tieback will carry the design load for the service life of the structure and that the unbonded length has been established. Performance tests examine the tieback's load-carrying capacity, location of load transfer, elastic behavior of the anchor soil system, and movement with respect to time. The proof test is a simple method of examining load-carrying capacity, total movement under load, and movement with respect to time. Creep tests are performed in cohesive soils to examine long-term behavior of the anchor.

Although corrosion of tiebacks has not been a problem, corrosion protection of both the unbonded length and the anchorhead area is recommended. Where conditions indicate a strong potential for corrosion, the use of more costly encapsulated tiebacks should be considered.

When the height of the bulkhead is such that the bending stress in the sheet piles becomes excessive, usually in the range 10 to 12 m, it may prove to be cost-effective to utilize a combi-wall (i.e., a bulkhead made up of a combination of steel sheet piles and either steel H-piles or steel pipe piles). As discussed earlier, the H-piles may be fabricated from conventional rolled shapes, with sheet pile interlocks welded along the edges of one flange and used with conventional Z-shaped sheet piles. Alternatively, they may be patented H-sections rolled with interlocks on the flanges and used with special compatible sheet pile sections. In situations where even greater strength is required, the intermediate sheet piles may be omitted and the patented H-piles interlocked together.

In some circumstances, particularly if the bulkhead is to be constructed where the existing grade on both sides of the bulkhead is at or near the finished grade elevation on the upland side of the bulkhead, a tied-back reinforced concrete wall constructed by the slurry wall method may be an attractive alternative to a sheet pile bulkhead. Slurry walls are constructed by excavating panel sections of trench to the required line and grade, and filling the trench with bentonite slurry during the excavation process. The slurry within the panel is cleaned with special desanding screens and centrifuges prior to the placement of reinforcement and concrete. The concrete is placed by the tremie method. Individual panels can vary from 600 to 1500 mm in thickness and 2 to 10 m in length. Practical depths for structural walls, where multiple tiers of tiebacks can be utilized, are usually in the range 45 to 50 m. However, if only one tier of tiebacks can be installed, the bulkhead height is usually limited to about 10 m.

Numerous wharf and bulkhead structures have been constructed using various combinations of slurry wall sections. In many examples, tiebacks of various types have been used to provide lateral support or resistance to overturning. Occasionally, a slurry wall deadman has been used. A comprehensive treatment of sheet pile bulkhead performance and design is found in Tsinker (1997).

2.8.3 Pile-Supported Wharves

A wide variety of pile-supported wharf concepts are in use. A few generic concepts are illustrated in Figures 2.24 through 2.27. There are in use a virtually unlimited number of variations on and combinations of those illustrated concepts.

In general, pile-supported wharf structures are particularly attractive where the upper layers of the foundation soil are weak and the required water depth is substantial. The piles are able to transfer the loading imposed on the structure to deeper strong soil strata or to bedrock. Also, in deeper water applications, the underdeck slope serves as a relatively cost-effective means of resisting at least some of the lateral earth pressure from the soil retained by the wharf, compared to gravity-stabilized bulkheads.

At locations exposed to wave action, the piles of a pile-supported wharf will allow virtually free passage of the waves, and the underdeck slope will dissipate some of the wave energy. In this case, pile-supported wharves improve the harbor performance as opposed to closed wharf structures, which reflect wave energy. Similarly, pile-supported wharves are less sensitive to local scour than closed wharf structures.

In regions where substantial seismic activity is a concern, the relatively light and flexible pile-supported wharves are generally more suitable than the relatively massive gravity stabilized structures. The majority of modern pile-supported wharves constructed in regions of high seismic activity feature vertical-pile-only construction, as illustrated in Figures 2.74 and 2.75. The piles and pile cap or deck act as ductile moment-resistant frames to resist both static and dynamic lateral loading. Some modern pile-supported wharves have been constructed with both vertical and batter piles, where the more rigid batter piles resist the lateral loading and where a "seismic fuse" is provided in the connection between the batter piles and the deck structure (Figure 2.76). In the event of a strong seismic event, the fuse will fail, preventing overstressing of the batter piles. The fuse is located and constructed in such a manner that it can be repaired readily.

The potential for soil liquefaction during a seismic event is reduced as compared to a gravity-stabilized type of wharf structure because the dynamic effects in the soil mass during pile driving will tend to densify granular soils. Also, the shear strength of the piles acting as dowels in the underdeck slope and the soil mass below the wharf will increase the global stability. Usually, the effect of the piles will be to drive the critical slip surface below the tip of the piles, into stronger soils, and increase the resistance to global failure.

The lateral load-resisting mechanism of pile-supported wharves may consist of a number of options, including vertical piles, batter piles, various types of tieback systems, or combinations of these options. When only vertical piles are used, without a tieback system, lateral loading is resisted by frame action and bending of the piles. Since the stiffness of a pile is a function of its unsupported length, the piles at the rear of the wharf, where the underdeck slope is at a relatively high elevation, will have a shorter unsupported length than the piles near

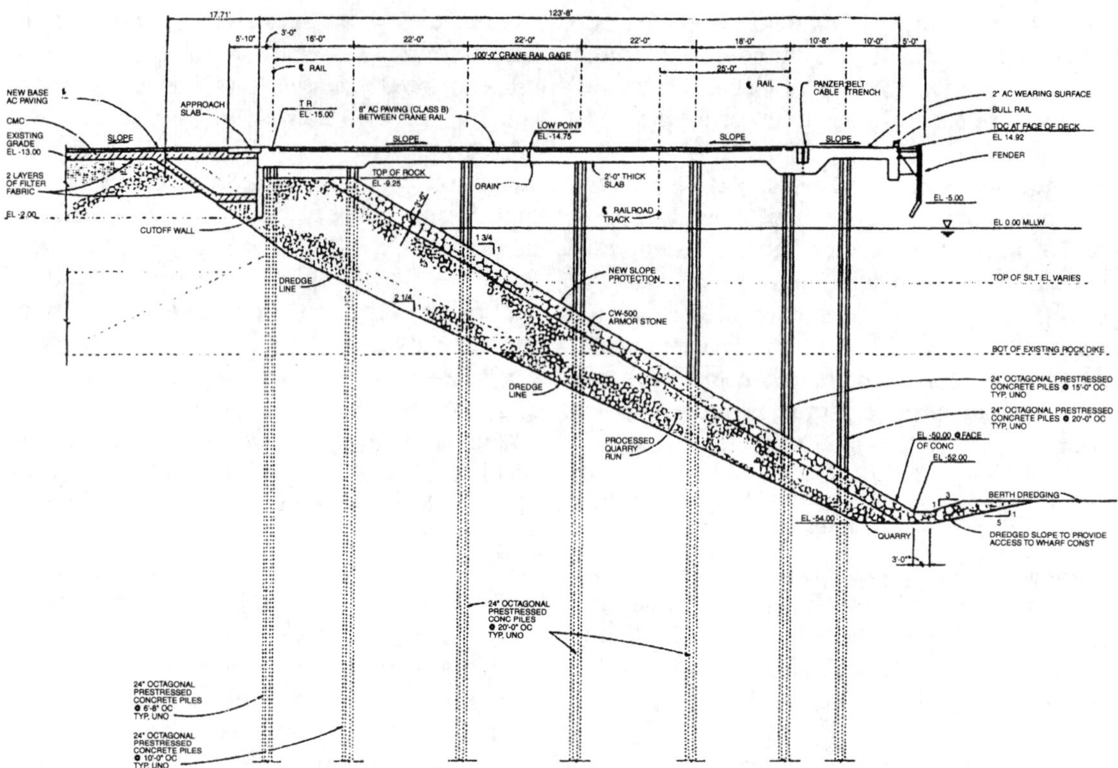

Figure 2.74 Vertical pile wharf at the port of Los Angeles, California. (Courtesy of Port of Los Angeles.)

the face of the wharf and will therefore be stiffer. Usually, for simplified conceptual analyses it may reasonably be assumed that the innermost one to three rows of piles will resist all the lateral loading on the wharf.

When a combination of vertical and batter piles is used (Figure 2.77), the vertical piles resist the lateral loading by bending and the batter piles resist the lateral loading by axial compression or tension. Since the axial stiffness of the piles is much greater than the bending stiffness, it may reasonably be assumed that the batter piles resist the entire lateral loading. Note that a single batter pile acting alone cannot resist a horizontal force by axial compression or tension. For a batter pile to resist a horizontal force, a mechanism must be provided within the structure to resist the vertical component of the batter pile force that accompanies the horizontal component. This vertical resistance is usually provided by an accompanying pile, either vertical or battered in approximately the same plane, but in the opposite direction. Often, dead weight, such as in the low-level platform type of wharf, is provided to resist some or all of the vertical uplift component of a batter pile loaded in compression.

When a combination of vertical piles and a tieback system is used, the stiffness of the tieback system is generally much greater than the bending stiffness of the vertical piles. Therefore, analysis of the structure may reasonably

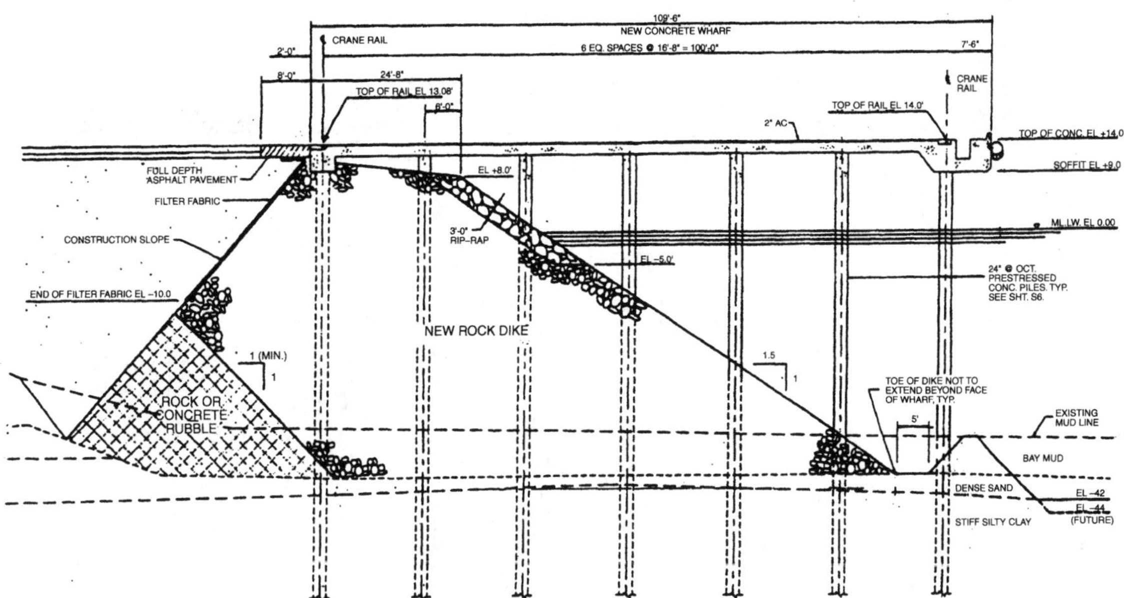

Figure 2.75 Vertical pile wharf at the port of Oakland, California. (Courtesy of Port of Oakland.)

and conservatively be based on the assumption that the tieback system alone resists the full lateral loading.

As for the other types of wharf concepts, care must be exercised to prevent contact between a berthing vessel's bulbous bow and the wharf structure. For pile-supported wharves, the first row of piles is normally set sufficiently far back from the face of the wharf to avoid contact during normal berthings at the maximum berthing angle.

The underdeck slope is an important element of pile-supported wharves. It serves as the transition from the elevation of the upland area to the elevation of the seabed at the wharf face. It must be designed to be stable under both static conditions and the dynamic conditions imposed by seismic events. The slope is designed in accordance with standard foundation engineering principles. It is usually constructed of some combination of dredged slope, fill slope, rock filter layers, and a rock armor layer. Often, the underdeck slope is originally constructed as a retention dike for dredged fill used to reclaim the upland area. The rock filter layers must be of an appropriate gradation to prevent loss of the soil beneath it, and the rock armor layer must be of sufficient size to remain stable when exposed to waves, the wakes of passing vessels, and the wash from propellers and thrusters. Often, a layer of geotextile is used as a filter between the soil and the rock layers, particularly above the low-water line.

The sequence for construction of the underdeck slope and installation of the piles can have a significant impact on the cost of the wharf. Usually, the most cost-effective scenario is to construct the underdeck slope first and then drive the piles through it. However, depending on the type and size of the piles, and the size

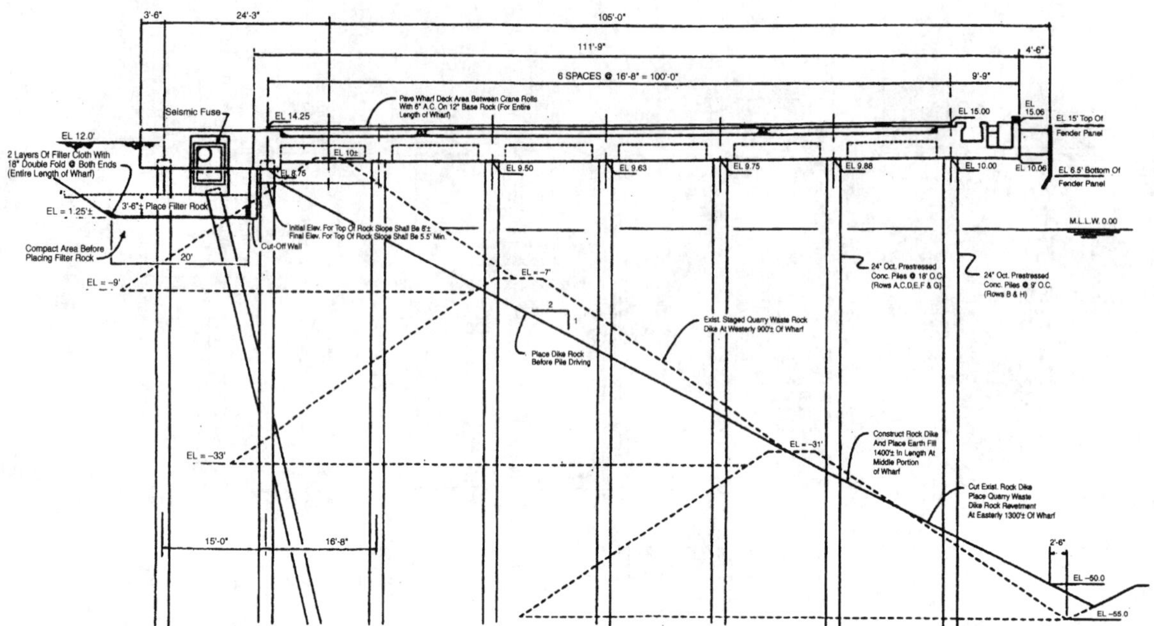

Figure 2.76 Vertical and batter pile wharf, with seismic fuse at the port of Long Beach, California. (Courtesy of Port of Long Beach.)

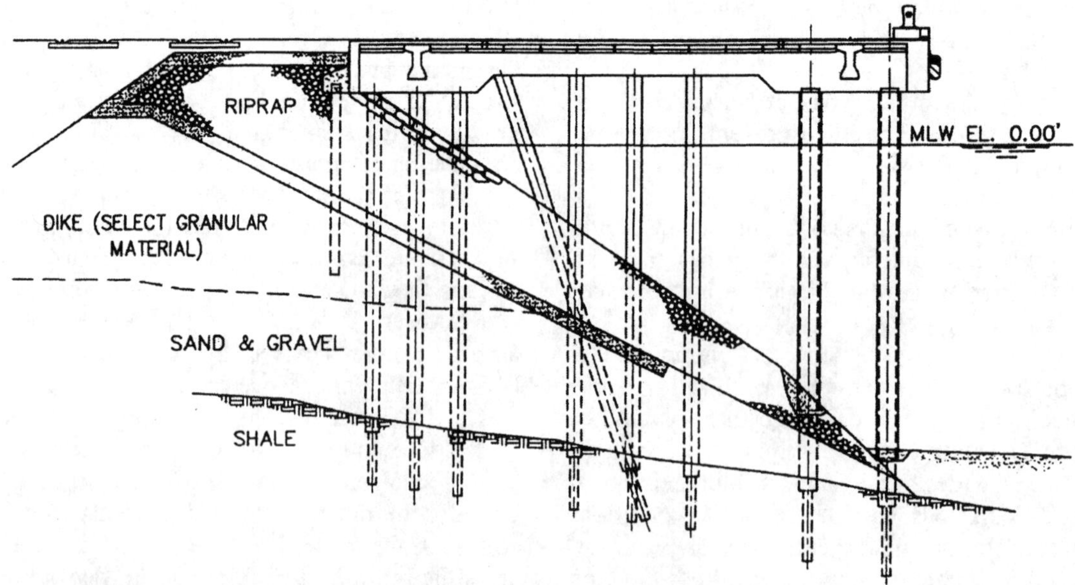

Figure 2.77 Vertical and batter pile wharf at Staten Island, New York. (Courtesy of Port Authority of New York and New Jersey.)

of the rock used for the underdeck slope, particularly the armor layer, it may not be practical to drive the piles after the armor layer is in place. In this case, the piles would be driven after the filter layer is installed, and then the armor rock would be placed around the piles prior to constructing the deck. Considerable care is required in placing the armor rock to avoid damaging the piles and to ensure that the piles are not displaced in the downslope direction by the additional lateral load imposed by the armor rock.

Once the basic configuration of the pile-supported wharf is established, it is necessary to evaluate alternative structural concepts for the deck system and pile system to determine the most cost-effective solution. Construction quality is extremely important for marine structures, and every attempt should be made to utilize precast concrete structural members for the deck to provide high-quality concrete for those members closest to the water and to eliminate costly over-water formwork. Where the deck is to be constructed of cast-in-place concrete, a flat slab configuration, without girders and beams, should be utilized. The flat slab presents the minimum surface area exposed to the penetration of corrosive elements, and eliminates corners, where the reinforcing bars are most susceptible to corrosion.

An economic analysis should be performed to compare the cost of various potential concrete deck framing systems. Preliminary designs and cost estimates should be prepared for each system for a range of pile bent spacings. Several potential concrete deck systems are illustrated schematically in Figure 2.78. The design of the concrete deck should comply with AASHTO (1996) code requirements or other code that may be applicable at the location of the structure. The cost of each system should then be plotted as a function of span length to determine the optimum deck system for any particular pile bent spacing. A typical plot showing the costs for various deck structural systems versus pile bent spacing is shown in Figure 2.79*a*, and these curves are combined into a single smoothed curve as shown in Figure 2.79*b*.

A number of different pile sections should then be investigated and compared on the basis of life-cycle cost, durability, ease of construction, and driving limitations. Steel pipe piles and prestressed concrete piles of various cross sections (Figure 2.80) are usually the most promising alternatives. Structural capacities of individual piles should be calculated on the basis of slender column behavior. Then, based on the soil characteristics, allowable axial capacity curves and, if applicable, p–y data tables should be generated for the different pile cross sections. For each type of pile section, the required number of piles, the pile penetration, and the pile weight are determined for a range of pile bent spacings. A cost analysis should then be prepared for the entire pile foundation system, for each pile type, and for varying bent spacings. The results for each individual type of pile should be presented as a series of curves of pile system cost versus pile bent spacing, as shown in Figure 2.81*a*. These various curves are then "smoothed" to create the single curve shown in Figure 2.81*b*.

After establishing the most cost-effective deck structural systems for a range of pile bent spacings and the most cost-effective pile systems for the same range of bent spacings, the optimum pile bent spacing is determined as shown in Figure 2.82, on which both the deck system cost and the pile system cost curves are plotted. The deck system cost increases with increasing bent spacing and the pile system cost decreases. By plotting the summation of those two curves, the U-shaped upper curve of Figure 2.82 results, with the low point representing the optimum pile bent spacing. Of course, since this analysis is usually not precise, a short range of pile bent spacings on either side of the low point of the curve may be considered as the optimum bent spacing and

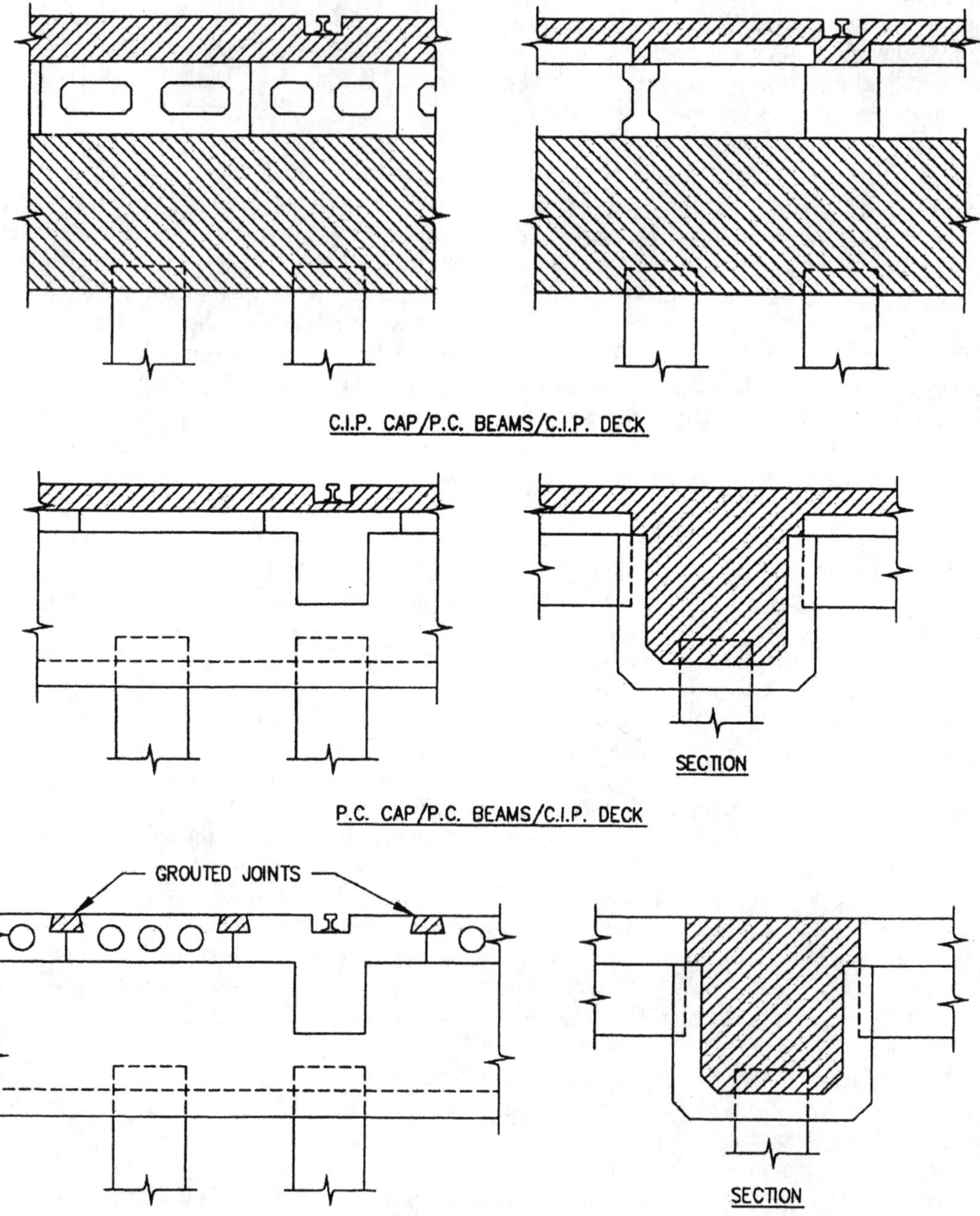

Figure 2.78 Typical wharf or pier deck structural systems.

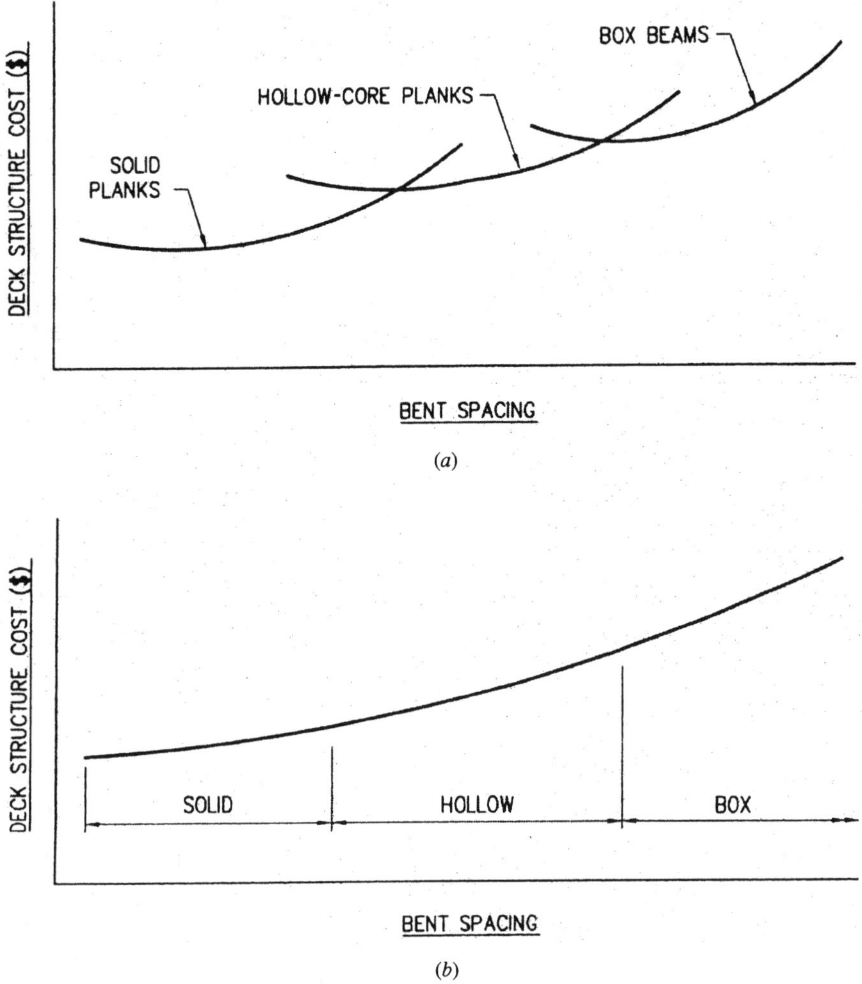

Figure 2.79 Deck cost optimization: (*a*) individual; (*b*) combined.

the final selection within this range may be based on other considerations. Then, going back to the deck system and pile system cost curves of Figures 2.79 and 2.81, respectively, the optimum deck structural system and pile type are identified. For more information on this subject matter the reader is referred to Tsinker (1997).

2.8.4 Pile-Supported Piers

As discussed in Section 2.4.2, pile-supported piers can be categorized as those supported by vertical piles only, those supported by a combination of vertical and batter piles, or those supported by batter piles only. Figures 2.83 and 2.84 illustrate vertical pile piers, Figure 2.85 illustrates a vertical and batter pile pier, and

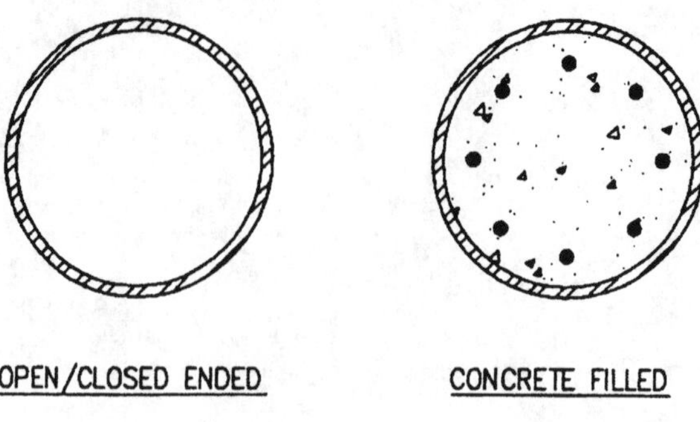

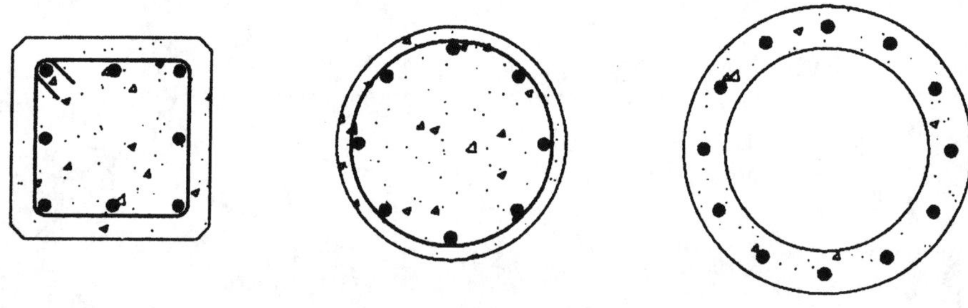

Figure 2.80 Typical pile sections.

Figure 2.86 illustrates a batter pile pier. The principles discussed for pile-supported wharves also apply for pile-supported piers, and the deck and pile optimization procedure for pile-supported wharves is also applicable.

One of the most significant factors affecting pier design and analysis is the impact force imparted by berthing vessels. It is often the primary lateral force experienced by the structure and is an extremely important design consideration. Normally, this force is transferred from the pier fender system to the deck structure, and finally is distributed among the pier's pile bents.

To optimize the design of pile-supported pier structures, it is essential to maximize the distribution of the berthing impact force among the supporting piles. The deck of most pier structures, when considered acting as a girder or truss in a horizontal plane, will distribute a concentrated horizontal force among a great many pile bents, with the actual extent of the distribution being dependent on the relative stiffness of the deck and piles. For piers supported by a combination of vertical and batter piles, it may reasonably be assumed that the batter piles alone will resist the horizontal loading on the pier.

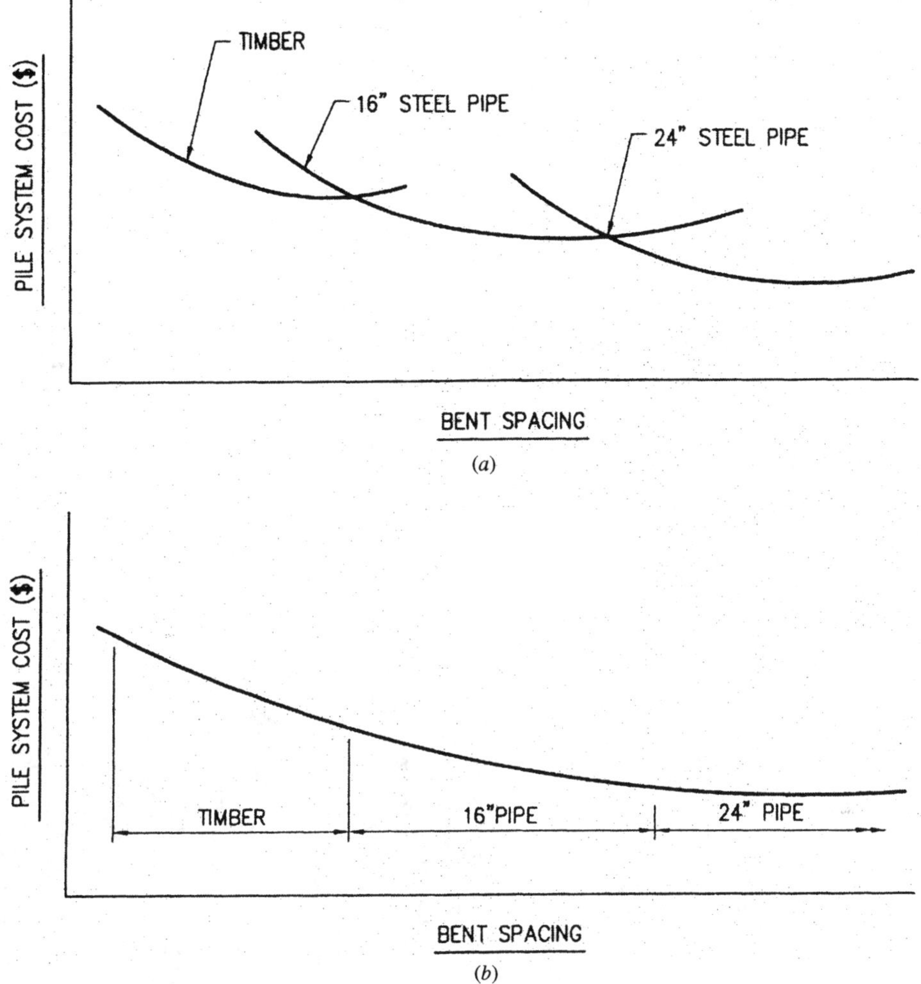

Figure 2.81 Pile cost optimization: (*a*) individual; (*b*) combined.

When designing piers, the engineer often has the option to vary many of the parameters that affect the stiffness of the deck and piles, and a number of alternatives must be evaluated to arrive at the optimum arrangement. The cost of the piles may constitute 50% or more of the total cost of pier structures capable of accommodating deep-draft vessels. By maximizing the distribution of the berthing impact force, the number and/or size of the piles, and consequently the cost of the structure, can be minimized.

Although computer analysis techniques are available to analyze virtually any deck structure and pile system configuration accurately, these analyses can be time consuming. To facilitate the rapid analysis of a great many deck and pile configurations, on a preliminary basis, a nu-

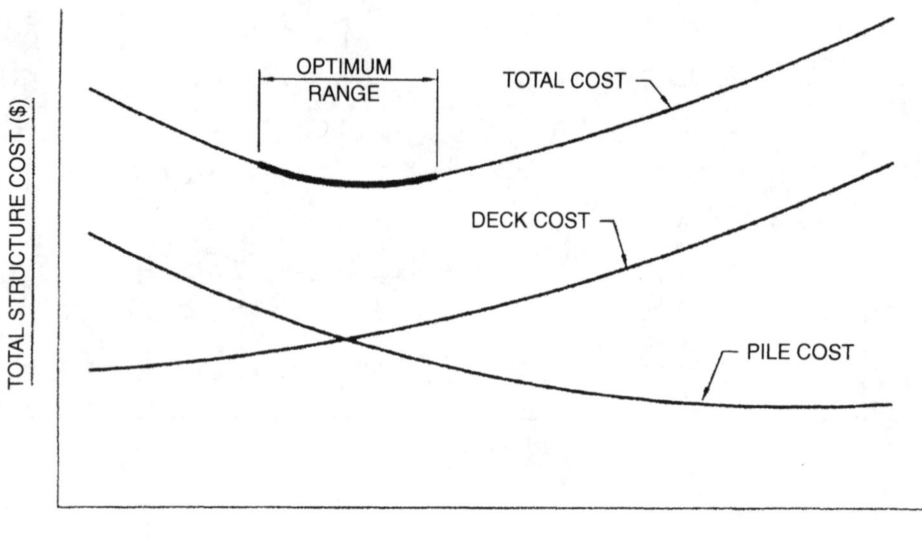

Figure 2.82 Pile bent spacing optimization.

Figure 2.83 Vertical pile pier at the Naval Weapons Station Earle, New Jersey. (Courtesy of Han-Padron Associates. Photograph by Brennan Photo.)

merical procedure (Padron and Elzoghby, 1986) and a series of curves for estimating the distribution of berthing impact forces have been developed (Padron and White, 1983). The curves are applicable for most deep-draft, pile-supported pier structures and can be used as a tool in optimizing alternative configurations at the preliminary design stage.

The actual distribution of the impact force among the pile bents depends on the relative stiffness of the deck and piles. The stiffness of the deck is a function of its width and thick-

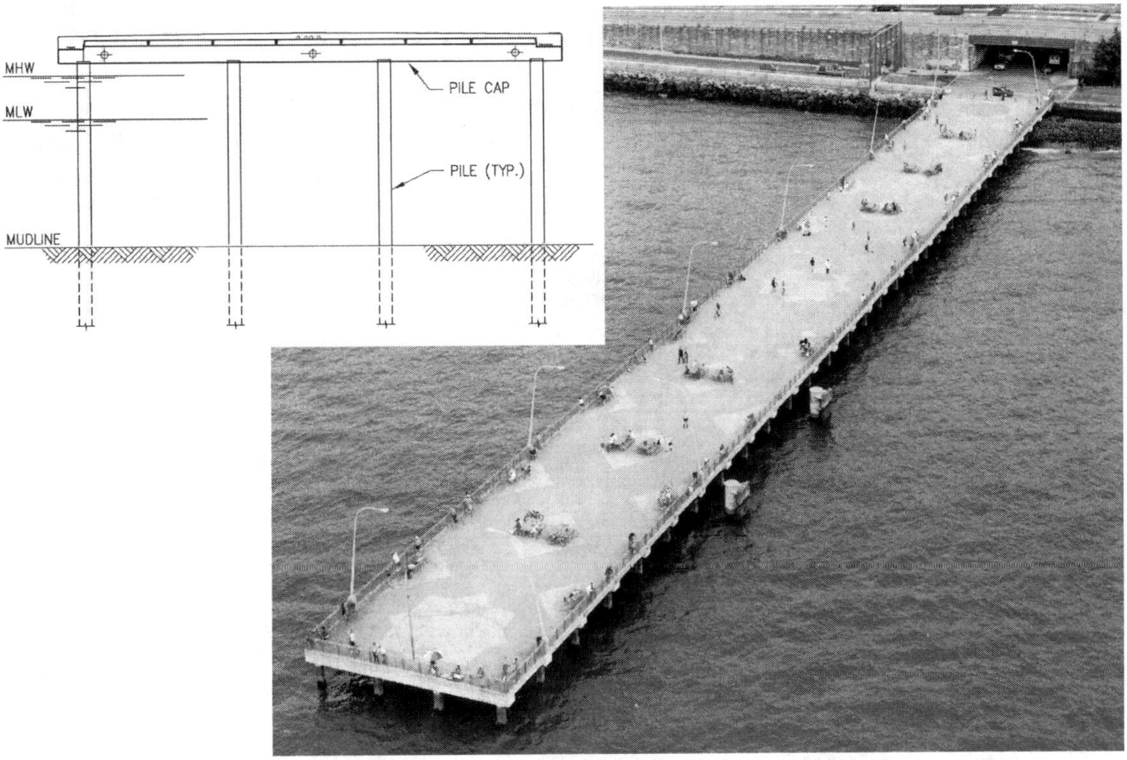

Figure 2.84 Vertical pile recreational pier, Brooklyn, New York. (Courtesy of Han-Padron Associates. Photograph by Highway Aerial Photography.)

Figure 2.85 Batter and vertical pile recreational pier, New York. (Courtesy of Han-Padron Associates.)

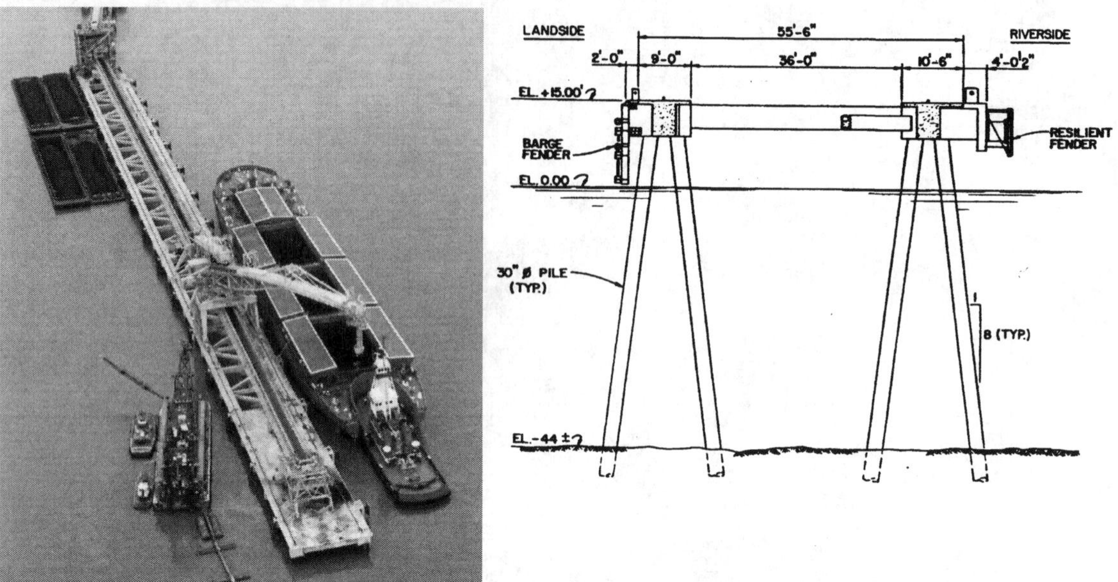

Figure 2.86 Batter pile at coal terminal on lower Mississippi River, Louisiana. (Courtesy of Han-Padron Associates and International Marine Terminals.)

ness, size and arrangement of deck beams, spacing of pile bents, deck material, and other special conditions. The stiffness of the piles in resisting horizontal forces perpendicular to the line of the pier (assuming that batter piles are used to resist these forces) depends on the number of batter piles per bent, spacing of the pile bents, angle of the batter, size and length of the piles, water depth, and the characteristics of the soil in which the piles are embedded.

Although there are many parameters that affect the distribution of the berthing impact force, some have a very significant effect and some a relatively minor effect. The curves shown in Figures 2.87 through 2.90 were developed by varying only the most important parameters, and a conservative combination of commonly used values were utilized for the less important parameters. The variable parameters are the width of the pier deck, the spacing of the pile bents, the number of batter piles per bent, and the angle of the batter piles; all other parameters are fixed. The curves shown in Figures 2.87 and 2.88 are for piers that have two batter piles per pile bent, and Figures 2.89 and 2.90 are for piers that have four batter piles per pile bent. Note that only the number of the batter piles per bent is significant and the actual arrangement of the batter piles is not important for the impact force distribution. The curves are used to determine the percentage of the total impact load taken by the most heavily loaded pile bent, referred to as the *primary bent*. When the impact occurs near the center of the pier, the primary bent corresponds with the location of the impact point, but when the impact occurs in the bent next to the end bent or next to an expansion joint bent, the end bent or the expansion joint bent is the primary bent. Figures 2.87 and 2.89 are for impacts near the center of the pier and Figures 2.88 and 2.90 are for impacts at the second bent. The second bent rather than the end bent or expansion joint bent is considered because when an impact occurs

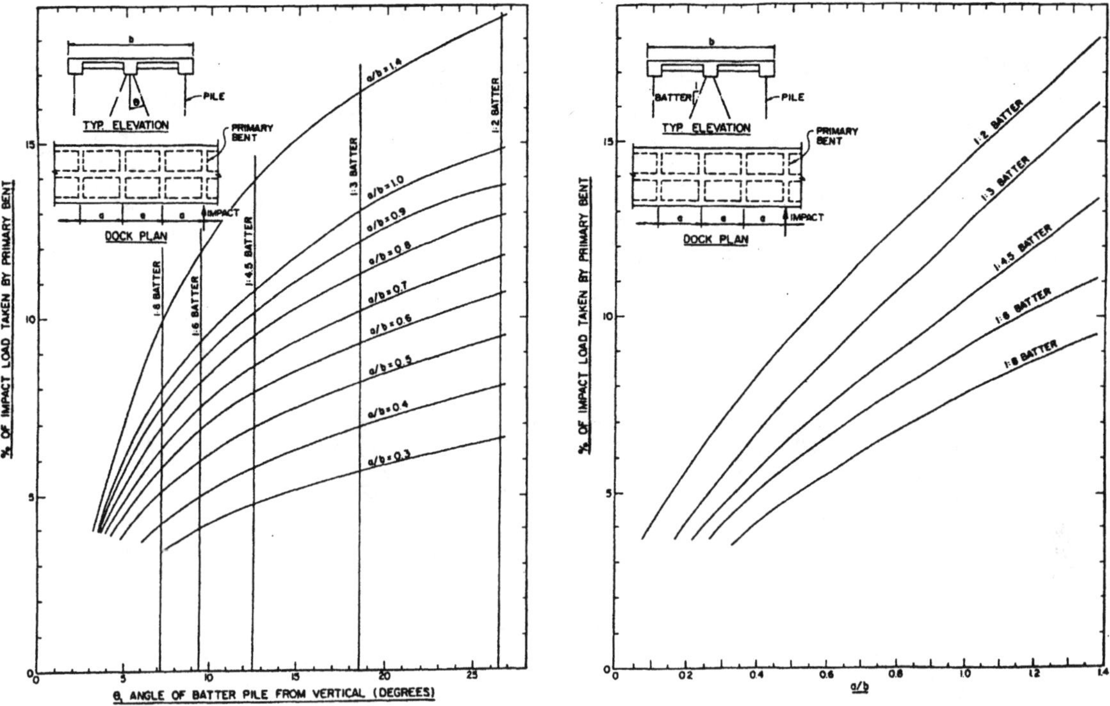

Figure 2.87 Primary bent loading, two batter piles per bent, midpier impact.

at the end bent or expansion joint bent the distribution of the force is very unfavorable, and this situation should be avoided, if possible.

In each figure the curves have been developed with either the bent spacing/deck width ratio or the angle of the batter piles as the abscissa, to simplify the investigation of alternative arrangements. Using either set of curves will provide the same results.

When using the curves, the designer should exercise some judgment and decide if adjustment of the values to suit the conditions for which the particular pier is being designed is warranted. For example, if the pier has an open, trussed deck, instead of a solid deck as assumed for the development of the curves, the deck will be more flexible and the primary bent will carry a greater percentage of the impact force. Conversely, if the soil on which the pier is supported is very soft and very long piles are utilized, the piles will be more flexible than assumed for developing the curves, and the primary bent will carry a smaller percentage of the impact force. Within the range of values normally encountered in the design of piers, these adjustments would typically be less than 15%.

2.8.5 Dolphins

Breasting and mooring dolphins may be designed as rigid structures, in which the horizontal load-resisting components are axially loaded batter piles, or as flexible structures, in which the horizontal load is resisted by one or more large-diameter vertical piles in bending. For these types of structures, the maximum loading condition is almost always the berthing

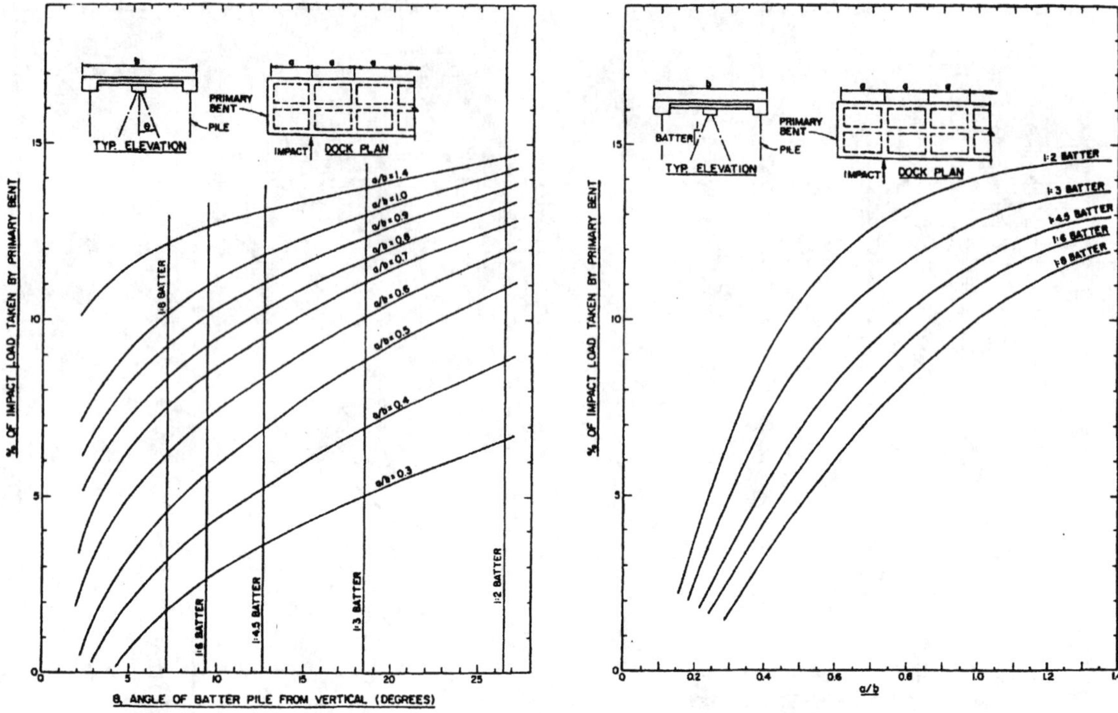

Figure 2.88 Primary bent loading, two batter piles per bent, end impact.

impact force for breasting dolphins and the maximum mooring line forces for mooring dolphins.

There is no universally accepted standard criteria regarding the design loading and corresponding allowable stress levels. It is recommended that the following criteria will provide the basis for a safe and reasonable design. For breasting dolphins, the most critical of:

- Maximum design berthing energy load, plus dead load and environmental loads, at normal allowable stresses
- Maximum design breasting force, plus dead load and environmental loads, at one-third increase in normal allowable stresses

For breasting or mooring dolphins, the most critical of:

- Maximum design mooring load, plus dead load and environmental loads, at one-third increase in normal allowable stresses
- One and one-half times the breaking strength of the largest mooring line, plus dead load and environmental loads, at 95% of yield stresses (This criterion allows a vessel with two mooring lines connected to a dolphin to have one line break and the second line loaded to 50% of its breaking strength without damaging the structure.)

2.8.6 Fender Systems

The first task in the design of a marine fender system is the calculation of the energy to be absorbed by the fender system. This is a critical step, of course. As explained in Padron (1986), there are three basic methods in use. The statistical method is based on actual energy mea-

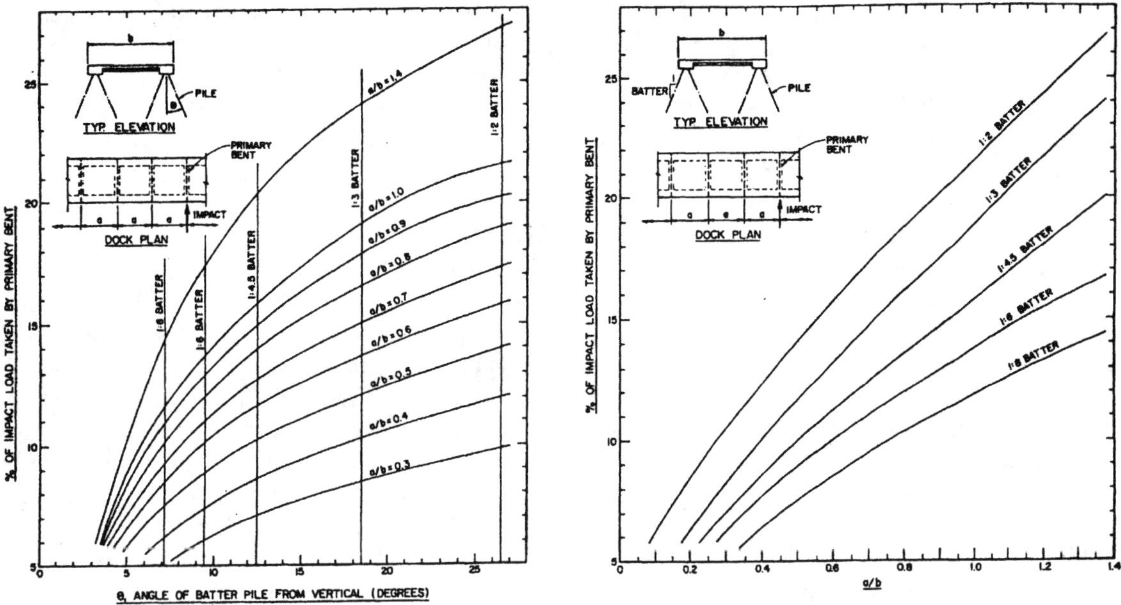

Figure 2.89 Primary bent loading, four batter piles per bent, midpier impact.

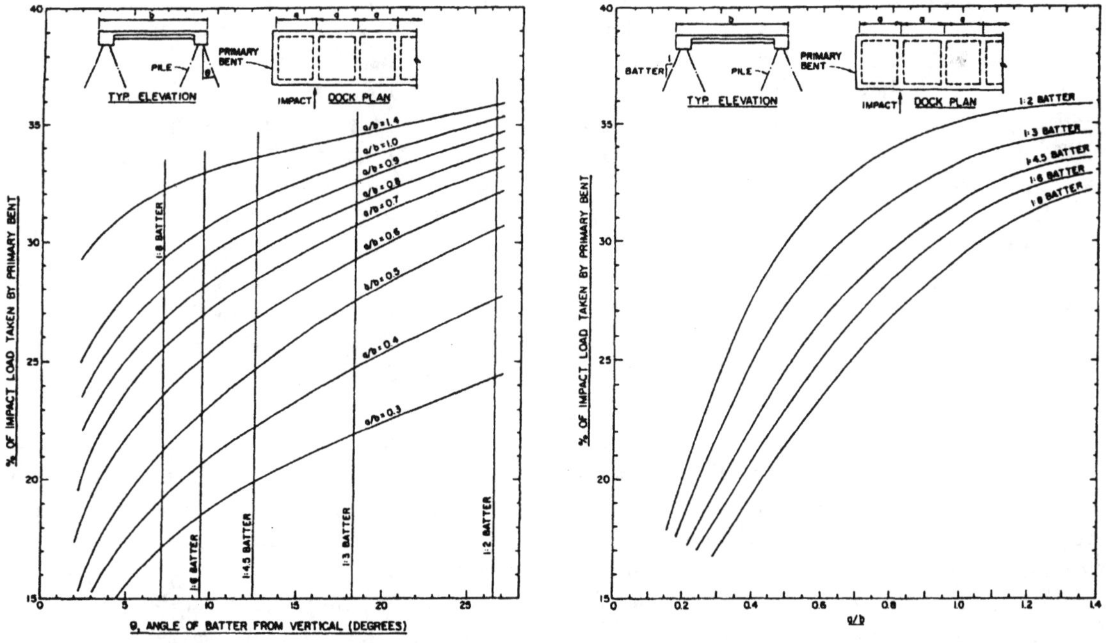

Figure 2.90 Primary bent loading, four batter piles per bent, end impact.

surements at existing berths. When suitable data are available, a statistical approach can be used to determine the design values of the energy to be absorbed by the fender system, taking into account the number and sizes of the vessels expected to use the berth during its design life. This method is a very logical approach to the problem, particularly when evaluating the consequences of a vessel exceeding the design energy level. The drawback of this method is that there are very few data available on which to base a valid statistical analysis, and the expenditure of effort and cost required to obtain such data can only be justified for facilities for very large vessels.

The mathematical modeling approach for calculating berthing energy is still in its infancy. This approach gives a good theoretical insight into the problem, but its practical application at present is very limited.

The oldest and most widely used method is the kinetic approach. This method is based on the equation for kinetic energy, which states that the kinetic energy of a body in motion, in this case the berthing vessel, is equal to one-half of the mass of the body times the square of its velocity. But the total kinetic energy of the ship does not have to be absorbed by the fender system. Usually, the required fender system energy is expressed as a factor, f, times the ship's kinetic energy. This factor is conventionally made up of four components: an eccentricity factor, C_e, an added mass factor, C_m, a softness factor, C_s, and a berth configuration factor, C_c.

$$f = C_e C_m C_s C_c \tag{2.59}$$

The eccentricity factor depends on the location of the point of impact with respect to the ship's center of gravity. A rather simple formula for C_e was presented a number of years ago and states that C_e is equal to the square of the radius of gyration of the ship divided by the sum of the squares of the radius of gyration and the distance between the ship's center of gravity and the point of impact. There is a more accurate analysis available, but the increased accuracy is usually not necessary. Typically, for a continuous fender system, C_e is taken between 0.5 and 0.6 and for a berth with individual breasting dolphins, C_e is taken between 0.7 and 0.8.

When a vessel hits the fender, not only does the mass of the vessel have to be decelerated, but also a certain mass of water surrounding and moving with the vessel. This is accounted for by the added mass factor. There has been a great deal of model testing and full-scale testing carried out to determine the value of C_m. Typically, designers have been using values of C_m between 1.2 and 2.0. Many different formulas for C_m have been proposed and nine different formulas are discussed in the PIANC (1984, 2002) reports. But added mass factor is a function of water depth, underkeel clearance, the distance to obstacles or walls, the underwater shape of the vessel, the berthing velocity, currents, the deceleration behavior of the vessel, and the cleanliness of the vessel's hull. Therefore, it is not surprising that there is not a single value or formula for C_m. It is recommended that the value of C_m be taken as 1.5 where there is a large underkeel clearance (i.e., the water depth is 1.5 times the draft of the vessel, or more). Where the water depth is only 1.1 times the draft of the vessel, C_m should be taken as 1.8. For intermediate water depths, interpolate between 1.5 and 1.8.

The softness factor accounts for the relation between the rigidity of the vessel and the fender. It is usually taken as 1.0 for "soft" fenders and 0.9 for "hard" fenders. But for large modern fenders that have a significant amount of deflection, a simple calculation shows that this factor is actually very close to 1.0. Therefore, it is recommended that C_s always be taken as 1.0.

The berth configuration coefficient is introduced to take into account the cushioning effect of the water that is squeezed between the berthing vessel and a solid quay wall. For a berth with a solid wall and a relatively parallel approach of the vessel, C_c is about 0.8. For an open structure or where the approach angle of the vessel is more than about 5°, C_c should be taken as 1.0.

Since the kinetic energy is a function of the square of the berthing velocity, the selection of a design value for the berthing velocity is critical. However, there are no firm rules governing the selection of the design berthing velocity because there are too many factors involved. The proper values have to be selected by experienced engineers familiar with the conditions at the site. As a general guide, the following values are recommended:

- Very favorable conditions 0.10 m/s
- General case 0.15 m/s
- Very unfavorable conditions 0.30 m/s

As a general rule, larger vessels tend to berth at lower velocities than smaller vessels. For piers intended for a range of vessel sizes, it would not be unusual to find the design berthing energy of the smaller vessel equal to that of the larger vessel.

Now that a method for calculating the required energy absorption capacity is defined, the designer must select a fender system that will provide the required capacity. However, the selection of the optimum fender system for a particular situation involves the evaluation of many factors. The first factor to be considered, of course, is the energy absorption requirement since the function of the fender system is to absorb energy. But there are usually many alternative fender systems that can satisfy the energy absorption requirement, and the designer must consider all the other factors in order to select the best system. These factors include:

- *Reaction force.* This is the force that is exerted on the vessel's hull and on the berthing structure during impact. The reaction force can have a significant effect on the cost of the berthing structure.
- *Standoff.* This is the distance between the face of the fender system and the face of the wharf or pier. Generally, fender systems with greater standoff require cargo transfer equipment with greater reach.
- *Deflection.* This is the distance that the face of the fender system moves in absorbing the vessel's kinetic energy. Generally, fender systems with greater deflection require greater standoff.
- *Hull pressure.* This is the pressure exerted on the vessel's hull by the fender unit and is derived by dividing the reaction force by the fender area in contact with the vessel. Hull pressure must be limited to levels that will not cause permanent damage to the berthing vessel.
- *Reaction–deflection relationships.* The nature of the reaction–deflection relationship determines the relative stiffness of the fender system.
- *Angle of impact.* The larger the angle between the vessel's hull and the berthing line, the less efficient some fenders can become.
- *Deceleration factor.* This is the rate at which the fender system causes the berthing vessel to decelerate. This is particularly relevant for fender systems for ferries.
- *Long-term contact.* This includes the changes in environmental conditions (i.e., wind, current, waves, and tide) while the vessel is moored. The fender system should not "roll up," tear, abrade, or be susceptible to other forms of damage when subject to long-term contact.
- *Coefficient of friction between fender system and vessel's hull.* This determines the resulting shear force when the vessel is

berthing with longitudinal and/or rolling motion and may have a significant detrimental effect on the energy absorption performance of the fender system. The magnitude of the shear force also may have a significant effect on the cost of the berthing structure.

- *Safety factor for vessel, berthing structure, and fender.* The more serious the consequences that would result should damage of the fender or supporting structure occur, the higher the safety factor should be. The mode of failure of a fender and its effect on the supporting structure should also be considered.
- *Costs.* Capital costs for both the fender system and the structure, as well as operation, maintenance, and repair costs, must be considered.
- *Capability of tug crews in assisting the berthing vessel and dock labor in maintaining the fender system.* Where maintenance is expected to be poor, a simple, possibly less efficient fender system may be preferable to a system requiring a higher degree of maintenance.
- *Repetition factor.* Fender types already used locally should be considered since their performance under actual conditions is known. Also, there may be an advantage in having interchangeability of spares, particularly if the number of new fenders required is small.
- *Frequency of berthing operations.* A high frequency of berthings normally justifies greater capital expenditures for the fender system.
- *Range of vessel sizes expected to use the berth.* While the energy absorption capacity of the fender system may be selected for the largest vessel expected to use the berth, the fender system must be suitable for the full range of vessels that the berth will accommodate.
- *Shape of vessel's hull in contact with the fender system.* Where vessels with unusual hull configurations or protrusions may be expected, special attention must be paid to the selection and arrangement of the fender system.
- *Range of water levels to be accommodated.* The fender system must be suitable during the full range of water levels that may occur at the berth. The design must consider both the largest and smallest vessels, in both the loaded and light conditions, at high and low water levels.
- *Degree of exposure.* Where the berth is exposed to severe wind and/or wave action, the fender selection may be governed by mooring conditions rather than berthing conditions.

When a continuous fender system is to be used, the contact length of a vessel during berthing is an important parameter. The contact length depends on the spacing of the fender piles and rubber fender units, and the connection details of the chocks and wales to the fender piles. The contact length can be approximated by the chord formed by the curvature of the bow and the berthing angle as follows:

$$L_c = 2r \sin \alpha \qquad (2.60)$$

where L_c is the contact length, r the bow radius, and α the berthing angle.

In lieu of detailed analysis, Table 2.21 may be used to estimate the contact length.

2.8.7 Mooring Fittings

Mooring fittings include bollards, cleats, and quick-release hook assemblies, as described in Section 2.5.6. A certificate issued by the manufacturer normally defines the working capacity of the mooring fitting. Base bolts are normally supplied by the mooring fitting man-

Table 2.21 Contact length estimation

Vessel Size (dwt)	Contact Length (m)
1,000–2,500	10
5,000–26,000	12
35,000–50,000	15
60,000–75,000	18
100,000–125,000	21

ufacturer and will retain the fitting from being pulled out if the supporting member is capable of resisting the bolt forces. The base bolts are subjected to both shear and uplift. Mooring fittings are often grouted into a recess in the structure to eliminate the shear acting on the bolts. The bolt force should be calculated based on the following factors:

- Mooring line force from mooring analysis
- Location of load application with respect to the fitting base
- The vertical angles based on the actual mooring angles for the highest and lowest tide levels for the various sizes of vessels that will be moored
- The horizontal angels based on the mooring line layout for the various sizes and positions of vessels that will be moored

The design of the installation of quick-release hooks should be based on withstanding the minimum breaking load of the largest mooring line, with a minimum safety factor of 1.2 on yield.

REFERENCES

AASHTO, 1996. *Standard Specifications for Highway Bridges,* American Association of State Highway Transportation Officials, Washington, DC.

ACI, 1995. *Building Code Requirements for Structural Concrete* (ACI 318-95) *and Commentary* (ACI 318R-95), American Concrete Institute, Detroit, MI.

AISC, 2000. *Manual of Steel Construction: Allowable Stress Design,* American Institute of Steel Construction, Chicago.

API, 2000. *Recommended Practice for Planning, Designing and Constructing Fixed Offshore Platforms: Working Stress Design,* 21st ed., API RP 2A, American Petroleum Institute, Washington, DC.

ASCE, 1996. *Design of Sheet Pile Walls,* American Society of Civil Engineers, Reston, VA.

———, 1998a. *Minimum Design Loads for Buildings and Other Structures,* Rep. 7-98, American Society of Civil Engineers, Reston, VA.

———, 1998b. *Seismic Guidelines for Ports,* Monogr. 12, American Society of Civil Engineers, Reston, VA.

Bowles, J. E., 1988. *Foundation Analysis and Design,* McGraw-Hill, New York.

Briand, J.-L., and Y. Lim, 1999. Tieback walls in sand: Numerical simulation and design implications, *ASCE J. Geotech. Geoenviron. Eng.,* Vol. 125, No. 2, pp. 101–110.

British Ship Research Association, 1973. *Research Investigation for the Improvement of Ship Mooring Methods,* BSRI, Northumberland, England.

British Standards Institution, 1984. *British Standard Code of Practice for Maritime Structures,* BS6349, BSI, London.

Cork, S., and T. Holm–Karsen, 2002. Vessels growth and the impact on terminal planning and development., *PIANC Proc. 30th Congress,* Sydney, Australia.

California State Lands Commission, 2002. *Marine Oil Terminal Engineering Standards,* CSLC, Sacramento, CA.

Das, B. M., 1993. *Principles of Soil Dynamics,* PWS-Kent, Boston.

Davies, T. G., R. Richards, and K. H. Chen, 1986. Passive pressure during seismic loading, *ASCE J. Geotech. Eng.,* Vol. 112.

Matsuo, H., and S. O'Hara, 1960. Lateral earth pressures and stability of quay walls during earthquakes, *Proc., 2nd World Conference on Earthquake Engineering,* Tokyo, Japan.

OCIMF, 1977. *Prediction of Wind and Current Loads on VLCC's,* Oil Companies International Marine Forum, London.

———, 1997. *Mooring Equipment Guidelines,* Oil Companies International Marine Forum, London.

Padron, D. V., 1986. Marine fender systems, *Dredg. Port Const.,* June.

———, 1998. Most advanced marine oil terminals, *Proc. Prevention First '98,* California State Lands Commission, Long Beach, CA.

Padron, D. V., and H. M. Elzoghby, 1986. Berthing impact force distribution on pile support structures, *ASCE Proc. Specialty Conference Ports '86,* Oakland, CA.

Padron, D. V., and S. M. White, 1983. Optimizing pier design by utilizing deck stiffness, *ASCE Proc. Specialty Conference Ports '83,* New Orleans, LA.

PIANC, 1984, *Report of the International Commission for Improving the Design of Fender System,* Suppl. Bull. 45.

———, 2001a. *Seismic Design Guidelines for Port Structures,* Suppl. Bull. 106, International Navigation Association, Brussels, Belgium.

———, 2001b. *Seismic Design Guideline for Port Structures,* Balkerna, Rotterdam, The Netherlands.

———, 2002. *Guidelines for the Design of Fender Systems, 2002.* International Navigation Association, Brussels, Belgium.

Schmertmann, J. H., 1955. The undisturbed consolidation behavior of clay, *ASCE, Trans.,* Vol. 120.

Seed, H. B., and I. M. Idriss, 1971. Simplified procedure for evaluating soil liquefaction potential, *ASCE J. Soil Mech. Found. Div.,* Vol. 97.

Seed, H. B., and R. V. Whitman, 1970. Design of earth retaining structures for dynamic loads, *ASCE Proc. Specialty Conference on Lateral Stresses in the Ground and Design of Earth Retaining Structures.*

Tsinker, G., 1995. *Marine Structures Engineering. Specialized Applications,* Chapman and Hall, New York.

Tsinker, G., 1997. *Handbook of Port and Harbor Engineering. Geotechnical and Structural Aspects.* Chapman and Hall, New York.

UNCTAD, 1987. *Operating and Maintenance Features of Container Handling Systems,* United Nations, New York.

U.S. Army Corps of Engineers, 1992. *The Seismic Design of Waterfront Retaining Structures,* USACE, Washington, DC.

U.S. Department of the Navy, 1984. *Harbors Design Manual,* DM 26.1, Alexandria, VA.

———, 1998. *Mooring Design Manual,* DM 26.4, Alexandria, VA.

Wang, S., 1975. Dynamic effects of ship passage on moored vessels, *ASCE J. Waterways Harbors Coast. Eng. Div.,* Vol. WW3.

Westergaard, H. M., 1933. Water pressures on dams during earthquakes, *ASCE Trans.,* Vol. 98.

Winterkorn, H. F., and H.-Y. Fang, 1991. *Foundation Engineering Handbook,* Van Nostrand Reinhold, New York.

3

SEISMIC DESIGN OF PORT STRUCTURES

Constantine D. Memos
National Technical University of Athens
Zografos, Greece

Alberto Bernal
B y A Estudio de Ingenieria
Madrid, Spain

3.1 INTRODUCTION

Many ports around the world lie on seismically active areas. In such cases port structures should be able to withstand earthquake loading and be designed accordingly. Guidelines for the seismic design of port structures are included in national codes and other relevant documents: for example, in Japan (Ministry of Transport, 1999), in Germany (EAU, 1996), in Spain (Puertos del Estados, 2000), in the United States (Ferrito, 1997; Werner, 1998); in the European Union (CEN, 1994), and in New Zealand (New Zealand Standards, 1992–1997). The International Navigation Association [(INA); formerly Permanent International Association of Navigation Congresses (PIANC)] focused international attention on the devastating effects of earthquakes on port structures by disseminating the work of its relevant working group published in concise form as a PIANC report (2001b) and in its full extent as a book (PIANC, 2001 a,b). The seismic design philosophy for port structures, suggested in these two publications, consists of a performance-based approach that is followed in this chapter. The approach includes the evaluation of deformations and permanent displacements on soil and structure under the design earthquake. This is considered an advance over current practice, for which methods based on a force-balance approach are used. In the latter methods the structures are designed to withstand a predetermined level of seismic loading specified primarily by seismic acceleration. However, for shakings outside the prescribed design-level conventional limit equilibrium-based methods do no provide information on the degree of damage, the loss of serviceability, and so on. In contrast, the recently introduced performance-based concept treated in this chapter, allows that port structures would be designed to maintain serviceability in frequently occurring earthquakes but to suffer a certain degree of damage in rare shakings. The accepted degree of damage in each level of earthquake depends on the importance and specific functions of a port structure. Estimation of the damage can be performed by a variety of analysis methods, ranging from simple to sophisticated, as will be seen in Section 3.5.1. The proposed guidelines are general enough to be used even when the required functions of port structures, the economic and social environment, and seismic activity may differ from region to region.

3.2 EARTHQUAKE MOTION

Earthquakes are complex natural phenomena, having their origin in the release of tectonic stress accumulated in Earth's crust. The seismic waves generated under a site propagate through crustal rocks to the surface of the bedrock. Then the ground motion propagates upward through the local soil deposits, reaching the surface and the existing structures (Figure 3.1). In this section we describe briefly the design earthquake parameters and make recommendations pertaining to the basic data to be collected and the analytical procedures to be followed for their evaluation.

3.2.1 Size of Earthquakes

The size of an earthquake is one of the fundamental factors that control the resulting ground movement. Some parameters are used to characterize the size of earthquakes, the most usual being intensity and magnitude. *Intensity* is a measure of the destructiveness of the earthquake as evidenced by human reaction

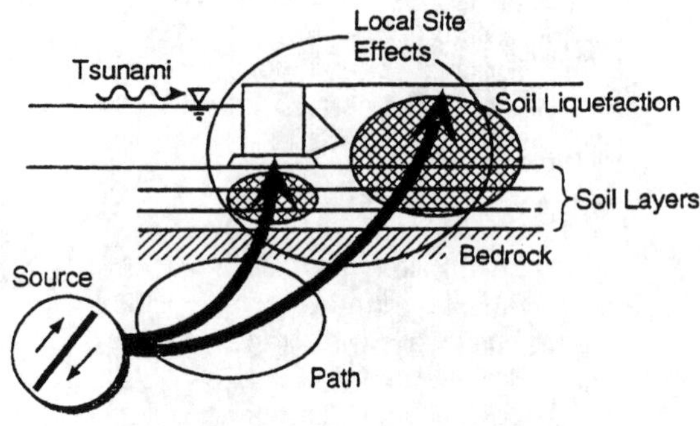

Figure 3.1 Propagation of seismic waves.

and damage observed. It varies from one location to another, depending on the size of the earthquake, the focal distance, and the local site conditions. Seismic damage and the corresponding intensity depend on characteristics of seismic motion (acceleration, duration, and frequency content) as well as the natural frequencies and vulnerability of the structures affected. Intensity is the best single parameter to use in defining the destructiveness of an earthquake at a given site, but it cannot be used as input for dynamic analysis. In many cases, especially for historic earthquakes, it is the only parameter available for characterizing earthquake motion.

Several different seismic intensity scales have been adopted in different parts of the world. The Modified Mercalli (MM) Scale (Wood and Neumann, 1931; Richter, 1958) is used widely in North and South America. The Medvedev–Sponheuer–Karnik (MSK) Scale (Medvedev and Sponheuer, 1969), recently revised as the European Macroseismic Scale (EMS) (Grünthal, 1998), is commonly used in Europe. These scales of 12 degrees (I to XII) are roughly equivalent. Japan maintains the Japan Meteorological Agency (JMA) Scale, which consists of eight degrees (0 to VII). Relationships between the various scales can be found in a publication by the Technical Committee for Earthquake Geotechnical Engineering (1999). Maximum intensity in the epicentral region, I_o or I_e, is a measure of the size of an earthquake.

Magnitude is a physical measure of the size of an earthquake, typically evaluated from the recorded data. There are several scales, based on the amplitude of seismograph records, each of which captures mainly the ground motion amplitude at a particular frequency. The Richter local magnitude (M_L) is based on waves of 1 to 5 Hz; the surface wave magnitude (M_S) is based on surface waves at a frequency of 0.05 Hz: the short-period body wave magnitude (m_b) is based on body waves at a frequency of 1 Hz. Moment magnitude (M_W) is calculated from the seismic moment, which is a direct measure of the factors that produce rupture along a fault. M_W is presently preferred by seismologists. The use of M_L for magnitudes up to 6 and M_S for magnitudes 6 to 8 does not differ significantly from M_W values. Earthquakes with a magnitude below 3 are considered microtremors; those measuring up to 5 are considered minor earthquakes with little associated damage. Maximum recorded magnitude is about $M_W = 9.5$ (e.g., Chilean earthquake of 1960).

3.2.2 Strong Ground Motion Parameters

The seismic input to be used for analysis and design depends on the type of port structure and on the type of analysis being performed. Principal parameters describing the seismic input are:

- *Peak ground horizontal acceleration* (PGA_H, or simply PGA). This is the maximum absolute value reached by ground horizontal acceleration during an earthquake. Similarly, there is a peak ground vertical acceleration, a peak ground horizontal or vertical velocity, and corresponding peak ground displacements.
- *Accelerogram $a(t)$*. This is the time history of acceleration. It can refer to its horizontal or vertical component. It is usually an irregular function of time. Figure 3.2 shows some examples of accelerograms from different types of earthquakes.
- *Acceleration response spectrum* [SA(T,D)]. This represents the maximum acceleration (absolute value) of a linear single-degree-of-freedom (SDOF) oscillator, with period T and damping D% of critical, when the earthquake motion is applied to its base. The SDOF oscillator is the simplest model of a structure. This spectrum represents a good approximation of the response of dif-

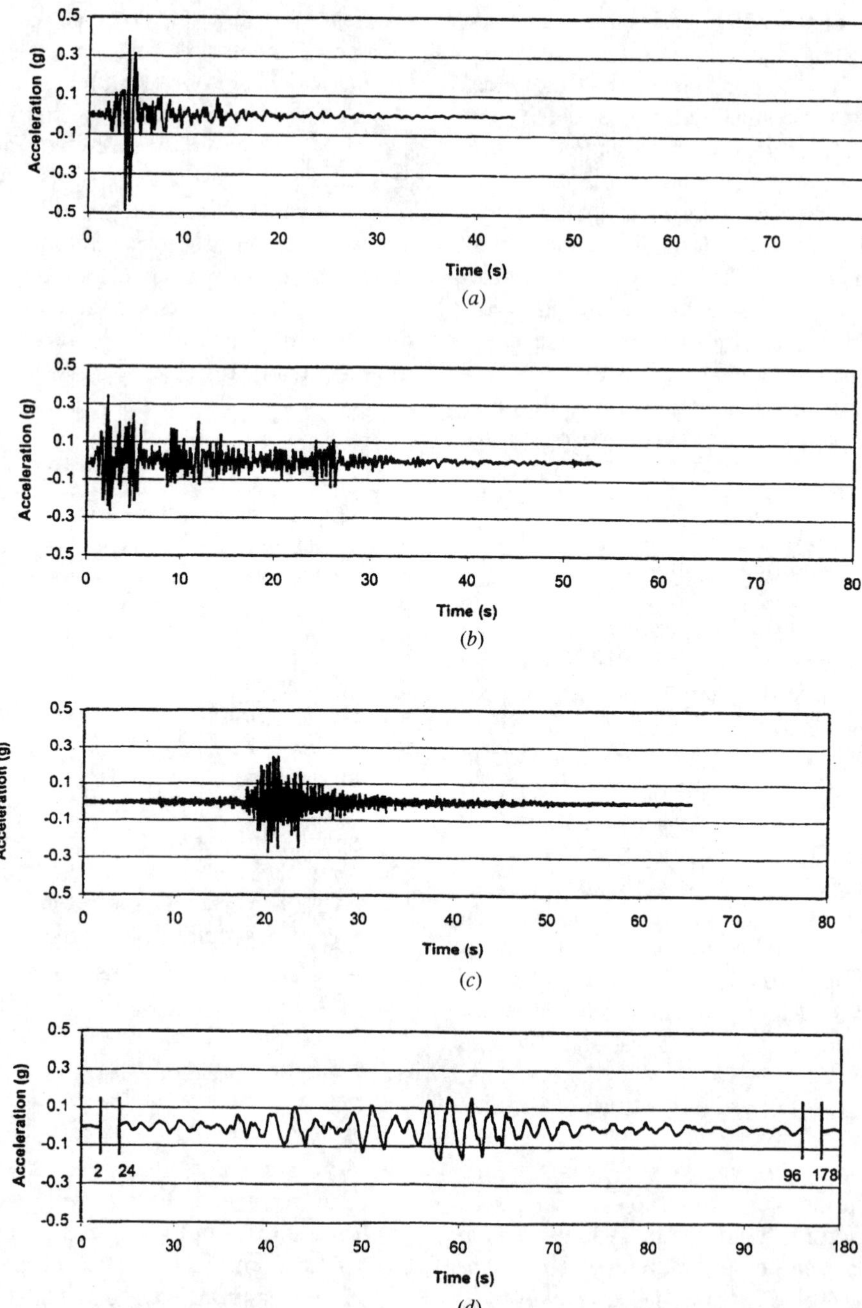

Figure 3.2 Examples of accelerograms from different types of earthquakes and site conditions: (*a*) Parkfield, California, June 28, 1966: record of almost a single shock; (*b*) El Centro, California, May 19, 1940: record of moderate duration with a wideband frequency content; (*c*) Lima, Peru, October 17, 1966: record with predominant high-frequency components; (*d*) Mexico City, Mexico, September 19, 1985: long-duration record with predominant low-frequency components.

ferent structures when they are subjected to an earthquake. Similarly, there is a velocity response spectrum, SV(T,D), and a displacement response spectrum, SD(T,D), where the velocity and the displacement are not absolute but relative to the base. The response spectra usually refer to horizontal movement but can also be defined for vertical movement. Figure 3.3 shows examples of response spectra of different types of earthquakes.

Ground motion reaching a site or a structure depends on several factors, which can be classified as seismic source, travel path, and local site conditions (Figure 3.1).

3.2.3 Seismic Source and Travel Path Effects

The tectonic mechanism in the seismic source zone influences the seismic waves generated. Depending on the position of the hypocenter and its relationship to the global tectonics processes, earthquakes can be classified as interplate or intraplate. *Interplate earthquakes* originate in contact zones between tectonic plates (the principal fragments into which

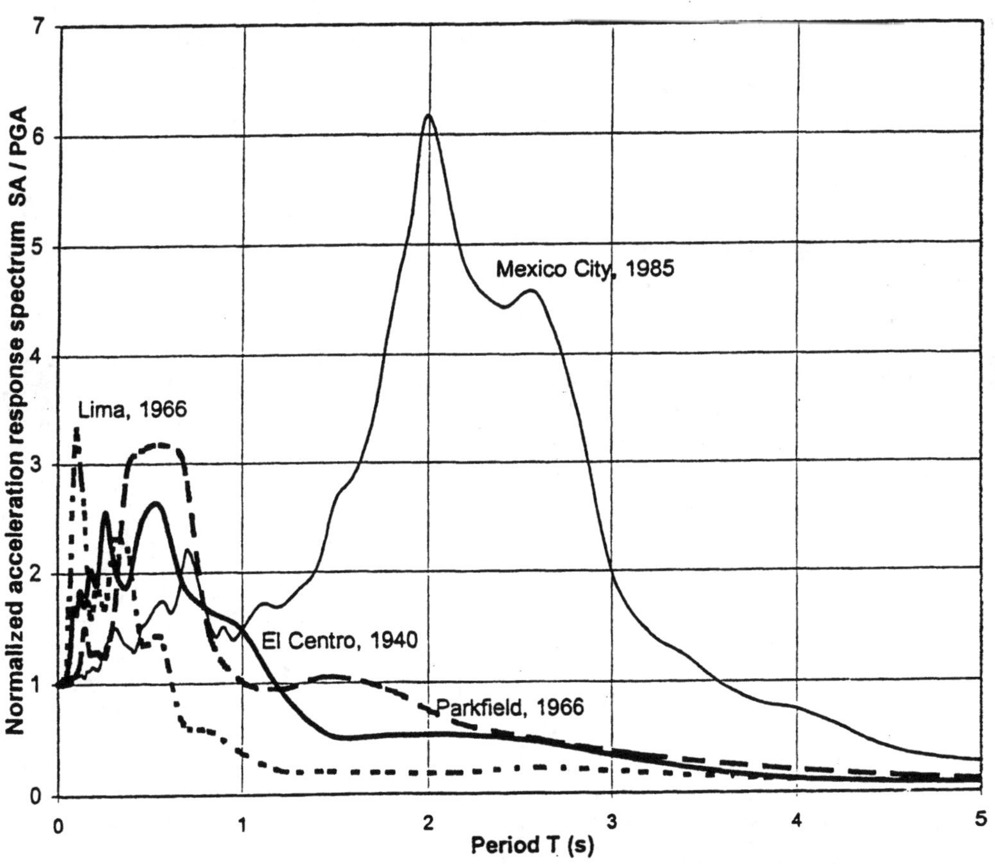

Figure 3.3 Examples of acceleration response spectra shapes.

Earth's crust is divided), where great stress is accumulated. Most destructive earthquakes, with magnitude above 7 to 8, belong to this category. Earthquakes that originate in the subduction zones are a particular type of interplate earthquake. *Intraplate earthquakes* originate in faults within a plate and away from its edges. Intraplate earthquakes typically do not reach such high magnitudes.

The types of faults causing the earthquakes are usually classified as *reverse, normal,* or *strike-slip.* Tectonic movement in reverse and normal faults occurs mainly in the direction of the dip, with a relatively high vertical component. Faults of strike-slip type are typically associated with horizontal movement. In general, reverse faults allow greater normal stress to be transmitted than normal faults and are capable of storing and, consequently, releasing in an earthquake a larger amount of energy. The type of fault also influences the frequency content of the vibratory motion and the intensity of the motions in the near field of the source (i.e., near-fault effects).

The influence of the travel path is related to the attenuation of ground motion with source distance. High frequencies attenuate faster than low ones. As a result, the frequency content of movement is different near the source than at longer distances. Thus, a conventional way of classifying earthquake motions is related to the distance between the hypocenter and the site where the earthquake is felt. Near field ground motions are traditionally distinguished from other types. The term *near field earthquake* is useful in characterizing earthquake motion close to a seismic source, as represented by their higher peak acceleration and greater content in high frequencies. However, a continuous variation of these parameters with epicentral distance is presently accepted.

In practice, the effects of seismic source and travel path are taken into account through magnitude and distance. The movements at the bedrock or at an outcropping rock have an amplitude that increases with magnitude and decreases with distance. Predominant periods are influenced by the same factors. Generally, the greater the magnitude or focal distance, the greater the predominant period. Figure 3.4 shows the influence of magnitude on the response spectra of several Mexican earthquakes. All the earthquake motions associated with these response spectra came from essentially the same source and were recorded at about the same distance. The variations in amplitude and frequency content, particularly in the long-period range, are apparent. Figure 3.5 compares the median response spectra for strike-slip events of magnitude 7 at several distances. The decrease in spectral ordinates and the displacement of predominant periods toward larger values are also apparent.

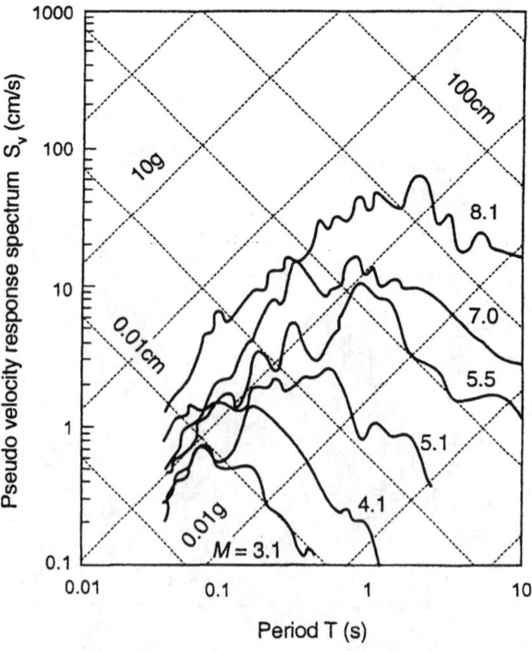

Figure 3.4 Effects of magnitude on response spectra. (After Anderson, 1991.)

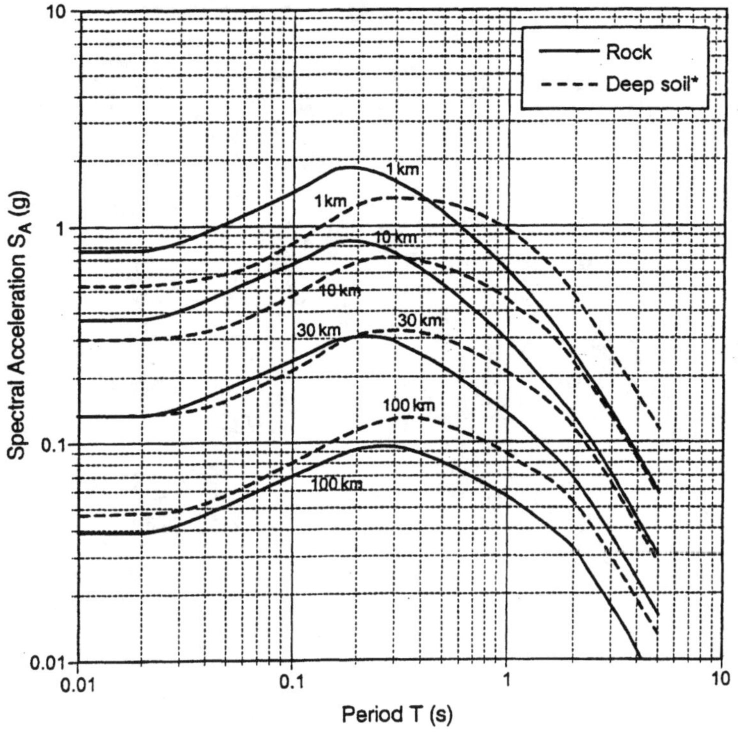

Figure 3.5 Effects of distance on response spectra.

3.2.4 Local Site Effects

There are important differences in ground motion, depending on the local site conditions. The main local factor that influences the movement of the surface during an earthquake is the presence of soil (especially when soft) over the bedrock. Effects of the topography (i.e., hill effects) and of the shape of the soil deposits (i.e., basin effects) should also be considered.

The effect of a soft soil layer is usually the amplification of the vibration, especially in some low frequencies. The main physical mechanism responsible for this amplification at soft soil sites is related to the resonance of the components of vibration with frequency near the natural frequency of the soil layer. The fundamental resonant period for shear wave, T_s, propagating with a shear velocity V_s through a soil deposit of thickness H, is given by $T_s \approx 4H/V_s$. At the surface, the up- and down-going waves of such a frequency are in phase, resulting in a wave of double amplitude. The resonance amplification could be modified significantly by damping and/or the nonlinear properties of the soil.

Figure 3.6 shows the differences in surface accelerations observed, depending on local site conditions. For low and medium PGA, acceleration in the soft soil surface is larger than acceleration in the rock. However, for high acceleration in the bedrock, the shear strength of soil limits the level of the acceleration, resulting eventually in deamplification.

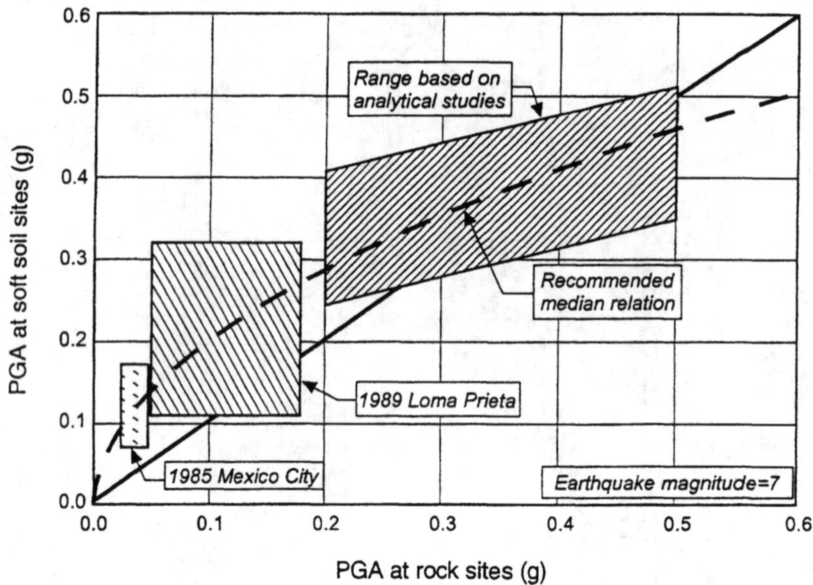

Figure 3.6 Comparison of peak horizontal acceleration at soft soil surface and at rock sites. (After Idriss, 1991.)

As shown in Figure 3.5, the frequency content of the motion at the surface of a soil deposit is different from that in a nearby outcrop rock or in the bedrock. Typically, the predominant periods in rock measure tenths of a second. Depending on the thickness and stiffness of the soil layer, the predominant period at the surface can be from several tenths of a second to a few seconds. Accordingly, the maximum seismic vibratory effects for structures founded on rock can occur for natural periods of tenths of a second (e.g., stiff buildings, gravity quay walls), while at the surface of a soil deposit, the structures most affected can be those with a natural period from several tenths of a second to a few seconds (e.g., high-rise buildings, offshore platforms, pile-supported wharves).

3.2.5 Design Earthquake Motion

Peak horizontal acceleration, response spectra ordinates, and accelerograms are commonly used parameters in the seismic design. The first two are computed from a seismic hazard analysis. There are two approaches to this, the seismotectonic approach and the direct approach based on seismic history.

3.2.5.1 Seismotectonic Approach. In the seismotectonic approach, there are two different but complementary methods:

- *Deterministic method.* The largest earthquake that could affect a given site is used. Commonly, the result of a deterministic analysis consists of the maximum credible earthquake (MCE), the maximum earthquake that appears possible under the known tectonic conditions, taking into account the seismic history of the region.
- *Probabilistic method.* Several values of earthquake motion parameters are used, generally acceleration or response spectra ordinates associated with annual exceedance probability.

The procedures for deterministic and probabilistic analysis are shown schematically in Figures 3.7 and 3.8. The process includes the following steps:

Step 1: Identification of Active Faults and Other Seismic Sources. All sources (both onshore and offshore) capable of producing significant ground motion at the site must be considered. In addition to hazards associated with the specific faults, broader seimotectonic provinces (i.e., regions with uniform tectonic and seismic conditions) are often defined. In light of the seismic hazard associated with unidentified sources, the inclusion of areal, or "random," sources is warranted in most regions of the world.

Step 2: Characterization of Each Seismic Source Activity

DETERMINISTIC METHOD. The maximum earthquake (M_i) and the minimum distance (R_i) from the source to the site are defined. The maximum earthquake is usually assessed from the regional seismic history with the help of regional tectonics. For R_i, epicentral, hypocentral, or fault plane distance may be used, depending on the attenuation model considered in step 3.

PROBABILISTIC METHOD. The parameters of the earthquake occurrence statistics are defined, including the probability distribution of potential rupture locations within each source and

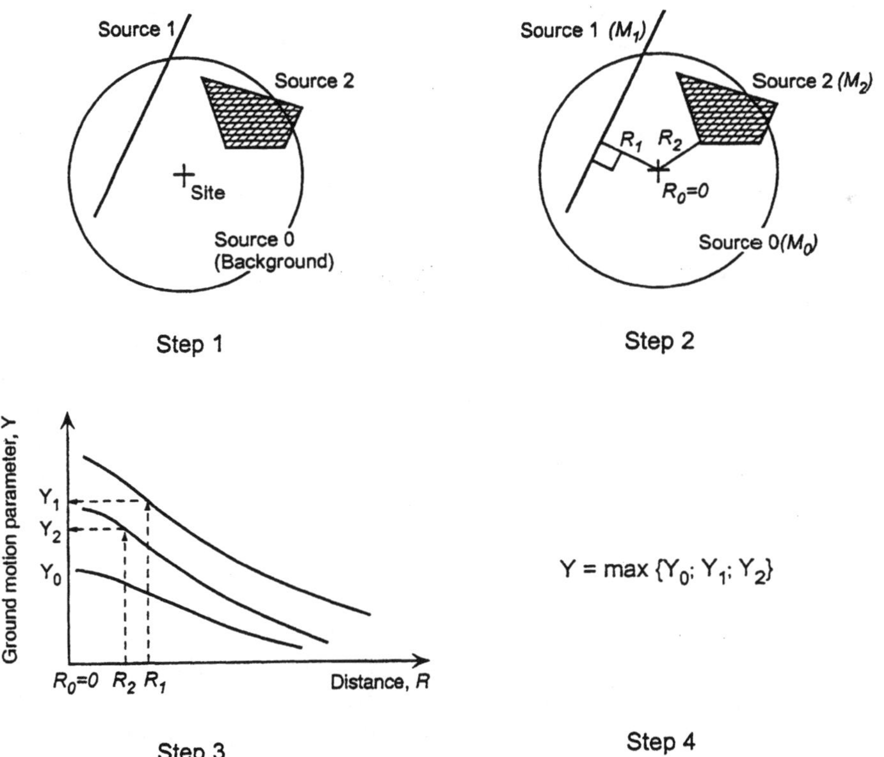

Figure 3.7 Principal steps in deterministic seismic hazard analysis.

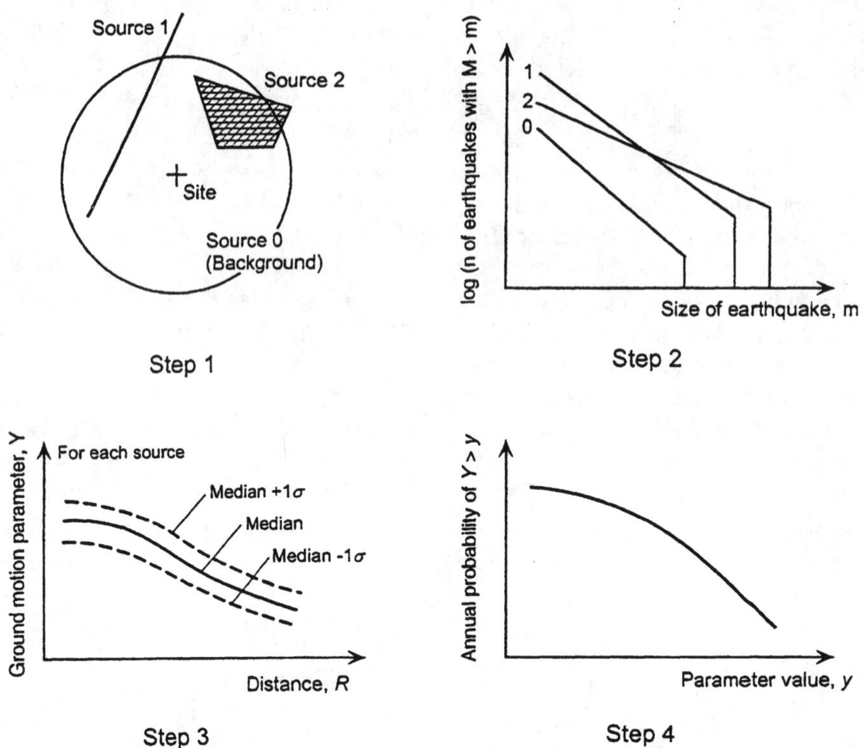

Figure 3.8 Principal steps in probabilistic seismic hazard analysis.

the recurrence relationship. Commonly, it is assumed that all points within the source have the same probability of triggering an earthquake. The recurrence relationship specifies the average rate at which an earthquake of given size will be exceeded, and also, the maximum earthquake, as discussed for the deterministic method.

For modeling the occurrence of earthquakes of different magnitudes, the conventional exponential model (Gutenberg and Richter, 1944) or any of its variants are commonly used. They are based on the Gutenberg–Richter relationship, which relates magnitude (or intensity) M with the mean annual number of events, n, that exceeds magnitude (or intensity) M:

$$\log n = a - bM \tag{3.1}$$

The coefficients a and b must be obtained by regression analysis of the data related to each seismic source. They could depend on the range of earthquake sizes used in the regression. If the seismic catalog is incomplete for small earthquakes, as usual, only earthquakes whose size is beyond a certain threshold level should be used.

The Gutenberg–Richter model was first derived for a set of regional data that included several seismic sources, as in seismotectonic provinces in regions of low seismicity, but it would not be appropriate for large earthquakes from a fault. For those cases, the elastic rebound theory predicts the occurrence of a large "characteristic" earthquake when the elastic stress in the fault exceeds the resistance of the rock. Characteristic earthquakes occur repeatedly, with a segment of a fault having the maximum possible source dimensions.

The temporal model commonly used for the occurrence of earthquakes is a Poisson model, in which each earthquake is a memoryless event, independent of the preceding and next events at the same location. Elastic rebound theory suggests that the occurrence of characteristic earthquakes on a particular fault segment is not independent of past seismicity. A renewal process better than a Poisson process would represent the occurrence of these large characteristic earthquakes. However, the Poisson model seems adequate, physically and observationally, to represent the smaller events, presumably those due to local readjustment of strain, secondary to main elastic rebound cycles. To overcome the limitations of simpler models, a hybrid model (renewal-Poisson) was proposed by Wu et al. (1995).

Step 3: Determination of the Attenuation Laws for the Acceleration, Response Spectra Ordinates, or Other Parameters of Interest

DETERMINISTIC METHOD. The attenuation relationship to be used should reasonably represent the "minimum attenuation" for the region. If a statistical analysis has been performed, a mean or median relationship would not be enough.

PROBABILISTIC METHOD. The mean attenuation function and the standard deviation should be computed through the statistical analysis of data from earthquakes of the same region or, at least, from earthquakes of similar tectonic environment, recorded in stations with local ground conditions similar to those of the site of interest.

Recently, it became common practice to use specific attenuation relationships for each of the spectral ordinates of different periods. A general expression for an attenuation relationship is

$$\log y = f_1(F_T) + f_2(M) + f_3(R) + f_4(S_T) + \varepsilon_\sigma \tag{3.2}$$

where y is the ground motion parameter or response spectrum ordinate; F_T a set of discrete variables describing the fault type; M = the magnitude; R a measure of distance; S_T a set of discrete variables describing the site subsoil conditions or a continuous variable, depending on the average shear wave velocity in the ground; f_i are functions (f_2 is often assumed linear in powers of M, f_3 depends on R, $\log R$, and sometimes, on M); ε_σ a random error term with zero mean and σ standard deviation.

This procedure supplies appropriate response spectra for rock. However, results for soil sites are averages of values from different soil conditions and do not necessarily represent any particular site. In many cases, it may first be preferable to obtain the ground motion parameters in rock and then to compute the seismic response at the soil surface. Selected attenuation relations for spectral ordinates can be found in *Seismological Research Letters* (Abrahamson and Silva, 1997).

Step 4: Definition of the Seismic Hazard. Ground motion, described primarily by PGA value and spectral ordinates, must be defined. Other parameters, such as intensity and duration, can be obtained similarly.

DETERMINISTIC METHOD. Calculate the values of the motion parameters at the site from each source of the earthquake and select the maximum.

PROBABILISTIC METHOD

1. Calculate the annual number of occurrences of earthquakes from each source which produce, at the site, a given value of the PGA_H (or other earthquake motion parameters).
2. Calculate the total annual number, n, of exceedance.
3. Calculate the return period of earthquakes exceeding the PGA_H value above (i.e., the reciprocal of the mean annual rate of exceedance):

$$T_R = \frac{1}{n} \tag{3.3}$$

4. Calculate the probability $P_R(a_{max}, T_L)$ of the PGA_H value being exceeded in the life T_L of the structure:

$$\begin{aligned} P_R(a_{max}, T_L) &= 1 - (1 - n)^{T_L} \approx T_L n \\ &= \frac{T_L}{T_R} \quad (\approx \text{ for } T_L/T_R << 1) \end{aligned} \tag{3.4}$$

If the attenuation relationships do not match the local site conditions (e.g., if attenuation relationships for rock are used and there is a surface soil deposit), the motion parameters resulting from the hazard analysis must be converted to the specific site conditions using empirical amplification ratios or numerical dynamic soil response models. For preliminary studies, or when data are scarce, the methodology discussed previously is applied to calculate the PGA_H values. Afterward, a response spectrum is scaled to this PGA_H. In this method, care must be taken in selection of the response spectrum shape, which depends on fault type, magnitude, distance to rupture zone, and local site conditions. Figure 3.9 shows the shapes of elastic response spectra proposed by Bea (1997).

3.2.5.2 Direct Approach Based on Seismic History. In some cases, a particular earthquake controlling the seismic hazard is defined directly from the seismic history. Seismic acceleration is also defined by historic data or derived from appropriate attenuation relationships. Accelerograms are selected from those recorded in earthquakes of fault type, magnitude, distance range, propagation path, and recording site conditions similar to those of the design earthquake, and are then used to define response spectra at various percentile levels suitable for design.

3.2.6 Design Accelerograms

Depending on the type of analysis, time histories can be defined, usually in the form of recorded or generated accelerograms. Recorded time histories must be selected from collections of accelerograms caused by earthquakes whose fault type, magnitude, and distance are similar to those of the design earthquake and can be scaled to the design PGA_H value. Commonly, it is necessary to use several accelerograms to cover the implicit contingency in its selection. When the strong ground motion parameters of the design earthquake are influenced by several seismic sources, it is not possible to find a single recorded accelerogram that incorporates all the characteristics of the design motion, including spectral ordinates and duration. In such a case, several records of each earthquake type should be used.

Typical sources of digitised accelerograms can be found at the National Geophysical Data Centre (www.ngdc.noaa.gov) in the United States and the K-NET (www.K-NET.bosai.go.jp) in Japan (Kinoshita, 1998).

For the modal analysis of structures, it is important that the frequency content of the recorded time histories match the spectral shape

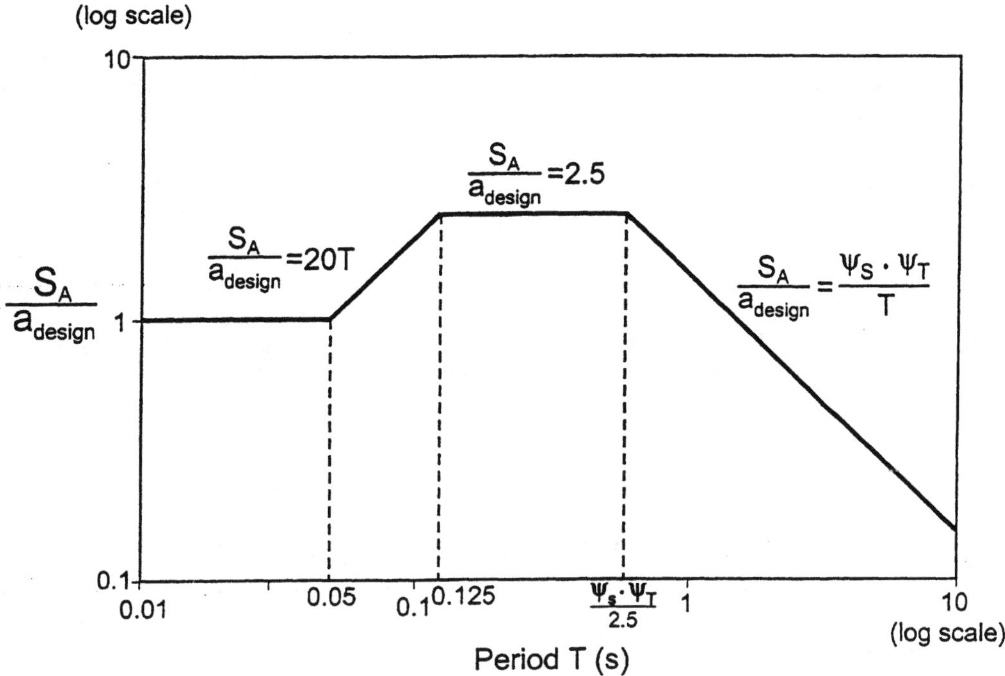

Soil condition	Ψ_s
A - Rock (V_s > 750 m/s)	1.0
B - Stiff to very stiff soils, gravels (V_s = 360 to 750 m/s)	1.2
C - Medium stiff to stiff clays (V_s = 180 to 360 m/s)	1.4
D - Soft to medium stiff clays (H = 1 to 20 m; V_s < 180 m/s)	2.0
E - Conditions other than above	Site-specific studies required

Tectonic condition	Ψ_T
Type A Shallow crustal faulting zones	1.0
Type B Deep subduction zones	0.8
Type C Mixed shallow crustal & deep subduction zones	0.9
Type D Intraplate zones	0.8
Default value	1.0

Figure 3.9 Design response spectra shape proposed by Bea (1997).

of the design earthquake. If the maximum recorded acceleration is different from the design PGA_H value, the recorded accelerogram can be scaled to this value. In some cases, a slight modification of the time scale in the recorded accelerogram is also acceptable to achieve better agreement between the frequency contents of the recorded and design acceleration time histories. For liquefaction analysis, it is also important that the duration of a recorded accelerogram match the design duration of the shaking.

Artificial time histories must match the PGA_H value, duration, and response spectrum of the design earthquake. Formulation of design earthquake motions for near-source regions may require adequate modeling of fault rupture process and travel path effects. Practice-oriented guidelines proposed by the Japan Society of Civil Engineers are summarized by Ohmachi (1999).

3.2.7 Ground Motion Input for Seismic Analysis of Port Structures

The seismic input to be used for analysis and design depends on the type of port structure and the type of analysis being performed, as discussed in Section 3.5. Simplified analysis of Earth or massive structures assumes a uniform field of seismic acceleration represented by a seismic coefficient. Therefore, the seismic coefficient should represent the average acceleration over the entire mass of the sliding block.

Simplified dynamic analyses of Earth or massive structures, such as Newmark-type rigid block sliding, requires a time history of acceleration averaged over the entire mass of the structure. For simplicity, the time history at a particular point scaled to the average of the peak acceleration over the sliding mass is often used.

Dynamic analysis of soil–structure interaction requires as input motion at the bedrock. If the seismic data consist of free field ground surface motions, it is necessary to compute subsurface–bedrock motion by numerical deconvolution techniques. On steel and concrete structures, the analyses are usually based on response spectra.

In some cases, the effects of elevation difference between the landward ground surface and the seabed can be an important issue that should be considered in the development of seismic acceleration, response spectra, and other earthquake motion parameters. At present there exist no well-established criteria referring to this issue, and therefore extra care should be given in design practice if increased reliability on the results of the simplified dynamic analyses is required.

3.3 DESIGN APPROACH

As stated in Section 3.1, deformations of both a structure and the surrounding soil are key design parameters for port as well as for many other structures. Conventional design methods based on a limit equilibrium approach are generally not suitable for an evaluation of these deformations, although some of them may be acceptable under specific conditions. The performance-based methodology presented here incorporates analyses that are capable of predicting the above-mentioned deformations and allows evaluation of the response of the structure and the associated damage for shakings above the limit equilibrium.

In performance-based design, two levels of earthquake intensity are typically used as design reference motions:

- *Level L1:* motions that are quite likely to occur during the lifetime of the structure
- *Level L2:* motions that are associated with rare events of very strong ground shaking

More specifically L1 shaking is typically defined as a motion with 50% probability of exceedance during the lifetime of the structure

Table 3.1 Level of damage in performance-based design

Level of Damage	Structural	Operational
Degree I: Serviceable	Minor or no damage	Little or no loss of serviceability
Degree II: Repairable	Controlled damage	Short-term or complete loss of serviceability
Degree III: Near collapse	Extensive damage in near collapse	Long-term or complete loss of serviceability
Degree IV: Collapse	Complete loss of structure	Complete loss of serviceability

Table 3.2 Acceptable level of damage in performance grades S, A, B, and C

	Design Earthquake	
Performance Grade	Level 1 (L1)	Level 2 (L2)
Grade S	Degree I: Serviceable	Degree 1: Serviceable
Grade A	Degree I: Serviceable	Degree II: Repairable
Grade B	Degree I: Serviceable	Degree III: Near collapse
Grade C	Degree II: Repairable	Degree IV: Collapse

under consideration, while L2 is typically defined as a motion with a corresponding probability of 10%. If the life span of a port structure is, say, 50 years, according to the foregoing definitions, the return period of L1 and L2 should be 72 and 475 years, respectively.

In regions of low seismicity, L1 may thus be relatively small and of minor engineering significance. In such cases only L2 may be taken into account, coupled with the appropriate damage criteria, assuming that performance under L1 is ensured to be adequate. This approach may seem similar to the conventional methodology (limit equilibrium and one-level earthquake); however, it should be noted that in the performance-based approach, some damage criteria associated with design earthquakes must be established.

In regions of moderate and high seismicity, the dual-level approach should be applied, since meeting the criteria for L2 does not necessarily imply that the criteria for L1 are met, and vice versa. In such cases, the design of the structure may be dominated by either L1 or L2 motion.

The damage caused by a specific earthquake can be classified according to PIANC (2001) in four levels, based on structural and operational considerations, as shown in Table 3.1. The acceptable level of damage associated with a port structure should be defined as above for all design earthquake motions, taking into account factors such as protection of human life and property, protection from spilling hazardous materials, functioning as an emergency transportation, and economy. This leads to assigning the performance grade of the structure. According to PIANC (2001b) four such grades are introduced: S, A, B, and C. For the typical dual-earthquake-level approach, the performance grades are provided through the matrix of Table 3.2.

Broadly speaking, grade S structures are those with a potential for extensive loss of human life and property, key structures needed to be intact for recovery purposes, installations handling hazardous materials, or those with a potential for devastating economic and serial impact following a strong earthquake. Grade A structures have less serious effects than grade

S structures or, if damaged, are difficult to restore. Grade B structures are ordinary structures, and grade C structures are small, easily restorable structures.

Once the damage criteria for the design motions have been defined (see Section 3.4), evaluation of an existing or newly designed structure can be undertaken. This can be achieved by employing a suitable method, as described in Section 3.5.1. It is noted that the performance grade of a structure is specified by its lower performance at any earthquake motion tested.

3.4 DAMAGE CRITERIA

A key element in performance-based design is identification of the deformation and failure modes pertinent to the port structure under examination and the main parameters to quantify these modes. The specific limiting values of the parameters that quantify the performance of a structure constitute the damage criteria for the design. These values must be expressed in engineering terms: for example, as displacements, tilting, stresses, and strains. Performance-based design is a relatively new methodology and the damage criteria are not fully developed. PIANC made an effort to specify values for these criteria, most of which will be followed in what follows.

The actual values of the damage criteria to be adopted in each case depend on the function and the seismic response of the port structure. They can be established based on Table 3.1, taking into account the needs and views of the owners and users of the facility. In general, the criteria for damage level degree I (serviceable structure) are dictated by operational considerations, whereas for higher degrees, structural aspects provide the primary design criteria to be adopted.

In what follows, general guidelines to determine damage criteria, together with some specific values of the latter, are proposed for the various types of port structures under examination. It is noted that the level of damage of a construction is defined by the highest degree of damage associated with its elements.

3.4.1 Gravity Quay Walls

Figure 3.10 shows the principal parameters for specifying damage criteria for gravity quay walls. Tilting is more frequent when the width/height ratio of the structure is less than about 0.75. In cases of poor quality of the foundation soil and of the backfill, more pronounced tilting and displacements may occur.

To increase the stability of gravity quay walls, several options are available, such as:

- To provide backfill material with a high angle of internal friction
- To provide the foundation soil with high resistance and low deformability by means of deep compaction techniques
- To cater for a cross section and fill material of the wall that will draw landward the center of gravity of the structure

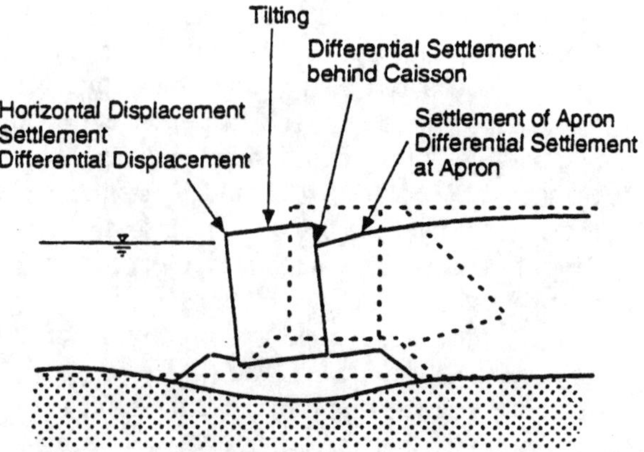

Figure 3.10 Parameters for specifying damage criteria for gravity quay walls.

- To increase the friction between the wall and the foundation soil in various ways (e.g., by giving the foundation surface of the wall a special antiskid shape)
- To design for inclined rather than horizontal contact surfaces between the precast blocks of a quay wall

It is noted that in many cases of gravity quay walls, with no cranes on rails, displacements can be acceptable up to a certain limit dictated by operational factors. However, tilting cannot be accepted in most circumstances. Parameters for specifying damage criteria include horizontal displacements, tilting settlements, differential settlements at the back of the wall, and so on. To specify the criteria for degree I damage, serviceability considerations should be taken into account (e.g., safe berthing, safe loading and unloading of vessels, ramp operations). For higher degrees of acceptable damage, structural stresses and deformations should be considered (e.g., in the case of gravity quay wall with cells formed by sheet piling, the stresses of the sheet piles, especially at the joints).

Indicative values of damage criteria for gravity quay walls are given in Table 3.3. Note that the values listed in Table 3.3 and in Table 3.4) have been specified independent of one another thus the horizontal displacement appears more stringent than the tilting in virtually all cases.

For cellular quay walls with cells formed by sheet piles, damage criteria proposed for gravity quay walls are applicable. Moreover, for degree I it is proposed that sheet piles and cell joints both work in the elastic region, degree II requires that cells work in the elastic region while cell joints can work in the plastic one but below the appropriate strain limit, degree III can tolerate cells working in the plastic state but below the appropriate strain limit and cell joints anywhere in the plastic region, and degree IV signifies cells and cell joints working anywhere in the plastic region.

3.4.2 Sheet Pile Quay Walls

A sheet pile quay is normally composed of interlocking sheet piles embedded in the foundation soil and of continuous or independent anchors and tie rods. For details consult Chapter 2. Typical failure modes are associated with the foregoing components. Taking into account the secondary effects of a local failure and the difficulty of restoration, the following sequence of yielding is usually preferred: (1) sheet pile above mudline, (2) sheet pile below mudline, (3) tie-rod, and (4) anchor system, e.g., sheet pile wall, concrete deadman, vertical or baffer piles.

Figure 3.11 shows the main parameters for specifying damage criteria of sheet pile quay walls. When based on serviceability, they in-

Table 3.3 Proposed damage criteria for gravity quay walls

	Degree I: Serviceable	Degree II: Repairable	Degree III: Near Collapse	Degree IV: Collapse
Residual horizontal displacement d/H^a (%)	<1.5	1.5–5	5–10	>10
Residual tilting toward the sea (deg)	<3	3–5	5–8	>8
Differential settlement behind caisson (m)	<0.1	—	—	—

[a] d, displacement at top of wall; H, height of wall.

Table 3.4 Proposed damage criteria for sheet pile quay walls

	Degree I: Serviceable	Degree II: Repairable	Degree III: Near Collapse	Degree IV: Collapse
Residual horizontal displacement d/H^a (%)	<1.5	—	—	—
Residual tilting toward the sea (deg)	<3	—	—	—
Differential settlement on apron (m)	<0.1	—	—	—
Peak response stresses and strains on sheet pile above mudline	Elastic	Plastic (allowable ductility not exceeded)	Plastic (allowable ductility not exceeded)	Plastic (allowable ductility exceeded)
Peak response stresses and strains on sheet pile below mudline	Elastic	Elastic	Plastic (allowable ductility not exceeded)	Plastic (allowable ductility exceeded)
Peak response stresses and strains on tierod	—	Elastic	Plastic (allowable ductility not exceeded)	Plastic (allowable ductility exceeded)
Peak response stresses and strains on anchor	Elastic	Elastic	Plastic (allowable ductility not exceeded)	Plastic (allowable ductility not exceeded)

[a] *d*, displacement at top of wall; *H*, height of wall.

clude, as in the case of gravity quay walls, horizontal displacements, tilting, and settlements. Parameters for damage criteria based on structural damage involve stresses on sheet piles, anchors, and tie rods. Proposed design criteria for sheet pile walls, in the case of the anchor being more difficult to restore than the wall itself, are included in Table 3.4.

3.4.3 Pile-Supported Wharves

A pile-supported wharf is composed of a deck supported by piles and a sloping dike with a small wall on its top. The deck and the top of the dike are connected to provide horizontal continuity to the platform. Taking into account the secondary effects of a local failure and the difficulty of restoration, the following sequence of yielding is usually preferred: (1) pile caps, (2) pile tops, (3) deck, (4) piles below dike surface of mudline, and (5) dike. Figure 3.12 shows the main parameters for specifying damage criteria for pile-supported wharves. Those referring to structural components are related to yielding of piles and deck, for which proposed design criteria are included in Table 3.5. The performance of the wall at the top of the dike should also be evaluated, using the damage criteria proposed for quay walls in preceding paragraphs. Damage criteria for dikes are not fully developed at present. They should be defined by the designer, depending on the displacements of the dike, which can damage the piles according to the criteria of Table 3.1.

3.4.4 Quay Walls with Cranes on Rails

A crane usually consists of a supporting structure with four legs and an upper structure for handling cargo. Both parts are generally made of steel-braced frames. Figure 3.13 shows the main parameters for specifying damage criteria

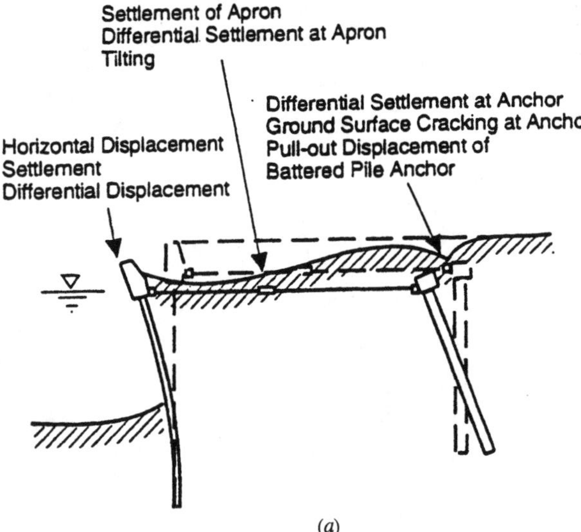

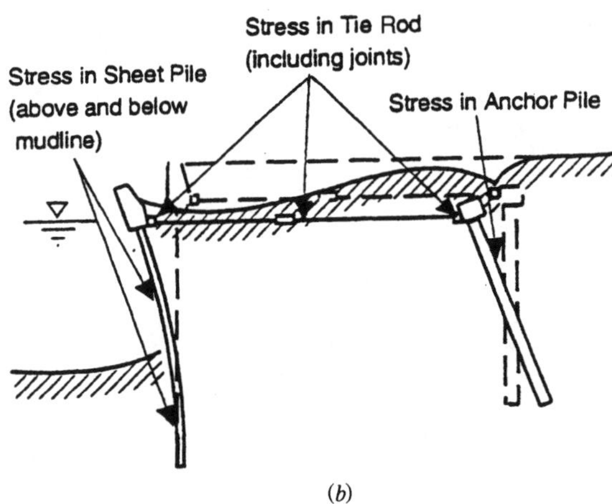

Figure 3.11 Parameters for specifying damage criteria for sheet pile quay walls with respect to (*a*) displacements and (*b*) stresses.

for cranes. They refer to the serviceability and stability of a crane due to rail displacement and to the structural performance of the crane itself. Damage criteria should be established by the designer based on the criteria of Table 3.1. Table 3.6 provides some guidance to this end.

3.4.5 Breakwaters

Breakwaters are usually either of the rubble-mound type or have a vertical face. Also, combinations of these two main types can be constructed (i.e., walls with vertical faces resting on rubble mounds or protected by rocks; Japanese-type composite breakwaters). Examples are found in Chapter 9. Breakwaters are designed to withstand severe storms, but only wave loads from a moderate sea state should be considered together with the design earthquake.

The design criteria to be adopted depend largely on the function of the particular breakwater under consideration. For example, in cases where the lee of the structure is used for berthing or for other operations, the allowable damage should be lower than the corresponding simple breakwater. Damage criteria have not been fully developed at present. They should be adopted by the designer, aiming at achieving the structural and operational behavior defined in Table 3.1 for the acceptable degree of damage for the performance grade assigned to the specific breakwater. According to PIANC (2001b), the following performance grades are proposed for breakwaters of various functions:

- *Structure to reduce wave penetration:* grade C
- *With access for people:* grade A, B, or C, depending on risk to human life
- *With berthing and cargo handling facilities, including conveyor belts:* grade B
- *With pipelines for oil and liquid gas:* grade A or S, depending on the risk of explosion

3.4.6 Special Consideration to Soil Liquefaction

Liquefaction of the foundation soil or earth fill is one of the main causes of damage to port

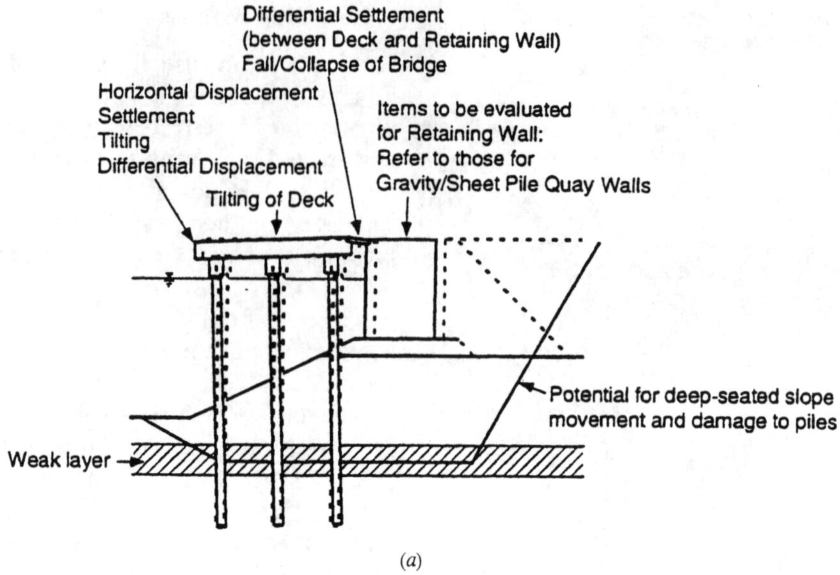

(*a*)

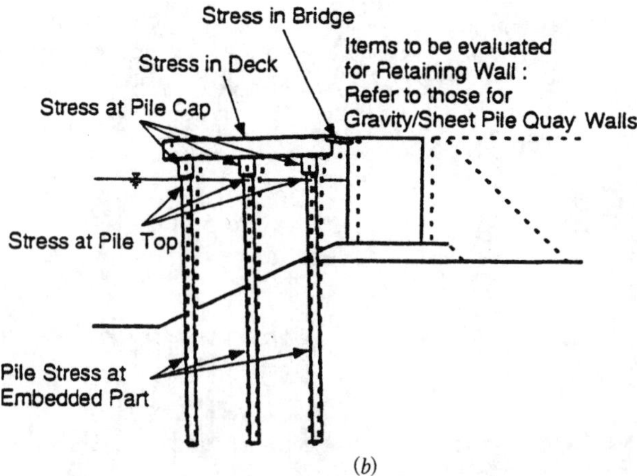

(*b*)

Figure 3.12 Parameters for specifying damage criteria for pile-supported wharves with respect to (*a*) displacements and (*b*) stresses.

structures during earthquakes. In the present state of practice, in most cases the evaluation of liquefaction potential is carried out by means of direct procedures, independent of the behavior of the entire structure. In such cases it is not easy to compute the effects of liquefaction (displacements, stresses, etc.) on the main structure and hence the expected degree of damage, especially for degrees II and III. Consequently, damage criteria are often reduced to the condition of no liquefaction in both design earthquakes.

Table 3.5 Proposed damage criteria for piles and decks of pile-supported wharves

	Degree I: Serviceable	Degree II: Repairable	Degree III: Near Collapse	Degree IV: Collapse
Residual tilting toward the sea (deg)	<3	—	—	—
Differential settlement between deck and land behind (m)	<0.1–0.3	—	—	—
Peak response stresses/strains on piles	Essentially elastic response with minor or no residual deformation	Controlled limited inelastic ductile response and residual deformation keeping the structure repairable	Ductile response near collapse (double plastic hinges may occur at only a limited number of piles)	Beyond the state of degree III
Displacement of dike (cm)	7–15	15–30		

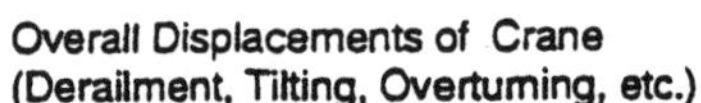

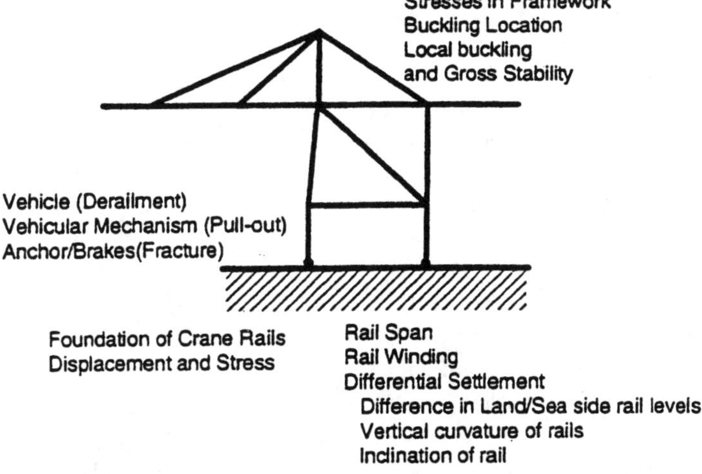

Figure 3.13 Parameters for specifying damage criteria for cranes.

Table 3.6 Proposed damage criteria for cranes on rails

		Degree I without Derailment	Degree II with Derailment	Degree III without Overturning	Degree IV with Overturning
Peak response stresses/ strains	Upper structure	Elastic	Elastic	Plastic (less than the ductility factor/strain limit for upper structure)	Plastic (beyond the ductility factor/ strain limit for upper structure)
	Main frame-work of supporting structure	Elastic	Plastic (less than the ductility factor/ strain limit for main framework)	Without collapse	Collapse
	Toe	Elastic	Damage to toe (including pullout of vehicle, fracture of anchor/brakes)	Damage to toe (including pullout of vehicle, fracture of anchor/ brakes)	Damage to toe (including pullout of vehicle, fracture of anchor/ brakes)

3.5 SEISMIC ANALYSIS

3.5.1 Types of Analysis

Before embarking on the seismic analysis of a port structure, it is necessary to establish near-field boundary conditions (i.e., the design earthquake motions and the geotechnical conditions of foundation ground and retained fill). Earthquake motions should be defined at a point characteristic of the geometry of the problem to be analyzed (bedrock, base of the structure, free field ground surface). This is achieved by regional seismic hazard analysis and by evaluation of the local site effects. The required knowledge of geotechnical conditions includes the geometry of soil layers and their dynamic deformability and resistance, including the effects of possible pore pressure buildup (liquefaction).

In general, three levels of sophistication are distinguished in seismic analysis of port structures:

1. *Simplified analysis,* where the stability limit state can be evaluated and an order of magnitude of residual displacements can be estimated
2. *Simplified dynamic analysis,* by which it is possible to evaluate displacements and stress/strains based on expected failure modes
3. *Dynamic analysis,* which is capable of estimating both failure modes and the associated displacements and stresses and strains

Selection of the type of analysis to be used depends on the required performance grade as well as on the required accuracy of the design. Table 3.7 shows the appropriate type of analysis with regard to these two factors.

3.5.2 Site Response

Site response analysis refers to the evaluation of ground motion modification due to local site conditions. Usually, it relates to the case in which one-dimensional analysis is adequately representative of the ground horizontal layering. Such analyses are necessary when seismic input data refer to the ground surface, and movement at bedrock is needed or when movement at the surface must be obtained from data at bedrock.

The main procedure in *simplified analysis* of site response is associated with the use of some amplification coefficients to obtain seismic acceleration/response spectra at the surface from regional or bedrock acceleration/response spectra resulting from seismic hazard analysis. These amplification coefficients depend mainly on the thickness and stiffness of soil over bedrock.

One-dimensional total stress procedures using the hysteretic equivalent linear model of the soil should be considered as *simplified dynamic analysis*. It allows estimation of the histories of acceleration and of shear stresses and strains at any point inside the soil profile when the history of acceleration at another point is known.

Dynamic analysis of site response is made by means of the effective stress (nonlinear) method, or in cases where degradation of resistance or elastodynamic properties are not demonstrated, by means of a total stress method employing a hysteretic equivalent linear model of the soil. In Section 3.5.8, the total stress and effective stress methods of analysis are discussed in somc detail.

3.5.3 Liquefaction

The assessment of liquefaction potential is usually based on a comparison between the seis-

Table 3.7 Types of analysis related to performance grades

	Accuracy for Performance Grade:			
Type of Analysis	C	B	A	S
Simplified analysis: Appropriate for evaluating approximate threshold level and/or elastic limit and order-of-magnitude displacements.	[a]	[b]	[b]	[b]
Simplified dynamic analysis: Of broader scope and more reliable. Possible to evaluate extent of displacement/stress/ductility/strain based on assumed failure modes.	—	[a]	[b]	[b]
Dynamic analysis: Most sophisticated. Possible to evaluate both failure modes and extent of displacement/stress/ductility/strain.	—	—	[a]	[a]

[a] Standard/final design.
[b] Preliminary design or low level of excitations.

mic load (effective or equivalent shear stress) and the capacity of the soil to withstand that load (resistance to liquefaction in terms of shear stress). Only in the most sophisticated dynamic analysis is the pore pressure buildup during seismic motion evaluated in detail as part of the response of the soil–structure system.

In simplified analysis the equivalent-to-earthquake shear stress (τ_{eq}) is computed from the cyclic stress ratio, CSR (τ_{eq}/σ'_{vo}), and the soil resistance to liquefaction (τ_{rl}) is deduced from the cyclic resistance ratio, CRR (τ_{rl}/σ'_{vo}). CSR is evaluated by the equation (Seed and Idriss, 1971)

$$\text{CSR} = \frac{\tau_{eq}}{\sigma'_{vo}} = 0.65\,\frac{a_{max}}{g}\,\frac{\sigma_{vo}}{\sigma'_{vo}}\,r_d \qquad (3.5)$$

where α_{max} is the peak horizontal acceleration at the ground surface, g is the acceleration due to gravity, σ_{vo} and σ'_{vo} are, respectively, the total and effective vertical overburden stresses, and r_d is a function of depth z that accounts for stress reduction due to soil deformability and for differences in phase of vibration at different depths. Detailed discussion of the values of r_d (z) can be found in Youd et al. (2001). They vary from 1.0 at the surface to 0.6 to 0.9 at a depth of 15 m. Average values for design purposes can be obtained from the expressions:

$$z < 9.15\text{ m:} \qquad r_d = 1.0 - 0.00765z$$

$$9.15\text{m} < z < 23\text{ m:} \qquad r_d = 1.174 - 0.0267z \qquad (3.6)$$

CRR is evaluated using empirical charts (Figures 3.14 through 3.16), depending on index properties resulting from some in situ tests (SPT N-values; q_c, cone penetration resistance; V_s, shear wave velocity). SPTs and CPTs are generally preferred because of the more extensive past experience, but V_s should be preferred in sites of gravelly soils that are difficult to test by SPT or CPT.

These charts are developed for earthquakes with magnitude $M = 7.5$, for horizontal or gently sloping ground surface, and for liquefaction occurring at low depths (less than 15 m). The values of CRR for earthquakes with magnitude other than 7.5 need to be corrected by a scaling factor, C_M, that accounts for the different number of cycles during the seismic movement:

$$\text{CRR}(M) = (C_M)\text{CRR }(7.5) \qquad (3.7)$$

Other scaling factors for high overburden stress, static shear stress, and age of deposit are discussed by Youd et al. (2001).

Several sets of values are proposed for C_M. Youd et al. (2001) recommend the expression

$$C_M = \left(\frac{7.5}{M}\right)^{2.56} \qquad (3.8)$$

as a lower bound for C_M values. Index properties (N, q_c, or V_s) to enter the charts for CRR show, in practice, large variability due to the state of stresses in the soil, and in the case of SPT, due to some operational details. The main cause of N variability is the difference in energy efficiency of SPT procedures. It could vary from as low as 30% for rope and pulley manual devices to about 80% for automatic free-fall devices. As a consequence, measured values of N need to be normalized to the one associated with the empirical correlation to be used. The energy efficiency associated with the N-value used in computing CRR in Figure 3.14 is about 60%, a value to which the measured N-value must be corrected. Other corrections to be effected relate to borehole diameter, rod length, and lining on the sampler.

The reference blow count $(N_1)_{60}$, normalized to an overburden pressure of 100 kPa ($\simeq$ 1 kilopoise (kp)/cm^2 $\simeq$ 1 atm) and corrected for

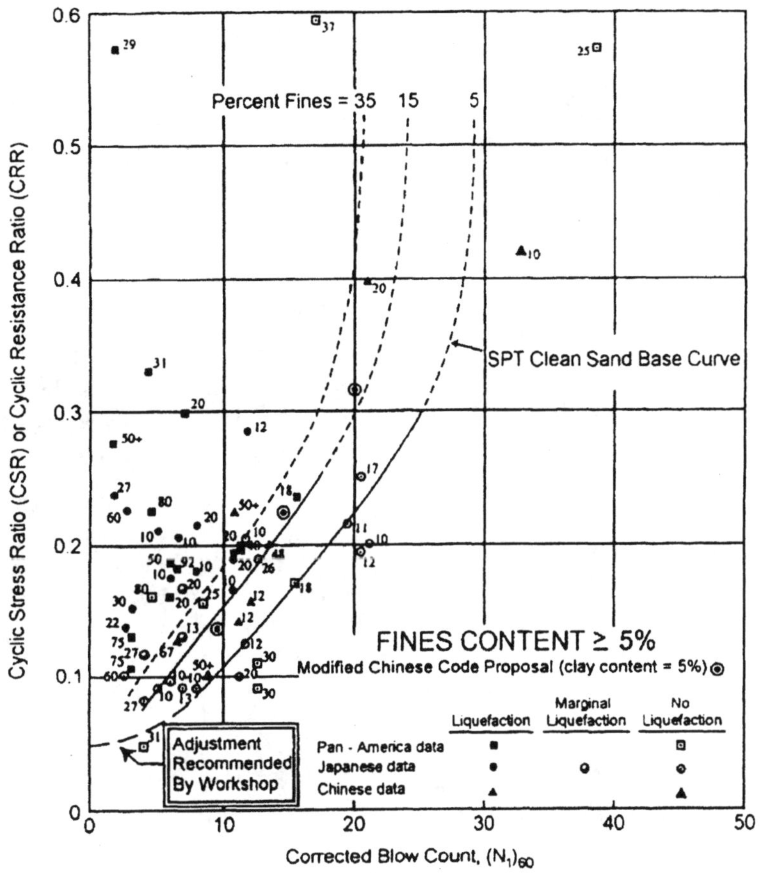

Figure 3.14 Empirical chart of CRR versus $(N_1)_{60}$ for magnitude 7.5 earthquakes. (After Seed et al., 1985; modified by Youd et al., 2001.)

the criteria described above can be expressed as

$$(N_1)_{60} = C_N C_E C_B C_R C_S N_m \tag{3.9}$$

Suggested values for these correction factors are listed in Table 3.8. CPT resistance should also be normalized to an effective overburden pressure of $p_a = 1$ atm ($\simeq$ 100 kPa $\simeq$ 1 kp/cm^2) through the expression

$$q_{c1} = C_q \frac{q_c}{p_a} \tag{3.10}$$

Youd et al. (2001) suggested use of the expression

$$C_q = \left(\frac{p_a}{\sigma'_{v0}}\right)^n \tag{3.11}$$

with n dependent on the grain characteristics of

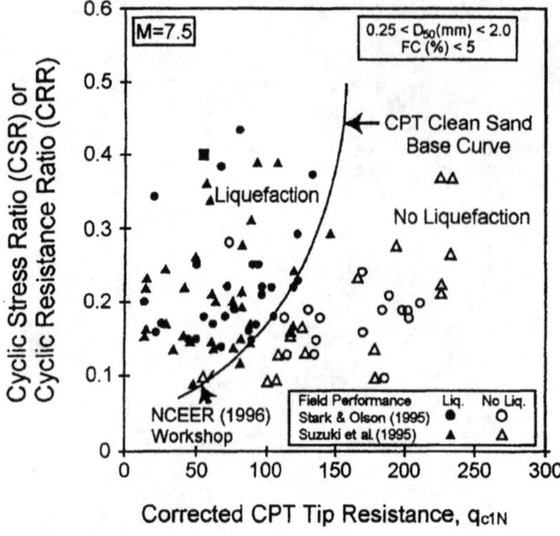

Figure 3.15 Empirical chart for calculation of CRR from CPT tip resistance. (After Robertson and Wride, 1998. See original publication for field performance references.)

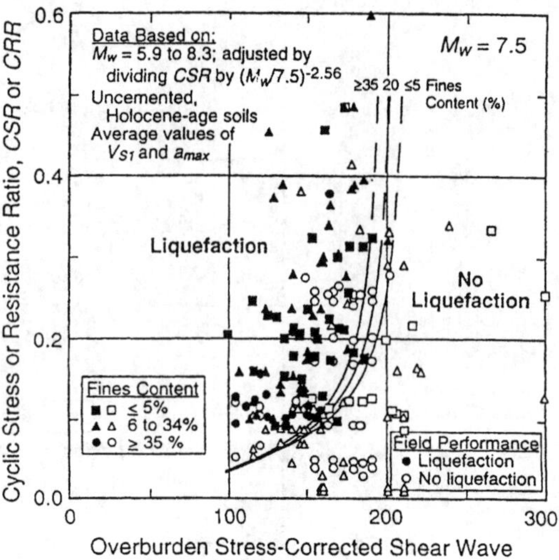

Figure 3.16 Empirical chart for liquefaction resistance versus shear wave velocity for uncemented sands. (After Andrus and Stokoe, 2000.)

the soil and ranging between 0.5 (clean sand) and 1.0 (clayey soils).

The measured shear wave velocity V_S should also be normalized to a reference effective overburden stress of p_a through the expression

Table 3.8 Correction factors for SPT blow count

Factor	Equipment Variable	Correction	Value
Overburden stress		C_N	$(p_a/\sigma'_{v0})^{0.5}$
			$C_N \leq 2$
Energy ratio	Donut hammer	C_E	0.5–1.0
	Safety hammer		0.7–1.2
	Automatic-trip donut hammer		0.8–1.3
Borehole diameter (mm)	65–115	C_B	1.0
	150		1.05
	200		1.15
Rod length (m)	3–4	C_R	0.75
	4–6		0.85
	6–10		0.95
	10–30		1.0
	>30		>1.0
Sampling method	Standard sampler	C_S	1.0
	Sampler without liners		1.1–1.3

Source: Robertson and Wride (1998).

$$V_{S1} = V_S \left(\frac{p_a}{\sigma'_{v0}}\right)^{0,25} \tag{3.12}$$

The influence of fines content can be included in the chart, as in Figure 3.14 or 3.16, or when the soil is characterized by the cone penetration q_c, it can be taken into account by the procedure described in Robertson and Wride (1998).

In simplified dynamic analysis, shear stress equivalent to an earthquake can be computed from total stress response analysis, and liquefaction resistance can be evaluated by the charts of Figures 3.14 through 3.16, which could be complemented by laboratory tests, especially on human-made fills of fine granular soils. In dynamic analysis, the responses of soil and structure are evaluated in a single model using finite element or finite difference methods and effective stress soil constitutive models. Pore pressure buildup during an earthquake is obtained directly from such an analysis.

3.5.4 Gravity Quay Walls

The gravity quay wall structure is being conventionally evaluated against seismic loading through simplified analysis using pseudostatic approaches. In these methods the dynamic loads are represented by additional static loads of magnitude proportional to the design seismic acceleration. The actual behavior of such quay walls, especially those of block work, are much more complex, and thus advanced methods have been proposed to simulate more accurately than the pseudostatic approaches the seismic response of the structure.

In the pseudostatic method the stability of a wall is usually assessed with regard to overturning, sliding, and the bearing capacity of the soil. When the width/height ratio of the structure is adequate, the critical check will be against overturning; otherwise, sliding will dictate the minimum required dimensions of the gravity quay wall. Tilting is often more serious than displacement; therefore, the safety factor associated with overturning is normally higher than the corresponding factor with respect to sliding. In the simplified approach, the earth pressures are usually estimated through use of the Mononobe (1924) and Okabe (1924) equation that modifies Coulomb's classical theory to account for the inertia forces due to shaking of the retained earth. The force vector is rotated through an angle ψ, defined by

$$\psi = \arctan \frac{k_h}{1 - k_v} \tag{3.13}$$

where $k_h g$ and $k_v g$ are, respectively, the horizontal and vertical accelerations of the excitation (Figure 3.17).

Introducing the notation above into the geometry of Coulomb's classical solution, the active pressure coefficient K_{ae} for a vertical wall *and a horizontal ground surface* is derived as follows:

$$K_{ae} = \frac{\cos^2(\phi - \psi)}{\cos\psi \cos(\psi + \delta)} \times \left[1 + \sqrt{\frac{\sin(\phi + \delta)\sin(\phi - \psi)}{\cos(\delta + \psi)}}\right]^{-2} \tag{3.14}$$

where ϕ is the angle of internal friction of the retained fill available during the earthquake, and δ is the friction angle between the backfill and the wall. The coefficient above involves

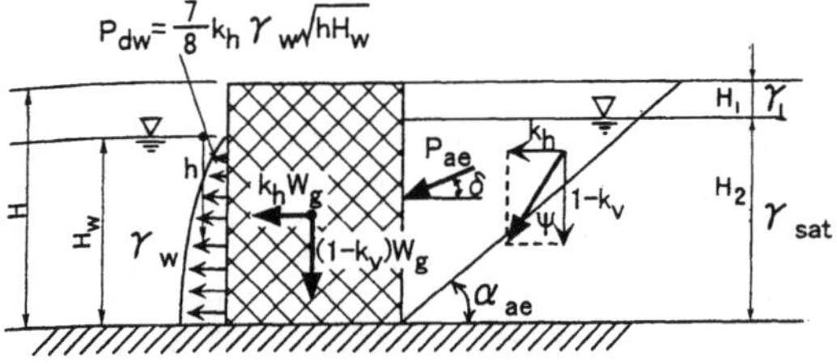

Figure 3.17 Nomenclature for seismic design of a gravity quay wall.

both static and dynamic effects, and it provides the total force acting at a level of about $0.4H$ to $0.45H$ from the base, H being the height of the fill.

A more recent approach (Seed and Whitman, 1970) distinguishes the static from the dynamic component of the total force in a solely horizontal shaking. In this method the coefficient associated with the dynamic component is given approximately by

$$\Delta K_{ae} = \tfrac{3}{4}k_h \qquad (3.15)$$

The dynamic component of the total thrust acts at a point $0.5H$ to $0.6H$ from the base of the wall.

In Mononobe–Okabe's expression, a complication arises due to the combination of static and dynamic components of the total force. The static component involves the unit weight of the buoyant fill since the hydrostic pressures are computed independently, whereas the dynamic component should be associated with the weight of the fill that participates in seismic movement. For saturated soil its is usually considered that pore water moves with soil particles. This discrepancy is accounted for in applying a modified horizontal seismic coefficient increased by the relation

$$k_h' = \frac{\gamma_{sat}}{\gamma_b} k_h \qquad (3.16)$$

and similarly for the vertical coefficient. It should be noted, however, that in most applications the vertical seismic excitation is ignored.

The relation (3.16) assumes uniform saturated fill along the total height of the wall. Since this is usually not the case, the modified seismic coefficient k_h' should take this into account. The corresponding modification of k_h is somewhat arbitrary, and therefore there are a few versions of expressing the above. One of these reads

$$k_h' = \frac{q + \gamma_1 H_1 + \gamma_{sat} H_2}{q + \gamma_1 H_1 + \gamma_b H_2} \qquad (3.17)$$

In this relation, which substitutes for the preceding one, q denotes the surcharge included in the seismic design, usually 50% of the surcharge used for static stability design, and γ_1, H_1, and H_2 are as shown in Figure 3.17. The modified seismic coefficient k_h' is used for calculation of the active pressure coefficient. In the case of combined static and dynamic components, as with Mononobe–Okabe's equation, the following relation is used to estimate the total Earth thrust:

$$P_{ae} = K_{ae}\tfrac{1}{2}\gamma_e H^2 \qquad (3.18)$$

where γ_e stands for an equivalent unit weight of a multilayered backfill. When two layers are present (Figure 3.17), the equivalent unit weight can be expressed by the weighted average of the unit weights γ_1 and y_b of the soil above and below the water table, respectively:

$$\gamma_e = \gamma_1 \left[1 - \left(\frac{H_2}{H}\right)^2\right] + \gamma_b \left(\frac{H_2}{H}\right)^2 \qquad (3.19)$$

In case it is required to include the effect of the vertical component of the seismic shaking, then the value of P_{ae} giving the total earth thrust through eq. (3.18) should be multiplied by the factor $(1 - k_v')$, where k_v' is a vertical seismic coefficient modified similarly to the horizontal coefficient.

It has been assumed in the discussion above that the retained backfill and the pore water act as a single body. Although this situation is quite common, there are cases of extremely coarse-grained fills in which soil skeleton and water can be considered as acting independently. It is quite complex to define the parameters and their values leading to such a state of independent movements. However, permeability is thought to be a key parameter, and for very high values the total earth thrust can be made of:

- A thrust from the soil skeleton computed by eq. (3.18) where k'_h is derived through eq. (3.17), in which γ_{sat} has been replaced by γ_1.
- A contribution from the dynamic water pressure, which can be estimated using Westergaard's expression. Due to the complicated nature of the cyclic motion of water into the voids of the soil skeleton and the fact that the two contributions to the total thrust do not act similtaneously, the part associated with the hydrodynamic pressure is taken to be reduced by, say, 30%.

Westergaard's expression can also be used to approximate the hydrodynamic suction in front of the wall. The total load is

$$P_w = \tfrac{7}{12} k_h \gamma_w H_w^2 \tag{3.20}$$

where γ_w is the unit weight of water and H_w is the height of the water column (Figure 3.17). Westergaard's load is applied at a distance $0.4H_W$ above the seabed.

Mononobe–Okabe earth pressures and simplified Seed and Whitman pressures can be computed using the static angle of internal friction of the backfill in cases where no significant earthquake-induced pore pressures are expected. In other cases, appropriately reduced resistance should be used. If liquefaction of backfill is expected, a conservative approximation of the pressures on the wall is that the backfill acts as a liquid having the unit weight of the saturated fill.

The dead loads of the wall will also be provided with a dynamic component. This is expressed as a product of the seismic acceleration by the mass of the shaken body (Figure 3.17). Additionally, other horizontal loads can be taken into account, such as a fraction of the horizontal bollard pull. The factor of safety F against earthquake hazard is defined as

$$F = \frac{k_t}{k_d} \tag{3.21}$$

where k_t the threshold seismic coefficient and k_d is the design seismic coefficient. The threshold seismic coefficient is the value of the seismic coefficient at which the limit equilibrium state occurs at the most critical failure mode. The design or effective seismic coefficient is the seismic coefficient used in the design of the quay wall. The latter value is normally lower than the one corresponding to the peak ground acceleration of the motion. An average relation between the design seismic coefficient and the peak ground acceleration a_{max} has been proposed on the basis of 129 case histories of gravity quay walls damaged ruring 12 earthquakes (Noda et al., 1975):

$$k_d = 0.6\left(\frac{a_{max}}{g}\right) \tag{3.22}$$

Similarly, Eurocode EC-8 (CEN, 1994) proposes a design seismic coefficient that depends on the acceptable wall displacements:

$$k_d = \frac{a_d/g}{r} \tag{3.23}$$

where a_d is the design acceleration and r is a reduction factor taking the values 2 for free gravity walls that can accept displacements up to $30a_d/g$ (cm), 1.5 for displacements up to $20a_d/g$ (cm), and 1.0 for other retaining structures allowing no displacements. Other intermediate values of the reduction factor can be adopted by the designer according to the prevailing conditions in each case.

Based on analysis of case histories in nonliquefiable sites, estimates of the horizontal displacement d normalized with respect to the wall height H and of the settlement s of a gravity quay wall were proposed by Uwabe (1983) as follows:

$$\frac{d}{H} = -7.0 + \frac{10.9}{F} \qquad (3.24)$$

$$s = -16.5 + \frac{32.9}{F} \quad \text{in cm} \qquad (3.25)$$

It is noted, however, that these predictions are too conservative and approaches based on simplified dynamic analyses should be preferred to estimate deformations of gravity quay walls due to seismic action. On liquefiable sites a rough estimation of wall displacement can be obtained from Table 3.9, which represents displacements observed in two levels of earthquakes: a threshold earthquake whose factor of safety F against earthquake hazard [eq. (3.21)] is approximately 1.0, and a severe earthquake whose factor of safety F [eq. (3.21)] is 0.5 to 0.67.

Evaluation of liquefaction potential can be done in a simplified manner by twice applying the procedure described in Section 3.5.3, once to the soil column in front of the wall and once to the backfill/natural soil column at the back of the wall, assuming in both cases horizontal soil layering and ground surface. In simplified dynamic analysis a more accurate picture of the seismic response of the quay wall is gained. However, more input information is required; this is usually furnished in the form of time histories of seismic acceleration representative of the site under examination. A sliding block analysis can then be appied to evaluate the permanent displacements, if any, of the wall (Newmark, 1965). Knowledge of the wall displacement is important in the design, since under several conditions, it might be acceptable as, for example, in cases where no damage to crane rails is induced by the wall displacement. A threshold acceleration a should be determined, given by the expression

$$\alpha_t = [V\mu - (\cos\delta - \mu \sin\delta)P_{ae} - P_w]\,\frac{g}{W} \qquad (3.26)$$

where μ is the coefficient of friction between the wall and the foundation material, W the (dry) weight of the wall, V the vertical stabilizing loads, P_{ae} the active earth thrust corresponding to a_t, and P_w the hydrodynamic load. Expression (3.26) requires an iterative procedure, since it contains two interrelated unknowns (a_t and P_{ae}). The threshold acceleration is then compared with the time history of excitation acceleration $a(t)$, and when the latter exceeds a_t the displacement relative to the rigid base can be obtained by double integrating the area between $a(t)$ and a_t. It should be empha-

Table 3.9 Normalized displacement of gravity walls at liquefiable sites.

	Normalized Displacement d/H (%)							
	During Threshold Earthquake				During Severe Earthquake[a]			
	0–5	5–10	10–20	20–40	0–5	5–10	10–20	20–40
Nonliquefaction	×					×		
Loose sand at backfill only		×					×	
Loose sand at both backfill and foundation			×					×

Source: PIANC (2001*b*).

[a] 1.5 to 2.0 times the threshold earthquake.

sized that to apply the sliding block analysis, several assumptions should be met: notably, rigid base, wall sliding without tilt, and rigid backfill movement.

Numerical studies based on the methodology described above have led to simplified expressions of the permanent displacement d (e.g., Richards and Elms, 1979), where $v_{\max}$ is the peak ground velocity and $a_{\max}$ is the peak ground acceleration:

$$d = 0.087 \frac{v_{\max}^2 a_{\max}^3}{a_t^4} \qquad \frac{a_t}{a_{\max}} \geq 0.3 \quad (3.27)$$

Another expression, recommended for design by Whitman and Liao (1985), is the following:

$$\frac{a_t}{a_{\max}} = 0.66 - \frac{1}{9.4} \ln \frac{d a_{\max}}{v_{\max}^2} \quad (3.28)$$

Some parametric graphs that can be used to estimate wall displacement have recently been developed based on results from dynamic analysis of a quay wall under various levels of shaking (Iai et al., 1999). The analysis was based on the multiple shear mechanism using the computer program FLIP under various simplifying assumptions.

In the simplified dynamic approach the evaluation of safety against liquefaction of foundation and backfill soils can be done basically by the procedures described in Section 3.5.3 but obtaining τ_{eq} from an analysis of the response by total stress procedures based on the hysteretic equivalent linear constitutive model of the soils. Soil resistance to liquefaction (τ_{rl}) can be deduced from the CRR described in Section 3.5.3, modified by considering the scaling factors: K_σ for high overburden stress and K_α for static shear stress. Youd et al. (2001) discuss these factors in detail. Liquefaction laboratory tests can help to assess the effects of the particular stress conditions under the wall.

Dynamic analysis generally uses finite element or finite differences techniques and involves soil–structure interaction. This analysis is site specific and the structure under examination can be idealized either as a linear or a nonlinear model, depending on the level of shaking relative to the elastic limit of the structure. Caisson gravity walls are usually modeled through linear models. Foundation soil and backfill are represented either through equivalent linear total stress or by means of effective stress models. In Section 3.5.8 some additional comments on these types of analysis are made. In dynamic analysis using the effective stress models, pore pressure buildup during earthquake and subsequent dissipation is obtainable directly.

3.5.5 Sheet Pile Quay Walls

Evaluation of the overall stability of anchored sheet pile walls under seismic loading permits a determination of the embedment length of the sheet piles and the anchor into the foundation soil, as well as the distance of the latter from the wall face. Evaluation of stresses induced in structural components of the wall permits the dimensioning of the sheet piles, the anchor, and the tie rod. The latter should be designed to fail last, whereas a balanced movement of the anchor can reduce the tie rod stresses considerably. However, there are design practices that allow for the wall to fail before the anchor, leading to large differences in the design forces applicable to tie-rods and anchors.

At the simplified analysis level, field data from nonliquefiable sites have been used by Gazetas et al. (1990) for estimating the degree of damage to anchored sheet pile walls due to earthquakes. The degree of damage was quantified according to the reported permanent displacement at the cap of the wall (Table 3.10). Two nondimensional factors were used to relate the damage incurred to the design characteristics of the sheet pile wall: the effective anchor index (EAI) and the embedment participation index (EPI). These are defined as follows:

Table 3.10 Degree of damage for sheet pile quay walls

Degree of Damage	Permanent Displacement at Top of Sheet Pile (cm)
0	<2
1	10
2	30
3	60
4	120

Source: Gazetas et al. (1990).

$$\mathrm{EAI} = \frac{D_{\mathrm{anc}}}{H} \tag{3.29}$$

$$\mathrm{EPI} = \frac{P_{pe}}{P_{ae}}\left(1 + \frac{D_f}{D_f + H}\right) \tag{3.30}$$

In general, earth and water pressures on the wall can be estimated following the procedures exposed in Section 3.5.4. The seismic active and passive earth pressures P_{ae} and P_{pe} can be calculated through relations (3.14) and (3.18), where for the passive pressure the minus sign should be placed in front of the square root. The remaining symbols are defined in Figure 3.18.

The depth of the effective point of rotation, D_f, can be approximated by

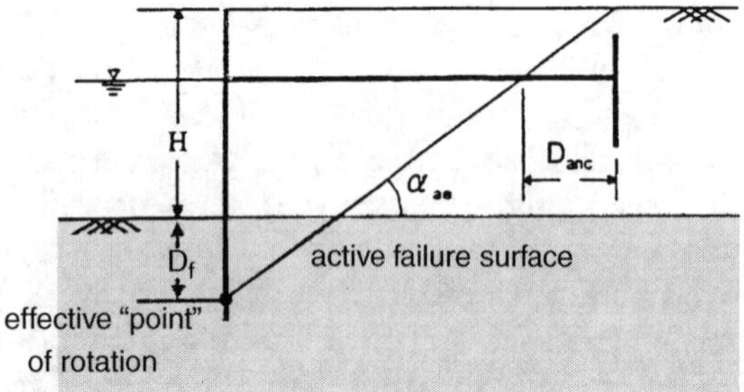

Figure 3.18 Definition sketch for anchored sheet pile quay wall.

$$D_f = [0.5(1 + k_h') - 0.02(\varphi - 20)]H \tag{3.31}$$

where φ is the angle of internal friction of the backfill soil in degrees and k_h' is the effective horizontal seismic coefficient. The angle of the active failure surface, α_{ae} (Figure 3.18), can be approximated by

$$\alpha_{ae}\ (\mathrm{deg}) = 45 - \frac{\varphi}{2} - 135(k_h')^{1.75} \tag{3.32}$$

Indices EAI and EPI can be used to provide a rough design chart related to the allowable displacement, as shown in Figure 3.19. On liquefiable sites a first approximation of the wall displacement can be obtained from Table 3.11, compiled from data referring to threshold earthquakes [i.e., with $F \approx 1.0$ in eq. (3.21)].

In simplified dynamic analysis the sliding block method can be used to evaluate the displacement of sheet pile quay walls due to seismic action. The concept is the same as that used in gravity quay walls (Section 3.5.4); however, there are differences related to the wall inertia and bottom friction forces that are not present in a sheet pile wall. To overcome these difficulties, Towhata and Islam (1987) assumed rigid-block motion of the pile–backfill system. Thus they obtained a threshold horizontal seismic coefficient, similar to eq. (3.26), reading

$$k_t = \frac{a \tan \alpha_{ae} - b + \tan(\varphi - \alpha_{ae})(1 + b \tan \alpha_{ae})}{1 + c \tan \alpha_{ae}} \tag{3.33}$$

where

$$a = \frac{mT_s + P_p + \frac{1}{2}\gamma_w(H_w + D_{\mathrm{emb}})^2 + \Delta U_p}{W_m}$$

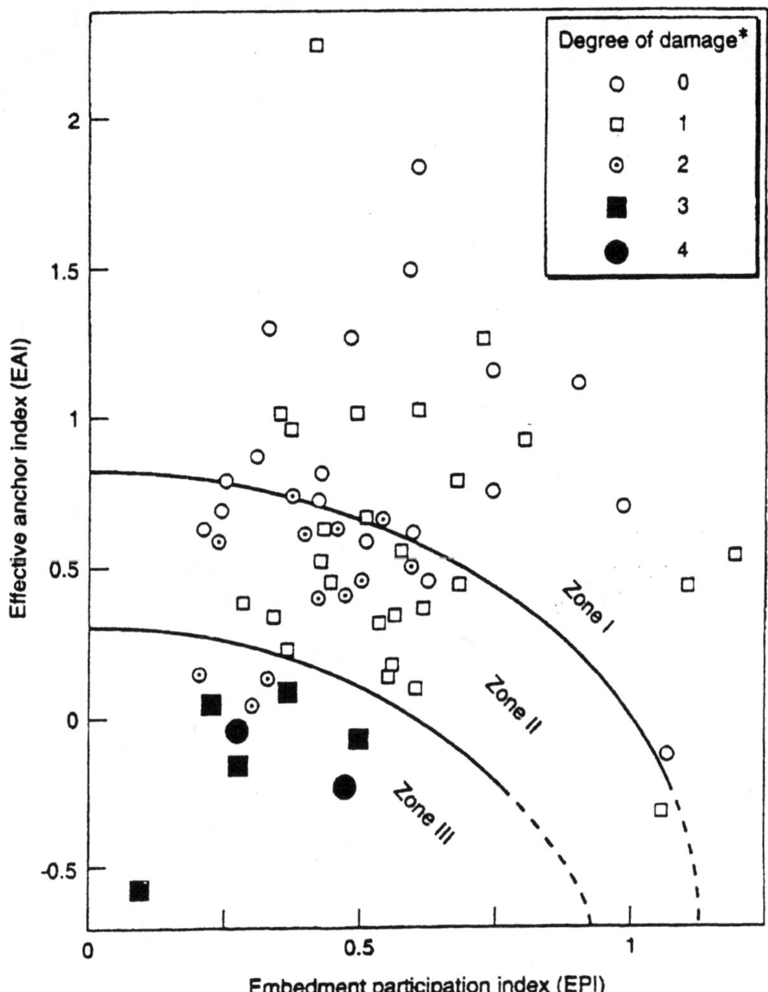

Figure 3.19 Rough design chart for sheet pile walls. (From Gazetas et al., 1990.)

Table 3.11 Displacement of sheet pile walls at liquefiable sites during threshold earthquake

	Normalized displacement d/H (%)			
	0–5	5–15	15–25	25–50
Nonliquefaction	×			
Loose sand behind the wall only		×		
Loose sand at backfill including anchor			×	
Loose sand at both backfill and foundation				×

Source: PIANC (2001b).

$$b = \frac{\frac{1}{2}\gamma_w(H_w + D_{emb})^2 \tan\varphi + \Delta U_a \sin\alpha_{ae}}{W_m} \times \tan\psi$$

$$c = \frac{1}{W_m}\left[\frac{23mnT_s}{8(K_p - K_a)} + \frac{17P_p\gamma_{sat}}{8K_p\gamma_b} + \frac{7}{12}\gamma_w H_w^2\right]$$

$$W_m = \tfrac{1}{2}[\gamma_{sat}(H_w + D_{emb})^2 + \gamma_{wet}(H - H_w)(H + 2D_{emb} + H_w)]$$

where m is a parameter depending on the anchor capacity (0 for no anchor capacity, 1 for full anchor capacity); P_p the static passive earth resistance; ΔU the excess pure water pressure due to cyclic shearing (subscript a for pressures within the active soil wedge, subscript p for pressures within the passive soil wedge); $n = 1$ when the anchor is above the water table and $n = \gamma_1/\gamma_b$ when the anchor is completely submerged, γ_b buoyant unit weight of soil; T_s the ultimate anchor resistance under static conditions; and K_a and K_p are the active and passive earth pressure coefficients under static conditions. The remaining parameters are as defined in Figure 3.20.

It was noted above that the passive earth pressure coefficient K_{pe} can be estimated by a similar expression to (3.14) due to Mononobe–Okabe. Alternatively, this coefficient can be decomposed into a static, K_p, and a dynamic component, ΔK_{pe}. Seed and Whitman (1970) proposed a simple expression for ΔK_{pe}, analogous to eq. (3.15):

$$\Delta K_{pe} = -\frac{17}{8}k_h \tag{3.34}$$

Also, the ultimate anchor resistance, T_e, during the earthquake can be calculated through

$$T_e = \frac{T_s(K_{pe} - K_{ae})}{K_p - K_a} \tag{3.35}$$

The seismic performance of sheet pile quay walls has been examined through a parametric study using a two-dimensional effective stress model (McCullough and Dickenson, 1998). It was found that the quality of backfill soil plays a crucial role in the seismic behavior of the wall, and that soil of improved characteristics, say, of 20 blows/30 cm, should extend roughly $2(H + D_{emb})$ behind the wall for minimal lateral displacements of the sheet pile wall. The same parametric study showed also that the displacements are more sensitive to earthquake motion than suggested by the graph of Figure

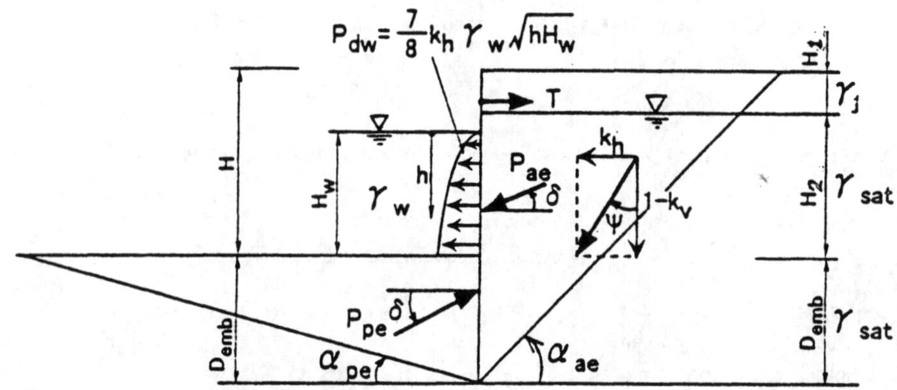

Figure 3.20 Seismic loading on a sheet pile quay wall.

3.19 and that the benefit of extending the sheet pile embedment and tie rod length beyond those normally specified by seismic design is not as substantial as indicated by the chart.

Dynamic analysis of sheet pile quay walls can be performed in a manner similar to that outlined in Section 3.4.1. For additional comments on dynamic analysis methods, refer to Section 3.5.8.

3.5.6 Pile-Supported Wharves

In this type of structure the primary interest in seismic analysis is the behavior of the pile–deck system under the inertia forces and its interaction with the slope dike. Structural deformations due to inertial effects develop during seismic movement. The worst situation occurs at the peaks of acceleration. Slope deformations develop progressively during an earthquake, but failure could happen some time after strong ground motions have ceased. This is particularly the case for liquefaction failures. Consequently, in most cases stresses and strains in the pile–deck system and those related to liquefaction at the dike slope can be evaluated independently. Structural configurations of wharves are frequently rather simple. However, complexity arises from soil–pile interaction, from significant torsional response resulting from the usual seaward increase in effective pile length, and from the behavior of shear keys between adjacent segments.

Detailed soil–pile system simulation would involve inelastic finite element modeling of the deck, the piles, and the foundation materials down to the bedrock, and laterally to a sufficient distance, as shown in Figure 3.21*a*. Two levels of simplification are used in practice to analyze the response of the pile–deck system. Figure 3.21*b* shows a model of piles with their points fixed to the ground and supported laterally by Winkler springs equivalent to soil. Seismic input is applied simultaneously and coherently to each spring fixed end and to each pile point. Deformability of lateral soil springs can be linear or defined by an appropriate set of nonlinear p–y curves. Figure 3.21*c* shows a more simplified model where no soil springs are considered but the effective pile lengths are selected to allow compatible horizontal displacements of the deck. This modeling can correctly predict elastic periods of vibration, horizontal displacements of the deck and stresses in the deck and in the pile cap. However, moments in the embedded part of the piles have to be evaluated afterward.

In simplified analysis, stresses in the pile–deck system are evaluated by the pseudostatic method or by single-mode spectral analysis. Stability of the dike slope can also be evaluated by a pseudostatic method. It is possible to take into account slope–pile interaction by introducing forces from piles to slope that help the stability of the latter and forces from slope to piles that introduce additional horizontal loads to the piles.

In *simplified dynamic analysis* a pushover analysis is used in conjunction with a single- or multiple-mode spectral analysis of a pile–deck system. The equivalent-depth-to-fixity or multiple-springs models shown in Figure 3.21 can be used to represent the soil–pile interaction.

Usually, the effective length of piles increases from land to sea. The stiffness of soil–pile system also varies, and consequently, the formation of plastic hinges in different piles will occur at different deck displacements. In such cases it is difficult to adequately represent the response to seismic input by an overall elastoplastic approach, which is the basis of spectral analysis with a force-reduction factor. Values of the elastic stiffness and force-reduction factor are not obvious. It is thus strongly recommended that one do a series of two-dimensional inelastic pushover analyses in which both transverse and longitudinal sections are subjected to incremental deck displacements, allowing an inelastic force-

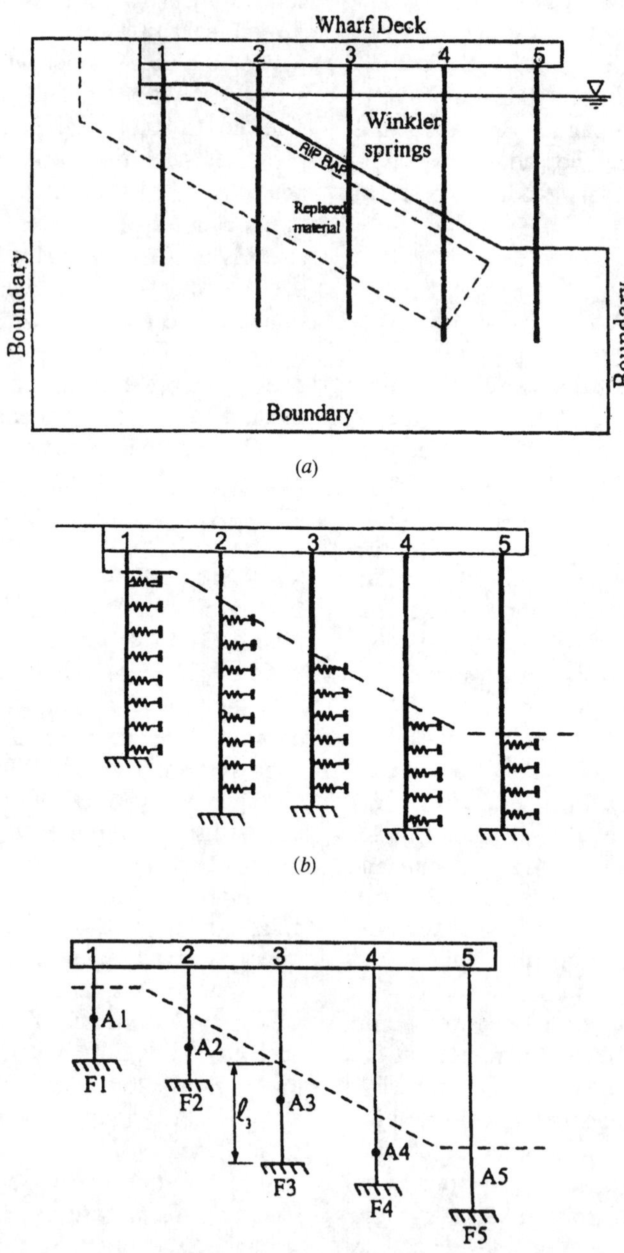

Figure 3.21 Modeling soil–structure interaction of a pile-supported wharf: (*a*) finite element modeling; (*b*) multiple-springs modeling; (*c*) equivalent-depth-to-fixity modeling.

displacements response (Figure 3.22). The results of such analysis can be used to determine the appropriate stiffness and damping in modal analysis.

In general, some iterations will be necessary to ensure compatibility between elastodynamic properties and results in modal analysis and in pushover analysis, but at the end of the process, a more realistic representation of peak response is achieved which eliminates the need to introduce force-reduction factors to convert to more realistic results the results of elastic analysis with initial stiffness.

The behavior of the dike slope for non-liquefiable sites can be evaluated by sliding-block, Newmark-type analyses, as described in Section 3.5.7 for slopes of breakwaters. Some comments on the evaluation of liquefaction potential of foundation and backfill soils are provided at the end of this section.

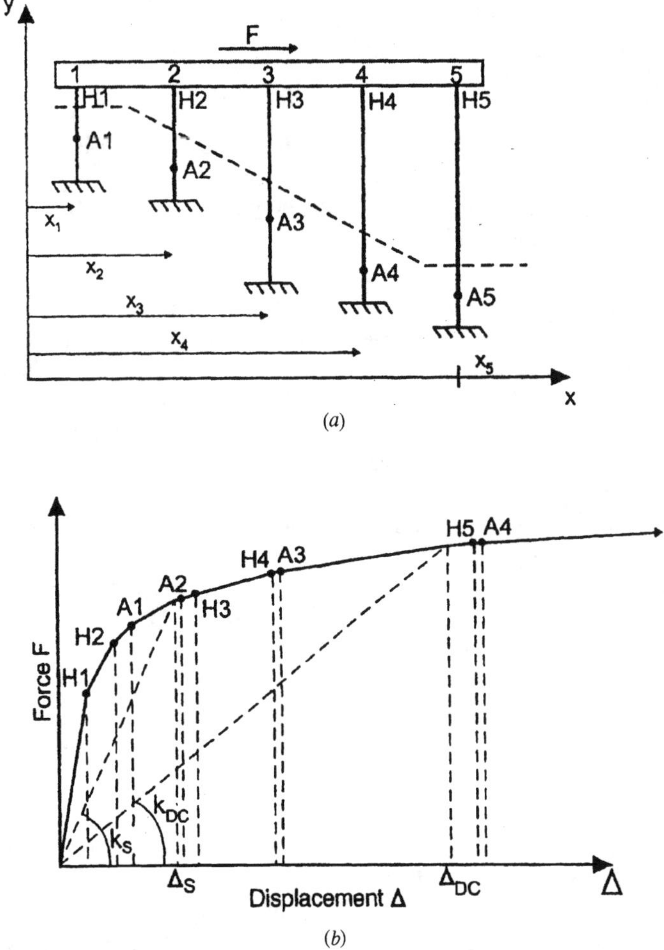

Figure 3.22 Pushover analysis of a wharf cross section: (*a*) section for analysis; (*b*) force–displacement response and hinge sequence.

Methods for dynamic analysis of pile-supported wharves include time-domain inelastic analysis of pile–deck system with the soil modeled by Winkler springs as shown in Figure 3.21*b*, and time-domain finite element analysis of overall model as shown in Figure 3.21*a*. The time-domain inelastic analysis of a pile–deck system is potentially the most accurate method for determining maximum displacements and inelastic rotations in plastic hinges. However, some difficulties arise from the large number of degrees of freedom to be included in the analysis, from the comparatively few three-dimensional time-domain computer programs available at present (January 2002) and from the rather limited experience in this type of analysis outside research-oriented applications.

The advantages and limitations of time-domain finite element analysis of an overall soil–structure system with the effective stress nonlinear constitutive relationships for the soil are as described in Section 3.5.8. This type of analysis allows the evaluation of the pore pressure buildup during an earthquake and of the liquefaction potential in foundation and backfill soils. If these are far enough from liquefaction, the structural analysis of a pile–deck system, eventually taking into account the effect on piles of permanent displacements of ground, produces an adequate description of wharf seismic performance.

The evaluation of safety against liquefaction of foundation and backfill soils is somewhat more complicated than in the case of horizontally layered soil with a horizontal surface, described in Section 3.5.3. Basically, the procedures described in Section 3.5.3 can be used, introducing in addition the effects of sloping geometry, and in some cases, of high overburden pressures. There is no simplified procedure equivalent to eq. (3.5) to compute the equivalent-to-earthquake shear stress (τ_{eq}). In simplified dynamic analysis, τ_{eq} can be obtained by total stress procedures using the hysteretic equivalent linear constitutive model of the soils. On the other hand, soil resistance to liquefaction deduced from the CRR described in Section 3.5.3 should be modified by considering the scaling factors K_σ for high overburden stress and K_α for static shear stress. Youd et al. (2001) present a detailed discussion of these factors. The effects of the particular stress conditions can be investigated similarly through liquefaction laboratory tests. In dynamic analysis using effective stress models, pore pressure buildup during earthquake and subsequent dissipation can be obtained directly.

3.5.7 Breakwaters

Breakwaters that are formed by precast gravity units can be designed seismically similarly to gravity quay walls. In most cases, however, no lateral earth load is exerted on this type of breakwater, in contrast with the corresponding load on quay walls. Rubble-mound breakwaters are designed to withstand wave loading, and their seismic analysis is often performed by simplified methods, due mainly to the simple repair practice of adding stones to restore the cross section to its previous shape. Due to the simplicity of the repair procedure, seismic evaluation of these structures is sometimes omitted. It is, however, recommended that in seismically active areas where, additionally, the foundation of rubble-mound breakwaters is made on weak soil, a proper seismic evaluation be performed even using simplified methods. Under the conditions described above, the most common mode of failure is sinking of the mound material into the weak subsoil and consequently, crest lowering and lateral spreading, associated in severe cases with characteristic toe bulging. If superstructure elements are present, these can possibly be tilted and/or displaced.

The simplified methods adopted in rubble-mound breakwaters are virtually those used in land embankments, with the modification of including an added mass due to hydrodynamic loading on the faces of the structure. Stability is assessed by a factor of safety obtained from limit equilibrium methods such as the Bishop or a similar method. To determine the minimum safety factor for the surface, gravitational and pseudostatic equivalent inertia forces are taken into account. The seismic acceleration along the height of the mound can increase from the acceleration at the toe a_t to that at the crest a_c. Usually, a_t is assumed to be half the maximum acceleration a_{max} specified for the particular site under consideration, while a_c includes the appropriate amplification of acceleration along the height of the mound. In absence of any other information, a_c can be taken equal to a_{max}. The effect of the added mass produced by hydrodynamic loading can be approximated in the simplified approach by Westergaard's formula applicable to a rigid body with vertical faces:

$$p(z) = \tfrac{7}{8}a\gamma_w(Hz)^{1/2} \tag{3.36}$$

where z is the depth of the layer under consideration, a the design horizontal acceleration, γ_w the unit weight of water, and H the total depth. Integration of the pressure over the height associated with the potential slip surface under examination, assuming a mean value of the design acceleration, gives the additional inertia load due to the hydrodynamic pressures. In simplified dynamic analyses, enhancement of the pseudostatic approach is made by including estimation of the permanent deformations that may occur due to seismic shaking.

For soils that do not develop large pore pressures and maintain most of their original resistance after earthquake shaking, a Newmark-type sliding-block analysis can be used (Newmark, 1965), assuming that the soil behaves as a rigid, perfectly plastic material. Makdisi and Seed (1978) have developed a procedure that has been used afterwards in practice. It contains the following main steps:

1. Search for the critical sliding wedges and the threshold seismic coefficients, k_y, in each wedge, for whose the safety coefficient against sliding equals 1.0. Values of k_y can be calculated using conventional limit equilibrium methods.
2. Determine the time history of earthquake-induced average acceleration or the equivalent seismic coefficient $k(t)$ in the sliding wedge. Two-dimensional equivalent linear total stress finite element algorithms or a modal analysis of a simple shear beam model can be used to compute the seismic response of the rubble mound.
3. Calculate the permanent deformations. The soil wedge slides with respect to its base when the equivalent seismic coefficient, $k(t)$, is larger than the threshold seismic coefficient, k_y. As shown in Figure 3.23, displacement of sliding wedge relative to its base is computed by double integration of $[k(t) - k_y]g$, where g is the acceleration due to gravity. Some corrections should be made to take into account the direction of seismic acceleration and the direction of slide.

As a first approximation, the residual displacement can be obtained from Figure 3.24. The use of Newmark analysis when large pore pressure buildup occurs is limited by the difficulty in assigning appropriate values to the resistance parameters of the soil. Also progressive failure in soils with strain softening cannot be represented easily by this type of analysis.

Dynamic analyses for rubble-mound breakwaters are based on numerical models that in-

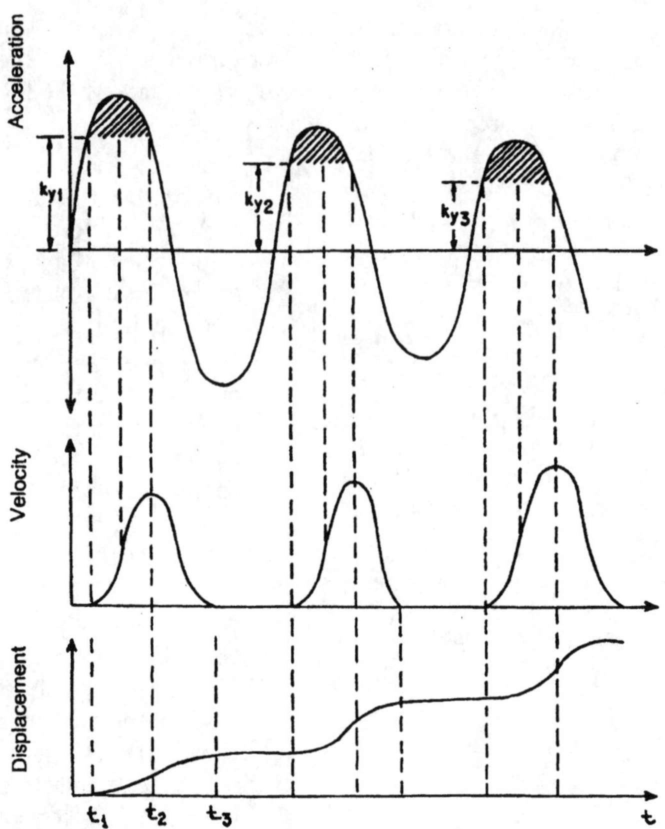

Figure 3.23 Residual earthquake-induced displacement obtained by double integration of acceleration. (After Seed, 1979.)

corporate calculation of the deformations of both the surrounding soil and the structure itself. Some comments on such types of analysis are presented in Section 3.5.8. In dynamic analyses, another refinement over simplified methods can be made related to estimation of the hydrostatic pressures and hence of the added-mass inertia component. Indeed, since rubble mounds are flexible rather than rigid structures inducing a height-wise modification of the horizontal acceleration, Westergaard's expression assuming constant acceleration is no longer valid. Instead, a site-specific numerical code, based, for example, on boundary element techniques, can be used to estimate more accurately the hydrodynamic pressures (Memos et al., 2001). Furthermore, this model can be coupled with the corresponding geotechnical model to account for interaction between the added mass due to hydrodynamic pressures and the resulting accelerations along the slopes of the breakwater.

3.5.8 Additional Comments on Dynamic Analysis Methods for Soils

Both total stress and effective stress constitutive relationships can be used for seismic anal-

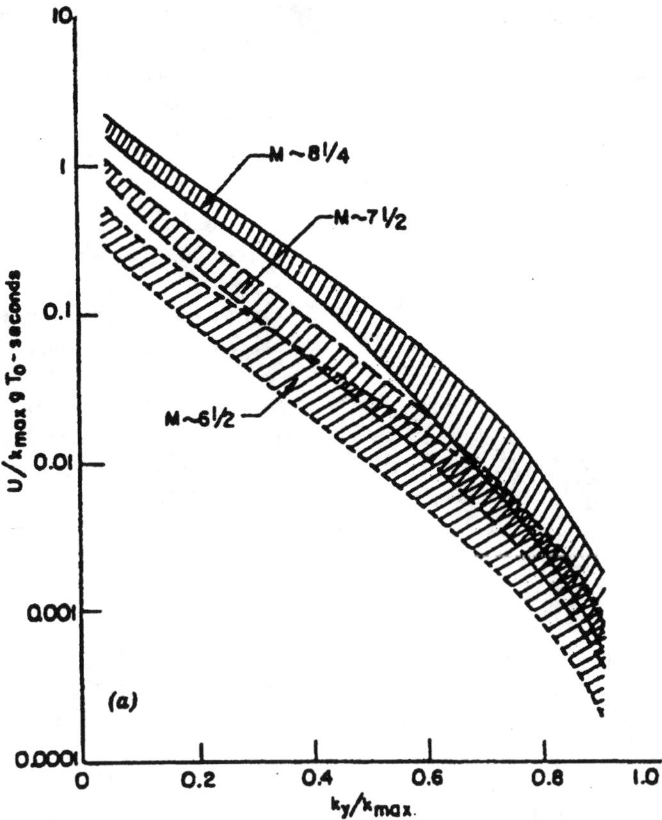

Figure 3.24 Normalized permanent displacement for earthquakes of different magnitudes. (After Makdisi and Seed, 1978.)

ysis of soils. Finite element or finite difference schemes are used to solve the elastodynamic problem. The principal concern in simulating soil behavior is its nonlinearity (i.e., the variation of shear modulus and damping with the level of shear strain) and also, the degradation of resistance if pore pressures develop during shaking. Total stress analysis has been the most widely used method. Computations are made preferably in the frequency domain. Soil is usually modeled by the equivalent linear hysteretic constitutive relationship, which does not include the changes in pore pressure and effective stress during shaking, and consequently, the changes in soil stiffness and strength. It allows the computation of histories of acceleration, stress, strain, and so on, but does not give any information about residual displacement or pore pressure change. The latter should be computed after obtaining the results of total stress analysis. In cases where low pore pressures develop, total stress analysis procedures supply the information needed to assess whether the stresses in the soil have reached the resistance envelope or to compute average acceleration in potentially sliding wedges. If the total analysis predicts clearly that liquefaction will occur, some remediation measurements should be taken without the need for more refined analysis.

When more precise information about the amount and time of pore pressure buildup is needed, a time-domain effective stress analysis is necessary. Soil is modeled by a nonlinear constitutive relationship with a simultaneous mechanism of pore pressure generation, which should be the one that better represents the physics of soil nonlinear stress–strain behavior and pore pressure buildup. Through the analysis above, pore pressure buildup and residual displacements after the earthquake can be obtained directly. Consequently, effective stress analysis methods are the most adequate for analyzing soil behavior during earthquakes. Many computer codes have been developed and utilized in practice incorporating an effective stress soil model. However, up to now, there has been limited experience on this soil constitutive model, and very few of the computer codes that implement it have been validated using well-documented case studies of the seismic performance of actual port structures. Therefore, the adequateness of their results in describing the actual behavior of port structures is not well established, and the uncertainty involved in their use is often unknown.

REFERENCES

Abrahamson, N. A., and W. J. Silva, 1997. Empirical response spectral attenuation relations for shallow crustal earthquakes, *Seismol. Res. Lett.*, Vol. 68, No. 1.

Anderson, J. G., 1991. Guerrero accelerograph array: seismological and geotechnical lessons, *Geotech. News,* Vol. 9, No. 1, pp. 34–37.

Andrus, R. D., and K. H. Stokoe, 2000. Liquefaction resistance of soils from shear-wave velocity, *ASCE J. Geotech. Geoenviron. Eng.,* Vol. 126, No. 11, pp. 1015–1025.

Bea, R. G., 1997. Background for the proposed International Standards Organization reliability based seismic design guidelines for offshore platforms, *Proc. Earthquake Criteria Workshop: Recent Developments in Seismic Hazard and Risk Assessments for Port Harbor and Offshore Structures,* Port and Harbour Research Institute, Yokosuka, Japan, and University of California, Berkeley, CA, pp. 40–67.

CEN, 1994. *Eurocode and Design Provisions for Earthquake Resistance of Structures;* Part 1–1: *General Rules: Seismic Actions and General Requirements for Structures* (ENV-1998-1-1); Part 5: *Foundations, Retaining Structures and Geotechnical Aspects* (ENV 1998-5), European Committee for Standardization, Brussels, Belgium.

EAU, 1996. *Recommendations of the Committee for Waterfront Structures, Harbours and Waterways,* 7th English ed. (English translation of the 9th German ed.), Ernst & Sohn, Berlin, 599 pp.

Ferritto, J. M., 1997. *Design Criteria for Earthquake Hazard Mitigation of Navy Piers and Wharves,* TR-2069-SHR, Naval Facilities Engineering Service Center, Port Hueneme, CA, 180 pp.

Gazetas, G., P. Dakoulas, and K. Dennehy, 1990. Empirical seismic design method for waterfront anchored sheetpile walls, *ASCE Proc. Specialty Conference on Design and Performance of Earth Retaining Structures,* ASCE Geotech. Spec. Publ. 25, pp. 232–250.

Grünthal, G. (ed.), 1998. *European Macroseismic Scale 1998,* Cahier du Centre Européen de Géodynamique et de Séismologie, Luxembourg, Vol. 15, 99 pp.

Gutenberg, B., and C. F. Richter, 1944. Frequency of earthquakes in California, *Bull. Seismol. Soc. Am.,* Vol. 34, pp. 1985–1988.

Iai, S., K. Ichii, Y. Sato, and H. Liu, 1999. Residual displacement of gravity quaywalls: parameter study through effective stress analysis, *Proc. 7th U.S.–Japan Workshop on Earthquake Resistant Design of Lifeline Facilities and Countermeasures against Soil Liquefaction,* MCEER-99-0019, Seattle, WA, pp. 549–563.

Idriss, I. M., 1991. Earthquake ground motions at soft soil sites, *Proc. 2nd International Conference on Recent Advances in Geotechnical Earthquake Engineering and Soil Dynamics,* St. Louis, MO, Vol. III, pp. 2265–2272.

Kinoshita, S., 1998. Kyoshin Net (K-NET), *Seismol. Res. Lett,* Vol. 69, No. 4, pp. 309–332.

Makdisi, F. I., and H. B. Seed, 1978. Simplified procedure for estimating dam and embankment earthquake-induced deformations, *ASCE J. Geotech. Eng. Div.,* Vol. 104, No. 7, pp. 849–867.

McCullough, N. J., and S. E. Dickenson, 1998. Estimation of seismically induced lateral deformations for anchored sheetpile bulkheads, in *Geotechnical Earthquake Engineering and Soil Dynamics III,* Geotech. Spec. Publ. 75, American Society of Civil Engineers, Reston, VA, pp. 1095–1106.

Medvedev, S. V., and V. Sponheuer, 1969. Scale of seismic intensity, *Proc. 4th World Conference on Earthquake Engineering,* Santiago, Chile, pp. 143–153.

Memos, C. D., A. Kiara, and C. Vardanikas, 2001. Hydrodynamic loading on rubble-mound breakwaters due to seismic shaking, *Proc. 24th IAHR Congress,* Beijing, China, Sept. 16–21, Theme E, pp. 134–140.

Ministry of Transport, 1999. *Design Standard for Port and Harbour Facilities and Commentaries,* Japan Port and Harbour Association, Tokyo, 1181 pp. (in Japanese); English edition (2001) by the Overseas Coastal Area Development Institute of Japan, Tokyo.

Mononobe, N., 1924. Considerations on vertical earthquake motion and relevant vibration problems, *J. Jpn. Soc. Civil Eng.,* Vol. 10, No. 5, pp. 1063–1094 (in Japanese).

Newmark, N. M., 1965. Effects of earthquakes on dams and embankments, 5th Rankine lecture, *Geotechnique,* Vol. 15, No. 2, pp. 139–160.

New Zeland Standards, 1992–1997. NZS 4203 (1992), *General Structural Design Loadings for Buildings;* NZS 3101 Part 1 (1995), *The Design of Concrete Structures;* NZS 3403 Part 1 (1997), *Steel Structures Standard;* NZS 3403 Part 2 (1997), *Commentary to the Steel Structures Standard;* Transit New Zeland (TNZ), *Bridge Design Manual.*

Noda, S., T. Uwabe, and T. Chiba, 1975. Relation between seismic coefficient and ground acceleration for gravity quay wall, *Rep. Port Harbour Res. Inst.,* Vol. 14, No. 4, pp. 67–111 (in Japanese).

Ohmachi, T., 1999. Formulation of level 2 earthquake motions for civil engineering structures, *Proc. 7th U.S.–Japan Workshop on Earthquake Resistant Design of Lifeline Facilities and Countermeasures against Soil Liquefaction,* MCEER-99-0019, pp. 497–506.

Okabe, N., 1924. General theory on earth pressure and seismic stability of retaining wall and dam, *J. Jpn. Soc. Civil Eng.*, Vol. 10, No. 6, pp. 1277–1323.

PIANC, 2001*a*. *Seismic Design Guidelines for Port Structures*. Rep. Work. Group 34, A. A. Balkema, Rotterdam, The Netherlands, 474 pp. 2001b.

———, 2001b. *Seismic Design Guidelines for Port Structures,* Suppl. Bull. 106, Permanent International Association of Navigation Congresses, Brussels, Belgium, 50 pp.

Puertos del Estado, 2000. *ROM 0.6: Acciones y Efectos Sismicos en las Obras Maritimas y Portuarias,* Madrid.

Richards, R., Jr., and D. Elms, 1979. Seismic behavior of gravity retaining walls, *ASCE J. Geotech. Eng. Div.,* Vol. 105, No. GT4, pp. 449–464.

Richter, C. F., 1958. *Elementary Seismology,* W. H. Freeman, San Francisco.

Robertson, P. K., and C. E. Wride, 1998. Evaluating cyclic liquefaction potential using the cone penetration test, *Can. Geotech. J.,* Vol. 35, No. 3, pp. 442–459.

Seed, H. B., 1979. Considerations in the earthquake-resistant design of earth and rockfill dams, 19th Rankine lecture, *Geotechnique,* Vol. 29, No. 3, pp. 215–263.

Seed, H. B., and I. M. Idriss, 1971. Simplified procedure for evaluating liquefaction potential, *ASCE J. Soil Mech. Found. Div.,* Vol. 97, No. SM9, pp. 1249–1273.

Seed, H. B., and R. V. Whitman, 1970. Design of earth retaining structures for dynamic loads, *ASCE Proc. Specialty Conference on Lateral Stresses in the Ground and Design of Earth Retaining Structures,* Ithaca, NY, pp. 103–147.

Seed, H. B., K. Tokimatsu, L. F. Harfer, and R. M. Chung, 1985. The influence of SPT procedures in soil liquefaction resistance evaluations, *ASCE J. Geotech. Eng.,* Vol. 111, No. 12, pp. 1425–1445.

Technical Committee for Earthquake Geotechnical Engineering (TC4), 1999. *Manual for Zonation on Seismic Geotechnical Hazards,* revised version, ISSMGE(TC4), International Society of Soil Mechanics and Geotechnical Engineering, Japanese Geotechnical Society, Tokyo, 209 pp.

Towhata, I., and S. Islam, 1987. Prediction of lateral movement of anchored bulkheads induced by seismic liquefaction, *Soils Found.,* Vol. 27, No. 4, pp. 137–147.

Uwabe, T., 1983. *Estimation of Earthquake Damage Deformation and Cost of Quaywalls Based on Earthquake Damage Records,* Tech. Note 473, Port and Harbour Research Institute, Yokosuka, Japan, 197 pp. (in Japanese).

Werner, S. D. (ed.), 1998. *Seismic Guidelines for Ports,* Monogr. 12, Technical Council on Lifeline Earthquake Engineering, American Society of Civil Engineers, Reston, VA.

Whitman, R. V., and S. Liao, 1985. *Seismic Design of Retaining Walls,* Misc. Paper GL-85-1, U.S. Army Corps of Engineers Waterways Experiment Station, Vicksburg, MS.

Wood, H. O., and F. Neumann. 1931. Modified Mercalli intensity scale of 1931, *Bull. Seismol. Soc. Am.,* Vol. 21, pp. 227–283.

Wu, S. C., C. A. Cornell, and S. R. Winterstein 1995. A hybrid recurrence model and its implication on seismic hazard results, *Bull Seismol. Soc. Am.,* Vol. 85, pp. 1–16.

Youd T. L., I. M. Idriss, R. D. Andrus, I. Arango, G. Castro, J. T. Christian, R. Dobry, W. D. Liam Finn, L. F. Harder, Jr., M. E. Hynes, K. Ishihara, J. P. Koester, S. S. C. Liao, W. F. Marcuson, G. R. Martin, J. K. Mitchell, Y. Moriwaki, M. S. Power, P. K. Robertson, R. B. Seed, and K. H. Stokoe, 2001. Liquefaction resistance of soils: summary report from the 1996 NCEER and 1998 NCEER /NSF workshops on evaluation of liquefaction resistance of soils, *ASCE, J. Geotech. Geoenviron. Eng.,* Vol. 127, No. 10, pp. 817–903.

4

LONG-TERM STRENGTH OF WATERFRONT STRUCTURES BUILT ON CREEP-SENSITIVE SOILS

Michael A. Gurinsky
Engineering Consultant
Fair Lawn, New Jersey

4.1 INTRODUCTION

Soils, especially cohesive soils, exhibit viscous creep behavior in which deformation and movement proceed under a state of constant stress or load. The theory of creep is a branch of the mechanics of soils treating the entire range of the phenomena resulting from changes in the state of stress and strain of a body as a function of time. The creep stress may ultimately lead to structural failure. The typical creep rate behavior of soils and deformation-dependent models have been discussed by Murayama and Thibata (1958), Scott (1963), Budin (1969, 1982, 1983), Budin et al. (1975), Budin and Demina (1979), Budin and Gurinsky (1985), Lacerda (1976), Pusch and Feltham (1981), Vyalov (1986), and others. Recent discussions on soil creep may be found in Kuhn and Mitchell (1993) and Gurinsky (1996, 2001, 2002).

The graphs in Figure 4.1 depict the way that creep deformation is developing in time. In this figure, the straight line 1, characterizing the development of deformation at a constant rate, represents a viscous flow; curve 3 corresponds to an attenuating creep at a decreasing rate, and curve 2 indicates a nonattenuating (progressive) creep. Since the rate of deformation in-

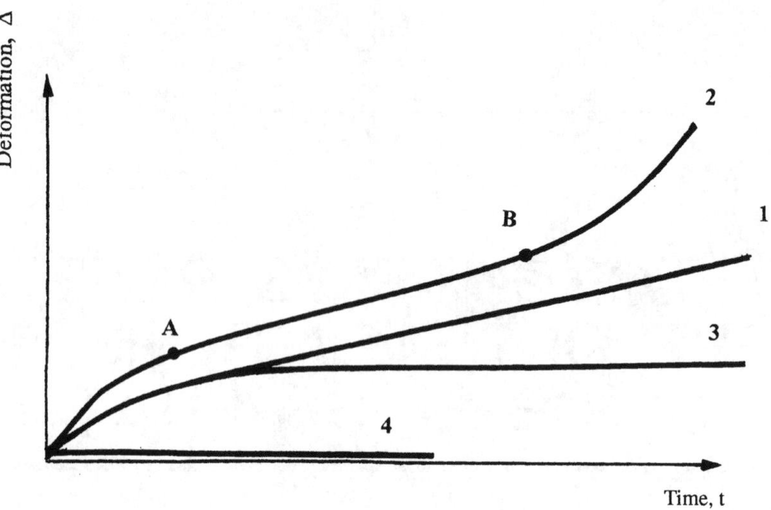

Figure 4.1 Development of creep deformation in soil with a time. 1, Steady-state creep; 2, nonattenuating creep; 3, attenuating creep; 4, creep process does not occur.

side segment *AB* of the latter curve is almost constant, this stage of the process can be regarded, by analogy with straight line 1, as one at which the flow is viscoplastic in nature. The configuration of creep curves normally depends on the magnitude of load (stress). If applied load $P < P_{th}$, creep process does not occur. Here, P_{th} is the initial creep threshold. If $P_{th} < P < P_{th1}$, creep process occurs and corresponds to an attenuating creep at a decreasing rate (curve 3). In the above, P_{th1} is creep threshold. By definition, P_{th1} is a constant threshold load (stress) beyond which steady-state creep process occurs (i.e., creep deformation develops at a constant rate). Therefore, if $P > P_{th1}$, a steady-state creep process (Figure 4.1, curve 1) is characterized by the development of deformation at a constant rate.

A point to be made is that unlike the theory of viscoplastic flow describing a steady-state process developing at a constant rate, the process of creep in soils is characterized by a variable rate of flow (as shown in Figure 4.1). Moreover, the latter process is associated not only with viscous deformations of soil but also with elastic deformations. This gives the reason to treat soils on the basis of the theory of elastoviscoplastic deformation, employing various rheological models.

Laboratory and theoretical studies of general behavior of creep soils under stress indicate the effects of their various viscous parameters on the rate of shearing deformations of the material that follows the applied stress. If the direct proportionality between the deformation velocity gradient normal to the sharing surfaces and applied shear stress is assumed as a quantitative description of the quality of the creep soil, the material is said to be proportionally constant.

The viscosity coefficient, η, is normally determined from tests of undisturbed soil samples using a special testing apparatuses and procedures. Descriptions of this apparatus as well as soil testing methods and procedures are found in most standard texts on soil mechanics. For preliminary calculations, the approximate values of η provided in Table 4.1 can be used. At $\eta \geq 10^6$ kN · day/m^2, the foundation is usually considered as noncreep.

Table 4.1 Approximate values of viscosity coefficient η of creep soil

Soil	η (kN · day/m²)
Clay of high plasticity	2.3×10^4
Clay of medium plasticity	2.5×10^4
Clay of low plasticity	2.3×10^4

For a detailed discussion on the rheological properties of creeping soils, the reader is referred to the fundamental works by Scott (1963) and Vyalov (1986) or any other relevant source. Peck (1948) reported that in the United States, long-term deformations of retaining walls constructed on creep soils resulted in an 18% failure of the slopes that these walls have been built to support. According to Peck, in 53 cases the walls were in the state of progressive displacement, 4% of the walls inspected showed significant displacement, and only 11% were in a state of stabilized displacement.

Similar phenomena were reported by Peterson (1953). Geuze and Tan (1954) reported a long-term displacement of a bridge pier resting on a 1.5-m bed of clay and peat. Haefeli et al. (1953) investigated deformations of the reinforced concrete viaduct caused by the creep of the slope-supporting abutment (Figure 4.2). The abutment had given in almost immediately on completion of the bridge and went on moving at a steady rate of approximately 37 mm per year. In 13 years the total displacement amounted to 480 mm, whereas the settlement was equal to only 14.9 mm, despite the fact that the layer of soil affected by the creep was 12 m deep. Peynircioglu (1957) provided details on displacements involving a 6.3-m retaining wall constructed in Devonian clays to hold back a steep slope in the Bosporus region. The slope that was cut off by the wall started to slide 2 years after completion of the construction and went on displacing at a daily rate of 1 mm until the wall has suffered substantial damage. Luga (1964) reported a case in which three bridge abutments constructed on stiff and hard clay have moved through horizontal distances ranging from 9.3 to 46 cm in 38 years (1916–1954). Mogilevskaya (1968) reported the case of long-term deformation resulting from soil creep at thc Farkhad hydropower station. There, a high frontal wall of the forebay erected on clay loam and sandy loam was displaced through 140 mm between 1947 and 1966.

4.1.1 Long-Term Deformation of Port Structures

Perhaps the most comprehensive investigation of soil retaining sheet pile walls built on creep-sensitive soils has been conducted by Budin and his colleagues (Budin, 1982). Two of these structures are depicted in Figure 4.3. In general, it was found that the creep of foundation soil caused a relaxation of the reaction stresses at the bottom of the sheet pile walls with the

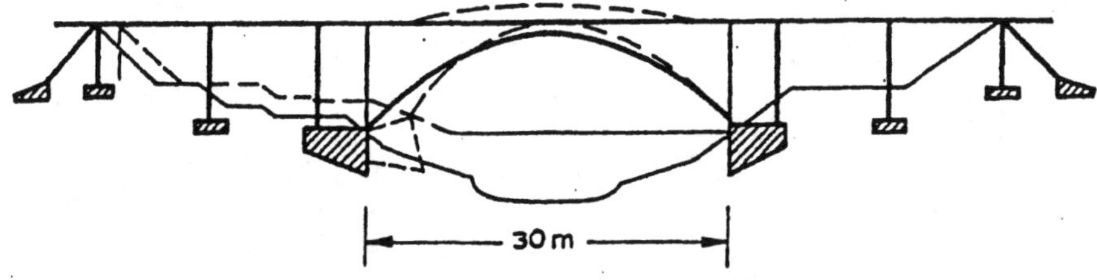

Figure 4.2 Displacement of a viaduct abutment in Switzerland due to creep of soil. (From Haefeli et al., 1953.)

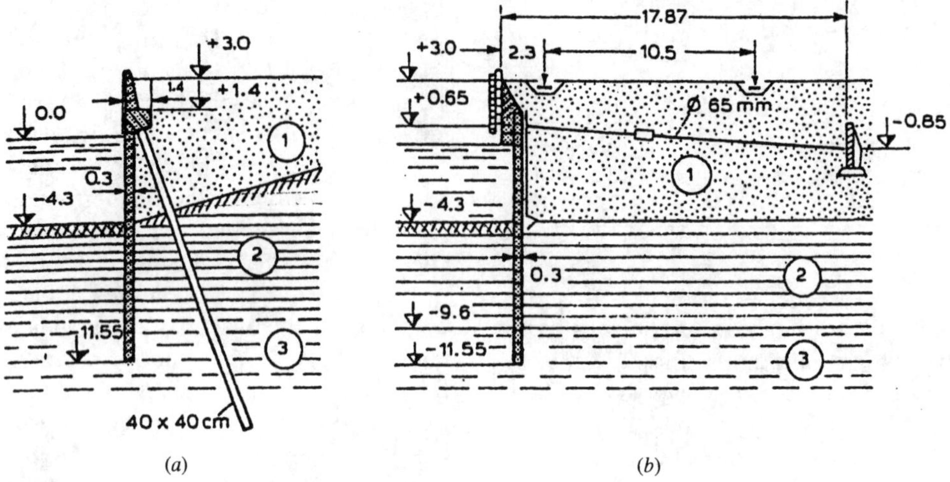

Figure 4.3 Sheet pile retaining walls constructed on clayey soils. 1, Sandback fill; 2, clay; 3, clay loam.

bending moments applied to the sheeting increased. For the structure depicted in Figure 4.3*a*, the field observations and measurements of deflections clearly illustrated the dependence of stress–deformation conditions on time. The top of this wall underwent a total displacement of 95 mm during the observation period (5 years and 5 months), of which 17 mm was registered almost immediately after the load was applied (Figure 4.4). For the structure depicted in Figure 4.3*b*, during the observation period, which lasted more than 4 years, the various sections of the wall deflected between 7 and 21.5 mm. This was followed by an increase in

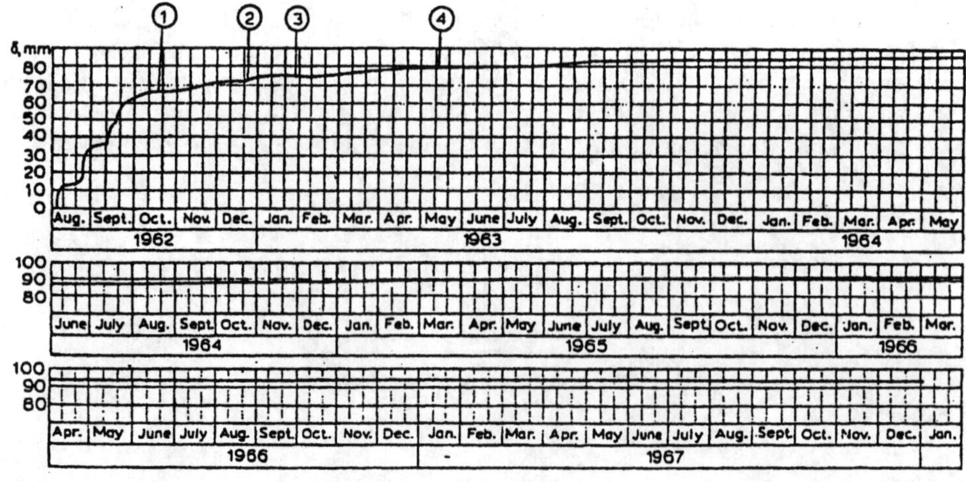

Figure 4.4 Displacements of sheet pile wall at the top: 1, Time of project completion uniformly distributed loadings; 2, $p = 0.16 \times 10^5$ Pa; 3, $p = 0.42 \times 10^5$ Pa; 4, $p = 0.5 \times 10^5$ Pa. (From Budin, 1982.)

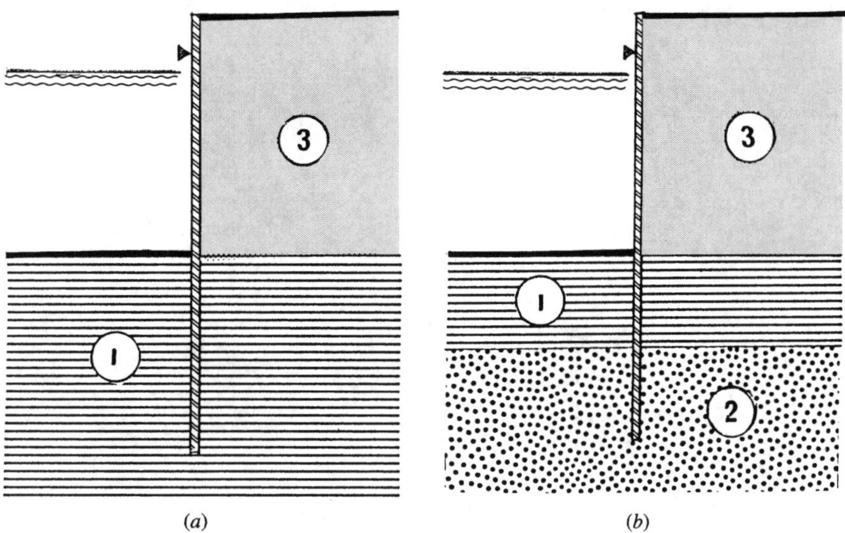

Figure 4.5 Practical cases of sheet pile wall construction in creep soil. 1, Creep soil; 2, underlaying noncreep soil; 3, granular backfill material.

anchor forces in the tieback system and substantial increase in bending stresses in sheet piling.

4.2 SHEET PILE BULKHEADS CONSTRUCTED ON CREEP-SENSITIVE SOIL

Numerous field investigations and observations of performance of sheet pile bulkheads built on creep-sensitive soils indicated the dependence of stress–deformation conditions of sheet pile bulkheads over time. The practice of using sheet pile bulkheads built on clayey soils usually shows steady increase in stresses in the sheeting over time under sustained loads. Sheet pile bulkheads constructed on soils with pronounced rheological properties are especially sensitive to postconstruction deformation of the foundation soil, which in some cases may lead to the failure of the structure.

Structural behavior of sheet pile walls built on creep-sensitive soils is best explained by relaxation of the soil passive resistance at the embedded part of the wall. The latter occurs due to soil "creeping off" the wall. As a result, the intensity of the passive soil pressure decreases, thus increasing stresses, bending moments, anchor forces, and deformations in the sheeting over time. The interaction between the sheet pile wall and the creeping foundation soil depends on the composition of the soil layers in which the wall is embedded. In practice, there are two basic scenarios of sheet pile bulkhead construction:

1. The embedded part of the wall is completely situated in the creep foundation soil (Figure 4.5*a*).
2. The embedded part of the wall extends through the top layer of a creep soil of limited depth and terminates embedded in a layer of a noncreep soil (Figure 4.5*b*).

In general, continuous creep soil foundations are not suitable for the long-term service

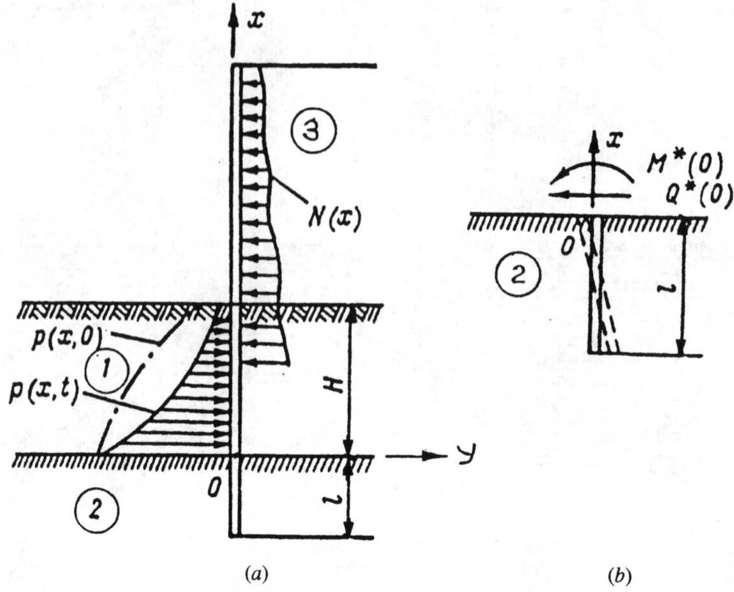

Figure 4.6 Cantilever sheet pile bulkhead installed in creep soil: (*a*) typical cross section; (*b*) embedded part fixed in creep soil. 1, Creep soil; 2, underlaying noncreep soil; 3, granular backfill. (From Budin, 1969.)

sheet pile walls because the strength of soft creeping soils usually does not provide sufficient passive resistance for wall long-term stability. In some practical cases, the sheet piling was driven into a lower layer of noncreep soil. This practice has, however, led to radical changes in time-dependent stress–strain conditions in wall components, manifested by overstress of the sheeting and overloading of the anchor system. The latter must not be overlooked by the designer.

On the basis of laboratory investigation and full-scale observation, Budin (1969, 1982) developed a conceptual theoretical approach to the design process of sheet pile bulkheads constructed on creep foundation soil. He has established that in the case of an interbedded creeping base, the process of relaxation of contact reactive pressures of the soil, $p(x,t)$, that is applied onto a flexible sheet pile wall structure with rigidity EI could be described by the following equation:

$$\frac{\partial^2 p(x,t)}{\partial x^2} = \frac{H\eta}{EI}\frac{\partial p(x,t)}{\partial t} \tag{4.1}$$

where x is the vertical coordinate, t the time, H the depth of the layer of creeping soil, η the viscosity coefficient of creeping soil (kN · day/m^2), E is Young's modulus of sheet pile material, and I the moment of inertia of the sheet pile.

The basic assumptions for eq. (4.1) are as follows (Figure 4.6):

1. The passive pressure diagram $p(x,0)$ is obtained from the conventional static analysis for the initial stage of soil–structure interaction ($t = 0$).
2. The load in active zone $N(x)$ is constant and is not time dependent.

Provided that all soil and structural data required are available, eq. (4.1) can be used to compute the distribution of bending moments $M(x,t)$, shear forces, and deflections in a sheet pile wall. Equation (4.1) does not, however, take into consideration the flexibility of the lower part of the wall embedded in noncreeping soil. This can be accounted for by using the expression:

$$\frac{EI}{H\eta}\int_0^x dx \int_x^H dx \int_0^t p(x,t)\,dt$$

$$= \int_0^x dx \int_0^x [M*(x,t) - M*(x,0)]\,dx$$

$$+ \frac{EI}{K}\{x\{F[p(x,t)] - F[p(x,0)]\}$$

$$+ f[p(x,t)] - f[p(x,0)]\} \tag{4.2}$$

where $M*(x,t)$ and $M*(x,0)$ are the summarized bending moments in a certain section of the wall at a certain postconstruction period of time, t, and at the initial time ($t = 0$), respectively, and

$$\begin{aligned} f[p(x,0)] &= Ky(0,0) \\ f[p(x,t)] &= Ky(0,t) \\ F[p(x,0)] &= K\theta(0,0) \\ F[p(x,t)] &= K\theta(0,t) \end{aligned} \tag{4.3}$$

where K is the horizontal spring characteristic of the underlying noncreeping soil and $y(0,t)$, $y(0,0)$, $\theta(0,t)$, and $\theta(0,0)$ are the displacement and angle of rotation of the wall at $x = 0$ at any given moment of time (t) and for the initial time ($t = 0$), respectively. Following is a discussion of the conceptual approach to the design of sheet pile walls to be constructed on creeping foundation soils that is based on the concept described by eq. (4.2).

4.2.1 Analysis of a Cantilever Sheet Pile Bulkhead

Consider the conventional approach to determining the passive pressure distribution on the wall structure. For cohesive soils, this will be associated with trapezoidal soil pressure diagram $p(x,0)$ (drained condition). This could be represented by its rectangular and triangular components. For the triangular component,

$$p(x,0) = \gamma K_p(H - x) \tag{4.4}$$

where γ is the soil unit weight, K_p the coefficient of reactive (passive) pressure, and H the height of the layer of creeping soil. The initial and boundary conditions can be described as follows:

$$\begin{aligned} p(x,t)|_{t=0} &= p(x,0) = \gamma K_p(H - x) \\ p(x,t)|_{x=0} &= 0 \\ p(x,t)|_{x=H} &= 0 \end{aligned} \tag{4.5}$$

For rectangular distribution of the initial reactive load, $p(x,0) = p(0) =$ constant, the boundary and initial conditions can be described as follows:

$$\begin{aligned} p(0,t) &= 0 \\ \frac{dp(H,t)}{dx} &= 0 \\ p(0,0) &= p(0) \end{aligned} \tag{4.6}$$

The initial relationship between the reactive soil pressure and the bending moment, $M*(0,0)$, and shear force, $Q*(0,0)$, acting in a certain section of the wall with coordinate $x = 0$ at the initial moment of the soil–wall interaction ($t = 0$) and at a given time are obtained on the basis of work by Snitko (1963) as follows:

$$f[p(x,0)] = \frac{24}{l^2}\,[M^*(0,0) + \tfrac{3}{4}\,Q^*(0,0)xl]$$

$$f[p(x,t)] = \frac{24}{l^2}\,[M^*(0,t)\,\tfrac{3}{4}\,Q^*(0,t)xl]$$

$$F[p(x,0)] = \frac{12}{l^2}\left[\frac{3M^*(0,0)}{l^2} + 2Q^*(0,0)\right]$$

$$F[p(x,t)] = \frac{12}{l^2}\left[\frac{3M^*(0,t)}{l^2} + 2Q^*(0,t)\right] \tag{4.7}$$

In the case in question, the right-hand side of eq. (4.2) can be rewritten as follows:

$$\int_0^x dx \int_0^x [M^*(x,t) - M^*(x,0)]\,dx + \frac{EI}{Kl^2}\{[M^*(0,t) - M^*(0,0)]\,f_1(x) + [Q^*(0,t) - Q^*(0,0)]f_2(x)\} \tag{4.8}$$

where

$$f_2(x) = 12\left(\frac{3x}{l} + 2\right)$$

$$f_2(x) = 12\left(2x + \frac{3l}{2}\right)$$

The summarized bending moments in eq. (4.2) can be expressed as follows:

$$\begin{aligned} M^*(x,t) &= [M(x,t)]_a - M(x,t) \\ M^*(x,0) &= [M(x,t)]_a - M(x,0) \end{aligned} \tag{4.9}$$

where $[M(x,t)]_a$ and $[M(x,0)]_a$ are components of bending moments resulting from the action of load $N(x)$ and $M(x,t)$ and $M(x,0)$ are components of bending moments resulting from reactive loads $p(x,t)$ and $p(x,0)$. The resulting bending moments and shear forces in section $x = 0$ can be expressed as follows:

$$\begin{aligned} M^*(0,t) &= [M(0,t)]_a - M(0,t) \\ M^*(0,0) &= [M(0,0)]_a - M(0,0) \\ Q^*(0,t) &= [Q(0,t)]_a - Q(0,t) \\ Q^*(0,0) &= [Q(0,0)]_a - Q(0,0) \end{aligned} \tag{4.10}$$

For $N(x)$ = constant, $[M(x,t)]_a = [M(x,0)]_a$, $[M(0,t)]_a = [M(0,0)]_a$, and $[Q(0,t)]_a = [Q(0,0)]_a$. On the basis of formulas (4.8) through (4.10), the basic equation (4.2) can be expressed as follows:

$$\frac{EI}{H\eta}\int_0^x dx \int_x^H dx \int_0^t p(x,t)\,dt = \int_0^x dx \int_0^x [M(x,0) - M(x,t)]\,dx + \frac{EI}{Kl^2}\{[M(0,0) - M(0,t)]f_1(x) + [Q(0,0) - Q(0,t)]f_2(x)\} \tag{4.11}$$

Subsequently, solution for $p(x,t)$ can be derived from eq. (4.11) as follows:

$$p(x,t) = p(x,0)\exp\left\{-\frac{EIKl^2t}{H\eta[F(x)Kl^2 + \Phi(x)EI]}\right\} \tag{4.12}$$

The functions $F(x)$ and $\Phi(x)$ normally depend on the geometry of the initial reactive load diagram $p(x,0)$. For the triangular initial reactive load, these functions are expressed as follows:

$$\Phi(x) = \frac{(3\alpha/\beta + \frac{9}{2}\beta + 6\alpha + 2)}{0.25\alpha(1 - \alpha - \alpha^2/3)} \tag{4.13}$$

$$F(x) = \frac{\alpha H^2(1 - \alpha + 0.5\alpha^2 - 0.1\alpha^3)}{6(1 - \alpha + \alpha^2/3)} \tag{4.14}$$

where $\alpha = x/H$ and $\beta = l/H$ and for the rectangular diagram of the initial load.

$$F(x) = \frac{\alpha H^2(1 - \frac{2}{3}\alpha + \frac{1}{6}\alpha^2)}{4(1 - \frac{1}{2}\alpha)} \tag{4.15}$$

$$\Phi(x) = \frac{6(3\alpha/\beta + 4\alpha + 3\beta + 2)}{\alpha(1 - /\alpha 2)} \tag{4.16}$$

Thus, by using the appropriate value of functions $F(x)$ and $\Phi(x)$ as described by eqs. (4.13) through (4.16) and assuming the initial load distribution $p(x,0) = p(0) =$ constant, the reactive soil pressures $p(x,t)$ at any time in question, t, can be determined from eq. (4.12).

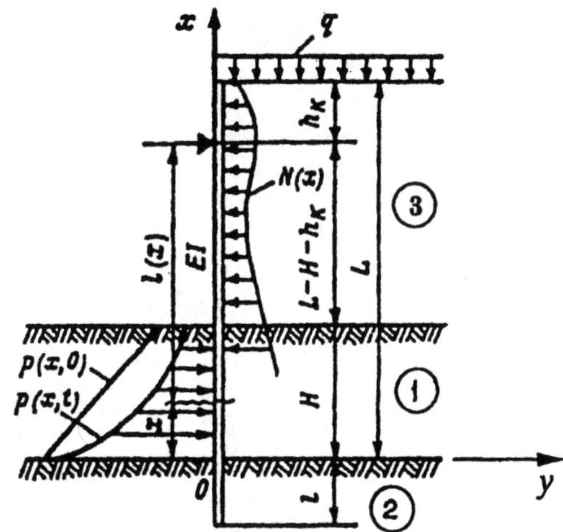

Figure 4.7 Design of single-anchored sheet pile bulkhead on creep foundation; typical soil pressures distribution diagram. 1, Creep soil; 2, underlaying noncreep soil; 3, granular backfill. (From Gurinsky, 1997.)

4.2.2 Analysis of a Single-Anchor Sheet Pile Bulkhead†

For a single-anchored sheet pile wall (Figure 4.7), the resulting bending moments can be obtained from the equations

$$M^*(x,t) = [M(x,t)]_a - M(x,t) - R^* l(x)$$
$$M^*(x,0) = [M(x,0)]_a - M(x,0) - R^* l(x) \tag{4.17}$$

where $R^*(t)$ and $R^*(0)$ are the anchor reactions at the moment of time t and at the initial moment of time, respectively; $l(x)$ is the arm of the force R^* relative to a certain specific section of the wall; and $l(x) = L - h_k - x$ (for definitions, see Figure 4.7). The expressions for bending moment and shear force in wall section with coordinates $x = 0$ and $y = 0$ are as follows:

$$M^*(0,t) = [M(0,t)]_a - M(0,t) - R_a(t)(L - h_k)$$
$$M^*(0,0) = [M(0,0)]_a - M(0,0) - R_a(0)(L - h_k)$$
$$Q^*(0,t) = [Q(0,t)]_a - Q(0,t) - R_a(t)$$
$$Q^*(0,0) = [Q(0,0)]_a - Q(0,0) - R_a(0) \tag{4.18}$$

By solving eq. (4.16) for eqs. (4.17) also using formula (4.8) and assuming that $[M(0,0)]_a$ and $[Q(0,t)]_a = Q(0,0)]_a$, the following expression is obtained:

†For analysis of more complex systems (e.g., double-anchored or multianchored bulkheads) the reader is referred to Section 6.8 by M. Gurinsky in Tsinker (1997).

$$\frac{EI}{H\eta}\int_0^x dx \int_x^H dx \int_0^t p(x,t)\, dt$$
$$= \int_0^x dx \int_0^x [M(x,0) - M(x,t)]\, dx$$
$$- \int_0^x dx \int_0^x \Delta R_a(t) l(x)\, dx$$
$$+ \frac{EI}{Kl^2}\{[M(0,0) - M(0,t)$$
$$- \Delta R_a(t)(L - h_k)]\, f_1(x) + [Q(0,0)]$$
$$- Q(0,t) - \Delta R_a(t) f_2(x)\} \qquad (4.19)$$

where $\Delta R_a(t)$ is the increment of anchor reaction for the specific period of time t, $\Delta R_a(t) = R^*(t) - R^*(0)$, and

$$f_1(x) = 12 \left| \frac{3x}{l} + \frac{2l}{2} \right|$$

and

$$f_2(x) = 12 \left| 2x + \frac{3l}{2} \right|$$

The initial anchor reaction force $R^*(0)$ and the reaction at a specific time, $R^*(t)$, are obtained from the formulations

$$R^*(t) = [R_a(t)]_a - R_a(t)$$
$$R^*(0) = [R_a(0)]_a - R_a(0) \qquad (4.20)$$

Solving eqs. (4.18) through (4.20), the expression for the distribution of the reactive (passive) pressure, $p(x,t)$, is as follows:

$$p(x,t) = p(x,0) \times \exp\left(-\frac{EIKl^2 t}{H\eta\{[F(x) - B(x)]Kl^2 + \phi^*(x)EI\}}\right) \qquad (4.21)$$

In eq. (4.21), the function $F(x)$ is obtained from eq. (4.15). The function $B(x)$ depends on the initial geometry of the reactive pressure diagram, described by $p(x,0)$. If distribution of the load $p(x,0)$ is triangular, $B(x)$ may be obtained from

$$B(x) = \frac{m^{tr}\alpha[(L - h_k)/H - \alpha/3]}{1 - \alpha - \alpha^2/3} \qquad (4.22)$$

where $\alpha = x/H$ and m^{tr} is a coefficient for determining the support reactions of the statically undetermined beams. In the case of a triangular diagram,

$$m^{tr} = \frac{H^4}{40(L - h_k)^2}\left(5 - \frac{H}{L - h_k}\right) \qquad (4.23)$$

In the case of a rectangular initial load $p(x,0)$ diagram,

$$B(x) = \frac{m^{re}\alpha^c H[(L - h_k)/H - (d/3)]}{2(1 - \frac{1}{2}\alpha)} \qquad (4.24)$$

where

$$m^{rec} = 0.055(L - h_k) \qquad (4.25)$$

The function $\Phi^*(x)$ is dependent on the geometry of the initial reactive load $p(x,0)$ diagram. In the case of a triangular reactive load distribution, $\Phi^*(x)$ is obtained from

$$\Phi^*(x) = \frac{24\{[H^2/6 - m^{tr}(L - h_k)](3x/l + 2) + (H^2/2 - m^{tr})(2x + 3l/2)\}}{x(H^2 - Hx + x^2/3)} \qquad (4.26)$$

In the case of a rectangular initial reactive load program, the formula for calculating $\Phi^*(x)$ is

$$\Phi^*(x) = \frac{12\{[H^2/2 - m^{\text{rec}}(L - h_k)](3x/l + 2) + (H + m^{\text{rec}})(2x + 3l/2)\}}{x(H - x/2)} \tag{4.27}$$

Thus, using the appropriate values of the functions $F(x)$, $B(x)$, and $\Phi^*(x)$ as described by eqs. (4.15) and (4.22) through (4.27) and assuming initial distribution of the reactive load $p(x,0) = p(0) =$ constant, the reactive soil pressure $p(x,t)$ at any time t can be determined by eq. (4.21).

***Example 4.1* (Budin, 1982).** Consider the following parameters of a single-anchor sheet pile wall (Figure 4.8): $L = 12.6$ m, $h_k = 2.35$ m, $H = 5.3$ m, and $l = 1.95$. The design depth of water is 4.6 m. The sheeting is driven into the stratum comprised of the creep clay (1) and noncreep loam (2) with the following engineering parameters:

- *Clay:* submerged unit weight $\gamma_c = 8$ kN/m^3; angle of internal friction $\phi_c = 16°$; cohesion $c = 0.01$ MPa; $\eta = 2.3 \times 10^4$ kN · day/m^2.
- *Loam:* submerged unit weight $\gamma_l = 10$ kN/m^3; angle of internal friction $\phi_l = 27°$; cohesion $c = 0.02$ MPa; spring characteristic $K = 0.32$ MPa

The wall is backfilled with granular material (sand) with the following parameters: unit weight in dry condition $\gamma_{s(d)} = 16$ kN/m^3; submerged unit weight $\gamma_{s(s)} = 10$ kN/m^3; angle of internal friction $\phi_s = 29.5°$. The wall is loaded by a uniformly distributed surcharge load $q =$

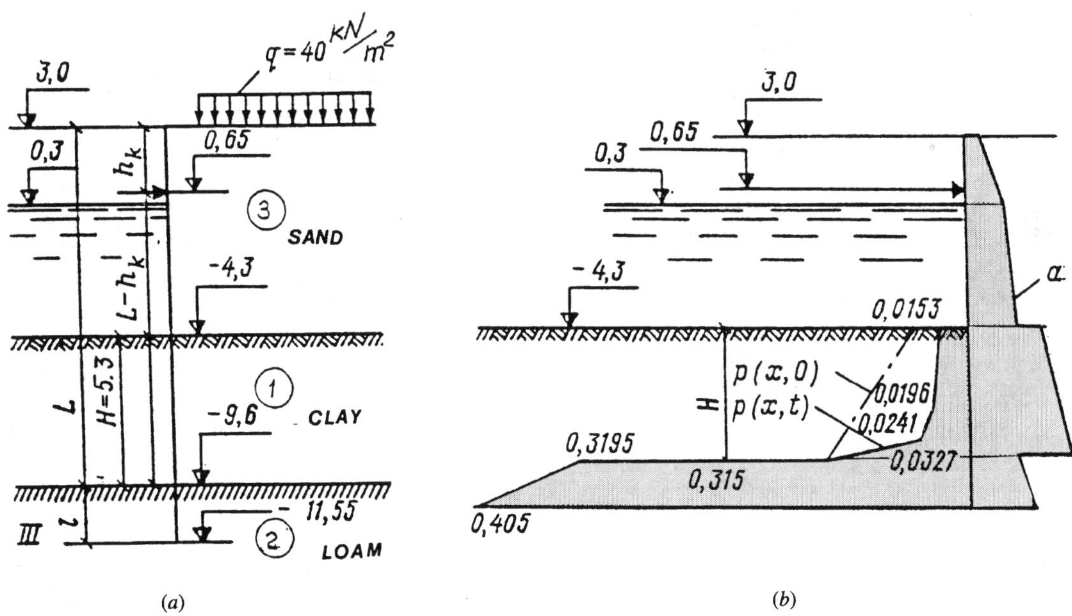

Figure 4.8 Design of single-anchored sheet pile bulkhead on creep foundation (Example 4.1): (*a*) design scheme; (*b*) soil pressures diagram (MPa). 1, Creep soil; 2, underlying noncreep loam; 3, granular backfill (sand). (After Budin, 1982.)

Table 4.2. Analysis of a single-anchored sheet pile bulkhead (Example 4.1)

	$p(x,0)$			$p(x,t = 40 \text{ years})$		
x/H	Triangular	Rectangular	Sum	Triangular	Rectangular	Sum
0.25	0.0715	0.0436	0.1151	0.0196	0.0131	0.0327
0.5	0.0476	0.0436	0.0912	0.0125	0.0116	0.0241
0.75	0.0238	0.0436	0.0674	0.0076	0.0120	0.0196
1.0	0	0.0436	0.0436	0	0.0153	0.0153

40 kN/m²; the wall rigidity EI = 63,400 kN · m².

Determine the bending moment in the wall structure after 40 years in service.

SOLUTION: First, determine the active and passive (reactive), $p(x,0)$ pressures for the initial period of the sheetpiling–soil interaction ($t = 0$). The results are presented in Figure 4.8*b* and Tables 4.2 through 4.4. The initial bending moment in the wall obtained from static analysis for $t = 0$ is $M(0) = 256$ kN · m. Next, the wall is analyzed for the long-term performance (t = 40 years). To calculate $p(x,t = 40)$ from eq. (4.21), determine functions $F(x)$, $B(x)$, and $\Phi^*(x)$ by using eqs. (4.22) through (4.27), where the coefficients m^{tr} and m^{rec} are determined from eqs. (4.23) through (4.25) and are as follows:

$$m^{tr} = 0.844 \qquad \text{and} \qquad m^{rec} = 0.56$$

The values of functions $F(x)$, $B(x)$, and $\Phi^*(x)$ that correspond to different values of x/H are shown in Table 4.3. Other values to be used in the design are

$$L - \frac{h_k}{H} = 1.94 \qquad \text{and} \qquad \frac{l}{H} = 0.368$$

The coordinates of diagram $p(x,t = 40)$ are determined from the formula (4.21). The results of these calculations are shown in Tables 4.2 and 4.4. The values of Kl^2 and t in this case are equal to 1215 kN and 14,600 days, respectively. The corresponding passive pressure diagram $p(x, t = 40)$ is shown in Figure 4.8*b*.

Table 4.3. Analysis of a single-anchored sheet pile bulkhead (Example 4.1)[a]

x/H	$F(x)$	$B(x)$	$\Phi^*(x)$
0.25	1.2/1.7	0.5/1.1	112.5/120.5
0.5	2.5/3.3	1.3/2.2	108.5/110
0.75	3.9/5.2	2.5/3.7	128/112
1.00	5.7/7.0	4.2/5.3	156.5/128

[a] Values indicated in the numerator and denominator are related to the triangular and rectangular components, respectively, of the generally trapezoidal diagram $p(x,0)$.

Finally, compute the new bending moment in the wall structure by using a new reactive pressure diagram, $p(x,t = 40)$. In this example, the bending moment in the bulkhead structure after 40 years in service obtained by a graphic method is equal to 310 kN · m, as compared to the initial bending moment of 256 kN · m.

4.3 GRAVITY WALLS ON CREEP-SENSITIVE SOILS

The practice of building gravity walls on creep-sensitive soil foundations reveals that they usually experience continuous horizontal displacements during their service life. Long-term

Table 4.4. Analysis of a single-anchored sheet pile bulkhead (Example 4.1)[a]

		$EIKl^2t$	
$\frac{X}{R}$	$[F(x) - B(x)]Kl^2 + \Phi^*(x)EI$	$H\eta\{[F(x) - B(x)] \times Kl^2 + \Phi^*(x)EI\}$	$\exp(-H\eta\{[F(x) - B(x)] \times Kl^2 + \Phi^*(x)EI\}$
0.25	770,000/825,000	1.29/1.20	0.275/0.301
0.50	743,000/752,000	1.34/1.32	0.262/0.267
0.75	875,000/766,000	1.14/1.29	0.32/0.275
1.00	1,070,000/946,000	0.93/1.05	0.395/0.350

[a] Values indicated in the numerator and denominator are related to the triangular and rectangular components, respectively, of the generally trapezoidal diagram $p(x,0)$.

displacements of gravity walls built on clayey soils in general and on a clay in particular have been investigated by many researchers, including Henkel (1957), Peynircioglu (1957), and Budin and Demina (1979). To prevent possible negative consequences of such displacements, which eventually may lead to structure failure, it is important to predict the long-term performance (e.g., of gravity walls, including their displacements over time).

In general, the character of gravity wall displacement over time is determined by maximum shear stresses in the soil at the wall base (τ_{max}). Under $\tau_{max} < \tau_{th}$, there will be no wall displacement. If $\tau_{th1} > \tau_{max} > \tau_{th}$, the creep has an attenuating character at a decreasing rate (Figure 4.1, curve 3). If $\tau_{max} > \tau_{th1}$, the wall displacement will take place continuously at a constant rate. Here τ_{th} is the initial creep threshold—a constant stress below which creep process does not occur—and τ_{th1} is the creep threshold at constant stress beyond which steady-state creep process occurs (Figure 4.1, curve 1). Some observations of long-term gravity wall displacements over time indicate that it could be substantial.

These displacements accumulated during the steady-state creep process (at a constant rate) may reach as much as 60 to 80 cm over the life of the wall (Budin and Demina, 1979). The displacement rate of a gravity wall built on creep-sensitive soil can be determined by a method developed by Maslov (1968):

$$V = \frac{h}{\eta}\left[\frac{2\tau_0}{\pi}\arctan\frac{B}{2h} - \left(\sigma_0 + \frac{1}{2}\gamma h\right)\tan\Phi - c_c\right] \quad (4.28)$$

where V is the shear stress at the base of the wall at depth h, h the depth of the soil active zone in the wall base, η the coefficient of soil viscosity, τ_0 the shear stress at the base of the wall, B the width of the wall base, γ the unit weight of the soil (tonnes/m^3), ϕ the angle of internal friction, c_c the structural cohesion, which is part of the soil cohesion (Budin and Demina, 1979).

For the sake of simplicity, the shear stresses are assumed distributed uniformly along the wall base and are given as $\tau_0 = E/B$, where E is the horizontal component of soil lateral thrust. The normal stress at the wall base σ_0 is also assumed as uniformly distributed and can be obtained by $\sigma_0 = (\sigma_1 + \sigma_2)/2$, where σ_1 and σ_2 are maximum and minimum values of normal stress at the wall base. Hence, the normal stress at the wall base, σ_0, is determined as equal to $\sigma_0 = Q/B + q_1$, where Q is weight of one linear meter of the wall length and q_1 is surcharge load on top of the wall (Figure 4.9).

The depth of the active zone h may be obtained from

$$\frac{2\tau_0}{\pi}\left[\arctan\frac{B}{2h} - \frac{Bh}{2[h^2 + (B/2)^2]}\right] = (\sigma_0 + \gamma h)\tan\phi + c_c \qquad (4.29)$$

In this equation the left part represents the shear stress τ at the base of the wall at depth h, and the right part represents the creep threshold of soil τ_{th}. The critical value of h can be determined graphically by choosing several values of h and creating graphs $\tau = f(y)$ and $\tau_{th1} = f(y)$ for the left and right parts of eq, (4.29) as indicated in Figure 4.10. According to Budin and Demina (1979), c_c can be determined from $c_c = \tau_{th1} - \tau_{th}$.

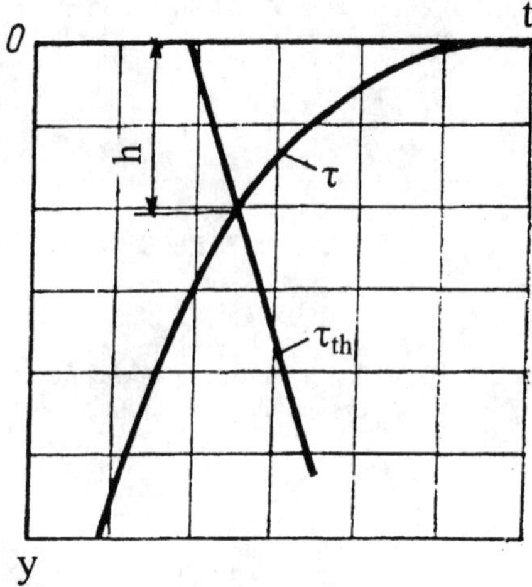

Figure 4.10 Determination of the active zone h in the wall base.

Example 4.2. The gravity wall indicated in Figure 4.11 (B = 8.5 m) is loaded with a uniformly distributed surcharge load, q = 2 tonnes/m^2. The soil at the wall base is creep-

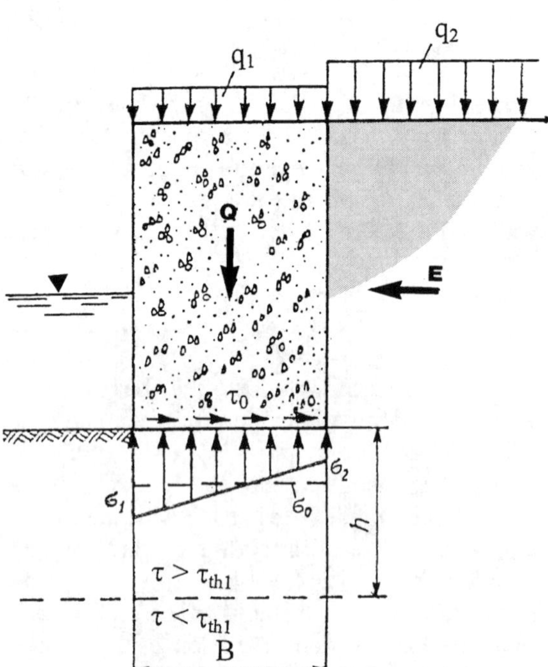

Figure 4.9 Design diagram for gravity wall built on creep-sensitive soil.

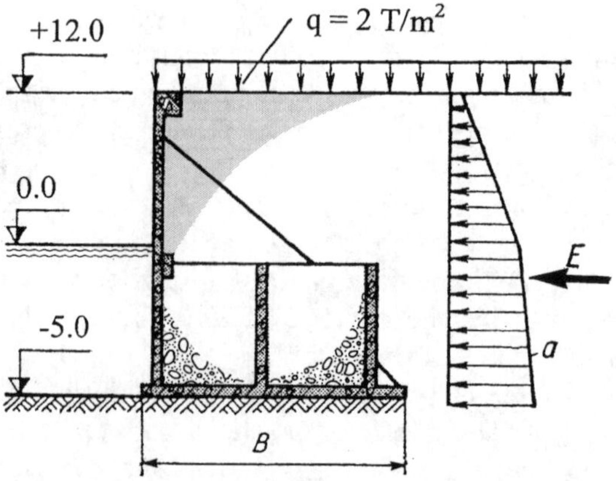

Figure 4.11 Design of gravity wall on a creep soil foundation (Example 4.2).

sensitive clay with an angle of internal friction $\Phi = 12°$, submerged unit weight $\gamma_s = 1$ tonne/m^2, and viscosity coefficient $\eta = 0.5 \times 10^4$ tonne · day/m^2. Assumed increase of soil creep threshold over initial creep threshold $\Delta\tau = \tau_{th1} - \tau_{th} = 0$.

The wall is backfilled with granular material (sand) having an angle of internal friction $\phi = 30°$, unit weight of dry soil $\gamma = 1.7$ tonnes/m^3, and submerged unit weight of soil $\gamma_s = 1$ tonne/m^2.

SOLUTION: The calculated value of horizontal load $E = 45.3$ tonnes. The latter corresponds to a value of shear stress at the wall base $\tau_0 = E/B = 45.3/8.5 = 5.33$ tonnes/m^2. Uniformly distributed normal stress at the base is equal to 18.9 tonnes/m^2. To find the depth of the soil active zone h, it is necessary to plot the curves $\tau = f(y)$ and $\tau_{th} = f(y)$, as suggested above. This resulted in a value of h that is approximately equal to 1.0 m. Finally, the wall displacement rate is found as equal to

$$V = \frac{1}{0.5 \times 10^4}\left[\frac{2 \times 5.33}{\pi} \arctan \frac{8.5}{2.1} - (18.9 + \tfrac{1}{2} \times 1) \tan 12°\right]$$

$$= 0.0001 \text{ m/day}$$

$$= 0.0365 \text{ m/yr} \approx 4 \text{ cm/yr}.$$

REFERENCES

Budin, A. Ya., 1969. Behavior of sheet pile retaining walls in soils susceptible to creep, *Soil Mech. Found. Eng.* (translation of *Osn. Fundam. Mekh. Gruntov*), No. 6.

———, 1982. *Thin Sheetpile Retaining Walls for Northern Regions,* Stroyizdat, Leningrad (in Russian).

———, 1983. Determination of rheological properties of soil, *Nauchn. Tr. Leningr. Inst. Vodn. Transp.* Vol. 176, pp. 140–146 (in Russian).

Budin, A. Ya., and G. A. Demina, 1979. *Quays Handbook,* Stroyizdat, Moscow (in Russian).

Budin, A., and M. Gurinsky, 1985. Computation of spring compensators for ground anchors, *Soil Mech. Found. Eng.* (translation of *Osn. Fundam. Mekh. Gruntov*), Vol. 22, No. 6, pp. 219–223.

Budin, A. J., W. M. Kirillov, and W. M. Kolga, 1975. Durability of hydro engineering structures on creeping soil foundation, *Proc. First Baltic Conference on Soil Mechanics and Foundation Engineering,* Gdansk, Poland, Vol. 3, pp. 327–336.

Geuze, E. C. W., and Tjong-Kie, Tan, 1954. The mechanical behavior of clays, *Proc. 2nd International Congress on Rheology,* Oxford.

Gurinsky, M., 1997. Sheet-pile bulkheads built on creep soils, in *Handbook of Port and Harbor Engineering,* G. Tsinker, ed., Chapman & Hall, New York.

———, 2001. Long-term strength of sheet pile bulkheads with ground anchors, *ASCE Proc. Specialty Conference Ports '01,* Norfolk, VA, Apr.

———, 2002. Long-term strength of prestressed ground anchors in creep-sensitive soils, *Proc. International Congress on Deep Foundations,* Orlando, FL.

Haefeli, R., Ch. Schaerer, and G. Amberg, 1953. The behavior under the influence of soil creep pressure of the concrete bridge built at Klosters by Rhaetiln Railway Company, Switzerland, *Proc. 3rd International Conference on Soil Mechanics and Foundation Engineering,* Vol. II, Zurich.

Henkel, D. J., 1957. Investigation of two long-term failures in London clay slopes at Wood Green and Northolt, *Proc. 4th International Conference on Soil Mechanics and Foundation Engineering,* Vol. II, London.

Kuhn, M. R., and J. K. Mitchell, 1993. New perspectives on soil creep, *ASCE J. Geotech. Eng.,* Vol. 119, No. 3.

Lacerda, W. A., 1976. Stress-relaxation and creep effects on soil deformation, Ph.D. dissertation, University of California, Berkeley, CA.

Luga, A. A., 1964. *The Effect of Creep in Clay Soils on the Displacement of Bridge Abutments.* Moscow (in Russian).

Maslov, N., 1968. *Long-Term Stability and Displacement of Retaining Structures,* Energia, Moscow.

Mogilevskaya, S. E., 1968. Results of studies of creep in loess soil in connection with deformation of water-development works, in *Studies of Rheological Properties of Soils,* Energia, Leningrad (in Russian).

Murayama, S., and T. Shibata, 1958. On the Reological Characteristics of Clays, Part 1, Bulletin N26, Disaster Prevention Research Institute, Tyoko, Japan.

Peck, R., 1948. Report on deformation of retaining structures in the USA, *Proc 2nd International Congress on Soil Mechanics and Foundation Engineering,* Rotterdam, The Netherlands.

Peterson, R., 1953. Discussion, *Proc. 3rd International Congress on Soil Mechanics and Foundation Engineering,* Zurich.

Peynircioglu, H., 1957. Earth movement investigations in a landslide area on the Bosporus, *Proc. 4th International Congress on Soil Mechanics and Foundation Engineering,* London.

Pusch, R., and P. Feltham, 1981. Computer simulations of creep of clay, *ASCE J. Geotech. Eng.,* Vol. 107, No. 1.

Scott, R. F., 1963. *Principles of Soil Mechanics,* Addison-Wesley, Reading, MA.

Snitko, N. K., 1963. *Static and Dynamic Pressures on Soils and Design of Retaining Walls,* Stroyizdat, Leningrad (in Russian).

Tsinker, G., 1997. *Handbook of Port and Harbor Engineering: Geotechnical and Structural Aspects,* Chapman & Hall, New York.

Vyalov, S., 1986. *Rheological Fundamentals of Soil Mechanics,* Elsevier, New York.

5

DESIGN OF REVETMENTS[†]

Krystian W. Pilarczyk
Dutch Public Works Department
Delft, The Netherlands

[†]This chapter is based partly on the paper by Klein Breteler et al. (1998b).

5.1 INTRODUCTION

The use of revetments such as riprap, blocks and block mats, various mattresses, and asphalt in civil engineering practice is very common. The granular filters, and more recently the geotextiles, are more or less standard components of the revetment structure [Permanent International Association of Navigation Congresses (PIANC), 1987a, 1992]. Within the scope of the research on the stability of open slope revetments, much knowledge has been developed about the stability of placed (pitched) stone revetments under wave load (CUR/TAW, 1995) and stability of rock under wave and current load (CUR/CIRIA, 1991; CUR/RWS, 1995).

Until recently, no or unsatisfactory design tools were available for a number of other (open) types of revetment and for other stability aspects. This is why the design methodology for placed block revetments has recently been extended in applicability by means of a number of desk studies for other (open) revetments: interlock systems and block mats; gabions; concrete mattresses; geosystems, such as sandbags and sand sausages; and other stability aspects, such as flow load stability, soil mechanical stability, and residual strength.

This chapter aims at giving a summary of the increased knowledge, especially that concerning the design tools that have been made available. Details may be found in Pilarczyk (1998).

5.2 THEORETICAL BACKGROUND OF WAVE LOADING

Wave attack on revetments will lead to a complex flow over and through the revetment structure (filter and cover layer). During wave run-up, the resulting forces by the waves will be directed opposite to the gravity forces. Therefore, the run-up is less hazardous then the wave run-down. Wave run-down will lead to two important mechanisms:

1. The downward-flowing water will exert a drag force on the cover layer, and the decreasing freatic level will coincide with a downward flow gradient in the filter (or in a gabion). The first mechanism can be schematized by a free flow in the filter or gabion with a typical gradient equaling the slope angle. It may result in sliding.
2. During maximum wave run-down there will be an incoming wave that a moment later will cause a wave impact. Just before impact there is a "wall" of water giving a high pressure under the point of maximum run-down. Above the run-down point, the surface of the revetment is almost dry, and therefore there is a low pressure on the structure. The high-pressure front will lead to an upward flow in the filter or a gabion. This flow will meet the downward flow in the run-down region. The result is an outward flow and uplift pressure near the point of maximum wave run-down (Figure 5.1).

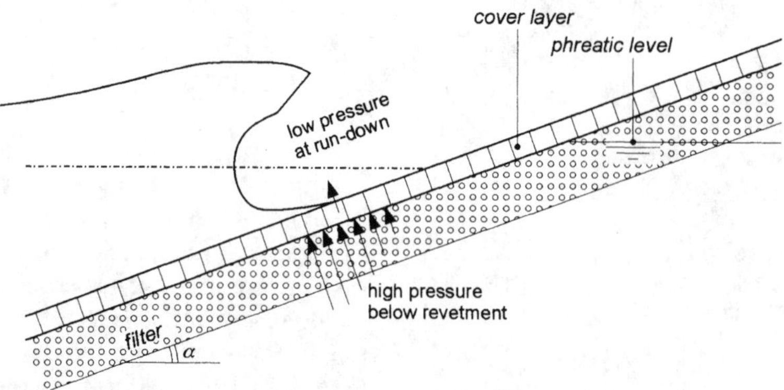

Figure 5.1 Pressure development in a revetment structure.

The schematised situation can be quantified on the basis of the Laplace equation for linear flow:

$$\frac{\partial^2 \phi}{\partial y^2} + \frac{\partial^2 \phi}{\partial z^2} = 0 \qquad (5.1)$$

where $\phi = \phi_b$ is the potential head induced in the filter or a gabion (m), y the coordinate along the slope (m), and z the coordinate perpendicular to the slope (m).

After complicated calculations the uplift pressure in the filter or a gabions can be derived. The uplift pressure is dependent on the steepness and height of the pressure front on the cover layer (which is dependent on the wave height, period, and slope angle; see Figure 5.2), the thickness of the cover layer and the level of the phreatic line in the filter or a gabion. In case of riprap or gabions, it is not dependent on the permeability of the cover layer if the permeability is greater larger than that of the subsoil. For semipermeable cover layers the equilibrium of uplift forces and gravity forces (defined by components of a revetment) leads to the following (approximate) design formula (Pilarczyk, 1998):

$$\frac{H_{scr}}{\Delta D} = f\left(\frac{D}{\Lambda \xi_{op}}\right)^{0.67} \quad \text{with} \quad \Lambda = \sqrt{\frac{bDk}{k'}} \qquad (5.2a)$$

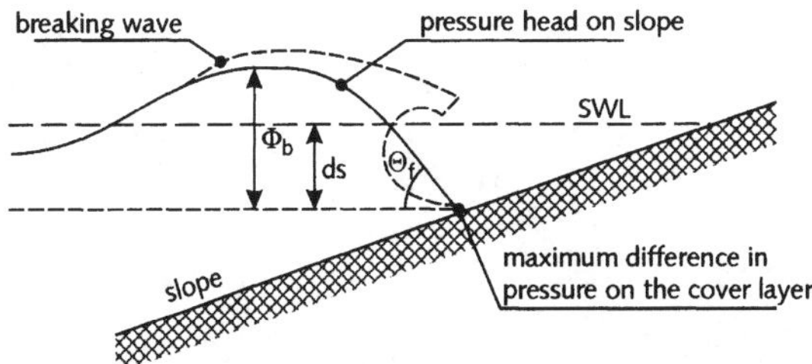

Figure 5.2 Pressure head on a slope.

or

$$\frac{H_{scr}}{\Delta D} = f\left(\frac{D}{b}\frac{k'}{k}\right)^{0.33} \xi_{op}^{-0.67} \qquad (5.2b)$$

or

$$\frac{H_{scr}}{\Delta D} = F\xi_{op}^{-0.67} \qquad (5.2c)$$

where H_{scr} is the significant wave height at which blocks will be lifted out (m); $\xi_{op} = \tan\alpha/\sqrt{H_s/(1.56T_p^2)}$, the breaker parameter; T_p is the wave period at the peak of the spectrum (s); Λ the leakage length (m); $\Delta = (\rho_s - \rho)/\rho$, the relative volumetric mass of cover layer; b the thickness of a sublayer (m), D the thickness of a top (cover) layer (m); k the permeability of a sublayer (m/s), k' the permeability of a top layer (m/s); f the stability coefficient, mainly dependent on structure type, tan α, and friction; and F the total (black-box) stability factor. The leakage length (Λ) and stability coefficient (F) are explained in more detail in the next sections.

5.3 STRUCTURAL RESPONSE

5.3.1 Wave Load Approach

There are two practical design methods available: the black-box model and the analytical model. In both cases, the final form of the design method can be presented as a critical relation of the load compared to strength, depending on the type of wave attack:

$$\left(\frac{H_s}{\Delta D}\right)_{cr} = \text{function of } \xi_{op} \qquad (5.3a)$$

For semipermeable cover layers, the basic form of this relation is

$$\left(\frac{H_s}{\Delta D}\right)_{cr} = \frac{F}{\xi_{op}^{2/3}} \quad \text{with maximum}$$

$$\left(\frac{H_s}{\Delta D}\right)_{cr} = 8.0 \quad \text{and } \cot\alpha \geq 2 \qquad (5.3b)$$

or, in more general form (also applicable for riprap and cot $\alpha \geq 1.5$), as defined by Pilarczyk (1990, 1998):

$$\left(\frac{H_s}{\Delta D}\right)_{cr} = \frac{F \cos\alpha}{\xi_{op}^{b}} \qquad (5.3c)$$

where H_s is the (local) significant wave height (m), Δ the relative density, D the thickness of the top layer (m), F the revetment (stability) factor, ξ_{op} the breaker parameter, and b = exponent ($0.5 \leq b \leq 1.0$). The approximate values of stability factor F are $F = 2.25$ for riprap, $F = 2.5$ for pitched stone of irregular shape, $F = 3.0$ to 3.5 for pitched basalt, $F = 4.0$ for geomattresses, $3.5 \leq F \leq 5.5$ for block revetments (4.5 as an average/usual value), $4.0 \leq F \leq 6.0$ for block mats (higher value for cabled systems), $6.0 \leq F \leq 8.0$ for gabions, and $6.0 \leq F \leq 10$ for (asphalt or concrete) slabs. Exponent b refers to the type of wave–slope interaction, and its value is influenced by the roughness and porosity of a revetment. The following values of exponent b are recommended: $b = 0.5$ for permeable cover layers (i.e., riprap, gabions, pattern grouted riprap, very open block mats), $b = 2/3$ for semipermeable cover layers (i.e., pitched stone and placed blocks, block mats, concrete- or sand-filled geomattresses), and $b = 1.0$ for slabs.

The relative density is defined as follows:

$$\Delta = \frac{\rho_s - \rho_w}{\rho_w} \qquad (5.4a)$$

where ρ_s is the density of the protection material and ρ_w is the density of water (kg/m^3). For porous top layers such as sand mattresses and gabions, the relative density of the top layer must be determined, including the water-filled pores:

$$\Delta_t = (1 - n)\Delta \qquad (5.4b)$$

where Δ_t is the relative density, including pores, and n is the porosity of the top layer material. D and Δ are defined for specific systems, such as:

- *Rock:* $D = D_n = (M_{50}/\rho_s)^{1/3}$ (= nominal diameter) and $\Delta_t = \Delta = (\rho_s - \rho_w)/\rho_w$.
- *Blocks:* D = thickness of block and $\Delta_t = \Delta$.
- *Mattresses:* $D = d$ = average thickness of mattress and $\Delta_t = (1 - n)\Delta$, where n is the bulk porosity of fill material and Δ is the relative density of fill material [for common quarry stone, $(1 - n)\Delta \sim 1$].

The breaker parameter is defined as follows:

$$\xi_{op} = \frac{\tan\alpha}{\sqrt{H_s/L_{op}}} \qquad (5.5)$$

The wave steepness S_{op} is defined as

$$S_{op} = \frac{H_s}{L_o} = \frac{2\pi H_s}{gT^2} \qquad (5.6a)$$

where

$$L_{op} = \frac{g}{2\pi} T_p^2 \qquad (5.6b)$$

α is the slope angle (°), L_{op} the deepwater wavelength at the peak period (m), and T_p the wave period at the peak of the spectrum (s).

The advantage of this black-box design formula is its simplicity. The disadvantage is that the value of F is known only very roughly for many types of structures. The analytical model is based on the theory for placed stone revetments on a granular filter (pitched blocks). In this calculation model, a large number of physical aspects are taken into account. In short, in the analytical model nearly all physical parameters that are relevant to the stability have been incorporated in the leakage length: $\Lambda = \sqrt{bDk/k'}$. The final result of the analytical model may, for that matter, again be presented as a relation such as eqs. (5.2c) or (5.3c), where $F = f(\Lambda)$.

With a system without a filter layer (directly on sand or clay and geotextile) the permeability of the subsoil (eventually with gullies/surface channels), not the permeability of the filter layer, is filled in. For the thickness of the filter layer which depth changes at the surface affect the subsoil are examined. One can assume 0.5 m for sand and 0.05 m for clay. The values for D and Δ depend on the type of revetment.

In the case of a geotextile situated directly under the cover layer, the permeability of the cover layer decreases drastically. Since the geotextile is pressed against the cover layer by the outflowing water, it should be treated as a part of the cover layer. Water flow through the cover layer is concentrated at the joints between the blocks, reaching very high flow velocities and resulting in a large pressure head over the geotextile. The presence of a geotextile may reduce k' by a factor of 10 or more.

To be able to apply the design method for placed stone revetments under wave load to other systems, the following items may be adapted:

- The revetment parameter F
- The (representative) strength parameters Δ and D
- The design wave height H_s
- The (representative) leakage length Λ
- The increase factor Γ (friction or interlocking between blocks) on the strength

Only adaptations such as these are presented in this summarizing review. The basic formulas of the analytical model are not repeated here. For these, readers are referred to (CUR/TAW, 1995).

The wave attack on a slope can be roughly transformed into the maximum velocity component on a slope during run-up and run-down, U_{max}, by using the formula

$$U_{max} = p\sqrt{gH_s}\xi_{op} \tag{5.7}$$

(For irregular waves and smooth slopes: $1 < p < 1.5$.)

5.3.2 Flow Load Stability

There are two possible approaches for determining the stability of revetment material under flow attack. The most suitable approach depends on the type of load:

- *Flow velocity*: horizontal flow, flow parallel to dike
- *Discharge:* downward flow at slopes steeper than 1:10, overflow without waves; stable inner slope

When the flow velocity is known or can be calculated reasonably accurately, Pilarczyk's relation (Pilarczyk, 1990, 1998, 1999) is applicable:

$$\Delta D = 0.035\,\frac{\Phi}{\Psi}\,\frac{K_T K_h}{K_s}\,\frac{u_{cr}^2}{2g} \tag{5.8}$$

where Δ is the relative density, D the characteristic thickness (m) (for riprap $D = D_n$ = nominal diameter as defined previously), Φ the stability parameter, Ψ the critical Shields parameter, K_T the turbulence factor, K_h the depth parameter, and K_s the slope parameter, u_{cr} the critical vertically averaged flow velocity (m/s), and g the acceleration due to gravity (g = 9.81 m/s^2). These parameters are explained below.

5.3.2.1 *Stability Parameter,* Φ. The stability parameter Φ depends on the application. Some guide values are:

Revetment Type	Continuous To Player	Edges and Transitions
Riprap and placed blocks	1.0	1.5
Block mats, gabions, washed-in blocks, geobags, and geomattresses	0.5–0.75	0.75–1.0

5.3.2.2 *Shields Parameter,* Ψ. With the critical Shields parameter Ψ, the type of material can be taken into account:

Riprap, small bags	$\Psi \approx 0.035$
Placed blocks, geobags	$\Psi \approx 0.05$
Blockmats	$\Psi \approx 0.07$
Gabions	$\Psi \approx 0.07$
Geomattresses	$\Psi \approx 0.07$

5.3.2.3 *Turbulence Factor,* K_T: The degree of turbulence can be taken into account with the turbulence factor K_T. Some guide values for K_T are:

- Normal turbulence:
  - Abutment walls of rivers $K_T \approx 1.0$
- Increased turbulence:
  - River bends $K_T \approx 1.5$
  - Downstream of stilling basins $K_T \approx 1.5$
- Heavy turbulence:
  - Hydraulic jumps $K_T \approx 2.0$
  - Strong local disturbances $K_T \approx 2.0$
  - Sharp bends $K_T \approx 2.0$ (to 2.5)
- Load due to water (screw) jet $K_T \approx 3.0$ (to 4.0)

5.3.2.4 *Depth Parameter,* K_h: With the depth parameter K_h, the water depth is taken into account, which is necessary to translate the depth-averaged flow velocity into the flow velocity just above the revetment. The depth parameter also depends on the development of the flow profile and the roughness of the revetment. The following formulas are recommended. For a fully developed velocity profile,

$$K_h = \frac{2}{\log(12\,h/k_s)^2} \tag{5.9a}$$

For a nondeveloped profile,

$$K_h = \left(\frac{h}{k_s}\right)^{-0.2} \tag{5.9b}$$

For very rough flow ($h/k_s < 5$),

$$K_h = 1.0 \tag{5.9c}$$

where h is the water depth (m) and k_s is the equivalent roughness: The equivalent roughness depends on the type of revetment/geosystem; for riprap, k_s is usually equal to 1, or twice the nominal diameter of the stones, for bags it is approximately equal to the thickness (d), for mattresses it depends on the type of mattress; k_s is about 0.05 m for smooth types and about the height of the rib for articulating mats. In the case of dimensioning the revetment on a slope, the water level at the toe of the slope must be used for h.

5.3.2.5 *Slope Parameter,* K_s. The stability of revetment elements also depends on the slope gradient under which the revetment is applied, in relation to the angle of internal friction of the revetment. This effect on the stability is taken into account with the slope parameter K_s, which is defined as follows:

$$K_s = \sqrt{1 - \left(\frac{\sin \alpha}{\sin \theta}\right)^2} \qquad (5.10a)$$

$$= \cos \alpha \sqrt{1 - \left(\frac{\tan \alpha}{\tan \theta}\right)^2}$$

or

$$K_s = \cos \alpha_b \qquad (5.10b)$$

where α is the transversal slope of the bank (°), θ the angle of internal friction of the revetment material, and α_b the slope angle of the river bottom (parallel along the flow direction) (°). The following values of θ can be assumed as a first approximation: 40° for riprap, 30 to 40° for sand-filled systems, and 90° for stiff and anchored mortar-filled mattresses and (cabled) blockmats ($K_s = \cos \alpha$). However, for flexible nonanchored mattresses and block mats (units without contact with the neighboring units), this value is much lower, usually about three-fourths of the friction angle of the sublayer. In case of geotextile mattress and block mats connected to geotextile lying on a geotextile filter, θ is about 15 to 20°. The advantage of this general design formula of Pilarczyk is that it can be applied in numerous situations. The disadvantage is that the scatter in results, as a result of the large margin in parameters, can be rather wide. With a downward flow along a steep slope, it is difficult to determine or predict the flow velocity, because the flow is very irregular. In such a case, formulas based on the discharge are developed (Pilarczyk, 1998).

5.3.3 Soil Mechanical Stability

The water motion on a revetment structure can also affect the subsoil, especially when this consists of sand. Geotechnical stability is dependent on the permeability and stiffness of the grain skeleton and the compressibility of the pore water (the mixture of water and air in the pores of the grain skeleton). Wave pressures on the top layer are passed on delayed and damped to the subsoil under the revetment structure and to deeper layers (as seen perpendicular to the slope) of the subsoil. This phenomenon takes place over a larger distance or depth as the grain skeleton and the pore water are stiffer. If the subsoil is soft or the pore water more compressible (because of the presence of small air bubbles), the compressibility of the system increases and large damping of the water pressures over a short distance may occur. Because of this, water under-tension and over-tension alternately may develop in the subsoil and corresponding to this, increasing and decreasing grain pressure. It can lead to sliding or slip circle failure (Figure 5.3).

The design method with regard to geotechnical instability is presented in the form of design diagrams. An example is given in Figure 5.4 [for more diagrams and details, see Pilarczyk (1998)]. The maximum wave height is a function of the sum of the cover layer weight (ΔD) and filter thickness (b_f).

5.3.4 Filters

Granular and/or geotextile filters can protect structures subjected to soil erosion when used in conjunction with revetment armor such as riprap, blocks and block mats, gabions and mattresses, asphalt or concrete slabs, or any other conventional armor material used for erosion control (PIANC, 1987a, 1992).

However, there is still a misunderstanding about the function of geotextiles in the total design of these structures, especially in comparison with the granular filters. In this section the general principles of designing revetments incorporating granular or geotextiles are reviewed. Attention is paid to replacing a granular filter by a geotextile, which may often lead to geotechnical instability. Furthermore, it appears that a thicker granular filter gives greater geotechnical stability but a lower cover layer

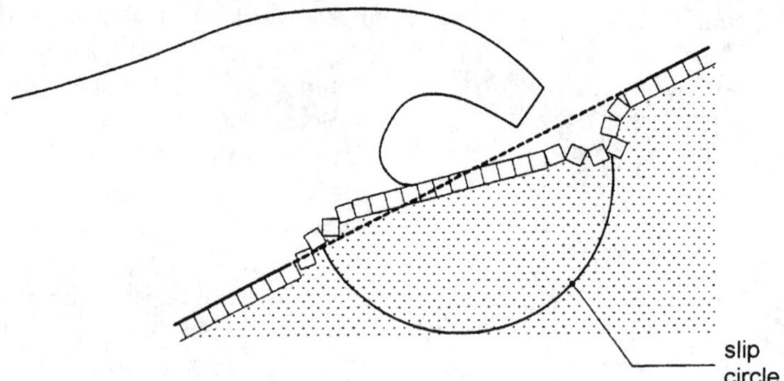

Figure 5.3 Development of an S-profile and possible local sliding in sand.

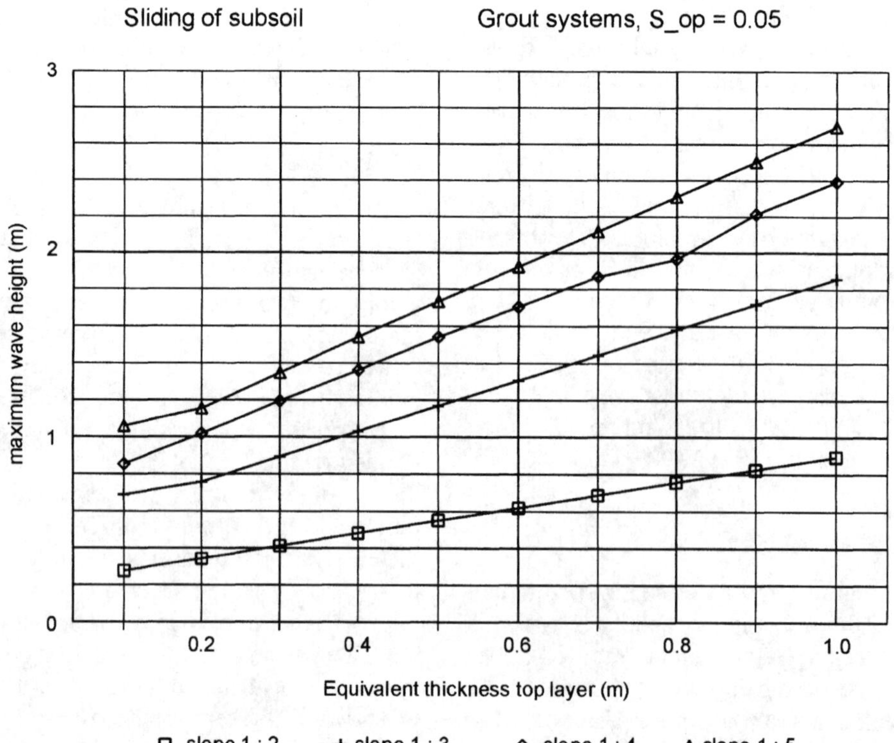

Figure 5.4 Geotechnical stability; design diagram for mattresses and $H_s/L_{op} = 0.05$.

stability (uplift of blocks). The conclusion is therefore that the wave loads must be distributed (balanced) adequately over the sand (shear stress) and the cover layer (uplift pressure). Too much emphasis on one failure mechanism can lead to another mechanism.

Filters have two functions: erosion prevention and drainage. Traditional design criteria for filters are that they should be "geometrically tight" and that the filter permeability should be larger than the base (soil) permeability. However, it results in a large number of layers, which are often unnecessary, uneconomical, and difficult to realize. In several cases a more economical filter design can be realized
using the concept of geometrically open filters (e.g., when the hydraulic loads/gradients are too small to initiate erosion). Recently, some criteria for geometrically open filters, including geotextiles, were developed (and are still under further development). However, application of these criteria requires knowledge or prediction of the hydraulic loads.

In the cases when the erosion exceeds an acceptable level, a filter construction is a proper measure for solving this problem. In revetment structures geotextiles are used primarily to protect the subsoil from washing away by the hydraulic loads, such as waves and currents. Here the geotextile replaces a granular filter. Unfortunately, merely replacing a granular filter by a geotextile can endanger the stability of other components in the bank protection structure. In the present section we show that designing a structure is more than just a proper choice of geotextile.

Filter structures can be realized by using granular materials (i.e., crushed stone), bonded materials (i.e., sand asphalt, sand cement), and geotextiles, or a combination of these materials. Typical filter compositions are shown in Figure 5.5. The choice between a granular filter, bonded filter, or geotextile depends on a number of factors. In general, a geotextile is used because of easier placement and relatively lower cost. For example, placement of a granular filter under water is usually a serious problem; quality control is very difficult, especially when placement of thin layers is required.

When designing with geotextiles in filtration applications, the basic concepts are essentially the same as when designing with granular filters. The geotextile must allow the free passage of water (permeability function) while preventing the erosion and migration of soil particles into the armor or drainage system (retention function).

In principle, the geotextile must remain more permeable than the base soil and must have pore sizes small enough to prevent migration of the larger particles of base soil. Moreover, concerning the permeability, not only the opening size but also the number of openings per unit area (percent open area) is of importance (Pilarczyk, 1999).

It has to be stressed that geotextiles cannot always replace the granular filter completely. A granular layer can often be needed to reduce (damp) the hydraulic loadings (internal gradients) to an acceptable level at the soil interface. After that, a geotextile can be applied to fulfill the filtration function.

In respect to the filters for erosion control (granular or geotextile), a distinction can be made between geometrically tight filters, geometrically open filters, and transport filters (when a limited settlement is allowed). Only geometrically tight filters are discussed. For other type of filters, readers are referred to Pilarczyk (1999).

5.3.4.1 Design Criteria for Geometrically Tight Granular Filters. In this case there will be no transport of soil particles from the base, independent of the level of hydraulic loading. That means that the openings in the granular filter or geotextile are so small that the soil particles are physically not able to pass the opening. This principle is illustrated in Figure 5.6 for granular filters. The main design rules (cri-

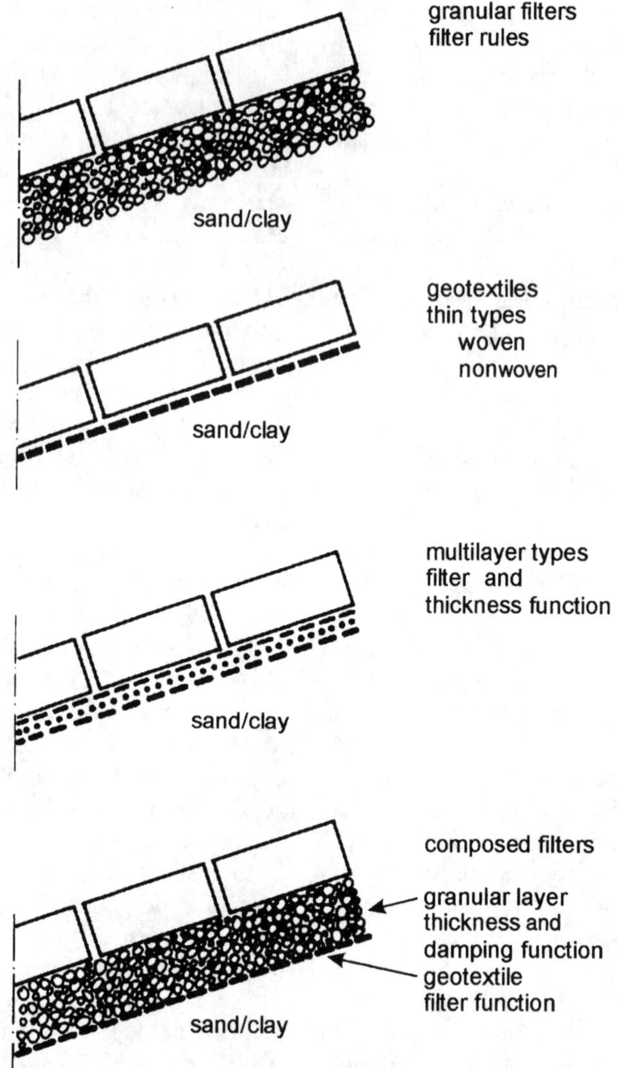

Figure 5.5 Examples of filters.

teria) for geometrically tight (closed) granular filters and geotextiles are summarized below. The more detailed information on design of geotextile filters is given in Pilarczyk (1999).

The soil tightness of the initial situation can be checked by means of the well-known criteria for granular filters. *Interface stability* (also called the *piping criterion*) is given by

$$\frac{D_{f15}}{D_{b85}} \leq 4 \text{ to } 5 \tag{5.11}$$

where D_{f15} is the grain size (m) of the filter layer (or cover layer) which is exceeded by 15% of the material by weight, and D_{b85} is the grain size (m) of the base material (soil)

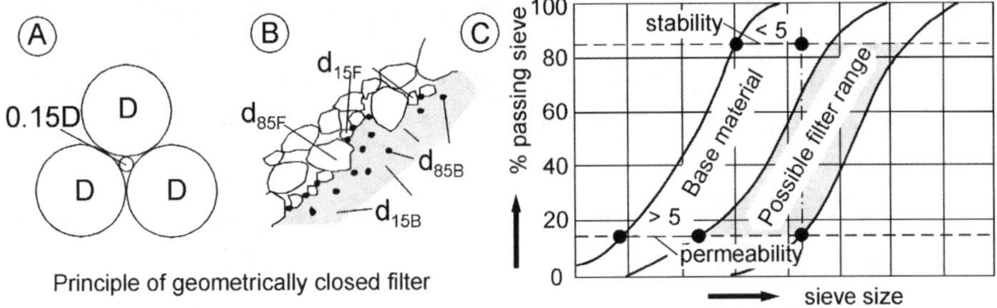

Figure 5.6 Principles of geometrically tight filters.

which is exceeded by 85% of the material by weight. The factor 4 in eq. (5.11) was given by Terzaghi and Peck (1967). The factor 5 is determined for normal wide-graded materials. Sometimes a similar equation is defined as

$$\frac{D_{f50}}{D_{b50}} < 6 \text{ to } 10 \tag{5.12}$$

However, eq. (5.12) is less general than eq. (5.11) and can be used for small gradations only. Therefore, eq. (5.11) is recommended for general use. However, in the case of very wide gradation, the situation requires an additional check with respect to the internal migration. In this respect, an important parameter is the *uniformity coefficient* Cu, defined by eq. (5.13) and the shape of the sieve curve:

$$\text{Cu} = \frac{D_{b60}}{D_{b10}} \tag{5.13}$$

Internal stability can be judged roughly by the following rules:

$$D_{10} < 4D_5 \tag{5.14a}$$

$$D_{20} < 4D_{10} \tag{5.14b}$$

$$D_{30} < 4D_{15} \tag{5.14c}$$

$$D_{40} < 4\ D_{20} \tag{5.14d}$$

The permeability criterion is given by

$$\frac{D_{f15}}{D_{b15}} > 5 \tag{5.15}$$

5.3.4.2 Summary of Design Rules for Geotextiles

Current Definitions for Geotextile Openings. There are a large number of definitions of the characteristic of geotextile openings. Moreover, there are also different test (sieve) methods for the determination of these openings (dry, wet, hydrodynamic, etc.) which depend on national standards. These all make the comparison of test results very difficult or even impossible. That also explains the necessity of international standarization in this field. Some of the current definitions follow.

- O_{90} corresponds with the average sand diameter of the fraction of which 90% of the weight remains on or in the geotextile (or 10% passes the geotextile) after 5 minutes of sieving (method: dry sieving with sand).
- O_{98} corresponds with the average sand diameter of the fraction of which 98% of the weight remains on or in the geotextile after 5 minutes of sieving. O_{98} gives a practical approximation of the maximum filter opening and therefore plays an important role in the sand tightness criterion for a

geotextile in strong cyclic loading situations. O_{98} is also referred to as O_{max}.

- O_f is the filtration opening size (FOS). O_f is comparable with O_{95} (hydrodynamic sieve method);
- AOS is the apparent opening size (according to the ASTM method), also called the EOS (the effective opening size). The AOS is determined by sieving spherical glass particles of known size through a geotextile. The AOS, also frequently referred to as O_{95} (dry sieve method), is defined as a standard sieve size, x (mm), for which 5% or less of the glass particles pass through the geotextile after a specified period of sieving.
- D_w is the effective opening size, which corresponds with the sand diameter of the fraction of which 10%, determined by the wet sieve method, passes through the geotextile. D_w is comparable with O_{95}.

The transport of soil particles within a grain structure is possible when there is enough space and a driving force (groundwater pressure, hydraulic gradients within the soil). In most cases it is the intention to prevent the transport of small soil particles in the subsoil, and therefore the term *soil tightness* is used and not the term *space for transport* or *pore volume* (in the case of the transport of water, the terms *pore volume* and *water permeability* are used). The relation between pore magnitude and grain diameter can be characterized by: pore diameter ≈20% of the grain diameter. Just as for the characterization of the performance of a grain structure with regard to the transport of soil particles, for geosynthetics, too, the term *soil tightness* is used.

As mentioned earlier (Figure 5.6), in a theoretical case when the soil is composed of spheres of one size diameter, all spheres can be retained if all apertures in the geosynthetic are smaller than the diameter of the spheres. Usually, the soil consists of particles with different diameters and shapes, which is reflected in the particle-size distribution curves. Smaller particles can disappear straight across the geosynthetic by groundwater current. In this case the retained soil structure can function as a natural filter (Figure 5.7). The better the soil particles are distributed, the better the soil tightness of the soil structure is effected. Smaller soil particles get stuck into the spaces between larger ones, and the soil structure prevents the flow of fine particles. When certain particle-size fractions are lacking, the soil structure is not stacked very well and cavities develop through which erosion can occur. The displacement of soil particles depends not only on the soil tightness but also on the hydraulic gradient in the soil structure. Morever, the dynamic effects due to heavy wave loading may not allow the forming of a natural filter, and the process of washing out may continue.

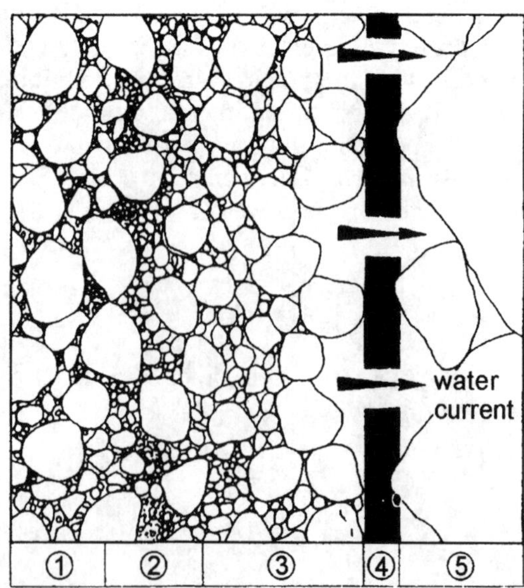

Figure 5.7 Natural filter with a soil-retaining layer. 1, Original soil structure; 2, filter zone; 3, filter cake; 4, geosynthetic; 5, revetment.

According to some researchers, the forming of a natural filter is possible only for stationary flow (CUR, 1993). However, this is also possible for nonstationary flow, for small values of the hydraulic gradients. For heavy wave attack (i.e., exposed breakwaters) this is usually not the case. In extreme situations, soil liquefaction is even possible. In such situations the soil particles can still reach the surface of a geotextile and be washed out.

To judge the risk of washout of soil particles through the geosynthetics, some aspects have to be considered. An important factor is the internal stability of the soil structure. In the case of loose particle stacking of the soil, many small soil particles may pass through the geosynthetic before a stable soil structure is developed near the geosynthetic. Also, proper compaction of soil is very important for the internal stability of soil. The internal stability is defined by the uniformity coefficient Cu [see eq. (5.13)]. It is defined as D_{b60}/D_{b10}. If this ratio is smaller than 6 (to 10), the soil structure is considered internally stable. In the case of vibration (e.g., caused by waves or by traffic), stable soil structures can be disturbed. To avoid such situations, the subsoil has to be compacted in advance and a good junction between geosynthetic and subsoil has to be guaranteed, and possibly, a smaller opening of geotextile must be chosen.

The shape of the sieve curve also influences the forming of a natural filter. Especially when Cu > 6, the shape of the base gradation curve and its internal stability must be taken into account (Pilarczyk, 1999). For a self-filtering linearly graded soil, the representative size corresponds to the average grain size, D_{b50}. For a self-filtering gap-graded soil, this size is equal to the lower size of the gap. For internally unstable soils, this size would be equivalent to D_{b30} to optimize functioning of the filter system. It is assumed that the bridging process involved would not retrogress beyond a limited distance from the interface.

Soil Tightness. With respect to the soil tightness of geotextiles, many criteria for geometric soil tightness have been developed and published in the past (Pilarczyk, 1999). An example of such design criteria, based on Dutch experience, is presented in Table 5.1. An additional requirement is that the soil should be internally stable. The internal stability of a grain structure is expressed in the ratio between D_{b60} and D_{b10}. As a rule, this value has to be smaller than 10 to guarantee sufficient stability.

However, in many situations, additional requirements will be necessary, depending on the local situation. Therefore, for design of geometrically tight geotextiles, the method applied in Germany can be recommended (see Heerten, 1982a and b; PIANC, 1987a; BAW, 1993). In this method a distinction is made between stable and unstable soils. Soils are defined as *unstable* (susceptible to down-slope migration) when the following specifications are fulfilled:

- A proportion of particles must be smaller than 0.06 mm.
- The soil must be fine, with a plasticity index (I_p) smaller than 0.15 (thus, it is not a cohesive soil).

Table 5.1 Design requirements for geosynthetics with a filter and separation function

Description Filter	Function/Soil Tightness
Stationary loading	$O_{90} \leq 1$ (to 2) D_{b90}
Cyclic loading with natural filter (stable soil structure)	$O_{98} \leq 1$ (to 2) D_{b85}
Cyclic loading without a natural filter (unstable soil structure)	
When washout effects acceptable	$O_{98} \leq 1.5D_{b15}$
When washout effects not acceptable	$O_{98} \leq D_{b15}$

- 50% (by weight) of the grains will lie in the range $0.02 < D_b < 0.1$ mm.
- The soil will be clay or silt with Cu < 15.

If I_p is unknown at the preliminary design stage, the soil may be regarded as a problem soil if the clay size fraction is less than 50% of the silt size fraction. The design criteria are presented in Table 5.2. More detailed information may be found in Pilarczyk (1999).

In the case of fine sand or silty subsoils, however, it can be very difficult to meet these requirements. A more advanced requirement is based on hydrodynamic sand tightness, that is, that the flow is not capable of washing out the subsoil material, because of the minor hydrodynamical forces exerted (although the apertures of the geotextile are much larger than the subsoil grains).

Requirements Concerning Water Permeability. To prevent the forming of water pressure (uplift) in the structure, causing loss of stability, the geotextile has to be water permeable. One has to strive for an increase of water permeability of a construction in the direction of the water current. In the case of riverbank protection, it means that the permeability of the geotextile has to be larger than the permeability of the soil on which the geotextile has to be applied. In the case of a dike slope or dike foundation, the geosynthetic is often applied on an impermeable layer of clay. Proper permeability of geotextiles is very important in respect to the stability of relatively less permeable cover layers, as, for example, block and block mats. When a geotextile lies directly under the cover layer, it reduces the open area of the cover layer considerably, and as a result, the uplift forces increase (see Example 5.1 in Section 5.4.2). The water permeability of woven fabrics and nonwovens may decrease in the course of time, owing to the fact that fine soil particles, which are transported by the groundwater flow from the subsoil, block the openings in the geotextile, or migrate into the pores of the geosynthetic (clogging).

To prevent mineral clogging, the pore size of the geotextiles has to be chosen as large as possible; but, of course, this pore size still has to meet the requirements for soil tightness. The danger of clogging increases when the soil contains more than 20% silt or in the case of gap grading of a soil. On the other hand, there usually is no danger of clogging when the total hydraulic gradient (over the subsoil and geotextile together) is less than 3, or when the subsoil is well graded. In all situations it holds that the soil must be internally stable. For less critical situations, no clogging can be expected if:

$$\text{Cu} > 3: \quad \frac{O_{95}}{D_{b15}} > 3 \qquad (5.16a)$$

Table 5.2 Design criteria for geometrically soil-tight geotextiles[a]

	$D_{b40} < 60$ μm		$D_{b40} > 60$ μm	
	Stable Soil	Instable Soil	Stable Soil	Instable Soil
Stationary loading	$O_{90} < 10D_{b50}$ $O_{90} < 2D_{b90}$	$O_{90} < 10D_{b50}$ $O_{90} < D_{b90}$	$O_{90} < 5D_{b10}\text{Cu}^{1/2}$ $O_{90} < 2D_{b90}$	$O_{90} < 5D_{b10}\text{Cu}^{1/2}$ $O_{90} < D_{b90}$
Dynamic loading		$O_{90} < D_{b90}$ $O_{90} < 0.3$ mm (300 μm)	$O_{90} < D_{b90}$	$O_{90} < 1.5\ D_{b10}\text{Cu}^{1/2}$ $O_{90} < D_{b50}$ $O_{90} < 0.5$ mm

[a] O_{90} is determined by the wet sieve method.

Cu < 3: criterion of internal stability of soil should be satisfied and/or geotextile with maximum opening size from soil-tightness criteria should be specified

(5.16b)

In respect to the water permeability of geosynthetics and geotextiles, a distinction should be made between "normal to the interface" and "parallel to the interface," For geotextile filters, permeability parallel to the interface is of the importance, while for drainage structures the permeability normal to the interface is of the most importance. As a general design criterion for flow normal to the interface, one can hold that the water permeability of a geosynthetic/geotextile has to be greater than that of the soil at the side from where the water flow comes. As a rule, one can keep to

$$k_{\text{geotextile (filter)}} = k_{\text{soil}} \times \text{factor} \tag{5.17}$$

where k_s and k_g are usually (basically) defined for laminar conditions. For normal (stationary) conditions and applications and clean sands, a factor of 2 is sufficient to compensate for the effect of blocking. If a geotextile is permeable with a factor of 10 more than the (noncohesive) subsoil, overpressure will usually not occur, neither below the geosynthetic nor in the case of reduced permeability caused by clogging or blocking. However, for special applications (i.e., for dam-clay cores with a danger of clogging), this factor can be 50 or more.

Detailed treatments with respect to the effect of hydraulic loading on permeability characteristics of the geotextile and possible interactions with the subsoil may be found in Van Santvoort (1994) and Pilarczyk (1999). However, basic information on this aspect as required for total design with geotextiles is given below.

Water Permeability Normal to the Interface. The function of the water permeability requirement in the total approach of designing with geotextiles is to bring the design of the filter into harmony with the subsoil. The requirement that excess pressures should not occur means that the eventual loss of stability at the filter occurs no sooner than the similar loss of stability does in the subsoil (i.e., migration of particles, softening of the subsoil and resulting sliding, etc.) as a consequence of groundwater flow (critical gradients). The basic (starting) requirement is that the gradient over the geotextile should maximally be equal to the gradient of the subsoil:

$$i_n \leq i_b \tag{5.18}$$

where i_b is the gradient in sublayer (or in subsoil, i_s) and i_n is the gradient normal to the geotextile.

The permeability of a geotextile can be characterized by the permeability coefficient $k_g = k_n$ (m/s) or by the permittivity ψ (s^{-1}) (see also Figure 5.8). Permittivity can be calculated directly from test results and expresses the rate of flow through the geotextile per unit area and per unit hydraulic head, and it is also defined as permeability per unit thickness of geotextile:

$$\psi = \frac{Q}{A\,\Delta h_g} = \frac{v_f}{\Delta h_g} = \frac{k_n}{T_g} \tag{5.19}$$

where ψ is the permittivity (s^{-1}), Q the flow rate through the geotextile (m^3/s), v_f the filtration velocity (m/s), A the surface area of geotextile (m^2), Δh_g the hydraulic head difference across the geotextile (m), k_n the permeability coefficient of the geotextile (k_g), normal to the interface (m/s), T_g the thickness of the geotextile (m). [*Note:* The term k_n/T_g in eq. (5.19) is often used for rough estimation of the permeability coefficient for other (not tested) thicknesses for the same type of geotextile.]

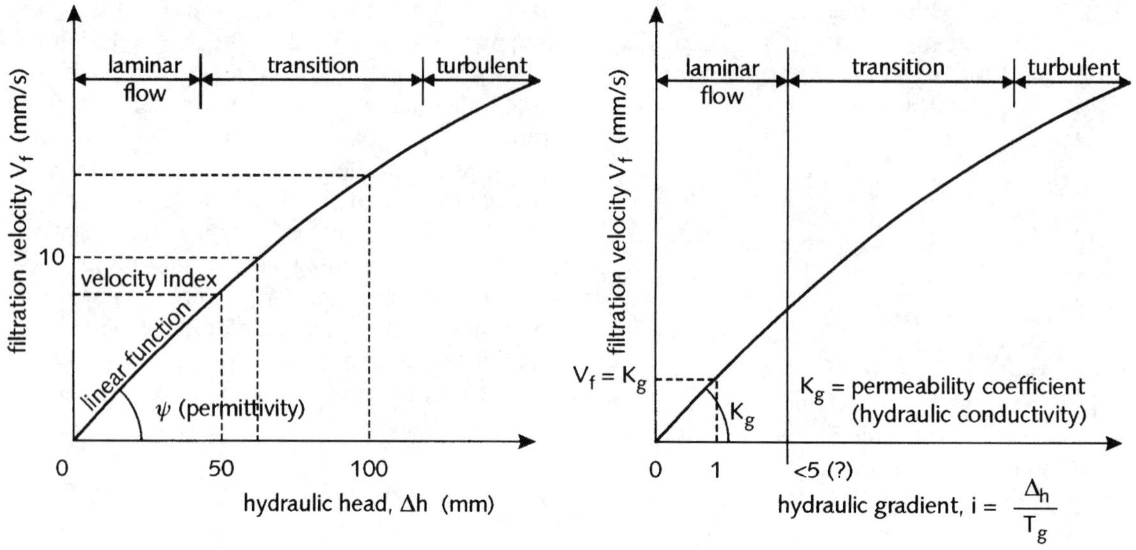

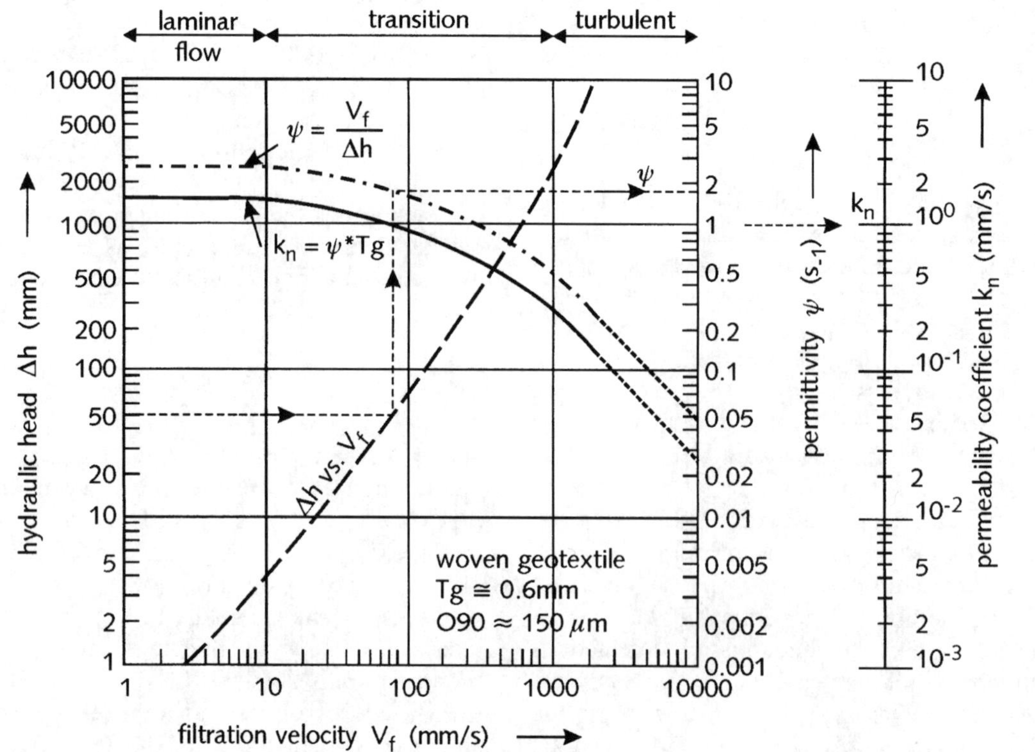

Figure 5.8 Example of the relationships between permittivity and hydraulic conductivity (permeability coefficient) as a function of hydraulic head.

The main problem of using permittivity is the definition of the thickness of geotextile. Usually, thickness under a normal stress of 2 kPa is used. Definitions and an example of test results concerning the determination of permittivity and permeability for geotextiles are presented in Figure 5.8. Combining formulas (5.17) and (5.18) with formula (5.19), and applying a continuity principle, yields

$$k_n i_n = k_b i_b \qquad (5.20)$$

where k_b is the permeability coefficient of the base material (subsoil, k_s) (m/s) and i_b is the gradient in the base material; it provides

$$\psi = c k_b \frac{i_b}{\Delta h_g} \approx c_\psi k_b \qquad (5.21)$$

In general, permittivity of a certain geotextile is a function of the hydraulic head. Only in the zone of laminar flow is the permittivity more or less constant (see Figure 5.8). When permittivity is defined outside this zone, the associated hydraulic head (Dh) should be mentioned.

The total safety factor c_y incorporates a number of uncertainties (i.e., permeability of soil, the loss of permeability due to the effects of clogging and stress, etc.) and, depending on application, can be equal to 10^3 for clean sands and up to 10^5 for critical soils and severe applications in dams (see CFGG, 1986). The last value seems to be rather conservative.

Holtz et al. (1997) propose to use, in addition to other permeability criteria, the permittivity criterion related directly to the soil type, defined by a certain percentage of passing a sieve of 0.075 mm. These criteria were established originally by the U.S. Federal Highway Administration (FHWA, 1995). These permittivity requirements are:

- $\psi \geq 0.5\ s^{-1}$ for soils with < 15% passing the 0.075-mm sieve.
- $\psi \geq 0.2\ s^{-1}$ for soils with 15 to 50% passing the 0.075-mm sieve.
- $\psi \geq 0.1\ s^{-1}$ for soils with > 50% passing the 0.075-m sieve.

The flow rate Q through the geotextile can be defined as

$$Q = v_f A = \psi\ \Delta h_g\ A = k_n A i_g \qquad (5.22)$$

Discussion. There are a large number of often very unclear and confusing definitions of permeability of geotextiles, especially when the permeability of the geotextile must match the permeability of a certain soil. The basic equation, for both soil and geotextile, is Darcy's equation; $v_f = ki$, which is valid for laminar flow conditions. The hydraulic gradient i is the average hydraulic gradient in the soil. For example, for the gradient along (parallel to) geotextile in revetments with a thick granular layer above, loaded by run-up and run-down, often a tangent of slope angle can be applied (tan α). However, for thin granular filters (layers) this gradient can be much larger than tan α. For other applications i can be estimated by using a conventional flow net analysis for seepage through dikes and dams or from a rapid drawdown analysis.The permeability of geotextiles is characterized by a number of different (national) index tests. Therefore, at this moment, the standard specification sheets provided by manufacturers include such definitions as: the permeability (filtration velocity, v_f) at the hydraulic head, equal to 50, 100, or even 250 mm; permittivity defined at the standard filtration velocity (v_f), equal to 10 mm/s (at a certain hydraulic head); the head loss index corresponding to a filtration velocity of 20 mm/s; and so on.

Actually, CEN/CR ISO (1998) prepared a European standard introducing only one index test, the *velocity index*, which defines the filtration velocity corresponding to a head loss of 50 mm across a specimen. The flow velocity

v_f, expressed in mm/s, equals the unit discharge q expressed in $l/m^2 \cdot s$. However, to be able to draw conclusions about the proper choice of permeability for various conditions and applications, it is necessary to perform (to measure) the full permeability characteristics and prepare a collective plot of the velocity v_f and head loss Dh for each specimen. The test range must be sufficiently wide to allow determination of the permeability parameters for laminar flow. If the full permeability characteristics of the geotextile product have been established previously, then for checking purposes it can be sufficient to determine the velocity index at a head loss of 50 mm only.

In case only one or two test data with standard index specifications (v_f, Δh, and T_g) are known, the approximate estimation of permeability can be done by using the equation $v_f = ki^m$, where $i = \Delta h / T_g$ and $0.5 \leq \text{m} \leq 1.0$ ($m = 1$ for laminar flow and $m = 0.5$ for turbulent flow). When only one point is available, the rough approximation can be obtained by applying $m = 0.7$.

By plotting a line through two points on log-log paper, the exponent m can be determined, and for $i = 1$, k can be approximated. By using this equation the permeability can be extrapolated roughly to the required conditions/definitions (e.g., $k_g = v_f T_g / \Delta h$ or $kg = \psi T_g$ at $\Delta h = 50$ mm, or k_g at $v_f = 10$ mm/s, or k_g or ψ for laminar conditions) and can be used as a first approximation. If in doubt, more data or additional tests can be requested.

The permeabilities defined outside the laminar zone can also be interpreted in the following way, for example, if k_g established at $v_f = 10$ mm/s is equal or larger than $k_{s(\text{soil})}$, the geotextile fulfills the requirement of permeability for filtration velocities (in the soil) lower than 10 mm/s. For larger filtration velocities a new estimation of k_g is needed, related to the higher v_f, to check the requirement $k_g > k_s$, because k_g will decrease for higher v_f in a transition or turbulent flow zone.

With regard to the hydraulic efficiency of the geotextile filter, full advantage should be taken on the permissible upper limits of the opening size, provided that the required mechanical filter effectiveness (soil tightness) is ensured (DVWK, 1993). The reason for this is that an open (and possibly, thicker) structure is generally superior to a dense structure with regard to the filter stability. Thus, when the permeability is decesive for the design, the largest admissible opening size resulting from the soil-tightness criteria should be used to ensure a permeability as high as possible. There are usually no problems with obtaining a sufficient permeability when $1 \leq O_{90}/D_{90} < 2$ is used as a soil-tightness criterion. Research results (Mlynarek, 1994) indicate that the permeability of a soil–geotextile system is defined primarily by the permeability of a soil. However, geotextile permeability will decrease in soil due to compression. It is of special importance for thick nonwovens. Therefore, the factor of safety should be increased accordingly to account for the critical nature of the application, type of geotextile, and the severity of the soil and hydraulic conditions.

This discussion indicates that there are still much uncertainties in proper application of the permeability criteria. However, for normal (stationary) conditions and stable soils it is usually not a real problem. For critical projects, performance tests simulating soil–geotextile interaction can be recommended.

5.4 STABILITY CRITERIA FOR PLACED BLOCKS AND BLOCK MATS

5.4.1 System Description

Placed block revetments (or stone/block pitching) are a form of protection lying between revetments comprised of elements that are disconnected, such as rubble, and monolithic revetments, such as asphalt/concrete slabs. Individual elements of a pitched block revetment are placed together tightly in a smooth pattern. This ensures that external forces such as waves and currents can exert little drag on the blocks and also that blocks support each other without loss of flexibility when there are local subsoil irregularities or settlement.

A (concrete) block mat is a slope revetment made of (concrete) blocks that are joined together to form a mat (Figure 5.9). The interconnection may consist of cables from block to block, of hooks connecting the blocks, or of a geotextile on which the blocks are attached with pins, glue, or by other means. The spaces between the blocks are usually filled with rubble, gravel, or slag. The major advantage of block mats is that they can to be laid quickly and efficiently and partly under water. Block mats are more stable than a setting of loose blocks, because a single stone cannot be moved in a direction perpendicular to the slope without moving nearby stones. It is essential to recognize that even with a small movement of an individual stone, a significant interactive force with the surrounding stones is mobilized. Large movements of individual blocks are not acceptable, because transport of filter material may occur. After some time, this leads to a serious deformation of the surface of the slope. Block mats are vulnerable at edges and corners. If two adjacent mats are not joined together, the stability is hardly better than that for pitched loose stones.

5.4.2 Design Rules with Regard to Wave Load

The usual requirement that the permeability of the cover layer should be larger than that of the underlayers cannot be met in the case of a closed block revetment and other systems with a low-permeable cover layer. Such a cover layer introduces uplift pressures during wave attack. In this case the permeability ratio of the cover layer and the filter, represented in the leakage length, is found to be the most important structural parameter, determining the uplift pressure. This is also the base of the analytical model.

The analytical model is based on the theory for placed stone revetments on a granular filter (CUR/TAW, 1995). In this calculation model, a large number of physical aspects are taken into account (see Figures 5.1, 5.2, and 5.6). In short, in the analytical model, nearly all physical parameters that are relevant to stability have been incorporated in the leakage length factor. The final result of the analytical model may, for that matter, again be presented as a relation such as eqs. (5.2) or (5.3), where $F = f(\Lambda)$. For systems on a filter layer, the leakage length Λ is given as

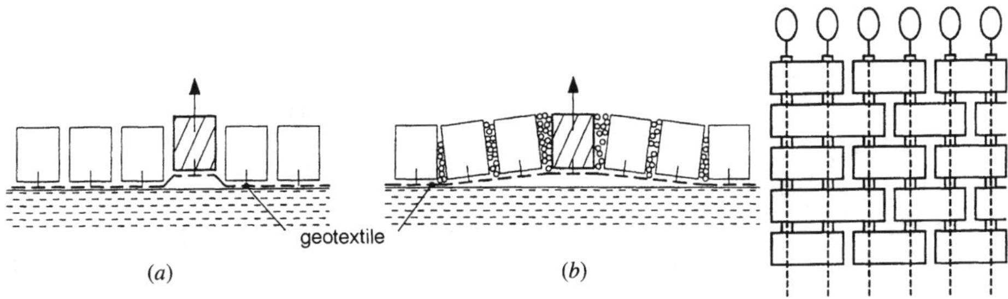

Figure 5.9 Examples of block mats: (a) standard block mat; (b) washed in with granular material strong interaction (interlocking).

$$\Lambda = \sqrt{\frac{b_f D k_f}{k'}} \quad \text{or} \quad \frac{\Lambda}{D} = \sqrt{\frac{b_f k_f}{D k'}} \qquad (5.23a)$$

where Λ is the leakage length (m), b_f the thickness of the filter layer (m), k_f the permeability of the filter layer or subsoil (m/s), and k' the permeability of the top (cover) layer (m/s).

In a system without a filter layer (directly on sand or clay, without gullies being formed under the top layer) the permeability of the subsoil (eventually, with gullies/surface channels), not the permeability of the filter layer, is filled in. To determine the thickness of the filter layer, it is necessary to know to what depth changes at the surface affect the subsoil. One can assume 0.5 m for sand and 0.05 m for clay. The values of D and Δ depend on the type of revetment. When schematically representing a block on a geotextile on a gully in sand, the block should be regarded as the top layer and the combination of the geotextile and the small gully as the filter layer (Figure 5.10). The leakage length can be calculated using

$$\Lambda = \sqrt{\frac{(k_f d_g + k_g T_g)D}{k'}} \qquad (5.23b)$$

where k_f is the permeability of the filter layer (gully) (m/s), d_g the gully depth (m), k_g the permeability of the geotextile (m/s), T_g the thickness of the geotextile (m), D the thickness of the top layer (m), and k' the permeability of the top layer (m/s).

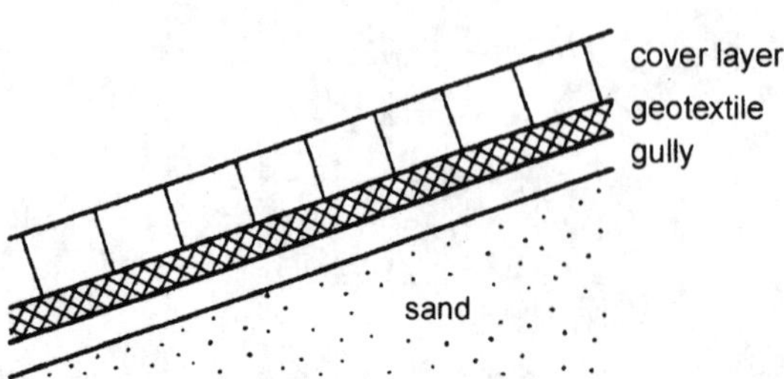

Figure 5.10 Schematization of a revetment with gully (cavity).

In the case of a geotextile situated directly under the cover layer, the permeability of the cover layer decreases drastically. Since the geotextile is pressed against the cover layer by the outflowing water, it should be treated as a part of the cover layer. Water flow trough the cover layer is concentrated at the joints between blocks, reaching very high flow velocities and resulting in a large pressure head over the geotextile. The presence of a geotextile may reduce k' by a factor 10 or more (Figure 5.11).

The leakage length clearly takes into account the relationship between k_f and k' as well as the thickness of the cover and filter layers. For the theory behind this relationship, reference should be made to literature (see CUR/TAW, 1995); Klein Breteler et al., 1998a and b). The pressure head difference that develops on the cover layer is greater with a large leakage length than with a small leakage length. This is due primarily to the relationship k_f/k' in the leakage length formula. The effect of leakage length on the dimensions of the critical wave for semipermeable revetments is apparent from the equations

$$\frac{H_{scr}}{\Delta D} = f\left(\frac{D}{\Lambda\,\xi_{op}}\right)^{0.67} = \frac{H_{scr}}{\Delta D} = f\left(\frac{D}{b}\frac{k'}{k}\right)^{0.33}$$

$$\xi_{op}^{-0.67} = \frac{H_{scr}}{\Delta D} = F\xi_{op}^{-0.67} \qquad (5.24)$$

where H_{scr} is the significant wave height at which blocks will be lifted out (m); $\xi_{op} = \tan\alpha/\sqrt{H_s/(1.56T_p^2)}$, the breaker parameter; T_p the wave period (s); Δ the relative volumetric mass of cover layer, $= (\rho_s - \rho)/\rho$; f the stability coefficient, dependent primarily on structure type and with minor influence of Δ, $\tan\alpha$, and friction; and F the total (black-box) stability factor.

These equations indicate the general trends and have been used together with measured

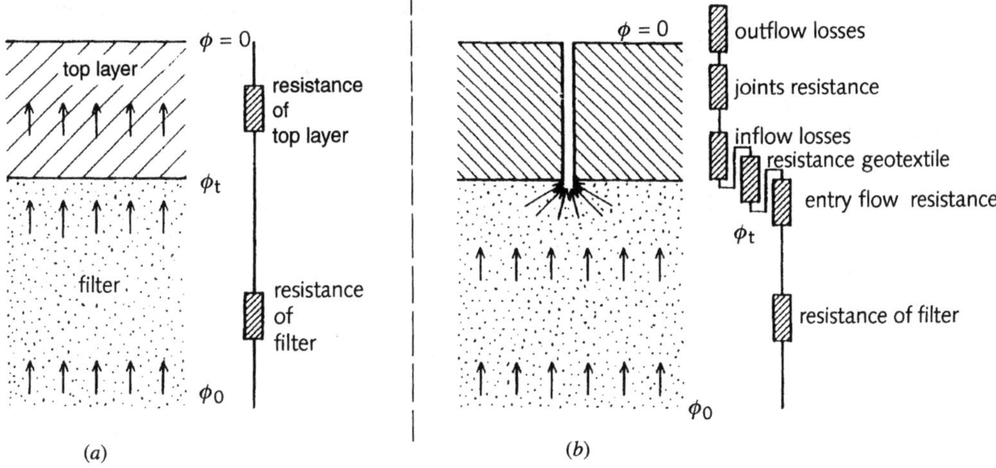

Figure 5.11 Combined flow resistance determining the permeability of a system: (*a*) homogeneous permeability; (*b*) unhomogeneous permeability.

data to set up the general calculation model (CUR/TAW, 1995; Pilarczyk, 1998). This method works properly for placed/pitched block revetments and block mats within the following range: $0.01 < k'/k_f < 1$ and $0.1 < D/b_f < 10$. Moreover, when $D/\Lambda > 1$, use $D/\Lambda = 1$, and when $D/\Lambda < 0.01$, use $D/\Lambda = 0.01$. The range of the stability coefficient is: $5 < f < 15$; the higher values refer to the presence of high friction among blocks or interlocking systems. The following values are recommended for block revetments:

- $f = 5$ for static stability of loose blocks (no friction between the blocks).
- $f = 7.5$ for static stability of a system (with friction between the units).
- $f = 10$ for tolerable/acceptable movement of a system at design conditions.

From these equations, neglecting the usually minor variations of f, it appears that:

- An increase in the volumetric mass, Δ, produces a proportional increase in the critical wave height. If ρ_b is increased from 2300 to 2600 kg/m^3, H_{scr} is increased by about 23%.
- If the slope angle is reduced from 1:3 to 1:4 (tan α from 0.33 to 0.25), H_{scr} is increased by about 20% (due to the breaker parameter, ξ_{op}).
- An increase of 20% in the thickness of the cover layer, D, increases H_{scr} by about 27%.
- A 30% reduction in the leakage length, Λ, increases H_{scr} by about 20%. This can generally be achieved by halving the thickness of the filter layer or by doubling the k'/k_f value. The latter can be achieved by approximation:
  - By reducing the grain size of the filter by about 50%, or
  - By doubling the number of holes in (between) the blocks, or
  - By making hole sizes 1.5 times larger, or
  - By doubling joint width between blocks.

Example 5.1. In 1983 the Armorflex mat on a 1:3 slope was tested on a prototype scale at

Oregon State University: closed blocks with thickness $D = 0.12$ m and open area 10% on two types of geotextiles and very wide-graded subsoil ($d_{15} = 0.27$ mm, $d_{85} = 7$ mm). In the case of a sand-tight geotextile, the critical wave height (instability of mat) was only $H_{scr} = 0.30$ m. In the case of an open net geotextile (opening size about 1 mm), the critical wave height was more than 0.75 m (maximum capacity of the wave flume). The second geotextile was 20 times more permeable than the first. This means that the stability increased by factor of $20^{0.33} = 2.7$.

In most cases the permeabilities of the cover layer and sublayer(s) are not known exactly. However, based on the physical principles described above, the practical black-box method has been established where parameter Λ and coefficient f are combined to one stability factor F. F depends on the type of structure, characterized by the ratios k'/k_f and D/b_f. With the permeability formulas from (CUR/TAW (1995) it is concluded that the parameter $(k'/k_f)A(D/b_f)$ ranges between 0.01 and 10, leading to a subdivision into three ranges of one decade each. Therefore, the following types are defined:

- *Low stability:* $(k'/k_f)(D/b_f) < 0.05$–0.1
- *Normal stability:* 0.05–$0.1 < (k'/k_f)(D/b_f) < 0.5$–$1$
- *High stability:* $(k'/k_f)(D/b_f) > 0.5$–1

For a cover layer lying on a geotextile on sand or clay, without a granular filter, the leakage length cannot be determined because the size of b_f and k cannot be calculated. The physical description of the flow is different for this type of structure. For these structures there is no such a theory as for the blocks on a granular filter. However, it has been proved experimentally that eqs. (5.3b) or (5.24) are also valid for these structures.

It can be concluded that the theory has led to a simple stability formula [eq. (5.24)] and a subdivision into four types of (block) revetment structure:

a1. Cover layer on granular filter, possibly including geotextile, low stability
a2. Cover layer on granular filter, possibly including geotextile, normal stability
a3. Cover layer on geotextile on sand
a4. Cover layer on clay or on geotextile on clay

The coefficient F is quantified for each structure type by way of fitting eq. (5.3b) to the results of a large collection of results of model studies from all over the world. Only large-scale studies are used because both the waves and the wave-induced flow in the filter should be well represented in the model. In the classification of structures according to the value of $k'D/k_f b_f$, the upper limit of $k'D/k_f b_f$ is 10 times the lower limit. Therefore, the upper limit of F of each structure type (other than a1.1) is assumed to be $10^{0.33}$ 2.14 times the lower limit, since $F = f(k'D/k_f b_f)^{0.33}$. A second curve is drawn with this value of F. In Table 5.3 all available tests are summarized, and for each type of structure a lower and upper boundary for the value of F is given (see also the example in Figure 5.12). The lower boundary gives with eq. (5.3b) a stability curve below which stability is guaranteed. Between the upper and lower boundaries the stability is uncertain. Whether the structure will be stable or not depends on various unpredictable influences. The upper boundary gives a curve above which instability is (almost) certain.

The results for structure type a3 (blocks on geotextile on sand) may be applied only if the wave load is small [$H_s < 1$ or 1.5 m (max.)] or to structures with a subsoil of coarse sand ($D_{50} > 0.3$ mm) and a gentle slope ($\tan \alpha < 0.25$), because geotechnical failure is assumed to be the dominant failure mechanism (instead

Table 5.3 Lower and upper values for *F*

Type	Description	Average Value		
		Low *F*	High *F*	Usual *F*
a1.1	Pitched irregular natural stones on granular filter	2.0	3.0	2.5
a1.2	Loose blocks/basalt on granular filter, low stability	3.0	5.0	3.5
a2	Loose blocks on granular filter, normal stability	3.5	6.0	4.5
a3	Loose blocks on geotextile on compacted sand/clay	4.0	7.0	5.0
a4	Linked/interlocked blocks on geotextile on good clay or on fine granular filter	5.0	8.0	6.0

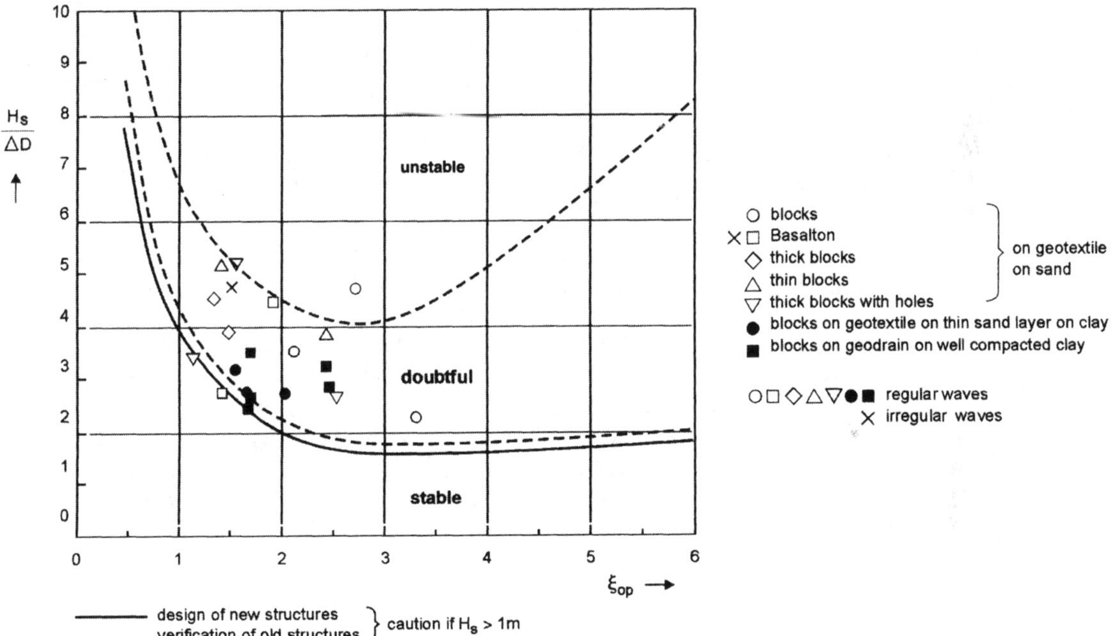

Figure 5.12 Example of a stability function for type a3 (loose blocks on geotextile on sand).

of uplift of blocks). Good compaction of sand is essential to avoid sliding or even liquefaction. For loads higher than $H = 1.2$ m, a well-graded layer of stone on a geotextile is recommended (e.g., layer 0.3 to 0.5 m for 1.2 m $< H <$ 2.5 m).

The results for structure type a4 can be used on the condition that clay of high quality with a smooth surface is used. A geotextile is recommended to prevent erosion during (long-duration) wave loading. The general design criteria for geotextiles on cohesive soils are given by Pilarczyk (1999).

In the case of loose blocks, an individual block can be lifted out of the revetment with a force exceeding its own weight and friction. It is not possible with the cover layers on linked or interlocking blocks. Examples of the second type are block mattresses, shiplap blocks, and cable mats. However, in this case high forces will be exerted on the connections between the blocks and/or geotextile. In the case of blocks connected to geotextiles (i.e., by pins), the stability should be treated as for loose blocks, to avoid the mechanical abrasion of geotextiles by moving blocks. In comparison with loose blocks, the lower boundary of stability of cabled mats can be increased by a factor of 1.25 (or 1.5, if additionally grouted.) Such an increase in stability is allowable only when special measures are taken with respect to proper connection between the mats. The upper boundary of stability ($F = 8$) remains the same for all systems. Application of this higher stability requires optimization of design. This optimization technique (including application of geometrically open but stable filters and geotextiles) can be found in CUR (1993) and CUR/TAW (1995).

To be able to apply to other semipermeable systems the design method for placed stone revetments under wave load, the following items may be adapted: the revetment parameter F, the (representative) strength parameters Δ and D, the design wave height H_s and the (representative) leakage length Λ. The basic formulas of the analytical model are presented in CUR/TAW (1995) and Pilarczyk (1998). Table 5.4 gives an overview of usable values for the revetment constant F in the black-box model for linked blocks (block mats). The terms *favorable, normal,* and *unfavorable* refer to the composition of the granular filter and the permeability ratio of the top layer and the filter layer (see CUR/TAW, 1995). In a case of a fine granular filter and a relatively permeable top layer, the total composition can be defined as favorable. In the case of a very coarse granular layer and a less permeable top layer the composition can be defined as unfavorable. For the case of blocks connected to a geotextile and concrete-filled mattresses on a filter layer, the construction can usually be defined as between unfavorable and normal, and the stability factor $F = 3.0$ to 3.5 (max. 4.0) can be used. For block mats and permeable mattresses on sand, $F = 5$ (max. 6.0) can be used. Higher values can also be used when extreme design loading is not very frequent or when the system is (repeatedly) washed in by coarse material providing additional interlocking. This wide range of recommended values for F gives only a first indication of a suitable choice. Furthermore, it is essential to check geotechnical stability using the design diagrams [see, e.g., Figure 5.4, and for a full set of diagrams, see Pilarczyk (1998, 1999)].

Table 5.4 Recommended values of revetment parameter *F* for block mats[a]

Type of Revetment	F (−)
Linked blocks on geotextile on sand	5–6
Linked blocks on geotextile on clay	
Good clay	5–6
Mediocre (sandy) clay	4.5–5
Linked blocks on a granular filter	
Favorable construction	5–6
Normal construction	4–5
Unfavorable construction	3–4

[a] The lower values are those for blocks connected to a geotextile; the higher values are those for cabled blocks.

5.5 STABILITY CRITERIA FOR CONCRETE-FILLED MATTRESSES

5.5.1 Concrete Mattresses

Characteristic of concrete mattresses are two geotextiles with concrete or cement between them. The geotextiles can be connected to each other in many patterns, which results in a variety of mattress systems, each having its own appearance and properties. Some examples are shown in Figure 5.13. The permeability of the mattress is one of the factors that determine the stability. It is found that the permeability given by suppliers is often the permeability of the

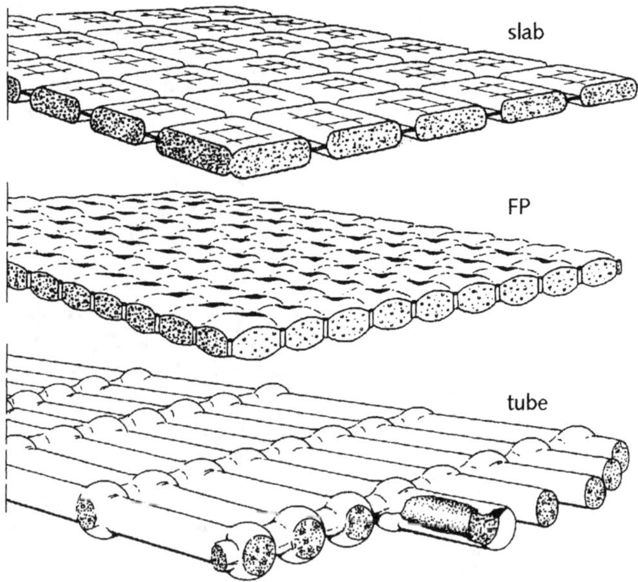

Figure 5.13 Examples of concrete-filled mattresses.

geotextile or of the filter points (Figure 5.14). In both cases, the permeability of the entire mattress is much smaller. High permeability of the mattress ensures that any possible pressure buildup under the mattress can flow away, as a result of which the uplift pressures across the mattress remain smaller. In general, with a subsoil of clay and silty sand, the permeability of the mattress will be higher than the permeability of the subsoil. Therefore, the water under the mattress can usually be discharged without excessive lifting pressures on the mattress. The permeability of the mattress will be lower than the permeability of the subsoil or sublayers if a granular filter is used or with a sand or clay subsoil having an irregular surface (gullies/cavities between the soil and the mattress). This will result in excessive lifting pressures on the mattress during wave attack.

5.5.2 Design Rules with Regard to Wave Load

The failure mechanism of the concrete mattress is probably as follows:

1. Cavities under the mattress will form as a result of uneven subsidence of the subsoil. The mattress is rigid and spans the cavities.
2. With large spans, wave impacts may cause the concrete to crack and the spans to collapse. This results in a mattress consisting of concrete slabs that are coupled by means of the geotextile.
3. With sufficiently high waves, an upward pressure difference over the mattress will occur during wave run-down, which lifts the mattress (Figure 5.1).
4. The pumping action of these movements will cause the subsoil to migrate, as a result of which an S-profile will form and the revetment will collapse completely.

It is assumed that local settlement of the subsoil will lead to free spans of the concrete mattress. Then the wave impact can cause the breaking of these spans if the ratio of H_s/D is too large for a certain span length (Figure

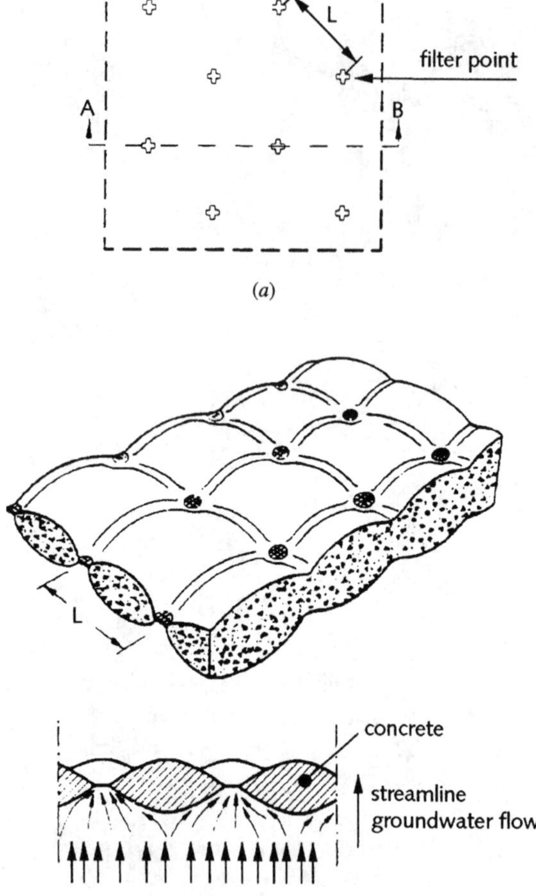

Figure 5.14 Principles of permeability of a filter point mattress: (*a*) top view; (*b*) cross section *AB*.

5.15). A calculation method is derived on the basis of an empirical formula for maximum wave impact pressure and the theory of simply supported beams. The collapsing of small spans (less then 1 or 2 m) is not acceptable, since these will lead to too many cracks. The empirical formula for the wave impact is (Klein Breteler et al., 1998a and b)

$$\frac{F_{\text{impact}}}{\rho g} = 7.2 H_s^2 \tan \alpha \qquad (5.25)$$

where F_{impact} is the impact force per meter of revetment (N).

Calculation has resulted in an average distance between cracks of only 10 to 20 cm for a 10-cm-thick mattress and wave height of 2 m. This means that at such an H_s/D ratio the wave impacts will chop the mattress to pieces. For a mattress 15 cm thick and a wave height of 1.5 m, the crack distance will be on the order of 1 m. Apart from the cracks due to wave impacts, the mattress should also withstand uplift pressures due to wave attack. These uplift pressures are calculated in the same way as for block revetments. For this damage mechanism the leakage length is important. In most cases the damage mechanism by uplift pressures is more important then the damage mechanism by impact. The representative/characteristic values of leakage length for various mattresses listed in Table 5.5 can be assumed.

Taking the foregoing failure mechanisms into consideration, the following design (stability) formula has been derived for the mattresses [eq. (5.3b)]:

$$\frac{H_s}{\Delta D} = \frac{F}{\xi_{\text{op}}^{2/3}} \qquad \text{with} \quad \left(\frac{H_s}{\Delta D}\right)_{\max} = 4 \qquad (5.26)$$

where Δ is the relative volumetric mass of the mattress (−) = $(\rho_s - \rho)/\rho$ [ρ_s the volumetric mass of concrete (kg/m^3)], D the mass per m^2/ρ_s (which can be called $D_{\text{effective}}$ or D_{average}), and F the stability factor (see below).

For an exact determination of the leakage length, one is referred to the analytical model (Klein Breteler et al., 1998a and b). However, besides mattresses of the tube mat (crib) type with relatively large permeable areas, for example, the other types are not very sensitive to the exact value of the leakage length. The following values of F can be recommended for use in design calculations:

- F = 2.5 or ($\leq$3) for low-permeable mattresses on (fine) granular filter

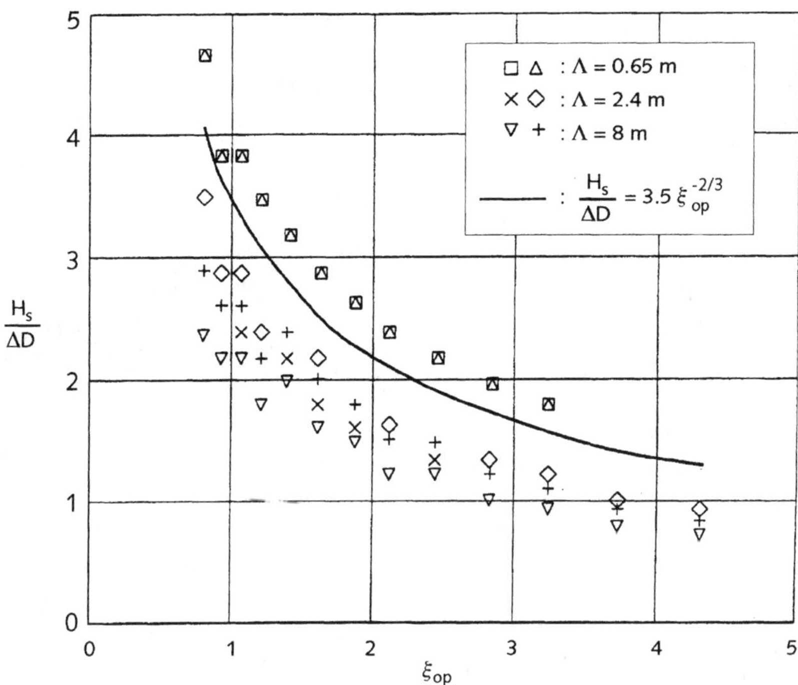

Figure 5.15 Calculation results for concrete mattresses ($H_s/\Delta D < 4$ because of acceptable crack distance due to impacts on spans).

Table 5.5 Leakage lengths for various mattresses

	Leakage Length Λ (m)		
Mattress	On Sand[a]	On Sand[b]	On Filter
Standard–filter point	1.5	3.9	2.3
Filter point	1.0	3.9	2.0
Slab	3.0	9.0	4.7
Articulated (crib)	0.5	1.0	0.5

[a]Good contact of mattress with sublayer (no gullies or cavities underneath).
[b]Pessimistic assumption: poor compaction of subsoil and presence of cavities under the mattress.

- $F = 3.5$ or (≤ 4) for low-permeable mattress on compacted sand
- $F = 4.0$ or (≤ 5) for permeable mattress on sand or fine filter ($D_{f15} < 2$ mm)

The higher values can be used for temporary applications or when the soil is more resistant to erosion (e.g., clay) and the mattresses are anchored properly.

5.6 STABILITY OF GABIONS

Gabions are made of rectangular baskets of wire mesh filled with stones. Waves and currents would easily wash away small stones, but the wire mesh prevents this. A typical gabion length is 3 to 4 m; width, 1 to 3 m; and thickness, 0.3 to 1 m. Gabions less then 0.5 m thick but with substantial length and width are usually called Renomattresses.

An important problem in this protection system is durability. Frequent wave or current attack can lead to failure of the wire mesh because of the continuously moving grains along the wires, finally cutting through. Another problem is corrosion of the mesh. Therefore, mesh with plastic coating or corrosion-resistant steel is used. On the other hand, the system is less suitable where waves and currents lead to frequent grain motion.

5.6.1 Hydraulic Loading and Damage Mechanisms

Wave attack on gabions will lead to complex flow over and through the gabions. During wave run-up the resulting forces by the waves will be directed opposite to the gravity forces. Therefore, the run-up is less hazardous then the wave run-down. As noted in Section 5.2, wave run-down leads to two important mechanisms: The downward-flowing water will exert a drag force on the tops of the gabions, and the decreasing phreatic level will coincide with a downward flow gradient in the gabions. During maximum wave run-down there will be an incoming wave that a moment later will cause a wave impact. Just before impact there is a "wall" of water, giving high pressure under the point of maximum run-down. Above the run-down point the surface of the gabions is almost dry, and therefore there is a low pressure on the gabions. The interaction of high and low pressure is shown in Figure 5.1.

A simple equilibrium of forces leads to the conclusion that the section from the run-down point to the phreatic line in the filter will slide down if:

- There is insufficient support from gabions below this section.
- The downward forces exceed the friction forces (roughly): $f < 2 \cdot \tan \alpha$ (f is the friction of the gabion on the subsoil and α is the slope angle.

From this criterion we see that a steep slope will easily lead to exceeding of the friction forces, and furthermore, a steep slope is shorter then a gentle slope and will give less support to the section that tends to slide down. Hydrodynamic forces such as wave attack and current can lead to various damage mechanisms. The damage mechanisms fall into three categories:

1. Instability of the gabions
   a. The gabions can slide downward, compressing downslope mattresses.
   b. The gabions can slide downward, leading to upward buckling of downslope mattresses.
   c. All gabions can slide downward.
   d. Individual gabions can be lifted out due to uplift pressures.
2. Instability of the subsoil
   a. A local slip circle can occur, resulting in an S-profile.
   b. The subsoil can wash away through the gabions.
3. Durability problems
   a. Moving stones can cut through the mesh.
   b. The mesh may be corroded.
   c. The mesh can be ruptured by mechanical forces (vandalism, stranding of ship, etc.).

5.6.2 Stability of Gabions under Wave Attack

An analytical approach to the development of uplift pressure in gabions can be obtained by applying the formulas for uplift pressure under an ordinary pitched block revetment, with as leakage length $\Lambda = 0.77D$. With this relation the stability relations according to the analytical model are also applicable to gabions. Substitution of values that are reasonable for gabions in stability relations according to (CUR/CIRIA, 1991) provides stability relations which indeed match a line through the measured points. After complicated calculations the uplift pressure in the gabions can be derived (Klein Breteler et al., 1998a and b). The uplift pressure depends on the steepness and height of the pressure front on the gabions (which is dependent on the wave height, period, and slope angle), the thickness of the gabions, and the level of the freatic line in the gabions. It is not dependent on the permeability of the gabions if the permeability is larger then the subsoil. The equilibrium of uplift forces and gravity forces leads to the following (approximate) design formula:

$$\frac{H_s}{\Delta D} = F\xi_{op}^{-2/3} \qquad \text{with } 6 < F < 9$$

$$\text{and slope of 1:3 } (\tan\alpha = 0.33) \qquad (5.27a)$$

or, using Pilarczyk's equation (5.3c) with $b = \frac{2}{3}$ and $F = 9$ (see Figure 5.16), yields

$$\left(\frac{H_s}{\Delta D}\right)_{cr} = \frac{F\cos\alpha}{\xi_{op}^{b}} = \frac{9\cos\alpha}{\xi_{op}^{2/3}} \qquad (5.27b)$$

where H_s is the significant wave height of incoming waves at the toe of the structure (m), Δ the relative density of the gabions (usually, $\Delta \approx 1$), D the thickness of the gabion (m), F the stability factor, ξ_{op} the breaker parameter [= $\tan\alpha/\sqrt{(H_s/1.56T_p^2}$, T_p the wave period at the peak of the spectrum (s)]. It is not expected that instability will occur at once if the uplift pressure exceeds the gravity forces. On the other hand, the result above turns out to be in good agreement with experimental results. Experimental verification of the stability of gabions is rather limited. Small-scale model tests have been performed by Brown (1979) and Ashe (1975) (Figure 5.16).

5.6.3 Motion of Filling Material

It is important to know if the filling material will start to move during frequent environmental conditions because it can lead to rupture of the wire mesh. Furthermore, the integrity of the system will be affected if large quantities of filling material are moved. During wave attack the motion of the filling material usually occurs only if $\xi_{op} < 3$ (plunging waves). Based on Van der Meer's formula for the stability of loose rock (CUR/CIRIA, 1991) and the assumption that filled gabions are more stable then loose rock, the following criterion is derived (Van der Meer formula with permeability factor $0.1 < P < 0.2$, number of waves $2000 < N < 5000$, and damage level $3 < S < 6$):

$$\frac{H_s}{\Delta_f D_f} = \frac{F}{\sqrt{\xi_{op}}} \qquad \text{with } 2 < F < 3 \qquad (5.28)$$

where H_s is the significant wave height of the incoming waves at the toe of the structure (m), Δ_f the relative density of the grains in the gabions (usually, $\Delta \approx 1.65$), D_f the diameter of grains in the gabion (m), F the stability factor, and ξ_{op} the breaker parameter [= $\tan\alpha/\sqrt{(H_s/1.56T_p^2}$; T_p is the wave period at the peak of the spectrum (s).

5.7 SCOUR AND TOE PROTECTION

Toe protection consists of the armoring of the beach or bottom surface in front of a structure that prevents it from scouring and undercutting by waves and currents. Factors that affect the severity of toe scour include wave breaking (when near the toe), wave run-up and backwash, wave reflection, and grain size distribution of the beach or bottom materials. Toe stability is essential because failure of the toe will generally lead to failure throughout the entire structure. Toe scour is a complex process. Specific (generally, valid) guidance for scour prediction and toe design based on either prototype or model results have not yet been developed, but some general (indicative)

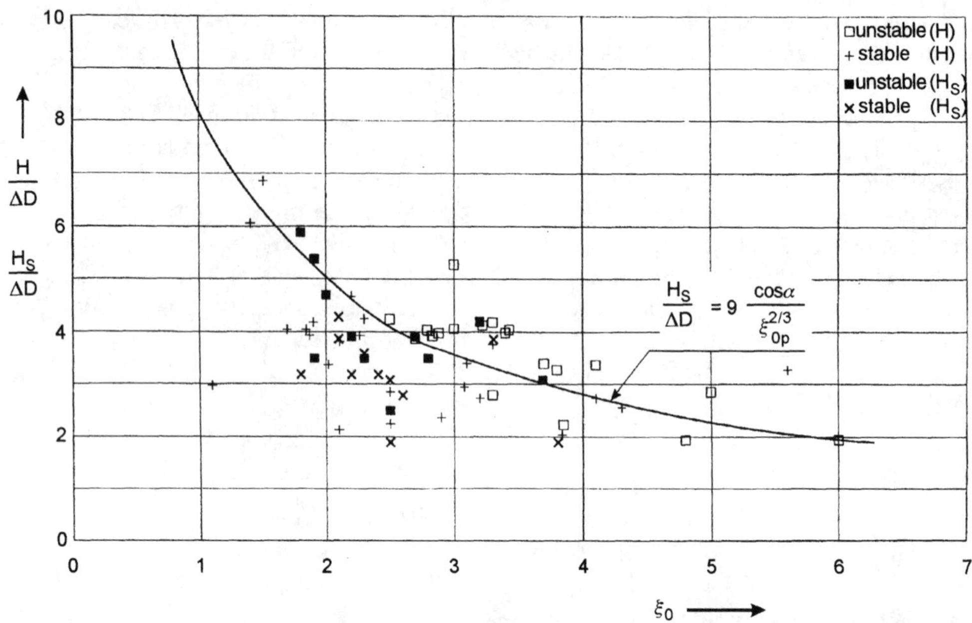

Figure 5.16 Summary of test results of Ashe (1975) and Brown (1979) and design curves.

guidelines for designing toe protection are given in U.S. Army Corps of Engineers (1984) and CUR/RWS (1995).

The maximum scour force occurs where wave downrush on the structure face extends to the toe and/or the wave breaks near the toe (i.e., shallow-water structure). These conditions may take place when the water depth at the toe is less than twice the height of the maximum expected unbroken wave that can exist at that water depth. The width of the apron for shallow-water structures with a high reflection coefficient, which is generally true for slopes steeper than about 1 in 3, can be planned based on the structure slope and the scour depth expected. The maximum depth of a scour trough due to wave action below the natural bed is about equal to the maximum expected unbroken wave at the site. To protect the stability of the face, the toe soil must be kept in place beneath a surface defined by an extension of the face surface into the bottom to the maximum depth of scour. This can be accomplished by burying the toe, when construction conditions permit, thereby extending the face into an excavated trench the depth of the scour expected. Where an apron must be placed on the existing bottom, or can be only buried partially, its width should not be less than twice the wave height. Some solutions for toe protection are given in U.S. Army Corps of Engineers, (1984), CUR/CIRIA (1991), and PIANC (1987a and b, 1992).

If the reflection coefficient is low (slopes milder than 1 in 3), and/or the water depth is more than twice the wave height, much of the wave force will be dissipated on the structure face and a smaller apron width may be adequate, but it must be at least equal to the wave height (minimum requirement). Since scour aprons generally are placed on very flat slopes, quarrystone diameter equal to $\frac{1}{2}$ or even $\frac{1}{3}$ of the primary cover layer probably will be sufficient unless the apron is exposed above the water surface during wave action. Quarrystone of primary cover layer size may be extended over the toe apron if the stone will be exposed in the troughs of waves, especially breaking waves. The minimum thickness of the cover layer over the toe apron should be two quarrystones. Quarrystone is the most favorable material for toe protection because of its flexibility. If a geotextile is used as a secondary layer, it should be folded back at the end and then buried in cover stone and sand to form a Dutch toe. It is recommended that an additional flexible edge (at least 1 m) consisting of loose material that can easily follow the scour at the toe be provided. The size of toe protection against waves can also be estimated roughly by using the common formulas on slope protection and schematizing the toe by mild slopes (i.e., 1 in 8 to 1 in 10). Some alternative toe protection designs are shown in Figure 5.17.

Toe protection against currents may require smaller protective stone, but wider aprons. The necessary design data can be estimated from site hydrography and/or model studies. Special attention must be given to sections of the structure where scour is intensified (i.e., to the head, the areas of a section change in alignment, the channel sides of jetties, and the down-drift sides of groynes). Where waves and reasonable currents (>1 m/s) occur together, it is recommended that the cover size be increased by at least a factor of 1.3.

Note that the conservatism of the apron design (width and size of cover units) depends on the accuracy of the methods used to predict the waves and current action and to predict the maximum depth of scour. For specific projects a detailed study of scour of the natural bottom and at similar structures nearby should be conducted at a planned site, and/or model studies should be considered before determining a final design. In all cases, experience and sound engineering judgment play an important role in applying these design rules.

5.8 PROTECTION AGAINST OVERTOPPING

If a structure (revetment) is overtopped, even by minor splash, the stability can be affected. Overtopping can (1) erode the area above or behind the revetment, negating the structure's purpose; (2) remove soil supporting the top of the revetment, leading to the unraveling of the structure from the top down; and (3) increase the volume of water in the soil beneath the structure, contributing to drainage problems. The effects of overtopping can be limited by choosing a higher crest level or by armoring the bank above or behind the revetment with a splash apron. For a small amount of overtopping, a grass mat on clay can be adequate. The splash apron can be a filter blanket covered by a bedding layer and, if necessary to prevent scour due to splash, by riprap, concrete units, or asphalt.

No definite method for designing against overtopping is known, due to lack of a proper method of estimating hydraulic loading. Pilarczyk (1990) proposed the following indicative way to design the thickness required for protection of the splash area (Figure 5.18):

$$\frac{H_s}{\Delta D_n} = \frac{1.5 \cos \alpha_i}{\Phi_T \xi^{2b} (1 - R_c/R_n)} \tag{5.29}$$

where H_s is the significant wave height; D the thickness of the protective unit ($D = D_n$ for rock); α_i is the angle of crest or inner slope; ϕ_T the total stability factor (1.0 for rock, 0.5 for placed blocks, and 0.4 for block mats); ξ the breaker index $\xi = \tan \alpha (H_s/L)^{-0.5}$; α is the slope angle and L is the wave length); R_c the crest height above still water level; and R_u the wave run-up on the virtual slope with the same geometry (see Figure 5.18).

The length of protection in the splash area, which is related to the energy decay, depends

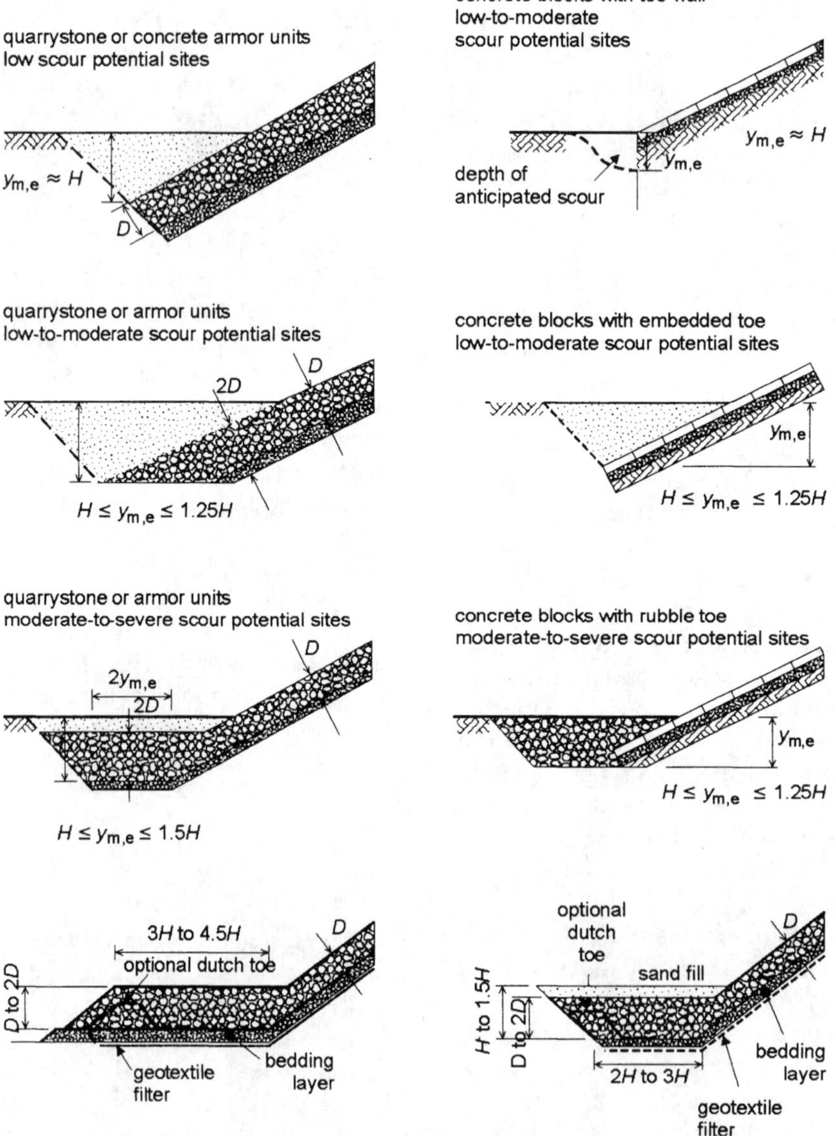

Figure 5.17 Alternative toe protections.

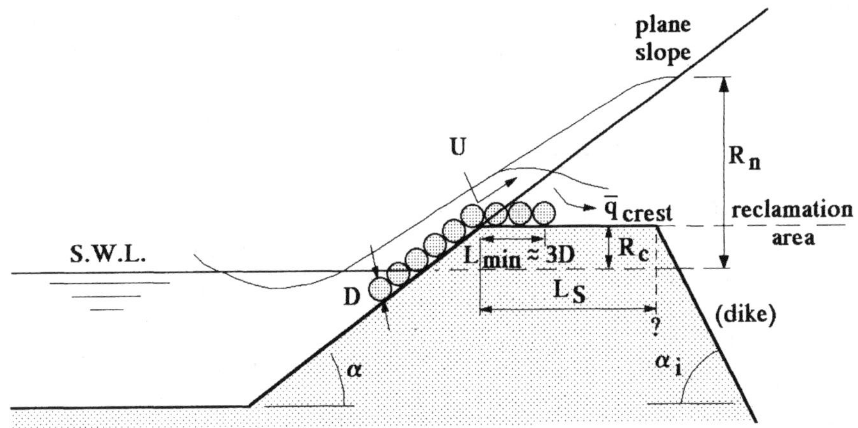

Figure 5.18 Definition of splash area.

on the permeability of the splash area. However, it can be roughly assumed to be

$$L_s = \frac{\psi}{5} T \sqrt{g(R_n - R_c)} \geq L_{\min} \quad (5.30)$$

with a practical minimum ($L_{\min}$) equal to at least the total thickness of the revetment (including sublayers) as used on the slope. ψ is an engineering judgment factor related to the local conditions (importance of structure), $\psi \geq 1$.

The stability of rockfill protection of the crest and rear slope of an overtopped or overflowed dam or dike can also be approached with the *Knauss formula* (Knauss, 1979). The advantage of this approach is that the overtopping discharge, q, can be used directly as an input parameter for calculation. Knauss analyzed steep shute flow hydraulics (highly aerated/turbulent) for the assessment of stone stability in overflow rockfill dams (impervious barrages with a rockfill spillway arrangement). This type of flow seems to be rather similar to that during high overtopping. His (simplified) stability relationship can be rewritten in the form

$$q = 0.625 \sqrt{g}\, (\Delta D_n)^{1.5}(1.9 + 0.8\phi_p - 3 \sin \alpha_i) \quad (5.31)$$

where q is the maximum admissible discharge ($m^3/s/m$); g the gravitational acceleration (9.81 m/s^2); Δ the relative density $= (\rho_s - \rho_w)/\rho_w$; D_n the equivalent stone diameter, $D_n = (M_{50}/\rho_s)^{1/3}$; ϕ_p the stone arrangement pacing factor, ranging from 0.6 for natural dumped rockfill to 1.1 for optimal manually placed rock (it seems to be reasonable to assume that $\phi_p = 1.25$ for placed blocks); and α_i the inner slope angle. *Note:* When using the Knauss formula, the calculated critical (admissible) discharge should be identified with a momentary overtopping discharge per overtopping fraction of a characteristic wave [i.e., the volume of water per characteristic wave divided by the overtopping time per wave, roughly (0.3 to 0.4)T (T is the wave period), 1, not with the time-averaged discharge (q).

5.9 JOINTS AND TRANSITIONS

Despite a well-designed protective system, the construction is only as strong as the weakest section. Therefore, special care is required when designing transitions. In general, slope protection for dikes and seawalls consists of a number of structural parts, such as toe protection, main protection in the area of heavy wave and current attack, upper slope protection (very often, grass mats), and berms for run-up reduction or as a maintenance road. Different materials and different execution principles are generally used for these specific parts. Very often, new slope protection has to be connected to an already existing protective construction which involves another protective system. To obtain strong, homogeneous protection, all parts of protective structures have to be taken into consideration.

Experience shows that erosion or damage often starts at joints and transitions. Therefore, important aspects of revetment constructions which require special attention are the joints and transitions, joints onto the same material and onto other revetment materials, and transitions onto other structures or revetment parts. A general design guideline is that transitions should be avoided as much as possible, especially in the area with maximum wave attack. If they are inevitable, the discontinuities introduced should be minimized. This holds for differences in elastic and plastic behavior and in the permeability or sand tightness. Proper design and execution are essential to obtain satisfactory joints and transitions.

When these guidelines are not followed, the joints or transitions may influence loads in terms of forces due to differences in stiffness or settlement, migration of subsoil from one part to another (erosion), or strong pressure gradients due to concentrated groundwater flow. However, it is difficult to formulate more detailed principles and/or solutions for joints and transitions. The best way is to combine the lessons from practice with some physical understanding of systems involved. Examples to illustrate the problem of transitions are given in Figure 5.19.

As a general principle one can state that the transition should be of a strength equal to or greater than the adjoining systems. Very often, it needs reinforcement in one of the following ways:

1. Increase the thickness of the cover layer at the transition.
2. Grout riprap or block cover layers with bitumen.
3. Use concrete edge strips or boards to prevent damage progressing along the structure.

Top edge and flank protection are needed to limit the vulnerability of the revetment to erosion continuing around its ends. Extension of the revetment beyond the point of active erosion should be considered but is often not feasible. Care should therefore be taken that the discontinuity between the protected and unprotected areas is as small as possible (use a transition roughness) so as to prevent undermining. In some cases, open cell blocks or open block mats (eventually, vegetated) can be used as transition (i.e., from hard protection into grass mat). The flank protection between the protected and unprotected areas usually needs a thickened or grouted cover layer, or a concrete edge strip with some flexible transition (i.e., riprap).

5.10 GENERAL CONSTRUCTION (EXECUTION) ASPECTS

Revetments are constructed in a number of phases, for example, (1) construction of the bank or dike body, (2) placement of toe structure, (3) placement of revetment sublayers (clay and/or filter layers), (4) laying the blocks or mattress, and (5) anchoring the mattress and, possibly, applying the joint filler. A well-compacted slope is important to produce a

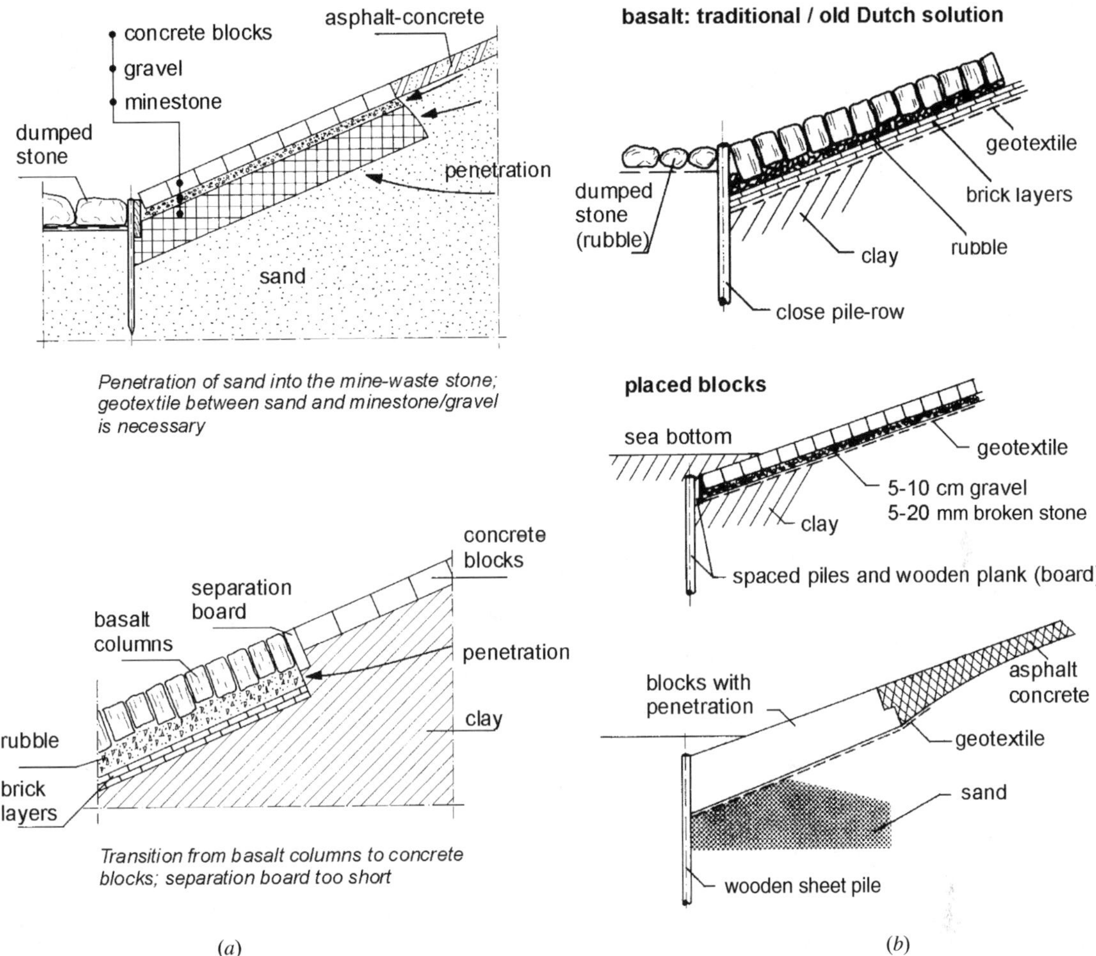

Figure 5.19 Transitions in revetments: (*a*) transition problems; (*b*) examples of transitions (toe protection).

smooth surface and thus ensure that there is a good connection between the mattress and the subsurface. When laying mattresses on banks it is strongly recommended that they be laid on undisturbed ground and that areas excavated too deeply be refilled carefully. Before using a geotextile, the slope must be inspected carefully for any projections that could puncture the material. When laying a mattress on a geotextile, care must be taken to ensure that extra pressures are not applied and that the geotextile is not pushed out of place. Geotextile sheets must be overlapped and/or stitched together with an overlap of at least 0.5 to 1.0 m to prevent subsoil being washed out. This is particularly important if the mattress is laid directly on sand or clay.

Block mattresses are laid using a crane and a balancing beam. The mattress must be in the correct position before it is uncoupled because it is difficult to pick up again and also time consuming. Provided that part of the mattress

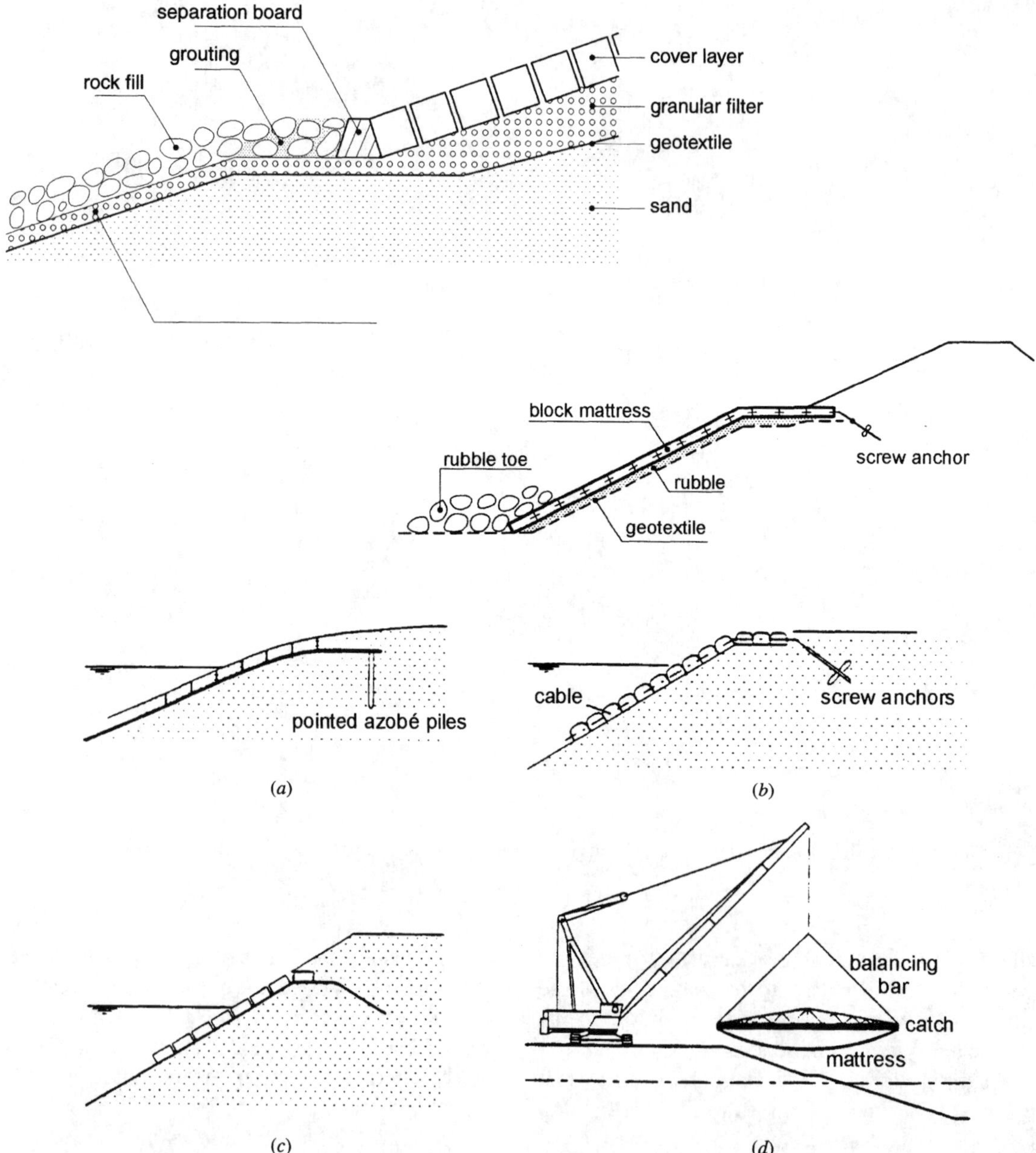

Figure 5.20 Construction aspects of revetments and methods of anchoring: (*a*) pointed azobé piles (wooden piles); (*b*) screw anchors; (*c*) buried geotextile; (*d*) placing a mattress.

can be laid above the waterline, it can generally be laid very precisely, and joints between adjacent mattreses can be limited to 1 to 2 cm. Laying a mattress completely under water is much more difficult. Nonetheless, the spacing between the blocks of adjacent mattresses should never be more than 3 cm.

Once in place, mattresses should be joined so that the edges cannot be lifted or turned up under the action of waves. Loose corners are particularly vulnerable. In addition, the top and bottom edges of the revetment should be anchored, as shown in Figure 5.20. In such a case, a toe structure is not needed to stop mattresses sliding. More information on execution aspects of revetments may be found in (CUR/RWS, 1995; CUR/TAW, 1995; Pilarczyk, 1998, 1999).

5.11 CONCLUSIONS

The newly derived design methods and stability criteria will be of help in preparing preliminary alternative designs with various revetment systems. However, there are still many uncertainties in these design methods. Therefore, experimental verification and further improvement of design methods are necessary. Also, more practical experience at various loading conditions is still needed.

REFERENCES

Ashe, G. W. T., 1975. *Beach Erosion Study; Gabion Shore Protection,* Hydraulics Laboratory, Ottawa, Ontario, Canada.

BAW, 1993. *Code of Practice: Use of Geotextile Filters on Waterway,* Bundesanstalt für Wasserbau, Karlsruhe, Germany.

Brown, C., 1979. *Some Factors Affecting the Use of Maccaferi Gabions,* Rep. 156, Water Research Laboratory, Australia.

CFGG, 1986. *Recommendations pour l'emploides geotextiles dans les systèmes de drainage et de filtration,* Comité Français des Geotextiles et Geomembranes, Paris.

CUR, 1993. *Filters in Hydraulic Engineering,* Civil Engineering Research and Codes, Gouda, The Netherlands (in Dutch).

CEN/CR ISO, 1998. *Guide to durability of Geotextiles and Geo-textiles Related Products,* European Normalization Committee (CEN), CR ISO, Paris.

CUR/CIRIA, 1991. *Manual on Use of Rock in Coastal Engineering,* Rep. 154, Gouda, The Netherlands.

CUR/RWS, 1995. *Manual on Use of Rock in Hydraulic Engineering,* Rep. 169, Gouda, The Netherlands.

CUR/TAW 1995. *Design Manual for Pitched Slope Protection,* Rep. 155, A.A. Balkema, Rotterdam, The Netherlands.

EAU, 2000. *Recommendations of the Committee for Waterfront Structures,* German Society for Harbour Engineering, Ernst & Sohn, Berlin.

DVWK, 1993. *Guidelines for Water Management,* No. 306, *Application of Geotextiles in Hydraulic Engineering,* German Association for Water Resources and Land Improvement, Bonn, Germany.

1984. *Proc. International Conference Flexible Armoured Revetments,* Thomas Telford, London.

FHWA, 1995. *Geosynthetics Design and Construction Guidelines,* FHWA-HI-95-038, Federal Highway Administration, Washington, DC.

Heerten, G., 1982a. Dimensioning the filtration properties of geotextiles considering long-term conditions, *Proc. 2nd International Conference on Geotextiles,* Las Vegas, NV.

———, 1982b. Geotextiles in coastal engineering, 25 years experience, *Geotextiles Geomembranes,* Vol. 1, No. 2.

Holtz, R. D., B. R. Christopher, and R. R. Berg, 1997. *Geosynthetic Engineering,* BiTech, Richmond, Quebec, Canada.

Klein Breteler, M., et al, 1998a, *Alternatieve Open Taludbekledingen* (*Alternative Open Slope Revetments*), H1930, Delft Hydraulics, Delft, The Netherlands (in Dutch).

Klein Breteler, M., K. W. Pilarczyk, and T. Stoutjesdijk, 1998b, Design of alternative revetments,

Proc. 26th International Conference on Coastal Engineering, Copenhagen, Denmark.

Knauss, J., 1979. Computation of maximum discharge at overflow rock-fill dams, *Proc. 13th Congress des Grand Barrages* (ICOLD), New Delhi, India, Q50, R.9.

Mlynarek, J., 1994. *Evaluation of Filter Performance of Geotextiles,* prepared for the Road and Hydraulic Engineering Division, Delft, The Netherlands.

PIANC, 1984. *Report of the International Commission for Improving the Design of Fender Systems,* Suppl. Bull. 45, Permanent International Association of Navigation Congresses, Brussels, Belgium.

———, 1987a, *Guidelines for the Design and Construction of Flexible Revetments Incorporating Geotextiles for Inland Waterways,* Rep. WG 4, PTC I, Suppl. Bull. 57, Permanent International Association of Navigation Congresses, Brussels, Belgium.

———, 1987b. *Special Issue on Propeller Jet Action, Erosion and Stability Criteria near the Harbour Quays,* Bull. 58, Permanent International Association of Navigation Congresses, Brussels, Belgium.

———, 1992. *Guidelines for the Design and Construction of Flexible Revetments Incorporating Geotextiles in Marine Environment,* Rep. WG 21, PTC II, Suppl. Bull. 78/79, Permanent International Association of Navigation Congresses, Brussels, Belgium.

———, 1997, *Guidelines for the Design of Armoured Slopes under Piled Quay Walls,* Suppl. Bull. 96, Permanent International Association of Navigation Congresses, Brussels, Belgium.

Pilarczyk, K. W. (ed.), 1990. *Coastal Protection,* A.A. Balkema, Rotterdam, The Netherlands.

——— (ed.), 1998. *Dikes and Revetments,* A.A. Balkema, Rotterdam, The Netherlands.

———, 1999. Geosynthetics and Geosystems in Hydraulic and Coastal Engineering, A.A. Balkema, Rotterdam, The Netherlands.

RWS, 1987. *The Closure of Tidal Basins, Rijkswaterstaat,* Delft University Press, Delft, The Netherlands.

Terzaghi, K. and R. B. Peck, 1967. *Soil Mechanics in Engineering Practice,* 2nd ed., John Wiley and Sons, New York.

U.S. Army Corps of Engineers, 1984. *Shore Protection Manual,* USACE Waterways Experiment Station, Vicksburg, MS.

Van Santvoort, G. (ed), 1994. *Geotextiles and Geomembranes in Civil Engineering,* rev. ed., A.A. Balkema, Rotterdam, The Netherlands.

6

REPAIR, REHABILITATION, MAINTENANCE, AND UPGRADING OF WATERFRONT STRUCTURES

Gregory P. Tsinker
Tsinker & Associates, Inc.
Niagara Falls, Ontario, Canada

6.1 INTRODUCTION

Aging and deterioration of many port-related waterfront structures, along with increasing demands on their operational capacities, impose requirements on their repair, rehabilitation, maintenance, and modernization. For example, the current demand of the U.S. Navy alone for repair and upgrading (modernization) of its infrastructure amounts to well over $1 billion and is increasing at an accelerated rate (Yao, 1998). The lifetime for properly designed, constructed, and maintained marine structures may be 25, 50, or more years. During their lifetime, these structures are susceptible to damage by ships, and deterioration due to the effects of marine environments, excessive use beyond that intended in the original design, aging of structural materials, and general obsolescence.

Different structural materials are affected in various ways by the marine environment. The most notable effects include corrosion of metals, degradation of concrete, attack on timber by marine organisms, fouling, and encrustation of virtually all materials. To meet today's service requirements, the load-carrying capacity of existing older marine facilities must be reviewed carefully. In general, modern dockside cargo handling and hauling equipment is much heavier than that used in the design of older marine structures. Hence, such equipment, when installed on older dock structures, may lead to a multitude of problems, such as settlement, distortion, sliding, and local and general instability.

Dramatic changes in ship size and shape, particularly introduction of bulbous and flared bows, created almost unique situations for piercing holes in sheet piling, damaging piles and solid wall constructions at an inaccessible low level. In some instances it was a cause of crane or even ship–train collision. Furthermore, larger ships with increased draft, and therefore smaller underkeel and propeller clearance, and with side thrusters, while approaching the berth without tug assistance, can cause considerable scouring effects, especially if the structure is built on an erodible foundation.

To evaluate the present capacity of an older structure or to select the most economical repair or rehabilitation scheme, the designer must have knowledge of its present physical condition. This is obtained through abovewater and underwater inspection. Basic guidelines on marine structures inspection are found in Permanent International Navigation Association of Navigation Congresses (PIANC, 1990). Basic information on the causes of deterioration of structural materials such as concrete, steel, and wood in the marine environment and a cost-effective approach to an evaluation of the load-carrying capacity of the existing older marine structures is provided in the following sections. We also deal with particular problems associated with ship propeller–induced scour in waterfront marine structures and with the remedial repair and/or rehabilitation work required to restore a structure's ability to carry the design loads.

6.2 DETERIORATION OF STRUCTURAL MATERIALS IN A MARINE ENVIRONMENT

Marine structures are subjected to various deteriorating agents throughout their service lives. The degree of deterioration depends on properties of ambient water, for example, whether it is sea or fresh water, its seasonal fluctuation, its tide range, its climatic conditions, and the chemical composition of construction materials. Concrete, steel, and wood are the primary materials used in construction of marine structures (as they are on land); therefore, most of the discussion of marine environment effects is devoted to these materials. Deterioration of marine structures also depends on the types of cargoes handled at the particular area of the port (e.g., salt, chemicals, etc.), the degree of protection from ship impact during the berthing and departing maneuver, the type of cargo handling technology used, and of course, the maintenance program in place.

6.2.1 The Marine Environment

The marine environment is harsh and typically produces adverse effects on marine structures.

The presence of soft bottom sediments, persistently hot and humid climates, the presence of dissolved chemicals in seawater, marine organisms, fouling, and last but not least, ice effects are responsible for high rates of steel corrosion, deterioration of concrete, and rapid degradation of wooden components of marine structures. Winds, waves, and currents, as well as weather factors such as rain, snow, fog, spray, and atmospheric icing, add greatly to the deterioration of marine structures.

6.2.1.1 Seawater and Fouling. Generally, seawater is viewed as a solution containing a greater number of chemical elements in different proportions. Concentrations of these elements are typically given in parts per million (ppm) by weight. The combined concentration of these elements in water or the total amount of dissolved solids in water is defined as *salinity*. The latter can also be assumed as the amount of dissolved solids in a water sample in parts per thousand (‰) by weight. The salinity of seawater usually varies from 31 to 38‰ and on average is approximately 35‰; it can be considerably less for nearshore coastal waters because of the freshwater effect, especially in the vicinity of large river mouths. Gaythwaite (1981) provides details on seawater composition.

Marine fouling is an accumulation of various marine growths and animal organisms on immersed and partly immersed surfaces of marine structure. Fouling may increase the weight of the structure, and more important, may substantially increase the surface area and its roughness. The latter may result in substantially increased drag forces due to current imposed on the immersed part of a structure (Tsinker, 1997). Additionally, fouling may increase the rate of steel corrosion, due to its destructive effects on protective coatings and oxygen concentration cell effects in areas where certain organisms (e.g., barnacles) exist. Furthermore, fouling creates difficulties for inspection and maintenance of the structure due to the presence at times of a quite (up to 300 mm) thick and tenacious growth. For detailed information, consult Gaythwaite (1981) and Tsinker (1995, 1997).

6.2.1.2 Variation in Water Levels. Variation in water levels as part of the marine environment may affect the deterioration of marine structures in a variety of ways. First and foremost, it may impose heavy cyclic loading upon retaining structures of miscellaneous designs due to tidal variation. The latter may result in overstress and premature deterioration of structural components due to fatigue or distress of foundation material. It also affects adversely the structural materials due to permanent wetting and drying process. From the viewpoint of exposure to a seawater environment, the face of a structure can be divided into several characteristic zones (Figure 6.1):

1. The uppermost zone exposed to the atmosphere. It tends to contain some

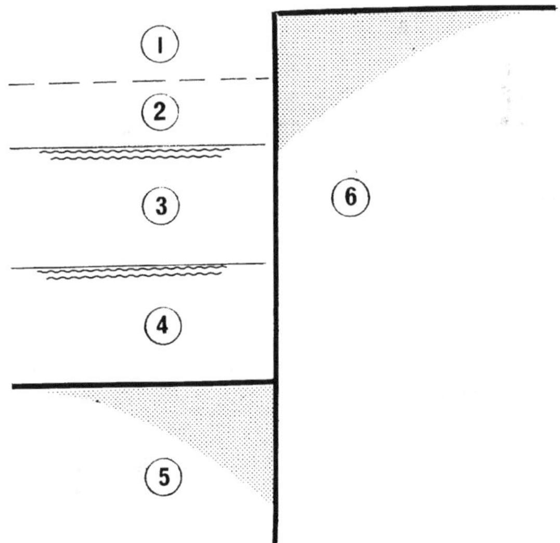

Figure 6.1 Distinct zones of material deterioration in a marine environment. 1, Atmospheric zone; 2, splash zone; 3, tidal zone; 4, zone of continuous erosion; 5, seabed zone; 6, backfill zone.

amount of salt, which increases the rate of atmospheric corrosion of metals and the deterioration of concrete.

2. The splash zone, which is exposed to atmosphere, wave impact, wind-driven spray, frost action, solar radiation, and drying winds which produce rapid evaporation. In general, it is characterized by intermittent wetting and drying.
3. The tidal zone, which is exposed to repeated cycles of wetting and drying as well as to freezing and thawing in a saturated condition and to the impact of waves. These waves can contain pieces of floating ice, gravel, or sand, which cause an abrasive action.
4. The zone that is continuously submerged in seawater.
5. The seabed zone.
6. The backfill zone. Below the mudline and below the low-tide level in the backfill material zone, the structure's elements are buried in the foundation and backfill materials. There all structural materials are relatively well protected, as the lack of oxygen prohibits oxidation and the existence of most organisms.

6.2.1.3 Weather Factors. Weather factors that may affect the integrity of structural materials are rain, snow, fog, spray, and atmospheric icing. For example, if the drainage system does not function properly, the rainfall is not removed from the dock surface expeditiously. The latter may result in overloading of the structure due to the buildup of hydrostatic pressure in the backfill material of the soil-retaining structures (e.g., gray wall, sheet pile bulkheads). In regions where snowfall is heavy and snow is not removed fast enough, it may accumulate and freeze. It is usually melted by spraying salt. This, in turn, may affect the corrosion rate of metal parts of the structure and increase the rate of concrete deterioration. Condensation of moisture due to fog, rain, or water spray, if not removed quickly, may also result in increased rate of metal corrosion and concrete and wood deterioration. Spray is created when waves break against a vessel or a structure. Wind hurls spray into the air, and during cold weather it may accumulate on a structure in the form of a great deal of ice. The heavy accretion of ice on vertical and horizontal surfaces of a dock, if not removed promptly, may result in structural damages and/or affect the stability of the structure.

6.2.1.4 Ice. Ice could be a major factor of structural damages when the dock operates in a cold region. Ice-induced forces and resulting damage to dock structures are discussed in detail in Tsinker (1995).

6.2.2 Concrete Deterioration in the Marine Environment

Despite general evidence of long-term durability of concrete structures in marine environment and their generally outstanding performance in sea and fresh water, in some cases serious concrete deterioration has been reported. The multitude of examples from the practice are found in specialty periodicals, technical journals, and conference proceedings. The reader is also referred to works by Gaythwaite (1981), PIANC (1990), and Tsinker (1995).

The general cause of concrete deterioration was noted to be cracking, resulting in the corrosion of embedded reinforcing steel. Corrosion occurs in the presence of oxygen, moisture, and electrolyte. Salt intensifies the electrolytic properties of concrete, thereby creating a corrosion cell, resulting in corrosion of steel. The deterioration of concrete is usually limited to a certain zone of marine environment, as indicated in Figure 6.1 (i.e., atmospheric, splash, tidal, continuous immersion, sealed and backfill zones). The atmospheric zone is generally characterized by concrete cracking due to corrosion of reinforcement.

The splash and tidal zones are the most vulnerable. They are characterized by cracking and spalling of concrete due to wetting and drying, frost action, corrosion of reinforcement, chemical decomposition of the hydration products of cement, dynamic effects of wave action, and abrasion by floating and piled-up ice floes. The part that is constantly submerged in seawater is basically vulnerable to loss of concrete strength due to chemical reaction between seawater and the hydration products of cement. Corrosion of reinforcement steel bars is seldom a problem in this zone. Concrete distress and deterioration may be the cause of more than one mechanism. Deterioration of concrete is a very complex phenomenon. It would be too simplistic to suggest that it is possible to identify a single, specific cause of deterioration for every symptom detected during evaluation of a concrete structure. In most cases, the damage detected will be the result of more than one mechanism. For example, corrosion of reinforcing steel may open cracks that allow greater access of moisture to the interior of concrete. This moisture could lead to additional damage caused by freezing and thawing.

With respect to concrete deterioration in marine environments, the causes of deterioration can be divided into two basic groups:

1. Corrosion of the reinforcing steel through carbonation or/and chloride added or penetrated
2. Deterioration of concrete through chemical reactions, construction errors, corrosion of embedded metals, design errors, erosion, freezing and thawing, settlement and movement, shrinkage, and temperature changes

6.2.2.1 Chemical Reactions. In general, deleterious chemical reactions may be classified as those that occur as a result of external chemicals attacking the concrete (e.g., acid attack, aggressive water attack, miscellaneous chemical attack, sulfate attack) or those that occur as a result of internal chemical reactions between the constituents of the concrete (e.g., alkali–silica and alkali–carbonate rock reactions). Each of these chemical reactions is described briefly below.

Acid Attack. The deterioration of concrete by acids is primarily the result of a reaction between an acid and the products of hydration of cement. In most cases the chemical reaction results in formation of water-soluble calcium compounds that are then leached away. In the case of sulfuric acid attack, additional or accelerated deterioration results because the calcium sulfate formed may affect concrete by the sulfate attack mechanism. If the acid is able to reach the reinforcing steel through cracks or pores in the concrete, corrosion of the reinforcing steel will result. The latter, in turn, will cause further deterioration of the concrete. Visual examination of the affected structure will show disintegration of the concrete (i.e., loss of cement paste and aggregate from the matrix). If reinforcing steel is affected by acid, rust staining, cracking, and spalling may be seen.

Aggressive Water Attack. Some waters have been reported to have extremely low concentrations of dissolved minerals. These soft or aggressive waters leach calcium from cement paste or aggregates (Holland et al., 1980). From the few cases that have been reported, there are indications that this attack takes place very slowly. For aggressive water attack to have a serious effect on marine structures, the attack must occur in flowing water, which contains a constant supply of aggressive water in contact with the concrete and which washes away aggregate particles that become loosened as a result of leaching of the paste. Visual examination will show concrete surfaces that are very rough in areas where the paste has been leached. If the aggregate is susceptible to leaching, holes where the coarse aggregate has been dissolved will be evident.

Alkali–Carbonate Rock Reaction. Certain aggregates of carbonate rock have been found to be reactive in concrete. The results of these reactions have been characterized as ranging from beneficial to destructive. The destructive category is apparently limited to reactions with impure dolomitic aggregates and are a result of either dedolomitization or rim-silicification reactions. If not detected, alkali–aggregate reactions may impair concrete durability to such an extent that the load-carrying capacity of a structure could be dangerously jeopardized. Alkali–aggregate reactions produce cracks through which oxygen, water, and chlorides travel to attack reinforcing steel.

Chlorides can amplify the adverse effects of alkali–aggregate reactions and increase concrete expansion significantly. Chloride salts react with the products of cement hydration to generate additional or secondary alkalis. Oddly enough, alkali–aggregate reactions do not decrease compressive strength of concrete appreciably. Because structural adequacy of concrete in situ is often determined based on compressive core strength, the data obtained from cores can be misleading. Therefore, when evaluating the structural adequacy of concrete affected by alkali–aggregate reactions, it is advisable to determine compressive and tensile strength, as well as the modulus of concrete elasticity. Visual examination of those reactions will generally show map or pattern cracking and a general appearance that indicates that the concrete is swelling. A distinguishing feature that differentiates an alkali–carbonate rock reaction from an alkali–silica reaction is the lack of silica gel exudations at cracks [American Concrete Institute (ACI), 1985a].

Alkali–Silica Reaction. Alkali–silica reaction is an expensive reaction, however, with the difference that the destruction is not preceded by weakening of the cement matrix but by gel formation in the aggregate. Under the influence of a high alkali content, silica gel is formed by the reaction from the silica oxide in the aggregate, which leads to the formation of cracks. Some aggregates containing silica ($Na_2SiO_3 \cdot nH_2O$ or $K_2SiO_3 \cdot nH_2O$), which is soluble in highly alkaline solutions, may react to form either a solid nonexpansive calcium–alkali–silica complex or an alkali–silica complex that can imbibe considerable amounts of water and then expand, disrupting the concrete. Visual examination of those concrete structures that are affected will generally show map or pattern cracking and a general appearance of concrete volumetric expansion. Damage through alkali–silica reaction can also be recognized by the typical crack pattern, which indicates flaking with structure and visually in the shape of a rough craquelure (map cracking).

Spencer and Blaylock (1997) pointed out that the alkali–silica reaction that occurs below the water surface usually proceeds faster than that in a dry condition. The alkali–silica reaction in salt water can cause significant damage to the underwater part of a relatively new marine structure (Spencer, 2001), while the above-water portion of the structure may remain in good condition. Engineers must be aware of the possibility of the concrete volumetric expansion below the mudline (Spencer, 2001). Hence, where the alkali–silica reaction in concrete is detected, the condition of concrete below the mudline must be investigated. For more information on alkali–silica reactions, interested readers are referred to Hobbs (1988), Acres International Ltd. (1989), Okada et al. (1989), Nieswaag et al. (1998), and Stark et al. (1999).

Miscellaneous Chemical Attack. Concrete will resist chemical attack to varying degrees, depending on the exact nature of the chemical. ACI (1985b) includes an extensive listing of the degrees of concrete resistance to various chemicals. To produce a significant attack on concrete, most chemicals must be in solution form and above a certain minimum concentration. Besides, for maximum detrimental effect, the chemical solution needs to be circulated in contact with the concrete. Concrete subjected to aggressive solutions under positive differential pressure may be vulnerable to their effects because pressure gradients tend to force the aggressive solutions into the matrix. If the face of the concrete is exposed to evaporation, a concentration of salts tends to accumulate at that face, resulting in increased chemical attack. In addition to the specific nature of the chemical involved, the degree to which concrete resists attack depends on the following: the temperature of the aggressive solution, the water/cement ratio of the concrete, the type of cement used, the degree of consolidation of the concrete, the permeability of the concrete, the degree of wetting and drying of the chemicals on the concrete face, and the extent of chemically induced corrosion of the reinforcing steel (ACI, 1985c).

Visual examination of concrete that has been subjected to chemical attack will usually show surface disintegration, spalling, and the opening of joints and cracks. There may also be swelling and general disruption of the concrete mass. Coarse aggregate particles are generally more inert than the cement paste matrix; therefore, aggregate particles may be seen as protruding from the matrix. Laboratory analysis may be required to identify the unknown chemicals that are causing the damage.

Sulfate Attack. Naturally occurring sulfates of sodium, potassium, calcium, or magnesium are sometimes found in soil or in solution in groundwater adjacent to marine structures. Sulfate ions in solution will attack concrete. There are usually two chemical reactions involved in sulfate attack on concrete. First, the sulfate reacts with free calcium hydroxide which is liberated during the hydration of the cement to form calcium sulfate (gypsum). Next, the gypsum combines with hydrated calcium aluminate to form calcium sulfoaluminate. Both of these reactions result in an increase in volume. The second reaction is responsible for most of the disruption due to a volume increase in the concrete (ACI, 1985a). In addition to the two chemical reactions above, there may also be a purely physical phenomenon in which the growth of crystals of sulfate salts disrupts concrete. Visual examination of concrete exposed to sulfate attack will show map and pattern cracking as well as a general disintegration of the concrete. Laboratory analysis can verify the occurrence of the reactions described.

6.2.2.2 Construction Errors. Failure to follow specified procedures and good practice, or outright carelessness, may lead to a number of conditions that may be grouped together as construction errors. Typically, most of these errors do not lead directly to failure or deterioration of concrete. Instead, they enhance the adverse impact of other mechanisms previously identified. Each error is described briefly in the following paragraphs. It should be noted that errors of the type described in this section are equally likely to occur during repair or rehabilitation projects as during new construction.

Adding Water to Freshly Mixed Concrete. This practice will generally lead to concrete with lowered strength and reduced durability. As the water/cement ratio of the concrete increases, the strength and durability will decrease and its shrinkage and permeability will increase.

Improper Consolidation. Improper consolidation of concrete may result in a variety of defects, the most common being bugholes, honeycombing, and cold joints.

Improper Curing. Unless concrete is given adequate time to cure at proper humidity and temperature, it will not develop the characteristics that are expected and that are necessary to provide for adequate durability. Symptoms of improperly cured concrete can include various types of cracking and surface disintegration. In extreme cases where poor curing leads to failure to achieve anticipated concrete strengths, structural cracking may occur.

Improper Location of Reinforcing Steel. This may result in reinforcing steel that is either improperly located or is not adequately secured in the proper location. Either of these faults may lead to two general types of problems. First, the steel may not function structurally as intended, resulting in structural cracking or failure. Second, the concrete cover over steel is reduced, which makes it much easier for corrosion to begin.

Movement of Formwork. Movement of formwork during the period while the concrete is going from a fluid to a rigid material state may induce cracking and separation within the concrete.

Premature Removal of Shores or Reshores. If shores or reshores are removed too soon, the concrete affected may become overstressed and cracked. In extreme cases there may be major failures.

Settling of the Subgrade. If there is any settling of the subgrade during the period after the concrete begins to become rigid but before it gains enough strength to support its own weight, cracking may occur.

Vibration of Freshly Placed Concrete. Most construction sites are subjected to vibration from various sources, such as blasting, pile driving, and from the operation of construction equipment. Freshly placed concrete is vulnerable to weakening of its properties if subjected to forces that disrupt the concrete matrix during setting.

6.2.2.3 Corrosion of Embedded Metals. When carbon steel reinforcement is embedded in concrete, the surface of the steel oxidizes to form a very thin surface film of ferric oxide (Fe_2O_3). This film is known as passive film because it is extremely stable when embedded in the highly alkaline cement matrix (pH normally greater than 11). Provided that the alkaline environment is sustained and the passive film remains intact, the reinforcement will undergo virtually no further oxidation over an indefinite period and the reinforced concrete structure will therefore exhibit none of the problems associated with corrosion of the reinforcement. However, if chloride ions reach high volumes in the reinforcing bar (rebar), the bar will form the anode, and its passivated part will form the cathode. At the anode, the iron goes into solution, which releases the electrons ($Fe = Fe^{2+} + 2e^-$). These electrons flow to the cathode, where water and oxygen reach to form hydroxide. The hydroxide ions react with the iron that has been released, whereby rust-creating products develop. A distinction must be made between carbonation and chloride-induced corrosion. Corrosion as a result of carbonation usually occurs evenly distributed in the area of the carbonated concrete. In corrosion initiated by chloride, the diameter of a rebar can be sharply reduced in places before volume increase will push off its cover. This pitting corrosion can be extremely detrimental because at this point of development, the structure will not display any early signs of distress. This must not be overlooked during inspection into possible corrosion. It should be noted that a commonly accepted chloride threshold value for depassivation of steel is about 0.2% of the total chloride ion by weight of cement, or about 0.8 kg/m^3 for typical high-strength concrete. The volume of the rust is several times the volume of original iron, and when produced it tends to burst the cover over reinforcement and/or to expand the original crack (Figure 6.2). As summarized by Holmes and Brundle (1987), the

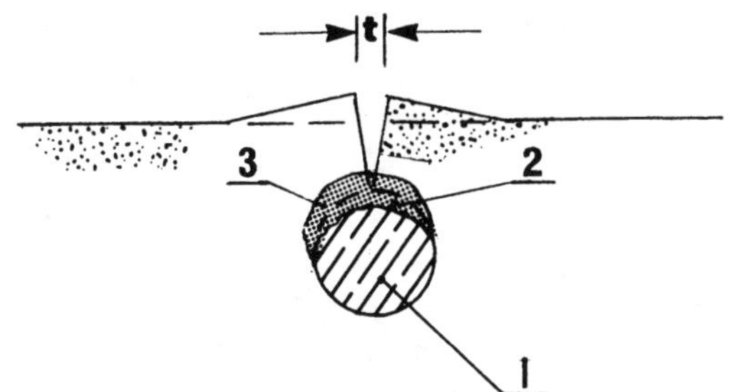

Figure 6.2 Concrete cracking due to corrosion of steel reinforcement. 1, Left metal; 2, steel area lost to corrosion; 3, product of steel corrosion.

factors affecting the corrosion rate are as follows:

1. The chloride concentration at the surface of the reinforcing bar
2. The initial integrity of the passive layer on the surface of the reinforcement
3. The electrical resistance of the concrete
4. The availability of oxygen to complete the cathodic reaction

High concentrations of chloride ions within the concrete matrix are either derived from the concrete mix constituents themselves or enter through the surface of the hardened concrete from the external environment. From the above standpoint, according to Nieswaag et al. (1998), the slag cement has a certain advantage in protecting the embedded steel of marine structures. In addition to the better sulfate resistance, this cement also retards the penetration of chloride into the concrete matrix.

The concentration of chlorides on concrete surfaces due to the effects of the external (marine) environment can vary significantly from one part of the structure to another. For example, in the splash zone, solar radiation and drying winds can produce rapid evaporation which could lead to deposition of salt crystals. Gradual buildup of these crystals can increase the chloride in this zone. In the tidal zone, salts are built up on a concrete surface during a falling tide, and as the tide level increases, the concrete surface is washed and the surface salt concentration is reduced. In the submerged zone, the chloride concentration at the surface of the concrete is relatively constant and the maximum concentration approximates the concentration of salts in seawater.

Review of the available large body of literature on steel corrosion in marine environments clearly indicates that the parts of concrete structures most susceptible to steel corrosion are those most saturated and/or exposed to intermittent wetting and drying. The saturated tidal zone of marine structures has more potential than other parts of the structure for cracking due to cyclic wetting/drying conditions, freeze and thaw attack, and other factors described in this chapter. It is also the most vulnerable to corrosion of reinforcing steel. In many cases the splash zone has been reported as being vulnerable to the corrosion of reinforcing steel.

In the case of embedded metal corrosion, visual examination of the existing marine structure will typically reveal rust staining of the concrete. This staining will be followed by cracking. Cracks produced by corrosion generally run in straight, parallel lines at uniform intervals corresponding to the spacing of the reinforcement. As deterioration continues, spalling of the concrete over the reinforcing steel will occur with the reinforcing bars becoming visible. One area where laboratory analysis may be beneficial is in determination of the chloride content in the concrete. This procedure may be used to determine the amount of concrete to be removed during a rehabilitation project.

6.2.2.4 Design Errors. Design errors may be divided into two general types: those resulting from inadequate structural design, and those resulting from lack of attention to relatively minor or very specific design details. In the case of inadequate design, the failure mechanism is simple—the concrete is exposed to greater stress than it is capable of carrying, or it sustains greater strain than its strain capacity. Visual examinations of failures resulting from inadequate structural design will usually show spalling and/or cracking. To identify inadequate design as a cause of damage, the locations of damage should be compared to the types of stresses that should be present in the concrete. If the type and location of the damage and the probable stress are in agreement, a detailed stress analysis will be required to determine whether inadequate design is the cause. In general, poor detailing may not necessarily lead directly to concrete failure. However, it may be the cause of concrete deterioration. For example, abrupt changes in section may cause stress concentrations in structural member. This may result in cracking.

Insufficient reinforcement at openings also tends to cause stress concentrations that may cause cracking. Poor attention to the details of draining a structure may result in the ponding of water. This ponding may lead to leakage or saturation of concrete and may result in severely damaged concrete if the area is subjected to freezing and thawing. Inadequately designed expansion joints may result in spalling of concrete adjacent to the joints. The use of materials with different properties (modulus of elasticity or coefficient of thermal expansion) adjacent to one another may result in cracking or spalling as the structure is loaded or as it is subject to daily or annual temperature variations.

In some instances, the incorrect design philosophy used for the selection of structure type may lead to premature deterioration of concrete components of marine structure. For example, in contemporary North American design practice there is a consistent trend to make the waterfront structure as flexible as possible. The driving force behind this trend is the desire to increase seismic resistance of the structure. This is normally achieved through avoidance of the batter piles and construction of all vertical pile concrete platforms. Examples are provided by Wallace and Mallick (1986), Johnson et al. (1995), Morley et al. (1995), Tsinker (1997), and many others. In the case where the batter piles are included, the overstressing cracks that encircle the top of batter piles as a result of several lateral forces (e.g., seismic, or ship impact) may be expected. In the case of all vertical pile platforms, seismic resistance is provided by the last rows of piles, driven into the underdeck slope (sometimes referred to as a *retaining dike*). These piles resist lateral load by frame action. The usual problem with these piles is that they are not readily accessible for inspection.

In most recent practice, the top 0.7 to 1.0 m of rear piles are left exposed at the top of the underdeck slope to provide for the ability to inspect the pile's top and the pile-to-deck connection. Furthermore, the-would-be economical concrete ribbed decks often are replaced with flat cast-in-situ concrete slabs of uniform thickness. The net result of the latter approach may be summarized as follows:

1. Flat deck reduces the possibility of high concentrations of chlorides and its susceptibility to salt spray effects and direct wave action that is characteristic for the ribbed deck.
2. Reinforcement of pile caps and beams of ribbed deck require use of stirrups, which is a notorious potential source of corrosion damage (Buslov, 1998). This usually occurs because of frequent violation of the minimum concrete cover requirements over stirrups during their installation; also, occasionally, they are moved sideways or bent on purpose to accommodate the reinforcement bars, lap splices, or embedded steel or just in-

stalled without proper care. Spalling of the concrete at stirrups triggers corrosion of the main reinforcing steel.

3. At their lower corners, beams of the ribbed deck are exposed most to the hostile effects of marine environment. Hence, the rate of chloride penetration and the probability of it reaching the bars at the beam edges are higher than may be expected in just a flat slab. In reality, the corrosion of reinforcing steel at corners (edges) starts earlier and therefore, concrete cracking and spalling there usually occurs much earlier and develops must faster than at the flat soffit surfaces. According to Buslov (1998), spalling of concrete cover on deck soffits (if any) usually begins eight to ten years later than at beam edges. It is common belief that introduction of prestressed concrete components (e.g., piles in marine applications) result in substantially improved performance of concrete wharves. This is because of its generally better quality and better durability in the marine environment and its lesser susceptibility to cracking under overloading condition. However, as admonished by Buslov (1998), the prestressed concrete in marine environment should be regarded as a mixed blessing because when it is good, it is very good, but when it is bad, it tends to be very bad. The reason for this may be as follows. The amount of reinforcing steel in the conventional reinforced concrete components is substantially larger than that in the prestressed concrete elements. The product of corrosion that occurs in the conventional reinforced concrete, because of its much larger quantity, tends to destroy the cover over reinforcing steel much faster than this will occur in prestressed concrete. Observations indicate that, in general, in conventional reinforced concrete the rebars may stay well embedded in concrete for quite an extended period even after complete spalling of the cover (Browne, 1980). In other words, as discussed, concrete spalling does not necessarily indicate serious structural damage despite some loss of metal to corrosion. On the other hand, in prestressed concrete components the volume of corrosion product required to crack the concrete cover is approximately the same as in conventional components. However, the volume of this product is generated from a high-strength steel which has a much smaller cross-sectional area than that of a regular reinforcing bar of equivalent strength.

Hence, a substantially larger percentage of reduction in steel area is required to crack the concrete cover. This reduction in steel area may be of such magnitude that very little of the prestressed steel cross-sectional area will remain in place by the time the cracks due to corrosion have formed. At this point, cracks over prestressed steel may be indicative of serious structural damages.

6.2.2.5 *Erosion.* Erosion of concrete in marine structures is a collective term for all forms of deterioration. In maritime works, the most characteristic form of erosion is mechanical abrasion in flowing water, usually caused by the action of ice, sediments, or miscellaneous floating debris that are rolling and grinding against a concrete surface. Ice is typically the principal source of concrete erosion. Moving ice, which can have compressive strength as great as 20 MPa, has been known to remove all the concrete cover and near-surface layers of reinforcement in marine structures (Hoff, 1988; Haroske, 2001). Ice abrasion at or near a waterline is a typical result of the combined effects of ice impact (or repeated impacts) and sliding of ice floes along the structure, which creates friction or drag on the concrete surface.

Wind- and/or current-driven ice floes can possess significant kinetic energy, much of which is dissipated into the concrete during collision with the structure. Some kinetic energy is lost in the crushing of ice. As driving forces continue to drag the ice floe against and along the structure, a local failure occurs in both ice and concrete. The degree of failure of the ice–concrete system depends on ice and concrete characteristics as well as on the dynamic response of the structure to the repetitive ice dynamic loading. With time, this repetitive loading can affect the aggregate bond near the surface of the concrete, and cause or propagate microcracks in the concrete matrix. When eventually the integrity of the surface of the concrete has become impaired, the ice dynamic/abrasion action may cause particles of the surface to be removed. Oblique impact forces on exposed aggregates can be especially damaging. Some ice floes may contain grit, which provides more abrasive impact on exposed concrete. Furthermore, a swelling effect can take place on a concrete surface when salts crystallize as a result of the pores drying out. Erosion can also develop due to cyclic freezing and thawing. This builds up tension in concrete. Similar effects develop from variation in moisture content on the surface. Frost can also contribute to the damage. Basically, normal-weight concrete is more abrasion resistant if hard, tough aggregates are used. Visual examination of a concrete surface exposed to any kind of abrasion typically reveals local scratches and a surface that looks worn and sometimes polished.

6.2.2.6 Freezing and Thawing. As the temperature of critically saturated concrete is lowered during cold weather, the freezable water held in the capillary pores of the cement paste and aggregates expands on freezing. If subsequent thawing is followed by refreezing, the concrete is expanded further, so that repeated cycles of freezing and thawing have a cumulative effect. By their very nature, concrete marine structures are particularly vulnerable to freezing and thawing simply because there is ample opportunity for portions of these structures to become critically saturated. Concrete is especially vulnerable in tidal and splash zones. Exposure in such areas as the face of walls, piles, deck structures, and bank protections enhances the vulnerability of concrete to the harmful effects of repeated cycles of freezing and thawing. Seawater accelerates damage caused by freezing and thawing. Visual examination of concrete damaged by freezing and thawing may reveal symptoms ranging from surface scaling to extensive disintegration. Laboratory examination of cores taken from structures that show surficial effects of freezing and thawing often show a series of cracks spreading parallel to the surface of the structure.

6.2.2.7 Settlement and Movement. Because concrete structures are typically very rigid, they can tolerate very little differential movement. As the differential movement increases, concrete components can be expected to be subjected to an overstressed condition. Ultimately, these components will crack or spall. Situations in which an entire structure is moving or a single element of a structure is moving with respect to the remainder of the structure are caused basically by subsidence attributed to long-term consolidations, new loading conditions, or by a wide variety of other mechanisms. In these cases, dock operators should be concerned not so much with cracking and spalling of concrete but rather, with overturning or sliding. In situations in which overall structure movement is diagnosed as a cause of concrete deterioration, a thorough geotechnical investigation should be conducted. Visual examination of structures undergoing settlement or movement usually reveal cracking or spalling or misalignment of structural components.

Because differential settlement of the foundation is usually a long-term phenomenon, review of instrumentation data, if any, will be helpful in determining causes of the apparent movement.

6.2.2.8 *Shrinkage.* Shrinkage is caused by the loss of moisture from concrete. It may be divided into two general categories: one that occurs before setting (plastic shrinkage) and another that occurs after setting (drying shrinkage).

Plastic Shrinkage. During the period between placing and setting, most concrete will exhibit bleeding to some degree. Bleeding is the appearance of moisture on the surface of concrete; it is caused by settling of the heavier components of the mix. Usually, the bleed water evaporates slowly from the concrete surface. If environmental conditions are such that evaporation is occurring faster than water is being supplied to the surface by bleeding, high tensile stresses can develop. These stresses can lead to the development of cracks on the concrete surface. Cracking due to plastic shrinkage will be seen within a few hours of concrete placement. Typically, the cracks are isolated rather than patterned. These cracks are generally wide and shallow.

Drying Shrinkage. This is the long-term change in volume of concrete caused by the loss of moisture. If this shrinkage could take place without any restraint, there would be no damage to the concrete. However, the concrete in a structure is always subject to some degree of restraint by either the foundation, another part of the structure, or by the difference in shrinkage between the concrete at the surface and that in the interior of a component. This restraint may also be attributed to purely physical conditions such as placement of a footing on a rough foundation, chemical bonding of new concrete to earlier placements, or both. The combination of shrinkage and restraints causes tensile stresses that can lead ultimately to cracking.

Visual examination typically shows cracks that are characterized by their fineness and absence of any indication of movement. They are usually shallow, a few centimeters in depth. The crack pattern is typically orthogonal or blocky. This type of surface cracking should not be confused with thermally induced deep cracking which occurs when dimensional change is restrained in newly placed concrete by rigid foundations or by old lifts of concrete.

6.2.2.9 *Temperature Changes.* Changes in temperature cause a corresponding change in volume of concrete. Temperature-induced volume changes combined with restraint similar to that occur in a dry shrinkage before damage can occur. Basically, there are three temperature change phenomena that may cause damage to concrete. First, there are temperature changes that are generated internally by the heat of hydration of cement in large placements. Second, there are temperature changes generated by variations in climatic conditions. Finally, there is a special case of externally generated temperature change: for example, fire damage. The latter, however, is not typical in marine structures.

Internally Generated Temperature Differences. The hydration of portland cement is an exothermic chemical reaction. In large-volume placements, significant amounts of heat may be generated, and temperature of the concrete may be raised by more than 35°C over the concrete temperature at placement. Usually, this temperature rise is not uniform throughout the mass of the concrete, and steep temperature gradients may develop. These temperature gradients give rise to a situation known as *internal restraint,* during which the outer portions of the concrete may be losing heat while the inner portions are gaining heat. If the differential is sufficiently great, cracking may occur. Simultaneous with the development of this internal restraint condition, as the concrete mass begins to cool, a reduction in volume takes place. If the reduction in volume is prevented by external conditions (e.g., by chemical bonding, mechanical interlock, or by piles or dowels extending into the concrete), the concrete is said to be *externally restrained.* If the strains induced by the external restraint are great enough, cracking may occur. There is increasing evidence, particularly in rehabilitation work, that a relatively minor temperature difference in thin, highly restrained overlays can lead to cracking. Visual examination usually reveals relatively shallow isolated cracking resulting from conditions of internal restraint. Cracking resulting from external restraint usually extends through the full section. Thermally induced cracking may be expected to be regularly spaced and perpendicular to the larger dimensions of the concrete.

Externally Generated Temperature Differences. The basic failure mechanism in this case is the same as that in internally generated temperature differences. In this case, the temperature change leading to concrete volume change is caused by external factors, usually climatic conditions. If the structure is not provided with adequate space expansion joints, the externally generated temperature differences may result in concrete cracking and/or spalling at expansion joints.

6.2.2.10 *Evaluation of Causes of Concrete Distress and Deterioration.* Given a detailed report of the condition of concrete in a structure and a basic understanding of various mechanisms that can cause concrete deterioration, the problem becomes one of relating the observation of symptoms to the underlying causes. When many of the different causes of deterioration produce the same symptoms, the task of relating symptoms to causes is more difficult than it first appears. One procedure to consider is based on that described by Johnson (1965). Johnson recommends the following steps to be taken for evaluation of the present condition of concrete.

Step 1. Evaluate Structure Design to Determine Adequacy. First consider what types of stress could have caused the observed symptoms. For example, tension will cause cracking, whereas compression will cause spalling. Torsion or shear will usually result in both cracking and spalling. If the basic symptom is disintegration, overstress may be eliminated as a cause. Second, attempt to relate the probable types of stress causing the damage noted to the location of the damage. For example, if cracking resulting from excessive tensile stress is suspected, it would not be consistent to find that type of damage in an area that is under compression. Next, if the damage seems appropriate for the location, attempt to relate the specific orientation of the damage to the stress pattern. Tension cracks should be roughly perpendicular to the line of tension stress. Some typical forms of concrete damages and deterioration as related to marine structures are depicted in Figure 6.3. Shear usually causes failure by diagonal tension, in which case the cracks will run diagonally in the web of a beam. Visualizing the basic stress patterns in the structure will aid in this phase of the evaluation. If no inconsistency is encountered during this evaluation, overstress may be the cause of the damage observed. A thorough stress analysis is warranted to confirm this finding. If an inconsistency has been detected, such as cracking in a compression zone, the next step in the procedure should be followed.

Step 2. Relate the Symptoms to Potential Causes. For this step, Table 6.1 will be of benefit. Depending on the symptom, it may be possible to eliminate several possible causes. For example, if the symptom is disintegration or

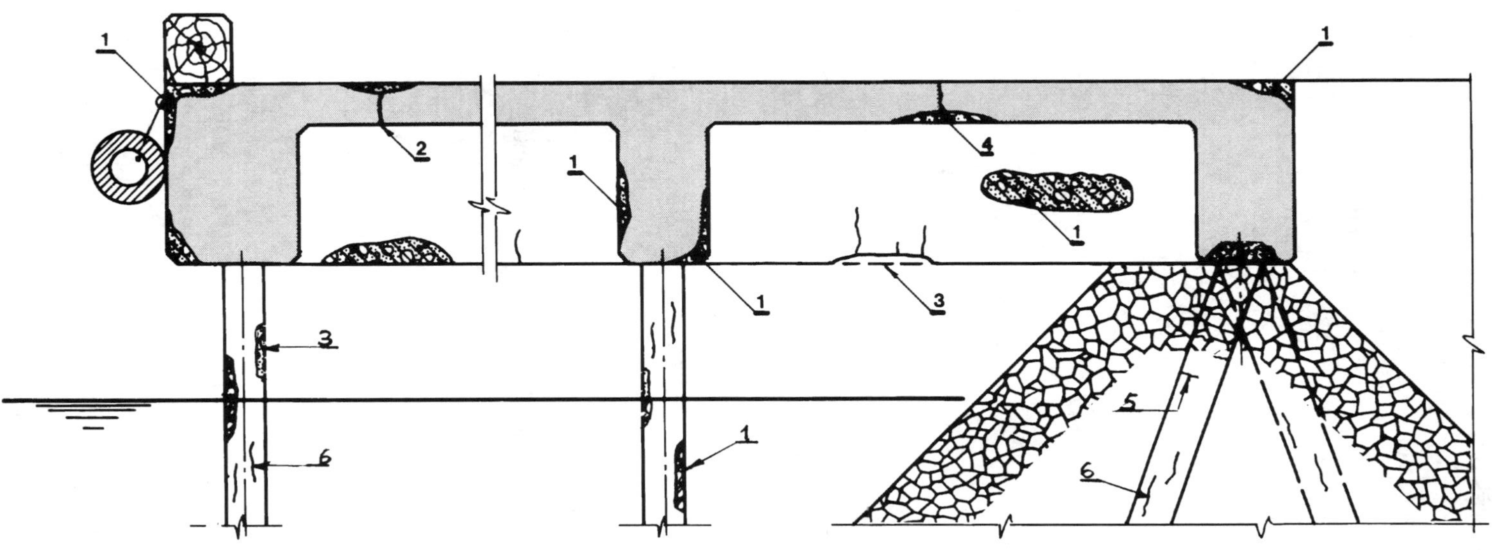

Figure 6.3 Typical forms of concrete damage and degradation in a seawater environment. 1 and 3, Spalled surface; 2, spalled surface and leaking crack; 4, underside of spalled concrete and leaking crack; 5 and 6, cracks.

Table 6.1 Relating symptoms to causes of distress and deterioration of concrete

Symptoms	Construction Faults	Cracking	Disintegration	Distortion/ Movement	Erosion	Joint Failures	Seepage	Spalling
Accidental loadings		×						×
Chemical reactions		×	×				×	
Construction errors	×	×				×	×	×
Corrosion		×						×
Design errors		×				×	×	×
Erosion			×		×			
Freezing and thawing		×	×					×
Settlement and movement		×		×		×		
Shrinkage		×						
Temperature changes		×				×		×

Source: U.S. Army Corps of Engineers (1986).

erosion, several potential causes may be eliminated by this procedure.

Step 3. Eliminate the Readily Identifiable Causes. From the list of possible causes remaining after relating symptoms to potential causes, it may be possible to eliminate two causes very quickly because they are relatively easy to identify. The first of these is corrosion of embedded metals. It will be easy to verify if the cracking and spalling noted are a result of corrosion. The second cause that is readily identified is accidental loading, as personnel at the structure should be able to relate the observed symptoms to a specific incident.

Step 4. Analyze the Available Clues. If no solution has been reached at this stage, all the evidence generated by field and laboratory investigations should be reviewed carefully. Attention should be paid to the following points:

1. If basic symptom is that of disintegration of the concrete surface, essentially three possible causes remain: chemical attack, erosion, and freezing and thawing. Attempts should be made to relate the nature and type of damage to location in the structure and to the environment of concrete in determining which of the possibilities above is most likely to be the cause of the damage.
2. If there is evidence of swelling of concrete, there are two possibilities, such as chemical reactions and temperature changes. Destructive chemical reactions such as alkali–silica or alkali–carbonate attack which cause swelling will have been identified during the laboratory investigation. Temperature-induced swelling should be ruled out unless there is additional evidence, such as spalling at joints.

3. If the evidence is spalling, and corrosion and accidental loadings have been eliminated earlier, the major causes of spalling remaining are construction errors, poor detailing, freezing and thawing, and externally generated temperature changes. Examination of the structure should have provided evidence as to the location and general nature of the spalling that will allow identification of the exact cause.
4. If the evidence is cracking, construction errors, shrinkage, temperature changes, settlement and movement, chemical reactions, and poor design details remain as possible causes of distress and deterioration of concrete. Each of these possibilities will have to be reviewed in light of available laboratory and field observations to establish which is responsible.
5. If the evidence is seepage and it has not been related to a detrimental internal chemical reaction by this time, it is probably the result of design errors or construction errors such as improper location or installation of a waterstop.

Step 5. Determine Why the Deterioration Has Occurred. Once the basic cause or causes of the damage have been established, there remains one final requirement: to understand how the causal agent acted on the concrete. For example, if the symptoms were cracking and spalling and the cause was corrosion of the reinforcing steel, what facilitated the corrosion? Was there chloride in the concrete? Was there inadequate cover over the reinforcing steel? Another example to consider is concrete damage due to freezing and thawing. Did the damage occur because the concrete did not contain an adequate air/void ratio, or did the damage occur because the concrete used was not expected to be saturated but for whatever reason was saturated? Only when the cause and its mode of action are understood completely should the next step of selecting a repair material be attempted.

6.2.3 Corrosion of Steel in the Marine Environment

The subject of metal corrosion in general and in the marine environment in particular is well covered in several fundamental works (Uhlig, 1948, 1971; Rogers, 1960; Laque, 1975; Dismuke et al., 1981). The interested reader is also referred to the abundance of literature on the subject matter presented in book form and published in numerous technical periodicals. In this work, only basic information is presented on metal corrosion in marine environment as related to maritime structures.

Corrosion of steel is an electrochemical process similar to that which takes place in a common flashlight battery; it occurs at the anode and is accompanied by a flow of electrons through the external "wire" to the cathode. Corrosion occurs because of small physical and/or chemical differences present in metals such as minor impurities or local composition variations, or environments; for example, changes in amounts of dissolved oxygen varying with the depth of immersion, nonuniform salt concentrations due to pollution, and so on. Two types of corrosion are recognized: dry and aqueous. The former may be described briefly as the metal directly oxidizing, thereby returning to a lower chemical energy level. This type of corrosion is slow and relatively uniform. Its rate is determined by temperature and diffusion of oxygen through the oxide. Thus the thickness and physical stability of the rust layer are significant. In practice, however, dry atmospheric conditions almost never exist in a marine environment.

A marine atmosphere where condensed moisture is very high (corresponding to 100%

relative humidity) is a very aggressive environment for metals. Under such a rather "wet" condition, the corrosion process is analogous to that of continuous seawater immersion, except that the thin wet electrolytic film has a marked effect on the corrosion pattern, the corrosion products, and the ease with which oxygen is transferred to the metal surface, resulting in accelerated corrosion.

Aqueous corrosion may cause the metal to go into solution at one point and to allow oxygen to be taken up at a second point while depositing the corrosion product at a third point. In fresh water, which is not as good a conductor as seawater, there is little corrosion because the corrosion rate there is directly proportional to the current. The salts dissolved in seawater greatly increase the water conductivity and hence its corrosiveness. To initiate the corrosion process there must be a complete electrical circuit in both the structure and the aquatic medium (electrolyte) such that negatively charged ions in the electrolyte flow from where they are produced, at the cathode, toward the anode. In the structure itself, therefore, the ions flow from the anode to the cathode, unless an opposing voltage is applied with the aim of suppressing this current. The presence of negative ions near the anode encourages positively charged metallic ions to dissolve into the electrolyte, where they combine with any available negative ions to form a corrosion product. Thus, the entire process continues unless the corrosion product itself forms a barrier to ionic movement. This "passive" coating re-forms and heals itself spontaneously provided that oxygen is available, but rapid corrosion can occur in crevices or under marine growth. The rate of corrosion in carbon steel is much higher than in other metals, including those that do not form a passive coating, because the magnetite produced in the process is a good conductor.

The chemical reactions that take place in iron (the principal constituent of steel) corroding in seawater are as follows. At the anode iron goes into solution:

$$Fe \rightarrow Fe^{2} + 2e \qquad (6.1)$$

The electrons flow to the cathode through the metallic circuit. At the cathode, oxygen converts hydrogen atoms into water:

$$2H^{+} + \tfrac{1}{2}O_2 + 2e \rightarrow H_2O \qquad (6.2)$$

or converts water to hydroxyl ions:

$$H_2O + \tfrac{1}{2}O_2 + 2e \rightarrow 2OH^{-} \qquad (6.3)$$

Adding (6.1) and (6.3) gives,

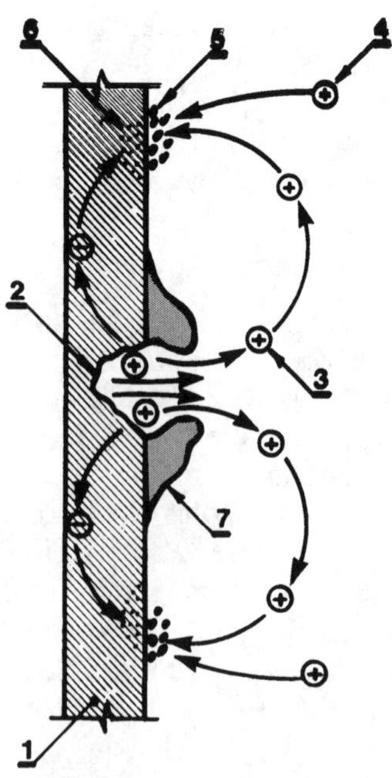

Figure 6.4 Mechanism of steel corrosion in a seawater environment. 1, Steel; 2, pit; 3, iron ion; 4, hydrogen ion; 5, hydrogen film; 6, impurity; 7, product of corrosion: $Fe(OH)_2$.

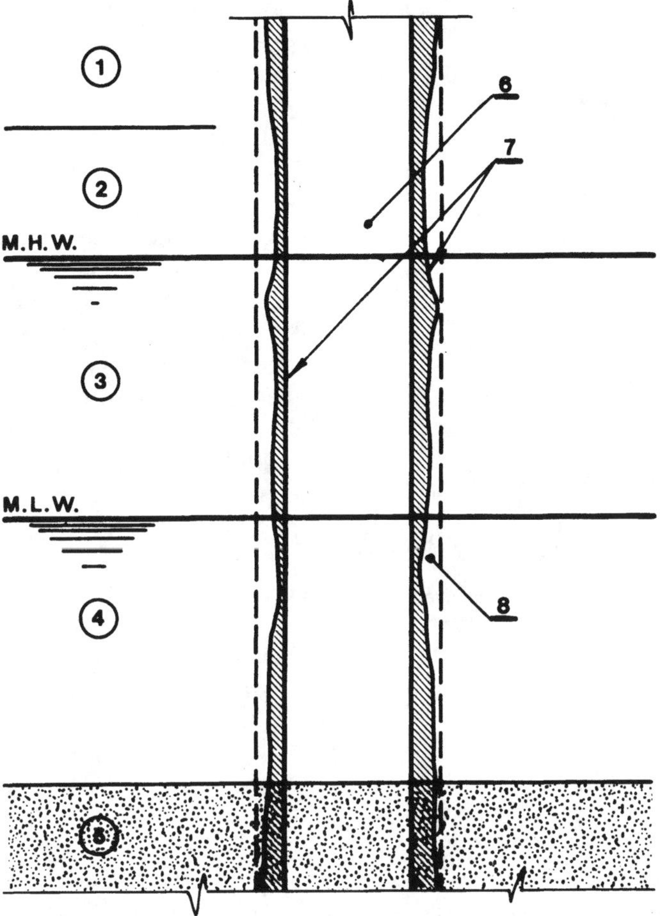

Figure 6.5 Typical vertical distribution of lost metal for representative H-pile in a marine environment. 1, Atmospheric zone; 2, splash zone; 3, tidal zone; 4, zone of continuous immersion; 5, seabed zone; 6, flange; 7, web; 8, metal lost to corrosion.

$$Fe + H_2O + \tfrac{1}{2}O_2 \rightarrow Fe(OH)_2 \qquad (6.4)$$

Iron is converted to ferrous hydroxide. Other reactions can occur, such as conversion of ferrous hydroxide to ferric hydroxide [$Fe(OH)_3$] by further reaction with oxygen. A typical mechanism of the corrosion process is shown in Figure 6.4.

Average corrosion rates are normally calculated by the total weight loss of metal or by metal thickness losses. The rate of corrosion typically varies depending on the zone of marine environment. The graphic representation of metal deterioration in marine environment is provided in Figures 6.5 and 6.6, and representative corrosion rate profiles and a cross section of steel sheet piling in a seawater environment is shown in Figure 6.7. As shown in the figures, the highest areas (rates) of metal loss to corrosion are located in the splash zone and im-

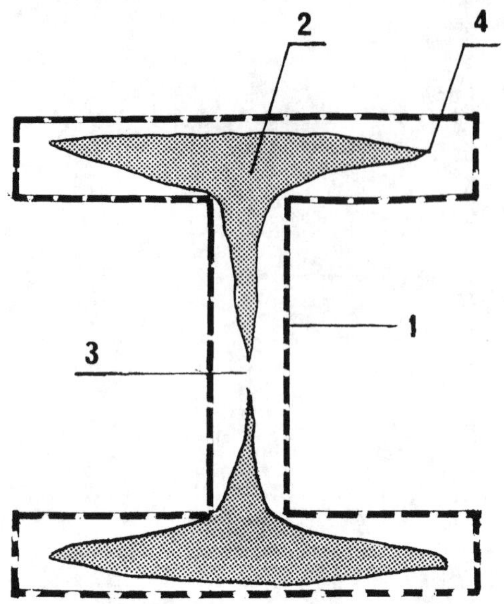

Figure 6.6 Loss of metal from a steel H-pile in a marine environment due to corrosion. 1, Original section; 2, corroded section; 3, typical perforation of web; 4, typical knife edge at flange end.

mediately below MLW (mean low water). This may vary markedly, however, depending on site conditions. For example, in some harbors with relatively large tide ranges the corrosion rate in the splash zone can be less pronounced, with the greatest metal loss just below MLW and just above the mudline. In the absence of local experience, the average rate of metal loss for mild steel in quiescent seawater can be considered as equal to approximately 5 mils per year (mpy; 1 mil = 1/1000 inch). However, studies of sheet pile bulkheads carried out by the U.S. Navy revealed the possibility of maximum corrosion rates, on the order of 8 mpy in temperate climates and up to 19 mpy in subtropical climates. Maximum rates were on the order of twice the average rates (Ayers and Stokes, 1961). Creamer (1970) reported corrosion rates up to 55 mpy in a splash zone on an offshore structure located in the Gulf of Mexico; however, more typical rates of metal corrosion in the splash zone are on the order of 25 to 40 mpy (Chellis, 1961).

In most cases, corrosion is not uniform over the surface of a structure. Severe pitting may form areas where structural stresses can concentrate. In some instances, pitting can be of more importance than uniformity. Pitting corrosion rates are generally greater than uniform rates, particularly for the first 10 years after installation. Pitting corrosion may result in complete perforation of sheet piling or steel piles, which eventually may lead to unacceptable reduction in structural strength, leak of backfill material, and so on. A practical example of steel sheet pile corrosion is illustrated in Figure 6.8.

Marine structures suffer severe and rapid corrosion in a splash zone because of the high availability of both oxygen and electrolytes and because any protection that might be afforded by the corrosion product itself is rapidly washed away. This is why the average rates of atmospheric corrosion usually are similar to those in the submerged zone. On other hand, the average rate of corrosion in the submerged zone is usually much smaller than in the splash zone, measuring about 0.15 mm per year in the first year and dropping to perhaps half this figure after about 15 years (Ractliffe, 1983). However, in warm climates the rate of steel corrosion in the submerged zone can be much higher (Buslov, 1979, 1983; Fukute et al., 1990)

Loss of metal due to corrosion is generally very low (sometimes even negligible) in areas of steel piling driven into undisturbed soils (Schwerdtfeger and Romanoff, 1972). The average corrosion rates there may be approximately 0.011 mm/yr (Ohsaki, 1982) in the first year and may decrease with time. According to Ohsaki, the corrosion rate is not affected by type of soil or its permeability. However, highly acidic or highly alkaline soils with a low

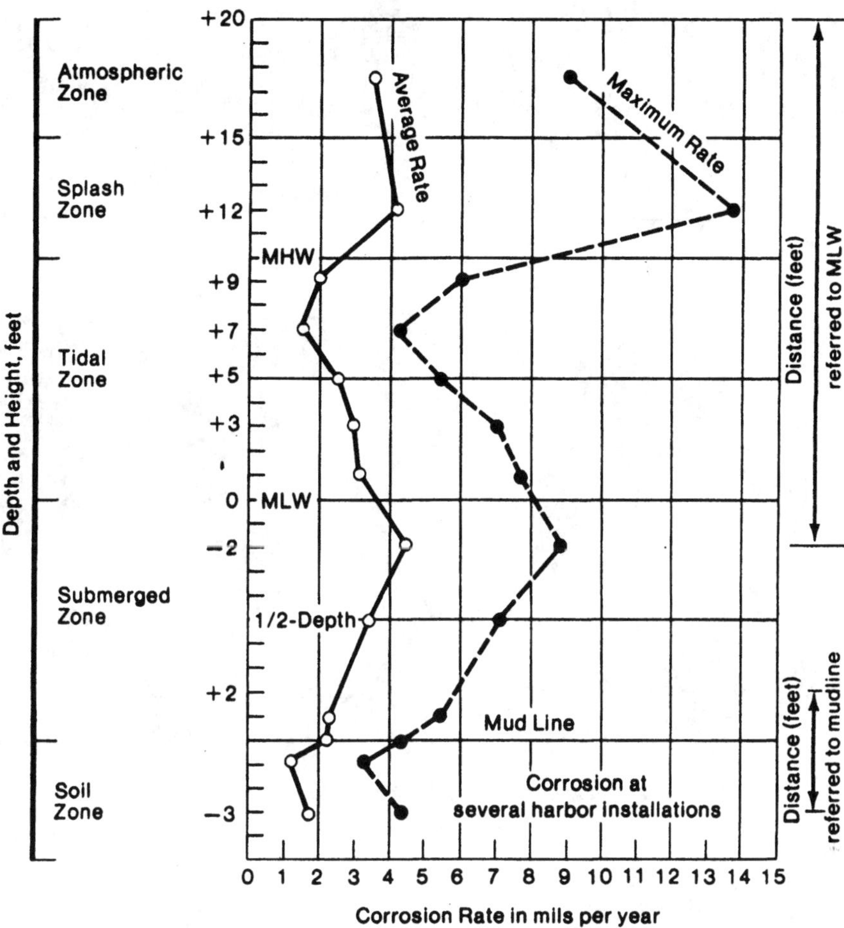

Figure 6.7 Corrosion rate profiles of steel sheet piling. (From Edwards, 1963.)

pH value can be corrosive. The accelerated corrosion may occur just below the mudline as a result of an abrupt change in the environment of the structure as it passes from the water to the soil. This may occur due to the removal of corrosion products by water flow, or sometimes, because of water (or soil) contamination.

The rate of corrosion varies from location to location and depends on the following environmental factors:

1. *Water temperature.* Chemical reactions, which include corrosion reactions, are in general accelerated in warmer water. However, the fouling rate is also accelerated in warmer water. The fouling provides a protective covering over the metal surface, which produces a retarding effect on the corrosion rate due to decreased access of oxygen to the steel. Therefore, contrary to expectations, cor-

Figure 6.8 Disintegration of steel sheet piling in a splash zone due to corrosion.

rosion rates in warm seawater (immersion zone) have not been found to differ significantly from those measured in cold water.

2. *Oxygen concentration.* The oxygen concentration is a corrosive agent of steel in seawater. The rate of oxygen concentration on the surface of steel depends on the water depth. Areas of low concentrations of oxygen are anodic to areas of higher concentrations. The increased corrosion rate in the submerged zone just below MLW as compared to the tidal zone is attributed to better aeration of water.
3. *pH value.* The pH value or degree of acidity or alkalinity generally varies from 7.2 to 2.2 (Dismuke et al., 1981). A pH value below 7 is acidic, and above 7 is alkaline. As alkalinity is increased above pH 9.5, iron tends to become passive and forms protective films that retard the diffusion of oxygen to the surface. Below pH 4, alkaline protective films are dissolved and the acid acts directly on metal, accompanied by the evolution of hydrogen. Hence, changes in pH values may affect the rate of corrosion.
4. *Water salinity.* Water in the open sea has a salt content of about 3.5%, consisting of various ions resulting from dissolved salts. The chloride ion is the most significant because of its high concentra-

tion. Chloride ions are able to penetrate the protective film formed by corrosion products to cause localized corrosion. The presence of chloride ions also affects the solubility of oxygen in water, hence increasing rates of corrosion.

5. *Water velocity.* Water velocity increases the rate of steel corrosion because of the increased availability and rate of diffusion of oxygen through the stagnant hydrodynamic sublayer of seawater on the surface of the metal. With zero velocity the overall corrosion rate of steel in seawater is generally lower than in flowing water.
6. *Marine organisms.* Organic matter in seawater has a marked effect on corrosion. The fouling community consists of a vast variety of marine plants and animals that attach themselves to marine structures. These marine organisms are generally found in fluidized mediums with pH values from 6 to 11 and can accelerate the corrosion rate in localized areas, which is a result of metabolic activity of living microorganisms. The normal metabolic activities of these microorganisms affect corrosion in seawater by altering anodic and cathodic reactions and creating a corrosion condition by producing differential aeration. Organically induced corrosion is a result of the activities of two classes of bacteria: aerobic, which require oxygen for their metabolic process; and anaerobic, which live and grow in an environment containing little or no oxygen (Cleary, 1969). Sulfate-reducing anaerobic bacteria are the most troublesome. These bacteria can affect deeply submerged steel components, such as piles. Microbial-induced corrosion in steel is discussed in Scott and Davies (1992). In a low-oxygen environment, bacteria can utilize the hydrogen formed cathodically at the steel surface to reduce sulfate from the electrolyte and to increase local corrosivity of the environment. The corrosion product related to marine organisms can be identified by its blackness. This type of corrosion typically results in the formation of pits on the steel surface.
7. *Pollution.* Pollution in harbors generally causes harm to the marine environment by its toxic effects or by depletion of the dissolved oxygen. The follow-up destruction of oxygen-dependent fouling organisms in seawater may alter corrosion rates unfavorably. In polluted water, however, the destructive effects of marine boring on wooden piles are much less pronounced than in clean water (Abood et al., 1995); (Tahal and Maltin, 1996).
8. *Wind.* Wind causes wave action, resulting in intermittent wetting in the splash zone. It also whips up the water surface and captures salt spray from breaking waves. The salt-laden spray evaporates, and the remaining salt crystals are deposited on metal surfaces. These salt crystals accelerate the corrosion of steel to which they adhere by attracting and retaining moisture and forming aggressive local cells.
9. *Rain.* Heavy rain can wash salt from a steel surface and thereby reduce the corrosion rate.
10. *Humidity.* Humidity forms a thin film of electrolyte on a steel surface in the atmospheric zone, and this promotes corrosion by diffusion of oxygen through this layer.
11. *Sun.* The sun affects the relative humidity, rate of evaporation, and temperature of the structure with subsequent impact on the corrosion of metal parts of the structure.

12. *Ice.* Winter ice conditions cause removal of all corrosion products and effectively expose totally bare steel every spring. In addition to the ice forces, sediments are often carried in the ice and they act as an abrasive and essentially polish the steel surface. Under such conditions, the loss of steel can reach more than 25 mils per year.

All the factors above, along with the damaging effects of ultraviolet exposure from the sun on the pigmentation and composition of many protective coating systems, affect the rate of steel corrosion in a complex way.

Additional factors that can affect the steel corrosion rate are the following:

1. *Steel composition.* The addition of copper to carbon steel with some inclusion of nickel and phosphorus is reported to provide superior corrosion resistance in the splash zone (Dismuke et al., 1981). Tests of other steel composition, such as ASTM A517 grade F and ASTM A242 type 1 (Cr–Si–Cu–Ni–P), have demonstrated even better performance (Schmitt and Phelps, 1969).
2. *Use of dissimilar metals.* Dissimilar metals coupled together can produce rapid corrosion in seawater because the galvanic current between different metals can be large unless steps are taken to insulate them from each other. As always, it is the anodic metal that corrodes. The most common anodic metals are magnesium, zinc, pure aluminum, aluminum alloy, clean steel, cast iron, chromium stainless steel, lead–tin solder, corroded steel, tin, mill scale on steel, brasses, copper, bronzes, copper–metal alloys, and chromium stainless steel. In marine pile construction, fasteners, welds, fittings, and so on, are sometimes fabricated from metals other than plain carbon steel. If these metals are cathodic to the pile and are much smaller relative to the pile surface area, little corrosion should be expected on the adjacent anodic pile surface. Similarly, cathodic metals would also be unaffected. In the reverse situation, small anodic fittings attached to the cathodic pile may corrode rapidly.
3. *Mill scale.* Mill scale, a tenacious deposit of iron oxide on the steel surface (a result of the hot rolling process) is more cathodic than steel. Exposed steel with mill scale present can be pitted about three times as deeply as descaled steels (Laque in Uhlig, 1948).
4. *Stray current.* Stray current from external sources around marine structures can cause rapid corrosion to the metal parts of a structure, especially at shipyards where improperly grounded welding equipment or electric service systems, for example, may result in localized electric currents. Faulty impressed-current cathodic protection systems, high-voltage cable crossings, berthed vessels, or other extraneous sources may also contribute to high localized corrosion rates due to stray electric currents. For comprehensive information on stray currents, consult Dismuke et al. (1981).

Corrosion deterioration limits a structure's service life. It may range from negligible to severe. Corrosion deterioration can develop into stress corrosion and/or corrosion fatigue. *Stress corrosion* is the cracking of metal under the continued effects of tensile stress and a corrosive environment. Tensile stress creates an anodic area that pits and increases the stress concentration. Corrosion at the tip of the notch causes a stress corrosion crack to start there. The cracks are very fine at the early stages of

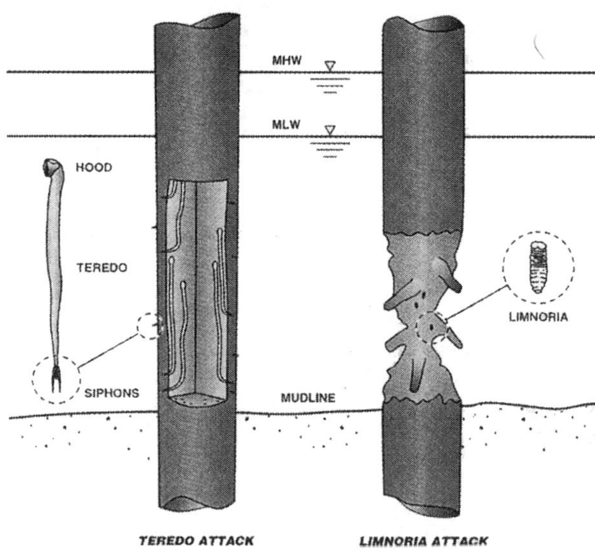

Figure 6.9 Typical pattern of damage to a wooden pile produced by marine borers.

growth, but become visible before failure. The failure appears like a brittle fracture.

Corrosion fatigue is a term that defines the reduction of the metal's air fatigue strength in a corrosive environment. In dry air, steel exhibits a fatigue limit defined as the stress below which the metal can sustain an unlimited number of load (stress) reversals. In a sea environment, fatigue damage can actually occur at all levels of stress, and fatigue strength in the high-cycle range is reduced to more than 40% of its value in dry air (Schmitt and Phelps, 1969). The rate of growth of fatigue cracks is about five times as rapid in seawater as in dry air, and cathodic protection can further accelerate crack propagation by encouraging the production of hydrogen at the crack root (Ractliffe, 1983).

6.2.4 Wood Degradation in the Marine Environment

Wooden structures in a marine environment are subject to mechanical damage and biological degradation. Above water, the wood can be damaged and destroyed by fungal (rot) and insect damage. Fungal spores and moisture can penetrate the heartwood by different mechanisms and destroy the wood, resulting in loss of structural member cross-sectional area. Furthermore, microbiological deterioration (bacteria and soft-rot fungi) can substantially reduce the compressive strength of the wooden structural members (e.g., piles) (Altiero, 1997). However, marine borers of all kinds are the primary culprits responsible for the destruction of marine structures constructed of wood in a seawater environment (Figures 6.9 and 6.10).

Destructive and fouling organisms are present at practically every seaport location, even in temperate climates. However, tropical and subtropical areas provide an environment that supports more species and greater activity than do more temperate and colder regions. It should be noted that in a freshwater environment, wooden structural components submerged below the lowest water level are normally well preserved and do not require any

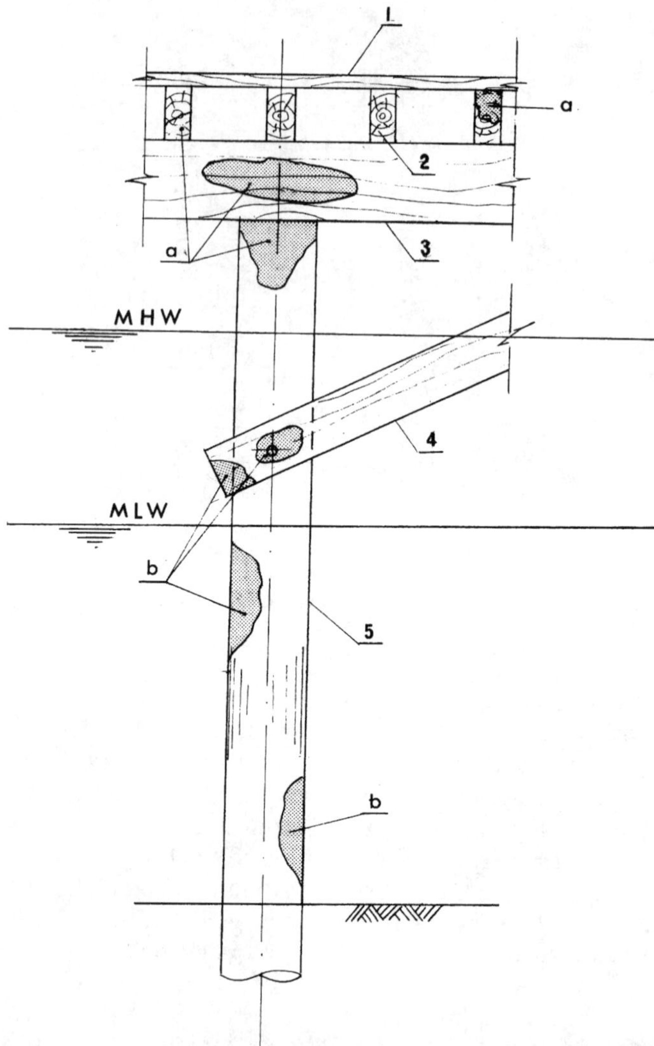

Figure 6.10 Degradation of wooden components of marine structures in seawater. a, Hidden fungal damage; b, hidden internal marine borer; 1, decking; 2, stringer; 3, pile cap; 4, brace; 5, pile.

treatment, and with time they may even gain some additional strength. Certain marine organisms in the seawater environment, such as mollusks and crustaceans, bore and destroy timber. Marine borers are found worldwide. Fortunately, no specific location contains all borers, and most areas have only a few kinds (Menzies and Turner, 1957). The presence of a particular species at a particular site is largely dependent on miscellaneous environmental factors, most of all on temperature and the salinity of water.

The most important mollusks found in North American waters are the teredinids and pholads. *Teredinids,* commonly known as *shipworms,* are perhaps the most treacherous marine borers, because the extent of damage they produce is largely unknown until a structure becomes unserviceable or collapses totally. Teredinids begin their life as free-swimming larvae less than 0.3 mm long. During their early period of life, which may last a few days to a couple of weeks, the larvae may settle and begin to bore if suitable unpreserved wood is found. Because the larvae are very small, the entry holes are almost invisible. On settlement the mollusk begins to metamorphose into a worm by using the wood as a sustained source of food. Within a year or so, the teredinid may grow to 25 mm in diameter and 120 cm in length (Atwood and Johnson, 1924). Clearly, if many such worms are present in a pile or in any timber member of a structure, it cannot retain much strength. Teredinids have been known to destroy untreated pine piling in 1 month. Although teredinid penetration is common within tidal zones, attack is generally heaviest near the mudline (Chellis, 1961).

Pholads, known as *rock borers,* resemble ordinary clams somewhat, as their bodies are entirely inside their shells. Pholads burrow for shelter and may penetrate soft rock, poor grade concrete, and clay as well as wood. Like teredinids, pholads are initially free-swimming and settle as larvae. Unlike teredinids, however, they do not burrow as deeply, often no farther than the length of their shell, about 50 to 65 mm. Their entrance hole (about 5 mm) is much bigger than that of teredinids.

Burrows from the crustacean group are related to crabs, shrimps, and so on. They are quite small but may be present in sufficient numbers to produce substantial damage in a relatively short time. The principal genera are *Limnoria* and *Sphaeroma*. The mode of attack of crustacean borers is somewhat different from that of the mollusks. The limnorians, for example, first penetrate the wood, then create tunnels just below and parallel to the surface with occasional perforations (for respiration) to the surface. The thin layers of their burrows are easily washed away, leaving a surface with myriad fine tunnels and exposing new wood for the next attack. According to Richards (1983), under heavy *Limnoria* attack, a pile radius in a tidal zone may diminish by 10 to 15 mm/yr—probably more in some cases.

Members of the genus *Sphaeroma,* known as *pill bugs,* often reach 10 to 15 mm in length. Damage by these borers is easily recognized by large (around 5 mm in diameter), round entry holes that are usually perpendicular to the wood surface and grain and about 10 to 20 mm deep. They attack wood predominantly in the tidal zone but have also been reported at the mudline (Atwood and Johnson, 1924). All types of borers respond to differences in environmental conditions such as water temperature, salinity, and the presence of a suitable substrate for boring. But other factors, such as water depth, current, pollution, siltation, oxygen content, turbidity, and suspended organic matter, may also be important (Turner, 1966). Changes in environmental conditions, such as prolonged droughts, heavy rains, or seasonal changes in water temperature, can also affect borers' activities mainly through regulation of reproduction. For example, the discharge of a power plant's warm water into a harbor can result in more severe teredinid damage (Bletchly, 1967). As noted earlier, borers are generally more active in the tropics and within normal ranges of oceanic salinity. Their activity is generally decreased by increase in water pollution turbidity, siltation, and by fast-moving currents.

In the past in many ports that were part of the large industrial cities, the problem of borers was almost nonexistent because of heavily polluted water. However, as stricter environmental regulations were introduced and the quality of water improved, the borers came back in mass. Their resurrection threatens many marine structures constructed from wood which now require protection from borers, or in some instances, complete replacement (Buslov and

Figure 6.11 Damage to a concrete caping wall inflicted by ship impact.

Padron, 1992; Abood et al., 1995; Gunas et al., 1993; Tahal and Matlin, 1996; Metzger, 1998). For more information on borers in marine environment and their effects on marine wooden structures, interested readers are referred to Gaythwaite (1981), Agi (1983, 1997), and Gaythwaite and Carchedi (1984).

6.3 PHYSICAL DAMAGE TO WATERFRONT STRUCTURES ATTRIBUTED TO DOCK OPERATION

These can be due to the following basic factors: ship hard docking, effects of cargo handling and hauling systems, ship propeller-induced scour, or a combination of all of the above.

6.3.1 Damage by Vessel and/or Cargo Handling Systems

Physical damage to the dock structure caused by a vessel can generally be categorized as either accidental or attributed to the method of ship handling. Accidental loadings are a short-duration, one-time event. These loadings can generate stresses higher than the strength of the structure, resulting in localized or general failure. The damage usually manifests itself in the form of loosening of the individual components of the wall, cracking, or in the extreme case, punching a hole in the wall (Figure 6.11). Unprotected by fender system(s) walls are clearly more vulnerable than those protected. Another ship-induced damage is the abrasion caused by the ship and its mooring lines. This usually manifests itself in severe concrete grooving, where the concrete is relatively soft due to soft aggregates used in the mix. Steel mooring lines can also penetrate joints in the coping, thereby causing further damage to the wall as vessel rolls. Determination of whether accidental loading caused damage to a structure usually requires knowledge of the events preceding discovery of the damage. Damage due to accidental loading is easy to diagnose.

Visual examination will usually show damaged steel or wooden structural components, and spalling or cracking of concrete that has been subjected to accidental impact. By their very nature, accidental loadings cannot be prevented. Accidental damage to a dock structure by a ship is usually attributed to ship mechanical problems such as power loss, excessive approach velocity, insufficient capacity of fenders, incorrectly installed fender system, and poor weather conditions. At times, poor coordination of berthing–unberthing operations creates an accidental situation when a ship approaches or leaves a berth at an unacceptable angle or with unacceptable speed.

Ship–structure collision under these conditions sometimes results in damage to the structure or its elements, such as the fender system, piling, and others, and/or to the ship. Traditionally, the problem of ship–dock structure damage due to collision impact has existed since the first primitive dock was constructed. Increase in ship sizes, particularly in the category of container ships, ferries, and oil and bulk carriers, substantially increased damage to port structures. Large ships with bulbous bows have added a new factor to potential damage to port structures.

At times, substantial damage to the berth structure caused by power failure, mechanical problems, or by errors in berthing of ships literally of all sizes, but mainly by large vessels with bulbous bows, has been reported. A bulbous bow is ideal for piercing holes in sheet pile bulkheads and in thin-walled concrete structures, damaging gravity vertical walls and catching the piling under piled platforms and all types of piled dolphins. Experience indicates that damage to the structure caused by bulbous bows usually occurs in the first row of piles nearest the face of the berth. In older structures these piles are centered as close as 0.3 to 0.5 m from the face of the berth. In some incidents, other support piling, the second row in particular, was damaged but not to the extent or frequency of the first (outer) row.

Damage to the structure usually occurred when the vessel had been approaching the wharf at an angle that exceeded 10 to 15° and/or with an unacceptably high velocity from the fender system energy-absorbing-capacity point of view. Hard docking may also be affected by weather conditions, for example, during heavy wind or high waves. On occasion, additional tugs may be required to combat the force of the winds to ensure safe docking of a ship with a large sailing area. Hard docking, or absence of or inadequate fendering, may result in greater wear and tear on dock walls. Incorrectly installed fender units (Figure 6.12) may result in damage to the edge of the dock structure and to the ship alike. Lack of a properly installed fender system and errors in berthing or unberthing maneuvers may also result in removal of mooring bollards and hand rails by flared bows. Flared bows have been responsible in a number of accidents involving cargo handling equipment: for example, container cranes. This author is aware of at least one case of vessel–train collision attributed to the ship's flared bow. In the latter case the train track was located about 2.75 m from the berth's face. Hard docking may result in damage or even complete destruction of the fender system (Thompson, 1977; Padron and Han, 1983; Tsinker, 1995 and 1997).

It must be noted that a fender system is, in general, a very vulnerable component of the berth. If not designed, installed, and maintained properly, it can be damaged severely or even

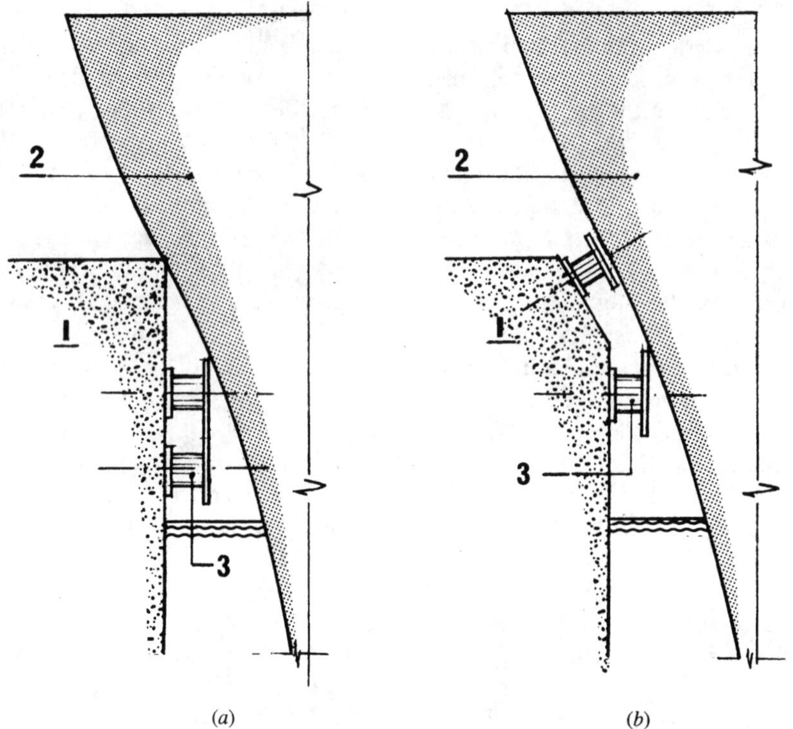

Figure 6.12 Fender installation: (*a*) incorrect; (*b*) correct. 1, Quay; 2, vessel; 3, fender unit.

torn off the structure completely by the ship. It is well known that the approach velocity and angle are the most important parameters that affect the berthing energy and ultimately, the ship impact load on a structure. Unfortunately, for a variety of reasons, in real life the actual vessel approach velocity, sometimes exceeds design values. This results in damage to fender units, dock structure, and sometimes to a ship. This is why the dock fender system should have sufficient redundancy and in most practical cases should not be designed with mathematical precision. The designer must be aware that the cost of the dock fender system in most practical applications contributes approximately 3 to 5% or less of the total cost of the dock. Hence, an insignificant reduction in the cost of a fender system will, in practice, not affect the total cost of the project; however, it may significantly affect the safety of dock operation. The most recent guideline on fender system design is found in the International Navigation Association (INA/PIANC, 2002). In some instances, the effects of new cargo handling systems installed on existing dock structures may result in substantial overloading. The latter, in turn, may affect a structure's stability and result in overstressing of some structural components (e.g., deck structure, piles). All this may also result in premature deterioration of the structure.

In general, structure in-service must cope not only with deterioration produced by new heavy loads generated by mobile, portal, and container cranes and by a variety of new heavy unit loads. Deterioration of the structure may also be caused by carelessly driven miscellaneous mobile equipment. Lift trucks can be

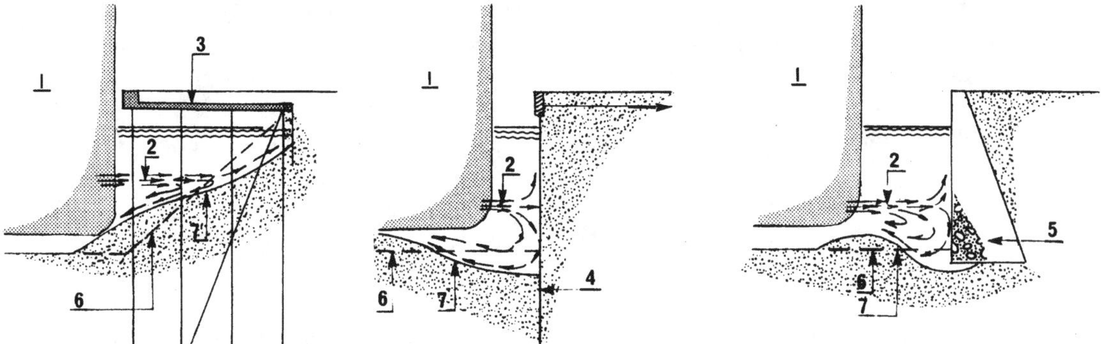

Figure 6.13 Propeller-induced scour. 1, Ship; 2, propeller (side unrusted)-induced water jet; 3, piled platform; 4, sheet pile bulkhead; 5, gravity quay wall; 6, original (designed) profile; 7, propeller jet–induced erosion.

driven with their forks catching the deck material, as can the goosenecks on tractors. Trailers and trucks can reverse into fences, handrails, lighting posts, mooring accessories, curbs, and other fixed objects.

Waterfront structures built in the past 60 years have been designed for about 25 kN/m^2 uniformly distributed load, whereas today we are looking at basic structures to handle unit and other loads on the order of two and more times that amount. Use of heavy cargo handling equipment that outstrips a quay's designed capacity to carry these loads creates material fatigue conditions resulting in heavy overstress and disintegration of structural components such as sheet piles, bearing piles, and deck structure. Cargo handling equipment today is very mobile and can be used anywhere in a port, and this is not always appreciated by their operators. The use of a heavy outrigger or wheel load sometimes results in local damage to a structure.

Normally, signs of structure overload are readily detected. It manifests itself in the form of permanent deformations, settlements, misalignment of individual components or the entire structure, localized damage to structural members, signs of fatigue, and so on. It should be noted that structural fatigue as a source of deterioration may also be attributed to the cyclic action of waves. This may be an especially important factor to look at in exposed (offshore) sites and in deep water.

6.3.2 Propeller-Induced Scour

A dramatic increase in vessel dimensions and installed engine power, the introduction of new types and special-purpose ships with bow thrusters, and the use of roll on–roll off ships in recent years are frequently reported as a source of dangerous scour, which in many cases threatens to undermine berth structures. Velocities at propeller jets at the exits of propellers, as well as at side thrusters, can easily be 11 to 12 m/s with resulting bed velocities from 3 to 4 m/s (Longe et al., 1987). Depending on circumstances, the average depth of scour could reach as much as 0.5 m per month and more. The theory of the propeller jet and its effects on erodible soils is summarized in Tsinker (1995).

In some cases, damage produced by propeller jets has extended to scour of the bottom in front of the quay wall, washing out grout seals between joints in gravity walls and in concrete sheet pile bulkheads (Chait, 1987; PIANC, 1995; Blokland and Smedes, 1996; Raes et al., 1996; Hamil et al., 1999; Belyaev, 2001; Vorobyov, 2001). Examples are presented in Figure 6.13. Disintegration of the grout seals caused sand backfill to leak through the joints, producing sinkholes in the surfacing of the quay. If allowed to progress, crane rails, rail tracks, buildings, and so on, can be undermined.

In general, scour is dangerous to any kind of marine structure, particularly to those relying on a designed bottom level to support the structures against horizontal and vertical loads: for example, all kinds of piled platforms, sheet pile bulkheads, gravity structures built on erodible foundation material, and riprap-protected slopes under open platform structures.

The main factors that affect the extent of damage done to a dock structure are as follows:

- Power of propulsion unit or side thruster
- Draft of vessel when berthing or unberthing
- Position of propeller or side thruster in relation to height above keel
- Position and shape of ship's rudder
- Shape of the vessel hull and beam
- Underkeel clearance
- Distance of quay and direction of thrust
- Duration of exposure to effects of a jet
- Type of bottom soil

Propeller-induced scour has proved to be a very serious factor to be considered in the design of a new port-related marine structures or reevaluation of the performance of existing structures in service. New and existing structures can be protected from propeller-induced damage to the structure in a variety of ways. Among these are the installation of different types of bed protection systems and/or the imposition of different types of operational constraints, such as reducing jet velocities by reducing the number of propeller revolutions or by increasing the underkeel clearance of the vessel. A decision is usually made on the basis of a cost–benefit analysis that takes into account the initial capital cost and maintenance cost. According to Verhey et al. (1987), in some cases it may be more cost-effective to install a cheaper bed protection system that requires regular maintenance work instead of applying an expensive but maintenance-free protection.

State-of-the-art review and a comprehensive list of relevant references related to subject matter is found in Tsinker (1995) and in recent works by Chin et al. (1996), Brazkiewicz (1997), Maynard (1999), Alhimenko and Belyaev (2001), Hamil et al. (2001), and Römisch and Hering (2002). In addition to propeller jet–induced scour, a variety of other factors can cause bed erosion, which may affect the performance of waterfront structures. These are wave attack associated with ship movement, underkeel current associated with a ship being moved laterally by tugs against a quay wall, or ship motion, for example, surging or rolling caused by long waves in the harbor.

6.4 INSPECTION OF WATERFRONT STRUCTURES

Despite some deterioration and sometimes, overloading conditions (compared to the original design criteria), some older and newer waterfront structures function without being overstressed and visibly deformed. On the other hand, others bear obvious signs of deterioration, overstress, and fatigue, and if reduced structural capacity is not properly addressed in time, this may eventually result in increased risk of consequential damage to the facility and to ships moored alongside the dock.

There are many good reasons for some waterfront structures to be in sound physical condition despite frequent overloading. The most common of these include the following:

1. The original design was based on conservative assumptions including assumed geotechnical soil parameters, combination of design loads, ignoring some specifics of soil–structure interaction, and so on.
2. The design may have included a substantial allowance for corrosion and excessive factors of safety.
3. Concrete had gained some additional strength over the value assumed originally.
4. Original timber or rubber-tire fenders were replaced by a new sophisticated fender system which reduced ship impact forces substantially.

The factors above may contribute to the fact that at times even substantially deteriorated or damaged structures do not compromise the safe use of the facility. Nevertheless, these structures should be repaired and properly maintained to ensure their prolonged life. In such cases, a "judicious neglect" method of repair [a term suggested by the U.S. Army Corps of Engineers (1986)] may be used. Judicious neglect is a repair method that involves taking no drastic action. It does not suggest ignoring situations in which damage to the structure has been detected. Instead, after a careful review of the circumstances (i.e., "judicious"), the most appropriate action may be to take no action at all.

Judicious neglect would be suitable for those cases of deterioration in which damage to the structure material is not causing any current operational problems for the structure and that will not contribute to future deterioration of the structure. For example, cracks in concrete, such as those due to shrinkage or some other one-time occurrence, are frequently self-sealing. This does not imply an autogenous healing and gain of strength, merely that the cracks clog with dirt, grease, or oil, or perhaps a little recrystallization occurs, and so on. The result is that the cracks are plugged and problems that may have been encountered with leakage, particularly if leakage is due to some intermittent cause rather than to a continuing pressure head, will disappear without repair work having been done. Hence, before decisions are made as to costly maintenance work, the use of existing dock facilities for servicing larger ships, or the use of much heavier cargo handling and hauling equipment are made, it is prudent to establish the actual capacity of the facility and possible load limitations associated with the facility's present and expected use for the design period of time.

A similar approach is advocated by Buslov (1992), who suggested that a cycle of about five years to repair only well-developed deteriorated concrete could be appropriate for most marine structures operated in a saltwater environment. The decision-making process also has to be based on the results of structure investigation. This must be carried out economically on an "as required" basis. As reported by Bakun (1986), in some cases, contractors conduct unfounded, expensive, and wide-ranging inspections of deteriorated marine structures. Also, structures often are restored to their original condition without prior evaluation of the reasons for the deterioration. Hence, for successful cost-effective decisions on structure repair and rehabilitation, a well-balanced, cost-effective inspection followed by engineering evaluation should be carried out. As a result, it may be concluded that the structure has served its design life and should be either rehabilitated for additional service life, redesigned for alternative use, or in the extreme case, should be removed altogether.

To avoid large-scale rehabilitation or repair work as a consequence of neglected systematic maintenance, systematic planning and budgeting of maintenance activities is necessary. The International Navigation Association (INA/PIANC, 1998), has developed guidelines to life-cycle management (LCM) aimed at harmonizing the decision-making process of budgeting and funding of systematic inspection and repair activities during the lifetime of port marine structures. This is discussed in the following section.

6.4.1 Systems Approach to Inspection, Maintenance, and Repairs of Port Structures†

6.4.1.1 Life-Cycle Management Concept. Maintenance and rehabilitation of port structures is a fairly complex process which due to the substantial costs associated, requires a high level of optimization. The traditional and proven way to optimize the multiparametric activities of this kind is to develop a system approach to their implementation. The most recent trend with regard to the maintenance of port structures is a commitment to consider seriously a life-cycle management (LCM) approach to the problem. Life-cycle management is based on a consideration of four major aspects of infrastructure maintenance: financial, technical, environmental, and safety, versus the four phases in the service life of facilities: planning/design, fabrication/construction, operation management, and removal (Figure 6.14). All aspects are evaluated for each phase in various reasonable combinations with relevant cost estimates of the present worth at the assumed discount rate. The end product of this work is a sequence of design/construction activities which results in a least possible sum of expenses over the life of the structure. The procedure is sometimes called *minimum whole-life costing* (WLC) *analysis*.

It is obvious that this series of evaluations requires a significant amount of information. The tests for environmental and safety requirements in each aspect–phase combination are relatively simple because these parameters are defined in the regulations. However, evaluation of the financial consequences of the specific technical decisions requires a significant amount of data, such as the service life specified for the structure, the existing condition of the main structural components, the rates of deterioration in various exposure zones, and forecasts for possible modifications and/or expected changes in loading and operational conditions. In the regular day-to-day life of port authorities, this information usually comes from technical personnel involved with the maintenance of structures. The inevitable personnel changes make this process highly unreliable. The implementation of LCM necessitates the creation of an information system that can provide reliable input. The results of all completed stages of LCM should be fed back to the LCM information system for future use (see Figure 6.14).

It is clear by the definition that LCM is most efficient when conceived and implemented at the planing and design phase. However, it can be applied at any other stage of the service life as well. The truth is that the majority of owners begin to think in LCM terms (sometimes without actually realizing it) when major maintenance problems become evident and multiple decisions and choices are at stake. It is when trucks on the operational deck area begin to fall into the sinkholes resulting from the leakage of fill through corrosion perforations in the steel sheet piles that the owner begins to ask: Shall we just weld the holes shut, or should we install cathodic protection as well? Had these questions been asked 20 years earlier, at least five or six design solutions and provisions would have alleviated the problem. The most

†This section is an expanded authorized reprinting from V. Buslov, *Proceedings of the 5th International Seminar on Renovation and Improvements to Existing Quay Structures*, Gdansk Technical University, Gdansk, Poland, 2001.

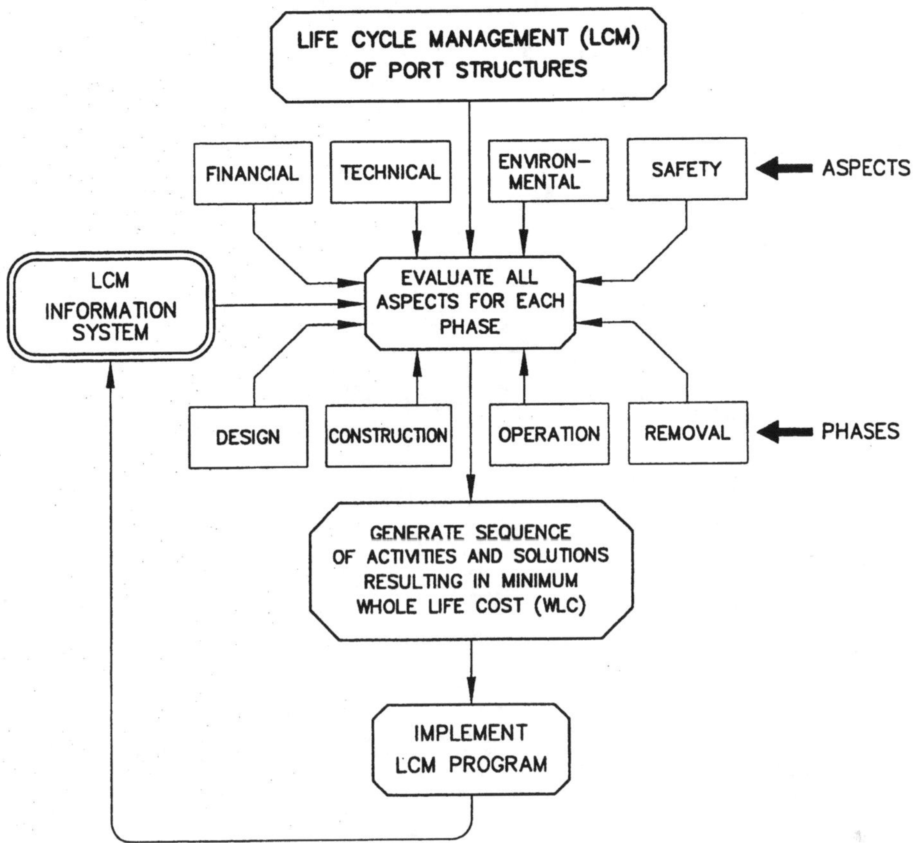

Figure 6.14 Principal components and activities of an LCM program. (From Buslov, 2001.)

comprehensive source of information on LCM of port structures is a report by INA/PIANC (1998). A new INA/PIANC group was recently formed to prepare an LCM implementation manual.

6.4.1.2 Maintenance and Inspection Procedures. The system approach at the foundation of LCM should be extended into all aspects of LCM, particularly into organization and implementation of the maintenance/inspection program. A typical inspection program operation is shown in the flowchart in Figure 6.15. Straightforward and almost self-evident as it seems at first glance, the sequence of activities shown in this chart was, in fact, developed as a result of many years of wharf surveys. The centerpiece and most active element of the program is maintenance inspection. Maintenance inspections are performed on a regular basis. In the program operation illustrated in Figure 6.15, the need for preconstruction inspections, condition surveys, and in many cases, special inspections is determined based on the results of maintenance inspections. Maintenance inspection is the only type

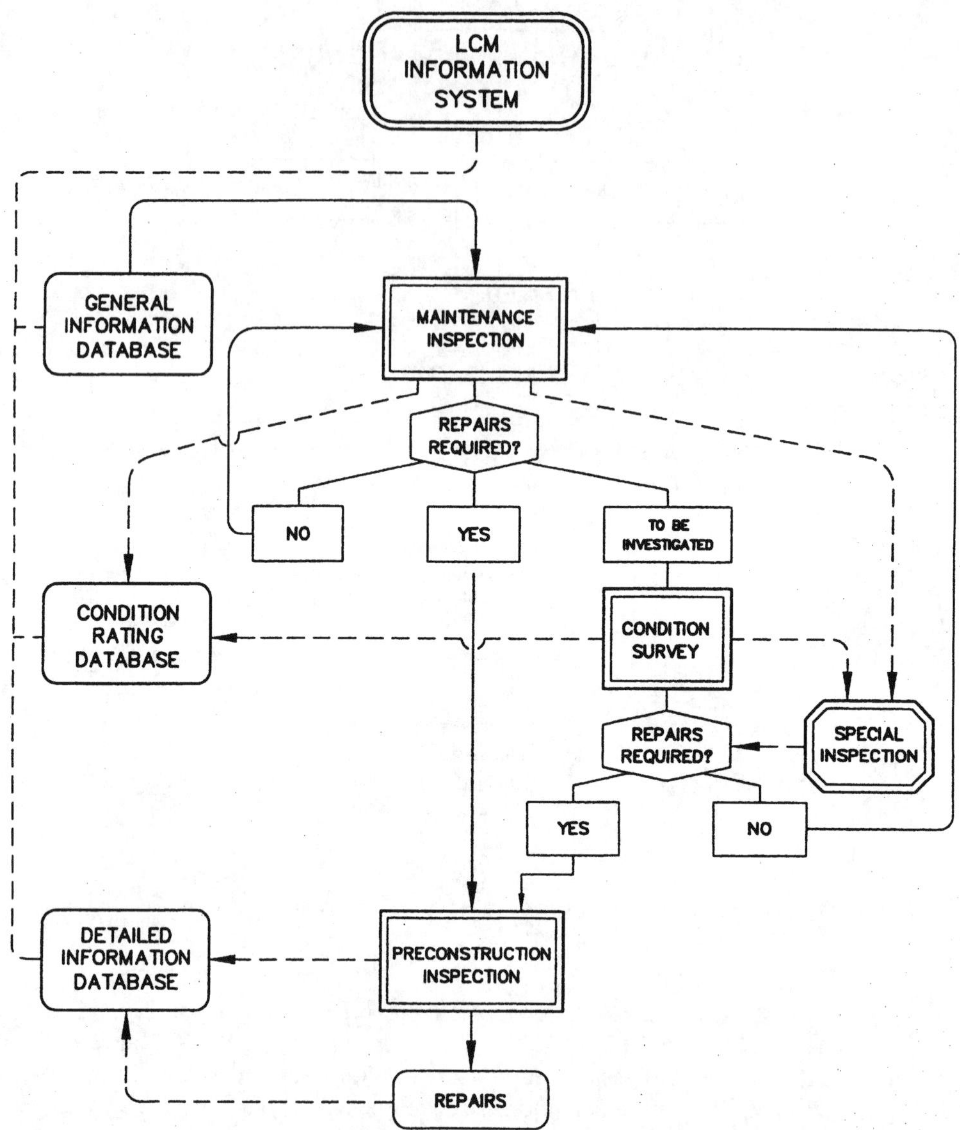

Figure 6.15 Inspection and maintenance procedure of an LCM program. (From Buslov, 2001.)

of survey that is scheduled and executed at fixed intervals regardless of other current activities on the wharves.

Although intended as brief surveys, these inspections generally include observation of each component. Maintenance inspections are geared toward obtaining the following information:

- Approximate number or percentage of components affected by each type of defect

- Location of isolated major defects, such as breakage and overstressing cracks, which may require emergency action
- Locations of corrosion damage concentrations
- General order of magnitude of defect sizes, as well as trends in location of specific defect types (e.g., the predominance of a type of defect on a particular corner or side of a component)

Analysis of the information collected is then performed to fulfill the goal of the inspection program: to determine whether repairs are required. Maintenance inspection may provide three alternative answers to this question:

1. The wharf is in good condition, and no repairs are required. The activities return to the regular cycle of maintenance inspections.
2. The intensity and scope of damage observed during maintenance inspection necessitate repairs. The next step is a detailed preconstruction inspection, performed to collect information necessary for the execution of repairs by the owner, or for the preparation of a repair bid package. During preconstruction inspections, the location and size of each defect are determined. This may be a time-consuming and, consequently, a costly activity. Therefore, detailed inspection should not be conducted unless a decision to do the repairs has been made.
3. Information collected during a maintenance inspection is insufficient to decide whether repairs are necessary. Additional investigation (e.g., structural damage analysis) is required to arrive at this decision. In port maintenance practice, investigations of this type are called *condition surveys,* in-depth engineering evaluations that should provide a final answer to questions as to the necessity of repairs. If repairs are not needed, activities return to the normal maintenance cycle. If repairs are considered necessary, the preconstruction inspection described in answer 2 is performed as the next step.

Special Inspections are initiated whenever some unexpected, unusual, or catastrophic damage has occurred on the structures: ship impact, storm damage, earthquake, damage from chemical products handled on the structure, and so on. The need for special inspections is frequently identified during maintenance inspections and condition surveys.

6.4.1.3 LCM Information System. As mentioned earlier, the input for the LCM implementation is derived from the LCM information system (Figure 6.14), which in a large port authority would probably be a computer database or several linked or interrelated databases. The capabilities (and hence the complexity) of the databases depend on the owner's needs and degree of readiness to commit time and resources for development of the system as well as for the work required to establish and maintain a database. Simplest and very useful is a database containing general information on port structures (top left corner, Figure 6.15). This is an "all about the wharf" databank, including main dimensions, allowable loads, structural data on basic components, mooring/berthing hardware, crane operational parameters, modifications, inspections and repairs (brief summary reports), and so on. The U.S. Navy inspection program uses a database system that maintains information on the dates, location, and brief scope of inspections preformed. Several inspection database programs for bridges have been developed in the United States.

A more advanced database can be designed to include the condition rating system (left middle portion of the diagram in Figure 6.15).

If a large number of structures are included in the program and many are recommended for repairs, the owner will need to establish the priority of repair work. The procedure for maintenance inspections may include assigning a condition rating based on the results of these brief surveys. For example, in the port of Los Angeles, a five-grade rating system for evaluation of the structural condition of concrete wharves is used. The structures are rated as being in good, fair, poor, serious, or critical condition. The Port Authority of New York and New Jersey has a similar rating system for preliminary structural evaluation, but wharf maintenance is driven by a three-grade rating based directly on the repair requirements: immediate, priority, or routine. The input for the condition rating databases is provided by the results of maintenance inspections and condition surveys (Figure 6.15).

When a database is designed to include input from detailed (preconstruction) inspections (lower left corner, Figure 6.15), it becomes quite complex, but the information capabilities of this relational "defect-level" database are almost limitless. The most sophisticated database program of this kind was developed by the port of Los Angeles for the inspection and maintenance of concrete wharves (Lindner et al., 1995). The database is called AIRIS (Automated Inspection and Repair Information System), and it manages a vast amount of information. The program operation, basically, is similar to that described above (Figure 6.15). The initial prioritization, at the level of maintenance surveys, is performed using the condition rating system. However, AIRIS goes one significant step further. At the level of preconstruction inspections, all of the characteristics of every defect observed on the structure are recorded. This allows the user to prepare bid documents very rapidly and with great flexibility, using preprogrammed reports. The database also allows the user, at the touch of a key, to analyze thousands of records spanning years of inspections, to perform trend and statistical analyses to evaluate such issues as deterioration rates and repair effectiveness.

The heart of the database, and the key to the maintenance process, is the inspection and repair module, where all information on inspection and repairs can be accessed. In addition, the database has a general berth information module which stores detailed information on the physical characteristics of the wharf, original design criteria, construction, fender system, cranes, and current tenant. The general berth information can be used in conjunction with the inspection and repair module to perform queries such as providing a listing of all wharves within the port which exhibit extensive seismic-related overstressing cracks where the design seismic load(s) are less than a specified value. Another possible query is to link all piles of a specific age and type, which exhibit specific defects, to the original construction contractor and/or precast concrete manufacturer. This provides a powerful analytical tool for use by engineers to improve future designs and construction quality control.

The following benefits are viewed as the primary justifications for the program:

- Preventive approach to maintenance of concrete wharves
- At-a-glance structural condition overview
- Repairs targeted at the root cause of the deterioration
- Speed of preparing repair contract documents
- Accuracy in estimating repair quantities

6.4.1.4 Implementation of LCM Program.

Implementation of the LCM program for a new and existing structure follows basically the same path, illustrated by the flowchart of Figure 6.16.

Step 1. Determine the Service Life Requirements. The most important input for LCM analysis is a service life requirement. For a new

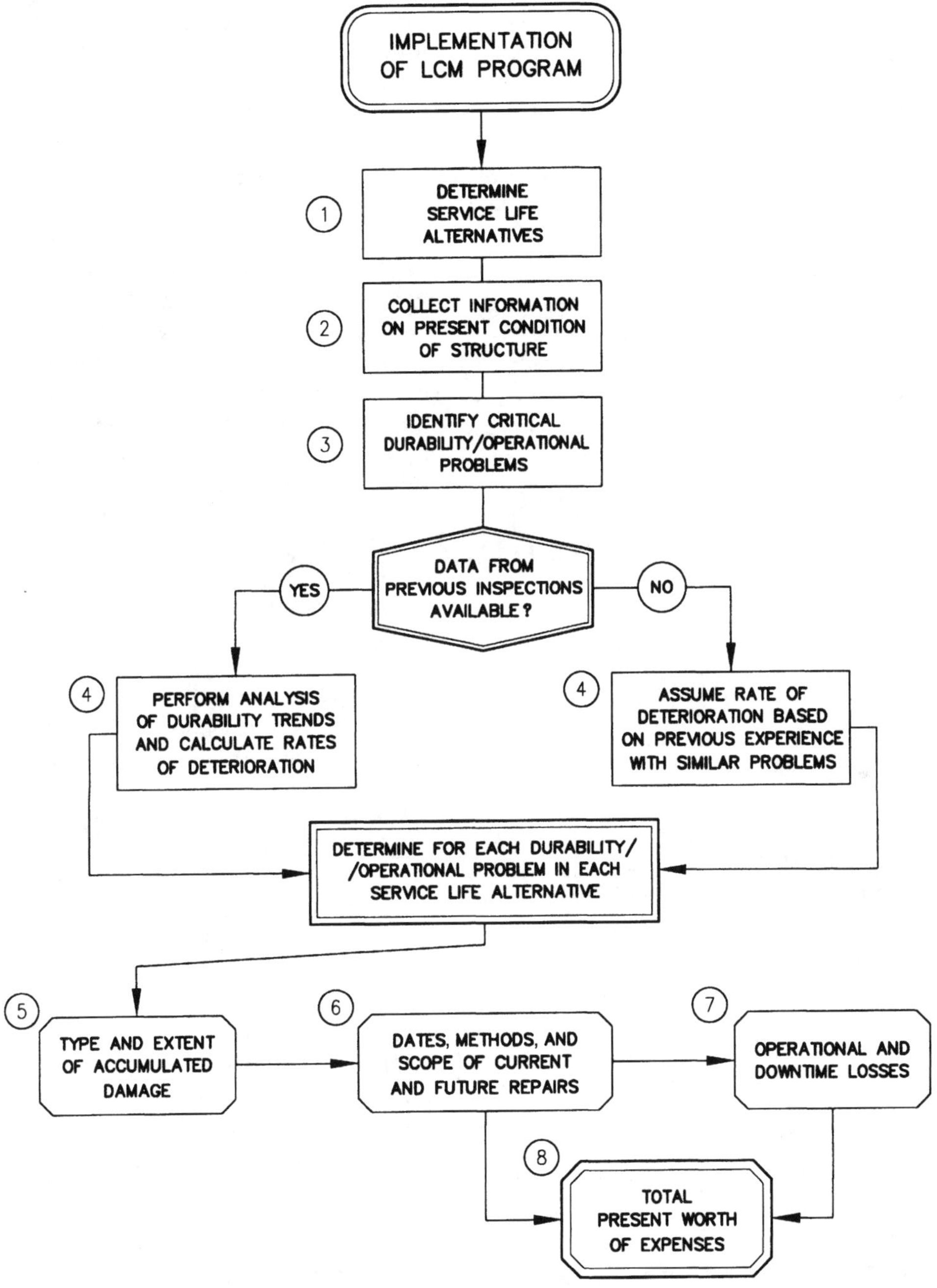

Figure 6.16 Implementation of an LCM program. (From Buslov, 2001.)

structure this parameter is the design service life (DSL). Presently, a 50-year DSL is frequently assumed for port structures. However, it is becoming more and more obvious that many wharves stay in operation for a much longer time, up to 80 years or more. In addition, recent studies in the general area of transportation infrastructure indicate that the current trend is for the DSL to be increased to 100 years of more (Freyermuth, 2001). There are several major factors responsible for this trend. One of them is the ever-increasing size and cost of facilities (a typical new post-Panamax container terminal is a good example). The other factor is a tremendous backlog in the rehabilitation of aging waterfronts around the world with limited funds available for this work. However, the most fundamental consideration is a need to support sustainable development (i.e., to satisfy the current rehabilitation requirements while leaving enough funds for new construction). The economics of sustainable development inevitably leads to longer life-cycle demands. A 100- to 120-year DSL concept is a present-day requirement for the design of major bridges. It is reasonable to assume that the design of new port structures will follow the same road.

The subject of service life requirements remains a highly debated issue and goes well beyond the brief discussion presented in this chapter. For instance, the general picture would be incomplete without mentioning that an analysis based only on economic considerations does not always support a long DSL requirement (Freyermuth, 2001). At the present typical U.S. net discount rate of 8% (commercial rate of return minus inflation), it is economically justifiable to replace facilities fully after 30 to 60 years in service. The controversy becomes less acute if normal fluctuations in the discount rate are considered (a 2 to 3% value is fairly compatible with a 100- to 120-year service life). Most important, though, is the fact that this "economics only" strategy directly contradicts the realities of present-day transportation infrastructure development, which point firmly to extended DSL for the new facilities.

The "determination of service life alternatives" indicated as a first step in LCM implementation (Figure 6.16) relates to an existing structure that has already been damaged in service. In this case, the owner is facing a choice (or range) of reasonable alternatives to extend the remaining life of the structure to a certain number of years (10, 20, 30, or more). When the owner has little or no latitude in the decision (e.g., the lease for the area expires in 10 years, the source of main specialized cargo will be shut down in 15 years), the service life requirement is firmly set. As a result, the LCM evaluation is reduced to the minimum whole-life costing (WLC) analysis for a single service life alternative (i.e., the procedure is identical to that for the DSL of a new structure).

When no obvious operational or administrative constraints exist, cost may become a major modifying factor in the owner's choice of a service life alternative. The prospect of a longer service life is always attractive, but not at any cost. This means that in the most general case, the full range of alternatives are to be considered and presented to the owner with their "price tags" attached. When the results of the LCM analysis are presented to the owner, he or she should be advised that the service life alternatives are based on the preplanned strategies. They can be fully efficient only if executed from "ground zero" to the end of the service life specified. The owner may always take a 10-year strategy and at the end of this period make a decision to extend the life of the structure for an additional 10 years. However, typically, this will result in considerably higher expenses than those of taking a 20-year strategy at the start of the LCM program.

Step 2. Collect Information on the Present Condition of the Structure. Once a decision on the range of service life alternatives to be evaluated is made, information should be acquired on the condition of the structure at the start of the LCM program. The results of the most recent inspection are used, but if such data are not available, the owner should be advised on the need to perform an inspection. The results of regular maintenance inspection may not be sufficient for an LCM analysis. Typically, a condition survey (see Figure 6.15) which yields quantitative data on the extent of existing damage may be required (e.g., remaining thickness, residual or locked stresses, long-term deformations).

Step 3. Identify Critical Durability and Operational Problems. In many ways, the LCM for a new structure is easier to plan and perform, due to the "unrestrained" range of possible technical solutions. After 20 to 30 years in service, various steel and concrete components of marine structures usually exhibit multiple and highly variable accumulated damage. The first test of the personnel involved in the execution of the LCM analysis is their ability to identify critical durability (or operational) problems. From the wide range of existing defects, several most critically important for the service life of a structure are to be singled out for further durability evaluation. For instance, general (area) corrosion loss of thickness at the top of the submerged zone of steel sheet pile bulkheads is not critical compared with pitting corrosion, which causes perforation of sheet piles and loss of fill. Relatively minor corrosion cracking on prestressed concrete components is much more threatening to the residual service life of a superstructure than is extensive spalling of concrete cover over mild steel reinforcement; breakage of armor concrete units on the breakwater is more critical than settlement of the crown wall; and so on. Frequently, several critical problems are identified. The time factor should be watched particularly closely. A problem that is not critical in the 10-year alternative may become critical after 20 years in service (this is why these strategies typically cannot be "added up" and should be set out and executed from beginning to end).

Step 4. Evaluate Durability Trends and Calculate Rates of Deterioration. The best way to forecast future damage for various service life alternatives is through an analysis of data on the past performance of a particular structure under the specific exposure conditions. The crucial role of the LCM information system at this stage of evaluation is particularly obvious.

Rates of steel corrosion are commonly extrapolated based on the total thickness loss measured during unprotected exposure. Analysis of the long-term deterioration of concrete marine structures may be more complicated. Frequently, the results of several follow-up inspections are to be screened in order to identify the "durability trend." If the quality of concrete in the structure is generally good, long-term damage will show up only at locations of original construction defects. Since the number of such defects (in "generally good" concrete) is limited, the rate of deterioration will eventually slow down. When the quality of originally placed concrete is generally poor, the rate of deterioration may show an upward trend in service. Forecasting the extent of future damage without a full understanding of durability trends may be reduced to the level of guessing. Evaluation of durability trends and rates of deterioration are probably the most technically and professionally demanding activities of the LCM analysis. They should be performed only by people who have substantial experience in this field.

Step 5. Determine the Type and Extent of Accumulated Damage. At this stage, the damage associated with identified critical durability problems is extrapolated to the dates of designated service life alternatives using back-calculated or assumed rates of deterioration. This work is performed based on the vast information collected to date about the fundamentals of deterioration of steel, concrete, and timber in a marine environment. In addition to the time of exposure, the effect of a wide array of environmental and operational factors is to be considered. In many cases, this evaluation should be done not just for all key structural components, but for their specific parts as well. In a recent LCM study performed for the owner of a 30-year-old bulkhead, damage to the structure had to be forecast in 10- and 20-year service life alternatives for three critical durability problems:

1. Perforations of steel sheet piles at MLW caused by high-rate pitting corrosion
2. Reduction in structural reserves associated with the loss of thickness of sheet pile webs, due to their corrosion in the splash zone
3. Structural failures (ruptures) in the wall caused by corrosion of tension wale connection bolts

A complex multiparametric study was required to develop the necessary recommendations (Han-Padron Associates, 2001).

Step 6. Determine the Methods and Scope of Current and Future Repairs. After forecasts of the type and extent of damage at different service life alternatives have been prepared, a long-term plan for remediation of this damage can be compiled. Again, a major input from the LCM information system is required, particularly with regard to the parameters of damage that may affect the repairs (e.g., the degree of cracking on prestressed components may indicate that the prestressing strands have already been lost and the members affected must be replaced, or the level of contamination by chlorides may preclude the use of corrosion inhibitors).

When several service life alternatives are considered, an optimized sequence of repairs and preventive maintenance actions should be evaluated to achieve maximum effect at minimum cost. For instance, it takes approximately 10 years after the first corrosion perforations are found on a steel sheet pile bulkhead before the problem becomes unmanageable. If the structure is abandoned in 10 years, patch welding of corrosion holes is cheaper than installation of cathodic protection. However, a 20-year service life extension may be achieved only if cathodic protection is installed soon after the first corrosion perforations are recorded.

Step 7. Evaluate Operational and Downtime Losses. Current and future repairs may cause an interruption of operational activities on a wharf. The cost of these interruptions to the owner sometime exceeds the cost of repairs. The downtime losses may reshape completely maintenance strategies based on technical merits. The results of the LCM analysis of the steel sheet pile bulkhead in step 5 showed that the cost of repairs of occasional ruptures in the wall caused by corrosion failure of wale connection bolts was significantly lower than the cost of total bolt replacement. However, when the cost of downtime caused by repairs of the ruptures was added, the total cost of this alternative became prohibitive. Similar problems are frequently encountered in planning of all major repairs used for marine structures: replacement of broken first row and batter piles, shotcrete repairs of concrete components, installation of protective coatings and sleeves, regrouting of crane rails, and so on.

Step 8. Calculate the Total Present Worth of Expenses Associated with the Execution of All Service Life Alternatives. The final step of the LCM analysis involves the preparation of engineering cost estimates for the service life alternatives considered. In a properly prepared LCM study, there is a well-defined repair or rehabilitation strategy behind each of the alternatives. Sometimes, owners are more concerned with immediate direct expenses than with the total costs of the alternatives included in the analysis. This requirement is never a problem because the immediate costs can always be identified easily. However, the final results of the LCM analysis should be presented as the total cost of all expenses, including the repair work and downtime associated with future rehabilitation projects. Standard discount rate tables are used to calculate the present worth of these future expenses.

In 1995, the U.S. Congress passed the National Highway System Bill, which requires that life-cycle cost analysis be performed for all projects costing $25 million or more. So far, no similar decisions have been made with regard to port structures, although the costs of many of them exceed the limit set by Congress. The condition of the aging highway system and the astronomical costs required for its current and future maintenance are apparently more in the public eye than similar problems with port infrastructures. However, it would come as little surprise if one day the mandatory requirement to perform an LCM analysis were introduced for large port construction and rehabilitation projects. For more information on LCM, readers are referred to Shvartsman et al. (1999), Freyermuth (2001), and Han-Padron Associates (2001).

6.4.2 Inspection

Inspection by definition is the process of gathering information on the actual state of an inspected structure. Its ultimate goal is to determine and record the condition of port structures and to detect any changes (deterioration) that have occurred since the structure was commissioned for operation. A systematic inspection may also be used to monitor the dynamics of change in the appearance, position, and local and global deterioration of the structure. All the information obtained may form a database on the condition of the structure and be used by the owner for decision making on regular maintenance and/or repair and rehabilitation. The complete database will also include all documentation pertinent to the structure in question (e.g., drawings of the design loads as built; environmental data; operational data; records of previous accidents and resulting repair work; inspections conducted previously; measurements and tests; computer, video, and audio records; records of maintenance works). A well-planned and well-organized database will provide easy access to any piece of information that is requested. It musts be stressed that the inspection must be carried out by the qualified personnel who have the knowledge of structural design and performance in a marine environment, deterioration mechanisms, and so on. Misjudgment of structural conditions may result in structural failures.

To enhance the ability of inspecting personnel to carry out inspections of marine structures and to standardize the process, the American Society of Civil Engineers (ASCE, 2001) developed the *Standard Practice Manual for Underwater Investigations*. The manual enhances the ability of owners to request an inspection type to match their project needs, to standardize the terminology and procedures included, and to improve the efficiency and cost-effectiveness of underwater inspection by tailoring the scope of work to specific inspection objectives. The manual addresses such important tasks as inspection types, scope of work and frequency; classifies underwater structural conditions; and recommends a course of action to ensure safe operation of the structure. The manual also defines minimum training and experience requirements for personnel involved with underwater inspection of marine structures. In addition, it provides guidance in conducting routine inspection of repaired structures when protected by cathodic protection and different types of encasements, coatings, and wraps.

Usually, two types of inspections are implemented in ports: routine or periodic inspection, and special inspection. The former is executed in accordance with the long-term continuous activity of the facility to review its condition on a regular basis. Special inspection is usually warranted when unscheduled or sudden events (e.g., heavy overloading, ship collision, fire, toxic chemical spillage, storm, earthquake) result in structural damages that require immediate attention. It is imperative that specialists be engaged to determine the manner and scope of the inspection and then act in either a supervisory role or to carry out an inspection as required. Success of an inspection is largely dependent on the *inspection criteria* (specifications) and execution of the inspection. Inspection criteria typically include but are not limited to the following:

- Data required for inspection, that is, a general description of the structure, construction, and service history; operational requirements; and original design and "as-built" drawings
- The purpose of the particular inspection task, and where in the structure the inspection is to be performed
- The inspection method and types of equipment to be used
- Sets of logs and forms for manual records
- The scope of inspection work, which includes hydrographic, geotechnical, diving, and material survey programs
- The work schedule

Inspection execution is affected significantly by the inspection method selected as well as by the objectives to be achieved. For example, searching for cracking or material deterioration subjected to marine growth requires thorough surface cleaning, and searching for propeller-induced bottom scour requires divers equipped with specialized equipment to determine the depth and extent of scour. During inspection it is imperative that the particular spot being inspected be properly identified. The most common method is to relate these spots to established profiles or pile bents along the structure and at given elevations.

6.4.2.1 Routine (Periodic) Inspection. This inspection procedure is a basic component of a port maintenance program. It aims at assessing the general overall condition of a structure and determining the level of maintenance or repair activities that may be required to keep the structure in sound shape. This inspection program is normally a long-term continuous activity and is most efficient when executed on time and when reports are reviewed on a regular basis. The manner, extent, and frequency of inspections is usually influenced by the maintenance strategy selected for the particular structure. The practice of routine inspections in commercial ports is highly variable; it depends on the type of structures in question, the material, environmental conditions, the character of operation, and so on. Underwater inspections are usually carried out once every 3 to 5 years, and visual examination of above-water structural ports may vary between 3 and 12 months. The spectrum of the inspection methods is very broad, ranging from simple visual observation, to the use of sophisticated equipment for measuring displacements, to cutting out samples for testing or drilling to obtain information on structural material and/or soil properties.

Visual Inspection. Visual inspection includes a rapid scan of the structure (both above and underwater). It is conducted for preliminary assessment of the general condition of the structure and to identify potential problem areas that require follow-up action. The inspecting diver must be skilled both as an engineer and as a diver. All diving operations must conform with standards established by the appropriate authority (e.g., PIANC, 1990; ASCE, 2001). The preliminary underwater inspection is essentially limited to visual observation by divers (often obstructed by suspended sediments). The main objective of the visual inspection is to ascertain whether any obvious damage to the structure has occurred. The basic qualitative data obtained from visual inspection are generally inadequate to accurately assess the condition of the structure. Therefore, if any significant damage is found in the process of visual inspection, a detailed inspection is usually ordered.

A visual inspection is also aimed at identifying and defining potential areas of distress. It would normally include a mapping of the current state of various types of structural deficiencies, such as construction faults, distortion or movement, and material deterioration. Construction faults typically found during a visual inspection include the use of structural elements and materials that deviate from the original design and specifications. They may include local deficiencies such as honeycomb evidence of cold joints, exposed reinforcing steel, irregular surfaces caused by improperly aligned forms, and a wide variety of surface blemishes and irregularities in concrete, poorly welded or insufficiently bolted connections in steel structures, and improperly treated wooden components. These faults are typically a result of negligence, poor workmanship, or failure to follow accepted good practice.

As the terms imply, distortion and movement are changes in alignment of the structural components, including buckling, settling, tilting, warping, and differential settlement of adjacent sections of the structure. A review of historical data such as periodic inspection reports may be helpful in determining when movement first occurred and the apparent cause and rate of movement. Material deterioration typically includes corrosion of steel, decay of timber, cracking, disintegration (blistering, delamination, weathering, dusting), and spalling of concrete. During visual underwater inspection the presence and extent of propeller jet–induced erosion, such as washout of sealing grout between concrete blocks or concrete sheet piles, as well as the presence and extent of bottom scour, are established. Visual survey of the structure is normally formalized by a clear, concise report supplemented by surface mapping.

Surface mapping is a procedure to survey a structure in which deterioration of the structural material is located and described. Surface mapping may be accomplished using detailed drawings, photographs, or videotapes. Typically, mapping begins at one end of a structure and proceeds in a systematic manner until all surfaces are mapped. The use of three-dimensional isometric sketches showing offsets of distortion of structural features is occasionally desirable. Areas of significant distress should be photographed for later reference. During a visual survey particular attention should be paid to the condition of joints. Opened or displaced joints (surface offsets) should be checked for movement if appropriate; various loading conditions should be considered when measurements of joints are taken. All joints should be checked for potential defects such as spalling or cracking, chemical attack, evidence of emission of solids, and so on. If joint filer is present, its condition should be examined. More information is given in U.S. Army Corps of Engineers (1986), Milwee and Aichele (1986), PIANC (1990), Tsinker (1995), Agi (1997), Buslov et al. (1998), ASCE (2001), and Heffron and Childs (2001). In some instances, to inspect the suspected parts of a structure, visual inspection could be extended to limited detailed inspection.

Detailed Inspection. Regularly conducted detailed inspections both above and underwater can in the final analysis prevent costly, wide-ranging detailed inspections. Concise documentation of each such inspection is essential for the success of cost-effective decisions and systematic evaluation of a structure's physical condition, as well as for a structure's maintenance program and for repair or rehabilitation measures. The objective of a detailed inspection is to gather quantitative data so that an engineering evaluation of dock capacity or the actual strength of its elements can be made. Specialized inspection equipment and techniques are required to gather quantitative data accurately to assess the condition of underwater parts of a structure.

In-place tests, also called *nondestructive tests,* are normally used in a detailed inspection to obtain information about the properties of materials as they exist in a structure. These tests are usually performed for two basic reasons: evaluation of the existing structure, and to monitor strength development during new construction. Inspection methods commonly used for a detailed structure survey, both above and underwater, are ultrasonic material testing, concrete testing by rebound hammer or the probe penetration method, material coring, and sampling. Some recently developed techniques include photogrammetry and alternating-current potential, and surface potential testing. Traditional tools for concrete testing, such as rebound hammer and probe penetration, are still indispensable for material investigation.

All inspection techniques require some cleaning of the surface to obtain accurate observations and measurements. The degree of cleaning required depends on the inspection technique to be used and the structural material. Surface cleaning usually involves removal of marine fouling and corrosion. It is usually a time-consuming operation; therefore, a cost-effective approach to the selection of the number of samples, as well as improved methods of surface cleaning, are needed to decrease surface preparation time. Among the presently used techniques for surface cleaning are powered and hand brushes, scrapers, and grinders, and high-pressure water jet cleaning systems. The latter is an especially efficient tool. It can deliver a water jet to the cleaning surface at 82.5 MPa and above (Smith, 1987; Momber, 1998). It is usually available in the form of a handheld water jet pistol with various nozzle assemblies that can easily be changed at the work site to match requirements. Because of the high pressure, these devices must be used with great caution, as the pressure tends to remove sound material on wooden and concrete surfaces.

The nondestructive methods of material inspection can be supplemented by cutting samples of steel material or core drilling for further laboratory analysis. *Core drilling* to recover concrete for laboratory analysis or testing is the best method of obtaining information on the condition of concrete within a structure. However, because core drilling is expensive, it should be considered only when sampling and testing of interior concrete is deemed necessary. When drill hole coring is not practical or core recovery is poor, a viewing system such as a borehole camera, borehole television, or borehole televiewer may be used to evaluate the interior concrete conditions.

Once samples of concrete, steel, or other materials have been obtained, whether by coring or other means, they should be examined in a qualified laboratory. In general, the examination includes petrographic, chemical, or physical tests. Evaluation of concrete compressive strength is done by conducting tests on samples prepared from the test specimens in the laboratory. The same specimens could be used to determine the chloride content. Experience indicates that concrete with evidence of extensive deterioration usually shows higher values of chloride content than does concrete with minor or no deterioration. Concrete strength in structural components inaccessible for coring, for example, anchor piles buried in soil, or piled relieving platforms, can be estimated on the basis of concrete age. With age (usually within a few years after construction), concrete gains strength. Strength increase depends on numerous factors, such as type of cement, curing temperatures, and water/cement ratio, and can reach 100% and more of the concrete compressive strength in 28 days. The approximate strength of mature concrete can be obtained using the equation

$$f'_{c(n)} = f'_{c(28)} \frac{\log(n)}{\log(28)} \tag{6.5}$$

where $f'_{c(n)}$ is the strength of matured concrete in n days after placing and $f'_{c(28)}$ is the 28-day strength of concrete. Equation (6.5) is invalid for concrete that is too old, older than, say, 5 years, because the concrete cannot gain much strength once all the available cement has been hydrated. More information on the computation of concrete strength for various ages is given in Plowman (1956), Kee (1971), and Shroff (1988).

A material survey, such as noticeable loss of steel due to corrosion, spalled and crumbling concrete, and shear and diagonal cracks in concrete elements, and its remaining strength is best recorded as a percentage of the minimum remaining section of the component. In some cases of underwater concrete structures and timber piles and cribs below minimum water level, one may expect some gain in material structural strength. Special attention must be paid to material conditions in tidal and splash zones, in previously repaired areas of relatively thin concrete elements, such as pier decks, piles, and sheet piles. In addition, the condition of existing coating systems, cathodic protection, fendering systems, and different kinds of mooring accessories must be surveyed as part of a detailed inspection.

Furthermore, *hydrographic surveys* and *geotechnical investigations* are essential parts of a detailed investigation. The objective of a hydrographic survey is to reveal erosion or deposition of material in the vicinity of the structure, and to identify areas where propeller-induced scour has taken place, and to detect if substantial undermining of the piling or a solid wall foundation has occurred. Different types of acoustic mapping systems are used for evaluation of the seafloor. These systems use the sonar principle, transmitting acoustic waves and receiving reflections from underwater surfaces. These can be used to perform rapid, accurate surveys of submerged horizontal surfaces in water depths of 1.5 to 10 m and produce survey results with an accuracy of ±50 mm vertically and ±300 mm laterally, and are very useful in fixing and investigating large scour holes and sediment buildup. It should be mentioned, however, that accuracy of these systems will decrease at depths above 10 m (Garlich and Chrzastowski, 1989).

Geotechnical investigations are usually performed to determine if the foundation and fill materials have undergone changes that might affect the strength of the structure. At this stage, previously collected geotechnical data should be reevaluated creatively. As a result of a routine detailed investigation, a structure can be rated as being in good, satisfactory, or poor condition. A good condition is usually defined as such when no invisible or just minor deterioration and/or damage are observed and no overstressing of structural elements is noted. In this case, the structure does not require any repair. A satisfactory (or fair) condition is usually limited to minor defects; however, all primary structural components are in sound condition, although localized areas of moderate to advanced deterioration may be present. These do not, however, reduce the load-bearing capacity of the structure significantly. In this case, some repair work may need to be carried out. A poor condition is characterized by advanced deterioration, overstressing of a widespread portion of the structure, and breakage of some structural components all of which separately or in combination may significantly reduce the load-bearing capacity. Depending on the degree of deterioration or damages, the repairs may need to be carried out with varying degrees of urgency.

6.4.2.2 Special Inspections. Special inspection is usually warranted when as a result of unusual overloading (e.g., accident, earthquake, ship collision) the structure sustained significant overstressing or failure of some primary structural components. In this case, detailed inspection of each significant structural component should be carried out followed by engineering evaluation of the load-bearing capacity of the structure. In most cases, operation of the badly damaged and/or deteriorated structure must be restricted as required and repair/rehabilitation work ordered with urgency to restore the structures safe function. For more information on detailed inspections, readers are referred to Scola (1989), PIANC (1990), Agi (1997), ASCE (2001), and Heffron and Childs (2001).

6.4.2.3 Cost-Effective Detailed Inspection. A cost-effective detailed inspection would concentrate on areas that may control the overall structural strength or stability (i.e., points of maximum deterioration, damaged elements, points of maximum stress, structural joints, the foundation, or the condition of the drainage system behind all kinds of soil-retaining structures, sheet pile bulkhead anchor systems, etc.). Inspections, especially underwater inspections, may be costly, and therefore, to be cost-effective routine inspections, especially underwater inspections, should be carried out at intervals as appropriate based on the extent of deterioration observed in a structure, the rate of additional deterioration anticipated, the importance of the structure, or other factors. The intervals may be increased as appropriate for nontypical cases (e.g., when special deterioration-resistant construction, materials such as special hardwoods, piles made from plastic materials, and so on, are involved). Essentially, the intervals between inspections depend on the structure's rated condition (e.g., good, satisfactory, poor) and the marine environment [e.g., fresh water with low or moderate currents, or aggressive, including brackish or salt water, polluted water, or waters with moderate to swift currents (more than 0.75 km/s)]. Based on the factors above, ASCE (2001) recommends that intervals of underwater inspection vary from 3 to 6 years, and in case of a serious (critical) structural condition, at intervals from $\frac{1}{2}$ to 2 years. General visual inspection as part of a maintenance program is usually applied to 100% of the components above the water level. Its obvious purpose is to identify signs of structural damage. Underwater inspection usually involves the removal of marine growth and testing, sampling, and measurement of damaged or deteriorated parts to determine the cause of damage or deterioration, its typical pattern, and its magnitude. Because of the costs involved in underwater inspection, it is usually performed on a limited number of structural components. Since the degree of reliability depends on the number of observations, the scope of cost-effective work should define the minimum number of elements to be inspected in detail. Often, these requirements are based on the personal experience of the inspector(s), the owner's requirements, and local conditions, and generally vary between 5 and 10%. Sometimes other criteria, such as at least one pile in each bent, and/or other criteria are used.

The main problem with the arbitrary criteria above is that they have no reliable basis and in some cases may result in an insufficient number of samples being inspected, while in others substantial funds are wasted on unnecessary detailed inspections (Barquett and Childs, 1983). Buslov et al. (2001) attempted to solve the problem based on a statistical approach to determining the representative sample size. Their approach is based on the idea of determination of the reliable minimum number of structural elements to be surveyed. On the basis of a survey carried out among colleagues involved in planning and execution of underwater survey, the authors attempted to establish the acceptable accuracy range and confidence level for several types of damaged components most frequently encountered in underwater inspections. They stated that most of respondents considered a 95% confidence level as the sufficient and minimum required parameter. It appears, though, that 10 to 15% tolerance is considered acceptable when the survey parameters are to be used for the evaluation of structural capacity; for the parameters used in the evaluation of durability and service life, 25% accuracy appears to be acceptable. Accuracy of data used for analysis depends on dispersion of the data itself, and the distribution parameters of the data to be collected constitute that part of the input required for calculation of the minimum number of observations; therefore, they are to be established.

As noted by Buslov et al. (2001), any statistical analysis that involves determination of the required minimum number of tests is possible only if the type of data distribution is established. The experience with surveys on marine structures shows that normal distribution may be assumed in most cases, provided that:

- The number of components is sufficiently large (hundreds).
- The type of damage surveyed is not associated with accidental (i.e., strictly local) activities such as damage by ship impact, overloading on decks, and so on.

Both the survey to be conducted and the sample (pilot) inspection to establish the distribution parameters should be performed for groups of elements with similar service conditions (e.g., concrete piles vs. steel piles, front row piles vs. back row piles, piles vs. sheet piles). Hence, the analysis should be performed separately for each group of components that may be considered as having distinctly different service conditions. The general statistical approach for determination of the minimum number of observations to satisfy the set accuracy requirements can be applied to any inspection task, including reliable determination of quantitative parameters such as remaining thickness of steel components, remaining reinforcing bar diameter in concrete structures affected by corrosion, diameter of timber piles affected by borers, and so on. However, when the quantitative parameters are surveyed, the procedure for statistical evaluation involves a series of additional factors, such as accuracy of the measuring tools, individual inspector's errors, mode and format in which the parameter is used in the capacity formula, and particularly, the pattern of field data distribution.

One possible approach to determining a structurally representative sample size could be to base the sample size on the type of damage observed. An example is a concrete structure with components exhibiting various stages of corrosion distress, chemical attack, freeze–thaw scaling, and so on. The complexity of establishing a sample size that addressed the required reliability of each type of damage would get quite complex and impractical. For this reason, the methodology developed by Buslov et al. (2001) presented below is focused on reportable defects. In underwater inspections of most structures, with a relatively short time allocated for the work and limited visibility, the most frequent and efficient mode of investigation is when the main task is to determine the quantity of elements affected by certain type of defects (broken piles, piles affected by corrosion or infested by borers, sheet piles with perforations, broken or dislocated armor units on breakwaters, cracked and/or scoured supports, etc.).

The reportable defects can be recorded using a simple scale, which allows us to describe the condition numerically. For this application, a scale of "damaged" and "not damaged" with a numerical representation of 1 and 0, correspondingly is used. Two additional necessary conditions are applied:

- Whenever significant differences exist in the structural configuration, geometry, or loading on exposure conditions, the structure should be divided conditionally into separate groups (populations) of uniform components.
- For a specific defect, normal distribution for the mean value of the sample throughout the population is assumed within each group of uniform components.

Several basic statistical definitions used in the derivation process are:

Sample: $$x = \{x_1, x_2, x_3, \ldots, x_n\} \tag{6.6}$$

Sample mean: $$\bar{\mu} = \frac{1}{n} \sum_i x_i \tag{6.7}$$

Sample standard deviation:

$$\bar{\sigma} = \left[\frac{1}{n-1} \sum (x_i - \bar{\mu})^2\right]^{1/2} \tag{6.8}$$

Thereafter, three main parameters are used as a basis for evaluation if the data from previous inspections or the pilot inspection conducted prior to the main survey describes the condition of the entire facility with sufficient accuracy. They are described briefly below.

Confidence interval is employed for evaluation of the mean value obtained from previous inspection data or from a small-scale random pilot inspection conducted prior to the main inspection project. With this parameter, it is required to achieve a probability of $1 - \alpha$ (or in other words, a confidence level of $1 - \alpha$) that the mean value calculated from inspection of the sample is located within the range illustrated in the following equation:

$$\bar{\mu} - \frac{\bar{\sigma}}{\sqrt{n}} t_{n-1}\left(1 - \frac{a}{2}\right) \le \mu \le \bar{\mu} + \frac{\bar{\sigma}}{\sqrt{n}} t_{n-1}\left(1 - \frac{a}{2}\right) \tag{6.9}$$

For example, to obtain a 95% confidence interval, $\alpha = 0.05$ should be used, and for a confidence level of 90%, $\alpha = 0.10$. Values for coefficient t are determined from Table 6.2.

The *mean value range* (accuracy level) shows the accuracy with which the calculated mean describes the condition for the population. As shown in eq. (6.9), it depends on several variables, and is expressed as a mean value $\pm X\%$. The *number of inspections, n,* is the parameter in question. It should be determined from eq. (6.9) prior to beginning the main inspection project, whereby the other variables are calculated or established as follows:

- *Confidence level* and *mean value range* from the considerations noted above
- *Mean* and *standard deviation* from the available (previous) survey data, or random small-scale pilot inspection

It should be pointed out that the procedure might involve a large number of iterations to come up with a solution; however, it can be simplified with the introduction of a small computer program. As input, the program will need only the number of damaged and undamaged piles, available from previously performed inspections or from the random small-scale pilot inspection mentioned earlier, as well as the accuracy criteria set for a specific inspection project. Alternatively, a set of charts can be created. Using these charts an inspector can graphically pick up the number of necessary inspection locations that will satisfy estab-

Table 6.2 Quantities of the *T*-distribution with ν degrees of freedom

	$p = 1 - a/2$								
$\nu = n - 1$	0.6	0.7	0.8	0.9	0.95	0.975	0.99	0.995	0.999
1	0.325	0.727	1.376	3.078	6.314	12.706	31.821	63.657	318.309
2	0.289	0.617	1.061	1.886	2.920	4.303	6.965	9.925	22.327
3	0.277	0.584	0.978	1.638	2.353	3.182	4.541	5.841	10.215
4	0.271	0.569	0.941	1.533	2.132	2.776	3.747	4.604	7.173
5	0.267	0.559	0.920	1.476	2.015	2.571	3.365	4.032	5.000
6	0.265	0.553	0.906	1.440	1.943	2.447	3.143	3.707	5.208
7	0.263	0.549	0.896	1.415	1.895	2.365	2.998	3.499	4.785
8	0.262	0.546	0.889	1.397	1.860	2.306	2.896	3.355	4.501
9	0.261	0.543	0.883	1.383	1.833	2.262	2.821	3.250	4.297
10	0.260	0.542	0.879	1.372	1.812	2.228	2.764	3.169	4.144
11	0.260	0.540	0.876	1.363	1.796	2.201	2.718	3.106	4.025
12	0.259	0.539	0.873	1.356	1.782	2.179	2.681	3.055	3.930
13	0.259	0.538	0.870	1.350	1.771	2.160	2.650	3.012	3.852
14	0.258	0.537	0.868	1.345	1.761	2.145	2.624	2.977	3.787
15	0.258	0.536	0.866	1.341	1.753	2.131	2.602	2.947	3.733
16	0.258	0.535	0.865	1.337	1.746	2.120	2.583	2.921	3.686
17	0.257	0.534	0.863	1.333	1.740	2.110	2.567	2.898	3.646
18	0.257	0.534	0.862	1.330	1.734	2.101	2.552	2.878	3.610
19	0.257	0.533	0.861	1.328	1.729	2.093	2.539	2.861	3.579
20	0.257	0.533	0.860	1.325	1.725	2.086	2.528	2.845	3.552
21	0.257	0.532	0.859	1.323	1.721	2.080	2.518	2.831	3.527
22	0.256	0.532	0.858	1.321	1.717	2.074	2.508	2.819	3.505
23	0.256	0.532	0.858	1.319	1.714	2.069	2.500	2.807	3.485
24	0.256	0.531	0.857	1.318	1.711	2.064	2.492	2.797	3.467
25	0.256	0.531	0.856	1.316	1.708	2.060	2.485	2.787	3.450
26	0.256	0.531	0.856	1.315	1.706	2.056	2.479	2.779	3.435
27	0.256	0.531	0.855	1.314	1.703	2.052	2.473	2.771	3.421
28	0.256	0.530	0.855	1.313	1.701	2.048	2.467	2.763	3.408
29	0.256	0.530	0.854	1.311	1.699	2.045	2.462	2.756	3.396
30	0.255	0.530	0.854	1.310	1.697	2.042	2.457	2.750	3.385
35	0.255	0.529	0.852	1.306	1.690	2.030	2.438	2.724	3.340
40	0.255	0.529	0.851	1.303	1.684	2.021	2.423	2.704	3.307
45	0.255	0.528	0.850	1.301	1.679	2.014	2.412	2.690	3.281
50	0.255	0.528	0.849	1.299	1.676	2.009	2.403	2.678	3.261
55	0.255	0.527	0.848	1.297	1.673	2.004	2.396	2.668	3.245
60	0.254	0.527	0.848	1.296	1.671	2.000	2.390	2.660	3.232
65	0.254	0.527	0.847	1.295	1.669	1.997	2.385	2.654	3.220
70	0.254	0.527	0.847	1.294	1.667	1.994	2.381	2.648	3.211
75	0.254	0.527	0.846	1.293	1.665	1.992	2.377	2.643	3.202
80	0.254	0.526	0.846	1.292	1.664	1.990	2.374	2.639	3.195
85	0.254	0.526	0.846	1.292	1.663	1.988	2.371	2.635	3.189
90	0.254	0.526	0.846	1.291	1.662	1.987	2.368	2.632	3.183
95	0.254	0.526	0.845	1.291	1.661	1.985	2.366	2.629	3.178
100	0.254	0.526	0.845	1.290	1.660	1.984	2.364	2.626	3.174

Table 6.2 ***(Continued)***

	$p = 1 - a/2$								
$\nu = n - 1$	0.6	0.7	0.8	0.9	0.95	0.975	0.99	0.995	0.999
105	0.254	0.526	0.845	1.290	1.659	1.983	2.362	2.623	3.170
110	0.254	0.526	0.845	1.289	1.659	1.982	2.361	2.621	3.166
115	0.254	0.526	0.845	1.289	1.168	1.981	2.359	2.619	3.163
120	0.254	0.526	0.845	1.289	1.658	1.980	2.358	2.617	3.160
∞	0.253	0.524	0.842	1.282	1.645	1.960	2.326	2.676	3.090

Source: Devore and Farnum (1999).

lished *confidence level* and *mean value range* criteria.

In each case, the engineering personnel responsible for planning and execution of the survey should make their own decisions as to the acceptable accuracy range and the minimum confidence level, depending on the type of damage and the goal of the inspection. However, the following general guidelines are recommended by Buslov et al. (2001) for underwater inspection of port structures.

Mean value range:

Used for structural evaluation ±15%

Used for assessment of durability problems, preventive maintenance, and service life predictions ±25%

Confidence level: 95%

Example 6.1 (Buslov et al., 2001) This example clarifies the methodology proposed for determining the sample size for an inspection. The example does not directly follow the methodology, but rather, illustrates different scenarios to make the methodology clearer. In this example, the number of randomly inspected piles is 10. The results of inspection for a specific defect are:

Damaged: 7 (denoted 1)

Undamaged: 3 (denoted 0)

Mean value:

$$\bar{\mu} = \frac{1}{n}\sum_i x_i = \frac{7 \times 1 + 3 \times 0}{10} = 0.7$$

Standard deviation:

$$\bar{\sigma} = \left[\frac{1}{n-1}\sum (x_i - \bar{\mu})^2\right]^{1/2} = \left[\frac{1}{10-1}(7 \times (1 - \bar{\mu})^2\right]^{1/2} = 0.2646$$

For a 95% confidence level, $\alpha = 0.05$, which means that a T-distribution coefficient of

$$t_9\left(1 - \frac{0.05}{2}\right) = t_9(0.975) = 2.262$$

should be used in eq. (6.9):

$$0.7 - \frac{0.2646}{\sqrt{10}}\, t_9\left(1 - \frac{0.05}{2}\right) \leq \mu \leq 0.7 + \frac{0.2646}{\sqrt{10}}\, t_9\left(1 - \frac{0.05}{2}\right)$$

After numerical simplifications the following mean value range was obtained:

$$0.511 \leq \mu \leq 0.889$$

The mean value range variation expressed in percentages is calculated as

$$\left(\frac{0.511}{0.7} - 1\right) \times 100 = -27\% \qquad \text{and}$$

$$\left(\frac{0.889}{0.7} - 1\right) \times 100 = +27\%$$

Thus, in this example, the mean value of 0.7 varies $\pm 27\%$ with a confidence interval of 95%. If this range is too large compared to the acceptable parameter, a new sample size should be determined by solving eq. (6.9). Assuming that at this stage mean and standard deviation can be constant for the entire population, and using the already calculated mean and standard deviation, eq. (6.8) is solved for n using several iteration cycles. For more details, interested readers are referred to the work of the aforementioned authors.

Finally, it should be pointed out that the method above is recommended for detailed underwater inspections. The visual inspection should be performed on 100% of accessible underwater structural elements. For more information on investigation and investigation techniques, interested readers are referred to Dunnicliff (1988), Bray and Tatham (1992), special publications by ASCE and PIANC, and other relevant sources.

6.5 ENGINEERING EVALUATION

At this stage, two basic tasks are to be performed:

1. Evaluation and interpretation of detailed inspection survey data
2. Structural analysis to evaluate a structure's actual load-bearing capacity.

6.5.1 Evaluation and Interpretation of Inspection Survey Data

Evaluation and inspection are important, sometimes, controversial, and often, very difficult phases of an assessment process. These are sometimes more art than science and require involvement of highly experienced personnel. In the case of structures that are constructed from, or include, wooden components exposed to seawater, it is imperative to have a clear idea of the spread of hidden fungal damage and damage inflicted by marine organisms, especially under fouling and in turbid conditions. In general, evaluation of wooden structures is carried out on the basis of available inspection data, which include as-built plans and the actual physical condition of structural members. The latter include yield or failure data, information on superstructure members for mechanical and fungal damage, and information as to mechanical and marine organism damage of underwater members.

Evaluation of concrete structures is carried out on the basis of the following data obtained from a detailed inspection:

- Presence of a yielding or failed element
- Degree of abrasion, cracking, and spalling of concrete
- Impact damage induced by floating debris, ice, or ships
- Degree of corrosion and deterioration of steel reinforcement
- Scour at the mudline

In an evaluation of concrete components, it must be recognized that the function of concrete in a reinforced concrete structure is to provide adequate compressive strength to protect reinforcing steel against corrosion. The following criteria are considered in evaluation of existing concrete:

- Actual compressive stress, usually obtained from cores.

- Ability to prevent a corrosive environment forming at the level of reinforcing steel due to the presence of chloride. For example, the chloride content threshold for active corrosion of steel is approximately 0.8 kg/m^3 of free chloride at the steel reinforcement level.
- Ability of concrete to prevent exposure of the reinforcing steel to water and to the atmosphere.

Furthermore, an overall evaluation of concrete and other structural materials and components should be evaluated on the basis of their performance in the past. For example, on occasion, contractors pull misplaced piles into their design position, thus inducing in them initial bending stress. In some instances, this may be the cause of premature pile deterioration or overstress and should therefore be taken into account in structural analysis.

Johnson (1989) and DiCastro et al. (1998) indicated that spalling concrete at the soffit of a beam or slab may produce a completely different effect on the remaining portion of beam or slab, depending on the kind of reinforcement used in these structures. In the case of regular reinforcement, the spalling concrete will result in increased tension in the bottom reinforcement; on the other hand, in the case of bottom-located prestressed reinforcement, the spalling concrete will produce additional tension in the beam/slab upper zone due to arch action.

The inspection data used for evaluation of metal components of marine structures is very similar to that used for evaluation of concrete structures. It includes the following:

- Evidence of yielding, failure, buckling, abrasion, and impact damage
- Oxidation/rusting in air loss of steel
- Web cracking and rusting
- Pitting corrosion
- Damages to riveted and bolted connections and damages to interlocks in sheet piling
- Condition of protective coating
- Presence or absence and condition of the cathodic protection system(s)
- Loss of steel to galvanic corrosion

Conversion of data provided by the inspection report into design parameters for evaluation of load-carrying capacity of the marine structure requires a knowledge of inspection methods and capabilities used, as well as knowledge of structural engineering.

6.5.2 Structural Analysis

It is intended to define the actual strength of an in-service structure and to compare it to the capacities required. This analysis should be carried out based on a thorough understanding of the mechanism of load–structure interaction. The required capacities are based on the present or on new short- or long-term operating conditions and suitable factors of safety. In the case of a damaged or badly deteriorated structure, the purpose of structural analysis is to determine the degree of redundancy present, that is, structural evaluation of reserve strength. The accuracy of such analysis depends on the engineer's ability to establish the actual strength of each subcomponent of the structure, realistic load criteria, and properly modeled load–structure interaction. A representative, examples of strength assessment of corrosion-damaged reinforced concrete slabs and beams is provided by Eyre and Nokhasteh (1992).

An important part of capacity evaluation is the correct assessment of the energy absorption capacity of the existing fender system and the capacity of the dock's mooring accessories (i.e., bollards, bitts, etc.). The actual or expected long-term bearing capacity of a structure is determined on the basis of projected rates of concrete or timber component deterioration and/or corrosion of steel components of the structure. The typical structural analysis is carried out in the following steps. Initially, the actual strength of each structural subcomponent is evaluated. This is carried out based on the original or as-built drawings and material survey data obtained from the inspection report. Next, the average strength of each structural component is established. Finally, based on the assumed factor of safety, the acceptable horizontal and vertical loads are established. If these loads are below required values, an engineer has the following options:

1. To reduce the impact force from berthing a ship, a new or improved fender system can be considered.
2. To increase the capacity of structural components such as piles and deck, additional new piles and/or an additional bracing system to reduce the free height of piles can be considered.

Furthermore, the size of the ship that can be used at the facility will be governed not just by the actual or improved strength of the waterfront structure, but also by specific site features, such as the length of the wharf, depth of water, geotechnical conditions of the seafloor, and berthing and mooring conditions. Again, if the structure under consideration cannot sustain impact from the theoretical ship, a new fender system can be installed and/or the structure must be reinforced accordingly. If the latter is considered to be an uneconomical solution, the use of the structure will be limited to receiving smaller ships. Evaluation of the load-carrying capacity of gravity-type soil-retaining quay walls usually does not present significant problem. In most cases these structures of miscellaneous designs are constructed on a bedrock foundation, and therefore, little displacement of the wall is expected. In this case the lateral soil pressure against a wall is determined as "pressure at rest," which exceeds normal active soil pressure. The accuracy of soil pressure analysis depends on the quality of available soil parameters as well as a thorough understanding of the mechanism of soil–structure interaction. The latter could be quite a complicated task for the case of a wall constructed on soft or creeping foundation materials. For details, consult Tsinker (1997).

A much more complicated task is establishing the actual capacity of sheet pile bulkheads. Sheet pile bulkheads are considered to be "flexible" structures because of their ability to yield slightly under the soil-induced pressure. Problems involved in the analysis of "flexible" retaining structures are more complicated than those applying to regular gravity retaining walls. The soil pressure is the main force acting against the sheet pile wall. The magnitude of this pressure depends on physical properties of the soil and character of soil–structure interaction.

Movement (deflection) of a structure is a primary factor in the development of earth pressures. However, this problem is highly indeterminate. Earth pressures may be affected by the time-dependent nature of soil strength, such as consolidation due to vibration, groundwater movement, soil creeping, and chemical changes in the soil. Despite various theories that have been developed to derive analytical solutions for the bending moment reaction force and sheet pile submerged depth, most designers in the field still rely heavily on past practice, good judgment, and experience. For details, consult Tsinker (1997).

6.6 MAINTENANCE, REPAIR, AND REHABILITATION OF MARINE STRUCTURES

6.6.1 Maintenance

In general, maintenance is a restoration of defects regularly carried out in structural elements aimed to arrest unacceptable deterioration of existing materials and to keep the structure operational. As a first step, maintenance includes periodic inspection of the structure. Again, the inspector must be familiar not only with the material but also with the structures intended performance. The inspector must, above all, be able to determine the actual reason for material deterioration (e.g., poor material quality, poor execution of construction work, overstress, poor maintenance, accidental damage). The inspector must fully establish the extent and causes of the problem; otherwise, the problem may recur, so that repeated work will be necessary. Actually, there is no sharp distinction between major maintenance work and minor repair work. Properly planned maintenance work is designed to reduce the rate of structure degradation to such a level as to avoid major repair work through the life span of the facility. It is commonly achieved through regular inspections and installation and proper maintenance of miscellaneous protective systems against corrosion of metal parts of the structure, biodegradation of wooden components, and degradation of concrete. As pointed out earlier, a cost-effective inspection program includes mandatory monitoring and recording of the condition of structure. It also includes planning and budgeting the maintenance work. To be inspectable, particularly in splash, tidal, and submerged zones, the surface of structural components must be cleaned from marine growth, loose materials, dirt, and so on. Above water, it is usually done by sandblasting, and under water, by use of high-pressure jet. Pneumatic tools can also be used in both cases (i.e., above and under water).

6.6.1.1 Timber. Essentially, the maintenance and repair of structures made from wood can be minimized by treating the timber before it is installed in the work. In some instances, the wooden marine structures can be protected "naturally" by miscellaneous fouling organisms (e.g., mussels, barnacles) adhering to structural elements. These organisms may prevent a borer attack by covering the exposed wood and feeding borer larvae. Natural protection can also be provided by the use of high-density tropical hardwood accredited with great durability in seawater. Ironically, the wooden structures usually last longer in heavily polluted water because it prevents the existence of borers. If marine borers are a problem that need to be remediated, the most common and economical remediation involves wrapping of timber components (e.g., piles with a tightly drawn and attached PVC or polyethylene sheet). Piles are usually wrapped with the jacket extended 50 to 100 cm into the solid mud and above the maximum water level. If extended up to the underneath of the concrete deck, then for better adherence, a splash zone epoxy or hydraulic cement mortar is applied at the interface. The purpose of timber warp basically is to seal off the oxygen that effectively kills all the parasites in the wood by suffocating them, thus preventing further deterioration. In case the top of the wrap is below the waterline, it is very important that there is a seal so that no more oxygenated seawater can get in by pumping action that may occur due to variation in tide level.

The most effective wrap system is one that has a positive closure. In the latter, the plastic wrap folds on ratchets over itself to form a tightly closed seal along its vertical seam. The horizontal seams, which contain a closed seal foam, are cinched tight with a band. Then the horizontal and vertical seams are nailed to adhere the wrap tightly to the piles and to close the seam firmly. Other forms of preventive repair work include mesh-style bags filled with concrete or fiberglass and epoxies that also seal the wood from oxygen. The latter types can also be used in deteriorated structures to add strength to timber piling. For details on pile protection from the effects of marine parasites, readers are referred to Baileys (1995) and Webber and Yao (2001).

6.6.1.2 Stone, Masonry, and Unreinforced Concrete Structures. Maintenance and small repair work of these structures is generally limited to fixing or replacing the grout that was used to fill joints. That softened by the action of seawater grout should be replaced. The integrity of the drainage system behind the wall should be inspected on a regular basis. It must be kept clear to allow the escape of water from behind the wall. If drainage became blocked, attempts should be made to clear them. Where the structure was originally faced with protective blocks (e.g., granite plates), any facing that has been lost should be replaced. Where the structure was intended to be watertight, where necessary, the new grout should be made watertight by using injection techniques. Alternatively, grout "socks" made of geotextile material can be installed.

6.6.1.3 Reinforced Concrete Quay Walls. Good maintenance program of these structures includes regular inspection and, if required, some minor repair work. As part of maintenance, the surface of the structure should be cleaned of marine organisms that may adversely affect structural materials. In particular, they should be removed around metal fixings so that the fittings can be maintained properly. Life (durability) of the structure can be extended by applying miscellaneous surface coatings and/or cathodic protection. The former is applied to resist penetration of carbon dioxide, chloride, oxygen, and water. In general, the coating should possess a good bond with the protected surface, ability to transmit water vapor from the surface to be protected (poor water vapor transmission can lead to debonding), resistance to ultraviolet light, and so on. There are several brands of coating material available on a market. With respect to their characteristics (i.e., bond, durability, costs, etc.), the most frequently used are saline and siloxane, epoxies, bitumen, acrylics, and similar. The former two act as water repellent and therefore has a high chloride penetration resistance. They transmit water vapor readily but have quite poor carbon dioxide resistance. It must be pointed out that the materials above are effective only when applied to dry surfaces.

Epoxies of various types have a good record for resistance to chloride, carbon dioxide, and water penetration. They can be applied to wet surfaces and can be used effectively under water and in splash or tidal zones. They are rated as durable despite the fact that they do not resist ultraviolet light well. Bitumen-based materials are of the surface-seal type. They bond reasonably well with the surface, have excellent resistance to carbon dioxide, chloride, and water, and are extremely durable (life expectancy over 20 years) provided that they are not exposed to ultraviolet light. Both epoxies and bitumen have poor water vapor transmission properties, but nevertheless, have a good record in practice for bonding to the treated surface.

Acrylics used for structure protection are usually of three types: emulsion, resin in solvents, and unsaturated resin normally based on methylmethacrylate. All acrylics have good resistance to ultraviolet light and many chemicals, and bond well with protected surface. The emulsion can be applied to damp surfaces. Acrylics have good resistance to carbon dioxide penetration when applied in a multicoating form. They have some resistance to chloride and water penetration, but their ability to resist oxygen penetration is poor. Cathodic protection may be used when chloride ingress has been so severe as to reach the reinforcement. Minor repair work as part of maintenance may include those repairs intended to arrest the deterioration as it is first noted. It may include small cracks, initial stage of resting of steel reinforcement, initial stages of concrete dilapidation, and so on. The maintenance repair work usually includes superficial repair to deteriorated concrete and the application of protection systems.

6.6.1.4 Steel Structures. Maintenance considerations of these structures usually begin at the design stage by inclusion of miscellaneous protection systems. The latter includes consideration of cathodic protection system(s) and/or different types of protective coating systems. The protective systems are designed to halt steel corrosion, which is most severe in a warm seawater environment. In practice, the original protective systems installed because of their poor maintenance may fail and need to be reinstalled. The selection of any type of cathodic system and its components should be made on the basis of cost-effective selection of materials, fittings, and construction and maintenance procedures; selection of any product should be based on proven experience. The system should be monitored on a regular basis, its physical condition verified periodically, and sacrificial anodes replaced as required. Preventive maintenance also includes periodic inspection, spotting, and recoating surfaces as required. The coating system should be applied on a thoroughly prepared clean surface. Its application requires special care in the tidal and splash zones.

6.6.2 Repair and Rehabilitation Works

Normally, repair and rehabilitation work is preceded by detailed inspection and engineering evaluation of the actual structure load-bearing capacity, as discussed earlier. The consideration of repair and rehabilitation methods must include the owner–operator's expectations from the structure: for example, its expected lifetime, function, appearance, and the like.

All the considerations above may affect the choice of the appropriate actions. Broadly speaking, the options can be as follows: to make just some cosmetic repair by allowing the deterioration to continue, to repair the damaged part(s) but take no other action, to repair substantially to prevent future deterioration, to replace a damaged part of the structure with a new one as required, and so on. In some cases the maintenance and repair program may include a certain cycle of partial repairs, whereby only the most deteriorated or damaged structural components are repaired. Useful speculations on the above are found in Neville (1997) and DiCastro et al. (1998).

However, in general, after the actual capacity of the structure is established, cost-effective decisions on follow-up actions can be made. If in its present condition the structure can safely carry the design service loads, just regular preventive repair work to mitigate the effects of deterioration is necessary to maintain the structure's operability for the design period; thus, costly repair/rehabilitation is avoided. At this stage, provided that all major structural elements are in a good condition, repairs could be performed to restore only deteriorated components of the structure.

If structural analysis or stress–strain–deflection surveys indicate that a structure is overstressed, structural restoration (rehabilitation) or downgrading of a structure's capacity through operation limitations is required. This could include restoration of all or just major structural components to their original or required strength, installation of a more efficient fender system, or major reconstruction of the whole or part of a structure. In the case of piled structures, the actual solution to the repair, or rehabilitation, is usually quite straightforward. This generally includes repair of piles (seriously damaged piles are usually replaced by new ones), and repair and/or strengthening of the deck structure. In some cases, addition of new piles, replacement of the existing fender system by a new, more efficient system, and repair of the under-platform riprap dike is required. Repair and rehabilitation of soil-retaining structures usually presents more serious problems. As stated previously, these structures typically are gravity-type retaining walls or anchored sheet pile bulkheads.

Before any repair or rehabilitation method to repair these structures is selected, it is prudent to establish not just the required capacity of the facility, but also the length of time for which this load-bearing capacity will be valid. It is also necessary to establish the operational limitations, such as ship size and her allowable approach speed and approach mode (e.g., assisted by tugs, or direct approach) and acceptable load condition. Last, but not least, it should be understood that a lack of proper supervision might result in a poor quality of repair or rehabilitation work. This usually occurs due to human nature or incompetence of the personnel involved. Hence, proper supervision is essential to achieve good-quality work.

6.6.2.1 Planning. To be successful, the repair or rehabilitation of the marine structure should be planned carefully. The main stages of planning procedure include the following steps:

1. Determination of all available basic relevant repair procedures. This involves review of previous experience of similar repair or rehabilitation work and development of a new idea relevant for the case under consideration.
2. Examination of the technical feasibility of selected repair or rehabilitation procedures. This can be done either on the basis of previous experience or through specific analysis relevant for the case under consideration.
3. Economic evaluation of selected repair or rehabilitation methods and procedures and selecting the most cost-effective alternative. This must take into account the cost of repair or rehabilitation and the cost of possible disruption of dock operation.
4. Preparation of repair or rehabilitation specifications that will typically cover detailed procedures, sequence of operation, materials, quality control, tolerances, and other relevant subjects.

As noted in step 3, the repair and rehabilitation work may interfere with port operation. This may manifest itself in interruption in ship docking and departure procedures, the necessity of moving the existing cargo handling equipment to temporary location, and in the worst-case scenario, interruption of work on the entire berth. Last, but not least, environmental and social aspects may have to be considered. All of the above should be considered at the structure repair and rehabilitation planning stage. Structure rehabilitation work planned properly and executed in a timely manner to meet the requirements of contemporary vessels and cargo handling and hauling equipment may present a feasible cost-effective alternative to replacement of the existing structure by a new one.

6.6.2.2 Repair and Rehabilitation of Timber Structures. These structures usually are repaired either by replacement of the deteriorated load-bearing components or repair of the affected members by strengthening using splice plates (timber, tropical hardwood, or steel) or shoring members. Longitudinal splits are sometimes held together by stick bolts. Often, metal fastenings within the tide and splash zones deteriorate more rapidly than the timber itself. The source of this decay or rot is trapped moisture. It must be carefully addressed in the repair and rehabilitation work. Furthermore, where decay and/or rot is present, the source of trapped moisture should be eliminated before the new timber is placed. Fire-damaged members usually remain as they are, provided that sufficient sound material below the surface char is present. In general, all timber materials used in marine structure, especially in a seawater environment, should be treated with preservatives (in most cases by creosote), with the exception of naturally resistant dense tropical hardwoods. Cutoffs and holes should be field treated. General information on wood structures repair is given in ASCE (1982).

In the repair and rehabilitation of marine timber structures, the most significant ongoing cost is associated with pile repair and/or replacement. The repair options usually include, but are not limited to, the installation of replacement creosote piles, PVC wrapping of existing piles, use of plastic, steel or concrete piles, and so on. In most cases, the former is the most economical; however, the use of creosote piles is not always permitted by environmental agencies. Creosote timber used in construction of a new dock or repair of existing ones has the disadvantage that it must be disposed of, when required, at special facilities at significant cost. If severe deterioration (decay) is confined to its upper part above water level, this part can be cut off and a new piece of timber installed and secured at both the pile and deck structure, as required.

Pile wrapping, if done properly, may have an excellent effect on pile durability and performance in seawater environment (Asavarenngchai et al., 1995; Collins and Garlich, 1998; Webber and Yao, 2001). As pointed out earlier, PVC pile wraps are installed basically to act as a barrier between the marine environment and the pile. When installed properly, these wraps also create an anaerobic condition design to suffocate the marine borers that have already penetrated the pile substrate. This is achieved by cutting off the resupply of oxygen to the protected area. Most often, the wraps are used for timber pile remediation when exposed to borer attack.

On irregular surfaces, different kinds of jackets are installed. Jackets not only protect piles from marine organisms but also strengthen them. Pile jackets most often incorporate a stay-in-place form of either fabric, fiberglass, or other materials. Jackets can also be constructed by using removable forms similar to those used for conventional column construction. The space between the form and pile is generally filled with concrete, epoxy grout, or similar material. Because of their ability to include reinforcing materials to strengthen a pile, jackets are often used to repair piles exhibiting considerable deterioration. Jackets are most helpful where piles are subject to impact from floating debris and/or ice sheets and in areas of severe abrasion by moving sediments or ice.

Depending on the specific jacket material and purpose (e.g., protection or strengthening), jackets may substantially increase the weight of the pile and its diameter. In some cases it may result in a substantial increase in structure dead loads and lateral hydrodynamic and drag loads. While normally of minimal effect, these loads should be considered in repair schemes. Furthermore, in seismically active areas the increase in pile stiffness should be considered in a structure's response to dynamic loads. This also should be carefully addressed during the repair design phase. Two representative examples are shown in Figure 6.17. The pile jacket that is shown in Figure 6.17*a* is fabricated from reinforced fiberglass form placed around the pile. Noncorrosive grout spacers are used inside the jacket to maintain proper spacing around the piling. Before installation of the jacket, the pile shaft must be thoroughly cleaned by water jet, sandblasting, or other suitable method. The jacket can be installed at the tidal zone area or positioned below the

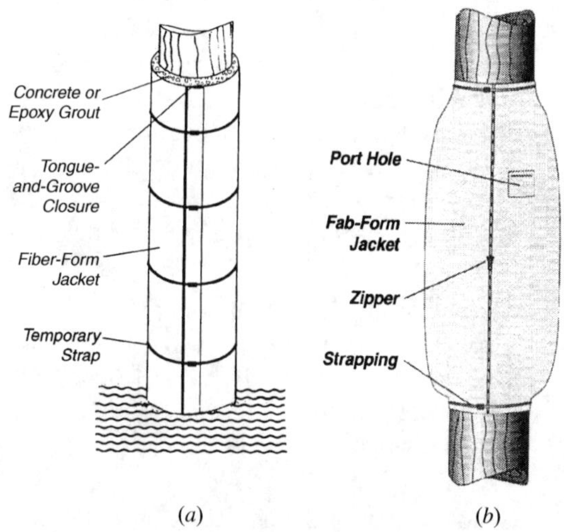

Figure 6.17 Synthetic forms for concrete encasement to restore and protect timber piles structurally.

mudline. In the latter case, the mud from within the jacket should be excavated. The reinforcing steel is used as determined by the engineer. After the jacket, which normally has a vertical tongue-and-groove closure positioned around the pile, it should be secured in place with a select strapping system at every 30 cm. Next, the bottom is usually scaled with epoxy grout, and the vertical seam is sealed by special tape and/or epoxy grout.

Finally, the jacket is filled with cementitious fill at a constant rate of placement but within allowable pressure against the jacket form. The top of the fill is sealed with the epoxy grout. In the example illustrated in Figure 6.17*b*, the pile is protected with a fabriform made from a nylon fabric. It is provided with a zipper closure system. This jacket is also installed on a thoroughly cleaned portion of a pile intended to be strengthened or protected. If steel reinforcement used to be installed, all ends of the rebars should be turned toward the pile to avoid damage to the fabric jacket. If required, spacers are installed to maintain the adequate mortal cover over reinforcing. First, the jacket is positioned around the pile and secured to the pile by straps. Then the zipper is closed from the top down, and the jacket is filled with cementitious mortar of required strength. The mortar is injected through the portholes in a uniform manner at a constant rate. In both cases illustrated in Figure 6.17, the mortar injection is similar to that used in a tremie technology. The latter is discussed in the following sections. In no case is the mortar dropped from top to bottom of the jacket, which could result in a segregation of mortar materials. Furthermore, the cementitious mortar normally should incorporate the antiwashout admixture and have the lowest possible water/cement ratio.

It should be noted that both types of jackets described are also used for repair or rehabilitation of concrete and steel piles. The procedure for repair of the deck structure usually involves replacement of damaged or deteriorated structural components by new ones. The deck structure could also be strengthened as required to handle the heavier loads generated by new cargo handling equipment. In this case the pile foundation may need to be reinforced by the installation of new piles. To function as intended, the latter must be properly incorporated into the dock deck system.

6.6.2.3 Repair and Rehabilitation of Gravity Walls. Erosion of structural material (in most cases, superficial) and erosion (scour) of a wall's toe protection mattress or foundation material are basic problems associated with gravity quay wall operations. Deterioration of structural material very seldom presents problems as to the integrity of the structure, but to prevent further deterioration (in some cases, severe dilapidation), large cavities in the wall face should be repaired.

Surface Repair and Rehabilitation. The depth of concrete or other material deterioration can range from surface scaling to 1 m and more in depth, which at times can reach several cubic meters in volume. In the latter case, to prevent further material deterioration that eventually may compromise the integrity of the wall, the cavity must be filled with concrete. Installation of new concrete usually does not present a problem in the zone above the water level. It must be noted that substitution of the lost concrete or other material with conventional concrete should not be used in situations where an aggressive factor that has caused deterioration of the original materials still exists. For example, if the deterioration has been caused by acid attack, aggressive water, or abrasion–erosion, it is doubtful that repair by the installation of conventional concrete will be successful unless the cause of deterioration is removed. Before the placement of new concrete, existing loose material must be removed until sound material is reached. To ensure a good bond between new concrete and existing material, the minimum depth of the cavity should be at least 150 mm. For better repair, the cavity should have vertical sides normal to the formed surface or keyed as necessary to lock the repair into the existing structure. The top inside face should be sloped up toward the front. Slope 1:3 is usually sufficient to prevent air pocket formation inside the repaired cavity.

Surfaces of existing material must be thoroughly cleaned by high-pressure water jet, sandblasting, or by another equally satisfactory method. Water jetting is an especially effective means of surface preparation. It is used for removal of all loosened and easily peeled layers and all foreign materials, cleanout, chloride-contaminated concrete, exposure and derusting the reinforcing steel, and so on. Concrete treatment by high-pressure water jet above and underwater is widely used (Momber, 1998). The basic advantages of the use of high-pressure water jets for concrete repair are as follows:

- High performance rates
- Prevention of damage to, or loosening of, built-in parts (e.g., reinforcing steel, clamping, sleeves)
- Environment friendliness through prevention of the formation of dust, gases, and so on
- Absence of percussion or vibrations
- Negligible reaction forces
- Wide range of application
- Wide range of tools
- Selective material removal
- Guaranteed high pull-off strength (more than 1.5 MPa)
- Good surface quality

Besides surface preparation, a high-pressure water jet may be used effectively to remove and cut the heavy concrete as required. The water jet penetrates the deteriorated concrete, breaks it, and flushes it away without causing damage to the reinforcement. High-pressure water jetting or sandblasting effects should be confined to the surface that is to receive the new concrete. If required, dowels and reinforcement can be installed to make repair self-sustaining and to anchor it to the underlying concrete. Before receiving new concrete, the repaired surface both under and above water is first carefully coated with a thin layer of epoxy-modified slurry not exceeding 3 to 5 mm and having the same water/cement ratio as the concrete to be used in the replacement. When the wall repair is carried out under dry conditions, to minimize strains due to temperature, moisture change, shrinkage, and so on, concrete for the repair should generally be similar to the old concrete in maximum aggregate size and water/cement ratio and should be vibrated thoroughly; internal vibration should be used if accessibility permits. The new concrete can be placed either by pumping it or by injecting it through strategically located ports (Figure 6.18*a*). An underwater concrete repair, although similar in technique to a dry repair, is, more complex to execute (Figure 6.18*b* and *c*).

All requirements and precautions normally applied to surface preparation for the water repair described above should be considered for underwater concrete installation. Additionally, underwater work requires removal of marine growth from repaired surfaces. All underwater activities are carried out by divers. The concrete mix used for placement under water usually contains an antiwashout admixture and requires the lowest possible water/cement ratio, preferably 0.45 (by weight) or less. The use of superplasticizers is necessary to promote the flow of concrete into the cavity. Below-water concrete should be placed in accordance with ACI (1989) or equal standards.

In general, underwater concrete (grout) must be proportioned basically for workability and usually not for a given strength level. Normally, the strength achieved by replacement concrete will be more than adequate. Concrete (grout)

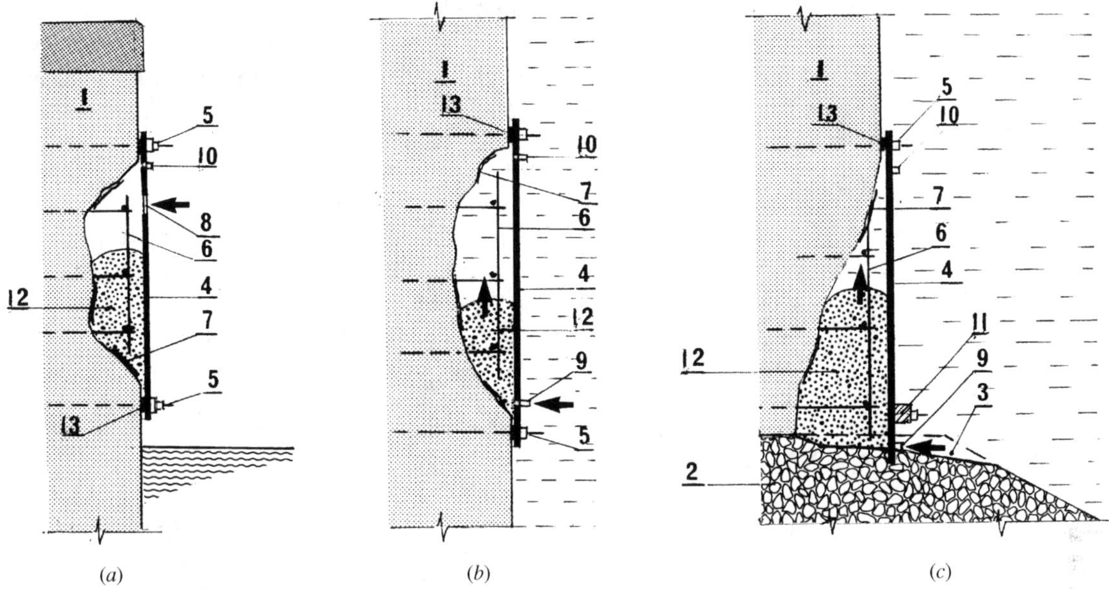

Figure 6.18 Gravity wall. Surface repair above (*a*) and below (*b* and *c*) water level. 1, Existing wall; 2, rubble mattress; 3, eroded part of rubble mattress; 4, formwork; 5, anchor bolt; 6, new reinforcement (installed where required); 7, bonding (epoxy) compound; 8, concrete placement port; 9, grout injection port; 10, water (grout) escape port; 11, cross bar for holding formwork; 12, new grout; 13, perimeter seal (foam).

placement equipment must be adequate to handle the designed concrete at a designed rate of placement. Underwater concrete (grout) is normally injected into the cavity from the bottom up. In the process of grouting, water escapes from the cavity through the upper port. To ensure good quality of repair, the concrete (grout) must continue to be injected some time after it starts escaping from the upper port. Installation of some reinforcement anchored inside the cavity can substantially enhance cohesiveness between new and existing concretes. The formwork is attached to the wall surface by means of anchor bolts. The size and number of these bolts must be proportioned in a way to take pressure from fresh concrete and weight of the form. In water colder than 10°C, formwork should be insulated to ensure proper curing of concrete. If the size of the cavity is substantial, several concrete injection ports can be installed in the formwork, and a diver will move the concrete carrying hose from port to port as required. To prevent concrete escape, the space between the formwork and the face of the wall, as well as potential voids between separate concrete blocks inside the cavity, must be thoroughly sealed.

A great many marine structures in North America were built 50 and more years ago. The concrete in most of the structures presently in service (most of them are nonreinforced mass concrete quay walls and chamber walls in navigation locks) does not contain entrained air and is therefore susceptible to freeze–thaw deterioration. The general approach to this wall repair or rehabilitation is to remove 0.2 to 1.0 m of concrete from the face of the wall, as required, and to replace it with new concrete. The deteriorated concrete can be removed by any suitable technique (e.g., blasting, cutting by high-pressure water jet). Any remaining loose or deteriorated materials that could inhibit the bond between old and new concrete ought to be cleaned out as discussed earlier.

Dowels 20 to 25 mm in diameter typically spaced at 1.0 to 1.5 m center to center in both directions are normally used to anchor the new concrete facing the existing wall. Dowel diameter, embedment, and spacing are proportioned to resist the load of fresh concrete or a potentially large hydrostatic pressure which could develop behind the reinstalled concrete. It is usually recommended that at least 0.25% of all dowels installed be field-tested. The embedment length is considered to be adequate when an applied pullout load is equal to or greater than the calculated yield strength of the dowel. Mats of reinforcing steel usually 15 to 20 mm in diameter (300 to 500 mm center to center) are hung vertically on dowels and fixed into position with provision for 75 to 100 mm of cover. Once the reinforcement and formwork are in position, replacement concrete is placed by using any conventional technique.

To minimize strain due to potential incompatibility of old and new materials, the new (replacement) concrete mix should be proportioned to simulate the existing concrete substrate; proper air entrainment in the new concrete can be obtained by use of admixtures that will ensure resistance to cycles of freezing or thawing. One of the most persistent problems in refacing gravity walls is cracking in the new concrete. These cracks, which may extend through the new concrete, are attributed primarily to restraint of volume changes resulting from thermal gradients and drying shrinkage. To reduce or minimize the effect of deterioration, concrete materials, mixture proportions, and construction procedures that will reduce concrete temperature differentials or minimize shrinkage of volume change should be considered.

For better quality (and appearance) a permanent formwork system of precast concrete panels can be used in wall repair and rehabilitation. Use of durable, high-quality concrete panels attached to the prepared concrete substrate has a significant potential for resolving the crack problem. Also, the use of precast formwork panels should shorten the time a quay wall must be out of service for repair or rehabilitation work. Permanent prefabricated concrete formwork panels have been used successfully to reface concrete wharves, navigation locks, and dams. Some practical examples and relevant references are given in Tsinker (1995).

Permanent formwork can also be constructed from steel sheet piles (Figure 6.19). They are installed in front of the existing deteriorated structure and fixed at the top and bottom. This follows by placing concrete fill between the sheeting and the prepared face of the deteriorated wall. Finally, the new caping is installed. A relevant example is reported by Martinez (1998). Replacement concrete can be installed by any conventional placement techniques; for underwater placement the tremie technique is normally employed. For concrete replacement up to 150 to 250 mm deep in dry conditions, the shotcrete method is generally more economical than placement of conventional concrete. Basically, this is because the former does not need costly formwork.

SHOTCRETING. This is the process of sand/cement mortar application at high velocity onto a concrete surface. It is done with the help of special equipment. As a material, shotcrete is a fine aggregate concrete with various sand/cement ratios, depending on the application. Shotcrete repair is typically used to reface large dry areas of deteriorated concrete. It is very effective for the repair of both vertical and horizontal surfaces that have deteriorated to variable depths. Shotcrete repair typically involves surface preparation, reinforcing, and concreting. A disintegrated area of concrete to be shotcrete treated is prepared in the way discussed

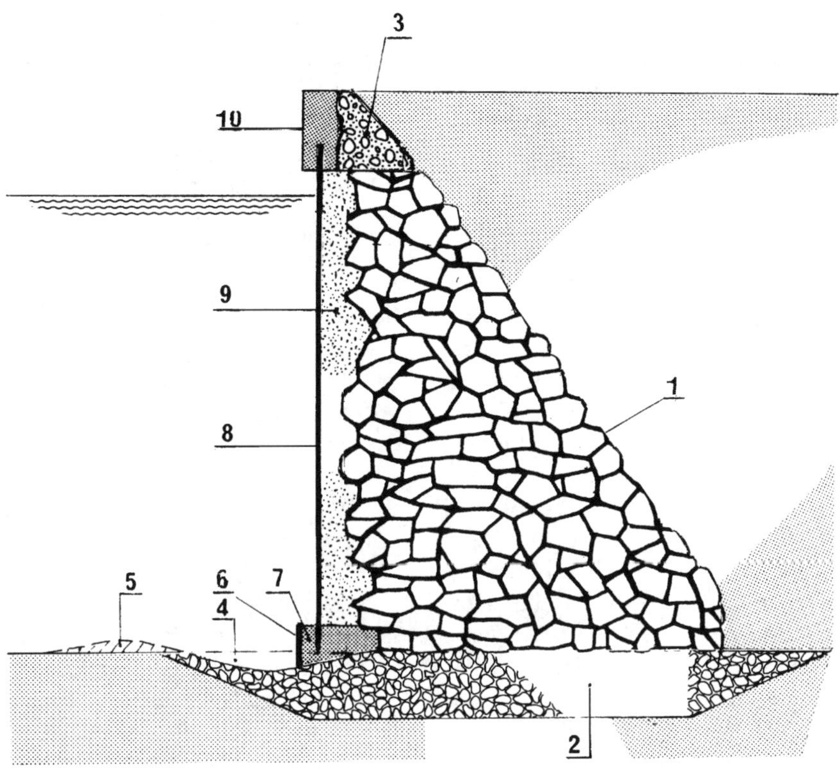

Figure 6.19 Underwater refacing of a cyclopean gravity wall. 1, Deteriorated wall; 2, stone mattress; 3, concrete caping wall; 4, eroded foundation in front of wall; 5, deposited material; 6, formwork; 7, underwater concrete for fixing of sheet pile panels; 8, steel panels fabricated from sheet piles; 9, tremie concrete; 10, new addition to wall caping.

earlier. Galvanized wire mesh is then attached to exposed reinforcing bars or dowels. A bonding agent is generally used to ensure good cohesiveness between the old and new concrete. Finally, well-graded concrete (sometimes, epoxy modified) with proper moisture content is applied to the repaired surface at required pressure. If done properly, the end result is a dense, high-strength, durable refacing capable of very good bond with the parent concrete. The resistance of shotcrete to cycles of freezing and thawing is generally good despite the lack of entrained air. This is attributed in part to the low permeability of properly proportioned and properly installed shotcrete. If applied properly, the shotcrete minimizes the ingress of moisture, thus preventing the new face from being critically saturated.

Use of steel fibers can greatly enhance the quality of shotcrete concrete repair (Hoff, 1987; Gilbride et al., 1988; Morgan, 1988). Low-permeability, steel fiber–reinforced concrete diminishes the depth of surface carbonization to no more than a few millimeters over a period of 10 years or more (Hoff, 1987). Although any fibers in a carbonized surface layer can be expected to corrode and disappear, the internal fibers beyond this zone will remain unaffected. If, for some reason, existing non-air-entrained concrete behind a shotcrete repair can become critically saturated by water (e.g., due to migration from beneath or behind the wall and the shotcrete is unable to permit the passage of water through it to the exposed surface), it is likely that the existing concrete will be more saturated during future cycles of freezing and thawing. If frost penetration exceeds the thickness of the shotcrete section under these conditions, freeze–thaw deterioration of the existing wall should be expected. Numerous examples of such effects are found in U.S. Army Corps of Engineers (1986).

UNDERWATER REFACING. It is normally accomplished by the installation of formwork under water and placing concrete, usually by tremie techniques. For better quality (and appearance), the permanently installed formwork systems were used successfully in the past. They were constructed in the form of locked-together, high-quality, durable concrete panels attached to the prepared concrete substrate, or panels, fabricated from steel sheet piles. Use of concrete panels resolve the cracking problem usually encountered in the refacing of deteriorated walls. Also, they cut the time of wall repair or rehabilitation, thus reducing the dock out-of-service period. For examples, consult Tsinker (1995).

A practical example of steel panels used for formwork system is illustrated in Figure 6.19. In this case, wall refacing proceeded in the following sequence. First, all soft and loose materials and marine growth were removed from the wall surface by divers using a high-pressure water jet; this was followed by the installation of panels fabricated from steel sheet piles. At the bottom, these panels were embedded in the cast-in-place underwater concrete foundation, and at the top, they were supported by a steel wale tied into the existing concrete caping wall by anchor bolts. The space between the wall and sheet pile panels was filled with tremie concrete, and the work was completed by construction of a new addition to the existing caping wall. Both the old and the new walls were posttensioned with each other by anchor bolts. Finally, the wall was outfitted with a fender system and mooring accessories.

The sheet pile form in front of the wall was designed to support the pressure imposed on it by the effective fluid pressure of the submerged weight of concrete backfill. To maintain the flexural stresses of steel sheet piling within allowable limits ($0.75f_y$), it was necessary to resist the height of the backfill lifts. To retard the corrosion process, the steel sheet piles were coated on the front face with epoxy polyamide coating. Because of the nonstructural nature of the sheet piling, no cathodic protection was required. However, to extend the service life of steel panels, some allowance for corrosion was considered.

The success of placing concrete under water is highly dependent on preventing mixing the fresh concrete with water, which may result in loss of fine cementitious particles from the mix. In modern practice, the mix design usually incorporates antiwashout admixtures and usually requires a minimum water/cement ratio equal to 0.45 (by weight) or less. The minimum dosage of antiwashout admixture should correspond to the allowable maximum loss of cementitious particles. Superplasticizers are used to promote better flow of concrete in forms. For more information on placing concrete under water, consult ACI (1989) and Tsinker (1997).

Wall Consolidation. A great deal of gravity quay walls presently in use have been in operation for quite a while; some of them were constructed in late nineteenth century. These walls, usually of blockwork, cyclopean, or mass concrete construction, have deteriorated significantly due to the effects of miscellaneous environmental factors, such as waves, current, ice, and so on, increase in live loads, cyclic loading effects, water seepage through the wall due to tidal effects, distress of foundation material(s), and others. Wall deterioration usually manifests itself in the form of a large void in the wall structure, internal and external cracking, uneven settlement of adjacent sections, loss of cement mortar, and so on. These walls are usually rehabilitated by cement mortar pumping in a variety of ways. The most commonly used method includes drilling boreholes vertically, or inclined in both directions, through which cement mortar is injected. For better consolidation, reinforcing bars are usually inserted into the boreholes before mortar injection. In some cases, if required, wall consolidation is carried out simultaneously with reinforcement of the distressed foundation soil.

A representative example is shown in Figure 6.20. The 1100-m-long wall depicted in this figure was constructed at the port of Providence, Rhode Island in the late nineteenth century late as a municipal wharf. It is a classic example of nineteenth-century granite masonry construction. For many years, this wall was used as a terminal for break-bulk, bulk, and LPG cargoes. In the mid-1970s the city of Providence undertook a major port modernization program to update its facilities to accommodate modern trends in maritime commerce and to provide expanded services for larger, deeper-draft vessels. For this, the seafloor in front of the wall had to be dredged down to 10.7 m. Such a deep dredge would, however, result in undermining of the existing wall. To prevent this from happening, the wall was first underpinned by small-diameter grouted piles drilled through the wall structure, which also consolidated the wall structure. Furthermore, to ensure safe dredging, steel sheet piles have been installed at the wall toe. As the work progressed, the survey revealed considerable outboard movement of the wall. This movement was apparently caused by the effects of vibration produced by the percussion drills. At the end of the construction period, wall movements gradually subsided as the new pile underpinning system assumed the foundation loads. In the process of wall rehabilitation prior to dredging, the contractor installed a temporary gravel and crushed stone berm at the wall's toe, which was in place until the foundation work was completed.

The underpinning grout piles were installed by drilling 15- and 20.5-cm-diameter holes vertically and sloped on a 1:2.5 batter through the granite courses and foundation soil to suitable foundation material; reinforcing steel was placed into the predrilled holes and grout was injected. Simultaneously, the casing was gradually withdrawn to form piles. The pile capacity was confirmed by conducting field load tests on special piles installed through and behind the wall. The initial construction specification called for the piles to be socketed approximately 1.5 m into the underlaying bedrock. Load tests, however, indicated that the piles could provide a suitable foundation without reaching bedrock. As a result, pile tip elevations were raised, resulting in considerable project savings. Pile design was based on the assumption that the wall would perform as a monolithic unit. It was therefore important that pile reinforcing be extended into the wall and that voids in the wall are consolidated during the grout installation process. Grout was installed at ambient hydrostatic pressure, and all precautions were taken to prevent formation of voids or discontinuities in the pile due to a sudden grout pressure drop, which could have resulted in negative pressure at the injection head. Grout injection proved to be difficult because of the presence of large voids encountered between granite blocks. The presence of these voids resulted in a loss of grout through the wall face. To prevent grout loss, the outside joints between courses were sealed by divers with packing material. This consisted of a mixture of cement, bentonite, and metal fibers

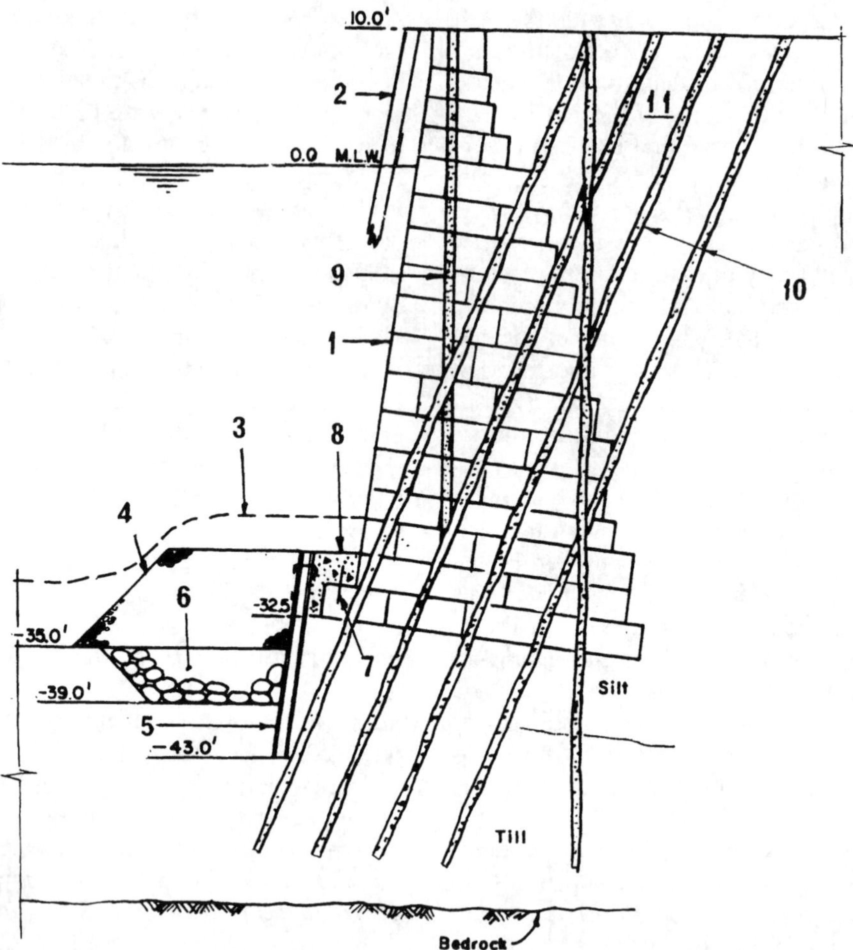

Figure 6.20 Rehabilitation of an existing gravity quay wall, berth 3 at the port of Providence, Rhode Island. 1, Granite blockwork wall; 2, old fender pile; 3, ground surface before modernization; 4, temporary (intermediate) rubble dam; 5, new steel sheet piling; 6, new scour protection; 7, steel dowel; 8, concrete plug; 9, grout-filled stitching hole; 10, underpinning piles; 11, miscellaneous fill. (From Pierce and Calabretta, 1986.)

made in the form of balls prepared on the surface and delivered under water to divers for wall packing. Additionally, vertical "stitching" holes were drilled and grouted close to the wall face. The line of these holes drilled at approximately 2.75 m center to center helped to consolidate the body of the wall and also helped to seal the wall face. Stitching holes were drilled through the cap stone to a depth of approximately 11.0 m and grouted with a thickened gravity-injected grout mixture. Most recently, similar work has been carried out for rehabilitation of a quay wall at the port of Ostend, Belgium (Van Damme et al., 1998). It is worth noting that dredging in front of any wall, a gravity wall in particular, must be carried out cautiously to avoid, or at least minimize, overdredging. The latter often is the cause of wall distress and instability, resulting in structural damage and excessive deflections.

Essentially, excavation below the level of a toe reduces the bearing capacity under the toe as well as a wall's resistance to sliding and overturning. Successive dredging operations may slowly reduce the level of soil at the wall face, thus reducing the chance of overdredging. The effect of this mode of dredging, however, may not necessarily be obvious, for a number of reasons; for example, the bottom-level investigation did not produce sufficiently accurate data (e.g., echo sounding can give a false reading close to a vertical wall), the presence of silt obscures the level of the firm material, and so on. If the bedrock bottom is excavated by use of explosives, the rock under the wall's base can be fragmented by adjacent blasting, resulting in degradation of the foundation. Hence, the amount of explosives used for rock excavation must be calibrated carefully to avoid the aforementioned effects. In concluding this section, it should be noted that to prevent leaking the backfill through the wall structure, the proper function of a wall drainage system is of paramount importance, and this must not be overlooked in a gravity wall repair and rehabilitation program. For more information on gravity quay walls repair and rehabilitation, readers are referred to Bray and Tatham (1992).

Rehabilitation of Distressed Foundation Material. A distressed foundation causing deformation to the gravity wall can be improved by one or a combination of the following methods.

GROUTING. Grouting is a technique by which foundation materials can be impregnated under pressure with cementitions or chemical grout that is allowed to set. The type and gradation of soil dictates the type of grout to be used. General information on soil grouting is found in most standard textbooks on geotechnical engineering.

JET GROUTING. Jet grouting is essentially a ground modification technology used to create in situ cemented formation of soil. Jet grouting is a rather remarkable technique now in use for well over 25 years. Although several techniques of jet grouting exist to date, the basic idea of jet grouting derives from a 1971 Japanese patent granted to Nakanishi. The method is used for modifying relatively soft soils to achieve general material improvement or to construct subgrade structural or load-bearing members without prior excavation. The chief use of jet grouting includes underpinning of existing structures threatened by subsidence. Jet grouting is radically different from traditional pressure grouting with cement or chemicals. It is essentially a partial soil replacement technique whereby a column is formed composed of a mixture of injected grout and in situ soil. The final product, called *soilcrete,* is made by complete, hydraulically induced mixing of cement slurry with native soil. It is usually done in the shape of a cylindrical column, and the column size and shape depend on the brand of equipment used and can be varied. The properties of soilcrete depend on the native soil and the way the process is applied. The typical phases of jet grouting are illustrated in Figure 6.21. In a typical jet grouting technique initially a pilot drill hole about 100 to 150 mm is installed to the required depth using standard drilling equipment. Then a grout-injecting stem capable of withstanding extremely high pressure is inserted. At the base of the stem is a monitor that typically has two nozzles: the lower for injecting grout and the upper, typically 2 mm in diameter, for a high-pressure water jet.

Water (sometimes a water–air mixture) jetting is usually operated at around a water delivery rate of ± 70 L/m at a pressure of about 400 bar. Grout is force fed via the lower nozzle at a pressure of about 40 bar. Portland cement and water mixed in a 1 : 1 ratio is the grout mix that is most commonly used. However, any

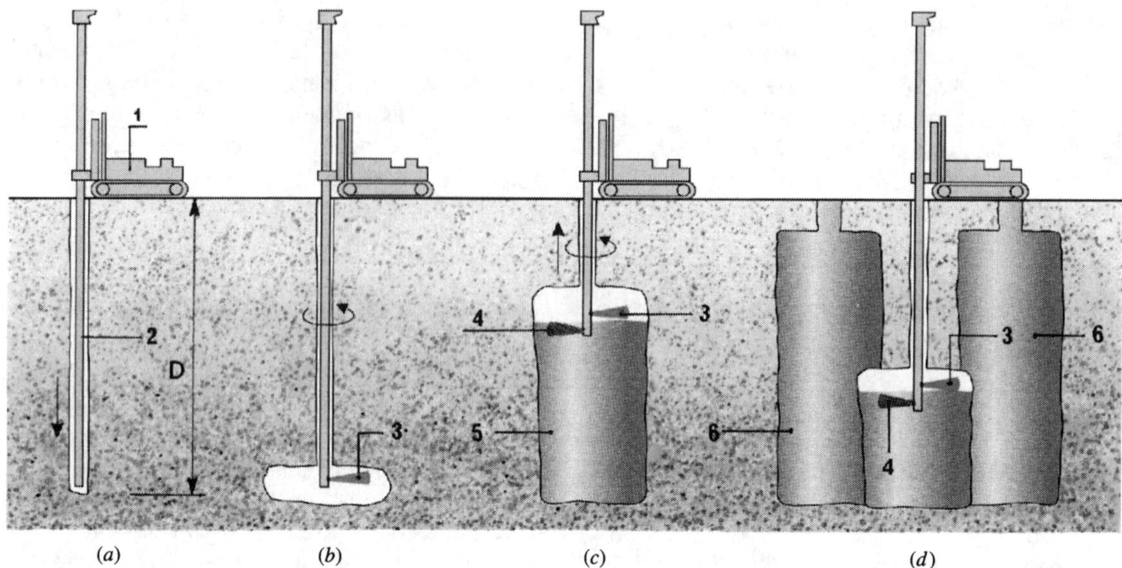

Figure 6.21 Typical sequence for soil jet grouting. 1, Mobile equipment (crane); 2, grout injecting stem; 3, high-pressure water jet; 4, grout injecting nozzle; 5, initial stage of soilcrete column formation; 6, complete soilcrete column.

pumpable material that solidifies when mixed with soil can also be used: for example, bentonite slurry, which is normally used for cutoff walls. Once at full depth, the stem is slowly extracted and rotated and grout is expelled radially from one or more base nozzles of the jetting monitor. The result of high-pressure jetting is that the in situ soil structure is destroyed and soil particles mixed thoroughly with grout. A continuous, full-height mortar column from the bottom up is formed at a single pass of the jetting head. When cement and water are mixed 1:1, the grout's typical compressive strength values can be about 7 MPa for clays and about 20 MPa for clean sand.

The diameter of the column formed can vary from 0.3 to 2 m, depending on the in situ soil conditions, jetting pressures, speed of rotation, and rate of extraction. The column spacing and alignment need to be controlled, although some tolerance is available when forming a wall by overlapping columns; also, some correction can be achieved by locally increasing the diameter of the column to give a connection to a misaligned column. The primary concern in the case of the cement grouts used with jet grouting techniques is the ability of the grout to set, cure, and develop appreciable strength at relatively low temperatures (say, below 5°C). Neat cement grouts consisting of normal portland cement and water will not set at all at about −1.8°C. Therefore, it will be necessary to use high early cement and/or accelerators. To date, cement grouts are the only grouts that have been used with different grouting techniques. An advantage of using cementious grouts is that seawater can be used for the mix, whereas fresh water is generally mandatory for chemical grouts. Cement grouts vary in terms of the types and/or proportions of portland cement, pozzolan admixtures, and water/cement ratio.

Chemical grouts are typically used in a pressure (injection) grouting technique. They vary considerably in terms of their chemical composition and end products (solids and gels). Practical aspects of the use of chemical grouts are discussed by Baker (1982), Gularte et al. (1992), Manfakh (1995), and Blakita and Cavey (1995). Grouts in each category, chemical or cement, have been used to grout different kinds of soils. The category of grout and the type of grout within the category, which can be used in a specific type of soil, will depend primarily on its rheological behavior (the manner in which it deforms and flows). The physical characteristics that will determine the rheological behavior of a grout are viscosity, rigidity, and granular content. Some grouts are nongranular and liquid (no rigidity) and will maintain the same viscosity with time until mass polymerization takes place. Because of the waterlike nature of these grouts, they are well suited for grouting sand and silts.

During the jetting process a mixture of grout and in situ soil may be expelled from the hole. The amount expelled depends on the void ratio of the in situ ground mass. In the case of sandy materials, little, if any material is expelled, whereas with consolidated clays, more can be expected. When the equipment is suitably modified, jet-grouted panels rather than columns can be formed. In this way considerable material savings can be obtained, as a relatively thin (300 to 500 mm) wall can be formed. Jet grouting can be stopped at any level below ground level, hence the technique can be used to underpin an existing structure or to connect a grout wall to the base of a structure. Jet grouting was developed using cementitious grouts, but the ambient conditions—for example, the retardation effects of the extremely low temperatures—may preclude the use of cement grouts. However, the technique can be adaptable to chemical grouts. The individual columns of soilcrete are the basic building units produced by jet grouting. They can be combined in various shapes and sizes to solve special problems related to new construction, or encountered in repair or rehabilitation projects.

Groups of interconnected columns can be so arranged as to make the group support required for vertical and horizontal forces. Typical layouts include interconnected rows, which can form cutoffs or retaining walls; a staggered grid pattern, which is used to improve the bearing capacity of foundation soil for a proposed structure; or for underpinning an existing structure or circular walls, which are used to provide a watertight barrier (cell) for local soil excavation.

Jet grouting offers several advantages. First, it is the only currently available technique for the relatively predictable treatment of clay, which is typically resistant to any kind of pressure grouting. Second, the process creates minimum vibration and therefore can be used safely in sensitive built-up areas. It requires no excavation prior to installation. It is a safe, maintenance-free, possibly much faster construction method than alternative techniques, requires only inert components, and has other advantages. The method's main disadvantage is that it can be applied only in relatively soft soils.

SOIL MIXING. In this technique the soil structure is destroyed by mechanical means, using an expanding bit type of device attached to the head of a drill string. As the drill is slowly rotated and extracted, a cement slurry is injected and a lean cement–soil column is formed. The technique was developed for improving ground conditions using cement stabilization techniques to increase various strength parameters. Originally developed in the United States, the techniques underwent substantial improvements in Japan (Bruce, 1996).

Repair of Scour Cavities at the Foot of a Wall. As pointed out earlier, a dramatic increase in vessel dimensions and installed power, and the introduction of bow (stern) side thrusters and special use vessels, has resulted in frequently reported scours that threaten to undermine (destabilize) all types of marine structures. Another source of scour may be the effects of heavy current and/or wave action. Where scour at a wall foot is found, it should be taken care of as soon as possible.

Depending on the size of the cavity at a wall's foot, the cavity could be filled by using one of the following methods or a combination:

- Tremie concrete
- Grout injecting into preplaced coarse aggregates
- Installation of concrete-filled bags with subsequently filling of the gaps between the base of the wall and bags by cementitious pressure grout

An example is depicted in Figure 6.22.

For good repair and rehabilitation before any of the concrete installation methods above are employed, the exposed surface of the existing concrete has to be thoroughly prepared; damaged and unsound concrete, as well as marine growth, have to be removed to ensure a good bond with the repair concrete. Practically all dry-land techniques, such as sand and grit blasting, water jetting, or jack hammering, can be used under water. Sand or grit blasting are also useful for removal of corrosion from rebars and other metal parts. It must be noted that marine fouling, which prevents bonding between old and new concrete, can present serious problems to structure repair. Hence, it is good practice to have all the marine growth removed just before placing new concrete. Steijaert and De Kreuk (1985) have pointed out that during the growing season (usually, June to September), concrete can be covered by tiny barnacles or mussels in 24 to 36 hours. Therefore, in areas where marine growth can present a problem, the new concrete has to be installed within 1 to 2 days after cleaning the repair area.

TREMIE CONCRETE. The basis of the tremie method is to pour concrete down a flexible or rigid tube 150 to 300 mm in diameter complete with a hopper into which fresh concrete is fed. At the end of the operation, the water in the void is replaced with concrete. For better results a chamfer should be provided at the high end at the feed side of the cavity. Some extra height should be provided for the outside wall to ensure complete filling. The best results in

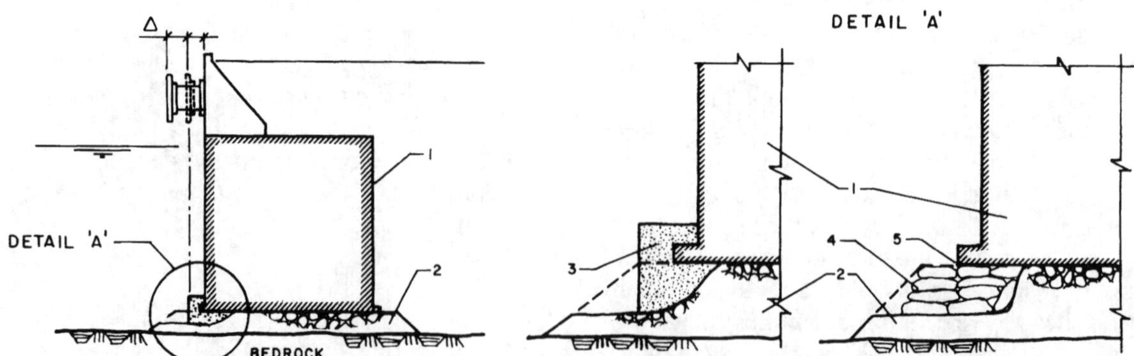

Figure 6.22 Repair of propeller-, current-, or other induced scour. Δ, Design compression of fender system; 1, gravity quay wall; 2, rockfill mattress; 3, tremie concrete, prepacked concrete, or special admixture; 4, concrete in bags; 5, pressure grout. (From Tsinker, 1989.)

filling voids are obtained with stable concrete mixes of high workability (slump 15 to 20 cm), rich in cement (minimum 350 kg/m^3), free of bleeding, and resistant to washing-out action. During concrete installation, free fall through a column of water should be avoided. For details on tremie concrete technology, consult Tsinker (1995, 1997).

GROUT INJECTING CONCRETING. Under this method, a void is initially filled with coarse aggregates; then the free space between these aggregates is filled by the injection of grout. In recent years, new, efficient antiwash admixtures have been developed that effectively prevent segregation of concrete or grout when placed underwater. A diver can simply distribute this type of concrete into a void using an underwater hose.

CONCRETE-FILLED BAGS. Under this method the diver initially installs 50 to 70% concrete-filled bags into the cavity. Because the cement paste is normally squeezed out through porous bags, a certain cementation within the bag structure takes place. Subsequently, for better interaction the space between bags and the wall in question is filled by pressure grout or tremie concrete.

Protection of Seafloor from Scour. The most severe erosion of the seafloor in front of any type of port-related marine structures (vertical wall construction in particular) is usually caused by propellers and side thrusters of large container ships, Ro-Ro (roll on–roll off) vessels, ferries, and, in some cases, very powerful tugboats. Propeller-induced flow of water near the seafloor causes bottom erosion, depending on the power of the ship's engines and the approach maneuvering techniques. Typical examples are illustrated in Figure 6.23.

Container vessels normally berth at a continuous quay where their bow thruster and main propeller(s) affect the stability of the seafloor. Ro-Ro vessels and ferries usually berth with their stern directed toward a transferring structure that can be constructed either in the form of a vertical wall or a piled system. Contemporary container ships usually have one bow thruster and a single main propeller, whereas Ro-Ro vessels and ferries may have a bow and a stern thruster and twin main propellers. On the other hand, bulk carriers usually do not have thrusters, except for relatively small coastal vessels of 5000 to 10,000 dwt.

A typical bow thruster ranges from 1.5 to 2.5 m in diameter. It may generate a maximum water jet thrust in a range 3 to 10 tonnes, with a power of 7 to 13% of main engine power for container ships and up to 17% in the case of ferries (PIANC, 1997). Because the jet velocity generated by the main propeller at the wall foot can be as detrimental to bottom stability as jet velocity produced by a side thruster, it is necessary to consider both in the design of seafloor protection.

It is usually assumed that in the most practical cases during ship berthing and unberthing operations the main propeller is used at its full power, except in the case of ferries. On the other hand, with some exceptions, the designer should assume that the full power of the bow thruster is used when berthing and/or unberthing. A ship berthing or unberthing at a continuous quay will at times use an angled rudder in conjunction with a main propeller ahead as well as its bow thruster. This combination generates a very complicated flow pattern that can result in severe bed erosion. In general, the depth of bed erosion depends on the rudder angle, and the final depth of bed erosion with a rudder present is significantly greater than that of a jet without a rudder (Hamil et al., 2001). Although theoretical solutions to calculating the effects of an angled rudder on jet velocities are available (e.g., McGarvey, 1996; Hamil et al., 2001), the best result can be obtained from physical mode tests. A simplified calculation method is given in PIANC (1997).

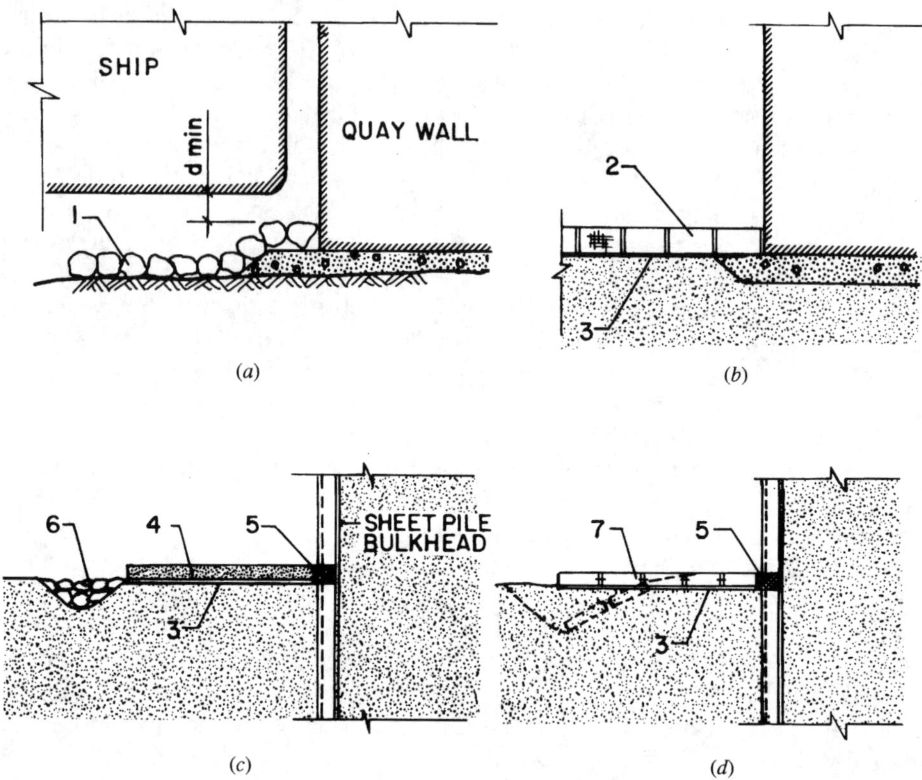

Figure 6.23 Typical scour protection systems; (*a*) stone rubble; (*b*) blockwork system; (c) prefabricated concrete; (*d*) flexible mattress. 1, Layered rock fill or rock armor placed on geotextile filter; 2, gabions (alternatively, concrete blocks, concrete in bags, etc.); 3, geotextile filter; 4, prefab concrete slabs; 5, cast-in-situ tremie concrete; 6, toe protection riprap; 7, cable-linked concrete blocks of various construction (alternatively, different types of fabric or bituminous mattresses; in the latter, no geotextile filter is required).

The level of bottom protection requires careful study because the protection installed could be damaged during maintenance dredging operations. Hence, from practical considerations the bottom protection should be installed somewhat below the lowest maintenance level. From the point of view of scouring effects, the greater the underkeel clearance, the lower the jet velocity at the level of protection, and therefore the jet thrust against the seafloor and the cost of protection are reduced. However, in the case of soil-retaining structures such as gravity quay walls and sheet pile bulkheads, lowering the level of bed protection will result in an increased retaining height of the backfill material behind the structure. The latter, in turn, will result in an increase in the cost of the structure. On the other hand, lowering the level of bottom protection will not significantly affect the cost of an open piled structure.

PIANC (1997) recommend the following criteria for placement of bed protection: 0.75 m below the lowest permitted dredging level for berths that are subject to maintenance dredging; if a berth is used regularly and the turnaround time is short, a minimum underkeel

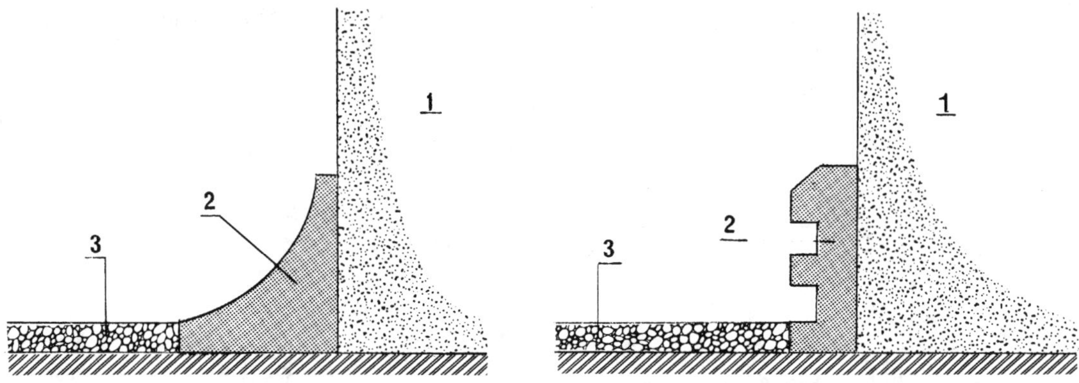

Figure 6.24 Typical arrangements of flow deflectors. 1, Quay wall; 2, flow deflector; 3, bottom protection system. (After Longe et al., 1987.)

clearance of 1 m when a ship is berthing or unberthing fully loaded is required (occasionally, when the maximum draft coincides with spring tides, this could be reduced to 0.6 m); a minimum underkeel clearance of 1.0 m should be adopted for the design of bottom protection at ferry and Ro-Ro terminals; where the ship has sufficient time for a relatively slow, hence careful approach with a very small underkeel clearance (e.g., bulk terminals), an underkeel clearance of 0.3 m may be adopted for the design of bottom protection. For more guidelines on the subject matter above, the reader is referred to PIANC (1997). Theoretical studies on bottom erosion by propeller jet action and its protection have been carried out all over the world. A list of the most important reports on these studies is given in Tsinker (1995), PIANC (1997), proceedings of specialized conferences and seminars (e.g., ASCE, PIANC), and most recently in Römisch and Hering (2002).

Theoretical studies usually have been followed by physical model tests and measurements of actual water velocities at bed level, to refine and better calibrate the theoretical methods for determining the propeller jet velocities for the design of bottom protection. Physical model studies are usually ordered where bed scour from propeller action or for other reasons may result in undermining of a new or existing important structure. The bed in front of the marine structure may be protected in a variety of ways. Some typical structural arrangements of seafloor protection are depicted in Figure 6.23. The seafloor in front of the wall can also be protected by the installation of flow deflectors of miscellaneous designs (Longe et al., 1987; Ducker and Miller, 1996b). Examples are shown in Figure 6.24.

Bottom protection in front of a marine structure is typically designed to be resistant to:

- The effects of jets generated by side thrusters and/or main propeller(s)
- Wave attack (if relevant)
- Currents due to the natural flow of water

The bottom protection should be designed to certain performance criteria as follows:

- Permeability to water flow caused by change in pore pressures
- Prevention of passing underlying materials through the system

- Capability of being installed, maintained, and repaired under water
- Flexibility to adjust to settlements
- Mechanical strength to resist accidental impacts
- Durability in service
- Economics in construction and maintenance

The basic structural types of bottom protection used in contemporary practice include installation of layered stones (riprap), concrete blocks of miscellaneous configurations, prefabricated concrete slabs, miscellaneous flexible mattresses made of gabions, prefabricated asphalt blankets, willow mattresses, grout-filled mattresses, and so on. Most forms of protection include different kinds of geotextiles for reinforcement, filtration, and containment of structural materials.

In conclusion, it should be mentioned that to exclude seafloor erosion due to propeller(s) action or at least to minimize its effects, some ports introduced restrictions on the use of bow thrusters during ship docking or departure maneuvers (Chait, 1987). These restrictions limit the power of ship bow thrusters or prohibit their use in close proximity to the face of the berth. In some ports, thrusters are permitted only in the case of emergency (e.g., breaking of mooring line or a sudden wind squall). Quays sensitive to potential bottom erosion must be inspected regularly and at no more than 6-month intervals so that preventive action can be taken if shown to be necessary (PIANC, 1997). The alternative to the restrictive use of bow thrusters is installation of reliable protection of the bottom at the face of the dock. The economic aspects of these alternatives are discussed by Van der Weijde et al. (1992).

BOTTOM PROTECTION SYSTEMS

Geotextiles. Geotextiles employed in geotechnical engineering in general and in bottom protection in particular are synthetic materials such as nylon, polyester, polypropylene, or similar. In their common form, geotextiles are flexible, permeable, durable sheet fabrics, resistant to tension and tear, and able to retain soils. Geotextiles are relatively new materials in marine engineering practices. Hence, when considering the use of geotextiles in a bottom protection system, or in any other marine applications, the designer must be thoroughly aware of exactly what geotextiles can and cannot do and how to select the best product currently available. It must be noted that misuse of geotextiles may have a detrimental impact on structure performance.

There are two ways of dealing with the problem of material selection: analytical and empirical. The first requires detailed knowledge of the technical properties of selected material, such as permeability, filtration characteristics, and strength. However, the analytical approach often does not cover various potential secondary aspects of design and therefore needs to be supplemented by empirical observations or by reference to relevant practical experience.

The basic data needed to describe a geotextile as a construction material are the type, size, and shape of its constituent elements, manufacturing technique, permeability, and soil retention properties. Typically, all these data are found in the manufacturer's literature. Geotextiles are materials made essentially from fibers or filaments. There are a great variety of geotextiles, but the most common ones are woven or nonwoven fabrics made from polymer fibers or filaments. Woven geotextiles are relatively homogeneous. Their strain at failure is usually between 15 and 40%. Nonwoven geotextiles are fabrics laid in a nonregular manner and bonded together by thermal, mechanical, or chemical processes. Nonwoven fabrics are relatively thick with a high porosity value (close to 0.9). Compressibility of nonwoven fabrics under load is noticeable, but the entanglement of fibers is such that the porosity usually remains as high as 0.8.

The properties above allow both woven and nonwoven geotextiles to be used for filtration, drainage, separation, containment, or protection purposes. When incorporated in bed protection system, geotextiles are used primarily as filters. The durability of the material selected and its long-term performance must be the designer's concern. A geotextile's properties and its performance may deteriorate with time. First, characteristics of the physical and/or chemical environments can lead to degradation of the fabric material. Second, performance of a geotextile filter may be reduced due to clogging, fatigue, and creep, which inevitably occur to any type of filter structure. Therefore, these must be considered at the design stage.

It is advisable that at the specification stage, recommendations by U.S. Army Corps of Engineers (1984), RILAM Technical Committee on Geotextiles (1985), Ingold and Miller (1988), PIANC (1992), U.S. Federal Highway Administration (1995), and other relevant recommendations be considered. Hydraulic properties of a geotextile fabric depend on its permeability and filtration characteristics. A geotextile filter's primary function within a bed protection system is to prevent migration of soil particles out of underlying soil.

To avoid soil erosion due to transportation of soil particles through a geotextile and/or clogging of the geotextile by soil particles, thus reducing the free flow of water through the geotextile, the geotextile filter must be designed to satisfy two important criteria: soil retention and fabric permeability. The discussion above assumes that the geotextile must have an opening small enough to prevent soil particle migration and must be permeable enough to allow free flow of water without inducing uplift load on a bed protection structure. Essentially, the soil retention criteria for the filter are dependent on the grain-size distribution of the soil.

To protect a geotextile from being damaged during installation of a cover layer, for example, when dumping riprap, a granular sublayer 0.2 to 0.3 m thick is generally used. Granular sublayers can greatly relieve downward hydraulic gradients in the subsoil. It is important, however, to realize that a sublayer can promote uplift load beneath cover layers of low porosity: for example, tightly placed concrete blocks. In such instances it is better to omit the sublayer and take all precautions necessary to protect the geotextile during installation of a cover layer. Otherwise, the latter must be designed to resist the consequential uplift load. Once the required geotextile properties are established, they must be specified properly to ensure that an appropriate geotextile is used in the bottom protection system.

Riprap. Riprap is a flexible bottom protection made from loose stones. It generally comprises randomly placed quarried rocks having narrow limits ranging from approximately $0.8W_{50}$ to $2W_{50}$. A minimum thickness of riprap protection equivalent to two layers of W_{50} stones is normally provided. This is equivalent to a minimum thickness of about 1.5 to 1.8 times D_{50}, depending on the shape of the rocks. Here, W_{50} and D_{50} are the average weight and diameter of stones included in the riprap system. The layer thickness should obviously be greater than the largest stone in gradation. The riprap used for bottom protection is made up of durable stone of sizes typically ranging from 100 to 500 mm. However, in general, the size of the stone included in a riprap system depends on the hydraulic conditions, and in some instances, much larger stones may be required. Stones used in riprap are usually specified by weight and typically range from 10 to 500 kg. They are normally placed in one or two layers. Riprap stability is dependent on the shape, size, and weight of stones as well as on their gradation. For greater stability the stones used in the riprap should have a blocky shape and uniform size and lie within the layer thickness. An oversized stone protruding from a layer can cause local weakness that may lead to a progressive failure of the system. It must be noted that if not properly proportioned, rocks included in the riprap cover could be swirled by a propeller jet and heavily damage the propeller's blades.

Because of its rugged nature, riprap can cause damage to a geotextile filter layer; hence it must be installed with great caution. The best way to protect a geotextile layer from a concentrated load from stones is to use a granular sublayer of smaller stones or to use a geotextile resistant to concentrated load. Where rocks are locally not available, precast concrete blocks can be used instead. These blocks should be designed for a specific minimum crushing strength with a minimum cement content and durable heavy aggregates. When used with an appropriate granular sublayer material and/or geotextile filter, concrete blocks provide a stable armor system. The smooth faces, however, reduce the interlocking capabilities of adjacent cubical concrete blocks.

Information on the design and selection of granular filters is provided in Chapter 7. It is also found in U.S. Army Corps of Engineers (1984), Sherard and Dunnigen (1989), and PIANC (1992). The stability of riprap cover composed of large stones can be enhanced by grouting the space between adjacent rocks or by granular material (PIANC, 1992; Raes et al., 1996).

With a small bed clearance the effect of propeller jet–induced bottom velocities on stone stability may be profound. It may be sufficient to move stones several tonnes in weight. Propeller jet–induced bottom velocities may be considered analogous to highly turbulent flow situations that exist in stilling basins of hydraulic structures such as spillways. Under such flow conditions, Prosser (1986) recommends the following equations for dimensioning of riprap stones:

$$D_{50} = \frac{U_b^2}{23} \tag{6.10}$$

$$W_{50} = \frac{\pi W_s}{6} D_{50}^3 \tag{6.11}$$

where D_{50} is the mean size of stones, W_{50} the mean weight of stones, and W_s the density of stone ($W_s \approx 2050$ kg/m^3). *Note:* Equation (6.11) assumes that stones are perfectly spherical.

Examples of stone size and weight as a function of bottom velocities as determined by eqs. (6.10) and (6.11) are presented in Table 6.3. Typical details of riprap bottom protection are shown in Figure 6.25. PIANC (1997) suggests that armor stones should be specified by adopting weight limits 0.75 to 1.25 times the weight that is calculated for the design. The range of size can be wider to suit the output from the particular quarry or, if desired, can follow the standard grading recommended in CIRIA/CUR (1991). A thickness of armor stone protection equivalent to two layers of rock is normally recommended. It is usual to specify that 50 to 75% by weight of the rocks should consist of those within the top half of the range specified. Useful information on riprap design is given in RÖmisch (1997).

Concrete Blocks. Many different designs use precast concrete blocks to protect the sea-

Table 6.3 Stone size to give no scour for given bed velocity [Eqs. (6.10) and (6.11)]

Velocity (m/s)	Mean Size of Stone, D_{50} (cm)	Mean Weight of Stone, W_{50} (kg)
1	4.5	0.10–0.15
2	17.5	7.5
3	40	90
4	70	475
5	110	1850
6	160	5700

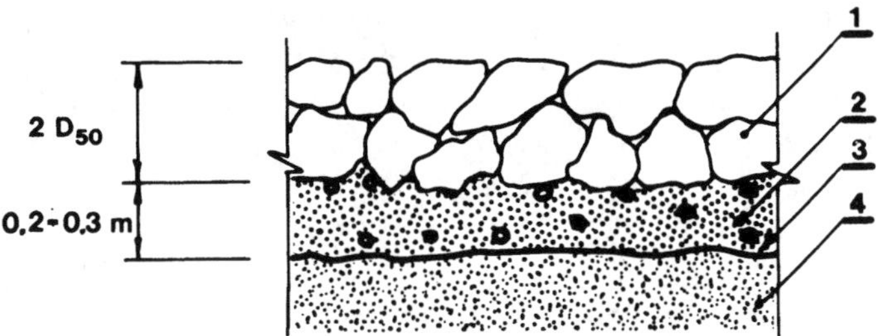

Figure 6.25 Detail of typical riprap protection. 1, Riprap; 2, fine stones (gravel); 3, synthetic fabric filter, layer; 4, natural bed.

bed in front of a quay wall from the effects of propeller-induced scour. To provide a stable armor system, these blocks are typically used with an appropriate sublayer and/or geotextile filter. Seafloor protection systems that utilize concrete blocks usually include loose noninterlocking blocks, cable-connected blocks, and blocks attached to geotextile.

Loose noninterlocking blocks of different geometric forms similar to armor rocks are placed directly on sublayers or geotextiles with no connection to each other. Similar to riprap, they stability of the cover layer formed by these blocks depends on the stability of the individual blocks. Therefore, if an unanticipated force would cause displacement of one block, this may be the start of a larger failure. Placement of loose blocks is a straightforward operation and is generally carried out by divers. Similar to riprap protection, stability of concrete blocks can be enhanced by filling the space between adjacent blocks with a granular material or grouting it with tremie concrete. This develops an interlocking effect between adjacent blocks and provides additional stability to the system by mobilizing the weight of adjacent blocks. When granular material is used, it is essential to have it remain in place, and it is therefore important to select blocks with inclined interfaces.

A loose interlocking block system requires a regular inspection and maintenance program. Cable-connected block systems are much more stable than loosely place concrete blocks. In these systems blocks are connected by cables running through the blocks in one or more directions. This ensures greater stability against block displacement due to propeller jet–induced forces or bed settlement. An additional advantage of cabling is that it reduces the risk of localized failure. Connecting cables can be either marine chains, galvanized wire ropes, or synthetic ropes. PIANC (1992) suggests that, in general, cabling should not be taken into account in the design for stability of cable-connected block systems against hydraulic loading and should be considered as the added factor of safety.

However, if a system were to rely on cabling for its stability under normal working loads, the risk of the blocks being repeatedly lifted off the sublayer or textile filter material and thus causing a pumping failure of a subsoil material or abrasion of cables and/or geotextile must be taken into consideration. The flexibility of the cabled system must also be considered. The flexural strength derived from the cabling means that blocks can bridge undesirable deformations in the subsoil, thus allowing some erosion in those areas. For this purpose the normal strength of a cable may be reduced by up to 40% when used as a sling and/or is subject to acute bending and poor splicing techniques.

The cabling facilitates construction work, as the blocks are made up into panels and placed under water using a crane and spreader frame.

Also, if repair is required, the full panel can be lifted out and replaced. Heavy armor rocks (2 to 4 tonnes) chained together, creating a cable-connected blocks system, has been used for bottom protection at the Frederikshavn, Denmark ferry terminal (Kristensen, 1987). The rocks were fitted with eye bolts used for placing and chaining the rocks together. A chain 20 mm in diameter was used to join every rock with another (Figure 6.26). The rocks were placed on a protection mat which consisted of woven sand pipes 5 cm in diameter. The mat was used as a sealing layer to protect subsoil material against waves and water flow effects. A mat made from propylene fiber fabric replaced a layer of a conventional granular sealer. It possesses great strength against mechanical stresses, which makes it suitable for use as a sublayer for rock armor. It weighs about 50 kg/m^2 and was made in the form of 5 × 5 m to 5 × 15 m covers. The total average thickness of this type of bed protection is about 1.2 m. The voids between rocks and sheet piles of a quay wall were filled with concrete. This type of bed protection system proved to be reliable against the propeller action of a ship with an engine power of 30,000 hp and more.

Gabions. By definition, a gabion is a steel-wire basket or polymer-mesh bag filled with loose stone. Gabions used for bed protection are normally fabricated in the form of box-shaped, compartmentalized containers made of polymer mesh or galvanized and/or polyvinyl chloride–coated steel wire mesh. The gabion container can be manufactured as a basket or as a mattress.

Boxes are commonly about 1.0 m high and 0.5 to 1.0 m wide. Gabions can also be fabricated in the form of a continuous mattress. The mattresses are typically 0.3 m thick and are generally used in a single layer to form a flexible free-draining protective cover to the bed surface. The mattress is divided into equally sized compartments that are assembled by lacing the edges together with steel wire or pol-

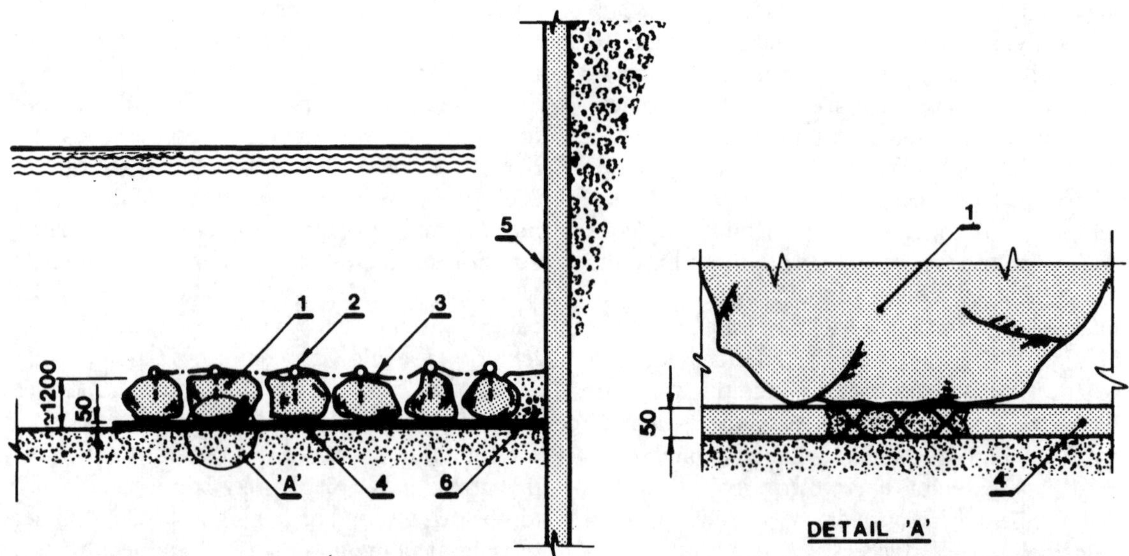

Figure 6.26 Chained armor rock system for bottom protection at the port of Frederikshavn, Denmark. 1, Rock (2 to 4 tons); 2, eye bolt; 3, chain, 20-mm diameter; 4, sealing woven sand pipe mat; 5, steel sheet pile bulkhead; 6, sealing tremie concrete. (After Kristensen, 1987.)

ymer rope. The individual units are then tied together and filled with stone. The lids are finally closed to form a large heavy mattress. To ensure best performance, properly sized filler rock should be specified. Flat stones and interior liners of any kind are not recommended in most cases.

For easy handling and shipping, gabions are supplied folded flat. They are readily assembled simply by unfolding and wiring the edges together and the diaphragms to the side. Then the adjacent gabions are filled to a depth of about 25 to 30 cm, and wire is placed in each direction and looped around two meshes of the gabion wall. This operation is repeated until the gabion is filled. The containers should be filled tightly to prevent any significant filler rock movement. They should be refilled as necessary to maintain tight packing. After the gabion is filled the top is folded shut and wired to the ends, sides, and diaphragms. Land prefabricated gabions are installed underwater by divers. An advantage of a gabion structure is that it can be built without heavy lifting equipment, using locally available stone.

A combination of the following features gives gabion structures some technical advantages over conventional rigid structures, particularly on sites where settlement or undermining is anticipated. Gabions are relatively flexible in the presence of unstable ground and/or moving water. This allows a gabion structure to settle and deform without failure and loss of efficiency.

Gabions are strong enough to withstand substantial flow velocity. They also are permeable and therefore usually do not require a sublayer drainage system. Gabions have quite high resistance to corrosion due to the well-bonded zinc coating on the wire (approximately 260 g/m^2). Gabions with polyvinyl chloride plastic coating are available for structures in locations where severe corrosion problems exist. Where the bottom is formed from soft soils, the bed protection may sink, due to the scour that may occur at the end of the protection system. In this case, the advantage of a flexible gabion system is that it will just settle without failure and subsequently adhere to the ground as scour occurs. To offer effective protection the gabion mattress must have sufficient length to reach the bottom of the expected scour. The projection of the mattress in front of the structure is generally estimated as equal to $1\frac{1}{2}$ to 2 times the expected scour depth. Normally, a thickness of 0.3 to 0.5 m is required so that the mattress can have enough flexibility and weight to remain adhered to the ground and remain stable even during the most severe propeller jet–induced velocities. Similarly to regular box-shaped gabions, gabion mattresses are fabricated on site near the site of installation and are then placed underwater.

The strength of a gabion mattress can be enhanced by asphalt mastic. An asphalt-grouted mattress retains its flexibility while the density of the fill is increased, with subsequent increase in the efficiency of the mattress performance. The thickness of a sand asphalt mastic consolidated mattress can be reduced to 0.15 m. Furthermore, the sand asphalt mastic offers additional corrosion and abrasion protection to the wire mesh. The minimum thickness of a regular gabion mattress (t_m) can be related to the stone size (D_{50}) and can be determined by formula (6.12); however, it should not be less than 0.3 m:

$$t_m(\text{min}) \times 1.8D_{50} \tag{6.12}$$

As recommended by PIANC (1992), the size of stone (D_{50}) can be determined by the formula

$$\frac{D_{50}}{h} = \left(\frac{U_b}{B_1(k'\psi_{\text{cr}}g\Delta_m h^{0.5}}\right)^{2.5} \tag{6.13}$$

where h is the depth of water, U_b the propeller-induced velocity at bed level, B_1 a factor that depends on flow condition $B_1 = 5$–6, k' a slope (α) reduction factor $= (1 - \sin^2 \alpha/\sin^2 \phi)^{0.5}$

(here ϕ is the angle of internal friction = 45°; for a horizontal bed, $k' = 1$), ψ_{cr} the Shields parameter (for stone-filled gabions, $\psi_{cr} = 0.1$), and Δ_m the relative density of fill material. It is obvious that in no case may the stone size be less than the size of the gabion mesh opening. If $U_b > 3.0$ m/s, a granular sublayer (0.2 to 0.3 m) should be incorporated into a gabion bed protection system. In all other cases it is sufficient to place the gabion system directly on the geotextile filter only.

The important advantages of gabion bottom protection are that it can be built without using heavy construction equipment, it is flexible, and it can easily be repaired or strengthened with impregnation of concrete or asphalt mastics. The basic disadvantage of gabion protection is its relatively high cost. Also, the long-term durability of the steel wire in severe marine environment may be in question. Even zinc or polyvinyl chloride–coated wire may be subject to deterioration at gaps in the coating which can be created by placement of rough angular stones. A damaged basket will eventually develop a hole and start losing stone.

Bergh and Magnusson (1987) reported the use of gabion mattresses for bottom protection at a ferry terminal in Stockholm, Sweden. There the design propeller jet velocity generated by propellers of large ferry vessels with an installed power of 31,000 to 36,000 hp was on the order of 10 to 12 m/s. Theoretically, these velocities would require the use of stones 4 to 6 m in diameter to protect the bottom from severe erosion. It is quite obvious that the use of rock fill of this magnitude is not practical. In fact, rock-fill protection is not likely to be a suitable solution if bottom velocities exceed 5 m/s.

At the aforementioned ferry terminal, several types of scour protection systems have been tested. One of them was a gabion mattress 0.3 thick (Figure 6.27*a*). Two years after gabion installation it was found that at free edges exposed to the direct propeller jet, the mattress had rolled up. Subsequently, at these locations the outer mattresses were completed with new mattresses placed underneath the previous ones and fixed to them to increase the weight at free edges. An inspection conducted 3.5 years later found the gabions in question in generally good condition. However, near the outer end of the mattress there still was a slight tendency of rolling up over a length of approximately 15 m.

Willow (Fascine) Mattresses. This type of bottom protection was used traditionally in inland ports and waterways. Its basic advantage is economy and usually local availability of all construction components: willow bushes and rock materials. The modern version of the traditional willow mattress is a flexible system comprising a bundle of young willow stems bound together to form a two-way square grid usually 1.0 × 1.0 m and approximately 0.5 to 0.7 m high. This is attached to a strong geotextile on the underside. The mattress is sunk into position on the bottom by loading it with quarry run up to 100 mm in diameter. A final layer of riprap is then added as required.

Cable-Connected Concrete Blocks. Numerous concrete block systems connected and/or anchored by cables running through the blocks in one or two directions have been developed and used for protection of the bottom from the effects of a ship propeller–generated water jet. Some of these systems are discussed in detail in PIANC (1992). The most crucial part of these systems are the connecting cables. They could be made from either steel or synthetic fibers. The latter has the advantage of not being susceptible to corrosion in a saltwater environment while having a strength comparable to that of steel wire ropes. A cabled concrete block system achieves stability from its own weight, and the cabling ensures greater stability against block displacement. Mattresses made of cabled blocks are installed on a carefully pre-

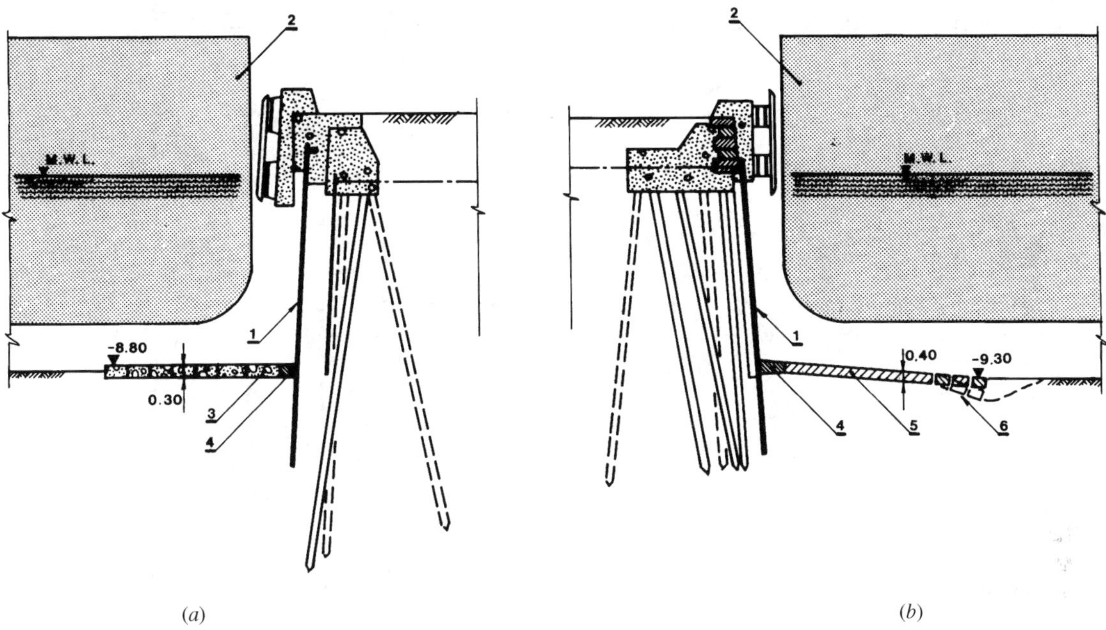

Figure 6.27 Port of Stockholm ferry terminal (Silja basin): (*a*) section through the north quay; (*b*) section through the south quay. 1, Quay wall; 2, ferry; 3, gabion mattress, 2 × 2 × 4 m; 4, tremie concrete; 5, precast concrete slab, 9 × 9.5 × 0.4 m; 6, concrete flap, 3 × 2.5 × 0.4 m. (After Bergh and Magnusson, 1987.)

pared layered drainage system, geotextile filer, or combination of both using a crane and a special spreader beam or frame.

Concrete Blocks Connected to Geotextile. Several structural system of this type have been developed (Kristiensen, 1987; Sawicki et al., 1998). In these systems, blocks may be either glued to the geotextile by a special type of adhesive or joined by mechanical means. Essentially, the geotextile fabric must possess adequate strength and durability as a connector and also perform as a filter. One practical example of this type of bottom protection is depicted in Figure 6.28. It was installed for protection of the seafloor in front of the Frederikshavn ferry terminal in Denmark. Concrete blocks were composed of double hexagonal concrete units ($f'_c = 40$ MPa) bolted together by means of eyebolts through the geotextile (polypropylene fiber) covered. The individual blocks were assembled in two parts; the upper part (50 kg per unit) is formed to reduce resistance to the flow of water, and the bottom of the lower part is concave, thus forming a vacuum that binds a block to the bed. The weight of the lower part is 83 kg. The average weight of the standard part of the cover is 420 kg/m^2. For better stability of the cover edges, they are made heavier than a standard cover. The weight of the cover along its outside edges is approximately 500 kg/m^2.

Standard sheets of cover 3.0 × 6.45 m were assembled on a leveled surface. The weight of one sheet was equal to 8 tonnes. Each cover unit was lifted by crane furnished with a special lifting device able to pick up the preassembled cover at each block. Finally, the adjacent standard sheets of cover were joined together as shown in Figure 6.28 and secured to the

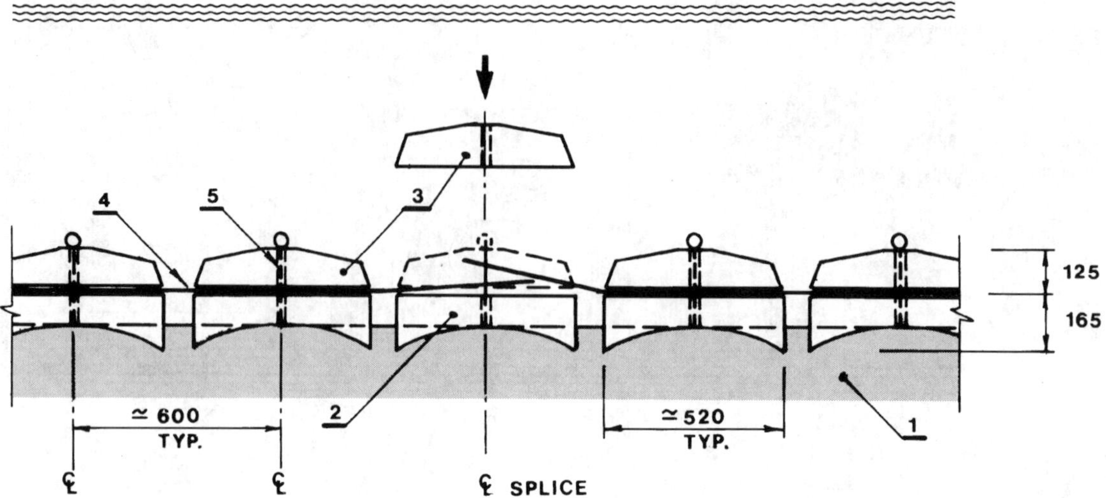

Figure 6.28 Detail of bed system comprising concrete blocks connected to geotextile as used for protection at the Frederikshavn ferry terminal, Denmark. 1, Seabed; 2, bottom concrete block; 3, upper concrete block; 4, geotextile; 5, eye bolt.

sheet pile wall by chains or hooks fixed to brackets welded to sheet piles. Design analysis of the foregoing system can be carried out based on recommendations by Sawicki et al. (1998).

Prefabricated Concrete Panels. This type of bottom protection is reliable, although not necessarily the most economical solution. The size of precast panels is controlled basically by the heavy lift equipment available. The panel reinforcement is usually designed to take stresses associated with handling during transportation and installation. The slab outer edge typically should be protected against undermining by the scour. For this purpose, riprap or a different kind of flexible cover attached to the outer edge is generally used. The space between the slab and the quay wall structure is normally sealed with tremie concrete. A typical example is illustrated in Figure 6.27*b*. There, precast concrete slabs 9.5 m wide, 9.0 m long, and 0.4 m thick with the outer end protected against propeller-induced scour by three concrete flaps 3.0 × 2.5 m have been installed in front of the sheet pile wall. To reduce uplift forces, flaps were furnished with drainage holes. Before placing slabs, the bottom was leveled by trailing a heavy steel beam over it. The slabs, each weighing 85 tonnes, were lifted by a floating crane and placed into their design position about 1 m from the sheet piling and about $\frac{1}{2}$ m apart. Subsequently, spaces between sheet piling and between adjacent slabs were sealed with tremie concrete. This type of construction proved to be satisfactory even in areas exposed to very high velocities. As expected, erosion had occurred at some of the flaps, which then moved down as anticipated, arresting further erosion.

Concrete-Filled Fabric Containers. High-strength, water-permeable fabrics made of synthetic fibers are being used as forming material for concrete in a wide variety of applications, particularly those involved in placement underwater. An early use of fabric as a form for concrete was proposed in a patent issued to J. Store

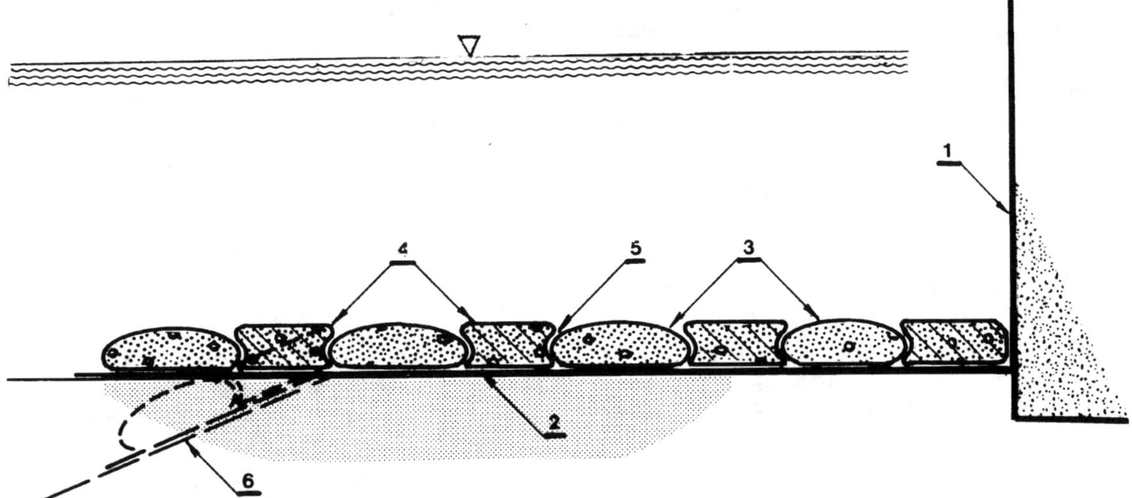

Figure 6.29 Bottom protection by concrete-filled geotextile tubes. 1, Quay wall; 2, geotextile filter; 3, tube laid first; 4, subsequent unit; 5, tongue-and-groove joint; 6, contour of potential scour.

of Norway in 1922. Since then it has been used in a variety of forms and applications. In contemporary marine and coastal engineering practices, fabric containers such as bags, tubes, and mattresses of different constructions, filled with concrete, are used effectively for bottom protection and underwater repairs of various kinds of marine structures. In bottom protection systems they are typically placed on a single or double layer of filter material.

Although theoretically, choice of fabric is independent of concrete mix proportions or placement methods, in practice, the fabric used for concrete forming is usually limited to specialized applications and the choice of mix and placement method should be appropriate to these applications (Lamberton, 1989). When concrete is injected in fabric forms underwater, the pressure within the delivery hose must be sufficient to expel the water through the water-permeable fabric and replace it with concrete. At the same time, the external water must not penetrate the system to contaminate the fresh concrete. Obviously, the conditions above apply only where the form is fully collapsed prior to the injection of concrete; if the form is suspended in an open manner, a normal tremie concrete technique can be employed. For reasons of accessibility, nearly all concrete is placed in fabric form by a pump. In North American practice, only woven fabric is regularly used as a concrete-forming material. If fully filled with concrete, the smooth, rounded contours of the concrete-filled bags may present an interlocking problem. To prevent bags from sliding upon each other and to ensure good contact with each other, they are usually filled between 50 and 70% full. Since the cement paste is normally squeezed out through the porous bags, a certain cementation within the bag structure takes place. The bags are normally placed in bond similar to block walling.

To provide for better and stronger performance, the diver can drive reinforcing bars through the bags. The bags also can be stabilized by cabling. An example of bed protection by concrete-filled tubes is shown in Figure 6.29. It is important that concrete or mortar "fluid" run to all points in the container readily without the aid of vibrators. For this reason the tubes are placed parallel to the face of the quay wall, being spaced continuously approximately one tube apart. Then the space between the tubes is filled with similar units. This makes a tongue-and-groove-like connection with those already in place. It should be noted that tubes at the edge could slump into a scoured trench. Therefore, it is necessary to continue the required protection far enough to account for potential scour at the protection end. For this purpose it is better to have edge tubes relatively shorter in length to cope with different scouring dimensions along the length of high-velocity jets. For better performance, edge tubes can be cabled; superplastisizers should be included in the mix, which reduces the water content required and hence maintains the strength. The use of ground blast furnace slag to replace a large proportion of the cement can make in situ concrete-filled tubes very economical. For short-term applications, nonsetting materials such as sand, bentonite combined with 10% cement, and others can be used as in-fill materials.

The equipment used for concreting is very simple; it usually comprises a mobile hopper or concrete pump and a boat or barge for feeding out the fabric sheath. The concrete mix can be supplied from ready-mix trucks that pour into the hopper continually. Woven textile bags and tubes are available in various diameters and lengths. Tubes can be placed on the bed and then grouted with concrete. They are usually placed on a woven filter cloth. A small tube, factory-stitched to the outer edge of the filter cloth, can provide toe protection.

Mattresses are usually designed for placement directly on a leveled bed. They are laid in place when empty, joined together, and then pumped full of concrete. This results in a mass of pillowlike, rather flexible concrete units with regularly spaced filter meshes for hydrostatic pressure relief. The filter is usually a normal geotextile material, as described previously. In general, polyester fabric must not be used for any kind of fabric containers that will be filled with concrete, as this is susceptible to attack by chemical reactions generated from hardened concrete. For strength and/or stability, concrete-filled geotextile containers used in marine applications must be designed to withstand the impact of different kinds of hydraulic loads (e.g., propeller jet, wave-induced currents and uplift forces).

In concluding this section, it should be noted that in the view of some experts, introduction of bottom protection in front of waterfront structures is not always justified economically; some port authorities prefer not to spend money on bottom protection, but rather, limit the use of bow side thrusters to minimize scour problems. However, in busy ports, reliable bottom protection is usually considered necessary (van der Weijde et al., 1992).

WIDTH OF BOTTOM PROTECTION SYSTEM. As discussed earlier, a ship propeller(s) and/or side thruster may induce high-velocity currents able to substantially undermine a berth structure. Essentially, where required the seafloor in front of the marine structure should be properly protected from erosion. The size of the protection depends on many factors, but mostly on the size and power of the maximum design ship, type of the structure (e.g., solid face walls or open construction), roughness of bottom protection, underkeel clearance, and the economy of bottom protection. For example, in some instances it may be more economical to construct a less expensive bottom protection system but carry out inspections and, if necessary, repair works at regular intervals, as opposed to installation of a maintenance-free but expensive protection system (Verhey et al., 1987). In all cases, however, the minimum size of the protection system must be sufficient to protect the structure from being adversely affected by bottom erosion. For example, in the case of a gravity quay wall, it must prevent formation of substantial scour in front of the wall that can adversely affect wall stability, and in the case of a sheet pile bulkhead, the amount of passive pressure near the toe must not be reduced to an unacceptable value. As a mater of fact, substantial scour in front of a sheet pile wall may result in a heavy irreversible overstress of both the sheet pile wall and its anchoring system. This is discussed further later in the chapter. According to Ducker and Miller (1996b), depending on wall design and underkeel clearance of the ship, the extent of the area to be protected in front of the wall is between three and four times the propeller (stern screw) diameter. The best result in determining the bottom protection width can be obtained from physical model tests.

Improvement of Quay Wall Stability. Stabilization of a gravity quay wall from sliding, overturning, or other excessive movements can be required due to miscellaneous wall conditions, (e.g., overstressed foundation material, overloading due to dysfunctional drainage system, change in operating fleet of vessels resulting in increased mooring, impact and related loadings, and/or use of heavier cargo handling and hauling equipment). In general, before wall stabilization measures are considered, the wall's structural integrity should be ensured as discussed earlier: the wall's drainage system functions as intended, the wall structure is consolidated and/or repaired as required, and last but not least, the seabed in front of the wall is protected from significant scour induced by the ship's propellers. Wall stability can be enhanced utilizing a variety of techniques. These include basically, but are not limited to, wall underpinning (alternatively, grouting of foundation soil), usually in combination with the installation of posttensioned ground anchors, reduction in soil lateral pressures against the wall, or a combination of both. These methods are discussed briefly below.

WALL UNDERPINNING IN COMBINATION WITH INSTALLATION OF POSTTENSIONED GROUND ANCHORS. A typical example is shown in Figure 6.30. In this case, the wall is stabilized by underpinning and the installation of ground anchors. The former helps to absorb the increased bearing stresses upon foundation soil, due to an increase in soil lateral pressures against the wall, and the latter produces increased righting moments, which results in overall enhanced wall stability. An effect similar to underpinning can be achieved by grouting the foundation soil with cementitious or chemical grout or installation of underpinning piles, as shown in Figure 6.30.

Recently, jet grouting columns have been used effectively in many ports worldwide for underpinning existing old gravity quay walls. One practical example is illustrated in Figure 6.31. In this particular case, 81 soilcrete columns 1.0 m in diameter were installed by a jet grouting technique to underpin an existing quay wall. It was achieved by drilling a 150-mm-diameter pilot hole to the required depth of about 2.8 m below the base of the wall, and inserting in this hole a 90-mm-diameter rotating stem through which a mixture of three parts pulverized fuel ash to one part cement was discharged into the soil, which had previously been cut by a water–air jet (the water delivery rate was 70 L/m at 400 bar pressure). The foundation soil comprised a stiff brown fine gravelly clay that became very sandy and silty locally (glacial till). When the rear row of columns was installed, they were redrilled, to take about 20-m-long rock anchors secured into the sandstone bedrock. Initially, these anchors were stressed to 1.25 times their design load of 630 kN, and subsequently distressed to the design load. The job was completed successfully with a minimum disruption of dock operations. More practical examples of the use of jet grouting technology for the enhancement of gravity wall performance are given in Van Damme et al. (1999) and later in this chapter.

REDUCTION IN SOIL LATERAL THRUST. In soil-retaining structures, in most cases the horizontal component of soil pressure is the principal

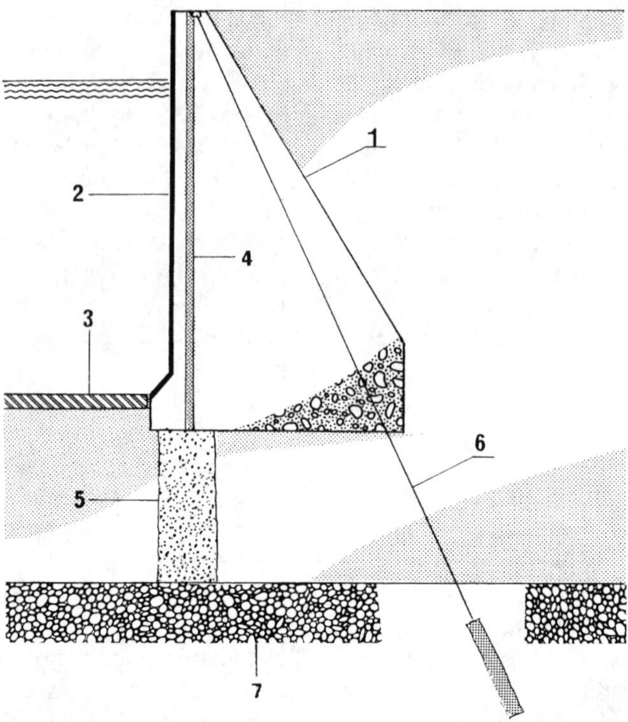

Figure 6.30 Typical arrangement for stabilization of gravity quay wall. 1, Consolidated body of quay wall; 2, repaired/rehabilitated face; 3, new (or existing bottom protection); 4, grouted jet grout pilot bore hole (approximately 150 m in diameter); 5, soilcrete column produced by jet grouting process; 6, ground (rock) anchor; 7, firm soil (bedrock).

load acting against the wall structure. In general, the soil pressure depends on soil geotechnical properties such as density, cohesion, and the angle of internal friction. Hence, where required, the wall stability may be enhanced by improving parameters of backfill material. The latter is usually achieved by replacement of poor-quality backfill material by good material (e.g., quarry run, rock fill, coarse-grained sand) or by in situ improvement of the geotechnical properties of the existing backfill material. The lateral soil thrust can be reduced by installation of miscellaneous pressure-relieving systems in a soil's active pressure zone, as shown in Figure 6.32.

Soil Replacement. Replacement of poor soil with good granular material such as selected rock fill, gravel, or coarse-grained sand is the easiest and in a way the most popular (however, not necessarily the most efficient) method of reduction in soil lateral pressure. To be effective, the new fill has to be extended far enough beyond the slip line (Figure 6.32*a*). If left within a potentially unstable slip wedge, new fill may result in even heavier lateral soil thrust against a wall. The replacement material should be placed from the edge of the excavation toward the wall. This will allow for maximum mobilization of shear forces within the body of a fill. Replacement of regular back-

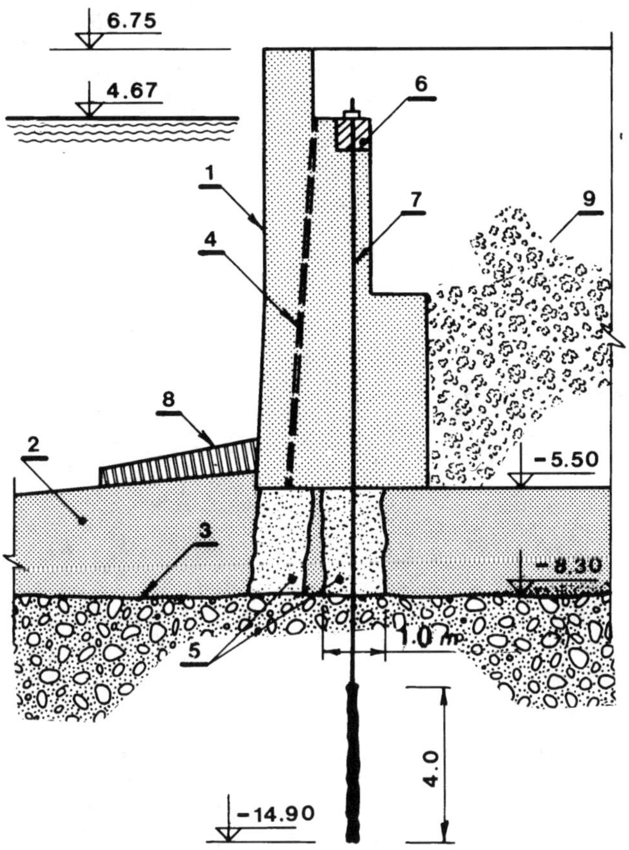

Figure 6.31 Restoration of the Brocklebank dock west quay wall. 1, Quay wall; 2, stiff brown fine gravelly clay becoming very sandy and silty locally (glacial till); 3, sandstone bedrock; 4, jet grout borehole, 150-mm diameter; 5, jet grout column (soilcrete); 6, cast-in-situ new concrete beam; 7, new posttensioned rock anchor; 8, antiscour mattress; 9, Miscellaneous backfill. (After Coomber, 1987.)

fill material with good-quality rock fill may result in substantial reduction in soil lateral pressure in the replacement zone. Excavation of poor soil from behind a gravity retaining wall will almost always result in reduction in soil-bearing stresses due to reduction in both vertical and horizontal soil pressures. For detailed information, consult Tsinker (1995).

In Situ Improvements to Existing Soils in Active and Passive Zones. This is usually achieved by vibration, installation of stone columns, soil mixing, and other methods. In general, soil vibro technology is readily available and is used to solve a wide range of static, dynamic, and seismic foundation problems through the use of various vibration systems to densify and/or reinforce the soil. Modern vibro systems are able to treat the soil in dry and underwater conditions to a depth of 65 m or more below working grade. Basically, vibro systems are used for soil vibro-compaction and/or vibro-replacement. The former is used to densify relatively loose granular materials

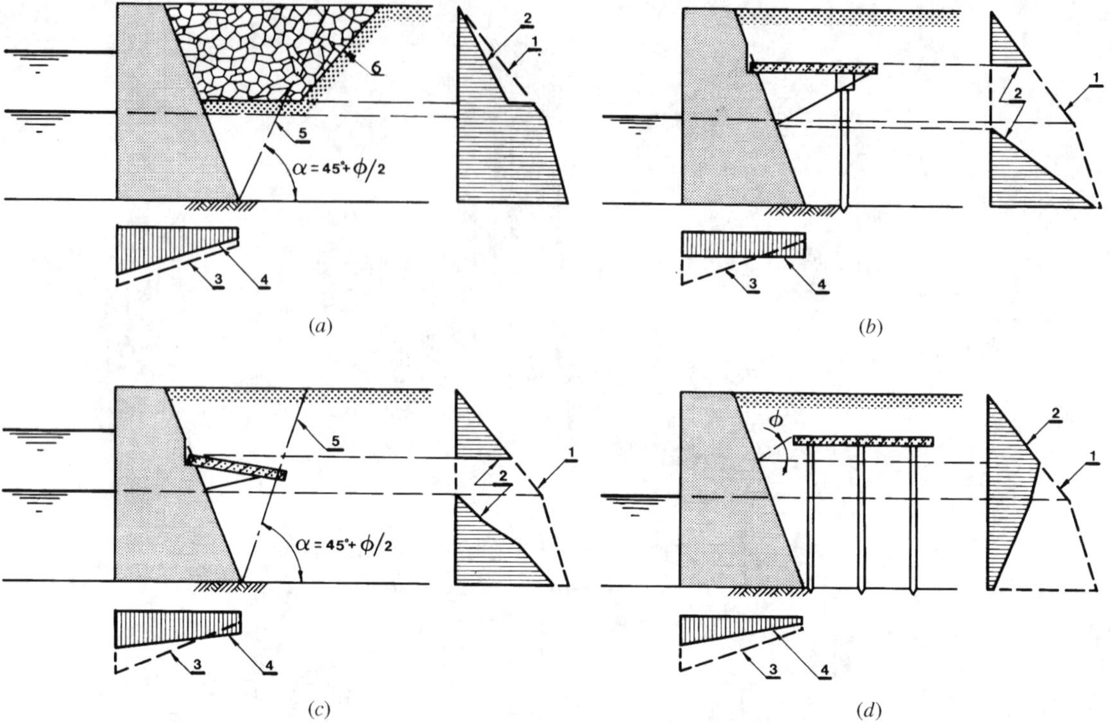

Figure 6.32 Methods of reducing lateral soil pressure on gravity retaining wall: (*a*) replacement of poor backfill soil with good-quality granular material; (*b*) pressure-relieving slab supported on wall interior and piles; (*c*) pressure-relieving slab supported on wall interior and backfill; (*d*) pressure-relieving piled platform placed behind a wall. 1, Original soil lateral pressure diagram; 2, new soil lateral pressure diagram; 3, original bearing pressure diagram; 4, new bearing pressure diagram; 5, slip line; 6, excavation. (From Tsinker, 1989.)

such as sands and gravel. In cohesive soils that do not respond well to vibration, improvement may be achieved by vibro-replacement of the cohesive soil by stone columns.

Vibro-Compaction (Figure 6.33*a*). This method of soil densification is usually accompanied by water jetting. It reduces intergranular cohesion, allowing the soil particles to move into a denser configuration, typically achieving a relative density of 70 to 80% and more. The improvement in soil characteristics depends on the type of granular soil (e.g., fine-, medium-, or coarse-grained sands), spacing between vibration points, and time of compaction. In most cases, the spacing between vibration points is between 2 and 4 m, with a center arranged on a triangular or square pattern. Vibro-compaction not only improves soil parameters such as angle of interval friction, but also mitigates its liquefaction potential.

The process of soil vibro-compaction usually starts with water jetting of the vibration probe to the desired depth. In the process some small particles are washed away. The vibrator is surged up and down as necessary to agitate the surrounding granular soil. Under the action of induced dynamic forces, the soil particles surrounding the base of the vibrator are rearranged to a denser state. The vibrator is raised

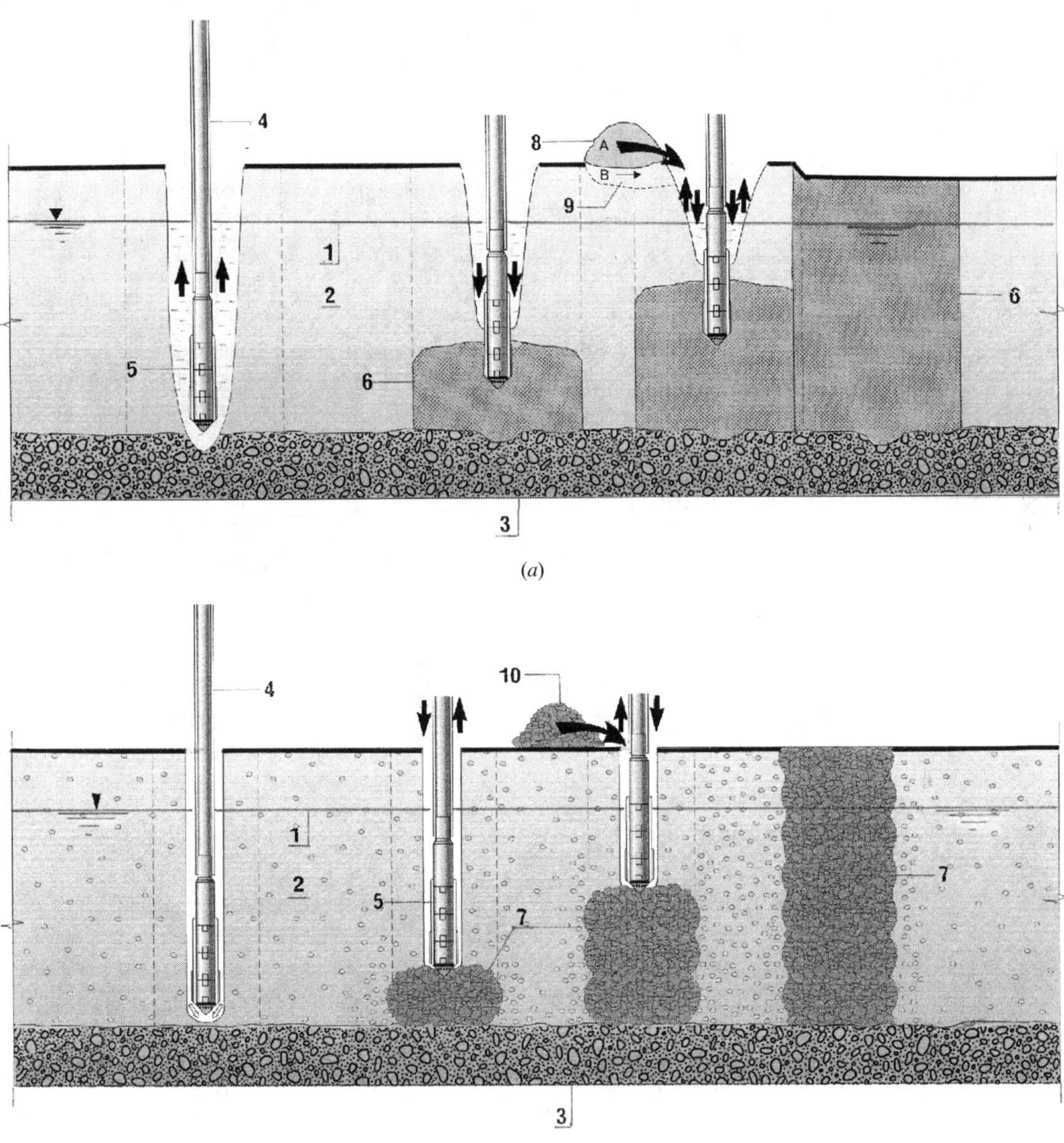

Figure 6.33 Standard methods of soil improvement: (*a*) vibro-compaction; (*b*) vibro-replacement of existing soil with stone columns. 1, Water table in backfill soil; 2, existing backfill soil; 3, bedrock; 4, probe; 5, vibrator; 6, densified soil; 7, stone column; 8, imported material; 9, in situ material used to compensate reduction in volume of soil due to its densification; 10, imported stone material.

incrementally as compaction of the underlying soil is achieved. As a result of soil densification, the original surface of the dock may drop by 5 to 15% of the treated depth. In this case, some additional granular material is brought in to compensate for the reduced volume. Finally, the surface of the treated area is densified by a surface compactor.

Soil densification in the active soil pressure zone of soil-retaining structures should be carried out with caution; compaction in too close proximity to a structure may result in a significant increase in the lateral soil pressures acting on the retaining structure (Tsinker, 1997). To prevent this from occurring, the relevant construction specification should stipulate the critical distance behind the structure where no compaction must be carried out. This can best be determined from field testing. In one specific case involving a steel sheet pile bulkhead, little or no increase in lateral soil pressures has been developed when deep vibratory compaction of backfill soils was carried out at a distance of 4 m from the bulkhead centerline (Castelli, 1991). For information on water jetting, readers are referred to Tsinker (1997).

In conclusion, it should be mentioned that a wide variety of soil compaction equipment and compaction procedures are readily available for vibratory compaction of granular soils. In general, specialized equipment, particularly high-horsepower, horizontally oscillating vibrofloats could be more efficient than probes powered by vibratory hammers. The equipment and procedures described above are especially effective for compaction of clean coarse-grained sands, gravel, and other coarse-sized materials.

Vibro-Replacement. This technique is used for improvement of cohesive, mixed layered, and other soils that do not densify easily when subjected to vibration along. A representative sketch of vibro-replacement procedure is depicted in Figure 6.33*b*. The vibro-replacement technology is used for replacement of soils that responded poorly to densification with columns made of dense coarse-grained materials (e.g., crushed stone, gravel, coarse-grained sand). The columns act as a vertical drain through which excess pore pressure dissipates readily. Column spacing can be arranged to suit varying combinations of loads, soil types, and performance requirements. Typically, columns are spaced at 2 to 3 m in both directions. Two basic methods are used to install the crushed stone (gravel) columns: the wet, top feed method and the dry, bottom feed method. In the former technique, jetting water is used to liquefy the soil material, stabilize the probe hole, and feed the stone material down into the hole. In this process it must be assured that the stone material reaches the tip of the vibration. This is the most commonly used and usually the most cost-efficient method of column installation. However, in some instances handling the spoil generated by the water jetting process may make this method more difficult to use on confined sites or in environmentally sensitive areas.

In the wet vibro-replacement method, a standard vibrator probe is used but with the addition of a hopper and supply tube, which is used to feed the stone material directly to the tip of the vibrator. Bottom feed vibro-replacement is basically a dry operation where the vibrator remains in the ground during the column construction process. Part or full elimination of flushing (jet) water eliminates the generation of spoil, thus extending the range of sites treated by this method. The treatment depth may reach 20 to 25 m. A combination of the wet and dry techniques is also used. In this case, the lower portion of a stone column is created by the dry vibro-replacement technique as described, and the upper part is installed by feeding the stone material into the probe's jet hole using the water-jet method.

In some instances where layers of stiff cohesive soils are present, each column hole can be preaugered to penetrate at least 0.5 m into the underlying firm soil. Stone backfill is then added and densified in a continuous operation, forming a compact granular (stone) column tightly interlocked with the surrounding soils, enhancing both shear and settlement characteristics. Stone columns placed behind a soil-retaining structure may substantially reduce the lateral thrust against the wall, increase soil's bearing capacity and substantially enhance its liquefaction potentials. A complete stone column system is normally overlaid with geotextile taped with a stone filter. The system is typically extended landward to cover the area designated for wick drains. For more information on potential benefits of the vibro-replacement method, readers are referred to Sonu et al. (1993) and Priebe (1995).

Soil Mixing. Soil mixing is a ground improvement technique involving the mechanical in situ blending of soil with cementitious materials using a hollow-stem mix tool. The injection of binder material may be in the form of either a slurry or a dry powder. Sets of one to six shafts, with mixing tools up to 2.5 m in diameter, are used to mix soft and loose soils to depths up to 30 m. As the tool is advanced into the soil, the hollow stem is used as a conduit to pump grout and mix it with the soil in contact with the tool. Penetration of the mix tool is set so that 100% of the soil to be treated comes into contact with the tool. Single column and/or panels are created with the process as the tools are worked in overlapping configurations. A range of unconfined compressive strengths between 70 and 6900 kPa is possible, depending on the native soil type. A characteristic case history involved soil mixing for improvement of backfill material is reported by Burke et al. (2001).

Pressure-Relieving Slabs and Piled Platforms. Pressure-relieving slabs placed between retaining structures could be very effective in the reduction of bearing stresses on foundation soil and increase a wall's sliding and overturning stability. If the water level permits, for better performance these slabs should be placed at a depth of about one-fourth the wall height. They should be of sufficient length to provide for the required shielding of the wall from the effects of soil located above the slab. The minimum length of such a slab is usually equal to $0.6H$ for gravity walls (H is the height of the wall from the dredge line to the top). The slab may be supported on the wall by one end while the other end is supported on either the backfill or piles (Figure 6.32*b* and *c*). In this case, a combination of reduced horizontal thrust and effect of the vertical load transmitted from slab to the interior face of the wall will result in smaller and better distributed bearing pressures upon foundation soil. If extended far enough beyond the slip circle, the pressure-relieving slab will also act as an anchorage to the wall. In the latter case, to reduce the value of the bending moment in the slab due to increased span, it may be broken into two parts jointed together through a hinge(s). In some cases, piled platforms independent of the wall structure or incorporated into it are used for the same purpose (Figure 6.32*d*). Practical examples of the use of pressure-relieving platforms and slabs are given in Tsinker (1995, 1997).

6.6.2.4 Repair and Rehabilitation of Sheet Pile Bulkheads. Most of the problems reported in new or relatively new sheet pile bulkheads are caused either by incorrect design or either poorly constructed anchor system. Less frequently, bulkhead failures are attributed to highly overstressed sheet piles due to undermining by overdredging or propeller-induced scour. However, the most typical cause of sheet pile bulkhead deterioration is corrosion of steel in a highly aggressive marine environment. Repair or rehabilitation of sheet pile bulkheads usually presents quite a complicated challenge. This is because stress and/or deflection conditions inherited by sheet pile bulkhead components as a result of overloading, material deterioration, scour, overdredging, or other reasons are practically irreversible. The latter is best illustrated by Figure 6.34. As can be seen, scouring or overdredging reduced the value of the original passive pressure in front of the anchored sheet pile bulkhead (E_p to $E_{p(1)}$), increased the bulkhead active span ($l_1 > l$), and produced additional deflection (Δ) and higher values of stresses in sheet piles and the anchor tieback rod ($R_{A(1)} > R_A$). Restoration of the foundation soil to its original level produces insignificant active pressure (E_A) in addition to remaining passive pressure ($E_{p(1)}$) and therefore cannot help much to restore a bulkhead's passive pressure to its original value (E_p). It also cannot bring the stress level in sheet piles and the value of anchor forces to their prescour level because it is practically impossible to change sheet pile deflection, that is, to bring it to the prescour position.

Practically, the situation as described could be improved by installation of an additional anchor system located below the existing anchors. In this case, new anchors must be tensioned to prevent or limit sheet pile deflection under design live load. This would minimize additional stress in a sheet pile structure. New ground anchors are typically drilled through sheet piles into the backfill. The problem with such repair or rehabilitation, however, is that in most practical cases this additional anchorage must be placed under water. Consequently, this type of construction may involve the use of divers or the installation of a special portable cofferdam around the working area. Examples are given in Tsinker (1995) and most recently in Alexeev (2001).

Typically, an additional anchor system is designed in the form of either posttensioned ground anchors or conventional, often posttensioned tieback rods secured at the existing anchor system or a new system, such as a sheet pile wall, concrete blocks (deadmen), or prefabricated concrete slabs. The representative examples are illustrated in Figures 6.35 and 6.36. The anchors themselves could be of conventional regular round-shaped steel bars or made of prestressed high-strength steel tendons. Both systems must be reliably protected from corrosion. Prestressed anchors made from high-strength tendons used in the construction

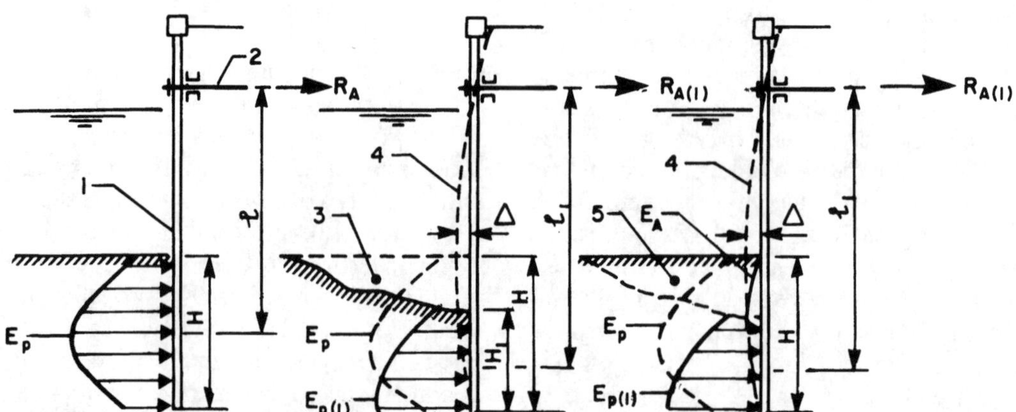

Figure 6.34 Changes in soil pressures in passive zone of sheet pile bulkhead due to scour effect. 1, Sheet pile; 2, anchor system; 3, scour; 4, deflection line; 5, restored foundation material. E_p, prescour passive pressure; $E_{p(1)}$, post scour passive pressure ($E_{p(1)} + E_A) < E_p$; E_A, Active pressure due to restoration of eroded foundation material; R_A, prescour anchor force; $R_{A(1)}$, postscour anchor force ($R_{A(1)} > R_A$); Δ, scour-induced deflection; l, prescour effective span; l_1, postscour effective span ($l_1 > l$). (From Tsinker, 1989.)

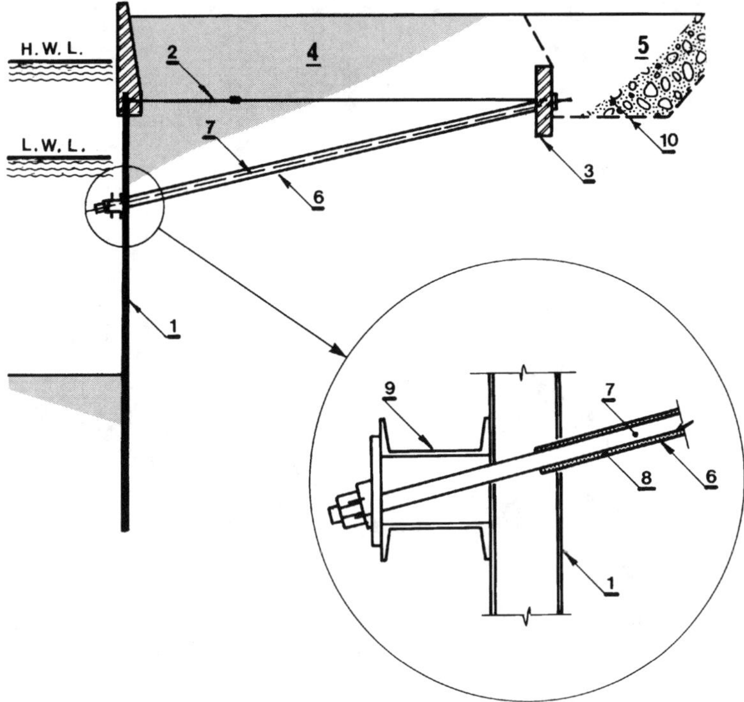

Figure 6.35 Secondary anchor system attached to existing anchor block(s). 1, Existing sheet pile bulkhead; 2, existing anchor tieback system; 3, existing anchor block(s) of miscellaneous constructions; 4, existing backfill soil; 5, existing backfill soil replaced with a good granular (or rock) material; 6, casing; 7, new tie rod; 8, grout; 9, new wale; 10, excavation of existing backfill soil.

of anchor sheet pile walls are discussed in Van Rooten and Zielinski (1996). Typical damage to sheet pile bulkheads involves the corrosion of steel sheet piles, deterioration of concrete sheet piles, and distress or failure of anchor systems. Flexural failure of the sheeting in general is a rare event. If corrosion of the sheeting is a problem at a given site, it is likely to progress most rapidly in the tidal/splash zone, where alternate wetting and drying occur. The presence of corrosion-induced holes may affect dock operation in a variety of ways. For example, the presence of holes in sheet piling may result in backfill loss, settlement of apron, damage to the crane truck, and so on.

Reinforced concrete sheet piles typically deteriorate due to corrosion of reinforcing steel ship impact, abrasive effects caused by waterborne debris or a combination of the above. The problem of sheet piling deterioration is particularly acute for bulkhead structures operating in a warm saltwater environment. Ultimately, sheet pile deterioration, which generally originates in the splash or tidal zone area, provides access of the water into the bulkhead's interior. This in turn may trigger severe corrosion of structural components and fasteners of anchor system (e.g., wale, tie rod, and connecting bolts and welds). It must be stressed that the wale is a very important part of a sheet pile bulkhead. It is typically located just above the mean high tide level, where in most cases the rate of steel sheet pile corrosion or concrete deterioration reaches its maximum value. In

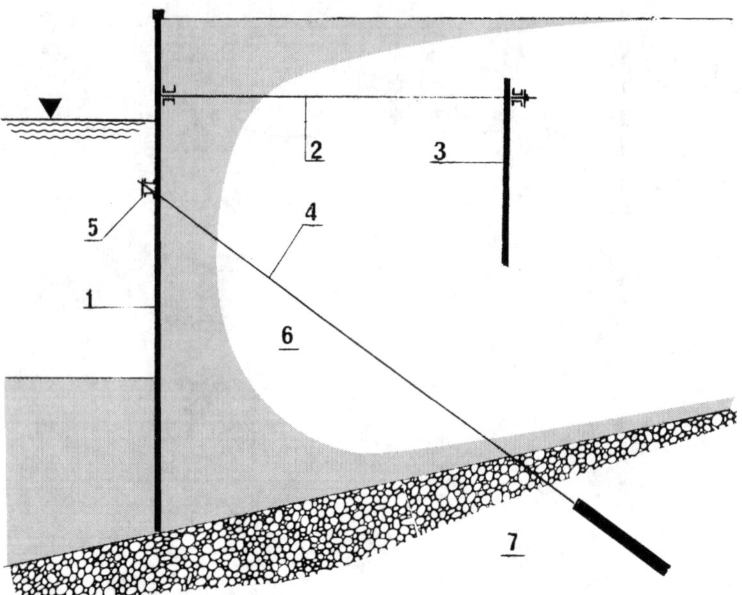

Figure 6.36 Ground (rock) anchors added to existing anchor system. 1–3, Existing sheet pile bulkhead tieback rod and anchor sheet pile wall; 4, new posttensioned ground (rock anchor); 5, new wale; 6, existing backfill material; 7, firm (bedrock) soil.

most cases of sheet pile bulkhead construction, the wale (usually composed of two channel sections placed back to back) is installed behind sheet piles. This is done primarily to protect it from being damaged by ship impact and for better corrosion protection. This makes the wale practically inaccessible for inspection, regular maintenance, and repair, however. It has been observed in a great many cases that seawater that has penetrated through holes in corroded steel sheet piles has been capable in a very short period of time of severely corroding the wale and adjacent portion of anchor rods.

As part of a sheet pile bulkhead anchor system, wales are vitally important to the integrity of the structure (the failure of a wale is essentially the failure of an anchor system), and its strength should therefore never be compromised. If any damage has been done to the wale structure, its remaining strength must be evaluated carefully and a decision must be made as to whether its present strength is sufficient to provide for safe service or it must be repaired or replaced by a new structure. If the remaining strength of the wale is sufficient to transfer service load to the tie rods, to preserve its status quo and eliminate sources of corrosion, which in most cases would be seawater penetration into the bulkhead interior, voids around the wale and in adjacent backfill material should be grouted. Large voids can be filled simply by placing tremie concrete through holes in sheet piling.

If material deterioration has not caused significant overstressing in bulkhead components, in most cases just a cosmetic repair might be sufficient. It would basically include plugging of holes in sheeting to prevent backfill material from being washed away. This is usually done by grouting voids behind sheeting or by constructing a new face in front of deteriorated sheet piles. Typical examples of this type of repair are given in Horvath and Dette (1983), Kray (1983), Porter (1986), Gaythwaite (1990), Taylor and Davies (2002), and Tsinker (1995).

An example of a typical repair of deteriorated concrete sheet pile is depicted in Figure 6.37. In this and similar cases, before receiving new concrete, the surface of the existing concrete must be thoroughly cleaned by any suitable method, as discussed earlier, and carefully coated with a thin layer of epoxy-modified slurry not exceeding 3 to 5 mm. The latter should have a similar water/cement ratio (by weight) to that of concrete used in sheet pile repair. Also, the replacement concrete should have properties similar to these of the original concrete. New concrete should be thoroughly densified after installation. A tighter repair will result if the concrete is placed through the chimney at the top of the form, as shown in Figure 6.37. It is good practice to place a pressure cap inside the chimney following completion of concrete placement. This operation should be repeated at 30-minute intervals until the concrete hardens. If located on the face side of the wall, the projected concrete left by the chimney should normally be removed on the second day. Placement of concrete under water is similar to that using the dry technique but is more difficult to execute. In the latter case, the concrete is normally pumped in bottom up, as discussed earlier.

Last, but not least, it is important to understand that mere replacement of steel lost to corrosion or installation of new concrete to replace deteriorated concrete cannot bring the stress level in a sheet pile system to its predeterioration level. Hence, where a wall is overstressed due to sheet pile weakening because of deterioration, to bring the stress level within allowable limits, it must be rehabilitated by one or a combination of the methods discussed earlier.

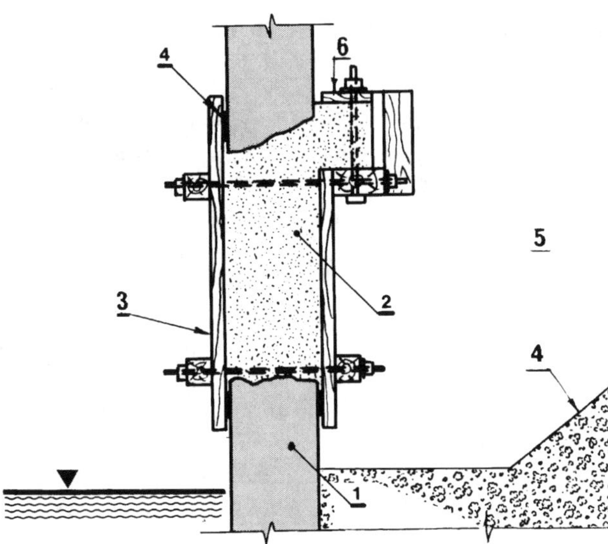

Figure 6.37 Repair of locally damaged concrete sheet pile. 1, Sheet pile; 2, new concrete; 3, formwork to be removed after concrete gains the desired strength; 4, excavation behind the wall; 5, replacement-quality granular fill; 6, pressure cap.

For the purpose of this work it is interesting to note the case of anchored sheet pile bulkhead repair described by Pettit and Wooden (1988). The wharf comprises 61-cm-wide precast concrete tongue-and-grove sheet piles driven into the mud and anchored with steel tieback rods secured to deadmen. Probably because of the absence or poor quality of the filter material behind the wall, the backfill material was washing out through sheet pile joints. Large gaps had formed, which eventually led to extensive pavement distortion. The repair was done by sealing joints between adjacent sheet piles by the jet grouting method. This procedure made it possible to carry out repair work between tieback rods safely without disturbing them. A special latex–cement slurry bonded the bulkhead sheet piles, and in 5 days over 150 seams were sealed. Similar work has been carried out in the ports of Ostend, Belgium and Klipeda, Lithuania (Anonymous, 1997; Habil et al., 2001). In the latter case, 525-m-long steel sheet pile bulkhead has been jet grout–treated not just for the purpose of rehabilitation, but also for establishing a deeper berth to receive a Panamax class vessel. The wall was also strengthened by the installation of additional anchor system.

In cases such as those above, a concrete wall constructed behind the sheet piling may obstruct free water movement in the backfill materials. This could result in excessive water pressures built up on the wall. Therefore, in cases where backfill and foundation materials do not provide for free water circulation, a reliable drainage system must be provided to the wall to relieve possible excessive hydrostatic pressure buildup on the wall due to the action of waves and/or tidal effects. Angled to the sheet pile wall, ground anchors may transmit a substantial vertical load to the sheeting. Hence, where an additional anchor system in the form of inclined ground anchors is to be installed, the effects of vertical components of anchor force should be considered in sheet pile wall flexural analysis. Furthermore, the embedded portion of the wall must possess bearing capacity (frictional resistance) adequate to resist the aforementioned vertical load with an appropriate factor of safety.

Experience indicates that most sheet pile bulkhead failures are attributed to the failure of its anchor system. This can occur due either to severe corrosion of anchor system structural components, heavy local overload, or a combination of both. In addition, heavy overstressing of both sheet piling and the anchor system can occur due to scour or overdredging of the seafloor. Failure of one or two anchor ties, however, may not necessarily be followed by failure of the entire bulkhead system; the system may just bulge in the area of failed anchors. Basically, this occurs due to the fact that a wale that is usually designed with substantial redundancy transfers the load from failed anchor(s) to adjacent anchors, although at the expense of being heavily overstressed and substantially deflected from its normal (original) position. Naturally, the adjacent rod on both sides of the failed anchor(s) will also be overloaded. In such an event, the failed anchors must be replaced.

To bring stresses in a wale system to an acceptable level, repaired or new tie rods must be tensioned. It is not practical, however, to expect that installation and tensioning of new or repaired rod(s) will significantly reduce the load in adjacent overstressed rods or in sheet piling. Again, this cannot happen, because tensioning of these rods cannot bring the backfill behind the sheeting into its original boundaries. As a matter of fact, an attempt to do this may cause the cantilever portion of a bulkhead to fail, due to the development of passive pressure behind it. Therefore, to prevent further overstressing of the existing rods immediately adjacent to a distressed or failed rod, new tie rods should be installed as close to the existing rods as required.

A wale/anchor tie system is usually analyzed as a continuous beam (wale) on elastic supports (anchorages). The elastic properties of each anchorage include elastic properties of the tie road and the anchor system (e.g., anchor block, anchor sheet pile wall, piles). The latter is best established by carrying out field or model tests. The stress level in overstressed sheet piles can be reduced by installation of an additional row of anchors as discussed earlier, or by replacing the existing backfill material with good-quality soil or rock fill (Figure 6.38). Porter (1986) reported a successful sheet pile bulkhead repair using lightweight expanded shale for partial replacement of existing fill. When poor backfill material is replaced with good-quality material, a reduction in earth pressure is achieved because of a smaller value of the coefficient of lateral pressure of the new fill. For example, the angle of internal friction of good-quality granular material could be equal to 40 to 45°. Therefore, replacement of say silty sand with a good-quality rock fill can reduce soil lateral pressure by a factor of about 2 (Tsinker, 1995).

However, one should not expect a significant reduction in stresses in a sheet pile bulkhead, due to the excavation of poor soil in its above-water zone. As stated earlier, the sheeting cannot rebound in its middle section after excavation is complete. Therefore, only the upper portion of the bulkhead will change its curvature after the excavation of part of the existing fill. Stress reduction in the most stressed middle portion of the sheeting can be achieved by distressing the anchor rod to allow for just slight sheet pile movement away from the fill. Practically this may be achieved by loosening the nut at the tie rod anchorage (e.g., concrete block, sheet piling).

In cases in which full restoration of the wall's original capacity is not warranted economically, the partially repaired dock area must be restricted to limited loading. A properly engineered sheet pile bulkhead repair or rehabilitation must consider maximum use of the existing material that is still in good condition. This is best illustrated by the relatively unusual example of steel sheet pile bulkhead rehabilitation shown in Figure 6.39. The job was carried out by Acres International Ltd. (1985). In this case, the steel sheet pile bulkhead, 12.7 m high, anchored with steel tie rods secured at the concrete anchor block was badly deteriorated, basically in the splash zone. In this case, a

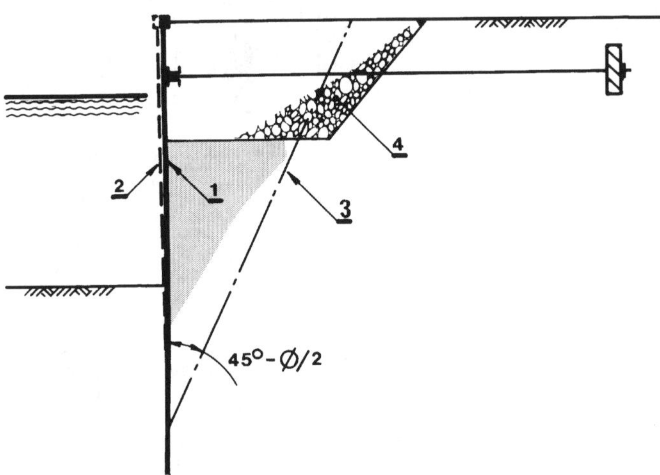

Figure 6.38 Replacement of poor backfill material with good-quality granular fill. 1, Original position of sheet piling; 2, deflected position of sheet piling due to failure of anchor system; 3, slip live; 4, good-quality replacement fill.

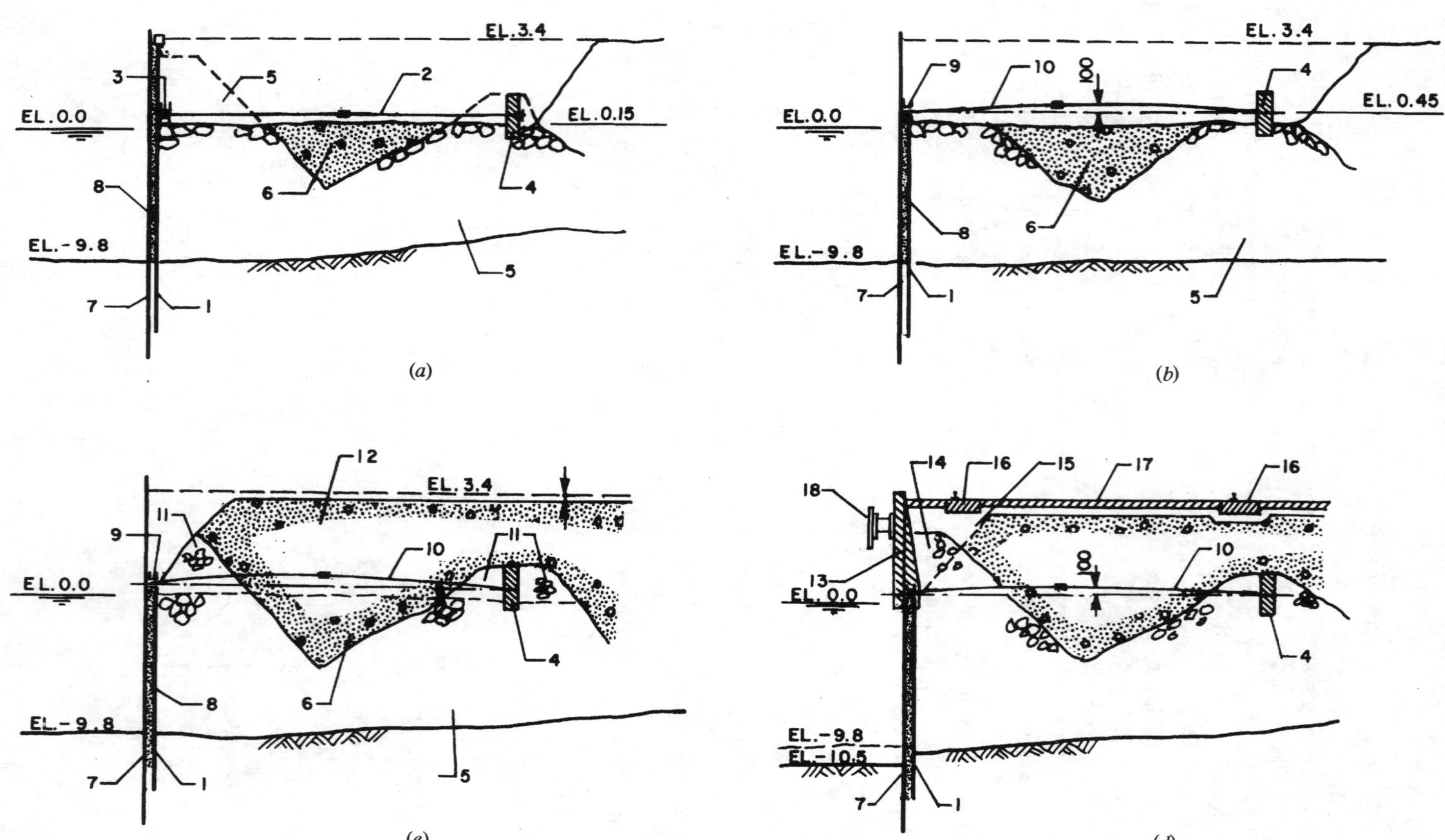

Figure 6.39 Example of steel sheet pile bulkhead rehabilitation. 1, Badly deteriorated sheet piles; 2, existing tie rod; 3, existing badly deteriorated wale; 4, existing solid concrete anchor block; 5, existing rubble; 6, existing fill; 7, new steel sheet piles; 8, gravel fill between new and existing sheeting; 9, new wale; 10, new and/or reconditioned tie rods; 11, partly reinstalled rubble; 12, new fill; 13, concrete cope; 14, complete rubble; 15, complete fill; 16, crane way; 17, new apron; 18, new fender system.

great deal of steel was lost from all structural components of the structure. The condition of the bulkhead forced the dock operator to take it out of service. Inspection of the structure has revealed that steel sheet piles below the minimum mean water level did not lose much steel to corrosion. It was also found that most of the anchor tie rods, with the exception of 1 to 1.5 m immediately adjacent to the sheet piles and anchor concrete block, could be reused. The terms of reference for dock rehabilitation called for an increase in water depth by 1.2 m.

The design rehabilitation work comprised the following stages (Figure 6.39).

Stage 1. New steel sheet piles (7) are driven as close as practicable to the existing sheet pile bulkhead (1), and the space between the old and new sheet piles is filled with gravel (8). Next, the existing backfill (6) and rubble (5) are excavated just below the existing tie rods (2); the excavation progresses from the face of the wall toward the concrete anchor block (4). Structural analysis indicted that in combined action new and existing sheet piles could resist the lateral thrust and the live load imposed by construction equipment with acceptable stresses and deflections while performing as a cantilever wall. It has been suggested that for the sake of safety the excavation (6) should expose at one time no more than five existing tie rods (2), which brought the total initial width of the excavation to about 7.5 m.

Stage 2. Existing tie rods (2) are cleaned, extended as required (in some cases tested), properly recoated, and certified for the purpose of reinstallation to support the new double sheet pile system. Because used rods' preparation and certification take time, five brand-new tie rods (10) are to be installed. At this stage, an existing concrete anchor block (4) is examined and if found to be in unsatisfactory condition, should be repaired or demolished as required, and a new anchor block of similar construction should be built. Next, existing sheet piles are cut right below the existing wales (3), and new wales (9) and refurbished existing tie rods (10) are installed.

Stage 3. Backfilling includes reinstallation of previously excavated rubble (11) and installation of new fill (12). At this stage, wall backfilling proceeds from the anchor block toward the face of the berth. Space in the immediate vicinity of the new sheet pile bulkhead is remained unfilled. This is required for development of uniform initial tension in all anchor tie rods before a concrete cope wall (13) is cast.

Stage 4. A concrete cope (13) is extended from its lowest level, located 60 cm below the minimum mean water level, up to 40 cm above the apron level and completed with a fender system (18). Finally, the space between the cope and fill is filled with rubble (14) and granular fill (15); an apron (17) and craneway (16) are built; and mooring accessories are installed.

Stage 5. The seafloor in front of the rehabilitated wall is dredged as required. As stated earlier, the dredging in front of the existing wall must be carried out cautiously to avoid or at least minimize the overdredging. The acceptable dredging tolerances must be addressed in the construction specifications.

The example above demonstrates a rational approach to bulkhead rehabilitation in which all useful structural components are reused with benefit to the owner.

6.6.2.5 *Repair and Rehabilitation of Open Piled Structures.* Open piled structures are normally built in the form of piers, piled dolphins, or marginal wharves (piled platforms). The latter usually have an underdeck slope protected by conventional riprap or with the other readily available protection systems. Typical damage to piled structures includes:

- Broken piles
- Pile and deck material deterioration
- Physical damage to deck elements caused by ship or cargo handling/hauling machinery
- Scour of the seafloor in front of the dock and/or underdeck slope protection caused by propeller-induced jet or wave action
- Damage or deterioration of elements of the fender system and/or mooring accessories

The repair and rehabilitation of open piled structures in most cases is a straightforward process.

Normally, broken piles that are beyond repair are removed from a structure and replaced with new ones. This is usually done by providing access to broken piles at the deck level. After a broken pile is removed, a new pile is driven through the window in the deck structure, after which the structure is integrated with the new pile by restoring the deck's original reinforcement or installing new reinforcement as required, subsequently filling the window with relevant material (most often with concrete). The deteriorated piles and deck structure are usually repaired by employing various basically conventional techniques. If eroded by propeller jet or wave action, the underdeck slope is evaluated for operation of vessels having side thrusters and/or powerful main propellers and restored accordingly.

Pile Repair. Deterioration of steel and concrete piles typically occurs in splash and/or tidal zones. There the piles are usually repaired or strengthened by employing a jacketing technique. When applied properly, the jacket will strengthen the repaired component and provide some degree of protection against further deterioration. Jacketing is especially useful where all or a portion of a pile has to be carried out under water. The pile jacketing technique is a process of pile repair by installation of a form around a deteriorated portion of a pile, followed by grouting the space between the pile and the form. The original pile can be concrete, steel, or wood.

Essentially, to ensure a good bond between the grout and the pile material left in place, the pile surface must be thoroughly cleaned. This involves removal of all loose materials, marine growth, dirt, and corrosion products developed on steel piles or reinforcing bars. Corrosion products should be removed down to bare metal. Furthermore, steel lost to corrosion should be compensated for if the loss might compromise the structure's integrity. This could be accomplished in a variety of ways, such as installation of reinforcement around piles, welding steel plates to steel piles, or installation of reinforced grout. In case a significant amount of removal is required, it may be necessary to provide temporary support to the structure during the pile repair operation.

In the past, forms for jackets fabricated from any suitable material have been used for pile repair. These forms were fabricated from different kinds of materials (e.g., fabric, steel, fiberglass, concrete). In North American practice, a variety of fabric, fiberglass, steel, and other proprietary form systems are available for pile jacketing. Reinforced concrete forms have been used widely to repair steel piles (Fukute et al., 1990). The forms can be manufactured either as a single unit or in two pieces to be joined together in the field. Forms can be round or square and can be installed one above the other, with the overlapping form having a molded open cavity to receive a bottom seal gasket.

Because of their light weight, fiberglass forms are used most frequently for pile repair. A typical example is illustrated in Figure 6.40. The thickness of a fiberglass form depends on forces and stresses that it may encounter during handling, installation, and grout injection. The minimum thickness of a typical fiberglass form could be in the range 3 to 5 mm. Most fiber-

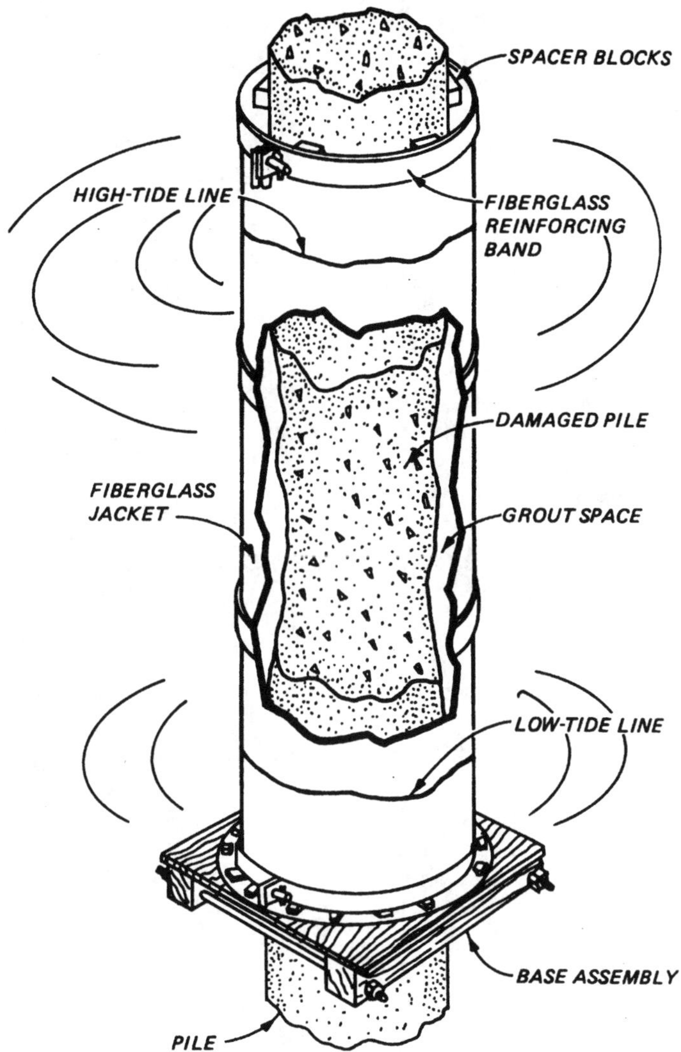

Figure 6.40 Typical repair of deteriorated concrete pile. (From U.S. Army Corps of Engineers, 1986.)

glass forms are relatively translucent, which provides some visual control of grout movement inside the jacket during the placement of grout.

Depending on the type of grout used in pile repair, the annulus formed by the form around a pile should be a minimum of 10 mm. This is secured by the use of different type of spacers. Once the form is in place, it may be filled using any suitable material. The choice of filling material is usually based on the environment in which it will serve, as well as knowledge of what caused the original material to fail. For example, if concrete pile deterioration was caused by its exposure to acidic water, jacketing with conventional portland cement concrete will not ensure against future disintegration. Grout is usually installed by tremie placement or by pumping in of epoxy-modified cementitious grouts. For details on tremie placement of concrete, consult Tsinker (1995, 1997).

The new concrete must adhere well to the previously cleaned surface of the existing pile and should attain the required level of strength and durability. This can be achieved by including in the concrete mix moderate amounts of silica fume and antiwashout admixtures (Khayat, 1992); silica fume increases strength and adhesion durability and provides a low chloride ion permeability value for the concrete, while the antiwashout admixtures, which are water-soluble polymers, are able to absorb some of the mixing water in concrete, thus reducing its migration from the cement paste, along with some suspended cement and fines, as it comes in contact with water.

The bond between a steel pipe pile and a concrete jacket system can be enhanced substantially by welding steel studs to the pile. This could be accomplished readily both in dry conditions and under water (Fukute et al., 1990). The grout, if required, could be reinforced with steel reinforcement placed in the annulus formed by the form around the existing pile. Also, grout could be strengthened by including a filter reinforcement into the mix. Khanna et al. (1988) reported the successful use of a steel fiber–reinforced concrete (SFRC) for the repair of distressed hollow-core concrete piles at Rodney Terminal in Canada. In this particular case, the distressed piles were repaired by pumping a steel fiber–reinforced concrete (SFRC) mix into a reusable steel form installed around the pile to construct a 150-mm-thick jacket of strong, freeze/thaw–resistant concrete intended to preserve (restore) the strength of existing piles and to act as a wearing coat (Figure 6.41). The steel fiber replaced conventional reinforcement in the concrete and provided a restraint against shrinkage and thermal cracking right out to the surface of the concrete. At this project, a typical concrete mix per cubic meter with a specified strength at 28 days of about 40 MPa included 415 kg of cement type 50, 105 kg of fly ash type F, 80 to 215 kg of sand, and collated and uncollated $l = 50/d = 0.5$ steel fiber. Typical fresh

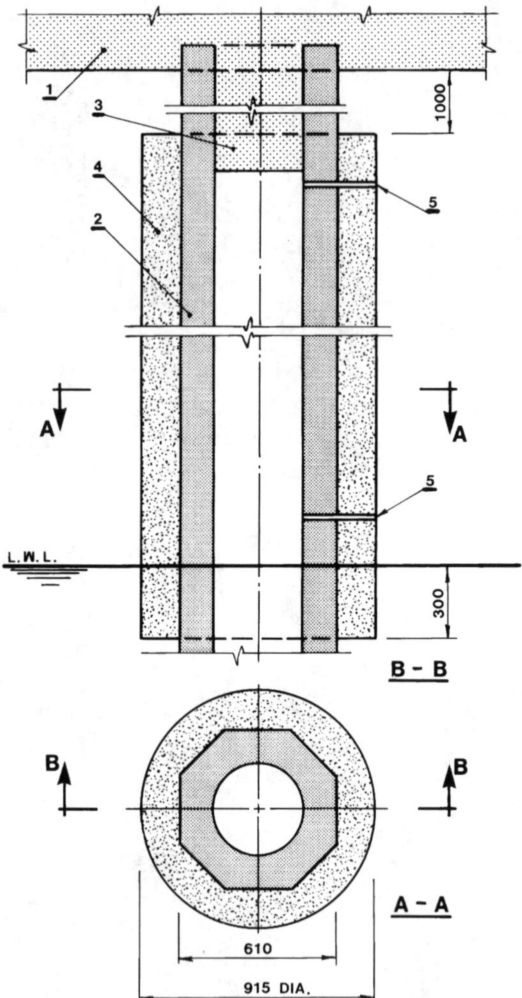

Figure 6.41 Restoration of distressed piles at Rodney Terminal, New Brunswick, Canada. 1, Deck structure; 2, hollow-core octagonal concrete pile; 3, concrete plug; 4, steel fiber–reinforced concrete jacket; 5, 25-mm-diameter drain hole. (After Khanna et al., 1988.)

concrete tests results showed the following average properties: slump, 158 mm; air content, 8.5%; temperature, 24°C; density, 2336 kg/m^3; and excellent workability.

A total of 362 piles were repaired there using SFRC. The piles were cleaned prior to installation of the form (usually a day or two ahead of concreting) by a high-pressure freshwater blaster operated in the range 28 to 35 MPa. The underwater part below the low level was washed by a diver who used the same wash nozzle gun as that used above water. Following cleaning, the clean formwork was assembled around the pile. The form was completely sealed around the pile and SFRC was pumped from the bottom up via a steel pipeline 125 mm in diameter. Three vibrators operated continuously until 4 to 5 minutes after completion of pumping. One of those vibrators was attached to the form just above the inlet nipple to help in avoiding blockage at this point. Two other vibrators were attached to the upper third of the form. The experience gained at Rodney Terminal indicated the following:

1. SFRC pile repair is a relatively simple method for jacketing marine piles.
2. When air entrainment is required for freezing and thawing resistance, the concrete strength and air void parameters specified must be met.
3. The SFRC method of marine pile jacketing is particularly attractive in tidal zones with a large range and strong currents because it minimizes work with a marine plant over water.
4. The quality of SFRC jackets may vary over the height of a jacket due to a nonuniform steel fiber distribution and a migration of air to the upper portion of the jacket. This was not found to be a serious problem, however, because the lower part of the jacket, where the SFRC quality is better, protects the pile zone, where freezing and thawing distress is at a maximum.
5. Collated steel fibers used on the project demonstrated a better toughness index than uncollated fibers for the same loading, partly because of the much larger number of collated fibers for the same weight.
6. SFRC can effectively replace conventional reinforcement, particularly where the SFRC is required for a purpose other than structural strength. This is mainly because installation of regular reinforcement, particularly by divers in a tidal zone, is expensive.

Usually, the most common concern regarding the use of SFRC in a marine environment is associated with potential corrosion of the steel fibers. However, a study carried out by Magnat and Gurusamy (1998) indicated that small fibers properly dispersed within a concrete mix do not compromise concrete integrity. The individual fibers located on the outside surface of SFRC usually just rust away without causing concrete deterioration. The success of SFRC pile jacketing at Rodney Terminal was affected significantly by the pumping equipment and setup, as well as a good maintenance program. Most recently, SFRC was used successfully in Israel for pile repair (Chernov and Buslov, 2001). More useful information on the subject is given in Boone et al. (1995) and Warren et al. (1995).

A jacket can be formed effectively by injecting epoxy-modified grout to fill the space between a form and a deteriorated pile. On some occasions, an epoxy grout suitable for marine applications was added to the prepact aggregates, pumped in bottom up via a hose of 25 mm internal diameter. In general, where epoxy grout is used for the construction of pile jackets, injection ports having a minimum internal diameter of 25 mm are normally spaced at intervals not exceeding 150 cm; to provide for even epoxy grout distribution, the injection ports are typically placed on alternate sides. To provide a better quality of jacket, the minimum space between the form and the pile should not be less than 15 mm.

As pointed out earlier, grout injection starts at the bottom injection port and proceeds upward. As the jacket is filled to each port, the lower port is capped off and the procedure repeated until the top of the jacket is reached. In the process, the injection hose is moved upward from port to port. Furthermore, deteriorated marine piles can be repaired by fabric-formed jackets filled with concrete. This type of repair has been used for timber, concrete, and steel piles in diameters up to 3.0 m and in lengths up to 17.0 m (Lamberton, 1989). After cleaning the pile and placing reinforcing steel, if required, the fabric forms are suspended from the deck structure, closed around the pile at the bottom of the jacket, and closed vertically, usually with a zipper. The sleeve is filled with concrete mortar through vertical injection pipes placed symmetrically down through the space between the pile and the jacket. Grout placement on one side only can cause the fabric-formed jacket to take a banana shape that is very difficult to straighten once formed. A factor of safety of 1.5 to 2.0 applied to 80% of the break load is generally used in fabric jacket component design. This applies specifically to the selection of zippers, which are available today in strengths up to 280 g/mm. Fabric sewing is also critical. The fabric at seams should be folded before stitching, and typical specifications usually call for the development of 80% of fabric break load at the seam.

When longer jackets are required, the fabric jackets may be reinforced with circumferential bands of steel strapping. The latter should be able to sustain the full hydrostatic pressure of concrete. Piles of a moderate batter (say, 1:10 and less) are filled from the high side, while jackets located in moderate current are filled from the upstream side. Fabric pile jackets are not appropriate for use in wave action such as that which occurs in a breaking wave zone or in flowing water with a current velocity above 1.0 m/s. Piles can also be repaired using miscellaneous polymers. Polymers in various forms have been used in marine environments for pile repair and protection for over 25 years. From simple coatings to plastic wraps, to membranes and composites, they have been used with widely varying degrees of success. Snow (1992) attempted to identify basic reasons for distress and failure of polymer systems that occurred in the past in order to present a workable solution for pile encapsulation. The practice has indicated that most jacket defects occur in the lower 10 to 15 cm. These defects are usually associated with concrete placement under water. Hence, to ensure a good quality of pile protection over its design length, a jacket should be extended by a minimum of 10 cm over its design length.

Both vertical and battered piles can also be repaired in dry conditions by the use of a special floated-in cofferdam. According to Alexeev (2001), cofferdams have been used for the repair of piles and sheet piles in Baltic and Black Sea ports and in the Russian Far East. Hand-applied epoxy-based compounds have also been used in underwater applications, but proved to be very difficult to handle, expensive, and in most cases did not provide for a durable, high-quality repair. In some instances, for enhanced protection of repaired steel piles and steel reinforcement of concrete piles, sacrificial anodes are added to the jacket system. These are fabricated in the form of a high-purity zinc mesh and reportedly are efficient and cost-effective (Leng, 2000). Anodes can be engineered for the expected life of a structure. For additional information on the use of sacrificial anodes for protecting steel in piles operating in corrosive marine environments, readers are referred to works by Uhlendorf (1998, 2001) and Uhlendorf and Janssen (2000).

Deck Repair and Rehabilitation. Basic types of concrete deterioration in deck structures supported on piles are shown in Figure 6.3. As pointed out earlier, deck deterioration could be caused by a multitude of adverse effects, such as ship impact–caused local damage to deck elements; corrosion of reinforcing steel, resulting in concrete cracking and spalling; overstressing caused by overloading, which may result in cracking and spalling; handling of cargo materials that are aggressive to deck structural materials; and others. All of the above, as well as in-place construction defects (e.g., high porosity of the concrete, which allows chloride intrusion and an alkali–aggregate reaction) will cause concrete deterioration.

Concrete repair must be accomplished such that further disintegration is arrested. This can be accomplished by one of the following methods: patching, grouting, shotcreting, strengthening, or by combinations of these.

PATCHING. Patching is the process of replacing loose, spalled, or crumbling concrete with new material (Figure 6.42). A good patch should restore structural integrity, be compatible with the parent concrete, and last as long as the structure does. Compatibility is defined by Emmons et al. (1993) as a balance of physical, chemical, and electrochemical properties and dimensions between a repair material and the existing substrate. This balance will ensure that the repair can withstand all the stresses induced by volume changes and chemical and electrochemical effects without distress and deterioration. For a discussion and references on the subject of electrochemical incompatibility of patches in reinforced concrete, readers are referred to Ping Gu et al. (1997). Concrete deterioration is symptomatic of many kinds of structural damage, so patching is part of almost every major concrete repair project. Suitable materials for patching dcteriorated concrete pier structures typically include epoxy-modified portland cement mortar–concrete for repair under dry conditions, and epoxy mortar for underwater repair. Sometimes, portland cement mortar–concrete or special cements are used. The choice of material depends on such factors as patch thickness, condition of placement (dry, wet, under water), shrinkage, and cost. Selecting the right material is essential. For example, epoxy mortar has high adhesive qualities but is expensive; it is thus a good

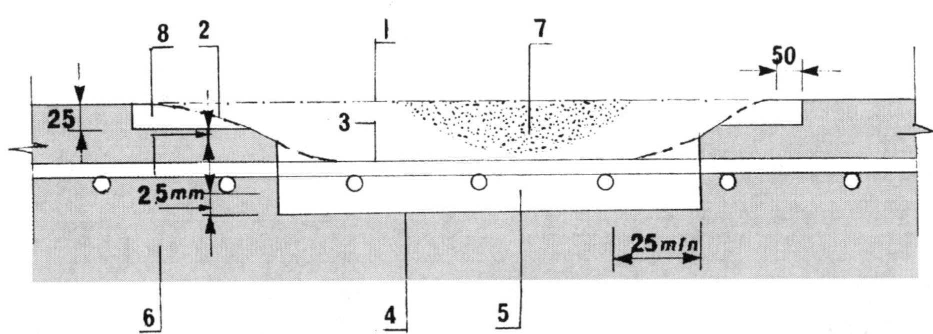

Figure 6.42 Surface repair (patch). 1, Original surface; 2, surface after deterioration; 3, existing rebar; 4, level of sound concrete; 5, concrete to be removed if deteriorated or if reinforcing is exposed; 6, sound concrete; 7, defective concrete replaced with good concrete; 8, shoulder.

choice for applications where high adhesion is important, such as freeze/thaw-resistant pier elements or a wear-resistant deck. For a deep patch, a better choice might be epoxy-modified portland cement mortar–concrete because it costs less and is more compatible with original concrete. Successful patching involves:

1. Good surface preparation; deteriorated concrete must be removed and the substrata must be taken down to a clean, sound finish. All foreign particles and materials, such as dust, laitance, grease, curing compounds, impregnations, and waxes must be removed by mechanical abrasion methods such as sandblasting. If a jackhammer is used for removal of deteriorated concrete, to limit microcracking of concrete substrate in the final removal stages, a light hammer should be employed. Normally, concrete should be removed a minimum 25-mm beyond the limits of the corroded rebar; a 25-mm shoulder usually is provided by sawcutting around the perimeter of the repair area (Figure 6.42). Exposed reinforcement must be cleaned to bare metal by sandblasting or another suitable method. Reinforcing that has lost a substantial area must be replaced.
2. Installing epoxy-modified portland cement mortar–concrete or epoxy mortar. To achieve the best results, the parent concrete should be moistened, allowed to dry until damp, and coated with epoxy-modified slurry. The latter is especially important when regular portland cement mortar–concrete is used for repair.

In general, new concrete installation is followed by tamping or ramming of the mortar in place to produce close contact between new and existing concrete. If portland cement mortar is used, the mortar should stand for half an hour after mixing to minimize its shrinkage in place, and then be remixed prior to use (ACI, 1984). It is good practice to place it in thin layers (about 10 mm thick). There is no need to wait between layers. An above-water repair is cured by using either water or a curing compound.

Again, for best performance the key criteria for cementitious materials used to patch a degraded concrete area are as follows: low shrinkage, for example, dry shrinkage not to exceed 0.05% at 28 days; high density, low permeability; minimum bond stress, 15.5 MPa at 28 days; and high strength, for example, a minimum compressive strength of 35 MPa and a minimum splitting tensile strength of 4.5 MPa, both at 28 days. Other requirements may include thermal and electrochemical compatibility with parent concrete, freeze/thaw criteria, and durability. When the entire surface of a structure is repaired, to retard ingress of chloride, it may be covered with an appropriate concrete sealant. For the sake of good maintenance, the frequency of sealant reapplication should be based on its in-place periodic testing.

GROUTING. This process is used for the repair of cracked concrete components. By definition, grouting is the process of placing materials into a surface fissure and/or interior crack. Typically, grouting is used to restore the monolithic nature of a structure and to restore the bond of reinforcing steel to concrete. Grouts are liquids that solidify after application. They are pumped into place, usually under substantial pressure, to ensure that the void is filled completely. The major grouting materials are chemical, epoxy, cement-based, and polymer grouts. Because of their excellent tensile quality, epoxy grouts are used more often than the others. They can be used to seal cracks as narrow as 0.05 mm. It must be kept in mind, however, that the presence of contaminants in cracks, including water, may reduce the effectiveness of the epoxy to repair structural cracks. To ensure grouting quality (e.g., to ensure that the grout travels the full depth of the concrete section), the crack is first flushed clean to assure port-to-port flow. During this operation, contaminants are removed from the concrete substrate. The grout specified should be capable of bonding to a moist substrate.

Typical procedures for concrete grouting include basically the following steps:

1. Cleaning the cracks of oil, grease, dirt, and fine particles.
2. Sealing surface cracks to keep grout from leaking before it has hardened. This is usually done by brushing an epoxy or similar material along the surface of the crack, then allowing it to harden. A strippable plastic may also be used to seal the crack if high injection pressure is not required. In the latter case, the crack should be cut out 15 mm deep and 20 mm wide in a V shape, then filled with epoxy.
3. Installing entry ports, which could take the form of special nipples or fittings placed flush with the concrete face over the crack. Typically, ports are installed at intervals that are greater than the thickness of the treated element to assure that the epoxy is traveling the full depth of the section. Grout injection begins at the bottom of the crack and an injection hose is moved to the next port only after the grout begins to flow from this port while injecting from below.
4. Injecting grout: this is usually done using hydraulic pumps, paint pressure pots, or air-actuated guns.

Typical examples of concrete repair are depicted in Figure 6.43 and Table 6.4. Where a large area of deteriorated concrete has to be treated, shotcreting technology can be employed economically for concrete restoration. Crack propagation can be arrested by strapping, that is, local reinforcing of the structure by installing exterior stitching dogs along the crack (Figure 6.44). The dogs should be anchored properly into holes predrilled on both sides of the crack. Nonshrink grouts or epoxy-based systems are used to anchor the stitching dogs into the holes. Strapping prevents further crack propagation and is typically combined with simultaneous crack repair. However, local stiffening of a deck structure by dogs may cause the concrete to crack at other parts of the deck. Hence, the effects of any changes to a deck structure should be evaluated thoroughly before any repair actions are considered. An example of the installation of additional reinforcement is shown in Figure 6.45. In this case, the shear capacity of the deck girder is enhanced by rebars installed across actual or expected cracks. Bars are installed into holes drilled at 90° (or close) to the shear crack. Holes and cracks are filled with epoxy grout pumped in under low pressure (0.4 to 0.6 MPa), after which reinforcing bars (typically, 15 to 20 mm in diameter) are forced into the hole. Bars should extend at least 500 mm on each side of the crack. The epoxy grout bonds the bar to the walls of the hole, fills the crack plane, and bonds the cracked concrete surfaces together in one monolithic form, thus reinforcing the section. Again, the crack(s) should be sealed in one way or another from outside the crack before grouting, to prevent the grout from escaping. In some cases, a deck structure can be strengthened by a posttensioning technique. For examples and a discussion of concrete strengthening, interested readers are referred to Lin and Burns (1981), Vejvoda (1990), and Tsinker (1995).

Repair and Rehabilitation of an Underdeck Slope. As discussed earlier, the introduction of large vessels using main propeller(s) and side thrusters in berthing and/or unberthing operations may result in significant erosion of an

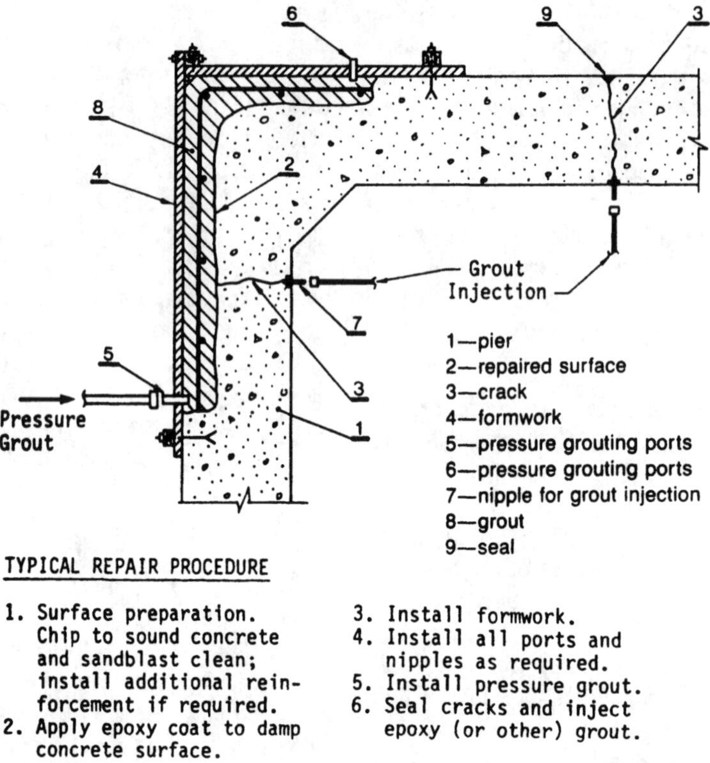

Figure 6.43 Typical repair of deteriorated concrete.

underdeck slope. However, the main threat of erosion of an underdeck slope comes from a vessel's main propeller jet when deflected by the rudder (PIANC, 1997). This produces very high water velocities on the adjacent slope during ship berthing and unberthing maneuvers. The effects of bow thruster(s) are less pronounced because the latter usually has a power of less than 20% of the main engine power. Damage to an underdeck slope as well as erosion of the seafloor in front of vertical-face structures inflicted by ship operations is a serious problem in modern port operation. This could be aggravated in the future, with the increased engine power and reduced underkeel clearance of larger vessels.

The basic problem in underdeck slope repair is the presence of the deck, which obstructs access to damaged area(s). This is why when a slope has to be repaired or reinforced by heavy armor blocks, the contractor may not have sufficient means to deliver the blocks directly under the deck. The heavy armor blocks can be replaced with grout-filled synthetic bags delivered to the damaged area in a small boat which could sail under the deck during low tide. After being preplaced by divers at the required location, the bags are filled under water with grout. As discussed earlier, the mound made by concrete-filled bags placed one upon another can be reinforced by stitching them together with rebars driven through bags while the grout is still fresh.

Fabriforms, grout-filled synthetic fabric mattresses, of miscellaneous designs, can also be used for slope repair. If required, they can

Table 6.4 Typical concrete deterioration and repair

Problem		Solution	
Description	Sketch	Description	Sketch
Leaking crack		Route and Clean crack, then install sealant	
Spalled surface		Remove unsound concrete, clean surface of concrete and reinforcement, install grout	
Spalled surface and leaking crack		Remove unsound concrete and rout crack, clean surface of concrete and reinforcement, install sealant and grout	
Underside spalled surface and leaking crack		Rout and clean crack, remove unsound concrete and clean surface of concrete and reinforcement, install sealant and grout	

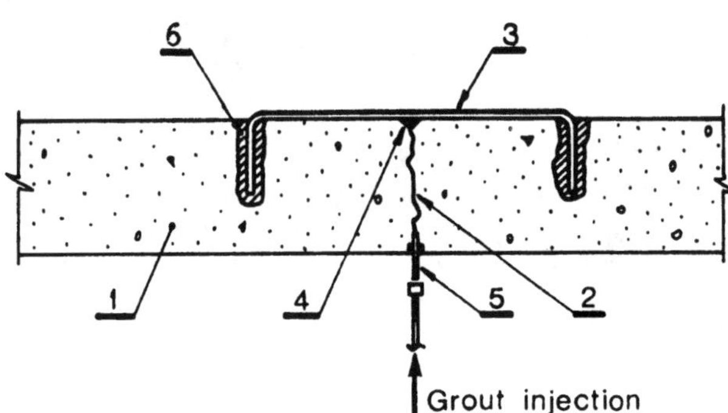

Figure 6.44 Concrete strapping. 1, Concrete element; 2, crack; 3, sticking dog; 4, seal; 5, grout injection port; 6, resin grout.

be placed in layers along and across the slope, creating a stable, thick layer. The system can be monolitisized further by sticking rebars through the layered mattresses. The leftover open space between the mattresses and piles can be filled by pumping in special concrete mixes modified with antiwash admixtures.

6.7 UPGRADING EXISTING WATERFRONT STRUCTURES

Nowadays, port authorities and terminal operators are increasingly seeking ways in which to adopt existing port infrastructure to meet the changing demands of their markets. This has

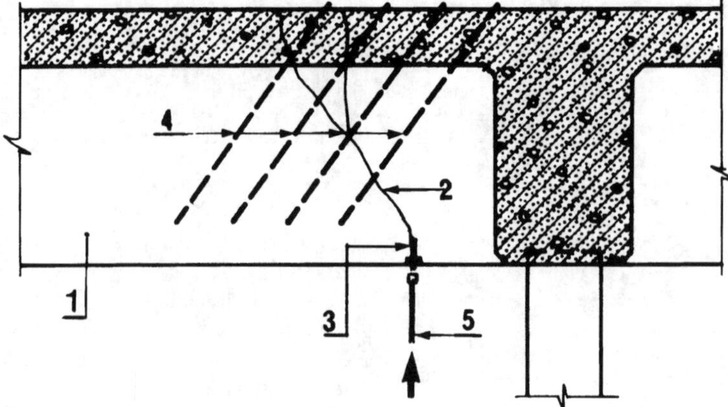

Figure 6.45 Concrete strengthening. 1, Concrete element; 2, shear crack; 3, grout injection port; 4, reinforcing bars; 5, grout injection hose.

turned ports into dynamic systems that are growing and changing as the purpose for which they were constructed has changed. From an engineering point of view, a port is a system that comprises miscellaneous facilities directed to economically efficient and safe handling of cargo. When the type of cargo or transportation mode changes, a port needs to be modernized accordingly to be adapted effectively to updated cargo handling and hauling equipment, new types of vessels and ground transportation, and so on.

The spectacular growth of ship sizes, especially in bulk transport, and the development of new ship types, such as container, roll on–roll off, car carriers, and large ferry ships, make many existing port facilities obsolete. Furthermore, larger and more powerful ships with bow thrusters, increased draft, and therefore less keel and propeller clearance approaching berths without tug assistance can cause considerable structural and scouring damage, especially if the structure is built on an erodible foundation. Consequently, to avoid the adverse effects of seafloor erosion on the performance of waterfront structures, a greater depth must be created in front of them. Alternatively, the use of side thrusters and/or main propellers could be restricted to a certain acceptable limit. Many ports presently in operation worldwide were built post–World War II, with facilities designed to service comparatively small vessels. At that time, port waterfront structures were designed basically for a maximum 40-kN/m^2 uniformly distributed load. Hence, to meet today's service requirements, the load-carrying capacity of these facilities must be carefully evaluated. Furthermore, most older facilities still in service struggle with problems such as those associated with insufficient water depth in front of a quay and antiquated cargo handling and hauling equipment and fender systems.

In most practical cases, all of the above is usually combined with structure physical deterioration, which reduces its ability to carry a load. To enable the existing waterfront structure to operate under conditions that are different from those for which it was originally designed, its load-carrying capacity must be evaluated and, if required, upgraded. Depending on local requirements, the latter may involve strengthening some structural components, enhancement of foundation and/

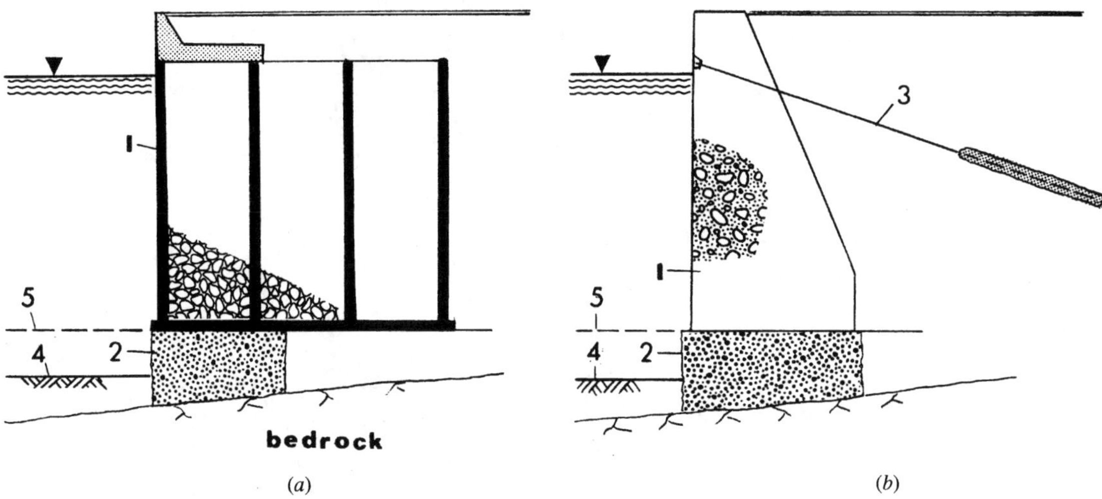

Figure 6.46 Increase in depth of water in front of existing gravity quay wall constructed on a foundation closely underlayered by bedrock: (*a*) underpinning; (*b*) base grouting. 1, Existing gravity quay wall; 2, grout; 3, ground anchor; 4, 5, new and old dredge line.

or backfill materials, addition of new structural components to existing structures, and other upgrades.

The approach to wharf upgrading is usually site specific and depends on the technical, economic, and operational specifics of a particular port. For example, construction work associated with port upgrading may interfere with the berthing and unberthing of vessels; the new construction in front of an existing berth may disrupt an existing single berth line; or the extension of a new berthing line farther into a harbor may necessitate replacement of the existing crane rail system with a new one for operation of new cargo handling equipment. Hence, any approach to port upgrading must plan to keep all interruptions to port operation to the minimum possible.

6.7.1 Gravity Quay Walls

To accommodate larger vessels and heavier cargo handling and hauling equipment, most older gravity quay walls should be upgraded substantially. The latter may include wall rehabilitation and/or increase in water depth at the wall's face. As discussed earlier, wall rehabilitation typically includes repair of its damaged and/or deteriorated parts and may also include improvements to foundation and backfill soils. The depth of water in front of a gravity wall can be increased in a variety of ways. Four basic methods are depicted in Figures 6.46 and 6.47. The choice of one of these schemes depends on the technical, economic, and operational conditions that exist at the particular port.

Structural modifications to an existing wall as depicted in Figure 6.46*a* may include wall underpinning utilizing the jet grouting technique, also referred to as a very high pressure (VHP) grouting method. Underpinning is carried out to the depth required to ensure wall stability after follow-up dredging of the seafloor in front of a new dock system. This method of structural upgrading is appropriate where the existing wall is sufficiently stable and is able to function safely under increased

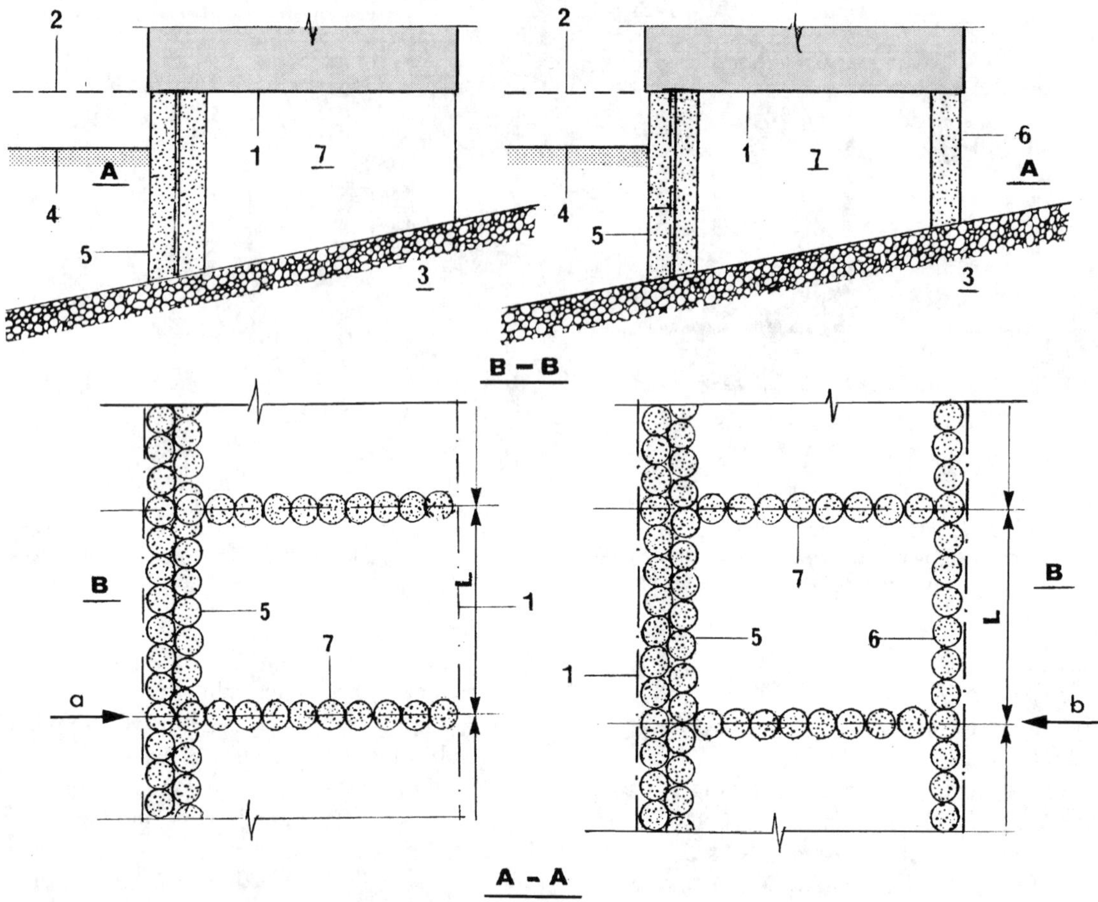

Figure 6.47 Underpinning of gravity wall by means of jet grouting technique. a, Water side of the wall; b, land side of the wall; 1, outline of existing wall's underside; 2, existing bottom; 3, competent foundation soil; 4, newly excavated bottom; 5, front grouted wall; 6, rear grouted wall; 7, transversal grouted wall.

load conditions. In this case, underpinning plays a dual role; it provides for increased depth of water in front of the wall and enables the foundation to accommodate the increased bearing stresses at the wall toe due to increased vertical and lateral loads. In some instances an increase in effective wall height can affect the wall's ability to resist the horizontal thrust of soil and overturning loads. In this case, additional wall support can be obtained using posttensioned ground anchors.

The wall modification depicted in Figure 6.46*b* comprises grouting the wall base and installing posttensioned ground anchors. This type of construction is generally used where the wall needs to be stabilized due to the increased depth of water in front of the structure and when the structure itself cannot safely accommodate increased loads (e.g., mooring forces and those generated by cargo handling and hauling equipment). Ground anchors are normally tensioned to prevent the wall from tilting toward the harbor due to elastoplastic deformation of the anchor system, and for more uniform distribution of bearing stresses at the wall's base. The installation level of ground anchors, and their size, angle, spacing, level of tension, and length, are based on wall stability analysis. In most practical cases, the double corrosion protection of anchors is considered.

Wall base grouting is the technique by which a foundation is impregnated under pressure with grout, which is allowed to set. In general, grouting is a process in which grout in liquid form is pumped into the voids of soil or rock and then hardens. As a result, the soil and rock are densified and gain additional strength. Grouting is generally used to alleviate difficult foundation problems and for remediation of existing foundation; it has also proved to be very effective in upgrading existing waterfront structures; it increases the shear strength and reduces the compressibility of soils under the wall's base. In marine applications, soil grouting is usually accomplished by injecting cement slurry or sand–cement mortars. Grout may be installed by the conventional injection grouting method, more commonly known as *grouting,* or by high-pressure grouting, known as *jet grouting*. The resulting solid foundation material effectively extends the wall height and safeguards the existing structure from being undermined by dredging (or overdredging).

A practical method of jet grouting was discussed earlier in the chapter and illustrated in Figure 6.21. The new foundation base can be created economically in a variety of ways. Two representative examples are depicted in Figure 6.47. The modified foundation includes a front screen wall normally created by jet grouting, and transverse walls spaced at a distance required to ensure all stability requirements (i.e., overturning, sliding, and bearing capacity requirements). In some instances, for the sake of wall stability, installation of the rear wall is also warranted. The latter has the advantage that the ground inside the closed caisson forms better interaction with the concrete wall system. In both cases, the double column front wall ensures the ground tightness of the system. If required, the existing to be rehabilitated wall could be posttensioned to the newly formed base. The total system could be further stabilized by installation of posttensioned ground anchors as required by stability analysis. Furthermore, the wall stability can be improved by reducing soil lateral pressures as discussed earlier and shown in Figure 6.32. Practical examples are given in Bray and Tatham (1992), Hallett et al. (1992), Ponnet et al. (1992), Thibant (1992), Tsinker (1995, 1997), and Van Damme et al. (1998, 1999).

Alternatively, the water depth in front of a gravity quay wall may be increased through constructing piled systems of miscellaneous designs in front of the wall. Typical examples are illustrated in Figure 6.48. The piled system depicted in Figure 6.48*a* typically comprises large-diameter king piles installed at a safe distance from an existing rehabilitated wall. This is necessary to prevent adverse effects of pile driving on the existing structure. The space between adjacent king piles in the direction parallel to the existing wall is determined by the pile's ability to carry out the design loads. The pile system is completed by driving a set of conventional sheet piles between the king piles. A concrete deck supported on piles and the existing wall can be assembled from precast components, be cast in situ, or be built as a combination of both. If the combined newly formed system is not able to resist mooring and ship impact forces safely, it may be further reinforced by the installation of posttensioned ground anchors. To protect the existing gravity wall from the effects of a ship's side thruster and main propeller jet, the old bottom between the wall and the new sheeting is protected in some way. Most recently, the system in question was used to upgrade ports in Belgium (Thibaut et al., 1992; Van Damme et al., 1998,

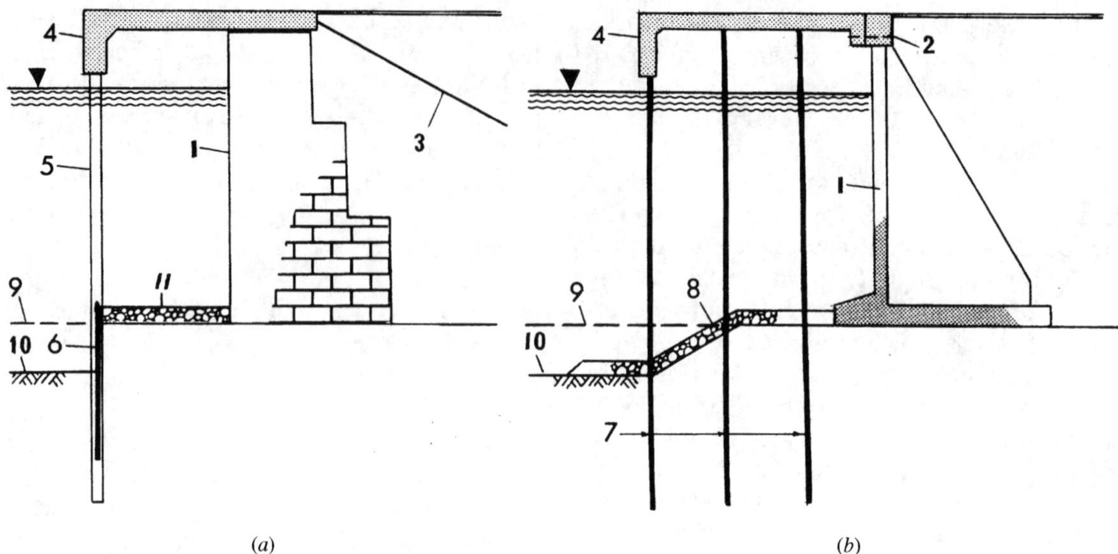

Figure 6.48 Increase in depth of water in front of an existing gravity wall constructed on soft foundation soils: (*a*) extension of gravity wall by construction of concrete deck supported on large-diameter piles; (*b*) use of pile platform. 1, Existing gravity quay wall; 2, posttensioned bar; 3, ground anchor; 4, concrete deck; 5, large-diameter king pile; 6, steel sheet piling; 7, steel or concrete piles; 8, slope protection; 9 and 10, old and new dredge lines; 11, protection from scour.

1999). Some specific design aspects of this system are discussed in DeMan et al. (1996).

The system illustrated in Figure 6.48*b* comprises a standard vertical pile platform constructed in front of an existing wall and joined to the latter by miscellaneous means. The platform is extended far enough from the face of the existing wall to have its underdeck slope safely constructed at the shortest possible distance from the wall's base. Again, depending on the platform–wall system's ability to resist mooring and ship impact forces, it may or may not be reinforced by the installation of ground anchors.

6.7.2 Piled Structures

Some representative solutions for upgrading piled waterfront structures are depicted in Figures 6.49 through 6.51. Piled waterfront structures are broadly classified as open pile constructions with a suspended deck and piled relieving platforms of miscellaneous designs (Tsinker, 1997). The former is constructed as an extension of the land-based terminal apron, where the structure is typically founded on a protected slope. In most practical cases, open piled structures carry little, if any, earth pressure and are designed to resist vertical loads generated by cargo handling and hauling equipment and horizontal forces due to ship impact and wind loading transmitted to the structure via a ship's mooring lines. Where a sheet pile wall is incorporated into the pile foundation on the back of the structure (Figure 6.51), a certain value of soil thrust against the structure is to be considered. Suspended deck structures are usually built in protected harbors and are extended seaward to cover the width of the underdeck slope. To prevent slope erosion, the

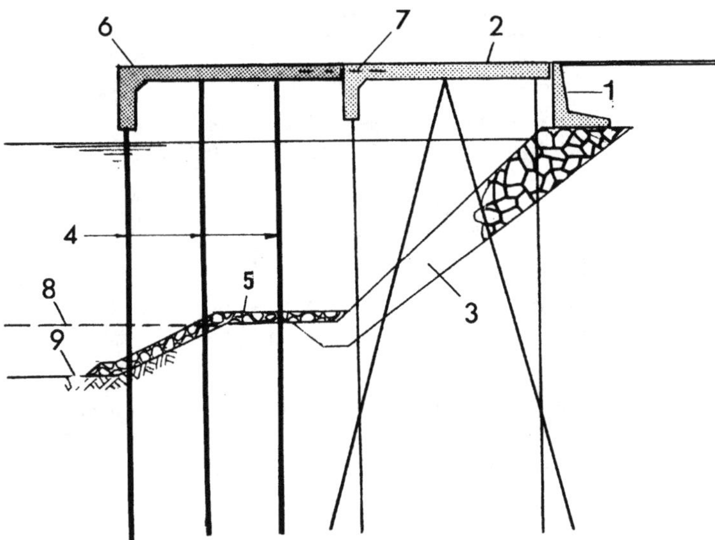

Figure 6.49 Upgrading (modernization) of open piled marginal wharf. 1, Retaining wall (cast in situ or prefabricated); 2, open piled structure; 3, underslope protection (riprap); 4, new piles; 5, new slope protection; 6, new deck; 7, posttensioned rebar; 8 and 9, old and new dredge lines.

latter is usually protected with armor stones, gabions, and so on, as discussed earlier. Relieving piled platforms include a sheet pile bulkhead at the front face of the structure. Thus, lateral soil pressure is considered in structure design. For detailed information on design and construction of all kinds of piled waterfront structures, readers are referred to Chapter 2 of this book and to Tsinker (1997).

As noted earlier, a structure needs to be upgraded basically when deeper water is required to receive larger vessels and/or where heavier or different types of cargo handling equipment have to be installed to handle different or increased volumes of cargo. In the case of marginal open piled structures, the depth of water can be increased by extending the existing structure further into the harbor. An example is shown in Figure 6.49. The addition should be integrated with the existing structures. This will increase the integrated structure's ability to resist larger horizontal forces. The projected length of the new structure depends on water depth requirements and on new and existing slope stability. An example of upgrading a crane rail system to accommodate heavier cranes having a larger gauge is depicted in Figure 6.50. In this particular example, a waterside crane rail beam is incorporated into an existing structure in the following steps:

1. The existing crane rail system is removed and the existing concrete deck structure is exposed.
2. Part of the existing deck and pile caps are demolished as required to accommodate the new beam.
3. New piles are installed and the new crane rail beam is cast-in-place and integrated with existing pile caps and deck structure.
4. The land-side rail system is built as required either in close proximity to the existing structure or at some distance.

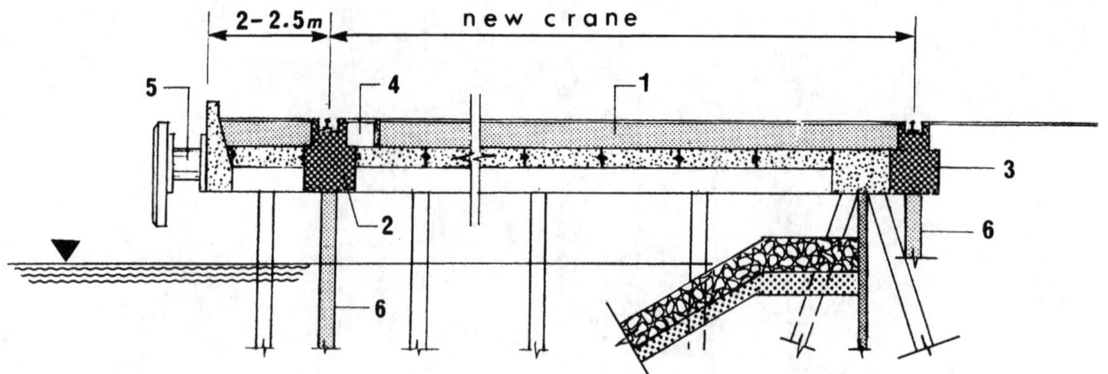

Figure 6.50 Upgrading the deck structure for the installation of a new gantry or portal crane. 1, Existing open piled structure; 2, cut-in new water-side crane rail beam; 3, new land-side crane rail beam supported on piles; 4, new power trench; 5, new fender; 6, new piles.

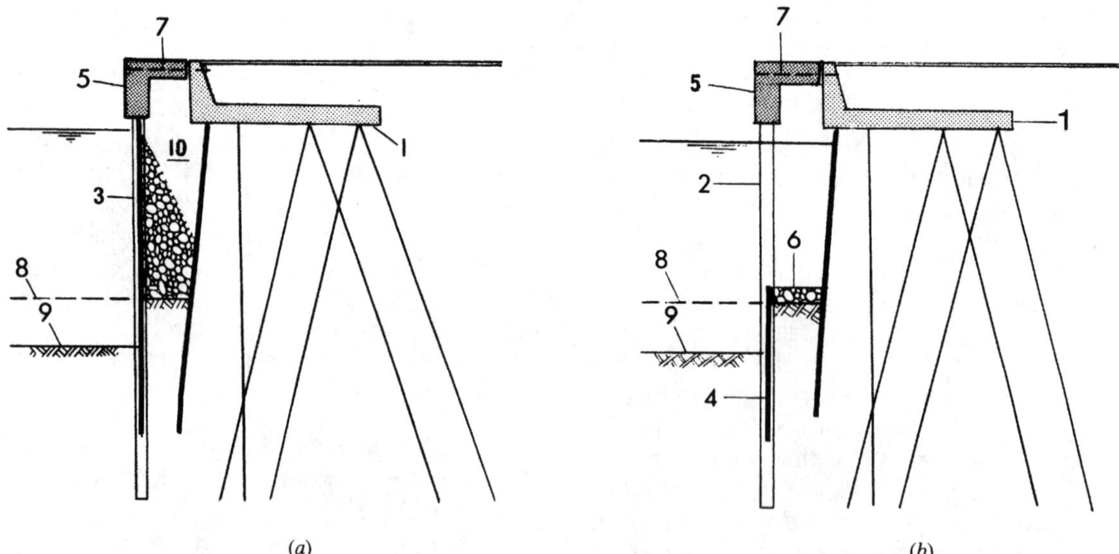

Figure 6.51 Upgrading of piled pressure-relieving platform. 1, Existing structure; 2, large-diameter king pile; 3 and 4, sheet piling; 5, concrete cap; 6, surface protection, 7, posttensioned rebar; 8 and 9, old and new dredge line.

5. The upper section of the deck structure is restored to its original shape, or as required.

6. Finally, a new fender system is installed, as required, to accommodate larger vessels.

Piled relieving platform structures can be upgraded as indicated in Figure 6.51. In case (*a*), a new sheet pile wall is installed in front of an existing structure and integrated with the existing structure. Depending on the vertical load, the sheet piling can be made from a regular sheet piling or built in the form of a combi-wall that includes vertical load-carrying king piles and regular sheet piles driven between the king piles. The space between the new sheet piling and the existing piling is filled with granular material (e.g., gravel, coarse sand). Finally, the seafloor in front of the wall is dredged as required. In case (*b*), the wall is upgraded by the installation of a combi-wall in front of the existing structure which eventually allows deepening of the seafloor in front of the existing wall. A number of practical examples of upgrading a piled structure are given in Ritchie and Watson (1983), Hofmann (1989), Simoen et al. (1990), Bray and Tatham (1992), Prucz et al. (1992), Smith et al. (1992), Ducker and Miller (1996a), Tsinker (1997), Ozolin and Soike (1998), and Van Damme et al. (1998).

6.7.3 Sheet Pile Bulkheads

In most cases, upgrading existing sheet pile bulkheads is a complicated task that requires special attention and thorough knowledge of the nature of the interaction between flexible soil-retaining structures and soil. Details were provided earlier in the chapter and in Tsinker (1995, 1997). The complexity of the sheet pile wall–soil interaction stems from the fact that the stress–deflection condition that is inherited by a flexible wall from previous loadings (overloadings), scour or overdredging effects, material deterioration, or other factors is practically irreversible.

If the sheet pile wall strength must be increased, or the seafloor in front of a sheet pile wall needs to be dredged, one of the methods illustrated in Figures 6.52 and 6.53 may be considered. These methods include installation of an additional anchor system(s) aimed at reduction in the wall's span (Figure 6.52*a* and *b*), soil replacement (Figure 6.52*c*), use of soil pressure-relieving systems (Figures 6.52*d* and *f*), installation of brand-new sheeting with partial or full use of remaining useful components of the old wall (Figure 6.52*e*), or construction of a new system immediately behind the existing bulkhead (Figure 6.53). As stated earlier, before attempting any upgrade, the existing sheet pile bulkhead must be repaired or rehabilitated as required. Following is a brief discussion on the systems noted above.

6.7.3.1 Installation of Additional Anchors (Figures 6.52a and b). Because the original anchors are usually placed very close to the mean water level, additional anchors needed to reduce the effective span of existing sheet piling are installed below the existing system and therefore under water. As discussed earlier, this necessitates the use of special underwater techniques or installation of a portable cofferdam around the working area to carry out the work under dry conditions. To be effective, new anchors must be tensioned to prevent or limit deflection of the sheet pile under the design live load. Because anchor tensioning cannot reduce the overall deflection of a sheet pile wall significantly, it also cannot reduce the stresses inherited by this wall from loads applied previously. Additional anchors are usually designed either in the form of ground anchors drilled through the sheeting into the backfill, or in the form of a conventional tieback rod secured at either the existing (sometimes strengthened) anchorage (e.g., concrete deadman, anchor blocks or plates, or piled structures of miscellaneous construction) or at a brand new structure located some distance behind the existing one.

If the existing anchor system cannot safely resist the combined load of existing and new anchor tie rods, it must be strengthened. This can be achieved in a variety of ways; for ex-

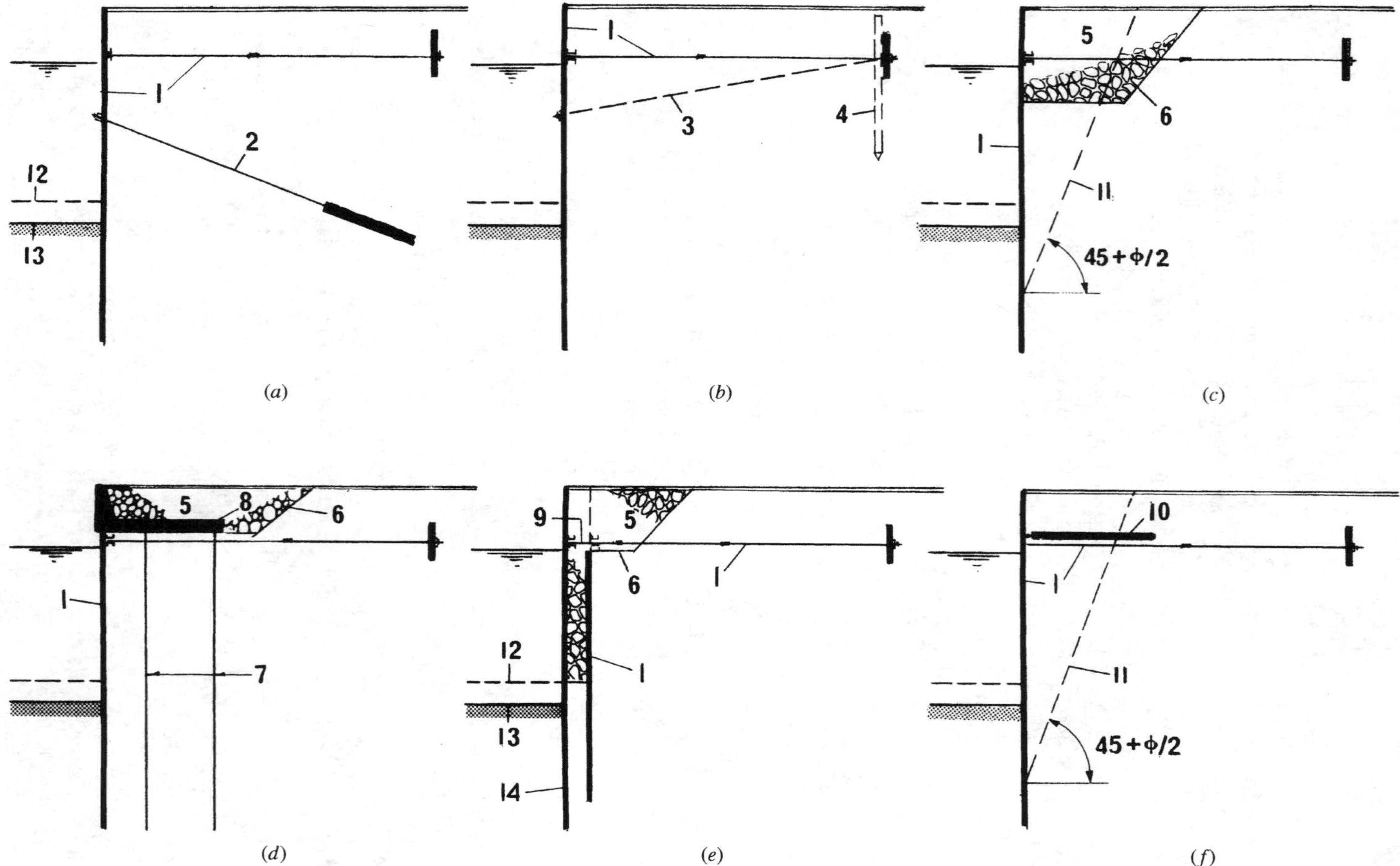

Figure 6.52 Upgrading of anchored sheet pile bulkhead; typical structural arrangements designed to increase the depth of water in front of the wall: (*a*) addition of new ground (rock) anchors; (*b*) installation of additional tieback secured at strengthened existing anchoring system; (*c*) replacement of existing backfill by good-quality granular material; (*d*) adding pressure-relieving system; (*e*) installation of new sheet piling with partial use of some useful components from the old system; (*f*) use of pressure-relieving (anchor) slabs. 1, Existing components of sheet pile bulkhead: sheeting and anchor system; 2, new ground (rock) anchor; 3, new tie rod; 4, new piles driven in front of existing anchor wall; 5, good-quality granular fill; 6, excavation; 7, new piles; 8, new pressure-relieving superstructure; 9, added piece of anchor rod; 10, new pressure-relieving (anchor) slab; 11, slip surface; 12 and 13, old level of seafloor and new dredge line, respectively; 14, new sheeting.

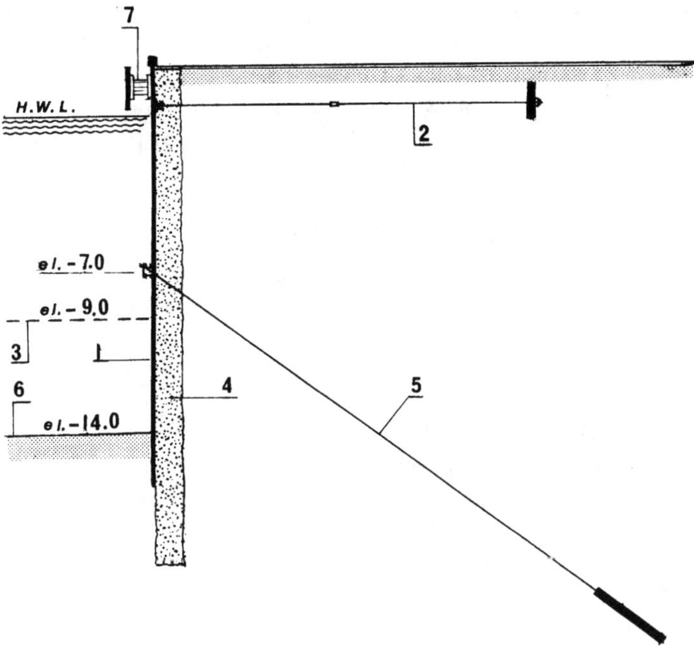

Figure 6.53 Upgrading of steel sheet pile bulkhead at Port of Klipeda, Lithuania. 1 and 2, Existing steel sheeting and anchor system; 3, premodernization level of bottom; 4, new jet grouting reinforced wall; 5, new ground anchor system; 6, new level of bottom; 7, New fender system.

ample, piles can be driven in front of anchor blocks, regular soil in front of anchors can be replaced with well-compacted rock fill, and so on. New tie rods are usually installed from the water side of the wall. Ishiguro and Miyata (1988) reported on new technology developed in Japan for underwater installation of additional anchor tie rods. This technology allows for installation of new tie rods through an existing or new deadman and backfill toward sheet piles and then at desired locations through sheet piles. Information on this technology is provided in Tsinker (1997).

An obvious disadvantage of the installation of additional anchors is that it requires the installation of external wale(s) to distribute the anchor force among individual sheet piles. This wale system can easily be damaged by a ship if not protected properly. On the other hand, it may pierce holes in a ship's hull. Where an external wale is used, the face of the wall superstructure (capping) must be extended far enough seaward to prevent dangerous contact between the wall and a ship.

In the ports of Ostend, Belgium and Klipeda, Lithuania, the sheet pile walls have been upgraded by construction of a reinforced concrete wall adjacent to the existing sheeting and the installation of additional ground anchors (Anonymous, 1997; Habil et al., 2001). At Klipeda, 525 m of existing sheet pile bulkhead has been upgraded by installation behind the existing sheeting of a new concrete wall that allows the water depth to be increased substantially (from 9 to 14 m) in front of the wall (Figure 6.53). The concrete wall was constructed by employing high-pressure grouting technology. The new bulkhead system was provided with an additional anchor system composed of post-tensioned ground anchors angled to the wall. The project was completed by the installation of a new fender system and bollards capable of handling Panamax class vessels.

In many cases, new gantry or portal cranes have to be installed to handle cargo. In the case of sheet pile bulkheads, because of the crane's potential to be located too close to the water-side edge of the dock, the load generated by a heavy crane can be prohibitively high. Hence, a water-side crane rail should be installed on a pile-supported beam. However, driving piles in close proximity to a bulkhead due to displacement of backfill soil toward the sheeting may result in excessive soil pressures on sheet piling and overloading of its anchor system. For details, consult Tsinker (1997). To prevent the above from happening, the bored piles have to be installed instead of conventional piles. An interesting technology for pile installation in close proximity to existing structures has been developed in Japan (Figure 6.54). In this case, the pile is installed in the following sequence:

1. A hollow-core pile is prepositioned at its design location.
2. The pile is driven in sections by a conventional hammer with simultaneous removal of soil from the pile interior; a soil-removing auger is structurally integrated with the jet grouting system.
3. When the pile reaches competent foundation soil, a concrete support block is formed at the pile tip by jet grouting.
4. The auger is removed and the pile is grouted as required.

6.7.3.2 Soil Replacement. As stated earlier, the effect of soil replacement on sheet pile bulkhead performance is less pronounced than it may be in the case of a gravity wall. This is because in practice, the sheeting cannot rebound at its middle section (the location of maximum stress) even after excavation in this area as shown in Figure 6.52*c*. Hence, only the upper part of the wall is free to change its curvature after excavation of the upper part of the existing fill. Some stress reduction in sheet piling can also be achieved by distressing the anchor rod to allow for some slight wall movement away from the fill. Practically, this may be accomplished by loosening the nut at the anchor block or by exposing the turnbuckles and loosening the tension in the rods. Porter (1986) reported a successful sheet pile bulkhead repair using lightweight expanded shale for partial replacement of the existing fill.

6.7.3.3 Pressure-Relieving Systems. These include piled systems as indicated in Figures 6.52*d*, or pressure-relieving slabs as shown in Figure 6.52*f*. In the former case, the existing sheet piling is actually converted into the part of a new structure that is pressure-relieving wall; it can carry much heavier loads, both vertical and horizontal, than could the original sheet pile bulkhead. Anchorage for the new system is provided by the existing anchor system. Again, if necessary, the existing stresses in the sheet-piling can be relieved by moving the sheeting slightly away from the fill. It must be noted that this system is appropriate only when the sheet pile system can carry some vertical load. Normally, sheet pile bulkheads that include large-diameter king piles can carry substantial vertical loads in combination with lateral soil thrust. Again, it must be remembered that pile driving in close proximity to the existing sheeting can be detrimental to the system. Hence, if these piles are required, they can be installed either in a form of conventional bored piles or as discussed in the preceding section. Alternatively, the possibility of installing low-displacement piles (e.g., H-piles or open-ended pipe piles) can be considered. A relevant practical example is provided in Barker and Chattaway (1990).

Incorporation of pressure-relieving slabs into a sheet pile bulkhead system may result in a reduction in soil lateral pressure and better

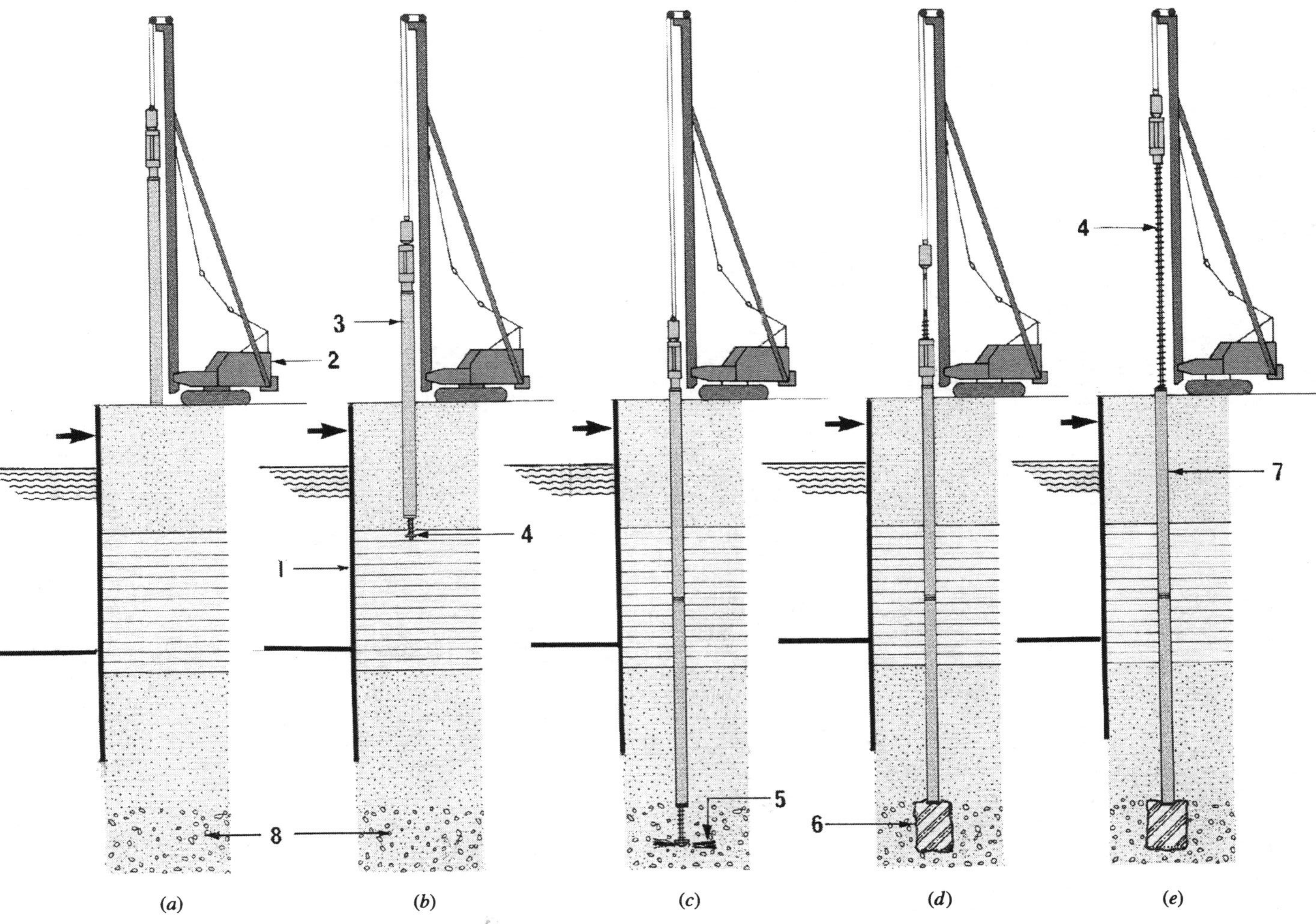

Figure 6.54 Pile installation in close proximity to existing sheet pile wall: (*a*) pile repositioned at design location; (*b*) pile is driven in sections with simultaneous removal of soil from its interior; (*c*) pile has reached competent soil and the soil removal auger, integrated with a jet grouting system, is extended to the depth required for the formation of concrete pile supporting block; (*d*) the block is formed; (*e*) the auger is removed and the pile is grouted as required. 1, Sheet pile bulkhead; 2, pile-driving system; 3, pile section; 4, auger–jet grouting system; 5, jet-grouting process; 6, concrete block; 7, complete pile.

anchorage of sheet piling. When used in sheet pile bulkheads, the pressure-relieving slabs have to have a minimum length equal to no less than $0.7H$, where H is the height from the dredge line to the wall top. This is required to ensure that the slab will intersect a slip surface behind the wall and therefore have support from a stable portion of the backfill at its rear end. These slabs are hinged to the sheet pile system and therefore transmit a vertical load to the sheeting. Hence, the new structural systems that incorporate a pressure-relieving system such as a platform or slab must be reviewed for the combined effect of a reduced lateral soil thrust combined with a vertical load.

Naturally, sheet pile penetration into the foundation soil must be adequate to resist the vertical design load. Sheet pile resistance (Q_R) to vertical loads due to the effects of a pressure-relieving slab (Figure 6.55), $Q_1 + Q_S$, should conform to equilibrium $Q_R = 1.5(Q_1 + Q_S)$, where Q_S is the weight of sheet piling, Q_1 is part of the weight of soil plus any type of live load transmitted through the slab to sheet piling, and 1.5 is a safety factor. Sheet piling resistance to vertical load in general will be $Q_R = Af$, where A is the area of embedded sheet piling and f is the soil friction resistance. The reaction $Q_1 \approx 0.5l(\gamma h + q)$, where γ is the soil unit weight, and q is the live surcharge load. Hence $Af = 1.5[0.5l(\gamma h + q) + Q_S]$, from whence the required minimum length of the soil pressure relieving slab, $l_{\min}$, is formulated as follows:

$$l_{\min} \leq \frac{1.33Af - 2Q_s}{\gamma h + q} \tag{6.14}$$

For detailed information on the design of waterfront structures provided with pressure-relieving slabs, readers are referred to Tsinker (1995).

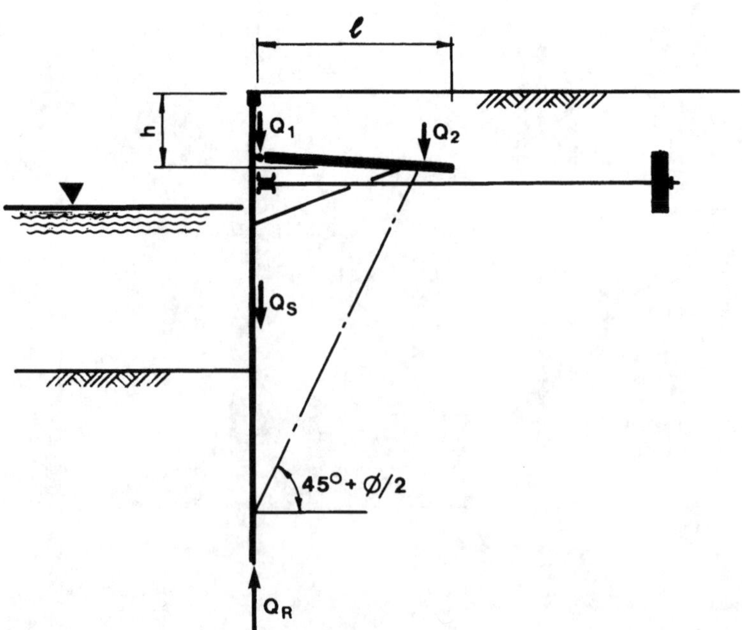

Figure 6.55 Design approach for a sheet pile bulkhead with pressure-relieving slab.

6.7.4 Retrofit of Existing Structures for the Effects of an Earthquake

Past experience has shown that an earthquake can cause serious damage to ports, with substantial loss of operation, resulting in negative economic impacts. For example, the devastating Hyogoken–Nanbu earthquake in 1995 resulted in the loss of virtually all 240 berths in the port of Kobe, Japan's largest port, which before the earthquake handled about 30% of the country's international trade in containers (Matso, 1995; Werner et al., 1996, 1997; Noda et al., 1998). The cost of reconstruction of the port (approximately $150+ billion) dwarfed earthquake damage to ports in California: the Loma Prieta earthquake in 1989 ($10 billion) and the Northridge earthquake in 1994 ($30 billion). Some of the damages experienced at Kobe is shown in Figure 6.56.

The primary cause of earthquake-related damage to port infrastructure is the liquefaction of loose, saturated granular foundation and backfill soils caused by violent ground shaking. Landslides and heavy waves generated by earthquake-induced ground motion may also be a source of hazards that could lead to severe damage to port-related structures. The heavy damage to port infrastructure and loss of operations promoted port operators to initiate massive retrofit programs intended to prevent earthquake-inflicted damages to older structures. Following is a concise discussion of some basic measures that have been developed to make existing structures more earthquake resistant. These developments are based on new and improved national and local standards and regulations concerned with structural design in earthquake-affected areas.

6.7.4.1 Piled Structures. In most cases, older piled structures (e.g., piers and suspended deck platforms) are supported on vertical and batter piles. From many reports it follows that batter piles at open piled structures have often performed poorly during earthquakes. This manifested itself in a form ranging from simple cracks and spalls in piles, to pile tops being shattered and having significant vertical separation from deck structures and horizontal offsets. Vertical piles and deck structures usually suffered much less damage than that of batter piles. This is because the large lateral stiffness of batter piles mobilizes much more seismic force at decking connections than more flexible vertical piles normally do.

The use of batter piles at waterfront structures is typically not advisable because of their generally poor seismic performance. However, where the use of batter piles is mandatory due to nonseismic considerations, basic seismic requirements for pile foundation analysis should be considered (i.e., including estimation of moments, shear force lateral displacement, axial loads, and liquefaction-induced soil movement). Subsequently, all of the above should be included in the design of piles and their connections to the deck structure, to be sufficiently strong and ductile to resist earthquake-induced forces. All the aforementioned usually leave the designer with two basic alternatives for upgrading open pile structures: the addition of new batter piles to an existing system or replacement of existing batter piles with a vertical pile, as required. In general, vertical piles, which are more flexible laterally than batter piles accommodate lateral displacement much better than do the latter. Records indicate that vertical piles, in general, performed satisfactorily during major earthquakes in California and Japan (ASCE, 1998).

Many older piled wharves were constructed under poor geological conditions utilizing nonprestressed piles. These piles typically do not have the ductility of the reinforcing steel normally required in modern piled waterfront structures. The seismic resistance of such structures may be improved by the addition of prestressed concrete piles or cylindrical steel piles. It should be understood, however, that the ad-

Figure 6.56 Port of Kobe, Japan. Damage to port waterfront inflicted by soil liquefaction due to effects of the Hyogoken–Naubu earthquake in 1995.

dition of new piles aimed at stiffening an existing structure while reducing the structure's lateral displacement may, in fact, increase the seismic forces acting against it. The reduction in lateral displacement will be beneficial in reducing the level of damage to existing piles; however, the new piles must be strong enough to resist the increased lateral seismic forces. The upgraded pile system should provide adequate ductility for the structure. For practical examples, the reader is referred to Birdy et al. (1989), Fotinos et al. (1992), and Johnson et al. (1995). For more information on seismic retrofitting of piled waterfront structures, readers are referred to Priestly (1997) and ASCE (1998).

6.7.4.2 Soil-Retaining Structures. The seismic performance of soil-retaining waterfront structures is influenced primarily by soil liquefaction in adjacent backfill material and underlying foundation soil. By definition, liquefaction is a process wherein loose, saturated granular soils subjected to shaking tend to decrease in volume due to collapse of the soil skeleton. This volume reduction is restricted by the rate at which pore water can flow out of the soil matrix, resulting in a dramatic increase in pore pressure and a temporary loss of stiffness and shear strength at the soil. Soil liquefaction causes an increase in lateral pressure against a quay wall that, in turn, may cause lateral, vertical, and rotational movement of the wall. Soil liquefaction phenomena are quite complex; for a detailed discussion, interested readers are referred to Wong et al. (1975), Prakash (1994), Vahdani et al. (1994), Evans and Zhon (1995), and Kramer (1996), Port and Harbour Research Institute (PHRI), Japan (1997), and ASCE (1998). As noted earlier, soil liquefaction and its impact on retaining structures manifested itself in the most spectacular way during the Hyogoken–Naubu (1995) earthquake that struck the port of Kobe. In the process of planning retrofitting measures concerned with the earthquake performance of soil-retaining quay wall, the liquefaction potential of backfill materials and foundation soils must be evaluated thoroughly. Soil liquefaction potentials can be predicted from the gradation and SPT N value or by performing cyclic triaxial tests on undisturbed samples. The latter method is based fundamentally on a procedure proposed by Seed and Idriss (1971) which evaluates soil liquefaction potentials by comparing the liquefaction strength of soil and the cyclic shear stress generated in soil layers during an earthquake.

Remediation measures against soil liquefaction should conform to the extent of seismic risk appropriate for the particular port. In the process, any potentially liquefiable materials in backfill and foundation soils must be identified and fully analyzed and the expected effects of liquefaction on the wall predicted. For this, special attention should be paid to the acceptability of the amount of settlement that can be tolerated and the effects of the increase in soil lateral thrust against the wall. If potentially unstable soils are identified, remedial strategies should be developed. This is especially important for sheet pile bulkheads, which are sensitive to severe overloading of the sheeting and loss of support at the anchoring system (e.g., blocks, wall). Details are provided in Tsinker (1997).

In general, the performance of soil-retaining structures exposed to seismic conditions must be analyzed with a focus on the dynamic behavior of the foundation and backfill soils, with an emphasis on the overall stability of the structure (e.g., sliding, overturning, bearing capacity, deep-seated instability). For the relevant guidelines, readers are referred to Kramer (1996), Prakash (1994), ASCE (1998), PHRI (1997), and Tsinker (1997). Remedial methods used to date typically include soil compaction, soil solidification by mixing in cementations materials, soil replacement, and reducing the pore pressure in backfill material during an

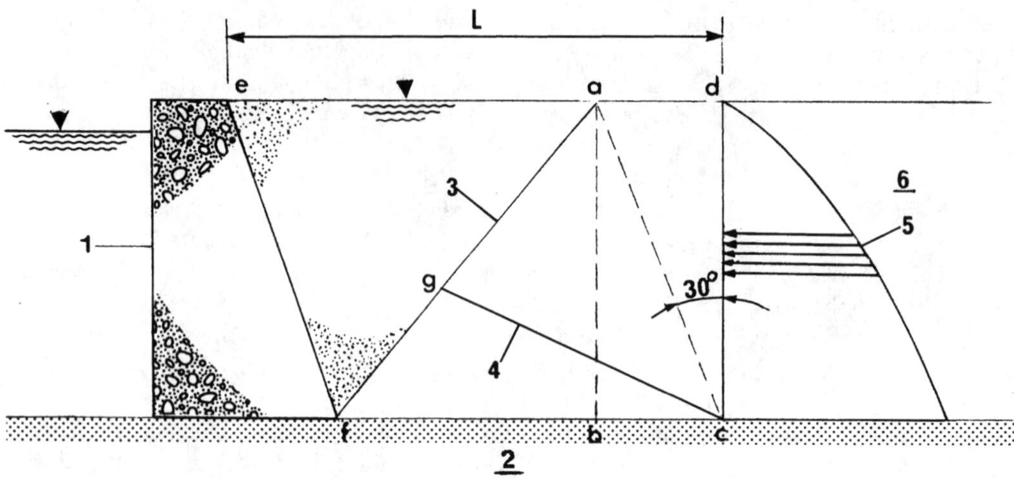

Figure 6.57 Investigation of stability of a gravity wall with respect to pressure from liquefied backfill material. *L*, Length of soil improvement area; 1, gravity wall; 2, nonliquefiable foundation soil; 3, active failure surface; 4, passive failure surface; 5, lateral pressure diagram from liquefied soil; 6, liquefied soil. (After PHRI, 1997.)

earthquake by installation of vertical drains made of coarse-grained materials (e.g., gravel, stone). All these methods were discussed earlier. Installation of vertical drains for liquefaction remediation is efficient and the most frequently used method for upgrading soil-retaining structures and slopes. Drains can be formed using a variety of methods. One of these is illustrated in Figure 6.33. It should be noted that to reduce the chances of drains being clogged by the penetration of fine particles into the column, they should be constructed from selected crushed stone or gravel. The design of drains (e.g., their size and installation pattern) is based on a balance between the rate of excess pore water pressure increase due to earthquake motions and the rate of dissipation of pore water pressure due to drainage. Essentially, for better results the design should be based on detailed information on the permeability of surrounding soil and its liquefaction strength. For a detailed design procedure, readers are referred to PHRI (1997). Useful guidelines are also provided in ASCE (1998). Based on extensive research on seepage flow analysis, PHRI (1997) recommends that for a gravity wall, the backfill soil should be improved at least to the distance indicated in Figure 6.57. In the figure, the lateral pressure at the boundary *ef* is normal active earth pressure, whereas pressure at the boundary *cd* (i.e., pressure from liquefied soil) is much greater than active soil pressure.

The pseudostatic approach to system equilibrium indicated in Figure 6.57 requires that sufficient shear resistance be mobilized in the body of backfill soil along the passive failure surface *gc*. For detailed information on miscellaneous methods used for determining the required extent of the soil improvement area (*L*) and a comprehensive list of relevant references, readers are referred to PHRI (1997) and ASCE (1998). The relevant practical example is found in Sonu et al. (1993) and most recently in Noda et al. (1998). As confirmed by Sonu et al. (1993), in some instances the danger of soil liquefaction can be precluded successfully by the installation of properly spaced gravel drains. Such drains may help to reduce the amount of pore water pressure due to earthquake effects by providing an escape route for excess pore water via vertical drains. As pointed out by Sonu et al., the soil densification could be detrimental to the effectiveness of gravel drains because the reduced pores space would retard the draining speed of the ground pore water. The investigators indicated that several marine facilities in Japan protected by gravel drains suffered no damage due to earthquake-induced soil liquefaction. This is best illustrated by a structure constructed at Fishermen's Wharf in the port of Kushiro, Hokkaido, Japan. The wharf's bulkhead is made of steel cylinders 9.17 m in diameter and approximately 9.0 m high. These cylinders are filled with dredged sand. Gravel drains 400 mm in diameter and approximately 6.0 m long were installed within this cellular bulkhead at a space of 1.2 m center to center. After installation the drains were topped with layers of gravel and paved over. On January 15, 1993, the port of Kushiro was struck by a major earthquake (magnitude 7.8 on the Richter scale) and a postearthquake inspection of port facilities demonstrated decisively that gravel drains can effectively prevent soil liquefaction. Fishermen's Wharf suffered no damage; there were no indications of soil liquefaction in the gravel-drain-protected area. In sharp contrast, adjacent areas immediately outside the protected structure suffered severe damage from soil liquefaction.

The seismic performance of sheet pile bulkhead, in general, has been reported as poor (Iai

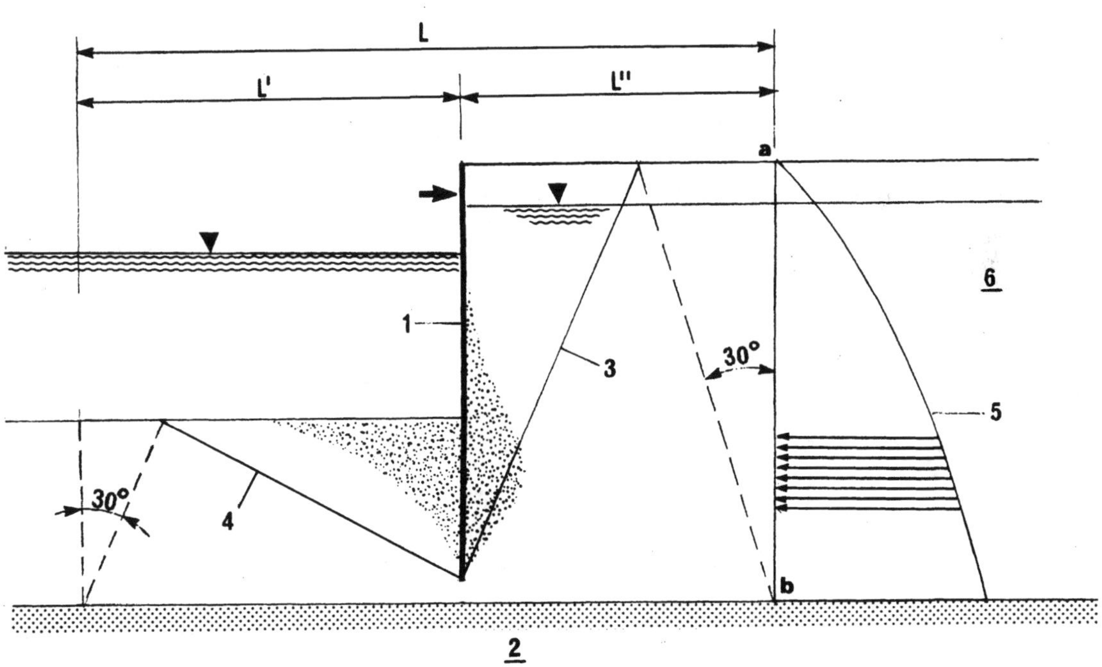

Figure 6.58 Basic diagram for investigating the stability of sheet pile bulkhead with respect to pressure from liquefied backfill material. *L*, Soil improvement area; 1, sheet pile bulkhead; 2, nonliquefiable soil; 3, active failure surface; 4, passive failure surface; 5, lateral pressure diagram from liquefied soil; 6, liquefiable soil. (After PHRI, 1997.)

et al., 1994; Vahdani et al., 1994; Mejia and Yueng, 1995; Tsinker, 1997). It usually manifested itself in excessive wall deformation, tie rod/wale/anchorage system failure, failure of the soil passive resistance zone, global instability, and so on. Loss of passive soil resistance at the anchorage and below the dredge line due to soil liquefaction in some cases resulted in catastrophic failures of bulkheads. The seismic performance requirements for anchored sheet pile bulkheads are discussed in Ferritto (1997), PHRI (1997), ASCE (1998), and Dickenson and McCullough (1998). Similar to gravity retaining structures, permanent lateral displacement at the top of the sheeting should be limited to values that are consistent with the

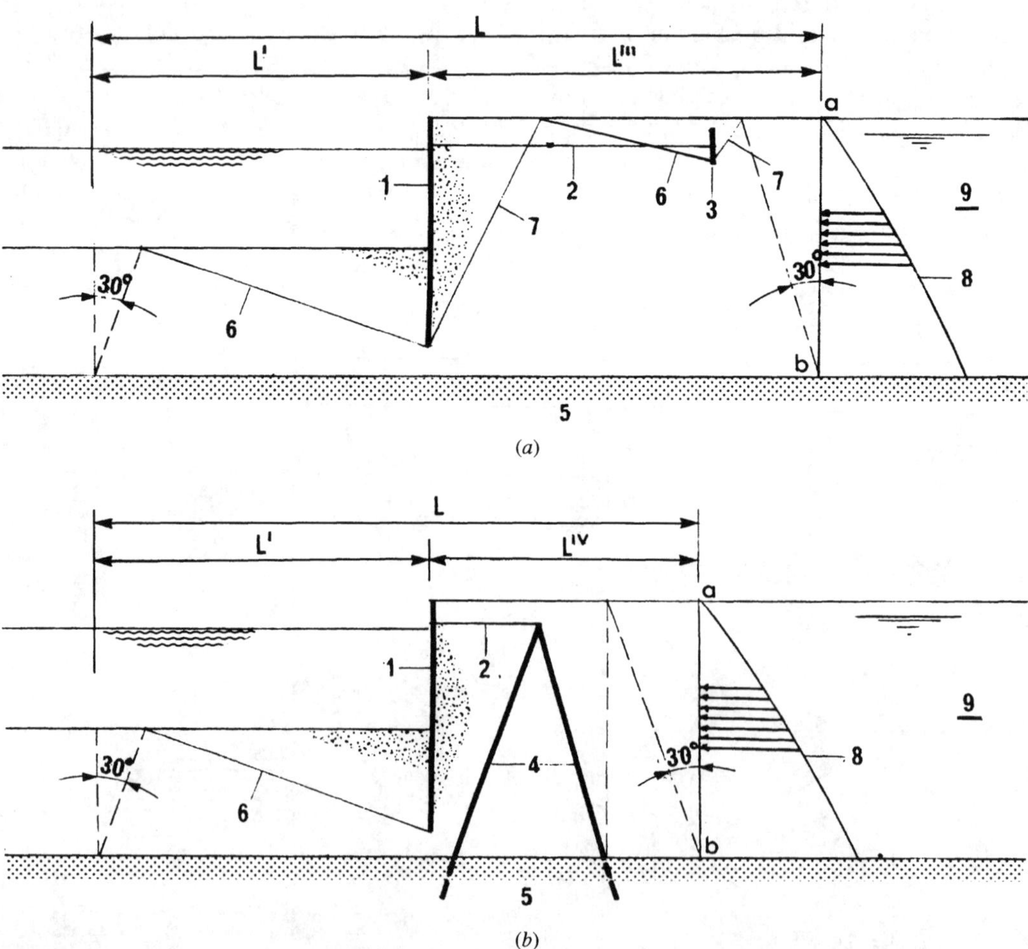

Figure 6.59 Influence of existing anchor system on the extent of soil improvement on the land side of the sheet pile bulkhead: (*a*) sheeting anchored by means of a tie rod secured at a deadman, anchor plate, etc.; (*b*) sheeting anchored by means of a tie rod secured at an A-frame pile system. 1, Sheet piling; 2, tie rod; 3, anchor block; 4, anchor piles; 5, nonliquefiable soil; 6, passive failure surface; 7, active failure surface; 8, lateral pressure diagram from liquefied soil; 9, liquefiable soil.

displacement tolerances of port components such as crane rails or other vital port elements located near the bulkhead. To prevent all of the above from happening, the seismic behavior of existing (as well as new) sheet pile bulkheads should be reviewed with emphasis on the dynamic behavior of the foundation and backfill soils and the increase in lateral soil thrust and decrease in soil passive resistance below the dredging line due to earthquake effects (Ebeling and Morrison, 1993; U.S. Army Corps of Engineers, 1994; Kramer, 1996).

Liquefaction-related phenomena are the primary seismic hazards to sheet pile bulkheads. Hence, to reduce the risk of failure of existing bulkheads, the water-end land-side soils adjacent to the bulkheads should be improved accordingly. The extent of a soil improvement area depends largely on the depth of sheet pile penetration into the foundation soil and its anchor system type (e.g., deadman anchorage, A-frame batter pile system, sheet pile wall). Examples are shown in Figures 6.58 and 6.59. Figure 6.58 illustrates a general approach to determination of a soil improvement area; it is divided into two parts: water- and land-side areas. Two examples are shown in Figure 6.59. The extent of soil improvement on the water side are (L^{1}) depends basically on sheet pile penetration in the foundation soil, while soil improvement (L^{111} and L^{IV}) on the land side largely depends on the type of bulkhead anchoring system. In any case, the extent of soil improvement on the land side (e.g., L^{111} and L^{IV}) should not be less than L^{11}, as indicated in Figure 6.58. The latter could be especially relevant when wall anchorage is designed in form of an A-frame batter pile system (Figure 6.59*b*). Seismic reevaluation of existing sheet pile bulkheads may indicate that in addition to the soil improvement on both the water and land sides of the wall, the system is strengthened structurally by the rehabilitation or repair of damaged structural components, installation of an additional anchoring system, and so on.

REFERENCES

Abood, K. A., M. J. Ganas, and A. Matlin, 1995. The Teredos are coming! The Teredos are coming! *ASCE Proc. Specialty Conference Ports '95,* Tampa, FL.

Acres International Ltd., 1985. Rehabilitation wharf no. 7 at Port-of-Span, Trinidad, unpublished report.

———, 1989. Alkali aggregate reaction in hydraulic structures, *Proc. 20th Annual Seminar,* Niagara Falls, Ontario, Canada, Apr. 14–15.

Agi, J. J., 1983. Structural evaluation of marine structures, *ASCE Proc. Specialty Conference Ports '83,* New Orleans, LA.

———, 1997. MAST: A marine structures maintenance system, *Proc. 4th International Seminar on Renovation and Improvements to Existing Quay Structures,* Gdansk Technical University, Gdansk, Poland, May.

Alexeev, I., 2001. New technologies and experience of repair of quays in Baltic Sea ports, *Proc. 5th International Seminar on Renovation and Improvements to Existing Quay Structures,* Vol. 1, Gdansk Technical University, Gdansk, Poland.

Alhimenko, A. L., and N. D. Belyaev, 2001. Application of propeller induced scour calculation method to various physical processes, *Proc. 5th International Seminar on Renovation and Improvements to Existing Quay Structures,* Vol. 1, Gdansk Technical University, Gdansk, Poland.

Altiero, K., 1997. Engineers ignore significant deterioration, *Civil Eng.,* January.

ACI, Committee 224, 1984. Causes, evaluation and repair of cracks in concrete structures, *ACI J.,* May–June.

ACI, 1985a. Guide to durable concrete, in *Manual of Concrete Practice,* ACI 201.2R, Part 1, American Concrete Institute, Detroit, MI.

———, 1985b. *Guide to the Use of Waterproofing, Dampproofing, Protective, and Decorative Barrier System for Concrete,* ACI 515.1R, Part 5, American Concrete Institue, Detroit, MI.

———, 1985c. *Guide for Making a Condition Survey of Concrete in Service,* ACI 201.1R, Part 1, American Concrete Institute, Detroit, MI.

———, 1989. Concrete placed under water, Chapter 8 in *Guide for Mixing, Measuring, Transporting and Placing Concrete,* ACI 304R-89, American Concrete Institute, Detroit, MI.

Anonymous, 1997. HSS quay works commence at Osfend and Klipeda, *Dredg. Port Constr.,* May.

Asavarenngchai, S., S. Kulsa, and D. Torseth, 1995. 1994 Port of Seattle timber study, *ASCE Proc. Specialty Conference Ports '95,* Tampa, FL.

ASCE, 1982. *Evaluation, Maintenance and Upgrading of Wood Structures: A Guide and Commentary.* American Society of Civil Engineers. Reston, VA.

———, 1998. *Seismic Guidelines for Ports,* S. D. Werner, ed., Monogr. 12, American Society of Civil Engineers, Reston, VA.

———, 2001. *Standard Practice Manual for Underwater Investigations,* American Society of Civil Engineers, Reston, VA.

Atwood, W. G., and A. A. Johnson, 1924. *Marine Structures: Their Deterioration and Preservation,* National Research Council, Washington, DC.

Ayers, J. R., and R. C. Stokes, 1961. Corrosion of steel piles in salt water, *Proc. ASCE,* WW-3, Vol. 87, Aug.

Baileys, R. T., 1995. Timber structures and the marine environment, *ASCE Proc. Specialty Conference Ports '95,* Tampa, FL.

Baker, W. H., 1982. Planning and preparing structural chemical grouting, *ASCE Proc. Conference on Grouting in Geotechnical Engineering*. New Orleans, LA, Feb.

Bakun, G. B., 1986. Cost-effective approach to pier inspection and maintenance, in *Maritime and Offshore Structure Maintenance: Proc. Institution of Civil Engineers Conference,* Thomas Telford, London.

Barker, J. E., and J. Chattaway, 1990. Deepening and developing quay walls in Felixstowe Port: examples of special design and construction technologies in difficult situations, *PIANC Proc. 27th International Congress,* Osaka, Japan.

Barquett, R. L., and K. M. Childs, 1983. Evaluation of publication analysis techniques for evaluation of the condition of waterfront structures, presented at the Non-destructive Engineering Annual Meeting, University of California–Santa Cruz.

Belyaev, N. D., 2001. Influence of ship propeller jet on formation of scour hole, *Proc. 5th International Seminar on Renovation and Improvement to Existing Quay Structures,* Vol. 1, Gdansk Technical University, Gdansk, Poland.

Bergh, H., and N. Magnusson, 1987. *Propeller Erosion and Protection Methods Used in Ferry Terminals in the Port of Stolkholm,* Bull. 58, Permanent International Association of Navigation Congresses, Brussels, Belgium.

Birdy, J. N., A. L. Fly, and W. E. Hurtienne, 1989. Effect of pile reinforcement on seismic behavior of concrete wharf, *ASCE Proc. Specialty Conference Ports '89,* Boston.

Blakita, P. M., and J. K. Cavey, 1995. Rest in peace, *Civil Eng.,* Dec.

Bletchly, J. D., 1967. *Insect and Marine Borer Damage to Timber and Woodwork,* Ministry of Technology, Her Majesty's Stationery Office, London.

Blokland, T., and R. H. Smedes, 1996. In-situ tests of current velocities and stone movements caused by a propeller jet against a vertical quay wall, *Proc. 11th International Harbour Congress,* Antwerp, Belgium.

Boone, W. M., D. B. Clark, L. Eddy, and E. I. T. Theisz, 1995. Marine concrete repair utilizing silica fume and polyfiber reinforced shotcrete, *ASCE Proc. Specialty Conference Ports '95,* Tampa, FL.

Bray, R. N., and P. F. B. Tatham, 1992. *Old Waterfront Walls: Management, Maintenance and Rehabilitation,* E & FN Spon, London.

Brazkiewicz, I., 1997. Remarks concerning determination of propeller jets behind ships, *Proc. 4th International Seminar on Renovation and Improvements to Existing Quay Structures,* Gdansk Technical University, Gdansk, Poland, May.

Browne, R. D., 1980. Mechanisms of corrosion of steel in concrete in relation to design, inspection, and repair of offshore and coastal structures, in *Performance of Concrete in Marine Environment,* SP-G5, American Concrete Institute, Detroit, MI.

Bruce, D. A., 1996. The return of deep soil mixing, *Civil Eng.,* Dec.

Burke, G. K., D. L. Lyle, A. L. Sehn, and T. E. Ross, 2001. Soil mixing supports a deepwater bulkhead in soft soils, *ASCE Proc. Specialty Conference Ports '01,* Norfolk, VA.

Buslov, V. M., 1979. Durability of various types of wharves, *ASCE Proc. Specialty Conference Coastal Structures '79.*

———, 1983. Corrosion of steel sheet piles in port structures, *ASCE J. Waterway Port Coast. Ocean Eng.,* Vol. 109, No. 3, Aug.

———, 1992. Marine concrete: when to repair, what to repair, *Concrete Int.,* May.

———, 1998. Durability and maintenance of concrete port structures: American experience, *Proc. 29th PIANC International Navigation Congress,* The Hague, The Netherlands.

———, 2001. System approach to inspection, maintenance and repair of port structures, *Proc. 5th International Seminar on Renovation and Improvements to Existing Quay Structures,* Gdansk Technical University, Gdansk, Poland.

Buslov, V. M., and D. V. Padron, 1992. Return of borers, *Military Eng.*

Buslov, V. M., R. E. Heffron, M. Rowghani, and T. E. Spencer, 1998. Underwater cracking of prestressed concrete piles, *ASCE Proc. Specialty Conference Ports '98,* Long Beach, CA.

Buslov, V. M., R. E. Heffron, and A. Martirossyan, 2001. Choosing a rational sample size for the underwater inspection of marine structures, *ASCE Proc. Specialty Conference Ports '01,* Norfolk, VA.

Castelli, R. J., 1991. *Vibratory Deep Compaction of Underwater Fill,* authorized reprint from ASTM Standard Tech. Publ. 1089.

Chait, S., 1987. *Undermining of Quay-Walls at South African Ports due to the Use of Bow Thrusters and Other Propeller Units,* Bull. 58, Permanent International Association of Navigation Congresses, Brussels, Belgium.

Chellis, R. D., 1961. *Pile Foundations,* McGraw-Hill, New York.

Chernov, V., and V. Buslov, 2001. Piles rehabilitation in the Port-of-Haita, *Proc. 5th International Seminar on Renovation and Improvements to Existing Quay Structures,* Gdansk Technical University, Gdansk, Poland, May.

Chin, C. O., Y. M. Chiew, S. Y. Lim, and F. H. Lim, 1996. Jet scour around vertical pile, *ASCE J. Waterway Port Coast. Ocean Eng.,* Mar./Apr.

CIRIA/CUR, 1991. Spec. Publ. 83, Rep. 154, Construction Industry Research and Information Association.

Cleary, H. J., 1969. On the mechanism of corrosion of steel immersed in saline water, *Proc. Offshore Technology Conference,* Houston, TX.

Collins, T. J., and M. J. Garlich, 1998. Pile jackets: issues in design and performance, *ASCE Proc. Specialty Conference Ports '98,* Long Beach, CA.

Coomber, D., 1987. Restoration of quay wall, *Civil Eng.,* Jan./Feb.

Creamer, E. V., 1970. *Splash Zone Protection of Marine Structures,* OTC Pap. 1274, Offshore Technology Conference, Houston, TX.

DeMan, M., W. Thibaut, R. Dedeyne, P. Depoorter, P. Bouquet, and L. Maertens, 1996. Some specific design aspects in quay wall renovation projects, *Proc. 11th International Harbour Congress,* Amsterdam.

Devore, J. L., and R. Farnum, 1999. *Applied Statistics for Engineers and Scientists.* International Thomson Publishing, Andover, Hampshire, England.

DiCastro, D., P. Harari, M. Radmir, V. M. Buslov, and J. A. Zwamboru, 1998. Service performance of marine structures in ports of Israel. *Proc. 29th PIANC Congress,* The Hague, The Netherlands.

Dickenson, S. E., and N. McCullough, 1998. Mitigation of liquefaction hazards to anchor sheet pile bulkheads, *ASCE Proc. Specialty Conference Ports '98,* Long Beach, CA, Mar.

Dismuke, T. D., S. F. Coburn, and C. M. Hirsh, 1981. *Handbook of Corrosion Protection for Steel Pile Structures in Marine Environments,* American Iron and Steel Institute, Washington, DC.

Ducker, H. P., and C. Miller, 1996a. Comparison of technical and cost factors regarding different construction designs for the building of a container berth, *Proc. 11th International Harbour Congress,* Antwerp, Belgium.

———, 1996b. Harbour bottom erosion at berths due to propeller jet, *Proc. 11th International Harbour Congress,* Antwerp, Belgium.

Dunnicliff, J., 1988. *Geotechnical Instrumentation for Monitoring Field Performance,* Wiley, New York.

Ebeling, R. M., and E. E. Morrison, 1993. *The Seismic Design of Waterfront Retaining Structures,* Tech. Rep. ITL-92-11, NCEL TR-939, U.S. Naval Civil Engineering Laboratory, Port Hueueme, CA.

Edwards, W. E., 1963. Marine corrosion: its cause and cure, *Proc. Annual Appalachian Underground Short Course,* Tech. Bull. 69, West Virginia University, Morgantown, WV, Oct.

Emmons, P. H., A. M. Vaysburd, and J. E. McDonald, 1993. A rational approach to durable concrete repairs, *Concrete Int.,* Vol. 15, No. 9, Sept.

Evans, M. D., and S. Zhan, 1995. Liquefaction behavior of sand–gravel composites, *ASCE J. Geotech. Eng.,* Vol. 121, No. 3.

Eyre, J. R., and M. A. Nokhasteh, 1992. Strength assessment of corrosion damaged reinforced concrete slabs and beams, Pap. 9851, *Proc. Institute of Civil Engineers Structures and Buildings,* No. 94, May.

Ferritto, J., 1997. *Design Criteria for Earthquake Hazard Mitigation of Navy Piers and Wharves,* Tech. Rep. TR-2069-SHR, U.S. Naval Facilities Engineering Services Center, Port Hueneme, CA, Mar.

Fotinos, G. C., G. M. Servenbi, and L. L. Scheibel, 1992. Earthquake damage repair and retrofit of the 7th Street terminal Port-of-Oakland, *ASCE Proc. Specialty Conference Ports '92,* Seattle, WA.

Freyermuth, C. L., 2001. Life-cycle cost analysis for large segmental bridges, *Concrete Int.,* Feb.

Fukute, T., O. Kiyomiya, and K. Minami, 1990. *Steel Structures in Port and Harbour Facilities: Actual Conditions of Corrosion and Counter-Measures,* Bull. 68, Permanent International Association of Navigation Congresses, Brussels, Belgium.

Garlich, M. J., and M. J. Chrzastowski, 1989. An example of sidescan sonar in waterfront facility evaluation, *ASCE Proc. Specialty Conference Ports '89,* Boston, May.

Gaythwaite, J. W., 1981. *The Marine Environment and Structural Design,* Van Nostrand Reinhold, New York.

———, 1990. *Design of Marine Facilities for Berthing, Mooring and Repair of Vessels,* Van Nostrand Reinhold, New York.

Gaythwaite, J. W., and D. R. Carchedi, 1984. *Rehabilitation of Waterfront Structures: Design Criteria and General Considerations,* presented at Seminar on Rehabilitation of Waterfront Structures, Boston Society of Civil Engineers, Section of ASCE, Boston.

Gilbride, P., D. R. Morgan, and T. W. Bremner, 1988. Deterioration and rehabilitation of berth faces in tidal zones at the Port of Saint John, *Proc. 2nd International Conference on Concrete in Marine Environment,* SP-109, St. Andrews, New Brunswick, Canada.

Gularte, F. B., G. E. Taylor, and R. H. Borden, 1992. Temporary tunnel excavation support by chemical grouting. *ASCE Proc. Conference on Grouting, Soil Improvements and Geosynthetics,* New Orleans, LA, Feb.

Gunas, M. J., M. P. Hunnemann, and D. R. Goulet, 1993. Marine borer activity on the rise in New York harbor, *Public Works.*

Habil, W. J., and K. Van der Eecken, 2001. VHP method for the quay wall renovation, *Proc. 5th International Seminar on Renovation and Improvements to Existing Quay Structures,* Gdansk Technical University, Gdansk, Poland.

Hallett, D., D. Pritchard, and M. Meletion, 1992. Innovative underpinning technique permitting alongside deepening at existing gravity quay walls, *Proc. 10th International Congress,* Antwerp, Belgium.

Hamil, G. A., H. T. Johnston, and D. P. Stewart, 1999. Propeller wash scour near quay wall, *ASCE J. Waterway Port Coast. Ocean Eng.,* July/Aug.

Hamil, G. A., J. A. McGarvey, and D. A. B. Hughes, 2001. *The Effect of Rudder Angle on the Scouring Action Produced by the Propeller Wash of a Maneuvering Ship,* Bull. 106-2001, Permanent International Association of Navigation Congresses, Brussels, Belgium.

Han-Padron Associates, 2001. Waterfront bulkhead life cycle assessment study, unpublished report, H-P, New York.

Haroske, G., 2001. *Wear Resistant Concrete in Hydraulic Structures,* Bull. 108, International Navigation Association, Brussels, Belgium, Sept.

Heffron, R., and K. M. Childs, 2001. New ASCE standard practice manual for underwater investigations, *ASCE Proc. Specialty Conference Ports '01,* Norfolk, VA.

Hildebrand, F. B., 1968. *Finite Difference Equations and Simulations,* Prentice Hall, Upper Saddle River, NJ.

Hobbs, D. W., 1988. *Alkali–Silica Reaction in Concrete,* Thomas Telford, London.

Hoff, G. C., 1987. Durability of fiber reinforced concrete in a severe marine environment, presented at the Catharine and Bryant Mather International Conference on Concrete Durability, Atlanta, GA, Apr.

———, 1988. Resistance of concrete to ice abrasion: a review, *Proc. 2nd International Conference on Concrete in Marine Environment,* SD-109, St. Andrews, New Brunswick, Canada.

Hofmann, K. F., 1989. *New Methods for the Construction of Quay Walls, in the Port-of-Hamburg,* Bulletin 64, Permanent International Association of Navigation Congresses, Brussels, Belgium.

Holland, T. C., T. B. Husbuands, A. D. Buck, and G. S. Wong, 1980. *Concrete Deterioration in Spillway Warm-Water Chute, Raystown Dam, Pennsylvania.* MP SL-80-19, U.S. Army Corps of Engineers Waterways Experiment Station, Vicksburg, MS.

Holmes, C. W., and S. G. Brundle, 1987. The effect of an arid climate on reinforced concrete in marine environment, *Proc. Conference on Coastal and Port Engineering in Developing Countries,* Vol. 1, China Ocean Press, Beijing, China.

Horvath, J. S., and J. T. Dette, 1983. Rehabilitation of failed steel sheet-pile bulkheads, *ASCE Proc. Specialty Conference Ports '83,* New Orleans, LA, Mar. 21–23.

Iai, S., Y. Matsunaga, T. Morita, M. Miyata, H. Sakurai, H. Oishi, H. Ogura, Y. Audo, Y. Tanaka, and M. Kato, 1994. Effects of remedial measures against liquefaction at 1993 Kushiro–Oki earthquake, *Proc. 5th U.S.–Japan Workshop on Earthquake Resistant Design of Lifeline Facilities and Countermeasures against Soil Liquefaction,* National Center for Earthquake Engineering Research, Buffalo, NY.

INA/PIANC, 1998. *Life Cycle Management of Port Structures: General Principles,* Rep. Work. Group 31, Suppl. Bull. 99. International Navigation Association, Brussels, Belgium.

———, 2002. *Guidelines for the Design of Fender Systems: 2002,* International Navigation Association, Brussels, Belgium.

Ingold, T. S., and K. S. Miller, 1988. *Geotextiles Handbook,* Thomas Telford, London.

Ishiguro, K., and Y. Miyata, 1988. *Reinforcement of an Anchored Sheet Pile Wall with Additional Lower Tie-Rods,* Bull. 61, Permanent International Association of Navigation Congresses, Brussels, Belgium.

Johnson, S. M., 1965. *Deterioration, Maintenance and Repair of Structures,* McGraw-Hill, New York.

———, 1989. Design of marine structures: life cycle cost factor, *ASCE Proc. Specialty Conference Ports '89,* Boston, May.

Johnson, R., G. England, and D. Baska, 1995. Port-of-Seattle terminal 5; practical seismic design for a concrete pile-supported wharf, *ASCE Proc. Specialty Conference Ports '92,* Tampa, FL.

Kee, Chin Fung, 1971. Relation between strength and maturity of concrete, *ACI J. Proc.,* Vol. 68, No. 3, Mar.

Khanna, J., P. Gilbride, and C. R. Whitcomb, 1988. Steel fiber reinforced concrete jackets for repairing concrete piles, *Proc. 2nd International Conference on Concrete in Marine Environment,* SP-109, St. Andrews, New Brunswick, Canada.

Khayat, K. H., 1992. In-situ properties of concrete piles repaired under water, *Concrete Int.,* Mar.

Kramer, S. L., 1996. *Geotechnical Earthquake Engineering,* Prentice Hall, Upper Saddle River, NJ.

Kray, C. J., 1983. Rehabilitation of steel sheet pile bulkhead walls, *ASCE Proc. Specialty Conference Ports '83,* New Orleans, LA, Mar. 21–23.

Kristensen, M., 1987. *Seabed Erosion in Ferry Berths,* Bull. 58, Permanent International Association of Navigation Congresses, Brussels, Belgium.

Lamberton, B. A., 1989. Fabric forms for concrete, *Concrete Intl.,* Dec.

Laque, F. L., 1975. *Marine Corrosion: Causes and Prevention,* Wiley, New York.

Leng, D. L., 2000. Zinc mesh cathodic protection systems, *Mater. Perform.,* Aug.

Lin, T. Y., and N. H. Burns, 1981. *Design of Prestressed Concrete Structures,* 3rd ed., Wiley, New York.

Lindner, J. T., V. M. Buslov, R. R. Hernandez, V. Hall, and M. Rowghani, 1995. Concrete wharf inspection and maintenance program port of Los Angeles, *ASCE Proc. Specialty Conference Ports '95,* Tampa, FL (originally published in *Port Technol. Int.,* No. 2, London).

Longe, M. J. P., M. P. Hergert, and M. R. Bylk, 1987. *Problems d'érosion aux ouvrages de quai existants causes par les propulseurs d'étrave et les helices principles des novires lors de deurs accostages aou appareillages,* Bull. 58, Permanent International Association of Navigation Congresses, Brussels, Belgium.

Magnat, P. S., and K. Gurusamy, 1998. Corrosion resistance of steel fibers in concrete under marine exposure, *Cement Concrete Res.,* Vol. 18.

Manfakh, G. A., 1995. *Deep Chemical Injection for Protection of Old Tunnel,* Tech. Publ. 1089, American Society for Testing and Materials, Philadelphia.

Martinez, M. N., 1998. Cyclopean seawall repair and berth deepening, 10th Avenue marine terminal, San Diego, California, *ASCE Proc. Specialty Conference Ports '98,* Long Beach, CA.

Matso, K., 1995. Lessons from Kobe, *Civil Eng.,* Apr.

Maynard, S. T., 1999. *Inflow Zone and Discharge through Propeller Jet,* Bull. 102, International Navigation Association, Brussels, Belgium.

McGarvey, J. A., 1996. The influence of rudder on the hydrodynamics and the resulting bed scour of a ship's screw wash, thesis submitted to the Queen's University of Belfast for the degree of Ph.D., Belfast, Ireland.

Mejia, L. H., and M. R. Yueng, 1995. Liquefaction of Caoralline soils during the 1993 Guam earthquake, in *Earthquake-Induced Movements and Seismic Remediation of Existing Foundations and Abutments,* Geotech. Spec. Publ. 55, American Society of Civil Engineers, Reston, VA.

Menzies, R. J., and R. Turner, 1957. The distribution and importance of marine wood borers in the United States, in *Proc. Symposium on Wood for Marine Use and Its Protection from Marine Organisms,* Spec. Tech. Publ. 200, American Society for Testing and Materials, Philadelphia.

Metzger, S. G., 1998. The *Limnoria* has landed, *Proc. ASCE Specialty Conference Ports '98,* Long Beach, CA, Mar.

Milwee, W. I., and W. F. Aichele, 1986. Modern inspection techniques in port maintenance, *ASCE Proc. Specialty Conference Ports '86,* Oakland, CA, May 19–21.

Momber, A., 1998. The case for waterjetting, *Concrete Eng. Int.,* Apr.

Morgan, D. R., 1988. Recent developments in shotcrete technology: a materials engineering perspective, presented at World of Concrete '88, Las Vegas, NV, Feb.

Morley, M. A., G. M. Serventi, and J. A. Ejan, 1995. Design considerations for the expansion of Charles P. Howard Terminal, Port of Oakland, *ASCE Proc. Specialty Conference Ports '95,* Tampa, FL.

Neville, A., 1997. Maintenance and durability of structures, *ACI Concrete Int.,* Nov.

Nieswaag, B., P. Hagenaars, P. deJoug, A. Pruijssers, P. van den Berg, A. Siemes, and W. deSitter, 1998. The durability of concrete structures in major marine works in the Netherlands, *Proc. 29th PIANC Congress,* Sec. II, Subject 3, The Hague, The Netherlands.

Noda, S., M. Ingeki, and T. Sonoyame, T., 1998. Damage to port facilities by the 1995 great Haushin–Awaji earthquake and reconstruction in the Port of Kobe, *Proc. 29th PIANC International Congress,* The Hague, The Netherlands.

Ohsaki, Y., 1982. Corrosion of steel piles driven in soil deposits, *Soils Found.* (Japanese Society of Soil Mechanics and Foundation Engineering), Vol. 22, No. 3, Sept.

Okada, K., S. Nisthbayashi, and M. Nakamura, 1989. *Proc. 8th International Conference on Alkali–Aggregate Reaction,* Tokyo.

Ozolin, E. W., and D. K. Soike, 1998. Planning and implementing an upgrade for a pile-supported wharf to serve post-Panamax crane operation, *ASCE Proc. Specialty Conference Ports '98,* Long Beach, CA.

Padron, D., and H. Y. Han, 1983. Fender system problems in U.S. ports. *ASCE J. Waterway Port Coastal Ocean Eng.,* Vol. 109, Vol. 3, Aug.

Pettit, P., and C. Wooden, 1988. Jet grouting: the pace quickens, *Civil Eng.,* Vol. 58, No. 8.

PHRI (Port and Harbour Research Institute), Japan, 1997. *Handbook on Liquefaction Remediation of*

Reclaimed Land, A. A. Balkema, Rotterdam, The Netherlands.

PIANC, 1990. *Inspection, Maintenance and Repair of Maritime Structures Exposed to Material Degradation Caused by a Salt Water Environment,* Rep. Work. Group 17, PTC II, Permanent International Association of Navigation Congresses, Brussels, Belgium.

———, 1992. *Guidelines for the Design and Construction of Flexible Revetments Incorporating Geotextiles for Inland Waterways,* Suppl. Bull. 78/79, Permanent International Association of Navigation Congresses Brussels, Belgium.

———, 1995. *Port Facilities for Ferries Practical Guide,* Suppl. Bull. 87, Permanent International Association of Navigation Congresses, Brussels, Belgium.

———, 1997. *Guidelines for the Design of Armoured Slopes under Open Piled Quay Walls,* Suppl. Bull. 96, Permanent International Association of Navigation Congresses, Brussels, Belgium.

Pierce, F. C., and V. V. Calabretta, 1986. Rehabilitation of gravity wall by underpinning, *ASCE Proc. Specialty Conference Ports '86,* Oakland, CA.

Ping Gu, J. J. Beaudoin, P. J. Tumidajski, and N. P. Mailvaganam, 1997. Electrochemical incompatibility of patches in reinforced concrete, *Concrete Int.* No. 8, Aug.

Plowman, J. M., 1956. Maturity and the strength of concrete, *Mag. Concrete Res.,* Vol. 8, No. 22, Mar.

Ponnet, L., L. Van der Eecken, R. Dedeyne, and W. Thibaut, 1992. Deepening of gravity quay walls by means of the very high pressure grouting technique: a pilot project in the Port of Antwerp, *Proc. 10th International Harbour Congress,* Antwerp, Belgium.

Porter, D. L., 1986. Innovative repairs to steel sheet pile structures, *ASCE Proc. Specialty Conference Ports '86,* Oakland, CA, May 19–21.

Prakash, S., 1981. *Soil Dynamics,* McGraw-Hill, New York.

———, (ed.), 1994. *Analysis and Design of Retaining Structures against Earthquakes,* proceedings of sessions sponsored by the Soil Dynamics Committee of the Geo Institute of ASCE, Washington, DC, Nov.

Priebe, A., 1995. The design of vibro replacement, *Ground Eng.,* Dec.

Priestley, M. J. N., 1997. Myths and fallacies in earthquake engineering! Conflicts between design and reality, *Concrete Int.,* Feb.

Prosser, M. J., 1986. *Propeller Induced Scour,* RR2570, British Port Association, London.

Prucz, Z., B. T. Martin, and J. L. Richstein, 1992. Modification to coal pier 6 made necessary by a deeper channel, *ASCE Proc. Specialty Conference Ports '92,* Seattle, WA.

Ractliffe, A. T., 1983. *The basis and essentials of marine corrosion in steel structures, Proc. Institute of Civil Engineers,* Part 1, No. 74, London, Nov.

Raes, L., F. Elskens, K. Römisch, and M. Sas, 1996. The effects of ship propellers on bottom velocities and on scour: near berth and protection methods using flexible revetments, *Proc. 11th Harbour Congress,* Antwerp, Belgium, June.

Richards, B. R., 1983. Marine borers, *Proc. American Wood Preservers' Association.*

RILAM, 1985. *Synthetic Membranes,* Draft Recomm. 47, SM Technical Committee.

Ritchie, W. D., and W. W. Watson, 1983. Modernization of a container terminal, *ASCE Proc. Specialty Conference Ports '83,* New Orleans, LA.

Rogers, T. H., 1960. *The Marine Corrosion Handbook,* McGraw-Hill, New York.

Römisch, K., 1997. Bottom attacks caused by shopmaneuvers of ships, *Proc. 4th International Seminar on Renovation and Improvements of Existing Quay Structures,* Gdansk Technical University, Gdansk, Poland.

Römisch, K., and W. Hering, 2002. *Input Data of Propeller Induced Velocities for Dimensioning of Bed Protection near Quay Walls,* Bull. 109, International Navigation Association, Brussels, Belgium.

Sawicki, A., M. Kulczykowski, W. Robakiewicz, J. Mierczynski, and J. Hauptmann, 1998. New type of bottom protection in harbors: design method, *ASCE J. Waterway Port Coast. Ocean Eng.,* July/Aug.

Schmitt, R. J., and E. M. Phelps, 1969. Corrosion performance of constructional steels in marine applications, talk prepared for presentation at the First Annual Offshore Technology Conference, Houston, TX.

Schwerdtfeger, W. J., and M. Romanoff, 1972. *NBS Papers on Underground Corrosion of Steel Piling, 1962–1971,* Monogr. 127, National Bureau of Standards, Washington, DC, Mar.

Scola, P. T., 1989. U.S. Navy underwater inspection of waterfront facilities, *ASCE Proc. Specialty Conference Ports '89,* Boston, May.

Scott, P. J. B., and M. Davies, 1992. Microbiologically induced corrosion, *Civil Eng.,* May.

Seed, H. B., and I. M. Indriss, 1971. Simplified procedure for evaluating soil liquefaction potentials, *ASCE J. Soil Mech. Found. Eng.,* Vol. 97, No. SM9.

Sherard, J. L., and L. P. Dunnigan, 1989. Critical filters for impervious soils, *ASCE J. Geotech. Eng.,* Vol. 115, No. 7, July.

Shroff, A. C., 1988. Evaluating a 50-year old concrete bridge, *Concrete Int., Des. Constr.,* No. 5.

Shvartsman, D., P. Harari, and Z. Rubin, 1999. Green book of port hydrotechnicial facilities as the tool for improving its maintenance activities and planning its development, *Proc. 12th International Harbour Congress,* Antwerp, Belgium.

Simoen, R., D. Gunst, J. Serras, P. Mortier, D. VandenBossche, J. De Regge, D. Goosens, and A. Bernard, 1990. Achievements of some port renovation projects planned, designed and realized in Belgium, *PIANC Proc. 27th Congress S-2,* Subject 4, Osaka, Japan.

Smith, A. P., 1987. New tools and new techniques for the underwater inspection of waterfront structures, *Proc. Offshore Technology Conference,* Houston, TX.

Smith, G. W., C. H. Evans, and M. A. Knott, 1992. Naval homeport facilities at Pensacola, Florida, and Mobile, Alabama, *ASCE Proc. Specialty Conference Ports '92,* Seattle, WA.

Snow, R. K., 1992. Polymer pile encapsulation: factors influencing performance, *Concrete Int.,* May.

Sonu, C. J., K. Ito, and H. Oishi, 1993. Harry Seed, liquefaction and the gravel drain, *Civil Eng.,* Dec.

Spencer, T. E., 2001. Underwater deterioration of concrete piles, *ASCE Proc. Specialty Conference Ports '01,* Norfolk, VA, Apr./May.

Spencer, T. E., and A. J. Blaylock, 1997. Alkali–silica reaction in marine piles, *Concrete Int.,* Jan.

Stark, D. C., G. Horeczko, and M. Rowghani, 1999. Cracking in prestressed concrete in seawater, *Concrete Int.,* Jan.

Steijaert, P. D., and J. F. De Kreuk, 1985. Underwater repairs with cement-based concretes, in *Behavior of Offshore Structures,* Elsevier Science, Amsterdam.

Tahal, V., and A. Matlin, 1996. Marine borers are back, *Civil Eng.,* Oct.

Taylor, D., and K. Davies, 2002. *Engineering the Rehabilitation of Reinforced Concrete Port Structures, Proc. 30th Congress,* PIANC, Sydney, Australia.

Thibant, W., J. Himpe, and E. Van Celst, 1992. Renovation of the north quay of the third harbour dock in the Port-of-Antwerp, *Proc. 10th International Harbour Congress,* Antwerp, Belgium.

Thompson, W. J., 1977. Damage to port structure by ships with bulbous bows, *ASCE Proc. Specialty Conference, Ports '77,* Los Angeles, CA.

Tsinker, G. P., 1986. *Floating Ports: Design and Construction Practices,* Gulf Publishing, Houston, TX.

———, 1989. The dock-in-service: evaluation of load carrying capacity, repair, rehabilitation, *ASCE Proc. Specialty Conference Ports '89,* Boston.

———, 1995. *Marine Structures Engineering: Specialized Applications,* Chapman & Hall, New York.

———, 1997. *Handbook of Port and Harbor Engineering: Geotechnical and Structural Aspects,* Chapman & Hall, New York.

Turner, R. D., 1966. *A Survey and Illustrated Catalogue of the Teredinidal,* Harvard University Press, Cambridge, MA.

Uhlendorf, H. J., 1998. Preparation of the surface, *Hansa,* No. 11.

———, 2001. New procedures for preparation of surfaces in heavy corrosion protection, *Proc. 5th International Seminar on Renovation and Im-*

provements to Existing Quay Structures, Gdansk Technical University, Gdansk, Poland.

Uhlendorf, H. J., and W. Janssen, 2000. Corrosion protective measures on harbour structures in the Jade, *Hansa,* No. 8.

Uhlig, H. H. (ed.), 1948. *The Corrosion Handbook,* Wiley, New York.

———, 1971. *Corrosion and Corrosion Control,* 2nd ed., Wiley, New York.

U.S. Army Corps of Engineers, 1984. *Shore Protection Manual,* Vols. I and II, Coastal Engineering Research Center, USACE, Washington, DC.

———, 1986. *Evaluation and Repair of Concrete Structures,* EM 1110-2-2002, USACE, Washington, DC.

———, 1994. *Engineering and Design: Design of Sheet Pile Walls,* EM-1110-2-2504, U.S. Department of the Army, Washington, DC.

U.S. Federal Highway Administration, 1995. *Geotextile Engineering Manual,* FHWA, Washington, DC.

Vahdani, S., R. Pjke, and U. Siriprusanen, 1994. Liquefaction of calcareous sand and lateral spreading experienced in Guam as result of 1993 Guam earthquake, *Proc. 5th U.S.–Japan Workshop on Earthquake Resistant Design of Lifeline Facilities and Countermeasures against Soil Liquefaction,* National Center for Earthquake Engineering Research, Buffalo, NY.

Van Damme, L., L. Bols, L. Van der Eecken, and F. Aerts, 1999. *Rehabilitation of Harbour Infrastructure: The Belgium Experience,* Bull. 101, International Navigation Association, Brussels, Belgium.

Van Damme, L., E. Bosschem, J. De Regge, and L. Bols, 1998. Port-of-Ostend: renovation of the outer harbour into a modern fast harbour for the 21st century, *PIANC Proc. 29th Navigation Congress,* The Hague, The Netherlands.

Van der Weijde, R. W., P. J. M. Heijndijk, A. C. Noordijk, and J. T. Kleijheeg, 1992. Scouring by ships' propellers: tracing, repair, prevention, *Proc. 10th International Harbour Congress,* Antwerp, Belgium.

Van Rooten, C., and A. J. Zielinski, 1996. Adoption of post-tensioning techniques for anchoring quay walls, *Proc. 11th International Harbour Congress,* Antwerp, Belgium.

Vejvoda, M. F., 1990. Strengthening of existing structures with post-tensioning, *Concrete Int.,* Sept.

Verhey, H. J., T. Blokland, M. P. Bogaerts, D. Volger, and R. W. L. van der Weijde, 1987. *Experiences in the Netherlands with Quay Structures Subjected to Velocities Created by Bow Thrusters and Main Propellers of Mooring and Unmooring Ships,* Bull. 58, Permanent International Association of Navigation Congresses, Brussels, Belgium.

Vorobyov, Y. L., 2001. Approximate evaluation of jet velocities and pressures caused by a side thruster against vertical quay wall, *Proc. 5th International Seminar on Renovation and Improvements to Existing Quay Structures,* Vol. 2, Gdansk Technical University, Gdansk, Poland.

Wallace R. L., and B. C. Mallick, 1986. Tacoma terminals pier: design for economy, *ASCE Proc. Specialty Conference Ports '86,* Oakland, CA.

Warren, G. E., L. J. Malvar, C. Iuaba, and D. Hoy, 1995. Rehabilitating the navy's waterfront infrastructure, *ASCE Proc. Specialty Conference Ports '95,* Tampa, FL.

Webber, D., and J. Yao, 2001. Effectiveness of pile wraps for timber bearing piles, *ASCE Proc. Specialty Conference Ports '01,* Norfolk, VA.

Werner, S. D., and S. E. Dickenson (eds.), 1996. *Hyogoken Naubu Earthquake of January 17, 1995: A Post-earthquake Reconnaissance of Port Facilities,* report by the Committee of Ports and Harbors Lifeline of the Technical Council of Lifeline Earthquake Engineering of the ASCE, New York.

Werner, S. D., S. E. Dickenson, and C. E. Taylor, 1997. Seismic risk at ports: cases studies and acceptable risk evaluation, *ASCE J. Waterway Port Coast. Ocean Eng.,* Nov./Dec.

Wong, R. T., H. B. Seed, and C. K. Chan, 1975. Cyclic loading liquefaction of gravelly soils, *ASCE J. Geotech. Eng. Div.,* Vol. 101, No. GT6.

Yao, S., 1998. Modular concepts for rapid repair of navy waterfront structures, *Proc. ASCE Specialty Conference Ports '98,* Long Beach, CA, Mar.

7

FLOATING TERMINALS

Gregory P. Tsinker
Tsinker & Associates, Inc.
Niagara Falls, Ontario, Canada

7.1 INTRODUCTION

From experience it is evident that for locations where waterways are obviously unstable, possess a wide range of water levels, and lack good morphological and hydrological data, a floating wharf arrangement is an ideal approach to port design (Tsinker, 1986). The conditions above may exist at many port locations, but they are most typical at inland navigational routes. Typically, an alluvial river never takes a straight course from its source to the sea. Normally, geological conditions cause the river to align with one of several general geometric forms, among which meandering or divided flow is most common. Analysis has shown that the main factors affecting the geometry of river channels are discharge, bed slope, and sediment load. If the discharge and sediment load entering a channel were to remain constant, an equilibrium state would result. However, in actuality, these factors do not remain constant or even totally cyclic in behavior, so irregular changes in the river regime must be expected and accommodated at port locations.

A coastal port designer knows that when normal sediment transport along a coastline is obstructed to accommodate harbor works, a change in sediment transport usually results. Similarly, intervention in the natural conditions of a river will usually cause scour or deposition that affect river stability. An inland port designer does not usually have to be concerned with the complexity of the total length of a river, but with a limited reach only, where the primary objective is to obtain a dynamically stable channel at the port site.

In considering the importance of navigation, the magnitude of scour and/or deposition at port location might be the cause for costly improvement schemes, and the port authorities have to wrestle with the economics of navigation benefits before being able to reconcile the cost of these improvement schemes. Subsequently, because of the instability of alluvial rivers, the inland port designer usually evaluates two basic alternatives for dock construction: bottom-fixed docks in combination with river channel stabilizing schemes, or floating docks. The latter would normally be more adaptable to changes in the geometry of a river channel and, in most cases, eliminating the need for costly channel-stabilizing structures (e.g., jetties, dikes, bank revetments).

Although floating docks have long been used in port construction, waterfront structures traditionally are built in the form of bottom-fixed structures. In the past two to three or so decades, however, there has been quickening interest by engineers and port operators in floating docks to resolve some port construction and operation problems. Port development practice has shown that floating terminals can be implemented successfully at inland ports and protected seaports and harbors as well as offshore. In general, for given specific conditions (e.g., deep water, strong current, short period suitable for construction, difficult and/or uncertain soil conditions, environmental concerns), construction of floating rather than bottom-fixed terminals may be advantageous. A floating terminal can be an excellent solution for construction of permanent and temporary wharves in remote areas with no means of access other than by water. The pier and other dock components, such as access bridges and mooring systems, can be built off site at well-established yards and floated in to the site. By keeping just the on-site portion of a dock, construction costs can be reduced considerably. In recent engineering practice this method has been used successfully for construction of different kinds of industrial plants in remote areas from the arctic to South American jungles.

Where water levels vary substantially, a floating dock has a definite advantage over a fixed dock in that it moves up and down along with the ship, and changes in water level do not hamper loading and unloading operations. Soil condition is another important factor: Where the bottom soils are poor and unsuitable for fixed mooring structure foundations, the floating dock may be the only solution. In this sense the floating dock solution also presents a much more reliable alternative than the fixed dock in highly seismic regions, as it is less affected by earth movement during an earthquake. Furthermore, where construction of a bottom-fixed facility is found environmentally objectionable, a floating dock could present an acceptable solution.

Floating docks are also very important in military applications, where speed of construction is the most challenging engineering tasks. Last but not least, construction of temporary floating wharves for delivery of construction materials and equipment can be used effectively for developing other water-related industries. However, probably the most important application of floating docks is in developing countries, where the availability of waterborne transportation encourages the establishment of industries in remote communities along naturally migrating rivers.

The mobility of floating docks makes them highly suitable for use on untrained and unstable rivers, as they can be moved when necessary due to bank erosion or sediment accretion with minimum disruption to port operations. Despite some obvious advantages of floating ports, in some cases they may have higher operating and maintenance costs than those for fixed facilities. This may be especially true at locations of heavy ice movements and in corrosive environments. Also, in areas exposed to significant wave action, the operating stability of a dock may be reduced, thus increasing its downtime.

7.2 EXAMPLES FROM PRACTICE

Floating structures for miscellaneous purposes have been long used by engineers as alternatives to bottom-fixed structures. This has occurred because in many cases floating structures proved to be more economical and reliable, to offer operational advantages, and appear to be more acceptable from an environmental point of view. Floating structures in general have been used for the construction of floating houses, storage for liquid and dry bulk materials, bridges, airports, golf courses, and miscellaneous industrial facilities. Floating piers have been used over a long period of time, however, recently built floating terminals represent a real technological breakthrough for ports and harbor planners, designers, and operators.

As stated earlier, in the past several decades a great number of outstanding floating marine terminals have been built elsewhere in the world. They have been constructed for oil, containers, general cargo, passenger, and other special-purpose terminals. Experience in the construction of floating wharves in Peru, Brazil, the United States, the United Kingdom, Saudi Arabia, Russia, Japan, and many other countries, not to mention a great number of floating piers constructed for military application, has clearly demonstrated that under specific site conditions, as discussed above, floating marine terminals may represent a competitive solution to the fixed waterfront concept.

7.2.1 Simplest Form of Floating Dock

The simplest form of a floating dock is a small or medium-sized steel or concrete pontoon or barge, such as may be seen at many lakes, marinas, and small river terminals throughout the world. The pontoon is usually moored in one of the ways shown in Figure 7.1. The mooring system is designed to keep the pontoon in position against the action of environmental loads as well as berthing and mooring forces imposed on the dock system. The pontoon is free to rise and fall with the tide. Access to the pontoon is usually obtained through an articulated bridge or special ramp. The floating dock depicted in Figure 7.1 could be found at almost any place adjacent to a water recreational facility. It is usually used by small recreational craft and is designed to accommodate small recreational boats and to provide access to and from the land. This type of docking facility is often used where the water level fluctuates between 0.5 and 1.5 m. Typically, small floating docks consist of three basic structural components: a deck made of steel or wood supported on floats of miscellaneous designs, a floating or articulated access ramp, and a primitive anchor system of piles, heavy concrete or metal blocks, and others. The ferry dock depicted in Figure 7.2 operated until recently at Sandy Hook, New Jersey. This facility consists of a rehabilitated steel barge anchored at its designated location by means of two steel piles and linked to the shore by an articulated span. The barge can rise and fall with the tide but is restricted from horizontal movements by anchor piles. The mobile passenger dock depicted in Figure 7.3 is in operation at Roko Island, Japan. It is moved from place to place as required to serve local passenger traffic. It consists of a steel pontoon, an anchor system, an access bridge, and a rather modern fender system. In

Figure 7.1 Simplest form of floating dock.

Figure 7.2 Floating ferry dock at Sandy Hook, New Jersey.

Figure 7.3 Mobile passenger dock at Roko Island, port of Kobe, Japan.

this case, the articulated access bridge is sitting permanently at the pontoon's corner and can swing about the fixing point toward the land.

7.2.2 Small Passenger Docks

A great number of efficient floating docks serving local passenger traffic have been constructed throughout the world in the past two to three decades. Three representative examples of such docks are described in the following sections. These docks represent the state of the art in handling heavy local passenger traffic.

7.2.2.1 Kobe, Japan. Most recently, several new passenger docks were installed to serve local passenger traffic in and around the city of Kobe. Details of these docks are shown in Figures 7.4 through 7.7. Docks of this type represent the latest development in floating dock technology. Each of these structures is designed to accommodate simultaneously two relatively small passenger boats. They are not just functionally efficient but also very pleasant aesthetically. These structures are constructed from the following basic components: a steel floating pontoon, an access bridge, guiding towers, and efficient fender and mooring systems (Figures 7.4 through 7.6). The towers allow the pontoon to rise and fall vertically due to variations in water level. To reduce friction between the floating part of the dock and the towers, the former is equipped with sophisticated roller systems which keep the pontoon between guiding towers with minimal movement in the horizontal plane (Figure 7.7). To reduce the noise that may be produced during pontoon movements, the rollers of the guiding system are made either from plastic material or are fabricated from steel covered with plastic material. The system allows smooth, substantial (up to approximately 3.0 m) vertical movement of the pontoon. The articulated bridge is hinged at the land abutment, and at the dock side it is suspended on cables. It can be lifted and lowered with the help of two winches installed on the top of the tower. For protection

Figure 7.4 Floating passenger dock, Kobe, Japan.

Figure 7.5 Floating dock access bridge, Kobe, Japan.

Figure 7.6 Floating dock, Kobe, Japan. View of the pontoon is completed with fender and mooring system, shed, and railing.

Figure 7.7 Pontoon guiding system for a floating dock, Kobe, Japan.

from ship impact the system is equipped with floating pneumatic fenders and V-type vertical rubber units installed on both sides of the dock (Figure 7.6). The boat is kept safely at a berth with the help of a mooring system comprising bollards and mooring rings (Figure 7.6). The system is completed with pleasant-looking sheds on both the pontoon and the bridge and with a passenger protection railing system (Figure 7.6).

7.2.2.2 Bangkok, Thailand. The floating dock constructed on the Chao Phy River to serve local passenger traffic at the city of Bangkok (Figures 7.8 and 7.9) is conceptually similar to that described in Section 7.2.2.1. The floating part of the dock, the steel pontoon, is installed between two guiding dolphins made of steel piles. Above water level, these piles are completed with steel struts that turn the pile-strut system into a rigid frame. This system is designed to resist operational loads such as ship impact and mooring forces. The pontoon is guided up and down smoothly along the guiding frames with the help of roller guides (Figure 7.9).

Figure 7.8 Floating dock on Chao Phy River, Bangkok, Thailand.

7.2.2.3 Kiev, Ukraine. The presence of ice in navigational channels, in general, presents a challenge for both navigation and dock operations. Ice movements make winter navigation not just difficult but also under certain conditions may result in significant forces acting on port marine structures. Naturally, ice forces acting on any type of marine structures are affected not only by miscellaneous environmental parameters such as wind, current, tide, and temperature but also by the size and configuration of a structure as well as ice movements around the structure. For a comprehensive discussion on ice forces and ice–structure interaction, the reader is referred to Tsinker (1995). Floating structures are especially vulnerable to impact by moving ice. The floating passenger dock that can resist both static and dynamic ice forces is shown in Figure 7.10. This structure was constructed on the Dnieper River at Kiev, Ukraine. This dock consists of the following basic components: two steel pontoons (floats) that are placed within cells constructed from flat reinforced concrete sheet piles, a trusslike steel deck that is rested on pontoons, and an articulated access bridge. To make the deck level of a passenger ship that is operating at the facility, the bridge is positioned approximately 70 cm above the water level.

The sheet pile cells are designed to accommodate the pontoons and to protect them from being affected by ice and other floating objects. The cells are surmounted by steel pylons that

Figure 7.9 Dock guiding system for a floating dock, Bangkok, Thailand.

are designed to guide both pontoons at a high water level in the river and to resist the associated forces generated by a berthing boat. In the case in question, the cell–pylon system allows for dock operation under fluctuation in water levels up to 8.0 m. The operating deck is constructed in the form of a space trusslike structure designed for a live load of 2 kN/m^2. It is fully protected against ship berthing impact by means of a special fender system, which transmits a berthing impact force directly to the cell–pillon system. The deck structure has built-in buoyancy that may help it to remain afloat in the case of accidental overloading of pontoons. The fender system is designed in the form a steel, horizontally placed framelike structure. It is connected telescopically with the deck structure and is provided with a standard cylindrical rubber fenders installed on its external and internal faces. The flexibility of the fender system also contributes to the dissipation of a ship's berthing impact energy.

Because of complete protection from outside forces, with the exception of a short period during high water levels (spring flood), the pontoons are built without watertight bulkheads. The articulated access bridge is hinged at the land-based abutment. Its outer end is provided with two rollers, which travel on a steel cantilever platform extended toward the land from the deck structure. The bridge is arranged such that it is horizontal at the highest water level in the river (spring flood). At this point its outer

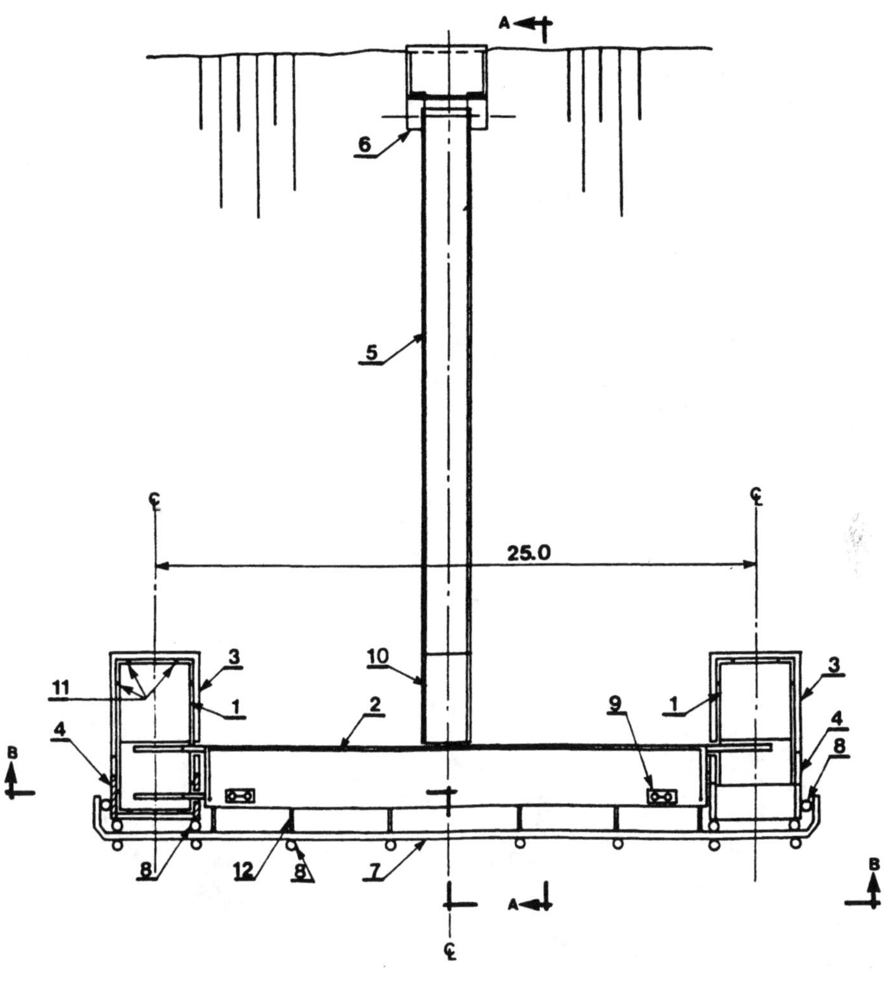

1—pontoon deployed into the dolphin
2—deck
3—bottom part of the dolphin
4—upper part of the dolphin
5—articulated access box-girder bridge
6—piled bridge abutment
7—fender system
8—tubular rubber fenders
9—double bitt
10—intermediate ramp
11—pontoon guides
12—telescopic support for the fender system
13—pivot
14—roller

(*figure continues*)

Figure 7.10 Passenger dock on the Dniper River, Kiev, Ukraine.

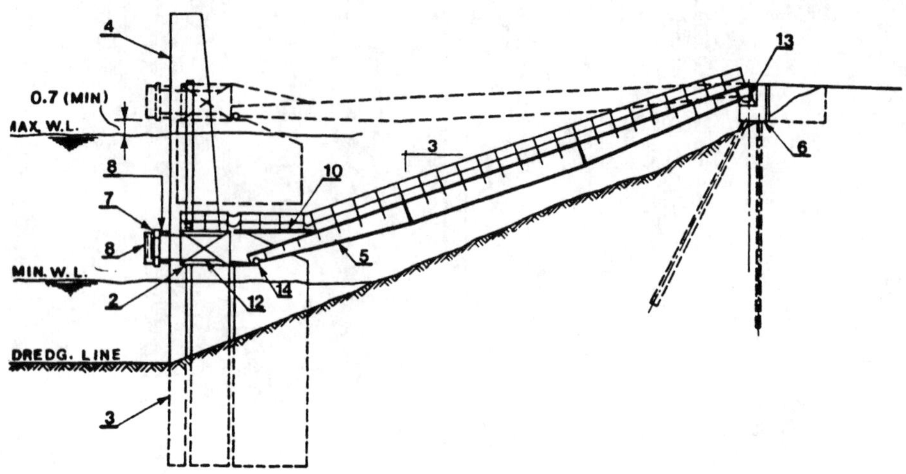

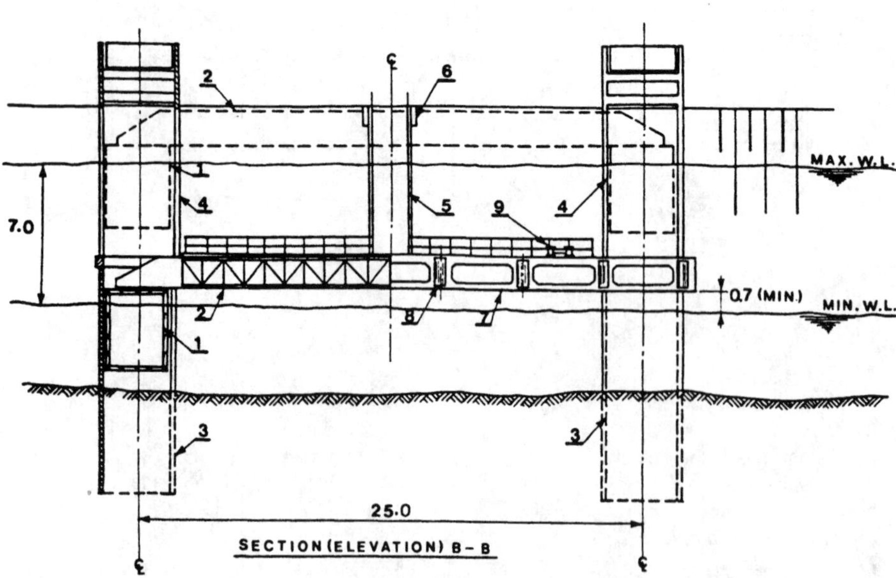

Figure 7.10 *(Continued)*.

end is at its closest position to the deck structure. Accordingly, at low water levels, the outer end of the bridge travels away from the deck structure. In all positions of the bridge relative to the deck, the gap between the deck and the bridge is bridged by the movable ramp (flap) that is hinged to the deck structure and slides on the bridge deck.

7.2.3 Large Floating Terminals in Europe

A great number of large and small floating terminals installed mostly on inland waterways are found elsewhere in Europe. Following are three characteristic examples.

7.2.3.1 Liverpool, England. This floating facility (Figure 7.11) has been installed in 1977 to service passenger traffic. It comprises six typical concrete pontoons 53.33 × 19.00 m with a molded depth of 5.0 m. Hence, the total length of a floating dock is 350.00 m. The pontoons were built in Dublin, Ireland, and towed some 230 km across the Irish Sea to Liverpool. For this, pontoons had to be designed for effects of wave action during the voyage (a design wave height of approximately 2.45 m). The design wave forces during the voyage period were greater than those expected at the permanent dock location in Liverpool, where the expected wave height was equal to approximately 1.25 m. The pontoons have been cast in situ. However, to facilitate the construction process, a number of internal walls were built in a form of open precast K-frames. It took 35 days to build one pontoon. The wall thickness of pontoons is 25 cm, and the thickness of the deck slab is 14.5 cm. Once in place, the pon-

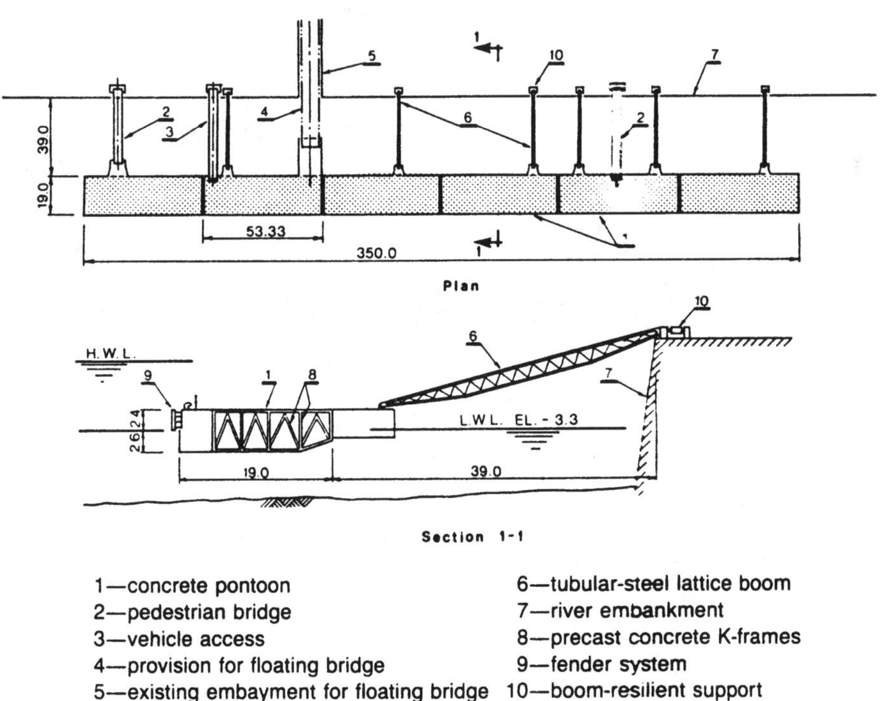

Figure 7.11 Plan and typical cross section, floating terminal at Liverpool, England. (After Hetherington, 1979.)

toons were joined together by means of prestressed 100-mm-diameter tendons composed of nineteen 15-mm-diameter strands (Figure 7.12). The tendons were placed in 125-mm-diameter flexible metal ducting sealed externally by surrounding concrete and filled internally with grease pumped in from the tendon anchorage. The operating stress in tendons is limited to 45% of their allowable strength as a precaution against fatigue loading. The presence of grease in the tendon's duct allows for tendon destressing to retrieve it, when necessary, for inspection and testing.

The total hinge force is approximately 16.6 MN. Six molded rubber pads, 120 × 60 × 10 cm, are placed between pontoons. They are designed to remain constantly in compression and to resist vertical shear forces and other load combinations. To prevent rubber deterioration, an anti-ozonant compound was included in the rubber compound. The dock is deployed 39 m from the edge of the existing river embankment and connected to it by articulated lattice steel booms, which allow the pier to rise and fall vertically on the tide. The lateral pier drift due to effects of tidal current is controlled by longitudinal anchor chains secured to the anchors. The booms transmit environmental loads (e.g., wind, wave and current, mooring and berthing forces) to supports installed on the top of the river wall using a special resilient hydraulic support system. This is required to accommodate geometrical distortion between adjacent booms due to possible (not uniform) motion of adjacent pontoons in the vertical plane. The support system is also designed to perform as a shock absorber to take a load of up to 2.3 MN resulting from ship berthing. Access to the pier is obtained by three articulated steel bridges, one of which is pivoted at both ends to the pier and to the abutment located at the river wall, and two others are pivoted at the river wall abutments and slide on the deck of the pier at the outer end. Pivoted at both ends, the access bridge also serves as a part of the pier mooring system. As shown in Figure 7.11, two bridges are for pedestrian use only, and the third one is used for access of vehicular traffic.

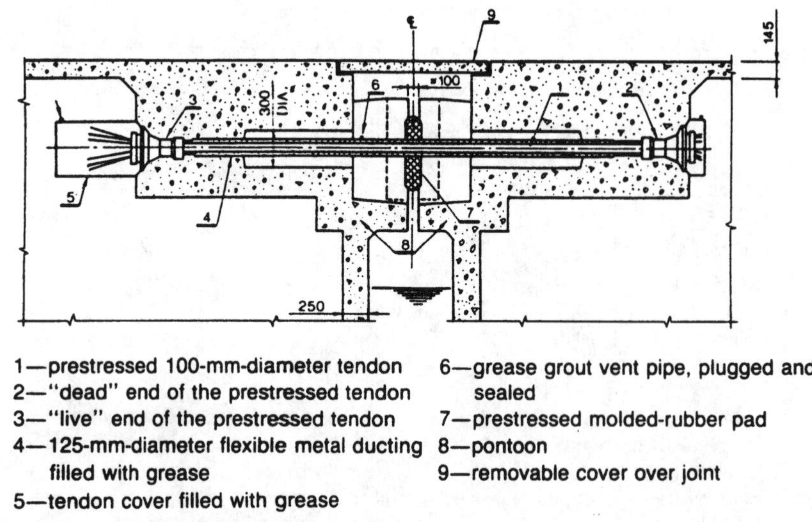

1—prestressed 100-mm-diameter tendon
2—"dead" end of the prestressed tendon
3—"live" end of the prestressed tendon
4—125-mm-diameter flexible metal ducting filled with grease
5—tendon cover filled with grease
6—grease grout vent pipe, plugged and sealed
7—prestressed molded-rubber pad
8—pontoon
9—removable cover over joint

Figure 7.12 Prestressed joint between adjacent pontoons, floating terminal at Liverpool, England. (After Hetherington, 1979.)

The performance of this dock is reported to be good despite the accidental damage to one pontoon caused by a ferry boat of 850 dwt (deadweight tonnage) which collided bow-on with the pontoon. The approaching speed of the boat at the time of accident was about 10 times the designed speed and a hole of about 80 × 30 cm was smashed in the 25-cm-thick pontoon berthing face. However, within a very short period of time, the damage was repaired successfully by employing a special concrete mix containing epoxy resin.

7.2.3.2 London, England. The floating dock, 348 m long and 24.5 m wide, installed at Tilbury (port of London) is designed to service oceangoing vessels and ferryboats. This dock is in service now for more than 60 years. The deck of the dock is composed of five steel continuous box girders designed to support transverse steel beams spaced up to 137 cm apart. The latter, in turn, supports timber decking made of 30- × 11.5-cm planks. The deck is supported by sixty-three 4.7-m-wide pontoons, 44 of which are 24.5 m long, 15 others 27.5 m long, and the remaining 4, 30.5 m long. All pontoons are spaced at 1.5 m (Figure 7.13). Larger pontoons are designed to support loads from bridges, booms, and so on. All the pontoons are provided with a midship and transverse bulkheads by means of which they are divided into several watertight compartments. The floating dock is retained in position by means of a mooring system of two dolphins, four booms, two pairs of breast mooring chains, three pairs of flood, and three pairs of ebb chains. The steel trusslike 48-m-long booms, being pivoted at the land abutments and the pier, rise and fall in a circular path. To fit this pattern of movement, the face of the dolphins is provided with rubbing timbers dressed to a suitable radius.

All mooring chains are secured at 0.9-m-diameter screw moorings or 1.4- and 0.9-m-diameter mushroom anchors. To prevent small craft from passing behind the dock, sets of upstream and downstream fender piles are provided. Access to the dock is provided by five single-span steel bridges, all except one being for passenger traffic. The fifth bridge is located at the ferry wharf and is designated to carry the road traffic. The pier carries a number of structures for accommodation of passengers and dock staff: one passenger bridge provides access to the upper floor of a passenger shelter. For more details on this floating terminal, readers are referred to Du Plat Taylor (1949) and Cornick (1958).

7.2.3.3 Antwerp, Belgium. Located on Schelde River in the heart of the city, this passenger terminal is serving heavy local passenger traffic, but it also has the capacity to receive relatively large seagoing passenger vessels. As shown in Figure 7.14, this facility includes a floating dock linked to the city embankment by an articulated bridge. The dock is placed within the recess provided in the embankment and therefore does not need special anchor system to keep it in place. The deck of the dock is covered with wooden planks. The land end of the articulated access bridge is hinged at the embankment, and its lower end slides on the deck surface. The bridge has the capacity to carry out the road traffic. The dock is fitted with fender and mooring systems. This type of installation is very typical for floating docks used in many European cities to serve primarily local passenger traffic.

7.2.4 Floating Navy Pier

An innovative concept for a floating pier has been developed by the U.S. Navy (Naval Civil Engineering Laboratory, 1982). (The actual design was carried out by T.Y. Lin International, San Francisco, California.) A plan and typical cross section of the structure are depicted in Figure 7.15. It should be pointed out that from an operational point of view, navy piers are

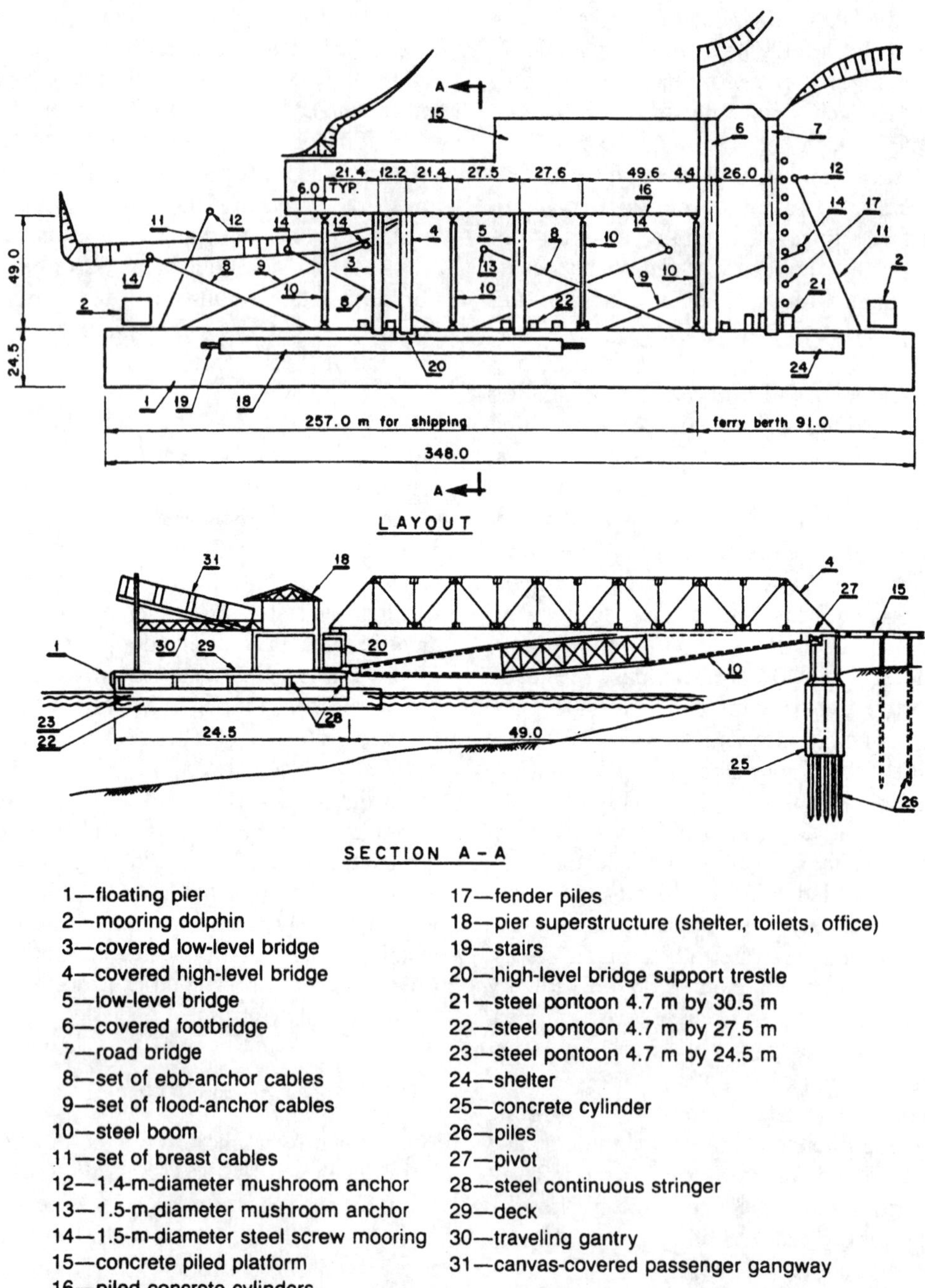

Figure 7.13 Plan and typical cross section, floating dock at Tilbury, port of London, England.

Figure 7.14 Passenger dock on river Schelde, Antwerp, Belgium.

quite different from ordinary commercial piers. The problems inherent in them have long been recognized (Chow and Haynes, 1983). The typical functions of navy docks is to provide support for minor vessel repair and resupply vessels with ammunition, food, potable water, and so on. To carry out its function, the navy pier must accommodate various services, such as electricity, steam, compressed air, potable water, and sewage. The pier also serves as a platform for contractors to refit vessels with improved equipment and/or weaponry. Finally, the pier has to accommodate a great number of sailors going on shore leave and returning to their ships. All of the above creates pier congestion. Furthermore, the pier deck may be cluttered by utility lines, vehicle parking, supply trucks, and electric substations. This usually results in a long turnaround time and inefficiency of operation. The vertical movement of a vessel along a fixed dock (due to tide effect) also presents some inconveniences in handling utility and service lines temporarily connected to the vessel. To overcome these problems, a double-deck floating pier solution has been proposed, and the conceptual design of such a structure has been developed for application to specific site conditions at the Charleston Naval Station.

The overall dimensions of the proposed pier are as follows: length 366 m and width 22.8 m. The pier's upper deck is connected to the shore by a centrally located 5-m-long and 6.1-m-wide articulated ramp that is designed to ad-

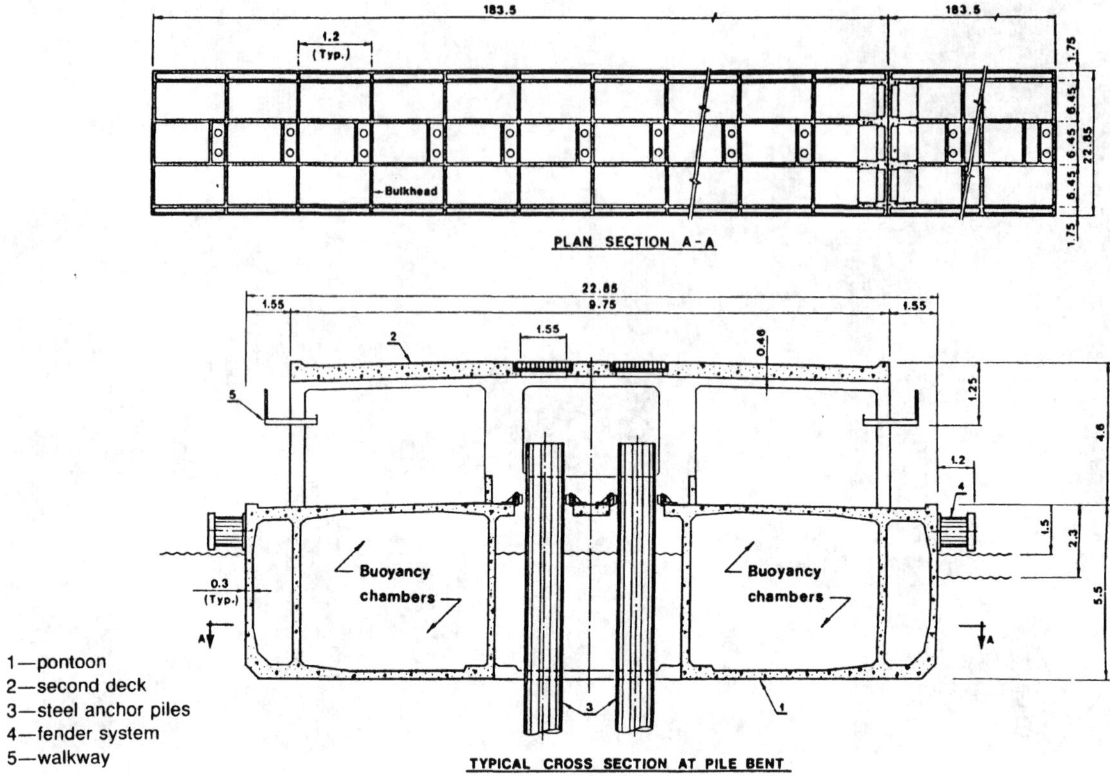

Figure 7.15 Plan and typical cross section of a U.S. Navy double-decked floating pier. (After Naval Civil Engineering Laboratory, 1982.)

just to water-level changes and the horizontal movements of the pier. The pier's lower deck is connected to the shore by two 4.6-m-wide articulated ramps that flank the upper ramp at each side. The maximum slope for all ramps is designed to be 1(V): 10(H) for an extreme tidal range of 3 m. On the shore side, ramps are pivoted, whereas at the pier they are free to slide. The double-deck pier structure consists of concrete pontoon surmounted by the upper deck structure. The upper deck structure is 19.8 m wide. A 1.5-m setback is provided on each side of the pier primarily to accommodate utility service lines and to provide for a walkway located below the upper deck and allow access to the utility lines. The upper deck is designed for a surcharge load of about 30 kN/m^2, truck loads of HS-20-44, a 200-kN forklift truck, and a 900-kN mobile crane. As might be expected, the outrigger load from a mobile crane constituted the controlling condition for determining the deck thickness of 46 cm. The pontoon is to be constructed from prestressed concrete. The longitudinal prestressing force for the upper deck is approximately 4.65 MPa. The loading criteria for the pontoon lower deck are a surcharge load of about 5 kN/m^2 and a wheel load of 200 kN from a forklift truck. The innovative design of the pontoon is controlled by operational as well as safety requirements. The pontoon hull, of double-wall construction, reflects the designer's concern for the stability and buoyancy of the pier in the case of puncture of its outer shell. The pontoon hull is designed for a hydrostatic pressure head of 6.7 m. The entire pontoon length of 366 m will be assembled from two or three typical prefabricated pontoons. It is assumed that these smaller pontoons will be constructed at the drydock and subsequently towed to the site of deployment, where they will be joined together in situ by posttensioning technique. The prestressing force will be applied longitudinally, transversely, and vertically at approximately 530 N/cm^2. Lightweight concrete (density of 2000 kg/m^3, compressive strength approximately 3.50 kN/cm^2) is considered for pier construction. As seen from the pontoon's cross section, depicted in Figure 7.15, the upper deck of the pier is placed at 4.6 m above the lower deck, which is located at the level that matches the average quarterdeck of a typical warship that will be served at the dock. The vertical distance between decks is sufficient to give the lower deck ample headroom for small vehicular traffic, and the lower deck area is sufficient to provide for parking, storage of equipment and materials, and utility service equipment, such as electrical transformers, saltwater pumps, and trash containers. The lower deck has a 1.5-m freeboard. The pontoon is outfitted with a modern rubber fender system that protects the pier from being damaged by docking vessels. The pier is anchored at the deployment site by vertical cylindrical piles driven through wells located along the centerline of the pier. Piles restrain the pier from excessive horizontal displacements but allow for free vertical movements. Fifty-eight vertical steel piles, 1220 mm in diameter with a wall thickness of 25 mm, are used to maintain the pier in place. A pair of vertical piles are located every 12 m along the pier length. Piles are designed to resist wind, current, and ship impact forces. The maximum wind force at 145 km/h, concurrent with a 6-knot current velocity applied to the pier with ships berthed along both its sides, resulted in a horizontal load of about 285 kN on each pile. As can be seen in Figure 7.15, piles contact the pier at a single point near the lower deck and restrain it against significant horizontal movement by cantilever action. The utility system (i.e., service lines) is located under the upper deck with outlets located at the service walkways. The latter clears the upper deck of most utility pipes, cables, hoses, and other obstructions. The utility pipes are suspended under the upper deck ramp at both the shore and pier sides. Pipes can accommodate three-dimensional movement with the help of swivel joints or flexible hoses. As follows from the above, the double-deck configuration of this navy floating pier increases the deck operational area without increasing pier width and dramatically improves its operational characteristics. It should be noted that a bottom-fixed version of the pier described above has recently been constructed at the naval base at Norfolk, Virginia.

7.2.5 Floating Terminals at Valdez, Alaska

Valdez harbor is a natural deepwater fjord located in the northeastern part of Prince William Sound in the Gulf of Alaska. The harbor, approximately 19 km long and 4 km wide, is protected on all its sides by steeply rising mountains. It remains ice-free throughout the year. The mean annual temperature there is about 2°C, with record extremes of −15.5°C and 30°C. The heaviest daily snowfall recorded there is 109 cm. Winds of speed greater than 30 km/h blowing from any direction occur there with a frequency of 3.5% throughout the year. Sudden gusts are frequent, and the maximum wind speed measured there is 156 km/h. The site is well protected from southwesterly winds but is fully exposed to the northeasterlies and to winds coming directly from the north. Because the harbor is protected from the open sea, waves within the harbor are normally limited to the wind-generated waves restricted by the available fetch. The maximum height of such waves is 2.9 m from the east. Waves of about 1.8 m could occur from the northwestern quarter. The maximum tidal range in the area is about 6.7 m. The tidal currents are weak and for dock design purposes have been assumed as about 0.4 m/s in any direction. However, this value is generally much less. The ice that forms in the shallow water does not interfere with ship operations, nor does it present a serious hazard to dock structures or its fittings. The special problem in the area is the possibility of an earthquake, and therefore, more than usual care must be taken in the design of land- and sea-based structures and equipment. Two important floating terminals have been constructed there. They are said to be the first of their kind in the world.

7.2.5.1 Alyeska Crude Oil Loading Port, Berth 1. The seabed at berth 1 consists of a very steeply sloping rock with little overburden (Brumitt and Dixon, 1980). Because of very deep water at the design location of berth 1, this site was not economically suitable for construction of a bottom-fixed marine structure and, on the other hand, was ideal for construction of a floating dock. The dock was designed and built there to receive tankers of up to 125,000 dwt. It comprises a floating pier and four bottom-fixed mooring dolphins (Figure 7.16). The pier is free to move vertically to accommodate tidal movements and wave action, including pitch and roll, but is restricted by a system of booms from movement on the horizontal plane. Restraint is provided by three-dimensional truss-type rigid struts hinged at the berth structure and at the land-based abutments (Figures 7.16 through 7.18). The four mooring dolphins are conventional deepwater steel jacket structures anchored to the bedrock of the seafloor. The floating pier is constructed in the form of a large space frame (118.87 × 21.34 × 16.8 m) (Figure 7.19) sup-

Figure 7.16 Layout of the Alyeska crude oil loading port at Valdez, Alaska. (Mooring dolphins are not seen.)

Figure 7.17 Side elevation of floating dock, Alyeska crude oil loading port at Valdez, Alaska.

Figure 7.18 Land-based foundation that is housing a universal hinge of the dock anchoring boom at the Alyeska crude oil load port at Valdez, Alaska.

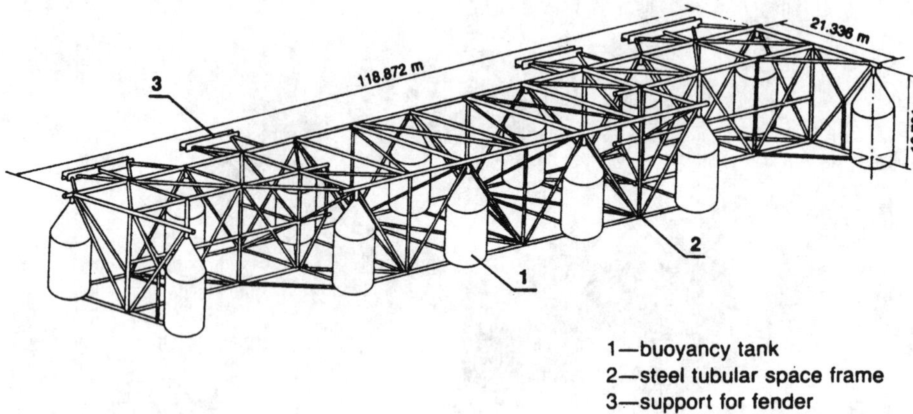

Figure 7.19 Floating dock space frame, Alyeska crude oil loading port at Valdez, Alaska.

ported on 13 large-diameter buoyancy tanks (d = 6.73 m) (Figures 7.19 and 7.20). For maximum safety each tank is double-shelled. Steel plates 12.7 mm thick were used to build the tanks, which weight 60 tons each. The total weight of structural steel used for the construction of space frame was 1040 tons, and the weight of the deck alone (including bracing) is 513 tons. The total structural weight of the floating dock, including the weight of its fender system and piping, is 2820 tons. The fender system comprising four Seibu H-type fenders 1700 m high by 1100 m long was designed to absorb the impact force generated by supertankers of 125,000 dwt at an approach speed of 10 cm/s. The dock was fabricated in Japan, put on a submersible barge, and towed across the ocean to the site of deployment, where it was installed permanently. The process of dock loading on a submersible barge is illustrated in Figure 7.21. This unique berth proved to be very reliable and the most economical solution for the particular location of berth 1. It should be noted that shortly after the dock was operational, it survived accidental impact by a supertanker but with no significant damage to the dock structure and its fender system.

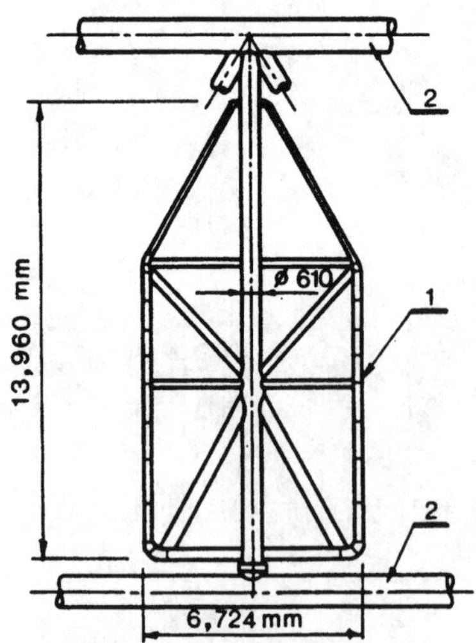

Figure 7.20 Buoyancy tank, Alyeska crude oil loading port at Valdez, Alaska.

7.2.5.2 Container Terminal. This facility was constructed near the northern bank of Valdez harbor. The project was first envisaged as a bottom-fixed structure, but placement of the fill to form the marshaling yard indicated that

Figure 7.21 Loading of the floating dock on a barge before moving it to site of deployment, Alyeska crude oil loading port at Valdez, Alaska.

a fixed structure could not be practical. As the 600,000 m^3 of fill settled to the bottom, the rock slope at the site of the pier slid into deeper water, providing a warning of what might happen during the earthquake. The follow-up studies indicated that the giant floating dock would cost no more than a fixed structure but would have a great many advantages, as discussed in Section 7.1. The floating dock was designed to handle a 40-ton container gantry crane and to accommodate container ships of up to 50,000 gross registered tons and barges of capacity up to 15,000 tons. It can also handle large ocean-going general cargo vessels and Ro-Ro (roll on–roll off) carriers. The dock's gantry crane can load and unload containers at a maximum outreach of 32 m. Trucks to haul containers can operate on either a one- or two-way loop from the marshaling yard onto the dock and back to the marshaling yard. The floating dock is comprised of three principal parts: a concrete floating pontoon, an anchor system, and two articulated access bridges (Figures 7.22 and 7.23). The pontoon was prefabricated off site and towed over 2520 km across the sea to the site of deployment. The delivery voyage involved crossing the Gulf of Alaska, a body of water known for sudden and intense storms.

Figure 7.22 Floating container terminal, port of Valdez, Alaska.

The delivery voyage design wave was determined as equal to 5.4 m. The 213-m-long, 30.5-m-wide, 9.15-m-deep (light draft 5 m) floating dock (gross displacement 32,000 tons) was towed in two equal pieces. This helped to reduce the design bending moment in the pontoon hull due to wave action by a factor of more than 5. The site environment data were established as follows:

- Normal wave height of 1.53 m with a corresponding length of 22.9 m
- Rare wave height of 2.4 m with a corresponding length of 32.2 m
- Extreme wave height of 2.1 m with a corresponding length of 37.2 m

Each of the two floating parts of the pontoon consisted of 16 watertight compartments to satisfy damaged stability requirements such that if any two cells are punctured below the waterline, the dock will remain afloat and stable. For ease and speed of construction, the pontoons were fabricated from a combination of precast concrete panels and cast-in-situ concrete (Figure 7.24). All the pontoon components were posttensioned together for strength and watertightness. The total 133 precast concrete pieces, ranging in weight from 40 to 50 tons, have been used for pontoon construction. Vertical joints between walls and the pontoon bottom slab were cast in situ. Each half of the pier also contains 234 prestressed concrete deck panels 1.83 m wide, spanning 7.62 m in width. There are two basic types of deck panels: those under the crane rail requiring heavier reinforcement and recesses into which crane rail bolts are grouted, and regular panels located elsewhere on the deck. The deck panels were used as a stay-in formwork to provide a work surface for a 30.5-cm-thick cast-in-situ concrete deck (total deck thickness, 46 cm). The typical pontoon framing system is depicted in Figure 7.25.

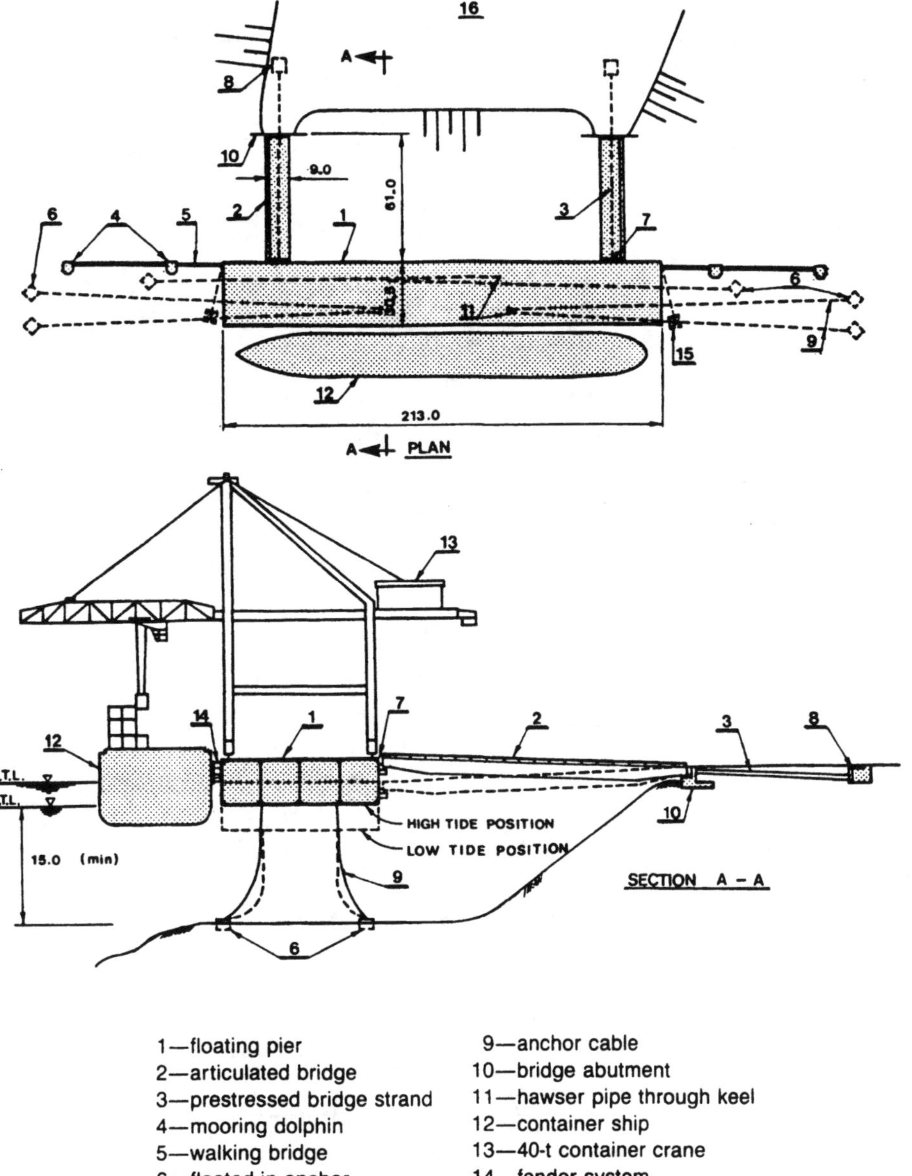

1—floating pier
2—articulated bridge
3—prestressed bridge strand
4—mooring dolphin
5—walking bridge
6—floated-in anchor
7—articulated joint
8—deadman anchor
9—anchor cable
10—bridge abutment
11—hawser pipe through keel
12—container ship
13—40-t container crane
14—fender system
15—emergency anchor
16—marshalling yard

Figure 7.23 Plan and typical cross section, port of Valdez, Alaska container terminal.

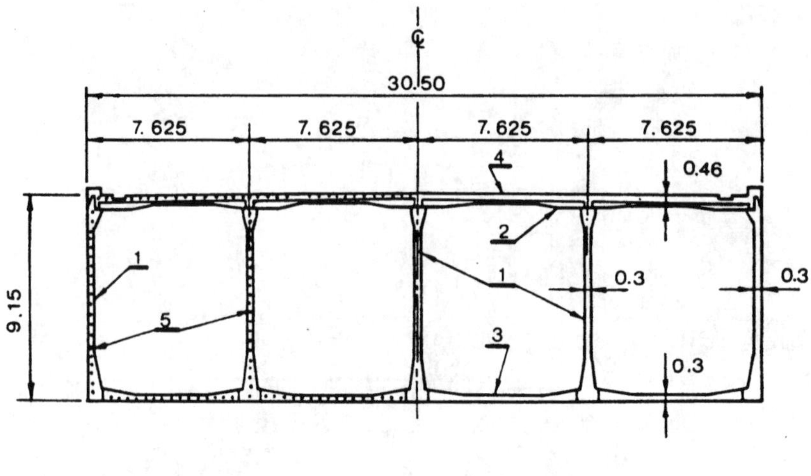

1—precast concrete wall
2—precast prestressed concrete deck
3—cast *in-situ* concrete slab
4—cast *in-situ* concrete deck topping
5—posttensioning tendons

Figure 7.24 Floating dock: typical cross section, port of Valdez, Alaska container terminal.

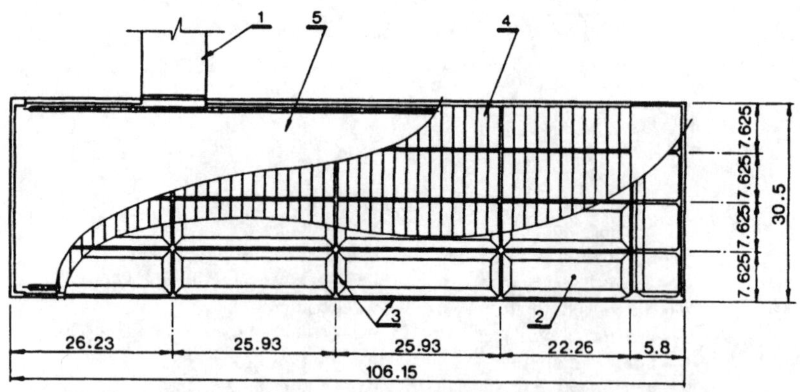

1—access bridge
2—haunched cast-*in-situ* bottom slab
3—precast wall panel
4—precast deck panel
5—cast-*in-situ* concrete deck topping

Figure 7.25 Floating dock: pontoon's typical structural framing, port of Valdez, Alaska container terminal.

At three locations inside each half of the pontoon, steel hawser pipe assemblies were cast in heavily reinforced concrete sections (Figure 7.23). They are used to accommodate six mooring cables. In addition, each half of the pontoon has a thickened section of the deck which serves as support for the access bridge. The pontoon is stressed by posttensioning of steel tendons placed vertically, transversally, and longitudinally throughout the hull. The detail is shown in Figure 7.26. Longitudinal and transverse posttensioning ensures hull watertightness and crack resistance. The exterior walls are posttensioned vertically with high-strength bars, which makes the hull more watertight and increases its ability to resist hydrostatic pressure, wave impact, and ship impact forces. Two large splicer tendons, each consisting of twenty-two 13-mm-diameter strands, are used to provide the moment resistance capacity when two pieces of the pontoon are joined together at a deployment site. These tendons are anchored 6.1 m back from each side of the joint. The anchor plates for these tendons are cast in a special thickened part of the pontoon hull. Access for stressing these splicer tendons was provided from inside the pontoon so that the dock integration can be carried out in dry conditions while the pontoon is afloat. To achieve a high degree of production repetition, all precast elements were standardized. The pontoon construction was carried out in two stages. During the first stage, the pontoon was constructed in the dry dock up to the

Figure 7.26 Installation of posttensioning tendons into the bottom slab of the pontoon, port of Valdez, Alaska container terminal.

deck level (including placing the precast deck panels). At this stage, the tendons below the waterline were posttensioned. Then the pontoon was floated out, having a nominal draft of about 3.6 m. During the second stage, the cast-in-situ concrete deck was completed and the tendons above the waterline were posttensioned. The total construction period for two units of floating pontoons (ready to tow to Alaska) was 6 months. The on-site pontoons integration procedure required good control to prevent damage to matching surfaces due to wave action or changes in the draft caused by shifting ballast. Initially, both pontoons were pushed together carefully at their bows until the pintels at the deck level engaged and the bottom bearing surfaces came in contact. Then the pontoons were partly posttensioned one to the other, and subsequently, the closure joint area was dewatered, cleaned out, and additional reinforcement was installed. Finally, the bottom plate, walls, and deck pours were completed, and after hardening of the concrete, the remaining tendons were posttensioned (Zinserling and Chichanski, 1982).

At the deployment site the floating dock was moored to six gravity anchors 6.1 × 6.1 × 4 m. These anchors have been constructed in the form of a hollow reinforced concrete shells each having a buoyant weight equal to approximately 205 tons. These shells were floated to their designed position, where they were filled with gravel to sink them to the seafloor. The anchor chains attached to these blocks can limit the dock movements along its longitudinal axis up to 0.9 m. Also, two "emergency" anchors have been added to anchor the system to limit accidental movement of the dock. Two 61-m-long 9.8-m-wide orthotropic steel box-girder bridges hinged at the floating pier and the land-based abutments link the dock to the marshaling yard; together, these bridges provide dock anchorage in a transverse direction. Each bridge consists of two box girders 3 m deep by 2.1 m wide joined together by a 9.8-m-wide orthotropic deck. The deck is overlaid by a 50-mm-thick asphalt riding surface. The bridge's maximum grades are +4.25 and −7.6% relative to the marshaling yard. Both bridges are secured to the land-based foundations by posttensioned tendons.

The floating dock is furnished with 42 units of Bridgestone Super Cell fenders and twelve 40-ton-capacity bollards placed uniformly along the face of the dock. It must be pointed out that a fender system is a very important (and costly) part of a dock system. If incorrectly designed or installed, it may be either ineffective, torn off the face of the dock, or both. The performance of a rubber fender system may be affected substantially by the ambient temperature. In cold regions, the low temperature can stiffen the rubber fender or result in ice buildup around the fender unit, both of which can seriously hamper the performance of the fender system. This occurs occasionally at the container terminal at Valdez (Figure 7.27). The floating part of the facility is linked to four bottom-fixed mooring dolphins by steel gangways. More information on this floating terminal is provided in *PCI Journal* and *Engineering News Record* articles (Anonymous, 1982a,b).

7.2.6 Floating Terminals in Peru

A number of floating docks have been built in recent years on the Amazon River and its tributaries. These facilities are part of the Trans-Andean Transportation Corridor, which is studying the construction of several inland ports and road network to facilitate the industrial development of the area. In this section three significant projects are discussed: the ports of Iquitos on the Amazon River, Pucallpa on River Ucayali, and Yurimaguas on the Huallaga River.

7.2.6.1 Port of Iquitos. The port of Iquitos is located on the headwaters of the Amazon

Figure 7.27 Floating dock: ice buildup around cell type fender unit, port of Valdez, Alaska container terminal.

River at latitude 3°47′S, approximately 3200 km from the Atlantic Ocean. In addition to being the major Peruvian port in the Amazon region, Iquitos is also the principal port of entry from the Atlantic. The port is designed and constructed to accommodate seagoing ships and is an important link in the trans-Andean transportation network system. The port is located on the left bank of an arm of the Amazon River called the Cauce Occidental. This arm carries the bulk of the flow in the river, and unlike many places along the river, the channel has remained relatively stable here for many years. Downstream from Iquitos the river has a minimum navigational depth of 8 m, which occurs in the months of September and October, when the water is at its lowest level. The draft and size of the vessels that are served at Iquitos are limited accordingly. The principal commodities passing through the port are lumber, plywood, sugar in bags, grain, flour, bottled beverages, machinery, jute, cement, and food items. The cargo is handled by mobile cranes, tractors, and forklifts instead of manually, as done previously. All cargo is carried onto and off the dock by trucks and trailers pulled by tractors. The new floating terminal (Figure 7.28), that replaced the outdated previously existed floating dock was commissioned in 1982. Part of the old dock was incorporated into the new facility at its upstream end. As shown in Figure 7.28, the new floating terminal comprises part of the old dock (87 m long and 9.3 m wide), five new steel pontoons, two steel end pontoons, two access bridges, and an anchor system. The total length of the floating pier is 274.5 m with available berthing face of 356 m on both sides of the pier. This made this facility one of the largest in the world. New principal steel pontoons (each 36 × 15.36 × 1.8 m) and end and bow small pontoons (each 1.8 × 15.36 × 1.8 m) are joined together 60 cm apart by a pinned connection. To eliminate end pontoon tilt when heavy loaded, they are provided with wooden separators placed between them and the principal pontoons. A 60-cm gap between the new pontoons is provided to accommodate the offshore anchor chains without interfering with pontoon movement. At the deck level, this gap is bridged by a steel cover plate. Each pontoon can easily be disconnected from the dock system and removed. The pontoons are designed in accordance with the requirements of the American Bureau of Shipping (1982). They are designed to load from mobile cranes of 10-ton capacity, HS-20-44 truck loading, and 6 kN/m^2 surcharge load. Structurally, pontoons are framed in both longitudinal and transverse directions and have watertight bulkheads as required. The pontoons are protected against ship impact by standard tubular rubber fenders with equally spaced bollards on each side. The dock rises and falls with changes in water levels without any significant lateral or longitudinal movement. It is retained in its design position by means of an inshore and offshore anchor system. A typical inshore anchor cable is comprised of wire ropes secured at the dock and the land-based piled abutment (Figure 7.29). The offshore part of the anchor system is constructed in the form of winch-operated cables secured at self-embedding gravity anchors placed on the riverbed. These cables are made of cast steel studlink chains (diameter 40 mm). The innovative self-embedding gravity anchors are designed to ensure maximum dock stability against movement toward the shore (Figures 7.30 and 7.31). Each particular group of anchor cables is designed to withstand the action of the most unfavorable combination of forces imposed on the dock by current, wind, and ship impact. Self-embedding anchors as used at Iquitos tend to dig into the soft riverbed when pulled by the anchor force. The anchor continues to dig in until sufficient resistance is developed and equilibrium is reached between the pulling and resistance forces. Two inshore anchor cables are installed at the dock downstream part (Figure 7.28, section 2–2). These cables are depressed by two concrete sinkers to provide for a minimum depth of water of 3.0 m and a navigable channel 30 m wide. This ensures sufficient docking conditions for the river barges that are served at the dock. Access to the dock is provided by two single-span 60-m-long 8.6-m-wide steel articulated bridges. These bridges are hinged to land-based concrete piled abutments and at their lower ends to support pontoons (26 × 8 × 1.8 m) (Figure 7.28). Both bridges are positioned to provide an additional berthing space 85 m long at the lower inshore part of the dock. The latter resulted in increased berthing space by approximately 30%. Space between the bridge and the dock is bridged by the articulated ramp (Figure 7.32). The 6-m-long ramp is hinged at the lower end of the lower part of the bridge and slides on the surface of the deck. The bridge system is prevented from any significant displacement in parallel to the riverbank by a

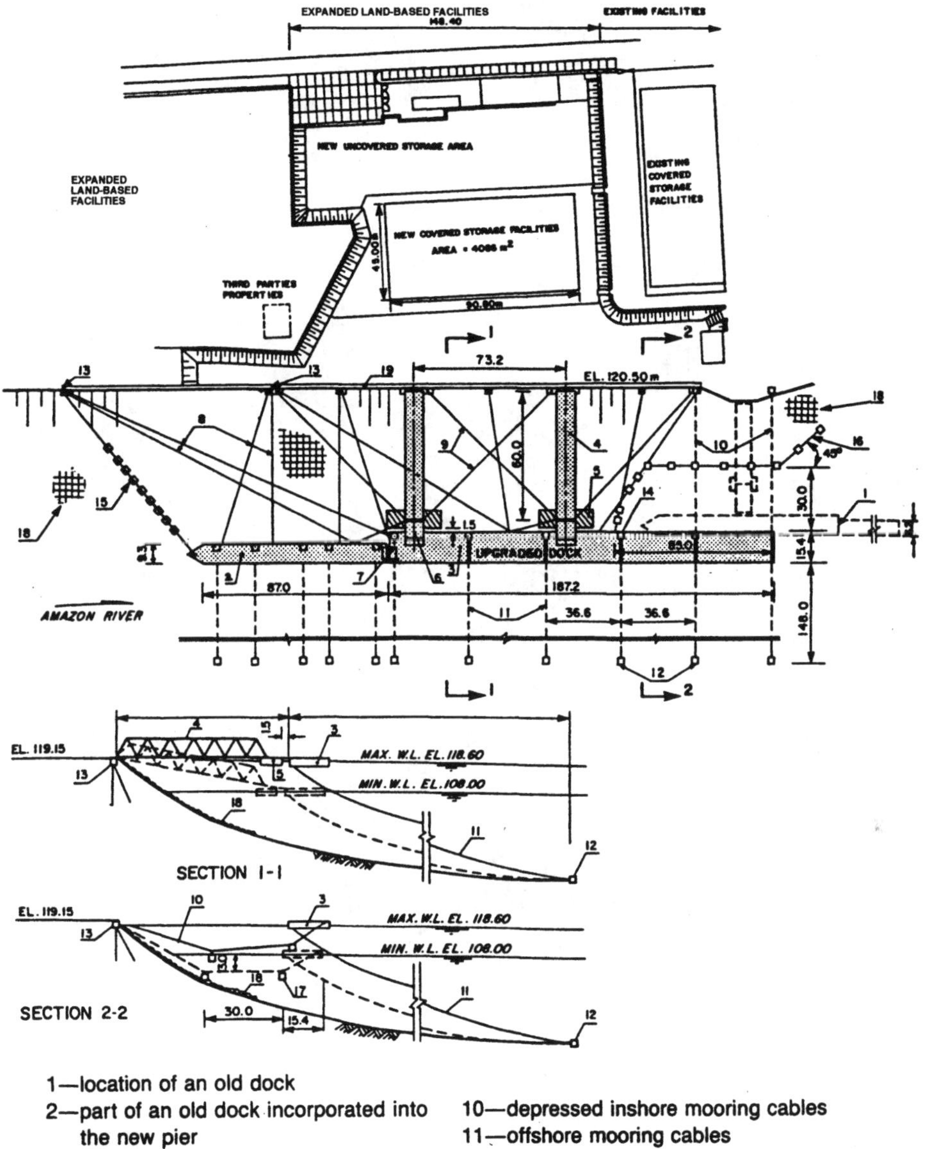

1—location of an old dock
2—part of an old dock incorporated into the new pier
3—new pier
4—access bridge
5—bridge support pontoon
6—ramp between bridge and pier
7—ramp between two parts of a new pier
8—inshore mooring cables
9—bridge keeping prestressed cables
10—depressed inshore mooring cables
11—offshore mooring cables
12—anchors
13—land-based piled deadmen
14—winch
15—upstream floating log
16—downstream navigational aids
17—concrete block
18—slope protection

Figure 7.28 Plan and typical cross sections, port of Iquitos, Peru Floating terminal.

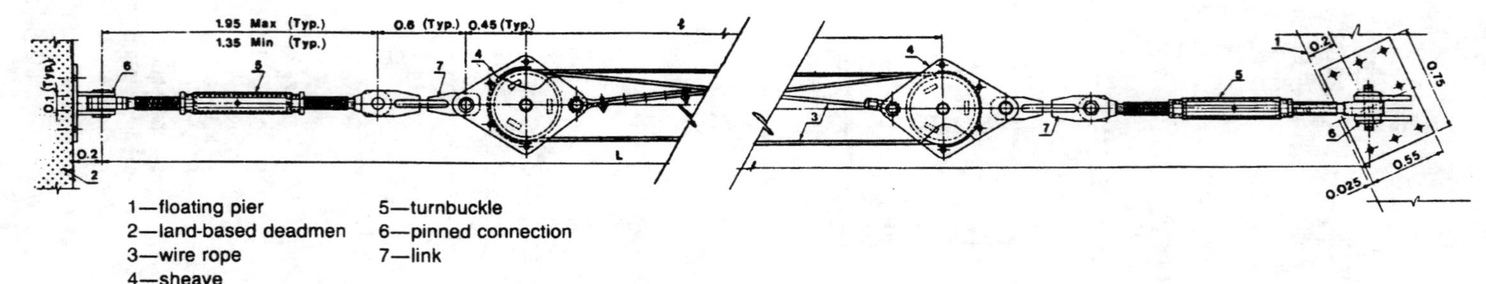

1—floating pier
2—land-based deadmen
3—wire rope
4—sheave
5—turnbuckle
6—pinned connection
7—link

Figure 7.29 Typical inshore mooring cable, port of Iquitos, Peru floating terminal.

Figure 7.30 Self-burying gravity anchor, port of Iquitos, Peru floating terminal.

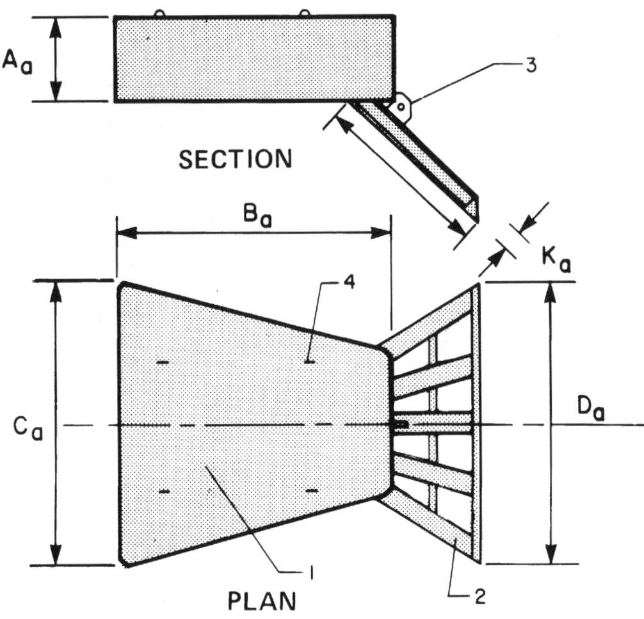

Figure 7.31 Structural arrangement for self-burying gravity anchor. 1, Concrete block; 2, steel plow; 3, attachment of anchor chain; 4, lifting lug. (*Note:* In case of port of Iquitos: A_a = 1.2 m, B_a = 4.0 m, C_a = 4.0 m, D_a =4.0 m, and K_a = 0.25 m.)

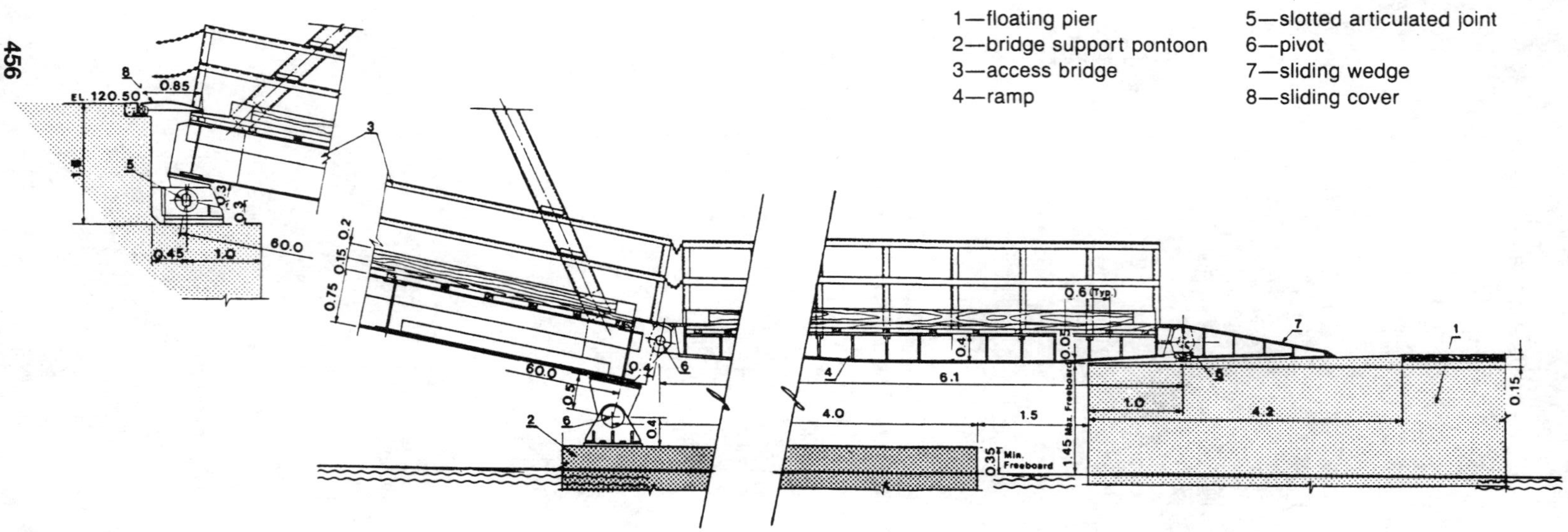
1—floating pier
2—bridge support pontoon
3—access bridge
4—ramp
5—slotted articulated joint
6—pivot
7—sliding wedge
8—sliding cover
EL.120.50
0.85
1.8
0.45
1.0
60.0
0.75
0.15
0.2
0.4
0.5
0.6 (Typ.)
0.05
6.1
4.0
1.5
1.45 Max. Freeboard
0.35 Min. Freeboard
4.2
0.15

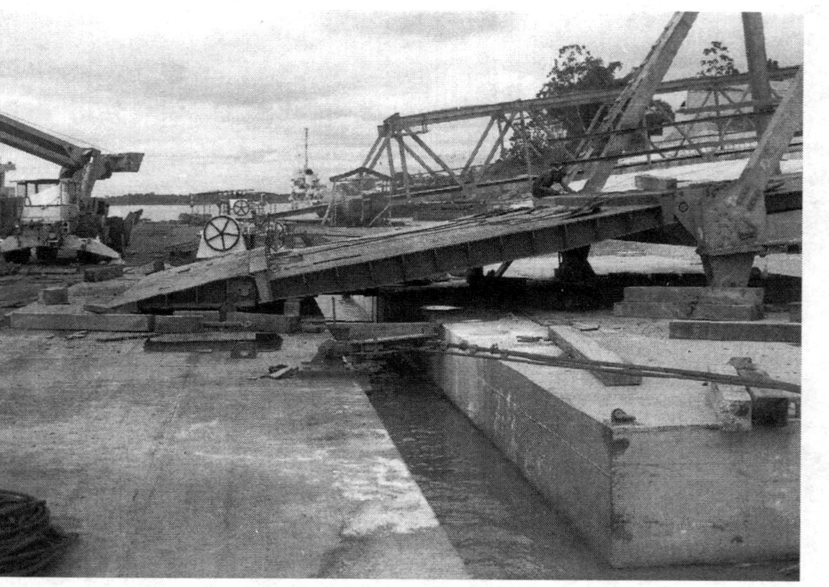

Figure 7.32 Articulated ramp, port of Iquitos, Peru floating terminal.

prestressed cable secured at the land-based abutment and the lower end of the bridge. The bridge land-side hinges are slotted slightly to permit some limited horizontal movement of the bridge without damage to its land-based bearings. As pointed out earlier, both bridges are separated from the floating dock system. This makes it possible to avoid tension in bridge structural elements due to potential pitching of dock pontoon when heavily loaded at one end. An additional merit of separation is that any dock pontoon can easily be withdrawn from the system for repair and/or maintenance. It should be pointed out that the safety of floating dock operations is to a great extent dependent on the safety and integrity of its anchor system, which can be damaged by floating debris and/or an errant vessel that may sail behind the dock and hit the anchor cables. At Iquitos, the principal anchor cables were protected by an upstream boom but also by a network of anchor cables of the old dock that was installed upstream (Figure 7.28). For some reason, in the early 1990s, port management moved the old section of the dock to a new location, thus exposing the new dock to potential disaster, which did not take long to happen. In 1993, an errant boat destroyed the upper bridge prestressed cable system, which resulted eventually in the loss of both bridges and their supporting pontoons. The dock was restored in 1995–1996. For more information on Iquitos, readers are referred to Delgado (1982) and Tanner et al. (1983).

7.2.6.2 Port of Pucallpa. Pucallpa is the second-largest Peruvian port in the Amazon region. It is situated on the left bank of the Ucayali River and is, like Iquitos, part of the Trans-Andean Transportation Corridor. This port is the only major Peruvian inland port which has a direct highway connection with the Peruvian Pacific Ocean coast. The hydrological conditions of the Ucayali River at Pucallpa are characterized by a maximum variation in water level equal to approximately 10 m and by a current velocity equal to 6 knots (3 m/s). At the port location, the Ucayali River is very unstable, and severe erosion and sediment deposition occur along the riverbanks. The meander at the Pucallpa appears to be migrating at an average rate between 150 and 200 m/yr (Fitzpatrick and Gamarra, 1983; Fitzpatrick et al., 1985). Under existing natural conditions, construction of a fixed dock was completely out of question and a mobile floating dock was proposed instead.

It was determined that this mobile facility would be moved to a new location every 5 years, subject to updated hydrological forecast. Accordingly, the dock and its two floating access bridges were fabricated as permanent components of the terminal. However, because its anchor system, consisting of anchor cables and concrete anchors, eventually would be buried deep in the heavily built-up sediment deposits and would thus be essentially nonrecoverable, it was constructed as an element designed to perform for a limited period of time. Thus, the floating terminal at Pucallpa consists of three principal parts: a floating dock, two floating access bridges, and an anchor system (Figures 7.33 and 7.34).

The dock consists of five steel pontoons each 36 m long, 18 m wide, and 2 m deep. The pontoons are joined together by means of articulated joints to form a the 180-m-long pier which provides 310 m of berthing space on both sides and about 3200 m^2 of operative area. The pontoons have longitudinal and transverse watertight bulkheads and are protected against ship impact by means of a fender system consisting of used rubber tires. Each pontoon has two bitts on both sides to provide convenient mooring for the wide variety of boats that use the port. The dock is designed to accommodate a 1500-gross registered ton barge and a variety of smaller boats. The deck of the pier is designed for a heavy truck load (H-20-44), mobile cranes with a lifting capacity of up to 10

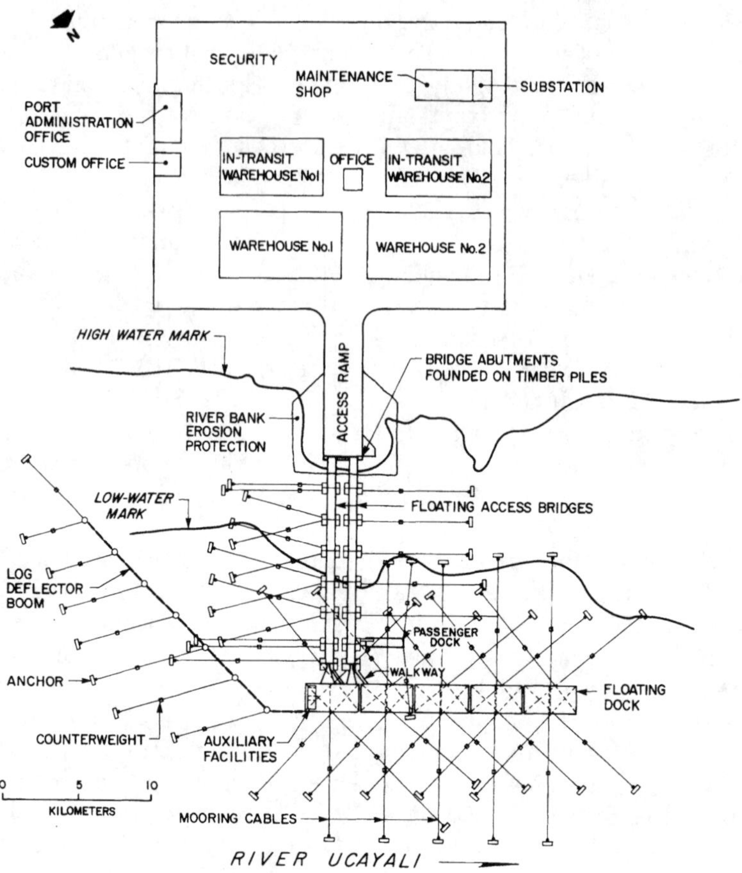

Figure 7.33 Plan, port of Pucallpa, Peru floating terminal.

tons, and a surcharge load 6 kN/m^2. The deck is covered with a hardwood floor 5 cm thick, placed on 5 × 10 cm timber planks, and has a full-perimeter curb 30 cm high. At six locations, inside each pontoon, steel hawser pipe assemblies are installed. These provide attachment for six anchor cables at the pier site. The head pontoon, on the inshore side, has a cantilever part designed to provide supports for two steel ramps of access bridges. The dock's anchor system consists of wire rope cables, concrete anchors, and hand-operated hoists. Cables are placed in upstream and downstream directions and perpendicular to the pier and the access bridges to resist current, wind, and ship impact forces. To provide the pier with the depth required for boats at its berthing faces, and to shorten the length of cables, the latter are depressed by means of concrete sinkers. The dock is linked to the shore by means of

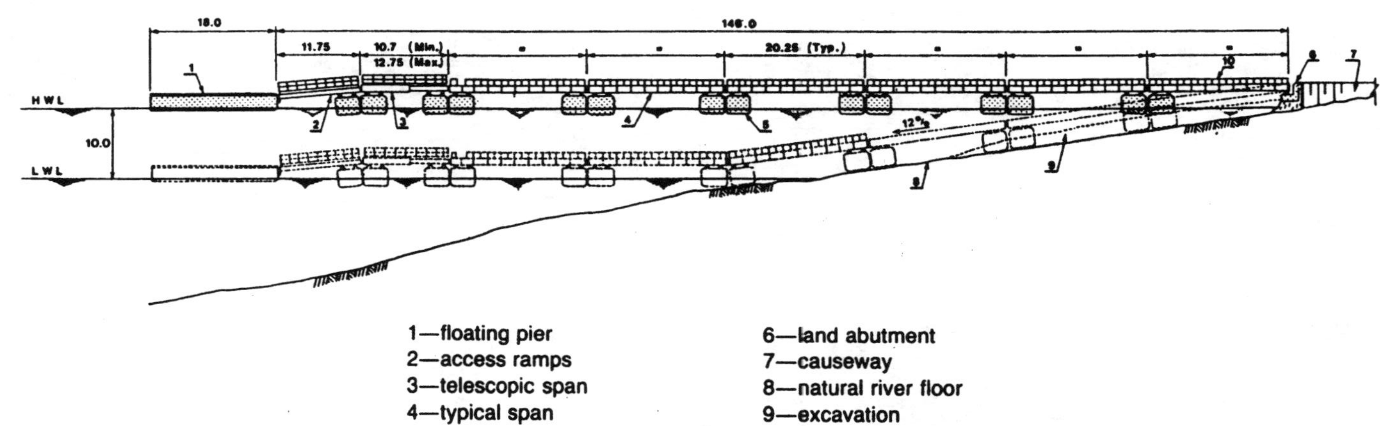

1—floating pier
2—access ramps
3—telescopic span
4—typical span
5—typical bridge support pontoon
6—land abutment
7—causeway
8—natural river floor
9—excavation
10—access ramp

Figure 7.34 Cross section, port of Pucallpa, Peru floating terminal.

two 146-m-long one-way floating bridges placed close to one another. Each bridge consists of five 20-m-long typical floating moduli, one modulus with a telescopic span and two ramps that link the bridge with the pier and the shore abutment (Figure 7.34).

The typical 20.25-m-long modulus consists of two steel pontoons 12 m long, 3.5 m wide, and 2.5 m deep connected rigidly to the 6.3-m-wide and 20-m-long span. Each pontoon has three watertight bulkheads placed in the longitudinal direction, provided with articulated joints for linkage with adjoining typical moduli. In addition, each modulus has an independent anchorage. The bridge span is designed as a regular single-span bridge for vehicular traffic and pedestrians. The bridge modulus with telescopic span is designed to lessen the bridge loading problem in the case of accidental pier movement toward the bridge. Ramps that link the bridges with the shore abutment and the floating dock are fixed to the pontoons at one end and hinged to the shore abutment or dock at the other.

A basic assumption of bridge design was that it would be entirely afloat at a high water level, but that as the water level recedes, as the pontoons would rest successively on the preleveled natural slope of the riverbank. This assumption proved not to be practical, however. In reality, the underwater preleveling of the slope proved to be a difficult task and has never been carried out by port authorities. This resulted in damage to the bridge due to excessive torsion and tension in articulated joints. As a result of bridge failure, the dock was out of service. This could be prevented if to maintain a steady slope during a period of low water, the bridges were provided with fixed resting platforms. Examples are provided in the following sections.

7.2.6.3 Port of Yurimaguas. This port is situated at the junction of the Paranapura River and the Huallaga River, one of the major tributaries of the Amazon River. Initially, the floating dock was operated on the Paranapura River side just about 400 m from the Huallaga River. However, because of the very high rate of sedimentation at the dock site in 1995, it was moved to the Huallaga River but still very close to all land-based port facilities. The dock relocation required complete replacement of the old anchor system with a new one. Because of a lot of small and large floating debris carried by the Huallaga River, some modifications to the dock structure were required. To prevent accumulation of such debris at the dock structure, these modifications basically included increased space between the bridge support pontoon and the dock and modification of the bridge pontoon. The layout of the relocated floating dock is shown in Figure 7.35.

The new dock anchorage system consists of the inshore and offshore wire rope cables secured at the dock and land-based piled concrete blocks and self-bearing concrete anchors very similar to those used at Iquitos, although much lighter. A typical inshore anchor cable is shown in Figure 7.36. To accommodate fluctuation in water levels in the Huallaga River of approximately 11.0 m, the length of the offshore cables and their tension force are controlled by six manually operated winches. The floating dock is comprised of 13 typical steel pontoons 6.1 m long, 3.65 m wide, and 1.6 m deep spaced at 92.5 cm and a wedge-type bow pontoon. The deck is rests on pontoons and is constructed in the form of three steel keelsons supporting steel transverse beams spaced at 76 cm.

The latter supports a wooden deck built of 5-cm-thick planks. The deck system is designed in such a way as to enable any single pontoon to be withdrawn from the system for repair and/or maintenance. Originally designed to handle a small truck and tractor–trailer loads in practice, however, the dock is occasionally used by very large, heavily loaded tracks and large mobile cranes (Figure 7.37). Access to the dock is obtained by means of a 38-m-long

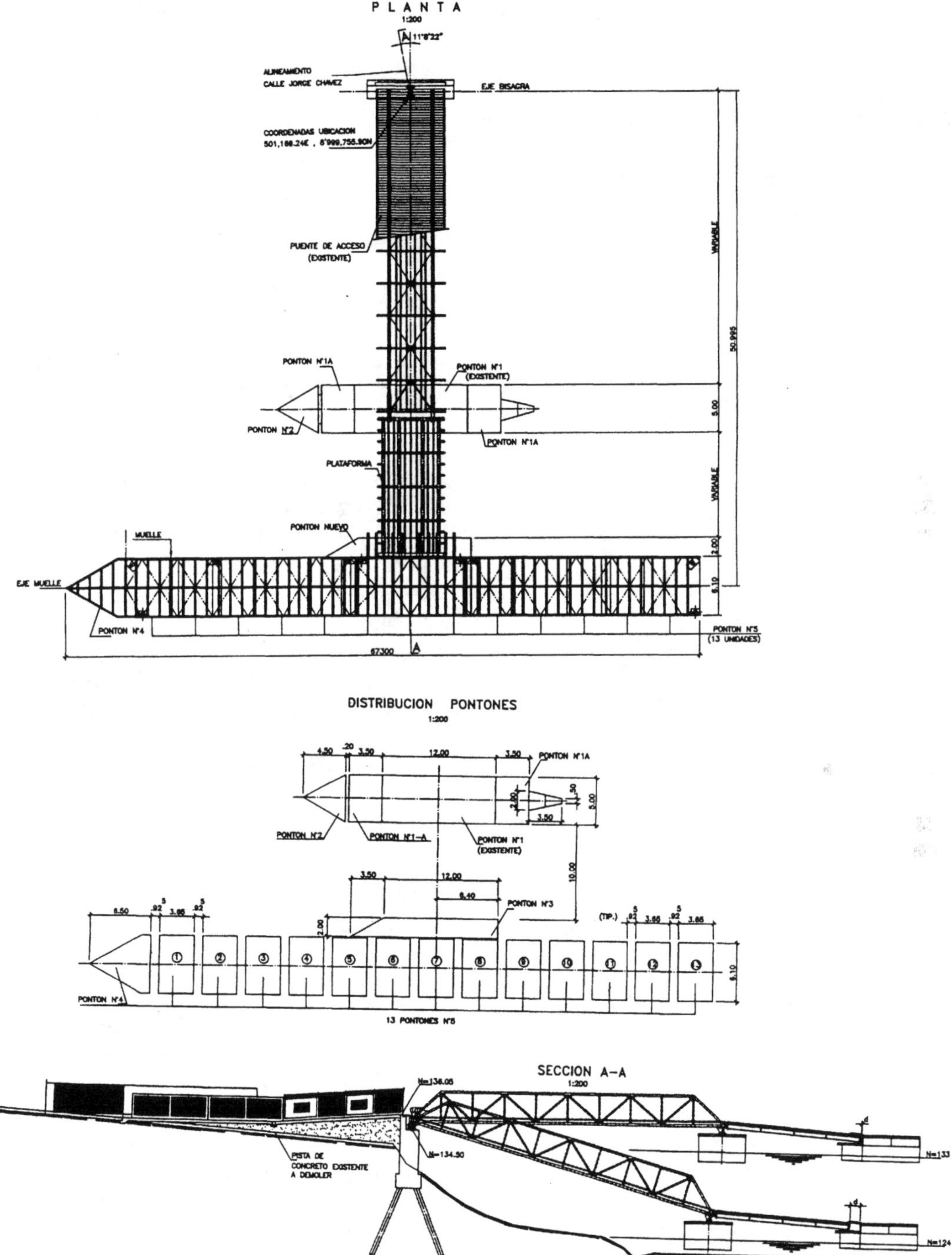

Figure 7.35 Plans and typical cross section, port of Yurimaguas, Peru floating terminal.

Figure 7.36 Typical inshore anchor cable, port of Yurimaguas, Peru floating terminal.

and 4.65-m-wide articulated bridge and an articulated platform 14.0 m long and 7.0 m wide that was added to the bridge. The bridge is hinged at the land-based piled abutment and at its lower end at the supporting pontoon. The platform is hinged at the lower end of the bridge while its lower end slides on a surface of a special pontoon that is added to the dock system and joined rigidly with it. Because of the added weight of the platform and the necessity to accommodate floating debris, some modifications of the bridge-supporting pontoon were required. These included an increase in pontoon size and a wedge-type pontoon placed upstream of the main pontoon (Figure 7.38). The purpose of the wedged pontoon is to deflect floating debris from the bridge-supporting pontoon and to ease its passage through the dock system. This pontoon is attached to the bridge-supporting pontoon by means of special guides that allow both of them to slide up and down against each other depending on the system loading condition. A free space approximately 10 m wide between the dock system and the bridge-supporting pontoon, in combination with the wedged pontoon, allow free movement of flooding debris which occasionally penetrates the floating boom system placed upstream of the terminal.

Similarly to Iquitos, the bridge–platform system is retained in place against all forces acting against it (e.g., drag due to current velocity and wind action, mooring forces, ship impact) by means of prestressed cables that angle from the lower end of the bridge and are secured at the land-based piled foundations. At the lowest water level in the river, the bridge can be inclined as much as 20° from the horizontal. At this angle, vehicles that are used in the dock operation to get in and out of the dock must be assisted by winches. The general view of the bridge–platform system is shown in Figure 7.39. The platform sliding end at the dock side is shown in Figure 7.40.

7.2.7 Flexiport at Port Stanley, Falkland Islands

The floating dock known as Flexiport was constructed by order of the British Ministry of Defence at Port Stanley, Falkland Islands to support military operations during the war between the United Kingdom and Argentina in the early 1980s (Figure 7.41). Moored offshore in approximately 6 m of water, the dock essentially consists of seven offshore industry barges link together to provide both storage and quay space. This system is moored at dolphins and is linked to the shore by a causeway. Barges 91.4 m long and 27.4 m wide can withstand a

Figure 7.37 Port of Yurimaguas, Peru floating terminal. Heavy mobile crane (25-tonne capacity) is moving toward the dock.

Figure 7.38 Port of Yurimaguas, Peru floating terminal. Wedged pontoons deflect floating debris from the bridge-supporting pontoon.

Figure 7.39 Bridge/platform system, port of Yurimaguas, Peru floating terminal.

Figure 7.40 Platform sliding end, port of Yurimaguas, Peru floating terminal.

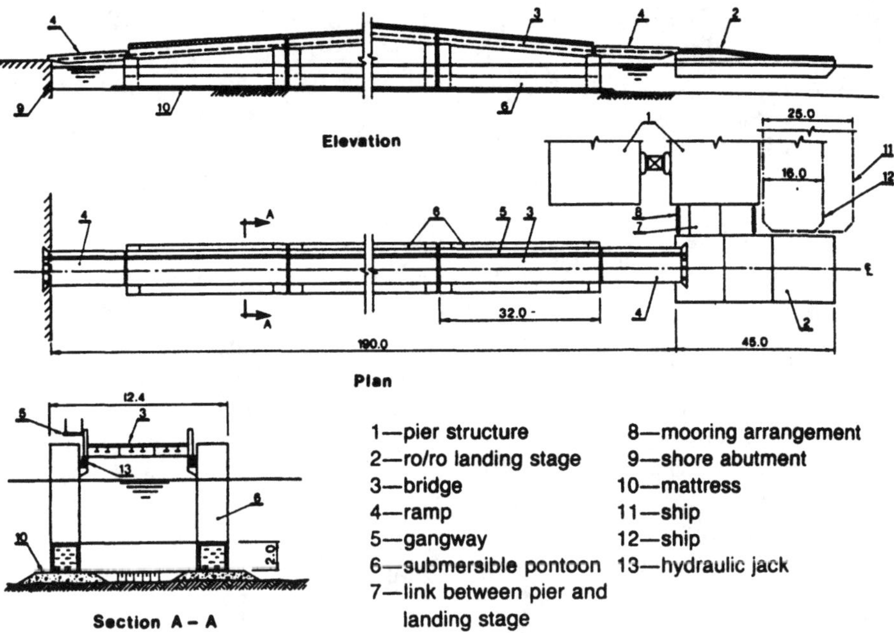

Figure 7.41 Flexiport, Falkland Islands (Islas Malvinas): plan and typical elevation.

deck loading of 15 t/m^2. Thus, the dock was able to sustain load from heavy mobile military and construction equipment. Each barge is also able to sustain a heavy impact load (e.g., when ships dock in the high winds that are a regular feature of Falklands weather). Pneumatic fenders have been installed between the floating dock system and the bottom-fixed dolphins to absorb the ship's impact forces. The total berthing face of the dock is 305 m. The link between the seaward end of the access causeway and the dock is built in the form of the floating landing stage 45 m long and 18.2 m wide projected at right angles to the dock. This provides a landing area for Ro-Ro ramps of vessels having beams between 16 and 25 m. The width of the landing stage is sufficient to allow for two-way vehicular traffic and convenient turning.

The access causeway is comprised of five framelike submersible pontoons 32 m long, 12.4 m wide, and 2 m deep. These pontoons are surmounted by four columns placed at the pontoon corners. Two 15-m-long ramps link the causeway to the shore abutment and to the Ro-Ro landing stage. The total length of the causeway is 200 m, which probably makes it the longest dock-to-shore cargo transfer link ever built for a floating dock. The pontoons that support the causeway's driveway span 7.3 m wide along with the gangway adjacent to the driveway. The gradient of the driveway is 1 : 10 and it is placed 3 m above the highest tide. The driveway is adjustable in height to ±0.5 m by means of a hydraulic jack. Submersible pontoons are installed on the mattress by flooding their internal compartments. A passive water ballasting system operated by submersible pumps and valves enables the trimming of each section and its evacuation in the case of need. It actually took only 48 hours to install the causeway pontoon upon its arrival at the site.

After installation, all pontoons were ballasted 100%. The completed access causeway could handle two-way traffic moving in both directions concurrently. Causeway spans are designed to carry vehicles having a gross weight of 30 tons, and the entire facility is designed to withstand wind forces of 100 knots and the force from waves 1.5 m high. In the postwar period the dock is used to service offshore oil rigs and to handle containers and miscellaneous bulk cargo. For more information on Flexiport, readers are referred to Anonymous (1983, 1985).

7.2.8 Offshore Floating Terminals

The worldwide demand for energy and raw materials has created a strong need for specialized facilities able to handle large, deep-draft vessels. The conventional practice indicated that in a great many cases, construction of an artificial deepwater harbor could be prohibitively expensive. The latter resulted in construction of offshore deepwater facilities able to accommodate the large deep-draft bulk carriers. These facilities are not protected from the effects of environmental loads (e.g., waves, wind, and currents). At these facilities, the low berth occupancy due to rough sea conditions, strong winds, and so on, is compensated for by a high rate of material handling. Naturally, all offshore terminals are site specific. In most cases, they are constructed in the form of bottom-fixed structures linked to the shore by trestles of miscellaneous design. In some instances, a better alternative to a long trestle is construction of an offshore transshipment island used as a transitional storage for the material delivered to or from the island to the port facilities by smaller ships rather than being transported directly to or from a large vessel. For a detailed discussion on port-related offshore terminals, readers are referred to Tsinker (1997).

In a great many cases, specifically for loading, discharging, or both, of liquid bulk materials (e.g., crude oil, liquid natural gas, miscellaneous petroleum materials), offshore terminals are constructed in the form of a single-point mooring (SPM) (Tsinker 1995, 1997). SPMs are fixed at the location one way or another and linked to the shore-based storage facilities by pipelines. They can operate efficiently in rough seas and are not sensitive to directional changes of wind, waves, and currents. With a tanker moored via bow lines only to a SPM, it is free to swing around the SPM and always stay head on to waves, wind, current, or a combination. SPM operation usually requires minimal help from tugs or other support vessels.

In some cases, floating bulk terminals (FBTs) are moored offshore at moorings that are similar to SPMs. These terminals are used to handle and store dry bulk materials (e.g., grain, coal, iron, ore, and other, and liquid bulks such as crude oil, liquid natural gas, and miscellaneous petrochemical products). An FBT usually consists of either a large vessel converted to a bulk storage facility or a specially designed floating facility. Typically, it is moored offshore to allow for handling of large vessels and is equipped with deck-mounted cargo handling equipment. Smaller bulk carriers are used to shuttle cargo continuously between the FBT and shore-located port facilities. Similarly to an SPM, an FBT system will always stay head on to waves, wind, and currents or a combination of some or all of them.

A bulk handling system is designed to transfer cargo effectively to or from a storage vessel and to or from an export (or import) carrier or shuttle carrier. At a correctly designed FBT, the available storage area, although limited, permits a continuous flow of bulk commodities between the FBT and shore-based port facilities. The FBT's mooring system is usually designed for the maximum adverse environmental conditions anticipated at the deployment site. It could be designed for permanent or temporary mooring. The latter is usually characteristic for areas susceptible to seasonal typhoons or hurricanes. It is designed to be disconnected from its mooring(s) as quickly as possible in advance of an approaching potentially damaging storm.

The advantage of an FBT is that it opens an existing port to a wide range of vessel sizes—actually to the largest vessels available. As noted earlier, an FBT also serves as a storage depot, which normally results in cost savings where additional land-based storage (e.g., grain silos) would be needed to accommodate an increase in exports and/or imports. The available floating storage may also reduce the cost of land transportation by locating the FBT closer to an exporter's shipping point, or alternatively, to an importer's receiving point. An additional advantage of an FBT is that it can serve as a fueling depot for both export–import carriers and shuttle vessels. Usually, FBTs are dedicated to handling one type of commodity only. However, they can be designed to handle and store several related commodities (e.g., corn, wheat, rice) to be stored in different dedicated holds.

The cargo handling equipment installed on an FBT typically allows for unloading cargo from a shuttle vessel and transferring it directly to an export carrier for shipment or to the storage located on the FBT in the event that there is no carrier there to receive the cargo. Similarly, cargo can be unloaded from an import carrier directly to a shuttle vessel with provision for temporary storage of the cargo in the FBT's storage in case a shuttle vessel is not available on a regular schedule. Normally, an FBT cargo management and control system allows for continuous monitoring, accounting, and recording of cargo movements during cargo handling operations. Furthermore, dust control, fire and explosion prevention systems, and so on, are requirements for FBTs.

Last, but not least for the effective and safe handling of vessels at an FBT, it is normally furnished with a sophisticated on-board fender system and mooring accessories. A very important element of an FBT is its mooring system. Since accurate environmental and geotechnical site conditions, usually cannot be obtained, the mooring system is designed with an adequate safety factor. Typically, for relatively shallow waters (e.g., 30 to 50 m), two types of mooring systems are used, referred to here as system A and system B. They are shown schematically in Figures 7.42 and 7.43. Essentially, both systems are single-point moorings. As shown in Figure 7.42, system A includes a mooring buoy equipped with a turntable to allow the FBT to swing head-on in prevailing environmental forces. The buoy is anchored in place by a minimum of four anchor chains shackled to the anchors. The FBT is attached to the buoy by a marine hawser.

The choice of anchors depends on the sea bottom soil conditions. To keep the FBT off the buoy during abrupt changes in wind direction, a tug usually stands astern of the FBT. It is also used to maintain a constant tension on the hawser between the FBT and the buoy. The size of the buoy depends on mooring forces applied to it. It is usually made from steel and is subdivided into watertight compartments by a bulkhead system. These compartments are typically filled with a fire-resistant polyurethane foam to provide reserve buoyancy. The buoy is furnished with fenders for protection against damage from possible collision with the FBT or a shuttle or export/import vessel. Mooring cables made of studlink marine chains are secured to the buoy and are attached to the anchor at the seafloor. The latter could be a single usually high capacity conventional anchor or large-diameter steel piles. If required, to limit displacement of the buoy, the mooring cables could be pretensioned.

The FBT's nylon mooring, however, is attached to a turntable placed on the buoy. The hawser must be long enough to keep the FBT at a safe distance away from the buoy. In a B-type mooring system as shown in Figure 7.43, the buoy is replaced with a chain-anchored buoyancy column, usually referred to as a *riser*. To maintain the riser's vertical position under

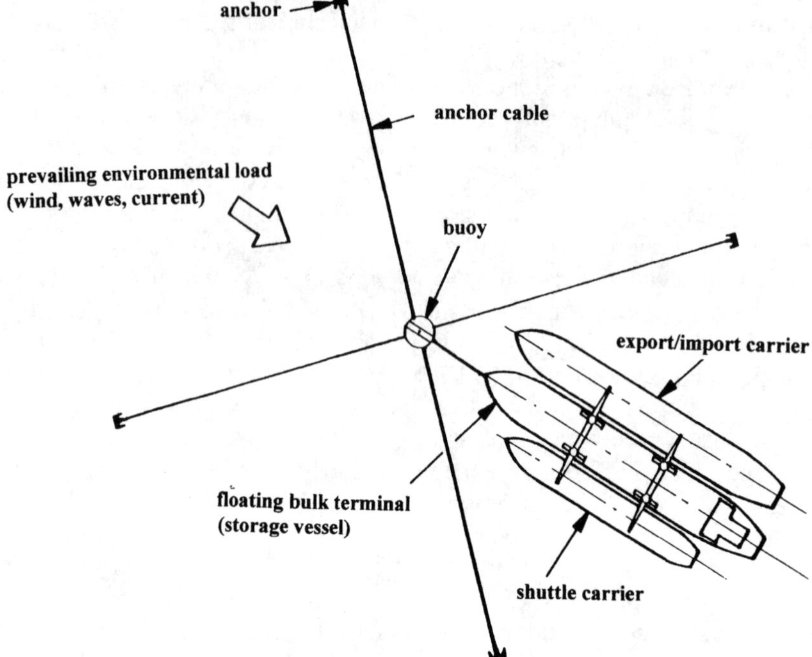

Figure 7.42 Floating bulk terminal, mooring system type A.

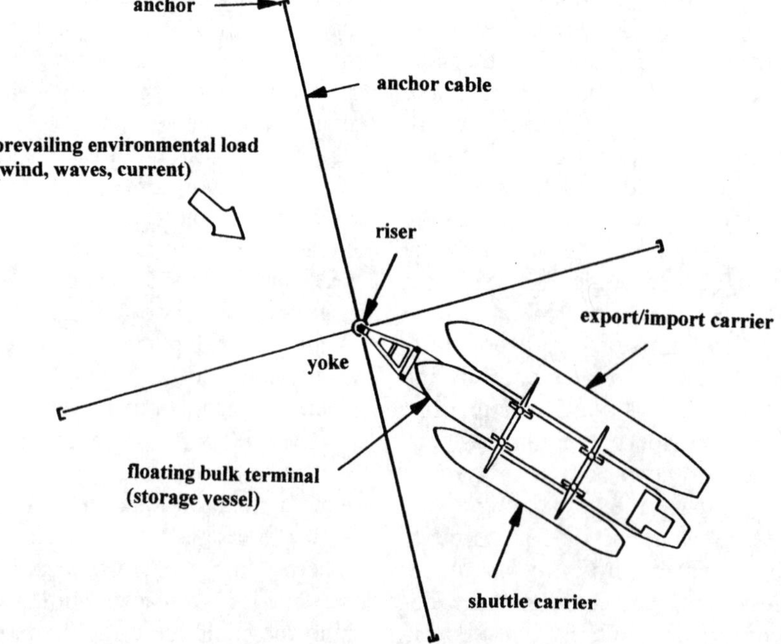

Figure 7.43 Floating bulk terminal, mooring system type B.

Figure 7.44 Floating liquid bulk terminal installed at Selat Island, Sumatra, Indonesia. (From IMODCO AMCA International commercial brochure.)

mooring loads, a system of anchor chains is attached to the top of the riser as required. The riser itself is fixed at the seafloor base via a universal hinge mounted to the top of this base. The FBT is anchored to the riser via a yoke that helps to keep the terminal fixed relative to the river position against the action of all environmental forces. Although B system is somewhat more expensive than system A, it does not require a standby tug. System B is also more adaptable to greater variations in wave height and tidal ranges.

Several representative examples of FBTs are presented below. The FBT illustrated in Figure 7.44 was installed offshore at Selat Laland, Sumatra, Indonesia. It was constructed on the basis of a modified 137,000-dwt tanker and is used to load processed crude oil from a fixed offshore production system at a rate of 15,000 barrels per day and to offload it, at similar rates, to export tankers up to 137,000 dwt. As shown in Figure 7.44, export tankers are using side-by-side berthing at the FBT. The processed crude oil reached the FBT via the undersea pipeline system and a single-point mooring (not shown in Figure 7.44). In a way, this kind of FBT operation is typical for loading and unloading of the liquid bulk materials. It can be found at a great many locations worldwide.

The next example (Figure 7.45) is an illustration of an innovative approach to construction of specialized FBTs. This facility is installed at Ardjuna Field, Java Sea, Indonesia. It is used for collecting, storing, and exporting liquefied natural gas (propane). Constructed from prestressed concrete at Tacoma, Washington, this 66,000-gross displacement (30,000-dwt) facility was towed all the way down to its deployment site located 20 miles offshore. The propane loading and discharge rates are 20,000 barrels per day. The concrete hull 140.5 long, 41.5 m wide, and 17.2 m deep has a fully loaded draft equal to 12.5 m. The total storage capacity of the 12 heavily insulated tanks, six of which are located below the deck, is 375,000 barrels. The facility is moored permanently at the type B single-point mooring. More examples of similar nature are given in VSL International (1987).

Figure 7.45 Floating LPG storage facility installed at Ardjuna Field, Java Sea, Indonesia. (Courtesy of Concrete Technology Corporation and Abam Engineers, Inc.)

Examples of FBTs designed to handle dry bulk cargoes are given in Figures 7.46 and 7.47. Dry bulk cargo handling technology, depending on the cargo's nature, typically include miscellaneous cranes, conveyor systems, and specialized loading and unloading equipment. Cargo handling systems are usually classified as either continuous or discontinuous, for example, grab-type unloaders are discontinuous, and those based on the use of specialized loading and unloading systems using a continuous process are continuous. The former system is typically used to handle large and coarse materials. It is characterized by low cargo handling rates. The spillage rate of these systems is usually high, and control of the dust emission is also difficult with a grab.

Pneumatic or mechanical unloaders are typically very efficient where their use is appropriate. Normally, two traveling gantry unloaders, each having a peak capacity of half the design FBT throughput, are used. They are usually designed for independent operation. For example, in the case of an export situation, one unit could be transferring cargo from the shuttle carrier to the export vessel while the other machine (not shown in Figure 7.46) could be retrieving cargo from the FBT's storage hold(s) and loading it to the export vessel. In general, two (or more) independent units give the terminal operator flexibility of operations

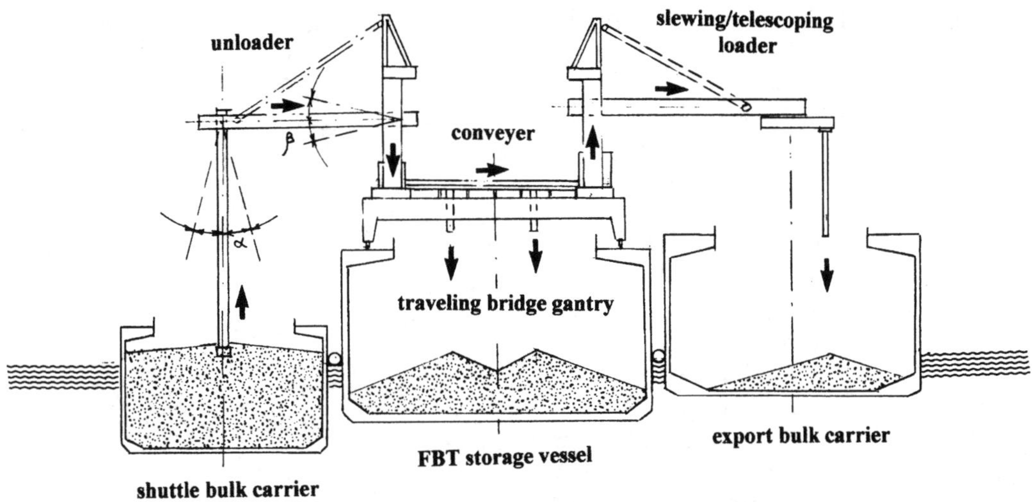

Figure 7.46 Typical cross section, floating dry bulk exporting terminal.

as well as backup needed to maintain continuous operation in case of the malfunction of one unit. The mobility of the gantry allows it to be positioned favorably in relation to the holds of any vessel to ensure effective transshipment of the cargo. Some form of mechanical conveying system is provided to move the cargo from unloading to loading unit. At the unloading side, a telescoping pipe is typically provided to limit the cargo's free fall.

An FBT is usually constructed from a modified active or inactive vessel. In most cases,

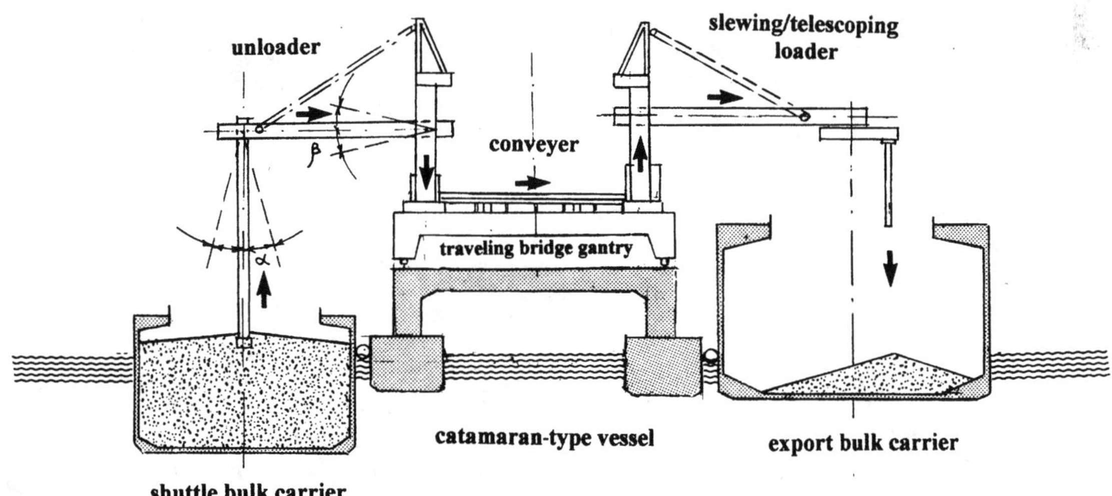

Figure 7.47 Typical cross section, catamaran-type floating dry bulk exporting terminal.

this is a bulk carrier in sound condition. Typical modifications to the vessel include structural strengthening of its deck and hull as required, refurbishing, and complete overhaul, if necessary, of the power, electrical, communication, and other utility systems, providing living quarters for stevedoring crews and marine personnel and furnishing the FBT with an efficient fender system and mooring accessories. Where personnel transfer by a helicopter is required, a new helideck is installed. Last, but not least, mooring fittings to keep the FBT safely offshore should be provided. If required, various holds of the FBT may serve as segregated storage for more than one variety of bulk commodity. Although FBT storage is designed for quick loading and unloading, it usually permits longer storage of cargo during situations such as a strike, or prolonged bad weather conditions. A sketch of a typical FBT and its operation is presented in Figure 7.46.

In the last 20 or so years, the new concept of FBTs has been developed and implemented at several ports around the globe. In this concept, the conventional vessel is replaced by a catamaran type of vessel (Figure 7.47). This vessel is designed for roll-resistance stability and is used at relatively shallow water locations (e.g., mouth of a river, shallow sea). The usually shallow draft allows this system close-to-shore operation. The system is highly mobile, can easily be moved from one location to another, and can become are "instant port" when necessary. The disadvantage of this system is its lack of permanent storage. However, where required the space between catamaran floats can be used for mooring barges, converted into temporary storage.

7.3 SITE CRITERIA AND DOCK LAYOUT

As pointed out in Section 7.1, selection of a location for the deployment of a floating dock is normally based on, but not limited to, the following conditions:

1. The site should be protected from the effects of high waves, strong current, moving ice, large pieces of floating debris, and so on. However, where an attractive site is exposed to the effects of the aforementioned agents, thereby prohibiting floating dock deployment and operation, the possibility of breakwater construction should be considered.
2. The port basin must be sufficiently large to allow the largest design vessel to enter and leave the harbor safely at a reasonable speed, and to approach and depart the terminal safely after being loaded and/or unloaded.
3. Prevailing wave, wind, ice, and other relevant conditions should be considered carefully in dock design and deployment.

Site environmental conditions and the mode of port operation affect the layout of the floating dock. As a result, it may take the form of a marginal wharf or finger pier (Figure 7.48). The choice of layout, as well as its overall dimensions, are determined by the mode of port operation. The dock overall dimensions are controlled by the number of berths required, as well as by the length of an average vessel to be handled at the dock and the cargo handling equipment to be used. Normally, a floating dock is oriented to minimize mooring forces and ship impact. It is usually done by aligning the dock's longitudinal axis with the current direction. At locations with a weak current, the floating pier is normally deployed parallel to the prevailing wind direction. As a general rule, however, a floating dock receiving primarily fully laden ships should be oriented parallel to the direction of the dominating current. This will reduce ship berthing impact. On the other hand, at a dock receiving primarily ships in bal-

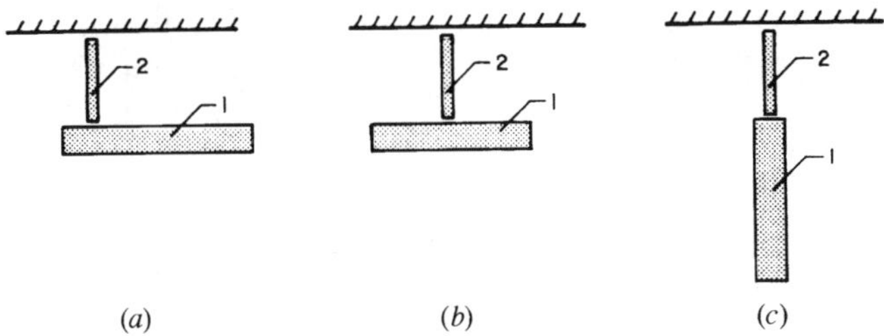

Figure 7.48 Typical layouts of floating docks: (*a*) L-shaped pier; (*b*) T-head pier; (*c*) finger pier. 1, floating pier; 2, access bridge (From Tsinker and Vernigora, 1980.)

last condition, wind forces may be more dominant than the current force. Hence, in this case, ship berthing impact may be reduced by orientating the dock parallel to the prevailing direction of strong winds.

Where these recommendations cannot be followed, good judgment should be exercised, and larger berthing energies should perhaps be considered.

The new port design is, by its very nature, concerned with the future. Therefore, port layout and its berthing facilities have to be tailored to the expected ship and cargo turnover requirements. Ships presently in use and ships that are known or projected to be built in the future have a great impact on the design of new and/or the modernization of existing port facilities. Ship sizes (e.g., length, beam, and draft) govern not only the pattern of their low-speed maneuvers while approaching the dock but also influence the size and depth of a port approach channel and basin, the length of docking facilities, and the layout of fenders and mooring accessories.

The type of cargo and ship cargo-carrying capacity influence a dock's operational land requirements and cargo handling and hauling equipment, such as cranes, conveyors, and pumps, and the dock layout. In the past, relatively small ships and low handling rates for cargo handling and hauling equipment enabled long, narrow piers to be built inside a harbor to maximize the length of berthage provided on both sides of a pier. With increased in port productivity, finger piers became less efficient than marginal wharves, and their use is now restricted to locations where there is no strong cross current and the available water frontage is of limited length. In the latter case, a maximum number of berths may be provided within narrow confines. This is usually characteristic for the municipal docks that serve local passenger traffic, cruise terminals, and especially for naval basis. In all these cases, piers are projected out from the shore to provide berths on both sides of the pier for as many ships as required. The floating part of the dock is maintained at the location by its anchor system and is linked to the shore or bank by an articulated bridge(s). The latter could be hinged at both ends or hinged at the land-based abutment and slide on the deck of the pier at its lower end.

Marginal floating docks are linked to the shore by one or more articulated access bridges to form a T- (in the case of one bridge only) or L- shaped configuration (Figure 7.48). The latter is usually preferred where the current flow exists in one direction only, so that floating craft can carry out berthing or unberthing maneuvers safely using both sides of the dock. Utilization of the land side of the dock permits simultaneous berthing of a greater number of vessels, resulting in increased dock effectiveness. It should be pointed out that using more than one access bridge generally improves the flow of traffic to and from the dock. The latter results in increased dock productivity but has the disadvantage that some of the land-side portion of the dock cannot be used as a berth.

For either a marginal wharf or a finger pier, when the pier length is not sufficient to provide effective mooring for a large ship, mooring dolphins (usually of bottom-fixed construction) are used in addition to the floating part of the dock. A practical example is illustrated in Figure 7.22. These dolphins are not designed to withstand the impact of the ship. The number of mooring dolphins and their layout, depending on the dock arrangement but largely on ship size, may vary from one to six (e.g., very large vessels may require up to six mooring dolphins). The dolphins' layout should provide for the most effective holding of the ship at the dock. Mooring dolphins are provided as required with bollards, quick-release hooks, and /or other means of moorings to handle a ship's mooring lines. In some instances, dolphins are outfitted with fenders for protection from accidental ship impact.

In general, a pier is required only along the ship hatches. Hence, the length of just one berth pier could be limited to the length required for cargo handling only, provided that additional moorings are installed to handle the ship mooring bow and stern lines. To determine the basic sizes of the dock layout, it is necessary first to ascertain the number of berthing points needed to keep a ship's waiting downtime to an economic level. Ideally, the dock is designed to handle a certain volume of a specific kind of cargo. In this case it will be designed to handle specialized ships only. Unfortunately, the latter is possible only in the case of a transport-industry chain operating under unified management, controlling the shipping fleet and the land distribution system. The width of an individual floating pier depends on the type of equipment used in cargo handling and whether one or both sides of the pier are used for berthing of vessels. Usually, for normal dock operation, 10 m is considered as the required minimum pier width. However, considerations of pier buoyancy and/or stability from a dock operation point of view may sometimes necessitate much larger width.

Finger piers are generally designed as two- or four-berth facilities. When several finger piers are used in a port, the width of the basin between adjacent piers must be sufficient for a vessel to maneuver in and out (usually with the aid of tugs). It should be noted that there is no definite rule for establishing basin width. It generally depends on the mode of port operation and varies widely. Although it is important for a port to accommodate the demands of future forecasted traffic, the basin's width should be kept reasonably narrow to decrease internal transportation discharges and to limit investment and maintenance costs. For more information on site selection and terminal layout, readers are referred to Tsinker (1986, 1995, 1997).

For bottom-fixed structures, the depth of water in front of the dock is usually determined based on the draft of the largest conceivable vessel to be operating at the port assuming the general trend in the shipbuilding industry toward increased vessel size and draft. This does not necessarily apply to floating docks, however, where the water depth at the berthing face must be deep enough to provide for safe operation of the vessels currently in use. If necessary, the basin can easily be deepened without interfering with the floating part of the terminal. Water depth in the port and port layout are discussed in detail in Chapters 1 and 2.

7.4 BASIC ELEMENTS OF FLOATING DOCKS

Generally, the typical floating dock consists of the following principal elements:

1. *Floating pier:* a relatively long floating system whose purpose is to accommo-

date vessels and cargo handling equipment and to provide space for cargo and passenger traffic at the dock and for the relevant facilities.

2. *Access bridge(s):* designed to provide convenient access to a pier for traffic and passengers at any water level. The bridge structural scheme depends basically on the amplitude of water fluctuation. Unlike regular bridges, the access bridges of floating docks have no permanent slope. Their angle of inclination changes depending on the level of water and the load intensity on both the bridge and the pier.
3. *Mooring system:* designed to prevent the floating pier and access bridge(s) from moving out of their design location.
4. *Fender system:* installed to prevent the ship and/or dock from being damaged during ship berthing and unberthing operations.
5. *Mooring accessories:* installed to provide convenient and reliable moorings to keep a ship safely at the dock during loading and unloading operation.

7.4.1 Floating Pier

There are four basic structural schemes generally used for construction of floating piers. They are as follows (Figure 7.49): (1) one long pontoon, (2) several large pontoons joined by pivots, (3) a series of smaller pontoons spanned by a number of single-span decks hinged with each other, and (4) a series of smaller pontoons surmounted by a continuous deck. From a capital cost point of view, the last two alternatives are the least economical because of the extra weight of the deck, which in turn necessitates enlarging the pontoons (Umansky, 1939). In general, because of better load distribution between pontoons, the structural scheme indicated in Figure 7.49*d* requires less structural steel for deck and pontoon construction than

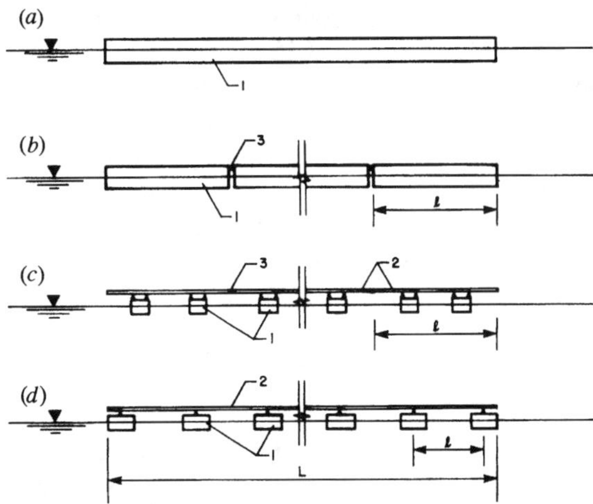

Figure 7.49 Typical structural arrangements of floating piers. 1, Pontoon; 2, deck; 3, articulated joint. (From Tsinker and Vernigora, 1980.)

the one shown in Figure 7.49*c* and therefore could potentially be more economical.

The most economical solution, however, and the one that ensures a pier's maximum floating stability, is shown in Figure 7.49*a*. Alternative (*b*), which consists of several large pontoons linked to each other by pivots is ranked from an economy viewpoint somewhere between alternatives (*a*) and (*d*). It should be pointed out that although alternative (*a*) could be more attractive economically, alternatives (*b*), (*c*), and (*d*) may be preferable from a dock operation and maintenance point of view because they allow the removal of one or more pontoons, or the entire section from the pier structure, for repair or maintenance with little interruption to dock operations.

7.4.2 Access Bridge

The access bridge is a very important part of a floating dock. For efficient dock operation, any means of access to the pier must provide for effective traffic circulation and the shortest possible distance from the pier to the storage area. This can be achieved in several ways, depending on local site conditions. Four practical schemes for bridging the space between the pier and the land are shown in Figure 7.50. The type of bridge to be used in floated dock construction is normally selected on the basis of economic, structural, and functional considerations; the cost of bridge maintenance and its possible replacement should also be taken into account. The bridge must be convenient and safe in operation. In some cases, aesthetic considerations may also affect the selection of bridge system.

7.4.2.1 Articulated Bridge. This type of bridge (Figure 7.50*a*) is generally used when the shore or riverbank at the terminal site is stable with no significant erosion or situation. A single-span articulated bridge is generally

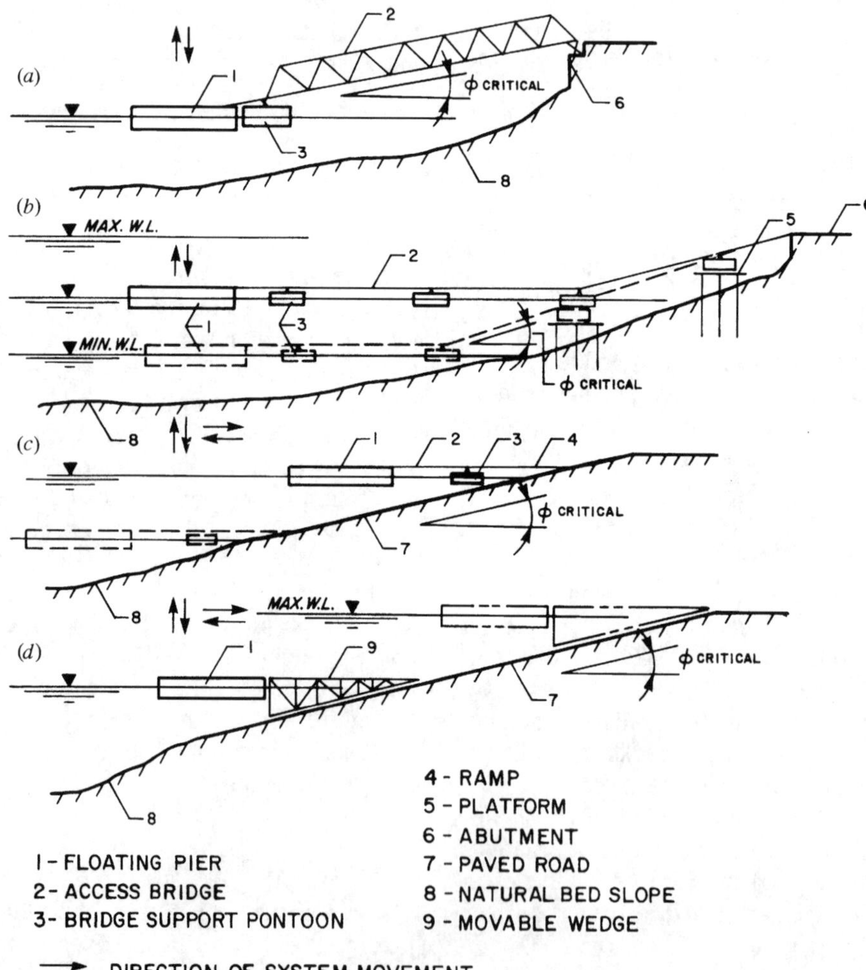

Figure 7.50 Typical means of access to floating pier: mechanism of pier–bridge interaction. (From Tsinker and Vernigora, 1980.)

used where the water-level fluctuation is less than 10 m. The length of the bridge depends primarily on the magnitude of variation in water levels and should be such that its inclination at the lowest water level was negotiable by vehicles or passengers. Generally, an articulated bridge is hinged at the land-based abutment at one end and supported on either a special pontoon or a pier deck at the other end. A practical example is shown in Figure 7.51. It illustrates the initial phase of installation of an access bridge during construction of the floating terminal at the port of Iquitos, Peru on the Amazon River. The outer end of the bridge is supported on a pontoon, then the bridge–pontoon system is stabilized through prestressed cables attached to the deadmen on land (Figure 7.51). To allow some horizontal movement of the bridge due to changes in load magnitude and resulting elastic elongation of cables, hinges at the land abutment are slotted. Structurally, depending on the length of the span, articulated bridges are very similar to the ordinary steel highway bridges and are designed following the same principles. The important considerations in articulated bridge construction should be given to use of a lightweight, free-draining deck with good traction and easy access to hinges for inspection and maintenance. In the case of vehicular traffic, the bridge slope should not exceed an 1:8 up and down gradient with the pier at its minimum normal draft at the lowest design water level or at its maximum normal freeboard at the highest design water level. The latter, however, seldom, governs the design of a conventional cargo handling floating terminal. Critical to the effectiveness and comfort of passenger bridges are the gradient and floor finishes. It is generally accepted that in the case of passengers

Figure 7.51 Installation of access bridge, port of Iquitos, Peru. In this operation, one of pier pontoons is used to provide temporary support to the bridge. 1, Amazon River; 2, bridge; 3, bridge abutment; 4, bridge-supporting pontoon; 5, one of five pier pontoons used for bridge installation; 6, upstream prestressed anchor cable for bridge stabilization; 7, existing floating dock that remained operational during new dock construction.

Figure 7.52 Access bridges system, port of Dover, U.K., ferry terminal. Two lower bridges are provided for access of vehicular traffic, and the upper one is a passenger access walkway.

serving a floating terminal, the former should not exceed 10° and that the most effective floor surfaces are either a bauxite–pitch epoxy system or indented rubber tiles.

For long-span bridges carrying heavy vehicular traffic, the lower end reaction could be very high. In this case it can be operated through bottom-fixed tower(s) with counterweight arrangements. In this case the gap between the bridge and the pier is usually bridged by a relatively small articulated flap. This scheme is allowed to keep the bridge reaction on the floating part of the dock to a minimum. The tower(s) above allow the bridge to be lifted clear of the pier when not in use, which may be beneficial for pier maintenance or other practical situations (e.g., severe weather conditions, security reasons). It should be noted that the adjusting tower-controlled access bridges are commonplace at Ro-Ro ship-to-shore connection. General guidelines for the design of such systems are given by the Permanent International Association of Navigation Congresses (PIANC, 1995).

A practical example is shown in Figures 7.52 and 7.53. These figures illustrate the access bridge system installed at the ferry terminal at the port of Dower, United Kingdom. The tidal range there is 7.2 m. Rebuilt completely in 1988–1989, the system comprises two twin-lane vehicular bridges each 53.0 m long and 6.8 m wide and a separate passenger access walkway 53.0 m long and 2.0 wide located above the upper roadway. The Dower bridges are provided with hydraulically operated finger flaps (short sliding ramps) at the

Figure 7.53 Hydraulically operated finger flaps. Note the hinged connection at the lower end of the articulated bridge.

lower (seaward) end (Figure 7.53). Each finger can be activated individually by its own hydraulic ram to cater for the wide variations in required access width. For safety reasons, the flaps have a minimum longitudinal overlap of 1.5 m to cater to vessel surges during loading operations. The bridges are raised and lowered by electrically operated winches. In some practical applications, the access bridge may be hinged at both ends to the land-based abutment and floating pier. An example is shown in Figures 7.22 and 7.23. In this case it is used as a part of the total anchor system, which exposes it to considerable longitudinal force. Similarly, where the bridge is separated from the pier and its outer (seaward) end is supported on a pontoon, the gap between the end of the bridge and the pier is bridged by an articulated ramp or platform ending with a wedge that can slide within limits on the pier deck. The ramp may overlap the deck by 1.0 to 1.5 m (minimum). The advantage of the use of an articulated ramp or platform is that the length and weight of the access bridge can be kept to a minimum. The bridge and ramp or platform decking is often built from timer planks. Several representative practical examples are shown in Figures 7.28, 7.32, 7.35, 7.37, 7.39, and 7.40.

7.4.2.2 Floating Bridge. Where the variation in water level exceeds 10 m, or the riverbank is flat or consists of very soft or loose soils not suitable for construction of fixed structures, the floating bridge could be used to provide access to the pier. Typically, a floating access bridge consists of pontoons interconnected by relatively short spans (Figure 7.50*b* and *c*). Because a floating bridge is normally placed across the current, the type of bridge arrangement above provides better protection from miscellaneous floating objects (e.g., trees, ice flows) which otherwise might accumulate against the bridge. At high water the bridge is entirely afloat, and as the water level begins to recede, the bridge pontoons begin gradually to settle on fixed platforms until at low water most of the pontoons are supported on fixed platforms at specific grade (Figure 7.50*b*). In general, this solution is suitable for locations where the river is relatively stable and scouring or accretion around the fixed platforms is not excessive.

At remote locations, where the cargo traffic is insignificant and the distance from the dock to the land is substantial, a floating bridge could be cut short as indicated in Figure 7.50*c*. In this case the floating pier and floating bridge must be moved frequently toward and away from land as the water level fluctuates. Vehicles gain access to the pier by a road built on a riverbank that is almost entirely submerged at high water. Operation of such dock and traffic circulation is more efficient at the high water level because of the dock proximity to land-based installations. This scheme is less costly to build and more flexible in operation than is a single long floating bridge. Being highly mobile, it is not affected dramatically by changes in the course of the river because it may easily be moved to another location. Operating costs of such systems, however, may be quite high, due to the necessity for frequent movement up and down the bank, expensive access road maintenance, difficulties with dock anchoring, protection from floating debris, and so on. Protection of floating bridges along with floating piers from floating debris, errant small boats, and so on, can be achieved in most practical cases by installing floating booms.

Structurally, floating bridges are very much similar to floating piers. There are three basic design alternatives for floating bridge construction:

- A chain of pontoons joined by articulated joints
- A chain of spaced-apart pontoons with individual deck sections joined by articulated joints
- A chain of spaced-apart pontoons with a continuous deck

Basically, floating piers differ from floating bridges by the shape of the pontoons. For example, pontoons used for construction of floating piers are usually of rectangular configuration. On the other hand, pontoons that are used for construction of floating bridges to reduce drag forces exerted on them by the current often have a streamlined shape. A floating bridge is usually linked to a shore abutment and to the pier by articulated ramps. The length of the ramp at the land-based abutment is usually equal to the length of the standard section of the bridge. This is required to provide the bridge with a uniform slope at low water levels, which must not exceed the critical angle for the design traffic. An alternative solution to a standard ramp could be an articulated bridge.

7.4.2.3 Mobile Wedge. This scheme (Figure 7.50*d*) works very much the same way as the short floating bridge system illustrated in Figure 7.50*c*. In this case, access to the pier is obtained via a mobile wedge which, depending on water level, moves together with the pier toward and away from land as the water level fluctuates. The wedge is usually moved on a track. Vehicles gain access to the pier–wedge system by a road that is mostly submerged at high water. This scheme is practical when the riverbank is gently sloped, relatively stable, and consists of firm soil. This type of construction has advantages and disadvantages similar to those for short floating bridges.

7.4.2.4 Vertical Lift Bridge. The vertical lift access bridge (Figure 7.54) requires a substantial amount of on-site work related basically to the construction of bridge support towers. Structurally, this type of construction is very similar to that illustrated in Figures 7.52 and 7.53. The main advantage of the vertical lift bridge is that it does not create a serious obstruction to river flow, but the capital cost for its construction could be prohibitive. The bridge consists of the following basic elements:

- Lifting span
- Support towers
- Counterweights
- Winches
- Access ramps(s)

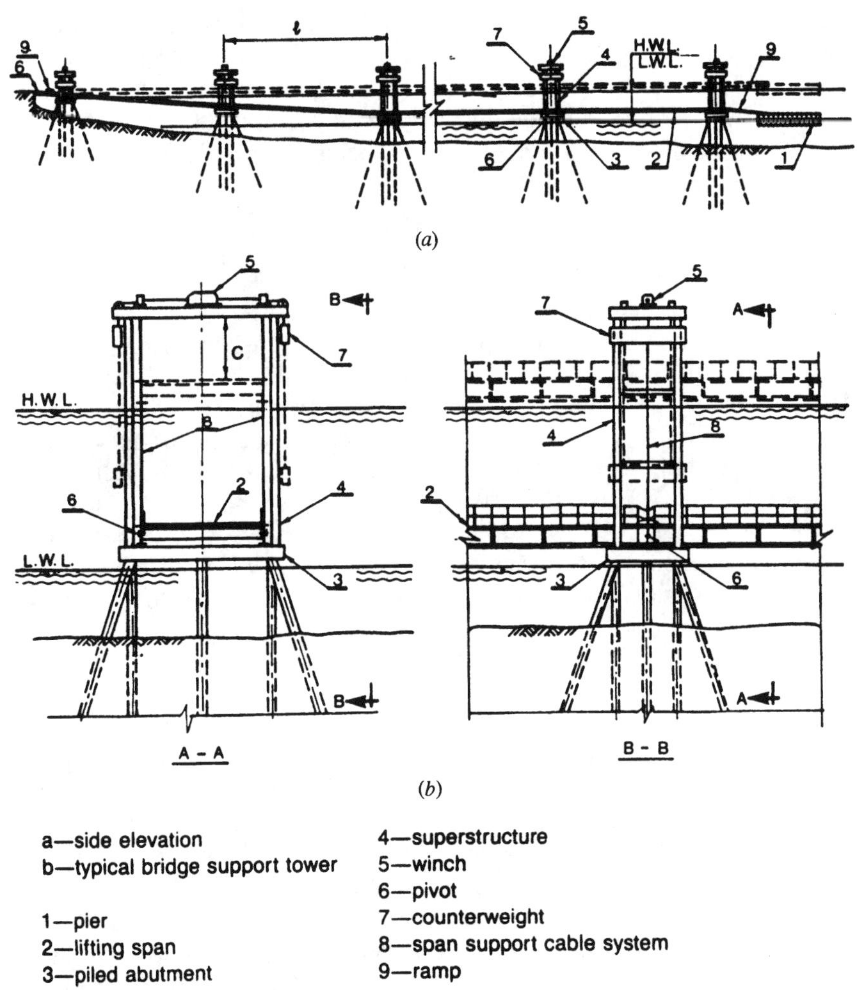

Figure 7.54 Vertical-lift access bridge.

A typical lifting span consists of several interconnected individual spans linked together by pivots. Each pivot nest is slotted to accommodate some horizontal movement. The structure of the span depends on its length. It can be fabricated in the form of steel trusses or steel girders of miscellaneous designs. The bridge is linked to the shore abutment and the floating pier via articulated ramps. The first one is pivoted to the bridge span and the land-based abutment. At pier side, the ramp is pivoted to the span and slides on the pier deck similar to conventional floating or articulated bridges. Similar to typical highway vertical lift bridges, the weight of the span is balanced by counterweights which slide on rails attached to the tower superstructure. Basically, all counterweights should be sufficient to balance the movable span and its attachments into any position, except that there should be small positive dead load reactions at the supports when the span is seated on "dogs." The unbalanced condition noted above is used in the design of lifting winches and power equipment. Counterweights are usually made of concrete enclosed in steel boxes.

The position of the bridge's span is controlled by winches installed on the top of the tower. Once positioned, the span is fixed in place by means of locking devices ("dogs") located at each end of the span. The position of each span is changed according to the height of the water level. Special attention should be paid to the design and construction of bridge support towers, which are designed to prevent significant settlement or movement. Where foundation piles are used, consideration is usually given to the use of batter piles to resist torsional and horizontal forces caused by movement of the span and other relevant horizontal forces. The lift span and counterweight are held in position during their movement by means of special adjustable guides. These guides may be of either the sliding or rolling type.

7.5 MOORING SYSTEMS

The basic design requirement for construction of a floating pier mooring system is to provide for safe and efficient dock operation. The reliability of a floating dock mooring system is particularly important with regard to access bridges, which are typically designed on the premise that the mooring system will keep the pier in place within an acceptable range of transverse and longitudinal motions. Environmental forces acting on the dock in combination with ship impact and mooring forces are considered in mooring system design. Extensive analysis is usually required to determine the design magnitude of mooring forces and their distribution between individual mooring system components and evaluation of the overall behavior of mooring system.

7.5.1 Basic Arrangements

Typical mooring systems usually consist of onshore and offshore parts. The onshore moorings terminate on land at deadmen, and the offshore moorings are placed underwater and secured to anchors of miscellaneous designs. Under a certain site conditions, onshore and offshore moorings can be replaced by mooring dolphins designed to retain the pier in place against all relevant loads. Basic types of mooring systems used in floating dock construction are presented in Figure 7.55.

7.5.1.1 System Consisting of Onshore Part Only. This system (Figure 7.55*a*) employs articulated booms (minimum of two) and onshore mooring lines angled to the land from the pier. The booms are hinged at both ends, and depending on their length and the importance of the structure, are made of timber or steel. Booms control pier motion normal to the land direction. Mooring lines (usually made of

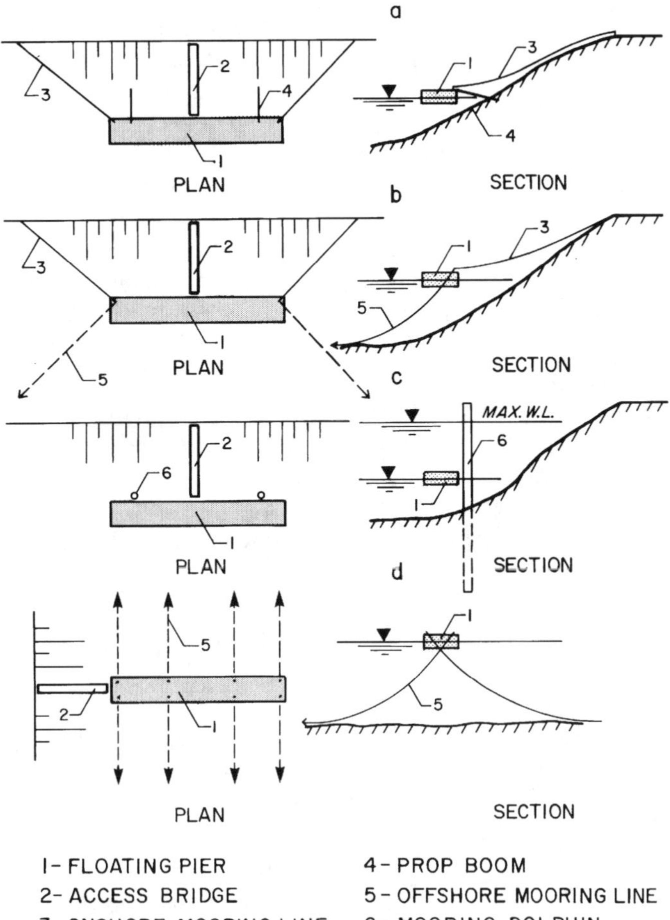

Figure 7.55 Typical mooring systems for floating piers. (From Tsinker and Vernigora, 1980.)

steel wire ropes) are secured at the pier and to the land-based gravity or piled deadmen. These lines control pier motion parallel to the shore and also keep it from moving away from the shore. Depending on pier size, the magnitude of forces involved, and the distance to the shore, the entire mooring system can employ rigid booms only to control pier motion in all directions. Examples are shown in Figures 7.16 and 7.18. The mooring system shown in Figure 7.55*a* is generally used when the distance between the pier and the shore is relatively short.

7.5.1.2 System Consisting of Onshore and Offshore Mooring Cables. The onshore part of this system (Figure 7.55*b*) is similar to that shown in Figure 7.55*a*, and the offshore part consists of underwater mooring cables (usually, studlink mooring chain or galvanized cables) secured at the pier and to the anchors. To prevent the system from being loose at low water levels or overstressed at high tide, the length of offshore mooring lines is normally controlled by winches located on the pier deck. The number of mooring lines and their layout depend on the forces exerted on the pier. This system is used when the distance between the pier and the land is appreciable.

7.5.1.3 System Consisting of Mooring Dolphins. A minimum of two dolphins are employed in this type of mooring (Figure 7.55*c*). The number of dolphins and their structures are normally controlled by the pier length and the magnitude of mooring forces involved. When the water level fluctuates, the pier slides up and down along the dolphins' face. In this case the pier-to-dolphin mooring connections are designed to allow for free vertical movement of the pier without significant displacements while permitting some degree of list and trim of the structure. The simplest form of connection is a hoop that could slide along the simplest form of the dolphin, which is the guide pile, or cluster of piles. In some instances, mooring dolphins could be constructed in the form of piled, or gravity-type structures. In this case, depending on the range of water-level fluctuation, these dolphins are typically furnished with mooring accessories, such as bollards, quick-release hooks, bitts, and so on, installed at the top of the dolphin or at two or more levels to provide for safe and convenient mooring of the pier. In some cases, dolphins can be used in combination with onshore mooring cables. This eliminates the necessity to use bollards or other mooring accessories installed on dolphins. The pier can be moored either in front of dolphins or behind them. In the latter case, the dolphins are used not only for pier anchorage but also as a conventional breasting dolphin protecting the pier from ship impact. A conventional dolphin mooring system is normally used in relatively shallow water (up to 12 to 15 m), where construction of bottom-fixed structures is usually economically feasible. In water deeper than 15 m with significant water-level fluctuation, the hybrid system shown in Figure 7.56 can be considered. The latter actually is a combination of dolphins with an articulated rigid mooring system.

7.5.1.4 System Consisting of Two-Way Underwater Mooring Cables. This system (Figure 7.55*d*) is generally used for mooring of finger piers projected seaward from the shore. In this case, mooring lines control a pier's motion parallel to the shore direction. If hinged to the pier, the access bridge controls a pier's motion normal to the shore. It should be mentioned that in all aforementioned mooring schemes the length of the underwater mooring lines may be reduced by the use of sinkers. In some cases sinkers are used to depress onshore mooring lines, to provide access to the pier from the inshore side. The practical example is shown in Figure 7.28. All of the schemes above, with some variations, have been used in practice to cater to existing physical conditions and operational requirements of floating terminals. For more information on anchor systems, readers are referred to Tsinker (1986).

7.6 ANCHOR SYSTEMS

Different types of anchors, from commercially available conventional steel anchors to specially designed steel or concrete anchors, are commonly used to provide anchorage for floating docks and other floating objects, such as ships, drilling rigs, floating bridges, and other. On the basis of their performance, anchors are usually classified as follows:

- Commercially available steel anchors, whose holding power depends primarily on their ability to penetrate the soil to develop passive resistance in the soil
- Concrete anchors jetted into the soil, whose pullout capacity is based on the value of passive resistance of the surrounding soil
- Gravity anchors, whose holding power depends solely on the weight of the anchor and the friction forces developed on contact between the anchor and the bottom

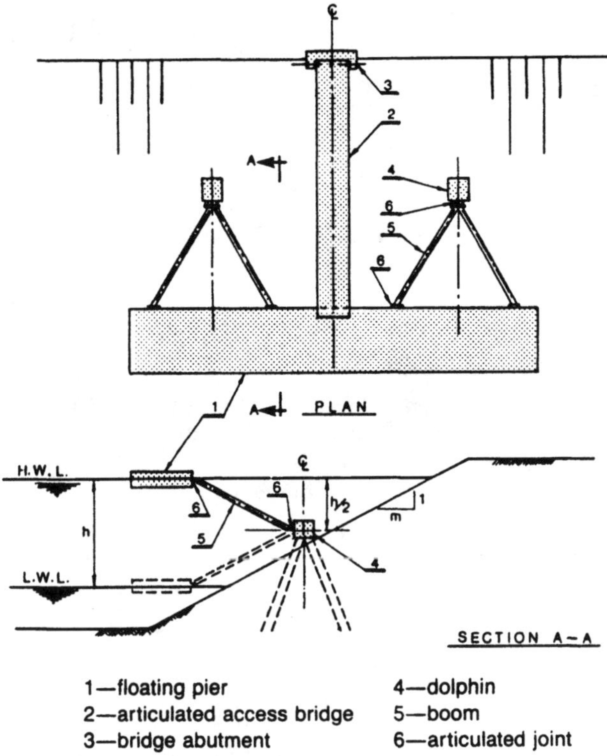

Figure 7.56 Hybrid mooring systems.

- Piled anchors of miscellaneous designs
- Special anchors, which combine properties of gravity anchors with buried anchors, suction piles, suction anchors, and so on

7.6.1 Commercially Available Conventional Steel Anchors

In general, the holding power of a conventional steel anchor depends on the mass of soil that is displaced by the anchor and physical properties of that soil. They generate most of their holding power by mobilizing the shear strength of the soil in which they are embedded. The weight of the anchor is usually not as imported as its fluke area and the depth of penetration into the soil. These two basic parameters determine the amount of soil displacement and hence the drift of the anchor in soil.

The shape of the commercial anchor is usually streamlined. This helps to minimize resistance when it penetrates the soil. Naturally, soil physical properties have a major influence on anchor holding power. In sands, anchors normally penetrate easily, but during a short moment of tripping, when the anchor rises on its fluke points, it may fall on its side. For this reason they are usually provided with stabilizers that are indispensable for anchor performance; they correct the anchor position to ensure its penetration. For better penetration in sand, anchor flukes have to be sharp. This helps to achieve the deepest possible penetration. In soft soils it is sometimes difficult for an anchor to trip, and it may slide over the bottom surface with its flukes pointing upward and neither trip nor penetrate.

Mud (clayey soils) is reasonably good anchoring soil because it allows the anchor to penetrate completely and covers the flukes. However, the holding power of a conventional anchor in mud is usually only about half as much as it is in sand. If the mud lies over the very hard layers into which the anchor cannot penetrate, it will probably fall on its side. Soft clay in combination with layers of mud is a variable anchoring soil because it can be disturbed prohibitively during penetration, and where the clay is of the unconsolidated type, the anchor holding power can be reduced significantly. Hard stiff clay is a good anchoring soil for anchors, which trip well, have sharp fluke points, and a sharp well-shaped shank which cuts through the soil without causing substantial resistance. Anchors that do not meet these requirements will not penetrate hard stiff clay at all and will have a holding power only equal to approximately their own weight. The fluke area is the major factor in conventional anchor holding power (Thorne, 1998). Doubling the weight of anchor while maintaining the same fluke area and penetration improves the anchor holding power by only 4%, whereas doubling the actual fluke area improves the holding power by 80% (Tsinker, 1986). In general, in any kind of soil, it is the fluke area combined with its penetration that determine the anchor holding power and its drift under the load.

The fluke–shank angle is the second most important factor in conventional anchor performance, as it determines the degree of penetration into the soil. It has been found that in hard soil, a fluke angle of 32° gives the best soil penetration. On the other hand, a 50° angle may obstruct an anchor's penetration into the soil and cause it to keel over at its stock stabilizers. (Without stabilizers it will slide on its side and will therefore neither reset nor penetrate.) In mud, it is the crown that enables the anchor to trip and penetrate. Here, anchors with a fluke–shank angle of 32° will cease penetrating far more rapidly than will anchors with a fluke–shank angle of 50°.

The pulling angle of anchor cable also has significant influence on anchor performance. If the anchor is pulling at an angle of zero (the anchor chain is parallel to the bottom), better anchor efficiency should be expected through deeper penetration. However, tests have indicated that it is not necessary to apply a pulling force at a zero angle which actually results in a very long anchor cable. This angle can be increased up to 6° with no significant reduction in anchor performance but with a substantial reduction in the length of the anchor chain (Van den Haak, 1983). It should be noted that some authorities recommend use of a maximum angle of 3° between the anchor chain and the bottom (Quinn, 1972). In conjunction with the above it must be pointed out that because of too many uncertain variables involved in an anchor system design, one should not attempt to determine the anchor holding capacity with mathematical precision. In all practical cases, common sense must be exercised when the type of anchor is selected and its holding power is analyzed.

For detailed information on conventional anchors presently in use, readers are referred to Tsinker (1986). In conclusion, the following guidelines may be helpful to the designer in selecting commercially available conventional anchors:

- The workload on the anchor should constitute approximately 60% of maximum anchor capacity. (Maximum anchor capacity by definition is load obtained at continuous anchor drag.)
- The workload on the anchor should be no more than 50% of the breaking load of the anchor chain; in turn, a chain-proof load has to be approximately 70% of its breaking load.

- When selecting the type of anchor to be used in a floating dock system, some fieldwork has to be carried out to determine the bottom condition and forecasting the anchor holding power. The latter can be achieved by testing the holding power of large-scale model anchors. The test results may be considered as reliable if the bottom soil condition is consistent over the area and if the test results are rendered to be accurate and consistent.
- Potential anchor drag must be evaluated and considered carefully in the total design of a floating dock system.

Finally, it should be pointed out that the commercially available conventional anchors should be used as designed. Sometimes, attempts to increase the fluke area, or to make an anchor heavier, or to introduce other modifications to the anchor structure may affect the balance and performance of the anchor. Furthermore, if used anchors with unknown service histories are to be employed, they should be inspected carefully for cracks or other indications of impending failure. It is good practice to inspect anchors after 5 years of service. Anchors that have been in service for 20 years or more should be inspected more frequently. After setting, the anchors should be pretested to the maximum load and reset if not holding.

7.6.2 Gravity Anchors

The performance of gravity anchors depends primarily on their weight and the grip on soil to resist load. They usually rest on a floor but could also be partly or completely buried into it. The primary advantage of gravity anchors stems from their ability to resist uplift forces, which permits the use of a shorter length of mooring cables. Gravity anchors are generally used where the floor cannot be penetrated by conventional anchors and where an unusually high holding power is required. Gravity anchors are usually constructed in the form of a heavy concrete block. This necessitates the use of heavy floating load handling equipment for transporting and installation of these blocks. A gravity anchor can also be fabricated in the form of a floated-in box sunk at a predetermined location, then filled with either granular materials or tremie concrete.

The underwater weight required for a gravity anchor may be determined by (Figure 7.57)

$$Q_u = k_s P \left(\frac{\cos \alpha}{f} + \sin \alpha \right) \qquad (7.1)$$

where Q_u is the underwater weight of the gravity anchor; k_s the factor of safety against sliding (depending on the importance of the structure, $k_s = 1.5$ to 2.0 for normal load design conditions and $k_s = 1.1$ to 1.3 for extreme load combinations); P the maximum pullout mooring force; f the friction coefficient between concrete gravity anchor and underlying soil ($f = 0.25$ for clay, $f = 0.4$ for sand, and $f = 0.45$ to 0.55 for rock; these values of f are approximate; accurate values can be obtained from the field or large-scale model tests; more information on friction coefficient is given in Chapter 2); α the angle between the direction of the pulling force (or mooring line) and horizontal. If the angle α is very small (e.g., does not exceed 3 to 6°), eq. (7.1) can be simplified as follows:

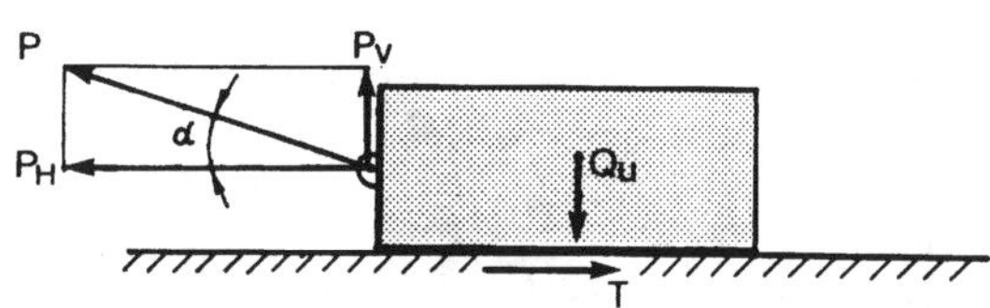

Figure 7.57 Gravity anchor design scheme.

$$Q_u = \frac{k_s P}{f} \quad (7.2)$$

For more information on gravity anchors, readers are referred to Tsinker (1986). The capacity of gravity anchors is usually limited only by their size and weight as well as limitations imposed by fabrication, transportation, and installation. Most recently, special large gravity anchors have been used for construction of the Hood Canal floating bridge near Seattle, Washington (Myint Lwin and Gloyd, 1984); the floating dock at the port of Valdez, Alaska; and elsewhere.

7.6.3 Piled Anchors

These anchors, depending on the mooring load, could be represented by a single pile driven into the foundation soil, a cluster of vertical piles, or a combination of vertical and batter piles brought together by a concrete cap. These anchors develop their holding capacity primarily by mobilizing the lateral soil pressure and skin friction in the surrounding floor material. Their advantage is an ability to resist both uplift and lateral loads. They required specialized installation equipment, however, and their underwater installation, particularly in deep flowing water, is very expensive. Additionally, piled anchor design requires knowledge of detailed geotechnical data to full pile depth, which is difficult to obtain and is therefore expensive. For information on piles and piled structures design, consult Chapter 2.

7.6.4 Special Anchors

Special anchors usually combine properties of conventional anchors with those of gravity anchors. Practical example of one of such an anchor is shown in Figure 7.31. This anchor develops its resistance against pulling-out force by digging into the soft strata. The anchor will continue to dig in until sufficient passive resistance is developed and the equilibrium between pulling-out forces (P_H) and soil passive resistance (E_p) is reached (Figure 7.58). As a result of digging into the soft strata, the anchor may tilt, but its inclination should not exceed 30°. Stability of the anchor is determined by the following equation, which is valid for a small angle between the pulling-out force and horizontal [e.g., $\alpha = 3$ to 6°, under which $P \approx P_H$ (see Figure 7.58)]:

$$k_0 = \frac{\sum Q + E_p h_3}{P_H h_4} \quad (7.3)$$

where k_0 is the factor of safety against overturning ($k_0 = 1.5$ for normal load combinations, and $k_0 = 1.2$ for extreme load combinations); Q, E_p, P (or P_H) and h_3 and h_4 are forces acting on the anchor and their respective locations from the tip of the anchor (Figure 7.58). A safety factor against sliding for self-burying anchors k_s of 1.2 for normal load conditions and 1.05 for extreme load conditions is considered as sufficient. Essentially,

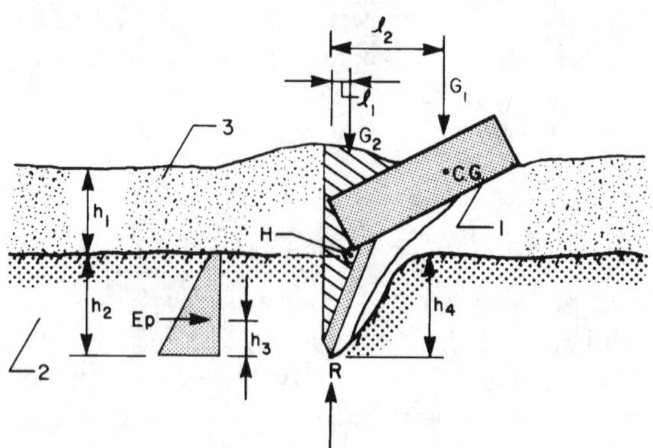

Figure 7.58 Self-burying anchor design scheme. (After Tsinker and Vernigora, 1980.)

the value of E_p should be determined on the basis of detailed knowledge of the physical properties of bottom soils.

7.6.5 Suction Piles

The suction pile concept recently introduced to the field of marine engineering is an exciting innovation in mooring system construction. It is particularly efficient for securing large floating structures in deep water with poor soil conditions. Most recently, a large number of suction piles was installed successfully in deep water at a soft clay site (Colliat et al., 1998). As reported, "the installation behaviour appeared to be easy and very smooth, as well as in good agreement with the prediction."

A typical suction pile consists of a hollow large-diameter steel cylinder with a closed top and open bottom (Figure 7.59). The pile is installed by creating reduced pressure within a pile's body, thus utilizing the available hydrostatic pressure. At the initial stage of pile installation, it is gently lowered onto the bottom and its sidewalls penetrate the bottom soil deep enough to form a seal around the lower edge of the pile. Then, by pumping out the water trapped inside it, the pile is forced by the hydrostatic pressure into the bottom. Once it has penetrated the bottom to its full design depth, the pump is stopped and the pore pressures in the soil around the pile regain their ambient values. Consequently, the system performs similar to a conventionally driven or drilled pile.

For successful installation, the pump capacity must be sufficiently high to create an adequate pressure difference between the ambient hydrostatic pressure and the internal pressure within the pile. The maximum available pressure on the pile corresponds to the head induced by the water depth plus atmospheric pressure. The suction pile technique is most effective with piles that are large in diameter and relatively short in length. Suction piles can be recovered by creating overpressure in the pile interior simply by reversing the flow of pump. The obvious advantage of suction piles compared to conventionally driven ones is that the former require no heavy floating pile-driving equipment for their installation. The only equipment that is required is a floating crane able to handle the weight of the pile and the standard pump.

The process of pile installation is shown in Figure 7.59. Similar to a conventional pile, the penetration resistance of the suction pile (R) is the sum of the skin and rim resistances:

$$R = R_s + R_r \tag{7.4}$$

where R_s is the skin resistance and R_r is the pile rim resistance. These could be determined by the formula

$$R_s = 2\pi D \int_0^h f \, dh \tag{7.5}$$

where D is the pile diameter, f the unit skin friction at the particular depth of a seabed (if a reliable value of f is not available, for preliminary analysis a theoretical value of f can be reduced by 10 to 20%), and h is the penetrated depth.

$$R_r = A_r q_r \tag{7.6}$$

where A_r is the projected area of pile rim and q_r is the soil bearing capacity at the pile tip. If a reliable value of q_r is not available, for the preliminary analysis the theoretical value of q_r can be reduced by 10 to 20%. When pile penetration resistance at a certain depth equals the available thrust capacity, the ultimate penetration depth may have been reached. The available thrust capacity (T) is calculated as

$$T = G_u + A_p \, \Delta P \tag{7.7}$$

where G_u is the underwater weight of a pile, A_p

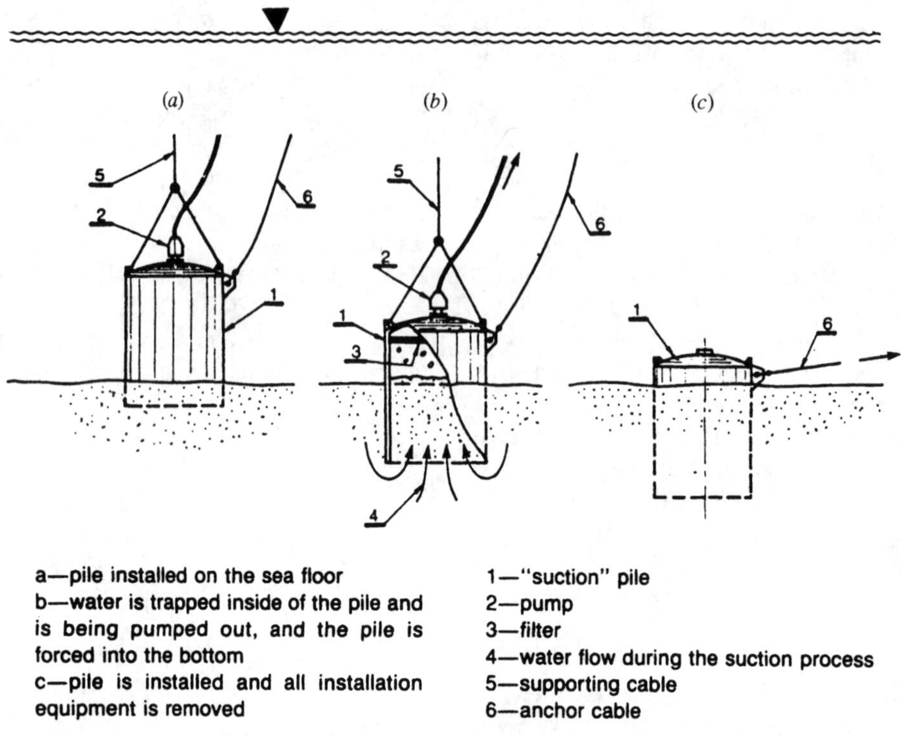

Figure 7.59 Installation phases, anchor suction pile.

the projected area of pile, and ΔP the pressure difference between the ambient and internal pile area.

Equations (7.4) and (7.7) are the base on which basic parameters of suction piles could be determined. Furthermore, the effects of groundwater flow during pile installation on soil parameters should be considered. The results of suction pile trials have been reported by Hogervorst (1980), Cuckson (1981, 1984), Epskamp and Hannink (1992), and Olberg et al. (1997). These trials indicated that the initial penetration of the suction pile into soil under its own weight can provide sufficient sealing for subsequent depressurizing of the pile inside. These tests also demonstrated that in all soils encountered, suction piles could be installed effectively and fast (20 to 60 minutes). Installation of test piles was not hampered by the presence of small obstacles in the ground or by nonvertical positioning of the pile. Furthermore, in sandy soils it was observed that the sand inside the pile could become liquefied, thus eliminating internal pile friction. In clays, the internal friction could be practically eliminated by applying internal friction reducers, which means that the pile skin resistance determined by eq. (7.5) could be halved. This could also apply to the resistance of a pile's rim.

As stated previously, a pile installed by suction effects is generally of large diameter (D) and of short length (H). Usually, the H/D ratio of such piles is less than or equal to 3. The bending stiffness of typical suction piles is very high, and they are generally regarded as being very stiff, moving in the soil as a rigid body. Field tests have indicated that for laterally loaded suction piles submerged into sandy soils, the allowable lateral pile load P_h can be determined by Broms's formula:

$$P_h = \frac{\gamma D H^3 K_p}{2(a + h)} \tag{7.8}$$

where γ is the underwater weight of the sand, K_p the coefficient of passive earth pressure, a the distance from the pulling point at the top of a pile to the floor surface, and h the part of the pile that is buried in the bottom.

During the aforementioned trials it was found that the pile center of rotation was located near the pile rim, which confirmed the practicality of the foregoing theoretical approach. Recently, the finite element method has been employed for determination of the suction pile's ultimate bearing capacity and resistance to lateral forces (Epskamp and Hanuink, 1992). It must be stressed that the results obtained for pile analysis are heavily dependent on input data (i.e., soil parameters). One should realize, however, that these parameters, obtained by standard methods in the field, can be altered substantially by the process of pile installation.

On the basis of very interesting, although limited experience in the use of suction piles and experimental findings to date, it can be concluded that suction pile technique may have great future in the field of marine engineering. For more information on suction pile anchors, consult Larsen (1989), Dyvik et al. (1993), Sparrevik (1996), and Colliat and Foulhoux (1998).

7.6.6 Selection of Anchor System

Essentially, each project is site specific; hence in choosing an anchor system for the particular floating dock, the following should be considered:

- Depth of water and properties of the bottom soil
- Magnitude of anchor force
- Drag potentiality
- Effects of cyclic loading on a bottom soil
- Character of the dock installation (e.g., is the dock installation permanent or temporary; in other words, do the anchors need to be recovered at a later date or not?)
- Costs of the anchor, its installation, and its maintenance

Obviously, the depth of water and bottom soil parameters are the most important factors in choosing the type of anchor. Hence, prior to deciding which type of anchor should be used, it is necessary to carry out soil investigations compatible to that carried out for the design of a bottom-fixed structure. Care should be taken to determine the presence and nature of boulders or rock outcrops that may be present in the vicinity of the anchor location, as this could influence the choice of anchor. Load magnitude and its character (e.g., static or cyclic) may influence selection of the type and sizes of anchor. Both laboratory tests and field observations have indicated that repetitious (cyclic) loading may have a strong effect on submerged sands, and to a lesser extent, on clays or cohesive silts. For more information on anchors, readers are referred to Tsinker (1986).

7.7 MOORING DOLPHINS

Mooring dolphins can be designed in the form of either piled or gravity structures. The simplest form of piled structure is a flexible dolphin comprised of wood or steel piles. The number and material of piles included in a flexible dolphin depend on the physical properties of bottom soil, the height of the dolphin, and the magnitude of the mooring force acting against the dolphin. The dolphin is designed primarily to sustain horizontal forces of miscellaneous nature (e.g., pull-through mooring lines, ship impact). At locations where bottom soil conditions permit pile driving, piled structures are commonplace. Depending on local conditions (i.e., depth of water, water fluctuation, soil condition, design load, etc.), the former may be constructed in the form of vertical pile(s) only or may comprise vertical and/or batter piles.

Sheet-pile cells are suitable for almost any soil conditions and make excellent dolphins which can sustain a very large mooring force. Cellular dolphins are usually capped with a heavy concrete slab to which mooring accessories such as bollards and bitts are anchored. It is obvious that diameter of the sheet pile cell driven into the penetrable bottom is smaller than that build on a bedrock foundation. Although a sheet pile cell is not always the most economical solution for dolphin construction, it is certainly the simplest. It can be constructed directly at the design location but could also be prefabricated and mounted in situ. The latter may considerably reduce the over-water construction time. The sheeting must extend to a sufficient depth below the bottom to prevent the cell from being undermined by erosion of the bottom. The minimum depth of penetration is usually equal to 3 m unless the bottom comprises rock or other hard material. Where dolphins have to be constructed on foundations that do not permit pile driving (e.g., rock, very hard clay), the gravity structures are usually considered. The latter are typically designed in a form of a floated-in concrete caissons, open-bottom concrete or steel cylinders, steel sheet pile cells, or in some special cases, steel jackets anchored to the floor. These structures have the advantage of considerably reducing the construction time over water. However, for their installation, they usually require construction of a thoroughly leveled rock-filled mattress.

The floated-in caissons are typically constructed in the form of cellular boxlike or cylinder units with a closed bottom and with diaphragm walls dividing the unit into several compartments. Caissons may be constructed from prefabricated elements or be cast in situ. After been launched into the water, caissons are towed to the site of deployment, where they are ballasted down to the prepared mattress and filled with a high-quality granular material. For detail discussion on caisson structures, the readers are referred to Tsinker (1986, 1997).

7.8 MOORING CABLES

As discussed earlier, the floating pier mooring cable system typically comprises onshore and offshore portions. The onshore moorings usually consist of steel wire ropes secured at the floating pier and the land-based deadmen. The offshore moorings are typically comprised from cast-steel studlink anchor chains. For temporary dock installation, or for short-term operation, synthetic ropes can be used. These can be of nylon, polyester, polypropylene, or other materials. Standard wire rope is made of very strong, tough, durable steel, combining great strength with high resistance to fatigue. Its minimum standard tensile strength varies from 1570 to 2000 Mpa, depending on wire diameter. Most wire ropes have either a fiber or a steel core, which supports the wire strands of the rope. In marine applications, protective coating of wire ropes is usually considered for permanent floating structures. For maximum corrosion protection of mooring cables it is essential that wire ropes be electrolytically zinc-plated with a coating two to three times the weight of a normal galvanized coating. Non-galvanized wire ropes, used for mooring cables for temporary floating structures, must be lubricated periodically to obtain maximum cable performance.

It has been found that in the case of direct contact between wire rope and galvanically unprotected steel pontoon (or ship), the wire rope could become anodic and protect the steel pontoon galvanically, just as would a sacrificial metal (Malloch and Kolbe, 1978). This results in significant weight loss and reduction in wire diameter. Therefore, protection of wire rope against the foregoing type of corrosion is of great importance. This could be done either by electrical isolation of the cable from the steel pier structure or by installation on active galvanic or impressed current protective systems. Furthermore, if wire rope cable is submerged into the water, the velocity of water could increase corrosion rates (Nachman and Duffy, 1974). Safety factors on wire ropes are generally more conservative than on chains because of corrosion potentials and because of strength reduction due to handling, radius bending, fittings, and so on.

In selecting a wire rope for a mooring line, a working load factor (LF) equal to 3 to 5 is usually considered. The smallest value of LF is applied for temporary anchorage, and the largest is used for permanent structures. Structurally, the typical onshore mooring cable consists of the following components:

- *Wire rope.* The size and number of ropes in one set depends on the magnitude of the mooring force acting on the mooring cable.
- *Sheave.* Where mooring line consists of several wire ropes, the balanced arrangement of the ropes included in the system is achieved by sheaves. The sheaves must be suitable for the particular wire rope (i.e., their diameter must suit the diameter of the rope). If proper sheaves are not used, the useful life of the rope will be lessened.
- *Turnbuckle.* One or two turnbuckles are generally used to make adjustments to the wire rope in cases where more than two onshore mooring lines are required to operate the floating pier. Where only two onshore mooring lines are used, turnbuckles normally are not required.
- *Connecting link.* This is used to connect the turnbuckles (if required) and the sheaves.
- *Miscellaneous.* The other components of mooring cables are shackles of different types, thimbles, and sockets. Similar to the components listed above, these may be obtained from manufacturers' catalogs.

Practical examples are seen in Figures 7.29 and 7.36. Cast-steel studlink anchor chains are commonly used for offshore mooring lines. The anchor chain is usually secured to the anchor by means of a shackle. The separate shots of the chain are linked to each other by means of connecting links. Generally, the offshore mooring lines are operated by an onboard windlass. A working factor (WF) equal to 3 to 3.5 is normally used for selecting the size of a chain for an offshore mooring line application.

Great care must be exercised in the design of offshore mooring cables. These cables are less accessible for inspection and maintenance than onshore ones, and therefore, reliability of their performance must always be ensured. According to Stern and Wheatcroft (1978), the worst anchor chain end failures are related to malfunction of the windlass, or operator error. The component most frequently involved in the failure of anchor chain is the connecting link. It has been found that connecting link failure comprises about 65% of all chain–accessory failures reported. Usually, metallurgical defects or stress concentration problems due to the welding process are responsible for failures of connecting link. Although present specifications provide relatively reliable anchor chain and accessories, closer control of the welding process, as well as verification that an appropriate metallurgical structure is attained throughout the cross section, especially for quenched and tempered materials, is required.

7.9 FENDER SYSTEM

A fender system is a very important component of dock system. It is used to prevent direct contact between the pier and the vessel and to absorb and dissipate berthing impact energy. Its principal function is to prevent the ship and/or dock from being damaged during berthing or unberthing operations. The latter is particularly important in the case of floating piers. For details, consult Chapter 2.

7.10 MOORING ACCESSORIES

The purpose of mooring accessories installed on the dock is to provide a convenient and reliable means of moorings for ship while loading or unloading at the pier. The basic requirements for mooring accessories are simplicity, reliability, fast operation, and the ability to operate with the least damage to mooring lines. The basic mooring accessories are fabricated in the form of bollards, bitts, cleats, and mooring rings of variable capacities. In some cases, quick-release hooks are used for mooring of vessels. For details, consult Chapter 2.

7.11 DESIGN LOADS AND FORCES

The design and construction of a satisfactory floating dock system first and foremost depends on establishing realistic (both normal and extreme) loads that may act against the dock structure(s). This means that the dock in general and its individual components in particular must be designed to withstand all the normal and extreme dead and live loads and load combination. The latter in general includes a pier's own weight and the weight of the cargo handling and hauling equipment, loads from cargoes and passenger traffic, environmental loads such as wind, waves, current, ice, and so on, thermal loads, ship impact, and others.

It must be noted that safety of a dock operation as a design concern must take precedence over all other design considerations. The safety of a floating dock depends on subsequent loadings. Since any structure is always loaded after it is built and not always in the mode or manner used in the design, the selection of design loads is a task of paramount importance. It must be noted that in some cases, neither the owner nor the designer is quite sure what maximum loading the dock may be required to sustain in the course of its useful life. In this case, selection of appropriate loading should be based on the experience. Being largely subjective, it must be based on commonsense approach and, of course, is heavily dependent on owner or designer team experience.

To avoid divergent designs and to formulate reasonable design loads and load combinations so that a dock would not be cost prohibitive nor become obsolete shortly after construction, the designer should use recommendations of relevant codes and standards, but most of all, exercise great care in the formulation of loads generated by cargo handling and hauling equipment of all kinds, environmental loads, and operational specifics related to the particular site. To establish the required size of the dock and its components, it is necessary to determine critical loads imposed on it. Loads used in dock design are distinguished as normal and extreme. The most severe load combination is usually selected as the nominal design load. Normal load combination is one that could exist under dock normal operation. This usually comprises the worst combinations of dead load, buoyancy, ship impact force, mooring forces, along with forces generated by cargo handling and hauling equipment, and/or appropriate surcharge load. The above, of course, assumes that in all load combinations, common sense and good judgment is exercised. To proportion the dock structural components exposed to normal load combinations, the normal allowable stresses are used. Normal loading in combination with unusual forces such as earthquake, operation under damaged condition, and so on, is considered to be an extreme loading. In this case, allowable stresses used in design would be increased appropriately.

Typically, loads and forces acting on a dock and its structural components are defined as permanent, temporary, and special. *Permanent loads* include the weight of the pier structure, the weight of permanent structures located on a pier (e.g., warehouses, sheds, offices), and the weight of the fixed equipment. This kind of load is usually referred to as *dead load. Temporary loads and forces* include the environmental forces (e.g., wind, waves and ice, and forces related to dock operation). The latter includes all kinds of *live loads* (e.g., surcharge, moving equipment, etc., ship impact, and mooring forces). *Earthquake forces,* which mostly affect bottom-fixed structures affiliated with a floating dock (e.g., an access bridge and its land-based abutment, mooring dolphin, gravity anchors) are usually considered as *special loads*. Design loads and force related to the design of all kinds of port-related marine structures are discussed in detail in Chapter 2. The following is a discussion are the loads most characteristic of floating dock design.

7.11.1 Environmental Loads

Wind, waves, current, and ice constitute the principal environmental loads acting on floating docks. Wind acts on projected above-water-level parts of a pier, moored vessel(s), and cargo handling and hauling equipment sitting on a pier deck and exposed to wind. As pointed out earlier, floating docks are usually deployed at a well or relatively well protected from wave locations. However, at some location, waves up to 1 to 2 m high could be a factor of major concern. Furthermore, when it is necessary to tow the pier across the large body of water from a construction yard to the deployment site, waves could constitute a major design load to be considered in pier design. At some locations currents may generate substantial load against a floating pier. Current acts on projected below-water parts of a pier and a ship(s) moored alongside the pier. Load generated by current depends greatly on pier orientation.

In some cases, basically in countries with a cold climate, ice may be a major environmental load to the dock if it is located in areas unprotected from moving ice. Also, ice buildup on a pier structure due to water spray, fog, or for other reasons may affect both pier buoyancy and stability and therefore must not be overlooked. The aforementioned environmental loads are normally obtained on the basis of oceanographic and meteorological data which are needed for the design of any kind of offshore structure, but of a floating pier particularly. For more information consult Chapter 8.

7.11.1.1 Wind. Wind load acting on a pier system depends on the velocity of the prevailing wind in the area where the port is located, pier orientation, and the exposure to wind of pier areas and of vessels laying alongside a pier (Figure 7.60). The design wind force is usually based on a storm having an average expected recurrence interval of 50 years. In floating dock design, two components of wind force are considered: one that acts perpendicular to the pier and another parallel to the pier direction.

The wind force (P_w in kN) may be obtained from the equation

$$P_w = k \sum A p_w C \qquad (7.9)$$

where k is the shape factor = 1.3; ΣA the area exposed to wind of vessel cargo handling and hauling equipment located on the deck of

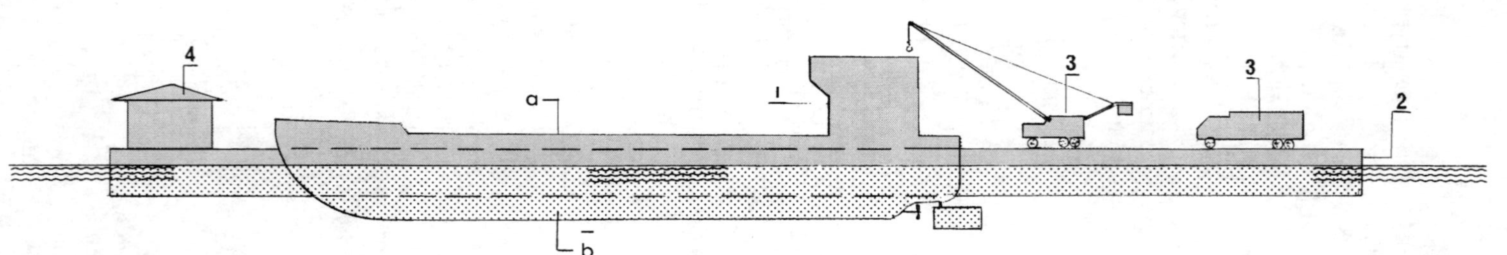

Figure 7.60 Floating pier. Wind and current pressure diagram on ship–pier system. *a*, Exposed to wind area; *b*, exposed to current area; 1, ship; 2, pier; 3, cargo handling and hauling equipment; 4, office–warehouse.

the pier and area of a pier projected above water (m^2; if the pier is occupied by vessels on both sides, the sheltering effect of the windward ship istaken into account; the wind force on the sheltered vessel is usually assumed to be 50% of would-be total value if this vessel had no shelter against the wind); and p_w the specific wind pressure (kPa; it varies with the square of the sustained wind velocity that is an average speed of wind during a time interval of 1 minute). The value of p_w is usually specified by local building codes, and if not available can be obtained from the equation

$$p_w = 4.74 \times 10^{-5} \times V_w^2 \qquad (7.10)$$

where V_w is the velocity of the wind (km/h). Wind speed is usually assumed to be the sustained wind speed at 10 m above the water surface.

C is the wind gust factor. The average value of the gust factor ranges from 1.35 to 1.45. The factor may be reduced based on the size of the design vessel(s) and/or pier. The reduction factor r is given as follows:

Length of Vessel and/or Pier (m)	Reduction Factor r
25	1
50	0.95
100	0.90
200 and more	0.85

In the case of wind blowing at an angle to the longitudinal axis of the pier, two components of wind force are to be determined: one acting parallel to the pier longitudinal axis and the other acting perpendicular to the pier. In calculating the maximum wind force, the vessel and the pier should be considered as being in light conditions. In floating dock design, it is usually assumed that cargo handling equipment such as cranes and loading towers would not operate under conditions where the wind is stronger than 25 km/h.

It is also customary to consider that a ship would not remain alongside the dock in light conditions in a severe storm; the ship would either take on ballast, or leave for the sea. Therefore, in dock design, the use of a unit wind pressure more than 1 kPa needs to be justified by special requirements. It should be mentioned that in some cases, docks are protected from being exposed to the seaward wind force by construction of special protective shields. In the past, such structures have been built in the ports of Le Havre, Marseilles, Dover, and at other locations (Minikin, 1963). For more information, readers are referred to Tsinker (1986).

7.11.1.2 Current. Three types of currents are normally considered in floating pier design: river, tidal, and wind-driven currents. The river and tidal currents are usually selected from available statistics. The value of wind-driven current at the still-water level is usually taken as equal to 1% of the sustained wind speed at a level of 10 m above the face of still water. The maximum design drag force on a pier or ship system due to current velocity is obtained on the premise that both the pier and the vessel that is moored at the pier are fully loaded. The force of current exerted on the ship–pier system (P_c) in kN could be obtained by

$$P_c = c \sum A\, V_c^2 \qquad (7.11)$$

where c is the empirical coefficient = 0.5 to 1.0 (smaller values are used for vessels and larger values for a pier structure; a maximum of c = 1.0 is used for square-shaped nosing piers; these values of c are conservative; the true value may be obtained from either model test or field measurements on similar structures); V_c the velocity of current (m/s); and $\sum A$ the underwater area of a ship–pier system exposed to current when the system is fully loaded (m^2).

In the presence of a strong current, the pier is usually oriented parallel to the stream. The design current velocity V_c is typically based on a current having an average expected recurrence interval of 5 years. In weak currents the pier could be placed at a certain angle to the direction of flow. In this case two components of drag force shall be considered in pier design: one that is acting parallel to the pier's longitudinal axis and the other acting perpendicular to the pier. If the pier is occupied by ships on both sides, than similar to the wind force, the force of current on a sheltered vessel could be taken as 50% of the total value of the drag force on a nonsheltered vessel. Current force acting on a ship is transmitted to the pier through her mooring lines and sometimes, although very seldom, by the ship bearing against the fender system.

7.11.1.3 Waves. Floating docks are usually deployed at locations protected from waves. However, at some locations, dock could be exposed to significant waves generated mostly by wind and/or passing large vessels. The size of wind-generated waves depends on the speed of the wind and the size and geometry of the harbor, and the size of the ship that originated the waves depends on the speed of the vessel and its proximity to the dock.

Because the speed of vessels within a harbor, especially near a dock, is usually limited, the height of shipborne waves usually does not exceed 0.6 to 1.0 m. However, the size of wind-generated waves could be much larger. Waves do not just produce a lateral impact force against the pier–vessel system, but also affect the pier structure by inducing "sagging" and "hogging" conditions on it. These are discussed in the following section. It should be noted that the wave-generated lateral load on a floating pier in a great many practical cases is secondary to that produced by wind, current, or ship impact. Hence, wave-generated lateral load typically does not control the design of a pier hall.

7.11.1.4 Ice Effects. Generally, in determining the ice loads against dock structures (e.g., pier, dolphins, mooring system), consideration is usually given to a number of ice effects. These depend on harbor–port location, port–dock layout, mode of harbor operation, type of ice control methods, and others. Typically, siting of a port is usually determined by the proximity of an industrial demand or settlement. In considering the effects of ice on port siting it is clear that maximum protection from moving ice is desirable while maximum accessibility to the port area must be provided. Protection is best achieved by locating the port in land-fast ice and with the entrance to the port basin oriented away from the direction of the prevailing wind or current. This will encourage ice floes to move out of the port basin during the breakup period.

If the port basin is exposed to moving ice, the ice loads on port marine structures could be severe, and vessel operations in port basin could be hindered significantly. Where ports must be located in moving ice, ice interaction with dock structures and its interference with vessel operations must be minimized. Where possible, a port should be located in the area that has a natural source of ice control. If the port is to serve an industrial complex, it should be located as close as possible to this complex so as to utilize any waste heat that may be available from it. If a port is located in a large harbor, the dock layout and its orientation should be such that movement of the wind- or current-driven ice had not produced a substantial load on marine structures and had minimum interference with harbor operations. The layout design would be of particular importance in a river port where moving ice could seriously hamper dock operations.

The intended operation of the port is obviously a basic input to the overall port and dock design. In that sense, allowances should be included in the system to mitigate the potential ice interference with port and vessel operations. The degree of tolerance of ice interference depends to some extent on port use. For example, if a vessel needs to be berthed immediately adjacent to the face of the dock for loading and/or unloading, it would be essential not to have pieces of ice jammed between the vessel and the dock. On the other hand, if the vessel uses loading arms, it might be able to tolerate some ice in that area; hence, in this case, ice control requirements might be lessened.

The degree of the ice problem also depends on whether the operation is supported by a dedicated icebreaking tug. If a tug were available to aid in berthing, ice problems facing the berthing vessel could be reduced. It is evident from the experience of winter port operation that passage of ships along the same track in a continuous ice sheet results in an accelerated formation of ice through repeated exposure of water to the atmosphere. After a certain number of passages, the accumulation of ice may reach the point where further passages will be precluded. The same ice buildup will occur in the pier area, and if ice accumulation proves to be excessive, ice control techniques should be required to permit a proper berthing procedure for the vessel.

The range of ice control techniques that might be employed to mitigate port operations could be broken into the following broad categories:

1. *Icebreaking.* This involves breaking, cutting, or destruction of ice by miscellaneous means, basically by use of icebreakers. In the vicinity of the dock, an icebreaking procedure usually associates with an ice removal procedure. This involves physical displacement of ice by lifting, submerging, or pushing it out of the dock berthing area.
2. *Ice suppression.* This is used to inhibit the formation of ice. It involves the use of various types of thermal discharge systems, which usually are effective in a relatively small, well-enclosed harbor. Among them are the bubbler systems used to move "warm" bottom water to the ice surface, ice dusting, which may assist in melting of ice during the breakup period, and other methods that are based on use of miscellaneous chemicals for ice melting. It should be mentioned, however, that all ice suppression methods that are based on the use of chemicals for ice melting could be objectionable from the point of view of environmental concerns.
3. *Prevention of ice formation.* This prevents ice formation in the vicinity of docking facilities by employing different kinds of insulating covers on a body of water. The effectiveness of this system could be greatly increased if combined with a warm-water discharge system or bubbler system. It should be noted that this technique is considered reliable and presents no safety or environmental problems.
4. *Ice diversion.* This includes deflection or arrest of moving ice to prevent a collision with dock structure or vessel. For ice diversion purposes, different types of structures, such as floating booms or bottom-fixed structures, are typically employed.

For better effectiveness, the ice control methods above can be used in different combinations. Evaluation of ice control techniques and/or combinations is thereof done on the basis of their effectiveness, costs, reliability, safety, and environmental considerations. Appropriate data collection usually precedes the determination of ice loads exerted on the dock. When the information on ice condition required for dock design is not available, it could be extrapolated from the nearest similar locations with relevant records. The collection of such data is the first step in the determination of ice loads exerted on dock structures. In general, a lateral ice load that may be exerted on the dock comes from wind- and/or current-driven ice, either in the form of individual floes, large sections of ice cover, or small pieces of ice jammed between the vessel and the dock during berthing maneuvers. They could also be attributed to forces of thermal origin. Ice buildup on vertical surfaces of a floating dock could affect its buoyancy and stability. Similarly, ice buildup on vertical surfaces of a bottom-fixed structure (e.g., mooring dolphins) could significantly increase foundation loads or create destabilizing buoyancy forces. In an earthquake zone, the "added mass" of the ice built up on a structure should be considered in an evaluation of its affect on the structure response to earthquake loading. The scope of this work does not allow for detailed discussion on ice, ice-related loads, and ice problems in ports. For detail discussions on all these subjects, interested readers are referred to Tsinker (1995).

7.11.2 Dead and Live Loads

General information on dead and live loads applied for the design of port-related marine structures is provided in Chapter 2. In this section, the only dead and live loads that are characteristic of floating dock design are addressed. These should be considered in conjunction with the loads discussed in Chapter 2.

7.11.2.1 Dead Load. The dead load is the weight in air of the overall pier structure, including the relevant weight of access bridge(s), passenger shed(s), office(s), storage, and so on, and permanently installed cargo handling equipment. For the purpose of preliminary design unit weight of a steel pier may be considered as equal to approximately 4 kN/m^2 of pier deck. For the same purpose, the weight of a pier constructed from concrete could be taken as equal to 25 kN/m^2 of pier deck. The weight of fixed mechanical equipment, piping and its liquid content, and structures located on the pier, such as offices, warehouses, and sheds, are obtained from appropriate sources.

7.11.2.2 Live Load. Since floating piers are rarely used as intermediate storage for cargo, they are commonly designed from crane and vehicular loads and for passenger traffic (5 kN/m^2) where applicable. In designing a floating pier for vehicular load, it is customary to assume that a 20-ton truck (traffic load class H-20-44 or HS-20-44) could, within reason, be spotted anywhere on the pier deck, and assumptions as to load distribution should comply with applicable requirements of the bridge construction regulations. However, the impact coefficients normally used for bridge design may generally be substantially reduced considering the relatively low speed of vehicular traffic on the deck of the dock. The latter also applies to design of the access bridge.

The vehicular design load that is used in pier design consists generally of individual wheel, axle, and truck loads as given by the appropriate specifications. In the design of a pier deck, the uniformly distributed live load rarely governs design because of the relatively short spans of deck structural members. The critical wheel load on deck elements depends on wheel contact area and load distribution through deck cover system. For details, consult Chapter 2. If there is likelihood that appreciable quantities of cargo can be shored temporarily on the pier deck, its actual load (but no more than 20 kN/m^2) should be considered in the pier design. A uniform live load of 15 kN/m^2 may be used for the design of floating docks that handle bulk materials (liquid or dry) by means of conveyors or pipelines, and where general cargo is of secondary importance.

At floating docks the general cargo typically is handled by mobile cranes. However, if required, portal cranes, container cranes, gantry cranes, or others are used. The full loads generated by these cranes are considered in pier design. Typically, the most unfavorable load combination is created by two portal cranes operating in close proximity to each other. It must be realized, however, that two neighboring portal cranes cannot produce maximum load in their adjacent legs simultaneously because, in this case, their jibs will interfere with each other. Typical loads produced by miscellaneous cranes used in port operation are discussed in Chapter 2.

In designing floating container terminals, where appropriate, provision for roll on–roll off operation has to be considered. The latter assumes two-way traffic for tractors delivering their trailers on wheels onboard (or offboard) the ship and returning to the shore (or ship) to pick up another trailer. In some instances loading and unloading operations could proceed simultaneously. Therefore, in such cases the deck of the pier and its access bridges have to be designed according to the rules applied to two-way bridge design, assuming that the heaviest truckload is HS20-44. As pointed out earlier, allowance for impact normally recommended for bridge design could be reduced at the designer's discretion assuming that traffic on the deck of a dock is moving not as fast as traffic on a regular highway bridge.

7.11.3 Thermal Load

Changes in the temperature of ambient water and air as well as solar radiation may cause the pier hull bend. The latter, in turn, may have a profound impact on stress distribution in the hull structure of a floating pier. Furthermore, the action of sunrays on a pier deck, or on one side of the hull, may result in substantial distortion of the hull. The latter can also result in stress development of substantial magnitude. Hence, the designer must be aware of the magnitude of these stresses and their effect on a hull structure.

7.11.4 Seismic Load

As mentioned earlier, earthquake forces have little effect on the floating pier itself. However, other structures included in a floating dock system—mooring dolphins, offshore gravity anchors, access bridges, and land-based bridge abutments and deadmen used for inshore mooring lines—must be proportioned to resist earthquake-induced forces. Seismic design considerations are discussed in Chapters 2 and 3.

7.11.5 Load Combinations

Dock elements must be designed with an acceptable and relatively uniform degree of safety under various load combinations. As dead load is practically constant through the life of a structure, a combination of dead load with any other load constitutes a basic combination in which safety factors are applicable. When dead load plus buoyancy is combined with two or more other loads (see the following load combinations), simultaneous occurrence of full design values of each load effect is less likely to occur than basic combinations. Therefore, an appropriate increase in permissible stresses in structural elements due to combinations of the dead load and buoyancy effect with two or more other load effects is justified. Because of the relatively short duration of some design loads, the probability of their simultaneous occurrence is very small. For example, it is usually considered that the seismic load does not need to be considered concurrently with maximum wind or wave loads. Generally speaking, all load combinations should be scrutinized by the designer on a very rational basis in consultation with the dock owner or operator.

In designing dock elements and their structural components, all potential loads should be considered to act in the following combinations, and whichever combination produces the most unfavorable effects on the pier, access bridge, mooring system, or any structural member concerned, should be selected.

1. D	7. $D + B + T$
2. $D + B$	8. $D + B + L + E$
3. $D + E$	9. $D + B + L + T$
4. $D + T$	10. $D + B + E + T$
5. $D + B + L$	11. $D + E + T$
6. $D + B + E$	12. $D + B + L + E + T$

In the groups of load combinations above: D is the dead load of the dock elements; B the buoyancy; E the environmental or seismic loads, whichever produces the most unfavorable effect; T the load produced by contraction or expansion due to temperature changes, shrinkage, or creep in component materials, or combination of the above; and L the live load, ship impact load, or vertical and horizontal loads related to the operation of cargo handling and hauling equipment, including the impact factor, hydrostatic pressure, and mooring forces. The following percentages of permissible stress are recommended for the load combinations above:

- 100% for group I, which includes load combinations 1 through 7
- 125% for group II, which includes load combinations 8 through 11
- 133% for group III, which includes load combination 12

Load combinations 1 through 7 and subsequent unit stresses as shown in this section are used in a working stress design method which is still often used in floating dock design.

7.12 FLOATING PIER DESIGN GUIDELINES

7.12.1 General Design Considerations

Following is basic information related to the design of a floating pier.

7.12.1.1 Design Stages and Design Requirements. Depending on project objectives and complexity, the overall pier design process may consist of the conceptual, preliminary, and final design stages. At the conceptual design stage, provided that project objectives have been defined in terms of certain engineering parameters, the structure's basic proportions and layout are to be established to satisfy such objectives. The *conceptual design process* may be carried out in either of two ways. The problem could be solved on the basis of accumulated experience and data for similar projects as required, or a novel concept could be developed. The former would focus on past practice and interpolate or extrapolate from similar projects. This method of conceptual design provides a quick and fairly reliable answer when adequate data are available. This method, however, would serve only as a rough guide if a novel concept is to be developed. In the latter approach, new ideas would require investigation in order to establish their technical and economical feasibility. During the conceptual design stage, the initial proportions of a pier are to be established.

The *preliminary design stage* is the next step of the process. During this stage more detailed structural analyses and calculations are required to modify earlier assumptions and obtain more accurate quantities for project economical evaluation. The pier structure is usually roughed out at this stage and its basic parameters (e.g., general arrangement, layout, weight, stability parameters) as well as some significant details are established. During the *final design stage,* detailed calculations and drawings, as well as technical specifications necessary to build the structure, are produced.

To carry out its function as part of a port or specific terminal, the floating pier should be designed to support the design loads and to provide an adequate berthing length. Furthermore, piers forming part of a floating terminal should accommodate traffic and equipment that is required to handle the designed volume of cargo and/or passengers. Pier general arrangement can be defined as the assignment of its internal and external space for all required functions and equipment. Generally, the internal space of a pier is used to provide sufficient buoyancy. However, in some cases, it may also be used successfully for storage, offices, and equipment required for dock operation.

The external space of a pier, namely its deck, is used for operation of cargo handling equipment and occasionally for temporary storage of cargo and/or passenger operations. The deck is also used for pier mooring accessories, engineering services, and sometimes for offices, rest rooms, and so on. The first step in solving the general arrangement problems of a pier is locating the main areas of operation. Space for cargo handling equipment with convenient access to, mooring accessories, entrances and exits for the traffic, and their boundaries within the pier must be defined. At the same time, certain important requirements must be met (e.g., adequate buoyancy and stability, structural integrity, and safety of operation). The latter includes considering subdividing a pier into watertight cells. The general pier arrangement is typically evolved by a gradual process of trial, check, and improvement. During this process several preliminary alternatives of pier arrangement are usually investigated, and selection of the most suitable one is finally made. Then the process continues into a more refined stage, simultaneously with pier structural development.

Safety of pier operation is a major general arrangement problem. Successful pier general arrangement must satisfy the following requirements.

1. *Safety of personnel.* All precautions must be taken with respect to the operation of cargo handling equipment and handling of dangerous cargoes which can be injurious to the safety of the working crew and/or passengers. The cargo handling pier deck must have a perimeter curb with a height of no less than 30 cm unless required otherwise, and a pier serving passenger traffic must have a railing of sufficient height.
2. *Stability of pier.* The pier must remain stable in various stages of loading and unloading conditions, including orderly and rapid operations.
3. *Stability under damaged conditions.* The pier must have an adequate fender system to be protected from being damaged by approaching vessels, but nevertheless, it must remain stable if accidental damage to its hull occurs and any two adjacent compartments are flooded. The maximum acceptable angle of a pier heel due to nonsymmetrical flooding is largely dictated by cargo handling equipment (permanently or temporarily) installed on its deck.
4. *Fire prevention.* Measures must be taken and firefighting equipment must be installed, especially on passenger and oil handling berths.
5. *Ventilation.* Provision must be made for ventilation of a pier interior, and if necessary, air conditioning should not be overlooked.

As stated earlier, the pier must be stable under all normal and extreme conditions of performance. This means that if the pier is displaced by an external force from its equilibrium position when floating in still water, it will return to that same position when the force is removed. The most important problem of pier stability is its transverse stability when it rotates about its longitudinal axis (rolling). As will be seen from further discussion, transverse stability of the pier is governed largely by the ratio of pier breadth to its draft. The pier is also capable of rotating about its transverse axis (pitching). For the ordinary pier, however, it is virtually impossible to make it unstable in this direction under normal conditions.

While in operation, the pier is exposed to hydrostatic pressure, hydrodynamic forces due to wave action, and gravitational forces arising from pier structural weight and the loads that it carries. All these forces cause the pier hull to deform and it bends like a beam predominantly in a longitudinal vertical plane. In addition, there could also be transverse and local deformations of the pier hull which arise from forces imposed on it. Therefore, it is necessary for the hull structure to have sufficient strength to resist this specific type of bending and relevant shear forces. However, as will be shown in further discussion, the pier longitudinal strength is of primary importance. This aspect of the pier strength to a large extent governs the ratio of the pier depth to its length. In general, the pier's molded depth depends on its buoyancy and freeboard requirements. The minimum freeboard depth of a conventional floating pier usually is no less than 1.5 to 2.0 m; however, in general, height of the pier freeboard should comply with the relevant size of an average vessel to be served at the dock. Furthermore, the height of a pier freeboard must also be sufficient to provide reserve buoyancy to enable a pier to stay afloat in the event of damage.

The depth to which a floating pier can be loaded should be specified and marked up visibly. To the best of this writer's knowledge, no such legislation exists concerning dock operation. However, in this writer's opinion, the hull of a floating dock should carry a *load line mark,* which in naval architecture is called the *plimsoll line.* Unlike an ordinary sailing vessel, the pier is not in constant motion from one place to another. Therefore, the shape of its hull should be as simple as possible while complying with requirements of buoyancy stability, carrying capacity, and operational requirements. A typical sequence of floating dock design is depicted in Figure 7.61.

7.12.1.2 Structural Materials. Floating pier structure should always be robust and tough to last in the hostile maritime environment for the required design life. It is usually constructed from steel, concrete, or a mixture of these (e.g., steel hull and concrete deck, or concrete hull and steel deck). Both materials have advantages and disadvantages. The main advantage as a structural material that steel enjoys over concrete is that for a given buoyancy, the weight of a steel pier is invariably less than the corresponding weight of a dock constructed from reinforced or prestressed concrete. However, the operation and maintenance costs of steel pier are relatively high.

It should be noted that steel is used more often than concrete for floating pier construction. This is basically because the great majority of established shipyards are reluctant to change their existing technology, which they have long used for steel shipbuilding; drastic changes in existing technology would inevitably increase the construction cost of a concrete pier. However, as indicated in Section 7.2, in recent years concrete floating piers have been used in ports successfully. The suitability of concrete for the construction of floating structures has been verified through successful design, construction, and maintenance operation of a great number of concrete floating piers, floating bridges, and floating dry docks and offshore oil production platforms.

Obviously, both steel and concrete or combinations will be used for floating dock construction in the foreseeable future, and

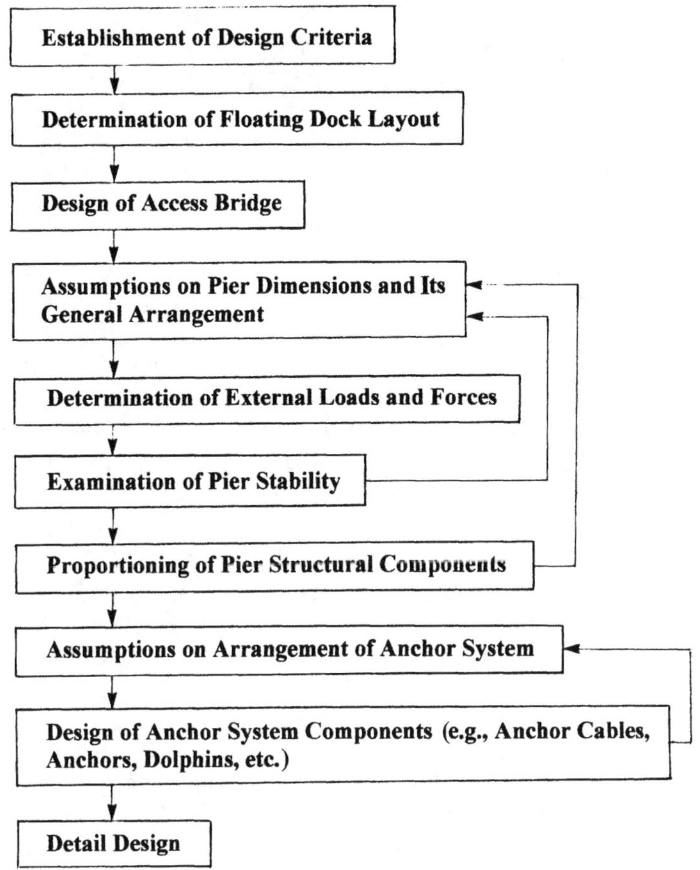

Figure 7.61 Logic diagram of floating dock development.

economy of construction will govern selection of the kind of structural material. Accordingly, a choice between a steel or concrete pier is usually based on the initial capital cost of pier construction and the cost of maintenance. Consideration of both factors could produce an accurate assessment of dock economy. In contrast to pier construction, access bridges are almost universally fabricated from steel.

7.12.1.3 Classification Societies and Governmental Authorities. Two important groups of organizations exert a considerable influence on design, construction, and safe operation of any kind of floating structures, including floating piers. These are different classification societies and governmental authorities. The oldest classification society known today is Lloyd's Register of Shipping, established more than 240 years ago (1760). Among others are the American Bureau of Shipping (United States), Register of Shipping (Russia), Det Norske Veritas (Norway), Bureau Veritas (France), Germanischer Lloyd (Germany), Registro Italiano Navale (Italy), and Nippon Kaiji Kyokai (Japan). These societies all have a long history of establishing standards and rules for constructing ships and offshore structures and have done a great deal to ensure the safety and well-being of all who sail on ships or work on different kinds of offshore floating installations.

Usually, the dock owner is not compelled to build a floating pier to the rules of the aforementioned societies, but the vast majority of major floating piers comply with the established rules, especially those that need to be towed across the sea to the site of deployment. Governmental authorities are concerned basically with the safety of dock operation. The work of both aforementioned bodies often overlaps to a certain extent. However, usually, government authorities will accept the strength of a floating pier as being adequate if it is built to the rules of a classification society. In recent years, classification societies have been instrumental in developing new methods of designing and analyzing ships and offshore structures such as fixed and floating oil rigs and others, and in so doing they are contributing greatly to the design of safe floating structures. On the other hand, legislation regarding safety of floating pier operation is the responsibility of the government. Some of the matters that governmental agencies are concerned with are load-carrying capacity, handling of dangerous cargoes, and lifesaving appliances. Floating piers designed to service passenger traffic are required to comply with especially stringent safety requirements. One such requirement is that a floating pier must possess sufficient buoyancy and stability even after being seriously damaged.

7.12.2 Steel Floating Pier

7.12.2.1 Basic Design Considerations. A typical steel pier used for floating dock and/or floating access bridge construction is the simplest form of vessel of ordinary steel, all-welded construction. Such a pier usually consists of a boxlike shell supported internally by trusses or frames and divided into several watertight compartments by system of bulkheads (Figures 7.62). In special cases, where a pontoon is reliably protected from being damaged by outside forces and also protected from being severely corroded during its design life, its framing could consist of trusses or frames only. An example is shown in Figure 7.62.

Floating pier pontoons typically are of a single-bottom construction and have a transverse framing system, which consists essentially of a series of closely spaced frames or trusses. These frames or trusses, along with transverse watertight bulkheads, stiffen the pontoon shell and its deck, on which the longitudinal strength primarily depends. Ultimately, they help support hydrostatic pressures against the pier hull and deck loadings from cargo handling equipment and other loads. Sometimes, to limit the effects of pier flooding due to a damaged condition, it may be provided with a double-sidewall hull. In this case its framing may consist of longitudinal and transverse trusses or frames only; no internal watertight bulkheads will be required.

Selection of the most appropriate structural alternative is usually based on thorough evaluation of site conditions and economy of pier construction. Typically, a steel pier includes the following basic structural components: sidewalls, deck, bottom, watertight bulkheads, and transverse and longitudinal frames and/or trusses. Steel plating is usually stiffened by keelsons and stringers. All of the components above form a tight and robust envelope that performs as a box girder in resisting shearing forces and bending moments imposed on the structure and provides for buoyancy that keeps the pier afloat.

The bottom plating constitutes the lower flange of the hull box girder. As a part of a watertight envelope it is subjected to hydrostatic pressures that are equal to the water head. Thus, it must withstand combined loads of tension or compression, depending on pier performance as a girder and bending moments due to hydraulic forces exerted on it. The keelsons and secondary girders make significant contri-

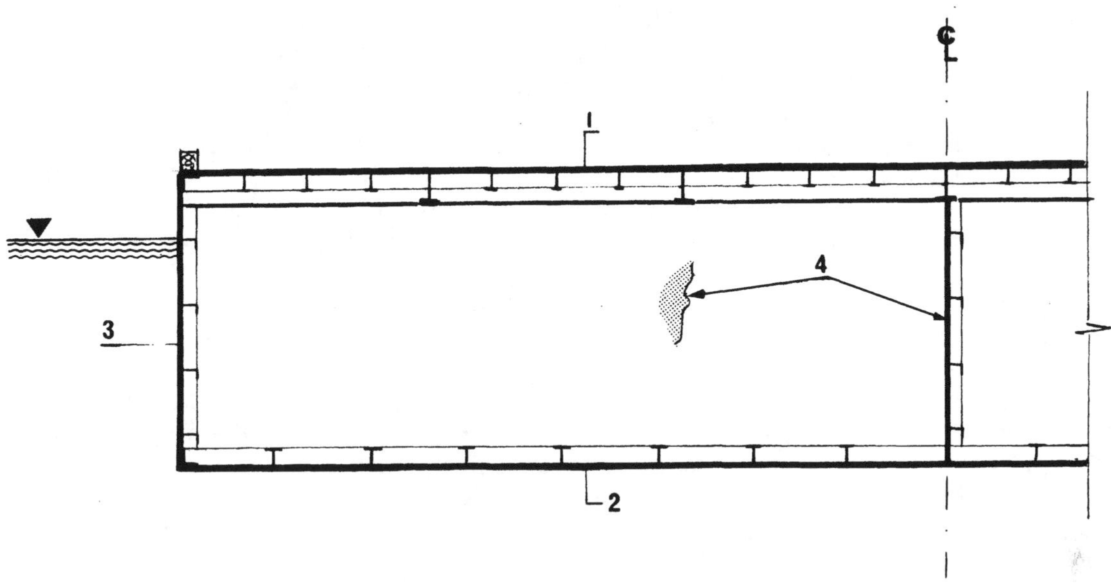

Figure 7.62 Typical framing of steel pontoons used in floating dock system. 1, Deck; 2, bottom; 3, sidewall; 4, transverse and longitudinal watertight bulkheads.

butions to the strength of the bottom plate by distributing the bending effects caused by hydrostatic loads to the main supporting boundaries (i.e., frames, bulkheads, or side plates). A pier deck forms the upper flange of a hull box girder and as such is subjected to tension or compression due to pier performance as a girder and locally to cargo handling equipment, cargo stored temporarily, loading from vehicular transport, and other relevant loadings. It could be constructed in the form of a plate supported on principal and secondary beams, or of orthotropic construction, similar to that used in steel bridges. The pier-side plating (along with longitudinal bulkheads) performs as a web for the hull girder and is an important part of the watertight envelope. It is exposed to various loads, such as hydrostatic pressure, dynamic water pressure due to wave action, potential ice impact force, and ship impact loading transmitted to the hull via the fender system.

The stringers and secondary beams make a significant contribution to the strength of the side plating by distributing all side loads and effects produced by these loads to transverse frames (or trusses) and bulkheads. For operation in ice, the pier hull must be strengthened locally as required. Watertight bulkheads are a major component of a pier's internal framing. They are provided to prevent progressive pier flooding under damaged conditions. The longitudinal bulkheads contribute to the strength of a pier hull girder nearly as effectively as does its side plating. The transverse frames and bulkheads basically act as internal stiffening diaphragms and resist in-plane torsion loads and do not contribute much to pier longitudinal strength.

Flat steel sheets used for side plating and watertight bulkheads are usually stiffened to perform their required function efficiently. The stiffening members typically are standard rolled sections with one edge welded to the plate they intend to reinforce. Obviously, all structural components of a pier, such as its sides, bottom platings, frames, and bulkheads being welded together interact with each other to provide overall edge restraint for each other. This, with each other interaction, causes a complex stress pattern that can be solved adequately by using standard numerical methods. Also, it can be represented very well by comprehensive, finite-element, three-dimensional mathematical models.

However, by developing his or her design scheme, the designer must be aware of potential misalignment of pier structural components during pier fabrication. Alignment is usually concerned with having two connected members continuously in the same plane, and in general, the ideal continuity of longitudinals can be obtained if the longitudinals themselves are continuous through cuts in bulkheads. Subsequently, this requires closely fitted slots in bulkheads with bulkhead plating continuously welded directly to longitudinals all around. This operation may be difficult and costly, however. It may be largely mitigated if heavy flat stiffening plates are used instead of rolled sections.

The minimum thickness of material used for pier construction should be based on recommendations of appropriate standards and guidelines. In general, thickness of pier shell plating depends on pier performance as a girder, but hydrostatic, hydrodynamic, and ship impact forces also have great effect on it. Furthermore, a certain margin is generally provided for corrosion. The latter depends on local water aggressiveness toward the steel in general and aims to ensure that during the design period of service, sufficient plate thickness must remain to ensure pier structural integrity and reliable performance. Depending on pier length and its intended performance, the thickness of its shell plating may vary. Midpier plate thickness, which is normally maintained through 40 to 75% of pier length, may gradually be reduced toward the ends. The operating pier should have special certification for such service. The shell plating of such piers is usually increased between the shallowest and the deepest winter service waterlines. The degree of hull strengthening depends on expected nature (e.g., static, dynamic, or both ice forces). Sometimes, hawser pipes break through the bottom plating. Normally, pipes welded to the bottom plating do not weaken it; they may be spaced as closely as working clearances permit. Similarly, pipes passing through bulkhead or deck plating when welded to plating do not affect their strength.

The pier transverse frames (or trusses) have two basic functions: first they help to resist all kinds of loads exerted on the shell and deck structure, and they also stiffens the hull against buckling. Similar functions are provided by transverse bulkheads, which extend from the bottom to the deck to provide for pier transverse watertight subdivision. The distance between transverse frames (or trusses) or watertight bulkheads is governed largely by pier length and its required stability under damaged conditions. Watertight bulkheads are usually located according to rules recommended by the relevant classification agency. As stated earlier, the pier structure is usually protected against ship impact by a fender system in which individual fender units are designed to absorb energy of a largest design vessel. In the case of a floating pier, it is highly advisable that the pier hull be protected by a secondary fender system from potential impacts by small boats such as barges, tugs, and service vessels, which could strike the pier between two adjacent principal fenders causing damage to a pier hull. The secondary fender system is usually installed in the form of wooden frames or strips, used tires, or small standard rubber units. Alternatively, steel sections such as half-pipes or corrugated plates can be used. If a wood fender is used, it is not recommended to use oak timbers placed directly against steel, because the tannic acid of oak may be corrosive. Pine is a suitable alternative.

7.12.2.2 Hull Materials and Welding Procedures. Steel ductility in either tension or compression, its durability, particularly with effective corrosion protection, and its widespread availability make it used almost universally in any kind of structure. Materials and construction procedures used for steel pier construction are the same or similar to those normally used for ordinary ship construction. Simplicity of floating pier structure stimulates the use of common-grade structural steel and automation of fabrication processes.

Common-grade structural steel is adequate in most cases for the service normally encountered by most floating piers. In some cases, however (e.g., a floating pier operating in a cold climate, a highly aggressive environment, or under special loading conditions), the use of special steel with enhanced properties may be required. For example, in cold-climate regions (−40°C or lower), use of improper steel could lead to development of brittle fractures in steel even at moderate stress levels, with eventual structural failure. Hence, it is imperative that the designer give careful consideration to selection of the proper grade of structural steel. Structural steel presently produced and means to improve notch toughness properties of steel materials provide the designer with a wide variety of steel grades that can meet design conditions in cold regions.

The greatest disadvantage of steel used in polluted or salt water is, of course, corrosion. Normally, the life of marine structures in the aforementioned environment is between 15 and 40 years. To extend the useful life of steel marine structures, special alloy steels [e.g., ASTM A620 (USS *Mariner*)] with up to three times greater resistance to saltwater splash corrosion are used. There are also other methods for extending the life of a steel pier in a marine environment. These include application of different kinds of protective coatings and cathodic protection. The latter is discussed briefly in the following section.

As noted earlier, the welding process today is a universally accepted method used for joining together metallic structural components to speed up process fabrication of steel structures. Welded joints are reliable for impact loads and severe vibration; a properly welded joint can be stronger than the materials joined. Other advantages of welded joints include the following:

- Less time is required for detailing, layout, and fabrication of steel structure, since fewer pieces are used.
- Punching or drilling, and reaming or countersinking of steel, are eliminated.
- The structure can be built in relative silence, thus preventing noise pollution.
- Because the core wire used in electrodes is of premium steel, and there is complete shielding of the molten metal during welding, properly deposited welds have a tremendous reserve of strength or factor of safety, far beyond what industry specifications usually recognize.

In today's shipbuilding industry, the trend is to build the structure on a subassembly basis, doing as much work as possible under ideal shop conditions where mass-production techniques can be fully employed. Similar construction technology applies to pier construction. The progress made in recent years in automatic and semiautomatic welding equipment and in the use of positioners and manipulators has made shop fabrication of pier modules and details for further assembly in the dry dock extremely attractive. This drastically reduces use of dry dock time, resulting in reduction in the cost of pier construction.

Developments in welding technology over the years have greatly increased the speed of the welding process while assuring high-quality welds. Much progress has been made in automatic manipulators, enabling the welding head to be put into proper alignment with the joint of the member in a matter of seconds. This alignment is maintained automatically along the length of the joint during welding.

Semiautomatic field welding is speeding up assembly of shop-prefabricated modules or details.

A correctly designed joint and properly made weld do not require special procedures to prevent cracks during welding or service. The need for special welding procedures increases, however, with heavy plate structural members. Most steels can be commercially arc welded with good results. Some steels are more weldable than others. The weldability of a metal refers to the relative ease of producing a satisfactory, crack-free, sound joint; and steel is said to be *weldable* if the weld joint required can be made without difficulty or with a resulting cost increase. It should be noted that basic structural steels used for floating pier construction usually do not require special precautions or special procedures. However, when welding thicker plates, the increased rigidity and restraint and the drastic quench effect makes use of the proper procedure vitally important. At the shipyard, the welding procedure is usually planned ahead. Preplanning welding procedure normally calls for one or all of the following:

- Joint design and plate preparation
- Minimized penetration to prevent dilution of the weld metal with the alloy elements in the plate
- Preheating, controlled interpass temperature, and sometimes even controlled heat input from the welding procedure to retard the cooling rate and reduce shrinkage stress

Additional requirements are found in special technical literature. The main objective of any welding procedure is to join the pieces with the most efficient, good-quality, crackless weld. A crack in a weld is never minor and cannot be tolerated. Cracks in the welds could be the result of a great many factors, some of which are as follows:

- Joint restraint, which causes high stress in the weld
- Inadequate bead shape, which could result in high tensile stress to cause a longitudinal crack or internal crack
- High carbon content of the base metal, which reduces weld metal through admixture
- Rapid cooling rate, which increases hardening ability and loss of ductility in the head-affected zone
- High impact loading under low temperatures
- Fatigue cracking due to a notch effect from poor joint geometry

To avoid crack formations due to the reasons cited above, the following basics should be considered:

- Create a bead with the proper width/depth ratio and surface (e.g., slightly convex).
- Keep joint restraint problems to a minimum.
- Avoid the use of inferior-welding-quality steels that have an excessively high percentage of those elements that always adversely affect weld quality (e.g., sulfur and phosphorus).
- Control input carefully. This may include preheating of basic metal, heating between weld passes, and postheating to control cooling rate. This lowers the shrinkage stresses and retards the cooling rate, helping to prevent excessive hardening in the heat-affected zone.

Because of their importance, the quality of welds must be monitored closely. There are several nondestructive ways to evaluate the weld quality. They are as follows:

1. *Visual inspection*. This assumes visual examination of the complete weld for

surface soundness, regularity, geometry, and alignment.

2. *Magnetic particle inspection.* In this method the base metal is magnetized as a superimposed electrical current is passed through, and finally, divided magnetic particles are applied to the plate surface. A flaw at or near the surface will form a pair of magnetic poles that will act as a magnet and attract the applied magnetic particles.
3. *Dye penetration inspection.* In this method, a liquid penetrant of low surface tension is used to penetrate a surface crack. The indication of cracks will appear as an accumulation of developer around a crack or fissure. Dye penetration inspection may be used in lieu of magnetic particle inspection.
4. *Radiographic inspection.* This is used to examine internal soundness of weldments and employs a source of electromagnetic radiation capable of penetrating the thickness of material under investigation. Radiographic inspection is mainly carried out primarily in highly stressed and sensitive structural areas.
5. *Ultrasonic inspection.* It is used as an alternative to radiography for the examination of welds.

Although ultrasonic and radiographic methods are both accepted for inspection of hull welds, each has distinct advantages and limitations. For example, the ultrasonic method is more sensitive than radiography to discontinuity (cracks, linear slag, and imperfection) and has an unlimited range of applicable thickness. Radiography is more sensitive in generalizing lack of fusion and slag inclusions, indicates shape and size of cracks, and is less dependent on subjective skill and judgment of technicians. However, it could be harmful in operating and requires greater precautions than the ultrasonic method.

7.12.2.3 Protection against Corrosion. Corrosion is the major problem facing engineers today; 40% of steel produced each year is used to replace the corroded metal (Staehle, 1975). Hence, protection of any kind of marine structures, and floating marine structures built of steel in particular, is a very important economic factor to be considered in floating dock design and operation. Deterioration caused by corrosion is a factor that could limit structure's service life and, therefore, when designing any type of marine structure that fabricated steel in particular, the corrosion potential of such a structure must be evaluated carefully; it may range from negligible to severe. If not properly controlled, deterioration caused by corrosion can develop into *stress corrosion.* The latter term is defined as the cracking of a metal under the combined effect of tensile stresses and a corrosive environment. Typically, a corrosion pit on a metal surface acts as a stress raiser. Corrosion at the tip of the notch causes a stress corrosion crack to start there. Another distinct deterioration caused by corrosion of steel is called *corrosion fatigue,* which is defined as the reduction in metal fatigue resistance caused by the presence of a corrosive medium. Fatigue cracks propagate faster in the presence of cyclic tensile stresses, which may have a pronounced effect on corrosion fatigue.

Obviously, where service life of a dock is limited, or where corrosion is not a factor of concern, corrosion control measures would be of limited nature or not needed at all. On the other hand, where corrosion potential is significant, then to avoid costly maintenance procedures, corrosion-protective systems must be considered to retard, or eliminate, corrosion-related deterioration of the structure. This is required to avoid excessive costly maintenance procedures, resulting in loss of service from prolonged downtime.

Many floating docks are expected to perform satisfactorily over a long period of time. Usually, a minimum service life of floating docks from 20 to 50 years or more is planned. Long service life requirements dictate that corrosion protection be provided for permanent floating piers from the outset. While in service, the structure must retain its integrity with an adequate margin of safety. Safety factors used in the design of a floating dock are usually necessary to compensate for excessive unpredictable load, potential defects in workmanship, improper service, and so on. Corrosion could alter adopted safety factors and increase safety hazards to operating personnel and equipment. Consequently, corrosion protection must be called for where these safety factors could appear in jeopardy over the intended service life. Sulfur, ammonium sulfate, salt, and raw sugar are examples of corrosive materials that can affect the dock structure, cargo handling equipment, and its maintenance. Handling of such bulks necessitates appropriate selection of basic construction materials and methods of their protection against corrosion.

Finally, corrosion control above the waterline is provided for aesthetic purposes of structure appearance. Methods of corrosion control vary from good design and fabrication to use of appropriate coatings, cathodic protection systems, or a combination of them. However, knowledge of corrosion fundamentals is necessary before deciding on the level of protection applied. In general, the rate of corrosion is influenced by factors such as water temperature, oxygen concentration, pH values, salinity, water velocity, marine organisms, pollution, wind, rain, humidity, sun, salt spray and other particles, stray electrical currents, and the presence of dissimilar metals.

It also depends on zone where corrosion may take place. There are three distinct zones of corrosion as related to floating marine structures:

1. *Atmospheric zone:* the area at the top of the floating pier accessible for visual inspection and maintenance.
2. *Splash zone:* the area from the water level and up to the bottom of the atmospheric zone. In this area a continuous film of water is maintained on the surface of the floating body exposed to the atmosphere. This area is also accessible for visual inspection and maintenance, although with some inconvenience.
3. *Submerged zone:* the area that is always submerged into the water. This area is not readily accessible for inspection without utilizing special equipment.

Note that in fixed marine structures (e.g., mooring dolphins), there are two more distinct zones of corrosion: tidal and soil zones. For details, consult Chapter 6. The rate of corrosion may vary considerably by zone; maximum loss of metal usually occurs in the splash zone immediately above the water level, however, significant corrosion of metal is also observed in the submerged zone. Where the steel structure is exposed to water on one side, the average rate of corrosion could vary from 1 to 4.5 mils/yr (1 mil = 0.0254 mm), reaching a maximum of 3 to 14 mils/yr (Ractlitte, 1983). Accordingly, if a steel structure is exposed to water on both sides, the total rate of corrosion is doubled. Corrosion may change considerably, however, depending on water pollution. The latter may include anything from domestic sewage and complex industrial wastes to spill of chemical materials handling at the port. Significant corrosion may also be caused by coupling of dissimilar metals and direct current from external sources such as improperly grounded welding generators, or ship and/or pier service systems. To prevent these causes of steel corrosion, use of bimetallic structures should be avoided in marine environments, especially in underwater and splash zones.

When rivets or bolts are used for fabrication, fastening components should preferably be cathodic to the parent steel structure. Welds should also be cathodic to the main structure. It is also good practice to avoid crevices and pockets when fabricating a pier. Crevices that could form at unsealed joints trap stagnant water. This causes cells to concentrate due to poor diffusion of oxygen into joints or the buildup of dissolved metal ions in the entrapped water. This may accelerate local corrosion attack. Similar results could be expected when water is trapped in different kinds of structural pockets. To prevent the accumulation of standing water, the dock must be designed so that proper water runoff occurs (i.e., camber on decks will allow water to run off, and all catchment areas have to have suitably located drain holes and slots to prevent the accumulation of water).

Furthermore, all structural members inaccessible for regular inspection and maintenance must be either corrosion resistant or be adequately protected against corrosion. For more information on corrosion and its effects on marine structures, consult Chapter 6. There are two basic methods of protecting structural steel against corrosion: application of protective coatings and utilization of cathodic protection, which we discuss next.

7.12.2.4 Protective Coating. Coating in general is an economical way to protect exposed carbon steel against corrosion. Its basic function is to reduce the area of metal exposed to the electrolyte (e.g., soil or salt waters). Good-quality coatings, in general, may have a service life of up to 15 years. In recent years, many coatings have been developed and evaluated for service in a marine environment. Recent information on the state of the art in this type of technology can be obtained from national associations of corrosion engineers or other relevant organizations, and from manufacturers. Typically, coatings used in submerged and splash zones in any type of marine structure are coal-tar epoxy, epoxy, metallized zinc, or aluminum with topcoats, phenolic mastics, and other.

The combination of a good service record and economy made coal-tar epoxies most often used in coating steel marine structures for immersion service. Epoxies, in general, have outstanding adhesion to clean metals. For good results before a coal-tar coating application, the surface of metal should be sand-blasted nearly white. Normally, epoxy-based coating forms a hard film with good resistance to penetration by fouling and good resistance to deterioration by cathodic action, thus offering good protection from corrosion in all zones of marine structure. The service life of coal-tar epoxies, as for many other coating systems, is related directly to the thickness of protective film. Usually, a minimum 20-mil dry film of coal-tar epoxies is recommended for submerged and splash zone thickness. However, some brands of coal-tar epoxies may suffer rapid deterioration and disbandment when exposed to ultraviolet light. Therefore, before selecting a particular brand of coal-tar epoxy, the engineer must be aware of the past record of the material. Straight epoxies normally are more expensive than coal-tar epoxies and generally are used to patch or apply new coatings to wetted and submerged portions of marine structure. For their successful performance, epoxies, too, require a nearly white sand-blasted surface. With proper application, epoxies may have 10 to 20 years of effectiveness in submerged zones.

Metallic coatings are also used to protect structural steel against corrosion in a marine environment. Steel metallizing is a process of applying a metallic coating to a substrata metal surface by spraying molten metal. For marine applications, in addition to steel metallizing, suitable sealers are used as topcoats to eliminate porosity of metallic coating. Adhesion of the applied metallic coating to the substrate is purely mechanical and is most effective when applied on a roughened, white-blasted metal surface. Flame-sprayed zinc coatings compare favorably with other high-ranking coatings when top-coated with an organic coating, except in the splash zone (Alumbaugh, 1966). The probable cause of poor performance of zinc in the splash zone is attributed to higher local velocities, which erode the protective films of zinc.

According to the American Welding Society (AWS, 1974), zinc with a sealer coat performs better than seal aluminum, but bare aluminum coating performs better than bare zinc. It should be pointed out that regardless of the merits of metallized coating, the cost of these systems with topcoats ranges from one-and-one-half to three times that of good coal-tar epoxy coating. Inorganic and organic zincs are most frequently used for protection of atmospheric zones of marine structures. A coat of vinyl red lead primer and a topcoat of silicone-alkyd copolymer resin is rapidly gaining acceptance in naval installations [American Iron and Steel Institute, (AISI), 1981]. The resistance of this coating to the action of destructive environmental agents such as sunlight, heat and cold, wind, rain and snow, and atmospheric contaminants is reported to be superior. The total film thickness for the system is about 5 to 8 mils.

Organic, zinc-rich coating systems have demonstrated good performance in marine atmospheric exposure. In general, these systems have better film-forming characteristics.

The ultimate life of any coating system is heavily dependent on surface preparation. Blast cleaning is commonly used in surface preparation of steel structures. A surface blasted to a near-white condition is expected to give the best results and is certainly of paramount importance for long-term service of coatings immersed in seawater. However, blast cleaning to near-white conditions may not be economical for less corrosive environments such as atmospheric exposure. Removal of oil, grease, rust, and other surface contaminants is usually considered to be adequate for many protective coating systems in atmospheric environments. Specifications and details concerning various metal surface preparation methods are normally provided in manuals prepared by coating manufacturers. As the purpose of coatings is to isolate the steel structure from its environment, it is very important that the coating be free of voids. The proper application of coating to marine structures and particularly to floating piers must be confirmed by an experienced and competent inspector who represents the owner.

Cathodic Protection. Cathodic protection is an electrochemical method of corrosion control. It effectively protects both bare and coated steel mooring structures that are totally immersed. In this method the protection of steel is achieved satisfactorily by the changing direction of flow of electric current by making the structure the cathode of an external electrochemical cell. The cathodic protection changes the direction of the flow of electric current to the steel structure and not from it, thus preventing iron ions from flowing out of the steel and causing its decomposition (galvanic action). Cathodic protection is achieved by establishing a direct current (dc) voltage between the protected metallic structure and an auxiliary anode so that the current flows through the water or moist soil to the structure. Marine structures could be protected by two types of cathodic protection systems: (1) a galvanic anode system and (2) an impressed current system. Both systems are illustrated in Figure 7.63.

GALVANIC ANODE SYSTEM. This system consists of a sacrificial anode connected to the structure electrically and immersed in the electrolyte (seawater). The protected surface acts as a cathode (hence, the name *cathodic protection* for the technique). Galvanic anode cathodic protection systems are frequently used for protection of marine structures. This is because properly installed, they require little or no attention for their design service life. Furthermore, galvanic anode systems require no

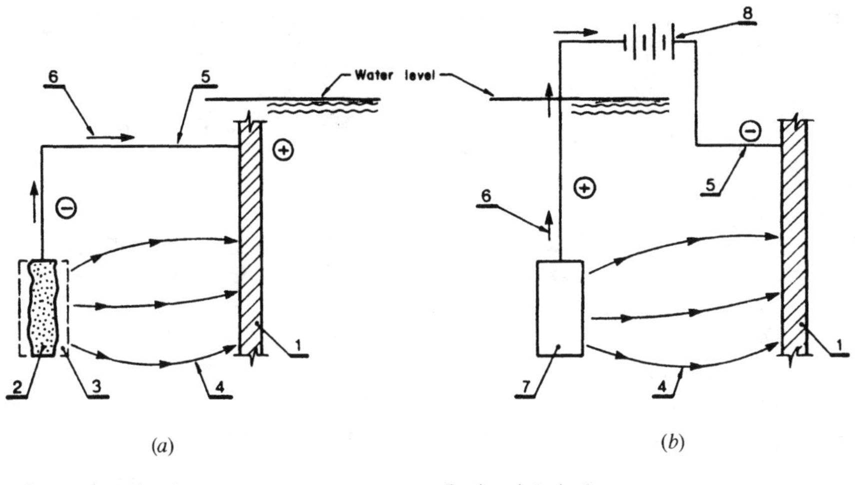

1—marine structure
2—sacrificial anode (zinc, aluminum, or magnesium)
3—metallic ions lost into solution
4—electric current flowing through electrolyte
5—insulated wire
6—electron flow in external circuit
7—inert anode, or graphite, lead alloy or other suitable material that will best discharge the immersed current
8—external source of direct current

Figure 7.63 Basic cathodic systems used for protection of metals from corrosion: (*a*) galvanic anode system; (*b*) impressed current system.

external source of power and are relatively easy to install and maintain. They are attractive under conditions where protective current requirements are not large and where low-driving voltages are adequate to obtain the current required. Typically, alloys used for sacrificial anodes are based on magnesium, zinc, or aluminum. The performance, and hence the suitability of a particular alloy for a specific application, depends on the composition of both the alloy and the electrolyte, the ambient temperature, and the anode current density.

The basic sacrificial anodes used in marine cathodic protection technology are as follows: zinc anodes, aluminum alloy anodes, and magnesium alloy anodes, and magnesium-clad zinc anodes among others. *Zinc* as an anode for cathodic protection of ship hulls was introduced over 140 years ago. However, over the past 30 years, zinc anode composition has changed from high-purity zinc to zinc alloys having small additions of aluminum (0.1 to 0.3%) and cadmium (0.025 to 0.06%) with allowable increases in iron (0.005% maximum) (AISI, 1981). Properly installed zinc anodes require no further attention. They are very efficient and can be designed for a life of 10 years or more.

Similar to zinc anodes, *aluminum alloy anodes* can supply current as long as the structure demands it. Various anodes of this type have been on the market since 1950; however, according to AISI (1981), they did not perform consistently in long-term marine service. In recent years, a propriety composition aluminum anode containing 0.45% mercury and 0.45% zinc has been marketed. Its efficiency is claimed to approach 95%, and its actual current output is about 2820 A · H/kg.

Magnesium alloy anodes are less efficient than zinc anodes because they waste some current on self-protection. If used in combination with protective coatings in seawater, magnesium anodes could impress too high a voltage gradient across the coatings. This may increase the probability of coating failure due to electroosmotic penetration by seawater and subsequent blistering of the film of coating by hydrogen gas evolution at the metal substrata.

Magnesium-clad zinc anodes are typical zinc anodes with approximately 5.3% magnesium cast as a jacket over all zinc faces. The basic advantage of this combination over a conventional zinc anode is that the magnesium, with its higher driving voltage, polarizes the structure much faster than zinc would and thereby permits a reduction in the current demand by the structure which is easily handled by the zinc after the magnesium jacket is consumed.

Some typical properties of galvanic anodes in cathodic protection are given in Table 7.1. The data decisively favor the use of aluminum alloy anodes. For more information on anode materials and their use in marine environments, readers are referred to British Standards Institution (1991).

IMPRESSED CURRENT SYSTEM. This system is normally used for the protection of marine structures where large quantities of current are required. It normally consists of the following basic components:

Table 7.1 Typical properties of galvanic anodes

Property	Zinc	Aluminum Alloy	Magnesium
Actual current output (A · h/kg)	780	2820	1100
Actual anode consumption (kg/A · yr)	11.3	3.1	8
Density (g/cm^3)	6.9	2.8	1.7

- Anodes and associated dc positive wiring
- Dc power supply and means of power regulation
- Negative return circuit from protected structure to dc power supply

Scrap iron and steel, graphite lead alloys, or iron–silicon materials are generally used as anodes in impressed current cathodic protection systems. Basic properties of these anodes are shown in Table 9.2.

The dissolution rate of iron or steel is approximately 9 kg/A · yr at high current, and this translates into enormous amounts of steel to be consumed. Therefore, the bulk of such material to be used as anodes would be very large, to avoid infrequent replacement. Graphite anodes are suitable for use in both freshwater and seawater and have been used successfully for years for cathodic protection of marine structures. Despite being insoluble material, graphite becomes decomposed in time, due primarily to electrochemical oxidation, and the graphite anode must be replaced in 5 to 15 years, depending on its size and location. The theoretical consumption rate for a graphite anode is 1 kg/A · yr. However, its actual rate of consumption is less than 10% of the theoretical rate at (low) anode current densities of about 10 to 20 A/m^2. At normal anode current densities in seawater, graphite deteriorates at approximately 0.35 to 0.55 kg/A · yr (Singleton, 1970). Assuming 0.4 kg/A · yr as a conservative value, a 35-kg graphite anode submerged in salt water would be completely consumed, providing 7 A continuously, in about 12.5 years (35/0.4 × 7). Practically, however, replacement must be made before the anode is greatly reduced in size, to avoid an excessive current density as the exposed surface of the anode diminishes. Graphite anodes can be suspended in water or mounted in insulated holding clamps attached to the structure. Because of their brittle nature, the

Table 7.2 Properties of anodes used in impressed current cathodic protection

Property or Characteristic	Graphite	High-Silicon Iron Alloy	Lead Alloy
Recommended anode current density range (A/m^2)	2.5–11	11–55	55–270
Consumption rate range ($g/A \cdot yr$)	365–1000	135–500	115–245
Density (g/cm^3)	2.22	7.0	11.3
Application limitations	Do not use in fresh water and do not bury in mud	Use lower consumption rates in fresh water	Do not use in fresh water and polluted water, and do not bury in mud
Special operating requirements	Do not exceed maximum current density of 110 A/m^2	Do not exceed maximum current density of 55 A/m^2	Use minimum current density of 30 A/m^2

graphite anodes must be protected against possible damage by floating objects or strong current. Relatively seldom, iron silicon materials (high-silicon alloys), lead alloys containing 2% silver, and other materials are also used as sacrificial metal for impressed current cathodic protection systems.

In the use of cathodic protection to submerged parts of steel marine structures, transformer–rectifiers are generally used to provide direct current for powering an impressed current cathodic protection system. A transformer–rectifier commonly consists of a step-down transformer and a rectifier stack (usually selenium, silicon, copper oxide, or copper sulfide) to convert alternating current (ac) to direct current (dc). Both transformer and rectifier stack are mounted inside weatherproof protective housing.

The design of a cathodic protection system (both a galvanic anode or an impressed current) typically include the following steps:

1. Determination of protective current requirements
2. Selection of the appropriate type of anode; determination of its size, number, and location as required for the assumed service life of the marine structure
3. Determination of the rectifier requirements for impressed current system
4. Design of a reliable wiring system and devlopment of the specifications for suitable mounting methods (i.e., securing of anodes to the structure)

A protective current, as required to protect a bare steel marine structure, depends on the rate of its average annual corrosion. It usually takes a current flow of approximately 21.5 mA/m^2 flowing continuously to corrode a steel surface at the rate of 1 mil/yr. Therefore, if the rate of corrosion in the submerged zone (of bare steel) is approximately 5 mils/yr, this means that an average corrosion current of $c = 21.5 \times 5 = 107.5$ mA/m^2 is flowing continuously. In this case, to stop corrosion by means of cathodic protection, a value equal to 107.5 mA/m^2 of current flow (with allowance for nonuniform current distribution) is required.

A coated structure naturally requires less current flow for protection against corrosion. The effectiveness of a sound coating may be assumed as 90 to 95%, and if similar to the previous example, the corrosion rate in the submerged zone is 5 mils/yr, the current density required for cathodic protection in this zone may be determined as equal to $c_c = 21.5 \times 5 \times (0.05 \text{ to } 0.1) = 5.35 \text{ to } 10.7 \text{ mA/m}^2$. Again, a margin for coating deterioration must be added to the current requirements on a coated marine structure. The total required current flow is estimated as the total area to be protected by means of cathodic protection times the flow requirements per square meter (c or c_c).

Since anodes for marine cathodic protection systems (both galvanic and impressed current) are standardized, their electrolyte resistance are known. Therefore, the total number of anodes required for cathodic protection may be determined by

$$N = \frac{cYI}{WUE} \tag{7.12}$$

where N is the number of anodes, c the consumption rate of anode material (kg/A · yr), Y the system's life expectancy (years), I the average total current requirement, W the mass of each anode (kg), U the utilization factor of anode material, and E an efficiency factor (used basically with galvanic anodes). The number of anodes required may be increased due to factors such as nonuniform current distribution between different structural components of a pier or nonuniform consumption of anodes.

In conclusion, it must be said that for design of satisfactory and economical cathodic protection systems additional factors such as water temperature and dissolved oxygen content in the water must also be taken into account. Because of the specialized nature of the design of cathodic protection systems and their importance for the extended life of a structure, the involvement of specialists in corrosion protection should be considered. For more details and practical examples, readers are referred to Tsinker (1986).

Combined Corrosion Protection. Sometimes a corrosion control method effective and economical for one zone of a structure might be unsuitable for another. Therefore, using one protective system only may have drawbacks. Ideally, a structure should be neither underprotected nor overprotected. The optimization of a total protective system could be accomplished by combining various protective systems. As noted earlier, cathodic systems provide protection for submerged steel but not for steel exposed to atmosphere. On the other hand, many coatings are economical and effective in both atmospheric and splash zones but less attractive for protection of the submerged zone, due to the high costs of maintenance. A combination of the two systems used concurrently generally offers economic benefits by providing effective and economic protection to the entire structure. It must be understood, however, that not all systems are compatible (e.g., some coatings may suffer disbandment when used in combination with an impressed current system).

Where combinations do work well, further economic benefits may be obtained by introducing a second system when the effectiveness of the first has decreased in a particular zone [e.g., utilization of an epoxy coating applied over a pier at the time of construction with subsequent (say, after 10 to 15 years) introduction of a galvanic cathodic protection system to be installed to protect the pier submerged zone]. However, the possible permutations and variations of various protective systems could be limited by economic practicalities. It should be noted that in some cases, the use of unprotected steel with an appropriate allowance for corrosion could be justified economically (e.g., when the useful life of the structure is limited by design constraints). Furthermore, the rate of corrosion can be reduced by providing proper ventilation of pier interior space. Natural ventilation relying on air density and temperature differences and wind usually is adequate for most uninhabited spaces. Ventilator openings are sized according to the cell volume and number of air exchanges per hour. In practice, this is achieved by providing the required number of ventilator pipes installed at the pier deck.

7.12.3 Concrete Floating Pier

7.12.3.1 Basic Design Considerations. As noted earlier, many engineers favored steel as a prime material for the construction of floating piers. However, because of the relatively high maintenance cost of steel structures in a corrosive marine environment, engineers more often chose low-maintenance-cost concrete, in particular, prestressed and prefabricated concrete for construction of miscellaneous floating structures and piers. The low maintenance cost of concrete is attributed basically to its durability and high resistance to corrosive effects of seawater. Concrete use for construction of floating structures is by no means a brand new idea. The record of construction and use of concrete ships goes back to 1848 Hetherington, 1920; (Bremner et al., 1996). However, most concrete vessels were built after World War I and during World War II. Although the total experience with concrete floating structures (including ships, pontoons, barges, etc.) is very limited compared to that of steel structures, the evidence presented by researchers and engineers, and the present condition of many floating concrete objects still operational after performing for several decades in seawater, clearly demonstrate the high durability and reliability of concrete as a construction material (Gerwick, 1975; Hetherington, 1980; Yee and Daly, 1984; Bremner et al., 1996). Floating piers, dry docks, offshore platforms, and other structures made of concrete are found elsewhere in the world. After being in service for extended periods of time, hundreds of them, except for isolated circumstances (e.g., hard berthing, exposed listed surfaces, overloading) are in good condition and exhibit reliable performance at low maintenance cost.

7.12.3.2 Concrete. The term *concrete* is often used as a general definition of a variety of materials used for construction of marine structures (e.g., plain concrete, reinforced concrete, prestressed concrete, ferrocement, and fiber-reinforced concrete). *Plain concrete* is an artificial conglomerate of cement and coarse and fine aggregates, including natural sand and gravel or crushed stone. Concrete that has been strengthened to resist the action of tensile stress by adding embedded steel bars is called *reinforced concrete. Prestressed concrete* is concrete that has been subjected to a permanent compressive stress of the same order of magnitude as the tensile stress expected to occur under load. The compressive stress is induced by action of tensioned high-tensile steel wires, strands, or bars anchored to the concrete. The steel (tendons) may be tensioned before placing the concrete, in which case, tension is transferred as compression to the concrete after it has hardened. This technique is known as *prestressing*. Alternatively, concrete may first be cast with tendons arranged either outside the concrete or in ducts. The tendons are tensioned and anchored at their extremities when the concrete has attained adequate strength. Finally, tendons are protected by grouting. This technique is known as *posttensioning*. The combination of both prestressing techniques is also possible and may be useful in prefabrication.

Ferrocement is a composite material usually defined as concrete consisting of cement and fine aggregate, reinforced by small-diameter steel wires distributed throughout a body of concrete. More practically, ferrocement consists of a number of layers of closely spaced steel meshes into which concrete mortar is forced. The strength of ferrocement is usually considered to come entirely from steel, with the concrete simply operating to keep water out (Turner, 1974). *Fiber-reinforced concrete* is a concrete containing a random dispersion of small fibers to reduce concrete intrinsic cracking and to increase its strength. Typically, the unit weights of plain and reinforced concrete, respectively, are 2300 and 2400 kg/m^3. The unit weight of ferrocement may reach 2900 to 3000 kg/m^3.

7.12.3.3 Concrete Cracking. Despite evidence of the long-term durability of concrete structures in marine environments and generally outstanding performance in the ocean, serious concrete deterioration has been reported. The general cause of deterioration was noted to be cracking, resulting in corrosion of the embedded steel. In marine environments corrosion occurs in the presence of oxygen, moisture, and an electrolyte. Salt, acting as a catalyst, intensifies the electrolytic properties of concrete, thereby creating a corrosion cell, resulting in steel corrosion. Steel corrosion, in turn, leads to the formation of large amounts of expansive products (about six to seven times the original volume) and further cracking of concrete in a direction parallel to the reinforcement. This process continues and two situations develop:

1. The reinforcing bars disintegrate, thereby reducing the flexural strength of the concrete structure.
2. Concrete spalling, delamination, cracks, and stains occur.

A comprehensive list of the physical and chemical causes of concrete deterioration is provided in works by Mehta (1980a,b) and Mehta and Gerwick (1982). The mechanism of concrete deterioration in marine environments is discussed in detail in Tsinker (1995).

Cracks that occur in concrete are classified in two basic categories:

1. Cracks caused by high stresses, such as thermal gradients while in service, tensile stresses, exposure of unprotected concrete to freezing and thawing, heating–cooling and wetting–drying cycles in service, impact forces, and a variety of expansive chemical reactions, such as those that occur in sulfate attack and in alkali–aggregate attack.
2. Self-imposed or intrinsic stresses. There are three types of intrinsic stresses that cause cracking: *plastic,* those occurring in the first few hours; *thermal contraction,* occurring in 1 day to 3 weeks and usually during construction of massive concrete structural members; and *drying shrinkage,* often occurring after several weeks and in some instances even after several months.

In modern engineering practice, various design codes and guidelines provide recommendations on limiting the crack width in concrete structures. For instance, the American Concrete Institute (ACI, 1980) recommends 0.15 mm as the maximum crack width at the tensile face of a reinforced concrete structure subjected to wetting and drying or spraying in seawater. The Fédération Internationale de la Precontrainte (FIP, 1985) recommends that crack widths at points nearest main reinforcement not exceed 0.004 times the nominal cover (e.g., 0.2 mm for a 50-mm cover, 0.3 mm for a 75-mm cover, and 0.4 mm for a 100-mm cover). The Russian Standard SNiPII-1I.14-69 calls for 0.05 to 0.1 mm as the maximum crack width for concrete in tidal zones or in areas exposed to seawater spray. The Japanese *Technical Standard for Port and Harbor in Japan* (OCDI, 1991) specifies permissible width of crack in seawater as a function of concrete cover (c) equal to

0.0035*c*. Most recently, Copola et al. (1996) pointed out that in cracked concrete with a maximum crack width greater than 0.2 mm, seawater has immediate access to the reinforcement.

7.12.3.4 Concrete Durability in a Marine Environment. The marine environment is notoriously aggressive with regard to material durability. Chloride-induced corrosion of the reinforcing steel, freeze–thaw attack on saturated concrete, alkali–aggregate reaction, sulfate attack on the aggregates, salt scaling, and abrasion from ice and sediments are the principal forms of attack. These are assisted and amplified by constant wetting and drying in the splash zone and by thermal strains. The durability of concrete in marine environments is dependent on the quality of materials used for concrete mix, as well as the mixing procedure and the volume and quality of water used.

Conclusions on the quality of materials derived from most reports and publications available to date are as follows. The most suitable cement for use in construction of durable marine structures seems to be either an ordinary portland cement, with C_3A ($3CaO \cdot Al_2O_3$) content less than 8%, or an ASTM type II cement, with a water/cement ratio (by weight) of 0.4 to 0.45. The cement content should be greater than 350 kg/m^3. The importance of a low C_3A content was demonstrated by some important studies of cement properties and their effect on concrete durability (Gjory, 1971). Gerwick (1975, 1976) proposed a lower C_3A content, giving as an acceptable range 3 to 8% for cement used for construction of marine structures.

Mehta (1988) stated that, theoretically, for sea concrete structures, use of ordinary portland cement having a C_3A content of 6 to 12% should be satisfactory, provided that cement is compatible with the admixtures used in concrete mix. With thick sections or large masses of concrete, blast furnace slag–portland cement may be employed (Barringer, 1997; Keck and Riggs, 1997). In this case, alkali content should be limited to 0.65% (Na_2O + $0.65K_2O$) (Gerwick, 1986). Where sulfate aggression is anticipated, pozzolanic additions to cement have been shown to increase durability. Inclusion of pozzolanic admixtures up to 30% by weight of cement, replacing a similar amount of portland cement, have been reported to produce more durable concrete (Tuthill, 1988; Gerwick, 1990). Normally, sulfate-resistant cements with a low C_3A content are used for marine applications. However, the acidity (pH) of cement paste must be maintained at a value high enough to prevent corrosion.

In North American construction practice, pozzolans, ASTM class F (fly ash) or N (natural) with limitations on free carbon, sulfur, and CaO, are used. In special cases where very high strength and impermeability are required, condensed silica fumes are used. In general, concrete sulfate resistance is virtually never a problem in seawater if cement-rich and impermeable concrete mixes are employed (Gerwick, 1986).

The corrosive action of external chlorides on embedded steel is one of the most severe and widespread problems of concrete structures in a marine environment. Due to the magnitude and severity of the problem, engineers should rely on protection methods to protect newly constructed structures against chloride-induced corrosion and other detrimental effects of a marine environment (Mehta, 1997). Corrosion protection measures are now a standard package provided for modern marine structures. This package typically includes appropriate design of the concrete mix and direct protection of the reinforcing steel. The former typically considers use of special corrosion-protection admixtures added to the concrete mix at the batch plant and mixed uniformly through the concrete.

Microsilica fume and a calcium nitrite corrosion inhibitor are two proven corrosion-protection products, which help to delay the onset of reinforcement corrosion (Berke et al., 1988; Lachemi et al., 1998). Microsilica fume reduces the permeability of concrete by slowing considerably the ingress of waterborne chlorides. A calcium nitrite corrosion inhibitor promotes stabilization of a natural passivating layer on the steel, thereby controlling the corrosion rate (Miller and Fielding, 1998). These two admixtures may be used individually or in combination for the most severe corrosion environments. Microsilica fume, which originated in Denmark and Norway, is a product also called condensed silica fume, with 85% or more silica (SiO_2). It has an extremely fine particle size, which allows microsilica to fill voids in the cement paste and between the cement paste and aggregate. This results in a far less permeable microstructure matrix. Usually, a specified microsilica quantity is based on the severity of the environment. In a typical marine environment, microsilica fume may amount to 10% by weight of cement. It is more effective with concretes of lower water/cement ratios, and when used in combination with superplasticizers, strength increases of 20 to 50% over that of conventional concrete can be achieved with the same water/cement ratios.

The highly alkaline environment of concrete creates a natural protective, passivating layer of iron oxide film on the steel in chloride-free concrete. Chlorides that eventually migrate through the concrete when reaching reinforcing steel can break down the passivating layer to initiate a steel corrosion process. It has been found that calcium nitrite can be used as a corrosion-inhibiting admixture; and when mixed directly into plastic concrete, it aborts the corrosion process chemically. Because the calcium corrosion inhibitor maintains an active corrosion-controlling system within a concrete mix, the admixture protects reinforcing bars from corrosion for the entire life of the concrete structure. Some studies have indicated that adding 2% calcium nitrite to concrete exposed to salt used for snow and ice melting can increase the concrete service life by 15 to 50 years (Peel, 1982).

The rate of chloride penetration into concrete depends on concrete *permeability*. The latter is the rate at which water or other liquid flows though the concrete. Low permeability is indeed the key to the overall good durability of concrete. It ensures better corrosion protection for the reinforcement and better protection of concrete structures from freeze–thaw damage in cold environments (Miller and Fielding, 1998). With time, concrete permeability in fresh water is usually less than that in seawater. Observed cccasionally, decreases in concrete permeability in seawater were explained by chemical reactions between the ions in seawater and hydrated cement, which produces crystallization products or chemical compounds that precipitate and decrease pore continuity in concrete (Harvey, 1980).

Use of cements with a low to moderate tricalcium aluminate (C_3A) content is beneficial in that cement combines with chloride ions to form an insoluble compound that blocks the pores. Because permeability in concrete occurs primarily along the interfaces between the cement matrix and coarse aggregate, selection of a mix with minimum bleed by use of aggregates having surface characteristics that promote physical or chemical bonding is required. A low water/cement ratio (0.4 to 0.45 by weight) is the principal factor to be considered in designing a mix of low permeability. The amount of water in concrete mix can be reduced by using water-reducing admixtures (superplasticizers). This is particularly useful where reinforcement is very congested or when high-strength or very dense concrete is required. Unfortunately, low concrete permeability is not specified explicitly in North American construction practice. As the industry becomes more concerned with the durability of concrete, it becomes apparent that permeability must be a specified design property.

Freeze–Thaw. Concrete used for construction of marine structures in cold regions has potential for spalling caused by the free–thaw cycle, by differential expansion, or by corrosion of steel reinforcing bars. To ensure concrete durability under freeze–thaw conditions, air entrainment agents are now invariably used to provide the required quantity and quality of the entrained air, pore size, and spacing. However, these agents must be controlled carefully to prevent any detrimental loss of concrete strength due to overdosage.

For ordinary concrete which has a water/cement raio of 0.4 or higher, air entrainment with a minimum spacing factor of 200 μm is usually recommended. Malhotra (1986) found that it is difficult to entrain sufficient air with the desired minimum spacing factor when the silica fume content is more than 20 to 30% by weight of cement. Mehta (1988), working with particular materials and test conditions, achieved a critical spacing factor of 400 μm for concrete with and without silica fume. Concrete durability depends heavily on mix preparation practices. The latter involves use of high-quality aggregates and water, mixing, placement, and cure.

Aggregates. It is desirable that aggregates which are likely to undergo physical or chemical changes should be avoided and that natural sand, gravel, or crushed rock that conform to ASTM C33, or a similar standard, should be used in concrete mix. Hard and strong aggregates are required particularly in tidal and splash zones, where resistance to heavy abrasion or erosion is usually required. The use of condensed silica fume in a concrete mixture improves the bond between aggregates and cement paste considerably, and consequently, the abrasion–erosion resistance of concrete. On the other hand, sandstones and quartz aggregates do not form strong chemical bonds with portland cement paste. Furthermore, coefficients of thermal expansion of sandstone and quartz are considerably higher than that of cement paste and therefore do not provide good abrasion–erosion resistance.

Water. In general, fresh water should be used for concrete mixes used for structural reinforced concrete. The chloride content of the water (and the mix) is an important factor in ensuring protection of reinforcing steel against corrosion.

Exothermic Reaction. The chemical reactions that cause concrete to cure are exothermic. Because concrete is not a good heat conductor, it is possible to generate high internal temperatures if the rate of heat generation exceeds the rate of heat dissipation. These high temperatures may impair concrete integrity. High temperatures may reduce the final achieved strength and cause concrete to lose its workability and encourages an excessive rate of evaporation from the surface. The latter way lead to plastic cracking. The temperature of mass concrete can be effectively reduced by partial replacement of portland cement with fly ash or granulated blast furnace slag (Bamfort, 1980). According to the FIP (1985), when the minimum thickness of concrete is greater than 600 mm and the cement content is more than 400 kg/m^3, use of a cement with a slow release of heat of hydration and other methods of controlling the temperature rise must be specified.

7.12.3.5 Concrete Preparation and Practices. The quality of concrete and its durability depends greatly on good construction practices, which involve batching, mixing, transportation, placement, consolidation, surface finishing, and curing. The batching sequence and methods should ensure the homogeneity of a concrete mixture. Properly mixed concrete may be transported to the site of placement by a variety of means (e.g., use of conveyors or trucks, or by pumping). If conveyed in trucks, a continuous agitation of concrete mix is usually required. When concrete is pumped, the forced pressure may in some instances cause the water to be absorbed into the aggregates, thus reducing the slump. High pressures may also affect entrained air. However, experience indicates that pumping, in general, is a reliable method of conveying concrete mixes. In all cases of concrete transportation, and especially in trucks or conveyors, it is important to preserve the homogeneity of the concrete during conveyance by avoiding segregation of its different composite materials.

When fresh concrete is placed against previously placed hardened concrete, to obtain a watertight cold joint or construction joint, the surface should be thoroughly prepared by sandblasting or using a high-pressure water jet. All dirt, foreign materials, and laitance should be removed and the coarse aggregate exposed. The indentation is considered sufficient when coarse aggregate is exposed to a depth of 6 to 10 mm. Subsequent concreting should start with a lift of a regular mix but not containing coarse aggregate, followed by regular mix. The two should be vibrated so that the first lift is well penetrated by the second. A good bond between previously placed and newly installed concrete is achieved by covering the hardened concrete with a thin layer of epoxy-modified concrete slurry just before placing new concrete. The latter provides a good watertight cold joint. The good-quality watertight joint can also be obtained through use of specially formulated mortar mixtures. Good-quality waterweight cold joints are especially important in reinforced concrete structures.

Concrete must be properly consolidated when placed. Even superplasticized "flowing concrete" should be vibrated, because nonvibrated concrete, particularly reinforced concrete, has a tendency to entrap 3 to 5% air, thereby reducing the bond strength and prompting settlement cracks around the reinforcing steel (Forssblad, 1987). Standard guidelines for concrete consolidation is proved in ACI 1201R, 311-4R, and 309R. When placing concrete in hot weather (above 30°C) or in cold weather (below 5°C), certain procedures must be followed. For details, the reader is referred to guidelines in ACI 305R, 306R, or similar codes.

Furthermore, it is worth noting that pozzolans and silica fume make the concrete adhere to forms. Therefore, if use of these admixtures is required, special measures should be taken to reduce adherence of the concrete to the forms. Surface finishing plays a significant role in good concrete performance; a smooth finish (i.e., a lack of voids and/or other surface defects) helps to improve concrete permeability and can greatly improve its resistance to abrasion. Freshly placed concrete should be cured properly. Curing is a process of maintaining proper humidity and temperature for a certain period of time after concrete mixture placement.

The curing process is essentially protection of fresh concrete against loss of moisture; it is associated with a supply of curing water externally. Where the ability to keep a concrete surface wet for the entire curing period is in doubt, or where there is a danger of thermal cracking due to too-cold curing water, a heavy-duty membrane curing compound should be used to seal the concrete surface (Mehta, 1988; Seanlon, 1997). In this regard, Gerwick (1986) warns that with membrane curing compounds; heat from the sun or cement hydration may degrade the curing compound; therefore, one or more additional applications may be required during the first day. For proper maintenance of humidity and temperature, FIP (1985) recommends that in cold weather the concrete should be at least 5°C at the time of placement and should be maintained above this temperature until it has reached a minimum strength of 5 MPa. The temperature may be raised by heating the mixing water and/or the aggregates. However, the temperature of water should not exceed 60°C by the time of contract with the cement. The exposed surface of fresh concrete should be moisturized with fresh water; seawater should not be used for curing of reinforced concrete, at least at the initial stage. Some data recommended by leading North American and European standards for design and construction of concrete offshore structures are presented in Table 7.3.

7.12.3.6 Protective Membranes. To minimize the ingress of chloride ions, the complete concrete structure operating in the marine environment may be protected by surface waterproofing membranes. Several traditional and relatively new types of protective membranes are currently used in marine structures engineering: hot-applied coal tar, coal tar epoxies, vinyl, polyurethanes and polybutadiene rubber coatings, chlorinated rubber, and others. Hot-applied coat tar and coal tar epoxies are probably the most common types of membranes used for the protection of metals and concrete, either with or without the aid of cathodic protection systems. Coal tar epoxies are known for their longevity and very good resistance to damage due to handling and abrasion.

Polyurethanes and polybutadiene coatings are typically specified where a concrete surface needs protection against abrasion, chemicals, and general abuse. Oshiro and Tanigawa (1988) reported that in Japan there is a growing tendency for use in marine environment of highly elastic coatings such as acrylic rubber. Any kind of protective membranes can be affected by ozone and ultraviolet rays. Hence, before specifying any of the protective membranes, the weathering effect of the foregoing agents and impermeability to chloride have to be examined carefully. When specified, membranes should be placed according to the manufacturer's instructions against a thoroughly prepared concrete surface. The selected material and method of application must minimize the development of blowouts and pinholes due to water vapor pressure from underneath the coating. However, if they occur, a repair procedure should be considered.

Finally, it should be noted that in addition to the challenges of nature, an environmentally conscientious engineer has to have a clear understanding of the potential toxic effect on the environment of different kinds of protective membranes. However, it should also be recognized that pressure by environmental concerns may sometimes be grossly overblown. Therefore, an engineer always needs to keep an open mind when designing marine structures, particularly when selecting protective membranes. For more information on the use of concrete in marine environments, readers are referred to Tsinker (1995, 1997).

7.12.3.7 Reinforcing Steel. Reinforcing steel can be conventional, prestressed, or posttensioned. Conventional reinforcing steel is best represented by regular deformed bars of miscellaneous grades. It may be used bare or coated in a variety of ways. For prestressed or posttensioned steel, high-tensile-strength bars or single- or multistrand seven-wire tendons are typically used; the former are generally used for short-length structural elements, and the latter for long-length structural elements. The use of posttensioned tendons has increased dramatically in all fields of reinforced concrete construction during the past several decades. The posttensioned tendon assembly is usually composed of steel strand(s), with anchors affixed to each end by toothed wedges, coated with corrosion-resistant lubricant, and encased in sheathing. The tendons are placed within sheathing, stressed and anchored, and then grouted. This assembly is typically combined with deformed reinforcing bars encased in concrete to complete the total reinforcing system. The ducts should be stiff enough to prevent local sag. They must also be tied to conventional reinforcing steel to prevent displacement from the design location by concrete mix during placement or consolidation. Ducts should be

Table 7.3 Concrete for marine structures

	ACI (1984)	FIP (1985)
	Material	
Cement composition	C_3A content of portland cement should not be less than 4%, to provide protection for the reinforcement. The maximum C_3A should be limited to 10% to obtain resistance to sulfate attack.	In the splash zone and atmospheric zone, portland cements with moderate C_3A content are recommended. Rapid-hardening cements should be used only for repair. Low-heat-of-hydration cements are preferred for structures with heavy dimensions.
Admixtures chlorides	No chlorides should be added intentionally. Total water-soluble chloride ion or a concrete mixture from all the component materials should not exceed 0.1% by weight of cement for normal reinforced concrete and 0.06% for prestressed concrete.	$CaCl_2$ or admixtures containing more than 0.1% chloride by weight of cement should not be used.
Chemical and air training	Where freeze–thaw durability is required, the concrete should contain entrained air as recommended by ACI 201.2R.	Air-entraining agents, workability aids, and retarders are often essential to obtain optimum mix design, but precautions should be taken to evaluate the side effects of each admixture type before use.
Pozzolanic	Pozzolans conforming to ASTM C618 (only natural pozzolans and fly ash are covered) may be used provided that tests are made to ascertain their relative advantages and disadvantages, especially with regard to sulfate resistance, workability, and corrosion of steel.	High-quality pozzolanic materials, such as special silica fumes, may be added to produce improved strength, durability, and workability.
Aggregates	Natural sand and gravel, or crushed rock conforming to ASTM C33, and lightweight aggregate conforming to ASTM C330. Marine aggregates may be used, provided that they have been washed to meet the chloride ions limits. No limits on maximum aggregate size are given.	Aggregates likely to undergo physical or chemical changes in concrete are to be avoided. Marine aggregates should not be used unless the chloride content is at an acceptable level and unless the aggregates have a sufficiently low seashell content.

	Specification	
28-day compressive strength minimum	35 MPa for all zones; 42 MPa where severe surface degradation is likely.	32 MPa for all zones, 36 MPa where abrasion resistance is required.
Water/cement ratio maximum	0.45 for the submerged zone, and 0.40 for the splash and atmospheric zones.	0.45 maximum, but 0.40 is preferred.
Cement content minimum	355 kg/m^3.	320 and 360 kg/m^3 for 40 and 20 mm maximum aggregate size, respectively; 400 kg/m^3 for the splash zone.
Consistency	No requirement.	No requirement.
Permeability minimum	No requirement.	No requirement.

protected against accidental entry of foreign materials (e.g., miscellaneous debris, concrete aggregates, dust).

Watertight galvanized sheaths are normally specified for the construction of posttensioned concrete marine structures, and in general, use of plastic sheaths should be avoided. Rigid metal ducts are generally used for long tendons, as the frictional characteristics of this type of sheath have proved to be favorable. The vents in sheathing are usually provided at all high points in tendon profile. In some instances, tendons may be installed into preformed holes. The latter may be used only on short tendons because during tendon friction, losses in concrete posttensioning are relatively high. Tendons should have smooth curves between designated high points and low points and have smooth sweeps around blockouts and sleeves.

Seven-wire strand (12.7 to 15.2 mm in diameter) typically is made with high-tensile steel. The most commonly used steel has a guaranteed ultimate strength of approximately 1860 MPa. Because stress–strain data are usually not available for each shipment of tendons, according to ASTM A416, one test should be considered for each 20-ton lot. The entrance to the ducts should be funneled, smooth, and abrasion-free. The strand is normally coated. The coating, sometimes referred to as *grease,* is a high-quality organic protective medium intended to provide lubrication and to form part of the corrosion protection system. Alternatively, a vapor-phase inhibitor powder may be dusted onto the strand. Anchor castings are made of durable iron and typically provide about 71 and 103 cm^2 of bearing area for use with 12.7- and 15-mm-diameter strands, respectively.

The tendon jacking procedure is carried out according to a predetermined sequence in order to equalize the effects of friction along the bench. Tendons are stressed with a hydraulic jack equipped with a calibrated pressure gauge. After the tendons are tensioned, they should be grouted as soon as possible. It is important to ensure that grouting tubes, hoses, ports, and valves do not leak during grouting. Leaks may result in plugging of the system and trapping of water in the sheath. Grout is usually introduced at a pressure of 0.5 to 1.0 MPa. During grouting, vents should be closed and the grout forced out through the strand end. The grouting mix should be selected according to standard practice for minimum bleed. The type of cement used in the grout should be the same type used in the basic concrete; the difference in the electrolytic properties of the cement in the grout and basic concrete may result in deterioration of the prestressing steel.

Shop prefabrication of reinforcing steel is a great benefit to the construction industry in general and to the construction of marine structures in particular. Prefabrication that is done under factory conditions and prefabrication of prestressed concrete elements in particular ensures better quality of steel fabrication while shortening construction time, requiring less on-site skilled labor and supervision, and ensuring easy inspection. It is extremely efficient where a great deal of repetition is required. However, whereas prefabrication gives all the advantages noted above, the cost of transportation can be high. Therefore, in some cases, well-organized on-site prefabrication may be a cost-effective alternative.

7.12.3.8 Concrete Cover over Reinforcing Steel. As pointed out earlier, corrosion of reinforcing steel in concrete structures exposed to seawater has often been observed to induce cracks and promote concrete deterioration. To prevent reinforcing steel corrosion, care must be exercised in not only in the design, mixing, and placing of the concrete mix, but also to provide the reinforcing steel with a proper concrete cover. In general, the concrete cover over reinforcement should be kept to a minimum because too thick a cover can lead to excessive cracking.

Current practice in the North Sea is to provide a 50-mm concrete cover over conventional reinforcement and a 70-mm cover over prestressed steel. In the United States, ACI 377R (ACI, 1984) stipulates the following concrete cover for reinforcing bars: 50 mm in the atmospheric zone not subjected to salt spray, 65 mm in the splash and atmospheric zones subjected to salt spray, and 50 mm in the submerged zone. The concrete cover over prestressed or posttensioned tendons must be increased by about 25 mm above these values. The Russian Code SNiPII-I.14 calls for a 70-mm cover over prestressed tendons, 50 mm over conventional bars, and a minimum 30 mm over secondary reinforcement. The code also allows for a 10-mm reduction in concrete cover for prefabricated concrete elements. For the tidal zone of offshore production platforms, FIP (1985) recommends a 75-mm cover over conventional reinforcement and 100 mm over prestressed steel. To date, concrete covers as low as 35 to 40 mm have been successfully used on many prestressed concrete piles. Even smaller concrete covers, as low as 10 mm, have been used successfully on concrete ships and pontoons, where special care was taken to achieve a dense cement paste cover.

7.12.3.9 Protection of Reinforcing Steel from Corrosion. A practical reality is that despite good design and construction practices, corrosion of reinforcing steel in concrete marine structures can occur. In current construction practice, to limit the chances of steel corrosion, reinforcing steel used in concrete marine structures is usually protected in one way or another. In the modern practice of marine construction, reinforcing steel is usually protected by fusion-bonded epoxy coatings, galvanization, and in some cases by cathodic protection. *Epoxy-coated steel* is one of the many innovative concepts initiated by the U.S. Federal Highway Administration in the 1960s to fight serious problems of reinforcing steel corrosion in bridges. In current construction practice, epoxy-coated reinforcement has become fairly well standardized. In the United States in 1981, the ASTM standard specification for epoxy-coated reinforcing steel bars was issued. A similar standard for epoxy-coated welded wire fabric was developed in 1988.

Epoxy coating is typically applied to the reinforcing steel by a fusion-bonding process. A range of thickness of fusion-bonded epoxy coating between 130 and 300 μm has been found to be very effective. Pullout tests for bond strength have indicated that the best performance may be achieved when the coating thickness does not exceed 300 μm (Read, 1987). Epoxy-coated bars should be handled with care to prevent the coating from being damaged during shop bending, installation, and placing of concrete. However, despite the good protection provided by epoxies, there are occasional reports on corrosion damage to coated rebars in the marine environment attributed to debonding of the coating (ACI, 1988).

It has been found that even a small slip of coated bar can result in disruption of epoxy coating. This implied that with any significant crack size, epoxy on either side of the crack may be subject to surface damage. More useful information on the performance of epoxy-coated rebars is given in Cairns (1994). Another way to protect the reinforcing bars from corrosion is galvanization. However, despite the fact that, in general, it does provide good protection from corrosion, it is not always satisfactory in a splash zone.

Galvanized reinforcing bars form a good bond with concrete. Zinc coatings are relatively ductile, and therefore, galvanized bars normally can bend without cracking. Damaged galvanized bar can be repaired easily by simple application of zinc-rich paint. Because of the sacrificial nature of zinc, neither uncoated reinforcing steel nor any other embedded metal dissimilar to zinc should be permitted in the same concrete element with galvanized bars, except as part of a cathodic protection system. For the same reason, galvanized bars must not be coupled electrically to uncoated bars. It is generally accepted that under relatively short-term (say, less than 35 years) exposure to a marine environment, galvanized and epoxy coatings are equally effective in corrosion protection. Therefore, a choice between the two is normally based mainly on cost and availability. From economic reasons, coated bars are usually installed at the exposed side of a structure in combination with uncoated reinforcement installed in corrosion-protected areas.

Fiber reinforcement is typically used as a secondary reinforcement system intended to mitigate problems associated with intrinsic cracking. Fiber reinforcement has been used commercially now for more than two decades. The greatest advantage of fiber-reinforced concrete (FRC) is its toughness and resistance to both shear and erosion. Fiber reinforcement, sometimes called *fibermesh,* provides an internal restraining mechanism that stabilizes intrinsic stresses, particularly during the first 7 days, when concrete is most vulnerable to shrinkage cracking. Unlike continuous regular or stressed reinforcing bars or tendons, fibers are discontinuous and generally are randomly distributed throughout the concrete mix. In general, fibers are not likely to replace conventional reinforcement. However, in cases such as precast fiber-reinforcement panels used for permanently installed formwork, rehabilitation, retrofit, and repair work of deteriorated marine structures and other relevant applications, fiber-reinforced concrete may be a feasible and economical alternative to regular reinforced concrete.

A great variety of fiber materials in various sizes and shapes have been developed for use in FRC. Currently, steel, glass, polymeric, and carbon fibers are commonly used. Fiber comes in many shapes, with circular, rectangular, semicircular, and irregular or varying cross sections. A convenient numerical parameter used for describing a fiber is the *aspect ratio,* defined as the fiber length divided by the diameter (or equivalent diameter in the case of a noncircular fiber). Typical aspect ratios range from 30 to 150 for length dimensions of 6 to 75 mm. Fibers may be straight or bent and come in various lengths, usually as fabricated bundles. The mixing action causes the engineered bundles to open and separate into millions of individual fibers distributed uniformly throughout the mix in a multidirectional pattern. As the concrete starts to harden, many microscopic cracks begin to form. At this stage, these cracks are intersected by the fibers, which halt their growth. Numerous studies have confirmed that a substantial reduction in fiber concrete cracking occurs due to drying shrinkage. Because failure of fiber-reinforced concrete is by fiber pullout, the use of deformed fibers increases the concrete's pullout strength and, consequently, its mechanical properties. The concrete used in a mixture with added fibers is of the common type, although the proportions should be such as to obtain good workability and take full advantage of the fibers. Corrosion of steel fibers in concrete with a high water/cement ratio may result in concrete deterioration. However, in actual applications, corrosion of carbon steel fibers has been found to be minimal.

A major disadvantage of glass fiber is its high vulnerability to deterioration in the alkaline cementations environment because it seriously damages the long-term properties of glass fiber–reinforced concrete. The application of glass fiber is quite complex. To date, the major application for glass fiber has been the spray-up process, in which glass fibers and a cement-rich mortar are sprayed on a surface simultaneously. The major shortcomings of polymeric fiber are low modulus of elasticity, poor bond with cement matrix, combustibility, and low melting point. A major difficulty in the use of FRC is the reduction in workability of concrete mix caused by the addition of fibers, especially when large quantities of fiber are required. This shortcoming may, however, be effectively eliminated by the use of superplasticizers.

7.12.4 Floating Pier Structural Design

7.12.4.1 Basic Design Considerations. A concrete floating pier is usually built in the form of a closed box type of cellular structure. Two basic structural systems of concrete pier hull are usually considered:

1. A cellular system with framing consisting of a modular arrangement of longitudinal and transverse watertight bulkheads (Figure 7.64*a*)
2. A system consisting of a double-walled hull with a stiffening of framelike longitudinal and transverse walls (Figure 7.64*b*)

In both cases, the pier structure must meet the following criteria:

1. It must be provided with a sufficient number of watertight cells to satisfy the damage stability requirements; the pier must survive and remain afloat and stable should any two adjacent cells be punctured below the waterline.
2. To minimize construction time, the amount of cast-in-situ concrete should be reduced to a minimum. Use of plant-built precast concrete panels should be maximized. To ease the erection process and speed pier construction, the number of

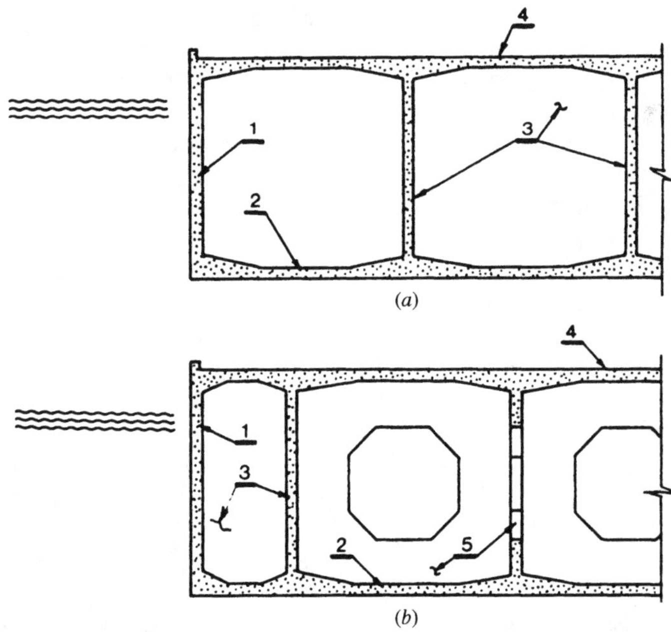

Figure 7.64 Typical framing systems of concrete pier. 1, External wall; 2, bottom slab; 3, watertight bulkhead; 4, deck; 5, framelike (or truss) bulkhead.

panels of different sizes and shapes must be kept to a minimum.

3. Construction must be simplified by allowing for as much repetitious use of formwork as possible. To promote simplicity and economy of construction, the hull structure has to be designed as a system of plate elements. It should be noted that in some cases, economy of material use is achieved by providing haunches in otherwise prismatic plate elements. These occur near the supports, which in effect distributes the load-carrying capacity of the plate elements in the same manner as for imposed moment distribution.
4. The hull shape developed must permit pier construction in the available drydock. It must possess sufficient buoyancy to permit its safe launch, delivery, and operation.
5. The pier must be robust, adequately stable, and require minimum maintenance.

Implicit in each of these criteria is a requirement of low cost. A low construction cost and minimum maintenance requirements must be the principal criteria in selection of a pier structural system.

There are two reinforcing systems that are normally used to provide a pier with sufficient strength and serviceability. The prime system of hull reinforcing is represented by posttensioning tendons. Posttensioned tendons usually control hull strength in the longitudinal direction (Figure 7.26) and can also be very useful for hull prestressing in the transverse direction. The secondary reinforcing system, in the form of regular deformed bars (ordinary or prestressed), is used in discrete areas where additional load capacity is required. The reinforcing steel must be protected from corrosion by proper construction details and construction practices. Reinforcement located in areas exposed to salt water should be given an appropriate concrete cover, which should comply with local codes and regulations. However, it should be no less than 50 mm for ordinary reinforcement and 70 mm for prestressed tendons. Adequate steel should be provided to control the temperature and shrinkage cracking. Most important of all is use of a good, sound, durable concrete mix. Minimum concrete strength should not be less than 30 MPa at 28 days and air entrainment should be provided.

For hull structural design the requirements stipulated in relevant codes and regulations, applicable to specific conditions, must be considered. Concrete floating piers usually are designed in accordance with the working stress method. However, a pier can also be designed according to the ultimate strength and serviceability design method. It is very important that the designer understand the kinematics of pier performance. Pier global bending creates tension and compression in its longitudinal and transverse members. If these exceed allowable limits, then tensile stresses may lead to through-thickness cracking of the pier hull. In general, the latter is unacceptable in a pier structure, especially in a pier's underwater zone.

Hull subdivision into watertight compartments (cells) is an important stage of pier design. Essentially, this is a trial-and-error procedure. The number and size of longitudinal bulkheads along with the shape and size of both the deck and bottom plate determine the hull section modules, which in the final analysis govern the hull's overall strength. On the other hand, transfer bulkheads are typically spaced to provide the most economical design of the pier shell (i.e., sidewalls, deck, and bottom plate). Transfer bulkheads perform a multitude of functions. They are designed to resist the torsional forces that may exist in a pier hull due to unsymmetrical load application to the pier deck and sidewalls. It is equally important that transfer bulkheads ensure a pier's structural integrity and its buoyant stability when damaged. The latter assumes that two adjacent cells formed by a transfer and longitudinal system of bulkheads are accidentally filled with water. Spaces between adjacent transfer bulkheads should also not result in excessively heavy sections of hull elements (e.g., sidewalls, deck, and bottom plate).

7.12.4.2 Hull Design

Forces. In general, a floating pier is subjected to the following loads:

- Deck load, which is generally similar to that used in design of fixed structures (for details, consult Chapter 2)
- Wave load of a dual nature (one that may exist at a deployment site or during transport of a pier from the site of construction to the deployment site)
- Local hydrostatic pressures, including dynamic effects on sidewalls and bottom plate
- Ship berthing impact transmitted to the hull via a pier fender system
- Loads of thermal origin
- Loads associated with pier construction, launch, and tow, if required, across a large body of water
- Environmental loads such as wind and waves; and where applicable, ice and current force must be considered

Pier structure must be sufficiently robust, able to resist the most adverse combination of the foregoing forces, as applicable, without failure and/or unacceptable deformations. The pier structural strength problem is both static and dynamic. Although floating docks usually operate at locations protected from heavy waves, they may sometimes be exposed there to rather significant waves (1.5 to 2.0 m in height). Hence, a complete study of pier structural strength should consider the effects of both static and dynamic forces, as required. If a pier has to be towed across the open sea, it can be exposed to heavy dynamic wave forces. These forces will cause the pier to bend in a longitudinal vertical plane like a beam. Hence, the pier must have sufficient structural strength to resist this type of bending and associated torsion. Furthermore, the pier can experience the transverse and local deformations that may arise from forces imposed on it during its voyage across the sea as well as at the deployment site.

The hydrostatic force on a floating body in still water is equal to the gravitational force, Mg, where M is the mass of the floating body and g is the acceleration due to gravity. For a common unloaded floating pier, the distribution of gravitational force per unit length is equal to the buoyancy force per unit length at every point on the pier. However, if the pier carries a load that varies along its length, the pier performance conditions may change. In this case the total hydrostatic force is still exactly equal to the gravitational force (this will now include the weight of the pier plus the live load); however, the distribution of hydrostatic force along the pier length is not necessarily uniform. The net force per unit length of the pier will be $bx - m_x g$, where m_x is the mass per unit length at any point on the pier and b_x is the buoyancy per unit length. Under these conditions, the pier structure may be represented by a beam carrying a load that varies along its length. Thus, the shear force for such a beam (Q) can be determined by integrating the load along its length:

$$Q_x = \int_0^L (b_x - m_x g)\, dx \tag{7.13}$$

where L is the pier length. On integrating a second time, the bending moment causing the pier to bend in a longitudinal vertical plane (M_x) will be obtained:

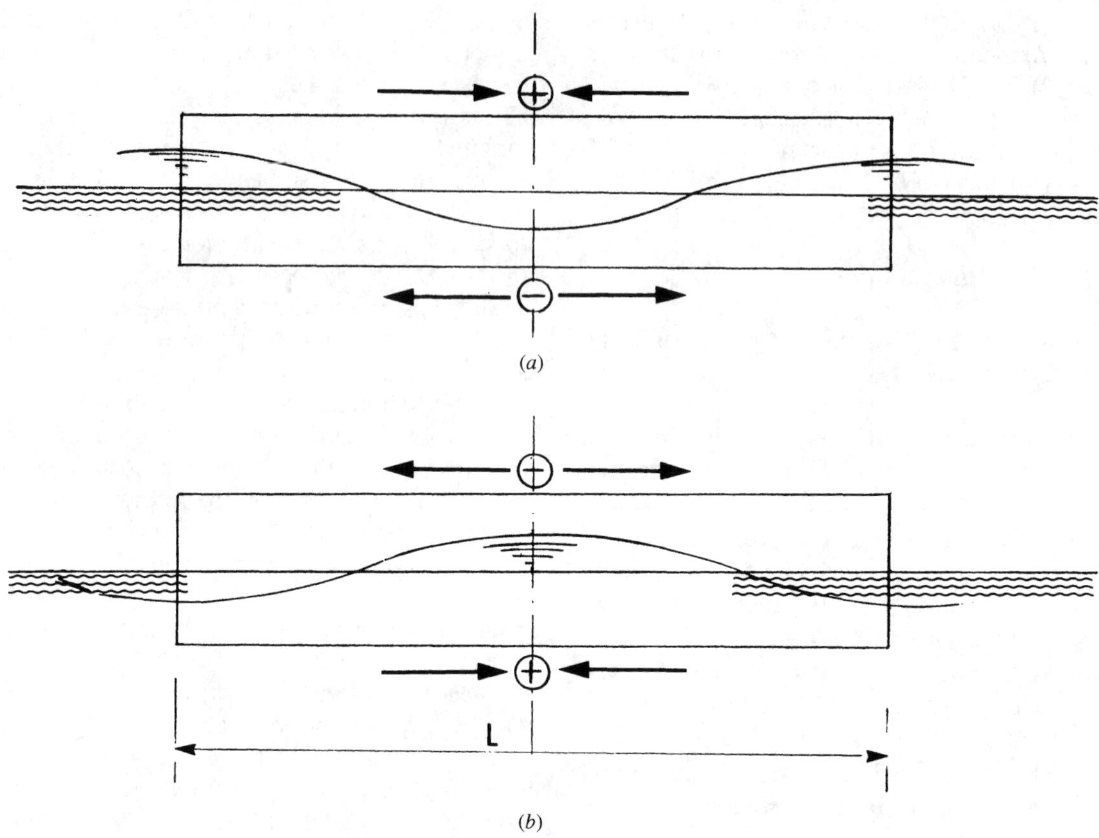

Figure 7.65 (*a*) Sagging and (*b*) hogging conditions of pier operation.

$$M_x = \int_0^x Q_x \, dx \qquad (7.14)$$

Since a pontoon's deadweight is usually distributed uniformly along the pier length and is balanced by the appropriate hydrostatic pressure, only live load on a pier deck and the corresponding hydrostatic pressure are used in the determination of Q_x and M_x. When a pier operates under wave conditions, the forces acting on it are different from those in still water. The alteration in pier static buoyancy due to variable immersion of its hull results in variable pressures. Pier motion in waves causes dynamic forces due to the accelerations involved. Despite the dynamic nature of pier–wave interaction, the traditional practice in naval architecture for the design of floating objects like piers is to reduce the quite complex dynamic problem to a much simpler quasistatic one.

The standard typically adopted procedure is to assume that a floating body is supported statically on a wave against both sagging and hogging conditions through calculation of the shearing forces and bending moments acting within a pier's hull. As illustrated in Figure 7.65, *sagging* occurs when the center of a pier (or any other floating object) tends to move vertically down relative to its ends, thus developing compression at the pier deck and tension at its bottom. *Hogging* is the reverse condition. Under each of these conditions, when there is a balance between the total downward forces and the total upward forces and the lines of action of these two forces coincide, the floating body is in equilibrium. Normally, an excess of buoyancy over weight at a certain portion of a floating body's length is counterbalanced by an excess of weight over buoyancy throughout the remaining portion. These unbalanced forces create the same conditions of shear and bending moments as in an ordinary beam.

Possible sagging and hogging conditions are usually checked and the greatest values of shearing forces and bending moments determined. In more refined investigations, calculations would be made with the crest of a wave passing a floating object located at regular intervals along its length. Then the peaks of the bending moment curves obtained for sagging and hogging conditions would be connected into two curves for maximum conditions. A complete description of the quasistatic approach to bending moment and shearing force calculations is given in MacNaught (1967).

As knowledge of wave-induced bending moments and their relation to total bending moments increased, a new design procedure was developed for ship design and adopted by classification societies (American Bureau of Shipping, 1982; Society of Naval Architects and Marine Engineers, 1980; and others). A number of empirical formulas with various degrees of accuracy are now available for the determination of still-water and wave-induced bending moments and shearing forces.

As noted earlier, a floating pier of rectangular shape has its weight uniformly distributed along the hull, and due to this condition, the bending moment and shear forces are zero in an unloaded pier hull in still water. Subsequently, wave-induced forces and live load only are considered in pier hull design. To simplify the design process, the quasistatic approach to sagging and hogging conditions in pier design is generally used. It is obvious that knowledge of wave dimensions such as length (L_w) and height (H_w) is very important in pier design.

In the shipbuilding industry, the design wave height (H_w) is calculated as the function of ship length (L). The simplest formula of this kind assumes the wave height to be $\frac{1}{20}$ of the ship length (L) (i.e., $H_w = \frac{1}{20}L$. According to Muckle (1975), $H_w = 0.607(L)^{.05}$ in meters. At one time, the latter equation was considered to be more-or-less representative practical proportions of wave height to ship length in actual sea waves. The current American Bureau of Shipping (ABS) rules recommend the following empirical equations for the determination of wave height as a proportion of ship length:

$$H_w = 0.0172L + 3.653 \quad \text{for } 61 \le L \le 150 \text{ m}$$

$$H_w = 0.081L + 3.516 \quad \text{for } 150 < L \le 220 \text{ m}$$

$$H_w = (4.5L - 0.0071L^2 + 103)10^{-2} \quad \text{for } 220 < L \le 305 \text{ m}$$

However, the ABS rules do not recommend that these values of H_w be used in quasistatic wave moment calculations.

For comparison, the three approaches to calculating wave height discussed above are plotted in Figure 7.66. Naturally, the wave heights (H_w) given in the formulas above represent the maximum values used in the shipbuilding industry. However, in floating pier design, use of the H_w values presented above may be justified only if a pier has to be towed across a large body of water. Where the pier is exposed to wave action at the deployment site, the actual wave height in the harbor usually is either known or can be established by standard methods. Under harbor conditions, wave height normally does not conform to the proportions presented above. In a harbor, a pier normally

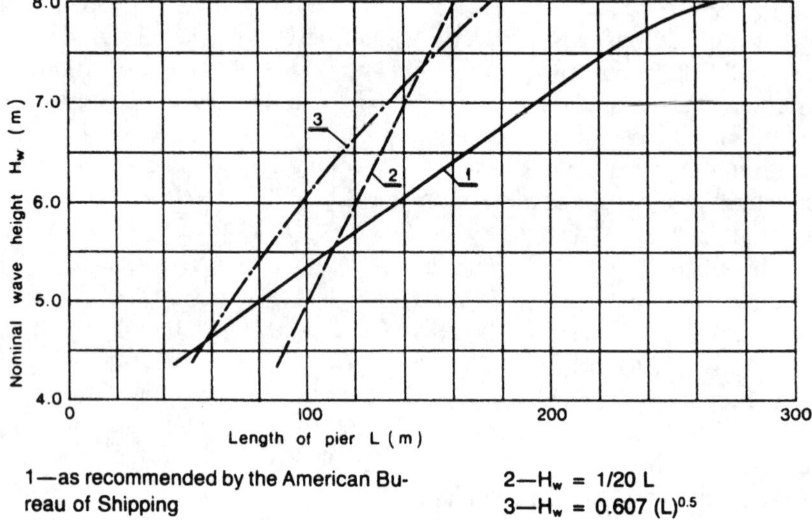

1—as recommended by the American Bureau of Shipping

2—H_w = 1/20 L

3—$H_w = 0.607 (L)^{0.5}$

Figure 7.66 Comparison of wave height calculated by different empirical methods as a function of ship (pier) length.

will be poised on a passing wave as a continuous beam rather than as a single span beam (a condition that may exist in the open sea). Both design situations are illustrated in the following examples.

Example 7.1 A concrete pier of constant rectangular cross section 90 m long, 15 m wide, and 5 m deep is to be towed across a sea (Figure 7.67). The total mass of the 3400-tonnes pier is distributed uniformly over its length and therefore constitutes a load of

$$\frac{3400}{90} \times 9.81 = 370.6 \text{ kN/linear meter of pier}$$

In this case the pier dead weight is the only load imposed on it during its journey across the sea. Assume that the pier is poised on a wave equal to its length ($L = L_w$). Therefore, the design wave height determined from $H_w = 0.607(L)^{0.5}$ is 5.76 m. To simplify the calculations, assume that the wave is of cosine form instead of being a trochoid, in which case the wave height (h_w) at any point above the still-water level will be given by the standard equation

$$h_w = \frac{H_w}{2} \cos \frac{2\pi x}{L}$$

or for the conditions assumed by

$$h_w = 2.88 \cos \frac{\pi x}{45}$$

Then the buoyancy per meter of pier under wave conditions

$$b_x = 1.025 \times 15 \times 2.88 \times 9.81 \times 10^{-3} \times \cos \frac{\pi x}{45}$$

$$= 0.434 \cos \frac{\pi x}{45} \quad \text{MN/m}$$

In this equation, 1.025 stands for the density

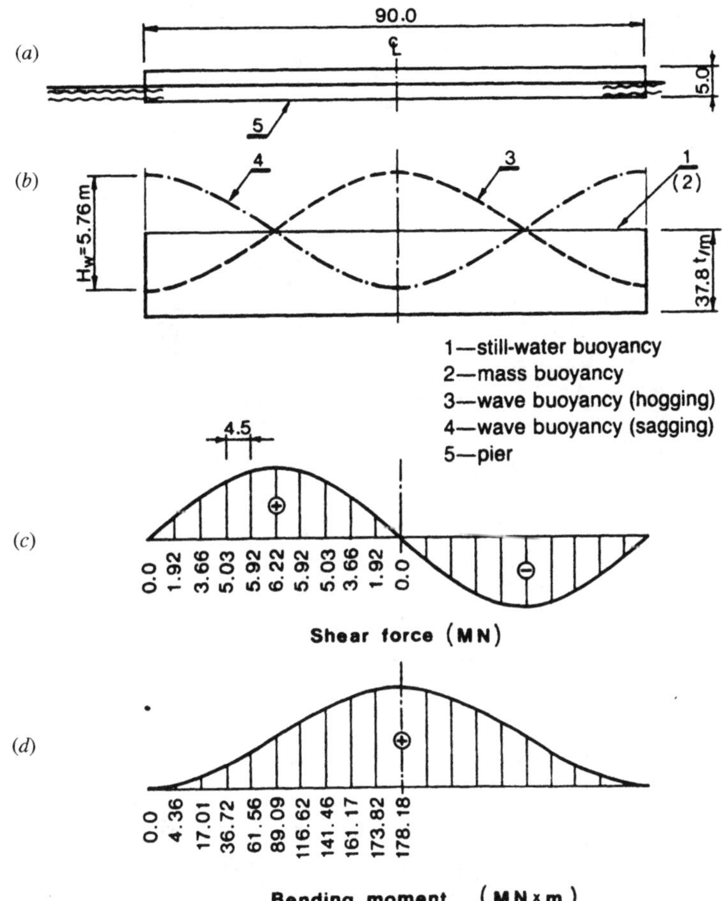

Figure 7.67 Example 7,1. Wave-induced sagging shearing force and bending moments. (*Note:* Wave-induced hogging shearing force and bending moments numerically are equal to sagging conditions, but, with the opposite sign.)

of seawater in tonnes/m³. Therefore, the wave shearing force will be determined from the equation

$$Q_{w(x)} = \int 0.434 \cos \frac{\pi x}{45}\, dx$$

$$= \frac{0.434 \times 45}{\pi} \sin \frac{\pi x}{45} + A$$

Because the shearing force is zero at $x = L/2$, $A = 0$. Hence,

$$Q_{w(x)} = 6.222 \sin \frac{\pi x}{45}$$

The wave-induced bending moment

$$M_{w(x)} = \int 6.222 \sin \frac{\pi x}{45}\, dx$$

$$= -\frac{6.222 \times 45}{\pi} \cos \frac{\pi x}{45} + B$$

Because the bending moment is zero at $x = 0$,

$$B = \frac{6.22 \times 45}{\pi} = 89.09$$

Therefore,

$$M_{w(x)} = 89.09 \left(1 - \cos \frac{\pi x}{45}\right)$$

At $x = 45$ m, the wave-induced bending moment at midpier is $M_w = 178.18$ MN · m sagging. The hogging bending moment in this case would be the same numerically but with opposite sign. Diagrams of shearing forces and bending moments are shown in Figure 7.67.

Example 7.2 The concrete pier in Example 7.1 is deployed at a protected location (Figure 7.68). The maximum height of the design wave in the harbor is 1.5 m. Assume that the length of the wave is $L = 20H_w$, which corresponds to a wave length $L = 30$ m. The pier is loaded with concentrated loads as shown in Figure 7.68. In this example, the midpier shearing force and bending moment will consist of load- and wave-induced forces. The buoyancy distribution curve due to a concentrated load only has an ordinate of 7.1 tonnes/m (see Figure 7.68). The bending moment at midpier due to the concentrated loads is obtained as follows. The load-induced bending moment, M_L is

$$M_L = -(200 \times 40 + 120 \times 30)$$
$$= -11{,}600 \text{ tonnes/m}$$

The buoyancy-induced bending moment, M_B, is

$$M_B = \frac{7.1 \times 45^2}{2} = 7188.75 \text{ tonnes/m}$$

Hence, the bending moment is −4,411.25 tonnes/m, which must be multiplied by $g = 9.81$ m/s^s to convert it to SI units. Hence, the net bending moment, M, is

$$M = -4{,}411.25 \times 9.81 \times 10^{-3}$$
$$= -43.27 \text{ MN/m}$$

This is a hogging moment.

Again, as in Example 7.1, assume that the wave is of cosine form, in which case the height at any point above the still-water level is given by the expression

$$H_w = \frac{1.5}{2} \cos \frac{2\ \pi x}{30} = 0.75 \cos \frac{\pi x}{15}$$

Then, the buoyancy per meter of pier under wave conditions is

$$b_x = 1.025 \times 15 \times 0.75 \times 9.81$$
$$\times 10^{-3} \times \cos \frac{\pi x}{15}$$
$$= 0.113 \cos \frac{\pi x}{15} \text{ MN/m}$$

Here, 1.025 represents the density of seawater in tonnes/m^3. By integrating the buoyancy, the wave-induced shearing force will be obtained as follows:

$$Q_{W(x)} = \int 0.113 \cos \frac{\pi x}{15}\, dx$$
$$= \frac{0.113 \times 15}{\pi} \sin \frac{\pi x}{15} + A$$

Here again, $A = 0$ and $Q_{w(x)} = 0.54 \sin (\pi x/15)$. The wave-induced bending moment is calculated as

$$M_{w(x)} = \int 0.54 \sin \frac{\pi x}{15}\, dx$$
$$= -\frac{0.54 \times 15}{\pi} \cos \frac{\pi x}{15} + B$$

At $x = 0$ ($M_w = 0$),

(a)

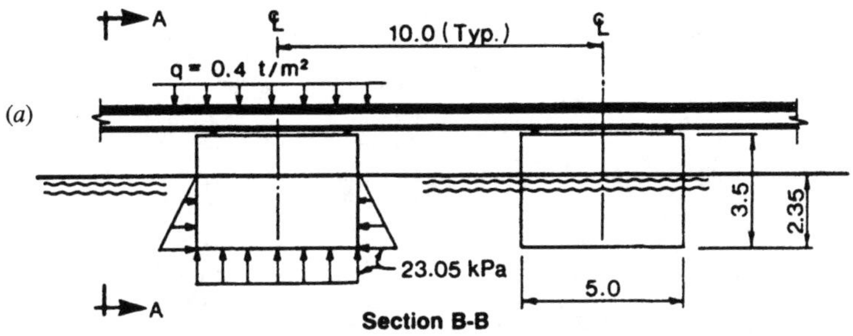

(b)

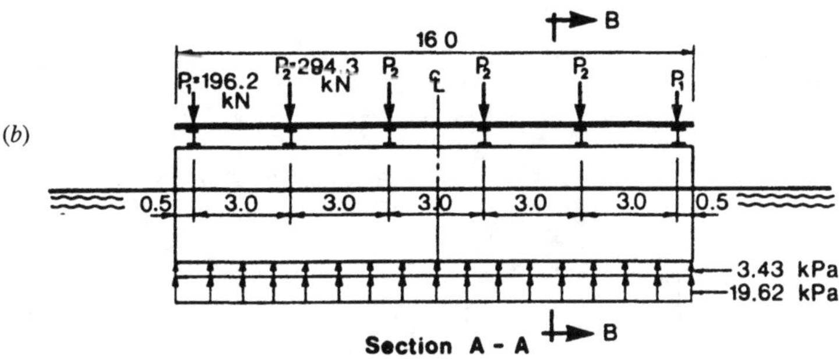

(c)

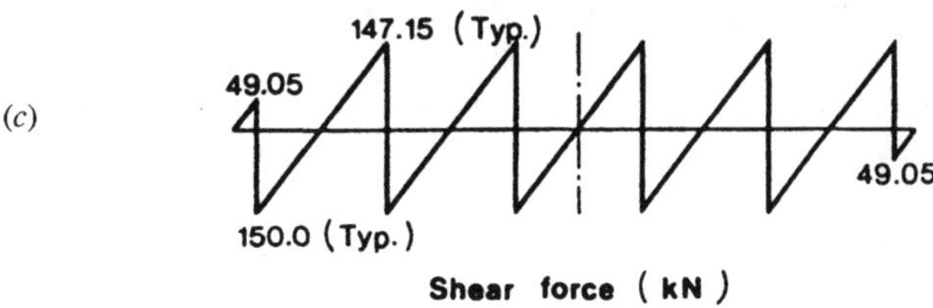

(d)

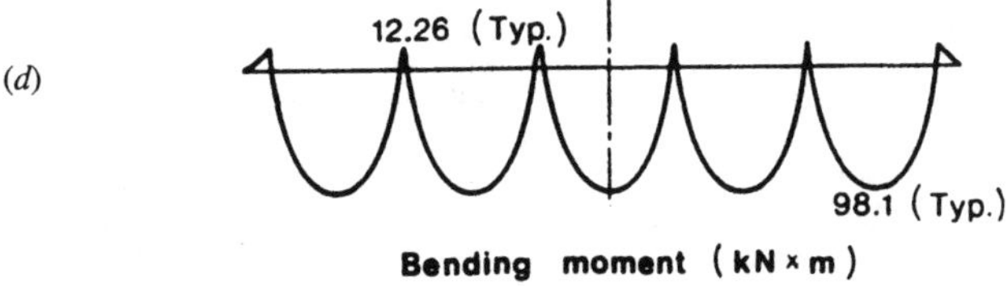

Figure 7.68 Example 7.2. Wave- and live load-induced bending moments and shearing forces in a pier deployed in a protected harbor.

$$B = \frac{0.54 \times 15}{\pi} = 2.58$$

and therefore,

$$M_w \times 2.58 \left(1 - \cos \frac{\pi x}{15}\right)$$

At midpier ($x = 45$ m), the wave-induced bending movement, $M_{w(x)}$, is $M_w = 5.16$ MN/m. Finally, the total bending moment at midpier, M is $M + M_w = 43.27 + 5.16 = 48.43$ MN/m.

Bending moments at any pier location could be investigated according to the procedure described above. From both examples it follows that in this particular case, voyage conditions rather than pier operation conditions govern pier design.

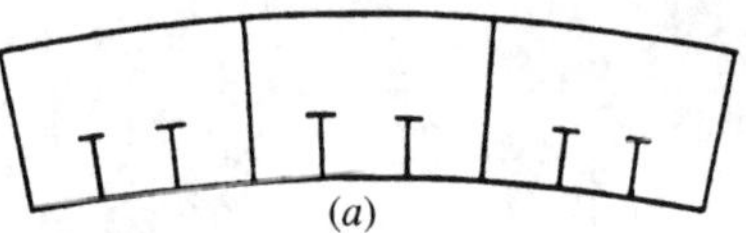

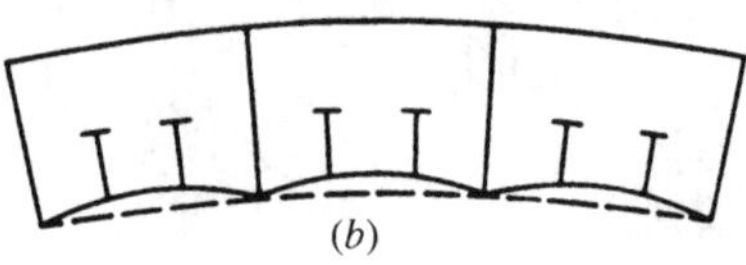

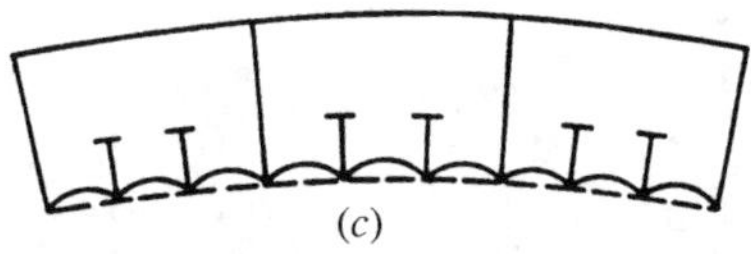

Figure 7.69 Pier structure response to outside forces: (*a*) primary; (*b*) secondary; (*c*) tertiary.

Pier Response to Outside Forces. A pier's structure is designed to resist the forces imposed on it by responding as a hull–girder structure. The latter includes calculation of stresses in pier structural components as related to its overall and local deflections (deformations). For simplicity, pier hull structural analysis (response to outside forces) can be divided into three phases (Figure 7.69):

1. *Primary response:* the response of the entire hull when bending as a beam under a longitudinal distribution of load (Figure 7.69*a*)
2. *Secondary response:* the stresses and deflections of a panel of stiffened plating (e.g., the panels of the bottom plate or deck structure, contained between two adjacent frames or bulkheads) (Figure 7.69*b*)
3. *Tertiary response:* the out-of-plane deflections and associated stress of individual panels of plating between stiffeners (Figure 7.69*c*).

The bending and shearing stresses and deflections corresponding to these three categories of structural responses are normally considered in pier structural design. Hull global deflections (δ) due to bending are found by double integration of the moment curvature equation. It may be expressed semiempirically as

$$\delta = C_\delta \frac{ML^2}{EI} \tag{7.15}$$

where E and I are, respectively, the module of elasticity and moment of inertia of the hull girder. The coefficient C_δ is often taken as 0.09 (Taggart, 1980). Where appropriate, any corrosion allowance should be excluded to obtain an "effective" cross-sectional pier area for calculating of I. Furthermore, it should be recognized that in actual service, a pier may be subjected not only to longitudinal bending, but also to other forces, such as forces that may induce side bending or torsion in the hull girder, not to mention the dynamic effects resulting from pier motion when under tow from construction site to deployment site. In this regard, it must be pointed out that in selecting the load combination to be used in hull girder analysis, the designer must not be overly conservative. He or she must use judgment based on the literature and results of recognized experiments to decide which are the most appropriate load combinations to use in a particular design.

In a long pier design, it is customary to extend the effects of the maximum bending moment on a pier hull over its midpier section in both directions. Accordingly, the midpier scantlings should be extended over the midpier section by $\frac{4}{10}$ its length in both directions. This will help to maintain the midpier section area at its full value in the vicinity of maximum shear force and will also allow for possible variation in the precise location of the maximum bending moment. In short piers, the midpier section is usually maintained unchanged through the full length of the hull.

Normally, all floating piers during their service life are subjected to fatigue produced by cyclic loadings. Therefore, the fatigue characteristics of a pier structure should be examined during the design stage. Experience indicates that under normal service conditions, fatigue-related problems are not really important with respect to the primary stresses in a hull-girder structure. This is due basically to the fact that the waves and/or other loads that produce the stress peaks in a hull girder do not occur often during the service life of a pier structure. Therefore, the endurance limit of a pier hull girder will normally not occur during a pier's service life under normal working conditions. As discussed earlier, pier bending deflection could be determined as for a common girder by assuming its particular loading. In addition to load-inflicted bending deflection, the actual pier deflection could be affected by thermal effects due to exposure to sun or cold temperature, although the amount is usually small.

Finally, consideration of pier strength and stability under damaged conditions is essential. Damage to a pier hull may occur due to accidental impact by incoming or departing ships. Under damaged conditions, a pier's stability could be severely distorted from that normal design conditions and therefore full appraisal the pier's its strength and stability in this state must be carried out during the design stage. Pier performance when damaged is an unusual working condition and from a design stress point of view it should be treated as an extreme loading condition. Accordingly, the allowable stresses in pier structures under damaged conditions must comply with the requirements of the most recent standards and regulations. The stability of a damaged pier is discussed in detail in Section 7.15.2.

7.12.4.3 Deck Design

Loads. The nature and magnitude of floating pier deck design live loads are similar to those used for the design of a fixed structure. The latter is discussed in detail in Chapter 2.

Similar to fixed marine structures, floating pier deck design loads are classified as follows:

1. *Permanent loads* include the weight of the deck structure and of permanently installed structures and fixed cargo handling equipment.
2. *Temporary loads* and forces include those related to effects of the environment and dock operation, including all kinds of *live loads* (surcharge, moving equipment), ship berthing and mooring forces, and so on.
3. *Special loads* include accidental loads, earthquake forces, and other unusual loads.

In deck design of a conventional floating pier, the loads and their combinations are distinguished as normal and extreme. *Normal load combinations* are those that could exist under normal dock operations. The most severe, realistically expected combinations of dead load and ship impact and mooring forces, loads generated by cargo handling and hauling equipment, and appropriate surcharge loads are usually selected as the normal design load combination for a given structure. Normal loading in combination with unusual forces such as an earthquake, an accidental load of a miscellaneous nature, or operation under damaged conditions is considered to be an *extreme load condition.*

Safe operation of any structure depends on the loadings subsequent to completion of construction. Because a pier is usually loaded after it is built and not always as designed, the selection of design loads is a problem of statistics and assessment of probability. In some cases, neither the owner nor the designer is sure what maximum loading the structure may be required to sustain in the course of its useful life. In addition, the loading selected is often subjective and depends largely on the designer's experience. To avoid divergent designs and to formulate reasonable design loads and load combinations, the designer should follow relevant guidelines, codes, and standards but also exercise common sense. *Reasonable design loads* are those that do not result in either prohibitive construction costs or structure obsolescence shortly after commissioning. The designer should, in close collaboration with the owner or dock operator, exercise great care in formulating design loads, particularly those generated by mobile cargo handling and hauling equipment. In deck design, concentrated loads such as outrigger reactions from large mobile cranes should be evaluated thoroughly. If heavy lifts by such cranes are seldom required, it is generally more economical to use specially designated strengthened areas of the deck for loading and unloading of such heavy loads, or to spread the outrigger load out with mats, steel beams, or by other means, rather than designing the entire deck to sustain heavy (however rare) concentrated loads.

As noted earlier, the deck of a floating pier is almost never used as a storage area, so a uniform distributed load is used only as a potential load situation. In some instances, however, a pier deck may be used as a short-term storage area for storing specific cargoes (e.g., stacked pallets, packed metal products). The intensity of such loads could be substantial, occasionally reaching a value of 50 kN/m^2 and above. If reasonably it can be expected, such a possibility must be taken into account in both pier structural design and in buoyancy and stability analysis.

Structural and Design Considerations. The deck is an integral part of a pier, and as a rule, the material used for deck construction is similar to that used for construction of the hull. Sometimes, however, a composite concrete–steel pier structure is considered. A deck structure typically consists of longitudinal girders and transversal beams, both of which use the deck slab (plate) as their common top flange. Being an integral part of a pier, the deck structure (similar to the pier bottom) is exposed to three basic types of deflections, as indicated in Figure 7.69: primary, secondary and tertiary. Accordingly, the deck components are exposed to longitudinal forces (tension or compression) due to pier hull-girder performance in combination with the effects of external loads.

A deck is a system in which all structural components are closely interrelated. For example, in a common concrete or steel deck structure a concentrated load applied to the deck between transverse beams (ribs) is first transmitted to the nearest transverse beams through the deck plate (slab). The transverse beams react on the longitudinal girders, which, acting as a continuous member, react on the transverse frames, trusses, or bulkheads that carry their load to the hull girder. In steel piers the deck structure usually is of orthotropic construction. The principal structural distinction between a common deck structure and an orthotropic structure is that in the latter, the main longitudinal structural beams have disappeared.

In an orthotropic deck, the closely spaced grid structure of the stiffened steel plate deck, with the ribs continuous and rigidly connected to each other at intersections, has a good load-distributing capacity for concentrated wheel loads which usually govern the deck design. Normally, the amount of steel used in orthotropic deck construction is less than it is in a deck of conventional design. As a percentage, however, the cost saving is much smaller than the saving in weight of steel because fabrication of an orthotropic deck is more difficult and therefore costly. For more details on orthotropic decks in floating piers, consult Tsinker (1986). The theory of orthotropic deck analysis design considerations may be found in any standard textbook concerned with steel bridge design and construction.

In a typical concrete pier, the deck structure is usually built in the form of a solid slab supported on a system of bulkheads, beams, and girders. To facilitate the construction process, the deck can be constructed of concrete precast panels supported on transverse bulkheads. Eventually, panels are joined together, and with bulkheads and sidewalls, posttensioned, thus forming a robust, continuous slab system. An example is shown in Figure 7.70.

The longitudinal joints in a deck system are typically made in the form of a large grouted shear key. Usually, there is no need to place reinforcing steel across shear keys because connections between panels and supporting elements normally prevent any separation of panels at a shear key. The presence of a shear key results in a significant reduction in the transfer distribution of a concentrated load. However, the presence of a longitudinal shear key does not materially impair the torsional stiffness of a deck slab, and the lateral load distribution is quite similar to that in a monolithic slab (Roesli, 1955; Walther, 1956). However, some stress concentration could occur when a load is placed directly over a shear key. This is evaluated in a report produced by Abam Engineers, Inc. (1967). Abam also found that in practical terms, the effect of haunches in deck panels is relatively minor; they cause just a slight increase in the negative moments at the supports and a slight decrease in positive moments at the midspan.

The deck of a complete pier is normally covered with topping material which protects the basic structural material (i.e., steel or concrete) from wear. It must also ensure safe riding conditions for cargo handling and hauling equipment and long-wearing service characteristics for a variety of working conditions. Actually, the pier deck cover is similar to that used in conventional highway bridges. It should be added that in some cases, particularly when pier internal space is used for storage, office space, operating space, or for energy-generating equipment, decking is also provided for insulation and/or fire protection.

The pier deck surface should satisfy the following goals:

1. *Lightweight.* The aim is the maximum saving in pier dead weight.
2. *Skid resistance.* The deck surface should have high surface friction (i.e., it should be rough, to increase friction and to minimize the effect of icing).
3. *Stability and durability.* The deck cover should not deform while being exposed to friction and braking forces generated by vehicles. It should be resistant to deterioration and potholing and should have a high resistance to abrasion.
4. *Protection of the pier deck against corrosion.* In some cases, a poorly designed

Figure 7.70 Construction of floating dock at Concrete Technology, Inc. drydock. (Courtesy of Abam Engineers, Inc.)

and placed deck wearing surface may aggravate rather than relieve the potential of deck corrosion. This could happen if the deck cover develops cracks or is insufficiently well bonded to the deck structure, thus forming voids in which moisture may accumulate; the resulting corrosion may spread undetected under the deck cover. The deck cover must be impervious to water and chemical agents, have no cracks that enable moisture to penetrate down to the deck surface, and have a good bond to the deck. The latter is essential for steel decks.

5. *Easy maintenance and repair.* If the deck cover is damaged, it should be possible to make necessary repairs without prolonged disruption in dock operation.

Choice of the type of pier deck cover is usually based on the performance record of various materials as well as on local experience. The following deck covers are typically used in floating pier construction.

WOOD DECKING. Wood decking is used primarily on passenger docks. It is also used on cargo handling piers and access bridges. For a steel pier or steel deck it is essential to protect the steel against corrosion by the application of reliable paint prior to the installation of wood decking. The wood decking must be installed so as to ensure good drainage of surface water.

BITUMINOUS SURFACE. A bituminous surface typically includes a seal coating topped with a suitable bituminous-mix wearing surface course which generally does not exceed 50 to 60 mm. The seal coating is used to prevent moisture from accessing the steel deck surface. It also provides a good bond between the deck and the overlying surfacing. Before seal application, the steel surface must be thoroughly cleaned by sandblasting.

The following materials and/or their combinations are generally used for seal coating:

- Epoxy resins with embedded grit to provide a good bond with overlying surfacing
- Different kinds of bituminous or resinous varnishes and paints with or without such additives as rubber powders, red lead, or others
- Dense, voidless asphalt mastic coatings 3 to 5 mm thick
- Inorganic zinc paints and other suitable materials available on the market

For surface courses bituminous mixes are used most often. The mix used for deck wearing surfaces should be robust, durable, stable at high temperatures, and not brittle at low temperatures. Unlike ordinary pavement the pier deck surfacing acts compositely with the structure and should therefore, withstand tensile and/or compression stresses due to pier bending.

ASPHALT–CONCRETE AND ASPHALT–MASTIC SYSTEMS. These systems are generally used in steel deck construction. An asphalt–concrete wearing surface is typically placed in two courses; the lower course is somewhat softer than the upper one. The total thickness of this type of decking is usually up to 50 mm. The mix content (percentage of bitumen and other aggregates) varies depending on the performance required. The performance of these surfaces has been reported to be generally satisfactory. Defects such as crack development, attributed to pier performance, are not common and to the best of this writer's knowledge have never been a problem leading to steel deck corrosion.

Longitudinal and transverse joints are generally required in asphalt–concrete wearing surfaces. They are typically made 5 to 7 mm wide and are filled with an appropriate joint filler material. Asphalt–mastic surfaces typically consist of a layer of mastic into which crushed stone aggregate 10 to 20 mm in size is embedded. Filled joints are not required in asphalt–mastic wearing surfaces. It is essential, however, that in any kind of bituminous surfacing the cross section of the cover be provided with sufficient slope toward the pier edges for the escape of surface water. An asphalt–mastic wearing surface should be inspected carefully on a regular basis. If cracks occur, they should be sealed as soon as possible and wear-out spots should be filled promptly with the appropriate compounds.

EPOXY RESIN COVERINGS. Epoxy belongs to a family of synthetic resins. Epoxy compounds for use in wearing surfaces are known under various commercial names. The mix should be placed immediately because its hardening starts right after mixing of its components. Complete hardening of epoxy resin wearing surface requires up to 24 hours. When properly mixed, placed, and cured, epoxy compounds have excellent adhesion to thoroughly clean steel and are tough and resistant to water, oils, solvents, and deicing salt. The compressive strengths of various resin compounds is up to 35 MPa, with a modulus of elasticity of about 2800 MPa at lower temperatures to 7000 MPa at higher temperatures. Epoxy resins are usually applied in a liquid state by spraying over a protected surface.

Use of sprayed-on epoxy wearing surfaces is very advantageous where weight saving is important for pier performance. Such a surface also has very good skid resistance. Epoxy resin coverings are rather expensive compared with bituminous wearing surfaces, but, the additional cost of this type of deck cover would be at least partly offset by a saving of structural steel due to a reduction in the dead load.

EPOXY CONCRETE AND EPOXY ASPHALT WEARING SURFACES. These types of deck cover are up to 30 mm thick and represent a mixture of concrete or asphalt with epoxy. An epoxy–asphalt wearing surface does not soften under high heat. An epoxy–concrete surface is waterproof, resilient, very adhesive, and less expensive than a sprayed-on epoxy surfacr. The big advantage of epoxy–concrete or epoxy–asphalt is that it could be placed, screeded, and rolled to the desired deck profile.

7.12.5 Control of a Pier Interior Environment

The interior environment of a pier must be consistent with the use of pier interior space (e.g., storage, offices, machinery and equipment). It can be maintained and modified if necessary by means of ventilation, heating, cooling, dehumidification, or by any combination of the above. For ordinary cargo or passenger piers, however, only natural ventilation is usually required. Ventilation is the process used to provide fresh outside air to a pier interior. In the case of natural ventilation, fresh air is supplied by natural draft and is provided for the removal of heat and noxious vapors, and for other reasons. The quantity of air required for each space to be ventilated could be determined by heat transfer or by empirical calculations. In a natural ventilation system, air movement is created by the difference in temperature and the density of inside and outside air. A typical natural ventilation system consists of an inlet, an outlet, and if necessary, a duct to the area served.

If the internal space of a pier is used for offices, rest rooms, or mechanical equipment, forced ventilation or a combination of mechanical supply/natural exhaust or natural supply/mechanical exhaust systems are used. Special attention must be paid to ventilation in spaces that have a potential for containing hazardous vapors. Examples of such spaces include oil and/or gas pump rooms, which in some special cases may be located on a pier, or the compressor room on liquefied gas terminals.

It should be noted that in a pier ventilation system, the ducts should be placed in as close to a straight line as possible, avoiding sharp bends or abrupt changes in duct sizes or shapes, all of which may result in a reduction in system efficiency. Noncorrosive materials such as galvanized steel or aluminum are used for duct construction. Usually, a heavy section of ductwork is welded into the penetrated structure, where structural compensation is required, and access holes and portable sections are normally provided to permit inspection and maintenance. Typically, it is not necessary to make ductwork watertight because it is typically placed as close to a deck as possible and therefore cannot break the watertightness of a pier in case of accidental damage and flooding of one or two compartments.

7.12.6 Pier Fittings

There is a broad assortment of hardware attached to a pier deck and hull to perform various functions, described next.

7.12.6.1 Curbs and Railings. Curbs and railings are fitted on the deck of a pier to protect movable machinery, personnel, and/or passengers from falling into the water accidentally. A perimeter curb usually 30 to 40 cm high is used on a pier used solely for cargo handling operations. No railing is required in this case.

Structurally, the curb could be made from timber, steel, or concrete components, or combinations of these materials. It is normally discontinued in the vicinity of mooring accessories such as bollards, to prevent interference with mooring lines.

For safety of operation, passenger piers are fitted with railings of miscellaneous construction. The upper rail is normally located 1.05 to 1.10 m above the deck level. The passenger dock railing usually has several portable sections about 1.5 to 3.0 m in length, convenient for handling manually. These sections are normally installed at locations where passengers enter and leave a ship.

7.12.6.2 Ladders. Ladders are installed on a pier hull where required to provide suitable access to the pier from small service craft. Ladders are usually installed at both ends of a pier, and if required, at its offshore and/or inshore faces. To prevent a ladder from being damaged by a ship, it is placed either in a recess or as close as practical to the fender. A common ladder consists of round steel bars (20-mm diameter) spaced at 300 mm and welded to two stringers spaced at 400 to 600 mm. Similar ladders are used for access to a pier's interior.

7.12.6.3 Manholes. Manholes, provided for access to a pier's interior, are flush-fitted to the deck's wearing surface. Because of infrequent use, only one manhole per compartment is usually considered. Manholes can be round or square, but their covers must be watertight. The minimum clear opening for a round hole should not to be smaller than 500 mm in diameter.

7.13 PIER CONSTRUCTION AND DEPLOYMENT

A floating pier is a simple boxlike, cellural-type pontoon, typically comprising an external shell and a system of watertight bulkheads. Structural simplicity permits easy fabrication of steel pontoons, even at a primitive shipbuilding facility. Depending on their size, the pontoons can be launched in a variety of ways. For details, consult Tsinker (1995, 1997).

Construction of a concrete pier is somewhat more complicated. Typically, a contractor has several alternative options for pier fabrication, but the selection is usually limited to the following:

- In situ casting (depending on pier height, it could be done either by employing ordinary concrete casting techniques or by a slipform method)
- Construction from prefabricated elements
- Combination of the two methods described above

Depending on available construction facilities, voyage conditions, and its actual length, a floating pier can be built in one piece, or in several units, subsequently towed to a deployment site and joined together while afloat to form a pier of designed length and/or form. The pier assembly is usually done by employing a posttensioning technique, pinned articulated joints, or another method. Because of a pier cellular arrangement, the cast-in-situ method of pier construction, due to repetitious assembling and dismantling of the formwork, can be relatively expensive. The cost of a cast-in-situ pier can be reduced by employing slipform techniques, is due to the potential reduction in the cost of formwork, in production time, and in subsequent standing charges at the drydock. The main advantage of pier prefabrication is that concrete work can proceed simultaneously on two fronts: at the precasting yard, where prefabricated elements are cast, and in the dry dock. This expedites the pier fabrication process and reduces costly dry dock time.

In addition, horizontal rather than vertical casting of prefabricated concrete elements results in better quality concrete work and drastically reduces the cost of formwork. The floor of the pier (sometimes the deck as well) is usually cast in situ. The assembly of precast walls follows. The precast panels extend from the bottom slab to the pier deck, and all construction joints in between them are subject to full posttensioning. The thickness of pier external walls (minimum 25 cm) is governed largely by the effects of environmental loads and by the conditions of pier operation. The thickness of internal walls is usually determined by the placement of posttensioned and regular reinforcement. The concrete cover over reinforcement in cast-in-situ internal walls usually amounts to 25 mm.

Construction joints between all prefabricated panels are cast in situ. In large concrete piers the minimum thickness of prefabricated internal walls is usually no less than 20 cm. Posttensioned tendons are threaded through prefabricated panels erected in line and through cast-in-situ joints. To avoid problems with tendon threading, bends in ducts and duct misalignment in adjacent prefabricated units must be avoided. Hence, great care must be exercised in duct placement. Ducts must be rigid and watertight. The wall thickness of ducts should not be less than 2 mm. If required, the ducting should be prebent to required curvatures before installation. All construction joint splices in ducts must be taped with waterproof tape. To facilitate grout injection and for its better quality, the inside diameter of ducts should be at least 5 mm larger than the diameter of the posttensioning tendon.

After construction in the dry dock is completed, the pier is launched and towed to the site of its deployment. Depending on dry dock size and voyage conditions, the pier can be built in one piece, or in several units, assembled at the site of deployment. An alternative method of pier launch, which is particularly helpful in developing countries and where dry dock facilities are not readily available, is construction of the pier on a riverbank during the dry season, setting it afloat at a time of high flood. The disadvantage of this method is the limited period of time available for pier construction. Another method is to cofferdam the appropriate area adjacent to the sea or waterway and build the pier in an enclosed area. On completion of construction work the cofferdam is breached (or dismantled) to flood the area so as to enable the pier to be floated out.

In some cases, the pier structure could be assembled from small units launched by crane and put together while afloat. From the above it follows that it is important to know in advance how and where the pier will be constructed, because actual construction facilities may greatly affect the pier structure. Upon completion, and normally before towing out, the pier is furnished with all required mooring devices and accessories and fender system. Again, depending on voyage condition, the pier can be towed either in one piece or in smaller units assembled at the deployment site. An example of multipontoon assembly to form a passenger ferry dock is shown in Figure 7.71.

Twin floating facilities, one of which is depicted in Figure 7.71, operate as commuter ferry terminals on either side of Burrard Inlet, Vancouver, B.C. The "E" configuration of this facility is formed by four concrete floating units joined together at the terminal side. The hull of each unit is of cellural construction, manufactured and prestressed in a dry dock and floated out upon completion. While afloat, the hull of each individual pontoon was completed by installation of the deck structure, consisting of precast, prestressed planks and cast-in-situ concrete topping. At the deployment site the units were assembled in the final E configuration by placing the concrete joints and posttensioning all the units together across the joints. The assembled dock is maintained in place by its anchor system, which includes two fixed dolphins at its sides.

Another interesting passenger-only floating ferry terminal has recently been constructed to serve crossings between Seattle and Bremerton,

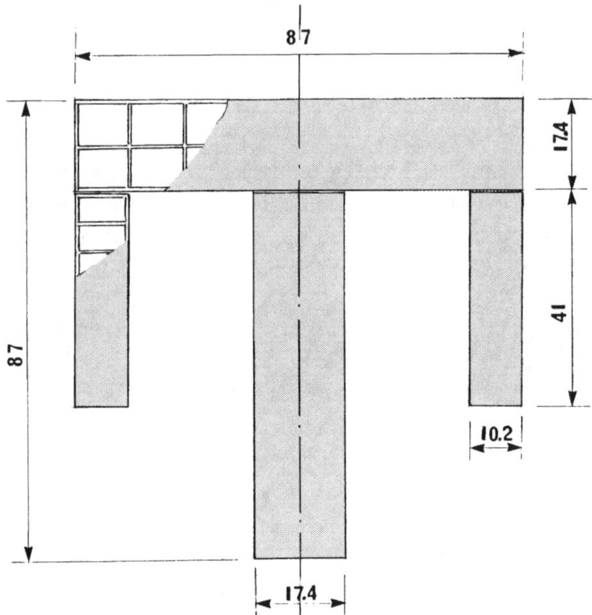

Figure 7.71 Floating ferry terminal, Burrard Inlet, Vancouver, British Columbia, Canada. (Courtesy of Abam Engineers, Inc.)

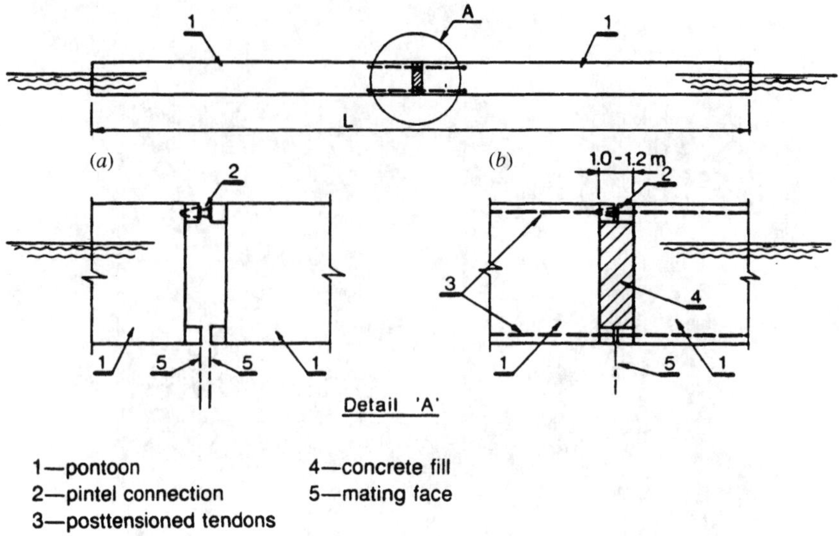

Figure 7.72 Stages of floating pier integration: (*a*) two pontoons are brought together; (*b*) integration is completed.

Washington (Cichanski et al., 2001). This facility comprises the following basic elements: a land-based piled trestle, articulated access bridge with enclosure, and two floats. The floating part of this facility is configured in an L-shaped layout with the landward floating unit serving as a ferry dock with an enclosure and foundation for the access bridge. The seaward unit connected to the landward float by the passenger ramp constitutes part of the floating breakwater. It is designed to sustain loads from large ferryboats. Both floats are maintained in place with the help of several bottom-fixed mooring dolphins.

Joining the separate units together to form the floating pier is an important construction stage. It requires good field control to prevent overstressing of separate units or damage to their matting faces. The latter may easily occur due to wave action, bumping, or sudden changes in draft caused by shifting ballast. Even though in many instances, the pontoons are identical, the construction tolerances in overall dimensions and weight could result in variations in draft and trim between adjoining units. Normally, during the assembling stage the separate pontoons are ballasted to compensate for differences in draft and to align the matting surfaces. The latter are usually furnished with steel bearing plates and/or adjustable pinned or other types of connections. Typical stages of pier integration are shown in Figure 7.72.

Floating docks are seldom towed long distances from the site of fabrication to the deployment site. Most typically, they are towed at no heavy wind and/or wave condition. The effects of environmental agents on dock towing are evaluated on a case-by-case basis. The minimum required power for the tow boat, N_t, can be computed from the following empirical formulation:

$$N_t = \frac{F_h V}{75} \quad \text{hp} \tag{7.16}$$

where F_h is the resistance to pier tow (kg) and

V is the speed of tow in still water (m/s). The resistance to pier tow, F_h, includes the resistance of water R_w in combination with wind, current, and wave effects. The resistance of water (R_w) usually constitutes the major resistance factor. It can be determined from the following empirical formulation:

$$R_w = (k_1A_1 + k_2A_2)V^2 \qquad \text{tonnes} \qquad (7.17)$$

where k_1 is the resistance factor for the front submerged area ($k_1 \approx 0.06$ tonne · s²/m⁴); A_1 the submerged area of the front leading wall of the individual pier or cluster of piers (m²); k_2 the friction of pier submerged area ($k_2 \approx 0.00015$ to 0.0002 tonne · s²/m⁴); A_2 the pier submerged (wet) area, which includes both sidewalls and the bottom slab (m²); and V the speed of the tow in still water (m/s). The optimum length of a towline L_t, which provides the best control of a pier under tow, can be determined from the following empirical formulation:

$$L_t = \frac{0.67\alpha L(20 + N_p)}{120 + N_t} \qquad \text{m} \qquad (7.18)$$

where α, the empirical coefficient, = 9.5 to 10.5; N_p is the tow boat actual registered power (hp); and L is the length of individual pier (m). The tug must be equipped with a pump of sufficient capacity and be ready to pump water out of a leaking or damaged pier. For more information on pier towing, readers are referred to Arctec Canada Ltd. (1984) and Acres International Ltd. (1987).

7.14 PIER MAINTENANCE

The purpose of pier maintenance is primarily to protect it against corrosion, but it includes maintaining good appearance, protecting equipment, minimizing abrasion damage, and timely repair of damaged or worn-out parts of a structure. The prudent dock owner should start maintenance program planning during the dock design stage. This would include selection of the type of corrosion protection system; selection of reliable high-performance coatings, which would help to maintain pier appearance for longer period of time; measures to prevent fouling of hawser pipes; and others. The high cost of a pier out of service provides a strong reason for the use of better protective measures. Actually, good pier appearance is associated with its long life, because of regular preventive cleaning and repainting. Furthermore, steel pier preservation also involves the use of continuous welds in exterior areas and wet spaces, elimination of crevices, and as much as possible, use of flat structural members instead of rolled sections.

All the considerations above reduce edge corrosion and therefore eliminate the need for frequent maintenance. A very important factor in steel hull preservation from corrosion and fouling is initial surface preparation before the start of coating. A good practice in steel pier construction is blast cleaning followed immediately by a primer application. Inspection and quality control are an important part of a hull preservation program. Inspection is the responsibility of a number of parties, including the owner's representative, coating vendor, and fabrication yard. The inspection requirements should be specific in scope. The choice of when, how, and where maintenance work will be done is an important consideration. Because the dock operator often cannot afford a loss of dock time for maintenance, a very well planned maintenance program should be in place to carry out maintenance work (recoating, repair, etc.) during a normal availability period.

7.15 PIER STABILITY

While in operation, a floating pier must be capable of safely carrying all kinds of design loads. In principal, the pier carrying capacity depends on its size, which must be sufficient to provide the necessary buoyancy to support all design loads. Another important requirement is that the pier must be sufficiently stable during normal loading conditions. This means that if when floating in still water, the pier is displaced from its equilibrium position by external forces, it should return to that position when the forces are removed. The pier must also remain afloat and stable if damaged. In the latter condition it should maintain a certain minimum freeboard.

Naturally, the pier should possess sufficient buoyancy to support its own weight and the loads it is designed to carry. It must maintain a certain amount of freeboard to provide convenient berthing and cargo handling operations. The design freeboard should also provide the pier with the reserve buoyancy that enables it to remain afloat and stable in the event of damage. During the voyage across the seas the pier is watertight and does not need to maintain a sufficiently high freeboard to prevent water from coming on board.

7.15.1 Stability of a Pier Constructed in the Form of an Individual Pontoon

Buoyancy and stability are principal properties of a body immersed into a fluid. The amount of fluid displaced by the body and the upward force (which is equal to the weight of that volume of fluid) are commonly known as the *buoyancy,* and the volume of the immersed part of the body is known as a *volume of displacement.* However, in naval architecture the term *displacement* is generally used to describe the mass of water displaced rather than the volume. The opposing force to buoyancy is the weight of the floating body, which for equilibrium, must equal the buoyancy. These forces act as if they were concentrated at points known as the *center of gravity* (CG) and *center of buoyancy* (CB) (see Figure 7.73). Since gravity acts downward and buoyancy upward, for equilibrium the two centers must in the same vertical line. This line is called an *axis of flotation*. For the body to stay afloat, its immersed part has to have enough volume to develop sufficient buoyancy. From the condition of equilibrium follows

$$B = G = \gamma V \tag{7.19}$$

where B is the buoyancy of the floating body, G the weight of the floating body, γ the specific weight of the fluid in which the body is immersed, and V the volume of the immersed portion of the floating body.

The stability of a body immersed in a fluid is the body's ability to balance an action of moments that try to overturn it. Adding weight (P) applied to the axis of flotation will increase the floating body's buoyancy (crosshatched area in Figure 7.73*b*). Now the total shaded area is the body's new displacement. It must be matched by the new gravity force if the system is to remain in equilibrium. If instead of being added at the axis of flotation, the weight had been applied at one side of a floating body, the body would tilt. Since the total weight of the floating system will remain the same as in the case illustrated in Figure 7.73*b*, its buoyancy will also not change. However, the system's new center of gravity will shift toward added weight and, to meet conditions of equilibrium, so will the center of buoyancy (Figure 7.73*c*). An increase in weight naturally would increase the system's tilt and its buoyancy, and continued increases in weight could eventually turn the floating body over.

In designing a conventional floating pier, however, the pier deck gradient is very small and usually should not exceed 3 to 8%. Considering a small angle of pier tilt, which practically does not affect the position of the center of gravity, for the purpose of further discussion, assume that the axes of pier rotation AB and CD (Figure 7.73*a*) always run through the center of gravity of the pier water-plane area and therefore are the principal axes of this area.

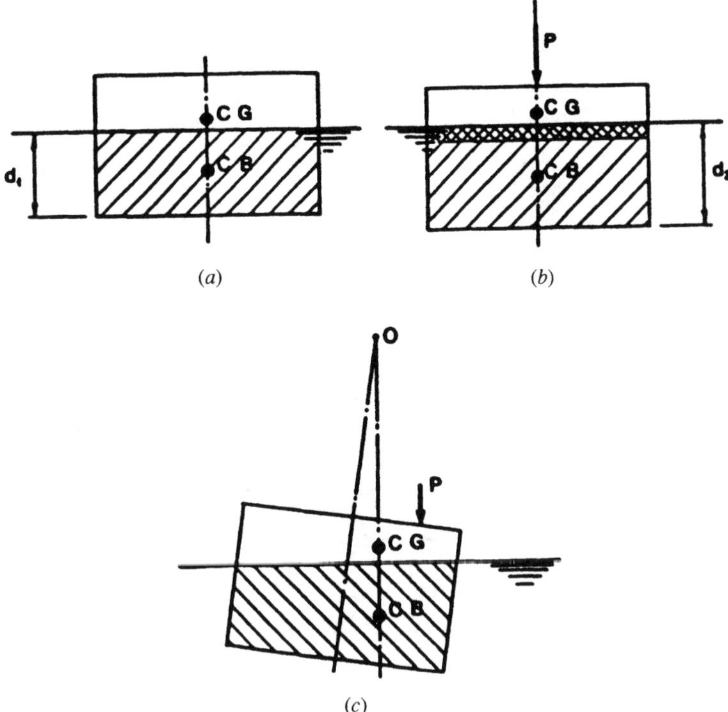

Figure 7.73 Definition sketch, buoyancy of a body immersed in fluid.

The pier tilt could be caused by the action of eccentrically applied force (P). Since in practice, pier inclination is limited and the angle of the tilt ϕ is rather small, the inclination could be expressed as

$$\phi = \tan \phi \qquad \text{or} \qquad \phi = \sin \phi \qquad (7.20)$$

Note that angle ϕ is the angle between the pier's original waterline W_0L_0 and waterline W_1L_1 after pier tilt (Figure 7.74*b*). The action of the force that tends to tilt the pier could be assumed to be the action of force P applied at the pier's center of gravity in combination with bending moment $M = Pl$, as shown in Figure 7.74*b*. There is now a leverage, or moment, trying to bring the pier back to an upright position. This moment is known as the *righting moment.*

When displaced from an upright position, the pier possesses the ability to return to an upright position after the load is removed; this capability is known as *stability.*

Assume that the distance between the pier's center of buoyancy (CB) and its center of gravity (CG) as equal to a. There are three practical positions of CG and CB:

1. CG is located above CB ($a > 0$). This is the most practical case of pier tilt.
2. CG is located below CB ($a < 0$).
3. CG and CB coincide ($a = 0$).

When pier tilts under the action of bending moment M, the forces of gravity and buoyancy are positioned on a new axis of flotation, and a new pair of forces (shaded wedges in Figure 7.74*b*) produce a stabilizing moment M'. Be-

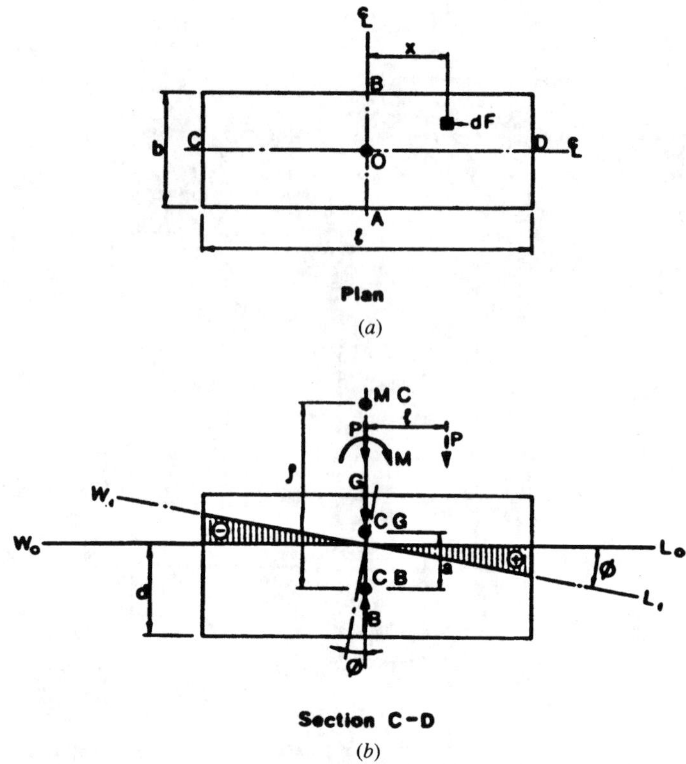

Figure 7.74 Definition sketch, pier stability.

cause the value of angle ϕ is small enough, the moment M' could be presented as follows:

$$M' = -\int^{F} \gamma \times \phi(dF) = -\gamma\phi \int^{F} x^2(dF) = \gamma\phi I \tag{7.21}$$

where $\gamma \times \phi(dF)$ is the buoyancy force and I is the moment of inertia of the water-plane area in relation to the axis of tilt AB (Figure 7.74*a*). The weight of the system together with the reaction of water also produces a pair of forces. Assuming that $\sin \phi = \tan \phi = \phi$, the moment of this pair M'' can be described as follows:

$$M'' = \gamma V a \phi \tag{7.22}$$

where V is the volume of displaced water. The sign of this moment depends on distance a. Under $a > 0$, this moment rotates the system in the same direction as moment M does, thus adding additional tilt to the system. Under $a = 0$, $M'' = 0$, and when $a < 0$, M'' produces a stabilizing effect. Under the condition of equilibrium, $M' + M'' + M = 0$, from which follows it that

$$M = \gamma\phi(I - Va) \tag{7.23}$$

Accordingly,

$$\phi = \frac{M}{\gamma V[(I/V) - a]} = \frac{M}{\gamma V(\rho - a)} \quad (7.24)$$

Equation (7.24) is known as the *metacentric formula of floating body stability*. Here

$$\rho = I \quad (7.25)$$

where ρ is known as the *metacentric radius* above the center of buoyancy. Thus the pier metacentric radius ρ above the center of buoyancy (CB) is found by dividing the moment of inertia of the water plane about its centerline by the volume of displaced water. Considering the rectangular form of the pontoon's waterplane area and its cross section, the transverse metacentric radius is equal to

$$\rho = \frac{\frac{1}{12}lb^3}{lbd} = \frac{b^2}{12d} \quad (7.26)$$

where l, b, and d are length, breadth, and draft of the pier.

If a pier has watertight compartments filled with ballast water, the metacentric radius of the system is expressed by the equation

$$\rho' = \frac{I - \sum i}{V} \quad (7.27)$$

in which $\sum i$ is the sum of moments of inertia of the water surface in each compartment. As can be seen from eq. (7.26), the height of the pier's metacenter above the center of buoyancy is affected by the pier's draft and its breadth; the pier length does not affect the height of the transverse metacentric radius. From eq. (7.26) it follows that with increasing draft the metacentric radius falls rapidly and could reach infinity under a draft value of zero.

The distance $\rho - a$ is often referred to as the *metacentric height of the pier above its center of gravity*. With eq. (7.24) one can judge the effect of the $\rho - a$ value on pier stability. A necessary condition of pier stability is that $\rho - a > 0$, which means that the pier metacenter must be located above its center of gravity. Under $\rho - a = 0$, infinitively large tilt is likely to occur. If $\rho - a < 0$, the floating object is unstable. The latter means that the smallest unbalanced force could cause the floating object to turn over.

A live load placed on a floating pier deck tends to raise the system's center of gravity, thus reducing the stability of the pier. Therefore, pier stability must be designed for the maximum possible live load placed on its deck. The minimum requirement for pier stability in still water is expressed as $\rho - a \geq 0.2$.

Equation (7.24) for $\phi = \tan\phi = \sin\phi$ can be rewritten as follows:

$$\phi = \frac{M}{\gamma I \alpha} \quad (7.28)$$

where

$$\alpha = l - \frac{aV}{l} \quad (7.29)$$

Pier sinking (draft) at distance x from the axis of tilt can be expressed by

$$d = \frac{Mx}{\alpha I \gamma} \quad (7.30)$$

Until now, the pier was assumed to rotate about one axis only. In reality, however, the freedom of a floating pier to rotate would allow it to tilt in various directions, depending on the position of the stability-disturbing force. However, for the sake of convenience in future discussions, it will be assumed that the pier rotates about two principal horizontal axes. Rotation about the longitudinal axis is known as *rolling*, and rotation about the transverse axis is known as *pitching*.

Assume now that the pier is exposed to a concentrated load P applied at any point on the pier deck (Figure 7.75). Equation 7.31 can be used to determine the amount by which the pier may sink at different locations due to the effects of load $P(l_x, l_y)$:

$$d_c = \frac{P}{\gamma}\left(\frac{I}{F} + \frac{l_x x_c}{\alpha_y I_y} + \frac{l_y y_c}{\alpha_x I_x}\right) \qquad (7.31)$$

where d_c is the pier sinking at point c due to the effect of concentrated load P only, and I_x and I_y are the moment of inertia of the pier at its water plane about axis x–x and y–y correspondingly.

In the case of a more complex shape of floating object, the same criteria as discussed before are applied. For example, take a catamaran pier, which could be part of a floating dock or floating bridge. This structure is composed of two pontoons joined together by a rigid deck (Figure 7.76). As in the case of a simple rectangular pier, the equilibrium of forces acting on a catamaran structure requires that the weight be equal to buoyancy, and the center of buoyancy and center of gravity must lie on the axis of flotation. A catamaran pier will sink uniformly if a load is applied along the system's initial axis of flotation, and it will tilt if a load is applied eccentrically. In the lat-

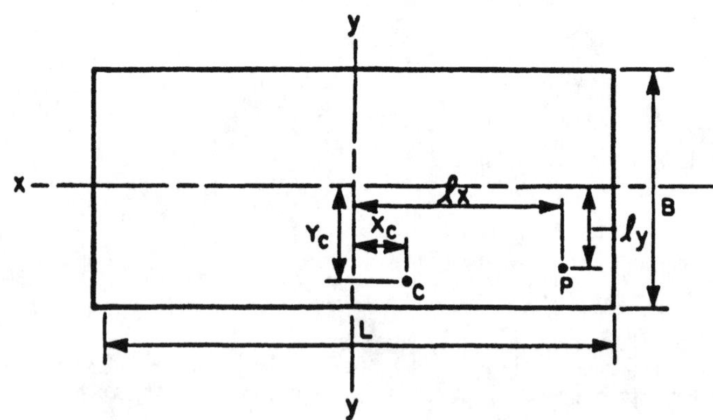

Figure 7.75 Plan of a pier exposed to concentrated load P.

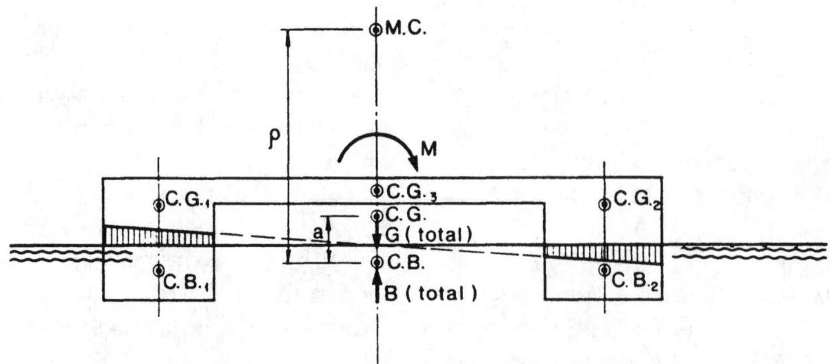

Figure 7.76 Stability of a catamaran pier.

ter case, each pontoon will tilt or heel and the buoyancy of each pontoon will change considerably, but the weight of the system will remain unchanged. The system's center of buoyancy will move out toward the immersed pontoon and a righting moment will appear that will tend to bring the catamaran back to an upright position. As before, the righting lever is actually proportional to the system's metacentric height, which is a function of the system's water-plane area moment of inertia and its buoyancy.

7.15.2 Stability of a Damaged Pier

In the event that the pier is damaged below its waterline, and if unrestricted flooding is permitted, the pier will eventually sink. To prevent this from happening, or at least to reduce the probability, the hull of a conventional pier is usually divided into a series of watertight compartments by means of watertight transverse and longitudinal bulkheads.

The compartmentalization of the pier's hull can localize and contain hull flooding in the case of accidental damage. Two important effects of pier flooding must be considered in floating pier design:

1. The pier must sustain a loss of buoyancy of limited magnitude.
2. The pier must remain stable when damaged.

In pier design it is usually assumed that the two adjacent compartments could be flooded and the subsequent effect of loss of buoyancy on pier stability must be evaluated. Depending on the system of watertight bulkheads, pier flooding could be either symmetrical or asymmetrical. Symmetrical flooding occurs if the pier has transverse watertight bulkheads only (Figure 7.77*a*). Asymmetrical flooding may occur in piers divided into a series of watertight compartments by means of transverse and longitudinal watertight bulkheads (Figure 7.77*b*).

Assume that the pier depicted in Figure 7.77 is damaged below the waterline at the area of the transverse bulkhead and its two adjacent compartments are flooded as indicated in Figure 7.77. Accordingly, the buoyancy of the pier between the bulkheads bounding these compartments is lost, and the pier must therefore sink in the water until it picks up buoyancy from undamaged compartments to maintain equilibrium. At the same time, because of the lost buoyancy, the overall position of the pier's center of buoyancy is altered. If pier flooding is symmetrical, the pier's center of buoyancy is shifted longitudinally only; if the flooding is asymmetrical, it alters the pier center of buoyancy in both longitudinally and transversely. To restore the center of buoyancy to its original position, the pier must be trimmed by ballasting of an opposite compartment or by repositioning the deck load.

The pier in question that was originally floating at the waterline W_0L_0 will now float at the waterline W_1L_1. It is usually assumed that the new waterline W_1L_1 should in no case be higher at any point than the pier deck. For determination of the position of the waterline W_1L_1, the *added weight method* is commonly used. In this method, the water entering the damaged compartments up to the original waterline can be regarded as an added weight $P = \gamma bld$. The values of d_1, d_2, d_3, and d_4 (Figure 7.77) can be determined with the help of eq. (7.31). However, because of the tilt toward its damaged part, the pier will be deeper in the water, so the *added weight P* will be somewhat greater than assumed originally. Therefore, additional approximations are necessary to obtain the desired degree of accuracy. These calculations can be programmed for the computer and any number of iterations could be carried out. Once the new buoyancy condition is established, the damaged pier stability should be investigated.

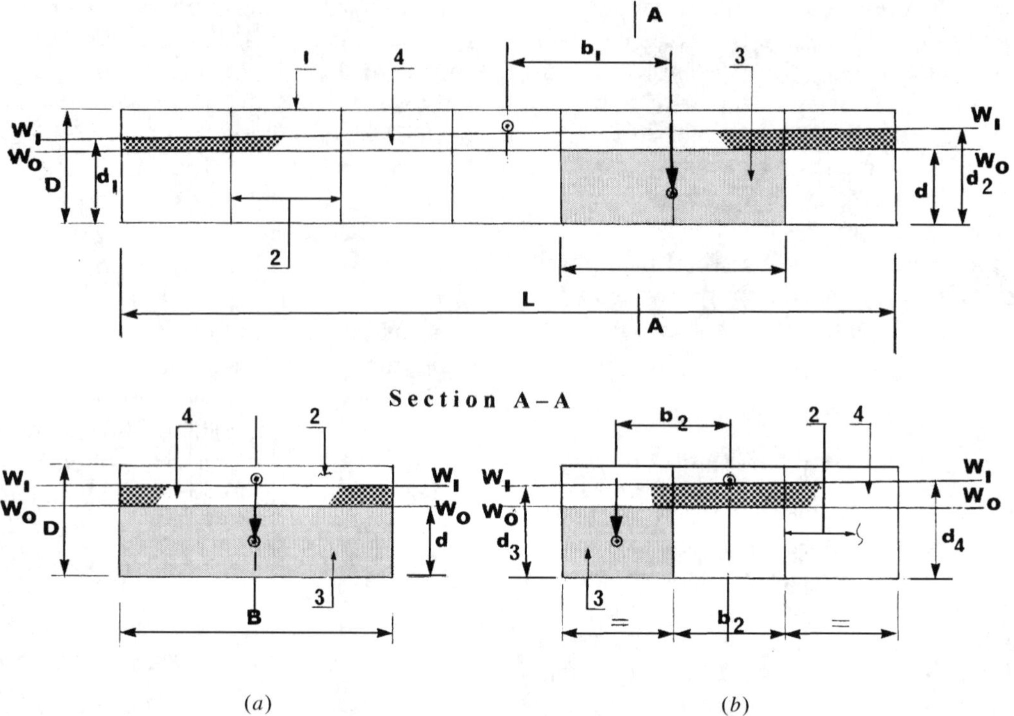

Figure 7.77 Stability of a damaged pier: (*a*) symmetric flooding; (*b*) asymmetric flooding. 1, Pier; 2, watertight bulkhead; 3, flooded area; 4, additional buoyancy to compensate for that lost.

Considering pier stability from the point of view of metacentric height ($\rho - a$), the effect of lost buoyancy from a position below the original waterline (W_0L_0) to the position above it would be that the center of buoyancy would rise. The magnitude of the pier's new metacentric radius above its center of buoyancy could be determined by eq. (7.25), in which the value of the water-plane area moment of inertia (I) must be corrected for the free surface effect of the water contained in the damaged compartments. The value of new metacentric height ($\rho - a$) has to be determined given the new position of the pier center of buoyancy corrected for the new waterline (W_1L_1). Asymmetrical flooding would cause the pier to heel. It is important to limit the angle of heel, and if it becomes excessive, the corresponding compartments on the opposite side of the pier should be flooded. While this crossflooding would make the pier sink deeper into the water, it would reduce or eliminate the angle of heel, which is considered more dangerous.

7.15.3 Pier Dynamic Stability

A freely floating unrestrained body possesses six degrees of freedom of motion that with respect to the body's center of gravity are defined as *surge, sway,* and *heave,* and *roll, pitch,* and *yaw*. The first three represent the body's translational movements, and the following three represent the body's rotation about its longitudinal, transverse, and vertical axes. In a real site condition, the floating pier represents the partly restrained floating body and is basically exposed to three degrees of motion: heave, roll, and pitch. These are depicted in Figure 7.78. In a real site condition all three motions can be generated simultaneously. All three are true oscillatory motions. If the pier is displaced from its equilibrium position by some force, when this force is removed, the pier will oscillate until its motion is damped out.

In practice, of those modes, heave and roll (i.e., vertical rise and fall and side-to-side rotation) are of particular significance in floating pier design. Pitching mode (i.e., end-to-end rotation) usually is not important because of the relatively long length of a pier structure compared to the length of the incident waves at the normal deployment sites of floating piers—protected from waves. The exception is locations exposed to long-period swell propagating along the longitudinal axis of the pier. The heave response normally is defined in terms of the ratio of heave amplitude to incident wave amplitude. The natural period of motion (i.e., time of one complete oscillation) of a pier of rectangular cross section in still water (T_H) can be determined as $T_H = 2.83(d)^{.05}$ in seconds, where d is the pier draft in meters.

The waves create a vertical force on the pier which can be shown to be a simple harmonic function of time. Heaving motion is thus generated and its magnitude depends basically on the ratio of the heaving period of the pier (T_H) to the period of encounter of the waves with the pier (T_E) (i.e., $n = T_H/T_E$). A resonant condition will occur when n approaches unity. Under resonant conditions, maximum amplitudes of heave will occur. Naturally, the pier should be designed to avoid a resonant condition that results in large amplitudes. Under given site conditions, the only way to avoid such a condition is to alter the pier draft so as to change the period of encounter.

This is very important in the design of floating access bridges and a pier consisting of pontoons placed apart with individual deck sections or with a continuous deck structure. Both systems above are usually designed to support a movable load (e.g., mobile cranes, trucks). To avoid large amplitudes or resonant conditions, the allowable speed of this traffic should be carefully specified. Basically, the acceptable limits on pier motion are dictated by the operation of certain cargo handling and hauling equipment and, at passenger terminals, the limits to pier motion are usually dictated by human response.

Both heave and roll are affected by the amount of damping present, but only heaving is affected significantly by the effect of added mass (a hydrodynamic mass of water that is moving along with a pier body). In some in-

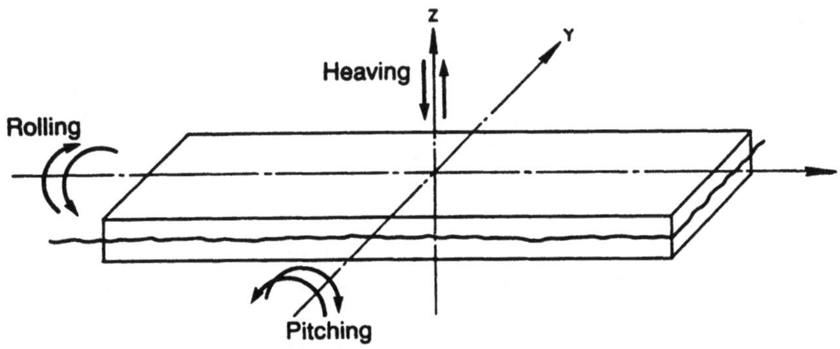

Figure 7.78 Motion of partly restrained pier (single pontoon).

stances, rolling may be affected by mooring restraint, which increases damping, thus reducing maximum amplitude. This does not significantly alter the natural period of pier rolling, however. Rolling and pitching phenomena are similar in that they are both rotations. However, for rolling, the added mass on the pier can be ignored, as this is likely to be relatively small. However, the added mass cannot be neglected for pitching. Rolling and pitching periods (T_R and T_p) can be determined by the equations

$$T_R = 2\pi \frac{r_t}{[g(\rho_t + d/2)]^{0.5}} \tag{7.32}$$

$$T_p = 2\pi \frac{r_l}{[g(\rho_l + d/2)]^{0.5}} \left(\frac{w_1 + w_2}{w_1}\right)^{0.5} \tag{7.33}$$

where r_t and r_l are the pier (pontoon) transverse and longitudinal radii of gyration, ρ_t and ρ_l the pier (pontoon) transverse and longitudinal metacentric radii above the pier's (pontoon's) center of buoyancy, g the gravitational acceleration, d the pier (pontoon) draft, w_1 the mass of the pier (pontoon), and w_2 the added mass (weight of the entrained water). As seen from eqs. (7.32) and (7.33), rolling and pitching periods are inversely proportional to the square root of ($\rho + d/2$). To keep periods long, and with small accelerations, this value should be small.

The designer has more control over the rolling period of a pier than over the periods of the other motions. In piers having a large value of $\rho_t + d/2$, reducing the value could give longer rolling periods. The designer has very little control over the natural pitching period, since the value of $\rho_l + d/2$ is very large, and since any relatively small changes that may be made in $\rho_l + d/2$ will have a negligible effect. Again, as floating piers are usually deployed in sheltered locations, pier pitching is unlikely to occur. Hence, the problem of pier pitching is more theoretical than practical. The three main motions likely to affect a floating pier in service have been discussed briefly here. It must be noted that during the voyage across a large body of water, the pier could be exposed to high waves, resulting in high amplitudes of these plus three more motions: surge, yaw, and sway (i.e., longitudinal transfer and rotation about the pier vertical axis, respectively). All aforementioned motions in different combinations may subject a pier to large forces. With pitching in particular, if the amplitude of the motion is substantial, the fore end of the pier may leave the water altogether. When it becomes immersed again, a large hydrodynamic force can be generated on the fore end of the structure. The combined effect of pitching and heaving, and to some extent rolling, could worsen the sail condition, resulting in the possibility of structural damage or failure. More information on the sail condition and requirements for pier design can be found in any naval architecture text. More information on the dynamics of floating bodies and a comprehensive list of references are given in Chapter 8 and Headland (1995).

7.16 PIER COMPOSED OF A CHAIN OF INDIVIDUAL PONTOONS

Examples of such systems are shown in Figures 7.49 and 7.79, where the multipontoon systems can be differentiated from each other basically by their deck system and pontoon sizes; the latter depend on the weight of the deck and the magnitude of the deck live loads. Normally, the systems illustrated in Figure 7.79*a* and *b* are the least economical. Both are usually employed in the construction of a relatively small pier (e.g., one used in a small craft marina). The systems depicted in Figure 7.79*c* through *f* are the most frequently used pier arrangements for the construction of floating piers. Selection of the pier system for a particular project is basically dependent on its economy.

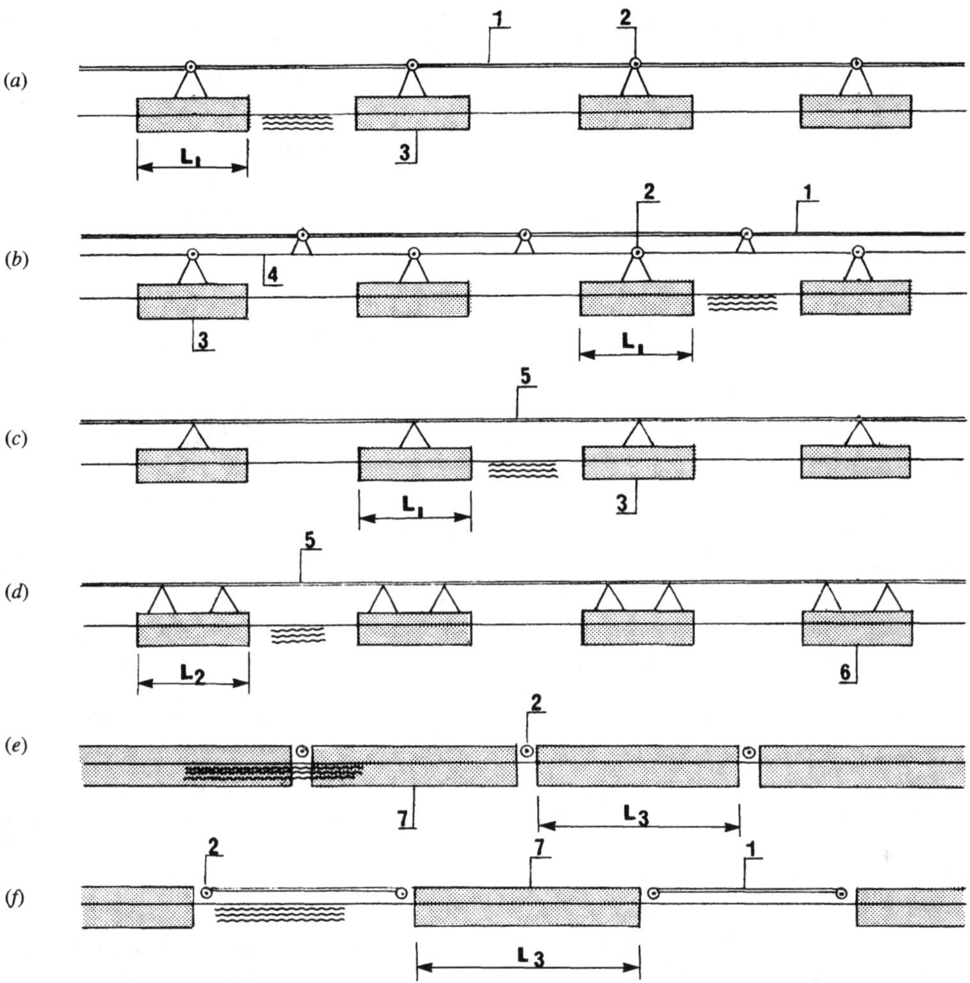

Figure 7.79 Structural arrangements of a multipontoon pier system: (*a*) system of individual pontoons (floats) bridged by short-span deck sections; (*b*) system that allows uniform load distribution between two adjacent pontoons (floats); (*c*) system of individual pontoons (floats) spanned by continuous deck; (*d*) system similar to (*c*) for large pontoons. (*e*) system of large pontoons hinged with each other. (*f*) system of large pontoons linked with each other through articulated deck sections. 1, Articulated deck section; 2, hinge; 3, small pontoon (float); 4, deck supporting system; 5, continuous deck; 6, middle-sized pontoon; 7, large pontoon.

The design process for a multipontoon system normally begins with the establishment of approximate sizes of support pontoons and the distance between them. This will be followed by analysis of the pontoon draft under dead and live loads and determination of the resulting gradient of the pier deck. These calculations must be reiterated as required until the desired result is achieved. In terms of analysis, the pier systems depicted in Figure 7.79*a* and *b* are statistically determinate. For the systems depicted in Figure 7.79*c* through *f* the resultant forces acting on an individual pontoon cannot be determined by applying equilibrium conditions only. These systems are described as being statically indeterminate, or redundant. For example, the systems depicted in Figure 7.79*c* and *d* can be analyzed similarly to a continuous beam resting on supports that settle under the load. In the pier systems depicted in Figure 7.79*e* and *f*, the deflection of any pontoon that is included in the system will result in position change in neighboring pontoons. The resulting shear forces at pontoon connections to the other pontoons in such a system(s) cannot be determined using static equilibrium conditions alone, as the buoyant forces acting on the pontoon are indeterminate.

To solve for the forces, it is necessary to make use of the relationship between loads acting on the pontoon and pontoon displacements. Theoretical bases for the analysis of a statically indeterminate structural system are given in Hetinyi (1946) and many relevant textbooks. The pier analysis can be carried out with the help of computers. In these analyses, the pier system is usually represented by simple spring-connected beams sitting on flexible supports.

The overall properties of the system are determined by the relatively simpler properties of its idealized parts. The simple components are interconnected at joints, or nodes. This joining corresponds to the imposition of displacement compatibility between elements. The stiffness properties of the individual elements relate the element nodal displacements and nodal forces. These are derived using element geometry and the stress–strain equation of the component material. External loads acting on the structure are replaced by statically equivalent nodal forces. In this case, the force and displacement terms are used in a general sense, implying moment and rotation components where necessary, in addition to simple loads and translations.

An analysis procedure based on the stiffness method regards the nodal displacements as the basic unknown quantities. Equilibrium relations are developed for the nodes, and these equate the external loads to the sum of the nodal forces acting on the elements that meet there. Substitution is made using the nodal compatibility conditions and the element nodal force–deformation relationships. The final result is a set of equations that relate the nodal displacements to the applied nodal loads. The equations are solved for the displacements and the element forces can then be recovered using the individual element stiffness.

Implicit in the procedure above is the assumption of linearity. The behavior of an element or structure is said to be *linear* if all displacements and internal forces vary in direct proportion to the loads applied. This assumption, which is reasonably accurate for pier structures subjected to normal working loads, is important for two reasons. First, the analysis is much simpler since the final set of simultaneous equations are linear and standard procedures are available for their solution. Second, it enables the superposition of solutions so that different load cases can be combined.

There are two basic causes of nonlinearity. The first is the behavior of the constituent material. The response of mild steel, for example, is linear until yielding occurs. The second cause of nonlinearity is associated with changes in geometry. If distortion of the structure causes changes in the equilibrium equations, nonlinearity will result. The pier analysis is usually based on two assumptions. The first is that individual pontoon floating in water will respond linearly to load (i.e., the pontoon's water-plane area and its geometry do not change with displacements). Second, the individual pontoon will rotate about its transverse axis only. The assumptions above substantially simplify the system analysis and do not result in unacceptable inaccuracy of calculations. Analysis of statically determinate and redundant pier systems and numerical examples are given in Tsinker (1986).

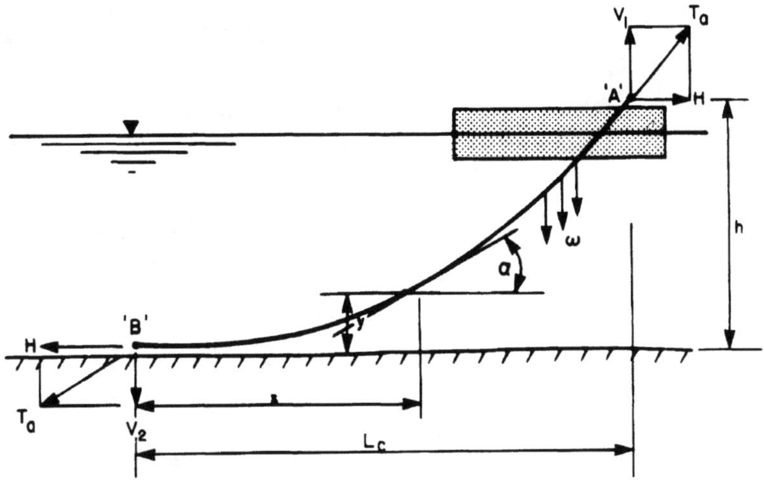

Figure 7.80 Catenary geometry definition sketch, offshore mooring line.

7.17 DESIGN OF A CABLE MOORING SYSTEM

The floating pier mooring system is designed to prevent it from being displaced from its design location. The basic design requirement for construction of a floating pier mooring system is to provide safe and efficient dock operations. The reliability of a floating dock mooring system is particularly important with regard to access bridges, which are designed on the premise that the mooring system will keep the pier in place, within an acceptable range of transverse and longitudinal motion. Miscellaneous mooring systems used for maintaining the floating docks of their design position are illustrated in Figure 7.55. They are basically of three categories: two-ways cable systems, a cable system combined with a rigid boom, and bottom-fixed mooring dolphins of miscellaneous design. The latter are discussed in Chapter 2.

7.17.1 Two-Way Cable Mooring System

A two-way cable system is illustrated in Figure 7.55*b* and *d*. The offshore mooring line is a cable freely suspended between its anchor points *A* and *B*, located, respectively, on a pier and at an anchor (Figure 7.80). The horizontal force *H* exerted on a pier is resisted by the weight of the anchor line with a wet unit weight ω. The designer's usual problem is to find the proper length and weight of the mooring cable and a reliable means of anchoring it at points *A* (pier) and *B* (anchor). As a reasonable approximation, assume that the cable weight ω is uniform per unit of horizontal projection of the mooring line. Since a cable can have no resisting moment, the anchor chain is analyzed as a flexible catenary, which may be expressed as

$$Hy - V_B x - \frac{\omega x^2}{2} = 0 \tag{7.34}$$

from which

$$y = \frac{V_B x}{H} + \frac{\omega x^2}{2H} \tag{7.35}$$

The variables are shown in Figure 7.80. From the condition of cable equilibrium,

$$V_A = V_B + \omega L \qquad (7.36)$$

$$V_A L = Hh + \frac{\omega L^2}{2} \qquad (7.37)$$

from which

$$V_A = \frac{Hh}{L} + \frac{\omega L}{2} \qquad (7.38)$$

$$V_B = \frac{Hh}{L} - \frac{\omega L}{2} \qquad (7.39)$$

Substituting these values in eq. (7.35), the equation for y may be rewritten as

$$y = \left(\frac{h}{L} - \frac{\omega L}{2H}\right) x + \frac{\omega x^2}{2H} \qquad (7.40)$$

The maximum force in the anchor cable will be at point A, and will be equal to

$$T_A = (H^2 + V_A^2)^{0.5} \qquad (7.41)$$

From these equations it may be deduced that with an increase in the vertical component of the reaction force, V_B will decrease. Therefore, if $V_B = 0$ ($\alpha_B = 0$, or is negligible), the dimension L will be equal to L_{max}.

$$L_{max} = \left(\frac{2Hh}{\omega}\right)^{0.5} \qquad (7.42)$$

As noted earlier, the length of the mooring line (L_{max}) can be reduced by assuming α_B equal to 3 to 6°. In this case, however, the result of an uplift force on anchor performance must be considered. From the catenary equation it follows that

$$f = \frac{\omega L^2}{8H} \quad \text{and} \quad H = \frac{\omega L^2}{8f}$$

where f is the cable sag. The equation for the anchor cable could be presented as follows:

$$y = x \tan \alpha - \frac{4f}{L^2} x(L - x) \qquad (7.43)$$

where

$$\frac{dy}{dx} = \tan \alpha_i = \tan \alpha - \frac{4f}{L^2}(L - 2x) \qquad (7.44)$$

or

$$\tan \alpha_i = \frac{h}{L} - \frac{\omega}{2H}(L - 2x) \qquad (7.45)$$

From eq. (7.45) it follows that

$$\tan \alpha_A = \frac{h}{L} + \frac{\omega L}{2H} \qquad (7.46)$$

$$\tan \alpha_B = \frac{h}{L} - \frac{\omega L}{2H} \qquad (7.47)$$

The minimum projected length of mooring cable ($\alpha_B \leq 6°$) could be determined by

$$L_{min} = -\frac{H \tan \alpha_B}{\omega} + \left(\frac{H^2 \tan^2 \alpha_B}{\omega^2} + \frac{2Hh}{\omega}\right)^{0.5} \qquad (7.48)$$

Finally, vertical components of anchor force at locations A and B are

$$V_A = H \tan \alpha_A \qquad (7.49)$$

$$V_B = H \tan \alpha_B \qquad 7.50)$$

Example 7.3 Assume the following conditions

of floating pier operation: $H = 200$ kN, $\omega = 1$ kN/lin m, and $h = 10$ m. Find the possible maximum and minimum projected length of mooring cable and reactions V_A and V_B. Assuming that $\alpha_B = 0$, from eq. (7.42) find

$$L_{\max} = \left(\frac{2 \times 200 \times 10}{1}\right)^{0.5} = 63.25 \text{ m}$$

Since in this case $V_B = 0$,

$$V_A = 63.25 \times 1 = 63.25 \text{ kN}$$

Assuming that $\alpha_B = 6°$ ($\tan \alpha_B = 0.105$) from eq. (7.48), we find

$$L_{\min} = -\frac{200 \times 0.105}{1} + \left(\frac{200^2 \times 0.105^2}{1} + \frac{2 \times 200 \times 10}{1}\right)^{0.5} = 45.64 \text{ m} \ll L_{\max} = 63.25 \text{ m}$$

$$\tan \alpha_A = \frac{10}{45.64} + \frac{1 \times 45.64}{2 \times 200} = 0.333$$

$$V_A = 200 \times 0.333 = 66.64 \text{ kN}$$

$$V_B = 200 \times 0.105 = 21.0 \text{ kN}$$

Note that V_A is acting downward and V_B upward.

As shown in this example, the assumed angle $\alpha_B = 6°$ offers a substantial reduction in anchor chain length. However, the uplift force V_B could be prohibitive for a certain type of anchor and may present a problem for anchor design. When a mooring system consists of two-way cables (i.e., offshore and inshore mooring lines), then to prevent collision between the pier and the bridge supporting pontoon, the system displacements must be carefully controlled. The degree of pier displacements under the effects of environmental and ship-induced forces depends on the initial tension in both offshore and inshore mooring cables and the magnitude of the forces acting on a pier.

In the absence of an external force H, forces H_1 and H_2 will be in equilibrium (Figure 7.81), where H_1 is the horizontal component of the initial tension in the inshore cable and H_2 is the horizontal component of the initial tension in the offshore cable. With the application of external force H, the previous stability of the system is upset and the pier moves the distance ΔL as shown in Figure 7.81.

$$\Delta L = L_2 - L_1 \tag{7.51}$$

The movement ceases when the new condition of equilibrium, $H + H_3 = H_4$, is established. Here H_3 and H_4 are the forces in the inshore and offshore cables, respectively. From examining the forces, it is seen that

$$H_4 = H_2 + \frac{H}{2} \tag{7.52}$$

$$H_3 = H_1 - \frac{H}{2} \tag{7.53}$$

To determine ΔL, the length of the offshore cable l_0, must be determined. It can be obtained from the equation

$$l_0 = L_1 \left(\sec \phi' + \frac{8}{3} \frac{S_1^2}{L_1^2 \sec^3 \phi'}\right) \tag{7.54}$$

where S_1 is the sag in offshore cable when $H = 0$, L_1 the horizontal distance between the two anchor points of the offshore cable when $H = 0$, and ϕ' the angle formed by a line connecting two anchor points of the offshore cable and the horizontal line through the lower anchor point

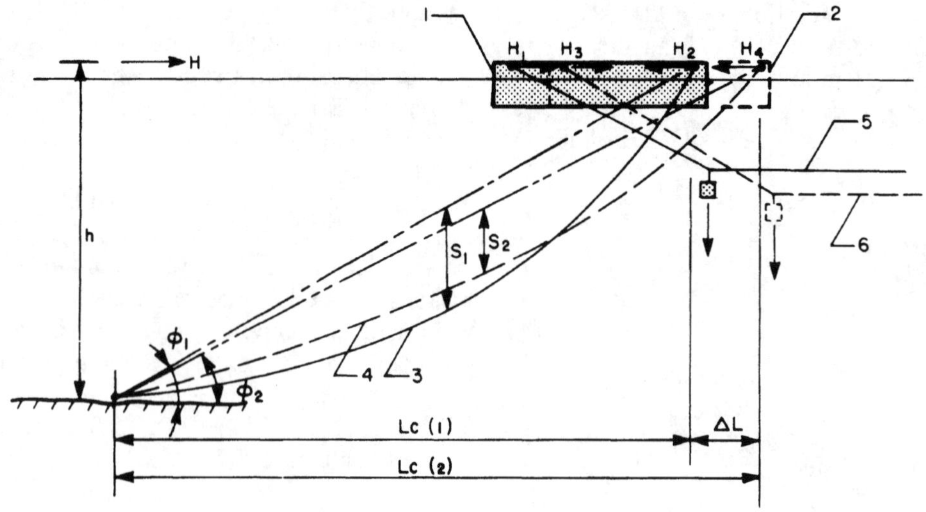

Figure 7.81 Displacement of floating pier system under effects of force *H*. (From Tsinker and Vernigora, 1980.)

when $H = 0$. The expression for sec ϕ' and S_1 may be presented in the forms

$$\sec \phi' = \frac{(L_1^2 + h^2)^{0.5}}{L_1} \tag{7.55}$$

$$S_1 = \frac{\omega L_1^2}{8H_2} \tag{7.56}$$

Using these expressions, equations for l_0 may be rewritten for L_1, H_2, L_2, and H_4 as follows:

$$l_0 = L_1 \left[\frac{L_1^2 + h^2)^{0.5}}{L_1} + \frac{\omega^2 L_1^5}{24H_2^2(L_1^2 + h^2)^{1.5}} \right] \tag{7.57}$$

$$l_0 = L_2 \left[\frac{(L_2^2 + h^2)^{0.5}}{L_2} + \frac{\omega^2 L_2^5}{24H_4^2(L_2^2 + h^2)^{1.5}} \right] \tag{7.58}$$

Equations (7.57) and (7.58) can be solved numerically to obtain the value of L_2. Thus, the distance moved by the pier may be determined by eq. (7.51). In the case of a two-way cable mooring system to allow the vessels use of the land side of the pier, it is necessary to depress the inshore cable(s). This is required to provide sufficient unobstructed depth at low water at the pier's land side. It may be achieved through attaching sinkers (weights) to the cable(s) on either side of navigation boundaries (Figure 7.82). A practical example of such a system is given in Tsinker and Vernigora (1980).

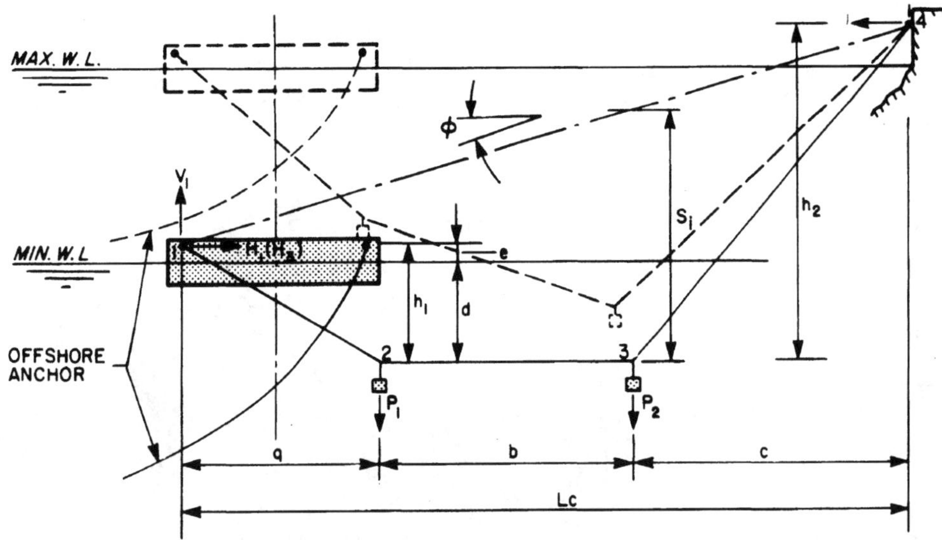

Figure 7.82 Movements of the inshore mooring cable depressed by two sinkers with changes in water level. P_1 and P_2, sinkers. (From Tsinker and Vernigora, 1980.)

To determine the weights of sinkers, the following loads acting on the staying chains should be examined. These loads consist of two parts:

1. V_1' and V_2' = weight of the cable
2. V_1'' and V_2'' = weight of the sinkers

Consequently, the follow expressions may be developed:

$$V_1 = V_1' + V_1'' \tag{7.59}$$

$$V_2 = V_2' + V_2'' \tag{7.60}$$

Similarly,

$$H_1 = H_1' + H_1'' = H_1' + H_2'' \tag{7.61}$$

where H_1' is the horizontal load created by weight of the chain, and $H_1'' = H_2''$ = part of H_1 dependent on the weight of the sinkers P_1 and P_2 (Figure 7.82):

$$V_1' = \frac{\omega l_i}{2} \tag{7.62}$$

in which l_i equals the length of the inshore cable, and

$$H_1' = \frac{\omega L^2}{8S_i} \tag{7.63}$$

in which S_i equals cable deflection above load P_2. Consequently,

$$H_1'' = H_1 - \frac{\omega L^2}{8S_i} \tag{7.64}$$

Therefore, the weight of sinkers P_1 and P_2 may be determined by

Figure 7.83 Typical small municipal passenger dock. Mooring system consists of two booms (steel pipes) and anchor cables.

$$P_1 = V_1'' = \frac{H_1''h_1}{q} \quad (7.65)$$

$$P_2 = V_2'' = \frac{H_1''h_2}{c} \quad (7.66)$$

7.17.2 Mooring System of Rigid Booms and Inshore Cables

The basic concept of this system is illustrated in Figure 7.55*a*. In this case pier movement toward the land or in the opposite direction is restrained by a stem of rigid booms of miscellaneous designs or by access bridges connected via hinges to both the pier and land-based abutments. Pier lateral movement is controlled by mooring cables that are angled from the pier toward the land-based anchor foundation. A typical example is shown in Figure 7.83, where a small passenger dock is restrained by a mooring system comprising of two steel booms and upstream and downstream steel cables angling from the dock to the land-based anchor rings. The booms are hinged at both ends. The land-based support of the boom is shown in Figure 7.84. This two-bracket system allows for free vertical movement and some lateral displacement of the boom outer end.

To reduce the adverse effects of a ship's hard approach to the dock, the booms can be equipped with energy-absorbing springs (Figure 7.85) or other energy-absorbing devices. In

Figure 7.84 Land-based articulated support for rigid boom. 1, Boom; 2, steel brackets; 3, hinge.

Figure 7.85 Boom equipment with energy-absorbing steel spring. 1, Boom; 2, spring; 3, hinge.

this case the boom is fabricated as a telescopic system with an energy-absorbing device placed between its movable parts. The anchor cables are designed as described in Section 7.17.1. In some instances where pier lateral displacement is very limited due to dock operation, the cables can be replaced by rigid booms that along with regular booms create a rigid pier guiding system. An example is shown in Figures 7.16 through 7.18. As pointed out earlier, the pier lateral displacement may also reduced by prestressing the mooring cables.

7.17.3 Mooring System of Bottom-Fixed Dolphins

Structurally, depending on foundation soils, mooring dolphins can be constructed as flexible or nonyielding dolphins. The former is typically represented by piled structures comprised of vertical piles. The simplest form of these dolphins is one consisting of a single vertical pile. This type of anchorage is used in small craft marinas. Nonyielding dolphins are built either in the form of piled structures that include battered piles, or as gravity structures of miscellaneous design (e.g., sheet pile cells, concrete caissons). Flexible dolphins allow for a certain elastic displacement when exposed to a horizontal load, while nonyielding structures do not. For details on mooring dolphins, readers are referred to Tsinker (1986, 1997).

The mechanism that keeps the pier from drifting away from the dolphins should allow for pier-free vertical movements. The simplest form of such a mechanism is the hoop, which can slide along the guide pile. Such a system is commonly used in small craft marinas. It can also be constructed in the form of a simple sliding system as shown in Figure 7.86, or in a rather complex manner as shown in Figures 7.7 and 7.9. More potential alternatives are given in Gaythwaite (1990).

If a mooring system consists of more than two dolphins, the potential misalignment of dolphins (Figure 7.87), uneven load distribution, or other factors may result in an unequal load distribution between individual dolphins. Because during the design stage the actual magnitude of a dolphin's misalignment is unknown, it is a common practice to increase the design lateral load acting against the dolphin by approximately 30%. Hence, where a mooring system comprises more than two dolphins of different stiffness, the total load (P_p) exerted on the system should be distributed among dolphins according to their relative stiffness. Accordingly, the lateral design load for each dolphin (P_d) can be expressed as

Figure 7.86 Sliding connection between a floating pier and a fixed structure.

$$P_d = 1.3 \frac{P_p}{\sum EI} (EI)_d \qquad (7.67)$$

where $\sum EI$ is the total stiffness of the system and $(EI)_d$ the stiffness of the particular dolphin within the system.

7.18 DESIGN OF AN ACCESS BRIDGE

An access bridge is a link between a pier and the land. It constitutes an important part of a floating dock system because efficient dock operation depends largely on the effectiveness of traffic circulation. There are several basic structural arrangements for bridging the space between the pier and the land. Some of them are depicted in Figures 7.50 and 7.54. They include articulated and floating bridges of miscellaneous designs, mobile wedges, and vertical lift bridges. The type of bridge to be used for a particular project depends on local site conditions and is selected on the basis of economical, structural and functional requirements. The cost of bridge maintenance and its possible replacement are also taken into account. The bridge must be convenient and safe to operate;

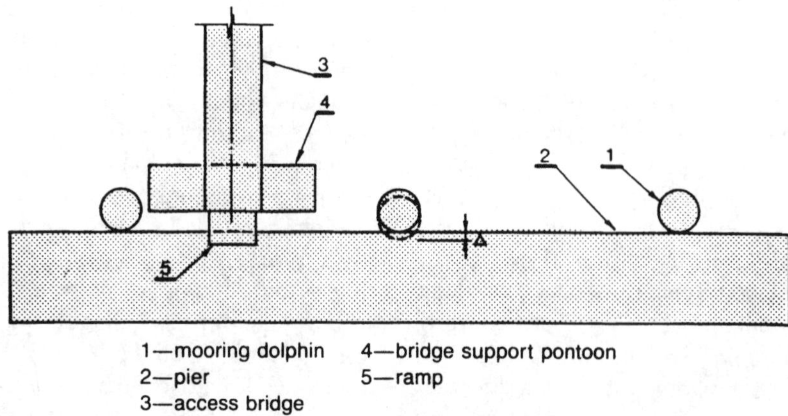

Figure 7.87 Mooring system comprising more than two dolphins.

sometimes aesthetic considerations may also affect selection of the bridge type. Any type of access bridge must be protected against damage from waterborne crafts and large floating objects such as trees and large logs. In practice, this is accomplished by placing a floating boom upstream from the dock. Navigation and warning lights and signs must also be installed to protect both the pier and the access bridge. Some basic design requirements for the most frequently used types of access bridges—articulated and floating bridges—are discussed below.

7.18.1 Articulated Access Bridge

Numerous examples of articulated access bridges are illustrated in Section 7.2. These bridges are generally used where the shore or riverbank is stable and there is no significant erosion or siltation. A single-span articulated bridge is used if the span range permits and if variations in water level are less than 10 m. The length of the bridge, which depends primarily on the magnitude of variation in water levels, should be such that its gradient at the lowest water level is negotiable by vehicles and/or passengers using it. In general, the maximum inclination of a bridge should not exceed a 1:8 gradient. To minimize the bridge load on a supporting pontoon or pier at its lower end, the bridge deck is usually designed as lightweight and free-draining, with good traction structure. The orthotropic deck structure conforms basically to the foregoing requirements (Wolchuk, 1963).

Structurally, articulated bridges are similar to regular highway bridges and are designed according to the same basic principles, where the governing loads include the same vehicles and cargo handling equipment under traveling conditions as used in pier design. Great attention should be paid to the design of the bridge end connection. Typically, articulated access bridges have a hinged connection at the land-based abutment, while its lower end either slides on the pier deck or is hinged at the supporting pontoon, separated from the pier structure, or at the pier.

Usually, a short transition structure in the form of a sliding wedge is added to the bridge lower end to span the gap between the decks of the bridge and the pier. A short cover plate is also required to cover the space created by the bridge pivot. If the bridge is hinged at its both ends, it could be heavily loaded axially, either in compression or in tension. To prevent bearings in hinges from being damaged due to some lateral displacements of the system, the bearings should be slotted to allow for some lateral movement. In long-span bridges where the load reactions are high, a separate pontoon is usually employed to support the lower end of the bridge in order not to affect the trim and capacity of the pier. In this case, the bridge–pontoon system is stabilized through cables that are angled from the lower end of the bridge to dedicated land-based anchorages. To limit the bridge's swing movements about its land-end hinges, these cables should be prestressed. The longitudinal compression load resulting from cable prestressing should be considered in bridge structure design.

Where the bridge is separated from the pier and its lower end is supported on a dedicated pontoon, the gap between the bridge end and the pier is spanned by an articulated ramp ending with a wedge that can slide on a pier deck within limits. The length of the ramp depends on the critical angle of the bridge slope and the free board of the pier and the bridge supporting pontoon (see Figure 7.88):

$$l_r = \frac{\sum D - \sum d}{\sin \phi_{cr}} - l_w \qquad (7.68)$$

where l_r is the length of the ramp between pivots; $\sum D = D_1 + D_2 - D_3$, where D_1 is the depth of the bridge supporting pontoon, D_2 the

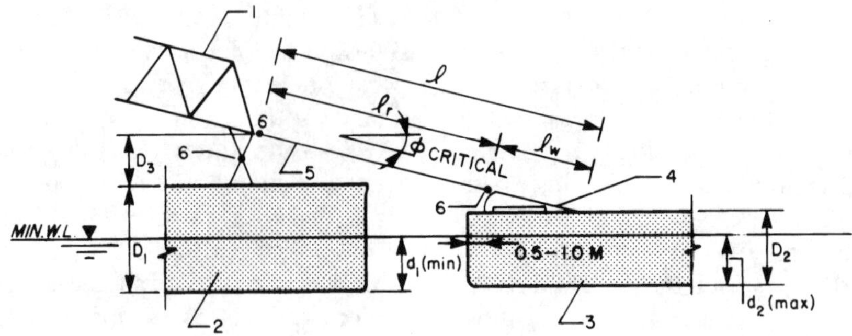

1 – Truss Bridge, 2 – Bridge Support Pontoon, 3 – Pier Pontoon,
4 – Sliding Wedge, 5 – Ramp, 6 – Pivot

Figure 7.88 Design scheme, articulated ramp between the access bridge and the pier. (From Tsinker and Vernigora, 1980.)

depth of the pier, D_3 the distance between the top of the bridge-supporting pontoon and the ramp pivot at the bridge; $\sum d = d_{1(\min)} - d_{2(\max)}$, where $d_{1(\min)}$ is the minimum draft of the bridge-supporting pontoon and $d_{2(\max)}$ is the maximum draft of the pier; ϕ_{cr} is the maximum allowable angle of the ramp (bridge) inclination to the horizontal plane acceptable for traffic operation; and l_w is the length of the sliding wedge.

By using a ramp, the length and the weight of the access bridge can be kept to a minimum. The bridge and the ramp decking are often made of timber planks no thinner than 5 cm. If the bridge inclination exceeds 5%, timber planks are normally placed across the bridge span to provide for better traction. Numerous practical examples of access bridges and articulated platforms of miscellaneous design are illustrated in Section 7.2.

A bridge-supporting pontoon must be strong enough to support the heavy load imposed on it by the bridge. To resist heavy concentrated loads, conventional bulkheads may be replaced by steel trusses placed in both longitudinal and transfer directions. An example is shown in Figure 7.89.

7.18.2 Floating Bridge

As pointed out earlier, floating bridges are typically used when water-level variation exceeds 10 m or when the slope of the riverbank or

Figure 7.89 Bridge-supporting pontoon during construction.

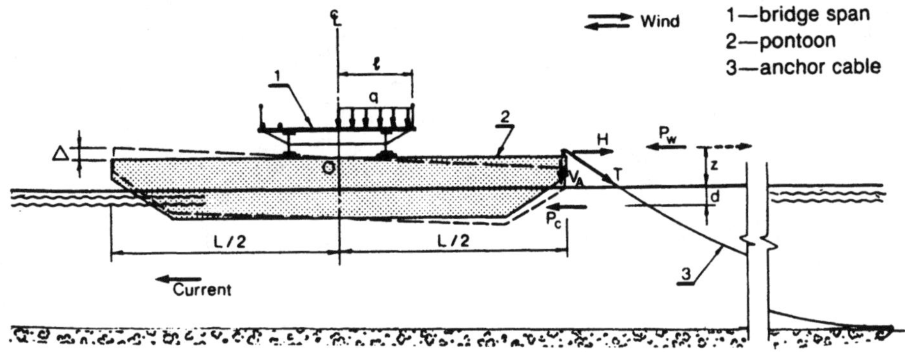

Figure 7.90 Pontoon loading diagram, floating bridge.

shore is flat or consists of a very soft or loose soil not suitable for construction of a land-based structures. A typical floating bridge consists of several pontoons linked with each other by short articulated spans. At a high water level, the bridge is entirely afloat, and as the water level begins to recede, the bridge pontoons gradually settle on fixed platforms until at low water most of the pontoons are supported on fixed platforms at a specific critical grade (Figure 7.50). Structurally, floating bridges are similar to floating piers, differing basically in the shape of the supporting pontoons. Pontoons used in floating piers are usually rectangular, while pontoons used in floating bridges are often streamlined, to reduce drag forces exerted on them by the current. The usual proportions of rectangular bridge pontoons are $L:B:D = 8:2:1$, where L is the length of the pontoon, B the width, and D the depth of the pontoon. As a rule, bridge pontoons require a minimum freeboard of 50 cm. The length of the pontoon must be sufficient for adequate stability normal to the bridge direction, and its maximum tilt in this direction should not exceed 3 to 5%. Pontoon width should be minimized to reduce drag load exerted on it by the current and to prevent major changes in the river channel due to the decreased cross section of the river channel at the bridge location.

Pontoon tilt depends on the magnitude of the current and wind forces exerted on the bridge system, and the magnitude and location of the live load on the bridge deck. Pontoon sinking (Δ) due to exposure to these forces and loads can be determined from the equation

$$\Delta = \frac{V_A}{A} + \frac{M}{\text{SM}} \tag{7.69}$$

where V_A is the vertical component of the mooring force, A the pontoon water-plane area, M the tilting moment, and SM the pontoon water-plane sectional modulus about its transverse axis. Here

$$M = P_c d \pm P_w z + V_A \frac{L}{2} + \frac{ql^2}{2} \tag{7.70}$$

where P_c and P_w are the current and wind forces exerted on the pontoon, q the live load (the most unfavorable combination) on the bridge deck, and d, l, r, z, and L are as shown in Figure 7.90. To balance the pontoons exposed to tilting forces, the bridge could be shifted out of the pontoon centerline.

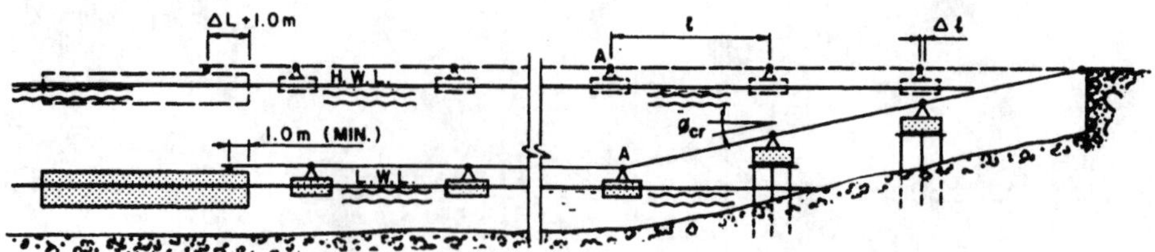

Figure 7.91 Change in floating bridge projected length due to variations in water level.

A floating bridge is usually linked to the land-based abutment and to the pier by articulated ramps. These ramps are designed according to the same rules and procedures as the design of a single-span articulated bridge. Alternatively, the floating bridge could be linked to the land via a regular articulated bridge. The up-and-down movements of the floating bridge due to variations in water level result in a change in the bridge's projected length (Figure 7.91). The difference between the minimum projected length of the bridge and its maximum (ΔL) is determined as follows:

$$\Delta L = n\,\Delta l = nl(1 - \cos\phi_{cr}) \quad (7.71)$$

where n is the number of typical sections in the bridge's articulated portion (note that in the scheme shown in Figure 7.91, $n = 3$), l is the length of a typical bridge span, and ϕ_{cr} is the angle of the bridge critical inclination from the horizontal plane. Subsequently, the length of the sliding ramp linking the bridge with the floating pier must allow for ΔL movement plus at least an additional 50 cm. Finally, if the floating bridge needs to be placed against a strong current in shallow water, hydrodynamic forces similar to ship motion in shallow water could cause rapid sinking (squat) of one end of the bridge pontoon in relation to another. In such a case, the pontoon can be equipped with a ballasting system to equalize the effect of hydrodynamic forces.

REFERENCES

ABAM Engineers, Inc., 1967. *Concentrated Loads on Haunched Deck Panels,* ABAM, Tacoma, WA.

ACI, 1980. *Control of Cracking in Concrete Structures,* ACI 224R-80, American Concrete Institute, Detroit, MI.

———, 1984. *Guide for the Design and Construction of Fixed Offshore Structures,* ACI 337R, American Concrete Institute, Detroit, MI.

———, 1988. *ACI workshop on epoxy-coated reinforcement,* Concrete Int., Dec.

Acres International Ltd., 1987. *Caisson: An Integrated Soft-Wave System for the Analysis and Design of Marine Caissons,* report prepared for Public Works Canada, Ottawa, Ontario, Canada.

AISI, 1981. *Handbook of Corrosion Protection for Steel Pile Structures in Marine Environments,* American Iron and Steel Institute, Washington, DC.

Alumbaugh, R. L., 1966. *Protective Coatings for Steel Piling: Results of 30 Months Tests,* Tech. Rep. R-194, U.S. Naval Civil Engineering Laboratory, Port Hueneme, CA.

American Bureau of Shipping, 1982. *Rules for Building and Classing Steel Vessels,* ABS, New York.

Anonymous, 1982a. Floating container terminal, Valdez, Alaska, *PCI J.,* July/Aug.

———, 1982b. Floating pier makes port debut, *Eng. News Record,* July 15.

———, 1983. *Link Span Design for Floating Falklands Port,* Dock and Harbour Authority, London, Dec.

———, 1985. *New Concepts for Port Construction Support,* No. 8, Port Construction International, Surrey, England.

Arctec Canada Ltd., 1984. *Study of Caisson Towing and Sinking Methods,* Vols. I and II, report prepared for Public Works Canada, Ottawa, Ontario, Canada, Sept.

AWS, 1974. *Corrosion Tests of Metalized Coated Steel: 19-Year Report,* AWS C2.14-74, Committee on Metallizing, American Welding Society, Miami, FL.

Bamfort, P. B., 1980. In-situ measurement of the effect of partial portland cement replacement using fly ash or granulated blast-furnace slag on the performance of mass concrete. *Proc. Institute of Civil Engineers,* Part 2, London, Sept.

Barringer, W. L., 1997. Before using fly ash, *Concrete Int.,* Apr.

Berke, N. S., T. G. Weil, and D. W. Pfeifer, 1988. *Concrete Int.,* Dec.

Bremner, T. W., T. A. Holm, and D. R. Morgan, 1996. *Concrete ships: lessons learned, Proc. 3rd CANMET/ACI International Conference,* SP-163, St. Andrews, New Brunswick, Canada.

British Standard Institution, 1991. Cathodic protection, part 1, in *Code of Practice for Land and Marine Application,* BS 7361, BSI, London.

Brumitt, E. W., and D. G. Dixon, 1980. Alyeska crude loading port, *ASCE Proc. Specialty Conference Ports '80,* Norfolk, VA.

Cairns, J., 1994. Performance of epoxy-coated reinforcement at the serviceability limit state, *Proc. Institute of Civil Engineers, Structures and Buildings,* Feb.

Chow, F., and H. Haynes, 1983. Innovative design for navy pier, *ASCE Proc. Specialty Conference Ports '83,* New Orleans, LA.

Cichanski, W. J., R. Easley, R. F. Henry, and G. Pool, 2001. Design/build project management of floating terminals: successful implementation in a complex environment, *ASCE Proc. Specialty Conference Ports '01,* Norfolk, VA.

Colliat, J. L., and L. Foulhoux, 1998. Suction piles vs. drag anchors for deep water moorings, *Proc. 8th International Conference on Offshore and Polar Engineering,* Montreal, Quebec, Canada, May.

Colliat, J. L., P. Boisard, P. Sparrevik, and J. C. Gramet, 1998. Design and installation of suction anchor piles at soft clay site, *ASCE J. Waterway Port Coast. Ocean Eng.,* July/Aug.

Copola, L., R. Fratesi, S. Monosi, P. Zaffaroni, and M. Colleperdi, 1996. Corrosion of reinforcing steel in concrete structures submerged in seawater, *Proc. 3rd CANMET/ACI International Conference,* SP-163, St. Andrews, New Brunswick, Canada.

Cornick, H. F., 1958. *Dock and Harbour Engineering,* Vol. 2, Charles, Griffin, London.

Cuckson, T., 1981. The suction pile finds its place, *Offshore Eng.,* Apr.

———, 1984. Alternative that works well in deep waters and poor soil. *Offshore Eng.,* Apr.

Delgado, R., 1982. Desarrollo portuario en la Amazonia, Puertos de Iquitos, Pucallpa y Yurimaguas, *Ing. Civil* (Peru), July/Aug.

Du Plat Taylor, F. M., 1949. *The Design, Construction and Maintenance of Docks, Wharves and Piers,* Eyre & Spottiswoode, London.

Dyvik, R., K. H. Andersen, S. Borg Hansen, and H. P. Christophersen, 1993. Field tests of anchors in clay, *ASCE J. Geotech. Eng.,* Vol. 119, No. 10.

Epskamp, G. J., and G. Hannink, 1992. Load test on suction piles in the port-of-Rotterdam, *Proc. 10th International Harbour Congress,* Antwerp, Belgium.

FIP, 1985. *Recommendations for the Design and Construction of Concrete Sea Structures,* Federation Internationale de la Precontrainte, London, Nov.

Fitzpatrick, J. B., and C. Gamarra, 1983. Bathimetry and its utilization to define short-term events on the upper Amazon, *Proc. Canadian Centennial Hydrography Conference,* Ottawa, Ontario, Canada.

Fitzpatrick, J. B., R. G. Tanner, and G. P. Tsinker, 1985. Alluvial rivers and their impact on layout and design of dock facilities, *Proc. 26th Congress of Permanent International Association of Navigation Congresses,* Brussels, Belgium.

Forssblad, L., 1987. *Need for Consolidation of Superplasticized Concrete Mixes,* ACI SP-96, American Concrete Institute, Detroit, MI.

Gaythwaite, J. W., 1990. *Design of Marine Facilities for Berthing, Mooring and Repair of Vessels,* Van Nostrand Reinhold, New York.

Gerwick, B. C., 1975. Durability of structures under water, *Proc. AIPC-PIP-CEB-RILEM-IASS Symposium on Behavior of Concrete Structures in Service,* Liege, Belgium, June.

———, 1976. Materials for concrete ships: durability and corrosion protection, coating, inspection, and repair, *Proc. Conference on Concrete Ships and Floating Structures,* Berkeley, CA, Sept.

———, 1986. *Construction of Offshore Structures,* Wiley, New York.

———, 1990. International experience in the performance of marine concrete, *Concrete Int.,* May.

Gjory, O. E., 1971. Long-time durability of concrete in seawater, *ACI J.,* Jan.

Harvey, H. H., 1980. Permeability of concrete in seawater, *Proc. International Conference on Performance of Concrete in Marine Environments,* St. Andrews, News Brunswick, Canada, Aug.

Hetherington, W. G., 1980. Floating jetties: construction, damages and repair, *Proc. Concrete Ships and Floating Structures Convention,* Nov. 1979, Thomas Reed Publications, Rotterdam, The Netherlands.

Hetinyi, M., 1946. *Beams on Elastic Foundations,* University of Michigan Press, Ann Arbor, MI.

Hogervorst, J. R., 1980. Field trials with large diameter suction piles, *Proc. Offshore Technology Conference,* Houston, TX.

Keck, R. H., and E. H. Riggs, 1997. Specifying fly ash for durable concrete, *Concrete Int.,* Apr.

Lachemi, M., G. Tagnit-Hamon, A. Li, and P.-C. Aitcin, 1998. Long term performance of silica fume concrete, *Concrete Int.,* Jan.

Larsen, P., 1989. Suction anchors as an anchoring system for floating offshore structures, *Proc. 14th Offshore Technology Conference,* Houston, TX.

MacNaught, D. F., 1967. Strength of ships, *in Principles of Naval Architecture,* Society of Naval Architects and Marine Engineers, New York.

Malhotra, V. M., 1986. *Mechanical Properties and Freeze/Thaw Resistance of Condensed Silica Fume Concrete,* ACI SP-91, Vol. 2, American Concrete Institute, Detroit, MI.

Malloch, R. D., and E. R. Kolbe, 1978. *Trowl Cable Corrosion,* Oregon State University, Corvallis, OR, Nov.

Mehta, P. K., 1980a. Durability of concrete in marine environments: a review, *Proc. International Conference on Performance of Concrete in Marine Environments,* St. Andrews, New Brunswick, Canada, Aug.

———, 1980b. *Durability of Concrete in Marine Environments,* ACI SP-65, American Concrete Institute, Detroit, MI.

———, 1988. Durability of concrete exposed to marine environment: a fresh look, *Proc. 2nd International Conference on Concrete in Marine Environment,* St. Andrews, New Brunswick, Canada.

———, 1997. Durability: critical issues for the future, *Concrete Int.,* July.

Mehta, P. K., and B. C. Gerwick, 1982. Cracking–corrosion interaction in concrete exposed to marine environments, *Concrete Int.,* Oct.

Miller, J. R., and D. J. Fielding, 1998. Durability by admixture, *Concrete Int.,* Apr.

Minikin, R. R., 1963. *Wind, Waves and Maritime Structures,* 2nd rev. ed., Charles Griffin, London.

Muckle, W., 1975. *Naval Architecture for Marine Engineers,* Newnes-Butterworth, London.

Myint Lwin, M., and C. C. Gloyd, 1984. Rebuilding the Hood Canal floating bridge, *Concrete Int.,* June.

Nachman, T. R., and E. R. Duffy, 1974. Effect of alloying additions on seawater corrosion resistance on iron–aluminum based alloys, *Corrosion,* Vol. 30, p. 10.

Naval Civil Engineering Laboratory, 1982. *Conceptual Design of Navy Floating Pier,* CR82.031, NCEL, Port Hueneme, CA.

OCDI, 1991. *Technical Standard for Port and Harbor in Japan,* Overseas Coastal Area Development Institute.

Olberg, T. S., T. Guttormsen, G. Molland, and J. Andersen, 1997. Full scale field trial of taut leg mooring using fibre rope and suction anchor attached to a semi-submersible drilling unit, *Proc.*

29th Offshore Technology Conference, Houston, TX.

Oshiro, T., and S. Tanigawa, 1988. Effect of surface coatings on the durability of concrete exposed to marine environment, *Proc. 2nd International Conference on Concrete in Marine Environment,* St. Andrews, New Brunswick, Canada.

Peel, J., 1982. Corrosion in concrete: a "chemical" solution, *Engi. Dig.* (Can. Eng. Publ.), Vol. 2, No. 10, Nov.

PIANC, 1995. *Port Facilities for Ferries: Practical Guide,* Rep. Work. Group 11, Suppl. Bull. 87, Permanent International Association of Navigation Engineers, Brussels, Belgium.

Quinn, A. F., 1972. *Design and Construction of Port and Marine Structures,* 2nd ed., McGraw-Hill, New York.

Ractliffe, A. T., 1983. The basis and essentials of marine corrosion in steel structures. *Proc. Institute of Civil Engineers,* Part 1, Pap. 8665, Nov.

Read, J. A., 1987. Rebar corrosion: coating a cure? *Civil Eng.,* Nov./Oct.

Roesli, A., 1955. *Lateral Load Distribution in Multibeam Bridges,* Rep. 223.10, Fritz Engineering Laboratory, Lehigh University, Bethlehem, PA.

Seanlon, J. N., 1997. Controlling concrete during hot and cold weather, *Concrete Int.,* June.

Singleton, W. T., Jr., 1970. Performance of various coating systems in a marine environment, *Mater. Protect. Perform.,* Vol. 9, No. 11, pp. 37–43.

Society of Naval Architects and Marine Engineers, 1980. *Ship Design and Construction,* SNAME, New York.

Sparrevik, P., 1996. Suction anchor piles: state of the art, *Proc. International Conference on Mooring and Anchoring,* Aberdeen, Scotland.

Staehle, R. W., 1975. *Corrosion,* Vol. 31, No. 1.

Stern, I. L., and M. F. Wheatcroft, 1978. Toward improving the reliability of anchor chains and accessories, *Proc. Offshore Technology Conference,* Houston, TX.

Taggart, R. (ed.), 1980. *Ship Design and Construction,* Society of Naval Architects and Marine Engineers, New York.

Tanner, R. G., G. P. Tsinker, and R. Delgado, 1983. Modernization and upgrading of the dock and associated facilities at Iquitos, Peru, *ASCE Proc. Specialty Conference on Ports '83,* New Orleans, LA.

Thorne, C. P., 1998. Penetration and load capacity of marine drag anchors in soft clay. *ASCE J. Geotech. Geoenviron. Eng.,* October, Vol. 124, No. 10.

Tsinker, G. P., 1986. *Floating Ports: Design and Construction Practices,* Gulf Publishing, Houston, TX.

———, 1995. *Marine Structures Engineering: Specialized Applications,* Chapman & Hall, New York.

———, 1997. *Handbook of Port and Harbor Engineering: Geotechnical and Structural Aspects,* Chapman & Hall, New York.

Tsinker, G. P., and E. Vernigora, 1980. Floating piers, *ASCE Proc. Specialty Conference Ports '80,* Norfolk, VA, May.

Turner, F. H., 1974. The contransfer concept for cryogenic carriers, *Ocean Ind.,* Nov.

Tuthill, L. H., 1988. Lasting concrete in a sulfate environment, *Concrete Int.,* Dec.

Umansky, A. A., 1939. *Floating Bridges, Transport Publishing House,* Moscow (In Russian).

Van den Haak, R., 1983. Choosing and using anchors properly, *Offshore Res.,* Vol. 1, No. 1.

VSL International Ltd., 1987. *Floating Concrete Structures: Examples from Practice,* VSL, Berne, Switzerland.

Walther, R., 1956. *Investigation of Multi-beam Bridges,* Report 223.14, Fritz Engineering Laboratory, Lehigh University, Bethlehem, PA.

Wolchuk, R., 1963. *Design Manual for Orthotropic Steel-Plate Deck Bridges,* American Institute of Steel Construction, New York.

Yee, A. A., and L. A. Daly, 1984. Concrete barges in the Pacific Area, presented at the American Concrete Institute Convention, New York, Oct./Nov.

Zinserling, M. H., and W. J. Cichanski, 1982. Design and functional requirements for the floating container terminal at Valdez, Alaska, *Proc. Offshore Technology Conference,* Houston, TX.

8

OFFSHORE MOORINGS†

John Headland and Eric D. Smith
Moffatt & Nichol Engineers
New York, New York

† This chapter includes some materials (including Figures 8.4 to 8.10 and 8.12 to 8.37) from the U.S. Navy's design manual *Basic Criteria and Planning Guidelines*, DM-26.5 (now included in the Department of Defense Handbook *Mooring Design*, MIL-HDBK-1026/4A, July 1, 1999). DM-26.5 was prepared for the U.S. Navy by John Headland.

8.1 INTRODUCTION

Offshore moorings comprise the assembly of anchors, chains, buoys, and associated hardware used to secure vessels at sea. These systems can be used to moor systems on a temporary or a permanent basis. Design conditions for the latter can be quite severe, especially in the open ocean. In this chapter we focus on the design of ship mooring systems in nearshore waters where water depths are 30 m or less.

8.2 OFFSHORE MOORING SYSTEMS

Moorings can generally be divided into *single-point moorings* (SPMs) and *multibuoy moorings* (MBMs). An MBM is also known as a *spread mooring*. Vessels are secured by a single line or structure in an SPM (Figures 8.1 through 8.3) and by multiple lines in an MBM (Figures 8.4 and 8.5). A vessel will swing, or weather-vane, around an SPM in order to align itself with prevailing wind, wave, and current conditions. This alignment tends to reduce the load on the mooring system. An MBM, on the other hand, holds a vessel in a relatively fixed position, and the vessel cannot move to align itself with prevailing conditions. As a result, an MBM can experience relatively high loads if wind, currents, or waves act at an angle to mooring. SPMs will involve considerably less hardware that an MBM for the same vessel and environmental conditions and are often less costly as a result. SPMs, however, require more

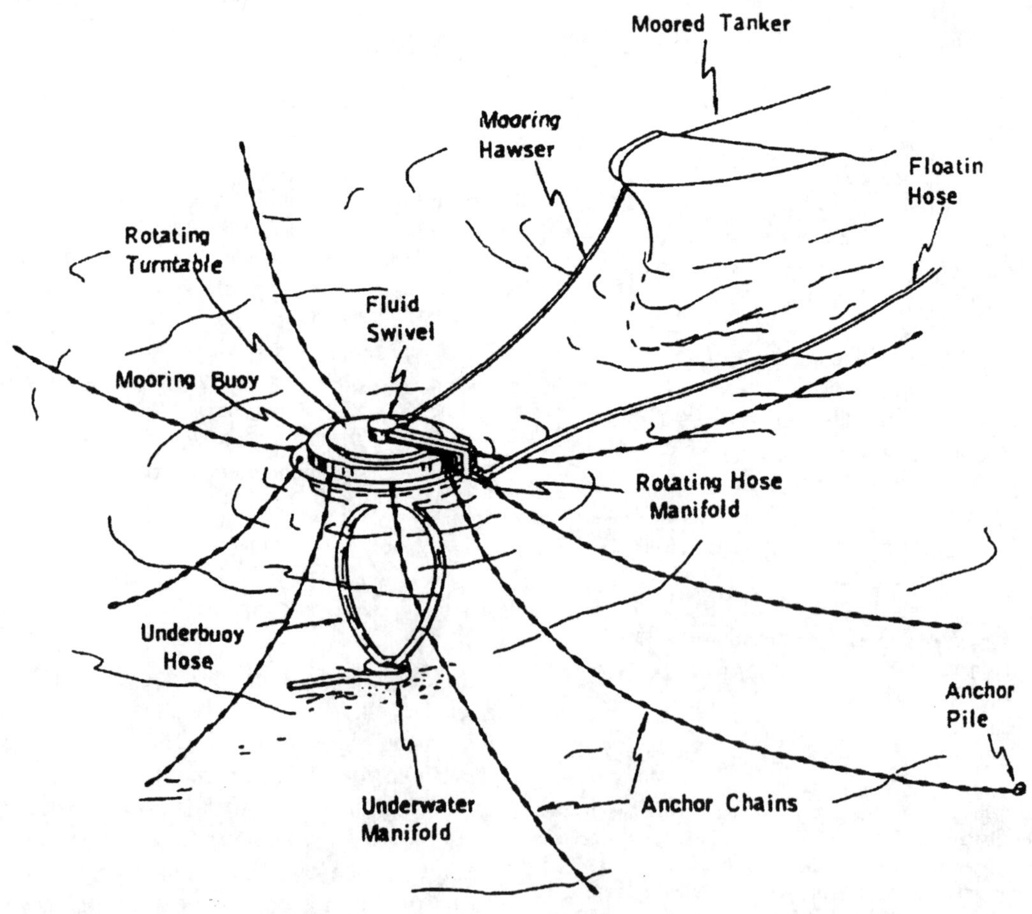

Figure 8.1 Catenary anchor leg mooring. (From Flory et al., 1977.)

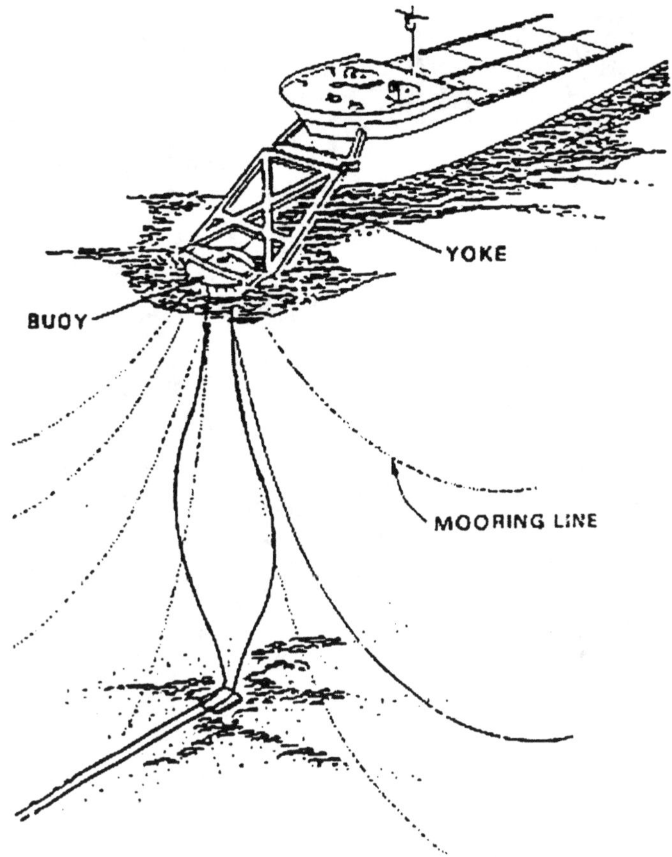

Figure 8.2 Catenary anchor leg mooring with rigid yoke. (From API, 1990.)

space than MBMs inasmuch as there must be room for a vessel to swing in a circle around the single mooring point.

8.3 MOORING SYSTEM COMPONENTS

Figures 8.1 through 8.5 show principal offshore mooring components. Components discussed below include anchors, sinkers, anchor chain, buoys, and mooring lines (hawsers).

8.3.1 Anchors

Drag, pile, dead-weight, and direct-embedment anchors are used in offshore moorings. Several types of drag anchors are shown in Figure 8.6. Figure 8.7 presents the important anchor features. The anchor shank transfers load to the anchor flukes that bear on the seafloor soil. The fluke tip is designed to penetrate the seafloor. Drag anchors resist horizontal loading efficiently but cannot resist much vertical loading. The ratio of mooring line length to water depth (mooring line scope) must be large to ensure a

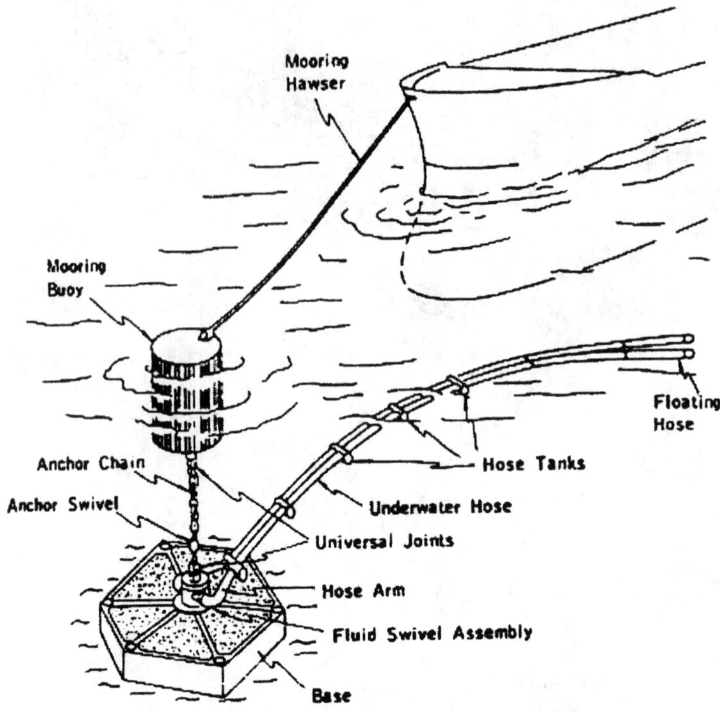

Figure 8.3 Single-anchor-leg mooring. (From Flory et al., 1977.)

near-zero line angle at the anchor. The design of drag anchors is presented in Section 8.6.4.

Pile anchors resist horizontal and vertical loading. Figure 8.8 presents some example pile anchor designs. Procedures for designing anchor piles are given in Rocker (1985) and by the American Petroleum Institute (API, 1987).

A dead-weight anchor also resists both horizontal and vertical loading. Figure 8.9 shows several types of deadweight anchor construction. Uplift loads are resisted by anchor weight, lateral loads are resisted by friction between the anchor and the seafloor. Design procedures for deadweight anchors are given in Rocker (1985). A direct-embedment anchor is driven, vibrated, or propelled vertically into the seafloor. The anchor is then expanded or reoriented to provide vertical and horizontal holding capacity. Rocker (1985) summarizes the design of this anchor type.

8.3.2 Sinkers

A sinker is a concentrated weight placed on the middle of a mooring line (Figure 8.10). Sinkers are used to promote horizontal loading at an anchor, to enhance mooring energy absorption, or both. A typical concrete sinker is shown in Figure 8.10.

8.3.3 Anchor Chains

Stud link chain is strong, highly durable, heavy, and has good shock-absorbing characteristics. Accordingly, chain is often used in mooring

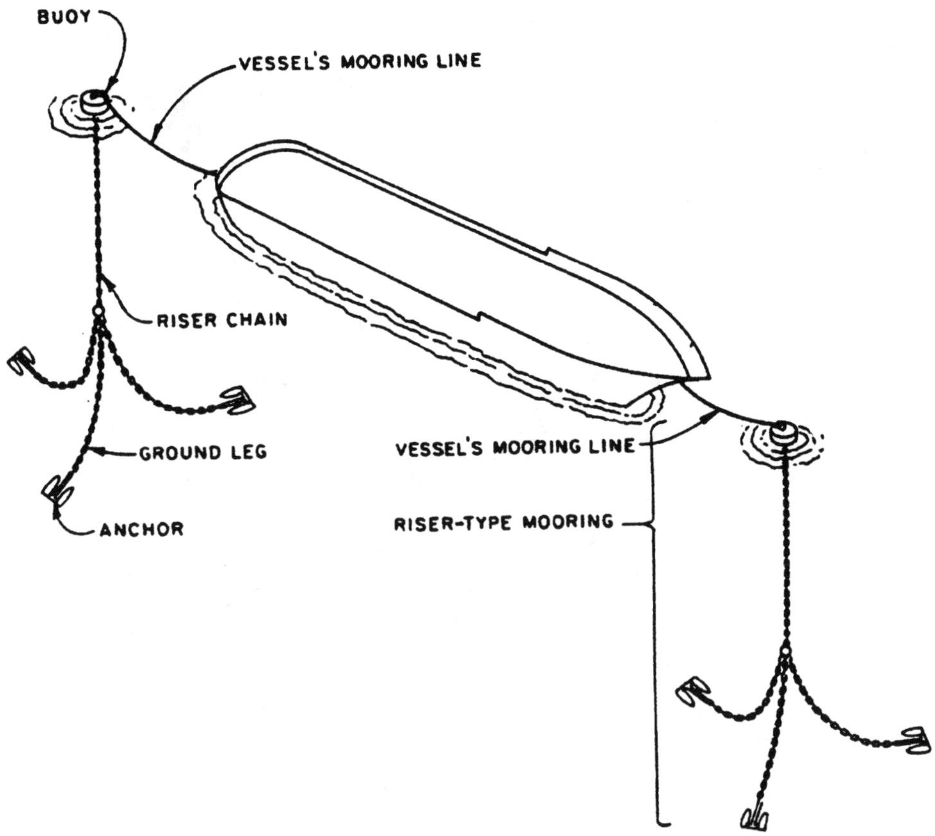

Figure 8.4 Bow-and-stern mooring.

systems. Other line types, such as wire or synthetic lines, are used less often in ship moorings. Chain links have a center stud that keeps the link from collapse under tension. The diameter of the bar used to make the link is the characteristic chain dimension. Chain length is generally reported in terms of shots. A shot is a 15-fathom (about 27.5 m) length.

8.3.4 Buoys

A buoy serves as a means of connecting the ship hawser to the mooring. The buoy also resists force in the case of a single anchor leg mooring (SALM). Buoys must have sufficient buoyancy to support the mooring. Berteaux (1976) and the American Bureau of Shipping (1975) provide buoy design guidance.

8.3.5 Mooring Lines or Hawsers

Ships are connected to moorings using mooring lines or hawsers. These elements are typically synthetic ropes, wire ropes, or chains. A discussion of mooring lines is presented by the Oil Companies International Marine Forum (OCIMF, 1997).

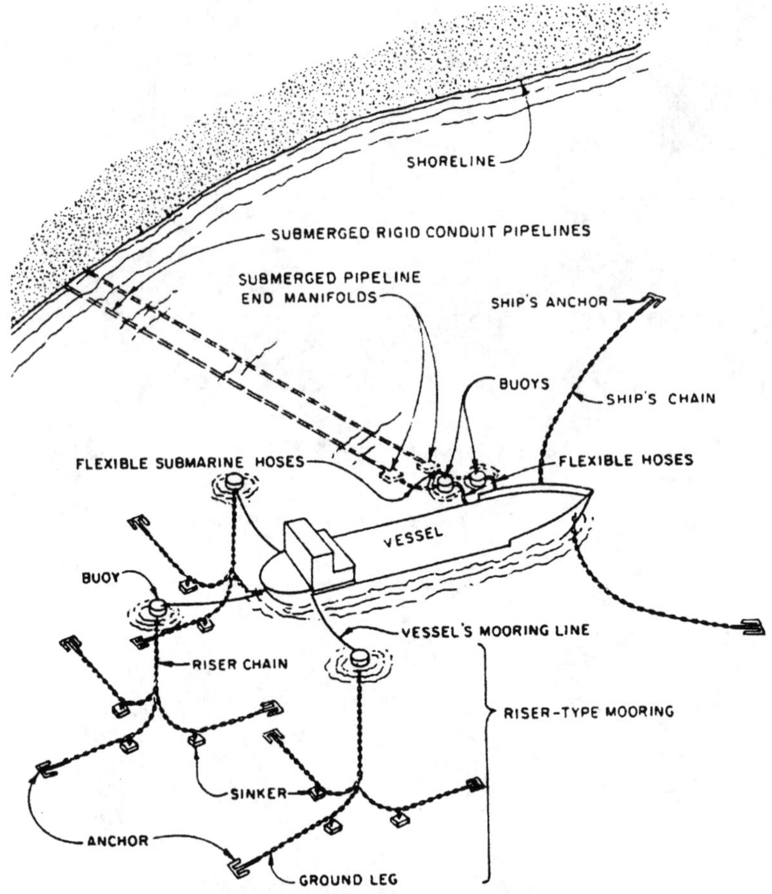

Figure 8.5 Typical spread mooring.

8.4 MOORING DESIGN PROCEDURE

Mooring design involves preliminary layout, selection of design criteria, evaluation of applied environmental loads, computation of mooring system loads, and design of mooring components. Figure 8.11 presents a flowchart outlining a general mooring design procedure and shows that several mooring design iterations are often needed to development a final design, especially for cases involving dynamic loading. A few of the critical design steps are discussed below.

8.4.1 Mooring Layout

The most important principle in mooring layout is to align the longitudinal axis of the moored vessel parallel to prevailing wind, waves, and current directions, as this practice will minimize the loads on the mooring. In many cases, winds will arrive from any direction; hence

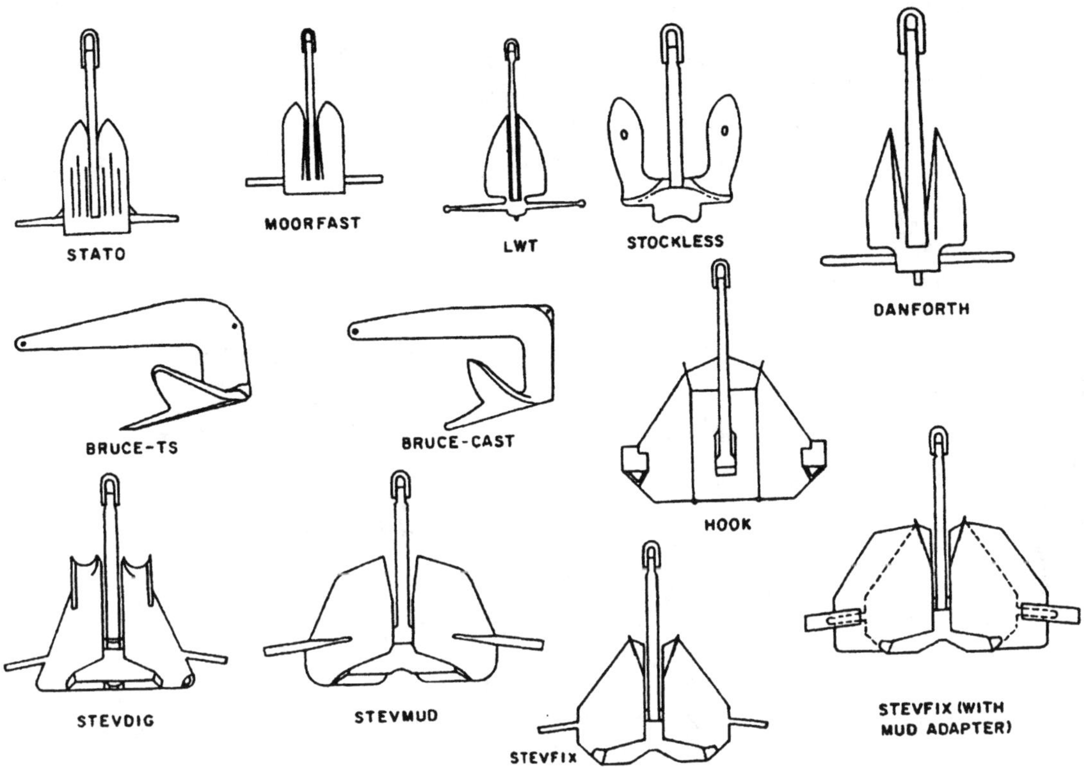

Figure 8.6 Drag-embedment anchors.

alignment with waves and currents is of great importance. Moorings located within harbors are generally protected from direct ocean wave attack, whereas those located outside a harbor are exposed directly to full ocean wave attack. Mooring layout may have to consider a range of vessels. Additionally, the layout will depend on mooring use, available space, allowable vessel movement, and the area required for safe maneuvering into the mooring.

8.4.2 Environmental Site Conditions

Environmental site conditions such as bottom soil characteristics, water depth, water levels, winds, currents, and waves are important to mooring design. Design criteria for water level, wind, current, and wave characteristics should be selected using probabilistic methods for safe/economical mooring design. The return period associated with an event, estimated using probabilistic principles, is defined as the average length of time between occurrences of that event. For example, a 50-year wind speed will occur, on average, once every 50 years.

Offshore moorings for offshore oil production are often designed using 100-year-return-period conditions (American Bureau of Shipping, 1975; API, 1990). The U.S. Navy (1985) recommends that ship moorings be designed using 50-year-return-period conditions, although these moorings are often unoccupied

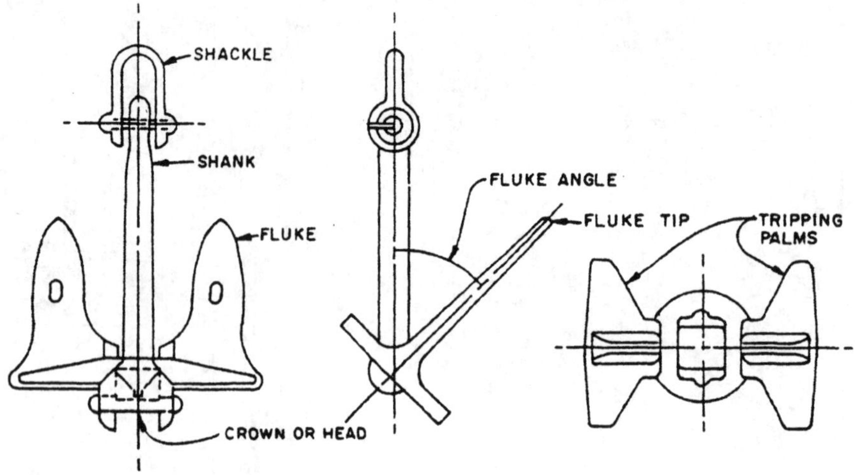

Figure 8.7 Elements of a drag-embedment anchor.

in severe conditions. The U.S. Navy (1986) provides additional guidance for selection of design conditions for a variety of mooring types. Some offshore moorings are designed for a vessel to exit the mooring when weather conditions exceed certain thresholds.

8.4.2.1 Soil Conditions. Anchor design is dictated by seafloor soil conditions. For example, pile or deadweight anchors can be designed for rock or hard soil conditions, whereas drag anchors generally cannot. Detailed soil investigations are required to develop a safe anchor design (see Rocker, 1985).

8.4.2.2 Water Depth and Water Levels. Water depth and water-level fluctuations have a direct bearing on the maximum allowable vessel draft, mooring line geometry, and area available for maneuvering. Although tide is the normal cause of water-level fluctuations, storm tides, seiche, and tsunamis can be a factor. Harbor sedimentation can reduce water depth and should be considered where appropriate.

8.4.2.3 Wind. Wind is a necessary design condition for nearly every mooring installation. Loads imparted by winds are idealized as static (i.e., steady state) in many design applications, but wind can impart considerable dynamic loading in some cases. Design wind speeds must be determined on the basis of local wind conditions, preferably from long-term wind measurements. Typically, recorded wind data must be adjusted for elevation, duration, and over-land/over-water effects to represent conditions at the mooring site (see U.S. Navy, 1985).

Design winds are normally corrected to a standard elevation of 10 m and standard wind duration. Wind speeds depend on duration, the largest gust associated with the shortest duration. Selection of the appropriate wind-speed duration has evolved on the basis of analytics and experience. A 1-minute duration is used in the design of single-point moorings for oil tankers [American Bureau of Shipping (ABS), 1975; Flory et al., 1977; API, 1990]. Wind loads on U.S. Navy vessels are based on a 30-

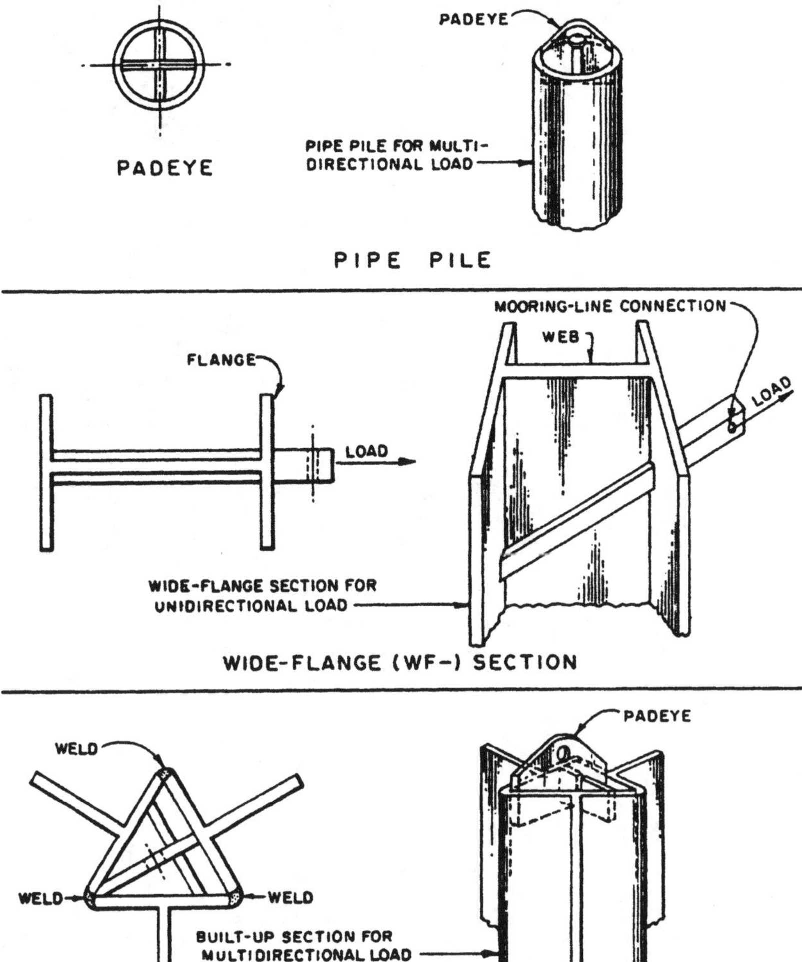

Figure 8.8 Types of pile anchors.

second duration (U.S. Navy, 1985). In reality, a wind field is characterized by a spectrum of wind speeds (i.e., a wind field with a range of speeds and durations). A vessel may or may not respond to individual wind gusts, depending on whether the frequency of the gust coincides with the natural period of the moored vessel system. The steady wind speeds above are intended to approximate a conservative equivalent of the dynamic loads experienced by a moored ship system in response to a wind spectrum. Section 8.7.3 presents methods for evaluating dynamic loads from wind and wave spectra. Additionally, it should be noted that

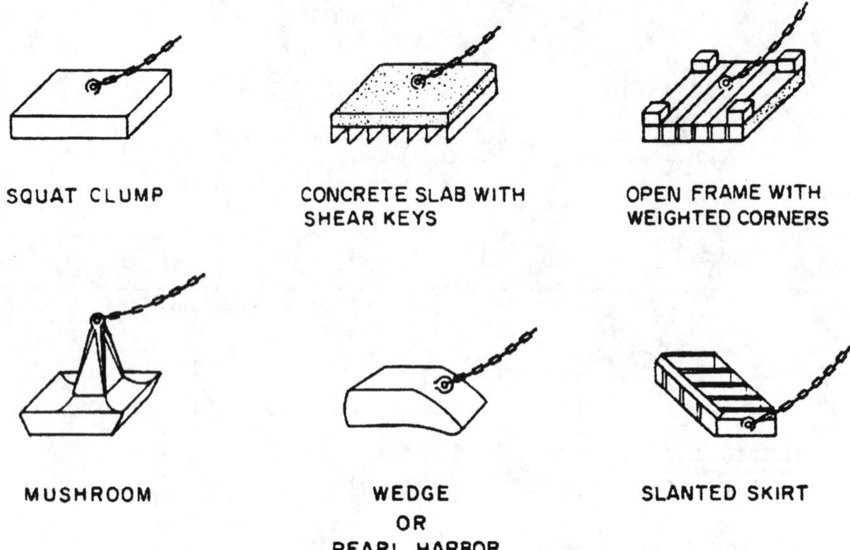

Figure 8.9 Types of dead-weight anchors.

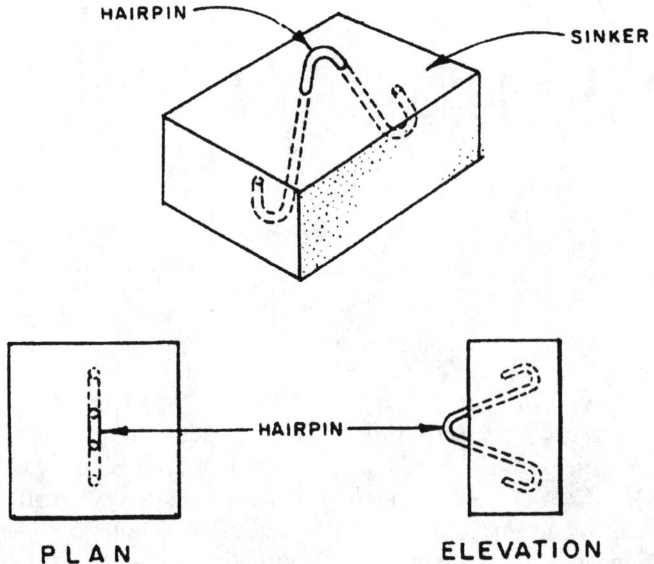

Figure 8.10 Concrete sinker block.

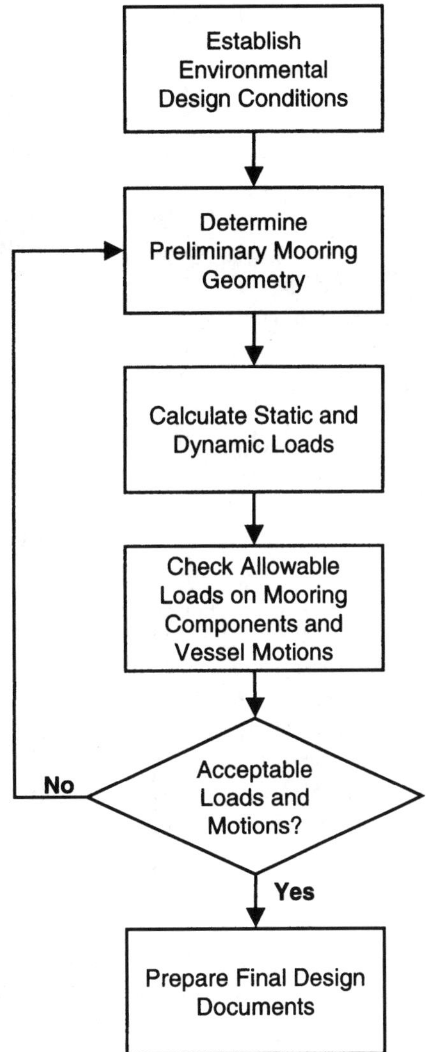

Figure 8.11 Mooring Design Process. (From Remery and Van Oortmerssen, 1973.)

rapid wind gusts and shifts in wind direction can produce very large loads on some mooring systems (Seelig and Headland, 1998). Simiu and Scanlon (1978) and the U.S. Navy (1985) present procedures for estimating design wind speeds using measured data. Batts et al. (1980) and Changery (1982a,b) give design wind speeds for various U.S. locations.

8.4.2.4 Currents. Currents of any appreciable magnitude can dominate mooring design; hence, it is highly desirable to align a mooring with prevailing currents. Currents can also dictate maneuvering into a mooring. Tidal currents are common in harbors and estuaries and are best estimated on the basis of direct measurements. Current speeds may be estimated using physical or numerical models where measurements are not available. Current speed and direction vary during the tidal cycle. Large eddies or macro vortices can form in some cases, imparting large mooring loads (de Kat and Wichers, 1990). Maximum tidal currents are generally used in design. Currents generated by river discharge are also important and are generally estimated from existing flow records. Tidal, oceanic, coastal, and wind-driven currents can be important at offshore moorings and are best estimated from measurements. Design current speed and direction should consider extreme conditions (e.g., storm tides, river floods), depending on the intended used of the mooring.

8.4.2.5 Waves. Wave loading can dictate mooring design, especially in the open ocean. A mooring located in a protected harbor may also be exposed to sea and swell. Design wave conditions are best estimated from long-term measurements taken at the mooring site. Such measurements, however, are seldom available and it is often necessary to resort to analytical wave hindcast studies. Waves generated by storm activity have periods in the range 6 to 20 seconds. Long waves, which may be a factor in mooring design, have periods ranging from 20 seconds to several minutes (i.e., seiche). Passing vessels can also impart significant dynamic loading to a moored vessel.

8.5 STATIC WIND AND CURRENT LOADS

Winds and currents impart a longitudinal load, lateral load, and yaw moment (Figure 8.12) to a moored vessel.

8.5.1 Wind Load

The following paragraphs summarize wind load formulas presented in U.S. Navy (1985). Alternative methods may be found in Remery and Van Oortmerssen (1973) and OCIMF (1994). OCIMF (1994) and U.S. Department of Defense (1999) should be consulted for design problems.

Lateral wind load is determined using the equation

$$F_{yw} = \tfrac{1}{2}\, \rho_a V_w^2 A_y C_{yw} f_{yw}(\theta_w) \qquad (8.1)$$

where F_{yw} is the lateral wind load, ρ_a the mass density of air, V_w the wind velocity, A_y the lateral projected area of ship, C_{yw} the lateral wind-force drag coefficient, and $f_{yw}(\theta_w)$ the shape function for lateral load where θ_w is the wind angle. The lateral wind-force drag coefficient depends on the hull and superstructure area of the vessel:

$$C_{yw} = 0.92\, \frac{(V_S/V_R)^2\, A_S + (V_H/V_R)^2 A_H}{A_y} \qquad (8.2)$$

where C_{yw} is the lateral wind-force drag coefficient, V_S the average wind velocity over a

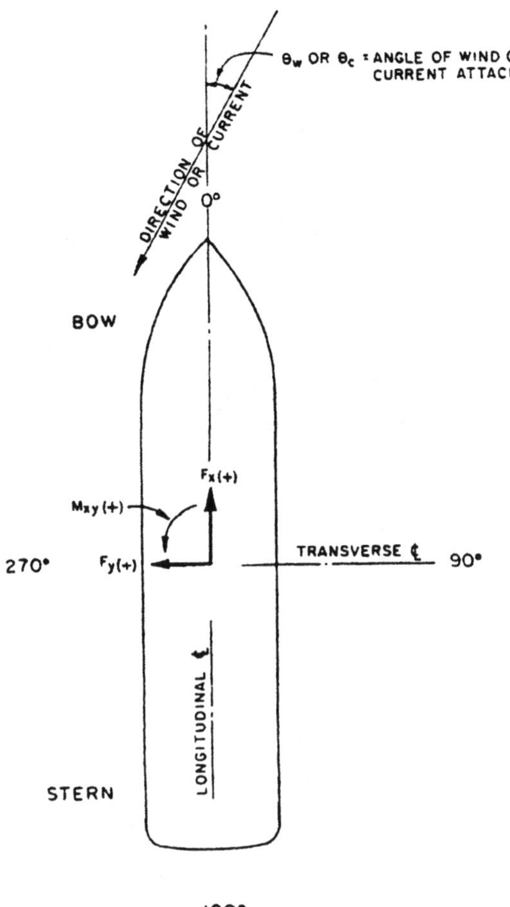

Figure 8.12 Static mooring load definition.

superstructure, V_R the average wind velocity at 10 m above sea level, A_S the lateral projected area of a superstructure, and A_H the lateral projected area of the hull.

The following formulas are also used:

$$\frac{V_S}{V_R} = \left(\frac{h_s}{h_R}\right)^{1/7} \tag{8.3}$$

$$\frac{V_H}{V_R} = \left(\frac{h_H}{h_R}\right)^{1/7} \tag{8.4}$$

The shape function for lateral load, $f_{yw}(\theta_w)$, is given as

$$f_{yw}(\theta_w) = \frac{\sin\theta_w - \sin(5\theta_w)/20}{1 - \frac{1}{20}} \tag{8.5}$$

Longitudinal wind load is determined using the following equation:

$$F_{xw}\ \tfrac{1}{2}\ \rho_a V_w^2 A_x C_{xw} f_{xw}(\theta_w) \tag{8.6}$$

where F_{xw} is the longitudinal wind load, ρ_a the mass density of air, V_w the wind velocity, A_x the longitudinal projected area of a ship, C_{xw} the longitudinal wind-force drag coefficient, and $f_{xw}(\theta_w)$ the shape function for a longitudinal load. The longitudinal wind-force drag coefficient varies according to vessel type and characteristics. Additionally, a separate wind-force drag coefficient is provided for a head wind (over the bow, $\theta_w = 0°$) and a tailwind (over the stern, $\theta_w = 180°$) conditions. The head wind (bow) wind-force drag coefficient is designated as C_{xwB} and the tailwind (stern) wind-force drag coefficient is designated C_{xwS}. The following longitudinal wind-force drag coefficients are recommended for hull-dominated vessels, such as aircraft carriers, submarines, and passenger liners:

$$C_{xwB} = 0.4 \qquad C_{xwS} = 0.4 \tag{8.7}$$

For all remaining types of vessels, except for specific deviations, the following are recommended:

$$C_{xwB} = 0.7 \qquad C_{xwS} = 0.6 \tag{8.8}$$

An increased head wind-force drag coefficient is recommended for center-island tankers:

$$C_{xwB} = 0.8 \tag{8.9}$$

For ships with an excessive amount of superstructure, such as destroyers and cruisers, the recommended tail wind-force drag coefficient is

$$C_{xwS} = 0.8 \tag{8.10}$$

An adjustment consisting of adding 0.08 to C_{xwB} and C_{xwS} is recommended for all cargo ships and tankers with cluttered decks.

The longitudinal shape function, $f_{xw}(\theta_w)$, differs over the head and tailwind regions. The incident wind angle that produces no net longitudinal force, designated θ_{wz} for zero crossing, separates these two regions. Selection of θ_{wz} is determined by the mean location of the superstructure relative to midships (see Table 8.1). For many ships, including center-island tankers, $\theta_{wz} \simeq 100°$ is typical; $\theta_{wz} \simeq 110°$ is recommended for warships.

The shape function for the longitudinal load for ships with single, distinct superstructures and hull-dominated ships is given below. (Examples of ships in this category are aircraft carriers and cargo vessels.)

$$f_{wx}(\theta_w) = -\cos\phi \tag{8.11}$$

where

$$\phi_{(-)} = \frac{90°}{\theta_{wz}}\theta_w \qquad \text{for } \theta_w < \theta_{wz}$$

Table 8.1 Selection of θ_{wz}

Location of Superstructure	θ_{wz} (deg)
Just forward of midships	80
On midships	90
Aft of midships	100
Hull-dominated	120

$$\phi_{(+)} = \frac{90°}{180° - \theta_{wz}}(\theta_w - \theta_{wz}) + 90° \qquad \text{for } \theta_w > \theta_{wz}$$

The value of $f_{xw}(\theta_w)$ is symmetrical about the longitudinal axis of the vessel. Therefore, when $\theta_w > 180°$, use $360° - \theta_w$ as θ_w in determining the shape function. For example, if $\theta_w = 330°$, use $360° - \theta_w = 360° - 330° = 30°$ for θ_w.

Ships with distributed superstructures are characterized by a "humped" cosine wave. The shape function for a longitudinal load is

$$f_{xw}(\theta_w) = \frac{-\sin\gamma - \sin 5\gamma/10}{1 - \frac{1}{10}} \tag{8.12}$$

where

$$\gamma_{(-)} = \frac{90°}{\theta_w}\theta_w + 90° \qquad \text{for } \theta_w < \theta_{wz}$$

$$\gamma_{(+)} = \frac{90°}{180° - \theta_w}\theta_w + \left(180° - \frac{90°\,\theta_{wz}}{180° - \theta_{wz}}\right) \qquad \text{for } \theta_w > \theta_{wz}$$

As explained above, use $360° - \theta_w$ for θ_w when $\theta_w > 180°$.

The wind yaw moment is calculated using the equation

$$M_{xyw} = \tfrac{1}{2}\rho_a V_w^2 A_y L C_{xyw}(\theta_w) \tag{8.13}$$

where M_{xyw} is the wind yaw moment, ρ_a the mass density of air, V_w the wind velocity, A_y the lateral projected area of a ship, L the ship length, $C_{xyw}(\theta_w)$ the yaw moment coefficient. Typical yaw-moment coefficient curves are presented in Figure 8.13.

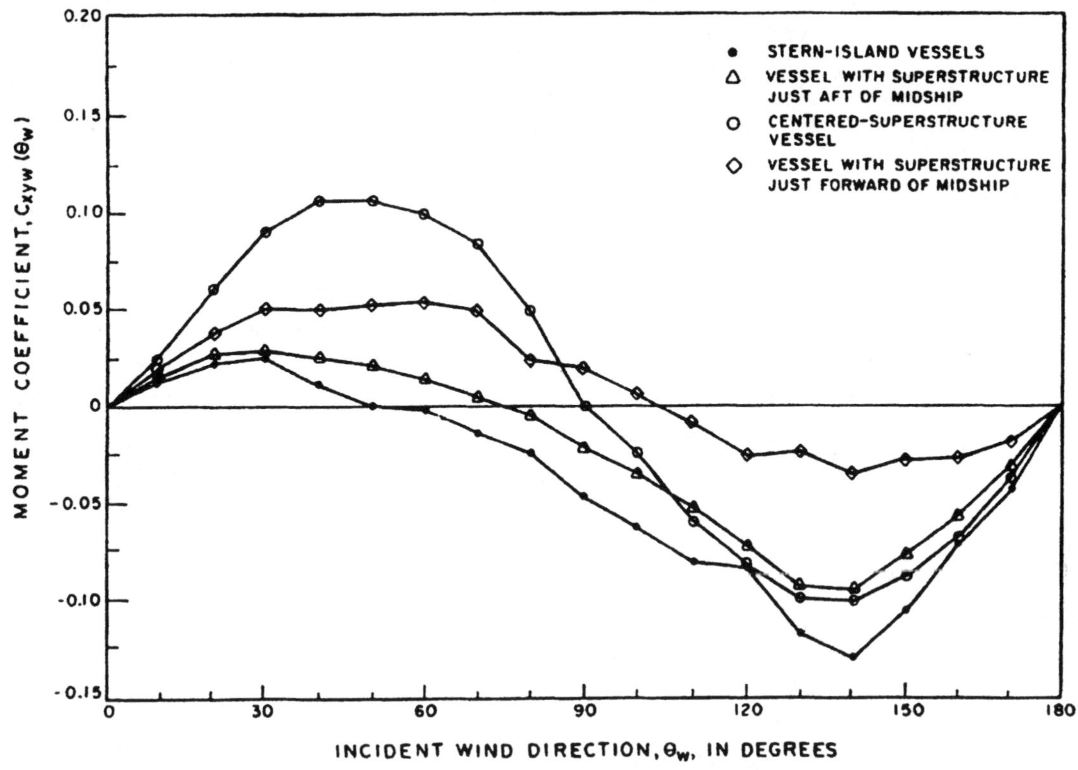

Figure 8.13 Wind yaw-moment coefficients.

8.5.2 Current Load

As in the case of wind, Remery and Van Oortmerssen (1973) and OCIMF (1994) present current load methods. The methods presented below are taken from U.S. Navy (1985) and U.S. Department of Defense (1999). Lateral current load is determined from the equation

$$F_{yc} = \tfrac{1}{2}\,\rho_w V_c^2 L_{wL} T C_{yc} \sin\theta_c \qquad (8.14)$$

where F_{yc} is the lateral current load, ρ_w the mass density of water, V_c the current velocity, L_{wL} the vessel waterline length, T the vessel draft, C_{yc} the lateral current-force drag coefficient, and θ_c the current angle. The lateral current-force drag coefficient is given by Seelig et al. (1992):

$$C_{yc} = C_{yc}|_\infty + (C_{yc}|_1 - C_{yc}|_\infty)\left(\frac{T}{\text{wd}}\right)^k \qquad (8.15)$$

where $C_{yc}|_\infty$ is the limiting value of C_{yc} for a large value of wd/T, $C_{yc}|_1$ the limiting value of C_{yc} for wd/T = 1, wd the water depth, and k a coefficient that ranges from 2 to 5. A value of 2 can be used for a wide range of ships and barges (see U.S. Department of Defense, 1999). The deepwater value, $C_{yc}|_\infty$, is given as

$$C_{yc}|_\infty = 0.22\sqrt{x} \qquad (8.16)$$

where

$$x = \frac{L_{wL}^2 A_M}{BV} \tag{8.17}$$

where L_{wL} is the length at the waterline, A_M the cross-sectional area at ship midsection, B the beam, and V the ship volume. $C_{yc}|_1$ is given as

$$C_{yc}|_1 = 3.2 \tag{8.18}$$

The longitudinal current load is determined using the equation

$$F_{xc} = F_{x\ \text{form}} + F_{x\ \text{friction}} + F_{x\ \text{prop}} \tag{8.19}$$

where F_{xc} is the total longitudinal current load, $F_{x\ \text{form}}$ the longitudinal current load due to form drag, $F_{x\ \text{friction}}$ the longitudinal current load due to friction drag, and $F_{x\ \text{prop}}$ the longitudinal current load due to propeller drag. Form drag is given by

$$F_{x\ \text{form}} = -\tfrac{1}{2}\rho_w V_c^2 BTC_{xcb} \cos\theta_c \tag{8.20}$$

where C_{xcb} is the longitudinal current-force form drag. Friction drag is given by

$$F_{x\ \text{friction}} = -\tfrac{1}{2}\rho_w V_c^2 SC_{xca} \cos\theta_c \tag{8.21}$$

where S, the wetted surface area of the hull, $= 1.7TL_{wL} + 35D/T$ (S in square feet with T, L_{wL} in feet and D in long tons), and C_{xca} is the longitudinal current-force friction drag coefficient. C_{xca} is computed as follows:

$$C_{xca} = \frac{0.075}{(\log R_n - 2)^2} \tag{8.22}$$

where $R_n = V_c L_{wL} \cos\theta_c/\nu$ is the Reynolds number [ν is the kinematic viscosity of water (1.4×10^{-5} ft^2/s)].

Propeller drag is the form drag of a vessel's propeller with a locked shaft. Propeller drag is given by the equation

$$F_{x\ \text{prop}} = -\tfrac{1}{2}\rho_w V_c^2 A_p \cos\theta_c \tag{8.23}$$

where A_p is the propeller expanded (or developed) blade area and C_{prop} is the propeller drag coefficient $= 1.0$. A_p is given by

$$A_p = \frac{A_{Tpp}}{0.838} \tag{8.24}$$

where A_{Tpp} is the total projected propeller area:

$$A_{Tpp} = \frac{L_{wL}B}{A_R} \tag{8.25}$$

Table 8.2 shows the area ratio, A_R, for six major vessel groups. (The *area ratio* is defined as the ratio of the waterline length times the beam to the total projected propeller area.)

The current yaw moment is determined using the equation

$$M_{xyc} = F_{yc}\frac{e_c}{L_{wL}} L_{wL} \tag{8.26}$$

Table 8.2 A_R for propeller drag

Vessel Type	A_R
Destroyer	100
Cruiser	160
Carrier	125
Cargo	240
Tanker	270
Submarine	125

where M_{xyc} is the current yaw moment and e_c/L_{wL} is the ratio of eccentricity. The value of e_c/L_{wL} is given in Figure 8.14 as a function of current angle, θ_c, and vessel type.

8.6 DESIGN OF MOORING COMPONENTS

8.6.1 Selection of Anchor Chain

Maximum mooring-chain tension includes both horizontal and vertical components. The maximum tension can be approximated from the horizontal tension, which is normally known, as follows:

$$T = 1.12H \tag{8.27}$$

where T is the maximum tension in the mooring chain and H is the horizontal tension in the mooring chain. This equation provides conservative estimates of mooring-chain tension for water depths of 30 m or less.

The U.S. Navy (1985) and U.S. Department of Defense (1999) allow a maximum chain working load in direct tension of

$$T_{\text{design}} = 0.35T_{\text{break}} \tag{8.28}$$

where T_{design} is the maximum allowable work-

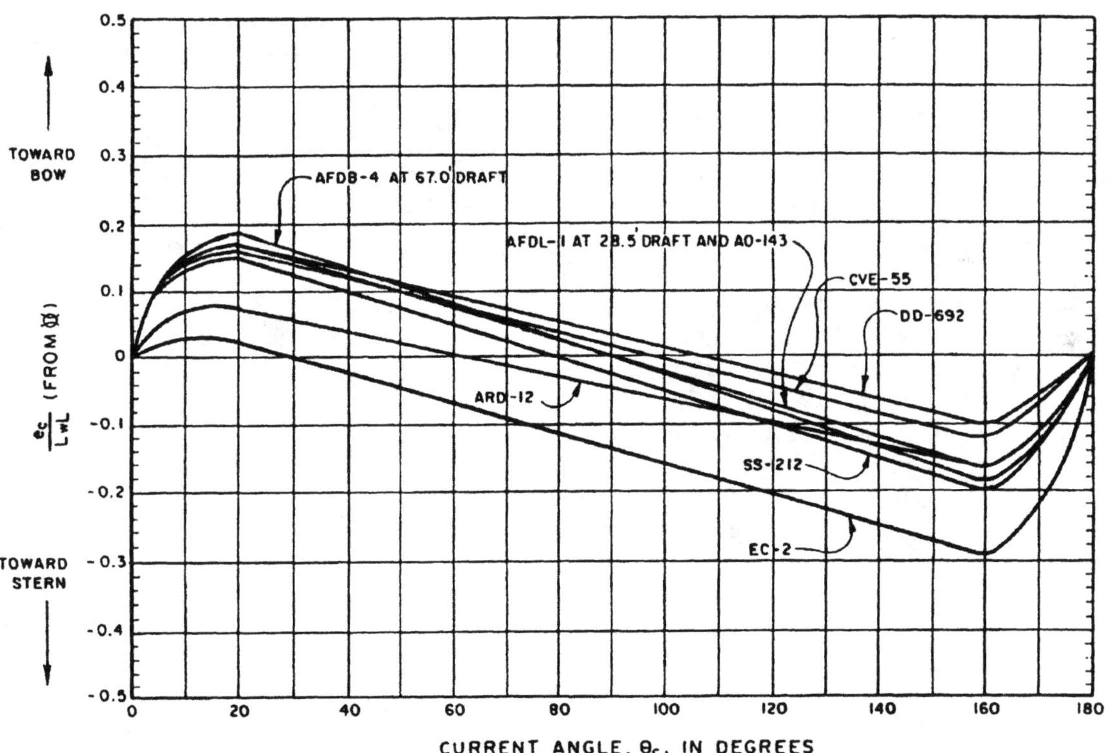

Figure 8.14 E_c/L_{wL} for various vessel types.

ing load on a mooring chain and T_{break} is the breaking strength of the chain.

For chain that passes through hawse pipes, chocks, chain stoppers, or other fittings which cause the chain to change direction abruptly within its loaded length U.S. Navy (1985) gives:

$$T_{design} = 0.25T_{break} \tag{8.29}$$

The ABS *Rules for Building and Classing Single-Point Moorings* (1975) calls for a factor of safety of 3:

$$T_{design} = 0.33T_{break} \tag{8.30}$$

The ABS *Rules for Building and Classing Mobile Offshore Drilling Units* (1997) specifies a minimum factor of safety of 2.7 for quasistatic mooring analysis and consideration of a smaller factor of safety for a dynamic mooring analysis.

Chains and fittings must be selected with a breaking strength equal or exceeding T_{break}. Chain breaking strength can be found in manufacturers' catalogs. Chain should be rounded up to the nearest 6 mm ($\frac{1}{4}$ in.). Larger chain may be desirable if excessive wear is expected.

The weight per shot of chain is presented in manufacturers' catalogs. The weight of chain in water is obtained by multiplying the weight in air by 0.87. When tables of actual chain weights are unavailable, the weight of a stud-link chain may be approximated as follows:

$$w_{air} = 9.05d^2 \tag{8.31}$$

$$w_{submerged} = 8.26d^2 \tag{8.32}$$

where w_{air} is the weight of chain (in air) in pounds per foot of length, d the diameter of chain (in.), and $w_{submerged}$ the weight of chain (in water) in lb/ft of length.

8.6.2 Computation of Chain Length and Tension

A chain mooring line behaves as a catenary. Figure 8.15 presents a definition sketch for use in catenary analysis. At any point (x,y), the following hold:

$$V = wS = T \sin \theta \tag{8.33}$$

$$H = wc = T \cos \theta \tag{8.34}$$

$$T = wy \tag{8.35}$$

$$c = \frac{H}{w} \tag{8.36}$$

where V is the vertical force at point (x,y), w the submerged unit weight of the chain, S the length of the curve (chain length) from $(0,c)$ to point (x,y), T the line tension at point (x,y), θ the angle of the mooring line with horizontal, H the horizontal force at point (x,y), and c the distance from the origin to the y-intercept.

The shape of the catenary is governed by the following equations:

$$y^2 = S^2 + c^2 \tag{8.37}$$

$$y = c \cosh \frac{x}{c} \tag{8.38}$$

$$S = c \sinh \frac{x}{c} \tag{8.39}$$

Equation (8.38) may be expressed more conveniently as

$$x = c \ln \left[\frac{S}{c} + \sqrt{\left(\frac{S}{c}\right)^2 + 1} \right] \tag{8.40}$$

Note that in the equations above, the horizontal load in the chain is the same at every point and

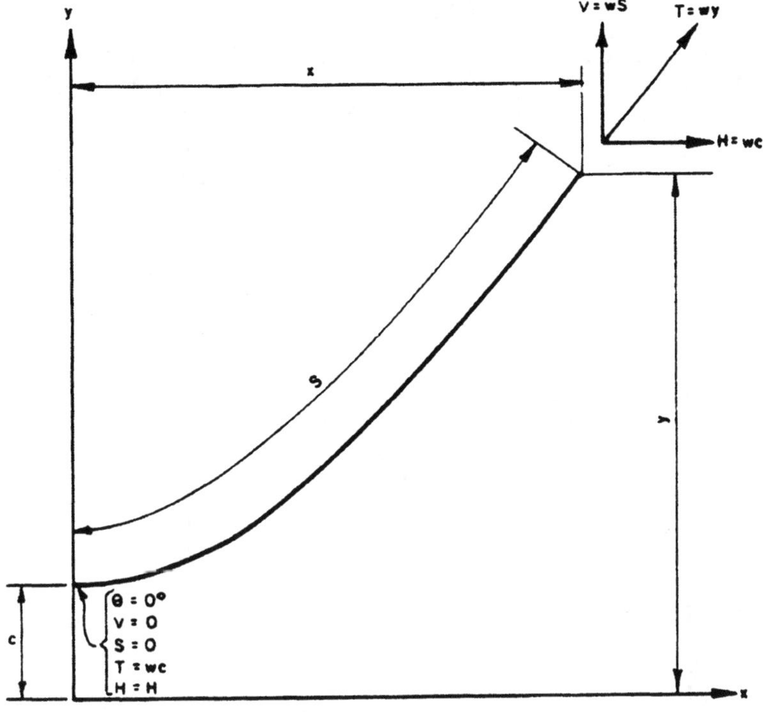

Figure 8.15 Catenary definition sketch.

that all measurements of x, y, and S are referenced to the catenary origin.

When catenary properties are desired at point (x_m, y_m), as shown in Figure 8.16, the following equations are used:

$$\sqrt{S_{ab}^2 - \mathrm{wd}^2} = 2c \sinh \frac{x_{ab}}{2c} \quad (8.41)$$

$$\frac{\mathrm{wd}}{S_{ab}} = \tanh \frac{x_m}{c} \quad (8.42)$$

$$x_m = x_a + \frac{x_{ab}}{2} \quad (8.43)$$

$$x_b = x_m + \frac{x_{ab}}{2} \quad (8.44)$$

where the terms in the equations above are as defined in Figure 8.16. Equation (8.41) is written more conveniently as

$$x_m = \frac{c}{2}\left[\ln\left(1 + \frac{\mathrm{wd}}{S_{ab}}\right) - \ln\left(1 - \frac{\mathrm{wd}}{S_{ab}}\right)\right] \quad (8.45)$$

8.6.3 Some Applications of the Catenary Equations

8.6.3.1 Case 1. The known variables are the mooring-line angle at the anchor, θ_a (which is zero: $\theta_a = 0°$); the water depth, wd; the horizontal load, H; and the submerged unit weight of the chain, w. The length of mooring line, S_{ab}, the horizontal distance from the anchor to the buoy, x_{ab}, and the tension in the mooring line at the buoy or surface, T_b, are desired. Pro-

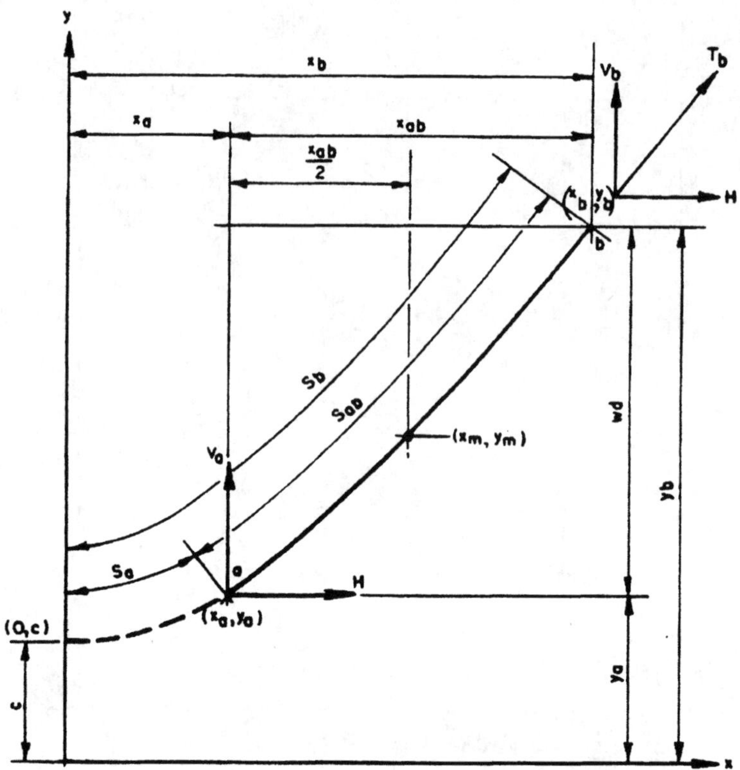

Figure 8.16 Catenary definition sketch.

cedures for determining these values are outlined in Figure 8.17. Check to determine if the entire chain has been lifted off the bottom by comparing the computed chain length from anchor to buoy, S_{ab}, to the actual chain length, S_{actual}. If the actual chain length is less than the computed length, case 1 cannot be used and case 5 must be used.

8.6.3.2 *Case 2.* The known variables are the mooring-line angle at the anchor, θ_a (or equivalently, a specified vertical load at the anchor, V_a); the water depth, wd; the horizontal load at the surface, H; and the submerged unit weight of the chain, w. This situation arises when a drag anchor is capable of sustaining a small prescribed angle at the anchor or an uplift-resisting anchor of given vertical capacity, $V_a = H \tan \theta_a$, is specified. The origin of the catenary is not at the anchor but is some distance below the bottom. The length of the chain from anchor to buoy, S_{ab}; the tension in the mooring line at the buoy or surface, T_b; and the horizontal distance from the anchor to the surface, x_{ab}, are desired. Procedures for determining these values are presented in Figure 8.18.

8.6.3.3 *Case 3.* The known variables are the horizontal distance from the anchor to the buoy, x_{ab}; the water depth, wd; the horizontal load, H; and the submerged unit weight of the chain, w. This situation arises when it is necessary to limit the horizontal distance from buoy to anchor due to space limitations. The

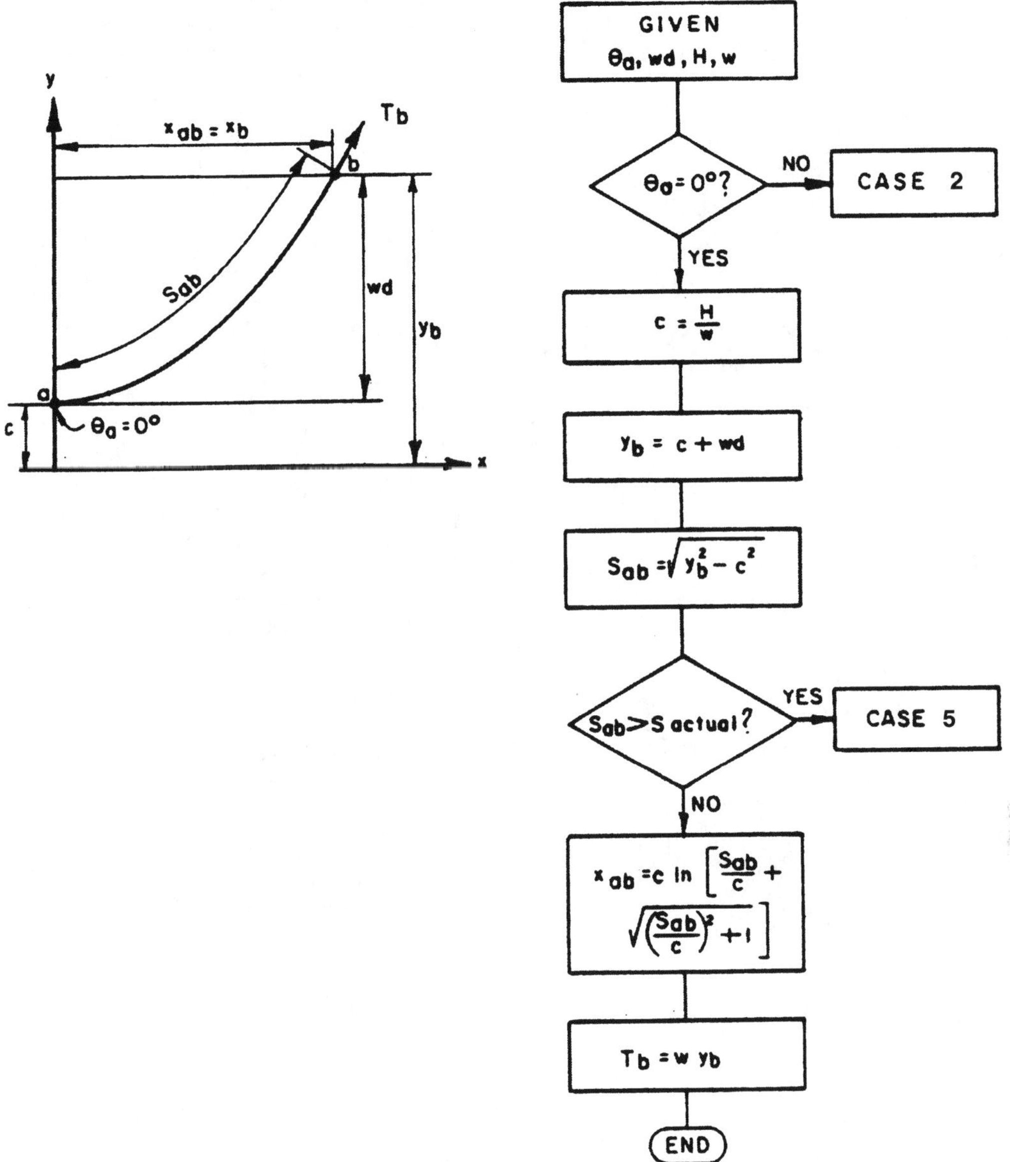

Figure 8.17 Case 1.

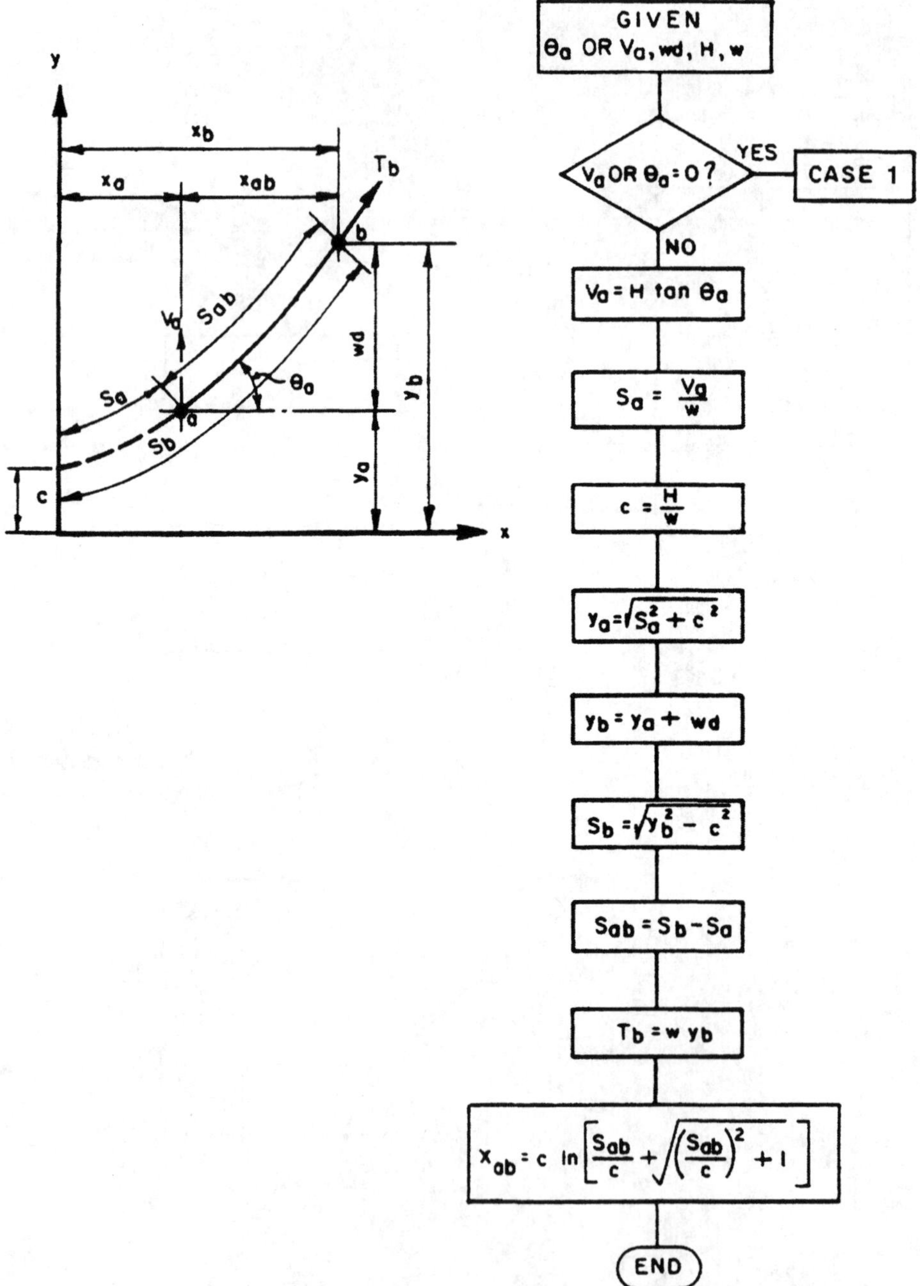

Figure 8.18 Case 2.

length of chain from anchor to buoy, S_{ab}, the tension in the mooring line at the buoy, T_b, and the vertical load at the anchor, V_a, are required. Procedures for determining these values are outlined in Figure 8.19.

8.6.3.4 Case 4. The known variables are the water depth, wd; the horizontal load, H; the submerged unit weight of the chain, w; the angle at the anchor, θ_a; the sinker weight, W; the unit weight of the sinker, γ_s; the unit weight of water, γ_w; and the length of chain from anchor to sinker, S_{ab}. The mooring consists of a chain of constant unit weight with a sinker attached to it. The total length of chain, S_{ac}; the distance of the top of the sinker off the bottom, y_s; and the tension in the mooring line at the buoy, T_{c2}, are desired. The solution to this problem is outlined in Figure 8.20.

8.6.3.5 Case 5. The known variables are the water depth, wd; the horizontal load on the chain, H; the submerged unit weight of the chain, w; and the length of chain from anchor to buoy, S_{ab}. The horizontal load, H, is sufficiently large to lift the entire chain off the bottom, resulting in an unknown vertical load at the anchor, V_a. This situation arises when one is computing points on a load–deflection curve for higher values of load. Solution involves determining the vertical load at the anchor, V_a, using the trial-and-error procedure presented in Figure 8.21. The problem is solved efficiently using a Newton–Raphson iteration method (Gerald, 1980); this method gives accurate solutions in two or three iterations, provided that the initial estimate is close to the final answer.

8.6.3.6 Load–Deflection Curve. A moored vessel will move in response to applied static loading (i.e., the vessel will move from an initial, preloaded position to a static equilibrium position). The force in the mooring chain will increase from a pretension value as the vessel moves. A load–deflection curve shows the relationship between load and deflection (Figure 8.22). The work required to move a vessel from initial to equilibrium position is equal to potential energy associated with raising the mooring lines and hardware. The work done is equal to the area under the load–deflection curve shown in Figure 8.22. Point A is the initial position of the vessel, point B, the static equilibrium position. Additional work will be done on the system if additional dynamic loads are applied and will move the vessel to point C. The work done on the vessel by dynamic loads is equal to the area under the curve from point B to point C. The additional work must be absorbed by the mooring system without exceeding the allowable load for the mooring line. This can only be the case if the mooring system is resilient [i.e., capable of absorbing work (or energy)].

Sinkers can be used to make a mooring more resilient (Figure 8.23). Curve 1 is the load–deflection curve without a sinker, curve 2 with a sinker. The vertical portion of curve 2 corresponds to lifting of the sinker off the bottom. Points A, B, and C are the initial, static equilibrium, and maximum dynamic position, respectively. The sinker increases energy absorption between points B and C as shown in shaded areas 1 and 2.

8.6.4 Anchor Design

Methods for designing drag anchors are presented below; pile, dead-weight, and direct-embedment anchors are discussed in Rocker (1985). The designer must determine anchor holding capacity, burial depth, and drag distance. Drag anchors can be used in mud or sandy seafloors, provided that the soil depth is sufficient for anchor embedment. Figures 8.24 and 8.25 provide maximum holding capacity for several anchor types in sand and clay/silt bottoms, respectively. The required maximum holding capacity, H_M, depends on the factor of safety, FS, and the horizontal load, H:

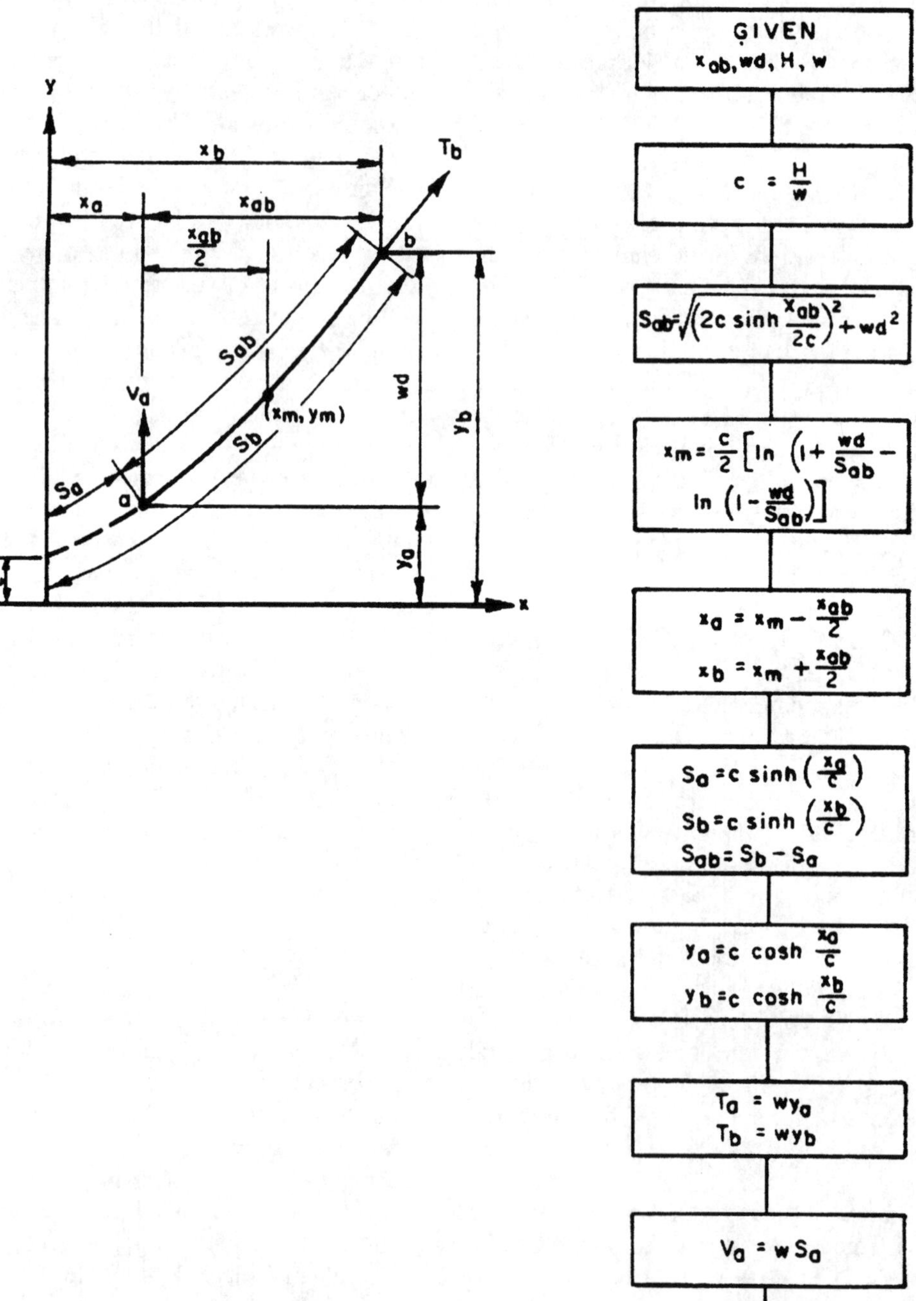

Figure 8.19 Case 3.

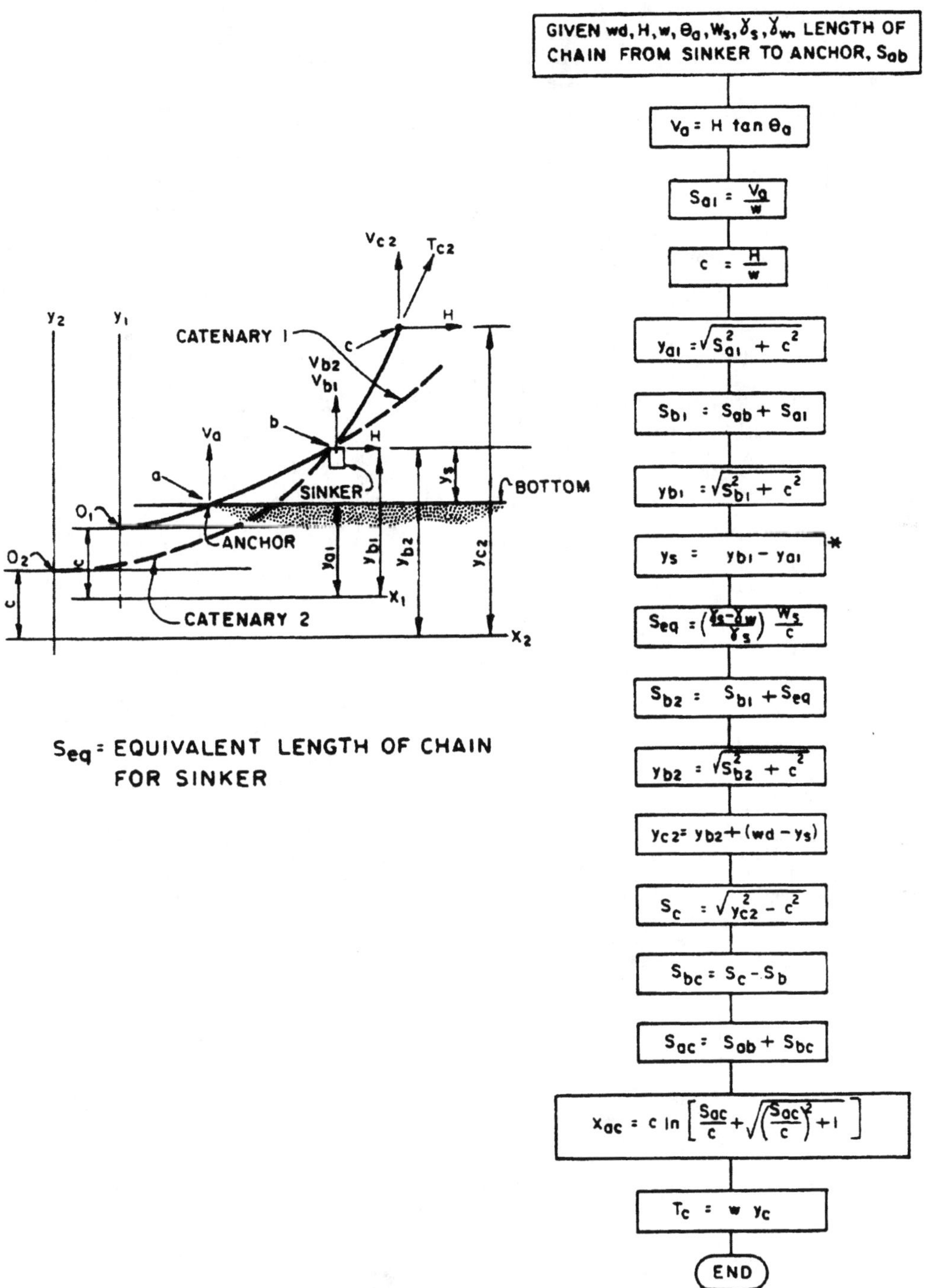

Figure 8.20 Case 4.

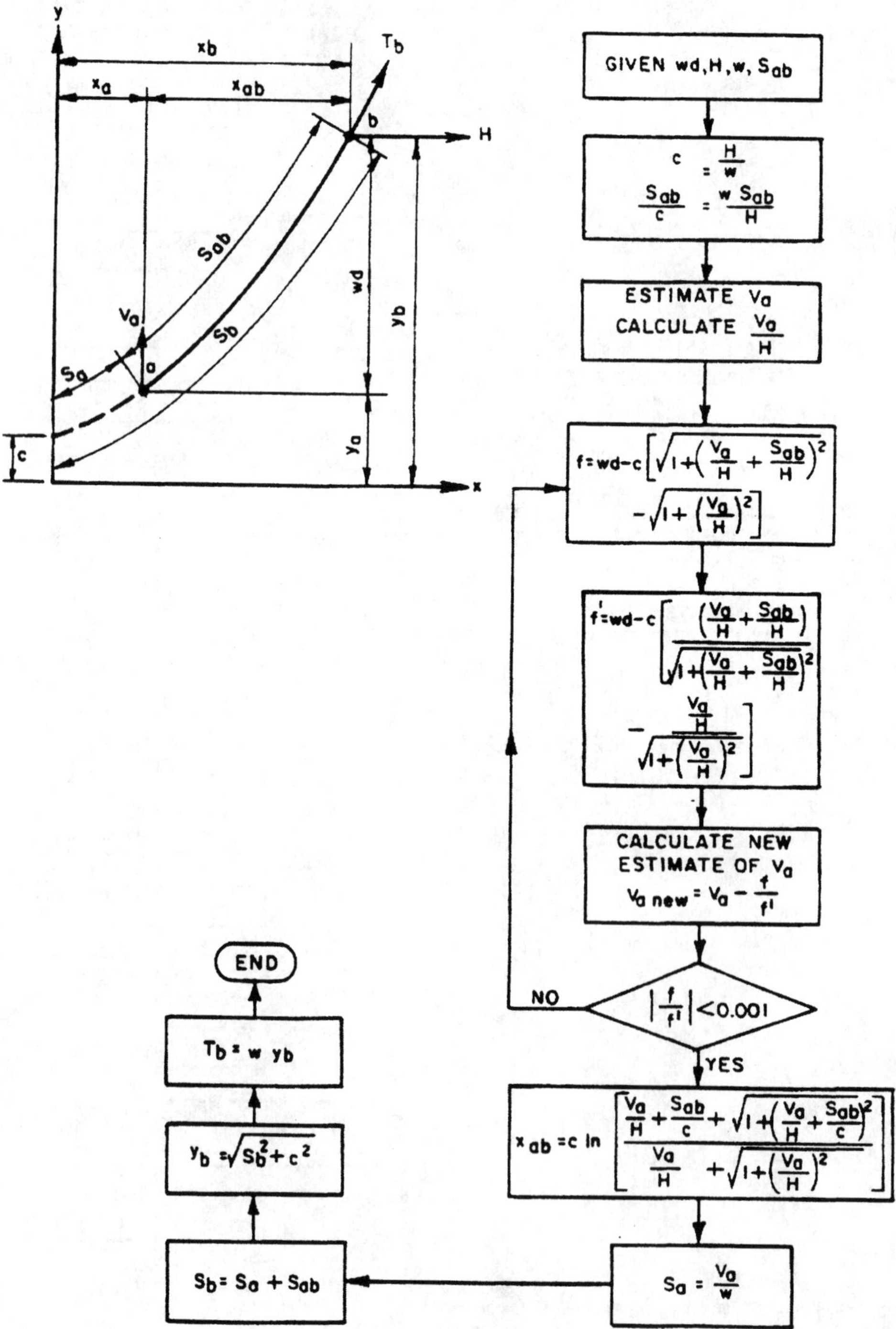

Figure 8.21 Case 5.

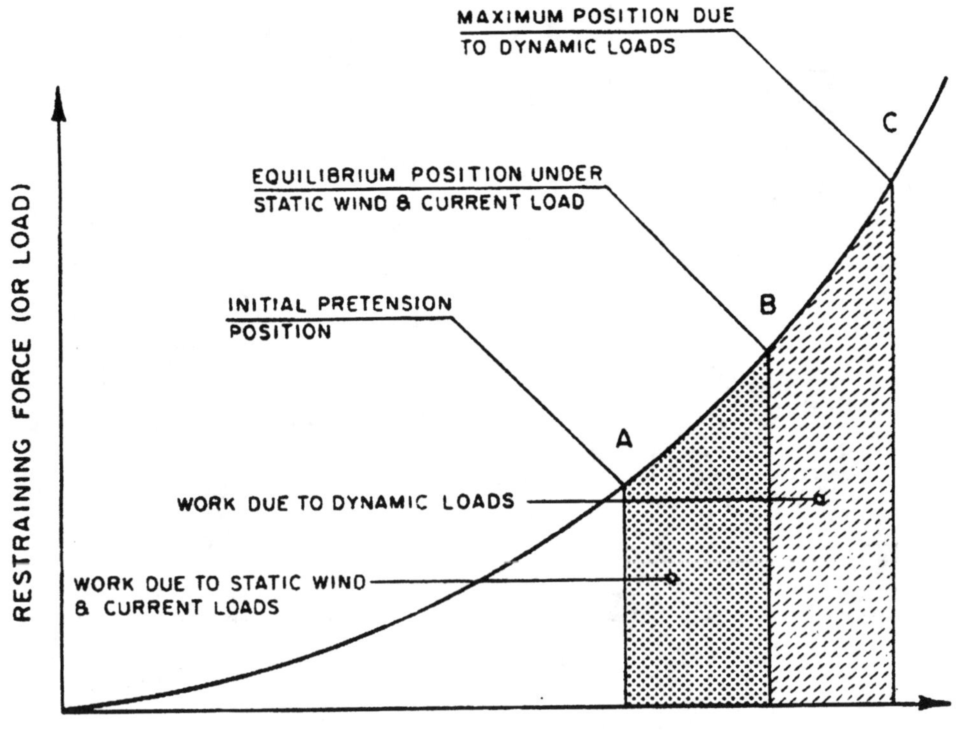

Figure 8.22 Load–deflection curve.

$$H_m = \text{FS} \cdot H \tag{8.46}$$

The normal factor of safety of 1.5 to 2.0.

Anchor holding capacities determined from Figures 8.24 and 8.25 assume that there is a sufficient depth of soil to allow for anchor penetration (see Table 8.3). The anchor capacity must be reduced when the soil depth is less than the penetration depth according to the following equation:

$$H_{Ar} = fH_a \tag{8.47}$$

where H_{Ar} is the reduced anchor capacity, f the anchor capacity reduction factor, and H_a the anchor capacity for full sediment depth. The correction factor, f, is determined using the equation

$$f = \frac{\text{actual soil depth}}{\text{required soil depth}} \tag{8.48}$$

Anchor drag distances in sand for Stockless and Stato anchors are given in Figure 8.26 based on a factor of safety of 1.5 and 2 for the Stockless and Stato anchors, resepectively. Anchor drag distances in sand for most anchors is about 3.5 to 4 fluke lengths (a factor of safety of 2). Anchor drag distances in mud can be determined from Figure 8.27. The drag distances for the Stockless (factor of safety of 1.5) and Stato (factor of safety of 2) anchors are

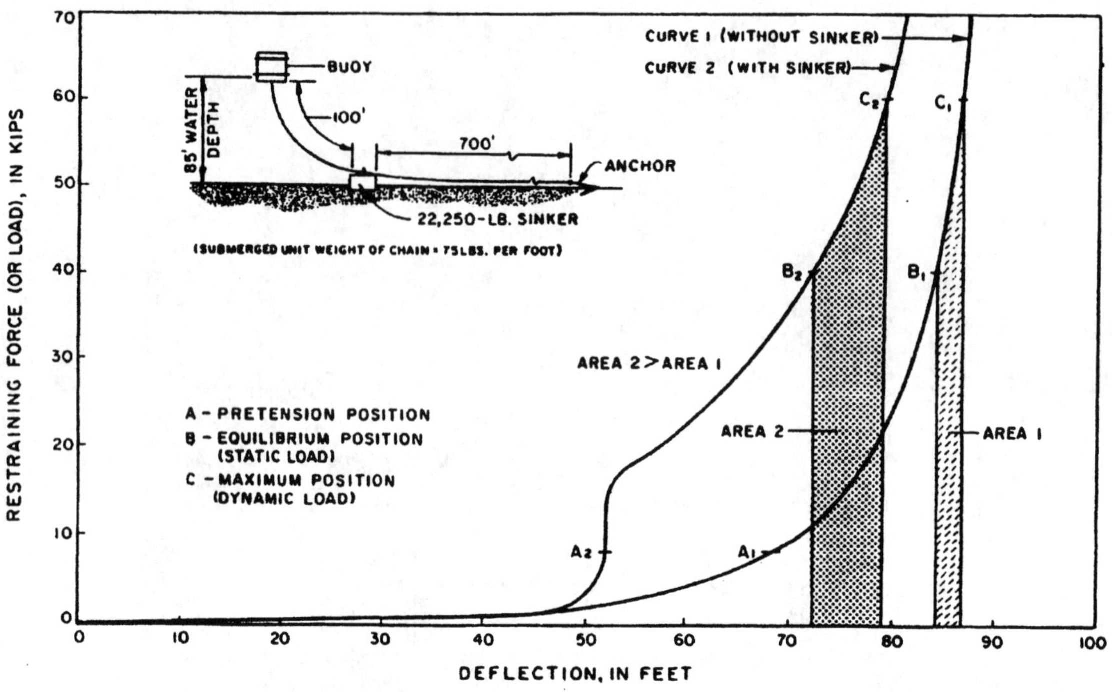

Figure 8.23 Load–deflection curves.

indicated in Figure 8.27. Figure 8.28 provides anchor drag distances for various commercially available anchors. Increased holding capacity can be achieved by using combinations of anchors. Methods for using multiple anchors are described in U.S. Navy (1985).

8.7 LOADS ON MOORING ELEMENTS

8.7.1 Static versus Dynamic Analysis

Mooring loads can be computed by means of static or dynamic analysis. Static analysis is appropriate when dynamic motions are not expected, dynamic analysis when they are. There are no hard-and-fast rules for determining when a ship-mooring system will experience dynamic motions. Strictly speaking, a dynamic analysis is required to determine if dynamic loads can be detected. On the other hand, many successful mooring have been designed without a dynamic analysis. Generally speaking, dynamic loads are likely when a mooring is exposed to ocean wave attack, large winds (50 knots or more), rapid windshifts, large currents, and current-induced eddies or macro vortices. SPMs are generally more vulnerable to dynamic wind and current loading than a MBM inasmuch as they tend to weather-vane in response to these environmental force. Large-wind-area vessels (e.g., cruise ships, dry docks, container ships, and tankers) exposed to high winds are vulnerable to dynamic loading. As stated previously, dynamic analysis is the only diagnostic for evaluating mooring vulnerability to dynamic loads.

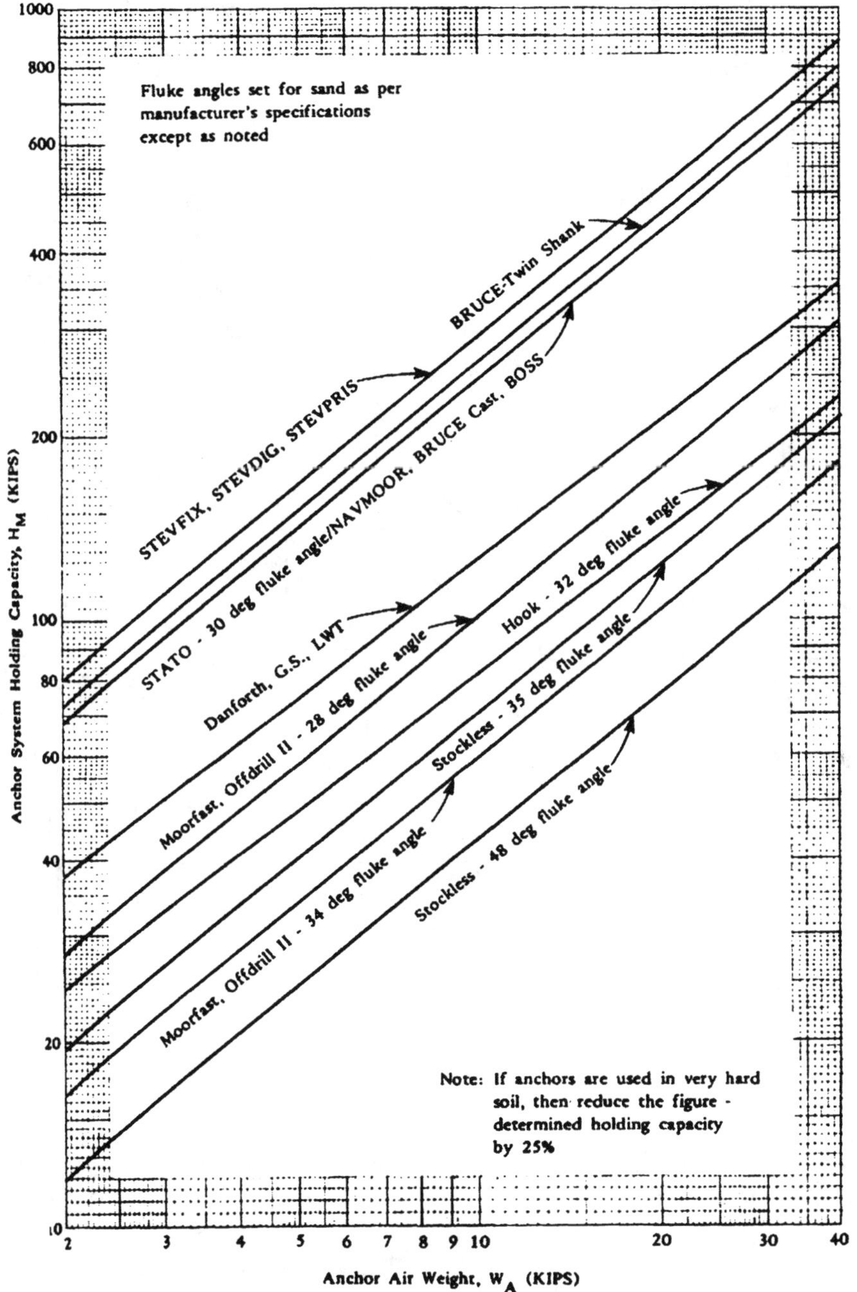

Figure 8.24 Drag-anchor holding capacity: sand bottom.

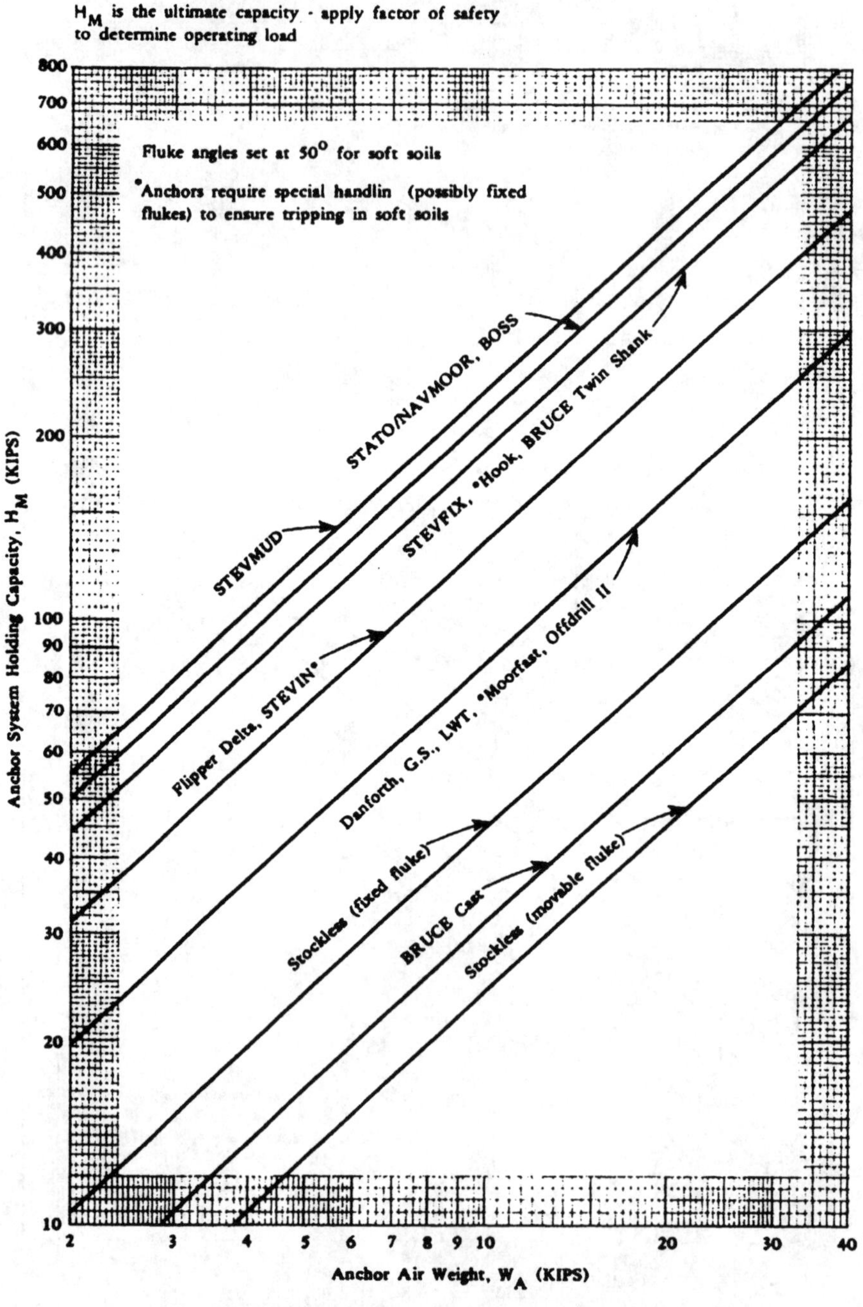

Figure 8.25 Drag-anchor holding capacity: mud bottom.

Table 8.3 Estimated anchor fluke tip penetration

Anchor Type	Fluke-Tip Penetration/ Fluke Length: Sand/ Stiff Clay	Mud (Soft Silts and Clays)
Stockless, Moorfast	1	3 (fixed fluke)
Offdrill II	1	4
Stato, Stevfix, Flipper delta, Boss, Danforth, LWT, GS (type 2), Bruce twin shank	1	4.5
Stevmud	1	5.5
Hook	1	6

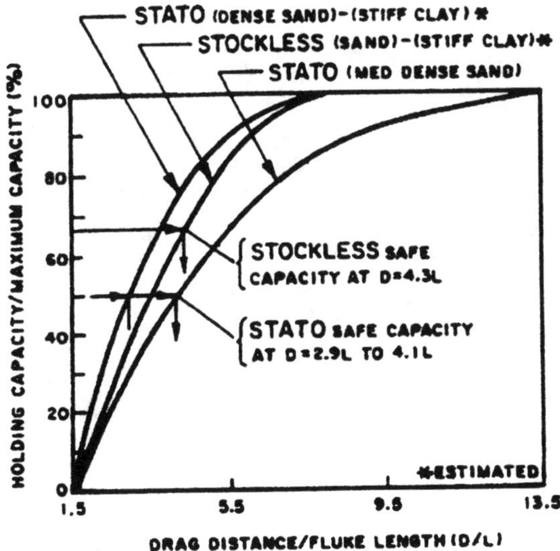

Figure 8.26 Anchor drag distance in sand.

8.7.2 Static Analysis

Winds and currents impose a longitudinal and lateral load and a yaw moment on a moored vessel. The vessel will move in response to these loads until it reaches an equilibrium position. The applied load is equal to the restraining load provided by the mooring system at the equilibrium position. The first step is to determine the total loads and moment:

$$F_{xT} = F_{xw} + F_{xc} \tag{8.49}$$

$$F_{yT} = F_{yw} + F_{yc} \tag{8.50}$$

$$M_{xyT} = M_{xyw} + M_{xyc} \tag{8.51}$$

where F_{xT} is the total longitudinal force, F_{xw} the longitudinal wind force, F_{xc} the longitudinal current force, F_{yT} the total lateral force, F_{yw} the lateral wind force, F_{yc} the lateral current force, M_{xyT} the total moment, M_{xyw} the wind moment, and M_{xyc} the current moment.

8.7.2.1 Single-Point Moorings. Figure 8.29 shows a vessel secured to an SPM exposed to wind and current. For static equilibrium, the applied loads must equal the restoring loads of the mooring system:

$$\sum F_x = 0 \tag{8.52}$$

$$\sum F_y = 0 \tag{8.53}$$

$$\sum M_{cg} = 0 \tag{8.54}$$

The vessel will weather-vane around the SPM until the equations of equilibrium above are satisfied. The forces and moment are a function of the angle between the vessel and both the wind angle, θ_w, and the current angle, θ_c which vary as the vessel rotates. Maximum hawser load is determined through a trial-and-error procedure where the angle of the vessel is adjusted until the point of zero moment is determined. The general procedure involves assuming a ship's position, calculating the sum of moments on the vessel, updating the position, and repeating this process until the sum of moments is zero (Figures 8.30 and 8.31). Accordingly, the point of equilibrium (zero moment) is best determined by computer or graphically.

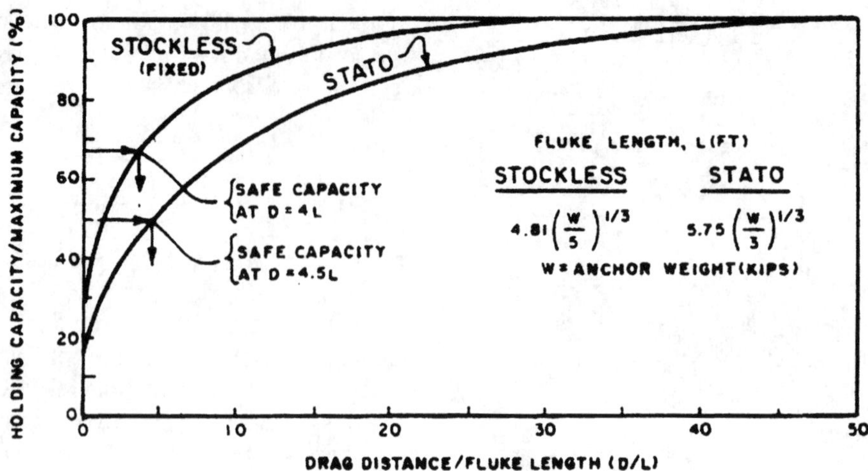

Figure 8.27 Anchor drag distance in mud.

$\sum M$ is determined as follows:

$$\sum M = M_{xyw} + M_{xyc} - F_{yT}\text{ARM} \quad (8.55)$$

where ARM is the distance from the bow hawser attachment point to the center of gravity of the vessel (ARM = 0.48LOA) and LOA is the overall length of the vessel. The horizontal hawser load, H, is determined using the equation

$$H = \sqrt{F_{xT}^2 + F_{yT}^2} \quad (8.56)$$

8.7.2.2 Multiple-Point Moorings. Mooring line stiffness influences the distribution of load among the mooring lines in an MBM. Figure 8.32 shows how a vessel reorients as the applied load is distributed among the mooring lines, which lengthen or shorten until they are in equilibrium with applied loads. As with the SPM, eqs. (8.51) through (8.53) must be satisfied for static equilibrium. The general problem is best evaluated using a computer program such as that described in Section 8.7.2.3. Simplified methods for several MBMs are presented in Figures 8.33 through 8.35. These simplified solutions are very approximate and should be used only for mooring layout. The computer methodology summarized below is required for final design.

8.7.2.3 Multiple-Point Mooring Computer Solution. U.S. Navy (1985) presents a computer program for MBM problems. The program determines forces in mooring elements and vessel displacements resulting from applied loading. Mooring elements may be comprised of hawsers and anchor chains. The hawsers are defined by load–deflection curves, while anchor chains are analyzed as catenaries. Solutions are obtained iteratively using the general Newton–Raphson method (Gerald, 1980).

Figure 8.36 summarizes the main program variables. The coordinate system is defined at the ship's initial position, with the origin, O, at the ship's center of gravity. When the ship

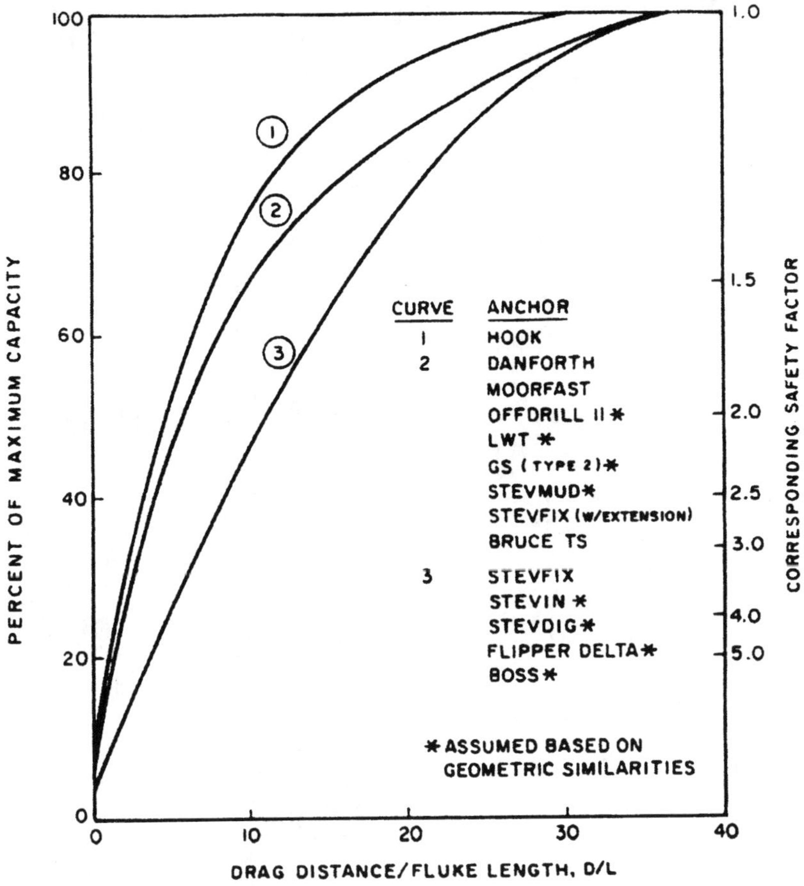

Figure 8.28 Anchor drag distance in mud.

moves as the result of unbalanced forces, the center of gravity moves to a new location, designated S. The new position is described by the surge and sway displacement, x and y, and the yaw angle, θ. The horizontal component of mooring line tension is computed from the horizontal distance between the attachment point on the ship and the fixed anchor point. Load–deflection curves are developed in a subroutine and are stored as look-up tables. The mooring-line length and direction are computed from the following equations:

$$x_2 = x_c \cos\theta - y_c \sin\theta \tag{8.57}$$

$$y_2 = x_c \sin\theta + y_c \cos\theta \tag{8.58}$$

$$x_3 = x_1 - x - x_2 \tag{8.59}$$

$$y_3 = y_1 - y - y_2 \tag{8.60}$$

$$r = \sqrt{x_3^2 + y_3^2} \tag{8.61}$$

$$\cos\theta_3 = \frac{x_3}{r} \tag{8.62}$$

$$\sin\theta_3 = \frac{y_3}{r} \tag{8.63}$$

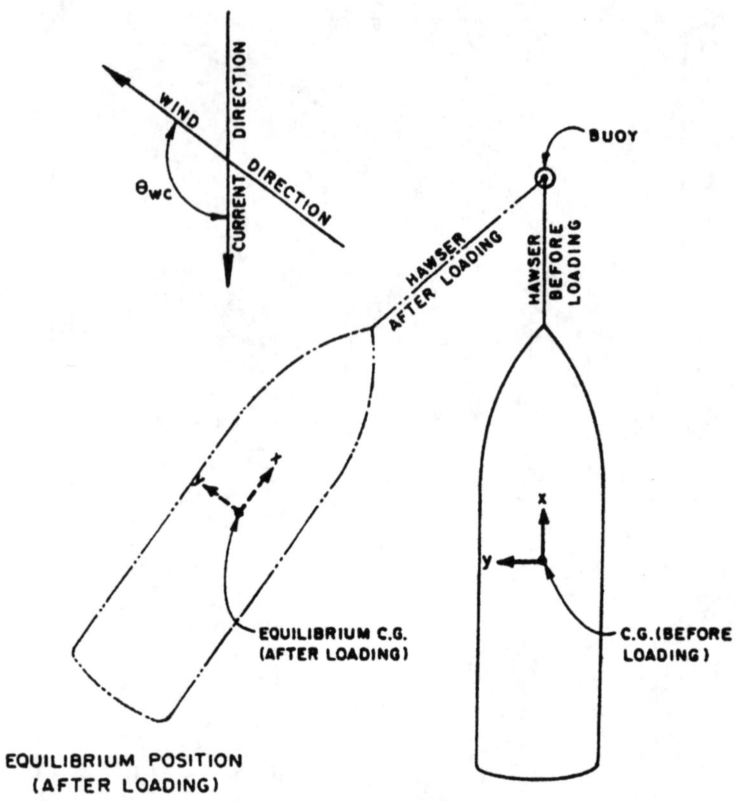

Figure 8.29 Static analysis of a single-point mooring.

Variables in the formulas above are as defined in Figure 8.36.

The x and y components of the force exerted by the mooring line on the ship and the moments (due to the mooring line) about the ship origin, S, are then given by

$$H = f(r) \tag{8.64}$$

$$F_x = H \cos \theta_3 \tag{8.65}$$

$$F_y = H \sin \theta_3 \tag{8.66}$$

$$M_{xy} = F_y x_2 - F_x y_2 \tag{8.67}$$

where H is the horizontal component of the mooring line force and the other variables are as defined in Figure 8.36.

The Newton–Raphson method requires the derivatives of the force components above with respect to x, y, and θ; these derivatives are presented in U.S. Navy (1985). The total forces and moment on the ship are computed by summing eqs. (8.64) through (8.66) for each mooring line and adding the applied static loads. The Newton–Raphson method is used to arrive at values of x, y, and θ for which the total force and moment on the ship are zero. In the expressions for the total differential of force components,

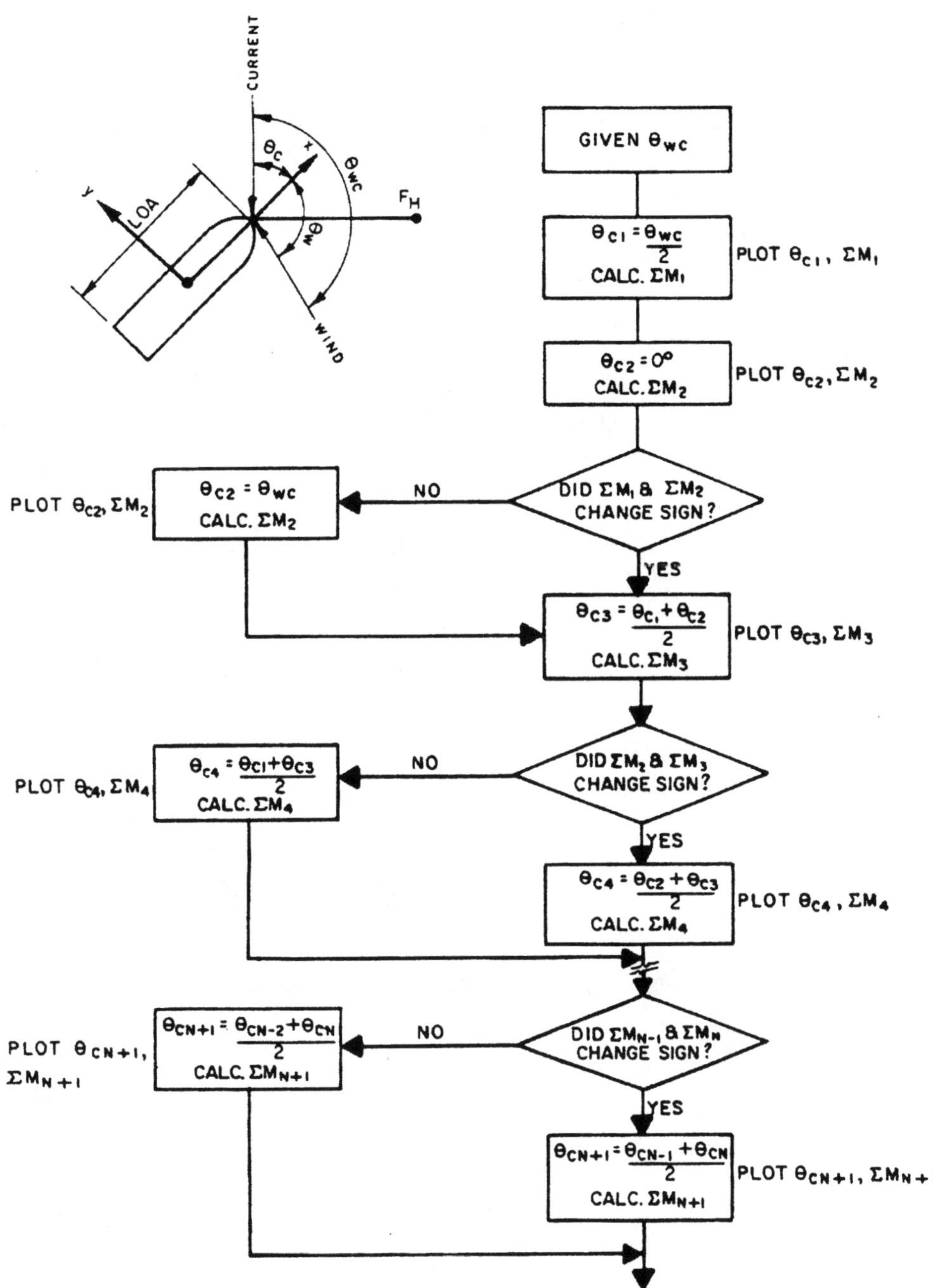

NOTE: CONTINUE PROCESS UNTIL DIFFERENCE BETWEEN θ_{CN+1} AND θ_{CN} IS SMALL OR UNTIL EQUILIBRIUM ANGLE CAN BE DETERMINED FROM GRAPH

Figure 8.30 Static single-point mooring analysis method.

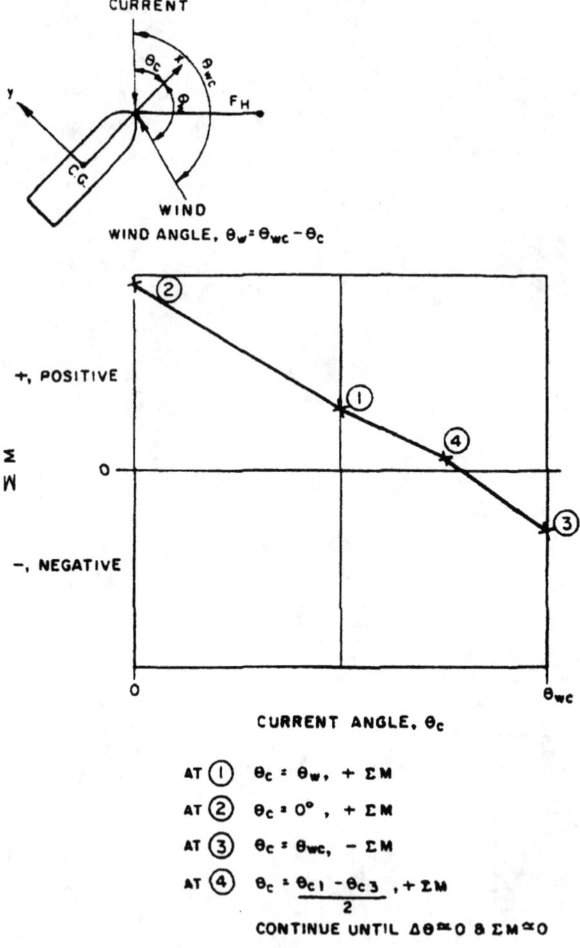

Figure 8.31 Example of static single-point mooring analysis.

$$dF_i = \frac{dF_i}{dx} dx + \frac{dF_i}{dy} dy + \frac{dF_i}{d\theta} d\theta \quad (8.68)$$

the differential motions forces are approximated by finite increments as follows:

$$\sum \frac{dF_x}{dx} \Delta x + \sum \frac{dF_x}{dy} \Delta y + \sum \frac{dF_x}{d\theta} \Delta\theta = - F_{xa} - \sum F_x \quad (8.69)$$

$$\sum \frac{dF_y}{dx} \Delta x + \sum \frac{dF_y}{dy} \Delta y + \sum \frac{dF_y}{d\theta} \Delta\theta = - F_{ya} - \sum F_y \quad (8.70)$$

$$\sum \frac{dM_{xy}}{dx} \Delta x + \sum \frac{dM_{xy}}{dy} \Delta y + \sum \frac{dM_{xy}}{d\theta} \Delta\theta = - M_{xya} - \sum M_{xy} \quad (8.71)$$

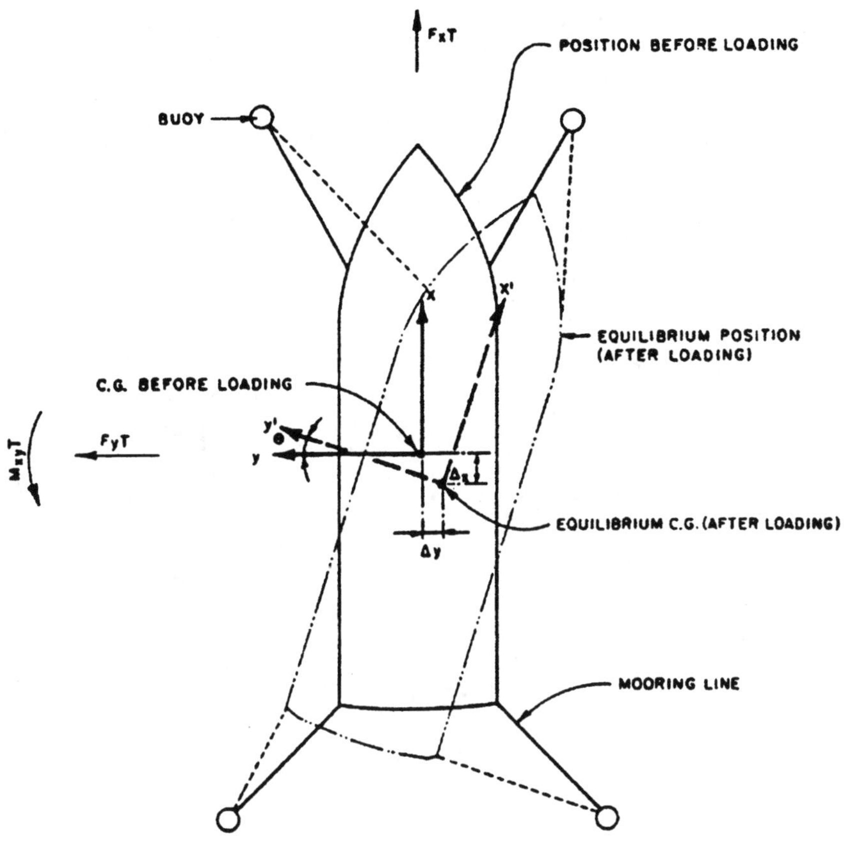

Figure 8.32 Spread mooring static analysis.

This set of equations is solved for Δx, Δy, and $\Delta\theta$; the ship is moved to

$$x + \Delta x \tag{8.72}$$

$$y + \Delta y \tag{8.73}$$

$$\theta + \Delta\theta \tag{8.74}$$

and the process is repeated until the total force components computed are all within the desired tolerance.

Load–deflection curves are computed for

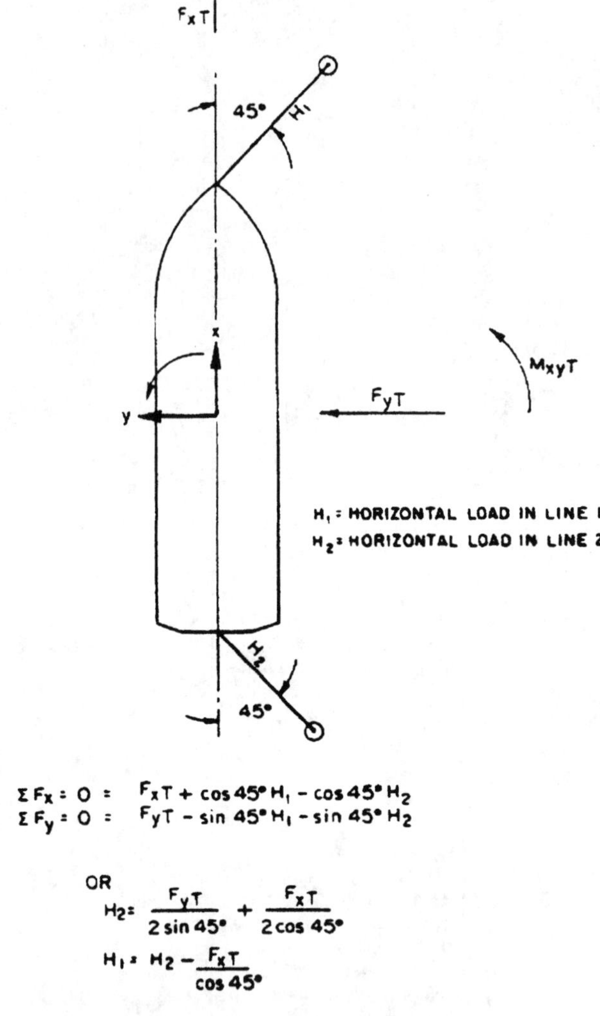

Figure 8.33 Bow-and-stern mooring analysis.

each mooring line and anchor chain. Mooring line characteristics are summarized in Figure 8.37. Anchor chain load–deflection curves are computed with the aid of catenary equations, as described in Section 8.6.2. The equations used are provided in U.S. Navy (1985).

8.7.3 Dynamic Analysis

The equation of static equilibrium (sum of forces and moment are zero) is solved for static analysis. For dynamic analysis, a form of the familiar equation of dynamic equilibrium

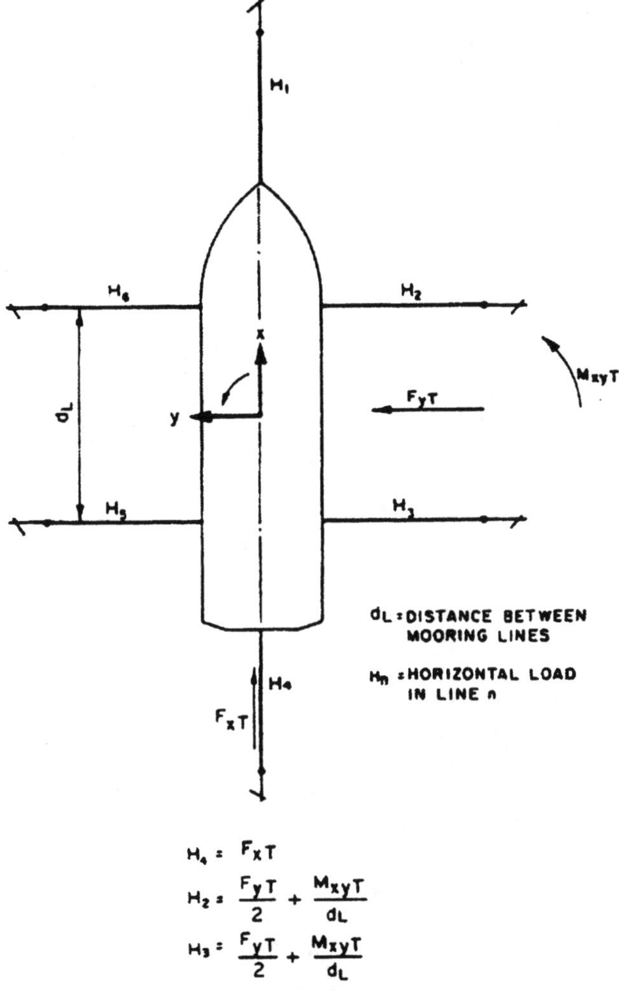

NOTE: DESIGNER MUST BE SURE THAT ALL SIGNS ARE CONSISTENT WITH APPLIED LOADS AND LINE ORIENTATIONS

Figure 8.34 Spread mooring analysis.

(force = mass × acceleration) must be solved. In its most general form, the equation is as follows:

mass × ship acceleration
+ damping coefficient × ship velocity
+ mooring or buoyancy force
= applied wind, current of wave force
on ship

In mathematical terms, this is

$$(m + a)\, x + bx + cx = F(t) \qquad (8.75)$$

where the coefficients a, b, and c represent added mass, damping coefficient, and restoring force, and the ship's mass is m. $F(t)$ represents a time-varying force due to waves.

There are many complications to eq. (8.75):

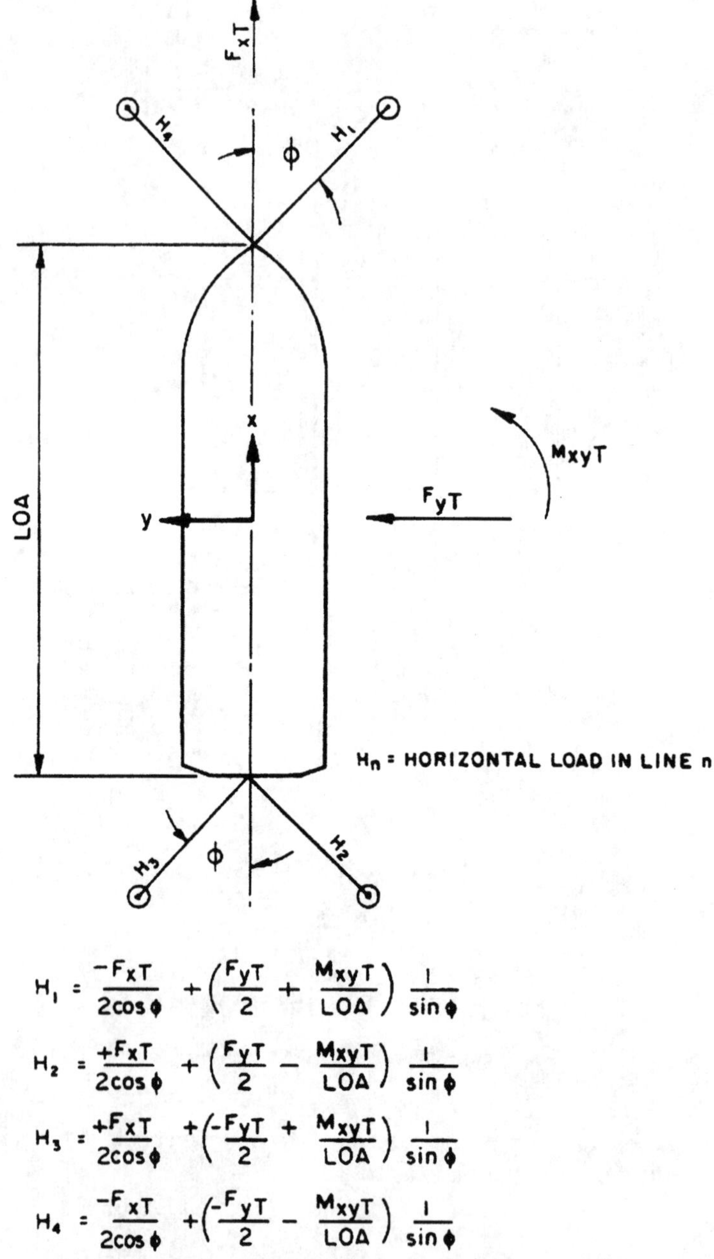

NOTE: DESIGNER MUST BE SURE THAT ALL SIGNS ARE CONSISTENT WITH APPLIED LOADS AND LINE ORIENTATIONS

Figure 8.35 Four-point mooring analysis.

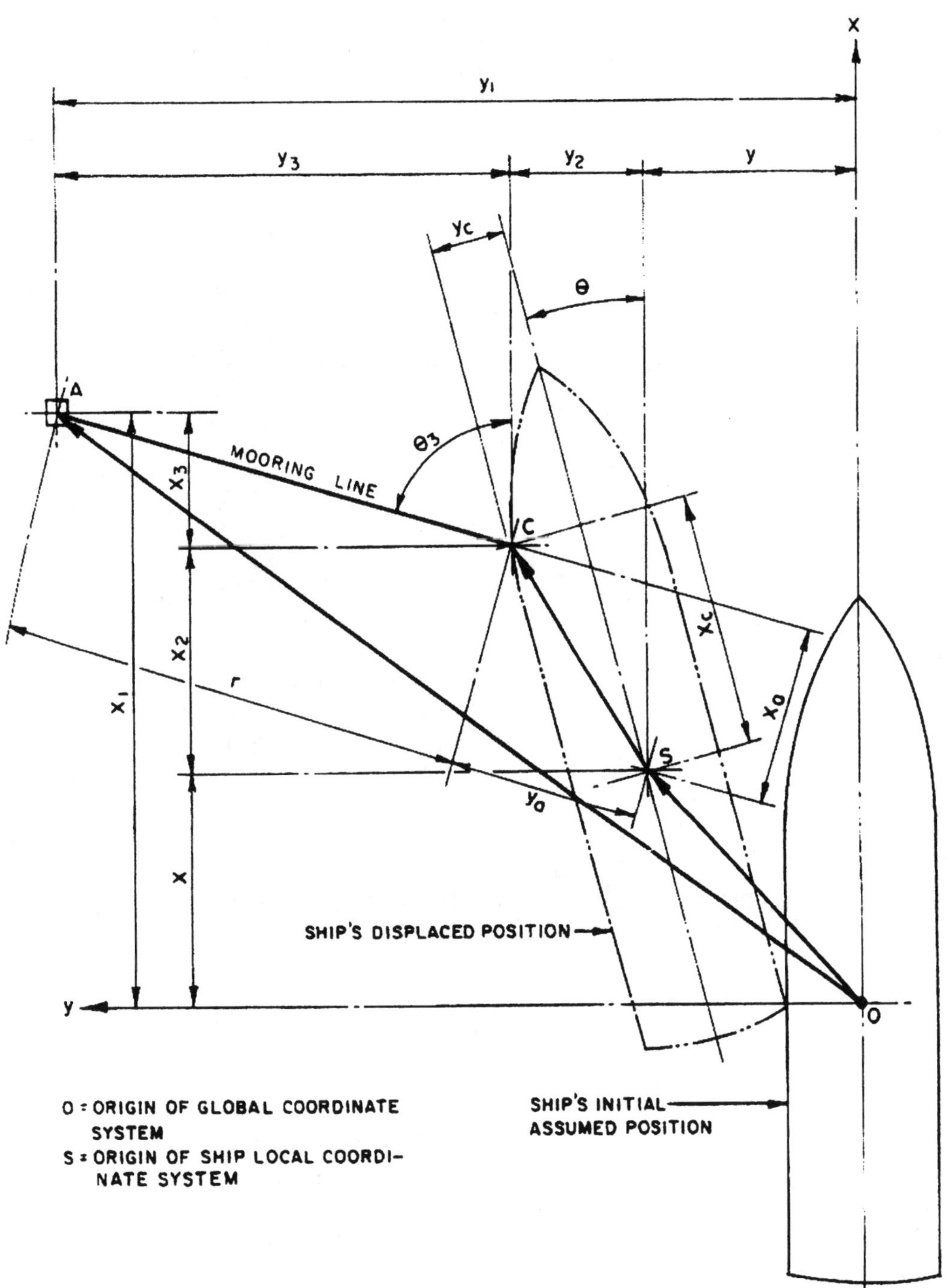

Figure 8.36 Definition sketch for static mooring computer model.

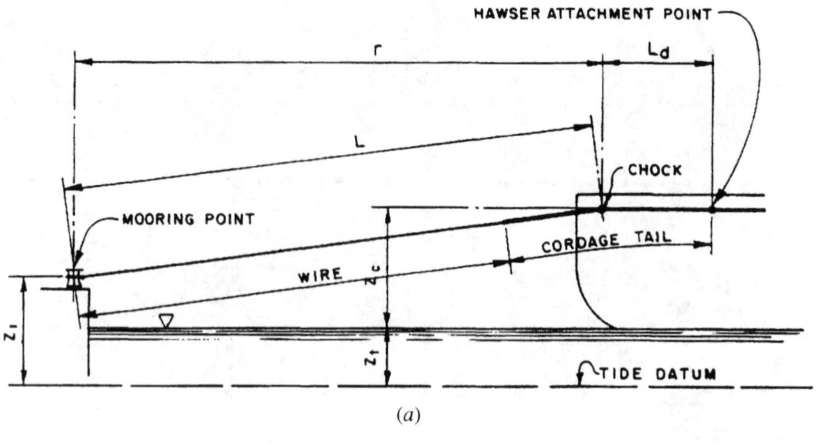

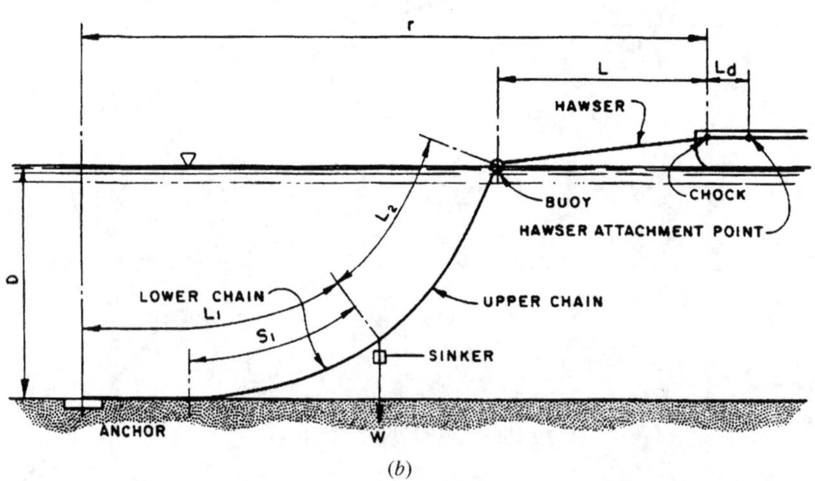

Figure 8.37 Definition sketches for (*a*) hawser and (*b*) anchor chain in static mooring computer model.

- The vessel can move in six modes (i.e., surge, sway, heave, roll, pitch, and yaw), so there must be a separate equation for each mode.
- The ship mass includes its physical mass as well as a mass generated as the vessel moves through the water (added mass).
- Damping can result from a variety of physical phenomena, including waves created by a vessel moving through the water and drag over the hull of the moving vessel (akin to a current force).
- The mooring forces tend to be nonlinear (i.e., the force–deflection curve is hyper-

bolic in shape rather than a straight line).

- The applied wave forces include both a linear first-order term proportional to wave amplitude and a second-order nonlinear term (second-order drift force) proportional to the square of wave amplitude.
- The applied wave forces associated with a typical ocean condition is characterized by a spectrum of waves having a range of wave height (or amplitudes) and period (or frequencies).
- The equations must be linearized and solved for each wave period (or frequency) to be solved using standard mass–spring and dashpot concepts of structural dynamics.
- The mooring forces on most SPM and MBM systems are dictated by horizontal surge, sway, and yaw motions.
- The natural periods of most SPM and MBM systems in surge, sway, and yaw are much longer than the periods of incident sea waves (i.e., less than 20 seconds). As a result, wave-induced loads on moorings tend to be dominated by second order drift forces, which act together with a typical wave spectrum to produce wave energy near the natural periods in surge, sway, and yaw.
- In a fashion similar to a wave spectrum, a typical wind spectrum has considerable energy near the natural periods in surge, sway, and yaw.

Interested readers should refer to Van Oortmerssen (1976) or Wichers (1988) for a more complete discussion of dynamic mooring analysis. The overall approach to a dynamic mooring analysis involves two basic mathematical models: (1) a hydrodynamic model, and (2) a mooring dynamics model. The hydrodynamic model is used to compute the added mass and damping coefficients associated with the vessel hull oscillating in the water. It should be noted that the resulting added mass and damping coefficients are therefore a function of frequency. The hydrodynamic model is also used to evaluate first- and second-order wave forces on the ship hull (assuming that the ship is at rest in the water). Wind and current loads are computed using methods like those presented for static analysis earlier in the chapter.

The mooring dynamics model takes the results of the hydrodynamics model and solves the governing equation of motion described above. There are two basic approaches to solving this equation: frequency domain and time domain. The dynamic equation of motion can be solved in the frequency domain in a manner analogous to the mass–spring–dashpot systems of structural dynamics, but only by linearizing the equations as mentioned above. This is a fatal limitation for most problems inasmuch as the mooring lines and the second-order drift forces are nonlinear. Moreover, frequency-domain analysis cannot simulate the slow "fishtailing" motions of single-point moorings.

Accordingly, time-domain analysis is preferred in design applications for SPMs and MBMs. Time-domain analysis involves integrating a time-domain version of the dynamic equation above and stepping the solution through time. This approach can handle any (including nonlinear) arbitrarily varying applied force from wind, waves, and currents and can also handle the nonlinear behavior of the mooring lines. A complete review of computational procedures for performing dynamic analyses of offshore moorings is beyond the scope of this chapter. Instead, in the following paragraphs we summarize the dynamic analyses of an MBM and an SPM.

8.7.3.1 Dynamic Analysis of a Spread Mooring. The following dynamic analysis example is for a 100,000-dead-weight ton (dwt) oil tanker secured to a six-legged MBM in a water depth of 20 m (Figure 8.38). The tanker

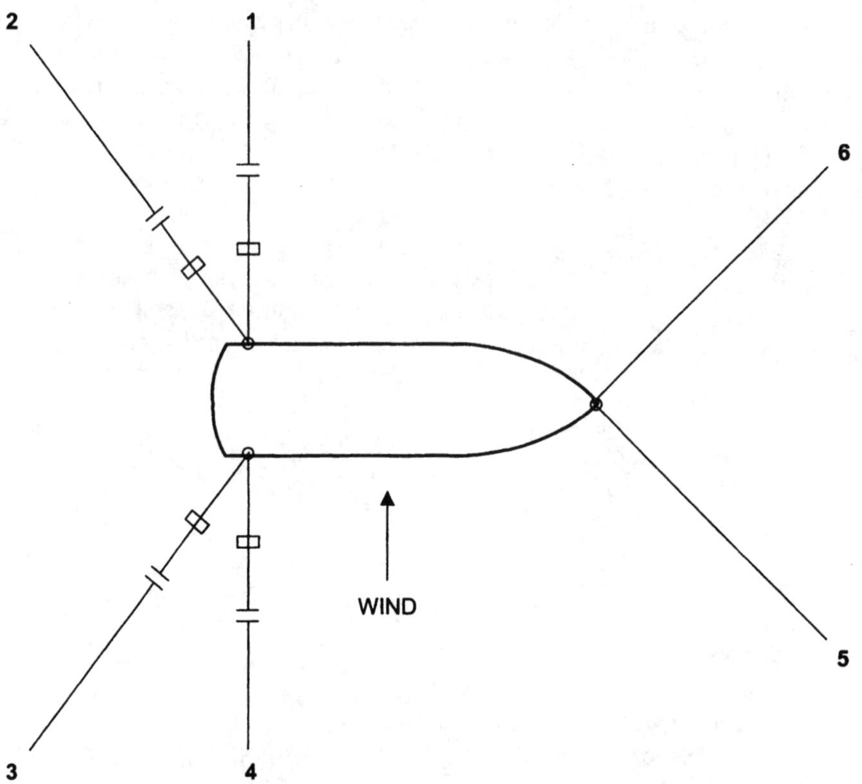

Figure 8.38 Six-legged mooring diagram.

is exposed to a range of beam-on winds in still water. The MBM mooring lines consist of wire rope hawsers and a 76- to 88-mm chain. Dynamic mooring analyses were performed using a three-degree-of-freedom time-domain model developed by the Maritime Research Institute in the Netherlands. The governing equations of the model are given below.

$$(m + a_{11})x_1 + a_{12}x_2 + a_{16}x_6 + b_{11}x_1 = F_1^{\text{wind}} + F_1^{\text{hydr}} + F_1^{\text{wave}} + F_1^{\text{moor}} \tag{8.76}$$

$$a_{21}x_1 + (m + a_{22})x_2 + a_{26}x_6 + b_{22}x_2 = F_2^{\text{wind}} + F_2^{\text{hydr}} + F_2^{\text{wave}} + F_2^{\text{moor}} \tag{8.77}$$

$$a_{61}x_1 + a_{62}x_2 + (l_{66} + a_{66})x_6 + b_{66}x_6 = F_6^{\text{wind}} + F_6^{\text{hydr}} + F_6^{\text{wave}} + F_6^{\text{moor}} \tag{8.78}$$

where m and l are the tanker mass and moment of inertia; $a_{i,j}$ the added mass coefficient; x_j, $\dot{x}_j$, and $\ddot{x}_j$ the acceleration, velocity, and displacement in the jth mode; $b_{i,j}$ the damping coefficient; F_i^{wind} the wind force in the i direction ($i = 1, 2, 6$); F_i^{hydr} the hydrodynamic reaction force in the i direction ($i = 1, 2, 6$); F_i^{wave} the wave drift force in the i direction ($i = 1, 2, 6$); and F_i^{moor} the mooring force due to a hawser in the i direction ($i = 1, 2, 6$).

Wind-force time histories were generated using spectral simulation techniques and the re-

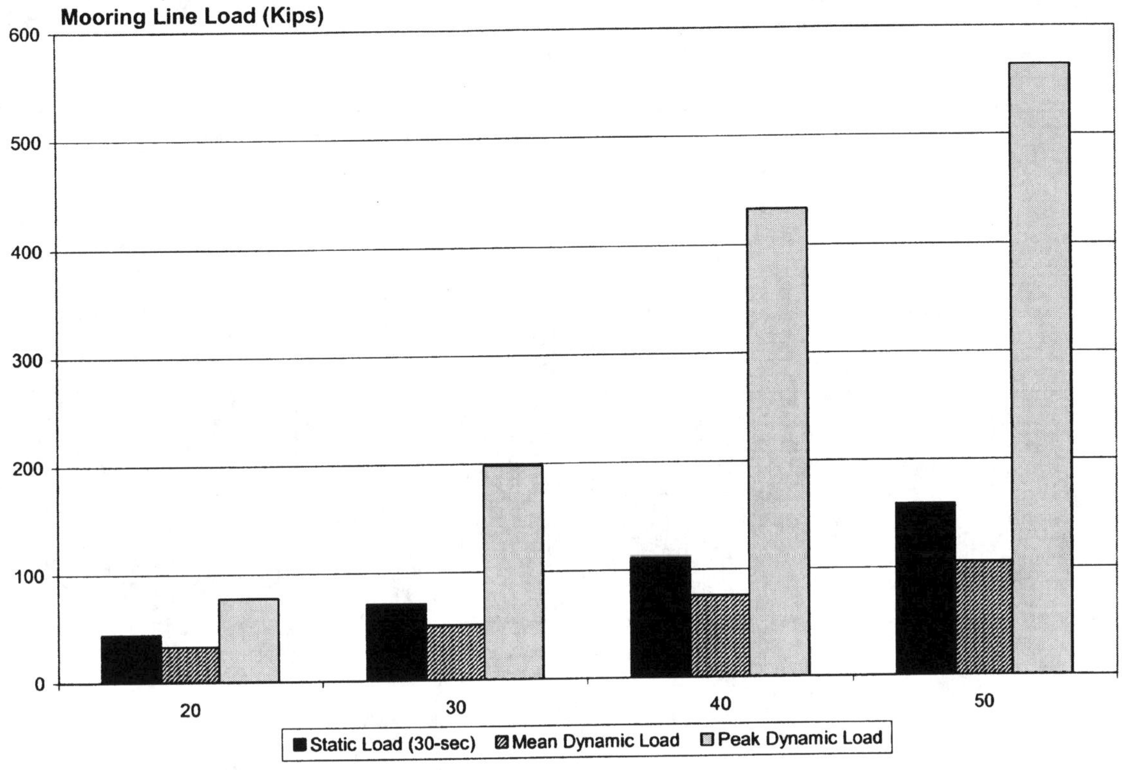

Figure 8.39 Dynamic versus static calculations for MBM.

cently developed wind spectrum reported by Ochi and Shin (1988). The Ochi and Shin spectrum is computed as follows:

$$S(f_*) = \begin{pmatrix} 583f_* & \text{for } 0 \le f_* \le 0.0 \\ \dfrac{420f_*^{070}}{(1 + f_*^{0.35})^{11.5}} & \text{for } 0.003 \le f_* \le 0.1 \\ \dfrac{838f_*}{(1 + f_*^{35})^{11.5}} & \text{for } f_* \ge 0.1 \end{pmatrix} \tag{8.79}$$

where

$$f_* = 10\,\frac{f}{V_w} \qquad S(f_*) = \frac{fS(f)}{C_{10}V_w^2} \tag{8.80}$$

where f is the frequency of oscillation (Hz), V_w the wind velocity, C_{10} the surface drag coefficient, and $S(f)$ the spectral density. The natural periods of motion of the moored dry dock system in surge, sway, and yaw are relatively long, and low-frequency motions dictate mooring line loads as a result. The initial mooring line pretension at the equilibrium position was 29.0 kN.

Static mooring analysis was performed using the techniques described with Sections 8.5

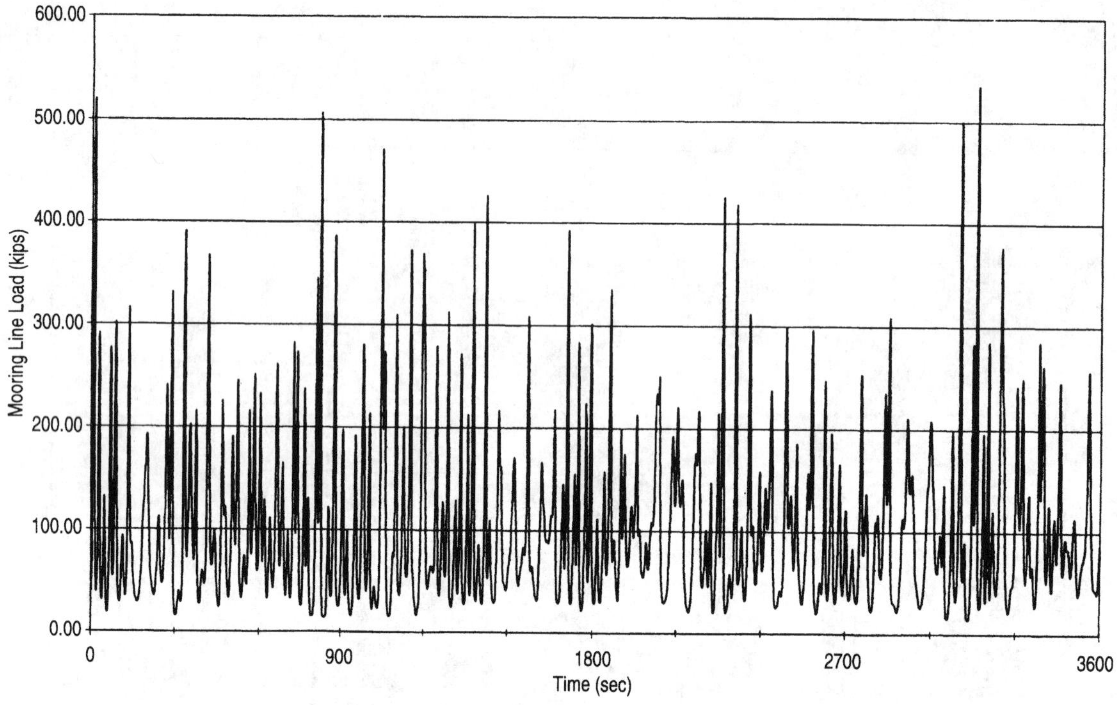

Figure 8.40 Mooring line load–time history.

and 8.7.2. Specifically, a steady 30-second wind speed was used to estimate the total wind load on the tanker. Dynamic analyses were performed for wind spectrums having mean hourly wind speeds of 20, 30, 40, and 50 knots. Each dynamic simulation was prepared using beam-on winds, a total simulation time of 3600 seconds (60 minutes). Figure 8.39 summarizes the dynamic analysis results in terms of the mean and maximum mooring line load and compares these to the forces obtained from static analysis results. Figure 8.40 presents a mooring line force–time history for a mean hourly wind speed of 50 knots. As shown in Figure 8.39, the peak dynamic mooring line loads greatly exceed mean mooring line loads. The static mooring line loads are between the dynamic mean and maximum mooring line loads. In summary, dynamic wind can be an important factor in the design of spread moorings. A similar analysis was prepared by Headland et al. (1989).

8.7.3.2 Dynamic Analysis of a Single-Point Mooring. SPMs let a moored vessel weather-vane or fishtail to align itself with prevailing environmental conditions, as stated previously. Although static analysis assumes that a vessel moored to a SPM will attain an equilibrium position, experience has shown that vessels slowly oscillate around an equilibrium position. The fishtailing phenomenon can result in large mooring hawser loads relative to those computed statically.

A model similar to that used in Section 8.7.3.1 was used for the present work. The governing equations follow.

$$(m + a_{11})x_1 + a_{12}x_2 + a_{16}x_6 + b_{11}x_1$$
$$= F_1^{wind} + F_1^{hydr} + F_1^{wave} + F_1^{moor} + PR \quad (8.81)$$

$$a_{21}x_1 + (m + a_{22})x_2 + a_{26}x_6 + b_{22}x_2$$
$$= F_2^{\text{wind}} + F_2^{\text{hydr}} + F_2^{\text{wave}} + F_2^{\text{moor}} \quad (8.82)$$

$$a_{61}x_1 + a_{62}x_2 + (l_{66} + a_{66})x_6 + b_{66}x_6$$
$$= F_6^{\text{wind}} + F_6^{\text{hydr}} + F_6^{\text{wave}} + F_6^{\text{moor}} \quad (8.83)$$

$$m_{\text{buoy}}x_{\text{buoy}} + b_{\text{buoy}}x_{\text{buoy}} = F_{\text{x buoy}} \quad (8.84)$$

$$m_{\text{buoy}}y_{\text{buoy}} + b_{\text{buoy}}y_{\text{buoy}} = F_{\text{y buoy}} \quad (8.85)$$

where m and l are the tanker mass and moment of inertia; $a_{i,j}$ the added mass coefficient; x_j, $\dot{x}_j$,

Table 8.4 Particulars of a 200,000-dwt tanker in ballasted condition

Property	Unit	Value
Length	m	310.00
Beam	m	47.17
Draft	m	7.56
Displacement volume	m^3	82,700
Longitudinal wind area	m^2	1,897
Lateral wind area	m^2	7,785

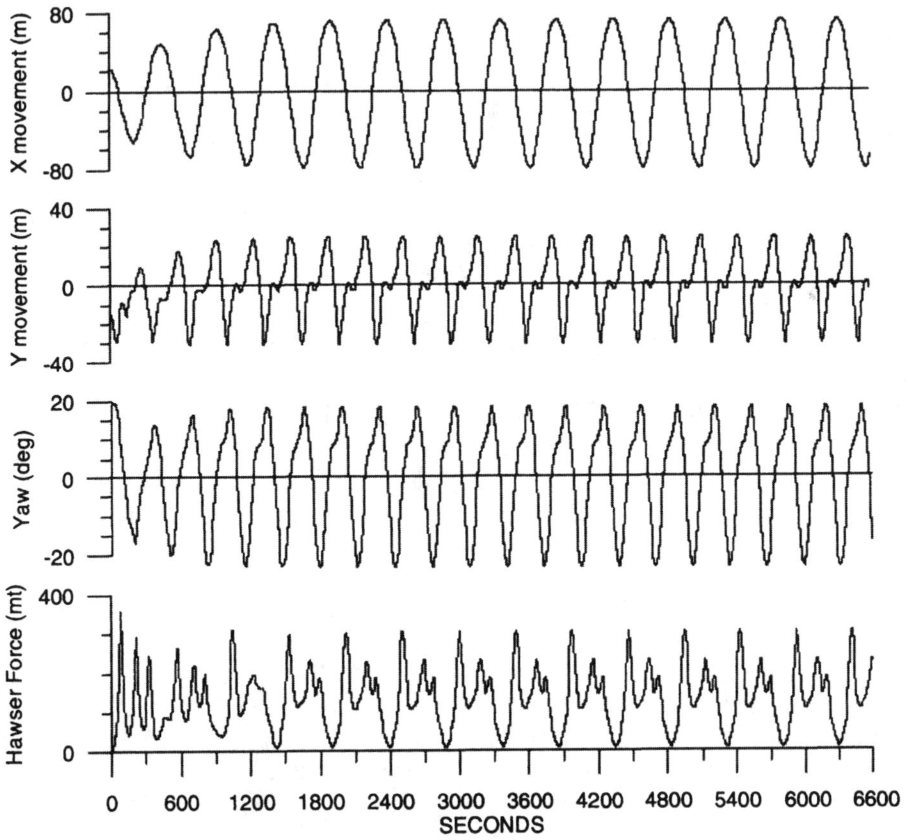

Figure 8.41 Dynamic analysis of a single-point mooring exposed to wind and current.

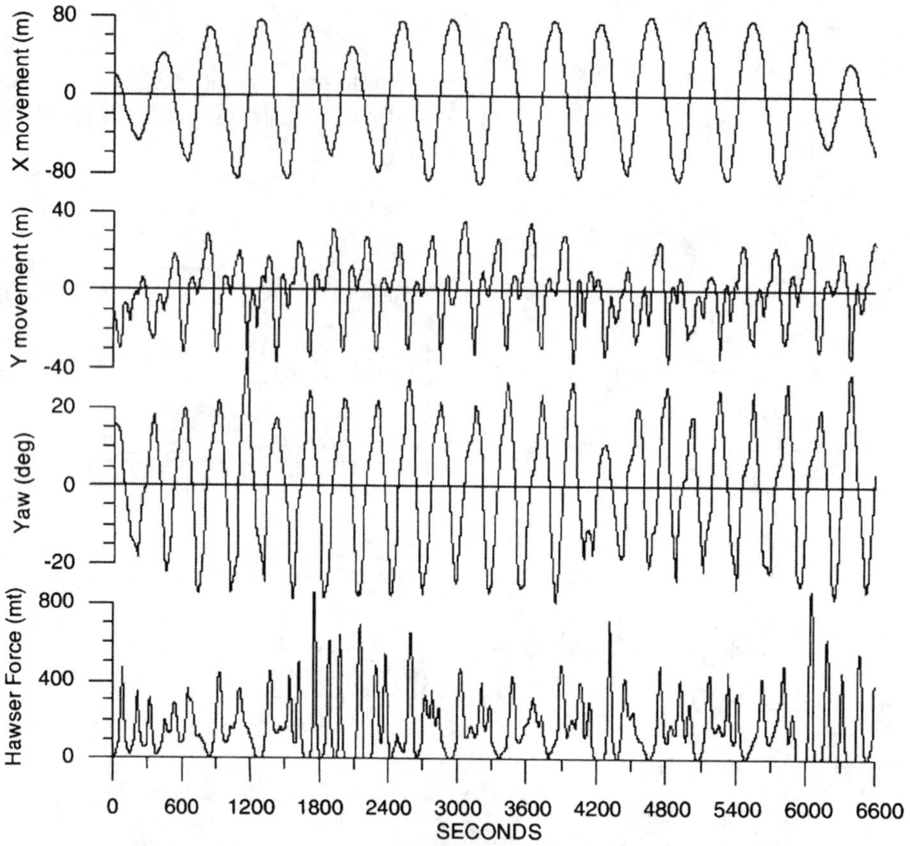

Figure 8.42 Dynamic analysis of a single-point mooring exposed to wind, current, and waves.

and $\ddot{x}_j$ are the acceleration, velocity, and displacement in the jth mode; $b_{i,j}$ the damping coefficient; F_i^{wind} the wind force in the i direction ($i = 1, 2, 6$); F_i^{hydr} the hydrodynamic reaction force in the i direction ($i = 1, 2, 6$); F_i^{wave} the wave drift force in the i direction ($i = 1, 2, 6$); F_i^{moor} the mooring force due to a hawser in the i direction ($i = 1, 2, 6$); PR the astern propulsion (negative); m_{buoy} the fixed mass of a buoy (fixed in program); b_{buoy} the damping of a buoy (fixed in program); and $F_{i\,\text{buoy}}$ the force acting on a buoy due to a hawser and chains. Steady wind and current drag forces in the equations above are computed using expressions similar to those presented in Section 8.4.3 but are computed using the relative velocity between the vessel and applied current. Viscous current drag forces are included in the model.

A dynamic mooring analysis was prepared for a 200,000-dwt tanker moored with an elastic hawser 90 m in length. This mooring problem was evaluated by Wichers (1988). The tanker characteristics are given in Table 8.4 for 25% loading conditions. The tanker was exposed to bow-on wind and current. The wind speed was 60 knots (30.9 m/s) and the current speed was 2 knots (1.03 m/s).

Simulation results are summarized in Figure 8.41, which shows an oscillatory fishtailing motion with a maximum mooring line load of

268 tonnes compared to a static value of 87 tonnes. Similar results are presented in Figure 8.42 for the same mooring configuration exposed to bow-on winds of 60 knots, bow-on currents of 2 knots, and a bow-on wave spectrum characterized by a significant wave height of 3.9 m and an average wave period of 10.2 s. In summary, the paragraphs above serve to illustrate that SPMs are prone to important dynamic motions not identified from static analyses.

REFERENCES AND RECOMMENDED READING

ABS, 1975. *Rules for Building and Classing Single-Point Moorings*, American Bureau of Shipping, New York, 1975.

———, 1997. *Rules for Building and Classing Mobile Offshore Drilling Units*, American Bureau of Shipping, New York.

API, 1987. *Recommended Practice for Planning, Designing and Constructing Fixed Offshore Platforms*, API RP 2A, 17th ed., American Petroleum Institute, Washington, DC, Apr. 1.

———, 1990. *Recommended Practice for Design, Analysis and Maintenance of Mooring for Floating Production Systems*, draft, American Petroleum Institute, Washington, DC, May.

Batts, M. E., M. R. Cordes, L. R. Russell, J. R. Shaver, and E. Simiu, 1980. *Hurricane Wind Speeds in the United States*, NBS Building Science Series 124, National Bureau of Standards, Washington, DC, May.

Berteaux, H. O., 1976. *Buoy Engineering*, John Wiley & Sons, New York.

Changery, M. J., 1982a. *Historical Extreme Winds for the United States: Atlantic and Gulf of Mexico Coastlines*, NUREG/CR-2639, National Oceanic and Atmospheric Administration, National Climatic Data Center, Asheville, NC, May.

———, 1982b. *Historical Extreme Winds for the United States: Great Lakes and Adjacent Regions*, NUREG/CR-2890, National Oceanic and Atmospheric Administration, National Climatic Data Center, Asheville, NC, Aug.

De Kat, J. O., and J. E. W. Wichers, 1990. *Dynamic effects of current fluctuations on moored vessel*, OCT 6219, *Proc. Offshore Technology Conference*, May, pp. 183–198.

Flory, J. F., F. A. Benham, J. T. Marcello, P. F. Poranski, and S. P. Woehleke, 1977. *Guidelines for Deepwater Port Single Point Mooring Design*, CG-D-49-77, U.S. Department of Transportation.

Gerald, C. F., 1980. *Applied Numerical Analysis*, Addison-Wesley, Reading, MA.

Headland, J. R., W. N. Seelig, and C. Chern, 1989. *Dynamic analysis of moored floating drydocks*, *ASCE Proc. Specialty Conference Ports '89*, Boston.

Molin, B., and G. Bureau, 1980. *A simulation model for the dynamic behaviour of tankers moored to SPM*, presented at the International Symposium on Ocean Engineering and Shiphandling, Gothenburg, Sweden.

Muga, B. J., and M. A. Freeman, 1977. Computer simulation of single point moorings, OTC 2829, *Proc. Offshore Technology Conference*.

Obokata, J., 1983. Mathematical approximation of the slow oscillation of a ship moored to single point moorings, presented at the Marintec Offshore China Conference, Shanghai, Oct.

Ochi, M. K., and Y. S. Shin, 1988. Wind turbulent spectra for design considerations of offshore structures, OTC 5736, *Proc. Offshore Technology Conference*, pp. 461–476.

OCIMF, 1994. *Prediction of Wind and Current Loads on VLCCs*, Oil Companies International Marine Forum, London.

———, 1997. *Mooring Equipment Guidelines*, Oil Companies International Marine Forum, London.

Owens, R., and P. Palo, 1982. *Wind Induced Steady Loads on Moored Ships*, TN: N-1628, Naval Civil Engineering Laboratory, Port Hueneme, CA.

Remery, G. F. M., and G. Van Oortmerssen, 1973. The mean wave, wind and current forces on offshore structures and their role in the design of mooring systems, OTC 1741, *Proc. Offshore Technology Conference*.

Rocker, K., 1985. *Handbook for Marine Geotechnical Engineering*, Naval Civil Engineering Laboratory, Port Hueneme, CA.

Seelig, W. N., and J. R. Headland, 1998. Mooring dynamics due to wind gust fronts, *ASCE Proc. Specialty Conference Ports '98*, Long Beach, CA, pp. 870–879.

Seelig, W. N., D. L. Kniebel, and J. R. Headland, 1992. Broadside current loads on moored ships, *ASCE Proc. Civil Engineering in the Oceans V Conference*.

Simiu, E., and R. H. Scanlon, 1978. *Wind Effects on Structures: An Introduction to Wind Engineering*, Wiley-Interscience, New York.

U.S. Department of Defense, 1999. *Mooring Design*, MIL-HDBK-1026/4A, U.S. Government Printing Office, Washington, DC, July 1.

U.S. Navy, 1985. *Fleet Moorings: Basic Criteria and Planning Guidelines*, DM-26.5, Naval Facilities Engineering Command, Alexandria, VA.

———, 1986. *Mooring Design: Physical and Empirical Data*, DM-26.6, Naval Facilities Engineering Command, Alexandria, VA.

Van Oortmerssen, G., 1976. *The Motions of a Moored Ship in Waves*, Publ. 510, Netherlands Ship Model Basin, Wageningen, The Netherlands.

Wichers, J. E. W., 1979. Slowly oscillating mooring forces in single point mooring systems, *Proc. Symposium on Behaviour of Offshore Structures*, London.

———, 1988. *A Simulation Model for a Single Point Moored Tanker*, Publ. 797, MARIN, Wageningen, The Netherlands.

9

Breakwaters

Gregory P. Tsinker

Tsinker & Associates, Inc.
Niagara Falls, Ontario, Canada

9.1 INTRODUCTION

The purpose of this chapter is to help those involved with port design, especially those who just recently joined the field of marine engi-

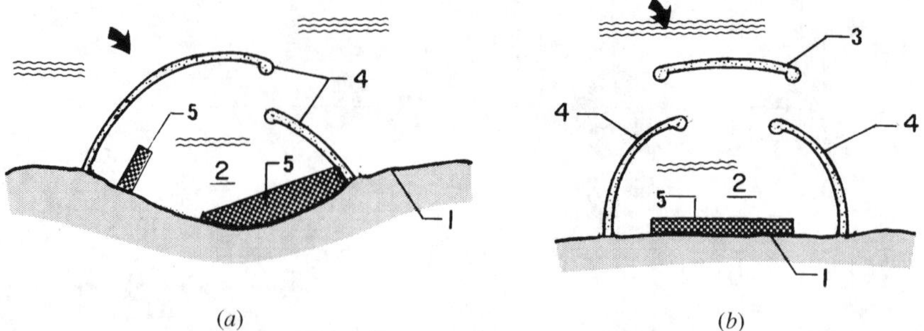

Figure 9.1 (*a*) Harbor protection by shore-connected breakwaters; (*b*) combination of shore-connected and offshore breakwaters. 1, Coastal line; 2, harbor; 3, offshore breakwater; 4, shore-connected breakwater; 5, port-related structures.

neering to better understand the phenomena of wave–structure interaction, as well as the role of breakwaters in port operation. The material provided in this chapter is basic and should not be treated as a guideline to detailed breakwater design. References to specific information that may be helpful in detailed breakwater design are found elsewhere in this chapter.

Harbor protection is normally achieved by the construction of breakwaters. They are constructed to protect exposed harbors otherwise lacking natural protection from adverse effects of waves, currents, migrated sediments, and drifting ice and to create a calm water area within the harbor which is required for safe and convenient ship operation. Breakwaters can be shore-connected (moles) or located offshore with no connection to the shore. The former are usually designed to protect the harbor water area from all the above-noted environmental phenomena. On the other hand, offshore breakwaters are designed primarily to provide protection from wave action to the area of shoreline located on the leeward side of the structure. The main purpose of using offshore breakwaters is to dissipate or reduce the enormous energy of a wave reaching the harbor in its lee. They may also provide additional protection for harbors, serving as a littoral barrier-sediment trap.

Depending on local side conditions, shore-connected and offshore breakwaters may be used separately or in combination with each other as depicted in Figure 9.1. Selection of the most suitable combination depends on the predominant direction of maximum waves and the mode of port operation. Essentially, the breakwater(s) layout should provide for the maximum protection of the harbor area.

9.2 BREAKWATER(S) LAYOUT

Ideally, breakwaters are placed so as to provide an area that is sufficiently large and protected from wave action so that it is adequate for safe entering and maneuvering of the largest design vessels and adequate functioning of the port. Where the harbor is also used to serve smaller vessels (e.g., fishing trawlers, pleasure boats) the additional interior breakwaters may be constructed inside the harbor to protect a fleet of smaller vessels from the effects of relatively large waves that do not affect the operation of large ships. Examples are shown in Figures 9.2 and 9.3.

As noted earlier, the basin size confined by breakwaters must be sufficiently large to provide safe and efficient port operation (i.e., must be adequate for safe and convenient ship entrance into the basin and the ship's approach to the terminal, either self-propelled or assisted by tugs). In general, the minimum harbor area is

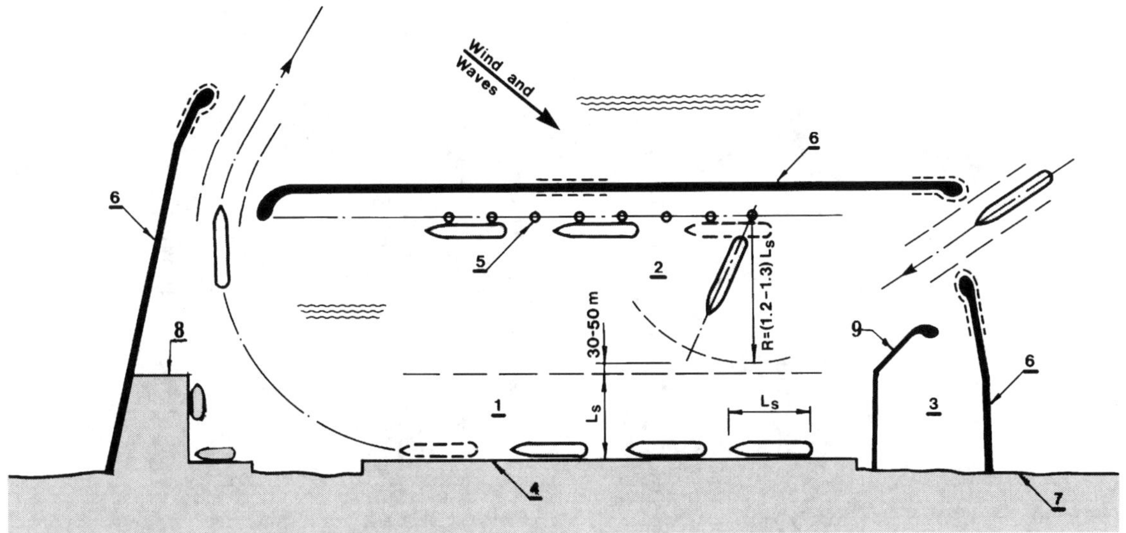

Figure 9.2 Typical layout of a middle-sized port placed at an artificially created harbor. 1, Berthing area; 2, anchorage area; 3, small-craft basin; 4, waterfront structure; 5, mooring dolphin; 6, breakwater; 7, coastal line; 8, tugboat docks; 9, in-harbor breakwater.

the space required to accommodate at least one ship berth with a turning basin, where a ship is turned by warping it around the turning dolphin or around the end of the pier. Depending on local conditions, the minimum harbor length can be taken as equal to five to six lengths of the largest ship, with a width of two to three lengths of the largest ship. For more information on harbor area requirements, readers are referred to Tsinker (1997).

Typical elements of a port basin usually include, but are not limited to, the ship stopping area, maneuvering area, turning basin, anchorage area, berthing area, and sometimes a special-purpose area (Figure 9.4).

9.2.1 Stopping Area

To provide for a ship's safe entrance into a port basin, the length of the basin adjacent to the entrance and immediately inside the basin should be sufficiently long to allow a ship entering the basin at a reasonable speed to stop safely. A preliminary ship stopping length (L_{st}) may be determined by (Dzhunkovski et al., 1964)

$$L_{st} = \frac{0.27 V_0^3 D}{N} \tag{9.1}$$

where V_0 is the speed of ship (knots), D the displacement (tonnes), and N the installed power (hp). In general, the stopping distance of a ship depends on such factors as ship mass (displacement), traveling speed, installed engine power, local environmental conditions, and on the assistance provided by tugs. No general rules exist for determining the length of this portion of the basin. It is usually taken as equal to 3.5 to 8 times the length of the longest ship expected to enter the basin, subject to ship mass and speed of entrée. The width of the necessary water area inside the basin beyond the port entrance, where the ship enters a calm water area and is no longer affected by possible cross-current and waves, could be

(*a*)

(*b*)

Figure 9.3 In-harbor breakwaters for protection of small-craft marinas: (*a*) port de Fontvieille, Monaco; (*b*) Osaka Bay, Japan.

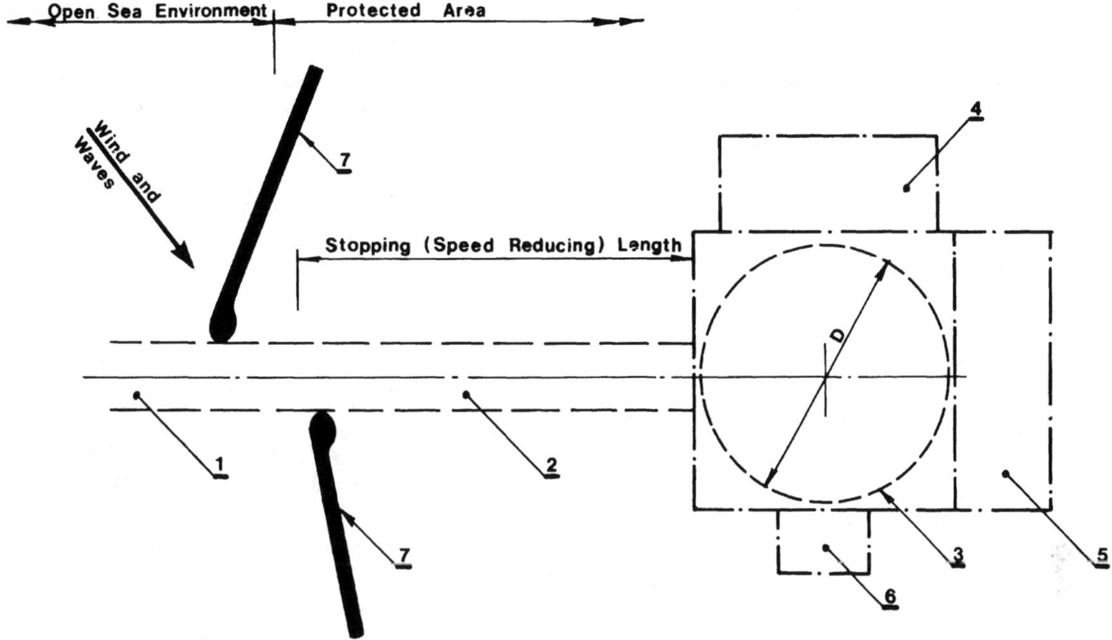

Figure 9.4 Port basin. 1, Access channel; 2, interior channel (stopping area); 3, maneuvering/turning area; 4, anchorage area; 5, berthing area; 6, special-purpose area; 7, breakwater.

smaller than the required width in the approach channel. However, in a transitional water area, where the ship bow enters calm water beyond the entrance to the protected area but the stern is still exposed to cross-current and/or swell action, the width of the water area must be adequate to account for the steering to which the ship may be subjected. In the modern science of port engineering, a rational stopping length and channel width are usually obtained by thorough planning and evaluation of a ship maneuvering into the basin. The latter is best obtained either by testing physical models or by using computer simulation models. More information is given in Dickson (1969), Permanent International Association of Navigation Congresses (PIANC, 1980), and Tsinker (1997).

9.2.2 Maneuvering Area

Before a vessel starts an actual berthing procedure, it usually stops in a maneuvering area. This area generally includes the necessary area to allow a ship to reduce speed and swinging area sufficiently. The width of the maneuvering water area must account for the potential drift during the stopping maneuver that may occur due to the engine of a single-screw ship running full speed astern. The water area needed for the movement of a ship during the turning maneuver is defined as a *turning basin*. A turning basin may be of various shapes, depending on the size of the area, the size and maneuverability of the ship, and the layout of berthing facilities. It is usually considered that during the turning operation the current in the turning area is less than 0.1 m/s and that if the turning ship is in a light condition, the wind speed should be less than 10 m/s. If these conditions are not met for any reason, the ship needs to be assisted by tugs adequate in number and power.

If the available water area permits the largest design ship to turn a complete circle without the aid of tugs or side thrusters, the minimum diameter of this circle should be a factor of 3.5 to 4 times the length of the ship, depending on the ship's maneuvering characteristics. Where the ship has tug assistance, the turning diameter could be reduced to two times the ship length or even less. Where the ship is turned by warping around a dolphin or around the head end of a pier, the turning diameter could be reduced to a minimum of 1.2 times the length of the ship. For practical examples, readers are referred to Tsinker (1997).

9.2.3 Anchorage Area

A large natural harbor protected naturally or by a breakwater system may offer sufficient water area for anchorages, where ships can safely be anchored while waiting their turn for berthing in favorable weather conditions or for any other reasons. If economically feasible, an artificial harbor created by a system of breakwaters could provide space for anchorage of relatively small and/or middle-sized vessels, whereas large ships can be anchored out in the open sea. An example is shown in Figure 9.2. Sometimes, the leeward part of a breakwater structure is used as a facility for mooring ships awaiting their turn to dock or as a specialized terminal (dock). Relevant examples are illustrated in Figure 9.5.

9.2.4 Harbor Entrance

In most cases, one entrance into a commercial harbor is sufficient for adequate functioning of a port. However, in some cases, particularly at coastal ports stretched along a coastline, two and sometimes more entrances are required. Ideally, a port entrance should be located on the lee side of a harbor formed by breakwaters. If it must be located on the windward end of the harbor, adequate overlap of the breakwaters should be provided so that the harbor interior is protected from wave action; at the same time the ship is able to pass through the restricted entrance and be free to turn with the wind before it is hit broadside by waves. Typically, a port entrance is formed by shore-connected or offshore breakwaters, or in most cases, by a combination of the two.

The most common desirable features of the entrance to a harbor formed by a breakwater area are as follows:

1. It must be oriented to protect the harbor from penetration by the most severe waves.
2. The width and configuration of the entrance must allow for the expected density of marine traffic and the frequency of use with a minimum of congestion.
3. To minimize the amount of dredging work, the entrance should be located as close as possible to natural channels.

Basic entrance parameters such as orientation and width and depth of water in general should provide safe navigation for the largest vessel expected to call at the port. However, it should not be wider than necessary; usually, selection of the entrance width is a compromise between safe navigation requirements and the degree of harbor protection. In general, an entrance should be located as far as possible from the shoreline. It must be designed such that a ship will not need to maneuver at the entrance; the ship can begin to maneuver only after passing the entrance and entering the protected area. To avoid the danger of having a ship grounded under the effects of heavy wind, waves, and current, the entrance channel should not be placed parallel to the shoreline. The angle between the channel axis and the shoreline, α_2 (Figure 9.6*b*), preferable should not exceed about 30°. On the other hand, from the standpoint of navigational safety, it is desirable to keep the angle α_1 (Figure 9.6*a*) as small as practical. This helps to prevent ship

(*a*)

(*b*)

(*c*)

Figure 9.5 Use of leeward part of breakwater for mooring of vessels or as specialized facility: (*a*) Peele Island ferry terminal, Canada; (*b*) and (*c*) port of Gibraltar's outer breakwater, used for mooring of vessels awaiting their turn to the dock, and the mole (breakwater linked to land) used as a passenger and navy facility.

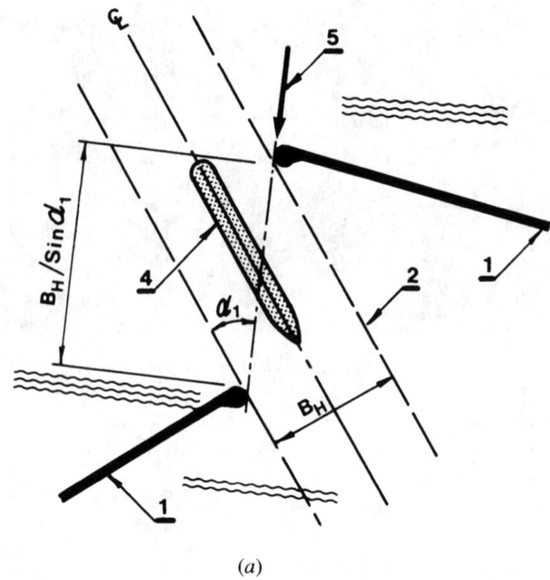

(a)

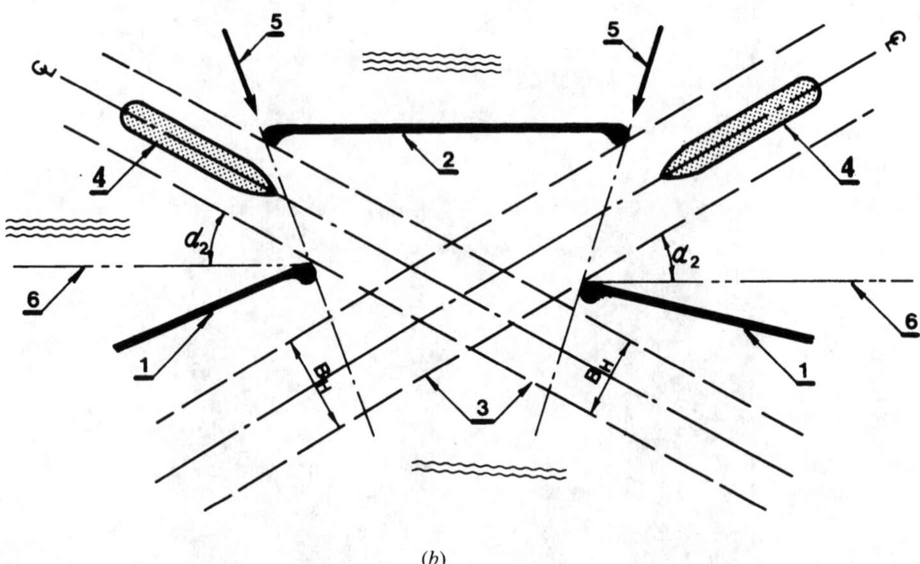

(b)

Figure 9.6 Port entrance orientation: (*a*) entrance formed by two shore-connected breakwaters; (*b*) entrance formed by two shore-connected breakwaters and offshore breakwater. 1, Shore-connected breakwater; 2, offshore breakwater; 3, access channel; 4, ship; 5, direction of major storms; 6, line parallel to the shoreline.

collisions with breakwaters under heavy wind and wave action. In practice, for safe navigation and better protection of the basin from waves, angle α_2 is normally considered equal to 30 to 35°. If shore-connected breakwaters cannot provide sufficient protection, an offshore breakwater should be placed to mitigate the wave situation (Figure 9.6*b*). An offshore breakwater can also improve ship negotiation into the basin.

Finally, where applicable, the entrance and breakwater layouts must provide protection against penetration into the harbor of wind- and/or current-driven ice floes. At the same time, ice movement away from the harbor should not be hindered. Structurally, the entrance geometry resembles the structure of breakwaters that form the entrance. Some typical entrances are presented in Figure 9.7. For further useful information, readers are referred to National Research Council (1981), Herbich (1992), and McBride et al. (1994).

9.2.5 Modeling and Testing Breakwater Layout and Harbor Entrances

The best design of breakwater layout and harbor entrances may be obtained through numerical modeling and physical model tests. Essentially, the success of numerical modeling and physical model tests depends on realistic input data that are included in defining harbor boundary conditions, breakwater structural and layout geometry, depth of water, properties of bottom contours, soil, wave, wind, and current climate, and so on. Being expensive, physical modeling is usually carried out to test the best results obtained on the basis of mathematical modeling. The latter normally includes use of computerized methods to analyze the wave climate inside the harbor and navigation conditions at the entrance.

9.3 BREAKWATER DESIGN

In general, breakwaters are classified as sloping or rubble-mound, vertical, composite, and special breakwaters. Selection of one type of breakwater structure or another for harbor protection depends first and foremost on local site conditions. The selection of a breakwater structure normally takes into account the following:

- Breakwater layout
- Site environmental conditions
- Utilization conditions
- Availability of construction materials and specialized floating construction equipment
- Cost of construction
- Cost of maintenance

Furthermore, a breakwater's effects on the surrounding topography and coastal line due to wave reflection and the quality of water inside the basin are taken into consideration when selecting the breakwater structure. As noted earlier, breakwaters are constructed to provide a calm basin protected from waves to allow safe ship navigation within the harbor and to protect the port facilities. They are also used to protect the port area from intrusion of littoral drift. Because sea waves have enormous power, the construction of breakwaters designed to mitigate such power is not easily accomplished, and the history of breakwaters records a great deal of damage and failure sustained by breakwaters. In the past 50 or so years, and especially since 1945, maritime technology has progressed a great deal. This has gradually made it possible to construct breakwaters with high stability against waves. As pointed earlier, in general, all breakwaters are classified as rubble-mound, vertical, composite, and special or unconventional breakwaters (Figure 9.8).

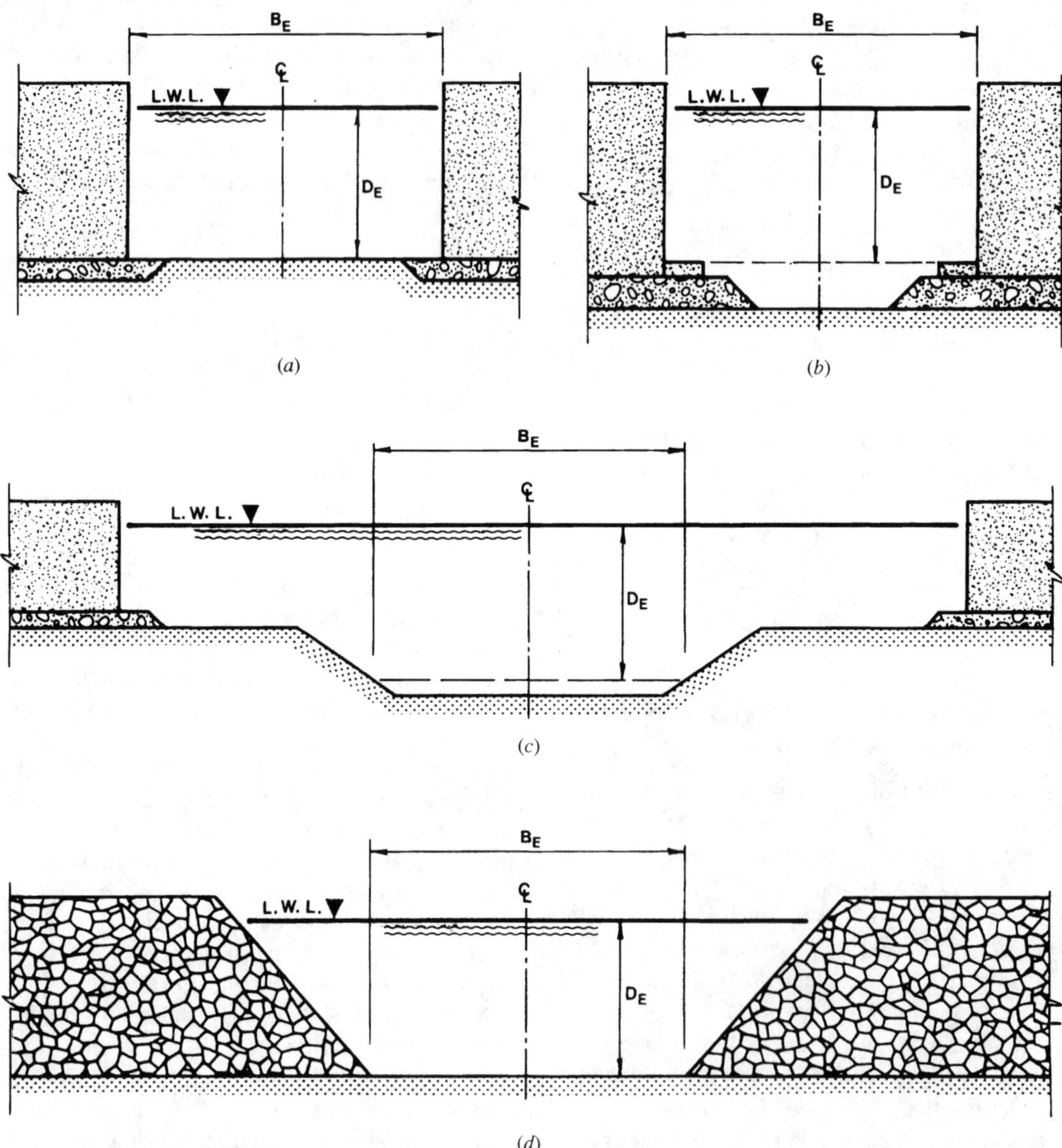

Figure 9.7 (*a*) and (*b*) Port entrance formed by vertical wall breakwaters; (*c*) dredged entrance protected by vertical wall breakwaters; (*d*) entrance formed by rubble-mound breakwaters.

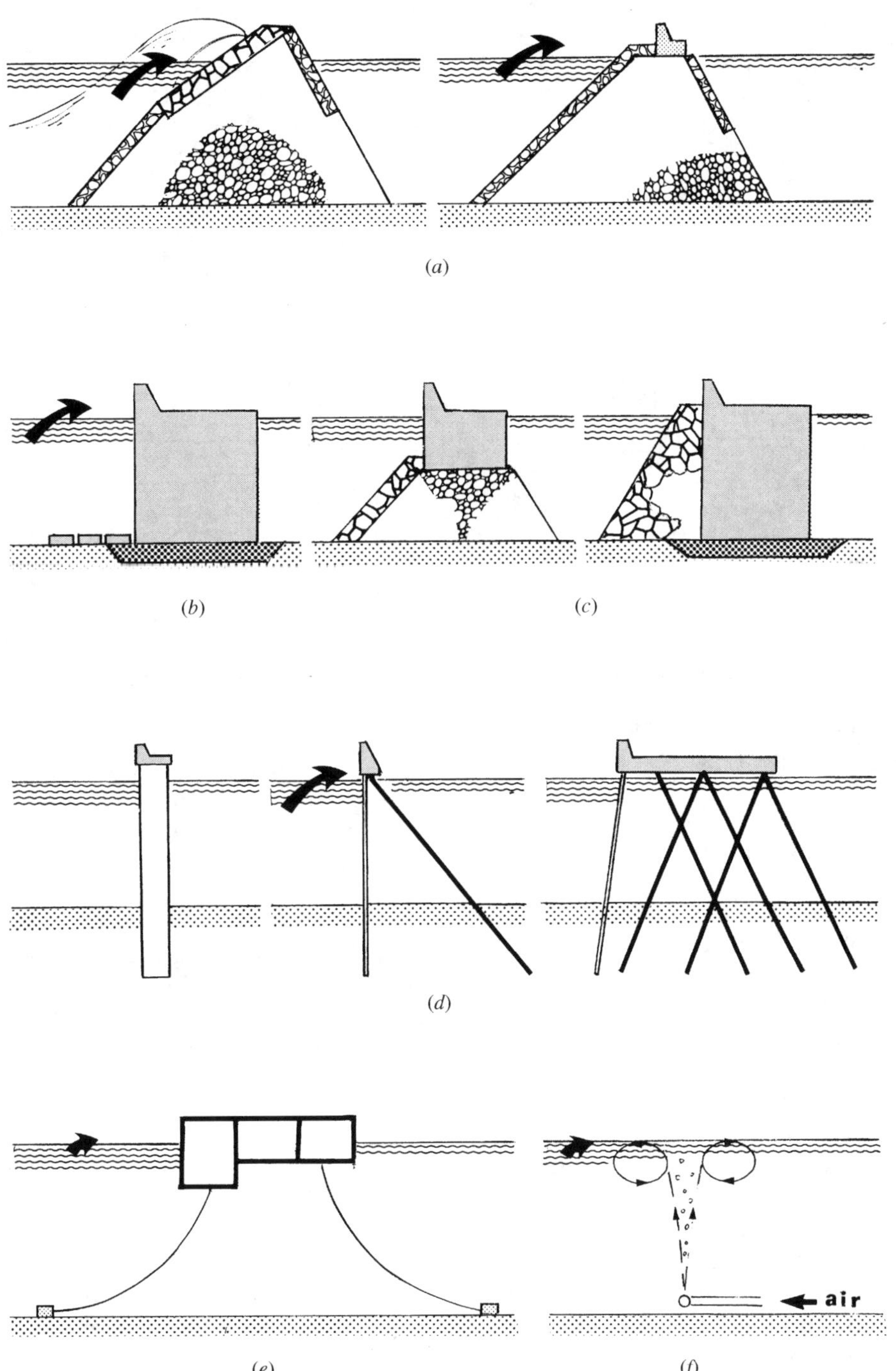

Figure 9.8 Basic types of breakwaters: (*a*) rubble-mound; (*b*) vertical face (gravity); (*c*) composite; (*d*) piled; (*e*) floating; (*f*) pneumatic.

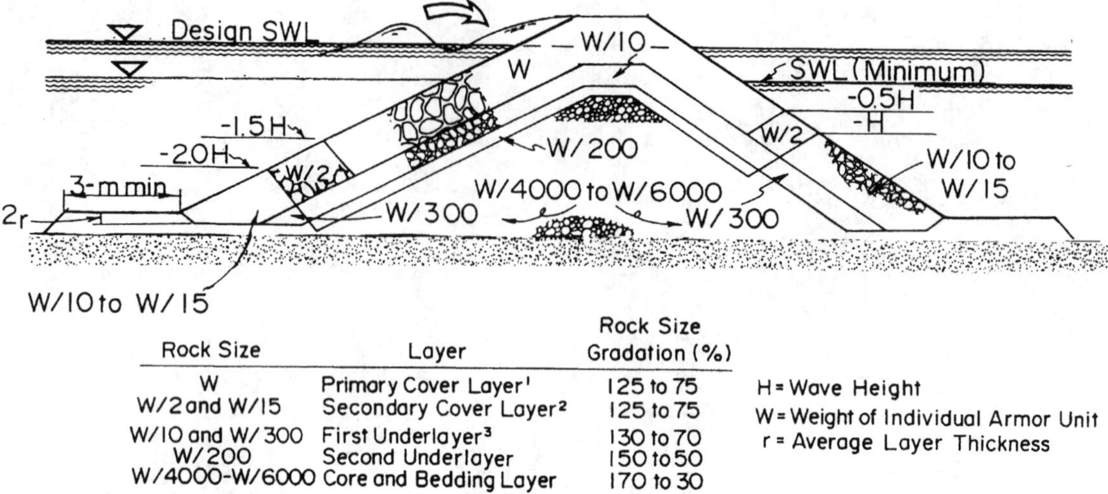

Rock Size	Layer	Rock Size Gradation (%)
W	Primary Cover Layer[1]	125 to 75
W/2 and W/15	Secondary Cover Layer[2]	125 to 75
W/10 and W/300	First Underlayer[3]	130 to 70
W/200	Second Underlayer	150 to 50
W/4000-W/6000	Core and Bedding Layer	170 to 30

Figure 9.9 Typical cross section, layered rubble-mound breakwater. (After U.S. Army Corps of Engineers, 1984.)

9.3.1 Rubble-Mound Breakwaters

Rubble mounds arranged in trapezoidal prismatic shape with gentle slopes represent the most common method to protect the harbors from the destructive effects of waves on port operation. The seaward slope of a rubble mound can effectively absorb wave energy, causing very little wave transmission and reflection. Further advantages of rubble-mound breakwaters are flexibility in shaping a breakwater's geometry and relative ease of construction and repair. The basic disadvantages of these structures lies in the need for a large volume of rock material for their construction and the large area of seafloor that they occupy. In some cases the latter could be objectionable from an environmental point of view. A typical rubble-mound breakwater has a rubble mound and an armor layer that usually consists of heavy rocks or shape-designed concrete blocks (Figure 9.8*a*). Use of shape-designed blocks has enabled modern-day rubble-mound breakwaters to resist the destructive power of waves effectively even in deep water.

A typical rubble mound is defined as a mound of randomly shaped and randomly placed stones protected from wave action by a cover layer of selected stones and/or specially shaped concrete armor units (armor units in a primary cover layer may be placed in an orderly manner or dumped at random). The rocks can also be placed in a layered manner, as shown in Figure 9.9. A multilayered rubble-mound breakwater ensures better stability and reduced wave reflection. This type of mound breakwater usually has a core composed from quarry run. The stability of the armor layer of the multilayered breakwater can be enhanced by using concrete blocks of miscellaneous geometry.

In terms of cross section, a rubble-mound breakwater may be classified in the following basic categories:

1. The conventional type, in which the rock material is placed in layers as required
2. A conventional structure provided with a parapet wall and concrete slab shaping the crest of the mound (Figure 9.10)

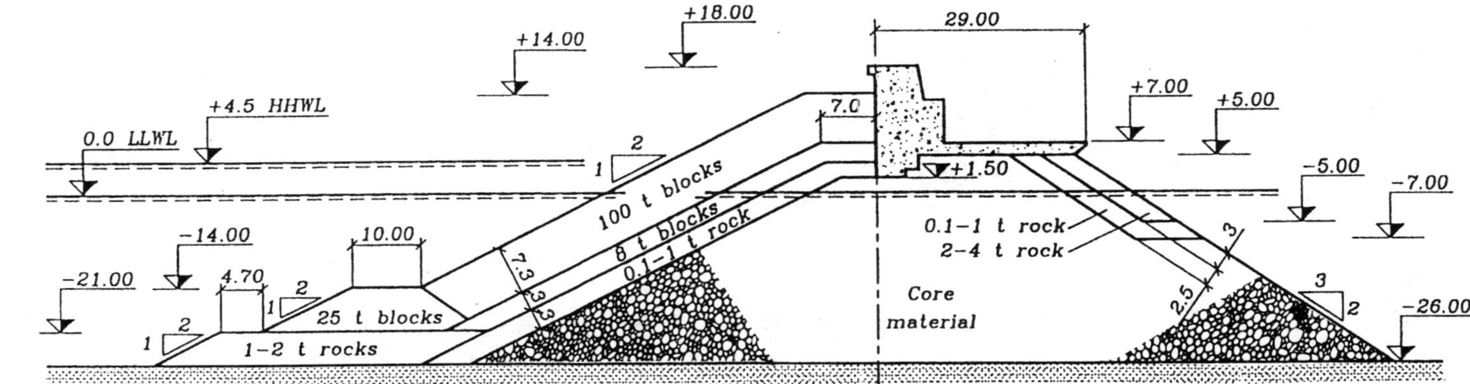

Figure 9.10 Typical cross section, rubble-mound breakwater with concrete parapet wall. (*Note:* All blocks are prismatic!) (Adapted from Burharth et al., 1995).

3. Berm breakwaters, structures with a horizontal section extending seaward and located close to the waterline (Figure 9.11)

Besides the three basic types of breakwaters above, there are structures composed completely from heavy rocks or artificial concrete blocks and "reef" or submerged breakwaters. The latter are not considered effective protection for a harbor operation but are useful in coastal protection. Reef breakwaters also do not create "visual" pollution and therefore are attractive from an environmental viewpoint in a heavily populated area.

Conventional Rubble-Mound Breakwater. It is normally composed of a bedding layer and a core of a quarry-run stones covered by one or more layers of larger stone and an exterior layer of large rocks or concrete armor units. An idealized cross section of conventional rubble-mound breakwaters is illustrated in Figure 9.9. The selection of stone size depends on site environmental conditions. For a detailed discussion and numerous examples, readers are referred to U.S. Army Corps of Engineers (1984), PIANC (1992), and Takahashi (1997).

Breakwater Complete with Parapet Wall. This is essentially the conventional type of breakwater in which a crest is formed by a parapet wall of miscellaneous construction.

Berm Breakwater. As stated by some authorities, berm breakwaters can be constructed and maintained at a considerable cost saving compared with conventional breakwaters with two layers of armor stone, especially at a considerable water depth and where the design waves are high (Baird and Woodrow, 1987; Blomberg and Green, 1988; Juhl and Jensen, 1993; Sigurdarson and Viggosson, 1994; PIANC, 2001; Jacobsen et al., 2002). The basic advantage of this type of breakwater is that the armor exposed to waves can be constructed from relatively small stones. A further advantage of berm breakwaters is their reliability. For example, the failure of a two-layer armor in conventional breakwaters tends to be abrupt. This may expose the core material to wave action, which may lead to a catastrophic failure of the breakwater. On the other hand, an abrupt catastrophic failure of berm breakwater does not occur if the design conditions are exceeded.

Although under such conditions some damage to or deformation of the structure should

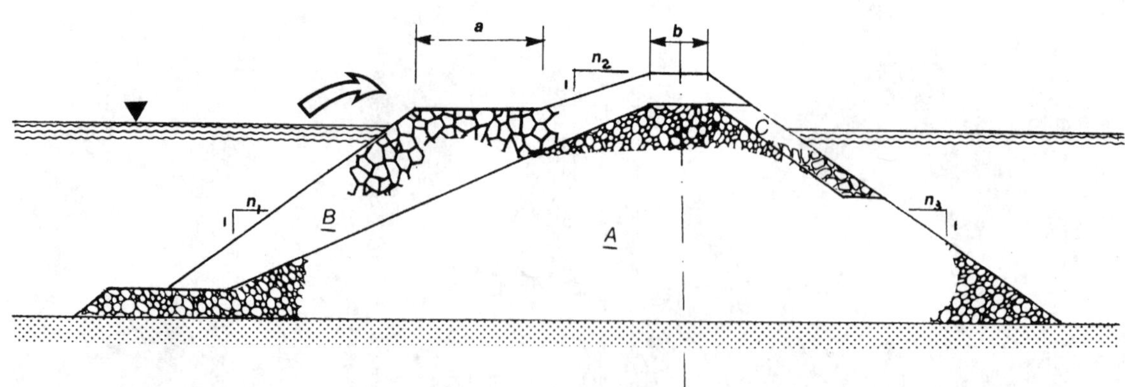

Figure 9.11 Typical cross section, berm breakwater. A, Core material; B and C, armor stones. n_1, n_2, n_3, a, and b variables depending on site condition.

be expected, but it can be speedily repaired. The other advantage of berm breakwaters is good utilization of a quarry-run rack. The latter is very important, especially in areas where large armor stones are not readily available. Use of regular quarry-run rock for breakwater constructions makes the construction procedures and follow-up maintenance of the breakwater much easier than for a conventional breakwater. Experience shows that even partly completely berm breakwaters may function satisfactory through winter storms and be repaired easily later (Viggosson, 1990).

Berm breakwaters do not represent a new idea in breakwater engineering. They have been long known as "reshaping" breakwaters that utilize the basic concept of dynamic equilibrium between the slope of a rubble mound and wave action, which in the final analysis results in the formation of an S-shaped slope stable against wave attack (Takahashi, 1997). The berm, essentially a horizontal platform built above the design water level, creates a large horizontal area where the waves can propagate among the armor rock mass, which has a high degree of porosity. The wave energy is dissipated there, resulting in a greatly reduced hydrodynamic force on the stone. This phenomenon allows smaller rocks to be used for breakwater construction. Over a period of time, the armor stone mass increases its stability, due to the process of consolidation, which results in an increase in the shear strength of the rock mass. For more information on berm breakwaters, readers are referred to Baird and Hall (1984); Burcharth and Brejnegaard-Nielsen (1986), Willis et al. (1987), Burcharth and Frigaard (1988), Fournier et al. (1990), Burcharth et al. (1995), and Bjordal (2003).

Discussion on state-of-the-art of berm breakwaters design and construction, as well as comprehensive list of references concerned with the subject matter is given in PIANC (2003a).

9.3.1.1 Design of a Conventional Rubble-Mound Breakwater. As noted earlier, a rubble-mound breakwater is a structure formed of superimposed orderly layers of natural stones of different size protected by large rocks or concrete blocks of miscellaneous designs. Important factors in effective and reliable breakwater design is a correct estimate of significant design parameters such as sea levels and characteristics of the design wave (i.e., height, length, and period) and understanding of potential failure modes. The latter is illustrated in Figure 9.12 and is discussed in detail by Burcharth (1994).

Breakwater failures may involve the following:

- Hydraulic instability of seaward face armor
- Instability of low-crested rock breakwaters
- Hydraulic instability of rock toe berm
- Wave run-up on seaward armored slopes

The conventional practice of breakwater design is deterministic in nature and is based on a concept of design loads that should not exceed the structure capacity to resist these loads. The latter is usually defined on a probabilistic basis as a characteristic value of the load (e.g., the expectation value of the design return period event preferably with consideration of the involved uncertainties). First and foremost, the latter is related to the determination of design wave height and the subsequent design load on a breakwater. This is because most relevant available formulas associated with breakwater design are either empirical or semiempirical in nature, based mainly on certain fittings of model test results or field observation of breakwater behavior under wave attack. An example is Hadson's and other formulas discussed briefly later in the chapter.

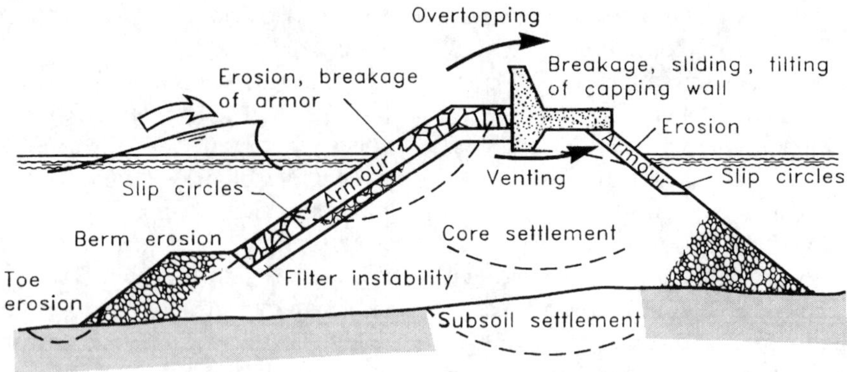

Figure 9.12 Failure modes of a rubble-mound breakwater. (From Burcharth, 1994.)

Wave Run-up, Overtopping, Transmission, and Reflection. These are important parameters and should therefore, be considered carefully in breakwater design.

WAVE RUN-UP AND OVERTOPPING. Wave run-up on a rubble-mound breakwater depends on the angle of seaward slope and its roughness, the depth of water in front of the slope exposed to waves, the slope of the bottom, and incident wave characteristics. In some cases the excessive run-up on breakwater is not acceptable because it may result in damage to the crest or rear face of the breakwater as well as in large waves in the lee of the breakwater. The latter may have an adverse effect on port operation in general and on vessel safety at berth in particular. Hence, selection of the breakwater crest elevation above sea level should be given proper consideration. There are a variety of methods to calculate wave run-up levels or wave overtopping discharges on simple with or without a crown wall or bermed slopes.

Following is an example from PIANC (1992), developed from Van der Meer's tests on armored slopes. The run-up levels above the static water level of exceedance level x, given by R_{ux}, that is, the run-up level relative to static level x, may be calculated using one of the following equations:

$$\frac{R_{ux}}{H_s} = \begin{cases} a\xi_m & \text{for } \xi_m < 1.5 \quad (9.2) \\ b\xi_m^c & \text{for } \xi_m > 1.5 \quad (9.3) \end{cases}$$

where ξ_m is the *Iribarren number*, $= \tan \alpha / s_m^{1/2}$, where $s_m = 2\pi H/gT_m^2$, where T_m is the mean wave period. For permeable structures (notional permeability factor, used in calculations of rock armor stability, $P > 0.4$) the upper limit on $R_{ux}/H_s = d$. The empirical coefficients a, b, c, and d can be obtained from Table 9.1.

Naturally, where the freeboard of the structure is smaller than the wave run-up R_u, overtopping of the breakwater will occur, producing waves in the sheltered area. For more information on wave run-up and overtopping, the

Table 9.1 Empirical coefficients *a*, *b*, *c*, and *d*

Exceedance Level (%)	*a*	*b*	*c*	*d*
0.001	1.12	1.34	0.55	2.58
0.02	0.96	1.17	0.46	1.97
Significant	0.72	0.88	0.41	1.35

Source: PIANC (1992).

reader is referred to U.S. Army Corps of Engineers (1984), Bruun (1985), and Van der Meer and Stam (1992).

WAVE TRANSMISSION AND REFLECTION. *Wave transmission* for rubble-mound breakwater can occur due to transmission by overtopping and transmission through the structure. The transmission coefficient, K_T, is usually expressed as follows:

$$K_T = (K_{To}^2 + K_{Tt}^2)^{1/2} \tag{9.4}$$

where K_{To} is a coefficient for waves overtopping the breakwater and K_{Tt} is a coefficient for wave transmission through the breakwater. K_T is a complex function of many parameters, including the size of the breakwater (i.e., width and height, composition of various layers of material, water depth, etc.) when wave penetration into the breakwater is more insignificant ($K_{Tt} \approx 1.0$) than $K_T = K_{To}$. For random waves a value of $K_{To} = (H_s)_t/H_s$ is generally used. In the above, $(H_s)_t$ is defined as a significant wave height of the sea state due to structure overtopping and H_s is a significant wave height that impinges on the structure. For more information readers are referred to Tanaka (1976), Seelig (1980), Allsop (1983), Sakamoto et al. (1984), U.S. Army Corps of Engineers (1984), and Takahashi (1997).

Wave reflection is usually represented by the coefficient K_R that is normally obtained from the laboratory experiments. To date, many studies have been carried out on the wave-reflection coefficient K_R (Losada and Gimenez-Curto, 1980; Selling, 1983; Allsop, 1990; Van der Meer, 1991; and others). The K_R value of a rubble-mound breakwater is generally small because attacking waves are usually effectively dissipated by this type of breakwater. However, as the slope angle increases and the wave steepness decreases, K_R increases. Furthermore, if the roughness of the slope is small, K_R is large. It should be noted that K_R is greatly reduced when wave overtopping does occur.

Wave Force on Armor Layer and Armor Stability. The armor layer stability is a very important requirement in breakwater design. If this layer is damaged, the breakwater's core can be opened up to direct exposure to wave attack, resulting in either substantial erosion or catastrophic failure of the structure. Waves cause water movement inside the breakwater body that in turn may cause displacements of the rubble stone on the slope (Figure 9.13). Damage to the armor layer due to the movement of water inside a breakwater's body usually occurs near the still-water level.

Forces acting on the armor stones are shown in Figure 9.14. There, P_h is the horizontal wave force, P_n is the uplift wave force, W' is the weight of the stone, and F_p is the frictional force that resists the two wave forces; F_p and W' are expressed as follows:

$$F_p = \mu_p(W' \cos \alpha - P_n) \tag{9.5}$$

$$W' = W\left(\frac{1 - 1}{S_r}\right) \tag{9.6}$$

where μ_p is the coefficient of friction and S_r =

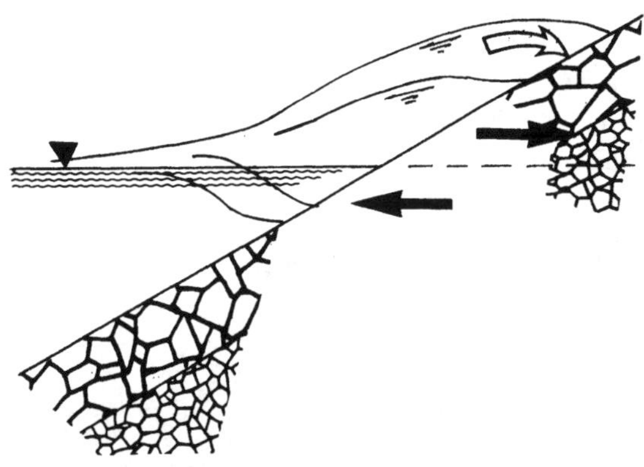

Figure 9.13 Wave action on rubble-mound breakwater.

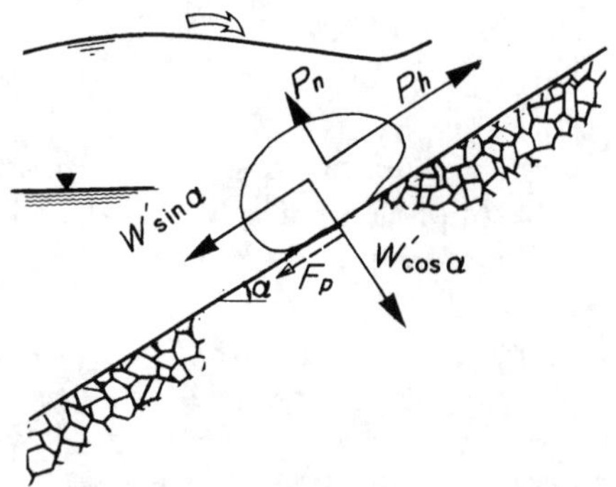

Figure 9.14 Wave forces acting on armor stone.

ρ_r/ρ_0 is the specific gravity of armor stones, ρ_r is the mass density of the armor stones, and ρ_0 is the mass density of water. The armor stone stability is determined from the formulation

$$P_h \pm W' \sin \alpha < F_p \tag{9.7}$$

Therefore,

$$W' > \frac{P_h + \mu_p P_n}{\mu_p \cos \alpha \pm \sin \alpha} \tag{9.8}$$

Forces P_h and P_n are due primarily to the drag and uplift forces generated by waves. Both of these forces are proportional to the square of the water–particle velocity; it is assumed to be proportional to $(gH)^{0.5}$. Because the drag and uplift forces are also proportional to the respective projected areas A_h and A_n of the stone, the forces P_h and P_n can be obtained from the following:

$$P_h = C_D P_0 g H A_h = C_D P_0 g H k_h \left(\frac{W}{\rho_r g}\right)^{2/3} \tag{9.9}$$

$$P_n = C_L P_0 g H A_n = C_L P_0 g H k_n \left(\frac{W}{\rho_r g}\right)^{2/3} \tag{9.10}$$

where C_D and C_L are the drag and uplift coefficients, respectively; k_h and k_n are coefficients that consider shape of the stone; and H is the wave height. Hence, the required weight of the stone, W, is determined from the following:

$$W = \frac{\rho_r g H^3}{N_s^3 (S_r - 1)^3} \tag{9.11}$$

in which the parameter N_s is called the *stability number*; N_s accounts for the effects of stone shape and wave conditions and is determined from the following:

$$N_s^3 = \frac{(\mu_p k_n C_L \pm k_h C_D)^3}{(\mu_p \cos \alpha \pm \sin \alpha)^3} \tag{9.12}$$

Equation (9.11) can be rewritten using the stone nominal diameter D_n as follows:

$$\frac{H}{\Delta D_n} = N_s \tag{9.13}$$

where

$$D_n = \left(\frac{W}{\rho_r g}\right)^{1/3} \tag{9.14}$$

$$\Delta = S_r - 1 \tag{9.15}$$

Equation (9.13) is very useful for understanding the scale of the structure in comparison to the wave height.

Evaluation of the required weight of armor blocks or stones is one of the most important aspects in the design of rubble-mound breakwaters. This is usually done either analytically or by conducting model tests. Also, the result of tests performed previously for the identical design conditions can be used for the design purpose. Many formulas have been proposed by different investigators for determination of the weight of the armor block (Iribarren, 1938; Larras, 1952; Hudson, 1958, 1959; Hedar, 1965; Price, 1979; Van der Meer, 1988b). Fundamentally, all proposed formulations are very similar to that expressed in eq. (9.10). The most popular one is the *Hudson formula*, which is a modified version of the Iribarren formula:

$$D_n^3 = \frac{H_s^3}{K_D \Delta^3 \cot \alpha} \tag{9.16}$$

where D_n is the nominal rock (block) diameter; H_s the significant wave height (the average of highest one-third of wave heights); K_D the empirical damage coefficient; $\Delta = \rho_s/\rho_w - 1$, where ρ_s/ρ_w in the ratio of the rock (block) and water densities; and α is the slope angle. *Note:* In some instances the Hudson formula is presented as

$$M = \frac{\rho_a H}{K_D \Delta^3 \cot \alpha} \tag{9.17}$$

where M is the mass of the armor unit and ρ_a is the mass density of the armor unit. The latter formula may be rewritten in terms of D_n and the stability number N_s as follows:

$$N_s = \frac{H}{\Delta D_n} = (K_D \cot \alpha)^{1/3} \tag{9.18}$$

Values of K_D are generally expressed as the volume of material removed out of the original cross-section profile expressed as a percentage of the total volume occupied by the primary armor. The U.S. Army Corps of Engineers (1984) gives values of K_D for no damage and higher degrees of damage (Table 9.2). It is recommended that $H = H_{1/10}$ should be used in the Hudson formula, where $H_{1/10}$ is the mean height of the highest one-tenth of the waves in a record.

The Hudson formula is still popular and has been used extensively in preliminary design in the past 30 years or so. The uncertainty inherent in the Hudson formula has been considered by Van der Meer (1988a, b), who proposed the following generalized formula:

$$\frac{H_s}{\Delta D_{n50}} = a(K_D \cot \alpha)^{1/3} S^b \tag{9.19}$$

where D_{n50} is the nominal diameter, calculated from the median mass derived from the mass distribution curve; a an empirical coefficient, = 0.7; and S^b, the damage number for (rock) armored slopes ($S = A_e/D_{n50}$, where A_e is the erosion area on the cross section and b, an empirical coefficient, = 0.15). Information on the reliability of data and its impact on breakwater design is given in Thomson and Shuttler (1976), Burcharth (1983, 1984, 1989, 1993, 1994), Van der Meer and Pilarczyk (1987a,b), Hall and Kao (1991), Goda (1994), PIANC (1994), Takahashi (1997), and U.S. Army Corps of Engineers (2001).

Several typical examples of armor systems used for slope protection are illustrated in Figure 9.15. More information on artificial concrete blocks used in breakwater construction and coastal engineering is given in Takahashi (1997). It should be noted that the invention of rectangular and shape-designed concrete blocks played a significant roll in promotion development and use of a rubble-mound breakwater. These concrete blocks are now used frequently where waves are high and where the natural stones are heavy enough to sustain wave forces safely are not readily available. The armor blocks are usually classified in terms of the following parameters:

- Shape (bulky or slender)
- Porosity (porous or solid as one unit and as a section)
- Interlocking (high or weak)
- Slope inclination (steep or gentle)
- Placement method [random or pattern placing (pitching), one layer or two layers]
- Wave energy dissipation (high or low)

Table 9.2 Recommended K_D values: no-damage criteria and minor overtopping

Armor Units	n[a]	Placement	Structure Trunk: K_D Breaking Wave	Structure Trunk: K_D Nonbreaking Wave	Structure Head: K_D Breaking Wave	Structure Head: K_D Nonbreaking Wave	cot θ
Quarrystone							
Smooth rounded	2	Random	*1.2*[b]	2.4	1.1	1.9	1.5–3.0
Smooth rounded	>3	Random	*1.6*[b]	3.2	1.4	2.3	[c]
Rough angular	1	Random[d]	[d]	2.9	[d]	2.3	[c]
					1.9	3.2	1.5
Rough angular	2	Random	2.0	4.0	1.6	2.8	2.0
					1.3	2.3	3.0
Rough angular	>3	Random	2.2	4.5	2.1	4.2	[c]
Rough angular	2	Special[e]	5.8	7.0	5.3	6.4	[c]
Parallelepiped[f]	2	Special[b]	7.0–20.0	8.5–24.0	—	—	
					5.0	6.0	1.5
Tetrapod and quadripod	2	Random	7.0	8.0	4.5	5.5	2.0
					3.5	4.0	3.0
					8.3	9.0	1.5
Tribar	2	Random	9.0	10.0	7.8	8.5	2.0
					6.0	6.5	3.0
Dolos	2	Random	15.8[g]	31.8[g]	8.0	16.0	2.0[h]
					7.0	14.0	3.0
Modified cube	2	Random	6.5	7.5	—	5.0	[c]
Hexapod	2	Random	8.0	9.5	5.0	7.0	[c]
Toskane	2	Random	11.0	22.0	—	—	[c]
Tribar	1	Uniform	12.0	15.0	7.5	9.5	[c]
Quarrystone (K_{RR})							
Graded angular	—	Random	2.2	2.5	—	—	

Source: U.S. Army Corps of Engineers (1984).

[a] n is the number of units comprising the thickness of the armor layer.

[b] *Caution:* The K_D values shown in *italics* are unsupported by test results and are provided only for preliminary design purposes.

[c] Until more information is available on the variation in K_D values with slope, the use of K_D should be limited to slopes ranging from 1:1.5 to 1:3. Some armor units tested on a structure head indicate a K_D-slope dependence.

[d] The use of a single layer of quarrystone armor units is not recommended for structures subject to breaking waves, and only under special conditions for structures subject to nonbreaking waves. When it is used, the stone should be carefully placed.

[e] Special placement with a long axis of stone placed perpendicular to structure face.

[f] Parallelepiped-shaped stone: long slablike stone with the long dimension about three times the shortest dimension (Markle and Davidson, 1979).

[g] Refers to no-damage criteria (<5% displacement, rocking, etc.); if no rocking (<2%) is desired, reduce K_D by 50% (Zwamborn and Van Niekerk, 1982).

[h] Stability of dolosse on slopes steeper than 1:2 should be substantiated by site-specific model tests.

[i] Applicable to slopes ranging from 1:1.5 to 1:5.

For example, tetrapods have a relatively slender shape and relatively high interlocking ability and therefore can be used on a steep slope. Tetrapods are usually placed randomly. This results in significantly large porosity of the armor section. Thus, the armor layer made from tetrapods dissipates the wave energy significantly. On the other hand, the Dolos blocks have a higher interlocking ability than the tetrapods, but they are structurally weaker because of their slender shape. The strength of the Dolos blocks can be increased if a large waist ratio ($r > 0.4$) is employed; r is the ratio of leg width to leg length. In recent years, several new bulky blocks with increased structural strength (e.g., accropods and others) have been developed and used successfully.

In the past 20 or so years, various investigators carried out research aimed at determination of the effectiveness of armor blocks of miscellaneous designs as well as determination of the strength of a block's structural components. A detailed account of these investigations is given in Takahashi (1997). More information on block stability is given in Chapter 5.

Armor blocks are normally installed on underlayers of filters. These, along with a toe section, are a very important part of breakwater structures. A breakwater's stability depends to a large extent on correct design of the layers above. The main purpose of inner layers is to reduce both transmission through the breakwater and the construction cost by using lower-cost materials for the breakwater core. These layers are designed to damp the wave penetration and to prevent pullout and washout of the fine material included in the breakwater core and from the seabed. Design of the underlayers and filter layers is quite complicated and is usually based on limited information and therefore must be carried out with care. Sublayer (filter) design is discussed in detail in Chapter 5.

Design Wave. Success of breakwater design depends greatly on a determination of the parameters of the design wave. Therefore, selection of the design wave used for breakwater design is of a paramount importance. In general, marine structures are designed to withstand the highest wave-induced force over their economic life. However, in the design of any structure subjected to nondeterministic and probabilistic environmental loads, such as waves, the ultimate design condition must invariably be weighed against economics that include capital (first) cost, maintenance and repair costs, and the consequence of structural damage with regard to the anticipated type of failure. Obviously, the possibility of a sudden

(*a*)

Figure 9.15 Armor blocks used for slope protection in rubble-mound breakwaters: (*a*) quarry rocks; (*b*) shape-designed concrete blocks; (*c*) acropod.

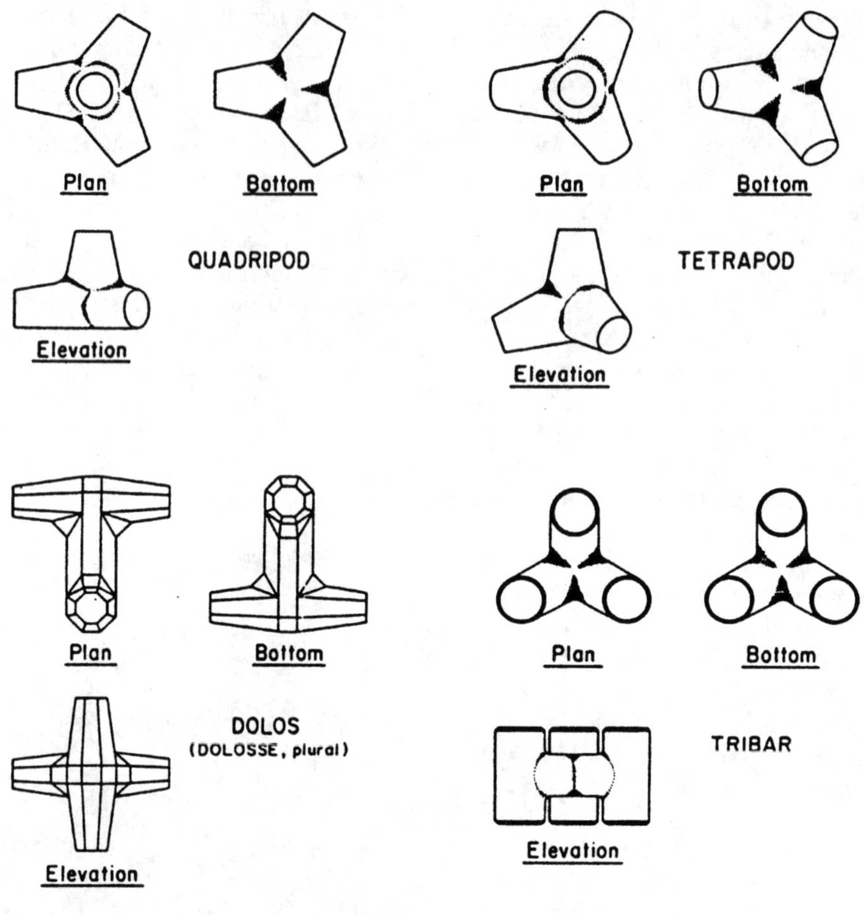

(*b*)

Figure 9.15 (*Continued*).

catastrophic failure that may involve loss of human lives as a result of a single high-wave action must be of serious concern. This is why when considering problems involving wave action in the design of ports and marine structures, engineers are always confronted with a dilemma regarding the selection of the design wave.

In general, the maximum wave that can occur at a site depends on many factors, among which the most important are depth of water, site sheltering, and underwater topography. In the design of marine structures exposed to wave action, consideration is usually given to structure type or function. For example, in the design of breakwaters and seawalls, some overtopping by very high waves may be accaptable. Also, rubble-mound breakwaters allow for some damage that can be repaired quickly and economically. The cost and extent of repairs of a structure, as well as potential delays in port operation, must be evaluated in detail, and the

(c)

Figure 9.15 *(Continued)*.

subsequent economics of construction, maintenance, and repair should be evaluated in relationship to the selection of design waves. This should also be supported by the appropriate risk analysis.

Current practice is first to obtain a long-term probability of wave distribution for the site under consideration, and the design wave is likely to be presented in terms of the significant wave height. The probability of other heights occurring in relation to the significant wave height is calculated through application of the Rayleigh distribution. From this distribution, a mean height, most probable height, highest of a given percentile, and so on, can be obtained. Oceanographers define the wave height as the average of the highest one-third of the *significant wave height* (H_s). This statistical term is related to one-third of the highest waves of a given wave group and defined by the average of their heights and periods. Composition of the higher waves depends on the extent to which the lower waves are considered. The practice indicates that an experienced observer who attempts to establish the character of the higher waves visually will record values that approximately fit the definition of the significant wave. Based on the cumulative Rayleigh probability distribution, the other statistical heights are related to the significant wave height approximately as follows:

- Average height $0.64H_s$
- Average highest 10% $1.29H_s$
- Average of highest 1% $1.68H_s$
- Highest $1.87H_s$

The long-term distribution can be obtained from hindcasting of historical storm records and/or can be predicted on the probability of the occurrence of meteorological conditions required for development of a maximum wave. The wave height and period can also be obtained through analysis of wave observations. When possible, both procedures should be used and the potential differences between results should be evaluated to ensure a more reliable outcome from the study. Both procedures are given in detail in the U.S. Army Corps of Engineers' *Shore Protection Manual* (1984). Wave forecasting is usually carried out by a specialist oceanographer or meteorologist familiar with the local climatic conditions.

In general, a 20- to 25-year-design wave, coupled with an annual extreme water level, is considered by some authorities as appropriate for the design of small to medium-sized projects. Again, selection of larger waves, having a recurrence interval more than 25 years, should be related to the economics of construction operation. Selection of the design wave (design conditions) for large and important projects requires a more detailed evaluation of all conditions mentioned previously. Actual storm wave data are quite scarce; hence, actual data reported are of particular interest. Some relevant information is summarized in Tsinker (1997).

Obviously, as the design wave height increases, the first cost of the structure also increases. As pointed out by NAVFAC DM-26.2 (U.S. Navy, 1982), the first cost must be related to an annual cost, which is accomplished by amortizing the first cost using an appropriate interest rate and time period. On the other hand, the annual maintenance cost will decrease if the structure is designed for a larger wave. This is usually based on some arbitrary assumptions made on a part of how many times in the life of the structure repairs will be carried out and how much it will cost. The experienced engineer can produce reasonable assumptions that enable him or her to select the design wave by adding the annual maintenance cost and the annual first cost to produce a curve that represents the annual cost of the structure. The engineer can then identify the wave height that represents the least annual cost. Because some of the assumptions involved in arriving at the "optimum" design wave height are arbitrary and not based on accurate data regarding the cost of maintenance, some latitude is permitted in selection of the design wave height when the optimization procedure above is used. If the cost varies by 5 to 10%, the optimum design wave would have a range of heights. The designer would use other factors, such as environmental, operational, and maintenance considerations to help select the proper design condition.

The economics of considering the smaller design wave (e.g., a wave with a recurrence interval of 25 years versus a wave having a 30-year recurrence interval) must be evaluated, and the physical and economic factors, such as design wave parameters versus the capital cost and costs of repair and maintenance, must be optimized. The principle of optimization is shown schematically in Figure 9.16 where the annual cost (first cost plus maintenance cost) is plotted as a function of design wave height. This plot is made by designing the structure for a range of wave heights. By definition, the typical wave system has heights that exceed H_s. Therefore, the ability of the structure to withstand the force generated by an occasional larger wave must be evaluated. In general, nonrigid structures such as ruble-mound breakwaters can sustain forces from waves in excess of H_s (sometimes with the acceptable damages). Hence, although in some cases, a single "maximum" design wave may not be critical, the concept of a "design wave," despite some differences in philosophy of design wave determination, remains central to the design of port-related marine structures. Practical examples of local damages inflicted to the breakwater by waves are illustrated in Figure 9.17. The detail information on wave field and wave impact on marine structures design is given in the appendix to this chapter.

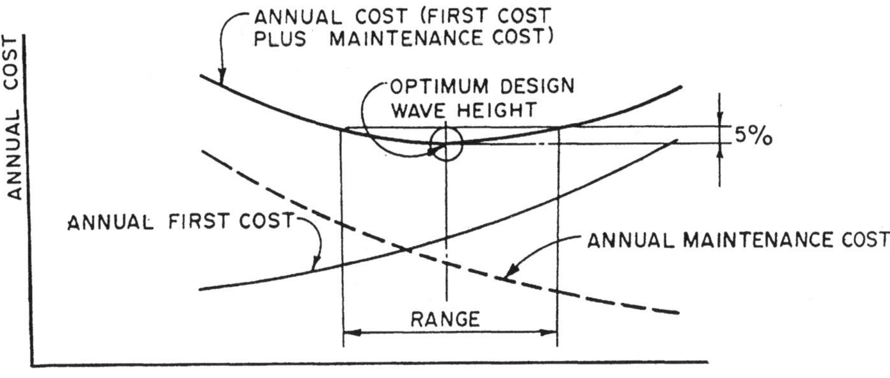

Figure 9.16 Selection of optimum design wave height. (From U.S. Navy, 1982.)

9.3.2 Vertical Breakwaters

Historically, breakwaters in general, and vertical breakwaters in particular, were built around the Mediterranean hundreds of years B.C. Ancient breakwaters were constructed of stone blocks and in the form of timber cribs filled with stones. A version of a caisson for breakwater construction was used in the Middle East at Caesarea around 20 B.C. The underwater part of these structures consisted of wooden forms filled with concrete mortar delivered underwater in baskets. A number of relevant examples are given in Bray and Tatham (1992), Franco and Verdesi (1993), Allsop and Bray (1994), Takahashi (1997), and PIANC (2003). The basic vertical breakwaters used in modern applications are depicted in Figure 9.18. A typical vertical breakwater consists of stone bedding placed on the seafloor and a gravity upright section (Figure 9.18*a*). The structures shown in Figure 9.18*b* and *c* are defined as composite breakwaters. The former comprises a (high or low) rubble-mound and gravity upright section, and the latter includes a regular upright section combined with a rubble-mound front section.

By definition the mound in a high-mound composite breakwater is higher than the low water level (LWL). In this case, waves can break on the mound that made it vulnerable to impulsive pressures and scouring effects. On other hand, a low-mound breakwater does not cause wave breaking on the mound. Because of its vulnerability to breaking waves and scouring effects, a high-mound structure should be adequately protected. For this reason, composite-type breakwaters with a low mound could be more economical.

Sometimes, armor blocks are placed in front of a breakwater. This is done to reduce both the wave reflection and the breaking-wave force on a vertical wall. This type of construction, termed a *composite breakwater*, is covered with wave-energy-dissipating large rocks or concrete blocks; it is also termed a *horizontally composite breakwater* (Figure 9.18*c*). Concrete blocks or large rocks have been used in the past to strengthen vertical wall breakwaters that have sustained damages from waves. These blocks or rocks have been placed in front of the vertical wall to dissipate wave energy and to reduce the wave force, especially that from breaking waves. In modern construction practice, shape-designed concrete blocks such as tetrapods and others are used more often.

Figure 9.17 Local damages to eastern breakwater inflicted by heavy waves, port of Callao, Peru.

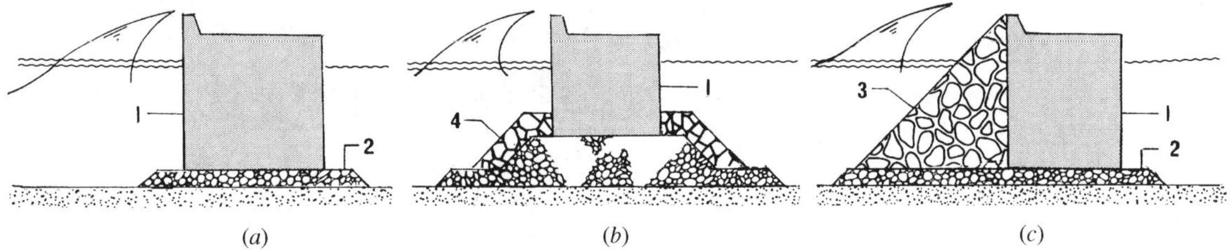

Figure 9.18 Vertical breakwaters: (*a*) low-mound breakwater; (*b*) high-mound composite breakwater; (*c*) horizontally composite breakwater. 1, Upright section; 2, stone bedding; 3, rubble-mound or artificial concrete blocks; 4, armor stone, or armor made of artificial concrete blocks.

Horizontally composite breakwaters are very similar to rubble-mound breakwaters armored with concrete blocks. Figure 9.19 shows how the cross section of a horizontally composite breakwater varies with mound height; as the mound height increases, a horizontally composed breakwater becomes very similar to a rubble-mound breakwater. This is particularly true for breakwaters with core stones placed in front of the vertical wall (Figure 9.19*d*) that are structurally nearly the same as rubble-mound breakwaters. Both, however, are conceptually different because the concrete blocks of a rubble-mound breakwater act as the armor for the rubble foundation, whereas the concrete bocks of a horizontally composite breakwater function to reduce the wave force and size of the reflected waves. In general, horizontally composite breakwaters are considered to be an improved version of vertical breakwaters.

Older vertical breakwaters have been built from large interlocking stone blocks generally laid dry or in lime or pozzolan mortar; later, for the same purpose, cement mortar was used. Concrete rather than stone blocks started to be used for breakwater construction sometime in the nineteenth century. At about the same time, iron-reinforced concrete caissons have been proposed for use in breakwater construction. Early in the twentieth century, concrete caissons have been used for breakwater construction in Europe, the United States, Japan, and other locations around the globe. The history of breakwater development is discussed extensively by Takahashi (1997) and in the *Proceedings of the International Workshop on Wave Barriers in Deep Waters*, held in Yokosuka, Japan in 1994.

9.3.2.1 Wave Transmission and Reflection. When waves act on vertical breakwaters, some energy of the incident waves is dissipated, and some of the remaining energy is reflected and generates reflected waves in front of the wall; the rest of the wave energy is transmitted and yields the waves transmitted behind the breakwater. Wave reflection sometimes presents a problem because of the additional agitation it may create in a port basin. The minimization of wave transmission is important in breakwater design because the principal function of a breakwater is to prevent wave propagation from occurring, thereby creating a calm water area behind the breakwater. The amount of wave reflection and transmission is usually determined by the wave reflection (K_R) and wave transmission (K_T) coefficients, which are defined as follows: $K_R = H_R/H_I$ and $K_T = H_T/H_I$, where H_I is the incident wave height, H_R is the reflected wave height, and H_T is the transmitted wave height, all of which usually correspond to the significant wave.

Wave Transmission. Transmitted waves are caused by wave transmission through the structure and overtopping. Transmission coefficients

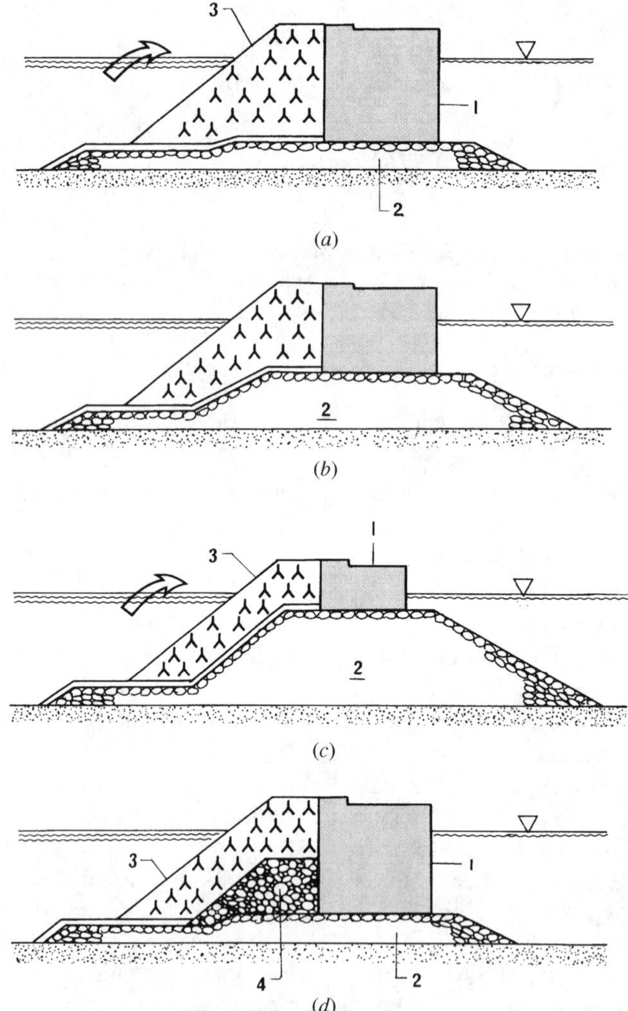

Figure 9.19 Horizontally composite breakwaters: (*a*) and (*d*) low rubble foundation; (*b*) relatively high rubble foundation; (*c*) high rubble foundation. 1, Upright section; 2, rubble foundation; 3, blocks; 4, inner core—quarry run. (From Takahashi, 1997.)

related to both causes are denoted as K_{Tt} and K_{To}, respectively, with the total transmission coefficient K_T being expressed as

$$K_T = (K_{T_t}^2 + K_{To}^2)^{0.5} \tag{9.20}$$

Transmitted waves created by overtopping waves have a complicated form with high-frequency components. They are produced by waves generated at the lee which result from the impact of the fall of the overtopping mass. Therefore, in general, the wave height and period transmitted are different from those of incident waves; that is, the wave period of transmitted waves is generally smaller.

Another phenomenon worthy of note is that transmitted irregular waves change characteristics as they propagate over a long distance; for example, distributions of wave height and period vary with the distance away from the breakwater. Wave transmission by vertical wall breakwaters is mainly by overtopping; therefore, the ratio of a breakwater's crest height h_c to the incident wave height H_I is the principal parameter governing the wave transmission coefficient. Based on regular wave tests, Goda and Kakizaki (1966) proposed the following equations for determination of the transmission coefficient for vertical breakwaters. For $\beta - \alpha < h_c/H_I < \alpha - \beta$,

$$K_T = \left\{0.25\left[1 - \sin\left(\frac{\pi}{2\alpha}\right)\left(\frac{h_c}{H_I} + \beta\right]^2 + 0.01\left(1 - \frac{h'}{h}\right)^2\right\}^{0.5} \tag{9.21}$$

For $h_c/H_I \geq \alpha - \beta$,

$$K_T = 0.1\left(1 - \frac{h'}{h}\right)$$

where coefficients $\alpha = 2.2$ and β is obtained from Figure 9.20; h' is the distance from the design water level to the bottom of the upright section of the breakwater.

Although according to Takahashi (1997), eq. (9.21) is based on regular wave tests, it is also applicable to the transmission coefficient of irregular waves with significant wave height. For example, most breakwaters in Japan are designed with a relative crest height $h_c/H_{1/3} = 0.6$, where $H_{1/3}$ is the design significant wave height. In this case the transmission coefficient calculated from eq. (9.21) is then equal to about 0.2 for the typical conditions $d/h = 0.6$ and $h'/h = 0.7$.

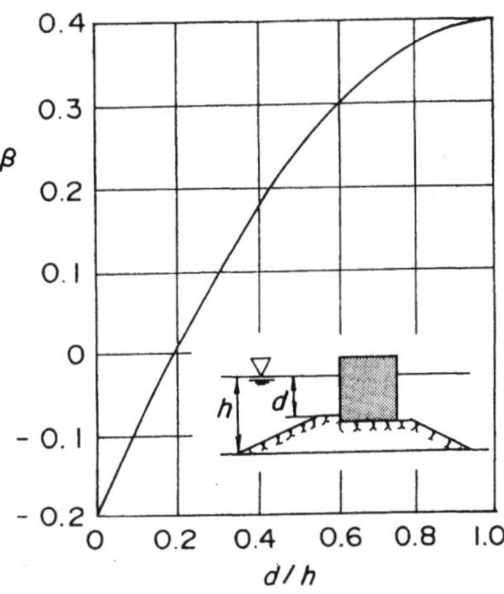

Figure 9.20 Nomograph for determining the factor $\beta = f(d/h)$ in eq. (9.21). (From Takahashi, 1997.)

Wave Reflection. Waves that usually exist in front of vertical breakwaters are standing waves, reflected by the wall. The reflection coefficient of vertical wall breakwaters is generally high, but is less than unity, usually due to some effects of the rubble-mound foundation and/or wave overtopping. The wave reflection coefficient K_R is considerable smaller when breaking waves act on the breakwater. For information on K_R, readers are referred to Tanimoto et al. (1987). Studies carried out by these investigators indicate that K_R tends to decrease with increases in $H_{1/3}/d$. They also found that K_R is affected by another important parameter, the relative crest height to the incident significant wave height ($h_c/H_{1/3}$), because it strongly influences breakwater topping.

9.3.2.2 *Wave Forces on Vertical Walls.* In an analysis of wave forces acting on vertical breakwaters, a distinction is usually made between the effects of nonbreaking and breaking waves. In some instances (e.g., when the breakwater is situated on a sloping seafloor), the effects of broken waves are also taken into consideration. In general, because of the effects of breaking waves, the slope of the sea bottom, wave interference with the rubble mound in composite breakwaters, overtopping, and other factors, determination of the wave pressure on vertical wall breakwaters is very difficult and in most cases is not straightforward. This is aggravated further by wave irregularity. This is why in most practical cases important breakwaters are designed based on results obtained from physical model tests or on the basis of empirical formulas formulated from model tests.

Until recently, the wave pressure on vertical breakwaters has been evaluated using the significant wave height, which essentially produces a smaller value of wave pressure than that produced by an actual maximum wave. At the present time, because of the spectacular failure of some important breakwaters [for examples see Takahashi (1997)], the maximum wave height is normally used to calculate the design wave pressure.

Wave direction is also an important factor in wave pressure calculations. If a wave is nonbreaking, the wave pressure in oblique seas can be calculated relatively easily. However, the effect of wave direction becomes especially pertinent when a wave is breaking; this phenomenon is usually evaluated experimentally. Last, but not least, the plane configuration of a breakwater may have a significant impact on wave pressure on vertical breakwaters.

Nonbreaking Wave Forces on Vertical Walls. These forces are primarily hydrostatic. Nonbreaking waves are usually expected where the fetch is limited and when the depth at the structure is greater than about 1.5 times the maximum expected wave height. The earliest proposed methods for nonbreaking waves are based on trochoidal wave theory. The *Sainflou* (1928) *method* simplified wave pressure theory and provided wave pressure distributions at the wave crest and trough. Accordingly, wave pressure at a wave crest is determined by the following equations (Figure 9.21):

$$p_1 = (p_2 + w_0 h)\frac{H + h_0}{H + h + h_0} \tag{9.22}$$

$$p_2 = \frac{w_0 H}{\cosh(2\pi h/L)} \tag{9.23}$$

whereas the pressure at a wave trough is determined from the following formulations:

$$p_1' = w_0(H - h_0) \tag{9.24}$$

$$p_2' = p_2 = \frac{w_0 H}{\cosh(2\pi h/L)} \tag{9.25}$$

$$h_0 = \frac{\pi H^2}{L}\coth\frac{2\pi h}{L} \tag{9.26}$$

In eqs. (9.22) through (9.26), H is the height of the original free wave (in water depth h) and h_0 is the height of the clapotis orbit center (mean water level at the wall) above the still water level (SWL). In general, the Sainflou formula correctly describes the standing-wave pressure and has been used all over the world for many years. The advantage of the Sainflou method has been ease of application, since the resulting pressure distribution may be reasonably approximated by a straight line. Experi-

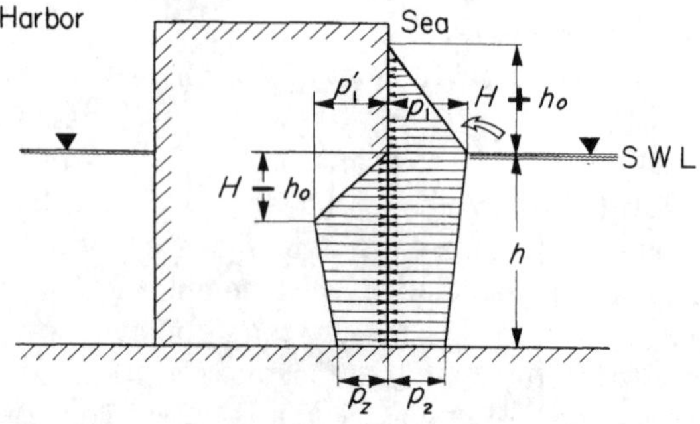

Figure 9.21 Simplified Sainflou wave pressure diagram.

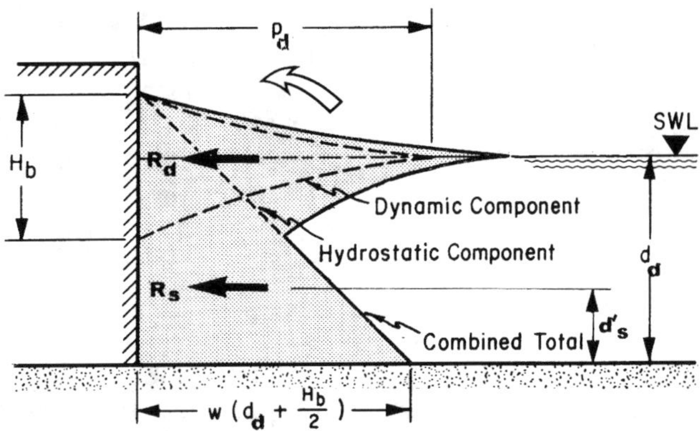

Figure 9.22 Minikin wave pressure diagram on vertical breakwaters.

mental observations by Rundgren (1958) have indicated that Sainflou's method overestimates the nonbreaking wave force for steep waves. The higher-order theory by Miche (1944), as modified by Rundgren (1958), to consider the wave reflection coefficient of the structure, appears to best fit forces measured experimentally on vertical walls for steep waves, while Sainflou's theory gives better results for long waves of low steepness. For a detailed discussion on Miche-Rundgren, readers are referred to U.S. Army Corps of Engineers (1984).

Breaking Wave Forces on Vertical Walls. Waves breaking directly against vertical-face structures exert high, short-duration dynamic pressures that act near the region where the wave crests hit the structure. Bagnold (1939) found that impact (impulsive) pressures occur at the instant that the vertical front face of a breaking wave hits a wall and only when a plunging wave entraps a cushion of air against the wall. Impulsive wave pressure is one of the most important problems in the design of a vertical breakwater, and it effects on breakwater performance must therefore, be evaluated thoroughly. A detailed account of the most recent and most important studies on wave impulsive pressures is given in Takahashi (1997). Based on his field observations and the results of Bagnold's studies, Minikin (1963) has developed a procedure for calculating breaking wave forces on vertical breakwaters. According to Minikin, the maximum dynamic pressure from a breaking wave is assumed to act at the SWL and could be determined from the formula

$$P_d = 101w \frac{H_b}{L_D} \frac{d_d}{D} (D + d_d) \tag{9.27}$$

where P_d is the maximum dynamic pressure, H_b the breaker height, L_D the wave length in water of depth D, d_d the depth at the toe of the wall, and D the depth one wave length in front of the wall.

The pressure distribution of dynamic pressure is shown in Figure 9.22. The pressure decreases parabolically from P_d at the SWL to zero at a distance of $H_b/2$ above and below the SWL. The force represented by the area under the dynamic pressure distribution is determined from

$$R_d \simeq \frac{P_d H_b}{3} \tag{9.28}$$

The hydrostatic contribution R_s to the dynamic force R_d must be added to determine the total force acting on a vertical wall.

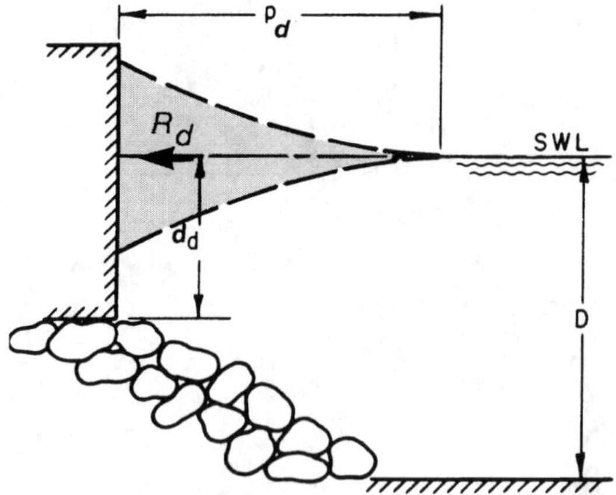

Figure 9.23 Minikin wave pressure diagram on composite vertical wall, dynamic component only.

It should be noted that Minikin's procedure as described can produce wave forces that are extremely high (e.g., 15 to 18 times those calculated for nonbreaking waves). Also, it should be understood that the high impact forces are very short in duration (on the order of hundredths of a second). Hence, they should be estimated carefully when used for determining a breakwater's stability against sliding or overturning. Furthermore, the Minikin formula (9.27) was originally derived for composite breakwaters composed of an upright section on a rubble substructure. In this case, D and L_D in eq. (9.27) are the depth and wave length at the toe of the substructure, and d_d is the depth at the toe of the vertical wall (i.e., the distance from the SWL down to the crest of the rubble substructure). For vertical walls where no or a very shallow substructure is present, formula (9.27) is adapted by using the depth at the structure toe as d_d, while D and L_D are the depth and wave length a distance one wave length seaward of the structure (Figure 9.23). Consequently, the depth D can be found from

$$D = d_d + L_d m \tag{9.29}$$

where L_d is the wave length in a depth equal to d_d and m is the nearshore slope. The forces and moments resulting from the hydrostatic pressure must be added to the dynamic force. For more information, readers are referred to U.S. Army Corps of Engineers (1984). Takahashi et al. (1992, 1994a,b) developed a simplified model of the impulsive wave pressures as a function of a wave's attacking angle, β, and the wave curvature angle, δ (Figure 9.24). These investigators found that the maximum average wave pressure intensity (p/w_0H) that appears in the transition region ($\delta > \beta > 0$) and its duration time, τ, can be approximated as follows:

$$\frac{p}{w_0 H} = \frac{0.4\pi k_m^2 k_l}{k_a} \frac{h + 075H}{h' + h_c} \tag{9.30}$$

$$\tau = \pi \left(\frac{\pi w_0 k_m^2 k_l^2 k_a H^2}{4\gamma g p_0} \right)^{0.5} \tag{9.31}$$

where w_0 is the specific weight of water; H the wave height; k_m a added mass correction factor (for practical calculations, k_m is usually as-

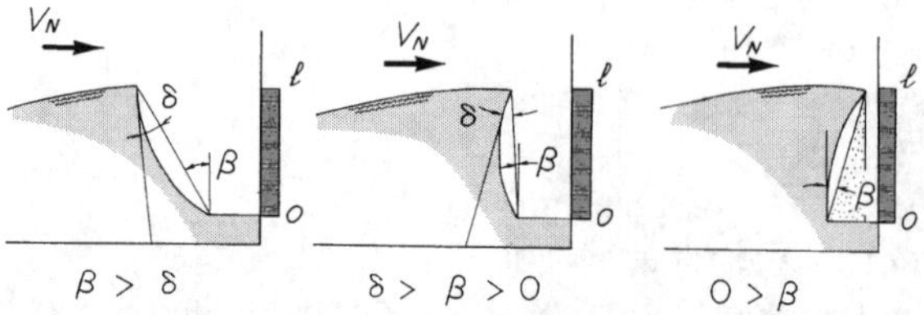

Figure 9.24 Three basic types of impulsive pressure. (From Takahashi et al., 1994.)

sumed equal to 0.83); k_l the impulsive height coefficient (k_l is the ratio of the wave-front height l to the wave height H; theoretically, it ranges from 0 to 1, although it is usually used as 0.4 to 0.9); k_a the air thickness coefficient (k_a is related to β and δ, and its minimum value is approximately on the order of 0.01 to 0.1); h the water depth; h' the water depth at the bottom of the wall; h_c the crest elevation of the wall; γ the specific heat ratio, = 1.4; g the gravity acceleration; p_0 the atmospheric pressure.

From eq. (9.30) it is obvious that the intensity of the impulsive pressure increases as the wave-front height increases and the amount of entrapped air decreases. As stated by Takahashi (1997), despite the fact that the impulsive pressure acts directly on the vertical wall, the total upright section of the breakwater responds dynamically to this pressure as a part of the elastic system, which includes the upright section, the rubble mound, and a seafloor soil. This results in a signifcantly reduced shear force that may cause the upright section to slide (Hayashi and Imai, 1964; Ito et al., 1966; Horikawa et al., 1975; Shimosako and Takahashi, 1994). For more information on the effects of breaking waves on vertical breakwaters, readers are referred to Oumerachi et al. (1993).

Goda's Formula. Goda (1972, 1973a,b) used results of his theoretical and laboratory studies to develop a comprehensive formula for calculation of wave forces acting on vertical wall breakwaters. It was later modified to account for the effect of the oblique waves and was used successfully for the design of numerous vertical breakwaters built in Japan. In the modified formula, the wave pressure acting along the vertical wall is assumed to have a trapezoidal distribution both above and below the still-water level, whereas the uplift pressure acting on the bottom of the upright section is assumed to have a triangular distribution as shown in Figure 9.22. The buoyancy is calculated using the displacement volume of the upright section in still water at the design water level. The theoretical elevation at which the wave pressure could be exerted η^*, and the representative wave pressure intensities p_u, p_1, p_3, and p_4 (Figure 9.25) in generalized form, are obtained from the following:

$$\eta^* = 0.75(1 + \cos\theta)\lambda_1 H_D \tag{9.32}$$

$$p_1 = 0.5(1 + \cos\theta)(\lambda_1\alpha_1 + \lambda_2\alpha^* \cos^2\theta)w_0 H_D \tag{9.33}$$

$$p_3 = \alpha_3 p_1 \tag{9.34}$$

$$p_4 = \alpha_4 p_1 \tag{9.35}$$

$$p_u = 0.5(1 + \cos\theta)\,\lambda_3\alpha_1\alpha_3 w_0 H_D \tag{9.36}$$

in which

$$\alpha_1 = 0.6 + 0.5\left[\frac{4\pi h/L_D}{\sinh(4\pi H/L_D)}\right]^2 \tag{9.37}$$

$$\alpha^* = \max\{\alpha_2, \alpha_I\} \tag{9.38}$$

$$\alpha_2 = \min\left\{\left(\frac{1-d}{h_b}\right)\frac{(H_D/d)^2}{3}, \frac{2d}{H_D}\right\} \tag{9.39}$$

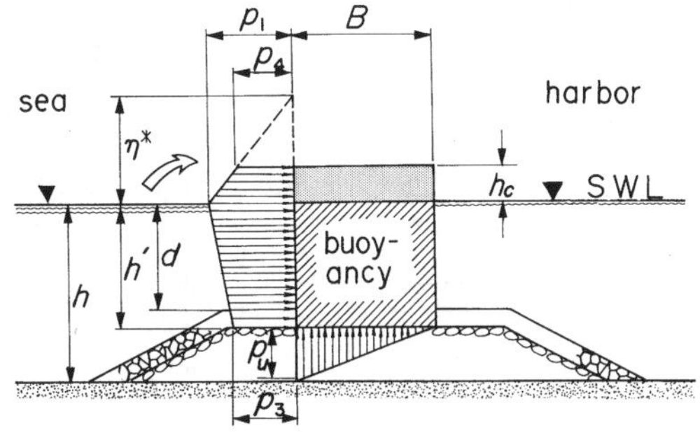

Figure 9.25 Goda's formula of wave pressure against a vertical wall.

$$\alpha_3 = 1 - \frac{h'}{h}\left[1 - \frac{1}{\cosh(2\pi h/L_D)}\right] \quad (9.40)$$

$$\alpha_4 = 1 - \frac{h_c^*}{\eta^*} \quad (9.41)$$

$$h_c^* = \min\{\eta^*, h_c\} \quad (9.42)$$

where θ is the angle between the direction of wave approach and a line normal to the breakwater; λ_1, λ_2, and λ_3 are modification factors depending on structural type; H_D and L_D are the design wave height and length, respectively; α_I is the impulsive pressure coefficient; w_0 the specific weight of seawater ($w_0 = \rho_0 g$); h_b the offshore water depth at a distance five times the significant wave height, $H_{1/3}$; min(a,b) the smaller value of a or b in eq. (9.39); and max(a,b) the larger value of a and b in eq. (9.38).

In Goda's formula the coefficient α_1 varies from 0 to 1.1 as the relative depth decreases, and the coefficient α_2 increases as d/h_b decreases, although it peaks and then decreases as d/h_b decreases; α_2 ranges from 0 to 1.0. The coefficient α_I was obtained by reanalyzing results of comprehensive wall sliding tests carried out by Takahashi et al. (1994b). It has a nondimensional value and represents the impulsive pressure component, which should be regarded as an additional effect of the slowly varying pressure component. This coefficient was introduced because the coefficient α_2 in Goda's formula does not estimate accurately the effect of the dynamic (impulsive) wave pressure and therefore the effective (quasi-static) pressure due to impulsive pressure under all practical site conditions. For detailed information on numerical values of α_I, readers are referred to Takahashi et al. (1994b) and Takahashi (1997).

Special Conditions That May Affect the Wave Force. These conditions include the elevation of the breakwater crest, the angle of wave approach to the structure, the breakwater alignment (e.g., concave shape), and others.

CREST ELEVATION. It is often not economically feasible to design a structure to provide for a nonovertopping condition by the design wave. Obviously, the wave force on a low-crest vertical wall is smaller than that on a high-crest wall, and sometimes this is a reason for the construction of low-crested breakwaters. However, wall overtopping may affect port operation negatively. Therefore, a reduction in the cost of breakwater construction should always be weighed against possible losses in port operation due to disruption in the process of ship handling at the berth. According to the U.S. Army Corps of Engineers (1984), the reduction factor in wave force due to wall overtopping r_R (Figure 9.26) can be defined as

$$r_R = \frac{R'_w}{R_w} \quad (9.43)$$

where r_R is given as follows:

$$r_R = \frac{H}{H_w(2 - H/H_w)} \quad \text{when } 0.5 < H/H_w < 1.0 \quad (9.44)$$

For more details on r_R, nomographs, and numerical problems related to the determination of r_R, readers are referred to U.S. Army Corps of Engineers (1984).

EFFECT OF ANGLE OF WAVE APPROACH. When breaking waves strike a vertical face breakwater at an oblique angle, α, the dynamic component of the force (R'_d) will obviously be less than that when waves strike perpendicular to the structure face (R_d). According to U.S. Army Corps of Engineers (1984), this force may be determined from

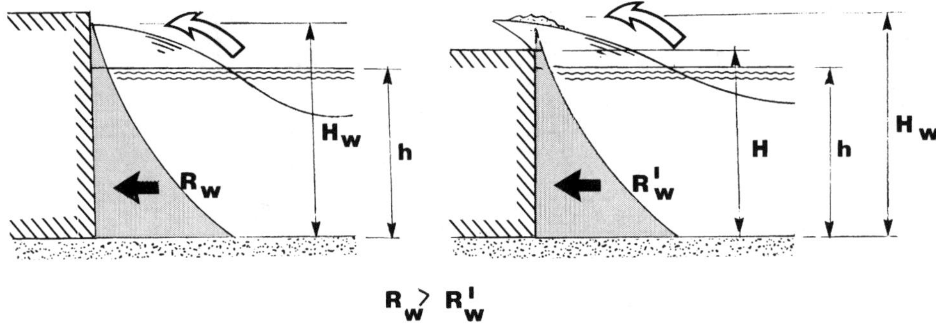

Figure 9.26 Effects of crest evaluation on wave force on a vertical breakwater.

$$R'_d = R_d \sin^2\alpha \qquad (9.45)$$

The reduction in breaking wave force is applied to the dynamic component only, not to the hydrostatic component. Furthermore, it must be understood that the maximum wave force does not act along the entire length of the wall simultaneously; consequently, the average force per unit length of wall will be smaller.

EFFECTS OF BREAKWATER ALIGNMENT. As pointed out by Kobune and Osato (1976), wave forces depend strongly on breakwater alignment. As discussed earlier, vertical wall breakwaters reflect waves, and therefore wave reflection and diffraction from this type of breakwater should be taken into account when computing wave forces. For example, in the case of concave alignment of a breakwater, the wave force could be amplified, reaching a value twice (or more) that of the regular force. Based on their experiments, Kobune and Osato (1976) stated that in the case of concave alignment, the maximum average amplification factor may be taken as 1.4. Naturally, the effect of amplification depends on the angle of concave alignment. Similarly, in the case of convex alignment, one may expect a reduction in the wave pressure on the vertical wall.

9.3.2.3 Stability of Upright Section. The upright section of a vertical breakwater must be stable against the overall (Figure 9.27) and local (Figure 9.28) failure modes. The former is concerned with a wall's stability against sliding, overturning, and overstressing the foundation material, as well as with the general stability of the rubble foundation, and the latter concerns mainly erosion beneath both the seaward and shoreward edges of the wall, seabed scour, and toe erosion and in some instances with the rubble punching failure at both edges of the wall.

Overall Stability. The standard quasistatic analysis of gravity wall stability is discussed in detail in Chapter 4. In most practical cases, in vertical wall breakwater analysis, the safety factor against sliding used is 1.4 to 1.6, and against overturning it is 1.5 to 2.0. A rather high safety factor against overturning is normally recommended to avoid the "hummering" effect of the rubble foundation. However, in many instances, wall sliding stability is more critical than overturning, especially for breakwaters with a low crown. The dynamic response and sliding stability of vertical breakwaters is discussed in Takahashi et al. (1994b), and PIANC (2003b). The bearing pressure on the rubble mound and underlaying subsoil is determined according to the procedures described in detail in Chapter 2.

The maximum allowable bearing pressure on a rubble mound, sometimes called the *toe*

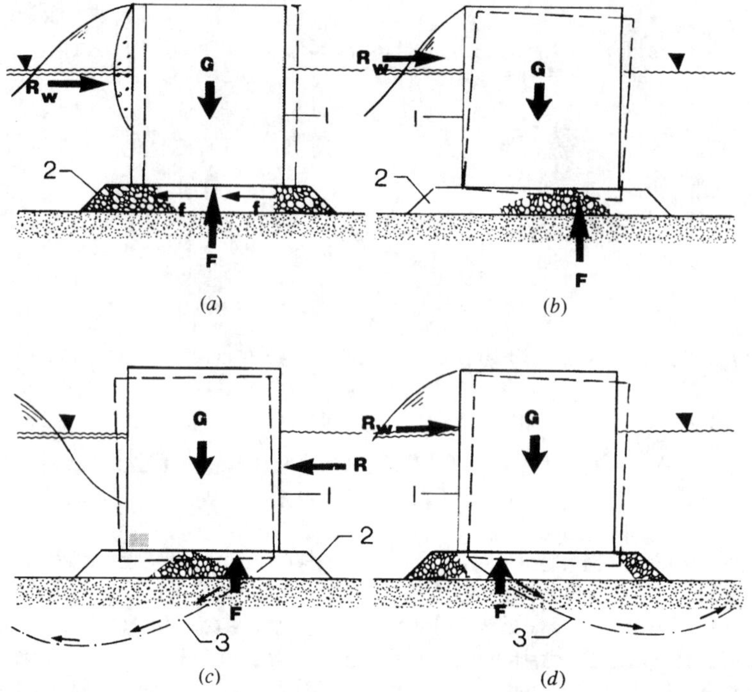

Figure 9.27 Overall failure modes of vertical breakwater: (*a*) sliding; (*b*) overturning; (*c*) and (*d*) settlement due to foundation failure. 1, Upright section; 2, rubble foundation (mattress); 3, slip surface.

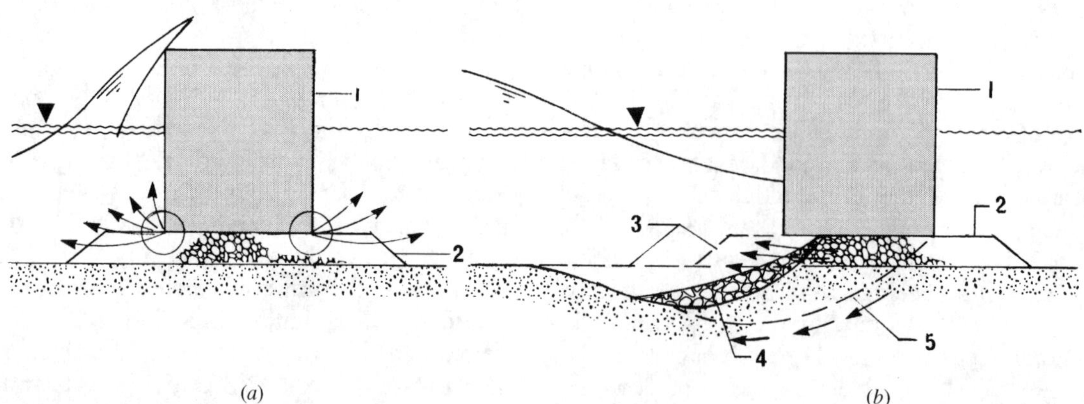

Figure 9.28 Local failure modes of vertical breakwater: (*a*) erosion and/or punching failure of rubble mattress at seaward and/or shoreward edges; (*b*) seabed scour and mattress erosion. 1, Upright section; 2, rubble foundation (mattress); 3, original profile of seabed and rubble foundation; 4, scour in front of the upright section; 5, potential failure plane.

pressure, is usually taken equal to 400 to 500 kN/m^2. In some instances the upper limit of this pressure was raised to 600 kN/m^2. Kobayashi et al. (1987) proposed a method for calculating the bearing capacity of a gravity structure that is resting on a rubble mound. In this method, the authors use the simplified Bishop method of circular slip failure for analysis of the rubble-mound foundation. For these analyses, Kobayashi et al. (1987) recommend a rubble cohesion coefficient c of 20 kN/m^2 and angle of shear resistance ϕ of 35°. These values have been obtained by the investigators from triaxial tests carried out on large samples.

9.3.2.4 Stability of a Rubble-Mound Foundation. The rubble-mound foundation of vertical breakwaters is vital for stability of the upright section; it must safely carry the wall's weight, be stable against wave forces acting against both the upright sections and the foundation itself, and must be protected against scouring by current and wave action. A rubble structure is usually composed of several layers of randomly shaped and randomly placed stones, protected by a cover layer of armor units selected from quarrystone or specially shaped concrete units. Armor units in a cover layer may be placed in an orderly manner to obtain good wedging or interlocking action between individual units, or they may be placed at random. These units may be displaced either over a large area of the cover layer, sliding down the slope en masse, or individual armor units may be lifted and rolled either up or down the slope. Empirical methods have been developed that if used with care will give a satisfactory determination of the stability characteristics of these structures when under attack by storm waves.

Design Factors. Bathymetry is a primary factor that is influencing wave conditions in the vicinity of the structure. The depths of the seafloor at the structure site partly determine whether a structure is subjected to breaking, nonbreaking, or broken waves for a particular design wave condition. Variation in water depth along the structure is also a significant factor to be considered as it affects the wave climate in the vicinity of the structure; it could be more critical where breaking waves occur than where the depth may allow only nonbreaking waves or waves that overtop the structure. A breakwater that is exposed to a variety of water depths, especially one perpendicular to the shore, should have wave conditions investigated for each range of water depth to determine the highest breaking wave to which any part of the structure will be exposed. The outer end of a breakwater might be exposed only to wave forces on its sides under normal depths, but it might be overtopped and eventually submerged as a storm surge approaches. The shoreward end might normally be exposed to lower breakers, or perhaps only to broken waves.

The character of wave–structure interactive is another very important factor to be considered. When waves impinge directly on rubble structures, they could do the following:

1. Break completely, projecting a jet of water roughly perpendicular to the slope
2. Break partially with a poorly defined jet
3. Establish an oscillatory motion of the water particles up or down the structure slope, similar to the motion of a clapotis at a vertical wall

According to U.S. Army Corps of Engineers (1984), the design wave height for a flexible rubble structure should usually be the average of the highest 10% of all waves (H_{10}). Damage from waves higher than the design wave can be progressive; however, the displacement of several individual armor units may not result in a complete loss of protection. The most severe wave condition for the design of any part of a rubble-mound structure is usually a combination of predicted water depth and extreme incident wave height and period that produces waves which would break directly on the part of interest. Last, but not least, the wave character and its effects on armor protection depends strongly on the approach angle. Kimura et al. (1994) found that as a wave approach angle decreases, the water velocity component parallel to the breakwater alignment increases, with the velocity near the vertical wall also increasing significantly. Special attention must be paid to the wave climate and wave effects around the breakwater head.

Armor for Protection of Rubble Foundation. Until about 1930, the design of rubble structures was based only on experience and general knowledge of site conditions. Empirical formulas that developed subsequently are generally expressed in terms of the stone weight required to withstand design wave impact. These formulas have to some extent been substantiated in model studies; they must be used with caution, however, exercising good engineering judgment. Local experience should also be taken into consideration and evaluated carefully. Physical modeling is often a cost-effective measure to determine the final cross-section design for the most costly rubble-mound structures.

The armor stones or concrete blocks should be heavy enough to withstand velocity-induced forces. Their required weight can be evaluated by the *Isbash formula* for stones embedded in the bottom of a sloped channel. This formula relates the stable weight of armor stones to the water particle velocity as follows:

$$W = \frac{\pi \gamma_r U^6}{48 g^3 y^6 (S_r - 1)^3 (\cos\alpha - \sin\alpha)^3} \quad (9.46)$$

where W is the required weight of armor stone; γ_r the specific weight of armor stone ($\gamma_r = \rho_r g$); U the water particle velocity on the stone; g the acceleration due to gravity; y the Ishbash number ($y = 1.2$ for embedded stones and $y = 0.86$ for stones placed on a flat bottom); S_r the specific gravity of stone; α the bottom slope.

Brebner and Donnelly (1962) proposed a method to determine the necessary weight of the armor unit directly from the wave height. In their method, the stable weight of the armor unit, W, can be obtained from the following formulation:

$$W = \frac{\gamma_r H_{1/3}^3}{N_s^3 (S_r - 1)^3} \quad (9.47)$$

where γ_r is the specific weight of the armor unit, $H_{1/3}$ the design significant wave height, and N_s the stability coefficient. The coefficient N_s depends on variables such as the shape of the armor unit, manner of placement, shape of the rubble-mound foundation, and wave parameters (e.g., height, period, and direction).

Takahashi et al. (1990) developed the following expression for the determination of N_s:

$$N_s = \max\left\{1.8,\ 1.3\left(\frac{1-\kappa}{\kappa^{1/3}}\right)\frac{h'}{H_{1/3}} + 1.8 \exp\left\{-1.5\left[\frac{(1-\kappa)^2}{\kappa^{1/3}}\right]\frac{h'}{H_{1/3}}\right\}\right\} \quad (9.48)$$

where

$$\kappa = \kappa_1 (\kappa_2)_B \quad (9.49)$$

$$\kappa_1 = \frac{2kh'}{\sinh 2kh'} \quad (9.50)$$

$$(\kappa_2)_B = \max\{\alpha_s \sin^2\theta \cos^2(kB_M \cos\theta),\ \cos^2\theta \sin^2(kB_M \cos\theta)\} \quad (9.51)$$

where $\max\{a,b\}$ denotes the maximum values

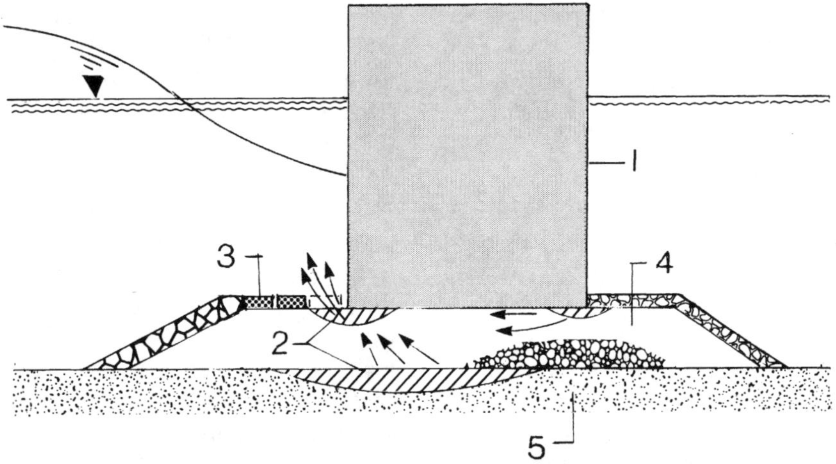

Figure 9.29 Rubble-mound and seafloor failure due to damage to tow protection blocks. 1, Vertical breakwater; 2, areas of potential erosion; 3, protection blocks; 4, rubble mound; 5, seafloor.

of a and b, h' is the water depth above the rubble-mound foundation, $k = 2\pi/L'$, L' the length of the wave that corresponds to the significant wave period at the depth h', B_M the berm width, and $\alpha_s = 0.45$ is a correction factor obtained from the wave tank experiments. Equation (9.48) can be also used for evaluation of the stability of the rubble-mound armor layer installed in the breakwater head. In this case, the term $(\kappa_2)_T$ is used instead of $(\kappa_2)_B$ to represent the water particle velocity at the breakwater head. Hence,

$$\kappa = \kappa_1(\kappa_2)_T \tag{9.52}$$

where

$$(\kappa_2)_T = \tfrac{1}{4}\,(\alpha_s \tau^2) \tag{9.53}$$

where τ is the ratio of the water particle velocity at the breakwater head to that of velocity incident wave; $\tau = 1.4$ is generally used for a wave angle of less than 45°. Sample calculations of the required weight of armor stones are given in Takahashi (1997). For more information on armor stability, readers are referred to U.S. Army Corps of Engineers (1984).

Foot Protection. The foot of an upright section is typically protected by concrete blocks or large rocks placed in front of the wall. The typical failure modes related to such foot protection are illustrated in Figure 9.29. Where some blocks have been removed, erosion of the rubble mound takes place near the foot of the wall. Also, through-wash (rapid current through the rubble mound) may cause scouring of the underlaying seabed. Foot protection concrete blocks are installed to prevent the direct intrusion of wave pressure into the rubble mound and subsequent pressure-induced current in the mound. These blocks work as a filter and also provide additional weight to the rubble mound. The critical force on a foot protection block is that due to differential pressure between the upper and lower faces of the block. The absolute value of the wave pressure under the block is found experimentally to be 5 to 40% less than that on the upper side. The differential pressure can be reduced and the stability of the block increased by the introduction of holes in the blocks. A block perforation ratio of 10% is recommended by Tanimoto et al. (1982). It should be noted that the holes should be small enough to prevent small stones and gravel from being washed away from the mound. The concrete foot protection blocks also act as armor for the rubble mound. Care should also be taken to prevent erosion of the seafloor underneath the thin rubble mattress. This could be accomplished by underlaying a rubble mound with a sheet of geotextile or installation of a specially designed layered stone filter.

Rubble-Mound Toe Protection against Scouring. In some cases the toe of a rubble mound could be vulnerable to scouring damage, which may lead to damage of the rubble mound and eventually failure of the breakwater structure. A number of design guidelines in the form of rules of thumb, or empirical formulas for determining the depth of scour have been proposed by Xie (1981), Irie et al. (1986), and Oumeraci (1994). Because scour protection is very important to breakwater performance it is usually included in breakwater construction where severe scouring is expected.

Typically, scour protection methods include use of gravel blankets, geotextile, or asphalt mats. These methods can limit scouring of the sea bottom to some extent; however, they cannot guarantee full protection from scour. Hence, at locations where a scour condition can be expected, a scour monitoring and breakwater repair and maintenance program should be in place. For more information, the readers are referred to Chapter 5.

9.3.2.5 Modified Traditional Vertical Breakwaters. As noted earlier, before the invention of reinforced concrete, vertical breakwaters were customarily constructed from heavy stone or concrete blocks installed under water. This was time-consuming, expensive work. The introduction of reinforced concrete and advances in concrete technology was followed by increased refinements in the construction of vertical breakwaters. Concrete floating-in caissons have become very popular in breakwater construction throughout the world. This happened because 90% or more of the work of building caissons is executed onshore under well-controlled conditions. Besides the better quality of construction, this also dramatically reduces the amount of downtime normally experienced offshore. Hence, the high cost of equipment and labor during periods of severe sea conditions is avoided. Caisson-type breakwaters have often been built where foundation soil precluded the construction of piled structures. However, caissons also have been found adaptable to loose alluvial soil formations at many locations in Europe and Japan.

In most cases, concrete caissons are built onshore where launching facilities such as shiplifts, marine rails, slip ways, dry docks, or flooded basins are available. On several known occasions, caissons have been constructed at locations with substantial seasonal water fluctuation (e.g., riverbanks). In eastern Canada, caissons built ashore have been launched using the *slipway launch method*, which includes construction of the caisson on a tilting platform adjacent to a fixed short slipway. There, the partly (or fully) completed caisson slips down the basin when the platform is tilted, then accelerates down-ramp (slipway), and finally splashes down in the basin (Tsinker, 1997).

Caissons can be of practically any height and do not require use of heavy lift equipment for their construction. They are normally constructed in close proximity to the project site. In some cases, however they can be fabricated at locations within a reasonable towing distance, where materials, labor, construction, and launching facilities are available. Usually, caissons are described as structures shaped like boxes. However, in modern engineering practice, the term *caisson* is also applied to floating cylinders used for the construction of breakwaters as well as other marine projects.

Numerous examples of the successful use of concrete cylindrical caissons in marine applications are given in Tsinker (1997). Caissons can be either slipformed or assembled from prefabricated components. A typical cross section of a breakwater constructed from a floating-in concrete caissons is depicted in Figure 9.30*a*. Normally, fabrication, launching, towing, and installation of large (long) caissons have presented no significant technical problems, and in most practical cases the design length of a caisson of conventional construction is influenced basically by tolerances specified for preparation of the bedding (rubble-mound mattress). Construction of the mattress, and specifically its densification and surface preparation, is a critical stage in the overall process of a breakwater construction; significant deviations from the horizontal plane and/or uneven settlements of the mattress may result in substantial global or local overstress of caisson structural components.

Therefore, success of caisson quay wall construction depends heavily on the quality of mattress preparation. Completed caissons are towed to a deployment site where they are ballasted down with water on the prepared bedding. Caisson towing is an important and sometimes complex marine operation. It is usually conducted when the environmental conditions, such as wind, waves, and currents, are suitable for the towing operation. Bathymetry along the tow route should be investigated carefully. The tow route selected should ensure that the caisson has an adequate underkeel clearance at all times. Due consideration must be given to the number and size of tugs required for caisson towing in the given environmental conditions. Extra care is required during caisson installation. Caisson installation and refloating are usually accomplished by careful ballasting of its compartments with water. For specific details concerned with rubble mattress construction caisson launch, tow, and installation of the mattress, readers are referred to Tsinker (1997).

Normally, vertical breakwaters are completed with a parapet that constitutes an integrated part of a structure. The parapets, sometimes curved in profile, are aimed at deflection of the water that is running up the wall and reduction in the chances of overtopping discharge. However, exposure of a parapet to breaking waves may result in a strong overturning moment (Benassai et al., 1974). For this reason the traditional straight parapet is sometimes made curved to better reflect the attacking wave, is set back (Figure 9.30*b*), or is substituted for with a sloping wall (Figure 9.30*c*). The advantage of a setback parapet is that forces acting on it are somewhat delayed with respect to those acting on a front wall, thus reducing the maximum force on the structure. Furthermore, vertical jets guided through the perforated front wall do not jump easily over the setback parapet, thus reducing wall overtopping.

With the dramatic increase in vessel draft, breakwater construction in open deep waters has become more prevalent in recent years. In deepwater locations, large waves generate tremendous forces on a breakwater, and a sloping-top breakwater (Figure 9.30*c*) has been found suitable for this application. A sloped superstructure is used to reduce wave forces on a breakwater structure. In addition, the downward component of wave force acting on the slope cancels at least part of the uplift pressure, thereby increasing breakwater stability. Sloping-top breakwaters have been used for many years, with the oldest structure of this type constructed at Naples, Italy in 1906. Another was built in the middle of the 1960s at Hanstholm Harbor, Denmark, where the overturning moment and total horizontal force act-

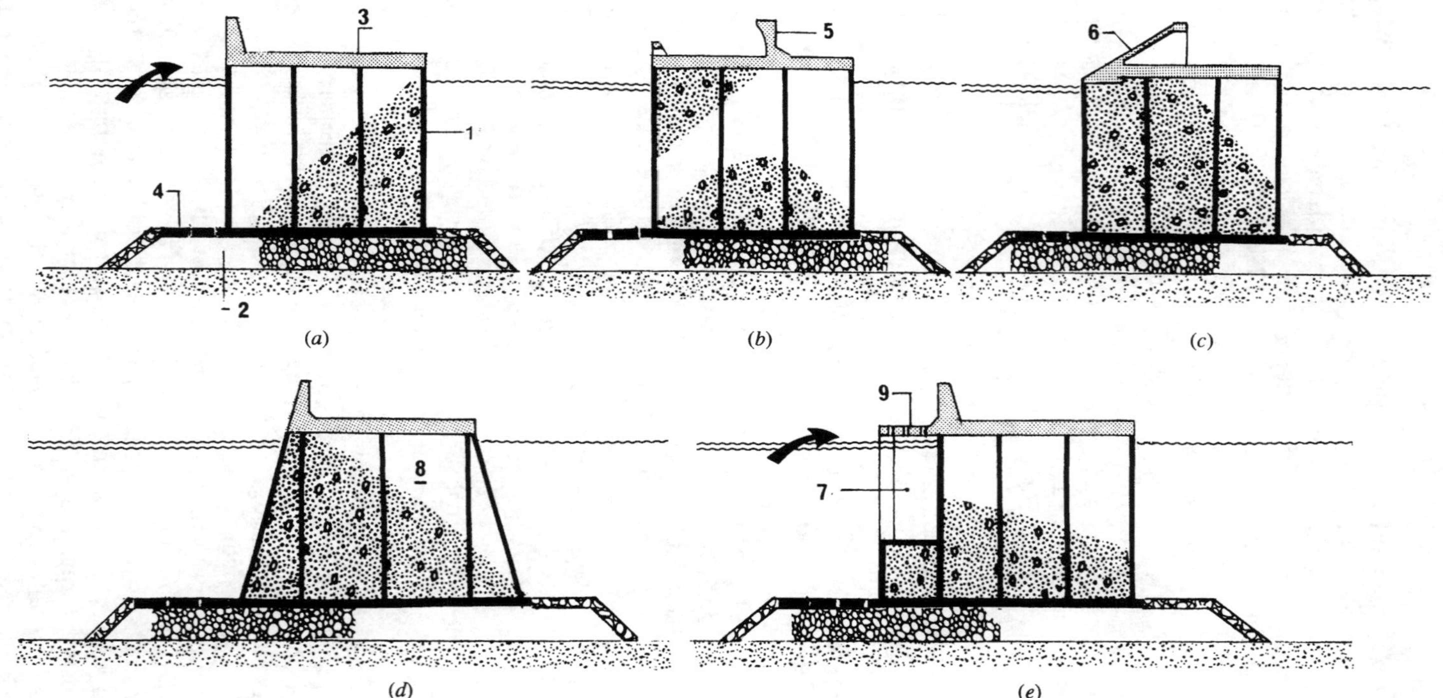

Figure 9.30 Vertical breakwaters: (*a*) conventional floating-in caisson breakwater; (*b*) breakwater with setback parapet; (*c*) sloping-top breakwater; (*d*) trapezoidal breakwater; (*e*) breakwater with slotted (alternatively, perforated) front wall. 1, Concrete caisson; 2, rubble mound; 3, concrete top complete with front parapet; 4, toe protection; 5, setback parapet; 6, sloping wall; 7, wave energy–dissipating chamber; 8, granular fill; 9, perforated top slab.

ing on a breakwater were reduced to about one-half the values of the vertical breakwater (Juhl, 1994; Ligteringen, 1994). Sloping-top caissons have also been constructed in deep-water areas in Japan, Taiwan, China, and Libya, where wave conditions are severe (Kuo, 1994; Xie, 1994; Takahashi, 1997).

For construction reasons the sloped part of a breakwater is usually located above sea level. However, the capability of a breakwater to resist wave attack is increased if the sloping part is extended below the still-water level (Takahashi et al., 1994b). A sloping-top caisson breakwater is very stable, yet the overtopping wave is significantly large. Hence, the crown height of a sloping breakwater must be somewhat higher than that of the ordinary vertical wall breakwater to obtain a similar transmission coefficient K_T value. For extensive discussions of sloping breakwaters, readers are referred to Tanimoto and Takahashi (1994) and Takahashi (1997).

Trapezoidal breakwaters (Figure 9.30*d*) are essentially vertical structures with enhanced resistance to wave forces due to their sloped walls. Tanimoto and Kimura (1985) found that the uplift wave pressure on the bottom of a trapezoidal breakwater is reduced due to upward water-particle velocity enhanced by the breakwater slope. These investigators recommended use of Goda's formula for determination of wave forces for this type of structure. A further advantage of a trapezoidal geometry for the upright section is less toe pressure applied on a rubble foundation. This is, of course, due to the greater width of the wall base.

In recent decades, in an attempt to reduce the wave forces against mainly vertical breakwaters, some new ideas have been introduced and new types of breakwaters have been developed. The latter include perforated and inclined wall breakwaters of miscellaneous design as well as semicircular structures (Takahashi, 1997). One example is shown in Figure 9.30*e*.

The idea behind front wall perforation is dissipation of wave energy, resulting in reduction of wave force against the structure. This type of breakwater usually incorporates a perforated front wall and wave energy dissipation chamber. The perforation is achieved in a variety of ways, but most often is in the form of circular holes or vertical or horizontal slots; curved slots are also employed. One of the earliest breakwaters with a perforated front wall was constructed at Comeau Bay, Canada (Jarlan, 1961). A similar structure was built later at the port of Kobe, Japan (Takahashi, 1997).

Because of their high wave-absorbing capacity and high stability against wave forces, this type of construction has been used increasingly worldwide for the construction of seawalls and breakwaters. In this type of construction, waves can enter and leave the wave chamber relatively freely. In the process, the wave energy dissipates through generating eddies. The larger difference in water levels inside and outside the wave chamber crates better conditions for dissipation of the wave energy. The size of the wave chamber depends strongly on the length of the design wave, and in general, the performance of a perforated wall breakwater is typically most efficient when the width of the wave chamber constitutes approximately 10 to 20% of the wavelength.

A wave chamber normally has a bottom slab, and to reduce the impulsive uplift pressures generated inside the wave chamber, the ceiling slab should be perforated. The magnitude of the wave force acting on a ceiling slab depends on its clearance from the still water level and degree of perforation. A ceiling slab is usually installed as a structural support to the front wall exposed to wave forces. If structurally acceptable, the ceiling slab could be removed all together. The wave forces on different components of a perforated breakwater reach their peaks at different phases of wave–structure interaction. Therefore, the wave pressure distribution at each of these phases must be evaluated. In fact, for a perforated breakwater, the peak load that may cause this breakwater to slide or overturn does not necessarily occur when the wave crest is located just in front of the perforated wall. The latter is discussed in details by Takahashi and Shimosako (1994). Numerous practical examples of most recent constructions of perforated breakwaters are given in Takahashi (1997).

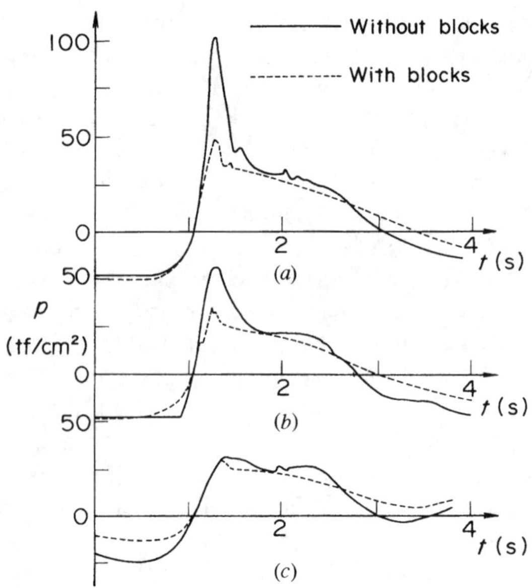

Figure 9.31 Change in wave pressure profile by wave energy–dissipating blocks: (*a*) heavy breaking wave; (*b*) regular breaking wave; (*c*) nonbreaking wave. (From Takahashi, 1997.)

9.3.2.6 Composite Breakwaters. Composite breakwaters are discussed briefly in Section 9.3.2.1, and some basic types of these structures are depicted in Figures 9.18*b* and *c* and 9.19. The rubblemound placed in front of the upright section of such breakwaters usually comprises large rocks or artificial concrete blocks. The purpose of placing these rocks and/or artificial blocks is to dissipate wave energy and to reduce the force of breaking waves. In most practical cases the top level of the rubble mound is at the top of the upright section. The wave transmission of composite breakwaters depends on the width of the rubble's berm, but primarily on breakwater elevation above maximum sea level. The wave reflection of composite breakwaters is very similar to that of regular rubble-mound breakwaters.

Wave Force on Upright Sections and Rubble-Mound Blocks. The blocks installed in front of an upright section act to reduce the wave force on the wall. Figure 9.31 illustrates the time history of wave pressure obtained from a model test, where the pressure profile with and without wave-dissipating blocks is indicated (Takahashi, 1997). The results depicted in Figure 9.31 demonstrate that the pressure component due to wave breaking is reduced significantly by the blocks, whereas the slowly varying wave pressure component changes only slightly. A reduction in wave force on the upright section depends in general on the size of the rubble section (the width of the berm) as well as the wave height/water depth ratio; when the latter ratio is close to unity, the wave reduction force is reduced to less than 0.8. Because the upright section in a composite breakwater is in fact the retaining wall, it is exposed to the lateral load produced by the rubble mound placed in front of it. This load, however, is very small compared with the wave force attacking the wall. It could be even smaller due to potential slight wall movements away from the rubble and therefore could be conveniently neglected in practice. However, in a deepwater situation it may be significant, and where relevant, its effects on the wall stability should be examined. As pointed out earlier, the rubble mound could be composed either of large rocks or of concrete blocks of miscellaneous designs. Examples are illustrated in Figure 9.15*b* and *c*.

To be efficient, a wave energy–dissipating system should cover the vertical wall completely to prevent the impulsive wave-breaking pressure on the wall. The top level of the blocks should therefore be at the same elevation as the vertical wall crown, whereas the berm of the rubble section at the top should be at least as twice as much as the height (diameter) of the block. The slope of the rubble section, α, is usually pretty steep (i.e., cot α = 1.34 to 1.5), and a gentle slope (e.g., cot α > 2) is seldom employed. The weight of the individual block included in the rubble mound could be evaluated by the Hudson or Van der Meer formula, as discussed in Section 9.3.1.1. In practice, the required weight of the wave energy–dissipating units is less than that calculated by the Hudson or Van der Meer formula. This is because the stability of such units placed in front of the vertical wall is inherently higher than the stability of similar armor blocks installed on a rubble slope, due to the higher permeability of the block section placed against a vertical wall. An additional factor is that both the Hudson and Van der Meer formulas have been developed to calculate the weight of armor blocks intended to protect rubble mounds. The most cost-effective solution could be obtained through physical large-scale model tests. This can help to determine the optimum shape and weight of the block section. For the same purpose, the results of experiments conducted previously in which the identical conditions have been employed can be taken into consideration.

9.3.3 Special (Unconventional) Breakwaters

Some special breakwaters are illustrated in Figure 9.8*d*–*f*.

9.3.3.1 Piled Breakwaters. Where the wave conditions are not severe and foundation soil is suitable for pile driving, piled breakwater structures (Figure 9.8*d*) may be cost-effective. In some instances (e.g., where a poor soil foundation extends to a significant depth), a piled structure can be the only alternative to breakwater construction. Piled breakwaters have been constructed in the form of a single row of steel or concrete pipes of various diameters driven to the depth required for adequate stability. These walls have been made watertight by use of miscellaneous interlocking systems. Examples are shown in Figures 9.32 and 9.33. Where required for better stability against wave forces, the breakwater is constructed in the form of sheet pile walls anchored to each other, with the space between them filled with a good-quality granular material. Relevant examples are presented in Figures 9.5 and 9.34.

For a poor soil foundation, elevated platform piled structures similar to that depicted in Figure 9.8*d* can be practical and economically feasible (e.g., a breakwater of this type was constructed successfully in Manfredonia, South Adriatic Sea, Italy in a water depth of 11 m) (Lamberti and Franco, 1994). The latter breakwater was installed on foundation soil that included a layered, 11 to 15 m, soft clayey silt, underlaid by a layer of medium-fine sand of 11 to 18 m. This structure was designed to sustain the force of a significant wave, H_s = 4.8 m, with a period T_s = 10 s. Practical design guidelines for piled marine structures are given in Chapter 2 and in greater detail in Tsinker (1997). The advantage of piled structures is that with the exception of seafloor protection from erosion in front of the breakwater, all civil works are carried out above sea level.

9.3.3.2 Floating Breakwaters. Interest in floating breakwaters (Figure 9.8*c*) was stimulated in the past 50 or so years either by the need to seek a means of rapid protection of shore areas in military operations, (e.g., construction of the Bombardon breakwater used by the allied forces during the D-day invasion in Normandy, France) (Lochner et al., 1948) or as a cost-effective alternative to a bottom-fixed structure in a deepwater area with relatively small waves. Experience has shown that the use

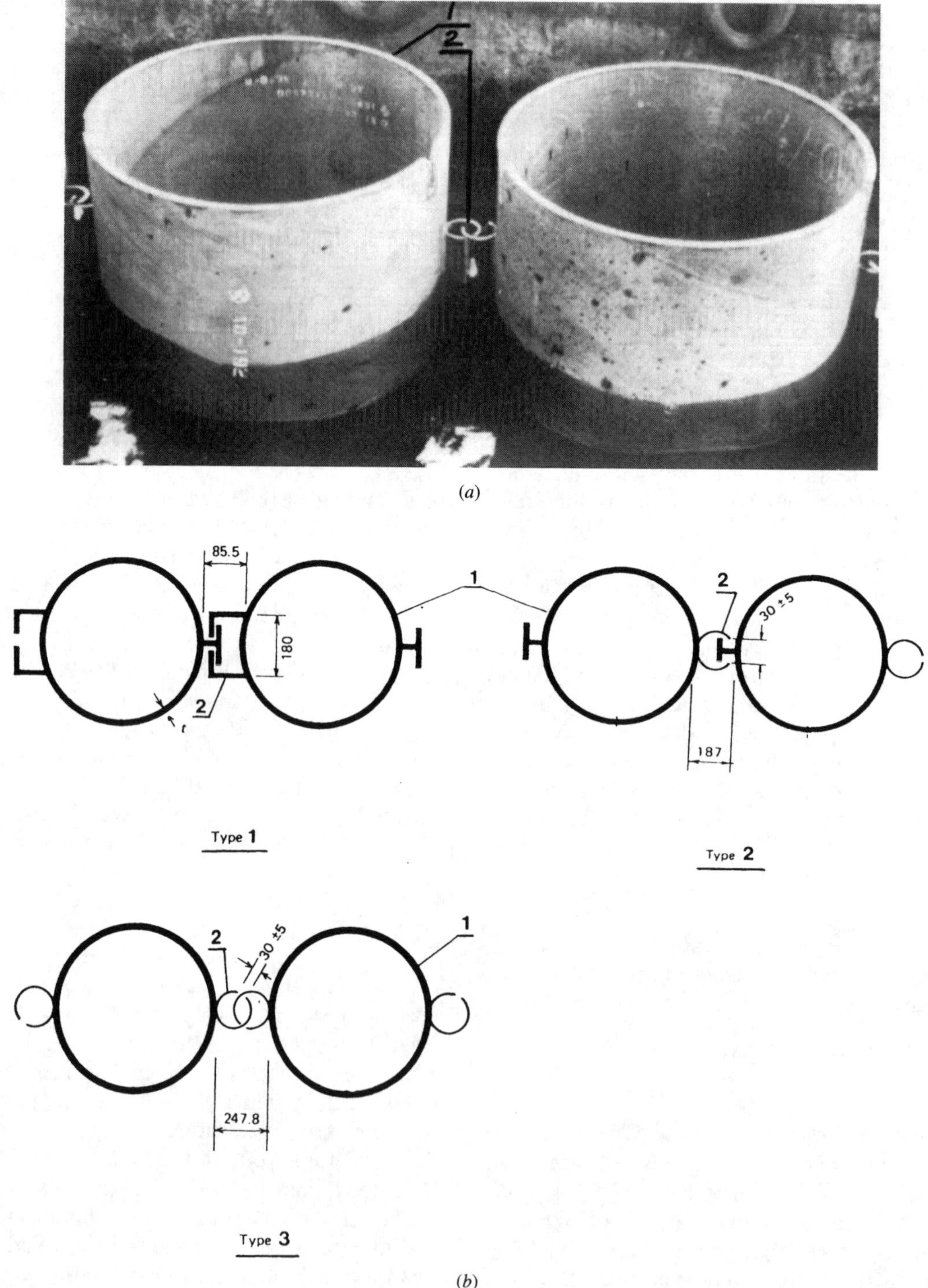

(*a*)

(*b*)

Figure 9.32 Breakwater comprises interlocked steel pipe piles. 1, Steel pipe pile; 2, interlocking system.

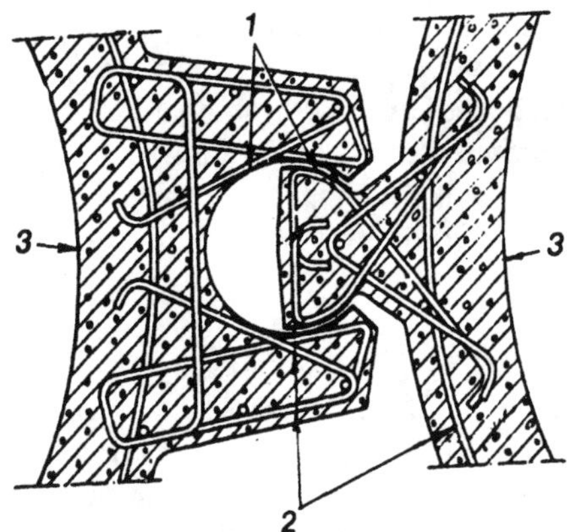

Figure 9.33 Interlocking system of concrete pipe piles. 1, Sheet metal; 2, steel reinforcement; 3, concrete pipe pile.

Figure 9.34 Breakwater constructed in the form of a double steel sheet pile wall.

of floating breakwaters can be structurally and economically feasible under certain conditions:

1. Deepwater conditions result in excessive construction costs for bottom-fixed structures.
2. Foundation conditions may present high costs for corrective action.
3. A bottom-fixed structure impedes the free interchange of water in a basin, resulting in minimized water quality parameters.
4. A bottom-fixed breakwater obstructs fish migration pathways.
5. A fixed structure causes "visual pollution" in the area.

Furthermore, the experience has confirmed the following:

1. Floating breakwaters offer a feasible and economic alternative to conventional breakwaters in deep water when exposed to moderate wave conditions (i.e., <2 m).
2. Floating breakwaters can be designed to reduce waves to acceptable levels as long as those waves are not long. A practical upper limit is on the order to 4 to 6 s; in most practical cases, floating breakwaters have been designed for wave periods of less than 4 s.
3. Floating breakwater structural units and their mooring systems are vulnerable to catastrophic failure during severe storms when they are needed the most.
4. Floating breakwaters require a relatively high level of maintenance compared to conventional fixed structures.

The advantages of floating breakwaters are that they can be fabricated off site and assembled rapidly where required. This makes them very useful, especially in military operations, as noted earlier. Conceptually, floating breakwaters are usually classified as reflective and/or energy-dissipative structures. By definition, the former are designed to reflect incident waves so that only a small amount of energy may pass through the system; the latter are those designed to dissipate wave energy, mostly by turbulence. Naturally, each structure acts partly in both ways.

A variety of floating breakwater types and their mooring systems have been developed for commercial use over the past 50 years. However, most floating breakwaters have been constructed in the form of rectangular pontoons moored in place by means of anchors and chains (soft catenary mooring) or by guide piles (rigid mooring). With regard to catenary mooring systems, a designer may select from a variety of anchor types, including drag anchors, dead-weight anchors, and pile anchors. Catenary computations, mooring design, and anchor selection are presented in Chapters 7 and 8. Early surveys of floating breakwaters are presented in Hales (1981) and McCartney (1985), and discussions of floating breakwater and mooring system design appear in Tsinker (1986), Oliver (1992), PIANC (1994, 1997), Headland (1995). Floating breakwater design is a complicated and iterative process, due to the interdependency of each design factor. For example, wave transmission performance depends on breakwater geometry, mass, and mooring properties. Similarly, mooring forces depend on breakwater geometry and mass. Finally, breakwater structural integrity depends on breakwater geometry, mass, and mooring forces.

There are four fundamental aspects of floating breakwater design: (1) buoyancy and floating stability, (2) wave transmission, (3) mooring forces, and (4) breakwater and structural integrity. A floating breakwater must possess sufficient buoyancy to support the weight of the breakwater and its moorings. Furthermore, the breakwater must be rotationally stable, in accordance with standard practices of naval architecture. The fundamental goal of a floating breakwater is to attenuate or reduce waves. Accordingly, evaluation of floating breakwater wave transmission constitutes a critical element of design. Moorings, whether constructed of piles or mooring lines and anchors, must hold a breakwater in place, and a careful assessment of mooring forces during design storm wave attack must be made to ensure the survival of the breakwater. In addition, the breakwater unit itself must sustain the stresses imposed by wave-induced hogging and sagging as well as those related to the moorings. With regard to the structural integrity of a breakwater unit, consideration must be given to both survival and fatigue-related stresses. Wave loading generally dictates the design of a floating breakwater, and wave-induced load is emphasized in this chapter. On the other hand, the designer should also evaluate other possible loads, such as those associated with currents, water-level variations, ice, wind, and vessel impact.

As for any typical civil or marine structure, a floating breakwater should be designed for both the strength and serviceability of its structural components (e.g., a floating section and its mooring system). As recommended by PIANC (1994), the following environmentally induced loads must be considered in floating breakwater design:

1. Wave-induced forces (EW)
   1a. Wave-induced forces–early return period (EW_0)
2. Current drag forces on a breakwater and boat(s) if moored alongside the breakwater (EC)
3. Wind drag (ED)
4. Ice loading (EI)
5. Dead loads (DL)

6. Live loads (*LL*)
7. Loads caused by temperature, prestress, shrinkage, and creep effects (*TL*)
8. Dynamic loads caused by the mass accelerations of the breakwater and mooring boats reacting to breakwater moorings

Despite the fact that wave forces constitute the predominant loads on a breakwater, the structure is designed for a combination of the loads listed above. The safety factors for the various load combinations depend on the probability of the simultaneous occurrence of all loads. With the exception of load 1a, PIANC (1994) recommends a recurrence interval of 50 years for environmental events. However, in practical designs, requirements in local and national codes and standards must be considered. In U.S. practice, the strength of a breakwater should meet or exceed the following load combinations (Oliver, 1992):

$$\mathrm{U} = \begin{cases} 1.2(DL + TL) + 1.6LL \\ \quad + 1.6(EW_0 + ED + EC) \\ 1.2(DL + TL) + 1.2LL \\ \quad + 1.3(EW + ED + EC) \\ 1.2(DL + TL) + 1.2LL \\ \quad + 1.3(EW_0 + ED + EC + EI) \end{cases}$$

Accordingly, serviceability should be investigated for the following unfactored load combinations:

$$\mathrm{S} = \begin{cases} DL + TL + LL + EW_0 \\ \quad + EC + ED + EI \\ DL + TL + LL + EW + EC + ED \end{cases}$$

Again, for detailed information on floating breakwater design, the reader is referred to PIANC (1994, 1997) and fundamental work by Headland (1995).

9.3.3.3 Pneumatic Breakwaters. Pneumatic (bubble) breakwaters (Figure 9.8*f*) have been known since the beginning of the century. These breakwaters are constructed from perforated pipes placed at a certain depth below the water surface and inflated with a compressed air supply system. Air delivered to perforated pipe rushes to the surface of the water, where it creates zones of circulating water, which in turn produce a surface-induced current directed against the waves. Interference with incoming natural waves by the induced current results in a dissipation of wave energy.

Perhaps because of their limited capacity, pneumatic breakwaters have never been used for the protection of major harbors. They have found, however, limited use in the protection of entrances to dry docks, navigation locks, and in some cases, as in-harbor secondary breakwaters. These systems have also been used to create an ice-free surface of water during winter navigation. A number of practical examples of pneumatic breakwaters are given in Tsinker (1997).

9.4 ECONOMY OF BREAKWATER CONSTRUCTION

As stated earlier, the primary purpose of breakwater system construction is establishing an area (basin) of calm water where a vessel can be safely moored without experiencing movements that can either jeopardize the safety of the vessel or disrupt the cargo transfer operations. In general, a ship's response to wave and wind excitation depends on many factors, including the port layout; however, first and foremost is a correctly designed and built breakwater system. Such a system is usually a multimillion dollar investment in construction of a part, so to be cost-effective, it must be neither underdesigned nor overdesigned. The middle ground is usually found as a result of realistic analysis of the types of vessels to be handled at the port (e.g., larger vessels need less protection from wave action than do smaller ones), as well as the kinds of services provided by the port (e.g., handling liquids or dry bulk, containers, lift on–lift off, roll on–roll off, or a combination of all or some of the above). For example, tankers generally can accept more movement at a berth than can dry bulk carriers or container ships. Hence, the port designer must establish acceptable criteria for a wave climate in a port area on the basis of a vessel's acceptable movements at berth. Some recommended practical limits of vessel movement are given in NATO (1988), Bruun (1990), and Tsinker (1997).

Ship movement requirements affect a breakwater system layout and its structural parameters (e.g., the breakwater's height above design sea level, the breakwater's ability to reflect and transmit waves). Alternative layouts and structural configurations are usually evaluated on the basis of potential delays in operation or even potential damage to a ship. In most port projects the number, length, and orientation of breakwaters is dictated by the size and geometry of the basin it is designed to protect. Along with breakwater stability, all of the requirements above play a roll in an economic evaluation. Basically, the latter involves evaluation of how much interruption in port operation could be tolerated economically due to installation of a less expensive structure(s). It is usually carried out based on the balance of the annualized cost of construction and maintaining the breakwater against the risk of operational losses due to unacceptable delays related to wave transmission.

In most cases this is not a simple task because port administrations usually oppose any risk in breakwater construction that may result in a loss of port operation. Limited damage to a breakwater system is usually accepted readily, however, when rare events such as severe hurricanes and typhoons pass through the port area. The severity of breakwater damage depends on both the characteristics of the original design parameters and the efficiency of maintenance. The former include the factors of safety, durability, and weight of the critical components, such as armor rocks and/or artificial concrete blocks, and size of the upright section. Effective maintenance includes regular survey and inspection of the structure for evidence of wear and tear or of damage following storms.

Hence, the process of optimization of breakwater design is a function of a careful assessment of a multitude of data, as suggested briefly above. For more practical information on the subject of economic optimization of breakwater construction, readers are referred to works of Van der Kreeke and Paape (1964), Smith (1987), Bruun (1990), Bruun and Smith (1990–1991) D'Angremond and Van Roode (2001), and Verhagen and Burcharth (2002), and to Section 9.3.1.1.

APPENDIX: DESCRIPTION OF THE WAVE FIELD†

9A.1 INTRODUCTION

Knowledge of the characteristics of the wave field in an area where port engineering works are to be executed is of paramount importance since wave attack produces the primary loading of such structures. Methods of predicting the wave field are available and can be employed to define the input to each case of port structure design. These methods range from the simplest parametric predictions of a single wave characterizing the sea state to more sophisticated probabilistic estimates and numerical models. Parametric predictions usually provide the wave height and period associated with the significant wave; stochastic analysis methods re-

†This appendix is contributed by Constantine D. Memos.

late the principal wave characteristics (i.e., wave height and period) of any wave with its probability of occurrence in the sea state under consideration; numerical models normally yield outputs in the form of either power spectra or time history of surface elevation at the required locations. The power spectra can be of the one-dimensional type (i.e., not displaying any information on the directional spread of wave energy) or of the two-dimensional type, incorporating such information. The latter can also be furnished approximately by simple expressions giving the percentage of each direction and the frequency band of the underlying spectra to the total amount of mechanical energy of the wave field at each particular point.

Historically, Sverdrup and Munk (1947) were the first to embark on developing parametric prediction methods, during World War II. Later, Bretschneider (1952, 1957) modified these methods, and Newmann (1953) and Neumann and Pierson (1957) introduced spectral analysis to describe ocean water waves. By their pioneering work it became possible to calculate representative parameters of the wave field, such as the mean or significant wave height, through spectral analysis. The stochastic approach to analyzing the wave field is based mainly on the renown work by Rice (1944) in information theory, whereby the spectral representation of a Gaussian process, such as the sea surface elevation in deep water, was associated with the probability density function of the excursion of its random variable. In 1952, Longuet-Higgins began publishing a series of articles related to the stochastic nature of a sea state. These pertained to such issues as the distribution of wave heights, wave directionality, broadband spectra, join density between wave heights and periods, and nonlinear and non-Gaussian waves.

In the following subsections a rough account of wave prediction methods and the corresponding results is given, focusing on the relevant probabilistic estimates that additionally can support risk analysis exercises for port structures. Also, some guidance is given on how the parameters required for design could be extracted from the results noted above, some of which have not yet been used in design practice.

9A.2 PARAMETIC WAVE FORECASTING

In general, wind-generated waves are the most important type of sea waves in the design of port structures. The conditions under which waves can be generated are classified into two categories: fetch-controlled or wind duration–controlled conditions. In the former, the duration of wind blowing is assumed adequately long; in the latter, the fetch over which waves develop and propagate is assumed unlimited. Parametric prediction models provide characteristic values of wave heights and periods of a sea state, usually defined by fetch, intensity, and duration of the wind field. The assumptions of these models generally include a uniform and constant wave field over the wave generating area, uniform water depth, and no wave breaking. Earlier models modified over the years are still being used in design practice. Some typical results are noted briefly below for water of any depth or for deep water only.

9A.2.1 Water of Any Depth

9A.2.1.1 Sverdrup–Munk–Bretschneider (SMB) Nondimensionalized Curves. These originated from the works of Sverdrup and Munk (1947) and Bretscheider (1952, 1958):

$$\overline{H} = 0.283 \tanh(0.53\overline{d}^{3/4}) \tanh \frac{0.0125\overline{F}^{0.42}}{\tanh(0.53\overline{d}^{3/4})} \tag{9A.1}$$

$$\frac{\overline{T}}{2\pi} = 1.2 \tanh(0.833\overline{d}^{3/8}) \tanh \frac{0.038\overline{F}^{1/3}}{\tanh(0.833\overline{d}^{3/8})} \tag{9A.2}$$

where nondimensionalization has been effected with respect to the constant wind speed u as follows. For wave height:

$$\overline{H} = \frac{gH_s}{u^2}$$

$$\overline{T} = \frac{gT_s}{u^2}$$

For wave period:

$$\overline{F} = \frac{gF}{u^2} \quad \text{fetch}$$

$$\bar{t} = \frac{gt}{u} \quad \text{wind duration}$$

$$\overline{d} = \frac{gd}{u^2} \quad \text{water depth}$$

In the relationships above, a bottom friction factor of 0.01 has been assumed.

9A.2.1.2 Nondimensional JONSWAP-Based Curves. These are based on Hasselmann et al. (1976):

$$H' = 200 \tanh(0.003877d'^{3/4}) \tanh \frac{0.0002129F'^{1/2}}{\tanh(0.003877d'^{3/4})} \quad (9A.3)$$

$$T' = 200 \tanh(0.07125d^{3/8}) \tanh \frac{0.00426F'^{1/3}}{\tanh(0.07125d'^{3/8})} \quad (9A.4)$$

The nondimensionalization is now made with respect to the friction velocity, u_*, as follows:

$$H' = \frac{gH_s}{u_*^2} \qquad T' = \frac{gT_p}{u_*}$$

where T_p is the period at the peak of the spectrum.

$$F' = \frac{gF}{u_*^2} \qquad d' = \frac{gd}{u_*^2}$$

The friction velocity is expressed in terms of the friction factor C_d and the wind velocity 10 m above the sea surface, u_{10}, as

$$u_* = \sqrt{C_d}\, u_{10} \quad (9A.5)$$

One of the several relations used for C_d is

$$C_d = 10^{-3}\,(0.75 + 0.067u_{10}) \quad (9A.6)$$

where u_{10} is expressed in m/s. It is noted that the wave predictive models above hold for fetch-controlled conditions, whereas for duration-controlled conditions, no widely accepted models exist for waters of any depth.

9A.2.2 Deep Water

For fetch-controlled conditions, eqs. (9A.1) through (9A.4) can be used. For deep water these relations simplify to the following SMB curves:

$$\overline{H} = 0.283 \tanh(0.0125\overline{F}^{0.42}) \quad (9A.7)$$

$$\frac{\overline{T}}{2\pi} = 1.2 \tanh(0.038\overline{F}^{1/3}) \quad (9A.8)$$

with the same notation as previously.

For duration-controlled conditions, the following JONSWAP-based curves can be used for deep water:

$$\overline{H}' = 5.767 \times 10^{-3}\, t'^{5/7} \quad (9A.9)$$

$$T' = 0.486t'^{0.411} \quad (9A.10)$$

where the nondimensional duration is expressed as $t' = gt/u_*$. The friction velocity u_* used in the nondimensionalization of the JONSWAP-based prediction curves can be estimated from wind measurements of land-based stations modified appropriately to cope primarily with altitude, surface relief, and temperature difference. The altitude correction can be effected by using the simple Prandtl–von Kármán boundary layer law:

$$u_{10} = u_z \left(\frac{10}{z}\right)^{1/7} \qquad (9A.11)$$

where u_z is the wind speed at z(m) above the surface.

The difference in boundary roughness between an overland wind and the corresponding wind blowing over the sea can be accommodated in simple terms as follows. If the ratio of the wind speed over the water to the corresponding wind speed over land is denoted by R, then for the wind intensities of interest to the design of breakwaters, it is roughly

$$R = \begin{cases} 1.2 & \text{for 4 Beaufort scale (Bf)} \\ 1.1 & \text{for 5 Bf} \\ 1.0 & \text{for 7 Bf} \\ 0.9 & \text{for 8 Bf or more} \end{cases}$$

After implementing the foregoing two corrections to u related to altitude and boundary roughness, a friction velocity can be calculated, and finally, the correction related to temperature difference can be introduced. This correction, required for wind velocities over the sea of less than 20 m/s, is expressed by a factor R_T, by which the previously calculated u_* has to be multiplied. For an atmosphere warmer than the sea by about 15°C or more, $R_T \approx 0.8$; for a negligible temperature difference, $R_T = 1.0$; and for an atmosphere colder by about 15°C, $R_T \approx 1.2$. An estimate of the fetch to be used in each case can be obtained by one of the methods presented in U.S. Army Corps of Engineers (1984, 2001).

9A.3 STOCHASTIC REPRESENTATION OF SEA WAVES

The inherent variability of the wind field as well as other nonhomogeneities of the parameters involved in a wave generating process are obviously transferred into the resulting wave field, which is thus supporting waves of a stochastic nature. Random waves can be represented in either time or frequency domain. In these representations it is assumed that the elevation of the free surface η at any particular point k in a time $\times$ space window of a sea follows a weakly steady-state random process. Then the ensemble mean,

$$M^{(t)} = \frac{1}{n}\sum_{k=1}^{n} \eta_k(t) \qquad (9A.12)$$

remains constant with time in the prescribed window. Also, the ensemble covariance,

$$C = \frac{1}{n}\sum_{k=1}^{n} \eta_k(t)\eta_k(t+\tau) \qquad (9A.13)$$

depends only on the time lag τ. Since $M = 0$ by definition, the autocorrelation function R obtained by the time history of the random variable η at any one given point,

$$R(\tau) = \lim_{T\to\infty} \frac{1}{2T}\int_{-T}^{T} \eta(t)\,\eta(t+\tau)\,dt \qquad (9A.14)$$

equals the ensemble covariance $C(\tau)$. This elucidates the ergodic nature of sea waves, which in practical terms is translated as indicating that a sea state meeting the weakly steady-state requirement can be adequately represented by a time history of sufficient length of the surface elevation at any point. It is also shown that $R(\tau)$ is bounded as follows:

$$-\sigma^2 + m^2 \leq R(\tau) \leq \sigma^2 + m^2 \quad (9A.15)$$

where σ and m are the standard deviation and mean, respectively, of the underlying process.

Moving to the frequency-domain representation, a spectral density function can be defined as follows:

$$S(\omega) = \lim_{T\to\infty} \frac{1}{2\pi T} |H(\omega)|^2 \quad (9A.16)$$

where ω is the angular velocity, and $H(\omega)$ stands for the Fourier transform of the variable $\eta(t)$:

$$H(\omega) = \int_{-\infty}^{\infty} \eta(t) \exp(-i\omega t)\, dt \quad (9A.17)$$

Then it is easily shown that the Fourier transform of the autocorrelation function $R(\tau)$ is identical to $S(\omega)$. Thus $S(\omega)$ and $R(\tau)$ are a Fourier pair and the following relations hold:

$$S(\omega) = \frac{1}{\pi} \int_{-\infty}^{\infty} R(t) \exp(-i\omega\tau)\, d\tau \quad (9A.18)$$

$$R(\tau) = \frac{1}{2} \int_{-\infty}^{\infty} S(\omega) \exp(i\omega\tau)\, d\omega \quad (9A.19)$$

The relations (9A.18) and (9A.19) constitute an important connection between the time- and frequency-domain representations of sea waves. Further, the mean value of the wave energy (divided by ρg) of the sea state, expressed as

$$\overline{W} = \lim_{T\to\infty} \int_{-\infty}^{\infty} \frac{1}{2T} \eta^2(t)\, dt \quad (9A.20)$$

is shown, using Parveval's theorem, to relate to the spectral density as

$$\overline{W} = \int_0^{\infty} S(\omega)\, d\omega \quad (9A.21)$$

Thus

$$\overline{W} = R(0) = \mathrm{E}[\eta^2(t)] = \mathrm{var}[\eta(t)] \quad (9A.22)$$

Wave directionality, an important design parameter, can be included in wave representations if the random variable of the process is conceived as the sum of the corresponding elevations of simple harmonic waves propagating in various directions. In the time domain the latter can be expressed in a coordinate system (x,y) on the sea surface, where x forms an angle θ with the wave direction as follows:

$$\eta(x,y,t) = \mathrm{a} \cos\left[\frac{\omega^2}{g}(x \cos\theta + y \sin\theta) - \omega t + \varepsilon\right] \quad (9A.23)$$

where α is the wave amplitude and ε is the phase at $(x,y,t) = (0,0,0)$. The summation of many such waves gives an elevation that can be written as

$$\eta(x,y,t) = \sum_i \alpha_i \cos(x \cos\theta_i + y \sin\theta_i) - \omega_i t + \varepsilon_i] \quad (9A.24)$$

where $i \to \infty$ and $0 \leq \theta_i \leq 2\pi$, $0 \leq \omega_i < \infty$. The phase angle ε_i is regarded as uniformly distributed in the range $-\pi \leq \varepsilon_i \leq \pi$. Amplitude α_i is also regarded as a random variable within $0 \leq \alpha_i < \infty$.

The total wave energy in a frequency × direction window,

$$[\omega - \Delta\omega, \omega + \Delta\omega] \times [\theta - \Delta\theta, \theta + \Delta\theta]$$

is expressed by extending the corresponding one-dimensional result as

$$W = \sum_{\Delta w}\sum_{\Delta \theta} \tfrac{1}{2}\, \alpha_i^2 \qquad (9A.25)$$

The two-dimensional spectral density function can then be defined by

$$S(\omega,\theta) = \frac{W}{\Delta\omega\ \Delta\theta} \qquad (9A.26)$$

for, $\Delta\omega$, $\Delta\theta \rightarrow 0$ for any $0 \leq \omega < \infty$ and $0 \leq \theta \leq 2\pi$. Using $S(\omega,\theta)$, Pierson (1955) expressed the profile $\eta(x,y,t)$ of a random sea by the following stochastic integral:

$$\eta(x,y,t) = \int_{-\pi}^{\pi}\int_{0}^{\infty} \cos\left[\frac{\omega^2}{g}(x \cos\theta + y \sin\theta) - \omega t + \varepsilon(\omega,\theta)\right] \sqrt{2S(\omega,\theta)\ d\omega\ d\theta} \qquad (9A.27)$$

The non-Riemann integral above can be transformed into a double series with appropriate limit values of the variables ω and θ.

It is obvious that a complete description of any sea state would be insurmountably involved, if not impossible. Therefore, only the main characteristics of the waves associated with their probability of occurrence are normally available for application. The most promising of these results are presented in the following sections.

9A.4 SHORT-TERM ESTIMATES FOR DEEP WATER

It is helpful to differentiate the processes of narrowband from those of wideband spectra. The former display simpler expressions of density functions of such wave estimates as wave height or wave period. A spectrum is said to be narrow when it is derived by a process of zero mean, which can be expressed in the form

$$\eta(t) = \alpha(t) \cos[\omega_0 t + \varepsilon(t)] \qquad (9A.28)$$

with $\omega_0 = \text{const}$. In a narrowband spectrum the broadness parameter ε tends to zero, where

$$\varepsilon = \left(1 - \frac{m_2^2}{m_0 m_4}\right)^{1/2} \qquad (9A.29)$$

and the spectral moments are given by

$$m_i = \int_0^{\infty} \omega^i\ S(\omega)\ d\omega \qquad (9A.30)$$

9A.4.1 Wave Heights: Narrowband Process

By assuming that the amplitudes $\alpha(t)$ of the process above are normally distributed and the maxima of the process are statistically independent, assumptions with a satisfactory approximation in deep water, and taking the marginal density of $\alpha(t)$ from the joint probability density function of amplitudes and phases, one arrives at the well-known Rayleigh distribution of wave heights H:

$$f(H) = \frac{H}{4\sigma^2} \exp\left(\frac{H^2}{8\sigma^2}\right) \qquad (9A.31)$$

where σ^2 is the variance of the amplitude of the underlying process, or equivalently, the zeroth moment of the corresponding spectrum, and H has been assumed to be double the amplitude α.

The cumulative probability, the mean, the variance, and the RMS value of the distribution of H above are, respectively:

$$F(H) = \Pr\{H' \leq H\} = 1 - \exp\left(\frac{-H^2}{8\sigma^2}\right) \qquad (9A.32)$$

$$E[H] = \sigma\sqrt{2\pi} \tag{9A.33}$$

$$\text{var}\,[H] = \left(1 - \frac{\pi}{4}\right) 8\sigma^2 \tag{9A.34}$$

$$H_{\text{rms}} = 2\sigma\sqrt{2}, \qquad \sigma = (m_0)^{1/2} \tag{9A.35}$$

From the Rayleigh distribution of wave heights, representative values can be derived, as, for example, the significant wave height in a sea state with amplitude variance $\sigma^2 = m_0$. By taking the probability

$$\Pr(H > H_*) = \tfrac{1}{3} \tag{9A.36}$$

it is found that $H = 1.485\sqrt{m_0}$ and

$$H_{1/3} = 2\{\sqrt{\ln 3} + 3\sqrt{\pi}\,[1 - \Phi(\sqrt{2\ln 3})]\}\sqrt{2m_0} \tag{9A.37}$$

where σ is the error function. The expression (9A.37) can be reduced to

$$H_S = 4.01\sqrt{m_0} = 4.01\sigma \tag{9A.38}$$

Generally, any representative wave height $H_{1/N}$ of the sea state can be expressed in a similar manner as

$$H_{1/N} = 2\{\sqrt{\ln N} + N\sqrt{\pi}\,[1 - \Phi(\sqrt{2\ln N})]\}\sqrt{2m_0} \tag{9A.39}$$

9A.4.2 Wave Heights: Wideband Process

Using results of Rice with respect to estimates of the number of points of given excursion from the mean as well as to the number of zero up-crossing points, the broadness of the spectrum derived by the process can be incorporated. Estimates of the two numbers of points mentioned previously are assumed statistically independent, and the probability density arrived at is expressed in terms of double amplitude H:

$$f(H) = \frac{2}{1 + 2E\sigma}\left\{\frac{\varepsilon}{\sqrt{2\pi}}\exp\left(\frac{-H^2}{8\varepsilon^2\sigma^2}\right) + E\frac{H}{2\sigma}\exp\left(-\frac{H^2}{8\sigma^2}\right) \times \left[1 - \Phi\left(-\frac{E}{\varepsilon}\frac{H}{2\sigma}\right)\right]\right\} \tag{9A.40}$$

where $E^2 = 1 - \varepsilon^2$ $(E > 0)$ and

$$\Phi(u) = \frac{1}{\sqrt{2\pi}}\int_{-\infty}^{u}\exp\left(-\frac{u^2}{2}\right)du$$

is the error function.

Expression (9A.40) shrinks to Rayleigh distribution for $\varepsilon = 0$ (i.e., for narrow spectrum), while for the other extreme value of ε, $= \varepsilon'$, a density function is found applicable to very wideband processes:

$$f(H) = \frac{1}{2\sigma}\sqrt{\frac{2}{\pi}}\exp\left(-\frac{H^2}{8\sigma^2}\right) \tag{9A.41}$$

In previous expressions the assumption $H = 2\alpha$ has been made despite the broadness of the process.

9A.4.3 Wave Periods: Narrowband Process

It is helpful for a mathematical analysis of this kind to define the wave period as the time between successive maxima in a surface elevation time history. Following such a definition for T, Longuet-Higgins found in 1975 the following expression for its density in a narrowband process:

$$f(T) = \frac{N}{1 + N}\frac{(\nu\overline{T})^2}{[(\nu\overline{T})^2 + (T - \overline{T})^2]^{3/2}} \tag{9A.42}$$

where $N^2 = 1 - \nu^2$ ($N > 0$); ν is an alternative to the ε measure of the broadness of the process, $\nu = [(m_0 m_2/m_1^2 - 1]^{1/2}$, which for narrowband processes is $\nu << 1$; and $\overline{T}$ is the mean value of T; $\overline{T} = 2\pi m_0/m_1$.

9A.4.4 Wave Periods: Wideband Process

In processes with broader boundwidths, similar to those in real-life situations, the wave period can be connected to the number of crossing points with a plane at any level y from the mean. This number of crossing points with a positive plane per unit time can be expressed in a general way as

$$\overline{N}_{y+} = \frac{1}{2\pi}\left(\frac{m_2}{m_0}\right)^{1/2} \exp\left(-\frac{y^2}{2m_0}\right) \quad (9A.43)$$

which gives the mean zero up-crossing period:

$$\overline{T}_{0+} = 2\pi \left(\frac{m_0}{m_2}\right)^{1/2} \quad (9A.44)$$

Longuet-Higgins (1983) produced the following density function of the wave period between two successive maxima:

$$f(\tau) = \frac{L\nu^2}{2\tau^2[\nu^2 + (1 - 1/\tau)^2]^{3/2}} \quad (9A.45)$$

where $1/L = \frac{1}{2}(1 + 1/\sqrt{1 + \nu^2})$ and τ is the nondimensionalized period with respect to the mean (i.e., $\tau = T/\overline{T}$, where $\overline{T}$ is as in the corresponding narrowband case).

Cavanié et al. (1976) used results referring to the expected number of positive maxima per unit time $\overline{N}$ due to Rice (1944):

$$\overline{N}_+ = \frac{1}{4\pi}\frac{1 + E}{E}\left(\frac{m_2}{m_0}\right)^{1/2} = \frac{1}{\overline{T}} \quad (9A.46)$$

where $\overline{T}$ is the expected time between two successive positive maxima. The density function they produced for the wave period T defined as previously reads

$$f(T) = \frac{\alpha^3\beta^2 T}{T^2\{[(T/\overline{T})^2 - \alpha^2]^2\alpha^4\beta^2\}^{3/2}} \quad (9A.47)$$

where $\alpha = \frac{1}{2}(1 + E)$ and $\beta = \varepsilon/E$. For a narrowband process (i.e., for $E = 1$), it is noted that $\overline{T}$ coincides with the corresponding value of Longuet-Higgins's expression as well as with the zero up-crossing period $\overline{T}_{0+}$ referred to previously.

9A.4.5 Wave Heights and Periods: Narrowband Process

Under the same assumptions as previously (i.e., for a Gaussian process with statistically independent maxima), Longuet-Higgins (1975) provided an expression for the joint density between wave heights and wave periods. Denoting by $\overline{T}$ the expected period between two successive maxima in the section dealing with the periods in a narrowband process, the joint density reads

$$f(\xi,\eta) = \frac{\xi^2}{\sqrt{2\pi}} \exp\left[-\frac{\xi^2(1 + \eta^2)}{2}\right] \quad (9A.48)$$

where

$$\xi = \frac{H}{2\sigma}$$

$$\eta = \frac{(T - \overline{T})}{\kappa \overline{T}}$$

$$\kappa = \frac{(\mu_2/\mu_0)^{1/2}}{\overline{\omega}}$$

$$\overline{\omega} = \left(\frac{m_2}{m_0}\right)^{1/2}$$

$$\mu_i = \int (\omega - \overline{\omega})^i S(\omega)\, d\omega \qquad \text{central}$$

moments of the spectrum

It is noted that the relation above is symmetric about the mean angular frequency, a fact not verified in nature, especially for the smaller waves.

9A.4.6 Wave Heights and Periods: Wideband Process

The corresponding results to the expressions quoted for the density of the wave period in a process of broad bandwidth are given below.

- Longuet-Higgins (1983):

$$f(r, \tau) = \frac{2}{\sqrt{\pi}} \nu \frac{r^2}{\tau^2} \exp\left\{-r^2\left[1 + \frac{(1 - 1/\tau)^2}{\nu^2}\right]\right\} L \tag{9A.49}$$

where $r = \rho/(2m_0)^{1/2}$ (ρ is the amplitude).

- Cavanié et al. (1976):

$$f(h\ \tau) = \frac{1}{4\sqrt{2\pi}} = \frac{\alpha^3}{\varepsilon E^2} \frac{h^2}{\tau^5 \mathrm{T}^{-4}} \exp\left\{-\frac{h^2}{8\varepsilon^2 \tau^4 \mathrm{T}^{-4}} \times [\tau^2 \mathrm{T}^{-2} - \alpha^2)^2 + \alpha^5\ \beta^2]\right\} \tag{9A.50}$$

where

$$\overline{T} = 4\pi \frac{E}{1 + E}\left(\frac{m_0}{m_2}\right)^{1/2},$$

has been nondimensionalized with respect to the corresponding mean value $h = 2\rho/(m_0)^{1/2}$.

Both joint densities above describe adequately the stochastic properties of sea states of average broadness, their main disadvantage being that they refer to the wave period between successive maxima. Thus the values of ε (or v) and T should be deduced from the record rather than by spectral analysis. For example, ε can be estimated by

$$\varepsilon = \left[1 - \left(\frac{\overline{N}_0}{\overline{N}_+}\right)^2\right]^{1/2} \tag{9A.51}$$

where $\overline{N}_0$ is the mean (zero-crossing) waves per unit of time and $\overline{N}_+$ is the mean positive maxima per unit of time.

Recent research has developed a model that under the usual assumptions of Gaussianity and ergodicity provide graphs for the joint probability density function of wave heights and periods, where the variables are defined in the common engineering way of up-crossings. Moreover, this model holds for any bandwidth, not just for small or average values of the ε parameter. Also, deepwater wave breaking can be taken into account as well as the exclusion of swell from wind waves. The required input consists of a pair of the mean (or other characteristic) values of wave height and wave period of the sea state under consideration, or equivalently of the correlation:

$$r(H,T) = \frac{1}{\sigma_H \sigma_T N} \sum_N (H_i - H_m)(T_i - T_m) \tag{9A.52}$$

where σ is the standard deviation, and the subscript m denotes mean value. A typical result of the joint density in dimensionless form is given in Figure 9A.1 for a value of $r(H, T) = 0.50$.

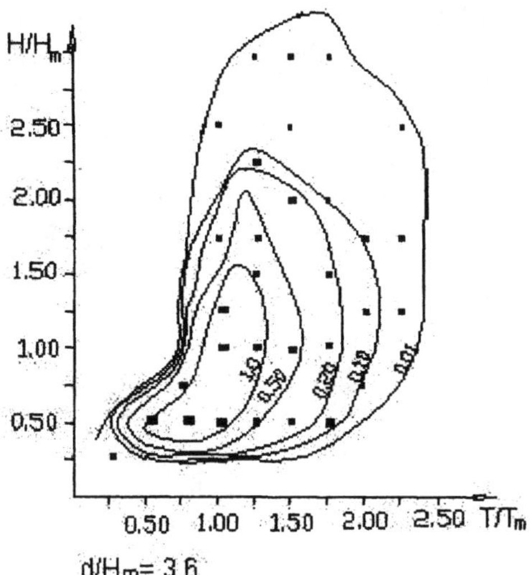

Figure 9A.1 Joint density of wave heights and periods in deep water. (From Memos and Tzanis, 2000.)

9A.5 SHORT-TERM ESTIMATES FOR SHALLOW WATER

Shallow water introduces complicated nonlinear processes in wave propagation, and the Gaussianity assumption on surface deviations no longer holds. Thus no analytical expressions for the density of wave heights or periods exist, in contrast with deep water conditions. The numerical model for the joint density of (H, T) referred to at the end of Section 9A.4 has received further treatment to cope with shallow-water conditions of negligible refraction effects. The phenomena of shoaling, depth-induced wave breaking, and wave re-forming after breaking have been included in a linear manner. Stokes third-order and cnoidal wave theories were used to describe the shoaling effect in shallow water. Required input parameters, in addition to those for deep waters, include the local slope and water depth at the point of interest. Typical results for two water depths at a bed slope of 5% are given in Figure 9A.2 for the deepwater waves corresponding to the graph of Figure 9A.1. Densities of wave heights or of periods can be obtained through Figures 9A.1 and 9A.2 by summing along the T or H axis, respectively.

9A.6 LONG-TERM PREDICTIONS

In Sections 9A.3 through 9A.5 a rough presentation has been made of the short-term representation of a seastate in deep or shallow water. Design of port structures often requires estimates of extreme events in the lifetime of the structure under study. Thus, long-term predictions of critical wave parameters, such as the wave height, are of paramount importance in engineering applications. These predictions should be based on local data spanning as long a time period as possible. The data are usually collected in 10- to 20-minute-long records typically every 3, 6, or 12 hours.

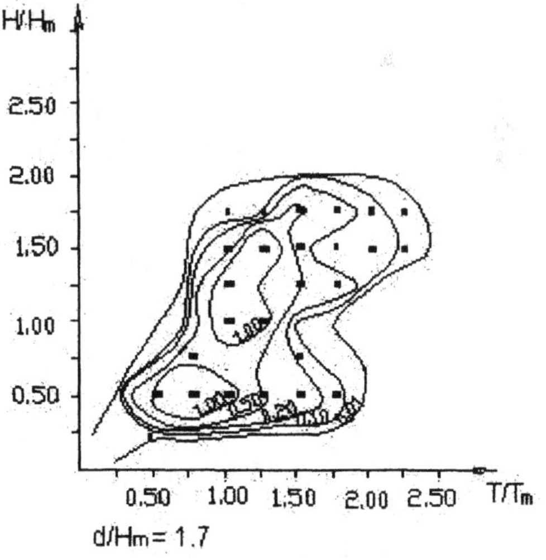

Figure 9A.2 Joint density of wave heights and periods in shallow water. H_m, deepwater mean wave height. (From Memos, 2002.)

It has been found that the significant wave height of storms follows the Weibull distribution quite closely. The two-parameter Weibull distribution can be written

$$f(x) = c\lambda^c x^{c-1} \exp[-(\lambda x)^c] \qquad 0 \le x < \infty \tag{9A.53}$$

where the parameters c and λ are to be determined by the measured data available. The value of c is usually close to unity; thus the one-parameter Weibull density is written

$$f(x) = \lambda \exp(-\lambda x) \tag{9A.54}$$

The density yielded by the Weibull distribution for small wave heights overestimates the corresponding value based on real data, while it compares favorably with the latter for high waves. The other main candidate to fit long-term wave height data is the log-normal distribution:

$$f(x) = \frac{1}{2\pi\sigma x} \exp\left[-\frac{(\ln x - \mu)^2}{2\sigma^2}\right] \qquad 0 \le x < \infty \tag{9A.55}$$

where μ and σ are parameters to be determined by the collected data.

The behavior of the log-normal distribution tends to be in the direction opposite that of the Weibull distribution: The densities at low wave heights are found to match the data well, whereas for large wave heights the log-normal densities overestimate measurements. It is not possible to say which of the two distributions better fits real-life data in general, since the latter depend on the geographical position of the study area, water depth, frequency of occurrence of extreme events, and many more physical parameters. However, the log-normal density function has an advantage that is sometimes helpful. This is that it can easily be combined with the corresponding density for the wave periods to provide the joint density of long-term wave heights and periods. Indeed, the distribution of wave periods can be described adequately only by a log-normal density. Hence, the joint density of, say, significant wave heights, also following a log-normal distribution, and zero-crossing periods can be expressed as follows, using a known property of normal densities:

$$\begin{aligned} f(H_s,T_z) &= \frac{1}{\sigma_H \sigma_T 2\pi\sqrt{1-\rho^2} H_s T_z} \\ &\quad \times \exp\left\{-\frac{1}{2(1-\rho^2)}\left[\left(\frac{\ln H_s - \mu_H}{\sigma_H}\right)^2 \right.\right. \\ &\quad - 2\rho \frac{\ln H_s - \mu_H}{\sigma_H}\frac{\ln T_z - \mu_T}{\sigma_T} \\ &\quad \left.\left. + \left(\frac{\ln T_z - \mu_T}{\sigma_T}\right)^2\right]\right\} \end{aligned} \tag{9A.56}$$

where μ and σ are the mean and standard deviation, respectively, and ρ is the correlation coefficient between H_s and T_z, and the subscripts refer to the variables.

There is an underlying link between the short- and long-term probabilistic estimates, f_S and f_L, respectively, of the main wave characteristics. This can be expressed formally by the following relationship for, say, the wave height:

$$f_L(H) = \frac{\sum_i \sum_j n_{ij} p_p p_j f_S(H)}{\sum_i \sum_j n_{ij} p_i p_j} \tag{9A.57}$$

where p_i is the frequency of occurrence of a sea state of given severity (denoted, e.g., by its significant wave height), p_j the frequency of occurrence of a given spectral shape (denoted, e.g., by its peak-shape parameter γ in JONSWAP (Joint North Sea Wave Project) spectra, for the same sea severity, i.e., for the same significant wave height), and n_{ij} is the number of waves in each sea state.

9A.7 WAVE SPECTRA

Wave spectra are a powerful alternative to the time-domain description of a sea state. They can also serve as input to spectral wave propagation models covering extensive sea areas. The characteristics of the observed one-dimensional power spectra depend mainly on the relevant geographical location, corresponding fetch, wind intensity and duration, presence of swell, and other parameters. Due to the scarcity of measured data for each location, researchers have developed typical spectral shapes by relating them with physical parameters. The main such parameters that are used include the wind friction velocity

$$u_* = \left(\frac{\tau_0}{\rho_\alpha}\right)^{1/2} \qquad (9A.58)$$

where $\tau\alpha_0$ is the tangential wind friction stress and ρ_α is the air density. The fetch is x, the frequency is ω, and the acceleration due to gravity is g. Thus, in general,

$$S(\omega) = F(\omega, g, u_*, x) \qquad (9A.59)$$

Typical spectra are related to either unlimited or limited fetch conditions. The main representatives of these two classes are the Pierson–Moskowitz and the Bretschneider spectra for the former conditions and the JONSWAP for the latter. Next, these three typical spectra are presented briefly for reference purposes.

9A.7.1 Pierson–Moskowitz Spectrum

This one-parameter spectrum was developed on data collected in the North Atlantic under unlimited fetch conditions and fully developed sea and wind speeds ranging from 20 to 40 knots. The proposed spectral shape is (Pierson and Moskowitz, 1964)

$$S(\omega) = \alpha\frac{g^2}{\omega^5}\exp\left[-0.74\left(\frac{g}{u\omega}\right)^4\right] \qquad (9A.60)$$

where α is a scale parameter equal to 0.0081 and u is the wind speed 19.5 m above the sea surface.

For narrowband spectra it was found that

$$H_s = \frac{0.21u^2}{g} \qquad (9A.61)$$

and the modal frequency is given by

$$\omega_m = \frac{0.87g}{u} \qquad (9A.62)$$

9A.7.2 Bretschneider Spectrum

This two-parameter spectrum is applicable both to fully and partially developed seas. It reads (Bretschneider, 1968)

$$S(\omega) = 0.313\left(\frac{\omega_m^4}{\omega^5}\right)H_s^2\exp\left[-1.25\left(\frac{\omega_m}{\omega}\right)^4\right] \qquad (9A.63)$$

This expression is of the general form

$$S(\omega) = \frac{A}{\omega^5}\exp\left(-\frac{B}{\omega^4}\right) \qquad (9A.64)$$

which is also applicable to the Pierson–Moskowitz spectrum. For such spectra the following values of moments are calculated: $m_0 = A/4B$, $m_1 = 0.306A/B^{3/4}$, $m_2 = \sqrt{\pi}A/4\sqrt{B}$, and $m_4 = \infty$. These values yield a very broad process in that they imply broadness parameters $\varepsilon = 1$ and $\nu = 0.43$. Despite that, the assumption is usually made of a narrowband spectrum in order to obtain values of the wave characteristics. Under this assumption the following results are found: $H_s = (4A/B)^{1/2}$, $\overline{H} = (\pi A/2B)^{1/2}$, $\omega_m = (4B/5)^{1/4}$, and $\overline{\omega} = m_1/m_0 = 1.23B^{1/4}$.

9A.7.3 JONSWAP Spectrum

In the widely known Joint North Sea Wave Project (JONSWAP), extensive measurements were performed which supported a spectral shape for fetch-limited conditions. This spectrum has the following expression (Hasselmann et al., 1973):

$$S(f) = \alpha \frac{g^2}{(2\pi)^4} \frac{1}{f^5} \exp\left[-1.25\left(\frac{f_m}{f}\right)^4\right] \times \gamma^{\exp[-(f-f_m)^2/2\sigma^2 f_m^2]} \qquad (9A.65)$$

where for the mean spectrum it was found that

$$\alpha = 0.076(\overline{x})^{-0.22}, \quad \overline{x} = \frac{gx}{u^2}, \quad (x \text{ is the fetch})$$

$$\sigma = \begin{cases} 0.07 & \text{for} \quad f \leq f_m \\ 0.09 & \text{for} \quad f > f_m \end{cases}$$

$$f_m = 3.5\left(\frac{g}{u}\right)(\overline{x})^{-0.33} \qquad \text{modal frequency}$$

where $\gamma = 3.3$ is the peak shape (or spectral peakedness) parameter, f the circular frequency, and u the wind speed 10 m above the sea surface.

Mitsuyasu showed that for single peaked spectra α and γ are both dependent on the dimensionless modal frequency $\overline{f_m} = uf_m/g$ as follows:

$$\alpha = 3.26 \times 10^{-2}(\overline{f_m})^{0.857} \qquad \gamma = 4.42(\overline{f_m})^{0.429}$$

He also proposed a slightly different relation for f_m, leading to

$$\overline{f_m} = 2.92(\overline{x})^{-1/3}$$

In the JONSWAP spectrum it is possible to replace the wind speed u by the significant wave height through the relation $u = Kx^{-0.615}H_s^{1.08}$ (u in m/s, x in km, H_s in m), where K is a factor taking the following values depending on γ:

γ	1.75	2.64	3.30	3.96	4.85
K	96.2	88.3	83.7	80.1	76.4

Parameter γ can formally be defined as the ratio of the spectral density at its peak in the JONSWAP spectrum to the corresponding ratio in the Pierson–Moskowitz spectrum for the same values of α and f_m. It is noted that higher γ values relate to narrower spectra and lower γ values to broader spectra. Values of γ in the range 7 to 10 are associated with the only-swell content of the spectrum.

Wave energy does not travel in a single direction, but it displays a directional spread associated with the intensity of the wave generating wind field. Wave energy spead, an important factor in engineering applications, is not taken into account by the spectra presented previously. To include information on wave directionality, a directional spectrum $S(f,\theta)$ is defined, θ the angle around the principal direction of propagation, as follows:

$$S(f,\theta) = S(f)G(f,\theta) \qquad (9A.66)$$

where $G(f,\theta)$ is a directional spread function dependent on both frequency and direction of the spectrum components.

Various expressions for the function G have been proposed over the years. One of the most complete, yet easy to use description of G due to Longuet-Higgins (1961) as modified by Goda and Suzuki (1975) is the following:

$$G(f\theta) = G_0 \cos^{2s}\frac{\theta}{2} \qquad (9A.67)$$

where

$$s = \begin{cases} s_{max}\left(\dfrac{f}{f_p}\right)^5 & f \leq f_p \\ s_{max}\left(\dfrac{f}{f_p}\right)^{-2.5} & f \geq f_p \end{cases}$$

and s_{max} is a factor depending on the concentration of the wave energy around the principal direction of wave propagation. Thus s_{max} should increase with increasing distance from the generating area of the wind waves. The following two extreme values of s_{max} have been proposed: $s_{max} = 10$ for wind waves in the generating area, $s_{max} = 75$ for swell well away of that area. In-between values of s_{max} can, of course, be adopted.

9A.8 DESIGN WAVE CHARACTERISTICS

In previous sections a brief account was given of the ways that a wave field can be represented along with those results related to engineering design. Below, some hints are given as to how to exploit these results in practical applications. One point to remember is that parametric wave forecasting methods predict the wave characteristics at a location in the open sea, which is usually offshore the study area. Therefore, these waves should be modified accordingly to accommodate the relevant processes, such as refraction, diffraction, reflection by the inclined seafloor, wave breaking, and so on. In cases of important constructions or lack of wind data, it is advisable to conduct wave measurements by wave riders, pressure sensors, or other equipment. The records collected are analyzed by a suitable method (e.g., zero up-crossing).

A very important parameter associated with the wave field is its directionality. This can be deduced from measurements by special wave riders, arrays of common wave riders, radars, or derives of infrared radiation. Directionality information can be fed in the one-dimensional spectrum to yield the directional spectrum of the wave field. From this information, coupled with a density function of the wave height, usually a Rayleigh distribution, the design parameters required can be deduced as described earlier. Summarizing:

- The significant wave height can be calculated from the zeroth-order spectral moment, $H_s = 4\sqrt{m_0}$. Various other characteristic wave heights can be calculated from the spectrum. For example, $H_{1/10} = 5.1(m_0)^{1/2}$ and $H_{max} \approx 7.5(m_0)^{1/2}$. The latter value corresponds to a 1% probability of exceedance in narrowband spectra, while in applications it is advisable to take $H_{max} = 2H_s$. Other values of wave heights based on the Rayleigh distribution are

$$\overline{H} = 0.64\, H_s \qquad \text{mean wave height}$$

$$H_{1/10} = 1.29\, H_s$$

$$H_{1/20} = 1.37\, H_s$$

$$H_{1/100} = 1.68\, H_s$$

- The average (zero-crossing) period is expressed through $T_z = (m_0/m_2)^{1/2}$, while the period T_p at the peak of the spectrum depends on the form of the latter. In the JONSWAP spectrum with $\gamma = 3.3$, it is $T_p = 1.286T_z$.
- The information above cannot in general provide long-term joint distribution of wave heights and periods. When the latter is required, it is proposed that several period values be taken in the calculations. Det Norske Veritas suggests the following range of T_p: $3.6\sqrt{H_s} < T_p < 5.5\sqrt{H_s}$ (T_p in seconds, H_s in meters). These values correspond to wave steepness from $\frac{1}{24}$ to $\frac{1}{10}$ in a Pierson–Moskowitz spectrum. The wave steepness of the maximum wave can be taken approximately as $\gamma_{max} = 1.5\,\gamma_s$, where γ_s is the steepness of the significant wave.

- The maximum deviation ε_{max} of the sea surface from the mean is quite helpful for determination of the crown level of breakwaters and other port structures. This can be estimated in deep water by assuming a Rayleigh distribution of wave heights as

$$\xi_{max} = H_s\left(\log \frac{N_z}{8}\right)^{1/2} \quad (9A.68)$$

where N_z is the number of (zero-crossing) waves, which can be calculated on the basis of the time length of the design storm and the corresponding average period. For quite long storms, ξ_{max} can be taken as $0.9H_s$.

- Knowledge of the short-term joint density of wave heights and periods (cf. Sections 9A.6 and 9A.7) can be exploited advantageously in a number of design issues. These include calculations pertaining to the primary armor units in a rubblc-mound breakwater, the stability of toe protection, wave transmission through the mound, loading of the crown structure, the dynamic equilibrium of nonarmored breakwaters, the displacement of armor units in berm breakwaters, overtopping estimates, and so on. The design aspects above require information in the form of values of the pair (H,T) for various probabilities of occurrence so that a full risk analysis be performed. Also, situations that evolve in time (e.g., armor units displacement, overtopping) require the probabilistic structure of the loading parameters to allow for the estimation of reliable results (see, e.g., Section 9A.4).

9A.9 RETURN PERIOD

A prerequisite to designing construction projects is determination of the design life of the relevant structures. The conditions applicable to each case are considered, in particular the physical factors and processes bearing on the project's efficiency and durability. Such factors could be fatigue of the project elements due to cyclic loading, feasibility for repair, corrosion of metal components, and so on. Table 9A.1 lists indicative minimum design life values.

Following the determination of the structure's design life, a risk analysis in case of failure has to be conducted. It is obvious that structures are not designed to face adequately every external load of any magnitude whatsoever. In the majority of cases that would be impossible, and in the remainder it would be uneconomical. Therefore, adopting an optimizing method, the probability of failure p of the structure which produces the smallest total cost is determined. To enable the drawing of conclusions, various factors are introduced into this analysis, of a social, economic, and technical nature, which in many cases are difficult to quantify. For instance, a much smaller value of p should be used in cases of rigid quay wall construction than would be used for riprap projects, where repairs are routinely undertaken. Moreover, the degree of hazard to human life in important installations sets the probability p of failure to very low levels. The designer's contribution in such cases plays a decisive role in determination of the optimum probability of failure of the structure. It is the designer who

Table 9A.1 Minimum design life of port structures

	Lifetime (years)
Quaywalls	60
Open piers (open piled structures)	45
Superstructures	30
Dry docks	45
Breakwaters	60
Shore protection works	60
Flood controls	100

Source: British Standards Institution (1988).

has a feeling for the overall safety that results from consideration of the constituent safety factors involved, of the weaknesses of the methods of calculation, and so on.

Having determined the probability p *of* exeedance of the design criteria and also the design life T of the structure, we are in a position to calculate the return period R of those conditions for which an exeedance of the criteria occurs in accordance with the relationship

$$p = 1 - e^{-T/R} \qquad (9A.69)$$

It should be noted that probability p refers to the design life T and includes all the events that exceed a given value. It is evident that for customary values of p the return period of the natural phenomenon should be significantly longer than the project's design life. For instance, if we assume that $R = T$, the probability of exceedance of the design conditions reaches a high 63%.

Inevitably, therefore, we resort to seeking extreme situations of natural phenomena for which usually no measurements of such duration are available. To this effect, methods of predicting values of extreme events in extended periods of time are adopted. These methods, however, should be applied very cautiously, by adhering to pertinent rules of statistical reliability. For instance, the time expansion above a certain limit associated with the values of a sample of given duration is not allowed. Usually, the Weibull distribution is applied for wave heights, the Fisher–Tippet II distribution for wind speed, and so on.

If the condition of exceedance of the design criteria requires the simultaneous occurrence of $\nu > 1$ extreme events, p is determined after evaluating the statistical correlation between the ν events and the probability of their simultaneous occurrence. For independent events it is $p = p_1 \times p_2 \times \ldots \times p_\nu$, and for totally dependent ones it is $p = (p_1 + p_2 + p_\nu)/\nu$. An example of the combination of extreme events may be the simultaneous occurrence of a maximum sea level and maximum wave height.

The relationship (9A.69) is also related to a risk level or exceedance probability of extreme events calculated in the following manner. We assume storms to be statistically independent phenomena with a Poisson distribution of the time intervals between successive events. Then the probability of n exceedances within a return period for which on average we have one exceedance ($m = 1$) is determined by

$$\frac{m^n}{n!} e^{-m} = \frac{1}{en!} \qquad (9A.70)$$

whereupon the probability is

• For no exceedance ($n = 0$)	0.368
• For one exceedance ($n = 1$)	0.368
• For two exceedances ($n = 2$)	0.184
• For three exceedances ($n = 3$)	0.061
• For over three exceedances	0.019
Total	1.000

We thus see that for $n \geq 1$ the probability is 0.632; in other words, we arrive at the same result as that noted previously. For return period R, different from the design life $T(m = T/R)$, we may calculate the probabilities for $n \geq 1$, or risk level, using the relationship (9.70). We thus have

m	0.05	0.1	0.5	1	2	5
Probability ($n \geq 1$)	0.049	0.095	0.393	0.632	0.865	0.993

REFERENCES

Allsop, N. W. H., 1983. Low-crest breakwater: studies in random waves, *ASCE Proc. Conference on Coastal Structures '83.*

———, 1990. Reflection performance of rock armoured slopes in random waves, *Proc. 22nd*

Coastal Engineering Conference, Delft, The Netherlands.

Allsop, N. W. H., and R. N. Bray, 1994. Vertical breakwaters in the UK: historical and recent experience, *Proc. International Workshop on Wave Barriers in Deepwaters*, Yokosuka, Japan.

Bagnold, R. A., 1939. Interim report on wave pressure research, *J. Inst. Civil Eng.*, Vol. 12.

Baird, W. F., and K. R. Hall, 1984. The design of breakwaters using quarried stone, *Proc. 19th International Conference on Coastal Engineering*, Houston, TX.

Baird, W. F., and J. Woodrow, 1987. The development of a design for a breakwater at Keflavik, Iceland, *Proc. Seminar on Berm Breakwaters: Unconventional Rubble-Mound Breakwater*. Ottawa, Ontario, Canada.

Benassai, E., F. Grimaldi, and A. Scotti, 1974. Interazione fro moto ondóso e la diga a paréte verticale fondata su pali del nuovo Pòrto di Manfredonia, *Atti XIV Convegno di Idraulica e Costruzióne Idràuliche*, Naples, Italy.

Bjordal, S., 2003. Laukvik fishing harbour rehabilitation of the breakwater-model tests. *PIANC Bull. 113*. International Navigation Assocation, Brussels, Belgium.

Blomberg, R., and P. Green, 1988. Berm walls tame violent seas, *Eng. News Record*, June 23.

Bray, R. N., and P. F. B. Tatham, 1992. *Old Waterfront Walls: Management, Maintenance and Rehabilitation,* CIRIA/E&FN Spon, London.

Brebner, A., and D. Donnelly, 1962. Laboratory study of rubble foundations for vertical breakwaters, *Proc. 8th International Conference on Coastal Engineering*, New Mexico City, Mexico, pp. 406–429.

Bretschneider, C. L., 1952. Revised wave forecasting relationships, *Proc. 2nd Conference on Coastal Engineering*, Council on Wave Research, Engineering Foundation, Berkeley, CA.

———, 1957. Review of *Practical Methods for Observing and Forecasting Ocean Waves by Means of Wave Spectra and Statistics*, U.S. Navy Hydrographic Office, Publ. 603, *Trans. Am. Geophys. Union*, Vol. 38, No. 2.

———, 1958. Revisions in wave forecasting: deep and shallow water, Proc. 6th Conference on Coastal Engineering Council on Wave Research, American Society of Civil Engineers, Reston, VA.

———, 1968. Significant waves and wave spectrum, *Ocean Ind.*

British Standards Institution, 1988. *Maritime Structures*, BS 6349, BSI, London.

Bruun, P., 1985. *Design and Construction of Mound for Breakwater and Coastal Protection,* Elsevier, Amsterdam.

———, 1990. *Port Engineering,* Gulf Publishing, Houston, TX.

Bruun, P. and Smith, O., 1990–1991. *Economic Optimization of Breakwaters in Port Engineering,* No. 824, Dock and Harbour Authority, London, Dec./Jan.

Burcharth, H. F., 1983. *Material, Structural Design of Armour Units: Seminaral Rubble Mound Breakwaters,* TRITA-VBI-120, Royal Institute of Technology, Stockholm, Sweden.

———, 1984. *Fatigue in Breakwater Armour Units,* Gulf Publishing House, Houston, TX.

———, 1989. *Uncertainty Related to Environmental Data and Estimated Extreme Event,* Rep. Subgroup B, Work. group 12, PTC 11, Permanent International Association of Navigation Congresses, Brussels, Belgium.

———, 1993. The design of breakwaters, in *Coastal and Harbor Engineering Reference Book*, M. B. Abbott, and W. A. Price, eds., E&FN Spon, London.

———, 1994. Reliability evaluation of a structure at sea, *Proc. International Workshop on Wave Barriers in Deep Waters*, Yokosuka, Japan.

Burcharth, H. F., and T. Brejnegaard-Nielsen, 1986. The influence of waist thickness of Dolosse on the hydraulic stability of Dolosse armour. *Proc. 20th ICCE*, Taipei, Taiwan.

Burcharth, H. F., and P. Frigaard, 1988. On 3-dimensional stability of reshaping breakwaters, *Proc. 21st International Conference on Coastal Engineering*, Malaga, Spain.

Burcharth, H. F., P. Frigaard, J. M. Berengher, and B. G. Madigal, 1995. Design of the Ciervana breakwaters, Bilbao, *Proc. Coastal Structures and Breakwaters '95*. Institution of Civil Engineers, London.

Cavanié, A., M. Arhan, and R. Ezraty, 1976. *A statistical relationship between individual heights and periods of storm waves, Proc. First International Conference on Behavior of Offshore Structures*, BOSS '76, Vol. 2, Norwegian Institute of Technology, Trondheim, Norway.

D'Angremond, K., and F. Van Roode, 2001. *Breakwaters and Closure Dams,* University Press, Delft, The Netherlands.

Dickson, A. F., 1969. Trends in vessel type and size: liquid cargoes, *Proc. Conference on Tanker and Bulk Carrier Terminals*, Institution of Civil Engineers, London.

Dzhunkovski, N. N., A. A. Kasparson, G. N. Smirnov, and A. G. Sidorova, 1964. *Ports and Port Structures,* Vol. 1, Stroyisdat, Moskow (in Russian).

Fournier, C. P., O. J. Sayao, and F. Caldas, 1990. Berm breakwater contamination study, Sergipe Marine Terminal, Brazil, *Proc. 22nd International Conference on Coastal Engineering*, Delft, The Netherlands.

Franco, L., and G. Verdesi, 1993. Ancient Mediterranean harbors: a heritage to preserve, presented at the Med-Coast Conference, Antalya, Turkey.

Goda, Y., 1972. Laboratory investigation of wave pressures exerted upon vertical and composite walls, *Rep. Port Harbour Res. Inst.,* Vol. 11, No. 2 (in Japanese); also, Experiments on the transition from nonbreaking to post breaking wave pressures, *Coast. Eng. Jpn.*, Vol. 15.

———, 1973a. Motion of composite breakwater on elastic foundation under the action of impulsive breaking wave pressure, *Rep. Port Harbour Res. Inst.*, Vol. 12, No. 3 (in Japanese); also, 1994, Dynamic response of upright breakwater, *J. Coast. Eng.*, Vol. 22.

———, 1973b. A new method of wave pressure calculation for the design of composite breakwater, *Rep. Port Harbour Res. Inst.*, Vol. 12, No. 3, pp. 31–70 (in Japanese); also, 1974, *Proc. 14th International Conference on Coastal Engineering*, Copenhagen, Denmark.

———, 1994. On the uncertainties of wave heights as the design load for marine structures, *Proc. International Workshop on Wave Barriers in Deepwaters*, Yokosuka, Japan.

Goda, Y., and S. Kakizaki, 1966. Study of finite amplitude standing waves and their pressure upon vertical wall, *Rep. Port Harbor Res. Inst.,* Vol. 5, No. 10 (in Japanese); also, 1967, Coast. Eng. Jpn., Vol. 10.

Goda, Y., and Y. Suzuki, 1975. *Computation on Refraction and Diffraction of Sea Waves with Mitsuyasu's Directional Spectrum*, Tech. Note 230, Port and Harbour Research Institute, Yokosuka, Japan (in Japanese).

Hales, L. Z., 1981. *Floating Breakwater: State-of-the-Art, Literature Review,* TR 81-1, U.S. Army Coastal Engineering Research Center, Fort Belvior, VA, Oct.

Hall, K. R., and J. S. Kao, 1991. The influence of armor stone gradation on dynamically stable breakwaters, *Coast. Eng.*, Vol. 15.

Hasselmann, K., T. P. Barnett, E. Bouws, H. Carlson, D. C. Cartwright, K. Enke, P. Muller, D. J. Olbers, K. Richter, W. Sell, and H. Walden, 1973. *Measurements of Wind-Wave Growth and Swell Decay during the Joint North Sea Wave Project (JONSWAP),* Deutsches Hydrographisches Institut, Hamburg, Germany.

Hasselmann, K., D. B. Ross, P. Muller, and W. Sell, 1976. Parametric wave prediction model, *J. Phys. Oceanogr.*, Vol. 6.

Hayashi, T., and T. Imai, 1964. Breaking wave pressure and sliding of caisson, *Proc. Japanese Conference on Coastal Engineering*, Vol. 11 (in Japanese).

Headland, J. R., 1995. Floating breakwaters, Chapter 5 in *Marine Structures Engineering: Specialized Applications*, G. Tsinker, ed., Chapman & Hall, New York.

Hedar, P. A., 1965. Rules for the design of rock-fill breakwaters and revetments presented at the 21st International Navigation Congress, Section 11-1, Stockholm, Sweden.

Herbich, J. B., 1992. Dredged navigation channels, in *Handbook of Coastal and Ocean Engineering*, Vol. 3, J. B. Herbich, ed., Gulf Publishing, Houston, TX.

Horikawa, K., Y. Ozawa, and K. Takahash, 1975. Expected sliding distance of high mound composite breakwater, *Proc. Japanese Conference on Coastal Engineering*, Vol. 19 (in Japanese).

Hudson, R. Y., 1958. *Design of Quarry Stone Cover Layers for Rubble Mound Breakwaters and Harbors,* Div. Res. Rep. 2.2, U.S. Army Corps of Engineers Waterways Experimental Station, Vicksburg, MS.

———, 1959. Laboratory investigation of rubble mound breakwater, *J. Waterways Harbors Div.*, Vol. 85, No. WW3, pp. 93–121.

Iribarren, C. R., 1938. *Una Fórmula para el Cálculo de Diques de Escollera,* Revista de Obras Pulicas, Madrid, Spain.

Irie, I., Y. Kuriyama, and H. Asakawa, 1986. Study on scour in front of breakwaters by standing waves and protection method, *Rep. Port Harbour Res. Inst.*, Vol. 25, No. 11 (in Japanese).

Ito, Y., M. Fujishima, and T. Kitatani, 1966. On the stability of breakwaters, *Rep. Port Harbour Res. Inst.*, Vol. 5, No. 14 (in Japanese); also, 1971, *Coast. Eng. Jpn.*, Vol. 14.

Jacobsen, A., J. E. Hagen, A. Andersen, C. Solberg, G. Viggoson, S. Vold, S. Bjordal, 2002. Berm breakwaters: an excellent wave breaker; some results from model tests, *PIANC Proc. 30th Congress*, Sydney, Australia.

Jarlan, G. E., 1961. A perforated vertical breakwater, *Dock Harbour Auth.,* Vol. 41, No. 488.

Juhl, J., 1994. Danish experience and recent research on vertical breakwaters, *Proc. International Workshop on Wave Barriers in Deepwaters*, Port and Harbor Research Institute, Yokosuka, Japan.

Juhl, J., and O. J. Jensen, 1993. Practical experience with berm breakwaters, draft worked out under the MAST I programme.

Kimura, K., S. Takahashi, and K. Tanimoto, 1994. Stability of rubble mound foundations of caisson breakwater under oblique wave attack, *Proc. 24th International Conference on Coastal Engineering*, Kobe, Japan.

Kobayashi, M., M. Terashi, and K. Takahashi, 1987. Bearing capacity of a rubble mound supporting a gravity structure, *Rep. Port Harbour Res. Inst.*, Vol. 26, No. 5.

Kobune, K., and M. Osato, 1976. A study of wave height distribution along a breakwater with a corner, *Rep. Port Harbour Res. Inst.*, Vol. 15, No. 2 (in Japanese).

Kuo, C. T., 1994. Recent researches and experiences on composite breakwaters in Taiwan, *Proc. International Workshop on Wave Barriers in Deepwaters*, Port and Harbor Research Institute, Yokosuka, Japan.

Lamberti, A., and L. Franco, 1994. Italian experience on upright breakwaters, *Proc. International Conference on Wave Barriers in Deepwaters*, Port and Harbor Research Institute, Yokosuka, Japan.

Larras, J., 1952. L'équilibre sous-marin d'un massif de materiaux soumis a la houle, *Gen. Civil.*

Ligteringen, H., 1994. Other European experience on deepwater breakwaters, *Proc. International Workshop on Wave Barriers in Deepwaters*, Port and Harbor Research Institute, Yokosuka, Japan.

Lochner, R., O. Faber, and W. G. Penny, 1948. The Bombardon floating breakwater, in *The Civil Engineer in War*, Vol. 2, *Docks and Harbours*, Institution of Civil Engineers, London.

Longuet-Higgins, M. S., 1952. On the statistical distribution of the heights of sea waves, *J. Mar. Res.*, Vol. 11, No. 3.

———, 1961. Observations of the directional spectrum of sea waves using the motions of a floating buoy, In *Ocean Wave Spectra*, Prentice Hall, Upper Saddle River, NJ.

———, 1975. On the joint distribution of the periods and amplitudes of sea waves, *J. Geophys. Res.*, Vol. 80, No. 18.

———, 1983. On the joint distribution of the periods and amplitudes in random wave field, *Proc. R. Soc. London, Ser. A,* No. 389.

Losada, M. A., and L. A. Gimenez-Curto, 1980. Mound breakwaters under wave attack, *International Proc. Seminar on Criteria for Design and Construction of Breakwaters and Coastal Structures*, Santander, Spain, Sec. II, pp. 127–238.

McBride, M. W., J. V. Smallman, and N. W. H. Allsop, 1994. Design of harbour entrances: breakwater design and vessel safety, *Proc. International Conference on Hydro-Technical Engineering for Port and Harbour Construction*, Vol. 1, Yokosuka, Japan.

Markle, D. G., and D. D. Davidson, 1979. Placed-stone stability tests, Tillamook, Oregon. U.S. Army Engineer Waterway Experiment Station, TR HL-79-16, Vickeburg, MS, April.

McCartney, B. L., 1985. Floating breakwater design, *ASCE J. Waterways Port Coast. Ocean Eng.,* Vol. 111, No. 2, Mar.

Memos, C. D., 2002. Stochastic description of sea waves, *J. Hydraul. Res.*, Vol. 40, No. 3.

Memos, C. D., and K. Tzanis, 2000. Joint distribution of wave heights and periods in waters of any depth, *ASCE J. Waterway Port Coast. Ocean Eng.*, Vol. 126, No. 3.

Miche, R., 1944. Mouvements ondulatoires de la mer in profondeur constante on décroissante, *Ann. Ponts Chaussees, Paris*, Vol. 114.

Minikin, R. R., 1963. *Winds, Waves and Maritime Structures.* Charles Griffin, London.

National Research Council, 1992. *Shiphandling Simulating: Application to Waterway Design*, National Academy Press, Washington, DC.

NATO, 1988. *Proc. Symposium on Advances in Berthing and Mooring*, Norwegian Institute of Technology, Trondheim, Norway. Kluwer Academic Publishers, Dordrecht, The Netherlands.

Neumann, G., 1953. *On Ocean Wave Spectra and a New Method of Forecasting Wind-Generated Sea*, Tech. Memo. 43, Beach Erosion Board, U.S. Army Corps of Engineers, Washington, DC.

Neumann, G., and W. J. Pierson, Jr., 1957. A detailed comparison of theoretical wave spectra and wave forecasting methods, *Dtsch. Hydrogr. Z.*, Vol. 10, No. 3, pp. 73–92.

Oliver, J. G., 1992. Floating breakwaters, *ASCE Proc. Specialty Conference Ports '92*, Seattle, WA.

Oumeraci, H., 1994. *Scour in front of vertical breakwaters: review of problems, Proc. International Workshop on Wave Barriers in Deep Waters*, Yokosuka, Japan.

Oumeraci, H., P. Klammer, and H. W. Partensky, 1993. Classification of breaking wave loads on vertical structures, *ASCE J. Waterway Port Coast. Ocean Eng.*, Vol. 119, No. 4, July/Aug.

PIANC, 1980. *Report on Large Ships*, Rep. Work. Group IV, Suppl. Bull. 35, Permanent International Association of Navigation Congresses, Brussels, Belgium.

———, 1992. *Analysis of Rubble Mound Breakwaters,* Suppl. Bull. 78/79, Permanent International Association of Navigation Congresses, Brussels, Belgium.

———, 1994. *Floating Breakwaters: A Practical Guide for Design and Construction,* Suppl. Bull. 85, Permanent International Association of Navigation Congresses, Brussels, Belgium.

———, 1997. *Review of Selected Standards for Floating Dock Designs,* Suppl. Bull. 93, Permanent International Association of Navigation Congresses, Brussels, Belgium.

———, 2001. *Guidelines for Design and Construction of Berm Breakwaters,* draft rep., Permanent International Association of Navigation Congresses, Brussels, Belgium, June.

———, 2003a. *State-of-the-Art of Designing and Construction Berm Breakwaters.* International Navigation Association, Brussels, Belgium.

———, 2003b. *Breakwaters with Vertical and Inclined Concrete Walls.* International Navigation Association, Brussels, Belgium.

Pierson, W. J., and L. Moskowitz, 1964. A proposed spectral form for fully developed wind tray based on the similarity law of S. A. Kitaigorodski. *J. Geophys. Res.*, Vol. 69, No. 24.

Price, A. W., 1979. *Static Stability of Rubble Mound Breakwaters,* Hydraulics Research Station, Wallingford, Berkshire, England.

Rice, S. O., 1944. Mathematical analysis of random noise, in *Noise and Stochastic Processes,* N. Wax, ed., Dover Publications, New York.

Rundgren, L., 1958. *Water Wave Forces,* Bull. 54, Royal Institute of Technology, Division of Hydraulics, Stockholm, Sweden.

Sainflou, M., 1928. Treatise on vertical breakwaters, *Ann. Ponts Chaussees, Paris* (translated from French by W. J. Yardoft, U.S. Army Corps of Engineers).

Sakamoto, Y., Y. Miyaji, T. Uenishi, and H. Takeda, 1984. *Experimental Study on Hydrodynamic Function of Rubble Mound Breakwaters,* Rep. 82, Civil Engineering Research Institute of Hokkaido Development Bureau (in Japanese).

Seelig, W., 1980. *Two-Dimensional Tests of Wave Transmission and Reflection Characteristics of Laboratory Breakwater,* TR 80-1, CERC, WES, U.S. Army Corps of Engineers Waterways Experimental Station, Vicksburg, MS.

Selling, W. N., 1983. Wave reflection from coastal structures, *Proc. Conference on Coastal Structures '83*, Arlington, VA.

Shimosako, K., and S. Takahashi, 1994. Determination of the sliding distance of composite breakwaters due to wave forces including breaking wave forces, *Proc. 24th International Conference on Coastal Engineering*, Kobe, Japan.

Sigurdarson, S., and G. Viggosson, 1994. Berm breakwater in Iceland: practical experience, *Proc. International Conference on Hydro-Technical Engineering for Port and Harbor Construction*, Yokosuka, Japan.

Smith, O. P., 1987. Cost-effectiveness of breakwater cross section, *ASCE J. Waterway Port Coast. Ocean Eng.*, Vol. 113, No. 5.

Sverdrup, H. U., and W. H. Munk, 1947. *Wind, Sea and Swell: Theory of Relations for Forecasting*, Publ. 601, U.S. Navy Hydrographic Office,

Takahashi, S., 1997. Breakwater design, chapter 10 in *Handbook of Port and Harbor Engineering*, G. Tsinker, ed., Chapman & Hall, New York.

Takahashi, S., and K. Shimosako, 1994. Wave pressure on a perforated wall, *Proc. International Conference on Hydro-technical Engineering for Port and Harbour Construction*, Port and Harbor Research Institute, Yokosuka, Japan.

Takahashi, S., K. Kimura, and K. Tanimoto, 1990. Stability of armor units of composite breakwater mound against oblique waves, *Rep. Port Harbour Res. Inst.*, Vol. 29, No. 2 (in Japanese).

Takahashi, S., K. Tanimoto, and K. Shimosako, 1992. Experimental study of impulsive pressures on composite breakwaters, *Rep Port Harbour Res. Inst.*, Vol. 31, No. 5.

Takahashi, S., et al., 1994a. Dynamic response and sliding of breakwater caisson against impulsive breaking wave forces, *Proc. International Workshop on Wave Barriers in Deepwaters*, Port and Harbor Research Institute, Yokosuka, Japan.

Takahashi, S., K. Tanimoto, and K. Shimosako, 1994b. *A proposal of impulsive pressure coefficient for design of composite breakwaters, Proc. International Conference on Hydro-technical Engineering for Port and Harbor Construction*, Port and Harbor Research Institute, Yokosuka, Japan.

Tanaka, N., 1976. Effects of submerged rubble-mound breakwater on wave attenuation and shoreline stabilization, *Proc. 23rd Japanese Coastal Engineering Conference,* (in Japanese).

Tanimoto, K., and K. Kimura, 1985. *A Hydraulic Experimental Study on Trapezoidal Caisson Breakwaters,* Tech. Note 528, Port and Harbour Research Institute, Yokosuka, Japan (in Japanese).

Tanimoto, K., and S. Takahashi, 1994. Design and construction of caisson breakwaters: the Japanese experience, *J. Coast. Eng.,* Vol. 22.

Tanimoto, K., T. Yagyu, and Y. Goda, 1982. Irregular wave tests for composite breakwater foundation, *Proc. 18th International Conference on Coastal Engineering,* Capetown, South Africa.

Tanimoto, K., S. Takahashi, and K. Kimura, 1987. Structures and hydraulic characteristics of breakwaters: the state of the art of breakwater design in Japan, *Rep. Port Harbour Res. Inst.*, Vol. 26, No. 5.

Thomson, D. M., and R. M. Shuttler, 1976. *Riprap Design for Wind Wave Attack: A Laboratory Study in Random Waves,* Rep. 61, CIRIA, London.

Tsinker, G. P., 1986. Floating ports, in *Design and Construction Practices*, Gulf Publishing, Houston, TX.

———, 1997. *Handbook of Port and Harbor Engineering.* Chapman & Hall, New York.

U.S. Army Corps of Engineers, 1984. *Shore Protection Manual,* USACE, Washington, DC.

———, 2001. *Coastal Engineering Manual,* Veri-Tech, Inc., Washington, DC.

U.S. Navy, 1982. *Coastal Protection Design Manual 26.2,* NAVFAC DM-26.2, Naval Facilities Engineering Command, Alexandra, VA.

Van der Kreeke, J. and Paape, A., 1964. On optimum breakwater design, *ASCE Proc. 9th International Conference on Coastal Engineering*.

Van der Meer, J. W., 1988a. *Rock slopes and gravel beaches under wave attack,* Ph.D. dissertation, Delft Technical University; published as Commun. 396, Delft Hydraulics, Delft, The Netherlands.

———, 1988b. Stability of cubes, tetrapods and accropode, *Proc. Breakwater '88*, Eastbourne, East Sussex, England.

———, 1991. *Stability and Transmission at Low-Crested Structures,* H-45, Delft Hydraulics, Delft, The Netherlands.

Van der Meer, J. W., and K. W. Pilarczyk, 1987a. Stability of breakwater armor layer: design formulae, *Coast. Eng.,* Vol. 2.

———, 1987b. *Stability of Breakwater Armor Layers: Deterministic and Probabilistic Design,* Commun. 378, Delft Hydraulic, Delft, The Netherlands.

Van der Meer, J. W., and C. M. Stam, 1992. Wave run-up on smooth and rock slopes of coastal structures, *ASCE J. Waterway Port Coast. Ocean Eng.,* Vol. 118, No. 5.

Verhagen, H. J., and H. F. Burcharth, 2002. Criteria for the selection breakwater types, *PIANC Proc. 30th Congress*, Sydney, Australia.

Viggosson, G., 1990. Rubble-mound breakwaters in Iceland, *J. Coast. Res.*, Vol. SI7.

Willis, D. H., W. F. Baird, and O. T. Magoon, 1987. *Berm Breakwaters,* American Society of Civil Engineers, Reston, VA.

Xie, S. L., 1981. *Scouring Patterns in Front of Vertical Breakwaters and Their Influence on the Stability of Foundation of Breakwaters,* Department of Civil Engineering, Delft University of Technology, Delft, The Netherlands.

———, 1994. *Recent research and experience on vertical and composite breakwaters in China, Proc. International Conference on Wave Barriers in Deepwaters*, Port and Harbor Research Institute, Yokosuka, Japan.

Zwamborn, J. A., and M. Van Niekerk, 1982. *Additional Model Tests. Doly Packing Density and Effect on Relative Block Density.* National Research Institute for Oceanology, Coastal Engineering and Hydraulic Division, Stellenbosh, South Africa, July.

10

NAVIGATION CHANNEL DESIGN

John Headland and Santiago Alfageme
Moffatt & Nichol Engineers
New York, New York

10.1 INTRODUCTION

Marine civil engineers are often called upon to prepare designs for shipping channels. The task generally involves selecting the appropriate channel depth and width, although it also involves selecting geometry for channel bends and areas for vessel turning (i.e., turning basins). Navigation channel design is complicated by the fact that the design engineer must provide channel dimensions that can safely accommodate large vessels, and the behavior of these vessels can only be quantified using sophisticated computer models and input from vessel pilots. Even then, the computer models can quantify the design process only to a certain extent. As a result, navigation channel design generally involves the use of design methods that have evolved over time and account for vessel behavior using rules of thumb. More sophisticated quantitative tools that simulate ship behavior have been developed in recent years and are becoming more customary for design engineers. In this chapter we provide a summary of design methods with a focus on simple design procedures to promote understanding on the part of readers. Quantitative tools are also discussed and illustrated through example applications.

10.2 SHIP CHARACTERISTICS

Navigation channel design starts with a solid understanding of the characteristics of the ships that are expected to use the channel. The basic ship characteristics include length, beam, draft, and displacement (Figure 10.1). Length overall (LOA) is the distance from vessel bow to stern. The vessel beam (B) is the maximum width of the vessel at the waterline. The draft (T) is the distance between the waterline and the bottom of the vessel keel. The length between perpendicular (LBP) may be defined as the distance between the stem and stern at the waterline. Displacement is the weight of the volume water displaced by the submerged vessel hull. Example vessel characteristics for typical design vessels are provided in the *Coastal Engineering Manual* (U.S. Army Corps of Engineers, 2001) and reproduced here as Table 10.1.

There are many other vessel characteristics that can be part of a navigation study, including vessel engine and rudder properties, vessel side and frontal wind areas, and sophisticated naval architectural properties. These characteristics are normally used in vessel maneuvering models and other sophisticated design tools.

10.3 DESIGN APPROACH

A basic understanding of ship behavior is helpful when designing a navigation channels. With respect to channel depth, there must be adequate water available during vessel transit. This can be accomplished by deepening the channel and/or transiting the channel during higher tidal elevations (provided that these conditions last long enough to permit transit). Factors affecting lateral channel dimensions include vessel speed, environmental conditions (winds, waves, currents), the extent of channel confinement, and whether more than one vessel will transit the channel at a time. Vessels are easiest to maneuver when traveling at high speed (5

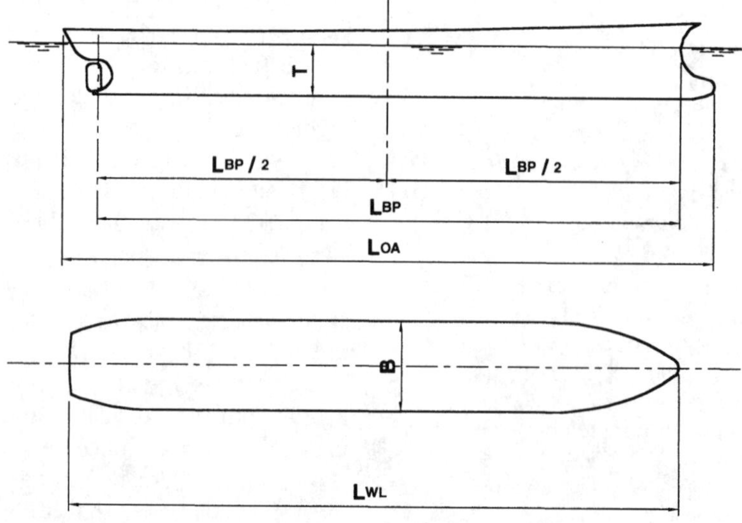

Figure 10.1 Basic ship characteristics.

Table 10.1 Characteristics of large ships

Name	Dead-Weight Tonnage	Length		Beam		Draft	
		m	ft	m	ft	m	ft
Tankers							
Pierre Guillamat	546,265	414.23	1,359.00	62.99	206.67	28.60	93.83
Nisseki Maru	366,812	347.02	1,138.50	54.56	179.00	27.08	88.83
Idemitsu Maru	206,000	341.99	1,122.00	49.81	163.42	17.65	57.92
Universe Apollo	114,300	289.49	949.75	41.28	135.42	14.71	48.25
Waneta	54,335	232.24	761.92	31.70	104.00	12.22	40.08
Olympic Torch	41,683	214.76	704.58	26.92	88.33	12.09	39.67
Ore carriers							
Kohjusan Maru	165,048	294.97	967.75	47.02	154.25	17.58	57.67
San Juan Exporter	104,653	262.00	859.58	38.05	124.83	15.44	50.67
Shigeo Nagano	80,815	250.02	820.25	36.86	120.92	13.23	43.42
Ore/oil carriers							
Svealand	278,000	338.18	1,109.50	54.56	179.00	21.85	71.67
Cedros	146,218	303.51	995.75	43.38	142.33	16.74	54.92
Ulysses	57,829	241.86	793.50	32.39	106.25	12.17	39.92
Bulk carriers							
Universe Kure	156,649	294.67	966.75	43.33	142.17	17.45	57.25
Sigtina	72,250	250.02	820.25	32.28	105.92	13.36	43.83
Container ships							
Sally Maersk	104,696	347	1,138	43	141	14.5	47.5
Mette Maersk	60,639	294.1	964.9	32.3	106.0	13.5	44.3
Korrigan	49,690	288.60	946.83	32.23	105.75	13.01	42.67
Kitano Maru	35,198	261.01	856.33	32.26	105.83	11.99	39.33
Encounter Bay	28,800	227.31	745.75	30.56	100.25	10.69	35.08
Atlantic Crown	18,219	212.35	696.67	27.99	91.83	9.24	30.33
Ocean barges							
SCC 3902	50,800	177.45	582.17	28.96	95.00	12.22	40.08
Exxon Port Everglades	35,000	158.50	520.00	28.96	95.00	9.60	31.50
Passenger/cruise ships							
Voyager of the Seas	142,000 (DT)	310.50	1,018.70	48.00	157.48	8.84	29.00
Grand Princess	101,999 (DT)	285.06	935.24	35.98	118.04	8.00	26.25
Imagination	70,367 (DT)	260.60	854.99	31.50	103.35	7.85	25.75

knots and greater). Under these conditions flow past the rudder is strongest and the ability of the rudder to influence vessel maneuverability is at a maximum. However, the ship must slow and eventually stop in order to berth at a port. Consequently, a vessel may have to travel at a less than desirable speed over large portions of the harbor. Tugs and/or thrusters are used to assist in navigation at lower speeds. Even so, it may not be possible to conduct safe maneuvers at low speed once wind, wave, and/or current conditions exceed critical values. Finally, closely passing vessels can impart large hydrodynamic loads to one other. These forces must be considered in the design width of the channel.

The overall design approach consists of the following steps:

1. Identify/select design vessel. The design vessel is normally selected on the basis of the overall economic interests of the port in question. At developed harbors, the design vessel may be one somewhat larger than the largest currently calling at the port. For example, a port servicing a post-Panamax container vessel may wish to examine the feasibility of accommodating a Suezmax container vessel. New harbors often cater to commercially popular vessels that import/export goods to/from the region in which the harbor is located. The design vessel for a new or existing harbor is normally defined in terms of the basic vessel characteristics summarized under item 2 above. Sometimes there are more than one design vessel (e.g., there may be a design container vessel and a design oil tanker).
2. Define environmental site conditions. Water depth, tide, current, wave, and wind conditions have a direct impact on the maneuverability of a moving vessel. Accordingly, these conditions must be established for each portion of the vessel maneuver into and out of a harbor. In this regard, it is common to divide the maneuver into a series of tangents or segments and to define environmental conditions for each segment.
3. Establish vessel speeds, tug assistance requirements, and other maneuvering procedures.
4. Estimate required channel depth.
5. Estimate channel width requirements.
6. Lay out channel over existing site bathymetry.
7. Confirm channel dimensions with a ship maneuvering/simulation model.
8. Determine requirements for navigation aids.
9. Develop channel quantities, dredging costs, and economic and environmental feasibility.

10.4 ENVIRONMENTAL SITE CONDITIONS

The design and operation of navigation channels depend on (1) water depth (i.e., bathymetry), (2) water levels, (3) winds, (4) waves, (5) currents, (6) visibility, (7) salinity, and (8) sedimentation and dredging requirements. Each of these factors is discussed in the following paragraphs.

10.4.1 Water Depths

Navigation channels are usually developed in waterways that are naturally deep or at least deeper than alternative harbor locations. Examples include rivers such as the Columbia, the Mississippi, and the Rhine, or estuaries such as the Chesapeake, Galveston, and San Francisco bays or the Hudson River. The existing bathymetry along the pathway over which the design vessel travels has a large bearing on the plan geometry of the channel. In most cases, the designer will lay out a proposed navigation channel over the deepest areas to minimize initial and long-term dredging requirements. This approach is desirable as long as it does not result in a large number of channel bends or bends that are too sharp, both of which can result in difficult navigation conditions.

10.4.2 Water Levels

Water levels at a port site are often dictated by astronomical tides although storm tides, and river flood levels may also influence water levels. Typical tidal datums used in the United States are mean higher high water (MHHW), mean high water (MHW), mean sea level (MSL), national geodetic vertical datum

(NGVD), mean low water (MLW), and mean lower low water (MLLW). Tidal datums for various locations throughout the United States and the rest of the world can be obtained directly from bechmark sheets published by the National Oceanic and Atmospheric Administration–National Ocean Service (NOAA-NOS). These sheets are available online from the Center for Operational Oceanographic Products and Services (CO-OPS) at *www.co-ops.nos.noaa.gov*.

10.4.3 Winds

Winds have a large bearing on ship maneuvering. Depending on a number of factors, it may not be possible to transit a channel safely when winds exceed critical thresholds. Accordingly, it is normally necessary to evaluate local wind conditions and summarize the statistics of wind speed and direction. These data can be summarized in terms of a wind rose (Figure 10.2) or a probability of exceedence plot (Figure 10.3).

10.4.4 Waves

Waves can have a significant bearing on navigation channel design. This is often particularly true of the outer portions of an entrance channel, where vessels must transit without protection from ocean wave attack. Waves affect design channel depths and widths. Wave statistics are available for most locations throughout the world. Recorded and hindcasted data are available from the National Data Bouy Center (*www.ndbc.noaa.gov*) and the U.S. Army Corp of Engineers (*www.bigfoot.wes.army.mil*), respectively, in the vicinity of a number of U.S. harbors. The United Kingdom Meteorological Office (*www.meto.govt.uk*) has some 20 years of hindcasted wind and wave conditions for most of the world. These data can be summa-

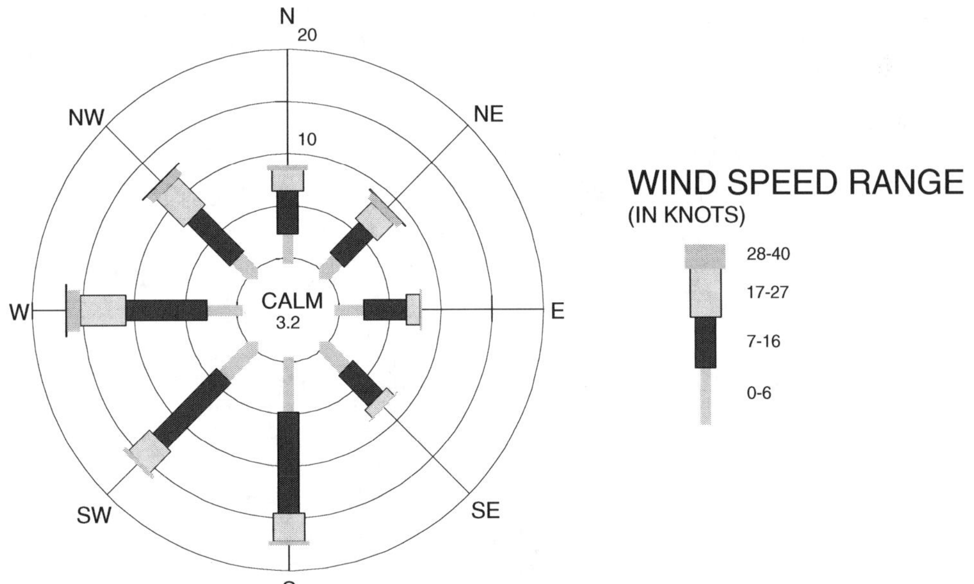

Figure 10.2 Wind rose diagram, offshore Long Island, New York.

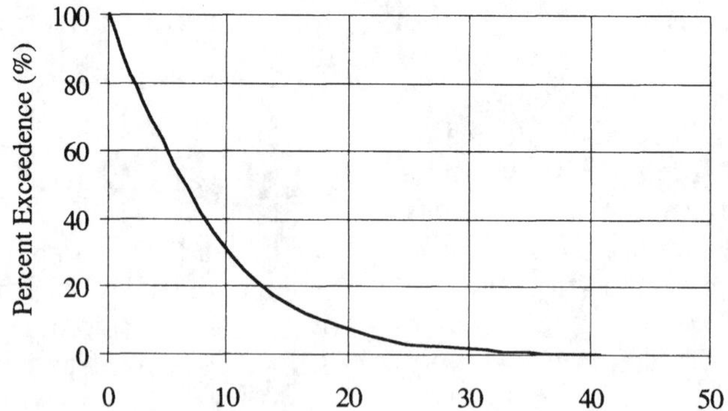

Figure 10.3 Wind speed-exceedence, offshore Long Island, New York.

rized statistically as shown in Figures 10.4 and 10.5.

10.4.5 Currents

Currents also have a considerable impact on channel design. Currents may be of tidal origin, may be the result of river flow, or both. Predicted tidal currents are published for most U.S. harbors and can be obtained easily using specialized software such as Tides & Currents. Tidal currents are normally available for specific locations within a harbor area. In many cases it is desirable to develop a numerical model of tidal currents throughout a harbor in order to provide local estimates of currents. An example result of such a numerical modeling exercise is provided in Figure 10.6.

10.4.6 Visibility

Visibility can inhibit safe navigation and may occasionally be limited by fog or other inclement weather. The persistence of poor visibility

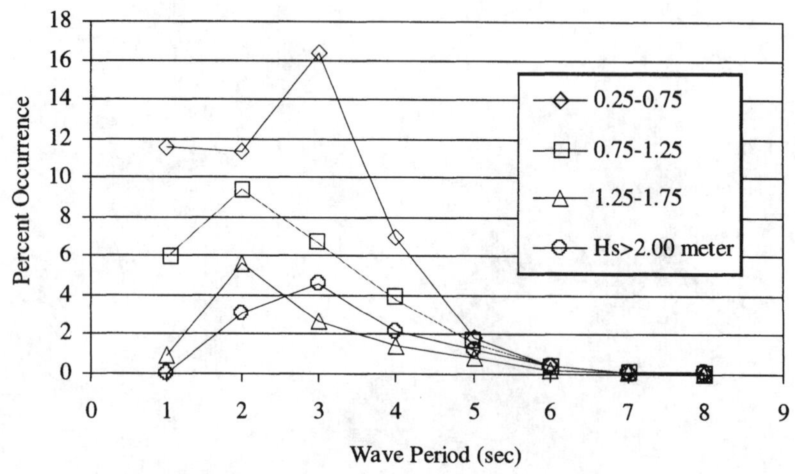

Figure 10.4 Wave height/period distribution.

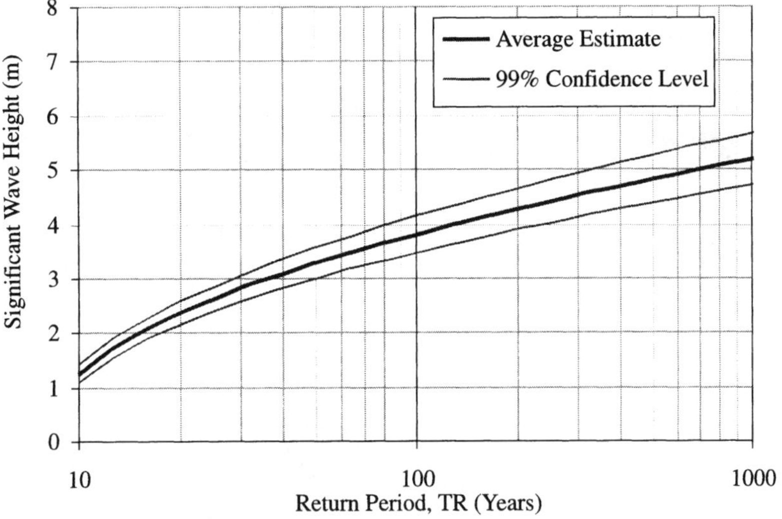

Figure 10.5 Design wave conditions.

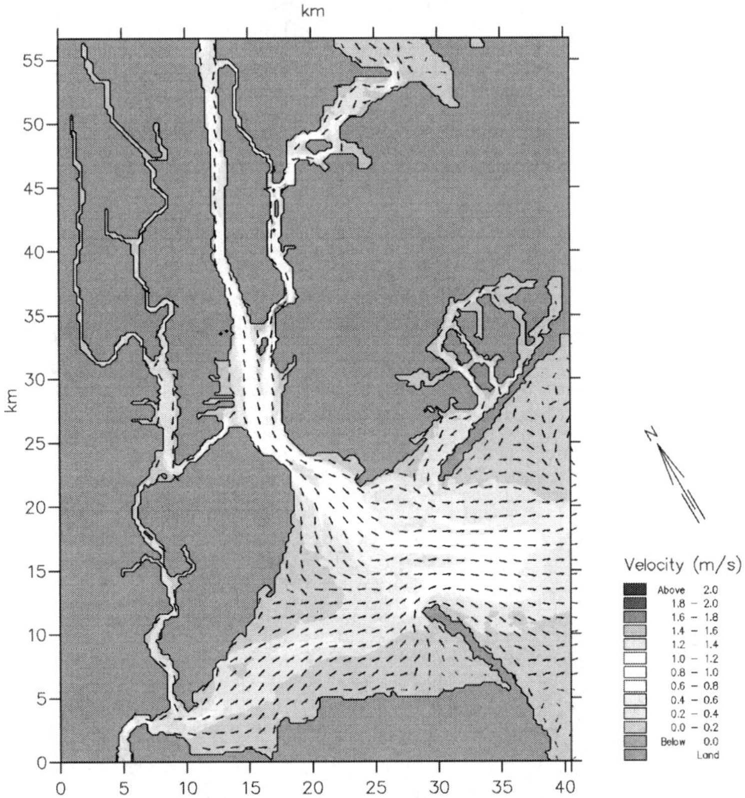

Figure 10.6 Modeled tidal currents, New York Harbor.

should be evaluated in areas vulnerable to lengthy periods of poor visibility.

10.5 DESIGN METHODS

A number of International channel design standards have been published in recent years. The approach summarized in this chapter is based on a synthesis of the following publications: U.S. Army Corps of Engineers (1965, 1983), National Ports Council (NPC, 1975), Permanent International Association of Navigation Congresses/International Commission for the Reception of Large Ships (PIANC/ICORELS, 1980), U.S. Navy (1981), Headland (1995), and Permanent International Association of Navigation Congresses/International Association of Ports and Harbours (PIANC/IAPH, 1997). These references provide the basis for channel depth, channel width, and turning basin requirements.

10.5.1 Channel Depth

Channel depth requirements are summarized in Figure 10.7, which shows that the required channel depth relative to a referenced water level must be based on (1) loaded vessel draft (including trim), (2) squat, (3) wave-induced motions, (4) safety clearance, (5) dredging tolerance, and (6) advanced maintenance dredging.

10.5.1.1 Loaded Draft. Vessel drafts vary according to vessel loading and can range from fully loaded to lightly loaded conditions. Normally, fully loaded or maximum vessel drafts are used in design. In some cases, however, it may be necessary to limit drafts to assure safe navigation.

10.5.1.2 Squat. The position of a vessel's keel relative to the channel bottom will lower with vessel speed. This phenomenon, known as *squat,* results when increased water velocities flowing past a moving ship hull produce a localized lowering of the water surface. In general, squat is a function of vessel speed, underkeel clearance, channel width and depth, and vessel dimensions. Squat varies as follows:

- Squat increases with the square of the forward speed.
- Squat increases as underkeel clearance is reduced.
- Squat is larger for width-constricted channels than for open areas.

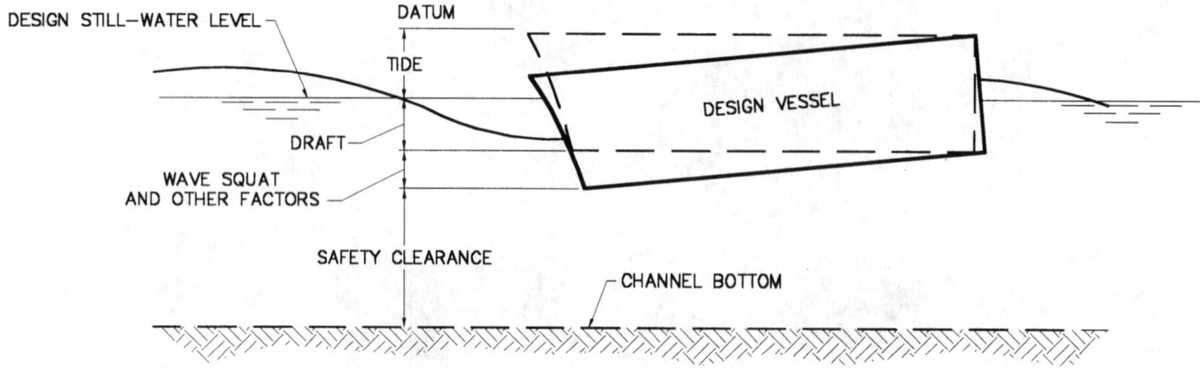

Figure 10.7 Channel depth requirements.

- Wide-beamed vessels tend to trim down by the bow and narrow-beamed vessels tend to trim down by the stern.

PIANC/IAPH (1997) presents the following:

$$\text{squat}(m) = 2.4 \frac{\nabla}{L_{pp}^2} \frac{F_{nb}^2}{\sqrt{1 - F_{nb}^2}} \qquad (10.1)$$

where ∇ is the volume of displacement (m^3) = $C_B L_{pp} BT$ (where C_B is a block coefficient, L_{pp} the length of the ship between perpendiculars, B the ship beam, and T the ship draft) and F_{nh} is the Froude depth number.

10.5.1.3 *Wave-Induced Motions.* A detailed examination of wave-induced motions has been computed using a methodology described in Headland (1995). A brief summary of this technique is presented here.

The wave response of a vessel is estimated using a six-degree-of-freedom ship motion model. The model accounts for both forward motion and shallow water effects. Frequency-domain computations are used to compute vessel motions in wave spectra. The model is used to evaluate the maximum vertical excursions of the bow of the vessel for various sea states, with the latter characterized by significant wave height, mean wave period, and wave angle relative to vessel heading. Maximum vertical bow motions are computed using a Gaussian distribution as follows:

$$P(z > z_{\max}) = e^{-z_{\max}^2/2m_0} \qquad (10.2)$$

$$m_0 = \int S_z(\omega)\, d\omega \qquad (10.3)$$

where $P(z > z_{\max})$ = probability that z exceeds the value $z_{\max}$ (where z is the vertical motion amplitude and $z_{\max}$ the approximate value of the maximum vertical motion amplitude), m_0 is the zero spectral moment of the vertical motion amplitude = RMS^2, and $S_z(\omega)$ is the vertical motion spectrum. The probability of exceedance used to estimate the maximum vertical bow motions is normally taken as 1/100 or 1%.

Computations using this method can be been used to determine the maximum vertical wave motion at the bow of a vessel transiting a navigation channel under a number of sea conditions. Example results are provided in Figure 10.8, which shows that the 10% exceedence bow motions of a Panamax container vessel for a particular site was computed as 0.25 m. These values will be exceeded only 10% of the time each year.

10.5.1.4 *Safety Clearance.* Safety clearance is the distance between the lowest calculated position of a vessel's hull and the channel bottom. When the channel bed consists of soft material, a safety clearance of 0.6 m (2 ft) is used. This allowance is increased to 1.2 m (4 ft) for hard rock or clay bottoms.

10.5.1.5 *Dredging Tolerance.* An additional 0.3 to 0.6 m (1 to 2 ft) beyond the design dredge depth is normally provided as a dredging pay item because of difficulties and costs associated with dredging a uniform and highly accurate depth.

10.5.1.6 *Advanced Maintenance Dredging.* In areas prone to siltation, an additional depth can be added to the channel to allow for the storage of accumulated sediment between maintenance dredging events. This approach prevents the premature loss of project depth and increases the time between dredging events. A detailed evaluation of advanced maintenance dredging normally requires the collection of field measurements and numerical modeling of sedimentation for various depths. Alternatively, periodic hydrographic surveys gain information necessary to evaluate the need for advanced maintenance dredging.

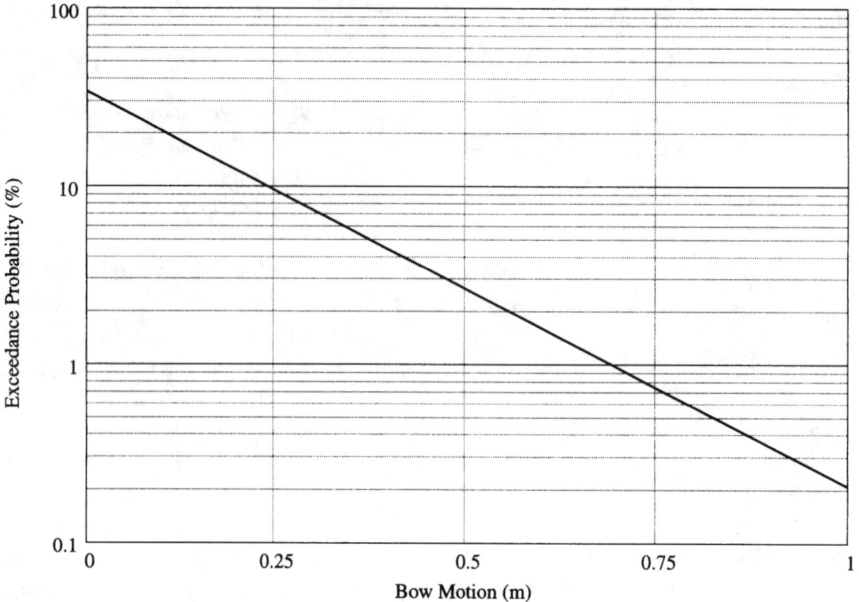

Figure 10.8 Probability of exceedence for bow motions.

10.5.2 Channel Width

The minimum width of a straight channel depends on the size and maneuverability of the vessel navigating the channel, type of channel bank, effect of other vessels in the channel, and effects of wind and currents. The width required consists of three distinct zones: (1) maneuvering lane, (2) ship clearance lane, and (3) bank clearance. Channel width design must account for the following factors: (1) vessel speed, (2) cross winds, (3) cross current, (4) longitudinal current, (5) significant wave height and length, (6) aids to navigation, (7) bottom surface, (8) depth of waterway, (9) cargo hazard level, and (10) traffic density.

The PIANC methodology for determining channel width is summarized is described in the following paragraphs. The bottom width of the waterway is given for a one-way channel by

$$W = W_{\text{BM}} + \sum_{i=1}^{n} W_i + W_{\text{Br}} + W_{\text{Bg}} \qquad (10.4)$$

and for a two-way channel by

$$W = 2W_{\text{BM}} + 2\sum_{i=1}^{n} W_i + W_{\text{Br}} + W_{\text{Bg}} + W_p \qquad (10.5)$$

where W_{Br} and W_{Bg} are the bank clearances on the "red" and "green" sides of the channel, W_p is the passing distance, W_i are the additional width requirements to allow for the effects of wind, currents, and so on, and W_{BM} is the basic maneuvering width that is required by the design ship to sail safely in favorable environmental and operational conditions. Maneuvering lane requirements are summarized as follows:

Ship maneuverability	Basic maneuvering lane, W_{BM}
Good	$1.3B$
Moderate	$1.5B$
Poor	$1.8B$

To the basic maneuvering lane width, W_{BM}, are added additional widths that give the maneuvering lane, W_m. The additional widths are given in Table 10.2.

10.5.2.1 Maneuvering Lane. The maneuvering lane must allow for the oscillating track of a maneuvering vessel. The oscillations occur as a result of a vessel's inherent directional stability characteristics. Generally, the width of the maneuvering lane will vary from 1.3 to 1.8 times the beam (B) of the vessel, depending on its controllability. Allowance must be made for the yaw of the vessel when there are cross currents or winds. Yaw angles of 5° are common, and yaw angles sometimes approach 10°. A yaw of 5° can add extra width to the maneuvering lane equivalent to half the beam for larger vessels. PIANC provides a method for computing the additional width needed when the vessel is exposed to cross currents and winds (see Table 10.2).

10.5.2.2 Ship Clearance Lane. In a multilane channel, it is necessary to separate maneuvering lanes by a ship clearance lane. This separation will avoid undesirable interaction between vessels traveling past one another. Ship clearance lanes are summarized in Table 10.3.

10.5.2.3 Bank Clearance. A vessel sailing close to a channel bank experiences a suction effect due to the asymmetrical flow of water around the vessel. To counteract this effect, helm must be applied to the vessel with a concomitant need for additional channel width. Bank clearance requirements are summarized in Table 10.4.

10.5.2.4 Bends. The minimum width of a channel will be larger in a bend than in a straight channel due to the additional width of maneuvering lane required for turning vessels. The ease with which a vessel can negotiate a bend depends on a number of factors, including bend radius, vessel length, vessel velocity, and bend deflection angle. There is a minimum bend radius which must be satisfied and is usually related to vessel length. Minimum bend radii tend to be in the range $8L$ to $10L$.

A number of methods have been proposed for estimating the required maneuvering lane in a bend. USACE (1965) suggests the average relationships for a 3810-m radius bend shown in Table 10.5. PIANC (1995) also provides guidance for channel bends.

The U.S. Army Corps of Engineers (1965) summarizes the Ghent Terneuzan method for determining channel bend widths. This method provides an estimate of the additional channel width required to safely negotiate the turn and is added to the base width to provide the total bend width. There are various geometrical means for achieving widening in a bend. Channels that have gradually widening cross sections can cause problems due to the variation in bank suction experienced throughout the bend. One of the most important factors in designing a bend is the ease with which the additional width at the bend can be marked with navigation aids. A widening of the inside of the bend is an established method to improve the navigability of the bend.

10.5.3 Basins and Maneuvering Areas

The following paragraphs summarize design criteria for stopping areas, turning basins, and berths.

Table 10.2 Additional widths for straight channel sections

Width, W_l	Vessel Speed	Outer Channel, Exposed to Open Water	Inner Channel, Protected Water
Vessel speed (knots)			
Fast >12		0.1B	0.1B
Moderate >8–12		0.0	0.0
Slow 5–8		0.0	0.0
Prevailing cross wind (knots)			
Mild ≤15 (≤ Beaufort 4)	All	0.0	0.0
Moderate >15–33 (> Beaufort 4–7)	Fast	0.3B	—
	Mod.	0.4B	0.4B
Severe >33–48 (> Beaufort 7–9)	Slow	0.5B	0.5B
	Fast	0.6B	—
	Mod.	0.8B	0.8B
	Slow	1.0B	1.0B
Prevailing cross current (knots)			
Negligible <0.2	All	0.0	0.0
Low 0.2–0.5	Fast	0.1B	—
	Mod.	0.2B	0.1B
	Slow	0.3B	0.2B
Moderate >0.5–1.5	Fast	0.5B	—
	Mod.	0.7B	0.5B
	Slow	1.0B	0.8B
Strong >3	Fast	0.7B	—
	Mod.	1.0B	—
	Slow	1.3B	—
Prevailing longitudinal current (knots)			
Low ≤1.5	All	0.0	0.0
Moderate >1.5–3	Fast	0.0	—
	Mod.	0.1B	0.1B
	Slow	0.2B	0.2B
Strong >3	Fast	0.1B	—
	Mod.	0.2B	0.2B
	Slow	0.4B	0.4B
Significant wave height H and length λ (m)			
$H_S \leq 1$ and $\lambda \leq L$	All	0.0	0.0
$3 > H_S > 1$ and $\lambda \approx L$	Fast	2.0B	
	Mod.	1.0B	
$H_S > 3$ and $\lambda > L$	Slow	0.5B	
	Fast	3.0B	
	Mod.	2.2B	
	Slow	1.5B	

Table 10.2 (*Continued*)

Width, W_I	Vessel Speed	Outer Channel, Exposed to Open Water	Inner Channel, Protected Water
Aids to Navigation			
Excellent with shore traffic control	0.0	0.0	
Good			
Average, visual and ship board, infrequent	$0.1B$	$0.1B$	
poor visibility	$0.2B$	$0.2B$	
Average, visual and ship board, frequent	$\geq 0.5B$	$\geq 0.5B$	
poor visibility			
Bottom surface			
If depth $\geq 1.5T$		0.0	0.0
If depth $< 1.5T$ then			
Smooth and soft		$0.1B$	$0.1B$
Smooth or sloping and hard		$0.1B$	$0.1B$
Rough and hard		$0.2B$	$0.2B$
Depth of waterway			
$\geq 1.5T$		0.0	$\geq 1.5T$ 0.0
$1.5T$–$1.25T$		$0.1B$	$<1.5T$–$1.15T$ $0.2B$
$<1.25T$		$0.2B$	$<1.15T$ $0.4B$
Cargo hazard level			
Low		0.0	0.0
Medium		$\geq 0.5B$	$\geq 0.4B$
High		$\geq 1.0B$	$\geq 0.8B$

Table 10.3 Additional widths for two-way traffic

Passing Distance, W_P	Outer Channel, Exposed to Open Water	Inner Channel, Protected Water
Vessel speed (knots)		
Fast >12	$2.0B$	—
Moderate >8–12	$1.6B$	$1.4B$
Slow 5–8	$1.2B$	$1.0B$
Encounter traffic density		
Light	0.0	0.0
Moderate	$0.2B$	$0.2B$
Heavy	$0.5B$	$0.4B$

Table 10.4 Additional widths for bank clearance

Width for Bank Clearance, W_{Br} or W_{Bg}	Vessel Speed	Outer Channel, Exposed to Open Water	Inner Channel, Protected Water
Sloping channel edges and shoals	Fast	0.7*B*	—
	Mod.	0.5*B*	0.5*B*
	Slow	0.3*B*	0.3*B*
Steep and hard embankments, structures:	Fast	1.3*B*	—
	Mod.	1.0*B*	1.0*B*
	Slow	0.5*B*	0.5*B*

Table 10.5 Maneuvering lane in a bend

	Width of Maneuvering Lane	
Controllability of Vessel	Deflection Angle 26°	Deflection Angle 40°
Very good	3.25*B*	3.85*B*
Good	3.70*B*	4.40*B*
Poor	4.15*B*	4.90*B*

10.5.3.1 *Stopping Distances.*

Vessels transiting navigation channels must maintain a minimum velocity in order to navigate safely. Before maneuvering to the berth, however, the speed of the vessel must be reduced to nearly zero. This is achieved by putting the vessel's engine astern until the forward movement has been arrested. The distance required to stop depends on the speed and size of the vessel. For moderate speeds an average stopping distance of five times the vessel length is considered a minimum. A stern engine tends to cause a lateral movement of the vessel, which must be accommodated by an increase in channel width.

10.5.3.2 *Turning Basins.*

Most vessels are turned either just before berthing or when leaving the berth. The minimum diameter required for turning will depend on whether the vessel has tug assistance. The following minimum diameters of a turning basin have been suggested and are generally consistent with PIANC guidelines:

	Diameter of Turning Basin
With tug assistance	2.0*L*
Without tug assistance	4.0*L*

The diameters might be reduced to 1.5*L* for tug assistance and 3.0*L* without tug assistance for good conditions, but these figures should be considered a lower limit.

10.5.3.3 *Berths.*

When dredged basins fronting a berth are constructed adjacent to shallow water, vessels may overshoot the berth, and an allowance should be provided for that possibility. The following guidance is suggested for favorable conditions:

	Length of Basin
Tug-assisted berthing	1.25*L*
Berthing without tugs	1.50*L*

Underkeel clearances at berths in sheltered areas will normally be 1 to 1.5 ft for soft bottoms.

10.5.4 Side Slopes

Dredging creates a side slope at the edge of the channel area and it is economical to have as steep a side slope as practicable. Most non-cohesive soils will not stand at a slope angle greater than 45°, while cohesive soils will initially stand at much higher angles. Over a period of time, however, the cohesive soils tend to flatten. For example, there are cases where dredged areas persisted for 6 months at slopes between 1:3 and 1:6 but flattened to 1:10 after two or three years and finally reached an equilibrium of 1:30 to 1:60 twenty years later. Table 10.6 shows the side slopes that are often adopted for various soils, which have been found to be satisfactory over long periods of time. The characteristics of muds and silts depend to a great extent on the extent of consolidation that is achieved. In some cases, muds and silts may act more like a fluid than a soil. A side slope of 1:3 has been used for the present work and is consistent with previous practice in the area.

10.5.5 Simulation Models

Real-time or man-in-the-loop and fast–time ship maneuvering simulation models are often used in preliminary and final channel design. These models reproduce the maneuvering behavior of a ship based on a mathematical representation of the environmental forcings and the vessel's reaction.

Table 10.6 Typical side slopes below water level for various soil types

Soil Type	Side Slope Vertical:Horizontal
Rock	Nearly vertical
Stiff clay	1:1
Firm clay	$1:1\frac{1}{2}$
Sandy clay	1:2
Coarse sand	1:3
Fine sand	1:5
Mud and silt	1:8 to 1:60

10.5.5.1 Fast-Time Simulators. Fast-time simulators typically include an autopilot algorithm that navigates a vessel along a channel by adjusting the rudder angle in response to wind, wave, and current actions. Therefore, fast-time simulation is a very useful tool for evaluation of navigation channel designs. Numerous models are available, and the following paragraphs summarize a model developed by the Maritime Research Institute of the Netherlands. Rather than present a lot of text about the model, a reasonable understanding of the utility of fast-time maneuvering models can be gained by presenting a hypothetical example. The model is capable of simulating the maneuvering behavior of ships, taking into account the ship's maneuvering characteristics, the desired track, rudder and engine actions, tug assistance, wind, wave, currents, shallow water, and bank suction. The model computes the track and course angles of the ship, the action of the rudder, the speed of the engine, and the force of the tug(s) on time steps during the maneuver. A flowchart describing the input/output data flow of the mathematical maneuver model is shown in Figure 10.9.

The following paragraphs summarize an example application of the model. Navigation conditions along a channel were investigated with the model for several operational and environmental conditions. Specifically, bank suction effects and wind forces on two design vessels transiting the channel under tug assist were determined. Required action of the rudder, the engine, and the tugs during the maneuver were investigated to determine the feasibility of the maneuver under each set of conditions tested.

Several model runs were completed based on design vessel characteristics, design criteria, and anticipated maneuvering. Transit within a

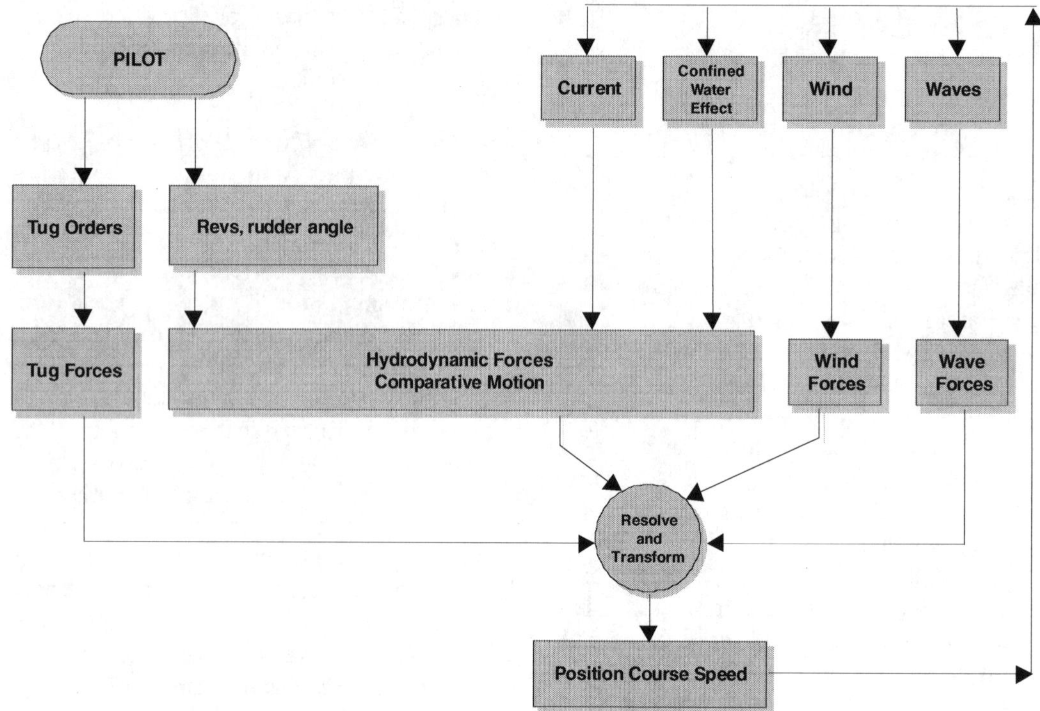

Figure 10.9 Fast-time ship maneuvering model flowchart.

harbor channel was considered taking into account similar vessels at adjacent berths. This configuration results in a minimum available channel width of approximately 330 ft for this particular example. Several runs were made to study the maneuverability of the vessel under various wind conditions.

The following figures illustrate representative results from the model. Figure 10.10 presents bank forces and resulting moment for the same maneuver and a 20-knot sustained cross wind (west). As depicted in Figure 10.11, wind forces are significant and an order of magnitude greater than bank suction effects. Figure 10.12 shows model results for a 25-knot cross wind and a sailing speed of 2 knots. As illustrated by the figure, the ship alignment is not parallel to the centerline of the channel even with tug assist. Although the vessel appears to be able to transit the channel by offsetting some of the wind forces through rudder application, the tugs and the vessel are nearly overpowered, and additional tug assist would probably be required.

This is typical use of a fast-time ship-maneuvering model and illustrates how such models can be used to examine channel dimensions developed using the design guidelines outlined in this chapter. In critical cases it may also be advisable to perform real-time or person-in-the-loop ship simulation. Such efforts rely on a mathematical maneuvering model such as the one described above. These models, however, are coupled with a simulated

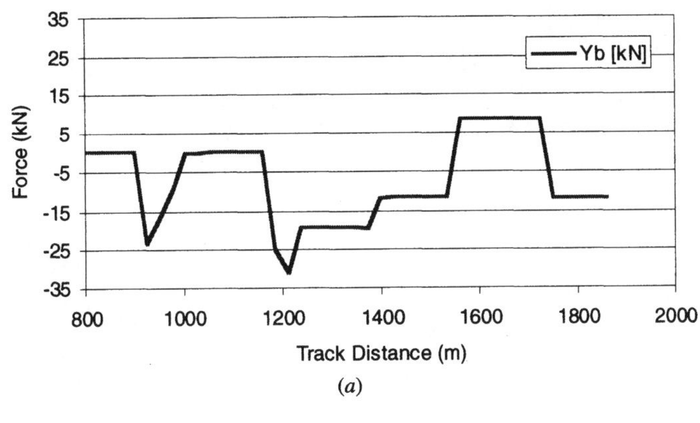

(*a*)

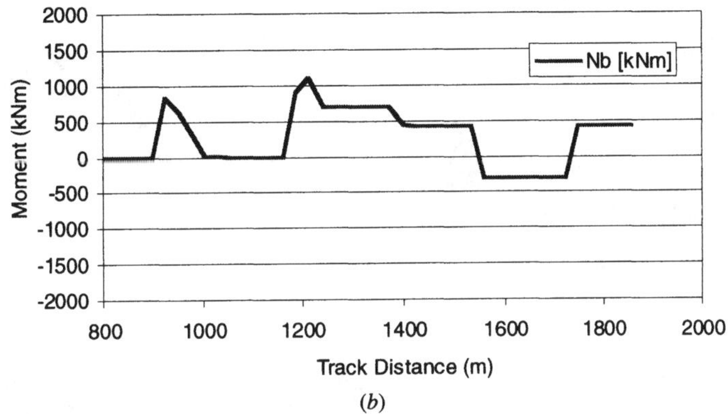

(*b*)

Figure 10.10 Bank-suction actions: (*a*) lateral bank-suction force (+, to port; −, to starboard); (*b*) bank-suction-induced moment. Vessel speed, 2 knots; wind speed, 25 knots.

vessel bridge and harbor environment. An actual harbor pilot "drives the ship" through the simulated environment. These models can be very realistic and useful in finalizing the design of navigation channels.

10.6 AIDS TO NAVIGATION

Markers and signals are used to aid safe navigation in channels. Navigation aids are important to safe transit of a channel, and the quality of the aids can have a direct bearing on channel dimensions as described above in the context of channel width design. A number of standard aids to navigation are listed below:

- *Beacons*. These lighted aids are placed on pile-supported foundations and are used to mark channel boundaries, turns, and hazard areas. Beacons are used in relatively shallow water depths.
- *Buoys*. These aids are used to mark channel boundaries, turns, and hazards and are held in place by a chain and anchor mooring.
- *Ranges*. Ranges are fixed structures aligned with a straight section of channel

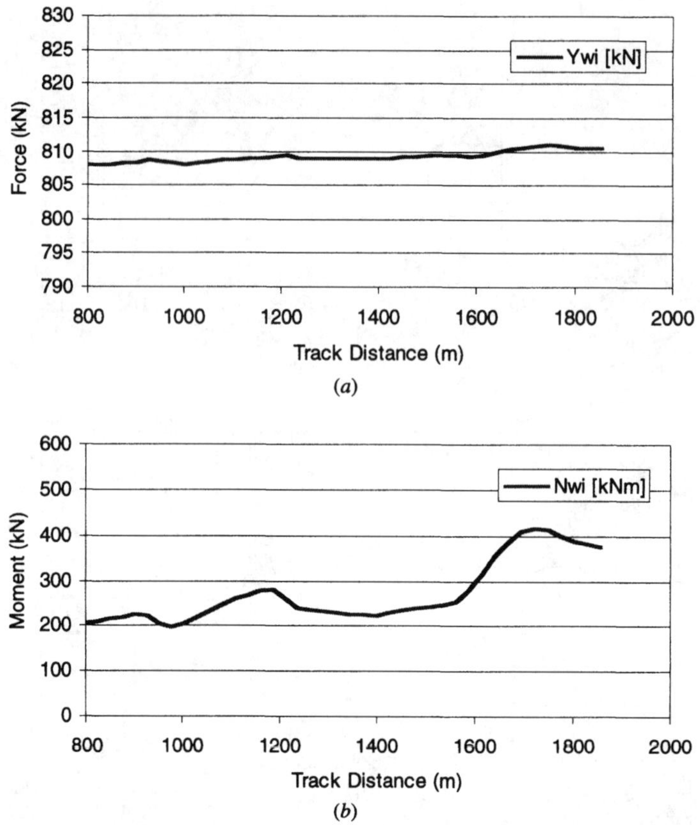

Figure 10.11 Wind actions: (*a*) lateral wind force (+, to port; −, to starboard); (*b*) wind-induced moment. Vessel speed, 2 knots; wind speed, 25 knots.

and are used by pilots to determine the position of a vessel in the channel.

- *Major lights*. These are high-intensity lights placed on high-elevation towers. They can be seen from a considerable distance.
- *Sea buoys*. These are large buoys marking the ocean entrance of a channel.

The U.S. Coast Guard provides additional design guidance for aids to navigation (U.S. Coast Guard, 1981, 1988a,b).

10.7 CHANNEL DEPTH DESIGN EXAMPLE

The following paragraphs present a channel depth design example for a 200,000 deadweight-ton (dwt) tanker for a sheltered interior channel and an entrance channel exposed to waves. The example illustrates the use of a methodology that accounts explicitly for vessel motions due to waves. Moreover, statistical uncertainties associated with inaccuracies in input data are taken into account through probability distributions of those variables. Results are

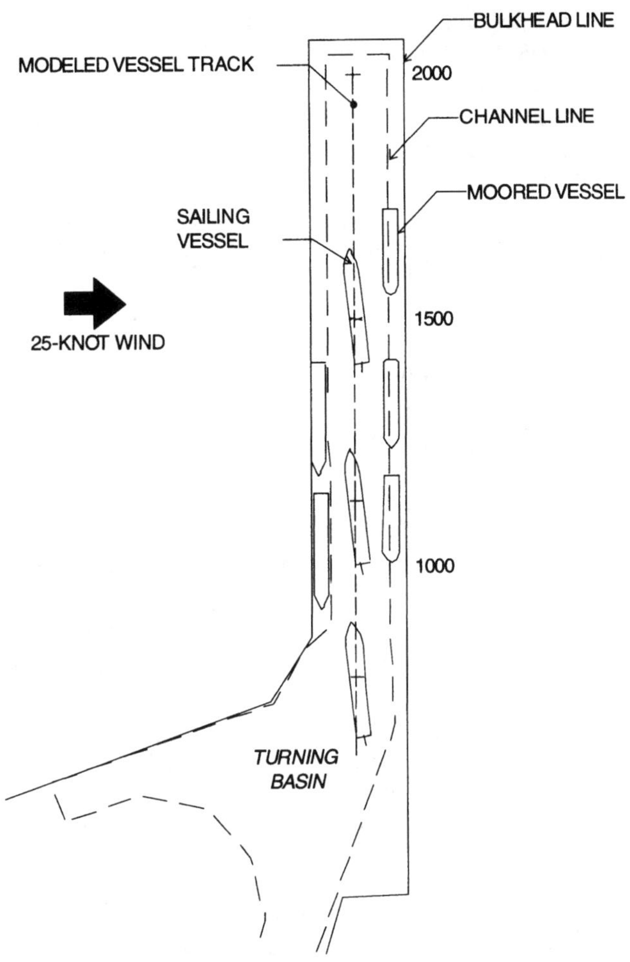

Figure 10.12 Vessel track for vessel speed, 2 knots, wind speed, 25 knots.

computed on the basis of Monte Carlo simulation techniques and are presented in terms of statistical distributions of underkeel clearances.

The U.S. Army Corps of Engineers (1983) and the U.S. Navy (1981) provide summaries of design criteria for navigation channel depths. The criteria summarized in these design handbooks are useful for quick assessments of channel depth requirements and are often used in feasibility studies. Relatively speaking, the handbooks provide a reasonably objective means for evaluating channel depths for sheltered navigation channels where ocean waves are not a factor. The handbooks do not, however, offer direct guidance as regards depth requirements for entrance channels exposed directly to ocean waves.

Depths for navigation channels are normally specified so as to allow for adequate underkeel clearance (UKC) for the largest vessel or group of largest vessels expected to operate in the waterway. Roughly speaking, underkeel clearance is the difference between the depth available in the waterway and the lowest instantaneous vertical position of the vessel hull experienced during transit. Following the terminology published by Kimon (1982), underkeel clearance depends on the following factors: (1) static draft (T) and static draft uncertainty (σT), (2) squat (S) and squat uncertainties (σS), (3) charted depth (h) and depth uncertainty (σh), (4) tide (t) and tide uncertainty (σt), (5) undercut (u) (dredging in excess of minimum), (6) siltation (st) and siltation uncertainty (σt), and (7) wave-induced motions (W).

Kimon (1982) presented a semiempirical methodology for determining UKC in ports. The approach included a graphical methodology for examining wave-induced vessel motions. Although Kimon (1982) presented a thorough and statistically consistent method for examining UKC, including the effects of waves, the method relied on an empirical calibration factor based on experience at an existing port in order to make a final assessment of UKC. The need to rely on empiricism stemmed from a lack of confidence regarding the behavior of uncertainties which were expressed in terms of normal distributions and estimated standard deviations. The basic equation suggested by Kimon for steady UKC requirements (i.e., in the absence of waves) is as follows:

$$\text{UKC} = k_1\sqrt{\sigma_T^2 + \sigma_S^2 + \sigma_h^2 + \sigma_t^2 + \sigma_{st}^2} + S \qquad (10.6)$$

where k_1 is an empirical coefficient which represents the number of standard deviations required to limit groundings to extremely low values. Kimon states that k_1 would have a value of 3.718 if the probability of grounding were limited to a value of 10^{-4} (i.e., 1 grounding per 10,000 transits) and each of the uncertainties were known to have a normal probability distribution with known mean and standard deviation. Instead, Kimon suggests that k_1 be determined for ports that allow a determination of UKC based on analytical computations and empirical experience at a specific port. The methodology was applied to the ports of Rotterdam (Netherlands), Fawley (United Kingdom), and Milford Haven (United Kingdom), and provided k_1 values of 5.31 (no waves), 3.95 (no waves), 2.66 (waves included).

10.7.1 Methodology

While the Kimon methodology offers a rational and quantitative process for evaluating UKC, recent advances in computer technology provide an opportunity to extend his approach. In the present chapter we extend Kimon's approach in two ways: (1) vessel response to waves is determined using a numerical ship motion model, and (2) statistical uncertainties are computed on the basis of Monte Carlo simulation. In this way, the basic calibration procedure suggested by Kimon can be avoided. It is emphasized, however, that site-specific experience should always be used to guide the selection of input parameters and the statistical distribution of those parameters. Furthermore, basic quantitative methods such as squat computation and vessel motion response should be verified as far as is practicable with physical model test results. The basic approach used here is to estimate actual UKC_a for a given set of input conditions as follows:

$$\text{UKC}_a = h - (T + S + t + st + W) \qquad (10.7)$$

All of the variables above are stochastic and depend on mean input parameters and assumed probability distributions. In general, the depth, tide, and siltation variable have a zero mean and an assumed range of values around the zero mean. Reliability computations can be estimated on the difference between the actual UKC_a (load) and the available UKC_f (strength) through the limit state function (i.e., failure or Z function):

$$Z = \text{UKC}_f - \text{UKC}_a \qquad (10.8)$$

The probability of failure (i.e., ship grounding) is determined on the basis of Monte Carlo simulations. Random values are drawn from the stochastic input variables, which are used to compute Z. A failure occurs when Z has a negative value.

10.7.2 Tanker Properties

The vessel chosen for the example computations is a 200,000-dwt (dead-weight tonnage) tanker (see Table 10.7). This tanker was chosen because there is considerable information published about the vessel and it offers an opportunity to demonstrate the extent to which the methods employed here are corroborated by experimental data (especially in waves). There is nothing unique about the choice of the 200,000-dwt tanker as regards the applicability of the probabilistic design procedure proposed; virtually any vessel type or size could be examined (e.g., container vessels, coal carriers, fishing boats, etc.). A complete series of physical model test data for the selected tanker is presented in van Oortmerssen (1975). Computations were prepared for a fully loaded tanker in an unrestricted water depth of 22.68 m, which corresponds to a water depth/draft ratio of 1.2. Furthermore, wave computations focused on head waves, which corresponds to a fully loaded tanker heading to sea through a shallow coastal entrance channel after loading at a protected port.

Table 10.7 Tanker characteristics

Item	Value
Length between perpendiculars	310.000 m
Breadth	47.200 m
Draft	18.900 m
Volume of displacement	235,000.000 m^3
Block coefficient	0.850
Midship section coefficient	0.995
Prismatic coefficient	0.855
Distance CG to midship section	6.610 m
Height of CG	13.320 m
Metacentric height	5.780 m
Longitudinal radius of gyration	7.500 m
Transverse radius of gyration	17.000 m

10.7.3 Computation of Vessel Squat

Blaauw and van der Knaap (1983) provide an excellent summary of the methods that can be used to compute squat for vessels sailing through restricted water. For tankers they suggest the use of a formula published by Eryuzlu and Hausser (1978). The Eryuzlu and Hausser formula, based on physical model tests, is used in this chapter and is applicable to h/T ratios from 1.08 to 2.78 for tankers traveling in channels of unrestricted width. Although this particular approach has been used in the example presented here, any of the other squat computation methods summarized by Blaauw and van der Knaap (1983) can be used in the probabilistic methodology. The Eryuzlu and Hausser formula is as follows:

$$S_{\max} = 0.113B \left(\frac{T}{h}\right)^{0.27} \left(\frac{V}{\sqrt{gh}}\right)^{1.8} \qquad (10.9)$$

where V is the velocity of the vessel and g is the acceleration due to gravity. A plot of eq. (10.9) is presented in Figure 10.13 for a range of vessel speeds. The stochastic nature of vessel squat can be taken into account by providing statistical distributions for vessel draft, water depth, and vessel speed.

10.7.4 Wave Response

Wave response of the tanker was estimated using a six-degree-of-freedom ship motion

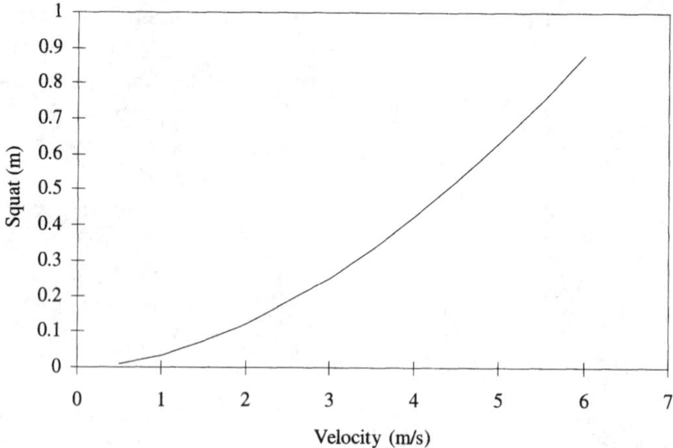

Figure 10.13 Computed squat for a 200,000-dwt tanker. *B*, 47.2 m; *T*, 18.9 m; *h*, 22.68 m.

model. The model accounts for both forward motion and shallow water effects. Frequency domain computations are used to compute vessel motions in wave spectra. For purposes of the present work, the model was used to evaluate the maximum vertical excursions of the bow of the vessel for various vessel motions sea states, with the latter characterized by significant wave height, mean wave period, and wave angle relative to vessel heading.

An example comparison of the model output to the physical model tests of van Oortmerssen (1975) is presented in Figure 10.14 for pitch and heave motions in quartering seas. The data are presented in terms of dimensionless response amplitude operators (RAOs) (i.e., ψ/a for pitch and X/a for heave). An RAO is the ratio of vessel motion to wave amplitude for a given wave period. Dimensionless wave frequency, presented on the abscissa of Figure 10.14, is defined as $\omega\sqrt{L/g}$ where L is ship length. The heave and pitch motions are presented inasmuch as these motions generally dictate the vertical motions of the keel near the vessel bow in head wave conditions. It should be noted that the computation results presented in Figure 10.10 correspond to the motions of the vessel at zero forward speed. Similar results may be available for forward speed; however, they are not readily available in the literature. It suffices to say here that the motions predicted for zero forward speed correspond to the maximum case because motions generally decrease with forward speed. The model does not account for wave current interaction. For the present time, the effect of current must be accounted for on the basis of relative vessel speed. Figure 10.15 presents a response amplitude operator for vertical motion at the vessel bow in head waves for various vessel speeds.

The RAOs presented in Figure 10.3 were combined with the following two-parameter spectrum for a significant wave height of 1 m and a range of mean wave periods:

$$S(\omega) = (0.25H_S^2)\left[0.817\left(\frac{2\pi}{T_m}\right)\right]^4 \omega^{-5} e^{[0.817(2\pi/T_m)]^4\omega^{-4}} \tag{10.10}$$

where $S(\omega)$ is the spectral energy density, T_m the mean wave period, H_s the significant wave

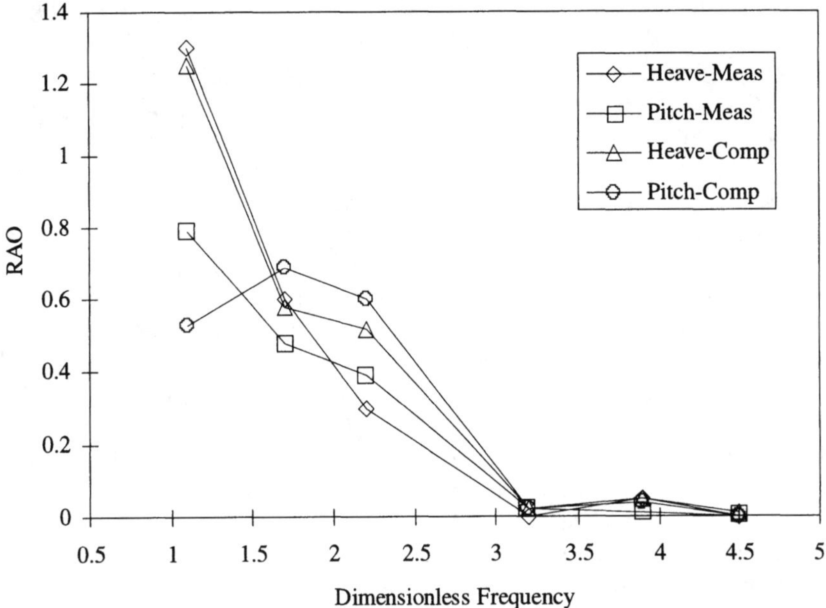

Figure 10.14 Comparison of computed vessel RAOs with measured results from physical model tests.

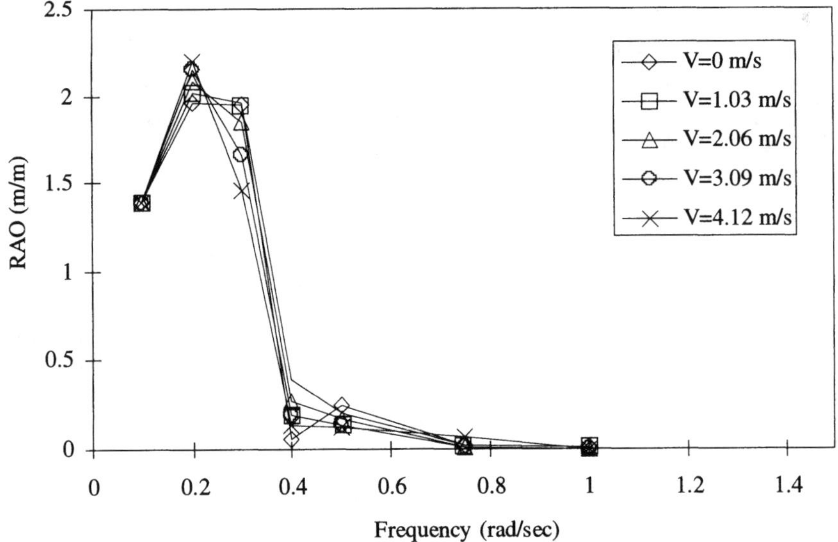

Figure 10.15 Vertical motion response amplitude operators for vessel bow at various forward speeds.

height, and ω the circular wave frequency ($2\pi/T_m$).

The spectral shape above is comparable to that published by ISSC. Consideration was given to use of the Jonswap and Goda spectral shapes, but the spectrum above provides a more conservative assessment of vessel motions. The wave spectrum above was combined with the RAOs presented in Figure 10.15 to provide a plot of root mean square amplitudes (RMS) of vertical bow motion for a range of mean wave periods a significant wave height of 1 m and a range of vessel speeds (i.e., 2, 4, 6, and 8 knots). The results of these computations are presented in Figure 10.16, which indicates that RMS vertical motion amplitudes increase with mean wave period and decrease with vessel speed. The values presented in Figure 10.12 are slightly smaller than those presented by Kimon (1982) but are considered reasonable insofar as the basic RAOs have been verified by physical model tests for the specific vessel examined.

The RMS response for a significant wave height other than 1 m can be obtained by multiplying the values presented in Figure 10.16 by the desired significant wave height. Kimon presented a relatively complicated and perhaps more accurate methodology for evaluating the extreme wave motions in a given sea state. The approach taken in this chapter is to use the following formula recommended by Price and Bishop (1974), which more readily allows for Monte Carlo simulation:

$$P(z > z_{\max}) = e^{-z_{\max}^2/2m_0} \tag{10.11}$$

$$m_0 = \int S_z(\omega)\, d\omega \tag{10.12}$$

where $P(z > z_{\max})$ is the probability that z exceeds the value $z_{\max}$ (where z is the vertical

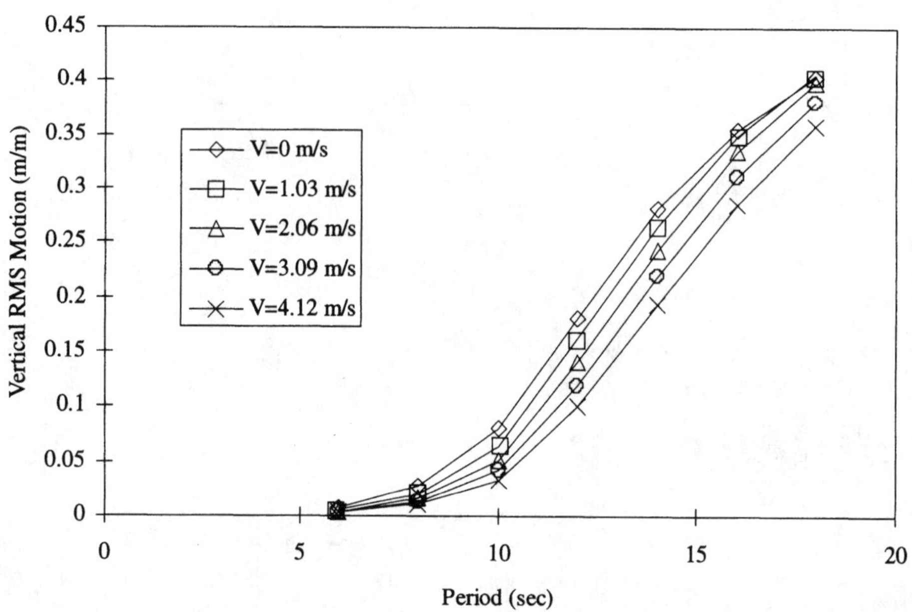

Figure 10.16 Vertical bow motion RMS values for H_s = 1 m and various vessel speeds.

Table 10.8 Variable assumptions used in squat and total vertical motion: 200,000-dwt tanker

Variable	Distribution	Mean	σ	Minimum	Maximum
Velocity (m/s)	Uniform	3.86	—	3	5
Tide (m)	Triangular	0.0	—	−0.25	0.25
Siltation (m)	Normal	0.0	0.1	−0.30	0.30

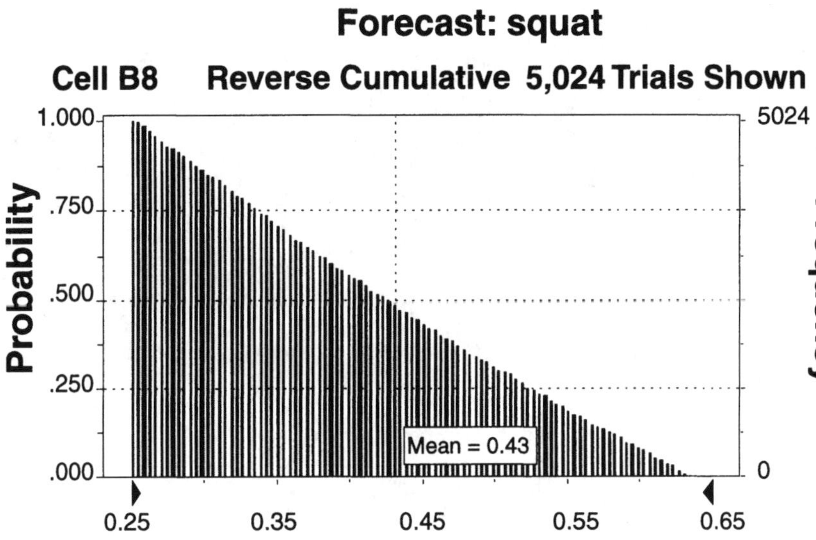

Figure 10.17 Cumulative probability distribution, squat.

motion amplitude and z_{max} is the approximate value of the maximum vertical motion amplitude), m_0 is the zero spectral moment of the vertical motion amplitude = RMS^2 and $S_z(\omega)$ is the vertical motion spectrum.

For the purposes of this chapter, the probability P was taken as 10^{-4}, as suggested by Kimon (1982), which corresponds to the vertical vessel motion, which will occur once in 10,000 for a given sea state. With regard to Monte Carlo computations, P is fixed in formula (10.11) and solved for values of z_{max} for various values of wave height, wave period, vertical RMS values, and vessel speed. As stated previously, the computations can be computed for various wave angles, but the results presented in this chapter correspond to head waves.

10.7.5 Results for Steady Conditions (No Waves)

Monte Carlo simulations for squat and total vertical motion (i.e., squat plus tide and siltation uncertainties) were computed for the conditions summarized in Table 10.8. Several distributions were examined for each of the stochastic values. The values of tide and siltation selected were consistent with the guidance suggested by Kimon (1982). Monte Carlo simu-

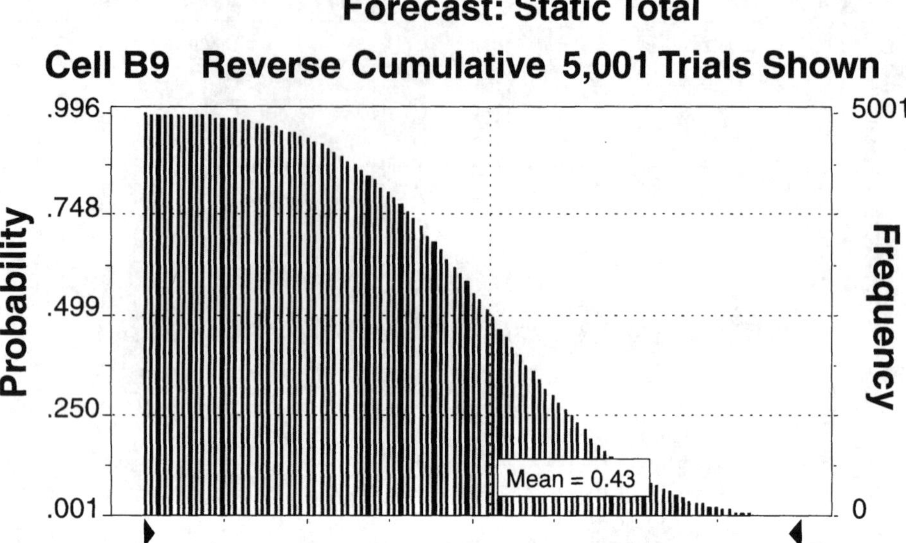

Figure 10.18 Cumulative probability distribution, total vertical motion (no waves).

Table 10.9 Variable assumptions used in squat and wave-induced total vertical motion: 200,000-dwt tanker

Variable	Distribution	Mean	σ	Minimum	Maximum
Velocity (m/s)	Uniform	3.86	—	3	5
Tide	Triangular	0.0	—	−0.25	0.25
Siltation	Normal	0.0	0.1	−0.30	0.30
Wave Height	Triangular	3.0	-	2.50	3.5
Wave Period	Triangular	16	-	14.0	18.0

lations were conducted for about 5000 trials, after which the simulated values of mean squat and total vertical motions had reached a more-or-less constant value. The simulated values of squat are summarized in Figure 10.17, and the total vertical motion of the vessel bow is summarized in Figure 10.18. With regard to the squat values computed, the mean squat value was 0.43 m, with maximum and minimum values of 0.25 and 0.64 m, respectively. These values can be compared to a deterministically computed value of 0.425 m for a vessel speed of 4 m/s. Accordingly, it can be concluded that the deterministic and mean values for the probabilistic values are similar; however, the range of possible squat values for the probability-based computations are much larger. A safe estimate of squat would consider the upper estimate of 0.64 m. The total vertical motion due to squat for deterministic computations would be 0.425 m, which neglects the stochastic nature of vessel speed and uncertainties due

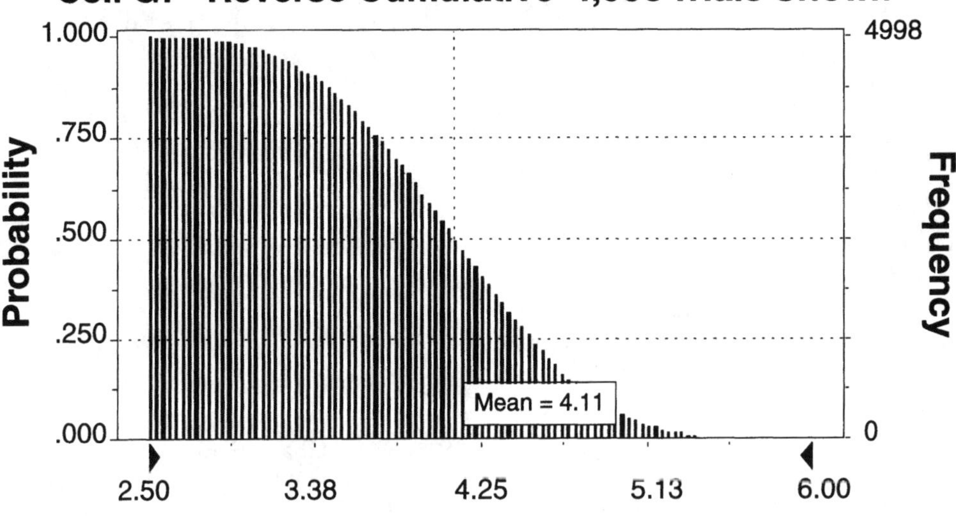

Figure 10.19 Wave-induced and squat total motions.

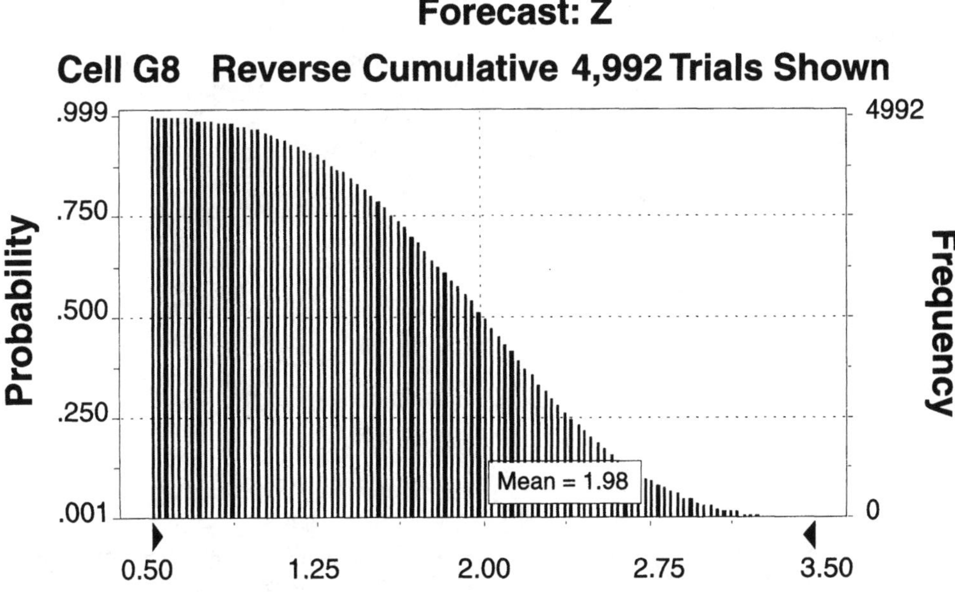

Figure 10.20 Simulation of failure function *Z*.

to variations in tide and siltation. Accounting for these variables, the probabilistic estimates indicate a range of predicted values extending from −0.18 m to a maximum of 0.9 m, which is roughly twice the mean and the value estimated deterministically.

10.7.6 Results for Wave Conditions

Monte Carlo simulations for squat and total vertical motion (i.e., squat and waves plus tide and siltation uncertainties) were computed for the conditions summarized in Table 10.9. In the absence of site-specific data, relatively simple probability distributions were estimated for the variations in wave height and period. Simulation values of total vertical motions are presented in Figure 10.19, which include the vertical motions due to combined wave-induced and squat excursions as well as uncertainties associated with variations in wave height, wave period, tide, and siltation. The maximum and minimum values of the total vertical motion is 5.64 and 2.42 m, respectively. The maximum value can be compared to the steady maximum value (without waves) of 0.9 m. Figure 10.20 summarizes the probabilistic-based estimates of the failure function Z. The maximum and minimum values of the Z function are 3.67 and 0.45, respectively. The lower Z value indicates that there is an adequate vessel UKC for the given vessel speed and wave conditions. A slight increase in wave height and/or period or a decrease in vessel speed would, however, probably result in a vessel grounding.

10.7.7 Summary

The proposed methodology provides a rational, quantitative, and statistically consistent method for determining UKC and, in turn, appropriate depths for navigation channels. Results using the methodology described in this chapter are generally comparable to those presented by Kimon (1982). Advantages of the present method, however, provide an easier means for accounting for the statistical uncertainty associated with the controlling stochastic variables. Applications of the methodology presented here can be used to make a rational assessment of the depth requirements for both interior channels (i.e., portions of the harbor free of waves) and ocean entrance channels exposed to large ocean waves. Further aspects of the model not discussed here include the impact of currents and directional wave spectra on model results and applications to other vessels types.

REFERENCES

Blaauw, H. G, and F. M. C. van der Knaap, 1983. Prediction of squat of ships sailing in restricted water, presented at the 8th International Harbour Congress, Antwerp, Belgium.

Eryuzlu, N. E., and R. Hausser, 1978. Experimental investigation into some aspects of large vessel navigation in restricted waterways, presented at the Symposium on Aspects of Navigability, Delft, The Netherlands.

Headland, J. R., 1995. Probabilistic design of navigation channel depths, *ASCE Proc. Specialty Conference Ports '95,* Tampa FL.

Kimon, P. M., 1982. Underkeel clearance in ports, presented at the Ship-Trans-Port Symposium, Rotterdam, The Netherlands.

NPC, 1975. *Port Approach Design: A Survey of Ship Behaviour Studies,* National Ports Council, London.

PIANC, 1995. Approach Channels, Preliminary Guidelines, PIANC Suppl. to Bull. 87, Permanent International Association of Navigation Congresses, Brussels, Belgium.

PIANC/IAPH, 1997. *Approach Channels: A Guide for Design,* Vol. 2, II-30(2), Permanent International Association of Navigation Congresses/International Association of Ports and Harbours, Brussels, Belgium.

PIANC/ICORELS, 1980. *Optimal Layout and Dimensions for the Adjustment to Large Ships of Maritime Fairways,* Rep. Work. Group IV, Permanent International Association of Navigation

Congresses/International Commission for the Reception of Large Ships, Brussels, Belgium.

Price, W. G., and R. E. D. Bishop, 1974. *Probabilistic Theory of Ship Dynamics,* Chapman & Hall, London.

U.S. Army Corps of Engineers, 1965. *Evaluation of Present State of Knowledge of Factors Affecting Tidal Hydraulics and Related Phenomena,* Rep. 3, Committee on Tidal Hydraulics, USACE, Washington, DC, May.

———, 1983. *Hydraulic Design of Deep-Draft Navigation Projects,* EM 1110-2-1613, USACE, Washington, DC.

———, 2001. *Coastal Engineering Manual,* EM 1110-2-1100, USACE, Washington, DC (6 vols.).

U.S. Coast Guard, 1981. *Aids to Navigation Manual-Administration,* COMDTINST M16500.7, USCG, Washington, DC.

U.S. Coast Guard, 1988a. *Aids to Navigation Manual-Seamanship.* CG-222.2, USCG, Washington, DC.

U.S. Coast Guard. 1988b. *Aids to Navigation Manual-Technical,* COMDTINST M16500.3, USCG, Washington, DC.

U.S. Navy, 1981. *Harbors,* DM-26.1, Naval Facilities Engineering Command, Alexandria, VA.

van Oortmerssen, G., 1975. *The Motions of a Moored Ship in Waves,* Publ. 510, Netherlands Ship Model Basin, Wageningen, The Netherlands.

11

DREDGING

Robert E. Randall
Texas A&M University
College Station, Texas

11.1 INTRODUCTION

11.1.1 Definitions

Dredging is the removal of bottom sediments from streams, rivers, lakes, coastal waters, and oceans, and the resulting dredged material is then transported by ship, barge, or pipeline to a designated disposal site on land or in the water where it is placed. According to Herbich (2000a and b), dredging is defined as raising material from the bottom of a water-covered area to the surface and pumping it over some distance. *Webster's* defines dredging as "(1) to dig, gather, or pull out with a dredge; (2) to deepen (as a waterway) with a dredging machine" and defines a dredge as "a machine for removing earth usually by buckets on an endless chain or suction tube." Dredged material is sediment that has been excavated by a dredge and has been or is being transported to a disposal site. The term *maintenance dredging* involves the removal of sediments that have accumulated since a previous dredging operation, and *new work dredging* involves the removal of materials that have not been previously dredged.

Dredging involves project planning, design, operation, and maintenance. Dredging, dredged material disposal, and other aspects of the overall navigation project should be considered as a total project. For example, the dredging and disposal equipment and procedures must be compatible. Navigation channel design dictates the dredging requirements, and refinement or modification to the navigation channels further influences the dredging requirements. Lake and cleanup dredging requires dredging over a dispersed area of localized shallow spots, precision removal of thin layers, and minimizing sediment resuspension during dredging.

Basic dredging requirements are determined by channel design and shoaling rates. The quantities of material to be dredged are determined from past records, and planning for dredging projects should be based on long-term requirements and hydrographic surveys. Horizontal positioning and depth measurements are conducted with electronic navigation and positioning equipment, and the data are usually reduced using computers. Accuracy and capabilities of positioning and surveying equipment using global positioning systems (GPS) have rapidly improved accuracy to within 1 m horizontally.

11.1.2 Brief History

Canal dredging was conducted as early as 4000 B.C. in the canals of Egypt using the labor of slaves, prisoners, and soldiers and with primitive tools such as spades and baskets. Huston (1970) indicates that the first dredger, called a spoon and bag dredger, scooped up the sediment with its spoon and placed the dredged material into the hold until it was full. The dredger then was poled or pulled to the shore, where the dredged material was again scooped out and placed on the shore. In 1435, the scraper dredge *Kraggelaar* loosened the bottom sediments and the water current carried the sediment out to sea. This type of dredging, called *agitation dredging,* is still used occasionally. The next dredge, called an *Amsterdam Mudmill* (Figure 11.1), was constructed of wood and operated by a treadmill turning an endless chain of buckets. The buckets excavated the sediment and placed it in a scow on the stern. Men operated the treadmill at first, and in 1620 a horse-driven Mudmill was used. The grab dredge or clamshell dredge was developed in Italy and Holland in the sixteenth century. The first hydraulic dredge was credited to the Basin of France in 1864. A centrifugal pump was built that had a 0.3-m (12-in.)-diameter discharge with a 0.61-m (24-in.) two-bladed impeller. Lebby conceived the first hydraulic hopper dredge in 1855, and it was named the *General Moultrie.* The 365-ton dredge had a wooden hull that was 46 m (150 ft) long, 3.1 m (10.3 ft) deep, and 8.1 m (26.7 ft) wide. It was equipped with a steam engine and a centrifugal pump with a 1.8-m (6-ft) impeller and a 0.5-m (19-in.)-diameter suction pipe. The production of this dredge was about 251 cubic meters (328 cubic yards) of dredged material per working day. Bazin introduced the idea of suction dredging in 1867 using a rotating harrow under the bow of a ship and suction pipes under the stern and applied it to dredging in the Suez Canal. Alexis von Schmidt designed the first pipeline suction dredge in 1874 and in 1876 obtained a patent for a three-spud pipeline suction dredge. Horace Agell developed the first two-spud dredge in 1884, and it contained most of the elements of a modern pipeline dredge. Modern dredges were developed in the twentieth century, and these dredges operate either mechanically or hydraulically, with most dredges using the hydraulic

Figure 11.1 Early dredge *Mudmill*. (From Herbich, 2000b. Reproduced with permission of The McGraw-Hill Companies.)

principle. These modern dredges have efficient pumps, heave compensating devices, electronic equipment for automatic controls, water jets, sophisticated navigation equipment, and advanced instrumentation. Land reclamation projects and environmental constraints continue to force the development of new advances in dredging and dredging equipment. In 2001, the pipeline dredges are as large as 95 cm (38 in.) in diameter, and ladder pumps have been added to the dredge ladder to extend the digging depth beyond 10 m (30 ft). Hopper dredges have two dragarms with submerged pumps to extend the digging depth to over 100 m (330 ft). The largest hopper dredge currently (2001) is the *Vasco deGamma,* which has a hopper capacity of 33,000 m^3 (43,000 yd^3).

11.1.2.1 Dredge Classification. Excellent sources of information related to dredging and dredging equipment include U.S. Army Corps of Engineers (USACE, 1983), de Heer and Rochmanhadi (1989), Bray et al. (1997), and Herbich (2000b). Dredges are classified as either mechanical or hydraulic as shown in Figure 11.2.

11.1.3 Characteristics of Dredged Sediments

11.1.3.1 Definitions. Dredged sediment is composed of solid particles, water, and gas. Sediments are commonly classified as gravel, sand, silt, and clay, depending on particle size. A sieve analysis is used to determine particle distribution if particles are sufficiently large. A sediment sample is shaken through a set of sieves with progressively smaller openings, and the results are illustrated in Figure 11.3.

Hydrometer analysis is conducted for particles finer than 0.04 mm. The diameter D_{10} means that 10% of the soil particles are finer than this diameter, and D_{30} is the diameter at which 30% are finer. D_{50} is the diameter at which 50% of the soil particles are finer than this diameter. Similarly, D_{60} is the diameter at which 60% of the soil particles are finer than this diameter. The uniformity coefficient (C_u) is an index of the particle size uniformity, and the coefficient of curvature (C_c) is

$$C_u = \frac{D_{60}}{D_{10}} \qquad C_c = \frac{(D_{30})^2}{D_{60}D_{10}} \tag{11.1}$$

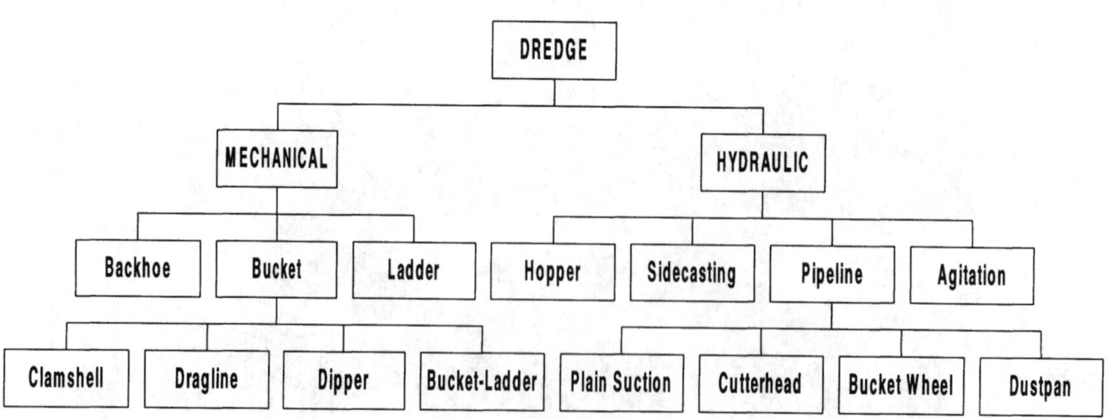

Figure 11.2 Classification of dredges.

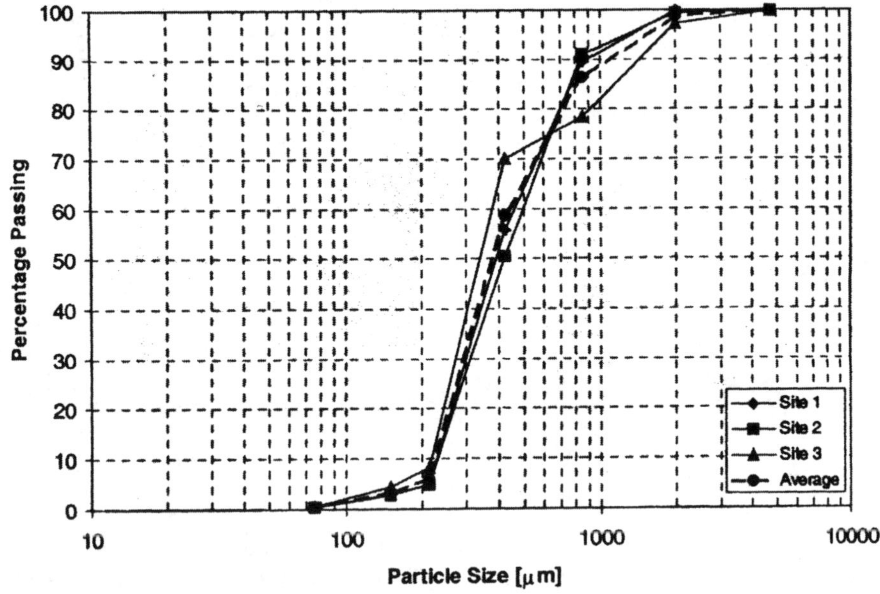

Figure 11.3 Results of example sieve analysis. (From Albar, 2001.)

If C_c is between 1 and 3 and if C_u exceeds 4 for gravels or 6 for sands, the soil is well graded (W). Otherwise, the sediment is poorly graded (P).

The porosity (n) is the ratio of the void volume (V_v) to the total volume (V) of the soil and is expressed as

$$n = \frac{V_v}{V} \qquad (11.2)$$

The void ratio (e) is the ratio of the volume of voids (V_v) to the volume of the solids (V_s) and is written as

$$e = \frac{V_v}{V_s} \qquad (11.3)$$

The water content (w) is the ratio of the weight of water (W_w) to that of the solid (W_s) and expressed as a percentage:

$$w = \frac{W_w}{W_s} \times 100 \qquad (11.4)$$

The degree of saturation (S_r) is the percentage of void space (V_v) that is occupied by water (V_w), expressed as

$$S_r = \frac{V_w}{V_v} \times 100 \qquad (11.5)$$

The unit weight (γ) is defined as the weight of material (W) divided by its volume (V), written as

$$\gamma = \frac{W}{V} = \frac{W_s + W_w}{V_s + V_w + V_a} = \frac{W_s + W_w}{V_s + V_v} \qquad (11.6)$$

The weight of dry material (W_s) divided by its volume is the dry unit weight (γ_d):

Table 11.1 Sediment classification according to IADC

Sand	Silt	Clay	Peat	Gravel
Weight	Weight by volume	Weight	Weight	Weight
Water content	Water content	Water content	Water content	Water content
Grain specific gravity	Grain size	Sliding resistance	Sliding resistance	Grain specific gravity
Grain size	Water permeability	Consistency ranges (plasticity)	Consistency ranges (plasticity)	Grain size
Water permeability	Sliding resistance (shear strength)	Organic content	Organic content	Water permeability
Frictional properties	Plasticity			Frictional properties
Lime content	Lime content			Lime content
Organic content	Organic content			Organic content

$$\gamma_d = \frac{W_s}{V} \tag{11.7}$$

The weight of solid particles (W_s) divided by the volume of dry solids (V_s) is defined as the unit weight of solids (γ_s):

$$\gamma_s = \frac{W_s}{V_s} \tag{11.8}$$

The Atterberg limits (shrinkage, plastic, and liquid) are the water contents at which soil consistency changes from one state to another. The plastic limit (PL) is the water content at which the soil crumbles when it is rolled down to a 4.6 cm (1.8 in) diameter thread. The liquid limit (LL) is the water content at which the soil on two sides of a groove flows together after the dish that contains the soil has been dropped 25 times through a distance of 1 cm. The shrinkage limit (SL) is the water content at which the soil no longer shrinks in volume upon further drying. The plasticity index (PI) is the difference between the liquid and plastic limits:

$$\text{PI} = \text{LL} - \text{PL} \tag{11.9}$$

11.1.3.2 Soil Classification. The International Association of Dredging Companies (IADC) system for identifying soil types and the parameters necessary for evaluating dredgeability are tabulated in Table 11.1. The classification system developed by the Permanent International Association of Navigation Congresses (PIANC) is as shown in Table 11.2. The United Soil Classification System (USCS) has been used in the United States. The soil is classified according to texture, plasticity, and engineering behavior. Table 11.3 illustrates the grain sizes recognized by the USCS.

11.1.3.3 Soil Investigations. The two main types of investigations are desk studies and field sampling. If sufficient historical and local information is available, a desk study can be sufficient to establish the geotechnical data. Field sampling is necessary for new work projects. This sampling includes drilling and boring in accordance with standards such as those established by the American Society for Testing and Materials (ASTM), British Standards Institution (BS), or German Standards (DIN). All dredging projects and sites are unique and the geotechnical site investigation must be tailored to the needs of the specific project. Open-

Table 11.2 Sediment classification system according to PIANC

Soil Type	Particle-Size Range (mm)	Sieve Size
Boulders, cobbles	>200	6 in.
Gravels		
Coarse	200–60	3–$\frac{3}{4}$ in
Medium	20–6	$\frac{3}{4}$–$\frac{1}{4}$ in.
Fine	6–2	$\frac{1}{4}$ in.–No. 7 sieve
Sands		
Coarse	2–0.6	No. 7–25
Medium	0.6–0.2	No. 25–72
Fine	0.2–0.06	No. 72–200
Silt		
Coarse	0.06–0.02	Passing No. 200 sieve
Medium	0.02–0.006	
Fine	0.006–0.002	
Clays	Below 0.002	Not applicable
Peats/Organics	Not applicable	Not applicable

water investigations are even more difficult, due to waves and currents. Soil properties for various types of equipment used in dredging are tabulated in Table 11.4.

Table 11.3 United Soil Classification System (USCS) definition of grain sizes

Component	Size Range
Cobbles	Above 76.2 mm (3 in.)
Gravel	76.2–4.76 mm (3 in. to No. 4 sieve)
Coarse	76.2–19.1 mm (3 in. to $\frac{3}{4}$ in.)
Fine	19.1–4.76 mm ($\frac{3}{4}$ in. to No. 4 sieve)
Sand	4.76–0.074 mm (No. 4 to 200 sieve)
Coarse	4.76–2.0 mm (No. 4 to 10 sieve)
Medium	2.0–0.42 mm (No. 10 to 40 sieve)
Fine	0.42–0.074 mm (No. 40 to 200 sieve)
Fines (silt or clay)	Below 0.074 mm (No. 200 sieve)

Drilling methods obtain soil samples down to the maximum dredging depth. Drilling methods include wash borings, augering, and rotary drilling. Rotary drilling requires boats, jack-up rigs, or bottom-supported rigs. Surficial sampling methods include surficial samplers (grab samplers), projectile or impact tube samplers, thin-walled tube samplers, or vibrating core samplers and are illustrated in Table 11.5.

11.1.3.4 Dredged Material Bulking Factors. Dredging by hydraulic methods mixes about 20% solids with 80% water, and this produces what is called *bulking*. Thus, the dredged material being placed in a disposal area is larger in volume than the dredged material in-situ. This increased volume is accompanied by increases in void ratio and water content. For clays and silts, the bulking may be quite substantial. The bulking factor (B) of a particular soil is defined as the ratio of the volume of soil in a containment area (V_c) after dredging to that volume of the soil in situ (V_i):

$$B = \frac{V_c}{V_i} \quad \text{(saturated conditions)} \qquad (11.10)$$

Table 11.4 Soil properties for various types of dredging equipment

	Soil Properties[a]											
Dredging Equipment	1	2	3	4	5	6	7	8	9	10	11	12
Hydraulic excavation												
Plain suction	×			×	×	×	×	×				
Trailing draghead suction	×			×	×	×	×	×				
Dustpan suction		×		×	×	×		×				
Mechanical excavation												
Bucket, shovel, backhoe, dragline	×	×	×	×	×	×	×	×				
Cutter blades rotary/fixed	×	×	×	×	×	×	×	×				
Hydraulic removal												
Suction pipeline				×	×	×	×	×	×	×		
Mechanical removal												
Bucket, shovel, backhoe, dragline				×	×	×	×					
Bucket-ladder, bucket wheel				×	×	×	×					
Hydraulic transport												
Pumped slurry in pipeline				×	×	×	×	×	×	×		
Mechanical transport												
Hopper (own hold)						×	×	×			×	×
Barge, self-propelled or towed						×						×
Land-based trucks, belts, etc.						×						×
Hydraulic disposal												
Pipeline slurry—land disposal						×	×	×			×	×
Pipeline slurry—water disposal						×	×	×			×	×
Mechanical disposal												
Bottom discharge—hopper or barge						×	×					×
Grabs/scrapers for emptying barges						×	×					×

[a]Compactness of granular soils; 2, consistency of cohesive soils; 3, sensitivity of cohesive soils; 4, water content; 5, mass density; 6, grain-size distribution; 7, plasticity of fines; 8, organic content; 9, particle shape and hardness; 10, rheologic property of slurry; 11, sedimentation rate in water; 12, bulking factor.

The bulking factor can also be expressed in terms of the specific weight as shown in

$$B = \frac{\gamma_{d,i}}{\gamma_{d,c}} \tag{11.11}$$

and in terms of the density and specific gravity of the solids (G_s) as

$$B = \frac{w_c G_s + 100}{w_i G_s + 100} \tag{11.12}$$

The type of sediment, dredging method, and particle-size distribution affect the bulking factor. Mechanical dredges normally cause smaller bulking of the in situ dredged material, and hydraulic dredges cause the larger bulking. Bray et al. (1997) presented bulking factors for different types of sediments dredged by a mechanical dredge that are tabulated in Table 11.6.

11.2 DREDGING METHODS AND EQUIPMENT

11.2.1 Large Hydraulic Dredges

Hydraulic dredges are commonly classified as large when the diameter of the pump discharge

Table 11.5 Geotechnical sampling equipment

Device Name	Illustration	Weight (lb)	Description
Peterson grab		40–90	Collects sample covering area of 1 ft^2 to a depth of 1 ft, depending on sediment type.
Shipek	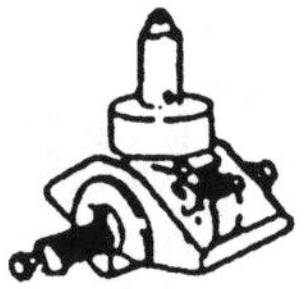	150	Collects sample covering area of 64 in^2 to a depth of 4 in.
Ekman	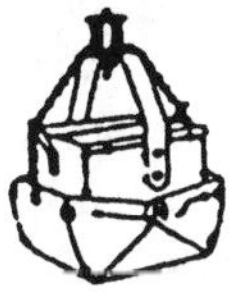	9	Works only in soft sediments covering an area of up to 64 in^2 to a depth of approximately 6 in.
Ponar		45–60	Collects sample covering area of 81 in^2 to a depth of 1 ft. Not effective in clay.
Drag bucket		Varies	Collects shallow sediment slice near the surface and comes in several sizes and shapes.
Phleger corer	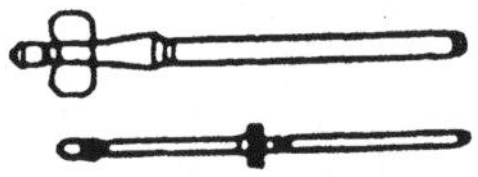	20–90	Core samples obtained by self-weight penetration or by pushing barrel into the sediment. Depth of penetration depends on sediment, with 2 to 6 ft being accomplished in soft sedimelnts.
Piston corer	[Picture not available]	200–300	Core samples obtained by self-weight and water pressure. Depth of penetration up to 80 ft.

and discharge pipeline are grater than 305 mm (12 in.). Hydraulic dredges perform both phases of the dredging operations (digging and disposing). Disposal is accomplished by pumping the dredged material through a pipeline that is floating, on land, or both to the disposal area or by storing the dredged material in hoppers that are emptied over a designated disposal area. Hydraulic dredges are more efficient, versatile, and economical. The dredged material is first loosened and mixed with ambient water by cutterheads or water jets and pumped as a fluid (slurry) through a long pipeline or to a hopper. The basic components of a hydraulic dredge are: dredge pumps, digging and agitation machinery, and hoisting and hauling equipment. Hydraulic dredges are categorized as hopper (trailing suction), pipeline (plain suction, cut-

Table 11.6 Bulking factors for different sediments using a mechanical dredge

Sediment Type	Bulking Factor, B
Hard rock (blasted)	1.50–2.00
Medium rock (blasted)	1.40–1.80
Soft rock (blasted)	1.25–1.40
Gravel, hard packed	1.35
Gravel, loose	1.10
Sand, hard packed	1.25–1.35
Sand, medium soft to hard	1.15–1.25
Sand, soft	1.05–1.10
Silts, freshly deposited	1.00–1.10
Silts, consolidated	1.10–1.40
Clay, very hard	1.15–1.25
Clay, medium soft to hard	1.10–1.15
Clay, soft	1.00–1.10
Sand/gravel/clay mixtures	1.15–1.35

Source: Bray et al. (1997).

terhead, dustpan), bucket wheel, and sidecasting.

11.2.1.1 Hopper Dredge. The development of self-propelled trailing suction hopper dredge revolutionized the dredging industry by reducing costs. These dredges are used extensively in Europe and the United States, and they can work in all but hard materials. Hopper capacities of several hundred to 33,000 m^3 have been built, and one of the largest is the *Vasco da Gama,* shown in Figure 11.4. The hoppers are usually unloaded through the bottom doors and some have pump-out facilities. The maximum dredging depth is 18 to 21 m (59 to 69 ft), but it can be increased to 40 m (131 ft) with the addition of a submerged pump at the draghead. The dragarms and dragheads extend from both sides of the hull, and each is lowered to the sea bottom. The dredge moves slowly over the area to be dredged while the dredge pumps move the sediment and water mixture (slurry) through the dragarm into the hopper bins. When these hopper bins are full, the dredge moves under its own propulsion system to a designated disposal area and empties the dredged material through the hopper doors in the bottom of the hull, or it uses inboard pumps to transport the dredged material through pump-out lines to shore for beneficial uses such as beach nourishment. Some of the modern dredges have a single hopper in the midsection, and the dredge is designed to split open and allow the dredged material to exit as the hull splits open (split-hull hopper dredge). The split-hull hopper dredge *Padre Island* is shown in Figure 11.5.

A hopper distribution system tries to minimize turbulence and allow for quick settling of the solid material. The slurry enters the hopper from some height entraining air that tends to keep material in suspension. Some distribution lines have been installed below the waterline to reduce entrainment and turbulence. Overflow weirs are located at opposite end of the hopper from where the slurry enters, to allow the sediment to settle to the bottom. The water is then allowed to overflow and return to the ambient water. Gratings are also installed in the hopper near the slurry inflow location to reduce entrainment and turbulence. The installation of submerged pump on the draghead or in the dragarm permits dredging at deeper depths. The addition of a submerged pump increases the dredging depth by reducing the chances of cavitation, and it permits the pumping of a higher-specific-gravity material. Most modern hopper dredges have pumps mounted on the dragarm. The draghead is an important part of the hopper dredge. Several types of dragheads are used, such as the California, Ambrose, Venturi, and IHC types. A grating is used to prevent large objects from entering the suction pipe. A new type of draghead uses a rotating cylinder with knives. A Venturi draghead creates a negative pressure just above the seabed that results in a 30 to 40% increase in the production of fine sand. An automatic draghead winch control system controls the movement of

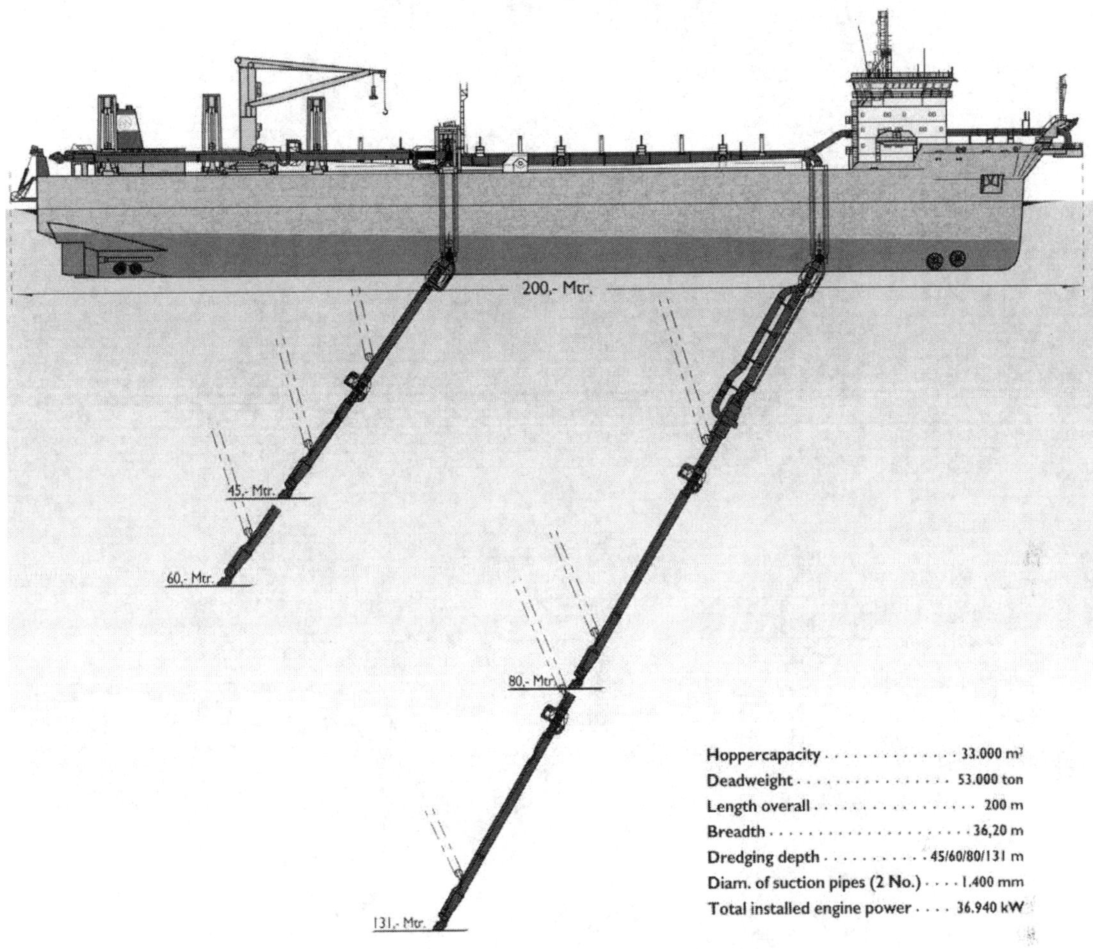

Figure 11.4 Hopper dredge *Vasco da Gama*. (Courtesy of Jan Van Den Nul.)

the suction pipe and draghead. It is coupled with a swell compensation system to maintain the correct pressure of the draghead on the bottom and the lateral position of the pipe hoist. The hopper dredge has the increased ability to operate in bad weather and minimizes the risk of damage to equipment.

11.2.1.2 Cutter Suction Dredge. The pipeline cutterhead dredge is a very versatile dredge, and its primary function is to excavate and move material hydraulically to a disposal location without rehandling. It is categorized by the size (diameter) of its floating discharge line. For example, a 61-cm (24-in.) dredge has a floating discharge line that has a diameter of 61 cm (24 in.). An operating cutterhead dredge is illustrated in Figure 11.6, showing pipeline, cutter, ladder pump, spuds, and the main cabin that houses the dredge pumps and crew facili-

Figure 11.5 Split-hull hopper dredge *Padre Island*. (Courtesy of Great Lakes Dredge and Dock.)

Figure 11.6 Cutterhead dredge *Alaska*. (Courtesy of Great Lakes Dredge and Dock.)

ties. During a dredging operation, the floating discharge and shore pipeline are connected to the dredge. Additional equipment to support the operation is required, such as a derrick, tugs, fuel and pipe barges, surveying boats, and other site-specific special equipment. A cutter is connected at the forward end of the ladder and connected to the shaft of the cutter motor. There are generally two types of cutters, classified as straight-arm or basket. Rotation of the shaft and cutter agitates soft or loose material and cuts hard material that is then picked up by the suction.

On a cutterhead dredge, the ladder supports the cutter, suction pipe, lubricating lines, and usually the cutter motor and reduction gear. The forward end of the ladder is supported by a framed structure with hoisting equipment to raise and lower the ladder. The length of the ladder determines the dredging depth and the dredging depth is typically considered to be 0.7 times the length of the ladder. Ladder lengths may be 7.6 m (25 ft) or less to 45.7 m (150 ft) or more and can weigh as much as 400 tons. The suction pipe supported beneath the ladder transports the dredged material to the dredge pump. The discharge line is connected to the floating pipeline by a ball joint. The diameter of the discharge line depends on the pump size, and the suction pipe diameter is usually 1.25 to 1.5 times the pump discharge diameter. Practically, the suction pipe is one size larger than the pump discharge, such as a 41-cm (16-in.) pump discharge and a 46-cm (18-in.) suction pipe.

The dredge pump is located forward in the hull with its center near the loaded waterline. A diesel engine, diesel electric motor, steam, or gas turbine can be used to drive the pump. In some cases, shore electric power may be used to drive the pump. The horsepower required varies from 186 to more than 11,186 kW (250 to more than 15,000 hp). The pump rotative speed varies from about 300 to 900 rpm. The dredge is moved and held in position with spuds. These devices (usually, two) are as big as 1.22 m (4 ft) in diameter, 30.5 m (100 ft) long, and weigh as much as 30 tons. Spuds are typically spaced not less than one-tenth the distance between the stern of the dredge and the cutter, so that moving ahead is not limited. A walking spud is a conventional spud placed in a groove such that it can move longitudinally along the centerline of the dredge.

The discharge line consists of three sections (pipe on dredge, floating line, and shore pipeline). The pipe on dredge runs along the dredge deck from the pump discharge to the stern, and it has a flap valve near the pump to prevent back flow. At the stern, it is connected to a swivel elbow or ball joint. The floating line extends from the stern of the dredge to the shore. It is made of 9.1 to 15.2 m (30 to 50 ft) sections, and floating pontoons supports each section. Strongbacks are used to connect the pontoons, and anchors are used where necessary to hold the pontoons in place. A walkway is often constructed on top of the pipeline. Floating pipelines are used more often when waves are expected. The shore pipeline consists usually of shorter and lighter sections with a Y-connection at the discharge end. A preliminary selection guide for cutterhead dredges is shown in Figure 11.7, based on production and length of discharge pipeline. Portable cutterhead dredges have been developed ranging from 20.3 cm (8 in.) to a 50.8-cm (20-in.) dredge. These are versatile and can undertake many small dredging projects.

11.2.1.3 Dustpan Dredge. A dustpan dredge is a hydraulic, plain suction vessel. The wide (about the same as the hull width) vacuum cleaner–like head is lowered to the bottom by winches. It has a high-velocity water jet to agitate and loosen material that is subsequently pumped through a floating pipeline to a disposal area. The dustpan dredge *Burgess* is shown in Figure 11.8.

11.2.1.4 Bucket Wheel Dredge. A rotating wheel equipped with bottomless buckets is used to cut or loosen soil which is then directed into the interior of the wheel and conveyed to the suction line. The bucket wheel is attached to the ladder as shown in Figure 11.9, but newer designs use a dual-wheel concept.

11.2.1.5 Agitation Dredge. An agitation dredge resuspends the dredged material and lets the natural currents move the sediments to another location in the dredged channel. The dredged sediments settle to the channel bottom, where the currents have slowed below the critical velocity that keeps the sediments in suspension. Another type of agitation dredge is called a water injection dredge. This dredge lowers its dredge head into the sediments and injects water into the sediment. If there is a slope in the channel, the liquefied mud

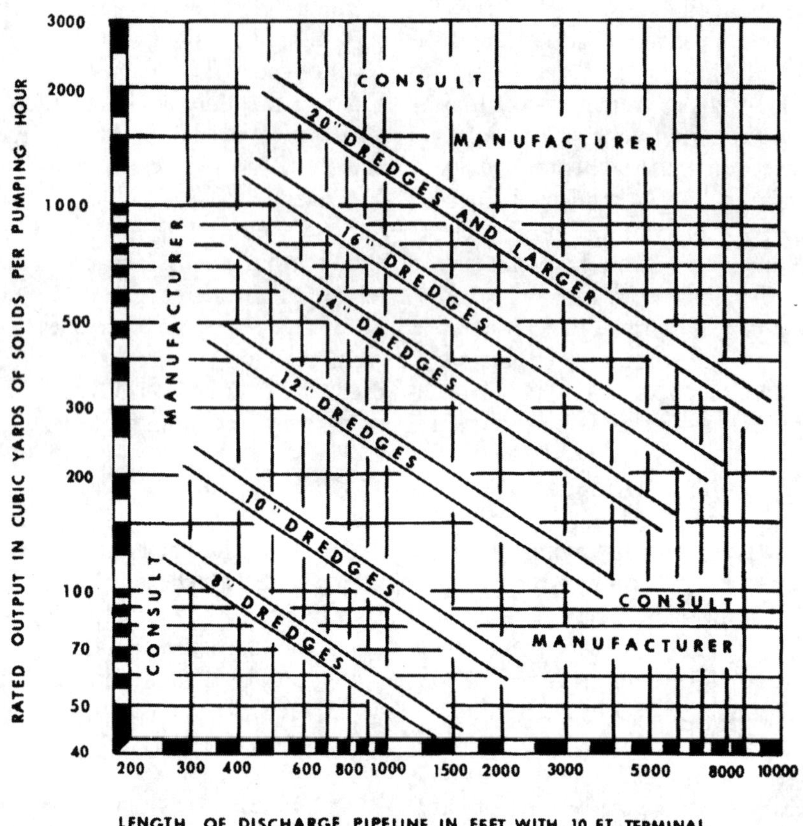

Figure 11.7 Preliminary selection guide for dredges. (From Herbich, 2000b. Reproduced with permission of The McGraw-Hill Companies.)

(dredged material) flows as a density current down the slope of the channel to a point of lower elevation.

11.2.1.6 Sidecasting Dredge. Sidecasting dredges discharge the dredged material to the side of the channel and allow for continuous dredging. Some sidecasting dredges are equipped with a hopper and a sidecasting boom as shown in Figure 11.10.

11.2.2 Small Hydraulic Dredges

Small hydraulic dredges are considered to have a discharge line that is 305 mm (12 in.) in diameter or less. These small hydraulic dredges, like large hydraulic dredges, perform both phases of the dredging operation (digging and disposing). The dredged material is first loosened and mixed with the ambient water by cutter heads or water jets and pumped as a fluid (slurry) through a long pipeline (cutter suction dredge) or to a hopper (hopper dredge). Disposal is accomplished by pumping the dredged material through a floating pipeline to a disposal area or by storing the dredged material in hoppers that are emptied over a disposal area. The basic components of a hydraulic dredge are dredge pumps, digging and agitation machinery, and hoisting and hauling equip-

Figure 11.8 Dustpan dredge. (Courtesy of U.S. Army Corps of Engineers.)

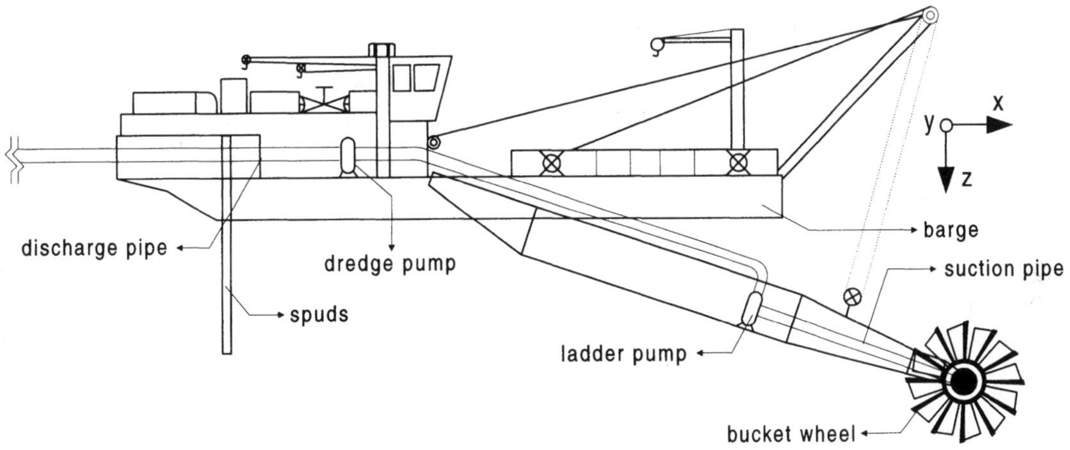

Figure 11.9 Bucket wheel dredge. (From Albar, 2001.)

ment. Hydraulic dredges are very versatile, efficient, and economical.

11.2.2.1 Cutter Suction. Similar to large cutter suction dredges, the small cutter suction dredge has a cutter that excavates and loosens sediment and makes it available to the suction pipe. The reduced pressure created by a centrifugal pump allows hydrostatic pressure to force the sediment and water through the suction pipe. The pump creates a pressure head on the discharge side that provides the energy to transport the sediment through the discharge pipe. Energy losses are caused by friction and minor losses. As long as the pump head is greater than the losses in the discharge pipe,

Figure 11.10 Sidecasting dredge *Schweitzer*. (Courtesy of U.S. Army Corps of Engineers.)

slurry will flow. A certain average velocity, called the *critical velocity,* is required to keep the sediment in suspension. If the average velocity in the pipeline drops below the critical velocity, the sediment begins to settle along the bottom of the pipeline and can clog the pipeline. A clogged pipeline can cause considerable downtime. Booster pumps placed along the pipeline are used to add energy to the pumping process and allow the delivery of the sediment through very long pipelines. Small hydraulic dredges (<0.3-m or 12-in. diameter) with pipelines longer than several thousand feet typically require booster pumps.

Cutter suction dredges have been shown to generate turbidity, but the turbidity is usually confined to near the bottom. Turbidity has been measured in the range of 5 to 200 mg/L (McLellan et al., 1989). Control devices placed around the cutter suction inlet have also been found to be effective in reducing turbidity (Hayes et al., 1988). These modifications include shielded cutters, cutterless dredgeheads, and different cutter designs such as the matchbox dredgehead. The pump location for a cutter suction dredge affects dredge production. Dredge pumps are limited by cavitation, and deep digging depths can severely limit the amount of sediment a dredge can pump before cavitation begins and greatly reduces pump efficiency. Therefore, many cutter suction dredges place a pump on the ladder that essentially eliminates cavitation.

Other cutter suction dredge limitations are the result of the advance mechanism used to advance the dredge through the dredging area. Fixed spuds typically result in the cutterhead being in the sediment only half of the time. The use of a spud carriage allows the dredge to increase its efficiency to approximately 75%. Having a bank height at least as large as the cutter diameter, which is two to three times the suction pipe diameter, allows the dredge to pump the maximum solids concentration. The cutter and swing winch power can also have an effect on dredging efficiency. An experienced operator can usually maintain an average slurry specific gravity of 1.2 to 1.3. Higher specific gravity can occasionally be attained but not for long periods of time. Thus, high solids content may be pumped for short periods, but the high solids content cannot be maintained. Several dredging companies have 203 to 305 mm (8 to 12 in.) small hydraulic dredges, and specialized dragheads could be developed that minimize turbidity and shield against debris entering the dredge system. Large debris entering the pumping system will cause the system to shut down. Underwater acoustic surveys of the dredging area are suggested to identify the location of the large debris and remove it prior to dredging.

Oil booms and silt curtains may be used to prevent the propagation of turbidity plumes and oil slicks. Water quality measurement programs need to be established to identify limits for turbidity levels. If the limits are exceeded, operations are suspended or operational modifications are made that bring the turbidity levels down to an acceptable level.

During 1993, the U.S. Army Corps of Engineer's Waterways Experiment Station conducted a survey of equipment that could be used for dredging contaminated sediment (Parchure, 1996). The survey was limited to equipment available in the United States. Sixty-four companies were contacted and 24 of these provided information on their products. Parchure (1996) determined that only seven companies offered equipment that appeared promising for dredging contaminated sediments, and the characteristics of these dredges and the Tornado 1 dredge with the Eddy pump are illustrated in Table 11.7.

Shark. The Dredging Supply Company, Inc. in Harvey, Louisiana manufactures the Shark cutterhead dredge (Figure 11.11) These small dredges use a centrifugal pump that has a 254-mm (10-in.)-diameter suction, 254-mm (10-in.)-diameter discharge, and a 813-mm (32-in.)-diameter impeller. The 254-mm (10-in.) pump has a continuous horsepower of 240 kW (322 hp) and the 305-mm (12-in.) pump has 300 kW (402 hp). The dredges are transportable by truck over the highway. The overall height from the water is 2.7 m (9 ft) with the spuds down, and the hull length is 12 m (39.5 ft). The maximum digging depth is 8.5 m (28 ft) and 9.5 m (31 ft) for the 254-mm (10-in.) and 305-mm (12-in.) dredges, respectively.

Versi-Dredge. The Versi-Dredge is manufactured by Innovative Material Systems, Inc. and has a shrouded cutterhead that minimizes turbidity. This small hydraulic dredge uses a 254-mm (10-in.) suction and discharge line, and the production rate is given as variable.

Super-Mudster. Dredgemaster International manufactures the Super-Mudster, and these dredges range in size from 203 to 406 mm (8 to 16 in.). They are designed to operate at various depths and production rates. Figure 11.12 is a useful guide for estimating the production rate for a given pipeline length. The 254-mm (10-in.) Super-Mudster operating with 1220 m (4000 ft) of pipeline is estimated to have a production of about 115 m^3/h (150 yd^3/hr). In order to pump light material a distance of 5 miles (8050 m or 26,400 ft), the pumping system would need booster pumps.

Matchbox. The matchbox dredge is a cutter suction dredge with a modified dredgehead called the matchbox (Figure 11.13). It consists of a cover that contains sediment and prevents the inflow of excess water. Side openings guide the sediment toward the suction intake as the dredge swings. These side openings are open in the direction of swing and closed on the opposite side to restrict the inflow of excess water. The dredge head has an angle control that keeps the dredge head parallel to the bottom contours regardless of the depth. Grates are located in the side openings to prevent ingestion of debris. The matchbox dredgehead was first used in First Petroleum Harbor in the Netherlands. It was tested in the United States in Calumet Harbor and New Bedford Harbor.

Table 11.7 Selected companies using small hydraulic dredges

Company and Location	Dredge Name	Dimensions (ft and in.)	Weight (lb)	Maximum Operating Depth (ft)	Suction and Discharge Pipe Dia. (in.)	Production Rate (yd^3/hr)	Pumping Distance (ft)	Cutter Type (C) Pump Type (P)
Keene Engineering Co., Northridge, CA	Nessie Model 8DX	*L*: 19 ft 1 in. *W*: 7 ft 2 in Draft: 16 in.	14,500	21.75	S: 8 D: 8	200	NA	C: bucket wheel P: centrifugal
Aquatics Unlimited, Martinez, CA	Aquamog Model PRX163	*L*: 30 ft 6 in. W: 10 ft Draft: 18 in.	18,500	20	S: 6 D: 6	60	5,000	C: auger, bucket, basket P: NA
Ellicott Machine Corp., Baltimore, MD	Mudcat Model 815	*L*: 31 ft 1 in. *W*: 8 ft Draft: 22 in.	13,200	15	S: 6 D: 6	50	NA	C: auger P: centrifugal
	Mudcat Model SP-915	*L*: 39 ft 5.5 in. *W*: 9 ft Draft: 21 in.	23,000	15	S: 8 D: 6	150	NA	C: auger P: centrifugal
	Mudcat Model MC-915	*L*: 39 ft 5.5 in. *W*: 9 ft Draft: 21 in.	22,000	15	S: 8 D: 8	400	10,000	C: auger P: centrifugal
	Mudcat Model MC 2000	*L*: 47 ft *W*: 8.5 ft Draft: 2.5 ft	28,300	20	S: 10 D: 8	125	1,400	C: basket, auger P: centrifugal

Dredgemasters International, Hendersonville, TN	Model HPC-8EC	*L*: 36 ft *W*: 10 ft Draft: variable	Custom designed	Custom designed	S: 10 D: 8	Variable	NA	C: basket, auger, dustpan P: centrifugal
	Super-Mudster HP-250-SM	NA	39,000	NA	S: 10 D: 10	200	3,000	C: basket, auger, dustpan P: centrifugal
Dredging Supply Company, Harvey, LA	Barracuda	*L*: 65 *W*: 17 ft 8 in. Draft: 3–4 ft	72,000	18	S: 10 D: 10	NA	NA	C: basket P: centrifugal
	Shark	*L*: 61 ft *W*: 11 ft 10 in. Draft: 4 ft	62,000	28	S: 10 D: 10	NA	NA	C: basket P: centrifugal
	Shark	*L*: 70 ft *W*: 11 ft 10 in. Draft: 4 ft 6 in.	74,000	31	S: 12 D: 12	NA	NA	C: basket P: centrifugal
Eddy Pump Corporation	Tornado 1	*L*: 82 ft *W*: 26 ft Draft: NA	NA	NA	S: 8 D: 10	NA	NA	C: none P: eddy (4-arm rotor creates vortex motion)
Innovative Material Systems, Olathe, KS	Versi-Dredge Model 4010	*L*: 30 ft 2 in. *W*: 9 ft 4 in. Draft: 20 in.	12,000	20	S: 10 D: 10	Variable	NA	NA

Figure 11.11 Shark dredge. (From Parchure, 1996.)

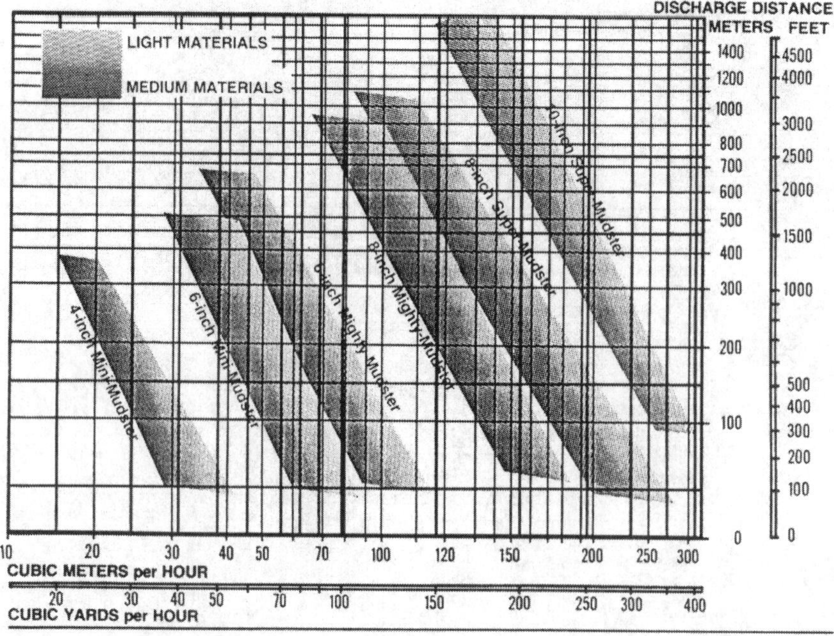

Figure 11.12 Estimated performance for Dredgemaster International cutter suction dredges. (Courtesy of Dredgemaster International.)

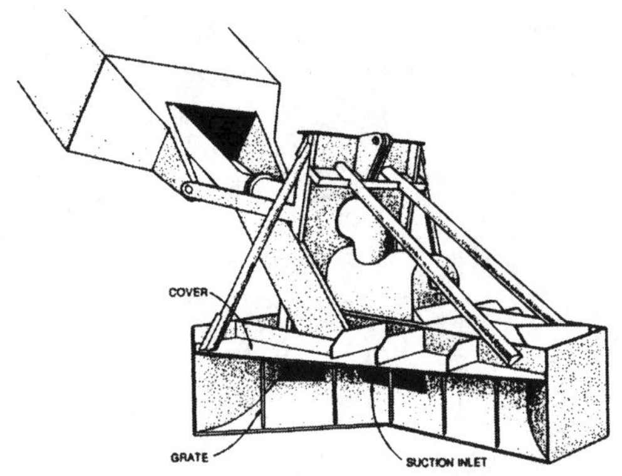

Figure 11.13 Matchbox dredgehead. (From Zappi and Hayes, 1991.)

Hayes (1986) reported the highest suspended sediment concentration was 31 mg/L above background. At New Bedford Harbor, Bean Dredging installed a matchbox dredgehead on one of their 305-mm (12-in.) cutterless dredges. Otis et al. (1990) reported average daily suspended solids at middepth of 79 to 609 mg/L, and background levels were observed 150 m from the dredge head.

11.2.2.2 Auger Dredge

Mudcat Auger Dredge. Buchberger (1993) reported results on Environment Canada's contaminated sediment removal program that demonstrated use of innovative technologies for the removal of contaminated sediment. It was shown that the Mudcat MC 915 ENV horizontal auger dredge caused minimal disturbance and adverse impact on the aquatic environment. Two of the performance standards were that the turbidity beyond 25 m (82 ft) from the removal location should not exceed ambient levels by more than 30% and total suspended solids should not exceed ambient levels by more than 25 mg/L. A similar Mudcat horizontal auger dredge is illustrated in Figure 11.14, showing the auger with cutting teeth and a shroud for minimizing sediment resuspension.

Ellicott International began production of the new Mudcat MC 2000 in 1999 (Seagren, 2000). The new dredge has a submerged centrifugal pump with 0.254 m (10 in.) suction and 0.203 m (8 in.) discharge on the ladder. The maximum digging depth is stated as 6.1 m (20 ft), but that depth can easily be extended. There is an automatic shroud that is used to control turbidity. The dredge fits on a single truck and does not require a permit for transport. The particle clearance for the dredge pump is 152 mm (6 in.), and pump performance curves are available. As an example, the dredge working with 427 m (1400 ft) of 203-mm (8-in.) discharge pipe and digging in 6.1 m (20 ft) of water with a terminal elevation of 3.1 m (10 ft) is estimated to produce 125 yd^3/hr for fine sand (d_{50} = 0.1 mm) with an in situ specific gravity of 2.1. The same production can be achieved through 915 m (3000 ft) for a 0.254-m (10-in.) discharge pipe. Booster pumps can be used to pump slurry longer distances.

Aquamog. Aquatics Unlimited manufactures the Aquamog dredge. It is a small hydraulic dredge capable of dredging to a depth of 0.254 m (20 ft). The suction and discharge diameters are 152 mm (6 in.) and the estimated production rate is 46 m^3/h (60 yd^3/hr). The dredge performs multiple functions from debris/oil cleanup to bucket/suction dredging using interchangeable attachments for the cutter (cutterheads to augers). The system minimizes turbidity by using shrouds around the cutter/auger heads and by the high suction at the intake.

11.2.2.3 Small Bucket Wheel Dredge. The Nessie dredge (Figure 11.15), manufactured by Keene Engineering Company, uses a small

Figure 11.14 Mudcat dredge. (Courtesy of Ellicott Machine Corporation.)

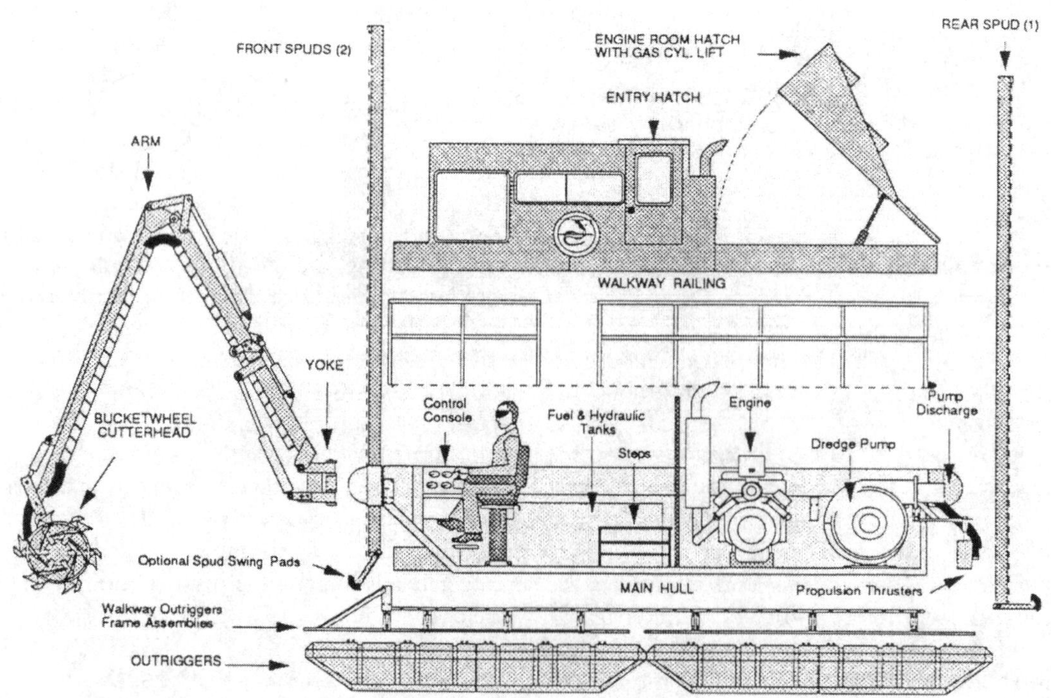

Figure 11.15 Small bucket wheel dredge, *Nessie*. (From Parchure, 1996.)

bucket wheel cutterhead to excavate the dredged material and a centrifugal pump (203 × 203 × 635 mm (8 × 8 × 25 in.). The pump produces a maximum head of 70.4 m (231 ft) of water and the maximum capacity is 290 L/s (4600 gpm). The maximum horsepower is 186 kW (250 hp). The length and width of the frame is 5.8 m (19.08 ft) and 4.22 m (13.83 ft), respectively, and the draft of the vessel is 0.43 m (17 in.).

11.2.3 Mechanical Dredges

11.2.3.1 Clamshell, Dragline, and Dipper Dredges. Dredges that remove sediments from waterways using various types of buckets and lifting mechanisms are classified as mechanical dredges. Example mechanical dredges are called dipper, clamshell, grapple, dragline, and bucket-ladder. These dredges lower a bucket to the bottom of the waterway and excavate the sediment. The sediment (dredged material) is hoisted or lifted to the surface and placed in a barge. A mechanical dredge using a conventional clamshell is illustrated in Figure 11.16. The barge is then moved to a disposal site where the dredged material is removed from the barge and placed in the disposal site. These mechanical dredges have limited ability to transport dredged material, no self-propulsion, and relatively low production. The advantages of a mechanical dredge are its ability to operate in restricted locations (e.g., docks, jetties, and piers) and its ability to treat and dewater dredged material in placer mining operations, and with special adaptations, such as watertight clamshell buckets, it can be used in working with contaminated sediments.

Figure 11.16 Clamshell dredge loading a barge. (Courtesy of Great Lakes Dredge and Dock.)

A clamshell dredge has a derrick mounted on a barge and is equipped with a clamshell bucket. It is best for working in soft sediments and works well in difficult-to-access areas. A dragline dredge has a steel bucket that is suspended from a movable crane as it is lowered to the bottom and then dragged toward the crane by a cable. A dipper dredge is a floating barge with a mechanically operated excavating shovel that works well in hard compact material. The Boston dipper dredge is shown in Figure 11.17.

Enclosed Watertight Clamshell Bucket. Thunder Bay Harbor in Canada is located on the shores of Lake Superior. The sediments in the harbor were contaminated with elevated levels of polynuclear aromatic hydrocarbons, chlorophenols, dioxins, and furans. Several technologies were considered for dredging and treatment (Ancheta et al., 1998). The dredging technology selected was the Cable Arm environmental clamshell bucket, which is an enclosed clamshell bucket, as illustrated in Figure 11.18. The bucket size used was a 5.5 m^3 (7.2 yd^3) and the cycle time was 2 minutes. The dredging operation was monitored for turbidity levels and a shutdown criterion of a 1-hour average of total suspended solids concentration of 25 mg/L above the ambient level 25 m (82 ft) from the dredging site. A total of 13,000 m^3 (17,030 yd^3) was dredged by the end of June 1998.

Figure 11.17 Dipper dredge. (Courtesy of Great Lakes Dredge and Dock.)

Figure 11.18 Cable Arm clamshell bucket. (Courtesy of Cable Arm, Inc.)

11.2.3.2 Conventional and Watertight Clamshell Buckets Comparison. Hayes (1999) reported on a cooperative effort by the Waterways Experiment Station, New England Division and Great Lakes Dredge and Dock (GLDD) to measure water quality data, while three types of dredge buckets were used to dredge sediments in Boston Harbor. The objective of the test was to compare three clamshell buckets (conventional, GLDD watertight, and Cable Arm). Turbidity sensors placed at the front of the dredge barge continuously measured the turbidity in formazin turbidity units (FTU). Four sensors were placed in a vertical array by the bow of the dredge and at depths of 1.4, 5.6, 8.1, and 10.5 m (4.5, 18.5, 26.5, and 34.5 ft). The results of the measurements are tabulated in Table 11.8. The results show the GLDD watertight bucket generated slightly less average turbidity than Cable Arm and both

Table 11.8 Average turbidity values for three different clamshell buckets

Bucket	Depth [m (ft)]	Observed Turbidity (FTU)	Background Turbidity (FTU)	Adjusted Turbidity (FTU)
Cable Arm	1.4 (4.5)	28.76	4.02	24.74
	5.6 (18.5)	9.30	5.69	3.60
	8.1 (24.5)	30.55	16.62	13.93
	10.5 (34.5)	66.48	15.64	50.85
	Average			2.25
Watertight	1.4 (4.5)	43.18	4.18	39.00
	5.6 (18.5)	4.62	3.67	0.96
	8.1 (24.5)	8.85	5.31	3.53
	10.5 (34.5)	51.82	24.28	27.54
	Average			18.50
Conventional	1.4 (4.5)	42.44	3.89	38.55
	5.6 (18.5)	24.23	3.12	21.10
	8.1 (24.5)	27.70	3.46	24.23
	10.5 (34.5)	60.40	20.22	40.18
	Average			31.05

Source: Hayes (1999).

Cable Arm and the GLDD watertight bucket produced less turbidity than the conventional open clamshell bucket. The GLDD watertight bucket produced higher turbidity in the upper part of the water column than the Cable Arm bucket.

11.2.3.3 Bucket-Ladder Dredge. A bucket-ladder dredge (Figure 11.19) has an endless chain of buckets, and each bucket is thrust into the material to be dredged and brings its load to the surface. When used in sand and gravel operations, the buckets discharge their contents onto vibrating screens to separate the sizes, and the separated material is placed on a barge for transport. The work cycle is continuous and more efficient than for grapple or dipper dredges. The buckets are attached to a chain that is guided and supported by a ladder. The lower end of the ladder is suspended from a hoisting gantry and lowered to the bottom for dredging and mining operations. The bottom sediment is cut by the rim of the bucket and then fills the bucket, which travels up the ladder and discharges to a barge alongside the dredge. The dredge is swung from side to side with the aid of anchors and mooring lines. Tugs are used to push or pull the loaded barges to a disposal or offloading area. Bucket-ladder dredges have been used in placer mining of gold and tin. Digging depth has increased from 15.2 m (50 ft) in 1905 to 53.4 m (175 ft) in 1973. Bucket sizes have increased from 0.2 to 1.5 m^3 (7 to 54 ft^3).

11.2.4 Pneumatic Dredges

A pneumatic dredge excavates bottom sediments and moves the dredged material to a barge or disposal area using compressed air. These dredges have been developed in Italy and Japan. The main advantage of pneumatic dredges is that no additional water is added to the in situ water content for transporting the material through the pipeline. In the case of dredging contaminated sediments, this means

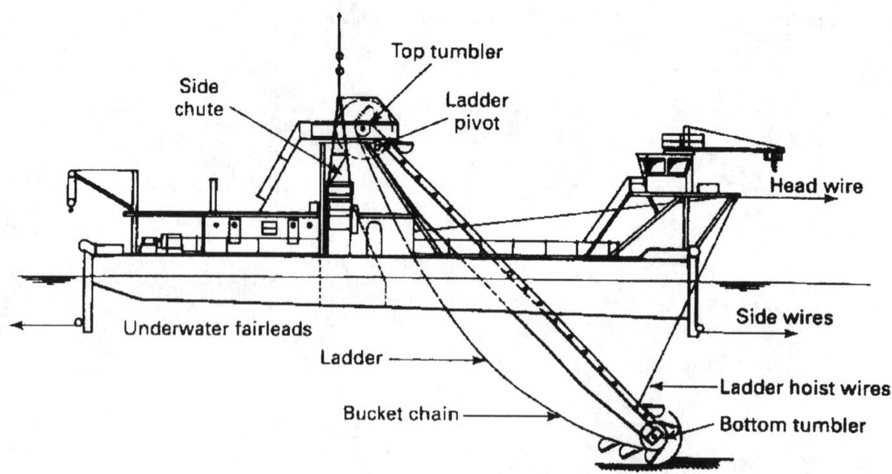

Figure 11.19 Bucket-ladder dredge. (From Albar, 2001.)

that the amount of water requiring treatment is much less than that required for hydraulic transport.

11.2.4.1 High-Density Dredge. A high-density dredge for removing contaminated sludge is described by Hara et al. (1998). The Shin–Ohmoto Maru dredge (Figure 11.20) uses a sludge pump to accurately remove contaminated sediment layers of 50 cm or less thickness. The dredge uses a global positioning system and is an automated dredge. The capacity of the dredge is 150 m^3/h (196 yd^3/hr) over a distance of 10,000 m (33,000 ft) and has a dredging depth of 1 to 8 m (3.3 to 26.4 ft). The sediment content is 50 to 80% by volume. The dredge has been used for projects in Japan and has been in operation since 1997.

11.2.4.2 Pneuma Dredge. The Pneuma pump (Figure 11.21) is a compressed air–driven displacement pump. The pump body incorporates three large cylindrical pressure vessels, each having a material intake on the bottom and an air port and discharge outlet on the top. Each intake and discharge outlet is fitted with a check valve, allowing flow in one direction only. Pipes leading from the three discharge outlets join in a single discharge directly above the pressure vessels. Different types of attachments may be fitted on the intakes for removal of various types of bottom material.

The Pneuma system was the first dredging system to use compressed air instead of a centrifugal pump to move slurry through the pipeline, and it can pump slurry of relatively high solids content with little generation of turbidity. It has been used in Europe and Japan and it has been evaluated in the United States (Richardson et al., 1982) and in Canada (Pelletier, 1992). During the dredging process the pump is submerged, and sediment and water are forced into one of the empty cylinders through an inlet valve. After the cylinder is filled, compressed air is forced into the cylinder, closing the inlet valve and simultaneously forcing the material out the outlet valve and into the discharge line. When the cylinder is empty, the air pressure is reduced to atmospheric pressure, the outlet valve closes and the inlet valve opens. The two-stroke cycle is then repeated. The distribution system controls the cycling

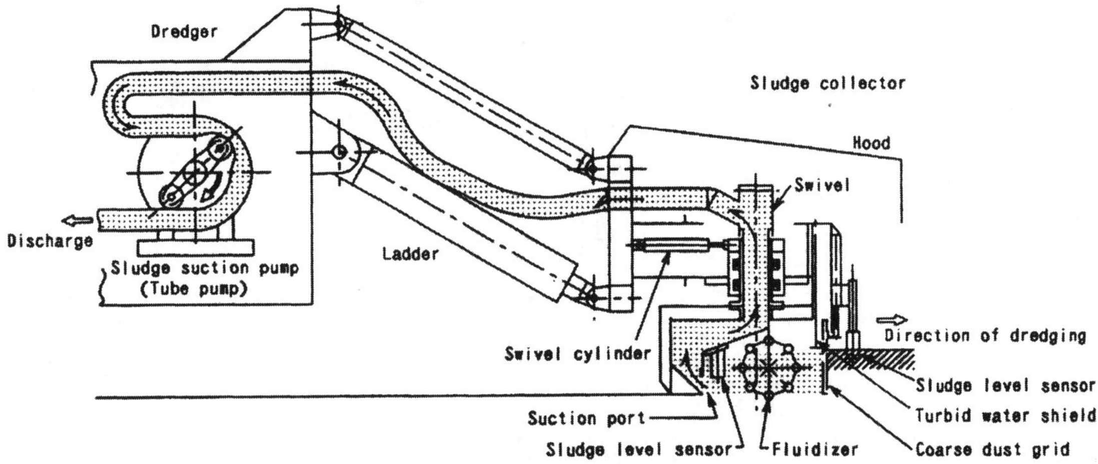

Figure 11.20 High-density dredge system. (From Hara et al., 1998.)

phases of all three cylinders so there is always one cylinder operating in the discharge mode. The system has been used in water depths up to 50 m (164 ft).

11.2.4.3 Airtight Bucket Wheel Dredge. An airtight bucket wheel dredging system developed by Iwasaki et al. (1992) is illustrated in Figure 11.22. The bucket wheel revolves slowly inside an airtight shroud that is submerged into the bottom sediments. The system reduces turbidity significantly and can dredge sludge while reducing or eliminating contamination of the surrounding waters. The dredged material enters a feed screw that agitates and fluidizes the material prior to delivering it to the pneumatic pump. A storage tank receives the material from the feed screw, and the pneumatic pump in conjunction with the simultaneous injection of compressed air delivers the material to the conveyor pipeline. The compressed air expands and forces the plug flow along the pipeline to the end discharge location.

11.2.5 Hybrid Dredge (Mechanical and Hydraulic)

A hybrid dredge combines mechanical excavation and hydraulic transport. As an example, a clamshell is used to excavate the sediment and the dredged material is placed in a bin or hopper. In this case water is not used to entrain the sediment during excavation. Next, the sediment is transported through a pipeline using a positive-displacement pump or other mechanism that uses a minimum amount of water.

11.2.5.1 Dry Dredge. Parchure and Sturdivant (1997) describe the development of an innovative dredge for removing contaminated sediments. The dredge has a sealed clamshell mounted on a rigid, extensible boom (Figure 11.23). The open clamshell is driven into the sediment hydraulically at a slow speed to minimize the resuspension of the sediments. Next, the clamshell is closed, and the sediment at its in situ density is contained inside the clamshell. The sediment is placed in a hopper of a positive-displacement pump on the dredge, and the pump forces the sediment through a pipe-

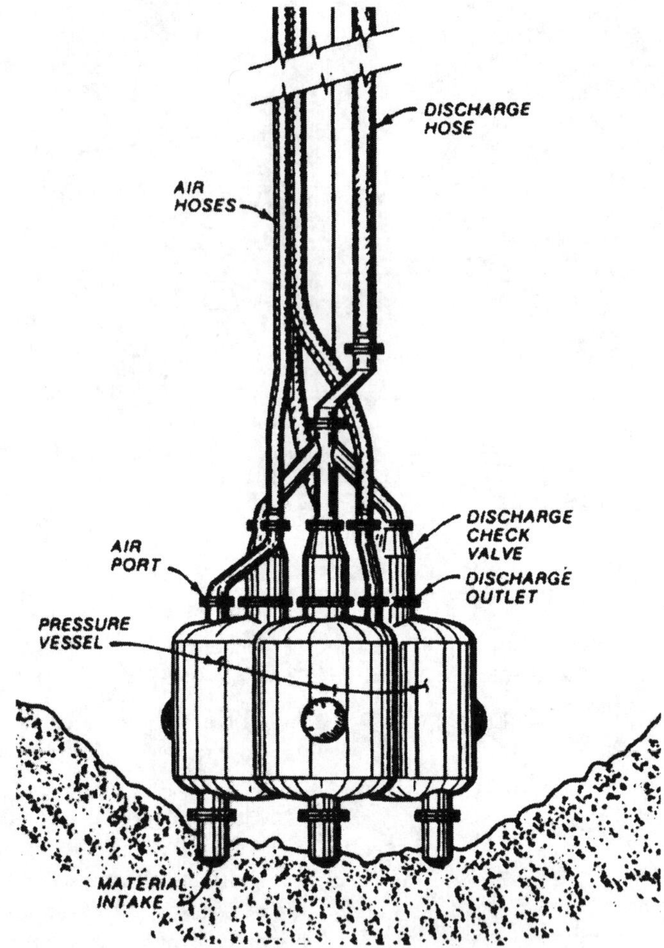

Figure 11.21 Pneuma pump system. (From Herbich and Brahme, 1991.)

line to the disposal site. The hopper can be equipped for debris screening, size reduction, vapor emission control, sediment homogenization, and blending of additives to modify flow properties. The pumped sediment has a toothpaste consistency. The discharge can be directed to a treatment process plant, an on-site disposal facility, or an enclosed transport vehicle. The dredge is portable and capable of being transported by truck. Field tests were conducted at the Waterways Experiment Station in Vicksburg, Mississippi. After the tests were completed, the dredge was modified and tested in Louisville, Kentucky. The production rate varied from 21.4 to 30.6 m^3/h (28 to 40 yd^3/hr). Parchure and Sturdivant (1997) concluded that the dredge was useful for shallow water less than 4.6 m (<15 ft), a low production rate, less than 30.6 m^3/h (<40 yd^3/hr), sediment with a high amount of fines and a low amount of sand, short pumping distances, and no wave or current action. The main advantage

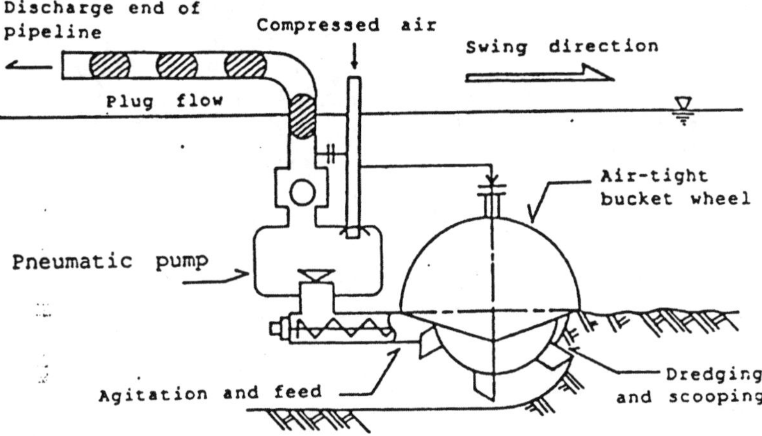

Figure 11.22 Airtight bucket wheel dredge. (From Iwasaki et al., 1992.)

Figure 11.23 Dry dredge working in lake. (From Parchure and Sturdivant, 1997.)

of the new dredge is the reduced volume of water to be handled and treated. The equipment is portable and small in size. The clamshell bucket capacity is 0.2 m^3 (7 ft^3), and the pipeline diameter is a 152 mm (6 in.).

11.2.5.2 Amphibex Dredge. The Amphibex dredge is a hybrid dredge that has been evaluated by Environment Canada and discussed by Santiago (2000). This dredge uses mechanical force (backhoe) initially to remove the sediment and then deposits the material in a hopper. A pump is used to transport the material to the disposal location (barge, disposal site). The production of this dredge is less than 76 m^3/h (100 yd^3/hr).

11.2.6 Comparison of Small Dredges

Several researchers (Hayes et al., 1988; Van Drimmelen and Schut, 1992; Randall 1992, 1994; Herbich, 2000a and b) have summarized equipment for removing contaminated sediments. A summary of the results is illustrated in Table 11.9. Comparisons of various dredge types working in nearly similar conditions have shown that conventional dredges have proven to be surprisingly successful (Hayes et al., 1984, 1988; Hayes, 1986; Otis et al., 1990). Turbidity measurements around conventional hydraulic suction dredges have shown resuspended solids concentrations of 5 to 200 mg/L.

Innovative technologies for dredging contaminated sediments were researched and reported by Zappi and Hayes (1991). The innovative hydraulic dredges included the Clean-up (Japan), matchbox, Refresher (Japan), modified dustpan, disk-cutter, bucket

Table 11.9 Comparison of resuspended sediment levels for various dredging equipment

Type	Production	Depth Limit [m(ft)]	Sediment Resuspension	Comments
Mechanical				
Open clamshell bucket	Low	9.1–12.2 (30–40)	High	
Watertight clamshell bucket	Low	9.1–12.2 (30–40)	Low	St. Johns River
Cable Arm clamshell bucket	Low	9.1–12.2 (30–40)	Low	Hamilton Harbor, Canada
Mechanical hydraulic				
Mudcat	Moderate	4.6–7.6 (15–25)	Low to moderate	Extensively used
Mudcat ENV	Moderate	4.6–7.6 (15–25)	Low	Sydney, Nova Scotia, Canada
Remote-controlled Mudcat	Low	4.6 (15)	Low to moderate	New development
Cleanup system	Moderate	21.3 (70)	Low to moderate	Japan
Cutterhead	Moderate to high	12.2 (40)	Low	New Bedford
Hydraulic-suction				
Refresher	Moderate to high	18.2–35 (60–115)	Low	Japan
Matchbox	Moderate to high	25.9 (85)	Low to moderate	Calumet Harbor
Wide sweeper	Moderate	30.5 (100)	Low	Japan
Pneumatic	Low to moderate	60.9 (200)	Low	Evaluated by USACE
Oozer	Moderate to high	18 (59)	Low	Japan
Mechanical–Hydraulic-pneumatic				
Screw-impeller	Low to moderate	6.1 (20)	Low	Japan (high density)
Airtight bucket wheel	Low to moderate	4.6 (15)	Low	Japan (high density)

Source: Hayes et al. (1988).

wheel, cutter suction, portable Mudcat, and IHC Roller Silt. The innovative pneumatic dredges investigated were the Pneuma, Oozer, and airlift dredges. Average suspended solids concentrations in Cape Fear River using the Pneuma dredge were less than 20 mg/L. Innovative mechanical dredges include the watertight clamshell bucket and the use of a turbidity barrier. Pennekamp and Quaak (1990) reported average above background suspended solids of 20 mg/L for a watertight bucket with a turbidity barrier, 35 mg/L for a conventional bucket with turbidity barrier, and 100 mg/L for a watertight bucket without a turbidity barrier. Small conventional hydraulic dredges (auger, plain suction, and cutter suction), mechanical dredges (watertight clamshell bucket), and hybrid dredges (DreDredge) are available in the United States.

Jaglal and McLaughlin (1999) discuss environmental dredging of polychlorinated biphenyl (PCB)-contaminated sediments for selected projects. These projects included Ruck Pond (Wisconsin), Sheboygan River (Wisconsin), Manistique Harbor (Michigan), River Raisin (Michigan, Waukegan Harbor (Illinois), and Lake Jarnsjon (Sweden). At Sheboygan River, 2905 m^3 (3800 yd^3) of PCB-contaminated sediment were removed using a modified clamshell dredge and backhoe. For this project the dredged sediment was placed on site and took two years to complete. A total of 87,929 m^3 (115,000 yd^3) of contaminated sediments were removed from three areas of Manistique Harbor using diver-assisted dredging and cutterhead dredges. The sediments were dewatered using sediment screens, settling tanks and basins, and mechanical dewatering equipment. The dewatered sediment was shipped off-site for disposal. Silt curtains and/or cofferdams were used to reduce resuspended sediments migration. In 1992, approximately 24,467 m^3 (32,000 yd^3) of PCB-contaminated sediments were hydraulically dredged from Manistique Harbor. Water depths were from 4.3 to 7.6 m (14 to 25 ft), and the depth of sediment dredged was from 0.3 to 2.1 m (1 to 7 ft). The dredged sediment was placed in a nearby containment cell and capped. The water generated from dredging was treated using sand filtration and carbon adsorption. In Sweden, 152,920 m^3 (200,000 yd^3) were dredged in 1993–1994 using a horizontal auger dredge. The dredged material was dewatered in filter presses and placed in a nearby landfill. A geotextile silt screen was used to contain migration of resuspended sediments.

Briot et al. (1999) summarize the several environmental dredging projects that include Sheboygan River, New Bedford Harbor (Massachusetts), Grasse River (New York), Ruck Pond (Wisconsin), and Manistique River and Harbor (Michigan). At New Bedford Harbor, a 305-mm (12-in.) cutterhead hydraulic dredge was used to remove 10,704 m^3 (14,000 yd^3) of contaminated sediments. Approximately 606 million liters (160 million gallons) of supernatant water from the confined disposal facilities were treated using settling, flocculation, sand filtration, micro filtration, and ultraviolet oxidation. The treatment plant design capacity was 1325 to 1514 L/min (350 to 400 gpm), and as a result the hydraulic dredging operations were limited. The Grasse River project used a horizontal auger dredge to remove 2676 m^3 (3500 yd^3) of sediments. A silt curtain system was used to control the resuspended sediment migration. Boulders and debris were removed prior to dredging activities. However, hidden rocks and cobbles still caused problems for the dredge. The water was treated using a specially designed treatment system.

11.3 CONTAINMENT BARRIERS AND UNDERWATER ACOUSTIC SURVEYS

Dredging of contaminated sediments that contain oil and grease requires the use of oil booms to contain the oil and grease that rises to the surface during dredging. Some oil booms can absorb the oil and grease, while others just

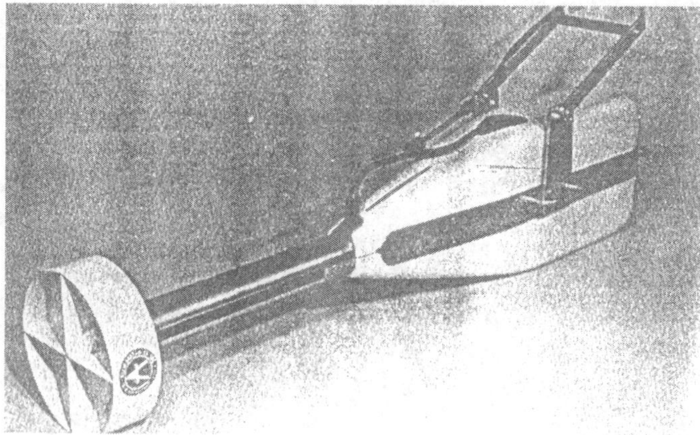

Figure 11.24 Combined side-scan and subbottom Sonar Towfish. (Courtesy of Klein Associates, Inc.)

contain the oil and grease within a confined area and other equipment is used to remove the oil and grease.

A concern while dredging contaminated sediments is resuspension of the contaminated sediments. Silt curtains are frequently used to prevent the propagation of the resuspended sediments beyond the dredging location. These curtains are either impermeable or permeable. They are commonly installed from the surface to the bottom, but the curtains are difficult to install and maintain in high currents and waves. Waves from passing ships, ship traffic, and river currents can also cause problems in the installation and maintenance of silt curtains.

Surveying waterways, lake bottoms, and the ocean floor is often accomplished using underwater acoustic equipment known as side-scan sonar and subbottom profilers. These systems have been used to locate sunken objects such as ships and airplanes. Side-scan sonar can map the bottom surface of a water body and can be used to locate debris such as cars, pipes, and other objects lying on the bottom of a waterway. The subbottom profiler has the ability to penetrate the bottom and locate similar debris buried in the sediments. A small (e.g., 6.7-m or 22-ft) survey vessel would have an acoustic system containing side-scan and subbottom transducers inside a housing, and either towed by the survey vessel (Figure 11.24) or attached to the side of the vessel. The survey vessel equipped with positioning equipment transects the area and locates the navigational position of the items. The output from the side-scan sonar and subbottom profiler requires an experienced operator to interpret the results and identify the location of objects or debris. Frequently, divers are used to further identify the objects. The identified debris locations can then be provided to a salvage contractor for removal using appropriate construction and salvage equipment.

A side-scan sonar sends acoustic beams sideways from the survey vessel and has selectable ranges of 30 to 150 m (100 to 500 ft). The frequency of the sonar ranges from 100 to 500 kHz. Better resolution of the bottom and any debris is obtained with the higher-frequency sonar. An example side-scan display in Figure 11.25 shows the detection of a wooden ladder and tires. The subbottom profiler operates at lower frequencies of 3.5 to 100 kHz and is capable of penetrating sediments. The depth of penetration depends on the sedi-

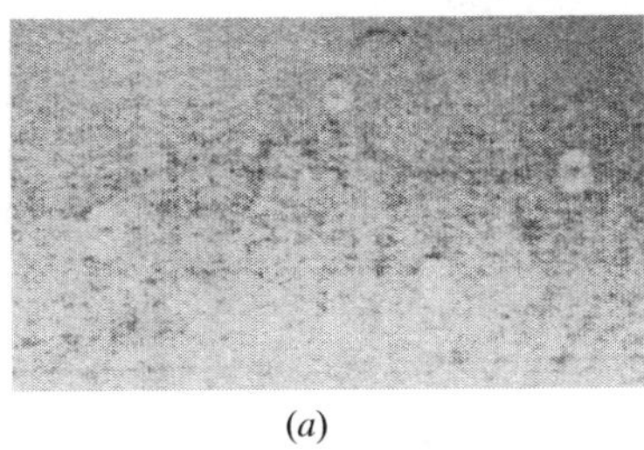

(a)

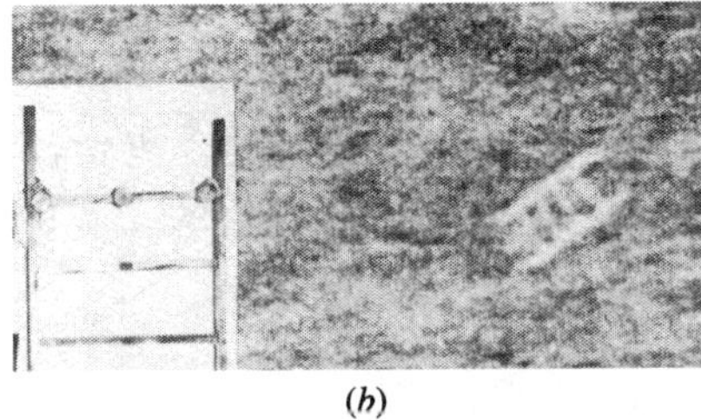

(b)

Figure 11.25 Side-scan sonar display showing and *(a)* tires and *(b)* wooden ladder. (Courtesy of Klein Associates, Inc.)

Figure 11.26 Example of HydroBat multibeam sonar survey system. (Courtesy of RESON, Inc.)

ments. For example, a sand layer can reflect the higher frequency and prevent penetration.

A multibeam depth sounder system covers more than just directly beneath the traditional single transducer depth sounders. The multi-element transducers form many individual beams electronically. These beams overlap and make it possible to insonify a wide swath of the sea bottom with one pass. Some multibeam systems are being used in shallow-water hydrographic surveys. These systems are expensive, but the improved quality and quantity of data are improving efficiency of survey and dredging operations. Some systems that are available include the HydroBat (Figure 11.26) and SeaBat systems by Reson Inc., the EM series from Kongsberg Marine/Simrad, the Sea Beam system from L-3 Communications, the Fansweep system from Krup Atlas, and the Echoscan from Odom Hydrographics. These systems are being used by the U.S. Army Corps of Engineers, consulting firms conducting contract survey projects, and dredging companies such as Great Lakes Dredge & Dock and Bean/Stuyvesant. As an example, the HydroBat multibeam system uses 120 beams and can cover a swath equal to as much as 7.4 times the water depth. The specifications include a 5-cm (2-in.) range resolution, accuracy of <10 cm (4 in.), wide coverage up to 180°, and capability to operate in water depths as shallow as 0.5 m (20 in.). Large amounts of data are collected with this system, so the computer systems for data logging must have a large storage capacity and be fast. A navigation positioning system (e.g., DGPS) is also interfaced with the system. The bathymetry data can be plotted as plan view charts or cross sections, and it is possible to export the data to other plotting software.

11.4 DREDGE PUMPS

Dredge pumps are designed with sufficient clearances to accommodate occasional gravel, rocks, and other debris passing through the pump without jamming. This typically limits the number of vanes contained in the impeller. Wear and abrasion due to sediment particles requires special materials and provisions for easy access for maintenance and repairs. The type of material pumped affects the performance of the centrifugal (dredge) pump. Dredge pumps must handle mud (silt–clay–water) mixtures as well as sands and gravels. The fluid being pumped is typically called *slurry*, which means that the water contains suspended solids (gravel, sand, clay, silt). Slurry has a variable specific gravity and a highly variable viscosity.

11.4.1 Viscosity

Viscosity is a function of concentration, temperature, past history, and rate of shearing stress. Figure 11.27 shows the variation of viscosity (μ) with the rate of shearing stress for a certain concentration and temperature. Different types of fluids have different relations between the shear stress (τ) and rate of strain (du/dy). For a Newtonian fluid, the relationship is $\tau = \mu(du/dy)$, as illustrated in Figure 11.27.

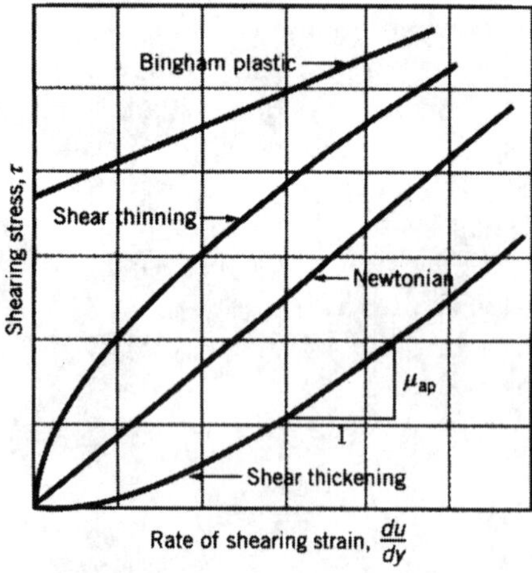

Figure 11.27 Shear stress as a function of rate of strain for Newtonian and non-Newtonian fluids. (From Munson et al., 1998.)

Solids–water mixtures display a variable proportionality between stress and rate of strain and are called *non-Newtonian fluids* (dilatant, pseudoplastic, or Bingham body). Sand–water slurries tend to behave similarly to Newtonian fluids, and the viscosity is the value of the carrier fluid, such as water or seawater. Silt–clay–water mixtures behave similarly to a Bingham body fluid, and the relationship between shear stress and rate of strain is defined by the equation above, where κ is the yield shear stress. The effects of viscosity are important in the design and analysis of dredge pumps, and it affects the velocity distribution in the pump passages, loss of head in passages, and the pump power.

11.4.2 Suction Inlet

The suction inlet design of a centrifugal pump is important if cavitation is to be avoided. Variables at the suction include the suction pipe diameter, curvature of inlet, eye diameter, and leading edge of vane. Ideal conditions require sufficient length of straight pipe without valves or disturbances preceding the entrance to the pump. *Prerotation* is the rotation of the fluid about the longitudinal axis of the pipe, and it may occur if the length of straight pipe is to short. Pumps are designed to operate near maximum efficiency, and prerotation is usually not a problem when operating near this point. When a pump must be operated at flow conditions other than those near maximum efficiency, guide vanes assist in maintaining maximum performance. However, guide vanes must be designed to prevent separation, cavitation, or high losses. The inlet section of the pump turns the fluid so that its relative velocity is along the axis of the impeller channels.

11.4.3 Impeller

Impeller performance is affected by inlet angle, vane shape, vane area, and exit angle. An impeller vane moves faster than the fluid so as to transfer energy from the vane to the fluid. The fluid is discharged at a mean angle relative to the impeller that is less than the vane angle. Observations indicate fluid flow into impeller is far from ideal at most flow rates, but it is close to ideal near the design flow rate. Figure 11.28 shows a large dredge pump on the left and the pump volute with an impeller on the right.

Head and efficiency increases as the number of vanes increase to a maximum point and then decreases due to losses caused by the increased number of vanes. Typical dredge pumps have three to five vanes, whose shapes are usually parts of a circular arc or a logarithmic spiral. The logarithmic spiral results in the best flow path, but many dredge pumps have circular arc vanes. The clearance between the impeller and the cutwater or volute tongue also affects the efficiency. Dredge pumps typically have a clearance about twice that of a water pump.

11.4.4 Modeling Dredge Pumps

Physical models of fluid flow phenomena are commonly used in various engineering disciplines. Engineers construct scale models of prototype systems and test them in laboratory facilities. These laboratory facilities include towing tanks, wave basins, wind tunnels, and pump test stands. In order to have complete similarity between the model and the prototype, the geometric, kinematic, and force ratios must be the same in both the model and prototype. Kinematic ratios include velocity and acceleration ratios, and force ratios include dimensional parameters such as Reynolds, Froude, Mach, and Weber, to name a few of the most common. For pumps, the most important similitude ratio is the ratio of the fluid velocity to the peripheral velocity of the impeller, which must be constant at all geometrically similar points. The most common dredge pump is the centrifugal pump, illustrated in Figure 11.29.

Figure 11.28 Example of a large centrifugal pump (courtesy of Mobile Pulley) and impeller (courtesy of Belesimo).

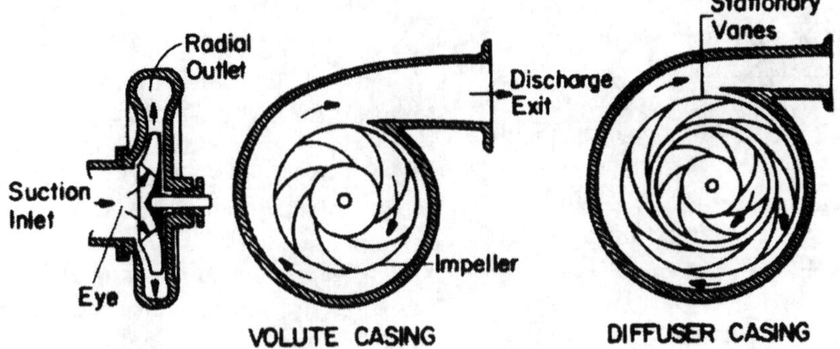

Figure 11.29 Major components of a centrifugal pump. (Courtesy of Allis Chalmers.)

11.4.4.1 Dimensional Analysis of a Pump. The head (H) produced, power (P) input and the efficiency (η) are the three important parameters for dredge pump selection and design. These are a function of the fluid density (ρ), angular velocity (ω), impeller diameter (D), fluid flow rate (Q), gravitational acceleration (g), and viscosity (μ). The head (H) depends on g, because H represents shaft work per unit weight of fluid. The product gH is independent of the acceleration of gravity (g) because the flow through the pump is totally enclosed and the fluid is incompressible. The efficiency (η) is the ratio of the output power to the input power. Dimensional analysis using the Buckingham Π theorem can be used to determine the dimensionless parameters important in modeling centrifugal pumps, and the analysis results in the pump affinity laws. The results of the dimensional analysis and neglecting the effects of viscosity are

$$\frac{gH}{\omega^2 D^2} = f\left(\frac{Q}{\omega D^3}\right) \tag{11.13}$$

$$\frac{P}{\rho\omega^3 D^5} = f\left(\frac{Q}{\omega D^3}\right) \tag{11.14}$$

$$\eta = f\left(\frac{Q}{\omega D^3}\right) \tag{11.15}$$

It is commonly assumed that the pump is operated at or near the maximum efficiency, and thus the efficiency is nearly constant. In this case, the equations above can be expressed as

$$\frac{gH}{\omega^3 D^2} = \text{constant} \tag{11.16}$$

$$\frac{P}{\rho\omega^3 D^5} = \text{constant} \tag{11.17}$$

$$\frac{Q}{\omega D^3} = \text{constant} \tag{11.18}$$

If the angular velocity (ω) is constant, the flow rate, head, and power of two similar pumps can be expressed as

$$\frac{Q_1}{Q_2} = \frac{D_1^3}{D_2^3} \tag{11.19}$$

$$\frac{H_1}{H_2} = \frac{D_1^2}{D_2^2} \tag{11.20}$$

$$\frac{P_1}{P_2} = \frac{D_1^5}{D_2^5} \tag{11.21}$$

Similarly, if the impeller diameter (D) is constant, then

$$\frac{Q_1}{Q_2} = \frac{\omega_1}{\omega_2} \tag{11.22}$$

$$\frac{H_1}{H_2} = \frac{\omega_1^2}{\omega_2^2} \tag{11.23}$$

$$\frac{P_1}{P_2} = \frac{\omega_1^3}{\omega_2^3} \tag{11.24}$$

These relationships are known as the *pump affinity laws*. Specific speed (N_s) is used in the selection of a pump to satisfy given operating conditions. The discharge rate (Q), head (H), and speed (ω) are normally required to have certain values for a particular pump installation. The commonly used specific speed is expressed as

$$N_s = \frac{NQ^{0.5}}{H^{0.75}} \tag{11.25}$$

where N is the pump rpm, Q the flow rate in gpm, and H is the head expressed in feet. The relationship between specific speed and the type of pump is illustrated in Figure 11.30, and the specific speed of centrifugal pumps ranges from 500 to 3500. In dimensionless form, the dimensionless specific speed is expressed as

$$n_s = \frac{\omega Q^{0.5}}{(gH)^{0.75}} \tag{11.26}$$

where ω is in rad/s, Q is in ft^3/s, g is in ft/s^2, and H is in feet; and in the SI system of units, ω is in rad/s, Q is in m^3/s, g is in m/s^2, and H is in meters.

11.4.5 Pump Characteristics

11.4.5.1 Dimensional Curves. Dredge pump characteristics are typically described by curves or graphs describing the variation of head, brake horsepower, and efficiency as a function of volumetric flow rate of water. An example of dimensional pump characteristics is illustrated in Figure 11.31 for 508 mm (20 in.) suction and discharge with a 107-mm (42-in.) impeller.

11.4.5.2 Dimensionless Curves. Dimensional characteristic curves require separate curves for each pump speed, impeller diameter, and specific gravity of mixture. The dimensionless characteristic curves have the advantage of requiring only one set of curves. The dimensionless parameters that are useful in pre-

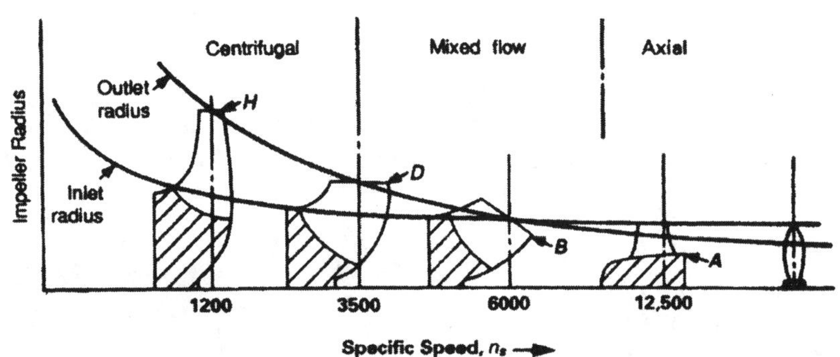

Figure 11.30 Suggested pump type as a function of specific speed. (Courtesy of Allis Chalmers.)

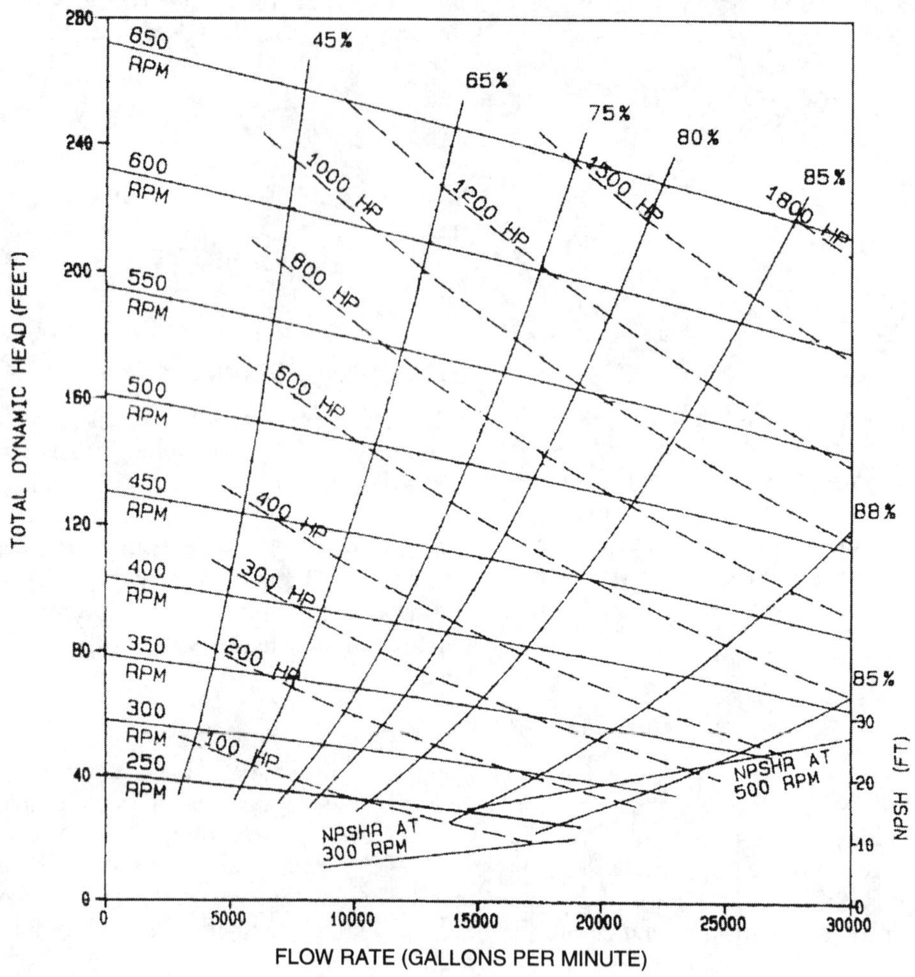

Figure 11.31 Example of pump characteristic curves for 20-in. suction and 20-in. discharge with a 42-in. impeller tested on clear water. (Courtesy of GIW Industries.)

senting dredge pump characteristics are (gH/ω^2D^2) versus dimensionless discharge $Q/\omega D^3$. Efficiency (η) is already dimensionless and is also plotted versus dimensionless discharge. For any given pump, there is just one set of curves for several sizes of impellers, pump speeds, and fluid densities. Discrepancies may occur for slurries of different rheological properties, such as silt–clay–water and coarse gravel–water mixtures. The complete characteristics for a given pump can be shown on a single graph, as shown in Figure 11.32.

11.4.6 Effect of Solid Water Mixtures on Pump Performance

The efficiency for pumping slurry mixtures is reduced below the efficiency when pumping

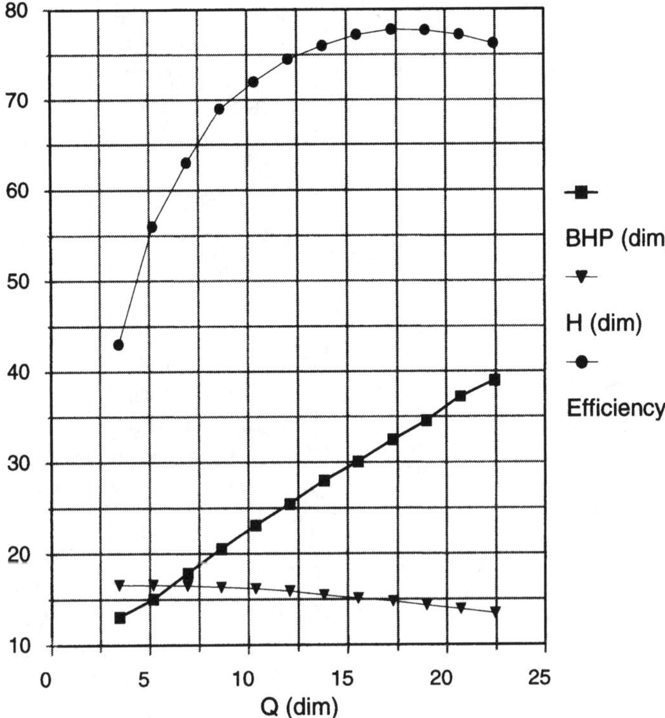

Figure 11.32 Dimensionless pump characteristics curves.

clear water by the ratio head of the mixture to that of clear water, as expressed by

$$\frac{H_m}{H} = \frac{\eta_m}{\eta} \tag{11.27}$$

where H_m is the head in meters of mixture, η_m is the efficiency when pumping slurry mixture, η is the clear water efficiency, and H is the head in meters of clear water. In the case of homogeneous mixtures and at a fixed capacity near the best efficiency point (BEP), the power increases as the ratio of specific gravity (SG) increases:

$$(\text{BHP})_m = (\text{BHP})\text{SG}_m \tag{11.28}$$

Figure 11.33 shows the effects of slurry on the head, power, and efficiency of a pump. Head (H_r) and efficiency (η_r) reduction factors can be defined as

$$H_r = \frac{H_m}{H} \qquad \text{and} \qquad \eta_r = \frac{\eta_m}{\eta} \tag{11.29}$$

Kazim et al. (1997) defines a head reduction factor (K_r) as

$$K_r = 0.13C_w\sqrt{(\text{SG} - 1)}\ \ln\frac{d_{50}}{20} \tag{11.30}$$

where C_w is the concentration of solids by weight, SG is the specific gravity of the solids, and the median particle size (d_{50}) and K_r are related to H_r as

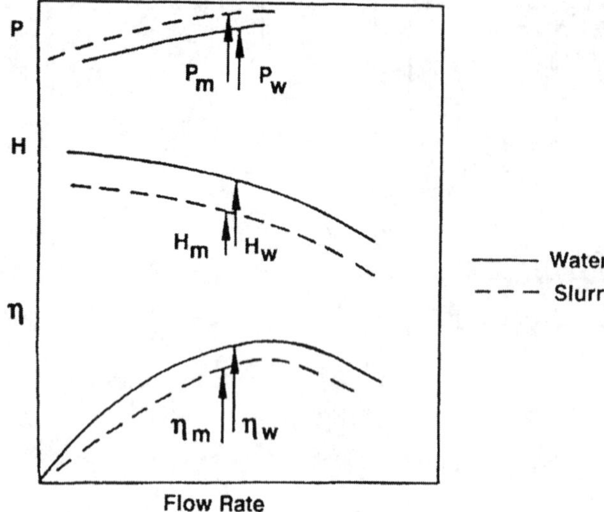

Figure 11.33 Effects of slurry on pump characteristics. (From Wilson et al., 1997.)

$$K_r = 1 - H_r \tag{11.31}$$

Wilson et al. (1997) discuss the effect of pump impeller size on the head reduction factor and show that the head reduction settling slurries depends of the impeller diameter (D) raised to the -0.9 power. Nomographs are presented in Wilson et al. (1997) to determine the head reduction effects as a function of concentration by volume, mean particle size, solids specific gravity, and impeller diameter.

11.4.7 Dredge Pump Systems Curves

The system head curve includes the static head and the friction loss. The friction-head loss is a function of pipe size, length, minor losses, slurry flow rate, and slurry specific gravity. When pumping water the system friction curve is zero when the flow rate is zero and then increases as the flow rate increases. For settling slurries that are common for dredging sand, the system friction curve has a minimum at the critical velocity that is normally based on the median grain size. The system head curve is superposed on the pump characteristics curves to determine the operating point, which should be near the best efficiency point (BEP). An example system head curve and the system head curve superposed on the pump characteristics curves are shown in Figure 11.34. The operating point is at the intersection of the system head curve with the pump head curve for 1000 rpm and the intersection is very near the best efficiency point.

Cutter suction dredges usually have long pipelines to transport the dredged material to the confined disposal facility. Therefore, the system head losses are due primarily to friction losses. Minor losses and elevation losses are generally small compared to friction losses. The hopper dredge does not have long pipelines, so the minor losses and elevation losses are more important than the friction losses, and the system head curve is much flatter than that shown in Figure 11.34.

Hopper dredges and cutter suction dredges frequently use ladder pumps that are mounted on the dragarm and ladder, respectively. Cutter suction dredges with very long pipelines (e.g., greater than a mile long) need to use booster pumps to transport the dredged material over the long distance. These ladder pumps and booster pumps are centrifugal pumps and are normally arranged in what is called *series operation*. When pumps are placed in series, the head of the two pumps is additive and the flow rate is the same in both pumps. If the pumps are placed in parallel operation, the flow rates are added and the head developed is the lowest head developed by the two pumps.

11.4.8 Dredge Pump Cavitation

Cavitation is defined at the formation and collapse of low-pressure cavities in a flowing liquid, often resulting in serious damage to pumps and ship propellers (*Reader's Digest Great En-*

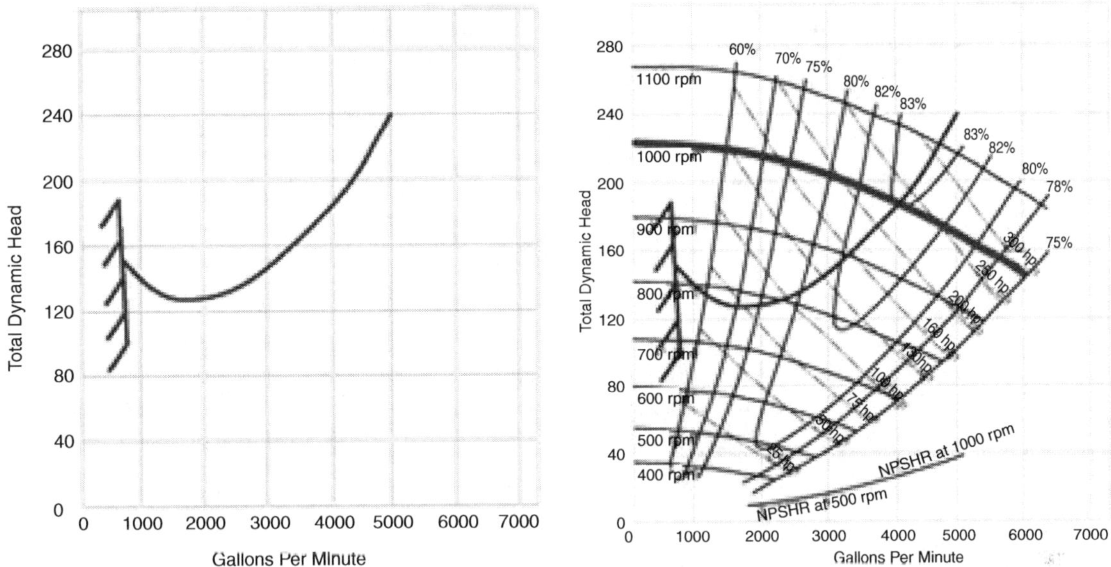

Figure 11.34 Example of system head curve and superposed on pump curves. (Courtesy of GIW Industries.)

cyclopedic Dictionary). As liquid fluid pressure is reduced at constant temperature below the vapor pressure by either static or dynamic means, vapor-filled cavities form. Cavitation bubbles form in areas of low pressure, expand, flow into high-pressure areas, and collapse or implode. The time between inception to implosion is typically only a few milliseconds. The collapse or implosion can severely damage the metal surfaces of a dredge pump. Cavitation occurs at the top of conduits or pump housing, in regions of high velocity (low pressure) such as impeller tips and in constricted passages, and where flow curvature exists, and it is often accompanied by eddies, vortices, or flow separation. Cavitation causes pitting of boundaries, reduction of pump efficiency, and vibrations.

When the pressure in a flowing liquid decreases to the vapor pressure of the liquid, the liquid begins to vaporize or cavitate. This cavitation condition is commonly called *incipient cavitation*. Herbich (2000a and b) describes *industrial cavitation,* which refers to the condition when the pump head is significantly reduced due to cavitation, as shown in Figure 11.35. It is also shown that a pump efficiency drop coincides with a reduction in pump head.

11.4.8.1 Net Positive Suction Head. The most common parameter used for evaluating cavitation in pumps is the net positive suction head (NPSH) produced for a given discharge or flow rate. NPSH is defined as the total head available to the pump above the vapor pressure of the liquid. The mathematical expression for NPSH is

$$\text{NPSH} = \frac{P_a}{\gamma} + \frac{P_s}{\gamma} + \frac{V_s^2}{2g} - \frac{P_v}{\gamma} \quad (11.32)$$

where P_a is the local atmospheric pressure, γ the specific weight of fluid or slurry being pumped, P_s the pressure at the suction inlet, V_s the average velocity at the suction inlet, g the gravitational acceleration (e.g., 32.2 ft/s^2 or 9.81 m/s^2) and P_v the vapor pressure of the carrier liquid (e.g., water). Consistent units

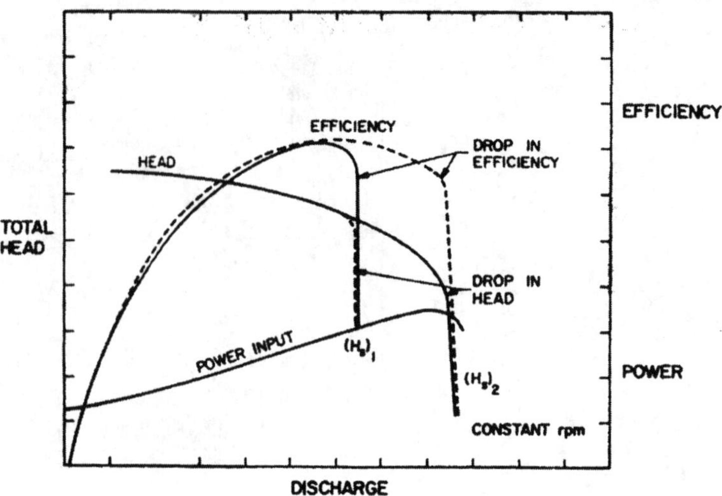

Figure 11.35 Pump performance showing abrupt cavitation. (From Herbich, 2000b.)

must be used, and each term has the units of head (e.g., feet of slurry, meters of slurry).

The minimum NPSH for incipient cavitation is determined from laboratory pump tests for different rates of discharge and is usually illustrated on the pump characteristics curves as a function of discharge and provided by the pump manufacturer. It is referred to as the *required NPSH* ($NPSH_r$). The actual operation of a pump occurs at some NPSH above $NPSH_r$. The available NPSH for a dredge pump can be calculated and compared to the required NPSH to determine if the pump system is expected to cavitate. The calculation of the available NPSH is relatively easy. In the case of a dredge pump the liquid is a mixture of water and solids and the pressure heads, NPSH, and so on, are expressed in ft (m) of mixture (not water).

The energy equation for the dredging application illustrated in Figure 11.36 is

$$\frac{P_1}{\gamma_m} + \frac{V_1^2}{2g} + z_1 = \frac{P_2}{\gamma_m} + \frac{V_2^2}{2g} + z_2 + H_L \tag{11.33}$$

where H_L is the friction and minor losses in the suction pipe and expressed in m (ft) of mixture. All the terms in eq. (11.33) are in ft or m of mixture and the subscripts 1 and 2 refer to the entrance to the suction pipe (e.g., drag head or cutter head) and suction inlet of pump, respectively. The pressure head in absolute terms is

$$\frac{P_1}{\gamma_m} = \frac{P_a}{\gamma_m} - \frac{P_v}{\gamma_m} + \frac{d}{SG_m} \tag{11.34}$$

where d is digging depth and SG_m is the specific gravity of the mixture (slurry) and is equal to γ_m/γ. The available NPSH ($P_2/\gamma_m - V_2^2/2g$) is now expressed as

$$\frac{P_2}{\gamma_m} + \frac{V_2^2}{2g} = \frac{P_a}{\gamma_m} - \frac{P_v}{\gamma_m} + \frac{d}{SG_m} - z_2 - H_L \tag{11.35}$$

or

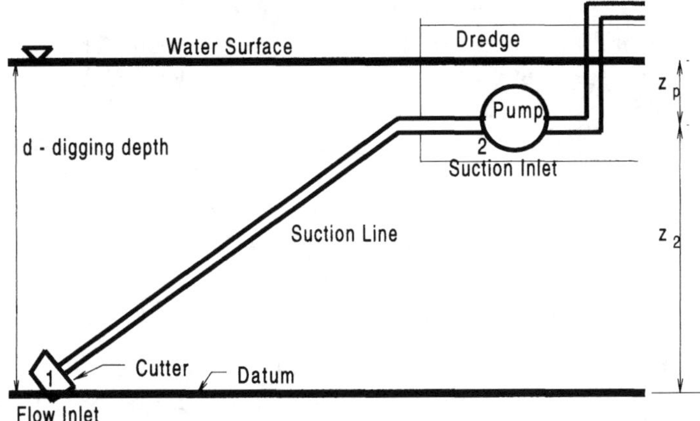

Figure 11.36 Dredge application for calculating net positive suction head (NPSH).

$$\text{available NPSH} = \frac{P_a}{\gamma_m} - \frac{P_v}{\gamma_m} + \frac{d}{\text{SG}_m} - z_2 - H_L \qquad (11.36)$$

Example 11.1 A dredge pump is located 5 ft below the water surface. The digging depth is 45 ft and the slurry specific gravity is 1.2. The estimated total head loss is 12 ft of mixture, and the vapor pressure at this location is 0.36 lb/in^2 at 70°F. The atmospheric pressure is assumed to be 14.7 lb/in^2.

SOLUTION:

$$\gamma_m = \gamma(\text{SG}_m) = 62.4(1.2)$$

$$= 74.8 \text{ lb/ft}^3 \text{ (11.75 kN/m}^3\text{)}$$

$$\frac{P_a}{\gamma_m} = \frac{14.7(144)}{74.8}$$

$$= 28.3 \text{ ft (8.6 m) of mixture}$$

$$= \frac{0.36(144)}{74.8}$$

$$= 0.69 \text{ ft (0.21 m) of mixture}$$

available NPSH

$$= \frac{P_a}{\gamma_m} - \frac{P_v}{\gamma_m} + \frac{d}{\text{SG}_m} - z_2 - H_L$$

$$= 28.3 - 0.69 + \frac{45}{1.2} - 40 - 12$$

$$= 13.1 \text{ ft (4.0 m) of mixture}$$

The available NPSH is compared to the required NPSH provided by pump manufacturer or from the pump characteristics curves. If the available NPSH is greater than the required NPSH, the pump will not cavitate. If the available NPSH is less than the required NPSH, the dredge pump is expected to cavitate. Most pump characteristic curves express the required NPSH in feet or meters of water. It is suggested that the available NPSH in feet or meters of mixture be used to compare with the required NPSH to give a margin or error in predicting cavitation.

11.4.9 Experimental Pump Test Facilities

Large laboratory test facilities are needed to test dredge pumps and develop the pump characteristic curves. In most cases the pump affinity laws are used to expand the results. To the author's knowledge, the largest pump test facility is located at the GIW Industries facility in Grovetown, Georgia. A schematic of the facility is illustrated in Figure 11.37. The facility can test pumps with suction and discharge sizes as large as 137 mm (54 in.). These large pumps are being used on the megahopper dredges such as the Vasco daGamma operated by Jan van deNul. Smaller facilities such the one at the Center for Dredging Studies on the campus of Texas A&M University (Figure 11.38) can be used to test small pumps (e.g., less than 254-mm or 10-in. pumps). Delft University and the Delft Hydraulics Laboratory in the Netherlands also have pump test facilities. Some dredge pump manufacturers also may have test facilities.

11.4.10 Effect and Removal of Gas in Dredge Pumps

Dredge pumps encounter mixtures of solids, liquids, and gases. Difficulties are not too troublesome for liquid–solid mixtures, except for plugging the pipe as a result of trying to pump too high-density material. For example, normal dredging operations commonly result in an average slurry specific gravity of 1.2. Laboratory experiments have accomplished a specific gravity of 1.45 when the slurry is silts and clays. In many estuaries, bays, harbors, and tidal rivers the sediments are gassy (hydrogen sulfide, methane, carbon dioxide, oxygen, and nitrogen. Dredging in these areas disturbs the sediments and the gases are released into the suction line. The gas collects in the pump and can severely reduce the flow of slurry and possibly cause the pump to lose its prime.

The Corps of Engineers developed the first gas removal system that was successful, on the dredge *Atlantic* in 1946. It consisted of a steam jet ejector attached on top of the suction side of the dredge pump. A common gas removal system consists of a gas accumulator and a source of vacuum. The accumulator is a vertical cylinder located on top of the suction pipe at the highest point in the suction line or near the dredge pump, as illustrated in Figure 11.39. The cylinder is about the same size as the suction pipe and is 2 to 3 diameters tall. The accumulator provides a means for the gas to separate from the solid–water mixture. The gas is then removed by the vacuum system. The vacuum is produced by either a piston vacuum pump or a steam- or water-driven ejector. Vacuum pumps can be damaged by water, but ejectors are not. Vacuum pumps can be simply controlled, but controls for ejector are more complicated. The vacuum pump have been found to be more effective with liquid-level control in the center of the accumulator, and ejectors more effective with the liquid level in the upper part of accumulator.

11.5 CUTTERS, DRAGHEADS, AND BUCKETS

The hydraulic pipeline dredge normally uses a cutter to excavate, loosen, or pulverize the material to be dredged and make it available to the suction inlet of the suction pipe. Basket cutters are the most common type of cutter, but other cutters are used, such as the bucket wheel cutter, auger, and disk cutters. Hopper dredges lower large dragheads attached to the end of the suction tube. Special dredges like the dustpan dredge use special dredge heads to remove maintenance material. Mechanical dredges such as grapple, clamshell, dipper, and endless chain buckets use different types of buckets to remove underwater sediments.

11.5.1 Cutters

11.5.1.1 Basket Cutter. The most common cutter used is the basket cutter, shown in Figure 11.40. It has multiple blades that rotate around

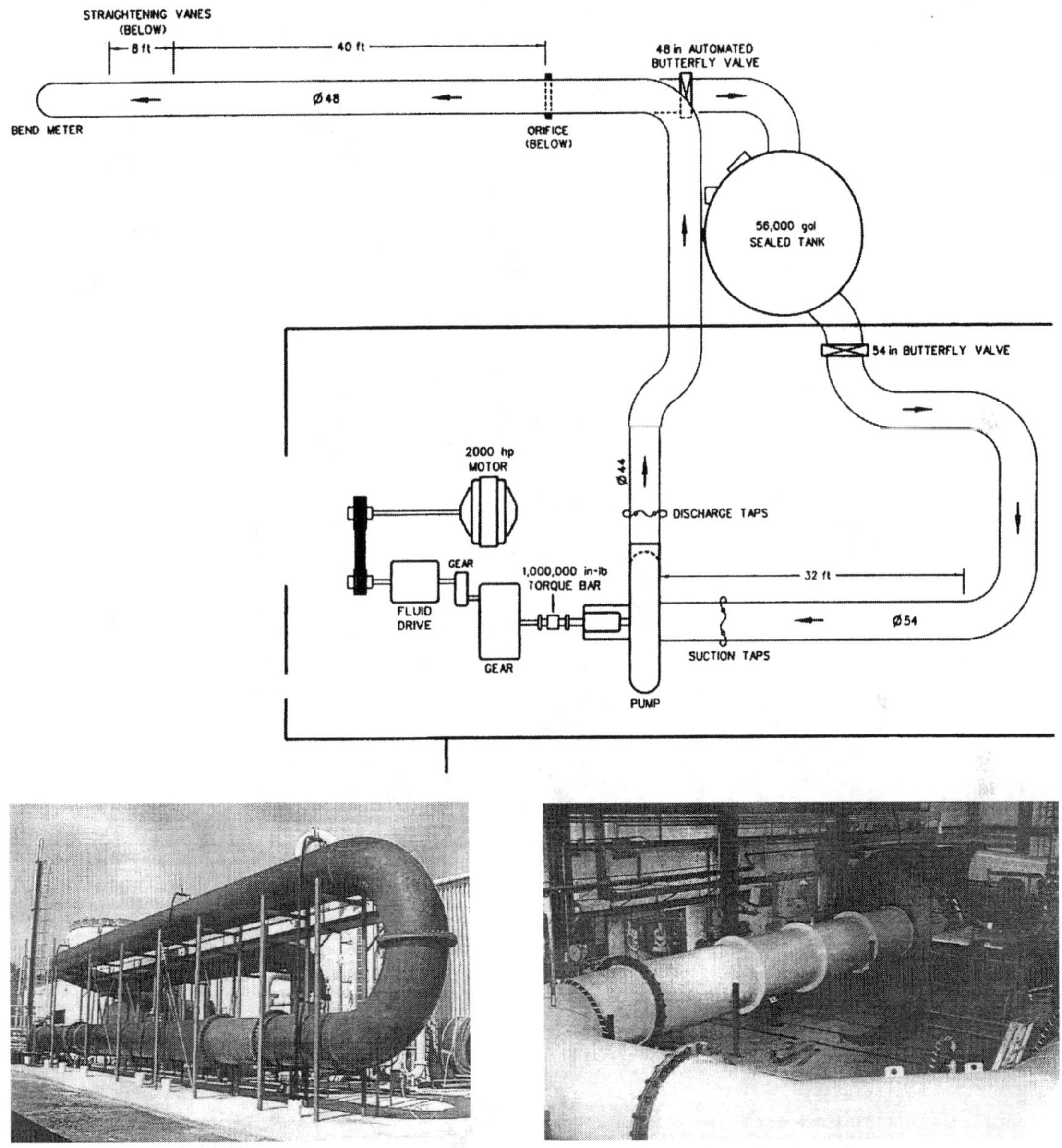

48-in. circulation pipe

54-in. inlet pipe

Figure 11.37 GIW pump test facility. (From Addie et al., 2001.)

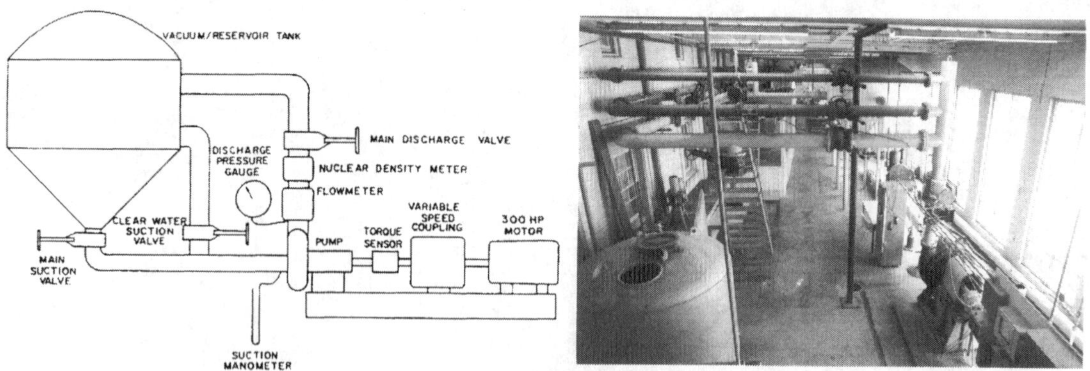

Figure 11.38 Dredge loop test facility at the Center for Dredging Studies at Texas A&M University.

Figure 11.39 Gas accumulator near suction inlet of pump. (From Adams and Herbich, 1970.)

Figure 11.40 Selected examples of cutters. (Courtesy of Ellicott, Div. of Baltimore Dredges, LLC.)

a longitudinally mounted shaft that is supported by a ladder. The ladder and dredge swing through an arc of a circle using swing winches and wire ropes attached to anchors. Basket cutters come in different shapes, number of blades, attachment methods, rake angles, and cutting edges. The cutter loosens the material in front of the suction inlet and prevents heavy stresses on the suction pipe that is supported in the lower half of the ladder. The cutter is commonly about three times the diameter of the suction pipe. The suction opening is strategically placed near the bottom of the cutter and very close to the cutter, to capture as much as possible of the loosened material under the influence of the high-velocity water flowing into the suction inlet. Because the basket cutter is open on the back side near the suction tube, some of the material will be left behind, and it is estimated that as much as half the material may be left on the channel bottom.

The geometry of a basket cutter is illustrated in Figure 11.41. The face angle should vary with the angle that the cutter shaft makes with the channel bottom. Large face angles make the cutter very conical and sometimes result in dragging of the ladder on the channel bottom, which creates large forces in the swing wires, which may lead to failure. Many baskets are square or trapezoidal in shape with relatively small face angles that are used best in shallow-water dredging. The large face angles are better for deeper dredging, with shaft angles of 25 to 35°. The rake angle is the angle between the tangent to the cutter circumference and the cutter blade or tooth on the blade, as shown in Figure 11.41. Turner (1996) indicates that the practical range of the rake angle is 25 to 30°. Very large rake angles require large amounts of power, and very small angles may result in limited excavation and low production.

The cutter drive used to rotate the cutter is either hydraulically or electrically driven. The cutter rotative speed and the cutter drive power must be compatible with the hydraulic transport capability of the dredge. Turner (1996) provides a relationship (Figure 11.42) between the strength of the channel sediments, production, and the standard penetrometer test (SPT) blow count, or the unconfined compressive strength of the channel sediments. There is much discussion on the accuracy of the data, but the dredger needs to have a method to determine the capacity of the cutter to ensure that the cutter is not limiting the production of the dredge.

11.5.1.2 Bucket Wheel Cutter. The bucket wheel cutter shown in Figure 11.43 rotates about an axis perpendicular to the ladder and

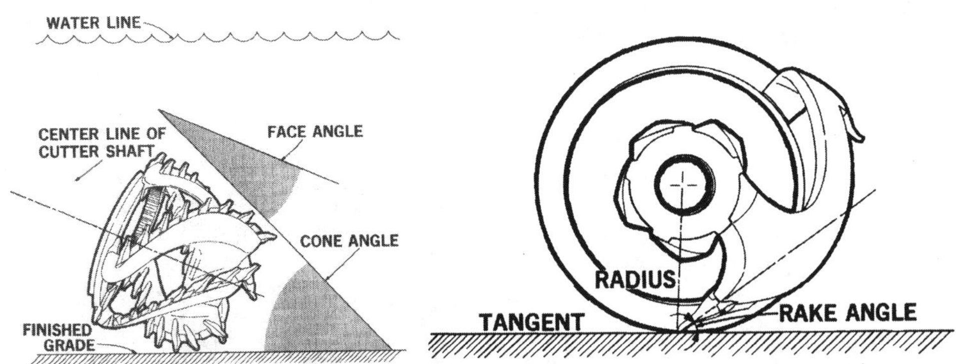

Figure 11.41 Cutter geometry and angles. (From Turner, 1996. Reproduced by permission of the publisher, ASCE.)

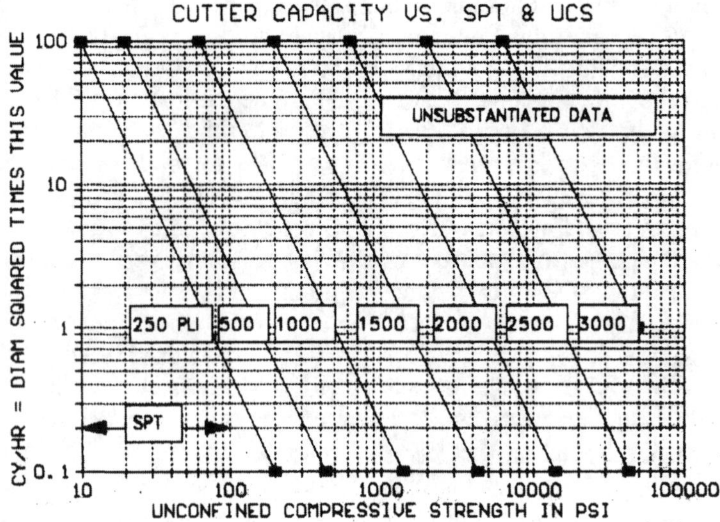

Figure 11.42 Estimation of cutter capacity as a function of SPT blow count and unconfined compressive strength of the sediments. (From Turner, 1996. Reproduced by permission of the publisher, ASCE.)

Figure 11.43 Bucket wheel cutter. (Courtesy of Lubecker.)

provides material to the suction pipe inlet. This cutter has bottomless buckets and teeth or blades of the bucket excavate the material. The advantages include positive cutting while swinging in either direction, greater cutting force per length of blade, capturing the excavated material without letting the material escape, handling heavy minerals, and better depth control. The biggest disadvantage of the bucket wheel cutter is its cost and the need to use a spud carriage to advance the dredge. Some of the increased cost is offset by the higher dredging efficiency of the spud carriage dredge.

11.5.1.3 Auger Cutter. Small hydraulic pipeline dredges (discharge diameter less than 305 mm or 12 in.) have used auger cutters for dredging in small water bodies such as lakes and quarries. The auger dredge (Figure 11.44) is composed of a cylinder 1.2 to 2.5 m (4 to 8 ft) long and 152 to 203 mm or 6 to 8 in. in diameter. The flighting on the cylinder forces the dredged material to the center of the dredge, where a submerged dredge pump is operating. The sediment enters the suction pipe and is pumped through the discharge pipeline to the disposal location. The Mudcat dredge developed by Ellicott Machine Corporation is

Figure 11.44 Auger cutter with cover. (Courtesy of Ellicott Machine Corporation.)

Figure 11.45 California and Fruhling dragheads. (From Bray et al., 1997.)

commonly used for shallow-water dredging in lakes and ponds and can be used to clear vegetation.

11.5.2 Dragheads

Trailing suction hopper dredges lower suction tubes, called *dragarms,* to the bottom of ship channels and remove channel sediments by pumping the sediments to the hopper. Once the hopper is filled with sediments, the dredge raises its dragarms and sails to a designated disposal area and places the material in the disposal area. A *draghead* is a large device that drags along the channel bottom in contact with the sediments. The bottom of the draghead normally has a grating that prevents large objects from entering the system. Several types of dragheads are used in the dredging industry, and these include the Fruhling, California, and Ambrose dragheads as examples. A schematic of the selected dragheads and a picture of the California draghead used on the Corps of Engineers' hopper dredge *Wheeler* are illustrated in Figure 11.45. In some cases, hopper dredging in ship channels encounters the presence of turtles that may be endangered. To protect the turtles, special devices are added to the draghead to prevent turtles from entering the draghead.

11.5.3 Buckets

Mechanical dredges use buckets such as those illustrated in Figure 11.46 to excavate channel sediments and place the dredged material into barges that are subsequently unloaded into dredge disposal areas. There are several types of buckets used. A clamshell bucket is one of the most common buckets used in dredging around docks and piers in ports and harbors. Backhoe buckets are also frequently used. The bucket sizes vary from as small as 0.76 m^3 (1 yd^3) to as large as 15.3 m^3 (20 yd^3). Recently, the dredging of contaminated sediments has necessitated making the buckets watertight to minimize resuspension of contaminants in dredged areas.

11.6 INSTRUMENTATION

Instrumentation is very important in all phases of dredging. Dredging equipment (e.g., cutter suction pipeline dredges, mechanical dredges, and hopper dredges) all perform more efficiently with instrumentation to measure density, flow velocity, pressures, and dredge position. Many dredges use instrumentation to provide information (data) to automate certain dredge operations. Differential global positioning systems (DGPSs) are used to record the position of dredges during the complete dredging operation (e.g., excavating, sailing, and placement). Survey vessels use DGPSs, depth sounders, and multibeam sonars to map the excavation and placement of dredged material.

11.6.1 Flow Meters

11.6.1.1 Magnetic Flow Meters. Magnetic flow meters (Figure 11.47) measure the average velocity of the fluid or slurry moving through the meter. These meters are normally installed in a vertical or inclined section of the suction or discharge piping near the pump. The average velocity is multiplied by the inside area of the pipe to obtain the average flow rate (e.g., gallons per minute, cubic feet per second, cubic meters per second). The magnetic flow meter operates on Faraday's principle, which says that the voltage induced in a conductor moving at right angles to the magnetic field is proportional to the velocity of the conductor through the field. The slurry is the conductor for dredging operations. Advantages of magnetic flow meters include: no obstruction to the flow, no additional head loss, electrical signal output linearly related to flow rate and can be input to computer data acquisition system, and the induced voltage not affected by turbulence, temperature, or viscosity. The flow meter does

clamshell

dipper

backhoe

chain of buckets

Figure 11.46 Examples of clamshell, dipper, backhoe bucket, and buckets on an endless chain for the bucket ladder dredge.

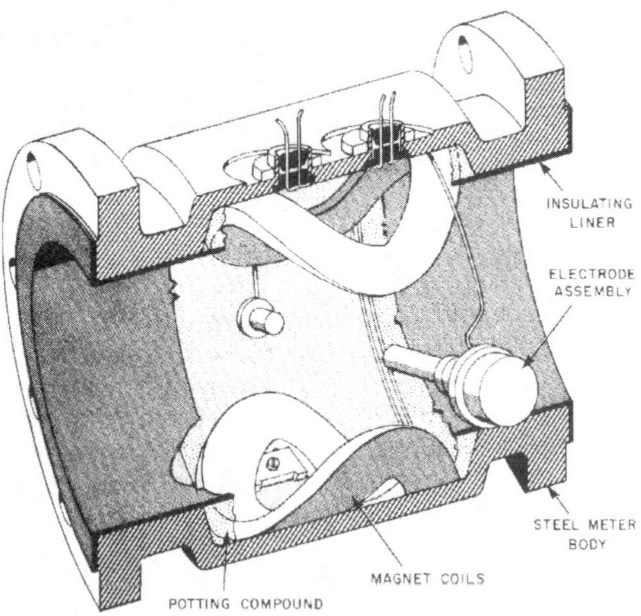

Figure 11.47 Magnetic flow meter. (Courtesy of Fischer Porter Co.)

require a liner that needs to be inspected periodically and the meter rotated to prevent uneven wear. The accuracy of a magnetic flow meter is usually within 1% of the range of flow. The instrument has proven to be reliable. It is relatively expensive and the unit cost depends primarily on the pipe size and can range from $10,000 to $30,000, but it can pay for itself in one dredging job by the improvement to production resulting from its use.

11.6.1.2 Sonic Flow Meters. The Doppler sonic flow meter is sometimes used in dredge installations. It is less expensive than the magnetic flow meter but less accurate. A transducer is mounted on the outside of the dredge pipe and a sound signal is introduced through the pipe wall. The signal propagates through the slurry and reflects off the moving solid particles and the returning frequency is shifted (Doppler shift) due to the velocity of the particle. The measured Doppler shift is directly related to the velocity of the particle in the pipe. Unfortunately, the velocity measured is a result of particles close to the pipe wall and does not give as accurate average velocity in the pipeline.

11.6.2 Density Meters

A nuclear density gauge (Figure 11.48) is commonly used to measure the density of the slurry flowing in a dredge pipeline. A source such as cesium-137 emits a stream of gamma rays that are absorbed by the material passing through the gauge typically located on the discharge side of the dredge pump. An ionization chamber located on the opposite side of the pipe measures the amount of energy reaching it and converts the radioactive energy into electrical energy. The system is calibrated to a known density, and the measured ionization during

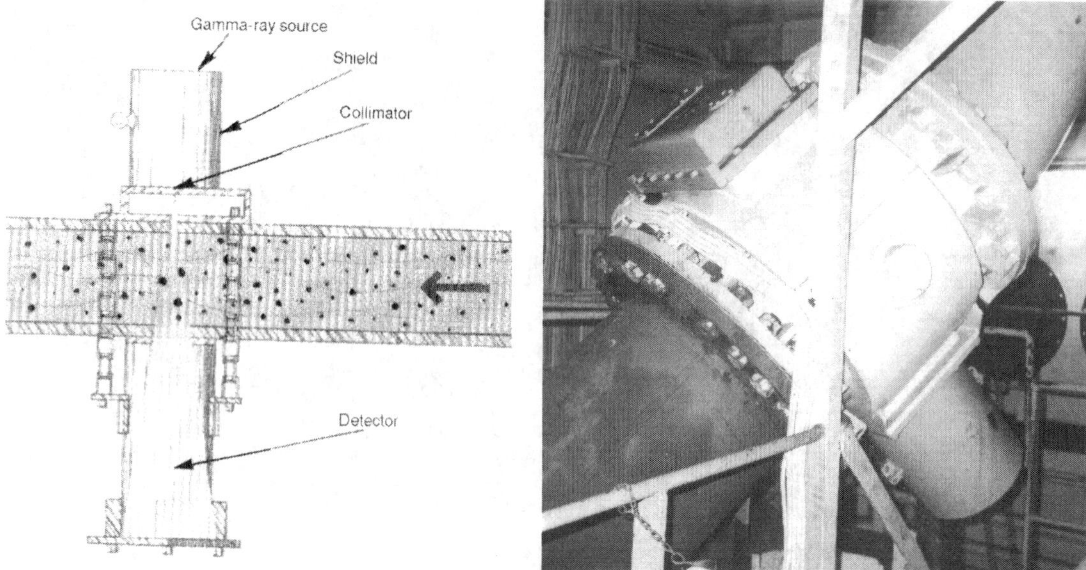

Figure 11.48 Nuclear density gauge. (Courtesy of TN Technologies.)

slurry transport is proportional to the density of the slurry being pumped. Strict safety procedures, construction and installation procedures, and regulatory procedures provide safe operation of the nuclear density gauge. A license is required for the installation and use of these instruments, and the user is not allowed to dismount or otherwise tamper with the source head. When the density measurement is combined with the velocity measurement of a magnetic flow meter, the amount of solids moved can be determined. This is an excellent method for helping the dredge operators maximize production.

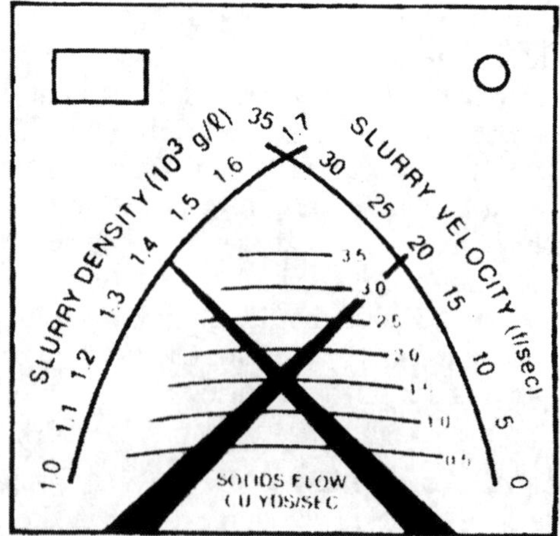

Figure 11.49 Cross-pointer display (Courtesy of TN Technologies.)

11.6.3 Production Meters

A production meter combines the outputs from the flow meter and the density gauge and displays the flow and density information on cross-pointer display (Figure 11.49). A scale on the cross-pointer display is produced behind the intersection of the pointers for density and velocity that indicates the solids pumped (solids production). The lever operator on the dredge uses this display to assist in maximizing the solids production by adjusting flow and density. In most cases, the dredge pumps are variable speed and the rotational speed can be adjusted to increase or decrease the velocity. The rate of swing and depth of the cutter can be adjusted to increase or decrease the density of the slurry.

11.6.4 Vacuum Relief and Bypass Valves

A vacuum relief valve is used to allow water to enter the suction pipe when the pressure in the suction pipe reaches a specified value. The purpose of this valve is to prevent cavitation. The bypass valve is used in the discharge line and also allows water to enter the suction line at a specified discharge pressure. A bypass valve is used to assist in the prevention of plugging the discharge pipeline. A commercially available valve for this purpose is the Hofer valve, which incorporates the vacuum relief and bypass valves. This system allows a metered amount of water to enter the suction pipe when set limits on suction and discharge pressure gauges are exceeded. The systems is meant to eliminate cavitation and pipeline plugging.

11.6.5 Automatic Light Mixture Overload Valve

The automatic light mixture overload valve (ALMO) is used on hopper dredges to divert slurry from the hopper when the density is to small for economic loading of the hopper. Input from a density gauge in the discharge pipe measures the density of the slurry, and when the value is below a set value (e.g., 1.2 specific gravity), the valve actuates gate valves that direct the slurry overboard to the water body.

11.6.6 Hopper Load Indicators

Hopper dredges incorporate sensors to determine the load of the dredged material added to the hopper during dredging. Once the desired load is reached, the hopper ceases dredging and lifts its dragarms and sails to the disposal location and places the material in the designated disposal area. These sensors are generally used to display the hopper loading visually to the dredge captain.

11.6.7 Cutterhead and Hopper Draghead Controllers

Cutterhead and draghead automatic control systems control the swing winches, ladder winch, spud carriage (if installed), and cutter speed for a pipeline dredge. For a hopper dredge, the controller automatically actuates the draghead winch to compensate for variations in the hopper dredge draft and the effects of tides and waves.

11.6.8 Flow Controller

The pipeline dredge requires velocity in the pipeline to transport the solids–water mixture and it must also swing and advance through the channel or dredging area. The solids normally require a minimum velocity to keep the solids from settling in the pipeline and plugging the pipeline. The dredge pump operates at different conditions when it is pumping slurry and when it is pumping water. Consequently, a flow (velocity) controller senses density and flow velocity and optimizes the velocity by controlling the pump speed (rpm) to ensure that sufficient velocity is maintained to carry the solids through the pipe. The flow controllers can save on fuel costs and increase the efficiency of the dredging operation.

11.7 PIPELINE TRANSPORT OF DREDGED MATERIAL

11.7.1 Slurry Transport in Pipes

Dredged material from cutter suction dredges is commonly transported by pipeline to dredged material disposal facilities or to beaches for beach nourishment. Hopper

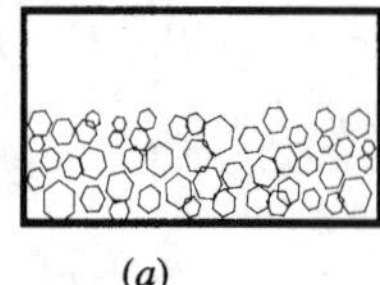
(a)

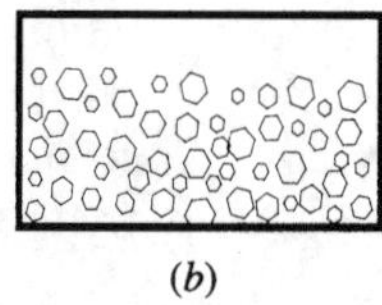
(b)

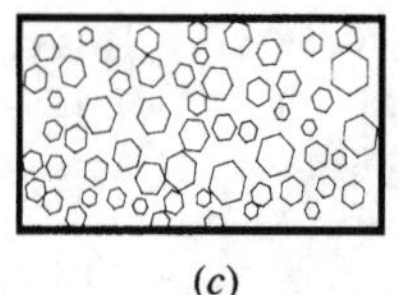
(c)

Figure 11.50 Sediment distribution in different slurry flow regimes: (a) fixed-bed; (b) heterogeneous; (c) homogeneous.

dredges also use pipelines to pump sand to beaches for beach nourishment. Even mechanical dredging operations may use hydraulic unloaders to unload barges hydraulically and transport the dredged material to a disposal location. The mixture of dredged material and water is called slurry. An important calculation in designing a dredging project is the determination of the amount of power and energy required to pump the slurry through the pipeline and to ensure that proper dredging equipment is available for the project. An equally important calculation is the estimation of production of solids that is eventually needed to estimate the time and cost to complete the dredging project.

The transport of sediments (dredged material) in pipelines depends on the water or seawater to support the sediments as it moves through the pipeline to the placement site. Typically, dredged material consists of sands, silts, and clays, and the settling velocity (v_t) of these sediments is important in calculating the energy (friction) losses in the pipeline. A simple equation (Schiller, 1992) for calculating the particle settling velocity is

$$v_t = 134.14(d_{50} - 0.039)^{0.972} \qquad (11.37)$$

where v_t is the settling velocity in mm/s and d_{50} is the median grain diameter in mm. This equation is good for sand particle sizes but not for gravel. Albar (2000) and Randall (2000a and b) compared additional settling velocity equations. Wilson et al. (1997) describe a very complicated settling velocity relationship that is recommended for use with the Wilson et al. (1997) friction loss expression.

There are three general regimes of slurry flow: fixed bed, heterogeneous, and homogeneous (Figure 11.50). The desirable regime for slurry transport is heterogeneous, which means that the sediment is just supported by the carrier fluid (water or seawater). Homogeneous flow means that the sediment and water are thoroughly mixed across the pipe cross section. Homogeneous flow requires more energy because the losses are greater because of the higher velocity required, and the higher velocity also greatly increases the wear in the pipeline, all leading to increased costs. Fixed bed means that the sediment is stationary on the bottom of the pipe, and eventually this situation can lead to plugging the pipeline, a very undesirable and costly occurrence.

Slurry transport in pipelines is also classified as settling and nonsettling flows. When the slurry contains sands, the sands can settle to the bottom of the pipe and this slurry is considered a settling flow. In some cases, dredging material may consist only of silts and clays, and this would be considered nonsettling. Wilson et al. (1997) address procedures for estimating energy losses in both settling and nonsettling flows. Most dredging projects involve the settling flows in the heterogeneous regime. Dredged material can also be classified as cohesive or noncohesive. Dredged material containing sands is considered noncohesive, and the viscous properties of the carrier fluid are used in the energy loss calculations. Dredged material containing almost all silts and clays (mud) is commonly cohesive and behaves as a non-Newtonian fluid that requires careful consideration. The majority of dredging projects involve dredged material that is considered noncohesive and is discussed herein; Wilson et al. (1997) and Shook and Rocco (1991) should be consulted for other types of flows.

Slurry composition is defined at the ratio of solids to the total amount of the slurry mixture by volume or by weight. The concentration of solids by volume (C_v) is commonly used in dredging and is expressed as

$$C_v = \frac{SG_m - SG_f}{SG_s - SG_f} \qquad (11.38)$$

where SG is specific gravity and the subscript *m*, *f*, and *s* represent mixture, fluid, and solids. The concentration of solids by weight (C_w) is the ratio of weight of solids to weight of mixture:

$$C_w = \frac{SG_s(SG_m - SG_f)}{SG_m(SG_s - SG_f)} \qquad (11.39)$$

The particle specific gravity for sand, silts, and clays is essentially 2.65, but the in situ specific gravity of dredged material typically ranges between 2.1 and 1.3. The in situ specific gravity is often used as SG_s in dredging calculations for the concentration by volume. In most cases, the highest concentration by volume that can be maintained is about 20%, which corresponds to approximately 40% by weight for typical dredging projects.

11.7.2 Determination of Critical Velocity

As a grain of sediment travels through a pipeline, there is a certain velocity that the fluid around it must maintain to prevent the grain from falling to the bottom of the pipe and becoming stationary. If this velocity is not maintained, much of the sediment will settle and eventually clog the pipeline. This condition is very undesirable, because it means shutting down the operation while the line is being unclogged. This critical velocity is a function of the specific gravity of the sediment, the grain size, and the diameter of the pipe. Wilson et al. (1997) presented a convenient method of determining the critical velocity (V_c) using a nomograph (Figure 11.51). Matousek (1997) developed an equation based on the nomograph to calculate the critical velocity. It is desirable to pump near the critical velocity based on the fact that head losses, power requirements, and pipeline wear are minimized at this value. The average slurry velocity is recommended to be 10 to 20% above the critical velocity.

$$V_c = \frac{8.8[\mu_s(S_s - S_f)/0.66]^{0.55} D^{0.7} d_{50}^{1.75}}{d_{50}^2 + 0.11 D^{0.7}} \qquad (11.40)$$

where μ_s is the coefficient of mechanical friction between the solid particles (typically taken as 0.44 or 0.55) and the pipe wall, d_{50} is the grain diameter (mm), and D is the inside pipe diameter (m) (Matousek, 1997). The mechanical friction coefficient of 0.44 agrees better with the results obtained from Figure 11.51.

The effect of pipe inclination on critical velocity is illustrated in Figure 11.52. These results show the critical velocity increases as the angle between the pipe and the horizontal increases up to an angle of approximately 35°. The increase in the critical velocity is determined using eq. (11.41). Thus, slurry pipe systems with inclined pipes must allow for the increased critical velocity, or else deposits may occur in the inclined sections and the pipe may

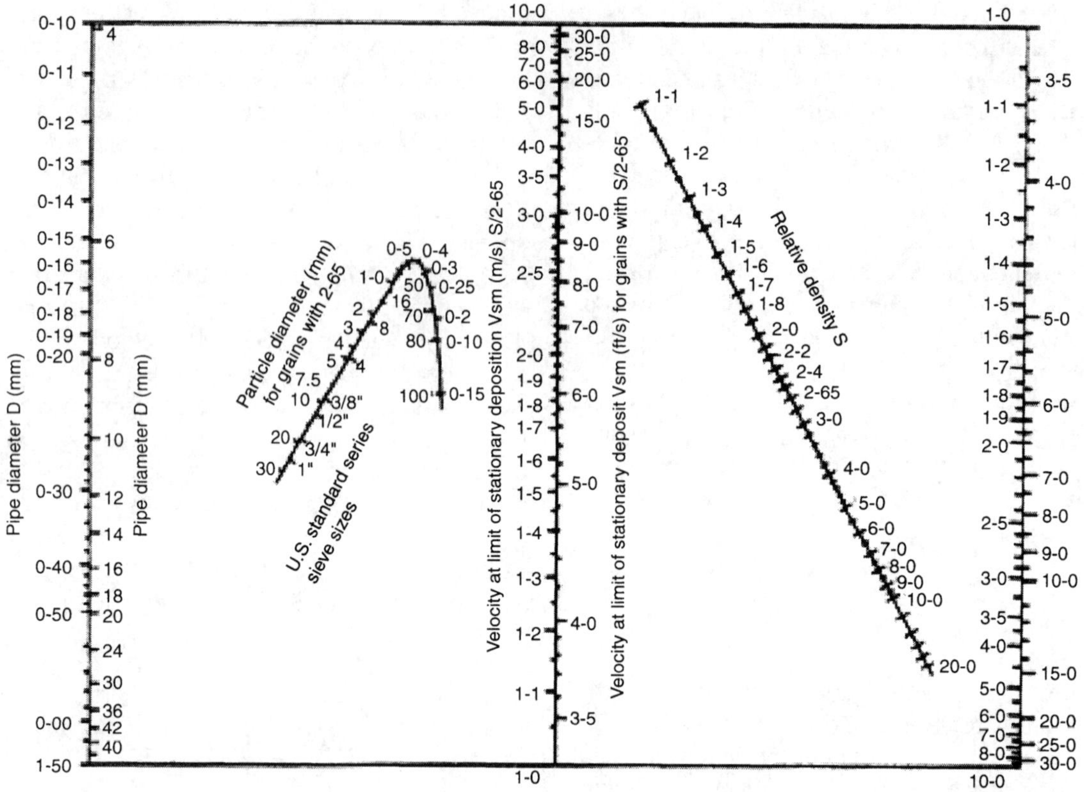

Figure 11.51 Nomograph for estimating the critical velocity in slurry pipelines. (From Wilson et al., 1997.)

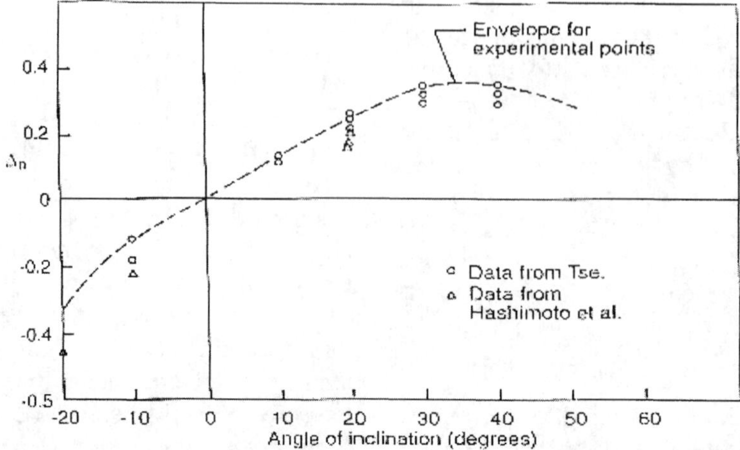

Figure 11.52 Effect of angle of inclination on critical velocity. (From Wilson and Tse, 1984.)

plug. It may also suggest that vertical sections may be preferable over short inclines to prevent deposits in the pipe.

$$V_c\text{ (inclined)} = V_c\text{ (horizontal)} + \Delta_D[\sqrt{2g(S_s - 1)D}] \quad (11.41)$$

11.7.3 Determination of Head Losses

It is important to calculate the head losses for the system because this is how the maximum pumping distance for a particular pump is determined. The major component of the head losses of a system is the frictional head loss. There are also minor losses due to pipe joints and bends, but when you are dealing with thousands of meters or feet of pipe, the frictional losses far outweigh the rest. Head losses due to friction are determined in the program by the procedure outlined by Wilson et al. (1997). The head losses per unit length expressed in meters of head per meters of pipe (feet of head per feet of pipe) are given by

$$i_m = \frac{fV^2}{2gD} + 0.22(S_s - 1)V_{50}^{1.7}C_v V^{-1.7} \quad (11.42)$$

$$w = 0.9V_t + 2.7\left[\frac{(\rho_s - \rho_f)g\mu}{\rho_f^2}\right]^{1/3} \quad (11.43)$$

$$V_{50} = w\sqrt{\frac{8}{f}}\cosh\frac{60d}{D} \quad (11.44)$$

where i_m is the head loss due to friction per unit length, d the median particle diameter, D the pipe inside diameter, ρ_s and ρ_f are the density of solid and fluid, respectively, V is the mean velocity of the mixture, f the friction factor for water, g the acceleration due to gravity, μ the dynamic viscosity of the fluid, S_s the specific gravity of solids, C_v the delivered concentration by volume, and V_t the particle terminal velocity. The value of 0.22 in the equation for i_m is a typical value for the ratio

$$\frac{i_m - i_f}{S_m - S_s} \quad (11.45)$$

where i_f is the hydraulic gradient of the fluid and S_m is the mean specific gravity of the mixture.

11.7.4 Evaluation of Friction Factor

In 1944, Moody developed the friction factor chart, and it is commonly used to determine the friction factor (f) as a function of Reynolds number (R) and the relative roughness (e/D). A convenient explicit expression for the friction factor developed by Swamee and Jain (1976) is

$$f = \frac{0.25}{[\log(\varepsilon/3.6D + 5.74/R^{0.9})]^2} \quad (11.46)$$

which is valid for the range of $5 \times 10^{-3} \leq R_D \leq 10^8$ and $10^{-6} \leq \varepsilon/D \leq 10^{-2}$.

11.7.5 Losses due to Fittings and Valves

Minor losses (h_m) in pipeline systems are energy losses caused by fittings, valves, elbows, contractions, swivel joints, pipe entrance, and exit conditions. In many cases these energy losses are calculated using the expression

$$h_m = K\frac{V^2}{2g} \quad (11.47)$$

where K is the minor loss coefficient. Selected values for these loss coefficients are tabulated in Table 11.10.

Minor loss coefficients can also be expressed as an equivalent length of pipe (L_e) that

Table 11.10 Selected minor loss coefficients for pipe systems

Pipe System Component	Minor Loss Coefficient, K
Suction entrance	
Plain end suction	1.0
Rounded suction	0.05
Nozzle	5.5
Oval	1.0
Funnel	0.1
Pear	0.02
Elbows	
Long radius 90° (flanged)	0.2
Long radius 90° (threaded)	0.7
Long radius 45° (flanged)	0.2
Regular 45° (threaded)	0.4
Regular 90° (flanged)	0.3
Regular 90° (threaded)	1.5
Stern swivel	1.0
Ball Joints	
Straight	0.1
Medium cocked	0.4–0.6
Fully cocked (17°)	0.9
End section	1.0
Valves	
Globe (fully open)	10.0
Angle valve (fully open)	5.0
Swing check (fully open)	2.5
Gate (fully open)	0.19
Gate (half open)	5.6
Gate (quarter open)	17
Ball (fully open)	0.05
Ball ($\frac{2}{3}$ open)	5.5
Ball ($\frac{1}{3}$ open)	210
180° return bend (flanged)	0.2
180° return bend (threaded)	1.5
Standard tee	1.8

has the same head loss for the same discharge. The equivalent length of pipe is defined as

$$f\frac{L_e}{D}\frac{V_2}{2g} = K\frac{V^2}{2g} \qquad (11.48)$$

where f is the friction factor, L_e the equivalent length of pipe, D the inside pipe diameter, and K the minor loss coefficient. The expression for the equivalent length of pipe in terms of the minor loss coefficient reduces to

$$L_e = \frac{KD}{f} \qquad (11.49)$$

Detailed minor losses expressed in terms of equivalent length of pipe are contained in a work of the Crane Company (1979). Additional general discussion of minor losses and general minor loss coefficients for expansions and contractions are contained in current fluid mechanics references such as Munson et al. (1998).

Example 11.2 The dredge pump centerline is located 5 ft below the water level (Figure 11.53). The cutterhead dredge is operating in 40 ft of water and has a 75-ft suction pipe and a 2600-ft discharge pipe, and each pipe has a diameter of 16 in. The ambient atmospheric pressure is 14.7 lb/in^2. The channel bottom material is fine sand (d_{50} = 0.2 mm), and the water has a specific gravity of 1.02 and average temperature of 60°F. The vapor pressure is 0.256 lb/in^2 absolute and the kinematic viscosity is 1.21×10^{-5} ft^2/s. Determine the friction head loss while the pump is operating with a slurry specific gravity of 1.24 and flow rate of 10,000 gpm. Neglect the effect of minor losses.

SOLUTION:

V_c = 2.75 m/s = 9.02 ft/s (from Figure 11.51)

$$i_m = \frac{f_w V_m^2}{2gD} + 0.22(S_s - 1)V_{50}^M C_v V^{-M}$$

where

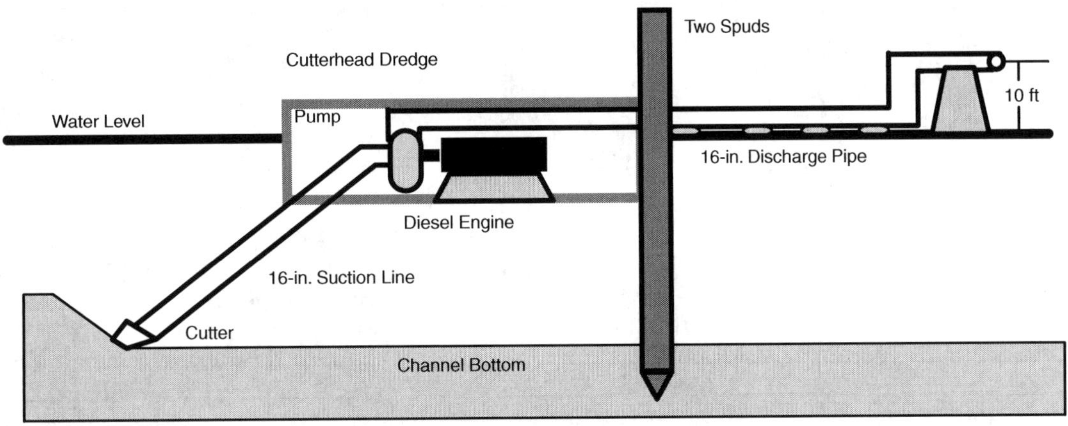

Figure 11.53 Dredge configuration for the example problem.

$$V_{50} = w\sqrt{\frac{8}{f_f}}\cosh\frac{60d}{\mathrm{D}}$$

$$w = 0.9v_t + 2.7\left[\frac{(\rho_s - \rho_f)g\mu}{\rho_f^2}\right]^{1/3}$$

$$= 0.9v_t + 2.7\left(\frac{\rho_s - \rho_f}{\rho_f}\,g\,\frac{\mu}{\rho_f}\right)^{1/3}$$

$$= 0.9v_t + 2.7\left(\frac{\mathrm{SG}_s - \mathrm{SG}_f}{\mathrm{SG}_f}\,g\nu\right)^{1/3}$$

$$w = 0.9v_t + 2.7\left[\frac{(\mathrm{SG}_s - \mathrm{SG}_f)g\nu}{\mathrm{SG}_f}\right]^{1/3}$$

$$= 0.9(0.0227) + 2.7\left[\frac{2.65 - 1.02}{1.02}(9.81)(1.17 \times 10^{-6})\right]^{1/3}$$

$$= 0.091 \text{ m/s}$$

$$V_{50} = w\sqrt{\frac{8}{f_f}}\cosh\frac{60d}{D}$$

$$= 0.091\sqrt{\frac{8}{0.0106}}\cosh\frac{60(0.0002)}{0.407}$$

$$= 2.5 \text{ m/s}$$

$$i_m = \frac{f_w V_m^2}{2gD} + 0.22(S_s - 1)V_{50}^M C_v V^{-M}$$

$$= \frac{(0.0106)(4.85)^2}{2(9.81)(0.407)} + 0.22(2.65 - 1)(2.5)^{1.7}(0.135)(4.85)^{-1.7}$$

$$= 0.0471 \text{ m of water/m of pipe}$$

$$= 0.0471 \text{ ft of water/ft of pipe}$$

$$h_L = 0.0471(2675) = 126 \text{ ft of water}$$

11.7.6 Estimating Production

Estimating production is important to determine the amount of time it requires to complete the dredging project and is also critical in the cost estimating process. The rate of production is the number of cubic yards of sediment per hour that are removed from the channel bottom by the dredge. From a cost estimation standpoint, this is an important factor simply because it determines how much time it takes to complete the project. More time on the job means more wear on the machinery, more fuel used, and more pay for the workers. It is therefore imperative that the estimate of the production be as accurate as possible. Turner (1996) estimates the production of a pipeline dredge using

$$P = 0.297QC_v \quad \text{or}$$
$$P = 0.297\, QC_{v\ \max} \cdot \text{DE} \qquad (11.50)$$

where P is the production (yd^3/hr), 0.297 is a conversion factor such that the production is in cubic yards per hour, Q the average flow rate (gpm), C_v the average concentration of solids by volume, $C_{v\ \max}$ the maximum concentration of solids by volume, and DE the dredging efficiency. Dredging efficiency is selected as 50% when using two fixed spuds and 75% when using a spud carriage.

11.8 DREDGED MATERIAL PLACEMENT

When dredged material is removed from waterways, it is typically placed in an open-water disposal site or confined disposal facility, or it is used beneficially (e.g., beach nourishment, habitat restoration, land reclamation, etc.). In the United States the USEPA (U.S. Environmental Protection Agency)/USACE (1992) discusses management of dredged material whether it is contaminated or clean, and whether it is new work and maintenance work. Placement of dredged material is regulated by the London Convention and local environmental laws.

11.8.1 Open Water

Disposal of dredged material in rivers, lakes, estuaries, and oceans is called *open-water disposal*. Open-water disposal is accomplished using pipeline dredges, hopper dredges, barges/scows loaded by mechanical or hydraulic dredges, and mechanical dredges. Open-water disposal sites are described as nondispersive or dispersive sites. In nondispersive sites the dredged material remains within the disposal site boundaries. In dispersive sites the dredged material is expected to leave the disposal site due to environmental forces such as ocean currents or other meteorological events (e.g., storms, frontal passages, hurricanes, etc.).

Pipeline dredges discharge slurry into the water body and the solids eventually settle to the bottom, forming a mound of dredged material. Some of the finer material may remain suspended, which is called a *turbidity plume,* and move out of the disposal area with the prevailing currents. Hopper dredges store the excavated dredged material in a large hopper within the hull of the vessel. The dredged material is a slurry that is a mixture of sediment and water. Overflowing can increase the amount of sediment in the hopper. The overflow procedure allows more sediment (i.e., sand) to settle out in the hopper while the water and fine material overflows back into the water body. When the hopper is full, the hopper dredge sails to the open-water disposal site and opens the bottom doors, or opens its hull to allow the dredged material to fall through the water column and form a mound of material on the water body floor. Most of the dredged material comes to rest on the bottom. Mechanical dredges may be used in shallow, protected bodies of water. The mechanical dredges (e.g., clamshell) remove the dredged material at nearly the in situ density and place it in a barge or scow. Once the barge or scow is filled, a vessel tows the barge to the open-water disposal site, where the dredged material exits the bottom of the barge or scow and falls to the bottom, as shown in Figure 11.54. Open-water dredged material disposal sites are commonly marked on navigational charts and are typically found just outside ship channels, where frequent dredging occurs to maintain the channel design depth. Rivers, bays, and estuaries also contain designated disposal sites. Local and international laws regulate these disposal sites.

Figure 11.54 Split-hull hopper barge placing dredged material in open water. (Courtesy of U.S. Army Corps of Engineers.)

11.8.2 Confined Disposal Facilities

Upland and nearshore areas that are diked to contain dredged material and allow clear water to return to the water body are called *confined disposal facilities* (CDFs) and are illustrated in Figure 11.55. Guidelines for design, operation, and management of CDFs are contained in USACE (1987a). The objective of the CDF is to have enough storage requirement for the dredging operation and to be capable of allowing the sediment to settle out of the water and remain within the diked boundaries while the water is allowed to flow over a weir structure and return to the water body.

Pipeline dredges are typically used to deliver dredged material in the form of slurry through a long pipeline to the confined disposal site. The dredged material enters the CDF at an inlet structure and the dredged material flows through the site and ponds. An outlet structure (e.g., weir) holds back the slurry for sufficient time to allow the solids to settle. Once the water clarifies as a result of the sediment settling, then the weir boards are removed and the water overflows back to the water body (e.g., river, bay, etc.). Eventually, all the water leaves the site through drainage and evaporation and the dredged material dries and consolidates (Figure 11.56).

Mechanical dredges sometimes load dredged material in barges or scows that are brought to the confined disposal site and unloaded onto trucks that are taken to the disposal site for emptying, or the dredged material is hydraulically slurried and pumped to the disposal site via a pipeline.

11.8.3 Beneficial Uses

Landin (2000) is an excellent reference that describes in detail the many beneficial uses of dredged material. U.S. Army Corps of Engineers programs such as the Dredged Material Research Program (DMRP) from 1973 through 1978, the Dredging Operations Technical Support Program (DOTS) from 1978 through the present, the Environment Effects of Dredging Program (EEDP) from 1982 through the present, the Dredging Research Program (DRP) from 1991 through 1996, the Dredging Operations and Environmental Research Program (DOER) from 1998 through the present, and the Wetlands Research Program (WRP) from 1990 through 1995 have contributed significant research results related to the beneficial use of dredged material. Conventional placement of dredged material at open-water and confined disposal sites continues to be the most economical placement procedure. However, there is strong desire and pressure to find funding and ways to use dredged material economically in a beneficial way.

11.8.3.1 Common Beneficial Uses of Dredged Material. Dredged material contains sands, silts, and clays, and these sediments can be considered a resource that can be used to benefit society. USACE (1987b) describes 10

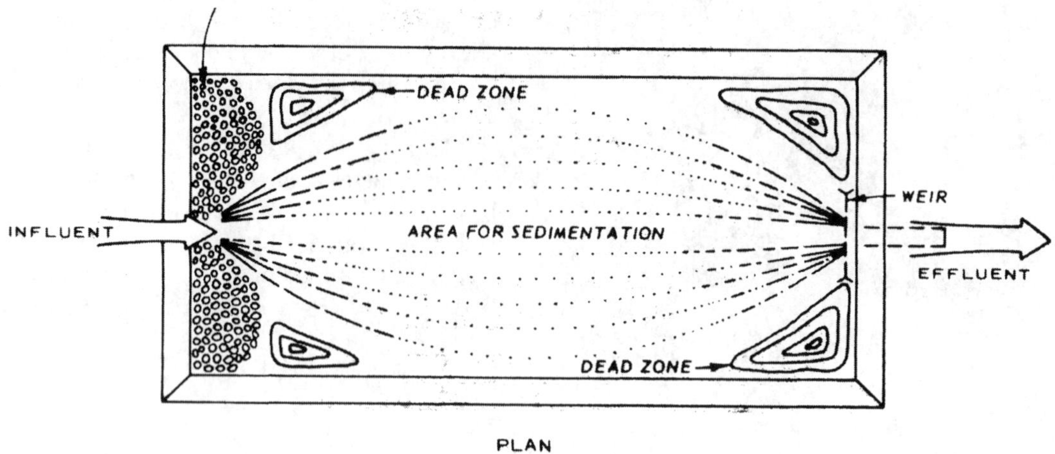

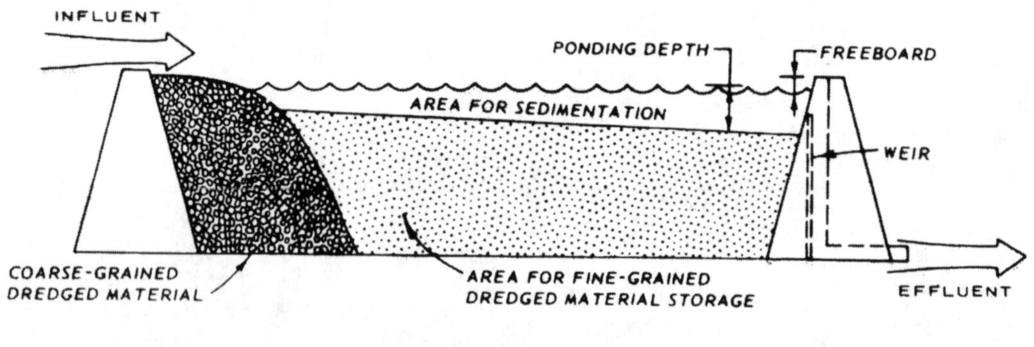

Figure 11.55 Confined disposal facility. (Courtesy of U.S. Army Corps of Engineers.)

Figure 11.56 Confined disposal facility after initial filling inlet (*a*), after filling (*b*), and after dewatering using trenches (*c*). (Courtesy of U.S. Army Corps of Engineers.)

broad categories of beneficial uses of dredged materials:

- Habitat restoration (e.g., wetlands, islands for birds and waterfowl)
- Beach renourishment
- Aquaculture
- Parks and recreation
- Agriculture, forestry, and horticulture
- Land reclamation and landfill cover
- Shoreline stabilization and erosion control (e.g., fills, submerged berms)
- Industrial use (e.g., port development, airports, residential development)
- Material transfer (e.g., fill, dikes, levees, parking lots, highways)
- Multiple purposes

Using dredged material beneficially is very desirable, but the cost of bringing the dredged material to the location where it can be used beneficially is often a difficult and costly endeavor. Examples of beach nourishment, habitat creation, and agriculture reuse are illustrated in Figure 11.57.

11.8.3.2 Newer Beneficial Uses of Dredged Material

Manufactured Soil. Graalum et al. (1999) suggest that a potential beneficial use of dredged material is to use it to manufacture topsoil. Previous manufactured soil projects include Toledo Harbor, New York/New Jersey harbor, Mobile Harbor in Alabama, and the Herbert Hoover Dike surrounding Lake Okeechobee in Florida. The manufactured soil helps reduce and recycle wastewater sludge and provides an additional alternative for the long-term management of dredged disposal sites by reducing the amount of land needed for the confined disposal facilities. Manufactured soil (MS) is created using dredged material (DM) and recyclable organic waste materials. The organic waste materials may be biosolids (BS) (e.g., sewage sludge, animal manure, yard waste), or biomass (BM) (e.g., cellulose or sawdust). The physical properties of the dredged material, such as sediment grain size and composition, are measured to determine the most appropriate use of the manufactured soil. The presence of salt, due to the brackish water in most areas, may pose a problem. Germination and growth tests for various combinations of the dredged material, biosolids, and biomass must be performed to determine the optimal percentage of each component (Lee and Sturgis, 1996). Biosolids are treated or reconditioned sewage sludge, but they can also be produced from reconditioned animal waste, such as cow manure. In sewage treatment facilities, the sewage sludge is reconditioned such that the pathogens

(*a*)

(*b*)

(*c*)

Figure 11.57 Examples of beneficial uses of dredged material: (*a*) beach nourishment (after and before); (*b*) habitat creation; (c) agriculture reuse. (Courtesy of U.S. Army Corps of Engineers.)

are reduced to appropriate levels, and the level is referred to as class B. USEPA restricts the use of class B biosolids. In some cases, the sewage may be reconditioned through chemical and thermal processes to obtain a class A level, and class A biosolids are unrestricted in how they are used.

In the manufacturing process, the volume of suitable dredged material at the confined disposal facility, or the estimated amount to be dredged, is determined to calculate the amount of topsoil to be produced. For a given amount of dredged material, the volume of biomass and biosolids is determined. Plant screening tests should be conducted to determine the required percentages of material. For example, studies using the N-Viro methodology by Lee et al. (1996) show ratios of 60% DM, 30% BM, and 10% N-Viro BS for the Toledo project. Ratios of 30% DM, 60% BM, and 10% N-Viro BS were used for the New York/New Jersey project. N-Viro soil is a patented reconditioned sewage sludge product from the N-Viro alkaline stabilization process. The amount of material to be processed is then dependent on the amount designated by the screening tests, as well as the amounts of dredged material and biosolids available. Once the volumes of the materials are determined, they can be used to calculate the costs associated with excavating and transporting the materials for the manufacturing process. Although each manufactured soil project is unique, the process as shown in Figure 11.58 requires certain elements that are common for all projects, pilot or full-scale.

The soil is processed using the estimated combination of dewatered dredged material, biosolid, and biomass. The mixing segment of the manufacturing soil process has two options: mixing by mechanical means or using agricultural equipment within the confined disposal facility (CDF). Mixing by mechanical means involves a compostlike blender in which the materials are combined (Lee, 1997). Mixing within the CDF can be accomplished using common agricultural equipment. Spreaders are used to distribute the reconditioned sewage sludge and biomass throughout the CDF. The materials are then mixed using a cultivator or rotor-tiller. The finished product is shipped to the intended use site or market. Contaminated dredged sediments may be converted to cover material for Superfund, mining, and landfill sites. If class A biosolids are used, uncontaminated material can be used as unrestricted topsoil for landscaping, roadside cover, parks, and bagged soil.

The issues affecting the feasibility of manufacturing soil include (1) finding a market or use for the converted topsoil, (2) determining the optimum site for the project, (3) deciding which biomass should be used and from where it will come, and (4) acquiring the biosolid or reconditioned sewage sludge. Graalum et al. (1999) discuss some of the anticipated problems, such as salinity and excavation of the dredged material, associated with converting dredged material to topsoil. A methodology is described for determining the costs associated with converting dredged material to topsoil and two examples of manufactured soil scenarios are discussed. Excavation, transportation, and manufacturing costs are applied to each of the materials used in the creation of the topsoil. The cost analysis is applied to two potential pilot sites along the Texas Gulf Intracoastal Waterway to demonstrate the methodology. The two sites, Matagorda Bay and the Bolivar Peninsula near Galveston Bay, show that manufactured soil for use in construction and landfill projects is feasible, with prices ranging from \$17 to \$26 per m^3 (\$13 to \$20 per yd^3) of soil manufactured. Although manufactured soil is more expensive than typical landfill cover and construction materials, it must be emphasized that the purpose of converting dredged material to topsoil is to reduce the volume of dredged material placed into a disposal area. In addition, use of dredged material from the disposal site saves costs associated with the

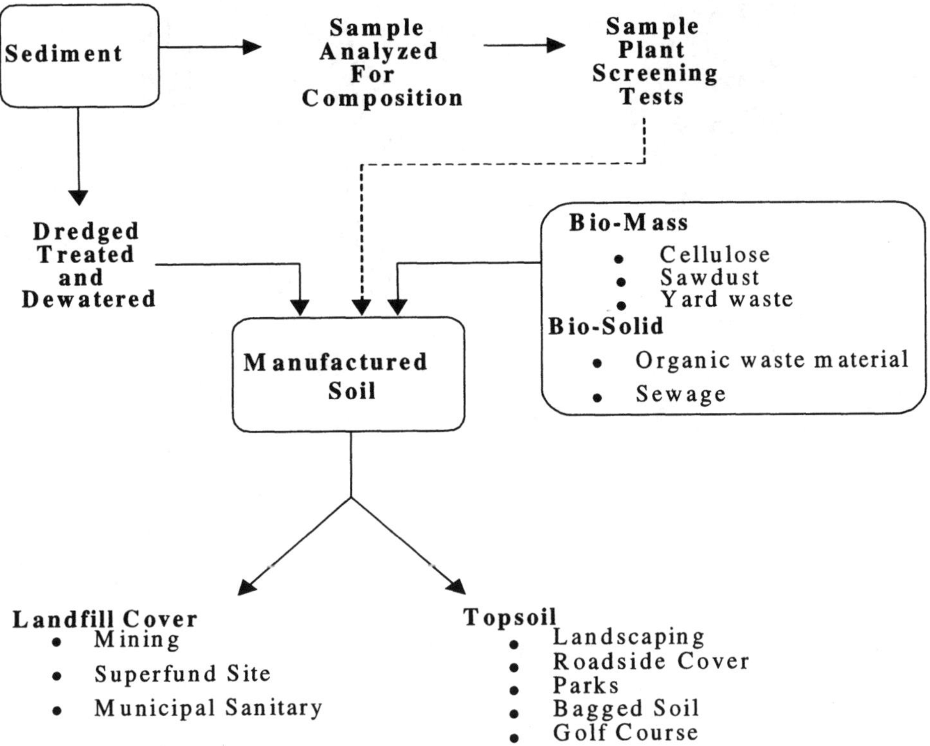

Figure 11.58 Process for creating manufactured soil. (From Graalum et al., 1999.)

future purchase of land for new disposal areas. It also uses sewage sludge and biomass that would otherwise have to be placed in a disposal location.

Geotubes. Areas along the Texas Gulf Intracoastal Waterway (GIWW) are currently widening in places due to erosion, and the loss of wetlands has become a concern (Randall et al., 2000). As long as there is no protection of the wetlands and erosion continues, the GIWW will continue to widen and inundate the wetlands at certain locations along the GIWW. A potential solution to the problem is the use of geotubes as erosion control structures. These structures can prevent inundation into the wetlands and provide long-term management of dredged material. Much of the eroding wetland sediments settles to the bottom of the GIWW and must eventually be dredged to maintain a navigable water depth of 3.7 m (12 ft). By using geotubes, it is expected that the frequency of dredging can be reduced, thus decreasing costs and the need for more confined disposal facilities (CDFs). Therefore, by placing dredged material–filled geotubes along the GIWW, the actual dredged material is used beneficially, erosion of the wetlands can be reduced, and the amount of material to be dredged decreases.

A geotube (Figure 11.59) is a coastal structure made from high-strength synthetics, which acts as a form to hold in soil and create an elongated structure to control or prevent ero-

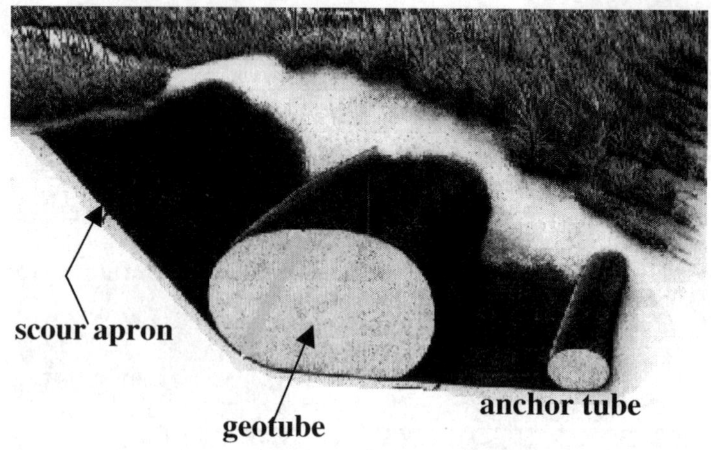

Figure 11.59 Example of geotube system. (Courtesy of Synthetic Industries.)

sion. It consists of two or more 4.57-m (15-ft)-wide geotextile industry standard sheets sewn together to form a shell. The most common geotube is a 9.14-m (30-ft) circumference tube, which usually yields a tube 1.83 m (6 ft) high and a 2.74 m (9 ft) wide. This size tube usually holds about 1.53 m^3 (2 yd^3) of dredged material per linear foot of the geotube (Wickoren, 1998). As described by Pilarczyk (1994, 1996), there have been many uses for geotubes, which include land reclamation, containment dikes, offshore breakwaters, dune reinforcement, and revetments.

Geotubes have been used in other coastal protection applications, such as groins, revetments, containment dikes, breakwaters, and dune reinforcement. Advantages of using geotubes include providing a beneficial use for dredged material, lower cost compared to traditional coastal erosion structures, decrease in work volume, use of local materials, and lower skilled labor required. On the other hand, resistance to punctures, fabric degradation, placement, height variation, and limited guidance are possible limitations of using geotubes. Geotubes are designed to withstand the pressures of hydraulic filling, provide sufficient permeability for dewatering, retain the dredged material, and resist erosion once in place. Geotechnical properties of the fill material are used to determine the appropriate geotextile material and the inlet spacing. The size of the geotextile mesh openings must retain the fill material and allow dewatering. High-strength geosynthetic material that is resistant to ultraviolet rays, oils, and chemicals are used. The spacing of the inlets is based on the fall velocity of the sediments and increases as the mean sediment grain size decreases. Two methods for hydraulically filling the geotubes are the small dredge method and the split pipe method. The small dredge method utilizes a small dredge that is directly connected to the filling tube. The small dredge is used only until the tube is filled. The split-pipe method uses the same dredge that is doing the maintenance dredging, and a branched pipe from the discharge line of the dredge is connected to the geotube for filling. Scour is a problem that must be addressed before a geotube is installed. Scour aprons are a standard method to prevent scour. Scour aprons can be installed using sandbags, sewn pockets, or a continuous small tube filled with dredged material that keeps the leading edge from floating.

11.9 CONTAMINATED SEDIMENTS AND CAPPING

11.9.1 Contaminated Sediments

Ports have been found to contain contaminants in the bottom sediments as a result of waste material for urban, industrial, river and navigation sources. Thus, dredged material from waterways, ports, and marinas must be tested to determine if the material is contaminated to some degree. In some cases the dredged material is contaminated and it requires special handling of the dredging and the placement of the dredged material. Special dredging equipment has been developed for dredging contaminated sediments. Placement options have also been developed to manage the dredged material placement in open-water disposal sites, confined disposal facilities, and for beneficial uses. The Marine Board (1997) and Herbich (2000) are excellent sources regarding information on contaminated sediments and their removal, especially in the United States.

National and international laws, regulations, and treaties regulate dredging and dredged material placement. Individual countries have their own regulations and laws that apply to inland and coastal water. In the United States, these regulations or laws include the Comprehensive Environmental Response, Compensation, and Liability Act of 1980 (CERCLA), National Environmental Policy Act (NEPA), and the Clean Water Act of 1972 (CWA). Many countries have similar regulations related to dredging and placing contaminated dredged material in an environmentally sound manner. The Marine Protection, Research and Sanctuaries Act (MPRSA), which is also called the London Convention, applies to the placement of waste in ocean waters outside the 12-mile limit, and except for fill, it applies to waters outside the 3-mile limit.

When contaminants are present in bottom sediments, benthic organisms are exposed and can uptake contaminants and pass them onto the marine life that feeds on these benthic organisms. The bottom sediments may be resuspended by dredging, fish trawling, and natural mechanisms (e.g., hurricanes, cyclones, and other meteorological events) that expose marine life in the water column to contaminants. As a result, marine life can ingest these contaminants that make them unsuitable for human consumption. Areas where contamination is probable include ports and coastal regions that are affected by urban and agricultural runoff, municipal and industrial waste streams, and other sources of pollution. The contaminants include heavy metals (e.g. lead, mercury, and cadmium), polynuclear aromatic hydrocarbons (PAHs), DDT, and polychlorinated biphenyls (PCBs).

The extent of contamination of water bodies varies around the world. In the United States, the Environmental Protection Agency estimates that approximately 10% of the nation's water is sufficiently contaminated to pose risks to fish and humans and wildlife that consume fish. It is estimated that 9.2 billion cubic meters (12 billion cubic yards) of total surface sediments are where benthic organisms live. About 229 million cubic meters (300 million cubic yards) of sediments are dredged from ports and ship channels annually, and about 2.3 to 9.2 million cubic meters (3 to 12 million cubic yards) are contaminated to the extent that special dredging and disposal techniques are required. Controls and technologies are required to handle contaminated sediments dredged from water bodies. In 1997, the U.S. Marine Board conducted a study regarding the status of contaminated sediments. A comparison of the controls) and remediation technologies used for the handling of contaminated sediments are tabulated in Tables 11.11 and 11.12.

11.9.2 Equipment

Special equipment for dredging contaminated sediments is discussed in Section 11.2. This special equipment includes watertight clamshell buckets, pneumatic dredges, dredges using positive-displacement pumps, and conventional dredges with covers to reduce resuspension. Several dredging companies have watertight clamshell buckets that do not allow the water or sediment to leave the bucket until it is placed in the barge. The Cable Arm bucket is watertight and makes a level cut on the bottom. Pneumatic dredges have been developed primarily in Japan for soft sediments, and the amount of water that must be treated is greatly reduced. In most cases, these special dredges have a much smaller production rates when compared to conventional equipment used for dredging clean sediments.

11.9.3 Placement Options

Contaminated sediment may be placed in open water provided that the sediment is capped with as layer of clean material. In some situations the contaminated sediment is found in a waterway, and the contaminated material is isolated from the water column by placing clean capping material over the contaminated sediments, and this is called *in situ capping*. Dredging of contaminated sediment by hydraulic and mechanical dredges may transport the contaminated sediments to a confined disposal facility

Table 11.11 Comparison of controls for contaminated sediment placement

Approach	Feasibility	Effectiveness	Practicality	Cost
Interim control				
Administrative	0	4	2	4
Technological	1	3	1	3
Long-term control in situ				
Natural recovery	0	4	1	4
Capping	2	3	3	3
Treatment	1	1	2	2
Sediment removal and transport	2	4	3	2
Ex situ treatment				
Physical	1	4	4	1
Chemical	1	2	4	1
Thermal	4	4	3	0
Biological	0	1	4	1
Ex situ containment	2	4	2	2
Scoring				
0	<90%	Concept	Not acceptable, very uncertain	$1000/yd^3$
1	90%	Bench		$100/yd^3$
2	99%	Pilot		$10/yd^3$
3	99.9%	Field		$1/yd^3$
4	99.99%	Commercial	Acceptable, certain	<$1/yd^3$

Source: Reprinted with permission from Marine Board (1997), courtesy of the National Academies Press, Washington, DC.

Table 11.12 Comparison of contaminated sediment remediation technologies

Feature Technology	State-of-Design Guidelines	Number of Times Used	Scale of Application	Cost/yd^3	Limitations
Natural recovery	Nonexistent	2	Full scale	Low	Source control, sedimentation storms
In-place containment	Developing rapidly	<10	Full scale	<$20	Limited technical guidance, legal/regulation uncertainty
In-place treatment	Nonexistent	2	Pilot scale	Unknown	Technical problems, few proponents, need to treat entire volume
Excavation and containment	Substantial and well developed	Several hundred	Full scale	$20–$100	Site availability, public assistance
Excavation and treatment	Limited and extrapolated from soil	<10	Full scale	$50–$1000	High cost, inefficient for low concentration, residue toxic, need for treatment train

Source: Marine Board (1997).

where the water and sediment must undergo treatment to decontaminate both the sediment and the water. Another possibility is to dewater the dredged contaminated sediment and use the contaminated sediment in a beneficial way, such as in construction bricks, fill material that is covered with a layer of clean material, in concrete aggregate.

11.9.4 Capping

Guidelines have been developed for capping contaminated sediments in open water, and these guidelines are described by Palermo et al. (1998). These guidelines contain case histories of capping projects and appendices that describe models used for modeling capping processes. In the United States a regulation that governs placement of the contaminated dredged material and subsequent cap material is the Marine Protection, Research, and Sanctuaries Act (MPRSA), also called the Ocean Dumping Act. Dredged material and/or fill material placed in U.S. waters (inland of the baseline to the territorial sea) is regulated by the Clean Water Act (CWA) of 1972. In addition to the CWA and MPRSA regulations, the requirements of the National Environmental Policy Act (NEPA) must be satisfied, and there are also some U.S. federal laws, executive orders, and others that must be considered dredging projects. Many other countries are signatories to the London Convention from which the MPRSA regulations were developed and therefore have similar requirements to meet regarding contaminated sediments and capping. These countries have similar national laws and regulations regarding the handling and placement of dredged material.

11.9.4.1 Definitions. *Capping* may be defined as the controlled placement of contaminated material at an open-water placement site, followed by a covering of clean isolating material called the *cap* (Palermo et al., 1998). There are two types of capping for open-water sites, called level bottom capping (LBC) and contained aquatic disposal (CAD), and illustrated in Figure 11.60. The purpose of level bottom capping is to place a mound of contaminated sediment on a flat or gently sloping seafloor and then cover it with clean isolating material such as sand. For a contained aquatic disposal (CAD) site, the contaminated material is placed where there is lateral confinement, such as a borrow pit, natural depression, or a subaqueous berm that confines the spread of the contaminated sediment, and a cap of clean material is placed over the contaminated sediment. The main advantage of the CAD site is the reduction in the amount of required capping material.

Another form of capping is *in situ capping,* in which the contaminated sediments are in place at the bottom of a waterway and there is a need to isolate the effects of the contaminants from the benthic organisms and the water column. In this case, clean material such as sand is placed over the contaminated sediments without dredging the contaminated sediments (i.e., the contaminated sediments are left in place). The cap isolates the contaminants from the benthic biota and prevents the contaminants from entering the water column. Palermo et al. (1996) provide guidance on in situ capping. The capping project is an engineered project that requires a team approach that includes the input from engineers, biologists, chemists, and dredge operators. The cap is designed to physically isolate the contaminated material from the benthic environment, reduce the flux of dissolved contaminants to the overlying water column, and prevent resuspension and transport of the contaminated material. The sequence of steps for designing capping projects are discussed in Palermo et al. (1998) and are summarized in Table 11.13.

11.9.4.2 Considerations for Selecting Capping Sites. Generally, capped disposal sites will be in designated disposal sites that are nondispersive, meaning that the sediments will remain inside the disposal site boundaries. If the site is located in ocean waters, the capping is regulated under the Marine Protection, Research, and Sanctuaries Act (MPRSA). If the site is in inland or coastal waters inland of the territorial sea, the capping is regulated by the Clean Water Act (CWA) for the United States. The selection of a capping site is guided by the same conditions as any other nondispersive open-water disposal site. However, the most

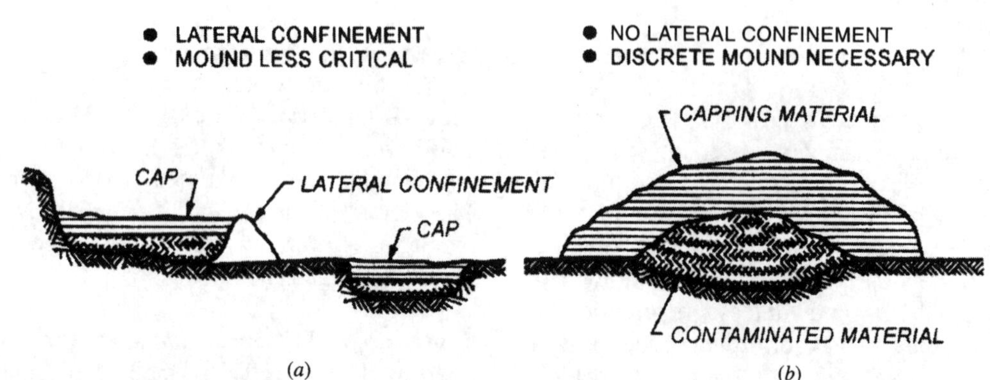

Figure 11.60 Capping contaminated sediment: (*a*) contained aquatic disposal; (*b*) level bottom capping. (From Palermo et al., 1998.)

Table 11.13 Summary of steps for designing capping projects

Contaminated Sediment	Common	Capping Sediment
Determine characteristics of contaminated sediments	Collect project data	Characterize and determine location of capping sediment
Decide on dredging equipment and placement procedures	Determine location and characteristics of potential capping site	Decide on dredging equipment and placement procedures
	Decide on navigation equipment and other positioning controls	
	Geotechnical compatibility must be determined, and this relates to the ability of the contaminated material to support the cap material considering the dredging and placement techniques	
Estimate mixing and dispersion (e.g., using a numerical model such as STFATE, MDFATE)	If unacceptable, controls such as a CAD, natural depression, or construction of berms for containment may be considered	Evaluate cap thickness and exposure time
Determine the mound geometry using a numerical model such as MDFATE	If unacceptable, controls such as a CAD, natural depression, or construction of berms for containment may be considered	Determine the cap geometry using a numerical model such as MDFATE
Estimate stability, erosion, and consolidation using a numerical model such as LTFATE	If unacceptable, controls such as a CAD, natural depression, or construction of berms for containment may be considered	Estimate stability, erosion, and consolidation using a numerical model such as LTFATE
	Define a monitoring program	

desirable capping site is one that is considered a low-energy environment (e.g., low wave and current), where there is low potential for erosion of the cap material. High-energy sites may be considered, but the erosion potential must be determined and the cap thickness may have to be increased to allow for cap erosion. Increased monitoring requirements and costs can be expected for high-energy environments.

11.9.4.3 Compatibility and Exposure Time. The compatibility of the contaminated dredged material and the cap material is important for a successful capping project. Some general statements regarding compatibility of the cap material and the contaminated dredged material are discussed. First, when the contaminated material is dredged mechanically and placed with split hull barges, the capping material can be placed by mechanical or hydraulic methods. On the other hand, when the contaminated material is fine-grained and placed by hydraulic methods, only hydraulic placement of the capped material should be considered because of the low strength of the contaminated material. The only exception is the slow, controlled placement of a sand cap by cracking hull openings, submerged spreaders, and hopper pump-out with a splitter plate.

The scheduling of the contaminated material placement and the subsequent cap placement must satisfy environmental and operational requirements. Once the contaminated material is placed on the sea bottom or in a contained aquatic disposal site, there is some necessary lag time preceding the placement of the cap. The important factors for determining the appropriate exposure time are the time required for benthic recolonization of the site, time required for self-weight consolidation, potential for effects prior to capping, and any monitoring requirements needed before the cap is placed. A period of 2 to 4 weeks is often allowed and is considered desirable for initial consolidation and monitoring surveys.

11.9.4.4 Contaminated Sediment and Cap Placement Techniques. For level bottom capping, the dredging and placement of the contaminated sediment needs to result in tight and compact mound that is easily capped. The compact mounds usually occur when the dredged material is near its in situ specific gravity. Mechanical dredging and point placement from barges are the most common technique for creating compact mounds. Contained aquatic disposal projects contain the lateral movement of the contaminated sediment when it is placed by either mechanical or hydraulic techniques. Consequently, the amount of cap material is minimized. When the contaminated material is placed hydraulically, sufficient time before capping is needed to allow for some settling and consolidation prior to placing the cap. This is necessary to prevent mixing of the contaminated material and the cap material. However, in some cases the cap material, usually sand, is sprinkled (falls at the particle settling velocity) over the contaminated dredged material to prevent mixing and/or displacing of the contaminated material. There are a number of different techniques and different equipment for dredging and placing the contaminated sediment and the cap material. Some of these techniques and equipment are summarized in Table 11.14.

11.9.4.5 Cap Thickness Design. The design of the cap requires selection of the material and thickness of the cap. The cap must physically isolate the contaminants from the benthic environment and the water column. The cap material for level bottom capping consists of relatively large volumes of clean dredged material (e.g., sand) that may come from other dredging projects or from a nearby source. Contained aquatic disposal sites and in situ capping normally require smaller volumes of clean sediments. Palermo et al. (1996) describe some guidance for nonsediment cap components for in situ capping.

The minimum thickness of the cap depends on the chemical and physical properties of the contaminated sediment and the cap sediment, water wave and current conditions, bioturbation of cap by benthic organisms, amount of consolidation, chemical isolation requirements, erosion, and other operational considerations, as illustrated in Figure 11.61.

A recommended sequence of steps for designing the cap thickness is described in Palermo et al. (1998) and is summarized in Table 11.15. The total cap thickness (t) consists of the total of the bioturbation thickness (t_b), consolidation thickness (t_c), erosion thickness (t_e), operational thickness (t_o) and the chemical isolation thickness (t_i) as defined by:

$$t = t_b + t_c + t_e + t_o + t_i \qquad (11.51)$$

In the recent history of capping projects the total cap thickness has been approximately 1 m (3 ft) of sand. When level bottom capping is used, the edge or apron of the contaminated

Table 11.14 Summary of selected equipment and techniques for placement of contaminated and cap material

Equipment and Placement Techniques	Comments
Surface discharge using conventional equipment	Hopper dredge (bottom discharge or pump out, barge or scow, pipeline.
Spreading by barge movement	Layer of cap material spread gradually by bottom dump barges whose bottom opening is controlled and the barge movement is controlled.
Hydraulic washing of coarse sand	Cap material is loaded on flat-top barges and washed overboard using high-pressure hoses.
Hopper dredge (cracked openings or pump down)	Hopper dredges crack open bottom doors or split hull and discharge the cap material slowly to create a thin layer of cap. Multiple layers are needed to establish the total cap thickness. Pump-out of the cap material is possible by pumping the cap material through pipes with end plates to direct the discharge in the direction of hopper movement and resulting in a sprinkling of the cap material. A discussion of this procedure is described in Randall et al. (1994).
Pipeline with baffle plate or sandbox	Spreading of cap material can be accomplished by attaching a baffle plate or sand box to the surface discharge end of the hydraulic pipeline. The baffle plate spreads the material radially and allows it to fall vertically into the water and it falls through the water column at its particle settling velocity. The sand box dissipates the energy with baffles and has holes such that the sand can fall through into the water and fall at the particle settling velocity. The spreader box is attached to a spud barge so that it moved about the spud with anchor lines (Sumeri, 1989).
Submerged discharge or diffuser	A discharge barge supports a vertical pipe with a submerged diffuser attached to the pipe near the bottom. The diffuser provides a larger pipe cross section that reduces the velocity of the material and directs the flow parallel to the bottom. The hydraulic pipeline is connected to the pipe fitting on the discharge barge to deliver the material for disposal or capping.
Sand spreader barge	Special equipment that was developed in Japan (Kikegawa, 1983; Sanderson and McKnight, 1986). Hydraulic dredge concepts are combined with a submerged discharge pipe that is controlled by a winch and anchor system to swing spreader side to side and forward in order distribute the sand over the area in need of cap.
Tremie pipe	The Japanese have developed the tremie technology for use with conveyor barges for the purpose of capping (Togashi, 1983; Sanderson and McKnight, 1986). The tremie pipe extends vertically from the barge to near the bottom and the telescoping characteristics allow the depth of the pipe to be adjusted as needed. Material may be placed in the pipe by either mechanical or hydraulic methods.

Table 11.14 (*Continued*)

Equipment and Placement Techniques	Comments
Geotextile fabric containers	Geotextile fabric can be placed in barges as a liner and the contaminated material is then placed in the barge mechanically or hydraulically. The fabric is brought together at the top and sewn together. A spilt-hull barge can then be moved to the disposal location, and the geotextile fabric container can be released through the bottom of the split-hull barge and fall to the bottom. These containers eliminate the spread of the contaminated material and reduce the amount of cap material. The cost of the use of geotextile material is more expensive that conventional bottom placement procedures.

Source: Palermo et al. (1998).

sediment mound often has only a thickness of 2 to 3 cm (1 in.) or less. Covering the apron-contaminated sediments with the total cap thickness is usually not done and a lesser thickness is placed. In some capping projects, intermediate caps may be desirable when multiple contaminated sediment disposals are anticipated. An example is a subaqueous borrow pit that in partially filled with contaminated sediments and will be used again is subsequent contaminated dredging projects. An intermediate cap can be placed, and its thickness is determined based on the risk associated with the time between the next placement events.

11.9.4.6 Monitoring of Cap Sites Level bottom capping, contained aquatic disposal, and in situ capping sites require monitoring to ensure that the cap is constructed and performing as designed and that the long-term integrity of the cap is maintained. Monitoring is necessary before, during, and after the placement of the contaminated sediment and cap material. Guidance for developing monitoring plans is found in Fredette et al. (1990a,b), Marine Board (1990), Pequegnat et al. (1990), and Palermo et al. (1998). Science Applicationis International Corporation (SAIC), (1995) describes a site-specific monitoring plan for the

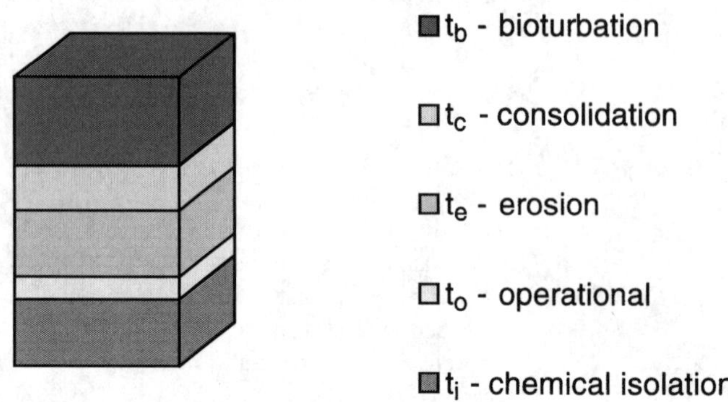

Figure 11.61 Cap thickness definitions.

Table 11.15 Summary of steps for determining the cap thickness

Steps	Additional Clarification
1. Determine the potential for bioturbation and define the appropriate thickness for the cap.	Benthic organisms colonize the disposal area and have the ability to burrow into the sediment and redistribute the sediment by bringing sediment from beneath the seabed surface to the surface. This is called *bioturbation,* and it is important to determine the bioturbation depth potential of the benthic organisms that naturally inhabit the disposal site. Knowledge of the bioturbation depth of the local species establishes the bioturbation thickness (t_b) for the cap. Bioturbation depths usually range between 0.3 and 0.6 m (1 to 2 ft).
2. Evaluate the consolidation that is likely to occur in the contaminated and cap material.	The contaminated dredged material that is placed on the seafloor (level bottom capping) or in a contained aquatic disposal site and subsequently undergoes some initial consolidation. It is important to determine through laboratory testing or numerical models the amount of consolidation that is expected. Cap material may also consolidate, although most cap material is sand that consolidates very little, if any. The consolidation thickness (t_c) is defined by the expected consolidation of the contaminated and cap material.
3. Determine the potential for erosion due to ocean currents and waves due to tidal circulation and historical storm events.	Ocean bottom currents and surface waves can induce water velocities over the cap that can erode the cap material. Tidal circulation and storm events drive the velocities that can erode a cap. Historical data and numerical models such as LTFATE can be used to estimate the amount of erosion that could occur as a result of these events. These estimates are used to establish an erosion thickness component for the total cap thickness.
4. Determine any operational considerations that would affect cap integrity.	Operational concerns can include ice gouging in cold-region areas, anchor dragging, ability to place thin layers, and unevenness of cap placement. It the capping site has potential for these or other operation concerns, an operational thickness (t_o) needs to be established as a component of the total cap thickness.
5. Evaluate the potential for short- and long-term contaminant flux through the cap and determine the cap thickness component necessary to chemically isolate the contaminants from moving through the cap.	The contaminated material may be found toxic to the benthic organisms and therefore is critical to ensure that the contaminated sediment does not move through the cap layer. The thickness required to isolate the contaminants chemically is called the chemical isolation thickness (t_i). The contaminants are normally tightly bound to the sediments, and prevention of the movement of sediments through the cap is needed. If the contaminants dissolve into the pore water, movement of the pore water through the cap must be prevented.

New England Division of the Corps of Engineers that has been used to evaluate capping operations for over 20 capping projects.

Multitiered monitoring programs are normally required for capping sites as described by Fredette et al. (1986), Zeller and Wastler (1986), Pearson (1987), and Palermo et al. (1998). Tiered monitoring programs have unacceptable environmental thresholds, sampling design, testable null hypotheses, and management options for conditions that exceed thresholds. Multidisciplinary advisory teams are normally employed to obtain the best technical advice in administering and completing the monitoring plan. Table 11.16 illustrates an example tiered monitoring plan for a hypothetical capping project. It should be noted that not all of the monitoring techniques in Table 11.16 would be used at every capping site.

11.10 DREDGING COSTS

Mechanical and hydraulic dredges conduct dredging of waterways and ports. A critical part of any dredging project is estimating the cost of the dredging project. In most cases, the production of a dredge is estimated and subsequently used to estimate the time required to dredge the volume of dredged material that must be removed. Knowledge of the cost of equipment, labor, mobilization/demobilization, and other costs are then used to estimate the dredging costs. Most dredging contractors have proprietary cost estimating software used to develop cost estimate for competitive bidding on the dredging projects. General approaches to developing cost estimates of dredging projects are described by USACE (1988), Bray et al. (1997), and Randall (2000a). The objective of this section is to summarize the general approach to estimating a dredging project.

The U.S. Army Corps of Engineers compiles data on the cost of dredging in the United States, and these data can be accessed through the Web site (www.usace.army.mil/ndc/cost.htm). Figure 11.62 shows the dredging costs attributed to the USACE, private dredging contractors, and the total (combined) costs for dredging in the United States from 1980 through 2000. It is shown in that the total cost of dredging in the United States in the last two years is just over $800 million. The total volume of dredged material removed by the Corp of Engineers and private contractors averages just over 300 million cubic yards during the 1980s and just under 300 million cubic yards during the 1990s (Figure 11.63). The average cost of dredging in dollars per cubic yard is shown in Figure 11.64, and these data indicate that the average cost in the 1980s was about $1.50/yd^3$. During the three-year period 1998–2000, the average cost has risen to about $3.00/yd^3$.

11.10.1 General Cost Estimation Scheme for Pipeline Dredge

11.10.1.1 Estimating Production. One of the key factors in determining the feasibility of a dredging project is the production rate that can be achieved. The *production rate* is defined as the volume of dredged material removed from the channel as a function of time, usually expressed as cubic yards per hour or cubic meters per hour. If a high production rate is maintained, the project can be completed more quickly and at a lower cost than if lower production rates are used. Many factors contribute to the production rate:

- *In situ sediment specific gravity.* High in situ specific gravity results in a higher delivered concentration of solids by volume (C_v). A higher concentration means a more efficient dredging operation because there is less water being pumped and more sediment.

Table 11.16 Example capping project monitoring plan

Monitoring Plan	Monitoring Frequency	Threshold	Management (Threshold Not Exceeded)	Management Options (Threshold Exceeded)
Consult site designation survey, technical advisory team, and Environmental Impact Statement (EIS) for physical and chemical baseline conditions				
Tier 1 • Bathymetry • Subbottom profiles • Side-scan sonar • Surface grab samples • Cores • Water samples	• Before placement • After placement • Annually	• Mound within 1.5 m (5 ft) of navigational hazard • Cap thickness decreased by 150 mm (6 in.) • Contaminant concentration exceeds limit in sediment or water sample	• Continue to monitor at same level • Reduce monitoring level • Stop monitoring	• Move to next tier • Cease using site • Increase cap thickness
Tier 2 • Bathymetry • Subbottom profiles • Side-scan sonar • Sediment profile camera • Cores • Water samples • Consolidation instrumentation	• Quarterly to semi-annually • Cap thickness decreases by 305 mm (1 ft) • Contaminant exceeds limit in sediment or water sample	• Continue to monitor at the same level • Reduce monitoring level	• Move to next tier • Replace cap material • Increase cap thickness • Cease using site	
Tier 3 • Bathymetry • Subbottom profiles • Side-scan sonar • Sediment profile camera • Cores • Water samples • Tissue samples	• Monthly to semi-annually	• Cap thickness degreases 305 mm (1 ft) • Contaminant exceeds limit in sediment or water sample • Contaminant exceeds limit in tissue sample	• Continue to monitor at same frequency • Reduce monitoring frequency	• Replace cap material • Increase cap thickness • Cease use of site • Change cap sediment • Dredge and remove

Source: Palermo et al. (1998).

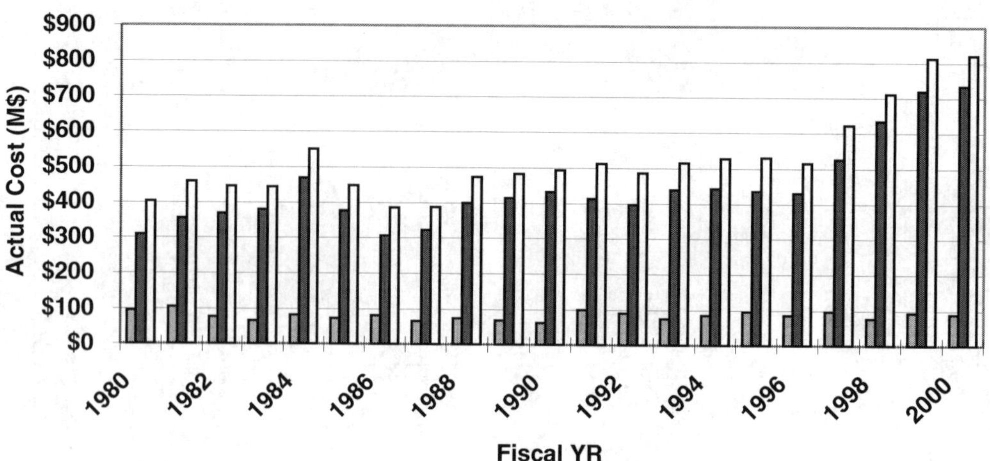

Figure 11.62 Total cost of U.S. dredging in millions of dollars for fiscal years 1980–2000.

- *Sediment grain size*. Small grain-size sediments are more easily transported and induce smaller frictional head losses.
- *Discharge lift*. This describes the elevation difference between the dredge and the disposal end of the discharge line.
- *Discharge pipe length*. It may sometimes be cheaper to lower the production rate rather than add a booster pump when the discharge line is too long for the main pump.
- *Minor losses*. Head losses that occur at discontinuities in the pipeline such as valves, joints, and bends.
- *Bank height*. Dredge cutters are about 1 m or several feet in diameter, and sometimes the depth of the cut (called the bank height) is smaller than the cutter diameter. In this situation, the optimum production is not achieved because not as much material is being taken into the suction line.
- *Dredge cycle efficiency*. Due to the nature of the dredge advance process, sediment is not being excavated 100% of the time. In fact, with a typical walking spud dredge, only about 50% of the time is the cutter in the sediment. Higher efficiencies (approximately 75%) can be attained using a spud carriage dredge.

The rate of production is the number of cubic yards of sediment per hour that is removed from the channel bottom by the dredge. From a cost estimation standpoint, this is an important factor simply because it determines how much time it takes to complete a project. More time on the job means more wear on the machinery, more fuel used, and more pay for the workers. It is therefore imperative that the estimate of the production be as accurate as possible.

As a grain of sediment travels through a pipeline, there is a certain velocity that the fluid around it must maintain to prevent the grain from falling to the bottom of the pipe and becoming stationary. If this velocity is not maintained, much of the sediment will settle and clog the pipeline. This condition is very un-

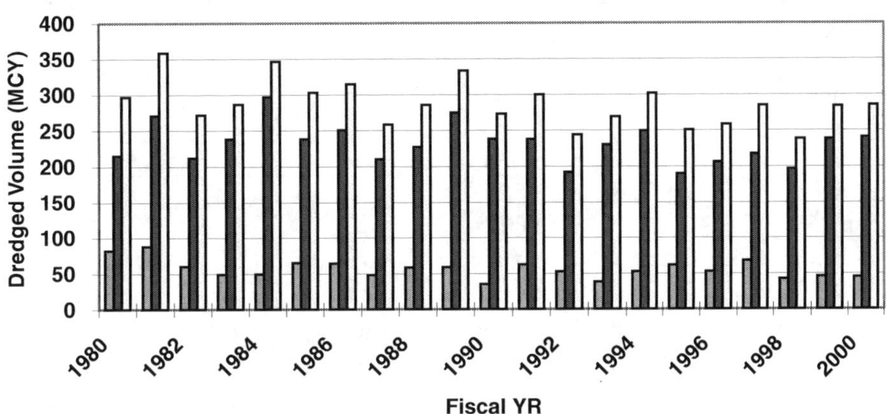

Figure 11.63 Total U.S. dredged volume in million cubic yards during fiscal years 1980–2000.

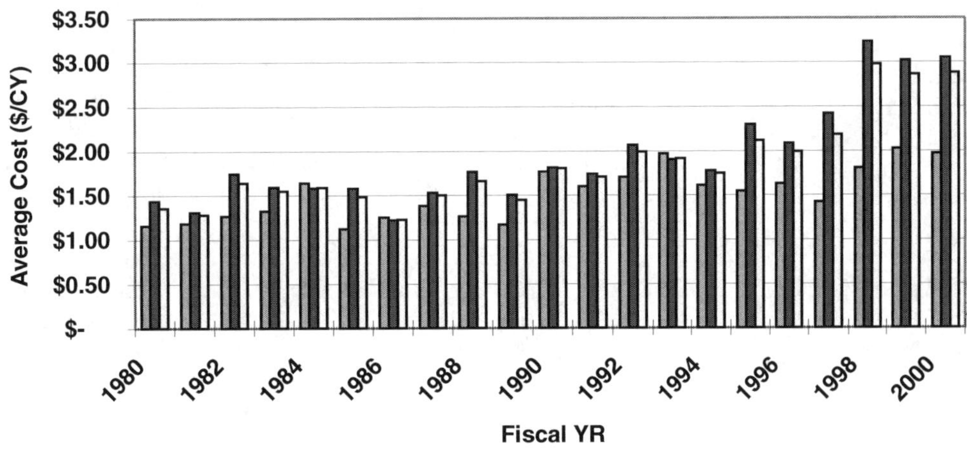

Figure 11.64 Average cost of U.S. dredging in dollars per cubic yard for fiscal years 1980–2000.

desirable, because it means shutting down the operation while the line is being unclogged. This *critical velocity* is a function of the specific gravity of the sediment, the grain size, and the diameter of the pipe. Wilson et al. (1997) present a convenient method of determining the critical velocity using a nomograph, and Matousek (1997) develop an empirical equation for the nomograph as described in Section 11.5. It is desirable to pump at an average velocity that is 10 to 15% above the critical velocity based on the fact that head losses, power requirements, and pipeline wear are minimized at this value.

The head losses must be calculated for the system to determine the maximum pumping distance for the installed pump. The major component of the head losses of a system is the frictional head loss that can be determined using the Wilson et al. (1997) method described earlier in this chapter. There are also minor losses due to pipe joints and bends that should be considered in the head loss calculations. These calculations are made for a range of average flow rates and the resulting curve, called the *system head curve,* is compared to the capability of the dredge pump as described by the pump characteristic curves.

A typical set of pump characteristics curves shows the pump head, brake horsepower (or shaft horsepower), and pump efficiency as a function of flow (discharge) rate. In a dimensional format, these curves are valid only for a pump with the same specifications and operating at the same speed. However, when translated into a dimensionless format, the values of head (H), horsepower (P), and discharge (Q) may be calculated for any similar pump operating at any speed.

The production rate of a cutter suction dredge can be calculated using the equation

$$P(\text{yd}^3/\text{hr}) = Q(\text{GPM})C_v \cdot 0.297\text{PF} \cdot \text{DE} \tag{11.52}$$

where DE is the dredge cycle efficiency (~0.5 for a walking spud dredge), P the production rate in cubic yards per hour, C_v the maximum concentration of solids by volume, PF the production factor that accounts for material differences, bank height and other factors, and 0.297 is a unit conversion factor. To calculate the production using the method described above, four values are needed: specific gravity of the mixture, impeller diameter, pump speed, and discharge rate.

The addition of a booster pump increases the head of the total system by the amount of pump head achieved by the booster. The process above is then repeated for the new system configuration. Once an acceptable number of boosters have been selected, the pump efficiency is obtained from the curves and the production is calculated using the formula stated above.

Since the characteristics of sediments vary widely and have a great influence on the actual production rates, the Corps has introduced the concept of a *material factor* to account for the difference (USACE, 1988). This factor takes into account different viscosities, grain sizes, and in situ consolidation to provide a description of relative ease of pumping. For instance, silt is easily excavated and, due to its tendency to remain in suspension, is easily transported. On the other hand, compacted clay is difficult to excavate and, once excavated, is difficult to pump through the line due to its tendency to consolidate and create "clay balls." Miertschin and Randall (1998) discuss the use of the bank factor, or bank height efficiency, that is a function of the ratio of the cutter diameter to bank height.

All of these factors are then applied to the production rate equation to yield the final production rate estimate.

11.10.1.2 Dredge Cost Components. Capital investment, operating costs, and mobilization and demobilization costs are some of the many costs that need to be considered. Dredges are expensive pieces of machinery, and for this reason, many dredges in use today are very old. Capital investment is very substantial for the contractors, and this investment is reflected in their bids for dredging work. Dredges are not the only large items of capital investment for contractors. Other large and expensive items include work tugs, crew and survey barges, derrick barges, fuel and water barges, work barges, and pipeline sections. Booster pumps are also very expensive, and therefore it is prudent to consider other options before adding a booster. The most complex part of determining a cost estimate is the dredge operating costs, and the major cost factors are:

- *Fuel costs.* Large quantities of fuel are consumed during a typical dredging project. The horsepower of the dredge and supporting equipment determines how much fuel is used. However, one also must consider the number of hours the dredge is in use and at what percentage of full power it is operating. Also, if thc dredge is not operating at high efficiency, excessive amounts of fuel may be consumed.
- *Lubricants.* With all the machinery on dredges and the supporting equipment, large amounts of lubricants are used. It is common practice to estimate the cost of lubricants at about 10% of the total fuel cost.
- *Dredge crew.* Dredging projects employ a large crew that includes the captain, officers, engineers, lever operator, winch operators, welders, deckhands, cooks, and electricians.
- *Land support crew.* There is also a crew that works away from the dredge that includes pipeline handlers, booster pump operators, fill foremen, and equipment operators.
- *Routine maintenance and repairs.* Many minor repairs are necessary during a project, as well as just routine maintenance. Replacing worn engine parts, damaged pipes and hoses, electrical consumables, and lubrication of equipment are regular maintenance repairs.
- *Major repairs and overhauls.* Major repairs usually do not occur during a project; they are included in contract bids as a small percentage of the cost of actual repairs.
- *Other considerations.* General overhead costs are usually taken at around 9% of the total cost of the project. Depreciation of the equipment is a fairly major item, especially for the pipelines that wear much more quickly than other components. Finally, the contractor has the right to make a profit on the job, so a reasonable percentage of the total cost must be added for this purpose.

One of the more abstract concepts when trying to quantify the cost of a dredging project is the cost of mobilization and demobilization. This amount varies greatly from contractor to contractor and job-to-job based on the distance from the site and readiness to mobilize. Before starting a dredging project, all the equipment involved must be inspected, some parts must be assembled, and everything must be prepared for transport to the dredge site. The move to the job site involves paying for tugs and relocating personnel. During the move the dredge and other equipment are sitting idle when they could be making money for the contractor somewhere else, so this is an indirect cost that must be considered. Once at the site, the equipment must be assembled and prepared for work and the discharge pipeline must be put in place. All of these tasks must be done in reverse for the demobilization process, unless the contractor has to begin work somewhere else immediately after finishing work at this site.

11.10.1.3 Difficulties Associated with the Development of Cost Estimating Software. Due to the large number of factors that must be considered when estimating the cost of a dredging project, it is quite a challenge to develop a general cost estimation program that yields accurate results. The most difficult aspect of creating such a program is the fact that each dredging contractor has different dredging equipment, different methods of completing a project, and different considerations for mobilization and demobilization. Also, each dredging project has its own unique set of problems associated with different aspects of its completion. For instance, some areas have strict environmental regulations that may prove costly to the dredging contractor. Problems such as these and numerous other unforeseen difficulties present a challenge to the person making the cost estimate. Although every effort may be made by the estimator to compensate for all possible scenarios, additional problems will inevitably arise. Hence, the cost estimator must pay careful attention to every dredging project detail.

11.10.1.4 Estimating Cost. Once the production rate is determined, the length of time required to complete the project is calculated. The operating and labor costs may then be determined based on the length of time that the equipment is in use and the workers are being paid. Other components of the final cost, such as mobilization and demobilization, are independent of the production rate and are calculated separately. A general flowchart for estimating the cost is shown in Figure 11.65.

Mobilization and demobilization costs are perhaps the most difficult to estimate. The main problem lies in the fact that no two dredges have to travel the same distance to arrive at a job site. Also, different dredges are always in different stages of readiness to mobilize. For example, contractor A may be completing a project 10 miles away from the site being considered, while contractor B may have a dredge not in use 100 miles away. In this example, contractor A has less distance to travel to the job site, but has more preparation to do before moving its operation. Also, contractor A's mobilization costs are basically included in the demobilization costs for the last project. Contractor B, however, has little preparation to do but must travel a greater distance. In this

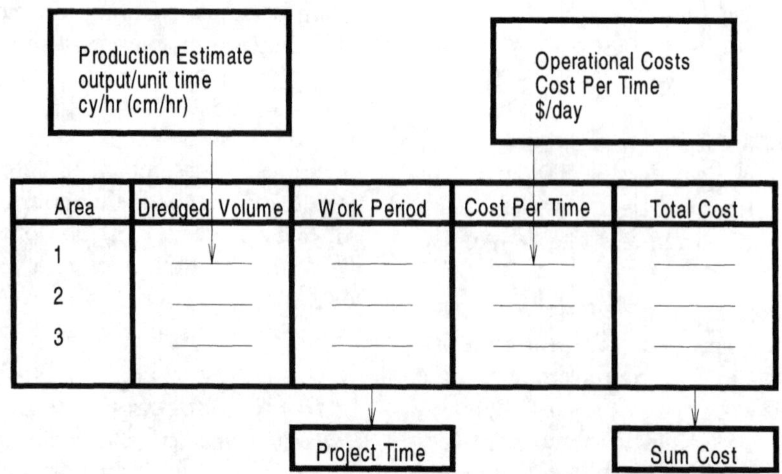

Figure 11.65 General flowchart for estimating dredging cost.

Table 11.17 Example of U.S. geographical variation of wage rates

Position	Louisiana, Mississippi, Alabama, Texas	New York,[a] New Jersey[a]	North Carolina,[a] South Carolina, Florida, Georgia	Washington[a]
Lever man	$6.10	$25.18	$17.31	$26.63
Welder	5.10	21.47	14.64	25.24
Mate	6.06	20.35	14.18	25.19
Deckhand	5.15	16.58	10.44	24.75
Engineer	5.15	21.83	11.15	25.24

Source: Belesimo (2000).

[a]Fringe benefits must be included.

scenario, contractor A would likely have substantially less cost associated with mobilization and would therefore give a lower bid price for the project. Many times, contractors are not compensated for demobilization costs and therefore do not include them in their bids. The mobilization/demobilization calculations are based on the assumption that all dredges are not in use immediately prior to transfer to the job site. The main factors in determining costs for this portion of the dredging project are therefore the distance to and from the job site, the size of the dredge, and the length of pipeline used on the job.

The largest component of the cost of a dredging project is the cost of hired labor, usually accounting for about 40% of the total cost. A dredging operation requires the services of many different personnel. A typical project requires a number of crewmen and laborers, both salaried and hourly. For a typical job, there is one each of the following salaried workers: a captain, officer, chief engineer, and an office worker. Depending on the size of the dredge being used, the number of the following hourly paid workers will vary: lever operators, dredge mates, booster engineers, tug crew, equipment operators, welders, deckhands, electricians, dump foremen, oilers, and shore crew. Data for the number of crew and their respective pay rates must be determined. Adjustments are then made to the wages using the values for fringe benefits, overtime pay, holidays, vacation time, taxes, and workers' compensation. The resulting total hourly rate is multiplied by the hours per day each crew member works to obtain the total daily cost of labor. These adjustments are not made to the salaried workers. The pay rates for the crew vary geographically, as illustrated in Table 11.17.

After the labor costs, the equipment costs are the largest part of the total project cost. The capital costs of the dredge, boosters, tugs, and barges are obtained from the database along with the number of each needed for the project. Bray et al. (1997) provide a method for computing routine maintenance and repairs, major repairs, insurance, fuel cost, lubricant costs, depreciation, and pipeline costs. Bray et al. (1997) recommended multiplying the capital value by 0.00014 to obtain the daily costs for minor repairs that can be completed while the dredge is operating.

For the average daily cost of repairs requiring the dredge or any other piece of equipment to be shut down, it is recommended to use a value of 0.0003 times the capital value. Insurance is calculated by multiplying the capital cost by 0.025 and dividing by the average number of working days per year. To calculate the fuel consumption, one must determine the effective hours per day that the dredge operates at 100% power. Options for entering the hours per day of the dredge operating at 100% power, 75% power, and 10% power are often used. From these values the effective hours at 100% are determined. The fuel cost per day is then taken as the item's horsepower times the hours per day at 100% times fuel cost per gallon times 0.048079. The cost of lubricants is estimated by using 0.1 times the fuel cost to obtain this value.

The depreciation is then calculated as the capital cost divided by the useful life and the number of days per year in operation. The total capital cost of the pipeline is determined by multiplying the total number of pipe sections by the cost per section obtained from the database. The same methods used above are used to calculate depreciation and repair costs, keeping in mind that the useful life of a section of pipe is much shorter than the equipment items, due to the constant abrasive wear of the material being pumped through it.

After the equipment and pipeline costs are determined, the overhead costs are then taken to be 9% of the total daily costs of equipment and pipeline. The time required to complete the project is calculated based on the production rate and the hours per month that the dredge is in operation. This value is then multiplied by the daily costs to obtain the total cost of execution.

11.10.1.5 Additional Considerations. In addition to the actual cost of the dredging project, the contractor's profit and bond are considered. These factors typically add up to roughly 10% of the total cost. Each dredging project is unique, and when determining the cost of the project, one must take into account the "big picture." There are aspects of each project that may reduce or increase the final price, and care must be taken to include these factors when making a cost estimate. For example, a particular project may necessitate additional effort to comply with local environmental restrictions.

11.10.2 Estimating for Mechanical and Hopper Dredge Projects

Bray et al. (1997) provides an outline for developing a cost estimating program for both the mechanical dredges and hopper dredges. USACE (1988) has done extensive research on the cost estimation procedure. USACE have developed software, but it is not been published in open literature. However, many of the procedures used in the software are published. The research conducted by USACE in the areas of labor cost estimation and mobilization/demobilization costs is particularly useful.

11.11 ENVIRONMENTAL CONSIDERATIONS

Regulation of the placement of structures, dredging of navigable waterways, and the disposal of dredged material within waters of the United States and ocean waters is a complex issue and is a shared responsibility of the Environmental Protection Agency (EPA) and the U.S. Army Corps of Engineers (USACE). MPRSA, CWA, and NEPA are the major federal statute/laws governing dredging projects, but a number of other federal laws and executive orders must also be considered. Jurisdiction of MPRSA and CWA are illustrated in Figure 11.66. Procedures for evaluating dredged material that is proposed for disposal in ocean waters is governed by USEPA/USACE (1991) and for disposal in inland or near coastal waters is governed by USEPA/USACE (1996).

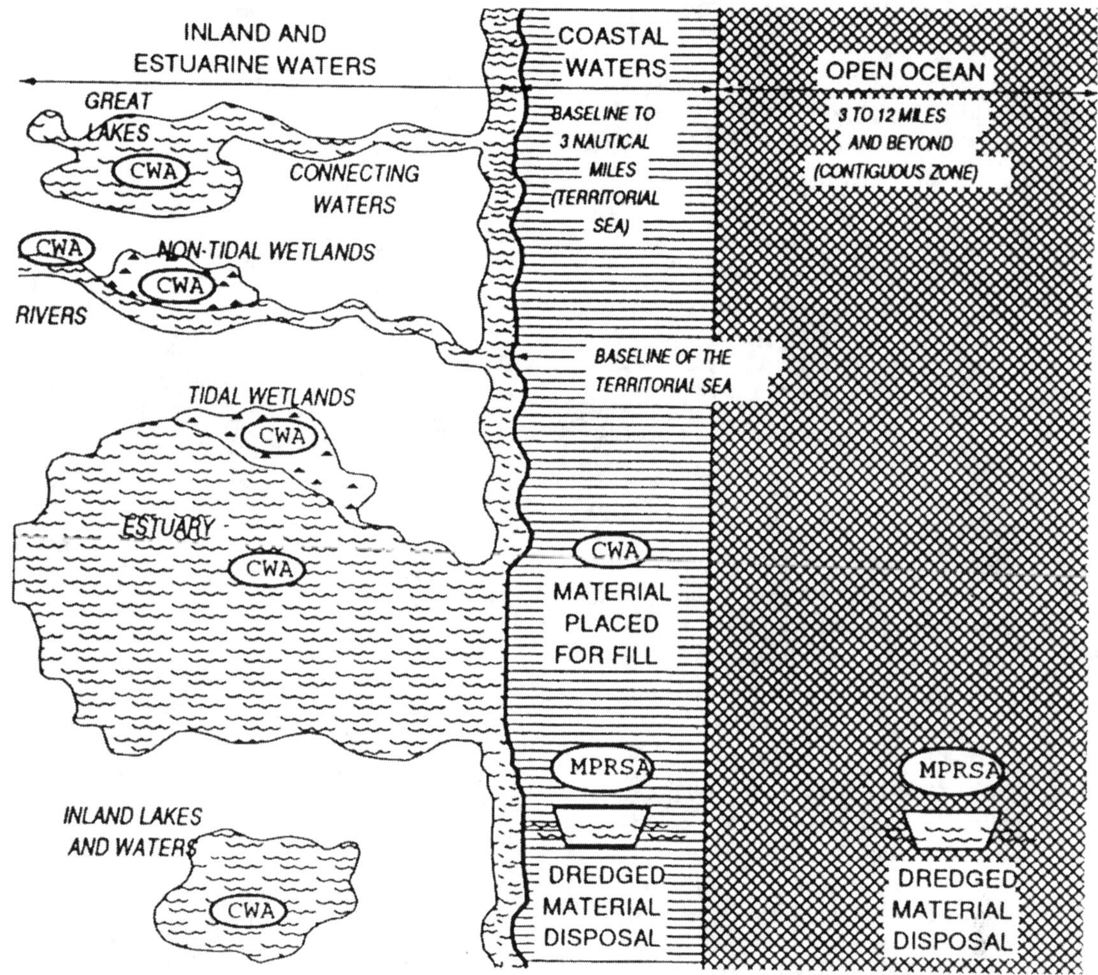

Figure 11.66 Jurisdiction boundaries for MPRSA and CWA. (From USEPA/USACE, 1992.)

11.11.1 Environmental Laws

A brief summary of U.S. environmental laws follows, and many of these laws can be found in the *Code of Federal Regulations* (CFR).

1. *Marine Protection, Research, and Sanctuaries Act–Ocean Dumping Act–London Dumping Convention.* This act is also referred to the London Dumping Convention and the Ocean Dumping Act. Section 102 requires the EPA to develop environmental criteria in consultation with the USACE. Section 103 assigns the USACE responsibility for authorizing the ocean disposal of dredged material. The USACE must apply the criteria developed by the EPA.

Section 102 gives authority to the EPA to designate ocean disposal sites. Section 103 authorizes the USACE to select ocean disposal sites for project specific use when an EPA site is not feasible or a site has not been designated.

2. *Clean Water Act.* The Clean Water Act is also the Federal Water Pollution Control Act and Amendments of 1972. Section 404 requires the EPA, in conjunction with the USACE, to publish guidelines for the discharge of dredged or fill material such that unacceptable adverse environmental impacts do not occur. Section 404 assigns responsibility to the USACE for authorizing all discharges and requires the application of USEPA guidelines. Section 401 provides the states a certification role for project compliance with the applicable state water quality standards.
3. *National Environmental Policy Act.* Dredged material disposal activities must comply with the applicable NEPA requirements regarding identification and evaluation of alternatives. Section 102(2) requires examination of alternatives to the action proposed, and these alternatives are analyzed in an Environmental Assessment (EA) or Environmental Impact Statement (EIS). For USACE dredging projects, USACE is responsible for developing alternatives for the discharge of dredged material, including all facets of the dredging and discharge operation, including cost, technical feasibility, and overall environmental protection. Compliance with environmental criteria of the MPRSA and/or the CWA guidelines is the controlling factor used by the USACE to determine the environmental acceptability of disposal alternatives. The NEPA process is finalized in one of two ways. First, a finding of no significant impact (FONSI) is the final decision document when an EA finds that the preparation of an EIS is not required. Second, an EIS is prepared, and the decision document, called a record of decision (ROD), specifies the recommended action and discusses the alternatives considered (USEPA/USACE, 1992).
4. *Resource Conservation and Recovery Act.* RCRA regulates the collection, generation, transportation, recovery, separation, and disposal of solid wastes and includes liquids, semiliquids, and contained gases. Under RCRA, a waste is hazardous if it is specifically listed as hazardous or if the waste has a hazardous characteristic. A hazardous characteristic means that the waste is ignitable, corrosive, reactive, or toxic. Anyone who generates, stores, treats, processes, or disposes of hazardous wastes must abide by the RCRA hazardous waste management provisions. Some wastes associated with offshore oil and gas exploration and production are exempt from RCRA, and the EPA lists exempt and nonexempt offshore oil and gas wastes (Butler and Binion, 1993).
5. *Comprehensive Environmental Response, Compensation, and Liability Act (Superfund) (1980) and Superfund Amendments and Reauthorization Act (1986).* CERCLA provided authority to publish a prioritized list of highly polluted locations or sites. Only hazardous substances are covered by CERCLA and petroleum is excluded from the definition of hazardous substances. Natural gas, natural gas liquids, and refined petroleum products such as gasoline are also excluded. However, substances associated with oil and gas exploration and production are hazardous, and these include methanol, caustic soda, and

many mud additives. The liability under CERCLA is very formidable because it is strict, retroactive, and joint. *Strict* means that it is immaterial whether the party involved is at fault or not. *Retroactive* means that a party may be liable for cleanup even before that party owned the location of the hazardous substance. *Joint* means each party that is potentially liable may be liable for the total cleanup costs and not just a proportionate amount. Also, a party that did not contribute at all to the pollution is not liable, but the party must prove that it did not contribute to the pollution and did not know of the polluted condition of the site. SARA provides a means and funding for quick, responsible cleanup of locations that threaten the environment or public health.

6. *Safe Drinking Water Act (1974)*. SDWA applies to the protection of drinking water from underground sources through underground injection.
7. *Oil Pollution Control Act (1990)*. The OPA-90 addresses marine oil spills for both onshore and offshore facilities.
8. *Toxic Substances Control Act (1976)*. TSCA is an act that requires that chemical manufacturers, importers, and processors must supply information related to chemicals handled by each organization. Crude oil and natural gas are naturally occurring substances and are excluded from the TSCA reporting requirements by EPA regulation.
9. *Occupational Safety and Health Act*. This act has wide applications and requires notification of users of hazardous substances. The notification includes the use of warning labels on containers and the issue of material safety data sheets (MSDSs). The organization is required to evaluate and inventory chemical hazards, properly label on-site containers, make the MSDS available to workers, train workers to protect themselves, and develop programs for communicating procedures for handling hazardous substances.
10. *London Dumping Convention (1972)*. The United States is a signatory to the International Treaty concerned with marine-waste disposal, which is the London Dumping Convention (LDC). LDC jurisdiction includes all waters seaward of the baseline of the territorial sea. MPRSA Section 102 criteria reflect the standards of the LDC.
11. *Coastal Zone Management Act*. The Coastal Zone Management Act requires the USACE to coordinate the permit review of all federal projects with all participating state-level coastal zone review agencies.
12. *River and Harbors Act of 1899*. This act requires a USACE permit for any work or structure in navigable waters of the United States. The act also requires permits for placement of fill material in navigable waters.
13. *Fish and Wildlife Coordination Act of 1958*. This act requires USACE to consult with federal and state fish and wildlife agencies to prevent damage to wildlife and provide for the development and improvement of wildlife resources for any proposed federal project in a stream or other body of water.
14. *Endangered Species Act of 1988*. This act establishes a consultation process between U.S. federal agencies and the Secretary of Interior or Commerce for conducting programs for the conservation and protection of endangered species. It protects threatened or endangered species of animal and plant life.

15. *Water Resources Development Act of 1986.* This act created a financing arrangement for dredging associated with navigation maintenance and improvement projects. Local sponsors finance one-half the cost of improvement and one-half the cost for additional maintenance dredging, and the federal government finances the other half.
16. *National Historic Preservation Act of 1966.* This act requires the consideration of the effects of the proposed project on any site building, structure, or objects that is or may be eligible for inclusion in the National Register of Historic Places.
17 *Migratory Bird Treaty Act.* Under this act an operator is responsible to place nets or other covers to keep migratory birds out of open pits and storage.
18. *Other Federal Statutes.* There are other federal statutes that are related to activities in the offshore and coastal waters, including the Comprehensive Environmental Response, Compensation, and Liability Act of 1980, Rivers and Harbors Improvement Act of 1978, Submerged Lands Act of 1953, Rivers and Harbors, Flood Control Acts of 1970, National Fishing Enhancement Act of 1984, Federal Insecticide, Fungicide, and Rodenticide Act (1972), the Marine Mammal Protection Act (1972), and the Hazardous Material Transportation Act (1990). In addition, there are numerous executive orders that affect ocean engineering projects such as Executive Order No. 11988, Flood Plains, which requires consideration of alternatives to incompatible development in floodplains. Another order is Executive Order No. 11990, Wetlands, which provides for protection of federally regulated wetlands. There are many other executive orders that may affect ocean engineering applications and these are found in the very voluminous *Code of Federal Regulations* (CFR), typically located in major libraries.

11.11.2 Permitting

The permitting system of the Corps of Engineers that is used in the United States is described briefly. Three types of permits exist: individual, nationwide, and general. An individual permit is required for locating a structure or for excavating or discharging dredged material in waters of the United States. Nationwide permits are issued for some smaller or minor water bodies, and general, permits are issued for certain regions that may require specific notification and reporting procedures. The typical Corps of Engineers review process is illustrated in Figure 11.67. The permit application process including an example is summarized in Herbich (2000b).

11.11.3 Environmental Impact Statements

An environmental impact statement is often required to assess the impact of the implementation of a new engineering system or change to an existing system. The cost of these studies is borne by the person, organization, or agency requesting the new or changed system. Herbich (2000a and b) discusses the preparation of an environmental impact statement.

11.11.4 Environmental Windows

The National Environmental Policy Act in 1969 started the requests for environmental windows in the United States. These environmental windows are periods set aside for no dredging in order to protect sensitive biological resources or their habitats from the effects of dredging. LaSalle et al. (1991) conducted stud-

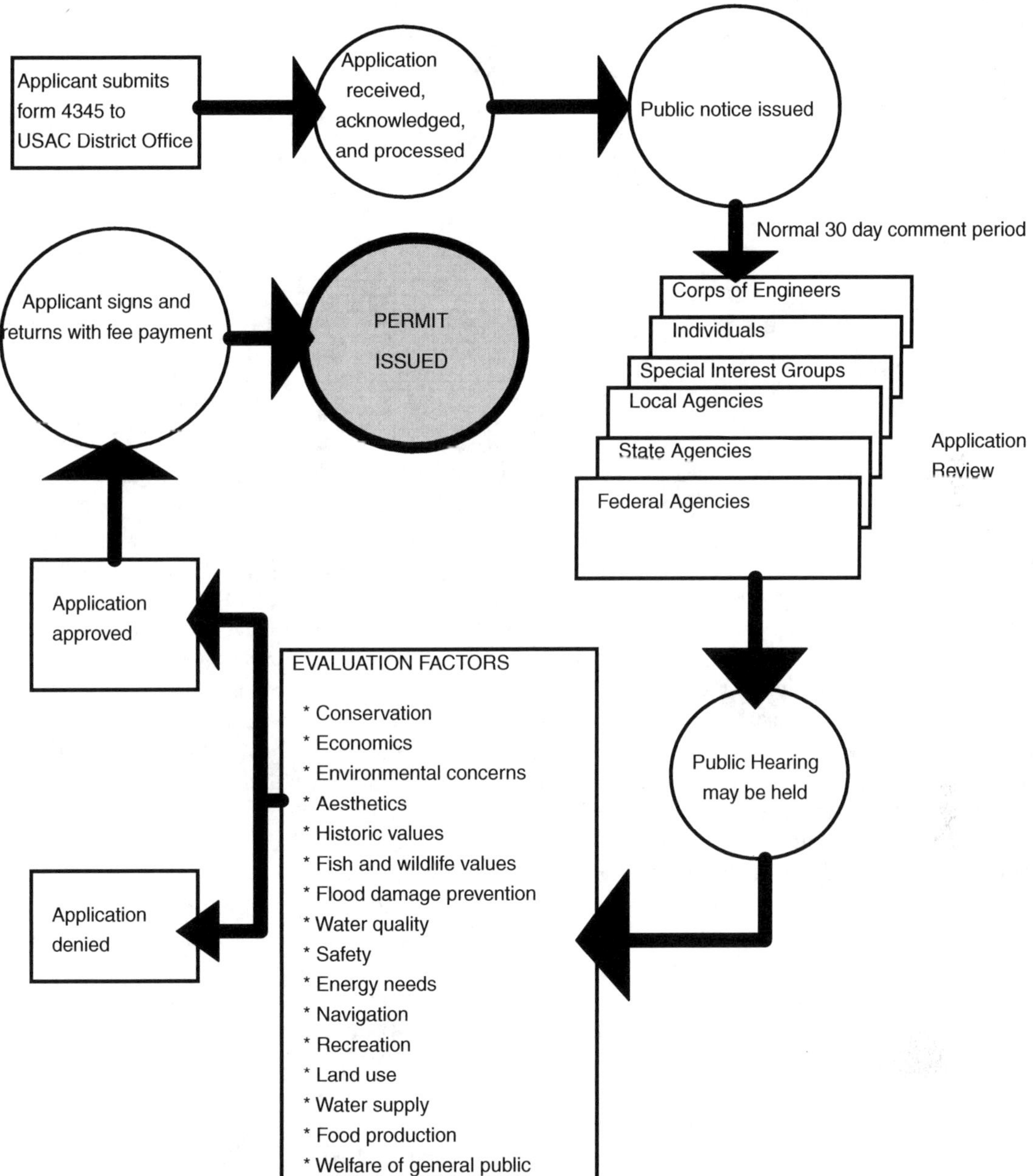

Figure 11.67 Flowchart for the Corps of Engineers permit review process.

Table 11.18 Example of multiple environmental windows for Hyannis Harbor, Massachusetts, dredging project

	Jan.	Feb.	Mar.	Apr.	May	Jun.	Jul.	Aug.	Sep.	Oct.	Nov.	Dec.
Environmental window												
Sea turtles						**	****	****	****	****	**	
Anadromous fish			**	****	****	**						
Winter flounder	**	****	****	****	****							
Shellfish spawning						**	****	****	**			
Shorebird mating and nesting			**	****	****	****	****	**				
Bathing and boating season					**	****	****	****	**			
Dredging period allowed	****								****	****	****	****

Source: Reine et al. (1998).

ies related to dredging operations in the Great Lakes area, and Sanders and Kilgore (1989) reported on seasonal restrictions related to dredging activities in freshwater areas. Dickerson et al. (1993) investigated the effects of dredging on sea turtles in the southeastern United States.

Reine et al. (1998) have updated these studies on environmental windows by conducting an extensive survey of districts and divisions of the U.S. Army Corps of Engineers. By type of dredge (pipeline, hopper, and mechanical), environmental windows affected hopper dredging about 83%, mechanical dredging 85%, and pipeline dredging 67% of the time. Agencies in the United States that most frequently request environmental windows include U.S. Fish and Wildlife, National Marine Fisheries, U.S. Environmental Protection Agency, state departments of natural resources, and state fish and game agencies.

Environmental windows are frequently imposed on dredging projects in both coastal and inland waterways. The physical disturbance to habitat and nesting is the justification for over 75% of the environmental windows in the United States. Other justifications include sedimentation and turbidity, entrainment/vessel strikes, suspended sediments, fish migration, dissolved oxygen reduction, and recreational activities (hunting and fishing). Marine resources that are frequently cited as justification for environmental windows are anadromous fishes (salmon and striped bass), colonial nesting water birds (pelicans and terns), and endangered species (whales and sea turtles).

In many situations the data are limited or not available to justify the environmental windows, but they are used anyway. The use of environmental windows increase the cost of dredging and often restricts the period of dredging to winter months when weather conditions are most dangerous for dredging operations.

Several environmental windows can be applied to a particular dredging project. Reine et al. (1998) illustrate one example of the effects of multiple applications of environmental windows on a dredging project for Hyannis Harbor, MA (Table 11.18). Various windows prevented dredging in all months except December and the first half of January. The weather conditions during these months are very difficult for dredging, and furthermore, the dredging could not be completed during the short six-week time frame. Consequently, an exemption was negotiated to allow dredging to occur between September and January.

In March 2001, the Marine Board of the National Research Council in the United States convened a two-day meeting to address the setting, managing, and monitoring on environmental windows, and the results are described in Marine Board (2001). The outcome of the meeting was a recommendation for a six-step process for setting, managing, and monitoring environmental windows. The recommendation included identifying stakeholders (individuals and industries that have a stake in the process), project sponsors, scientist team of experts (biology, chemistry, and resource experts), and an engineering team of experts (designers and constructors of project). Step 1 is to agree on procedures and timetable. Step 2 considers details of project and identifies resources of concern. Step 3 involves the science and engineering teams identifying availability and validity of data, recommending dredging technology, and prioritizing environmental windows. Step 4 consists of stakeholders and project sponsors reviewing the science and engineering teams' recommendations, consideration of socioeconomic implications, and setting windows. Step 5 involves conducting the dredging within the established windows, monitoring the project, and synthesizing the findings. Step 6 involves reconvening the stakeholders and project sponsors to review steps 1 through 5 and recommend improvements.

REFERENCES AND RECOMMENDED READING

Adams, J. R., and J. B. Herbich, 1970. Gas removal systems, *Proc. World Dredging Conference,* WODCON '70, Singapore.

Addie, G. R., L. Whitlock, and J. Dresser, 2001. Dredge pump testing considerations, *Proc. Western Dredging Association 21st Technical Conference and 33rd Texas A&M Dredging Seminar,* Houston, TX, June 24–27.

Albar, A., 2000. Effect of various terminal velocity equations on the result of friction loss calculation, *Proc. Western Dredging Association 20th Technical Conference and 32nd Texas A&M Dredging Seminar,* Warwick, RI, June 25–28.

———, 2001. Modeling of a bucket wheel dredge system for offshore sand and tin mining, Ph.D. dissertation, Ocean Engineering Program, Civil Engineering Department, Texas A&M University, College Station, TX.

Ancheta, C., R. Santiago, and G. Sherbin, 1998. An innovative approach to harbour remediation, Thunder Bay, Ontario, Canada, *Proc. 15th World Dredging Congress,* Las Vegas, NV.

Belesimo, F. J., 2000. Personal communication.

Bray, R. N., A. D. Bates, and J. M. Land, 1997. *Dredging: A Handbook for Engineers,* 2nd ed., Edward Arnold, London.

Briot, J. E., J. P. Doody, R. Romagnoli, and M. M. Miller, 1999. Environmental dredging case studies: a look behind the numbers, *Proc. Western Dredging Association 19th Technical Conference and 31st Texas A&M Dredging Seminar,* Louisville, KY, May 15–18.

Buchberger, C., 1993. Environment Canada demonstrations: Remediation technologies for the removal of contaminated sediment in the Great Lakes, *Terra et Aqua,* Vol. 50, January 1993.

Butler and Binion, L.L.P., 1993. *Environmental Law Simplified,* Penn Well Publishing, Tulsa, OK.

Crane Company, 1979. *Flow of Fluids through Valves, Fittings, and Pipe,* Tech. Pap. 410, Crame.

De Heer, R. J., and Rochmanhadi, 1989. *Dredging and Dredging Equipment: Dredging Equipment.* International Institute for Hydraulic and Environmental Engineering, Delft, The Netherlands.

Dickerson, D. D., D. A. Nelson, C. E. Dickerson, Jr., and K. J. Reine, 1993. Dredging related sea turtle studies along the southeastern U.S., presented at Coastal Zone '93, New Orleans, LA.

Fredette, T. J., G. Anderson, B. S. Payne, and J. D. Lunz, 1986. Biological monitoring of open-water dredged material disposal sites, *IEEE Oceans '86 Conference Proc.,* Washington, DC.

Fredette, T. J., D. A. Nelson, J. E. Clausner, and F. J. Anders, 1990a. *Guidelines for Physical and Biological Monitoring of Aquatic Dredged Material Disposal Sites,* Tech. Rep. D-90-12, U.S. Army Corps of Engineers Waterways Experiment Station, Vicksburg, MS.

Fredette, T. J., D. A. Nelson, T. Miller-Way, J. A. Adair, V. A. Sotler, J. E. Clausner, E. B. Hands, and F. J. Anders, 1990b. *Selected Tools and Techniques for Physical and Biological Monitoring of Aquatic Dredged Material Disposal Sites,* Tech. Rep. D-90-11, U.S. Army Corps of Engineers Waterways Experiment Station, Vicksburg, MS.

Graalum, S. J, R. E. Randall, and B. L. Edge, 1999. Methodology for manufacturing topsoil using sediment dredged from the Texas Gulf Intracoastal Waterway, *J. Mar. Environ. Eng.,* Vol. 1, pp. 1–38, www.oldcitypublishing.com/JMEE/JMEE.HTML.

Hara, S., Y. Hayase, H. Fujisawa, T. Hara, and S. Satoh, 1998. Automated sludge removal by using high density dredger, *Proc. 15th World Dredging Congress,* Las Vegas, NV.

Hayes, D. F., 1986. *Guide to Selecting a Dredge for Minimizing Resuspension of Sediment,* Environmental Effects of Dredging Tech. Note, EEDP-09-1, U.S. Army Corps of Engineers Waterways Experiment Station, Vicksburg, MS.

———, 1999. *Turbidity Monitoring during Boston Harbor Bucket Dredge Comparison Study,* project report, submitted to U.S. Army Corps of Engineers Waterways Experiment Station, Vicksburg, MS, Aug.

Hayes, D. F., G. L. Raymond, and T. N. McLellan, 1984. Sediment resuspension from dredging activities, presented at *Dredging 1984,* Clearwater, FL, American Society of Civil Engineers, Reston, VA.

Hayes, D. F., T. N. McLellan, and C. L. Truitt, 1988. *Demonstration of Innovative and Conventional Dredging Equipment at Calumet Harbor, Illinois,* MP EL-88-01, U.S. Army Corps of Engineer Waterways Experiment Station, Vicksburg, MS.

Herbich, J. B., 1995. Removal of contaminated sediments: equipment and recent field studies, in *Dredging, Remediation and Containment of Contaminated Sediments,* K. R. Demars et al., eds, (PCN) 04-012930-38, American Society for Testing and Materials, Philadelphia.

———, 2000a. *Removal of Contaminated Sediment by Dredging,* Dredging Engineering Short Course Notes, Center for Dredging Studies, Ocean Engineering Program, Civil Engineering Department, Texas A&M University, College Station, TX, Jan.

———, 2000b. *Handbook of Dredging Engineering,* 2nd ed., McGraw-Hill, New York.

Herbich, J. B., and S. B. Brahme, 1991. *A Literature Review and Technical Evaluation of Sediment Resuspension during Dredging,* Contract Rep. HL-91-1, U.S. Army Corps of Engineers Waterways Experiment Station, Vicksburg, MS.

Huston, J., 1970. *Hydraulic Dredging,* Cornell Maritime Press, Cambridge.

Iwasaki, M., K. Kuioka, S. Izumi, and N. Miyata, 1992. High density dredging and pneumatic conveying system, *Proc. 13th World Dredging Congress,* WODCON XIII, Bombay, pp. 773–792, Apr. 7–10.

Jaglal, K., and D. B. McLaughlin, 1999. Evaluation of large scale environmental dredging as a remedial alternative, *Proc. Western Dredging Association 19th Technical Conference and 31st Texas A&M Dredging Seminar,* Louisville, KY, May 15–18.

Kazim, K. A., B. Maiti, and P. Chand, 1997. A correlation to predict the performance characteristics of centrifugal pumps handling slurries, *J. Power Energy,* Vol. 211, No. A2, pp. 147–157.

Kikegawa, K., 1983. Sand overlying for bottom sediment improvement by sand spreader, *Management of Bottom Sediments Containing Toxic Substances: Proc. 7th U.S./Japan Experts Meeting,* prepared for the U.S. Army Corps of Engineers Water Resources Support Center by the U.S. Army Corps of Engineers Waterways Experiment Station, Vicksburg, MS, pp. 79–103.

LaSalle, M. W., D. G. Clarke, J. Homziak, J. D. Lunz, and T. J. Fredette, 1991. *A Framework for Assessing the Need for Seasonal Restrictions on*

Dredging and Disposal Operations, Tech. Rep. D-91-1, U.S. Army Corps of Engineers Waterways Experiment Station, Vicksburg, MS.

Lee, C. R., 1997. Manufactured soil from Toledo harbor dredged material and organic waste materials, presented at the International Workshop on Dredged Material Beneficial Uses, Baltimore, MD, July 28–Aug. 1.

Lee, C. R., and R. C. Sturgis, 1996. Manufactured soil: A productive use of dredged material, U.S. / Japan Expert Meeting, Oakland, CA.

Lee, C. R., T. C. Sturgis, H. C. Banks, Jr., and K. Johnson, 1996. *Evaluation of Toledo Harbor Dredged Material for Manufactured Soil,* Misc. Rep., USACE, Waterways Experiment Station, Vicksburg, MS.

Marine Board, 1990. *Managing Troubled Waters: The Role of Marine Environmental Monitoring,* National Research Council, National Academies Press, Washington, DC.

———, 1997. *Contaminated Sediments in Ports and Waterways,* National Academy of Sciences, National Academies Press, Washington, DC.

———, 2001. *A Process for Setting, Managing and Monitoring Environmental Windows for Dredging Projects,* Transportation Research Board and Ocean Studies Board, National Research Council, Washington, DC, Mar. 19–20.

Matousek, V., 1997. Flow mechanism of sand–water mixtures in pipelines, Ph.D. Dissertation, Delft University of Technology, Delft, The Netherlands.

McLellan, T. N., R. N. Havis, D. F. Hayes, and G. L. Raymond, 1989. *Field Studies of Sediment Resuspension Characteristics of Selected Dredges,* Tech. Rep. HL-89-9, U.S. Army Corps of Engineers Waterways Experiment Station, Vicksburg, MS.

Miertschin, M., and R. E. Randall, 1998. A general cost estimation program for cutter suction dredges, *Proc. 15th World Dredging Congress,* Las Vegas, NV, pp. 1099–1115.

Munson, B. R., D. F. Young, and T. H. Okiishi, 1998. *Fundamentals of Fluid Mechanics,* 3rd ed., Wiley, New York.

Otis, M. J., S. Andon, and R, Bellmer, 1990. *New Bedford Harbor Superfund Pilot Study: Evaluation of Dredging and Dredged Material Disposal,* U.S. Army Engineer Division, New England, Waltham, MA.

Palermo, M. R., S. Maynard, J. Miller, and D. D. Reible, 1996. *Guidance for In-Situ Subaqueous Capping of Contaminated Sediments,* EPA 905-B96-004, Great Lakes National Program Office, U.S. Environmental Protection Agency, Chicago.

Palermo, M. R., J. E. Clausner, M. P. Rollings, G. L. Williams, T. E. Myers, T. J. Fredette, R. E. Randall, 1998. *Guidance for Subaqueous Dredged Material Capping,* Tech. Rep. DOER-1, U.S. Army Corps of Engineers, Washington, DC, June.

Parchure, T. M., 1996. *Equipment for Contaminated Sediment Dredging,* Tech. Rep. HL-96-17, U.S. Army Corps of Engineers Waterways Experiment Station, Vicksburg, MS, Sept.

Parchure, T. M., and C. N. Sturdivant, 1997. *Development of a Portable Innovative Contaminated Sediment Dredge,* Final Rep., CPAR-CHL-97-2, Construction Productivity Advancement Research Program, U.S. Army Corps of Engineers Waterways Experiment Station, Vicksburg, MS, Sept.

Pearson, T. H., 1987. Benthic ecology in an accumulation sludge-disposal site, in *Ocean Processes and Marine Pollution,* Vol. 1, *Biological Processes and Wastes in the Ocean,* J. M. Capuzzo and D. R. Kester, eds., Krieger Publishing, Malabar, FL, pp. 195–200.

Pelletier, J. P., 1992. *Collingwood Harbour Sediment Removal Demonstration,* preliminary report on the Water Quality Monitoring Program, Environmental Protection, Environment Canada, Toronto, Ontario, Canada.

Pennekamp, J. G. S., and M. P. Quaak, 1990. Impact on the environment of turbidity caused by dredging, *Terra et Aqua,* Vol. 42.

Pequegnat, W. E., B. J. Gallaway, and T. D. Wright, 1990. *Revised Procedural Guide for Designation Surveys of Ocean Dredged Material Disposal Sites,* Tech. Rep. D-90-8, U.S. Army Corps of Engineers Waterways Experiment Station, Vicksburg, MS.

Pilarczyk, K. W., 1994. *Novel Systems in Coastal Engineering, Geotextile Systems and Other Methods: An Overview,* Rijkswaterstaat, Delft, The Netherlands.

———, 1996. Geotextile systems in coastal engineering, an overview, *Proc. International Conference on Coastal Engineering,* Vol. 2, pp. 2114–2127.

Randall, R. E., 1992. Equipment used for dredging contaminated sediments, *1992 International Environmental Dredging Symposium Proc.,* Erie County Environmental Education Institute, Buffalo, NY, Sept. 30–Oct. 2.

———, Dredging equipment alternatives for contaminated sediment removal, presented at the Conference on Environmental Dredging Technology, East Brunswick, NJ, sponsored by the Port Authority of New York/New Jersey, June 21–22.

———, 2000a. Estimating dredging costs, Appendix 9 in *Handbook of Dredging Engineering,* 2nd ed., J. B. Herbich, ed., McGraw-Hill, New York.

———, 2000b. Pipeline transport of dredged material, Chapter 7, Section 2 in *Handbook of Dredging Engineering,* 2nd ed., J. B. Herbich, ed., McGraw-Hill, New York.

Randall, R. E., J. E. Clausner, and B. H. Johnson, 1994. Modeling of cap placement at the New York mud dump site, *Dredging '94: Proc. ASCE 2nd International Conference of Dredging and Dredged Material Placement,* Orlando, FL, Nov. 13–16.

Randall, R., B. Edge, J. Basilotto, D. Cobb, S. Graalum, Q. He, and M. Miertschin, 2000. *Texas Gulf Intracoastal Waterway (GIWW): Beneficial Uses, Estimating Costs, Disposal Analysis Alternatives, and Separation Techniques,* Project Rep. 1733-S, Texas Transportation Institute, Texas A&M University System, College Station, TX, Sept.

Reine, K. J., D. D. Dickerson, and D. G. Clarke, 1998. *Environmental Windows Associated with Dredging Operations,* DOER Tech. Notes Collection (TNDOER-E2), U.S. Army Corps of Engineers Research and Development Center, Vicksburg, MS.

Richardson, T. W., J. E. Hite, R. A. Shafer, and J. D. Etheridge, 1982. *Pumping Performance and Turbidity Generation of Model 600/100 Pneuma Pump,* TR HL-82-8, U.S. Army Corps of Engineers Waterways Experiment Station, Vicksburg, MS.

SAIC, 1995. *Sediment Capping of Subaqueous Dredged Material Disposal Mounds: An Overview of the New England Experience, 1979–1993,* DAMOS Contrib. 95, Science Applications International Corporation Rep. SAIC-90/7573&c84 prepared for the U.S. Army Engineer Division, New England, Waltham, MA.

Sanders, L., and J. Kilgore, 1989. *Seasonal Restrictions on Dredging Operations in Freshwater Systems: Environmental Effects of Dredging,* Tech. Note EEDP-01-16, U.S. Army Corps of Engineers Waterways Experiment Station, Vicksburg, MS.

Sanderson, W. H., and A. L. McKnight, 1986. *Survey of Equipment and Construction Techniques for Capping Dredged Material,* Misc. Pap. D-86-6, U.S. Army Corps of Engineers Waterways Experiment Station, Vicksburg, MS.

Santiago, R., 2000. *Contaminated Sediment Removal General Management Options and Support Technologies,* Dredging Engineering Short Course Notes, Center for Dredging Studies, Ocean Engineering Program, Civil Engineering Department, Texas A&M University, College Station, TX, Jan.

Schiller, R. E., Jr., 1992. Sediment Transport in Pipes, Chapter 6 in *Handbook of Dredging Engineering,* J. B. Herbich, ed., McGraw-Hill, New York.

Seagren, E. H., 2000. Personal communications, Ellicott International, Mud Cat Division, Mar.

Shook, C. A. and M. C. Rocco, 1991. *Slurry Flow: Principles and Practice,* Butterworth-Heinemann, Boston.

Sumeri, A., 1989. Confined aquatic disposal and capping of contaminated bottom sediments in Puget Sound, *Proc. WODCON XII, Dredging: Technology, Environmental, Mining, World Dredging Congress,* Orlando, FL, May 2–5.

Swamee, P. K., and A. K. Jain, 1976. Explicit equations for pipe-flow problems, *ASCE J. Hydraul. Div.,* Vol. 102, No. HY5, pp. 657–664.

Togashi, H., 1983. Sand overlaying for sea bottom sediment improvement by conveyor barge, *Management of Bottom Sediments Containing Toxic Substances: Proc. 7th U.S./Japan Experts Meeting,* prepared for the U.S. Army Corps of Engineers Water Resources Support Center by the U.S. Army Corps of Engineer Waterways Experiment Station, Vicksburg, MS, pp. 59–78.

Turner, T. M., 1996. *Fundamentals of Hydraulic Dredging,* 2nd ed., ASCE Press, New York.

USACE, 1983. *Dredging and Dredged Material Disposal,* EM-1110-2-5025, U.S. Government Printing Office, Washington, DC, Mar.

———, 1987a. *Confined Disposal of Dredged Material,* EM-1110-2-5027, U.S. Government Printing Office, Washington, DC, Sept.

———, 1987b. *Beneficial Uses of Dredged Material,* EM-1110-2-5026, U.S. Government Printing Office, Washington, DC, June.

USACE, 1988. *Government Estimates and Hired Labor Estimates for Dredging,* EM 1110-2-1300, U.S. Government Printing Office, Washington, DC.

USEPA/USACE, 1991. *Evaluation of Dredged Material Proposed for Ocean Disposal (Testing Manual*), EPA-503/8-91/001, Office of Water, U.S. Environmental Protection Agency, Washington, DC.

———, 1992. *Evaluating Environmental Effects of Dredged Material Management Alternatives: A Technical Framework,* EPA 842-B-92-008, U.S. Government Printing Office, Washington, DC, Nov.

———, 1996. *Evaluation of Dredged Material Proposed for Discharge in Inland and Near-Coastal Waters: Testing Manual,* Office of Water, U.S. Environmental Protection Agency, Washington, DC.

Van Drimmelen, C., and T. Schut, 1992. New and adapted small dredges for remedial dredging operations, *Proc. WODCON XIII,* Bombay, India, pp. 156–169, Apr.

Wickoren, D. 1998. Personal communications.

Wilson, K. C., and J. K. P. Tse, 1984. Deposition limit for coarse-particle transport in inclined pipes, *Proc. Hydrotransport 9,* BHRA, The Fluid Engineering Centre, Cranfield, Bedfordshire, England, pp. 149–169.

Wilson, K. C., G. R. Addie, A. Sellgren, and R. Clift, 1997. *Slurry Transport Using Centrifugal Pumps,* Blackie Academic and Professional, New York.

Zappi, P. A., and D. F. Hayes, 1991. *Innovative Technologies for Dredging Contaminated Sediments,* Misc. Pap. EL-91-20, U.S. Army Corps of Engineers Waterways Experiment Station, Vicksburg, MS.

Zeller, R. W., and T. A. Wastler, 1986. Tiered ocean disposal monitoring will minimize data requirements, *Oceans '86,* Marine Technology Society, Washington, DC, pp. 3, 1004–1009.

12

ENVIRONMENTAL FACTORS IN PORT PLANNING AND DESIGN

Christopher M. Carr
Han-Padron Associates
New York, New York

12.1 INTRODUCTION

The objective of this chapter is to provide a practical overview of the range of environmental planning and design issues for port projects. Typical environmental control and mitigation measures are also outlined. The potential scope of environmental issues for port projects is enormous and the critical issues affecting the feasibility and cost of each project vary greatly. Many environmental issues and regulatory requirements need to be addressed depending on the nature and scope of the project and the local site conditions. In addition to the technical requirements of good engineering practice, the requirements of the environmental impact assessment (EIA) depend on the governing national, regional, and local environmental guidelines and regulations. Other practical considerations may include the degree of scrutiny and opposition that the project may receive from regulatory agencies and the public and the potential for litigation by environmental interest groups or others.

Port development is sometimes conceived and driven largely by macroeconomic and political considerations and often involves significant adverse environmental impacts. In the past, environmental issues have sometimes received minor or secondary consideration in port planning and design. However, this is no longer the case. Prudent environmental planning is essential and a comprehensive EIA is required for any major port project anywhere in the world. Even in remote areas, where environmental regulations might be limited in scope, funding by international sources such as the World Bank will require that the project be developed in an environmentally responsible manner, consistent with international standards.

Prudent planning requires detailed and unbiased assessment of a range of alternative site locations and design concepts for the port, taking into account the engineering needs, environmental impacts, and costs versus benefits. The full range of potential alternatives must be evaluated during the screening process and the economic need and financial feasibility of the project must be clearly demonstrated. The importance of these issues cannot be understated. Extended delays or cancellation of port projects during the regulatory review process is generally due to inadequate examination of feasible project alternatives and impact mitigation measures, or unconvincing analysis of project costs versus benefits.

This chapter includes overviews concerning available environmental guidelines for port projects, dredging-related environmental issues, regional environmental impacts, soil contamination, and environmental controls and mitigation measures for petroleum terminals, dry bulk terminals, and container terminals.

12.2 ENVIRONMENTAL GUIDELINES AND REGULATIONS

The World Bank has developed detailed guidelines for the environmental assessment of port projects (Davis et al., 1990) that provide an excellent starting point for environmental planning and the scoping of EIA studies for port projects. This includes a detailed series of environmental check lists plus narrative discussions of key environmental considerations. Additional guidelines on port-related environmental issues have been prepared by the International Association of Ports and Harbors (IAPH, 1991a,b, 1993), the American Association of Port Authorities (AAPA, 1993), the European Association of Parts Organization (ESPO, 1995), the Urban Harbor Institute of the University of Massachusetts (2000), the Permanent International Association of Navigational Congresses (PIANC, 1999), and various other reviews (e.g., Paipai, 1999; Abood and Metzger, 2000; Trozzi and Vaccara, 2000). International regulations include MARPOL and London Dumping Conference (LDC), described later in this chapter.

The appendixes to the AAPA guidelines include a number of fact sheets that are useful for environmental planning and best management practices (BMPs) for specific types of port facilities. The guidelines prepared by the

Urban Harbor Institute describe a number of case studies illustrating key environmental impact issues for recent port projects in the United States.

12.3 DREDGING, DISPOSAL, AND LAND RECLAMATION

Chapter 11 provides a detailed discussion of dredging in relation to port engineering. The present section supplements Chapter 11 and highlights key environmental considerations related to dredging operations, dredge material disposal, and land reclamation. More comprehensive reviews of environmental aspects of dredging, including the disposal of contaminated dredge material, are available [e.g., PIANC, 1996, 1998; National Research Council (NRC), 1997; Vellinga, 1997; Herbich, 2000].

12.3.1 Planning for Dredging Projects

The starting point for the planning of dredging projects is the characterization of the site conditions and detailed analysis of the dredging, disposal, and reclamation quantities. The site investigations may include bathymetric and subbottom profile surveys, side-scan sonar surveys, and geotechnical investigations and laboratory testing work based on soil borings and/or vibracores. The geotechnical data are used to evaluate the dredging characteristics of the material and to assess its performance as fill for reclamation or as dredge spoil. Historical bathymetric data and dredging records should also be reviewed, where available, to help assess the local sedimentation rates and estimate the long-term maintenance dredging requirements. This analysis may be supplemented by hydrodynamic and sediment transport numerical model studies to evaluate future sedimentation patterns and rates.

The calculation of dredging quantities is an iterative process related to optimization of the overall layout of a port and its breakwater system. For projects that involve extensive land reclamation, the relative balance of the dredge and fill quantities is of course a critical issue. Sometimes the layout and quantity estimates cannot be finalized until the later stages of the design effort, pending the results of physical hydraulic model tests of the harbor performance and navigation simulation studies to confirm the design of the entrance channel and turning basin. In calculating the dredging quantities an allowance should be included to account for overdredging beyond the nominal design dredge depth. This can be an important component of the total dredge quantity for dredging in relatively deep water where only a thin layer of sediment needs to be removed.

For the dredging of entrance channels in open-water conditions, the currents and operational wave climate at the site need to be evaluated carefully based on measured data or numerical modeling. The wave conditions can significantly affect the type of dredging equipment that can be used and the potential for operational downtime. Also, there may be large seasonal variations in a wave climate that can significantly affect the construction schedule. For channels that are prone to high-wave conditions, the depth requirements and associated dredging quantities may increase significantly.

12.3.2 Environmental Impacts for Dredging Projects

Dredging operations may involve a wide variety of physical, structural, chemical, and biological environmental impacts that need to be considered for port planning and EIA applications.

12.3.2.1 Physical Impacts. Dredging for port development generally requires large increases in local water depths and significant changes in the overall configuration of the seabed, over large areas. These changes can significantly modify the currents, waves, and water quality conditions in the project area. These effects can include, for example:

- Changes to circulation patterns and sediment transport processes
- Low mixing and poor water quality near the bottom of dredged basins and channels, resulting in depressed levels of dissolved oxygen
- Potential for increased salinity intrusion in estuaries
- Potential for local increases in wave heights due to changes in wave refraction patterns
- Potential for increased sedimentation rates and future maintenance dredging requirements
- Potential for beach erosion due to loss of offshore sand sources

12.3.2.2 Structural Impacts. Increased water depths can also present important structural and geotechnical concerns for any adjacent or nearby waterfront structures. Retrofitting or upgrading existing structures and facilities to accommodate the increased depths can be very costly or impractical, and presents major design challenges for harbor deepening projects. The scope of potential problems includes, for example:

- Reduced axial load capacity of pile-supported structures, due to reduced skin friction and increase in unsupported pile length, causing reduced buckling load resistance
- Reduced lateral load capacity for pile-supported structures due to reduced pile penetration depth
- Increased lateral loading of piles due to sloughing of slopes induced by dredging
- Reduced load capacity of bulkheads and other sheet pile structures, due to reduced toe (passive pressure) resistance
- Slope stability failure of adjacent reclamation areas and gravity retaining walls, such as crib walls and block walls
- Shoreline erosion caused by dredging-induced slope failure
- Damage to submarine pipelines and cables caused by dredging-induced slope stability failure or inadvertent overdredging

12.3.2.3 Chemical and Biological Impacts. Potential chemical and biological impacts and other water quality impacts caused by dredging activities have been investigated extensively. This includes a comprehensive series of research studies under the Dredge Material Research Program (DMRP) by the U.S. Army Corps of Engineers (USACE), Waterways Experiment Stations. Impacts may include:

- Temporary increase in turbidity and suspended sediment concentrations, potentially affecting marine life and visual quality of the water
- Release of contaminants to the water column due to sediment resuspension
- Temporary destruction of benthic habitat
- Long-term or permanent loss of high-value marine habitat (e.g., eelgrass)
- Fisheries impacts during spawning periods
- Fisheries impacts due to drilling and blasting operations

Dredging operations often do not cause severe chemical or biological impacts, provided that the work is performed in a responsible manner. Dredging-induced sediment resuspension and turbidity is generally not a severe problem, and the potential for release of contaminants to the water column (such as heavy metals) is generally low. Loss of benthic habitat may be a temporary condition, and recolonization often occurs within a time frame on the order of 6 to 12 months. However, such generalizations are difficult and the specific issues and impacts of each project must be evaluated on a case-by-case basis. Dredging-related impacts of particular concern include:

- Dredging in the vicinity of coral reefs or other highly sensitive environments
- Dredging of highly contaminated sediments
- Sediment plumes from excessive overflow from trailing suction hopper dredges when dredging fine-grained sediments

12.3.2.4 Dredging Impact Mitigation. Impact mitigation measures for dredging may be warranted in some cases or may be required by regulatory agencies as a matter of policy. Potential mitigation measures may include:

- Use of silt curtains or booms to contain sediment resupension
- Use of specialized dredging equipment to minimize sediment resuspension
- Seasonal restrictions to avoid fish-spawning periods or other critical periods

Depending on the site conditions, these measures may have limited effectiveness or may be difficult and costly to implement. For example, the use of silt curtains is generally impractical and ineffective in moderate-to-high current and wave conditions. The imposition of seasonal restrictions on dredging operations can have major cost implications for the project, increasing the overall construction schedule and incurring high costs for remobilizing dredging equipment.

12.3.3 Dredge Disposal Options

A number of dredge disposal and reuse alternatives may be considered for a given project. These include open-water disposal, confined aquatic disposal (CAD), unconfined upland placement, confined upland disposal facilities (CDF), and beneficial uses such as wetlands creation and beach nourishment. The range of potentially feasible disposal options depends largely on the grain size distribution and other physical characteristics of the dredge material, the level of chemical contamination, and the potential for adverse biological impacts.

During the evaluation of potential disposal options, cost and environmental factors typically govern the selection and quickly eliminate most options. However, other issues may also be important. These may include, but are not limited to, impacts to wetlands, threatened and endangered species, socioeconomic impacts, land use planning, political factors, and archaeological impacts. Not all of these issues apply to all options, so each option must be considered independently. It is the engineer and scientist's responsibility to identify the most cost-effective disposal option which has acceptable impacts.

Contaminated sediments that are not suitable for open-water disposal can be managed by various means, depending on whether the sediment is considered a hazardous waste from a regulatory viewpoint. In the United States, applicable regulations include the Resource Conservation and Recovery Act (RCRA) and the Toxics Substance Control Act (TSCA). Highly contaminated sediments are frequently disposed of in specially designed CDF facilities or at TSCA-approved landfills. Alternatively, the contaminated material may be cleaned using various remediation technologies and later reused.

12.3.3.1 Ocean Disposal. Ocean disposal is often the lowest-cost disposal option in cases where a certified offshore disposal site is available in proximity to the site, the material is not considered highly contaminated, and the material is deemed to be unsuitable for reclamation purposes or land reclamation is not required.

A systematic approach is required for the characterization and testing of the dredge material to determine its suitability for ocean disposal and other disposal methods. The London Dumping Conference (LDC) of 1972 (with subsequent annexes) is the major international guideline concerning the ocean disposal of dredge material and other types of wastes [International Maritime Organization (IMO), 1991]. The LDC guideline encourages ocean disposal as "often an economically and environmentally preferred disposal solution," under appropriate circumstances, for sediments that are not highly contaminated. Annex I of LDC includes blacklisted industrial wastes that are prohibited from ocean disposal, including compounds of organohalogens, mercury, and cadmium and various petroleum and other wastes. Annex II lists other industrial wastes that warrant special care. This includes, for example, arsenic, chromium, copper, lead, and zinc. Dredge material is not considered an industrial waste under LDC, and sediments that are contaminated with relatively low levels of Annex I or Annex II substances may be suitable for ocean disposal under certain circumstances. For example, sediments contaminated with Annex II substances may be suitable for ocean disposal provided that the contamination level is not considered "significant," which is generally defined by a threshold level of 0.1% by weight, or more. The provisions of Part B, Section 5.6 of LDC and Annex 28 [Resolution LDC23(10)] provide some specific guidance concerning the suitability of dredge material for ocean disposal. This includes a tiered approach for the examination and testing of dredge material, including physical, chemical, and biological/toxicological assessments.

The U.S. (USEPA/USACE, 1991) Environmental Protection Agency and the U.S. Army Corps of Engineers provide guidelines for the testing and characterization of sediments for dredging projects, including the *Green Book* for ocean disposal. Guidelines are also available for the characterization of sediment quality based on these test results (USACE, 1995). Potential water quality impacts from dredge disposal operations are generally evaluated using elutriate tests. The standard elutriate test is used to assess the potential for contaminant releases during open-water dredge disposal.

12.3.3.2 Confined Aquatic Disposal. Confined aquatic disposal (CAD) involves open-water placement of dredged material within a contained area and placement of a capping layer of clean material (typically sand). The dredged material is contained laterally by the construction of submerged dikes or by placement of the dredged material within natural holes or depressions. The cap layer is designed to prevent resuspension of the dredged material due to erosion, and to provide sufficient isolation against the migration of contaminants. CAD is a potentially feasible option for disposal of contaminated sediments that has been used or is being proposed on a number of recent projects. This approach is particularly attractive for the containment of moderately contaminated sediments in cases where adjacent upland areas for dredge disposal are not available or are very costly to develop and ocean disposal is not practical.

In theory, a relatively thin cap layer of sand can be quite effective in isolating contaminated sediments. A major design concern is the potential for erosion of the cap material during storms, due to waves and currents. There are also a number of construction and geotechnical engineering issues and concerns, such as:

- Availability of sand or other suitable cap material at reasonable cost
- Potential washout of contaminated sediments during construction prior to placement of the cap layer, due to erosion or the development of mud waves
- Accurate placement of the capping layer in a uniform, thin layer (as designed)

- Short- and long-term slope stability of the dikes (including seismic stability)
- Differential settlements leading to surface irregularities and increased erosion potential

12.3.3.3 Confined Upland Disposal Facilities. In the United States, environmental testing for upland disposal at confined disposal facilities (CDFs) is based on the manual for sediment testing (USEPA/USACE, 1994) and the modified elutriate test for effluent water quality. The results of elutriate and surface water tests are used to confirm whether the effluent from the CDF will conform to state water quality criteria.

The selection of upland dredge disposal sites is generally based on a three-phase process involving:

1. Initial screening of potential sites by examination of regional topographic maps, aerial photos, helicopter reconnaissance surveys, and site visits. A short list of potentially feasible sites is then developed based on the preliminary site analyses, considering environmental factors, landownership, constructability, and other factors.
2. Preliminary environmental and transportation assessments are performed and conceptual CDF facility designs are developed to prepare comparative construction costs. These results are used to rank the sites and select the final candidate site(s).
3. The final site selection stage includes follow-up site investigations (e.g., soil borings and topographic surveys), detailed environmental impact studies, and preliminary engineering studies to confirm the site feasibility and budget cost estimates.

An important factor in evaluating the storage capacity requirements for CDFs is the potential for increases in sediment volume due to bulking. The placed volume can be significantly larger than the original in situ volume, particularly for hydraulic dredging operations with fine-grained sediments.

12.3.3.4 Permit Applications. To meet the permitting challenges for major dredging and disposal projects, a multidisciplinary team may be required, including scientists, engineers, and planners. The scientific disciplines may include terrestrial, aquatic, and wetland biologists, environmental scientists, specialists on threatened and endangered species, chemists, toxicologists, geologists, and geochemists. Other disciplines may include civil, geotechnical, hydrologic, hydraulic, structural, and dredging engineers, economists, estimators, and planning specialists. For dredging projects in the United States, USACE is generally the lead agency, but the EPA and Fish and Wildlife Service (FWS) also have important roles in approving the necessary permits. Jurisdiction for these agencies includes, but is not limited to, the National Environmental Policy Act (NEPA), Clean Water Act (CWA), Marine Protection, Research, and Sanctuaries Act (MPRSA), Ocean Dumping Act, Endangered Species Act (ESA), and Fish and Wildlife Coordination Act (FWCA). Although the USACE has overall regulatory authority over dredging projects, it must follow EPA regulations and provide coordination among all interested federal and state agencies.

12.3.4 Land Reclamation

Port projects often require extensive land reclamation, and the economic viability of a project may depend largely on the feasibility of using dredged material for all or most of this reclamation work. The construction is generally performed by hydraulic filling operations:

by pumping dredged material into diked containment areas. This type of construction is similar in principle to the CAD and CDF facilities described previously, and presents similar technical issues.

In many areas of the world, reclamation is considered an accepted practice for port development and does not present major regulatory difficulties, provided that the work is accomplished in an environmentally responsible manner. However, in other areas, environmental regulations and public interests may strongly discourage land reclamation. Development options that minimize or eliminate reclamation may be preferred as a matter of general policy, regardless of construction cost considerations. The filling of large shallow- and open-water areas may involve extensive biological impacts such as the destruction of large wetland areas that may be impractical to mitigate or be unacceptable to the regulatory agencies (see Section 12.4.2).

In addition to environmental considerations, land reclamation can also present major geotechnical engineering challenges. Natural foundation conditions at reclamation areas are often very poor, often including thick layers of weak and compressible soils such as soft organic silts or clays. Ground improvement measures may be required to mitigate geotechnical problems due to slope stability and long-term settlement. This may include, for example, soil stabilization using preloading techniques and the installation of wick drains though soft compressible layers. Such remedial measures, when required, are very costly and can greatly affect the construction phasing and overall time frame for a project. Other important design concerns, for projects in earthquake-prone areas are the potential for liquefaction of hydraulically placed fill in reclamation areas and the need for costly remedial ground improvement measures such as vibro-compaction.

12.4 REGIONAL ENVIRONMENTAL IMPACTS

12.4.1 Infrastructure Impacts

The capacity and limitations of the existing infrastructure are major issues for the planning, design, and environmental assessment of port projects. The upland infrastructure includes numerous and diverse elements, such as the arterial and secondary road systems, rail services, electric power supply, gas supply, sanitary and stormwater sewer systems, wastewater treatment facilities, and firefighting and emergency response services. The capacity, suitability, and condition of these facilities need to be evaluated thoroughly, taking into account local and regional master plans for future improvements. The need for major infrastructure improvements must be identified at the initial planning and site-selection stage, since it can affect the overall feasibility of a project. Such improvements, if required, can have enormous costs or can be prohibitive due to adverse environmental impacts.

The potential infrastructure issues for port projects are wide-ranging and highly site-specific. Important issues can include, for example:

- The capacity of existing roads and bridges to accommodate the proposed construction and port operations and the need for new construction or repairs and upgrades to existing facilities. This includes the ability to adequately handle the projected increases in vehicular traffic, as well as structural capacity issues such as axle-load limitations and the seismic capacity of bridges servicing the port.
- The feasibility of using the existing municipal stormwater drainage system, and requirements for pretreatment facilities such as oil–water separators.

- The feasibility of using the existing sanitary sewer system, and requirements for on-site wastewater treatment.
- Requirements for electric power supply.
- Requirements and capabilities of local emergency management and security agencies, such as the local fire department, police department, and Coast Guard.
- Potential interference of tall structures such as communication towers and container cranes with flight path clearance requirements from adjacent airports or proposed airport development.

12.4.2 Biological and Wetlands Impacts

The avoidance or mitigation of adverse biological impacts is often a key factor affecting site selection and project feasibility. The potential scope of biological impacts is quite broad and includes, for example:

- Impacts on coastal and inland wetlands due to dredging and filling operations, wharf construction, effluent discharges, and other impacts
- Impacts on other types of high-value aquatic environments, including seagrass and coral reef areas
- Impacts on fisheries and fish spawning areas
- Impacts on threatened or endangered species of flora and fauna (regulated in the United States under the Endangered Species Act)
- Shade-related impacts at pile-supported platforms

12.4.2.1 Wetlands Impacts and Mitigation. Marine and freshwater wetlands are frequently located at or adjacent to sites of proposed port development. The avoidance and mitigation of impacts on these environmentally sensitive areas are often critical issues for port environmental planning.

Wetlands is a collective term for marshes, swamps, bogs, and similar areas found in flat vegetated areas, in depressions, and at shorelines of rivers, estuaries, and coastal regions. They are periodically or permanently inundated or saturated by surface or groundwater and serve numerous functions, including: marine food chain production, fish and wildlife habitat and diversity, water filtration and purification, nutrient removal and transformation, sediment stabilization and shoreline protection, and stormwater and flood control. They typically support a wide variety of flora and fauna, which may include threatened or endangered species. In the United States, activities such as dredging and the placement of fill in wetlands areas are controlled under the Clean Water Act (CWA) and various state and local water quality regulations. Wetlands that are regulated under Section 404 of CWA are termed *jurisdictional wetlands* and are identified based on several factors, including hydrologic and soil conditions and the nature of the plant life.

An important long-term goal of CWA is to achieve "no net loss" of wetlands. To achieve this goal, project plans should be developed in a manner that avoid wetlands-related impacts wherever possible and minimizes adverse impacts to the extent practical when they cannot be avoided. At sites that have jurisdictional wetlands, a sequential environmental review process is required whereby the applicant must first show that all available alternatives have been considered and that there are no practical alternatives that would have fewer adverse impacts on the aquatic ecosystem. Where significant impacts are deemed unavoidable, mitigation measures are generally required, which can be quite extensive, costly, and controversial. In some cases regulatory agencies may require that the wetland mitigation plans be implemented in advance of a project, as a prerequisite for the construction permit. Such requirements can present large additional costs, risks, and delays for the project.

A study by the NRC (2001) provides a useful overview of wetlands mitigation practices and trends in the United States within the context of the Clean Water Act. Requirements for compensatory wetlands mitigation plans are developed on a case-by-case basis for each project. Occasionally, regulatory agencies may accept user fees or grants in lieu of executing a wetlands mitigation project. Another potential option may involve the use of off-site multiproject mitigation banks that are developed either by the permitee or by a third-party agent. Environmentally responsible third-party agents may offer certain advantages in developing such banks, both for developers and for the regulatory agencies. Given sufficient financial and technical resources, an appropriate third-party agent may be able to provide the objectivity and commitment needed to administer, monitor, and maintain a major wetlands restoration or development project. From the viewpoint of the port developer, this approach might also reduce the initial risks of obtaining the construction permits in a timely manner, and later shift the responsibilities and risks for the ultimate performance of the mitigation project.

The scientific and engineering aspects of wetlands creation and restoration, together with numerous case studies, are described in Kusler and Kentula (1990), USEPA (1991), Hayes et al. (2000), Mohan (2000), and Zedler (2001). Regulatory preference is generally given to the restoration of existing wetlands rather than the creation of new wetlands. The creation of new artificial wetlands is problematic and generally presents significantly lower probability for success than does restoration of an existing deteriorated wetland. Historically, regulatory agencies have sometimes favored the construction of compensatory wetlands either on site or in an area contiguous to the wetland affected by the project. However, this strategy has tended to force on-site solutions that may not be optimal or even suitable. The natural development of marine wetlands requires highly favorable soil, topographic, and hydrologic conditions and generally occurs over a time frame of centuries or at least decades. In contrast, engineered wetlands renovation or development projects generally have a time frame of only a few years and may involve unsuitable soils, such as coarse-grained dredged material.

Wetlands impacts can present major problems and uncertainties for the long-range master planning of port facilities in environmentally sensitive locations. Historically, mitigation guidelines have sometimes included fixed ratios for compensatory wetlands, such as the "in-kind" replacement of 1 acre of impacted wetlands with 1.5 acres of compensatory wetlands. Such guidelines might be considered onerous from the viewpoint of the port developer, but at least provide a reasonable basis for long-range planning and real estate acquisitions. This is often no longer the case at a time when wetlands impacts are viewed with increasing regulatory scrutiny and the mitigation measures required are largely unpredictable and subject to an extensive review process on a case-by-case basis.

12.4.3 Water Quality Impacts

Water quality issues and regulatory requirements vary greatly depending on project location and scope. In the United States, the Clean Water Act (CWA) is the primary federal law related to water quality, but specific requirements are based primarily on state, regional, and local requirements. General requirements for wastewater receiving and treatment facilities at ports are described by IMO guidelines (1999).

Typical water quality issues for port projects include:

- Dredging and dredge disposal-related impacts (CWA Sections 401 and 404)
- Treatment of oil-contaminated stormwater drainage and wastes, with oil–water separators
- Treatment of storm drainage from dry bulk stockpiles
- Potential need for detention and treatment of storm drainage from yard drainage
- Development of operational storm water management plans, based on best management practices
- Development of site drainage and erosion and sediment control plans to mitigate construction-related impacts
- Development of oil spill response contingency plans
- Permit requirements, mixing-zone effluent limitations, and monitoring requirements for point discharges (CWA Section 401)
- Effluent limitations related to total maximum daily limitations (TMDLs) for the receiving water body, based on regional water quality requirements
- Special requirements affecting the degradation of water bodies designated as "impaired" [CWA Section 303(d)]
- Special requirements affecting the degradation of estuaries and coastal waters designated as having special significance (CWA Section 302)
- Effects of the proposed port facilities on local circulation patterns and flushing rates
- Effects of increased water depths (by dredging) on dissolved oxygen levels and other water quality parameters
- Potential for eutrophication of receiving water bodies due to increased nutrient loads
- Leaching of wood preservers, including creosote and chromium copper arsenate (CCA)
- Need for liners to prevent groundwater contamination at tank farms and dry bulk stockpiles
- Need for groundwater monitoring at petroleum bulk terminals, onshore dredge disposal facilities, and other facilities where groundwater contamination is possible

12.4.4 Erosion and Sedimentation Impacts

Breakwaters and other large-scale port structures can greatly affect local wave climate, currents, and sediment transport processes, and can cause significant changes in the configuration of the adjacent shoreline. Particularly severe effects can occur at open coast sites with energetic wave conditions and high rates of net longshore sand transport. In such cases the breakwater will block the longshore transport, causing rapid deposition on the updrift side and concurrent erosion of the coastline on the downdrift side.

Coastal erosion problems can be mitigated by periodic beach nourishment by dredging sand from offshore sources, bypassing the impounded sand on the updrift side of the port or providing sand from upland sources. Coastal structures such as groins and nearshore breakwaters may be used in combination with beach nourishment to reduce long-term maintenance costs. However, such structures tend to shift the location of the erosion problems and can cause additional adverse impacts that must be evaluated and mitigated.

12.4.5 Flood Impacts and Flood Control

Extreme flood conditions must be considered in establishing appropriate grade elevations for the port facilities. This is a critical design issue at low-lying sites in areas prone to high flood levels, since extensive fill volumes will be required from channel dredging or other sources.

Moreover, the placement of extensive fill can present geotechnical problems due to long-term settlements and slope instability, which may require costly remedial measures and further increase the fill volume requirements.

In coastal areas extreme flooding is generally due to high astronomical tide in combination with storm surge effects from hurricanes or other types of storms. Additional factors such as wave setup, wave overtopping, long-term rise in sea level, and regional ground subsidence may also be important. At river ports the facilities may encroach on the natural floodway, and increase upstream flood levels due to backwater effects. Such encroachments must also account for the long-term stability and migration of the river channel and the potential for structure-induced scour.

Whatever flood design criteria are used to establish the site grades, there is some risk that these conditions may be exceeded at some point during the life of the port. The potential for such flood damage should be assessed and may warrant provision of localized flood control methods for critical facilities.

12.4.6 Shipping and Navigation Issues and Impacts

Shipping studies should be performed to evaluate the range of vessels that will use the port and their respective berth occupancy requirements. Berth availability studies may be required to confirm the adequacy of port layout for the operational wave conditions. These studies should take into account seasonal variations in wave climate and the potential for long-period (infragravity) waves and seiche conditions, which can cause excessive vessel motions and unacceptable vessel downtime. Such studies normally involve numerical wave and hydrodynamic modeling together with three-dimensional physical hydraulic model tests of harbor wave agitation and vessel motions.

Monte Carlo system simulation studies may be warranted for the optimum design of key functional elements of the port, such as the number and types of berths, storage tank and stockpile capacities, number and capacity of units for cargo transfer and materials handling, and other facilities. These studies should consider the recent and anticipated trends in the shipping trades, including the potential need for servicing larger, deeper-draft vessels that are not yet available. Forecasts should be developed of the future vessel traffic and its impact on the shipping and boating traffic in the region.

A fundamental planning issue is the capacity of the existing and proposed port entrance channels to safely accommodate initial and future anticipated vessel traffic. PIANC channel design criteria guidelines are frequently used for this analysis and may be supplemented by detailed navigational and risk assessment studies using numerical ship navigation simulators. Other potential navigational impacts and hazards need to be checked, such as:

- Need for improved navigational aids
- Need for a new or improved vessel traffic control system
- Requirements for tug assistance and dedicated pilots
- Increased risks for shipping or boating accidents
- Potential for oil spills and other types of spills
- Requirements for offshore anchorage areas
- Increased queueing times at locks for barges and other vessels (for inland ports)
- Risk of collisions with bridge superstructure during extreme high water conditions
- Increased risks for ship collisions with bridge foundations, and the need for improved bridge fender systems or other protective systems

12.4.7 Vehicular and Rail Traffic Impacts

The surface transportation infrastructure presents major design and environmental impact issues for most port projects. The development of efficient highway and/or rail linkages is obviously essential for container terminal operations and for most dry bulk terminals. A high volume of fuel truck traffic is a potential concern for marine petroleum terminals. Traffic-related environmental and economic impacts can include increased traffic congestion, queueing at port entrances, air and noise pollution, higher maintenance costs for existing roads and bridges, and general degradation of quality of life and real estate values for the surrounding communities that are affected directly. Substantial new road construction to provide or improve access to the port may not be feasible except in relatively remote, undeveloped areas. In addition to construction costs for new roads and rail spurs, rights-of-way may be very difficult or impossible to obtain, and the environmental impacts may be unacceptable, such as the separation and isolation of commercial areas or residential communities. Axle-load limitations for existing roads and bridges can also be an important restraint.

Truck traffic generated during construction can be extremely high and can present unacceptable environmental impacts and construction logistics problems, depending on the nature and scope of the construction and the material supply sources. Unusually high traffic demands can be associated with:

- Land-based supply of core material and armor stone, for breakwater and seawall construction
- Land-based supply of fill for reclamation
- Land-based supply of cement and aggregates for projects requiring large quantities of concrete, such as concrete armor units for breakwaters or concrete block quay wall construction
- Off-site upland disposal of large quantities of dredge material, particularly if the material is considered contaminated or produces odors

In extreme cases, where impacts are severe or public and political opposition is keen, land-based construction supply operations may be greatly curtailed or eliminated. For example, for the construction of Massport's Deer Island project in Boston harbor, all vehicular access to the site was prohibited, due to adverse traffic impacts on adjacent residential areas. A dedictated ferry system was developed, at great cost, to ferry construction supplies, equipment, and construction workers to the job site.

For projects that involve significant traffic impacts, detailed surveys and assessments are needed for the existing transportation network and present traffic patterns. Future population growth and master plans for future growth should be considered in these analyses. Traffic increases associated with the port development need to be forecasted, and the impacts on the local and regional transportation system may be assessed using traffic simulation numerical models. Variations in existing and projected traffic also need to be considered, including hourly, daily, and seasonal variations. Peak traffic demands need to be evaluated as well as alternative traffic mitigation plans.

12.4.8 Air Quality Impacts

There are important air quality issues associated with most port development projects. These issues are most severe for ports that are located in urban and industrial areas where existing (background) air quality is already poor and the potential for adverse environmental impacts and regulatory concerns are more severe. At petroleum terminals, a major planning and cost consideration is the potential need for a vapor recovery system to mitigate the release of toxic and hazardous vapors during vessel loading operations. Air quality control systems are important design considerations for most types of dry bulk terminals, to limit emissions from stockpiles, conveyor systems, and other sources. At container ports, extensive air quality studies are generally required to access the project air quality impacts due to truck traffic and other sources.

Air quality studies generally require a detailed assessment of the site meteorology and the background (baseline) air quality conditions. Numerical model studies may be performed to assess the impacts of new emissions sources from the port, using dispersion models. Air quality control systems and mitigation measures are then designed to meet the appropriate regulatory criteria for pollutant concentrations and total emissions levels. Best management practice (BMP) techniques are often adopted, consistent with national and local guidelines and regulations.

12.4.9 Noise and Vibration Impacts

Noise impacts can be significant for container terminals and other types of port facilities. Noise studies typically involve field surveys to establish the existing background noise levels and sources, combined with accoustical numerical models to evaluate the effects of new noise sources. These sources may be analyzed as point sources, such as cargo handling equipment, or as line sources, such as roads and rail lines. Noise levels are simulated at the site and for surrounding areas, where the allowable noise criteria may be much more stringent. Noise impacts may also involve biological impacts such as the presence of nesting or feeding birds in the area.

Major sources of noise during construction include pile driving, dredging, compressors, and drilling operations. Mitigation measures may be needed to maintain noise levels within the allowable regulatory standards or to provide best management practice. The mitigation measure might include time restrictions on the construction operations (e.g., restrictions on nighttime construction), use of vibratory pile drivers in lieu of impact hammers, and various muffling devices for heavy construction equipment. In some cases, noise due to dredging operations during nighttime may exceed the allowable standards for adjacent areas. Since dredging operations must normally continue on a 24-hour basis, the mitigation of dredging related noise may not be practical.

For port operations, major noise sources include ships and support vessels, container and materials handling equipment, truck traffic, and rail services. Mitigation measures may include the use of muffler devices for equipment, designation of special high-noise areas, and the use of buffer zones and noise screens. As an example, for the new container port at Bayport, Texas, accoustical numerical model studies showed that operational noise levels would exceed the allowable level of 55 dBa for a nearby residential area. An unusual and costly solution is being adopted, involving the construction of a perimeter berm around much of the container yard, using dredge material. The berm will be vegetated and will provide multiple environmental objectives: noise barrier, visual screen, and beneficial use of dredge material.

A related issue is the potential for low-frequency vibrations caused by the construction or the port operations. Although this is not normally a problem, under certain conditions high-intensity ground vibrations can adversely affect adjacent structures and sensitive equipment.

12.4.10 Socioeconomics and Environmental Justice

The long-term economic viability of a project must be convincingly demonstrated, taking into account estimated construction, operating, and maintenance costs, anticipated sources and flow of income, taxes, finance charges, and proposed sources of financing. For major projects, extensive socioeconomic studies may also be required to assess the macroeconomic and social impacts of the project on the regional and national levels. Such studies are often required where funding is provided from international lending agencies.

Ports are often located in the vicinity of low-income residential areas. This can present environmental and political concerns regarding "environmental justice"—that the major burden of regional environmental impacts falls primarily on low-income and minority populations. In the United States, guidelines concerning environmental justice were developed in 1994 and need to be considered in port EIA studies.

12.4.11 Cultural Impacts

Ports are sometimes located in areas of major historical or cultural significance. Site surveys and historical and archeological assessments may be needed to identify structures or artifacts that have historical, religious, or cultural importance, as well as the presence of cemeteries or ancient burial grounds. For example, in the United States there are numerous old and sometimes decrepit structures that are deemed to have historical significance and are identified in the Historical Register. The proposed demolition or relocation of a structure that is included in this register would present significant (possibly insurmountable) difficulties for a project.

12.4.12 Visual Impacts

Adverse visual impacts need to be considered and can be critical for projects near residential or recreational areas. Although ports are generally located in areas that are zoned or otherwise planned for heavy industrial uses, the facilities and their lighting can often be viewed from considerable distances. Ports typically maintain 24-hour operations and can create nuisance problems due to "night glow" from area lighting and flashing lights from vehicles such as straddle carriers. Visual impacts from stockpiles, storage tanks, and other types of unsightly facilities can also be problematic, but mitigation may be difficult or impractical.

Waterfronts are viewed increasingly as important focal points and amenities for urban areas, and issues related to aesthetics and visual impacts are becoming more important for port planning. Redevelopment and enhancement of waterfront areas is being encouraged for commercial and residential redevelopment, tourism, parks and other recreational facilities, and public access. Port planners need to be sensitive to these issues and to the specific concerns of special-interest groups that may be affected directly by the proposed construction. Potential mitigation measures for visual impacts include:

- Strategic layout of the facilities to minimize lighting and other visual impacts
- Use of perimeter barriers or buffer zones
- Use of specially designed lighting fixtures

12.5 SOIL CONTAMINATION AND SITE REMEDIATION

Many waterfront sites throughout the world are heavily contaminated due to past industrial uses and dumping practices, where environmental controls were minimal or totally lacking. Contamination problems can be attributed to factors such as major spills or accidents, chronic leakage from pipelines or storage tanks, leaching from stockpiles, on-site disposal or outfalls of untreated or inadequately treated liquid wastes, and burial of chemicals and toxic wastes in drums or other containers. For example, a number of older marine petroleum at one time operated with minimal environmental controls and have required costly site remediation programs due to soil contamination or groundwater pollution problems. Occasionally, new port facilities are also located in areas that are highly contaminated, or plan to expand into adjacent areas that are contaminated.

In the United States, under the Super Fund Act, the EPA has identified and targeted a number of sites for cleanup and remediation. This includes the development of a priority list of 1000 Super Fund sites plus another 10,000 sites that require special consideration. Since the focus of these regulations is human health risks, groundwater contamination is a major concern at these sites. The presence of high contamination levels for in situ soils does not necessarily warrant costly remedial measures if the contaminants have low solubility and do not pose risks to groundwater or other drinking water supplies. In cases where site remediation is required, it can involve enormous costs, extensive litigation and negotiation, and take many years to complete. To the extent practical, EPA holds the parties that were originally responsible for the contamination responsible for the site remediation costs. However, in many cases the guilty parties are no longer available or do not have sufficient financial and insurance resources to incur these high costs.

The potential for soil contamination and the need for site remediation is an important planning considering for port development near areas of heavy industrial use (past or present). Frequently reported contaminants at port and other waterfront facilties include heavy metals and metalloids (e.g., arsenic), polychlorinated biphenyls (PCBs), pesticides, and polycyclic aromatic hydrocarbons. However, there is great variability in the types and levels of contaminants in sediments from region to region or harbor to harbor. Guidelines concerning threshold contamination levels have been developed by IAPH and are useful for preliminary screening studies and site assessments, supplementing available national and local regulations and guidelines.

12.5.1 Remediation Techniques

If pollution surveys demonstrate the presence of extremely high levels of contamination posing significant health risks, the site may require decontamination using isolation, immobilization, or various treatment techniques, as described in the following sections (IAPH, 1993; NRC, 1997).

12.5.1.1 Isolation. Where suitable, physical isolation by impermeable barriers is often the lowest-cost method for site remediation. Vertical isolation to control infiltration can be accomplished readily using pavements or synthetic or clay liner systems. Lateral isolation may involve the use of sheet pile systems, slurry walls, and other barriers. Special provisions may be required for external and internal drainage systems and vapor control.

12.5.1.2 Stabilization. Stabilization techniques involve mixing the soil with reacting agents to solidify the material and bind the contaminants permanently. It is generally accomplished on site, using reactor units. Reacting agents include lime, hydraulic cements, and other materials.

12.5.1.3 Thermal Cleaning. The contaminated soil material and water are removed and treated in a processing plant located on site. It is used primarily for soils contaminated with organic compounds, particularly hydrocarbons. The process generally involves a rotating furnace in which the soil and contaminants are burned at high temperature. Gases emitting from this process are subject to air quality control requirements.

12.5.1.4 Soil Washing. Soil washing involves the transfer of contaminants into the liquid phase, which is then decontaminated using water treatment processes. This method is applicable for soils contaminated by mineral substances (such as heavy metals) but certain processes can also be used for organic compounds. Soil washing was used at a demonstration project for a CDF in Saginaw, Michigan, for PCB-contaminated sediments.

12.5.1.5 Soil Vapor Extraction. In this method the contaminants are transferred into a gaseous phase, which is then collected and processed. This technique was used, for example, at Avila Beach, California for the remediation of the lighter hydrocarbon fraction of petroleum-contaminated soils.

12.5.1.6 Microbiological Degradation. Microbiological processes can be used in a variety of ways to destroy or break down the contaminants in polluted soil. It has been used for treating soils contaminated by petroleum hydrocarbons, often in situ. The biological reactions have an aerobic character and involve bacteria that are normally present in the soil and capable of breaking down the pollutants. The process can be stimulated by oxygenation, injection of suitable nutrients, and other techniques.

12.6 MARINE PETROLEUM TERMINALS

API 2610 provides general guidelines for planning and design of petroleum terminals, including marine terminals. Additional guidelines for design and equipment specifications at marine oil terminals have recently been developed by the California State Lands Commission.

12.6.1 Water Quality Control

Discharge of ballast water and oily wastes are regulated by the International Maritime Organization, under the MARPOL regulations. Additional guidance concerning port waste receiving facilities are described in other IMO publications and U.S. Coast Guard regulations.

In the planning of new marine petroleum terminals, ballast water treatment facilities are generally not required, since most tankers are now equipped with segregated ballast tanks, in accordance with MARPOL requirements. In the past, an important environmental and economic concern related to ballast water has been the potential for the introduction of alien biological species introduced from the discharge of ballast water obtained at remote coastal ports. This problem is now being managed by requirements to exchange ballast water in ocean waters under controlled conditions. Under these conditions the discharge of clean ballast water at port from ships equipped with segregated ballast tanks does not present significant problems.

Oil-contaminated stormwater and other oily wastewaters require treatment using oil–water separators. This includes, for example, the containment and treatment of drainage from spill-prone areas at loading platforms and from tank farm containment areas. Generally, oil water separator systems are designed to limit the oil and grease concentration to 15 ppm, unless local effluent requirements are more stringent.

Comprehensive oil spill control contingency plans are essential for all marine petroleum terminals. In the United States, these requirements are specified under Coast Guard regulations. The spill control plan must include detailed provisions for the use of oil spill containment booms and cleanup methods, including skimmers and dispersants [American Petroleum Institute (API), 1985; IMO, 1997, 1988]. In general, a tiered approach is required for oil spill management whereby the terminal must be suitably equipped and capable of handling small spills on an emergency basis. Larger spills are handled by mobilizing additional equipment, as needed, from designated remote locations by external agencies. For groundwater pollution control, mitigation measures may include the use of impermeable liners and periodic monitoring of water quality at monitoring wells.

12.6.2 Air Quality Control

A major air quality issue for marine petroleum export terminals is the potential need for costly vapor recovery systems to avoid the atmospheric release of toxic vapors during vessel loading operations. The vapor recovery system typically includes a vapor recovery loading arm, return pipeline to shore, and onshore vapor treatment facilities. Another design issue related to vapor control is the use of floating roof systems for storage tanks for certain types of petroleum products.

12.7 DRY BULK TERMINALS

Dry bulk terminals involve a number of environmental impact issues and may require extensive pollution control systems for water and air quality. The scope of these control systems depends greatly on the specific types of products being handled, the local site conditions, and the governing environmental regulations and guidelines. Descriptions of environmental issues and pollution control systems for typical bulk terminals are included in Mahr (1983), Soros (1992), and White (1992).

12.7.1 Water Quality Control

Important water quality issues at bulk terminals include the control and treatment of stormwater drainage, control of groundwater pollution, and pollution caused by spills during cargo transfer operations. For storm drainage, the principal concern is drainage from stockpile areas. Other areas of potential concern include truck wash-down areas, areas in the vicinity of conveyors and conveyor transfer stations, and other areas where materials are handled or transferred. As a minimum, primary treatment is generally required for storm drainage from stockpiles and other contaminated areas, for solids separation. This is generally accomplished using a gravity or pumped drainage system that directs flow into detention basins, tanks, or evaporation ponds (in arid regions). In some cases, pH adjustment of the wastewater may be required, and coagulants or flocculents may be added to accelerate the sedimentation process. The settled solids may either be recycled back into the stockpile or disposed off site in accordance with local regulations. Depending on the nature of the bulk product and the effluent discharge criteria, secondary wastewater treatment may also be required prior to effluent discharge.

An important design consideration is the need to preserve the quality (purity) of the stockpile product and avoid cross-contamination of different types of product or different product grades. This is a critical issue when solids are recycled after primary treatment. Cross-contamination can also present significant problems if secondary water treatment facilities are required, since the treatment of mixed wastewaters is often difficult or impractical. Isolated drainage and treatment systems are used to prevent this problem. Groundwater quality control mitigation may include the use of liners or pavements at stockpile areas to control groundwater infiltration. The installation of monitoring wells and the implementation of an operational groundwater monitoring program may also be warranted in some cases.

During cargo handling operations, accidental spillage of bulk products into the harbor can cause a variety of water quality problems. Mitigation of spillage is related primarily to the skill and care of operators and the type of equipment being used (e.g., the use of telescoping chutes). Spillage problems tend to be more acute at smaller bulk terminals where the material is transferred using grab buckets from shoreside cranes or from self-unloading vessels.

12.7.2 Air Quality Control

The control of dust and air pollutants is often a major concern at bulk terminals. This is particularly true for facilities located in urban or industrial areas where background air quality may already be poor. Potential sources of air pollution emissions include stockpiles, cargo transfer operations, conveyors, and conveyor transfer stations.

The range of potential air quality control systems and mitigation measures is broad and the requirements are again quite site specific. For stockpiles, mitigation methods may include use of spray systems (such as water canons), structural or vegetative perimeter screens, and (in rare circumstances) requirements for structural enclosures. Coating systems or liners may also be used to control emissions from inactive stockpile areas. For belt conveyor systems, air quality control mitigation measures may include:

- Installing partial enclosures, screens, and spray systems at transfer stations
- Minimizing the number of transfer stations
- Using covered conveyor systems
- Installing air quality control systems for covered conveyor systems, to collect and treat contaminated air

12.8 FACTORS AFFECTING PROJECT FEASIBILITY

As outlined in the preceding sections, numerous environmental factors need to be considered in the planning, permitting, design, construction, and operation of port facilities. However, for each project there are generally only a few fundamental technical and environmental issues that affect project feasibility, site location, and overall design. A partial list of potential key issues includes:

1. Environmentally acceptable dredge disposal options are available, at reasonable cost.
2. Dredge material does not include large quantities of rock that will require drilling and blasting and can be dredged using conventional equipment.
3. Dredge material does not include large quantities of highly contaminated sediments, requiring special treatment.
4. Suitable reclamation material is available at reasonable cost from the ongoing dredging operations or from other nearby sources.
5. Feasible mitigation options are available for impacts to high-value biological resources, including tidal marshes, seagrasses, and endangered species.
6. Permits for the dredging, disposal, and reclamation work can be obtained from regulatory agencies within a reasonable time frame.
7. The port development, including the breakwater construction, will not have major adverse effects on the adjacent coastline due to erosion, sedimentation, or water quality effects.
8. The infrastructure servicing the project, including roads and utility services, are generally adequate or can be improved with project budget limitations.
9. Environmental impacts related to transportation and other infrastructure improvements are acceptable.
10. Adequate upland area is available for environmental mitigation measures, such as wetlands mitigation, screens and buffer zones, and storm drainage detention ponds.
11. The upland area soils and groundwater are not excessively contaminated and do not require extensive cleanup and site remediation.
12. The project does not have major cultural resource impacts, such as important archeological artifacts or historically significant structures.

13. The project will not have major regional impacts that affect socioeconomics or cause environmental injustice.
14. All reasonable alternatives to the proposed project have been considered and evaluated in terms of environmental impacts, costs, and benefits.
15. The project is economically sound and can be adequately financed for initial construction, future construction phases, and future operations.
16. The project has a reasonable level of community and political support.

It is critically important that the key environmental and planning issues be clearly identified at the outset of the project and be addressed adequately throughout the initial planning, screening, and conceptual design phases. To the extent practical, it is also important that the project team work cooperatively and constructively with regulatory agencies, public interest groups, and all other interested and affected parties, as early as possible. Inadvertent alienation of such groups and attempts to force a single preferred solution can ultimately create major problems for the project.

REFERENCES

AAPA, (1988). *Environmental Management Handbook,* American Association of Port Authorities, Alexandria, VA.

Abood, K., and S. Metzger, 2000. Green ports: aquatic impact avoidance, minimization and mitigation for port development projects, *ASCE Proc. Specialty Conference Ports '01,* Norfolk, VA.

API, 1985. *Oil Spill Response: Options for Minimizing Ecological Impacts,* Publ. 4398, America Petroleum Institute, Washington, DC.

Davis, J., S., MacKnight, IMO staff, et al., 1990. *Environmental Considerations of Port and Harbor Developments,* Tech. Pap. 126, Transport and Environment Series, World Bank, Washington, DC.

ESPO, 1995. *Environmental Code of Practice,* European Association of Ports Organization, Brussels, Belgium.

Hayes, D. F., et al., 2000. *Wetlands Engineering Handbook,* U.S. Army Corps of Engineers Waterways Experiment Station, Vicksburg, MS.

Herbich, J. (ed.), 2000. *Handbook of Dredging Engineering,* 2nd ed., McGraw-Hill, New York.

IAPH, 1991a. *Guidelines for Environmental and Management in Ports and Coastal Area Developments,* International Association of Ports and Harbors, Tokyo, Japan.

———, 1991b. *Practical Guidelines for Ports on Environmental Issues-Water Pollution-A Concern for Port Authority,* International Association of Ports and Harbors, Tokyo, Japan.

———, 1993. *Practical Guidelines for Ports on Environmental Issues-Soil Pollution in Ports,* International Association of Ports and Harbors, Japan, Tokyo.

IMO, 1988. *Manual on Oil Pollution,* Section IV, *Combating Oil Spills,* International Maritime Organization London.

———, 1991. *The London Dumping Convention: The First Decade and Beyond,* International Maritime Organization, London.

———, 1997. *Field Guide for Oil Spill Response in Tropical Waters,* International Maritime Organization, London.

———, 1999. *Comprehensive Manual on Port Reception Facilities,* International Maritime Organization, London.

Kusler, J. A., and M. E. Kentula, 1990. *Wetland Creation and Restoration: The Status of the Science,* Island Press, Washington, DC.

Mahr, D., 1983. Systems approach to dust emission control at coal piles, *Power Eng.,* June, pp. 57–59.

Mohan, R. K., 2000. Design and construction of coastal wetlands using dredged material, in *Handbook of Dredging Engineering,* 2nd ed., McGraw-Hill, New York.

NRC, 1997. *Contaminated Sediments in Ports and Waterways: Cleanup Strategies and Technologies,* National Research Council, Washington, DC.

NRC, 2001. *Compensating for Wetlands Losses under the Clean Water Act,* National Research Council, National Academy Press, Washington, DC.

Paipai, E., 1999. *Guidelines for Port Environmental Management,* H. R. Wallingford Rep. SR 554, Nov.

PIANC, 1996. *Handling and Treatment of Contaminated Dredged Material from Ports and Inland Waterways,* Vol. 1, Rep. Work. Group 17, Suppl. Bull. 89, Technical Committee I, Permanent International Association of Navigation Congresses, Brussels, Belgium.

———, 1998. *Management of Aquatic Disposal of Dredged Material,* Rep. Work. Group I, Permanent Environmental Commission, Permanent International Association of Navigation Congresses, Brussels, Belgium.

———, PIANC, 1999. *Environmental Management Framework for Ports and Related Industries,* Rep. Work. Group 4, Permanent Environmental Commission, Permanent International Association of Navigation Congresses, Brussels, Belgium.

Soros, P., 1992. Technology: key to environmental successes, *ASCE Proc. Specialty Conference Ports '92,* Seattle, WA, pp. 189–202.

Trozzi, C., and R. Vaccara, 2000. Environmental impact of port activities, in *Maritime Engineering and Ports II,* C. Brebbia and J. Olivella, eds., WIT Press, pp. 151–161. Southampton, United Kingdom.

Urban Harbors Institute, 2000. *America's Green Ports: Environmental Management and Technology at U.S. Ports,* University of Massachusetts, Amherst, MA, Mar.

USACE, 1998. *Use of Sediment Guidelines (SQGs) in Dredged Material Management,* EEDP-04-29, U.S. Army Corps of Engineers, Waterways Experiment Station, Vicksburg, MS.

USEPA, 1991. *Restoring and Creating Wetlands,* U.S. Government Printing Office, Washington, DC.

USEPA/USACE, 1991. *Evaluation of Dredged Material Proposed for Ocean Disposal: Testing Manual,* U.S. Government Printing Office, Washington, DC.

———, 1994. *Dredged Material and Evaluation Manual: "Evaluation of Dredged Material Proposed for Discharge in Waters of the U.S.: Testing Manual,* U.S. Government Printing Office, Washington, DC.

Vellinga, T., 1997. *Dredge Marterial Management Guide,* Suppl. Bull. 96, Permanent Environmental Commission, Permanent International Association of Navigation Congresses, Brussels, Belgium.

White, P., 1992. "Waste Water Management at Bulk Terminals," *ASCE Proc. Specialty Conference Ports '92,* Seattle, WA, pp. 178–187.

Zedler, J. B., 2001. *Handbook for Restoring Tidal Wetlands,* CRC Press, Boca Raton, FL.

13

PORT SECURITY

Michael A. McNicholas
Phoenix Management Services Group, Inc.
Fort Lauderdale, Florida

13.1 INTRODUCTION

With few exceptions, the ports of the world are increasingly under threat of penetration, manipulation, and use by drug smugglers, stowaways, cargo thieves, pirates, and terrorists. Targets of these criminal elements include the terminals, vessels, cargo, containers, equipment, and personnel. In addition to concerns related to the rising number of incidents of violent piracy and waves of stowaways and refugees in the past several years, ports in these post–9/11/01 times must also face the looming specter of terrorist attacks involving weapons of mass destruction, or of a vessel being used by terrorists as a conveyance of an instrument of destruction. To deter or deny these threats effectively, ports must develop a security strategy that identifies the potential security threats, defines critical assets and information, integrates security resources and capabilities, and ensures successful design, implementation, and management of a world-class seaport security program.

The most comprehensive and effective seaport security program is one based on the military concept called *defense in depth.* Applied to a port, this concept involves the design and establishment of a series of security rings around and in the port, as well as encircling critical assets within the port (e.g., cranes, vessels). The number of security rings established and the specific components (security systems and measures) of each ring will vary and depend on the port's layout and operations, its assets, and the level, type, and duration of the threats. Although the rings themselves are permanent or long standing, the individual components within the rings may be long term or temporary, as in the case of measures implemented to address a crisis situation or a short-term threat. These security rings should be layered and integrated but be capable of functioning independently. Security ring components include physical security measures, procedural security standards, specialized assets, and personnel resources. It is important to appreciate that no one component can efficiently or effectively accomplish the overall task without the support of the others. As an example, while a security officer may be deployed at an entrance gate to control access, if there is no written or defined access control procedure or an identification badge system in use, he or she cannot perform this function effectively. Although many of the component systems and measures deployed in a security ring configuration may be permanent, others may be temporary, such as those enacted during a labor strike or terrorist alert. The temporary implementation or activation of special security components or procedures due to heightened threats should be preplanned and be part of an overall threat condition status system and detailed in a security standard operating procedures manual and emergency action plan. A world-class port security program, one that is designed for a multithreat environment, normally utilizes layered security rings and typically includes the following:

- External security ring
- Perimeter security ring
- Inner security ring
- Vessel security ring
- Site- and asset-specific security ring(s)

Essentially, security personnel employment and training are an integral part of a port security system that is directed and closely supervised by a port security director. Last, but not least, the port must have a comprehensive and realistic security policy carried out on the basis of a thoroughly developed security procedures manual.

13.2 SECURITY RINGS

13.2.1 External Security Ring

An external security ring normally includes intelligence operations outside and inside the port perimeter and close liaison with the government and law enforcement institutions.

13.2.1.1 Intelligence Operations. The continual tasking, collection, analysis, dissemination, and evaluation of strategic and tactical intelligence and information from confidential informants (i.e., persons associated with or inside criminal and terrorist organizations) and sources of information in the communities and regions surrounding and inside the port (e.g., truck drivers, warehouse laborers, documentation clerks, cargo surveyors, open press, news reporters) to provide advance indications and warnings of evolving and future criminal activities or threats targeting the port is one of the key tenets of a port security strategy. In many cases, the success of a port security program depends on the ability to receive advance knowledge of planned criminal/terrorist activities and to direct or manipulate events so that these situations are neutralized or contained outside the port or at one of the security rings.

13.2.1.2 Government and Law Enforcement Liaison. The establishment of active and ongoing relations with and support from national government agencies such as the police, customs, military, and intelligence services is a fundamental necessity. Also desirable are working-level contacts with international and non–host country law enforcement and intelligence agencies, such as Interpol and foreign customs services. These entities can provide vital information concerning activities by transnational criminal and terrorist organizations that may target the seaport or a vessel or its cargo.

13.2.2 Perimeter Security Ring

A perimeter security ring is an extremely important link in an overall port security system. It includes all kinds of physical security barriers, both on land and water, as well as measures for detection of perimeter intrusion by unauthorized persons and attempts to smuggling in or out narcotics, explosives, weapons of mass destruction, and so on. Perimeter security ring usually comprises security walls and/or fences of miscellaneous constructions, security towers, entrance and exit gates, and so on. Access to the port through the perimeter ring must be carried out strictly in accordance with the access control policy and procedures established for a particular port at a particular location.

13.2.2.1 Physical Security Barrier and Illumination. The entire land boundary of the port must be identified and protected by a wall or fence no less than 2.5 m in height and topped with three strands of barbed wire or baled concertina wire (Figure 13.1). The wire topping should be secured to arms that are angled outward at 45°. If the perimeter barrier is fencing, it should be constructed of either "climb-resistant" stretched steel or 9-gauge chainlink wire mesh with 50-mm openings, secured at the bottom with metal tubing or a concrete footing, to deter under-the-fence ingress.

The level of illumination along the perimeter barrier should be no less than 60-cm cd at ground level (similar to the level of lighting in a stadium), projecting 3 m inside the barrier and 6 m outside the barrier. This same lighting standard should be met in cargo and container staging areas, along berths, and on the exterior of buildings and warehouses. Good lighting is arguably the most effective and least expensive measure of deterrence against cargo pilferage, container theft, and similar violations.

13.2.2.2 Waterside Security Measures. A security launch with armed security officer(s) should patrol along berths and in nearby waters to deter or prevent unauthorized approach and access on the water side of the port by stowaways, smugglers, pirates, terrorists, and so on (Figure 13.2). Increasingly, narcotics trafficking organizations are using scuba divers to attach drug-laden torpedoes/boxes to the hulls and undersides of vessels. If this threat is suspected, the port security program should include the use of underwater security patrols

Figure 13.1 Sample perimeter barrier.

Figure 13.2 Security launch patrolling port waters.

(scuba), an antidiver system, or the installation of underwater close-circuit TV (CCTV) cameras. Similarly, drug smuggling and professional stowaway organizations utilize small boats or launches to transport their stowaways, drug couriers, and contraband to the water side of vessels for lading. Moreover, as demonstrated by the attack on the USS *Cole* in Yemen in 2001, terrorists utilize small launches and port service vessels to attack large vessels while in port. The 24-hour security patrol ensures that threats from the water side are deterred or prevented.

13.2.2.3 Perimeter Intrusion Detection. Generally, it is a common practice to deploy security officers at stationary and roving posts along the perimeter. These posts may include a security officer positioned in elevated towers along the perimeter (Figure 13.3), walking along the perimeter barrier and patrolling via mobile means. For maximum efficiency and effectiveness, a K-9 patrol team may be utilized to patrol the perimeter. Research by U.S. law enforcement has determined that the deployment of K-9 patrol teams is a force multiplier; one K-9 team is as effective as deploying three police officers. (*Note:* A K-9 team is a security force trained in detection of intruders, trucking, attack, building search, and so on.)

Perimeter intrusion detection can also be accomplished by or enhanced through the use of technological security systems; among them CCTV cameras, buried or taunt cable, microwave curtains, dual-technology PIR (passive infrared) motion detectors, and laser beams, all of which may be integrated into a manned central station.

13.2.2.4 Entrance and Exit Gates. The number of port entrances and exits should be reduced to a minimum and their purposes specifically defined. There should be separate gates for pedestrians and vehicles. Similarly, there should be separate gates for the entrance and exit of trucks transporting containers or cargo and vehicles driven by employees, vendors, clients, and visitors. Physically, the gates should be constructed so as to meet the same minimum standards as the chainlink perimeter barrier. These gates should lock with heavy-duty padlocks with the keys controlled by security personnel. A security gatehouse should be located at each primary access point. The gatehouse should have the basic items required to accomplish the tasks assigned: a fire extin-

Figure 13.3 Perimeter security tower.

Figure 13.4 Vehicle inspection for contraband.

guisher, first-aid kit, flashlight, rain gear, vehicle and visitor gate logs, 24-hour chronological security logbook, personnel authorization roster, telephone, emergency telephone notification list, security post orders, and a copy of the emergency action plan.

13.2.2.5 Access Control Policy and Procedures. All access points (gates) should be strictly controlled and there should be a comprehensive policy and specific written procedures that define the access of persons (employees, visitors, contractors, truck drivers, ship chandlers, etc.), vehicles (employee and visitor cars, trucks, etc.), and items (cargo, containers, trailers, ship's goods, spare parts, etc.) into and out of the port. "Authorized Personnel Only," "Identification Checkpoint," and "Subject to Search Upon Entry and Exit" signs should be posted and highly visible at all access points. Security officers posted at pedestrian gates should stop and challenge all persons, inspect their identification badges, and search any boxes, briefcases, or other items for contraband. Employees should present their ID badges to the security officer upon entrance and exit and wear their badges at all times while in the port. All visitors (clients, vendors, contractors, etc.) should be stopped at the gate, their visit confirmed with the sponsoring port employee, a temporary badge issued and visitor log completed, and any items opened and inspected for contraband. The interior and trunks of all vehicles should be checked visually for contraband (Figure 13.4). No privately owned vehicles should be permitted inside the terminal. All trucks entering the cargo gates should be stopped, the driver's license checked for validity, the cab inspected for contraband and unauthorized persons, container seals inspected, and relevant information recorded on a comprehensive gate log (Figure 13.5).

13.2.2.6 Access Control Badge System. Each person entering a port should be issued an identification badge. The ID badge program should be managed by a computer-based system that functions with proximity or magnetic strip badges, assigns zones of access, permits or denies a person's access into a specific zone, and records this activity into a database. The front of the employee ID badge should have a

Figure 13.5 Gate inspection of license and interior of cab.

color photo, the employee's name and signature, government identity document or passport number, position, and an expiration date. The back of the ID card should note the employee's date of birth, height, weight, color of hair and eyes, complexion, and the signature of the port director. Each employee's badge should be programmed to allow access to specific zones, this being based on his or her job or position requirements. Employees who have forgotten or lost their badges should be issued a temporary badge for the day or while a new badge is being prepared. Visitor badges generally are for one-day use, disposable, and should note the name of the visitor, government identity document or passport number, area or zones visiting, and the date issued. Nonemployees who work in the port frequently or temporarily, such as contractors, clients, and government representatives, should be issued a badge similar to the employee ID badge (but a different color). A permanent record of all nonemployee badges issued (with the captured data) should be maintained for at least two years.

13.2.2.7 Narcotics Control at Access Points. Attempts to smuggle drugs through access points and into the port may be conducted via hand-carried items, inside vehicles, and in containers/trailers and their cargo. Whereas hand-carried items such as briefcases and boxes can be inspected effectively by a hand search by a security officer, it is not as practical (in terms of time) or effective in the case of a loaded cargo container, empty trailer, or vehicle. In these cases, highly trained and certified narcotics detection K-9 teams should be utilized to inspect containers, cargo, and vehicles for narcotics (Figure 13.6). Alternatively, if financially possible, container x-ray stations should be positioned at vehicle and container entrance points to screen for narcotics (as well as other contraband).

13.2.2.8 Explosives Detection at Access Points. During times of heightened risks of terrorist attacks, bombing, or violent labor conflicts, extra security measures should be implemented to screen for explosive devices and weapons entering the port. In the event that there is a specific threat or reliable information of a planned attack, the security procedures should be enhanced further. The four primary means of searching and screening for explosive devices and weapons are: a visual and hand search, the use of a vapor analyzer to detect chemical odors from explosives, an x-ray machine (which vary in size from those used to screen letters and parcels to those that inspect vehicles and shipping containers), and an explosives detection K-9 team. These four measures may be used independently or in combination; this generally being determined by the level and type of threat. Special attention should be given to suspicious mail and delivery packages and unattended vehicles positioned at access points or near key assets or buildings.

13.2.2.9 Weapons of Mass Destruction Detection at Access Points. Ports must develop, test, and continually update contingency

Figure 13.6 Narcotics detection K-9 team at cargo gate.

plans for the rapid deployment of systems and measures for the detection of chemical, biological, and nuclear weapons (typically referred to as weapons of mass destruction). In many cases, a port will rely on the national government to provide such technical capabilities; however, it is critically important that the port security director develop policies, plans, and procedures that will ensure successful integration of these measures without significantly affecting the port's business or endangering the safety of its personnel. These contingency plans and procedures should be fully coordinated with the relevant government agencies and tested on a periodic basis. The contingency plans will interrelate with the port's disaster preparedness and recovery plan, which ensures business continuity and the safety and security of personnel.

13.3 INNER SECURITY RING

An inner security ring involves patrolling a port's interior area and is usually carried out in the form of mobile or foot security patrols. These are coordinated via an operation command center and supervised by the shift security supervisor.

13.3.1 Mobile Security Patrols

The interior areas of a port, such as container stacking zones, cargo staging areas, facility and maintenance buildings, equipment storage areas, and berths should be patrolled continuously by security officers in vehicles. These units should patrol in separate, overlapping zones. Security personnel should monitor general yard activities, restrict the movement of tractor drivers away from their vehicles, observe the transloading of cargo containers, and view stevedores and laborers working on the docks.

13.3.2 Foot Security Patrols

Periodic inspections and tallies of containers and seals throughout a yard by security officers are effective deterrents to cargo pilferage, drug smuggling, and container manipulation, as well as a means of establishing a specific time period of an incident. Foot security officers should be constantly vigilant that personnel are wearing valid ID badges and are in their authorized zones, that doors and windows of any structures and buildings are secured during nonoperational hours, and that drivers are not operating equipment at high rates of speed or in a dangerous manner.

13.3.3 Security Operations Command Center

Security systems specialist(s) should be deployed 24 hours a day in the security operations command center for the purpose of observing and operating the central station system (which manages and controls all perimeter intrusion detection measures; CCTV deployed in the patio, on the berths, and outside and inside buildings; building intrusion and panic alarms; access control system; fire alarm systems; etc.). All security systems should be fully integrated and support each other in the event of an incident.

13.3.4 Shift Security Supervisor

There should be one person designated as the overall shift security supervisor, who should direct, lead, and manage the terminal security officers and other deployed security resources (K-9 teams, security operations command center personnel, and vessel security teams). The shift security supervisor is a first-line management position and is a critical part of an overall security program. The shift supervisor should constantly patrol (in vehicle) the port, inspect/supervise security personnel, interact with other port managers, and respond to and take charge of incidents or potential security situations.

13.4 SITE- AND ASSET-SPECIFIC SECURITY RINGS

These security rings involve protection of specific buildings, such as administrative offices, selected warehouses, telephone station, and substations, essential equipment (e.g., cranes) and other critical assets.

13.4.1 Administrative Office Building

Dedicated resources should be deployed and procedures established so as to ensure the security of the administration building and its contents and the safety of its occupants. The number of entrances and exits to this building should be restricted to a minimum, with doors being secured with deadbolt locks when not in use. A security officer should be posted at each unlocked exterior-access door. Keys should be kept to a minimum and issued on a restricted basis by a designated key custodian. A computer-based key management system should be utilized. First floor–level windows (and those below) should be protected by bars or wire mesh. Lighting on the exterior of the building should be at the same level as that along the perimeter. At the main entrance, a security officer should screen all persons, check ID and visitor badges, and search all handbags, briefcases, boxes, and so on, for weapons and contraband. A segregated reception area should be located inside the entrance. All visitors should be escorted into the interior offices by the sponsoring port representative. The interior of the building should be divided into functional zones to establish access zones for employees. Access into each zone should be regulated through the ID badge system. The main entrance, secondary access points, and reception areas should be under constant surveillance by CCTV cameras, which are monitored and recorded by specialists in the security operations command center. Other sensitive areas, such as cashier windows and computer and telephone rooms, should be under observation and security monitoring by intrusion alarms and CCTV cameras.

13.4.2 Bonded and High-Risk Warehouses

Security measures and procedures similar to those noted in Section 13.1 should be implemented to ensure the security and integrity of cargo, buildings, and personnel.

13.4.3 Critical Assets and Essential Equipment

Security measures and procedures similar to those noted in Section 13.1 should be implemented, as appropriate, to ensure effective security and integrity of all equipment and assets, including cranes, electric plants, and telephone buildings.

13.5 VESSEL SECURITY RING

13.5.1 Basic Concept

Like other critical assets within a port, vessels must have their own security ring, which is a part of, but necessarily independent of, the terminal security apparatus. Effective vessel security (i.e., deterring and/or preventing stowaways, piracy, drug smuggling, pilferage, and terrorism) includes strict access control at the gangway. This involves a search of all items carried onboard and knowledge at all times of who is onboard. Maintaining a vessel security ring also includes providing security of its water side and conduct of postarrival and predeparture inspections. All vessels calling on the port should be assigned a vessel security team (VST). The VST should be deployed from time of arrival until time of departure. Upon each arriving vessel's clearance by government officials, the VST should immediately board the vessel and conduct a quick inspection of the deck and exterior of the superstructure. This inspection is to detect the presence of stowaways, terrorists, or narcotics, unlocked doors into the superstructure, possible hazard material emergencies (HAZMAT), and so on. All discoveries of undocumented persons, suspected narcotics, or HAZMAT situations should be reported immediately to the ship's captain. Other security discrepancies should be noted in the gangway logbook and reported to the ship's chief officer. Following this inspection, VST officers should deploy to their positions and continue with their duties.

13.5.2 VST Deployment for LO/LO Commercial Cargo and Container Vessels

The vessel security team typically should consist of no fewer than three security officers and one VST supervisor. One security officer should be posted at the gangway to control and document the entrance and exit of persons (stevedores, crew, visitors, vessel agents, government officials, etc.) and to search all parcels, bags, water coolers, and so on, carried on and off the vessel (Figure 13.7). At least one security officer should patrol the ship deck to monitor the activities of stevedores and ongoing cargo operations, and one officer should be posted to patrol the waterside area by scanning the waters for "swimmer" stowaways, scuba divers, drug smuggler launches, and so on (Figure 13.8). One VST supervisor should constantly patrol the vessel's decks and inspect and supervise the operations of security officers. Normally, all empty containers not inspected by port checkers should be inspected and sealed dockside at the hook by the VST supervisor.

13.5.3 VST Deployment for RO-RO Vessels

For Ro-Ro (roll on–roll off) vessels, the vessel security team should consist of no less than four security officers and one VST supervisor. One security officer should be posted at the top of the ramp to control and document the entrance and exit of all persons (stevedores, crew, visitors, vessel agents, government officials, etc.) and to search all parcels, bags, water cool-

Figure 13.7 Inspecting for possession of port ID badge of stevedore.

Figure 13.8 Waterside security patrol during threat alert.

ers, and so on, carried on and off the vessel. This officer should also scan the dockside constantly for unusual activity. Another security officer is posted on the ramp to inspect the undersides of trailers and inside vehicles (Otta-was) used to move trailers around a terminal, or embarking or disembarking trailers on Ro/Ro vessels, and to search and seal empty containers for the presence of stowaways, narcotics, and terrorists (Figure 13.9). One officer

Figure 13.9 Inspection of interior of Ottawa entering vessel.

should patrol the internal deck where loading trailers and equipment are being staged, and one more officer should patrol the open upper deck waterside, scanning the waters for "swimmer" stowaways, scuba divers, drug smuggler launches, and so on. The duty of the VST supervisor includes constantly patrolling the vessel decks, inspecting and supervising operations of the security officers, and taking charge of security situations.

13.5.4 Key Vessel Security Procedures

Noted below are key procedures proven to yield positive results in maintaining effective vessel security. Typically, they are as follows:

1. Posting a sign at the gangway that advises "Authorized Personnel Only—Present ID to Gangway Security—All bags, packages, etc. will be searched for weapons and contraband."
2. Having all stevedores and visitors relinquish their port ID badge or national ID card to the gangway security officer while onboard.
3. Using the visitor's log, stevedore list, shorepass log, and security logbook to document the entrance and exit of persons and all security incidents.
4. Maintaining all superstructure doors and cargo and deck hatches secured when not under guard.
5. Keeping deck maintenance and storage lockers and crane access hatches secured when not in use.
6. Installing rat guards on mooring lines.
7. Safely securing anchor chain cover while in port.
8. Restricting stevedores to immediate work areas.
9. Securing Jacobs ladder (rope ladder generally located midship) and pilot ladder.
10. Locking cargo bay access hatches when not in use.
11. Using water-side and dockside illumination at night.
12. Using plastic and paper seals on access points to assure minimum use.
13. Placing sawdust or flour around the anchor chain and mooring line holes and

Figure 13.10 The vessel inspection begins by checking each cargo bay.

in key crawl spaces (entry can be noted by hand- and footprints).

14. Keeping privately owned vehicles from parking on a dock next to a vessel.
15. Searching all ship's stores and ship chandler products dockside using a narcotics detection K-9 team.
16. Inspecting all empty containers for narcotics and stowaways and sealing them prior to loading onboard the vessel. The container and seal numbers for all containers and trailers loaded onboard should be recorded on a tally sheet.

13.5.5 Predeparture Search for Contraband and Stowaways

Upon completion of cargo operations, the VST supervisor should coordinate and lead the officers (with the exception of the officer posted at the gangway) in a systematic and comprehensive search of the vessel for stowaways and narcotics (Figures 13.10 through 13.15). The gangway security officer should restrict access to the vessel during this inspection. Upon termination of the vessel search, the VST supervisor should complete a vessel search

Figure 13.11 Inspection of ventilation shafts and crawl spaces between bays.

Figure 13.12 Security officer inspects the exterior of a ship's superstructure.

Figure 13.13 The VST team inspects a main deck (from stern to bow).

Figure 13.14 Inspection of the interior of fire hose box and storage areas.

Figure 13.15 A vessel search is completed following an inspection of the interior of the ship's superstructure, from the rudder room to the bridge.

certificate and provide signed copies to the ship captain, vessel agent, and shift security supervisor.

13.6 SECURITY PERSONNEL EMPLOYMENT AND TRAINING

Although not a component of a specific security ring, proper preemployment screening, training, and equipping of security personnel will have a direct impact on attaining the desired results from security personnel deployed in the various rings and in the success of the overall port security program.

13.6.1 Preemployment Screening

The screening of candidates for employment with the security department should typically include follow the following steps:

1. A candidate completes a detailed standard employment application form and provides a "good health" certificate, "no police record" certificate, and copies of all education and training documents.
2. The candidate is interviewed by a security supervisor.
3. The human resources group department verifies the candidate's prior employment and references.
4. An internal security investigator conducts in-person interviews of the candidate's neighbors and prior employers and checks the national police records.
5. The candidate undergoes a drug use test.
6. A candidates for a sensitive position (K-9 handler, investigator, supervisor, etc.) undergoes a polygraph test.
7. The candidate has a final interview with the security manager.
8. The human resources group advises the candidate of an employment offer, and

the candidate signs an employment contract.

9. The human resources group establishes a permanent personnel file, which includes the signed contract, completed application and photo, and all screening and investigation documentation.

13.6.2 Basic Security Training

All new security personnel, regardless of their permanent assigned position, should first be fully trained in the basics of seaport security (Figures 13.16 through 13.18). A comprehensive course for new security personnel should be approximately 200 hours in duration and typically include, but not be limited to, the following topics:

1. Security definitions in a port environment
2. Discipline, the chain-of-command, and ethics
3. Legal considerations
4. Uniform and equipment presentation
5. Personal defense tactics
6. Use of police batons, such as ASP or PR-24

Figure 13.16 Port security officer training.

Figure 13.17 Students learning firefighting techniques.

Figure 13.18 Executive protection team training.

7. Use of military/police grade gas (CS/CN)
8. Rules of access control of vehicles and persons
9. Security of installations, offices, and warehouses
10. Vehicle and foot patrols and static post rules
11. Report writing
12. Security forms and documentation
13. General and administrative orders
14. Post orders and special orders

15. Vessel security operations and post orders
16. Detection and search of persons and vehicles
17. Role of internal security and intelligence operations
18. Role of the shift security supervisor, K-9 units, and other special services
19. Leadership, motivation, and conflict resolution
20. First aid and firefighting
21. Use of radio and communication etiquette
22. The business and functioning of a commercial seaport
23. Use and qualification with a pistol and shotgun issued
24. Theft and cargo pilferage
25. Narcotics smuggling, stowaway, and piracy: trends and techniques
26. Terrorism awareness
27. Civil disturbances, protests, and labor strikes
28. Recognition of hazard materials and emergency response
29. Bomb threats and search procedures
30. Familiarization with the port security standard operating manual
31. Review of port emergency action plan
32. Port threat assessment
33. National and international port security regulations

13.6.3 Security Officer Equipment

All security officers and the shift security supervisor should wear a police or military-style uniform, hat, and black military boots. Each should be issued a safety vest and helmet and wear a black nylon military equipment belt which holds a three-D-cell Maglite flashlight, two flexcuffs, an ASP or PR-24 baton, a radio, CS/CN gas, a pistol, and extra ammunition (Figure 13.19).

13.7 PORT SECURITY DIRECTOR

All world-class port security programs require the assignment of a highly experienced, full-time security director and the support of a capable administrative staff. The security director should have a military and/or law enforcement background, extensive leadership and management skills, a solid understanding of the commercial maritime business and how a port functions, the ability to lead a large and multifaceted security organization, and broad knowledge of and experience in assessing and successfully confronting the various security threats faced by a commercial port. The secu-

Figure 13.19 Properly uniformed and equipped security officer.

rity director is charged with the development, implementation, leadership, and management of the overall port security program. In addition to managing all security operations and resources, the security director should define and establish all security policies, plans, and procedures, including the development of the port security operating manual, the port emergency action plan, and the disaster preparedness and recovery plan.

13.8 SECURITY OPERATIONS MANUAL AND CONTINGENCY PLANNING

There is an old military adage which states: Proper planning prevents poor performance. This is usually true for port security. The absence of comprehensive, realistic, and tested written security policies, plans, and procedures will ensure the failure of a port security program. There are three manuals that are essential to successful design, implementation, and management of a security program: the port security operations manual, the port emergency action plan, and the disaster preparedness and recovery plan. A brief description of each follows.

- The security operations manual defines clearly and in detail the policies, plans, and procedures for all security-related activities, operations, functions, responsibilities, processes, and personnel. The breadth of issues addressed in the manual is extensive and ranges from program vision to personnel security standards and training to physical security measures and access control procedures to computer security and competitive espionage, and from security post orders to serious incident management.
- The port emergency action plan identifies step-by-step plans to respond to various crisis situations. The latter include bombings, assaults, assassinations, kidnappings, terrorist attacks, civil disturbances, labor unrest and strikes, HAZMAT incidents, fire, and natural disasters.
- The disaster preparedness and recovery plan defines plans and procedures to ensure business continuity following natural and human-made disasters that affect the port significantly. These include earthquakes, tornadoes, hurricanes, HAZMAT incidents, and terrorist attacks using weapons of mass destruction and other devices.

To meet the increasing and ever-changing security challenges and threats to ports of the world, port management must design and implement a security strategy that is based on the concepts of defense in depth and layered, interrelational security rings. Such a well-planned strategy, when combined with the leadership of a highly professional security director, deployment of a well-trained security force, and preparation of a comprehensive security standard operating procedures manual and emergency contingency plans, will ensure the success of the security program in deterring or denying current and anticipated security challenges and meriting recognition as a world-class port security program.

INDEX